Springer Series in

MATERIALS SCIENCE 110

Springer Series in

MATERIALS SCIENCE

Editors: R. Hull R. M. Osgood, Jr. J. Parisi H. Warlimont

The Springer Series in Materials Science covers the complete spectrum of materials physics, including fundamental principles, physical properties, materials theory and design. Recognizing the increasing importance of materials science in future device technologies, the book titles in this series reflect the state-of-the-art in understanding and controlling the structure and properties of all important classes of materials.

99 **Self-Organized Morphology in Nanostructured Materials**
Editors: K. Al-Shamery and J. Parisi

100 **Self Healing Materials**
An Alternative Approach to 20 Centuries of Materials Science
Editor: S. van der Zwaag

101 **New Organic Nanostructures for Next Generation Devices**
Editors: K. Al-Shamery, H.-G. Rubahn, and H. Sitter

102 **Photonic Crystal Fibers**
Properties and Applications
By F. Poli, A. Cucinotta, and S. Selleri

103 **Polarons in Advanced Materials**
Editor: A.S. Alexandrov

104 **Transparent Conductive Zinc Oxide**
Basics and Applications in Thin Film Solar Cells
Editors: K. Ellmer, A. Klein, and B. Rech

105 **Dilute III-V Nitride Semiconductors and Material Systems**
Physics and Technology
Editor: A. Erol

106 **Into The Nano Era**
Moore's Law Beyond Planar Silicon CMOS
Editor: H.R. Huff

107 **Organic Semiconductors in Sensor Applications**
Editors: D.A. Bernards, R.M. Ownes, and G.G. Malliaras

108 **Evolution of Thin-Film Morphology**
Modeling and Simulations
By M. Pelliccione and T.-M. Lu

109 **Reactive Sputter Deposition**
Editors: D. Depla amd S. Mahieu

110 **The Physics of Organic Superconductors and Conductors**
Editor: A.G. Lebed

Volumes 50–98 are listed at the end of the book.

A.G. Lebed
Editor

The Physics of Organic Superconductors and Conductors

With 343 Figures

Springer

Professor Dr. Andrei Lebed
University of Arizona, Department of Physics
1118 East 4th Street, Tucson, AZ 85721, USA
E-mail: lebed@physics.arizona.edu

Series Editors:

Professor Robert Hull
University of Virginia
Dept. of Materials Science and Engineering
Thornton Hall
Charlottesville, VA 22903-2442, USA

Professor R. M. Osgood, Jr.
Microelectronics Science Laboratory
Department of Electrical Engineering
Columbia University
Seeley W. Mudd Building
New York, NY 10027, USA

Professor Jürgen Parisi
Universität Oldenburg, Fachbereich Physik
Abt. Energie- und Halbleiterforschung
Carl-von-Ossietzky-Strasse 9–11
26129 Oldenburg, Germany

Professor Hans Warlimont
Institut für Festkörper-
und Werkstofforschung,
Helmholtzstrasse 20
01069 Dresden, Germany

ISSN 0933-033X

ISBN 978-3-540-76667-4 Springer Berlin Heidelberg New York

Library of Congress Control Number: 2007938890

Springer is a part of Springer Science+Business Media.
springer.com

Typesetting: Data prepared by SPi using a Springer LaTeX macro package
Cover concept: eStudio Calamar Steinen
Cover production: WMX Design GmbH, Heidelberg

Printed on acid-free paper SPIN: 11559238 57/3180/SPi 5 4 3 2 1 0

This book is devoted to 25-th anniversaries of the discoveries of unconventional superconductivity, the field-induced spin-density-wave diagrams, and the "three-dimensional" quantum Hall effect in organic conductors. The editor dedicates his own contributions to the book to his wife, Natalia, for her loving support and patience.

Preface

Historically, quasi-low-dimensional superconductors were considered as the main candidates to observe high-temperature superconductivity. For a discussion of the related exotic mechanisms of superconductivity, suggested by W.A. Little and V.L. Ginzburg, see a chapter by D. Jérome in this volume. Unfortunately, high-temperature superconductivity has not been discovered yet in quasi-one-dimensional (Q1D) and quasi-two-dimensional (Q2D) organic materials. Nevertheless, very rich and, in many cases, unique physical properties of their metallic, superconducting, charge- and spin-density-wave phases allowed P.M. Chaikin to claim that the first organic superconductors, $(TMTSF)_2X$, are probably the most interesting electronic materials ever discovered.

Our book welcomes a reader to a fascinating world of exotic condensed matter physics, low temperatures, and high and ultrahigh magnetic fields. It is written by leading experts in the area from USA, France, Japan, Russia, United Kingdom, Germany, Canada, South Korea, Croatia, Hungary, and Switzerland. The book consists of six parts, subdivided into 27 chapters, which contain both the experimental results and their theoretical explanations. The majority of the chapters contain pedagogical introductions and all necessary illustrations to be read separately. Although we concentrate on physical phenomena, in the most chapters related chemistry and structural aspects of Q1D and Q2D organic materials are also discussed.

The goal of our book is to cover a broad range of physical effects, experimentally observed in organic conductors and superconductors. In Part 1, "Historical Surveys," the major experimental and theoretical discoveries, which have determined a development of the physics of organic compounds for more than two decades, are discussed from historical and pedagogical points of view. Among them are discoveries of the first organic superconductors, charge-density waves, charge ordering, field-induced spin-density waves, and three-dimensional quantum Hall effect (3D QHE). In Part 3, "Unusual Properties of a Metallic Phase," some novel types of quantum macroscopic phenomena in a magnetic field, which appear when electrons move along open orbits in the extended Brillouin zone, are described. Among them are very rich Fermi liquid and non-Fermi liquid angular magnetic oscillations and a novel type of the cyclotron resonance. Similar phenomena are discussed in Chaps. 5–9 of Part 2, "General Reviews." In Part 4, "Field-Induced Spin(Charge)-Density-Wave Phases," the corresponding phenomena and a related 3D QHE are described both from experimental and theoretical points of views. For these purposes, see also Chaps. 3 and 4 of Part 1 and Chaps. 5 and 7 of Part 2. In Part 5, "Unconventional Superconducting Properties," unconventional singlet d-wave, reentrant and possible triplet superconducting phases both in Q1D and Q2D organic superconductors are discussed. They are described also in Chaps. 1 and 3 of Part 1 and Chaps. 5–7 and 9 of Part 2. Electron correlations in reduced dimensions are considered in Part 6, "Electron Correlations in Organic Conductors," and Chap. 12 of Part 2. In Chaps. 10 and 11 of Part II, charge-density-wave like phases including charge ordering, solitons, and ferroelectric phases are considered.

Note that sometimes a historical term "organic conductor," which reflects only a chemical aspect of a compound, is misleading. As one can see from the contents of the book, this is particularly true in our case. Indeed, organic conductors and superconductors, discussed in this book, have a little common with typical organic polymers. We use the above-mentioned term to describe extremely clean crystalline conductors and superconductors with highly anisotropic electron spectra, demonstrating fundamental and, in many cases, unique solid-state physics.

It is our pride and not very often case in a modern materials science physics that significantly more than a half of the experimentally observed properties in organic conductors and superconductors have been successfully explained. Moreover, some of them were theoretically predicted. Therefore, one can describe the existing relationships between the experiment and theory

Fig. 1. Friendship between theory and experiment (drawing of Natalia Lebed)

in this area of a condensed matter physics as a friendship (see Fig. 1). This friendship may be transformed into a true love if some of its dark places are clarified. Among them are triplet–singlet controversy of the superconducting properties in $(TMTSF)_2X$ materials, the observations of the Giant Magic Angles Nernst effect, and some other still ill-understood phenomena. This work was partially supported by the NSF grant DMR-0705986.

Tucson, Arizona,
June 2007

A.G. Lebed

Contents

Part III Unusual Properties of a Metallic Phase

Part IV Field-Induced Spin(Charge)-Density Wave Phases

Part VI Electron Correlations in Organic Conductors

Contributors

E. Altman
Department of Condensed
Matter Physics
Weizmann Institute of Science
Rehovot, Israel
altman@wisemail.weizmann.ac.il

K. Biljaković
Institute of Physics
PO Box 304, 10001
Zagreb, Croatia

A. Bjeliš
Department of Physics
Faculty of Science
University of Zagreb, POB 162
10001 Zagreb, Croatia
bjelis@phy.hr

J.S. Brooks
National High Magnetic
Field Laboratory
Florida State University
1800 East Paul Dirac Drive
Tallahassee, FL 32310, USA
brooks@magnet.fsu.edu

C. Bourbonnais
Réseau Québecois sur les
Matériaux de Pointe
Département de physique
Université de Sherbrooke
Sherbrooke, Québec J1K-2R1
Canada
cbourbon@physique.usherb.ca

S.A. Brazovski
Laboratoire de Physique Theorique
et Modeles Statistiques
CNRS LPTMS
Bat.100A, Universite Paris-Sud
91405, Orsay-Cedex, France
brazov@lptms.u-psud.fr

S.E. Brown
Department of Physics
UCLA, Los Angeles
CA 90095, USA
brown@physics.ucla.edu

P.M. Chaikin
Department of Physics
New York University
New York, NY 10003, USA
chaikin@nyu.edu

S. Charfi-Kaddour
Laboratoire de Physique
de la Matière Condensée
Département de Physique
Faculté des Sciences de Tunis
Campus universitaire 1060
Tunis, Tunisia
samia.kaddour@fst.rnu.tn

R.W. Cherng
Department of Physics
Harvard University
17 Oxford Street
Cambridge, MA 02138, USA
rcherng@fas.harvard.edu

E. Demler
Department of Physics
Harvard University
Cambridge, MA 02138, USA
demler@cmt.harvard.edu

B. Dóra
Department of Physics
Budapest University
of Technology and Economics
H-1521 Budapest, Hungary
dora@kapica.phy.bme.hu

T. Giamarchi
DPMC-MaNEP
University of Geneva
24, Quai Ernest-Ansermet
CH1211, Geneva 4 Switzerland
Thierry.Giamarchi@physics.
unige.ch

S. Haddad
Laboratoire de Physique
de la Matière Condensée
Département de Physique
Faculté des Sciences de Tunis
Campus universitaire 1060
Tunis, Tunisia
shaddad@lagrange.physique.
usherb.ca

N. Harrison
National High Magnetic Field
Laboratory, TA-35, MS E536
Los Alamos National Laboratory
Los Alamos NM 87545, USA
nharrison@lanl.gov

M. Héritier
Laboratoire de Physique
des Solides d'Orsay
UMR 8502 CNRS - Paris XI
Universit Paris - Sud 11
91405 Orsay, France
heritier@lps.u-psud.fr

S. Hill
Department of Physics
University of Florida
PO Box 118440
Gainesville, FL 32611, USA
hill@phys.ufl.edu

D. Jérome
Laboratoire de Physique
des Solides UMR 8502
Université Paris-Sud
91405 Orsay, France
jerome@lps.u-psud.fr

W. Kang
Department of Physics
Ewha Womans University
11-1 Daehyun-Dong
Seoul 120-750, Korea
wkang@ewha.ac.kr

K. Kanoda
Department of Applied Physics
University of Tokyo
Hongo 7-3-1, Bunkyo-ku
Tokyo 113-8656, Japan
kanoda@ap.t.u-tokyo.ac.jp

M.V. Kartsovnik
Walther-Meissner-Institut
for Low Temperature Research
Bavarian Academy of Sciences
Walther-Meissner-Str. 8
D-85748 Garching, Germany
mark.kartsovnik@wmi.badw.de

A.V. Kornilov
Division of Solid State Physics
P.N. Lebedev Physics Institute
Russian Academy of Sciences
53 Leninskii prospekt
Moscow, 119991, Russia
kornilov@sci.lebedev.ru

J.-C. Lasjaunias
Centre de Recherches
sur les Très Basses Températures
CNRS, 25 rue des Martyrs, BP 166
38042 Grenoble cedex 9, France

A.G. Lebed
Department of Physics
University of Arizona
1118 East 4th Street
Tucson, AZ 85721, USA
lebed@physics.arizona.edu

K. Maki
Department of Physics
and Astronomy
University of Southern California
Los Angeles, CA 90089-0484, USA
kmaki@usc.edu

R.D. McDonald
National High Magnetic Field
Laboratory, TA-35, MS E536
Los Alamos National Laboratory
Los Alamos NM 87545, USA
rmcd@lanl.gov

P. Monceau
Centre de Recherches
sur les Très Basses Températures
CNRS, 25 rue des Martyrs, BP 166
38042 Grenoble, Cedex 9, France
pierre.monceau@grenoble.cnrs.fr

K. Murata
Graduate School of Science
Osaka City University
3-3-138 Sugimoto Sumiyoshi-ku
Osaka 558-8585, Japan
muratak@sci.osaka-cu.ac.jp

M.J. Naughton
Department of Physics
Boston College
Chestnut Hill
MA 02467, USA
naughton@bc.edu

F. Nad
Institute of Radioengineering
and Electronics
Russian Academy of Sciences
125009 Moscow, Russia

D. Podolsky
Department of Physics
University of California at Berkeley
Berkeley, CA 94720, USA
dpodolsky@berkeley.edu

V.M. Pudalov
Division of Solid State Physics
P.N. Lebedev Physics Institute
Russian Academy of Sciences
53 Leninskii prospekt
Moscow, 119991, Russia
pudalov@sci.lebedev.ru

C.A.R. Sá de Melo
School of Physics
Georgia Institute of Technology
837 State Street
Atlanta, GA 30332, USA
carlos.sademelo@physics.
gatech.edu

H. Shimahara
Department of Quantum
Matter Science
ADSM, Hiroshima University
Higashi-Hiroshima 739-8530, Japan
hiro@hiroshima-u.ac.jp

J. Singleton
National High Magnetic
Field Laboratory
TA-35, MS E536
Los Alamos National Laboratory
Los Alamos, NM 87545, USA
jsingle@lanl.gov

S. Takahashi
Department of Physics
University of Florida
PO Box 118440
Gainesville, FL 32611, USA

S. Uji
Graduate School of Pure
and Applied Sciences
University of Tsukuba
Tsukuba, Japan
uji.shinya@nims.go.jp

A. Virosztek
Department of Physics
Budapest University of
Technology and Economics
H-1521 Budapest, Hungary
and
Research Institute for Solid State
Physics and Optics
PO Box 49
H-1525 Budapest, Hungary
viro@szfki.hu

Si Wu
Department of Physics
University of Arizona
1118 East 4th Street
Tucson, AZ 85721, USA
wusi@physics.arizona.edu

V.M. Yakovenko
Department of Physics
University of Maryland
College Park, MD 20742-4111, USA
yakovenk@physics.umd.edu

H. Yoshino
Graduate School of Science
Osaka City University
3-3-138 Sugimoto
Sumiyoshi-ku
Osaka 558-8585, Japan
yoshino@sci.osaka-cu.ac.jp

D. Zanchi
Laboratoire de Physique Théorique
at Hautes Energies
Universités Paris VI Pierre
et Marie Curie - Paris
VII Denis Diderot
2 Place Jussieu
75252 Paris Cédex 05, France
drazen@lpthe.jussieu.fr

Wei Zhang
Department of Physics
University of Michigan
450 Church Street
Ann Arbor, MI 48109, USA
wzhangl@umich.edu

Part I

Historical Surveys

1

Historical Approach to Organic Superconductivity

D. Jérome

We give a brief account of how the search for superconductivity in organic matter, motivated by the quest for high T_c superconductors has developed since the 1960s. Even if the maximum T_c of organic conductors in 2005 does not exceed 13 K these are prototype systems which have contributed remarkably to the progresses in the physics of low dimensional conductors.

1.1 One-Dimensional Conductors

Superconductivity, i.e., the possibility for electrons to flow through a conductor without any energy loss has been the fuel for a very large number of industrial applications of physics in the twentieth century. To mention only a few of them, which are important in current life, there are the power-loss energy transfer, the production of very strong magnetic fields for medical imaging or for plasma confinement (the fusion problem), voltage standards, particle accelerators and magnetic flux sensors.

This new state of matter was discovered unexpectedly in 1911 in mercury with a critical temperature of $T_c = 4.15\,\mathrm{K}$ by Kamerlingh Onnes just after he had mastered the liquefaction of helium [1]. However, the phenomenon of superconductivity, in particular the value of its T_c, is strongly dependent on the material in which it is to be observed. Therefore, over the years a lot of new metals and intermetallic compounds have been discovered with increasing values of T_c reaching a plateau at about 24 K in the 1960s [2], see Fig. 1.1.

The absence of any satisfactory theory until 1957 did not prevent applications of superconductivity which developed after World War II but they were all bound to the use of expensive liquid helium as the cooling agent necessary for the stabilization of the superconducting state.

In the 1960s, when superconductivity had already used in applications, people were thinking about the possibility of discovering new materials to release the constraint of using liquid helium to reach the superconducting state. Then came the idea to look for materials out of the traditional series

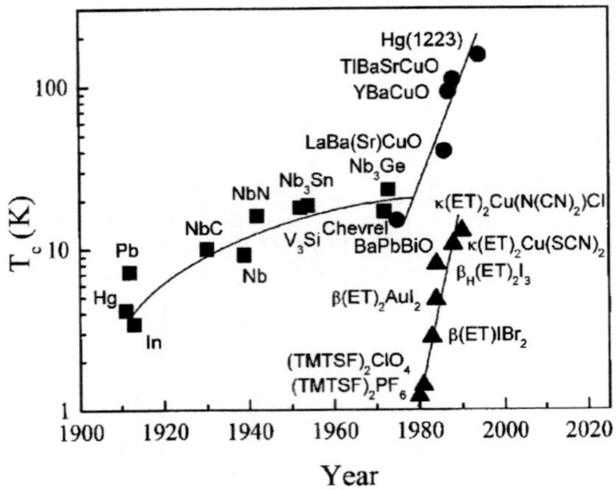

Fig. 1.1. Evolution of the superconducting critical temperature in metals and intermetallic compounds, cuprates and organic conductors (1 and 2D conductors)

of metals or alloys compounds namely new types of conductors in which the conduction would proceed through the transport of charge between molecular orbitals rather than atomic orbitals. Actually, the concept of a synthetic metal had already been launched by McCoy and Moore [3] simultaneously with the discovery of superconductivity in mercury when they proposed *to prepare composite metallic substances from non-metallic constituent elements.* As to the possibility of superconductivity in materials other than metals, Fritz London in 1937 [4] was the first to suggest that compounds with aromatic rings such as anthracene and naphtalene *might exhibit a current running freely around the rings under a magnetic field.*

The first successful attempt to promote metal-like conduction between open shell molecular species came out in 1954 with the synthesis of the molecular salt of perylene oxidized with bromine [5] although this salt was rather unstable. In the early 1960s, related to the quest for higher T_c Little made an important suggestion [6, 7]: a new mechanism for superconductivity expected to lead to a drastic enhancement of the superconducting T_c to be observed in specially designed macromolecules. The idea of Little was indeed strongly rooted in the isotope effect, one of the greatest successes of the theory proposed in 1957 by Bardeen Cooper and Schrieffer (BCS) [8] for the interpretation of superconductivity in materials known up to that date and based on the phonon-mediated electron pairing. In the BCS theory, the attractive interaction between electrons (or holes) which is a prerequisite for the Bose condensation of electron pairs into the superconducting state relies on the mass M of the ions which undergo a small displacement when the electrons are passing close to them namely, $T_c \,\alpha\, M^{-1/2}$. It can be noticed that Fröhlich proposed in 1954 a mechanism for superconductivity based on the 1D case [9]. Although the theory of Fröhlich has not been able to account for the superconductivity of metals, it turned out to be quite successful later for the interpretation of the transport properties in some 1D organic compounds, vide infra. In the excitonic mechanism of Little the charge carriers are moving along a conducting spine and it is an electronically polarizable medium which is used instead of the usual polarizable ionic lattice. Consequently, the small electronic mass m_e of the polarizable medium would lead to an enhancement of T_c of the order of $(M/m_e)^{1/2}$ times the value which is observed in a conventional superconductor, admittedly a huge factor. This has been the beginning of the concept of superconductivity at room temperature, still the Graal of scientists in the beginning of the twenty-first century. Simultaneously, Ginzburg [10,11] considered the possibility for the pairing of electrons in metal layers sandwiched between polarizable dielectrics through virtual excitations at high energy, see Fig. 1.3.

For 1D materials, the model of Little was based on the use of a long conjugated polymer such as a polyacetylene molecule grafted by polarizable side groups [12] (see Fig. 1.2). Admittedly, this formidable task in synthetic

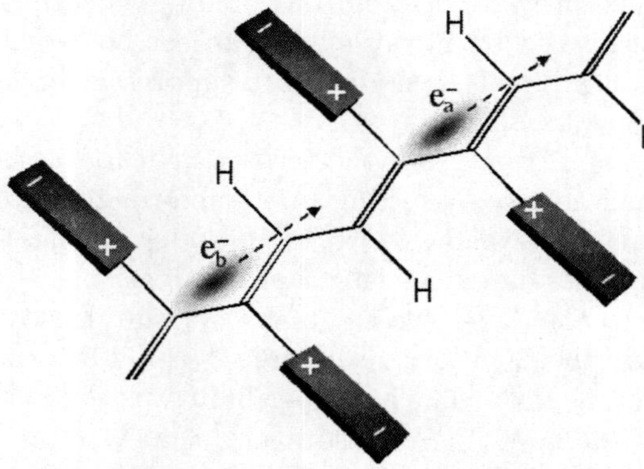

Fig. 1.2. Little's suggestion. Charge carriers *a* and *b* moving along a conducting spine are bound via a virtual electronic excitation of polarizable side groups

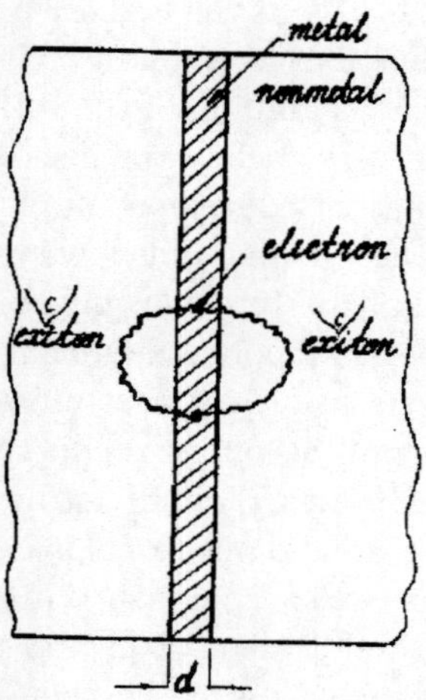

Fig. 1.3. 2D superconductor suggested by Ginzburg

chemistry has not reached its initial goal but the idea to link organic metallicity and one dimensionality was launched and turned out to be a very strong stimulant for the development of organic superconductors. Needless to say that a lot of basic physical problems had been overlooked in the seminal paper of Little [6] which was based on the popular Fermi liquid BCS theory. The first point is that 1D conductors had been regarded previously as textbook examples by Peierls [13], who stated a theorem according to which a gap should spontaneously open at the Fermi level in the conduction and lead to the stabilization of a dielectric insulating state at low temperature long before it has been established in the platino-cyanate chains [14]. Bychkov et al. [15] criticized the model of Little regarding its potentiality to lead to high temperature superconductivity. Following Bychkov et al., the 1D character of the model system proposed by Little makes it a unique problem in which there exists a built-in coupling between superconducting and dielectric instabilities. It follows that each of these instabilities cannot be considered separately in the mechanism proposed by Little in 1D and last but not least fluctuations should be very efficient in a 1D conductor to suppress any long range ordering to very low temperature.

At the beginning of the 1960s the term *high temperature superconductor* was already commonly used referring to the intermetallic compounds of the A15 structure, namely, Nb_3Sn or V_3Si [16]. The hidden 1D nature of the A15 structure provides an enhancement of the density of states at the Fermi level lying close to the van-Hove singularity of the density of states of the 1D d-band. Within the BCS formalism large T_c could be expected. They were actually observed (17–23 K) but an upper limit was found to the increase of T_c since the large value of $N(E_F)$ also makes the structure unstable against a cubic to tetragonal band Jahn–Teller distortion [17,18]. The theory showed that the actual compounds Nb_3Sn and V_3Si possess their optimum T_c [19]. In this context some papers at the beginning of the 1980s, based on metallurgical considerations, regarded 25–30 K as the highest possible value for T_c [2]. Last but not least, the possibility for superconductivity in organic structures was peremptorily rejected by the champion of the A15 materials [20].

Over the subsequent 16 years before the discovery of superconductivity in organic matter an intense activity developed in physics and chemistry communities. A lot of fascinating new compounds have been synthesized by chemists and studied by physicists. A first step following the ideas of Little has been the discovery of an inorganic polymer poly(sulphurnitride), which consists of alternating sulphur and nitrogen atoms covalently bonded in long chains, $(SN)_x$. The interest in this compound becoming a superconductor at 0.26 K [21] did not last very long, first because its T_c could not be significantly raised and second because the band structure did not reveal marked 1D effects. Nearly at the same time a major step has been accomplished in chemistry with the synthesis of the new molecule tetrathiafulvalene, TTF, by Wudl [22], which has deeply influenced the subsequent evolution of the chemistry of organic conductors. This molecule containing four sulphur heteroatoms in the fulvalene skeleton can easily donate electrons when it is combined to electron accepting molecules allowing the synthesis of the first stable organic metal, the charge transfer complex, TTF – TCNQ . The system is made up of two kinds of flat molecules each forming segregated parallel conducting stacks. This compound can be recognized as an organic conductor as the orbitals involved in the conduction (π-HOMO and π-LUMO for TTF and TCNQ, respectively) are associated with the molecule as a whole rather than with a particular atom with carriers in each stacks provided by an interstack charge transfer *at variance* with other organic conductors such as the doped conjugated polymers.

The announcement of a large and metal-like conduction in TTF – TCNQ was made in 1973, simultaneously by the Baltimore [23] and Pennsylvania [24] groups. The Pennsylvania group made a provoking claim announcing a giant conductivity peak of the order of 10^5 $(\Omega\,\mathrm{cm})^{-1}$ at 60 K arising just above a very sharp transition towards an insulating ground state at low temperature. This conductivity peak was attributed by their authors to precursor signs of an incipient superconductor. Unfortunately, the conductivity peak with such a giant amplitude could never be reproduced by other groups who anyway all agreed on the metallic character of this novel molecular material [25]. Besides, the Orsay group showed from X-ray scattering studies that the metal insulator transition was the consequence of the instability of a conducting chain predicted by Peierls [26] and also from the study of transport properties under pressure that the behaviour of the conductivity in the metallic regime could be attributed to collective Fröhlich fluctuations in a 1D regime [27]. Furthermore, the role of 1D electron–electron repulsive interactions has been recognized by the $4k_F$ signature in X-ray diffuse scattering experiments [28]. While high pressure appeared to be a parameter far more influential for organic conductors than for regular metals, it has failed to make TTF – TCNQ sufficiently 3D to prevent the onset of a Peierls instability and enable the stabilization of superconductivity [29].

Since decreasing the size of the Coulombic repulsion is expected to boost the conductivity of metals, other synthetic routes have then been followed. In the 1970s, the leading ideas governing the search for new materials likely to exhibit good metallicity and possibly superconductivity were driven by the possibility to minimize the role of electron–electron repulsions and at the same time to increase the electron–phonon interaction while keeping the overlap between stacks as large as possible. This led to the synthesis of new series of charge transfer compounds which went beyond the known TTF – TCNQ system, for example changing the molecular properties while retaining the same crystal structure. It was recognized that the electron polarizability is important to reduce the screened on-site e–e repulsion and that the redox potential $(\Delta E)^{1/2}$ should be minimized, [30, 31] to fulfill this goal. Hence, new charge transfer compounds with the acceptor TCNQ were synthesized using other heteroatoms for the donor molecule, i.e. substituting sulphur for selenium in the TTF skeleton thus leading to the TSeF molecule. Much attention was put on the tetramethylated derivative of the

TSeF molecule which, when combined to the dimethylated TCNQ gave rise to TMTSF – DMTCNQ. The outcome of this study has been truly decisive for the quest of organic superconductivity [32] since for the first time a metallic state could be stabilized under pressure in an organic compound as a ground state instead of the usual Peierls ground state observed in all the other compounds belonging to the TTF – TCNQ series [33]. In 1979, the Montpellier group synthesized a series of isomorphous radical cationic conductors based on TMTTF/indexTMTTF (the sulphur analogue of the TMTSF molecule) with an inorganic anion [34]. These materials were all insulating at ambient pressure but some of them did show superconductivity under pressure 20 years later and became quite important for the physics of 1D conductors.

At the same time, the Copenhagen group led by Klaus Bechgaard succeeded in the synthesis of a new series of conducting salts all based on the TMTSF molecule namely, $(TMTSF)_2X$ where X is an inorganic mono-anion with various possible symmetry, spherical (PF_6, AsF_6, SbF_6, TaF_6), tetrahedral (BF_4, ClO_4, ReO_4) or triangular (NO_3) [35]. All these compounds but the one with $X = ClO_4$ did reveal an insulating ground state under ambient pressure, Fig. 1.4.

The compound with $X = PF_6$ attracted much attention since the conductivity reaches the value of $10^5 \Omega^{-1} cm^{-1}$ at 12 K with still a strong temperature dependence at this temperature before the onset of an insulating ground state. This behaviour for the transport properties together with the absence of any lattice modulation [36,37], as precursors to the metal–insulator transition were new features in a field still dominated by the Peierls philosophy and stimulated further investigations under pressure suppressing the insulating state at liquid helium temperature under a pressure of about 9 kbar. The finding of a very small and still non-saturating resistivity at 1.3 K was a strong enough motivation to trigger further studies under pressure at even lower temperatures in a dilution refrigerator, which rapidly led to the discovery of a zero resistance state below 1 K. As this zero resistance state was suppressed by a magnetic field, superconductivity was claimed [38]. The discovery of organic

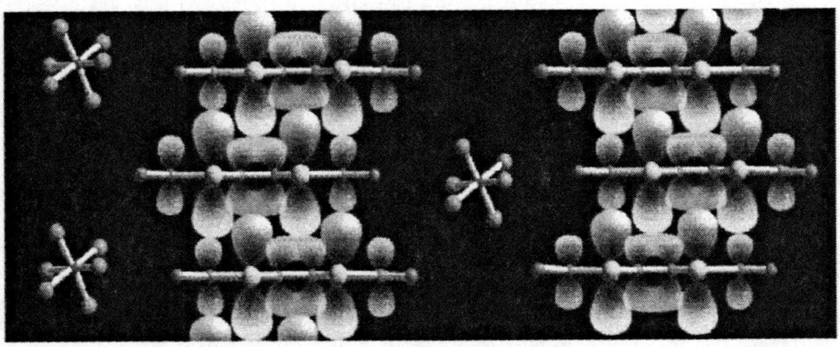

Fig. 1.4. Side view of $(TM)_2X$ conductors

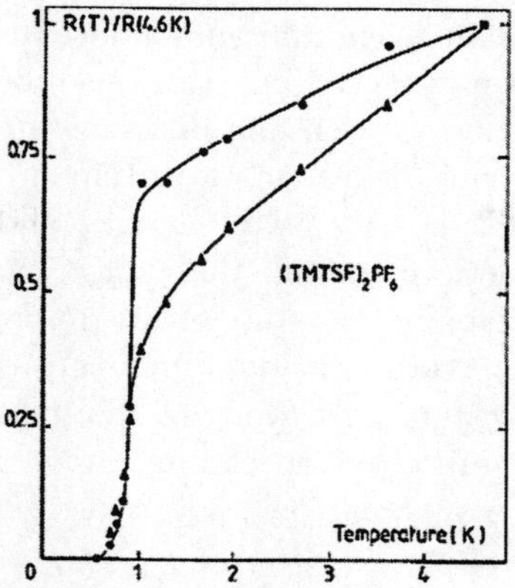

Fig. 1.5. First observation of organic superconductivity in $(TMTSF)_2PF_6$ under pressure [38]. The resistance of two samples is displayed

superconductivity in $(TMTSF)_2PF_6$ came out 15 years after the publication of Little and 10 years after the holding of an international symposium organized by Little at Hawai on the physical and chemical problems of possible organic superconductors [12]. Shortly after the discovery of superconductivity in $(TMTSF)_2PF_6$ [38], several other isostructural compounds with a variety of anions have been found superconducting in the vicinity of 1 K in the 10 kbar pressure domain [39]. Among them, $(TMTSF)_2ClO_4$ is the only exception becoming superconducting at low temperature without the need of a high pressure [40].

In addition, it has been realized that the highly insulating isostructural compound $(TMTTF)_2PF_6$, the sulphur counterpart of $(TMTSF)_2PF_6$, can be made to superconduct at low temperatures provided a pressure of 45 kbar is applied [41]. As a result, a unified $(TM)_2X$ phase diagram has been proposed experimentally [42]. This diagram suggests that apparently so different systems such as $(TMTSF)_2ClO_4$, superconducting under ambient pressure, and the strongly insulating $(TMTTF)_2PF_6$ belong to the same class of materials, i.e. the physical properties of the latter can be made equivalent to those of the former provided a large enough pressure is applied.

Moreover, according to their crystal structure these 1D conductors could potentially be prototypes of Luttinger conductors since the kinetic coupling between molecular stacks is very small compared to the longitudinal dispersion [43]. However, the commensurate filling of the conduction band, a consequence of the stoichiometry, makes the conduction band half or quarter-filled. Consequently, even a moderate value of the electron–electron repulsion of the order of the bandwidth opens a correlation gap at the Fermi level due to the half or quarter-filled electron–electron Umklapp scattering [44, 45] and all members of the $(TM)_2X$ family are actually Mott insulators with a gap decreasing from sulphur to selenium compounds. In addition, disproportionation reactions have been observed in several members of the $(TM)_2X$ series leading to a charge ordering with the onset of an insulating state clearly evidenced by NMR [46] and dielectric measurements [47].

When the Mott gap becomes small enough (under pressure) to compete with the transverse kinetic coupling, the charge carriers lose their 1D confinement and their behaviour in temperature approaches what is expected in Fermi liquids namely, a quadratic temperature dependence for the resistivity. However, the electron excitations of this "Fermi liquid" retain a low energy gap in the far infra red spectrum carrying most of the oscillator strength together with a very narrow and intense zero frequency peak in the conductivity [48, 97].

Among the questions which are still of great current interest for 1D organic superconductors is the nature of the superconducting coupling [50, 51]. Is it a spin singlet as for regular superconductors and high T_c compounds or a triplet coupling as this has been shown to be the case for the superfluidity of ^3He [52, 53] or some heavy fermion superconductors [54, 55], and ruthenates [56]. A recent investigation of the Knight shift and of the spin–lattice relaxation has shown that the superconducting phase of $(TMTSF)_2ClO_4$ is a spin-singlet state [57, 58]. Together with the existence of nodes for the superconducting gap [58] these recent data strongly suggest that the symmetry of the superconducting wave function is singlet-paired with a d-wave orbital symmetry.

It is also clear that a precursor superconductivity regime exists over an extended temperature domain above T_c, although the controversy about its origin (low-dimensional fluctuations, pseudo-gap, etc.) raised 25 years ago is still going on.

1.2 Two-Dimensional Conductors

In the beginning of the 1980s, a new derivative of the fulvalene molecule has been discovered and this opened a new series of organic superconductors. This is BEDT − TTF, an elongated version of TTF also called ET which has given a salt with ReO_4 becoming superconducting under pressure [59]. The crystal structure of $(BEDT-TTF)_2ReO_4$ bears much similarity with the molecular packing of the $(TMTSF)_2X$ series and shows a pronounced 1D character.

One year later, the same ET molecule contributed to the elaboration of half-filled band organic conductors with quite novel structures namely, materials displaying a two-dimensional (2D) conducting (or layer) structure. The first representatives of this 2D class of organic materials were the $(ET)_2X$ compounds where X is a linear triatomic anion such as I_3, IBr_2, AuI_2, etc. The salt with $X = I_3$ is particularly interesting as the β_H-phase, which can be stabilized at low temperature after a special pressure–temperature cycling, has provided a large increase of the superconducting T_c of the organics from 1–2 K in the $(TMTSF)_2X$ series up to 8 K [60–62]. When $(ET)_2I_3$ is cooled rapidly without any pressure treatment, random disorder of the ethylene groups of the ET molecules remains in the β_L-phase and strongly reduces the stability of the superconducting state with a T_c dropping down to 1.4 K [63] in agreement with the role of non-magnetic defects on the superconductivity of 1D organics [58].

In addition, $\beta_H(ET)_2I_3$ has provided a textbook example for Shubnikov–de Haas oscillations in a 2D conductor with extraordinarily large oscillations of the magnetoresistance [64]. Such a study has also enabled the determination of the overlap integral between conducting planes of 0.5 meV, a clear-cut illustration for the pronounced 2D character and the existence of angular magnetoresistance oscillations due to the c-axis warping of the Fermi surface in these ET conductors [65, 66].

A further increase in the T_c above 10 K has been accomplished in the 1990s with the discovery of the κ-phase salts namely, $\kappa - (ET)_2X$ with polymeric anions, X = $Cu(NCS)_2$ [67], $Cu(N(CN)_2)Br$ [68], and even up to 12.5 K in $Cu(N(CN)_2)Cl$ under a modest pressure of 0.3 kbar [69].

Most κ-phase salts show an isostructural face to face molecular dimers packing forming a 2D checker-board pattern. Very much like the compounds belonging to the $(TM)_2X$ family, most κ-phase systems can be gathered together in a generic temperature–pressure (chemistry) phase diagram [70]. The compound X = $Cu(N(CN)_2)Br$ is thus of particular interest since superconductivity at $T_c = 11.6$ K at ambient pressure makes NMR studies in the superconducting state accessible over a decade in temperature. ^{13}C Knight shift and relaxation studies in a κ-$Cu(N(CN)_2)Br$ single crystal have shown conclusively that the pairing in 2D organic superconductors must be spin singlet with the existence of nodes in the gap [71–73]. Among all compounds κ-Cl is also particularly interesting as it is the prototype in the series showing the complete sequence of states, namely, paramagnetic Mott insulating, antiferromagnetic, metallic and superconducting states within a few hundred bars. Magnetic [74], acoustic [75] and transport experiments [76, 77] have shown the existence of a first-order transition line in the T–P plane with a phase coexistence regime ending at a Mott critical point.

The study of this material has also enabled to illustrate the concept of bad metal, namely, a strongly correlated metal in the neighbourood of a Mott transition [78]. A recent reinvestigation of this compound suggests that the critical regime around the critical end point does not appear to obey the 2D Ising model universality class [79] *at variance* with a similar study of the 3D Mott transition performed in the vanadium sesquioxide [80].

Another aspect of the 2D organics is the experimental realization of the Jaccarino–Peter mechanism for superconductivity in systems comprising magnetic ions [81]. This mechanism is probably the one which allows the stabilization of superconductivity in $(BETS)_2FeCl_4$ (where BETS is the selenide parent molecule of ET) under intense magnetic field [82,83] compensating the exchange field of aligned Fe^{3+} ions.

1.3 Conclusion

This short presentation for the search for superconductivity in organic matter has been based on an historical approach because it can be instructive to see how the whole story has developed from quite provocative theoretical statements and led finally to the discovery of the first organic superconductor thanks to a very productive work between chemistry and physics communities over the past 35 years. It is worth citing Ginzburg in this conclusion [84]: *I believe that it was undoubtedly the discussion of the possible exciton mechanism of superconductivity that stimulated the search for such superconductors and studies of them* (organic superconductors and intercalated layered superconductors). In addition, the other statement of Ginzburg: *the organic superconductors are clearly interesting by themselves or, to be more precise, irrespective of the high T_c problem* cannot be more appropriate. While the actual mechanism behind organic superconductivity remains an enigma one can be confident saying that the model proposed by Little in 1964 is probably not the one applying to those materials.

Organic conductors (and superconductors) have shown their superiority compared to high T_c cuprates in terms of purity and in the variety of phenomena which can be studied in a single system keeping both structure and chemical purity constant thanks to the determining role of high pressure governing their physical properties. Recently, it was even believed that the entire physics of organics including superconductivity could be reproduced without the burden of high pressure techniques tuning the carrier concentration with a field effect technique. Unfortunately, the numerous articles published on this subject cannot be cited anymore. They do not represent the most brilliant days of condensed matter physics.

A perfect illustration for the scientific interest of organic conductors is provided by the amount of knowledge and new physical phenomena which came out of the studies of organic superconductors. To summarize, the optimized location for the superconductivity T_c close to the Mott localization resulting from a compromise between the interchain coupling and the strength of the correlations and the likeliness of an unconventional pairing mechanism which remains a challenging topic and awaits new experimental evidences. It is also important to mention the magnetic field confinement discovered in the 1Ds leading to the phenomenon of field-induced spin density wave phases [85–87] and the quantization of the Hall effect in these phases [88–90]. Furthermore, the angular-dependent magnetoresistance specific to these anisotropic conductors [91,92], the so-called Lebed oscillations in $(TMTSF)_2X$ provide a nice illustration for the new features of the quasi-1D conductors [93].

The 2D compounds have provided not only higher values of T_c and a textbook example for the 2D fermiology but also a playground for the study of the 2D Mott transition as well as the metallic phase in its vicinity.

At the end of this historical perspective of organic superconductors I wish to recall, in particular, some of my close colleagues who have played a pioneering role in this research, Friedel who, 40 years ago, pointed out the relevance of low dimensionality (1D and 2D) in superconductivity and has continuously supported the activity in 1D conductors at Orsay, Weger who introduced TTF – TCNQ to the physical investigations under pressure at Orsay in 1974, Bechgaard with whom the active cooperation established in the 1970s is still very much alive, Ribault who understood the importance of linking high pressure and very low temperature measurements. Brown has contributed to solve the controversy about organic superconductivity. Schulz, Giamarchi and Bourbonnais are all close theoreticians colleagues who made significant contributions to the 1D theory and who have been deeply involved in a tight cooperation with experimentalists.

References

1. H.K. Onnes, Proc. Akad. Wetensch., **14**, 113 (1911)
2. J. Hulm, B. Matthias, Science, **208**, 881 (1980)
3. H.N. McCoy, W.C. Moore, J. Am. Chem. Soc., **33**, 273 (1911)
4. F. London, J. Chem. Phys., **5**, 837 (1937)
5. H. Akamatsu, H. Inokuchi, Y. Matsunaga, Nature, **173**, 168 (1954)
6. W.A. Little, Phys. Rev. A **134**, 1416 (1964)
7. W.A. Little, Sci. Am., **212**, 21, 1965
8. J. Bardeen, L.N. Cooper, J.R. Schrieffer, Phys. Rev., **108**, 1175 (1957)
9. H. Fröhlich, Proc. R. Soc. A **223**, 296 (1954)
10. V.L. Ginzburg, J. Exp. Theor. Phys., **47**, 2318 (1964)
11. V.L. Ginzburg, Phys. Lett., **11**, 101 (1964)
12. W.A. Little, J. Polym. Sci. C **29**, 17 (1970)
13. R.E. Peierls, in *Quantum Theory of Solids.* (Oxford University Press, London, 1955)
14. R. Comès, M. Lambert, H. Launois, H.R. Zeller, Phys. Rev. B **8**, 571 (1973)
15. Yu.A. Bychkov, L.P. Gorkov, I.E. Dzyaloshinskii, Sov. Phys. JETP, **23**, 489 (1966)
16. G.F. Hardy, J.K. Hulm, Phys. Rev., **93**, 1004 (1954)
17. J. Labbé and J. Friedel, J. Phys., **27**, 153 (1966)
18. M. Weger, I.R. Goldberg, in: *Solid State Physics*, vol. 28 (Academic, New York, 1973)
19. J. Labbé, S. Barisić, J. Friedel, Phys. Rev. Lett., **19**, 1039 (1967)
20. B.T. Matthias, Phys. Today, **23**, 27 (1971)
21. R.L. Greene, G.B. Street, L.J. Suter, Phys. Rev. Lett., **34**, 577 (1975)
22. F. Wudl, G.M. Smith, E.J. Hufnagel, J. Chem. Soc. Chem. Commun., 1453 (1970)
23. J. Ferraris, D.O. Cowan, W. Walatka, J.H. Perlstein, J. Am. Chem. Soc, **95**, 948 (1973)
24. L.B. Coleman, M.J. Cohen, D.J. Sandman, F.G. Yamagishi, A.F. Garito, A.J. Heeger, Solid State Commun., **12**, 1125 (1973)
25. G.A. Thomas: et al. Phys. Rev., B **13**, 5105 (1976)
26. F. Denoyer, R. Comès, A.F. Garito, A.H. Heeger, Phys. Rev. Lett., **35**, 445 (1975)

27. R.H. Friend, M. Miljak, D. Jérome, Phys. Rev. Lett., **40**, 1048 (1978)
28. J.P. Pouget, S.K. Khanna, F. Denoyer, R. Comès, A.F. Garito, A.J. Heeger, Phys. Rev. Lett., **37**, 437 (1976)
29. B. Horovitz, H. Gutfreund, M. Weger, Phys. Rev., B **12**, 3174 (1975)
30. A.F. Garito, A.J. Heeger, Acc. Chem. Res., **7**, 232 (1974)
31. E.M. Engler, B.A. Scott, S. Etemad, T. Penney, V.V. Patel, J. Am. Chem. Soc., **99**, 5909 (1977)
32. J.R. Andersen, K. Bachgaard, C.S. Jacobsen, G. Rindorf, H. Soling, N. Thorup, Acta Cryst., B **34**, 1901 (1978)
33. A. Andrieux, P.M. Chaikin, C. Duroure, D. Jérome, C. Weyl, K. Bechgaard, J.R. Andersen, J. Phys. Paris, **40**, 1199 (1979)
34. J.L. Galigne, B. Liautard, S. Peytavin, G. Brun, J.M. Fabre, E. Torreilles, L. Giral, Acta Crystallogr. B, **34**, 620 (1978)
35. K. Bechgaard, C.S. Jacobsen, K. Mortensen, H.J. Pedersen, N. Thorup, Solid State Commun., **33**, 1119 (1979)
36. J.P. Pouget, Chem. Scr., **55**, 85 (1981)
37. J.P. Pouget, S. Ravy, Synth. Metals, **85**, 1523 (1997)
38. D. Jérome, A. Mazaud, M. Ribault, K. Bechgaard, J. Phys. (Paris) Lett., **41**, L95 (1980)
39. S.S. Parkin, M. Ribault, D. Jérome, K. Bechgaard, J. Phys., C **14**, L-445 (1981)
40. K. Bechgaard, M. Carneiro, M. Olsen, F.B. Rasmussen, Phys. Rev. Lett., **46**, 852 (1981)
41. H. Wilhelm, D. Jaccard, R. Duprat, C. Bourbonnais, D. Jérome, J. Moser, C. Carcel, J.M. Fabre, Eur. Phys. J., B **21**, 175 (2001)
42. D. Jérome, Science, **252**, 1509 (1991)
43. C. Bourbonnais, D. Jérome, Science, **281**, 1156 (1998)
44. V.J. Emery, R. Bruinsma, S. Barisić, Phys. Rev. Lett., **48**, 1039 (1982)
45. T. Giamarchi, Physica, B **230–232**, 975 (1997)
46. D.S. Chow, F. Zamborsky, B. Alavi, D.J. Tantillo, A. Baur, C.A. Merlić, S.E. Brown, Phys. Rev. Lett., **85**, 1698 (2000)
47. F. Nad, P. Monceau, C. Carcel, J.M. Fabre, Phys. Rev., B **62**, 1753 (2000)
48. N. Cao, T. Timusk, K. Bechgaard, J. Phys. (France), **6**, 1719 (1996)
49. A. Schwartz, M. Dressel, G. Grüner, V. Vescoli, L. Degiorgi, T. Giamarchi, Phys. Rev., B **58**, 1261 (1998)
50. A.G. Lebed, K.Machida, M. Osaki, Phys. Rev., B **62**, R795 (2000)
51. J.C. Nickel, R. Duprat, C. Bourbonnais, N. Dupuis, Phys. Rev. Lett., **95**, 247001 (2005)
52. D.D. Osheroff, W.J. Gully, R.C. Richardson, D.M. Lee, Phys. Rev. Lett., **29**, 920 (1972)
53. A.J. Leggett, Rev. Mod. Phys., **47**, 331 (1975)
54. R.H. Heffner, J.L. Smith, P. Birrer, J.O. Willis, C. Baines, F.N. Gygax, B. Hitt, E. Lippelt, H.R. Ott, A. Schenck, E.A. Knetsch, J.A. Mydosh, D.E. MacLaughlin, Phys. Rev. Lett., **65**, 2816 (1990)
55. L. Taillefer, J. Flouquet, G.G. Lonzarich, Physica (Amsterdam), B **169**, 257 (1991)
56. G.M. Luke, Y. Fudamoto, K.M. Kojima, M.I. Larkin, J. Merrin, B. Nachumi, Y.J. Uemura, Y. Maeno, Z.Q. Mao, Y. Mori, H. Nakamura, M. Sigrist, Nature, **394**, 558 (1998)
57. J. Shinagawa, Y. Kurosaki, F. Zhang, C. Parker, S.E. Brown, D. Jérome, J.B. Christensen, K. Bechgaard, Phys. Rev. Lett., **98** (2007)

58. N. Joo, P. Auban-Senzier, C. Pasquier, D. Jérome, K. Bechgaard, Eur. Phys. Lett. (2005)
59. S.S.P. Parkin, J.C. Scott, J.B. Torrance, E.M. Engler, J. Phys. (Paris) Coll., **44**, 1111 (1983)
60. V.N. Laukhin, E.E. Kostyuchenko, Yu. Sushko, I.F. Shchegolev, E.B. Yagubskii: JETP. Lett., **41**, 81 (1985)
61. K. Murata, M. Tokumoto, H. Anzai, H. Bando, G. Saito, K. Kajimura, T. Ishiguro, J. Phys. Soc. Jpn., **54**, 2084 (1985)
62. F. Creuzet, G. Creuzet, D. Jérome, D. Schweitzer, H.J. Keller, J. Phys. Lett., **46**, L-1079 (1988)
63. E.B. Yagubskii, I.F. Shchegolev, V.N. Laukhin, P.A. Kononovich, M.V. Kartsovnik, A.V. Zvarykina, L.I. Bubarov, JETP Lett., **39**, 12 (1984)
64. W. Kang, G. Montambaux, J.R. Cooper, D. Jérome, P. Batail, C. Lenoir, Phys. Rev. Lett., **62**, 2559 (1989)
65. M.V. Kartsovnik, P.A. Kononovich, V.N. Laukhin, I.F. Shchegolev, JETP Lett., **48**, 541 (1988)
66. K. Yamaji, J. Phys. Soc. Jpn., **58**, 1520 (1989)
67. H. Urayama, H. Yamochi, G. Saito, K. Nozawa, T. Sugano, M. Kinoshita, S. Sato, K. Oshima, A. Kawamoto, J. Tanaka, Cem. Lett., **55** (1988)
68. A.M. Kini et al., Inorg. Chem., **29**, 2555 (1990)
69. J.M. Williams et al., Inorg. Chem., **29**, 3262 (1990)
70. K. Kanoda, Physica C, **282**, 299 (1997)
71. H. Mayaffre, P. Wzietek, D. Jérome, C. Lenoir, P. Batail, Phys. Rev. Lett., **75**, (1995)
72. S.M. De Soto, C.P. Slichter, A.M. Kini, H.H. Wang, U. Geiser, J.M. Williams, Phys. Rev., **52**, 10364 (1995)
73. K. Kanoda, K. Miyagawa, A. Kawamoto, Y. Nakazawa, Phys. Rev., B **54** (1996)
74. S. Lefebvre, P. Wzietek, S. Brown, C. Bourbonnais, D. Jérome, C. Méziere, M. Fourmigué, P. Batail, Phys. Rev. Lett., **85**, 5420 (2000)
75. D. Fournier, M. Poirier, M. Castonguay, M. Truong, Phys. Rev. Lett., **90**, 127002 (2003)
76. P. Limelette, P. Wzietek, S. Florens, A. Georges, T.A. Costi, P. Pasquier, D. Jérome, C. Méziere, P. Batail, Phys. Rev. Lett., **91**, 016401 (2003)
77. F. Kagawa, T. Itou, K. Miyagawa, K. Kanoda, Phys. Rev. B, **69** 064511 (2004)
78. A. Georges, G. Kotliar, W. Krauth, M. Rosenberg, Rev. Mod. Phys., **68**, 13 (1996)
79. F. Kagawa, K. Miyagawa, K. Kanoda, Nature, **436**, 534 (2005)
80. P. Limelette, A. Georges, D. Jérome, P. Wzietek, P. Metcalf, J.M. Honig, Science, **302**, 89 (2003)
81. V. Jaccarino, M. Peter, Phys. Rev. Lett., **9**, 290 (1962)
82. S. Uji, H. Shinagawa, C. Terakura, T. Terashima, T. Yakabe, Y. Terai, M. Tokumoto, A. Kobayashi, H. Tanaka, H. Kobayashi, Nature, **410**, 908 (2001)
83. L. Balicas, J.S. Brooks, K. Storr, S. Uji, M. Tokumoto, K. Kobayashi, A. Kobayashi, H. Tanaka, V. Barzykin, L.P. Gorkov, Phys. Rev. Lett., **87**, 067002 (2001)
84. V.L. Ginzburg, Progress Low Temp. Phys. **12**, 1 (1989)
85. M. Ribault, D. Jérome, J. Tuchendler, C. Weyl, K. Bechgaard, J. Phys. Lett., **44**, L-953 (1983)
86. P.M. Chaikin, Mu-Yong Choi, J.F. Kwak, J.S. Brooks, K.P. Martin, M.J. Naughton, E.M. Engler, R.L. Greene, Phys. Rev. Lett., **51**, 2333 (1983)

87. L.P. Gorkov, A.G. Lebed, J. Phys. (Paris) Lett., **45**, L433 (1984)
88. J.R. Cooper, W. Kang, P. Auban, G. Montambaux, D. Jérome, K. Bechgaard, Phys. Rev. Lett, **65**, 1984 (1989)
89. S.T. Hannahs, J.S. Brooks, W. Kang, L.Y. Chiang, P.M. Chaikin, Phys. Rev. Lett., **63**, 1988 (1989)
90. M. Héritier, G. Montambaux, P. Lederer, J. Phys. (Paris) Lett., **45**, L943 (1984)
91. T. Osada, A. Kawasumi, S. Kagoshima, N. Miura, G. Saito, Phys. Rev. Lett., **66**, 1525 (1991)
92. M.J. Naughton, O.H. Chung, M. Chaparala, X. Bu, P. Coppens, Phys. Rev. Lett., **67**, 3712 (1991)
93. A.G. Lebed, JETP. Lett., **43**, 174 (1986)

2

From Sliding Charge Density Wave to Charge Ordering

P. Monceau

I will present a personal view how in the last 30 years a backward and forward movement in concepts and experimental results took place in quasi-1D inorganic and organic conductors, taking for examples sliding density waves, low energy excitations, metastability and memory effects, and dielectric response leading to the recent results of charge ordering.

Any graduate student starting the study of quantum theory of solids, as I did using the Kittel book, knows that the evaluation of crystal vibrations, periodic boundary conditions and electron density of states are the simplest for a linear chain, before being generalized in three dimensions.

However, the first synthesis of one-dimensional or quasi-one-dimensional (Q-1D) compounds was only realized in the beginning of the 1970s. Then, all concepts developed earlier could face their experimental proofs. Among these concepts, R.E. Peierls in 1955 showed the instability of a 1D metal interacting with the lattice towards a lattice distortion and the opening of a gap in the electronic spectrum – the so-called charge density wave (CDW), a concomitant lattice and electronic modulation with the same wave vector $q = 2k_{\mathrm{F}}$. Related to this CDW state, H. Fröhlich in 1954, just before the BCS theory, described a model in which the CDW can slide if its energy is degenerate along the chain axis, yielding thus a collective current without dissipation and leading to a superconducting state. In the 1960s Fermi instabilities were already considered by A.W. Overhauser to yield a SDW in the case of chromium.

The decade 1973–1982 can be considered as a "magic" decade for the 1D physics with the first synthesis of Q-1D materials and of many other families, discovery of the Peierls instability in many of these compounds, CDW and SDW sliding, organic superconductivity,.... Naturally I will not forget other fundamental discoveries in adjacent fields as the quantum hall effect (QHE) which is also connected with transport properties of some organic Q-1D.

Having been asked by the editor of this volume to write a historic survey of Q-1D materials, everybody will agree to consider impossible to survey so many research activities in a few pages. So, I will present my own point of view and how, among a great number of talentuous researchers, I have personally with many collaborators followed a research line along all these years. In the following, I will not make any reference to publications, except for the five figures.

After having obtained a Ph.D. in 1970 on microwave studies of type II superconductors, G. Waysand (Orsay) attracted our attention on publications of the research group of T.H. Geballe (Stanford) on superconductivity in layered compounds of transition metals dichalcogenides and on possible superconducting fluctuations at high temperatures (around 35 K) in intercalated 2H-TaS_2. Microwave is a good technique for searching such fluctuations and thus we started to interact with J. Rouxel, having recently settled his laboratory at Nantes. In the same period of time, crystals of 1D materials were just synthesized: first KCP, then TTF–TCNQ. Very soon by diffuse X-ray scattering and neutron scattering the nature of the metal-insulating transition was elucidated as being of the Peierls type. A huge peak in the conductivity just above the Peierls transition temperature of TTF–TCNQ raised a gigantic activity as being possibly due to some superconducting fluctuations. Among all the theories, J. Bardeen in 1973 revitalized the Fröhlich model of superconductivity.

Superstructures analysed as a Fermi effect with nesting properties of the Fermi surface were also discovered in the same year in transition metal dichalcogenides. By modification of the temperature conditions for growing $NbSe_2$, synthesis of a new fibre-like compound was realized at Nantes. The crystal structure made by A. Meerchaut (1975) revealed that this compound, $NbSe_3$, was formed of linear chains – six in the unit cell – running along the b-axis.

The first measurements we made on a bundle of fibres showed anomalies in the resistance at 145 and 59 K, change of the magnetic susceptibility at these temperatures, a peak in the specific heat and the decrease of the transition temperatures with application of pressure. These first results were accepted as a post-dead line publication at the 14th Low Temperature Conference held in Helsinki in the summer 1975.

At the end of 1975, I joined the group de A. Portis at Berkeley where N.P. Ong was measuring the Hall effect of TTF–TCNQ with a bimodal cavity to be rid off any contact effect which could disturb the measurements. The result of conductance of $NbSe_3$ we performed at 9.3 GHz is shown in Fig. 2.1. Astonishingly, the amplitude of the DC resistance peaks are nearly washed out. To corroborate these microwave studies, we performed DC and pulsed measurements with different current densities the results of which demonstrate the non-linearity in transport properties. The microwave results, in fact, make rememberings to similar results on type II superconductors in the vortex state at frequencies above the pinning frequency where the microwave resistivity is the same as that of the DC flux flow one.

It was J. Bardeen who, in 1978, assigned our results on $NbSe_3$ to the Fröhlich type conductivity resulting from sliding CDW (in his contribution to the book *Physical Properties in Highly One-Dimensional Conductors* edited

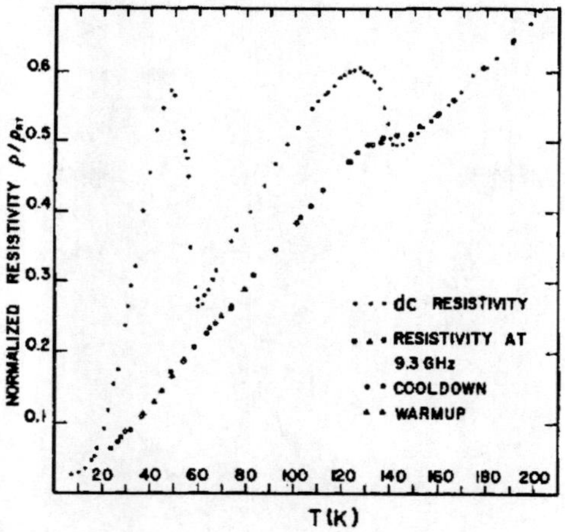

Fig. 2.1. Resistivity of $NbSe_3$ at 9.3 GHz compared to the DC resistivity as a function of temperature (from [1])

by T. Devreese, and in the concluding remarks in the Proceedings of the conference on 1D at Dubrovnick in September 1978).

In the mean time, CDW was also discovered in another trichalcogenide TaS_3 with two polytypes, CDW superstructures observed in $NbSe_3$ by electron diffraction. It was also shown that the superlattice peaks are not suppressed in the non-linear state, excluding the destruction of the CDW order towards the normal state (at least for the amplitude of the electric fields experimentally applied), as in the mixed state of type II superconductors where the normal state is recovered above H_{c_2}. Sliding occurs only for an applied electric field above a threshold value at which the CDW is depinned from impurities. Above this threshold, an AC voltage called NBN is generated.

Pursuing the analogy with superconductors, and inspired by the work on vortices on AC–DC interference showing Shapiro steps in the I–V characteristics, we performed similar experiments. Interference takes place when there is a locking between the RF field and the frequencies measured in the NBN and it is seen as an increase of the resistance of the sample for those frequencies, as shown in Fig. 2.2.

We associated these observations with the motion of the CDW. In the Fröhlich model, the lattice distortion moves with the electrons when the electrons are displaced in the k-space, so as to give a current flow $j = nev$, where n is the number of electrons condensed under the gap and v the velocity of the CDW. Assuming that the pinning forces are periodic with the phase of the CDW – thus the motion due to an external field is the superposition of a continuous drift and a modulation due to pinning at a recurrence frequency $\nu = (Q/2\pi)v$ with Q the CDW distortion vector – we showed the linear variation between the excess current carried by the CDW when sliding and the frequency ν where interference occurs, demonstrating the validity of the Fröhlich conductivity [2].

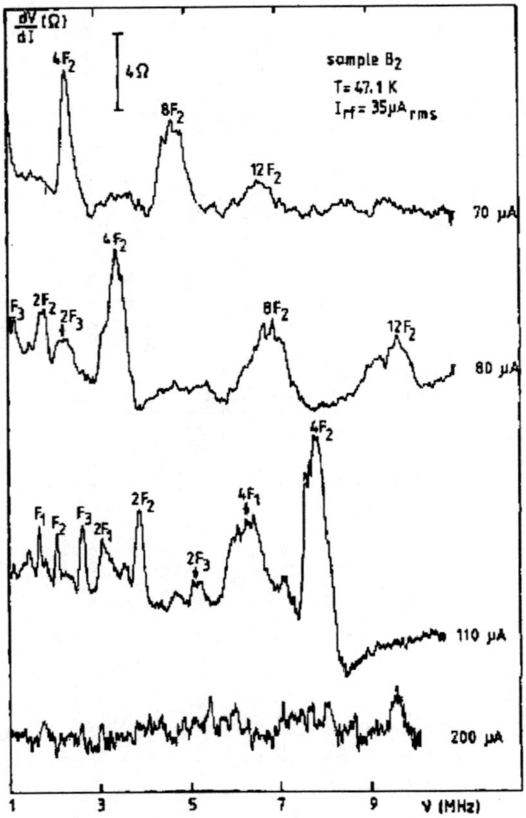

Fig. 2.2. Differential resistance dV/dI of $NbSe_3$ at 47 K in sweeping the frequency of a RF current of constant amplitude superposed to the DC current above the threshold value $-I_{th} = 59\,\mu A$ (from [2])

New CDW compounds were discovered: $(TaSe_4)_2I$, potassium and rubidium bronze $A_{0.3}MoO_3$ with A = K, Rb, $(NbSe_4)_{10}I_3$ with the Peierls transition above room temperature. A huge amount of theoretical and experimental works were performed for studying the electrodynamics of the CDW in a very broad frequency range in all these compounds, optical properties, metastability and hysteresis, screening effect, pinning effects by well-defined impurities, complete mode-locking, synthesis of high purity crystals, size effects,.... Similar depinning with very low threshold fields was also discovered in the SDW state of Bechgaard salts, $(TMTSF)_2X$, recovery of the SDW state by application of a magnetic field (field-induced SDW) on a pressured sample, plateaux in the Hall resistance,.... 1D materials are characterized by dominant valence forces between atoms along the chains. They act as restoring forces for bending. The deviation of the T^3 law in the specific heat at low T above T^* yields an estimation for the ratio between the bending force constant and the force constant between adjacent chains. Measuring this property in 1982, that was also the starting point for searching the deviation of tranverse acoustic mode dispersion using neutron scattering. At variance with $(TaSe_4)_2I$ and $A_{0.3}MoO_3$ where large crystals are available, the search for Kohn anomaly was performed only recently in $NbSe_3$ using X-ray inelastic scattering.

After many others, we measured the dielectric permittivity of CDWs, essentially TaS_3 and we started later with SDW. Analysis of the T dependence of the real part and imaginary parts of the low frequency conductance showed that the relaxation rate has two branches: a long time one diverges near some temperature while the short time relaxation increases monotically. We ascribed these two relaxations as α and β relaxations, providing evidence for a transition of the CDW and the SDW state into a glassy-like transition at low temperature.

An appropriate technique, among others, to study the disordered state of the charge and spin DW is very low T specific heat measurements. Thus, for CDW and SDW systems we showed that, as the temperature decreases, the total specific heat first follows a T^3 behaviour due to phonons, followed by a minimum and then at low temperatures by a T^{-2} behaviour. Energy relaxation has revealed aging in the heat response in the sense that the temperature signal $T(t_\omega, \tau)$ measured at a time $t_\omega + \tau$ depends on the duration of the heat perturbation or waiting time, as well as on the time τ elapsed since t_ω, as shown in Fig. 2.3 for TaS_3.

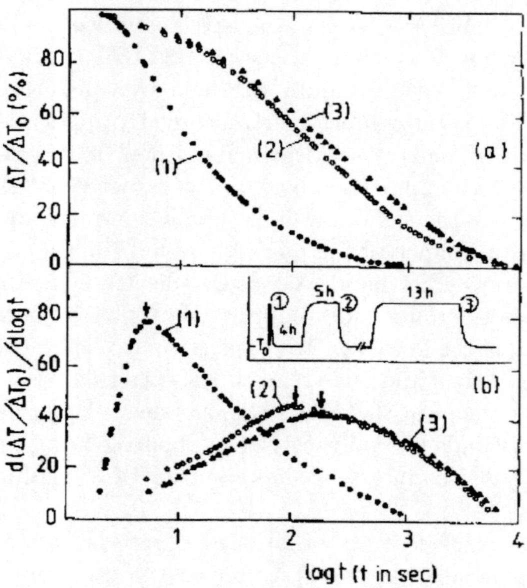

Fig. 2.3. (**a**) Variation of the temperature dependence $\Delta(T)/\Delta_0(T)$ as a function of log t for TaS_3 at $T_0 = 0.165$ K: after a heat pulse of 0.9 s (•), after that a heat flow has been applied during 5 h (13 h) and switched off (o and ▲, respectively). (**b**) Time dependence of the relaxation rate $d\Delta(T)/\Delta_0(T)/d \log t$ in the same conditions than in (**a**) (from [3])

Once the T^3 and the T^{-2} contributions are subtracted the residual specific heat follows a power law dependence. The amplitude of the T^{-2} term strongly increases when the waiting time increases. The spectrum of relaxation times shows a power-law distribution for intermediate waiting times, and interrupted aging for larger waiting times (in the sense that thermodynamical equilibrium is reached and that $T(t_\omega, \tau)$ does not depend on t_ω. Finally commensurate systems relax faster than incommensurate systems. Metastability due to bisolitons occurs at a sufficiently strong pinning potential from what it is possible to define an energy two-level system with a ground state separated from a metastable state by an energy barrier, explaining the T^{-2} contribution to the specific heat as the high temperature tail of a Schottky anomaly of the effective two-level system.

New techniques often developed for high T_c superconductors were applied to one-dimensional systems, in particular ARPES, high resolution diffraction using synchrotron radiation, ultra-vacuum STM, photo-induced effects, tunneling,....

The interlayer tunneling method has been very useful since the last decade in studies of layered HTS materials. Recently this method has been extended to other classes of layered materials as manganites and CDWs. In $NbSe_3$ and TaS_3, the elementary conducting layers are formed by elementary conducting chains assembled in a layer well separated from each other by a double barrier of insulating prism bases. By the interlayer tunneling technique on mesoscopic stacked junctions fabricated by ion focused beam, we have identified [4] the CDW gap and a zero bias conductance peak (ZBCP). The CDW gap values found are consistent with the data obtained by other techniques as STM, optics, ARPES and point contact spectroscopy. By application of a high magnetic field parallel to the layers which narrows the ZBCP, we found, in addition with the peak at 2Δ an additional peak in the gap at $V = 2\Delta/3$ [Fig. 2.4]. Due to the degeneracy of the CDW ground state, a non-uniform ground state can be realized by local change of the phase by π and simultaneous acceptance of one electron from the free band to conserve electro-neutrality. The resulting state is known as the amplitude soliton (AS). The AS energy $E_s = 2\Delta/\pi = 0.65\Delta$ is lower than the lowest energy of electron in the free band Δ. Therefore free electrons near the band edge tend to self-localized into AS states. Experimentally, the existence of the AS states has been reliably demonstrated only for dimeric compounds like polyacetylene. Then the additional peak in interlayer tunneling has been interpreted as resulting of transition of band carriers into AS levels.

Another remarkable feature is that onset of interlayer tunneling conductivity occurs above a sharp voltage V_t at low energies within the CDW gap. We ascribed V_t to the energy of the CDW phase decoupling between neighboring layers via the formation in the weakest junction of an array of dislocation lines (DL) – CDW phase topological defects. The first DL enters the junction at V_t and the staircase structure above V_t evidences the sequential entering of these CDW vortices in the junction area. There is a remarkable similarity

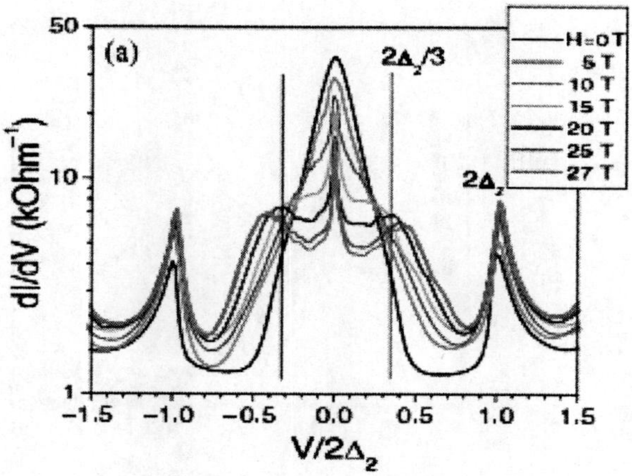

Fig. 2.4. Spectra of the tunneling conductivity dI/dV of $NbSe_3$ stacks at $T = 4.2$ K for the lower CDW at various magnetic fields applied parallel to the c-axis. The voltage V is normalized to the CDW gap $2\Delta_2$. The zero bias anomaly around $V = 0$ is strongly narrowed by the application of a high magnetic field, revealing the intergap peak at $(2/3)\Delta_2$ ascribed to amplitude solitons (from [4])

between layered superconducting and CDW systems that manifests itself in similar mechanisms of phase decoupling via formation of phase vortices. In both cases a threshold energy for phase decoupling associated with H_{c_1} for superconductors or V_t for CDW is much less than the value of the energy gap.

Low dimensional charge transfer compounds are known to be subject of strong electronic correlations. One of the best examples is found in the Bechgaard–Fabre salts. In the course of our thermodynamic measurements, we found that the magnitude of the anomaly in the specific heat at the SDW transition temperature of $(TMTSF)_2X$ salts was too large to be explained by the electron spin contribution alone, implying that the lattice is involved. Our original idea was to detect in the dielectric permittivity any contribution of the diffuse one-dimensional $2k_F$ scattering revealed previously in $(TMTTF)_2PF_6$ and $(TMTTF)_2Br$ which grows critically below 70 K. The huge surprise was to measure in the conductivity the opening of a new charge gap and a huge peak in the real part of the permittivity, ε' reaching 10^5–10^6 [Fig. 2.5]. Our results on $(TMTTF)_2PF_6$ and $(TMTTF)_2Br$ were presented at the ECRYS 1999 workshop, where we interpreted [5] this huge polarizability as reflecting the realization of a new charge ordered state of Wigner crystal type due to long range Coulomb interactions.

Charge disproportionation was then proved by NMR. The generality of this transition in the whole family of Fabre salts was established as well as the ferroelectric character of this charge ordered state. Many other compounds exhibit charge order, in particular two-dimensional organic quarter-filled compounds, for which charge patterns occur along stripes oriented along different axis depending on the anisotropy of the Coulomb interactions and of the

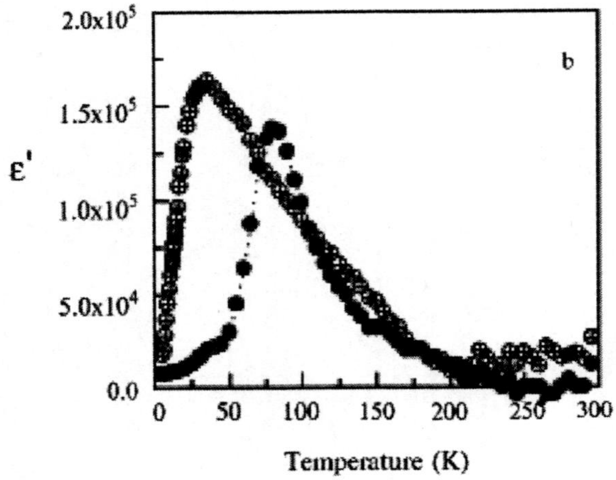

Fig. 2.5. Temperature dependence of the real part of the dielectric constant measured at 100 kHz in $(TMTTF)_2Br$ (⊕) and $(TMTTF)_2PF_6$ (from [5])

transfer integrals. Dielectric permittivity could be a nice tool for determination of the stripe orientation by applying the AC field parallel or perpendicular to the stripes.

I will not make any conclusions, because the future is unknown. However, the period of time covered by this short historical review has been exhuberant, totally exciting when dealing with collective properties having some similarity with superconductivity in a temperature range never reached before – till above room temperature. The occurrence of superconductivity with critical temperatures as high as 150 K has tremendously enlarged and opened these new perspectives. Even if CDW and SDW sliding do not yield a superconducting state and if, up to now, there is no increase of the conductivity when the DW slides above the extrapolation of the high temperature normal conductivity, quantum condensed states are known now to be stable in the high temperature range.

Nothing concerning this review would have been possible without the unvaluable collaboration of my colleagues, K. Biljaković, S. Brazovskii, J.M. Fabre, Yu. Latyshev, J.C. Lasjaunias, F. Levy, F. Nad, N.P. Ong, A. Orlov, M. Renard, J. Richard and T. Fournier whom I heartily thank.

References

1. N.P. Ong, P. Monceau, Phys. Rev. B **16**, 3443 (1977)
2. P. Monceau, J. Richard, M. Renard, Phys. Lett. **45**, 43 (1980)
3. K. Biljakovic, J.C. Lasjaunias, P. Monceau, F. Levy, Phys. Lett. **62**, 1512 (1989)
4. Yu. Latyshev, P. Monceau, S. Brazovskii, A.P. Orlov, T. Fournier, Phys. Lett. **95**, 266402 (2006); **96**, 116402 (2006)
5. F. Nad, P. Monceau, J.M. Fabre, J. Phys. IV (France) **9**, 361 (1999)

3

Field-Induced Spin–Density Waves and Dimensional Crossovers

A.G. Lebed

In this pedagogical and historical survey, we discuss how a discovery of a 2D $\rightarrow$ 1D crossover in a magnetic field by Gor'kov and Lebed and by Chaikin allows to explain an experimentally observed instability of a metallic phase with respect to a formation of the field-induced spin–density wave (FISDW) phases.

We introduce a momentum quantization law and describe in detail the physical meaning of the above-mentioned phase transition and also briefly review the most important theories and experiments.

For the first time, we present an expression for a free energy of the FISDW sub-phases, which is valid not only in the vicinity of the phase transitions but also at arbitrary low temperatures. It allows to explain such still not well understood phenomena as the first-order type of phase transitions between the different FISDW sub-phases and the non-trivial pressure and magnetic field dependences of the FISDW phase diagrams at low temperatures.

See also the chapters by Heritier, by Yakovenko, by Brown, Chaikin, and Naughton, by Lebed and Si Wu, by Kornilov and Pudalov, by Bjelis and Zanchi, by Maki, Dora, and Virosztek and by Haddad, Heritier, and Charfi-Kaddour.

3.1 Introduction

Experimental discoveries of the field-induced spin–density wave (FISDW) phase diagram by Chaikin's [1] and Ribault-Jerome's [2] groups in $(TMTSF)_2$ ClO_4 and its explanation in terms of the dimensional crossovers by Gor'kov and Lebed [3], Heritier et al. [4], Chaikin [5], and Lebed [6] opened a novel branch of solid sate physics. This branch, "Unconventional Magnetic Properties due to Dimensional Crossovers," has been very attractive for more than 20 years. It now includes: (1) FISDW phase diagrams, (2) field-induced charge–density wave (FICDW) phases, (3) 3D quantum Hall effect (3D QHE), (4) Fermi-liquid and non-Fermi-liquid Lebed magic angle phenomena,

(5) Yamaji–Kartsovnik oscillations, (6) Lee–Naughton–Lebed oscillations, (7) third angular effect and Danner–Kang–Chaikin oscillations, (8) cyclotron resonance on open orbits, (9) reentrant superconductivity phenomenon, and some other topics.

For detailed descriptions of the above-mentioned phenomena, see Chaps. 4, 5, 7, 16, 17, 19–21. Below, we restrict our discussion by a consideration of the FISDW sub-phases as the first historical examples of an application of the dimensional crossover concept [3] to magnetic properties of quasi-one-dimensional (Q1D) conductors.

3.2 Peierls Spin(Charge)–Density Wave Instability

It is well known that a pure 1D electron gas is unstable with respect to the spin(charge)–density wave (S(C)DW) formation since its electron spectrum

$$\varepsilon^{\pm}(\mathbf{p}) = \pm v_{\mathrm{F}}(p_x \mp p_{\mathrm{F}}) \tag{3.1}$$

possesses the so-called nesting property [7, 8],

$$\varepsilon(p_x) + \varepsilon(p_x + 2p_{\mathrm{F}}) = 0. \tag{3.2}$$

Here $+(-)$ stands for right (left) sheet of 1D Fermi surface (FS), and p_{F} and v_{F} are the Fermi momentum and Fermi velocity, respectively.

Using qualitative language, we can say that, in the 1D case, electrons and holes move along the conducting chains with exactly opposite velocities (see Fig. 3.1). The corresponding 1D quantum mechanical problem is

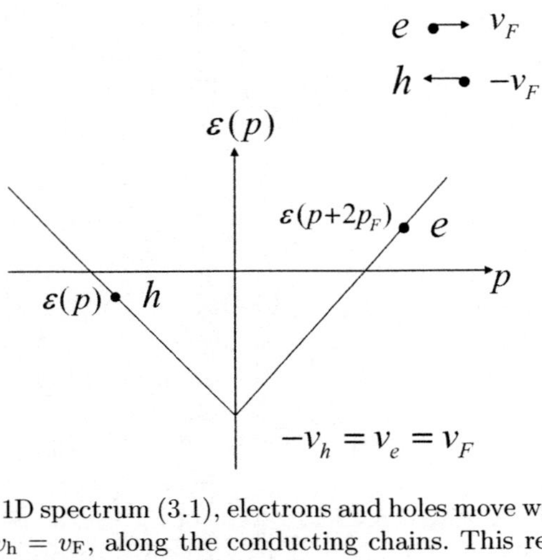

Fig. 3.1. For pure 1D spectrum (3.1), electrons and holes move with exactly opposite velocities, $v_{\mathrm{e}} = -v_{\mathrm{h}} = v_{\mathrm{F}}$, along the conducting chains. This results in the Peierls instability of a metallic phase (see the text)

always characterized by bounded states for an attractive potential in the Schrödinger equation. Therefore, in the case of attractive electron–hole interactions, electrons and holes create the Peierls electron–hole pairs, which correspond to a semiconducting S(C)DW phase with the order parameter

$$\Delta_{\mathrm{S(C)DW}}(x) = \Delta_{\mathrm{S(C)DW}} \; \cos(2p_{\mathrm{F}}x). \tag{3.3}$$

The finite degrees of "quasi-one-dimensionality" in real electron spectra can destroy the above-mentioned physical picture and can make a metallic phase to be stable with respect to the S(C)DW phase formation. Nevertheless, it is known that a Q1D spectrum, corresponding to the simplest variant of tight-binding model [3–8]

$$\varepsilon^{\pm}(\mathbf{p}) = \pm v_{\mathrm{F}}(p_x \mp p_{\mathrm{F}}) + 2t_b \cos(p_y b^*) + 2t_c \cos(p_z c^*) \tag{3.4}$$

preserves the nesting condition

$$\varepsilon(\mathbf{p}) + \varepsilon(\mathbf{p} + \mathbf{Q_0}) = 0. \tag{3.5}$$

As seen from Fig. 3.2, in this case, electron and hole move with exactly opposite velocities, $\mathbf{v}_{\mathrm{e}}(\mathbf{p}) = -\mathbf{v}_{\mathrm{h}}(\mathbf{p} + \mathbf{Q_0})$, along some direction. It is important that this property is valid for any electron momentum, $\mathbf{p}$, defining the electron position on the Q1D FS (3.4). This makes the electron–hole interactions to be effectively 1D and, therefore, in the case of attractive interactions, a ground state of the Q1D conductor (3.4) is the S(C)DW phase with the order parameter

$$\Delta_{\mathrm{S(C)DW}}(\mathbf{r}) = \Delta_{\mathrm{S(C)DW}} \cos(\mathbf{Q_0 r}). \tag{3.6}$$

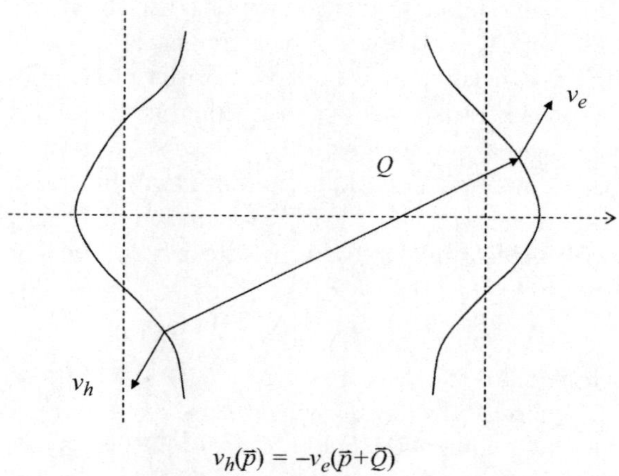

Fig. 3.2. For the simplest variant of the tight-binding spectrum (3.4), electron and hole, which organize the Peierls electron–hole pair, move with exactly opposite velocities, $\mathbf{v}_{\mathrm{h}}(\mathbf{p}) = -\mathbf{v}_{\mathrm{e}}(\mathbf{p} + \mathbf{Q})$

Note that the order parameter (3.6) corresponds to the so-called ideal nesting vector

$$\mathbf{Q_0} = (2p_\mathrm{F}, \pi/b^*, \pi/c^*). \tag{3.7}$$

Electron spectrum (3.4), corresponding to an electron jumping between the nearest neighboring molecular sites, naturally appears in a number of organic metals, where distances between the 1D conducting chains are large enough. This allowed Yamaji [8] to suggest that the SDW phase, experimentally observed in $(\mathrm{TMTSF})_2\mathrm{PF}_6$ conductor at ambient pressure, is due to the ideal nesting property (3.5) of its electron spectrum.

He also pointed out that the SDW phase in a sister compound $(\mathrm{TMTSF})_2$ ClO_4 is unstable due to the existence of some small (but important) corrections to its electron spectrum

$$\begin{aligned}\varepsilon^\pm(\mathbf{p}) = \; &\pm v_\mathrm{F}(p_x \mp p_\mathrm{F}) + 2t_b\cos(p_y b^*) + 2t_c\cos(p_z c^*) \\ &+ 2t'_b\cos(2p_y b^*) + 2t'_c\cos(2p_z c^*),\end{aligned} \tag{3.8}$$

where $t'_b \ll t_b$ and $t'_c \ll t_c$ are the so-called antinesting terms [3, 8].

Note that, if the main antinesting term in (3.8) is large enough, $t'_b \geq T_\mathrm{SDW}$, then electrons and holes move with quite different velocities. In this case, the corresponding quantum mechanical problem becomes 3D and the Peierls electron–hole pairs do not appear [3, 8] even in the case of the attractive electron–hole interactions.

3.3 Field-Induced Spin–Density Wave Instability

Nevertheless, in 1983, Chaikin's [1] and Ribault–Jerome's [2] experimental groups found that some phase transitions, which are roughly periodic in $1/H$, appear in $(\mathrm{TMTSF})_2\mathrm{ClO}_4$ conductor in a magnetic field. These transitions were interpreted by Gor'kov and Lebed [3], Heritier et al. [4], Chaikin [5], and Lebed [6] as the phase transitions between different FISDW sub-phases.

In this section, we describe how 2D→1D crossovers in a magnetic field, discovered by Gor'kov and Lebed [3], explain instability of the Q1D metallic phase (3.8) with respect to the FISDW formation. Let us consider a realistic Q1D electron spectrum (3.8) perpendicular to the conducting chains magnetic field

$$\mathbf{H} = (0, 0, H), \quad \mathbf{A} = (0, Hx, 0). \tag{3.9}$$

It is important that the nesting terms, t_b and t_c, preserve the nesting condition (3.5), whereas the antinesting terms, t'_b and t'_c, in (3.8) destroy it in the absence of a magnetic field. In fact, in the layered Q1D organic conductors, the parameter t'_c is very small, $t'_c \simeq 0.1\,\mathrm{K}$, and, therefore, we neglect it below. On the other hand, the parameter t'_b is typically of the order of 10 K and plays, as we show below, a crucial role. Therefore, instead of the

Q1D spectrum (3.8), we can consider the following 2D electron spectrum with one antinesting term

$$\varepsilon^{\pm}(\mathbf{p}) = \pm v_{\mathrm{F}}(p_x \mp p_{\mathrm{F}}) + 2t_b \cos(p_y b^*) + 2t_b' \cos(2p_y b^*). \tag{3.10}$$

Note that we also set $t_c = 0$ in (3.10), since t_c preserves the nesting condition (3.5) and, thus, is not important for the further consideration.

At first, we consider an instability of a metallic phase with respect to the FISDW formation using qualitative arguments related to the quasiclassical equation of motion for electrons

$$\frac{\mathrm{d}\mathbf{p}}{\mathrm{d}t} = \left(\frac{e}{c}\right)[\mathbf{v}(\mathbf{p}) \times \mathbf{H}], \quad \mathbf{v}(\mathbf{p}) = \frac{\mathrm{d}\epsilon(\mathbf{p})}{\mathrm{d}\mathbf{p}}, \tag{3.11}$$

and for holes

$$\frac{\mathrm{d}\mathbf{p}}{\mathrm{d}t} = -\left(\frac{e}{c}\right)[\mathbf{v}(\mathbf{p}) \times \mathbf{H}], \quad \mathbf{v}(\mathbf{p}) = \frac{\mathrm{d}\epsilon(\mathbf{p})}{\mathrm{d}\mathbf{p}}. \tag{3.12}$$

From (3.11) and (3.12), it follows that $\mathrm{d}\mathbf{p} \perp \mathrm{d}\mathbf{r}$, and, thus, a quasiparticle trajectory in a real space can be obtained from that in a reciprocal space by rotating it into $\pi/2$ (see Fig. 3.3). Note that the main feature of the electron spectra (3.8), (3.10) is that they are periodic in the extended Brillouin zone. Therefore, the quasiparticle trajectories, perpendicular to the conducting chains magnetic field (3.9), become periodic and restricted along the $\mathbf{y}$-axis in real space (see Figs. 3.3 and 3.4).

As it follows from Fig. 3.4, in a magnetic field, electrons and holes can move freely only along the conducting chains, whereas their trajectories along the $\mathbf{y}$-axis are localized. This means that electrons and holes cannot go to infinity along the $\mathbf{y}$-axis and, thus, the electron–hole pairing in a magnetic field (3.9) is "one-dimensionalized" [3]. 1D metal is known to be unstable against the S(C)DW phase formation, which is a reason why the Q1D metal (3.10) is unstable with respect to the FISDW phase formation in an arbitrary weak magnetic field [3–6].

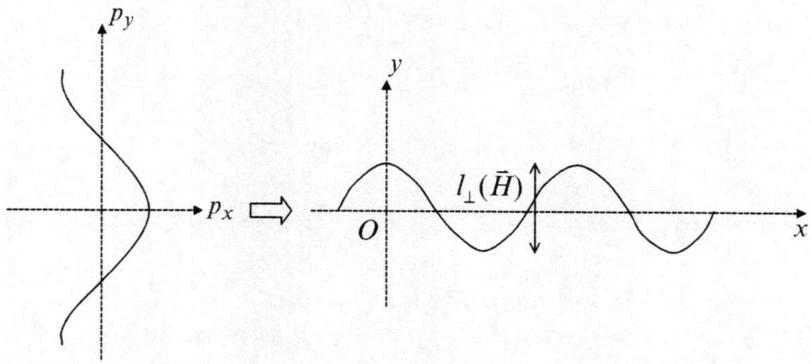

Fig. 3.3. Electron trajectory in a real (x, y)-space can be obtained from that in a reciprocal (p_x, p_y)-space by a rotation operation (see (3.11), (3.12) and the text)

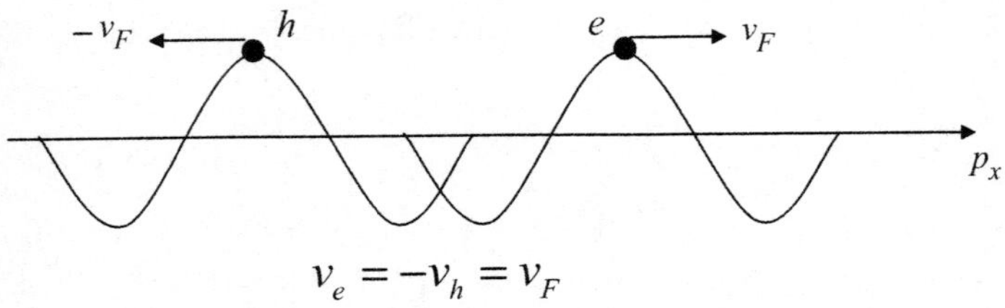

Fig. 3.4. Electron–hole interactions are "one-dimensionalized" in a magnetic field (3.9) (see the text)

Let us consider the 2D→1D crossover suggested above in some more detail. The quasiclassical equations of motion (3.11) for an electron, located on a right sheet of the Q1D spectrum (3.10), can be written in a magnetic field (3.9) as:

$$\frac{\mathrm{d}p_y}{\mathrm{d}t} = \left(\frac{e}{c}\right) v_{\mathrm{F}} H, \quad p_y b^* = p_y^1 b^* + \omega_c(H) t, \quad \omega_c(H) = \left(\frac{e}{c}\right) v_{\mathrm{F}} b^* H. \tag{3.13}$$

Therefore, the electron velocity component along y-axis is

$$v_y^e(t, p_y^1) = -2t_b b^* \sin[p_y^1 b^* + \omega_c(H)t] - 4t_b' b^* \sin[2p_y^1 b^* + 2\omega_c(H)t] \tag{3.14}$$

and, thus, the electron trajectory can be expressed as

$$y_{\mathrm{e}}(t, p_y^1) = \frac{2t_b b^*}{\omega_c(H)} \cos[p_y^1 b^* + \omega_c(H)t] + \frac{2t_b' b^*}{\omega_c(H)} \cos[2p_y^1 b^* + 2\omega_c(H)t], \tag{3.15}$$

where p_y^1 is an initial electron momentum at $t = 0$.

On the other hand, the quasiclassical equations of motion (3.12) for a hole, located on a left sheet of the Q1D spectrum (3.10), in a magnetic field (3.9), can be written as:

$$\frac{\mathrm{d}p_y}{\mathrm{d}t} = -\left(\frac{e}{c}\right)(-v_{\mathrm{F}}) H = \left(\frac{e}{c}\right) v_{\mathrm{F}} H, \quad p_y b^* = p_y^2 b^* + \omega_c(H) t. \tag{3.16}$$

Therefore, the hole velocity component along the y-axis is

$$v_y^h(t, p_y^2) = -2t_b b^* \sin[p_y^2 b^* + \omega_c(H)t] - 4t_b' b^* \sin[2p_y^2 b^* + 2\omega_c(H)t] \tag{3.17}$$

and, thus, the hole trajectory can be expressed as

$$y_{\mathrm{h}}(t, p_y^2) = \frac{2t_b b^*}{\omega_c(H)} \cos[p_y^2 b^* + \omega_c(H)t] + \frac{2t_b' b^*}{\omega_c(H)} \cos[2p_y^2 b^* + 2\omega_c(H)t], \tag{3.18}$$

where p_y^2 is an initial hole momentum at $t = 0$.

Let us consider an electron–hole pair which satisfies the nesting condition (3.5) and (3.7), where $p_y^2 b^* = p_y^1 b^* + \pi$. As it follows from (3.14) and (3.17), in the absence of the antinesting term (i.e., at $t_b' = 0$), the electron and hole velocity components along the $\mathbf{y}$-axis are opposite at any moment of time. Therefore, only the antinesting term, t_b', prevents the ideal electron–hole FISDW pairing with the ideal nesting vector (3.7).

Nevertheless, as it is seen from (3.15) and (3.18), the quasiparticle trajectories are localized along the y-direction in a magnetic field even in the presence of the antinesting term, $t_b' \neq 0$ (see Fig. 3.4). Therefore, electrons and holes always organize some nonideal FISDW pairs in an arbitrary weak magnetic field (3.9), whereas, in a high magnetic field, where the antinesting term, t_b', is not important, they organize the ideal FISDW pairs.

This point about the ideal and nonideal FISDW phases is clarified in Sect. 3.4.3. Here, we estimate a value of a magnetic field, above which the FISDW pairing becomes ideal. As seen from (3.15) and (3.18), the ideal FISDW electron–hole pairing is achieved at $2t_b'/\omega_c(H) \leq 1$, where the contributions of the antinesting terms to electron and hole trajectories are less than an interchain distance

$$\delta y \simeq \frac{2t_b' b^*}{\omega_c(H)} \leq b^* . \tag{3.19}$$

Equation (3.19) corresponds to some critical magnetic field

$$H_1 \simeq \frac{4t_b' c}{ev_{\mathrm{F}} b^*} , \tag{3.20}$$

which completely "one-dimensionalizes" the electron–hole interactions. (Note that, although the FISDW phase is stable in an arbitrary weak magnetic field [3], at magnetic fields lower than H_1, it is characterized by the nonideal nesting vector and, thus, by lower transition temperature [3–6])

The problem of FISDW phase formation, which is discussed above using qualitative arguments, was first considered in [3]. In particular, in [3], the above-mentioned problem is rigorously solved by the Green functions method. It is shown [3] that a generalized susceptibility of the Q1D electrons (3.10) diverges at low temperatures. This indicates FISDW instability and defines the FISDW transition temperature (see Fig. 3.5). (Note that, in [3], the FISDW problem is solved for the ideal nesting vector (3.7), which corresponds to the minimum of free energy at high enough magnetic fields, $H \geq H_1$ (see (3.20) and Fig. 3.6)).

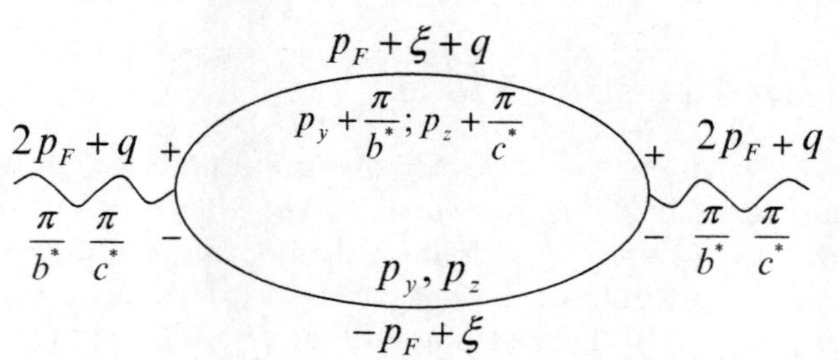

Fig. 3.5. A typical contribution to the mean field theory of a generalized electron susceptibility with respect to the FISDW formation [3–6]

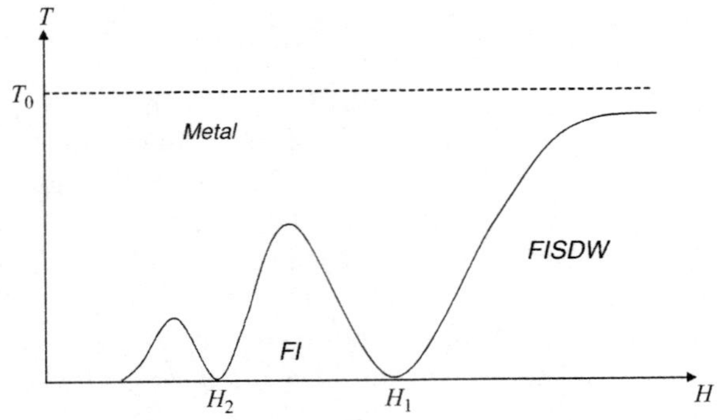

Fig. 3.6. *Solid line*: a solution of (3.21), which defines a transition temperature to the FISDW phase with the ideal nesting vector (i.e., for $q = 0$ in (3.22))

The main mathematical result of [3] is that a phase boundary between the metallic phase (3.10) and the FISDW phase (3.6) can be described as

$$\frac{1}{g} = \int_d^{\infty} J_0\left(\frac{4t'_b}{\omega_c(H)} \sin\left[\frac{\omega_c(H)x}{v_F}\right]\right) \frac{2\pi T \mathrm{d}x}{v_F \sinh\left(\frac{2\pi T x}{v_F}\right)}, \tag{3.21}$$

where $g \ll 1$ is a dimensionless coupling constant and d is a cut-off distance.

Note that the Bessel function, $J_0(\ldots)$, is a periodic function in integral (3.21) and, thus, the integral (3.21) logarithmically diverges at low temperatures at any value of a magnetic field. This means that (3.21) always has a solution and, therefore, the metallic phase (3.10) is absolutely unstable with respect to the FISDW phase (3.6) formation at low enough temperatures. This is the main physical result of [3], which is illustrated in Fig. 3.6, where (3.21) is graphically solved.

We stress again that (3.21) is written for the ideal nesting vector (3.7), which minimizes the free energy of the FISDW phase at high enough magnetic fields, $H \geq H_1$ (3.20). Although, as anticipated in [3], some other quantized nesting vectors may correspond to the minima of free energy at lower magnetic fields, $H \leq H_1$, the FISDW problem for arbitrary magnetic fields was fully solved only in [4,6].

3.4 Quantized Nesting Model

In this section, we discuss the so-called quantized nesting (QN) model, introduced by Heritier et al. [4] and developed by Lebed [6] and Virosztek et al. [9]. From the results of Sect. 3.3, it follows that only the antinesting term, t'_b, is responsible for possible deviations of the FISDW wave vector from its ideal nesting value (3.7). Let us suppose that $H \leq H_1$ and consider a phase transition to the FISDW phase, characterized by more general nesting vector

$$\mathbf{Q_0} = (2p_F + q, \pi/b^*, \pi/c^*). \tag{3.22}$$

3.4.1 Momentum Quantization Law

In this section, we introduce a momentum quantization law, which allows to consider the QN model using qualitative arguments. For a detailed analysis of the momentum quantization law, see the review by Lebed and Si Wu.

Let us consider Q1D electrons, which are characterized only by the major antinesting term

$$\varepsilon^{\pm}(\mathbf{p}) = \pm v_{\mathrm{F}}(p_x \mp p_{\mathrm{F}}) + 2t'_b \cos(2p_y b^*), \tag{3.23}$$

since only the antinesting term in (3.23) can be responsible for $q \neq 0$ in (3.22). We introduce electron wave functions near the right and left sheets of the Q1D FS (3.23) in the form

$$\Psi^{\pm}_{\epsilon}(x, y) = \exp(\pm \mathrm{i} p_{\mathrm{F}} x)\psi^{\pm}_{\epsilon}(x, y) \tag{3.24}$$

and expand the amplitudes $\psi^{\pm}_{\epsilon}(x, y)$ into the Fourier series with respect to a coordinate y

$$\psi^{\pm}_{\epsilon}(x, y) = \int_0^{2\pi} \psi^{\pm}_{\epsilon}(x, p_y) \exp(\mathrm{i} p_y y) \frac{\mathrm{d}p_y}{2\pi}. \tag{3.25}$$

In a magnetic field (3.9) perpendicular to the conducting chains, the Schrödinger equation for the electron wave functions (3.25),

$$\left(\mp \mathrm{i} v_{\mathrm{F}} \frac{\mathrm{d}}{\mathrm{d}x} + 2t'_b \cos\left[2p_y b^* - \frac{2\omega_c(H)x}{v_{\mathrm{F}}}\right]\right) \psi^{\pm}_{\epsilon}(x, p_y) = \delta\epsilon\ \psi^{\pm}_{\epsilon}(x, p_y), \tag{3.26}$$

can be obtained from (3.23) by using the Peierls substitution method [3],

$$p_x - p_{\mathrm{F}} \to -\mathrm{i}\left(\frac{\mathrm{d}}{\mathrm{d}x}\right), \quad p_y b^* \to p_y b^* - \left(\frac{e}{c}\right) A_y b^* \tag{3.27}$$

where $\delta\epsilon = \epsilon - \epsilon_{\mathrm{F}}$.

It is important that (3.26) can be solved analytically

$$\psi^{\pm}_{\varepsilon}(x, p_y) = \exp\left(\pm \mathrm{i} \frac{\delta\varepsilon}{v_{\mathrm{F}}} x\right) \exp\left[\pm \frac{\mathrm{i} t'_b}{\omega_c}\left(\sin\left[p_y b^* - \frac{2\omega_c x}{v_{\mathrm{F}}}\right] - \sin[p_y b^*]\right)\right]. \tag{3.28}$$

Let us show that (3.28) directly demonstrates two phenomena: the 2D→1D crossover in a magnetic field (3.9), described in the previous section, and a momentum quantization law. For this purpose, we calculate the Fourier component of the wave function (3.28) with respect to variable x,

$$\psi^{\pm}_{\varepsilon}(x, p_y) = \exp\left(\pm \mathrm{i} \frac{\delta\varepsilon}{v_{\mathrm{F}}} x\right) \sum_{m=-\infty}^{\infty} A_m(p_y) \exp\left(\mathrm{i} \frac{2\omega_c}{v_{\mathrm{F}}} m x\right), \tag{3.29}$$

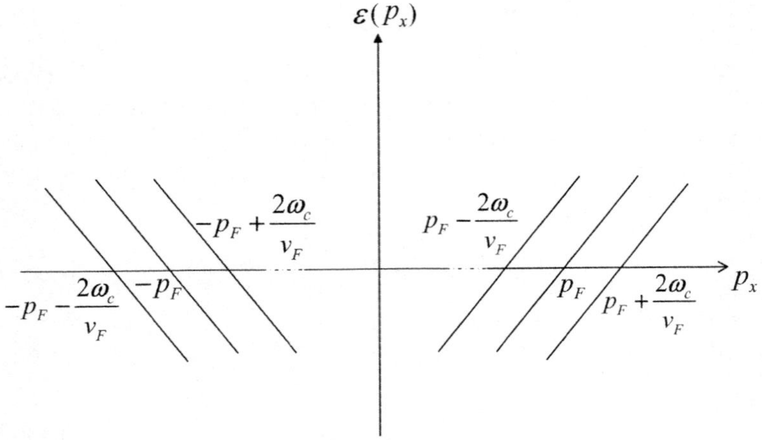

Fig. 3.7. Energy levels in the vicinities of $\pm p_{\mathrm{F}}$ (see (3.31))

where

$$A_m(p_y) = \frac{\omega_c(H)}{\pi v_{\mathrm{F}}} \int_{-\pi v_{\mathrm{F}}/2\omega_c}^{\pi v_{\mathrm{F}}/2\omega_c} \exp\left(\mp \mathrm{i}\frac{\delta\varepsilon}{v_{\mathrm{F}}}x\right) \exp\left(-\mathrm{i}\frac{2\omega_c}{v_{\mathrm{F}}}mx\right) \psi_\epsilon^\pm(x,p_y)\mathrm{d}x. \tag{3.30}$$

As shown from (3.28), electron energy, $\delta\epsilon$, does not depend on a momentum component, p_y, which is a direct indication of the 2D→1D crossover. Moreover, (3.29) introduces the following momentum quantization law: if an electron has a definite energy, $\delta\epsilon$, then its momentum component along the x-axis is quantized

$$p_x = \pm p_{\mathrm{F}} \pm \delta\epsilon/v_{\mathrm{F}} + [2\omega_c(H)/v_{\mathrm{F}}]\ n, \tag{3.31}$$

where $2\omega_c(H)/v_{\mathrm{F}}$ is a momentum quantum (see Fig. 3.7).

The momentum quantization law (3.31) introduced above plays a central role in the QN model. Indeed, as it follows from (3.31), the nesting condition (3.5) is obeyed in a magnetic field (3.9) for a number of wave vectors (3.22) with the quantized value of q

$$q = [2\omega_c(H)/v_{\mathrm{F}}]n, \tag{3.32}$$

where n is an integer.

3.4.2 Metal-FISDW Phase Transition Line

Let us introduce the QN model, which was first suggested by Heritier et al. [4]. To do this we calculate a general susceptibility of the metallic phase (3.10) with respect to the FISDW pairing corresponding to the nesting vector (3.22).

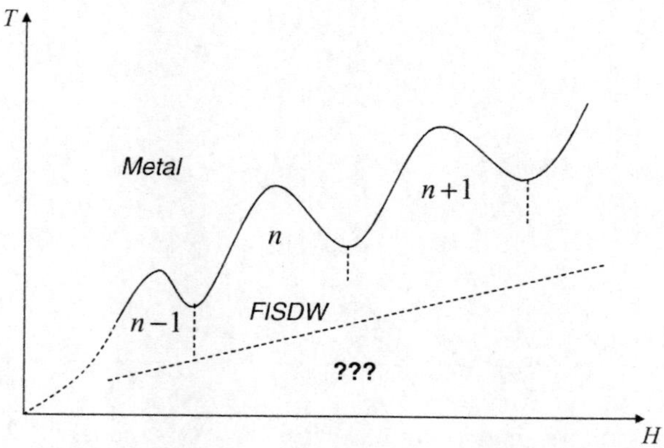

Fig. 3.8. Transition line between the metallic phase (3.10) and the FISDW sub-phases (3.22) and (3.32) [4]

As a result, we obtain the following equation [4]

$$\frac{1}{g} = \int_d^{\infty} J_0\left(\frac{4t'_b}{\omega_c(H)} \sin\left[\frac{\omega_c(H)x}{v_F}\right]\right) \cos(qx) \frac{2\pi T \, dx}{v_F \sinh\left(\frac{2\pi T x}{v_F}\right)}, \tag{3.33}$$

where there is an extra term, $\cos(qx)$, in comparison with (3.21).

From (3.33), it directly follows that the integral (3.33) is divergent as $T \to 0$ for a number of quantized nesting vectors (3.32). This means that (3.33), which defines a transition line between the metallic and FISDW sub-phases, has solutions for a number of quantized nesting vectors (3.22) and (3.32) at low enough temperatures.

To obtain the phase transition line, we need to maximize the transition temperature (3.33) with respect to a quantum number n (3.32). As a result, the following phase transition line is obtained by Heritier et al. [4] (see Fig. 3.8).

We stress that (3.33) is valid only in the Ginzburg–Landau (GL) region and, thus, defines only the transition line between the metallic phase and the FISDW sub-phases. To obtain the whole phase diagram, we need to calculate the free energy of the FISDW sub-phases out of the GL area, which was first performed by Lebed [6] and Vitosztek et al. [9].

3.4.3 Phase Transitions Between FISDW Sub-Phases

Let us discuss the main results of [6, 9], where the free energy of the Q1D electrons (3.10) are calculated in the FISDW phase and the phase transitions between different FISDW sub-phases are described (see Fig. 3.9).

We stress that the above-mentioned problem cannot be analytically solved for an arbitrary relationship between the FISDW order parameter, Δ_{FISDW},

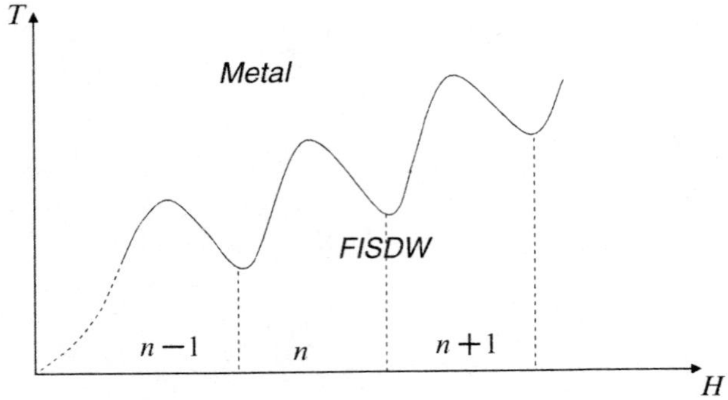

Fig. 3.9. The FISDW phase diagram within the QN model [4, 6], where *vertical dashed lines* correspond to the first-order phase transitions between different FISDW sub-phases, characterized by quantized nesting vectors (3.22) and (3.32)

and the magnetic field dependent cyclotron frequency, $\omega_c(H)$. Nevertheless, as pointed in [6], in many cases

$$\omega_c(H)/\Delta_n \geq 1, \tag{3.34}$$

where the order parameter

$$\Delta_{FISDW}(x,y,z) = \Delta_n \exp\left(\mathrm{i}\frac{\pi y}{b^*}\right)\exp\left(\mathrm{i}\frac{\pi z}{c^*}\right)\exp\left(\mathrm{i}\left[2p_\mathrm{F} + 2n\frac{\omega_c(H)}{v_\mathrm{F}}\right]x\right) \tag{3.35}$$

corresponds to the FISDW sub-phase with the quantized nesting vector (3.32). (Note that the condition (3.34) corresponds to the so-called magnetic breakdown phenomenon, which occurs through the FISDW gap, Δ_n. It is important that the QN model, introduced in the previous section, is also based on condition (3.34).)

As shown in [6], the free energy of the FISDW phase can be calculated at any temperature under the breakdown condition (3.34). In this case, it is possible to show [6] that an arbitrary term of the free energy expansion in the presence of a magnetic field can be obtained from the expression of the BCS free energy by the replacement

$$\Delta^2 \to \Delta_n^2 \; J_n^2[2t_b'/\omega_c(H)], \tag{3.36}$$

where Δ is the BCS gap, $J_n[\ldots]$ is the Bessel function of the nth order.

The terms of the second order in Δ_n are an exception. They, however, can be evaluated separately and correspond to magnetic field and pressure dependent effective coupling constant [6]

$$\frac{1}{g} \to \frac{\ln(t_b'/t_b^0)}{J_n^2[2t_b'/\omega_c(H)]}, \tag{3.37}$$

and a magnetic field dependent cut-off energy] [6]

$$\Omega = \gamma \omega_c(H)/\pi^2, \tag{3.38}$$

where t_b^0 is a critical value of the antinesting parameter, t_b', which destroys the SDW phase in the absence of a magnetic field, γ is the Euler constant. (In other words, as is shown in [6], it is possible to summarize an infinite number of the Feynman diagrams defining the free energy under magnetic breakdown condition (3.34).)

In this form the free energy in a magnetic field has local minima for sub-phases with the quantized wave vectors q from (3.22), (3.32) and the entire sum of the Feynman diagrams is transformed into the functional

$$\begin{aligned}\delta F(n,H) &= \Delta_n^2 \ln\left(\frac{t_b'}{t_b^0}\right) \\ &\quad - 4T\int_0^{\Omega} \mathrm{d}\xi \ln\left(\frac{\cosh\left[(\xi^2 + J_n^2[2t_b'/\omega_c(H)]\Delta_n^2)^{1/2}/2T\right]}{\cosh(\xi/2T)}\right).\end{aligned} \tag{3.39}$$

From here we must determine the phase diagram. If we use the condition that the variation of functional (3.39) with respect to Δ_n determines the equilibrium value of the gap for the nth sub-phase, Δ_n, then we can rewrite (3.39) as:

$$\begin{aligned}&\delta F(n,H) = \\ &\int_0^{\infty} \mathrm{d}\xi \left[\frac{J_n^2(2t_b'/\omega_c(H))\Delta_n^2}{(\xi^2 + J_n^2[2t_b'/\omega_c(H)]\Delta_n^2)^{1/2}} \tanh \frac{[\xi^2 + J_n^2(2t_b'/\omega_c(H))\Delta_n^2]^{1/2}}{2T}\right. \\ &\qquad \left. - 4T \ln \frac{\cosh\left[(\xi^2 + J_n^2[2t_b'/\omega_c(H)]\Delta_n^2)^{1/2}/2T\right]}{\cosh(\xi/2T)}\right].\end{aligned} \tag{3.40}$$

(Note that in the convergent integral (3.40) the upper limit is set to infinity.)

Since a free energy (3.40) of each FISDW sub-phase now depends explicitly only on a combination $\Delta_n^2\ J_n^2[2t_b'/\omega_c(H)]$, it follows that the optimum value of the wave vector q in (3.32) is reached at the maximum value of the Bessel function $J_n^2[2t_b'/\omega_c(H)]$. Thus, the phase diagram of a layered Q1D conductor consists of the FISDW sub-phases, determined by the wave vectors q from (3.22) and (3.32) for which $J_n^2[2t_b'/\omega_c(H)]$ is maximum.

The phase diagram of the Q1D conductor (3.10), obtained in [6], is shown schematically in Fig. 3.9. It is important that the transition temperature from metallic to the FISDW nth sub-phase, $T_n(H)$, is related to the energy gap, $\Delta_n(H)$, in the same way as in the BCS theory

$$T_n(H) = \frac{\gamma}{\pi}\Delta_n(H). \tag{3.41}$$

(Note that each FISDW sub-phase is accompanied by 3D QHE [10, 11]. For theories of the 3D QHE, see [12, 13].)

3.5 Beyond Quantum Nesting Model

We point out that the magnetic breakdown condition (3.34), which may be rewritten as

$$h = \omega_c(H)/\pi T_n(H) \geq 1 \qquad (3.42)$$

is well fulfilled in $(TMTSF)_2ClO_4$ conductor [14]. On the other hand, as stressed by Lebed [14], it is not fulfilled in a sister compound $(TMTSF)_2PF_6$ and some other materials.

Below, we discuss the results of [14], where the metal–FISDW phase transition line is theoretically analyzed for an arbitrary value of the parameter h. We also discuss in brief the experimental work by Kornilov et al. [15] performed on the $(TMTSF)_2PF_6$ conductor, where the main results of [14] are experimentally confirmed. In particular, in [14], it is shown that an account of a finite transition temperature, $T_n(H) \neq 0$, changes the main qualitative consequences of the QN model, described in the previous section.

In contrast to the QN model, it is shown [14] that (1) The longitudinal wave vectors of the FISDW sub-phases (3.22) and (3.32) are not strictly quantized (i.e., n is not an integer in (3.32) unless $n = 0$). (2) For small enough values of the parameter h, $h < h_c \simeq 1$, the FISDW phase diagram consists of two regions (a) a low-temperature region ("quantum FISDW") where there exist discontinuous (but noninteger) jumps of the FISDW wave vectors (i.e., the first-order phase transitions between different FISDW sub-phases) (see Fig. 3.10) and (b) a high temperature region ("quasiclassical FISDW") where the jumps and the first-order transitions disappear, but the FISDW wave vector (3.32) is still a nontrivial oscillating function of a magnetic field. (3) For large enough values of the parameter h, $h > h_c \simeq 1$, the FISDW phase diagram consists of a cascade of the first-order phase transitions between the different FISDW sub-phases. They are characterized by discontinuous (but noninteger) jumps of the FISDW wave vector (see Fig. 3.11).

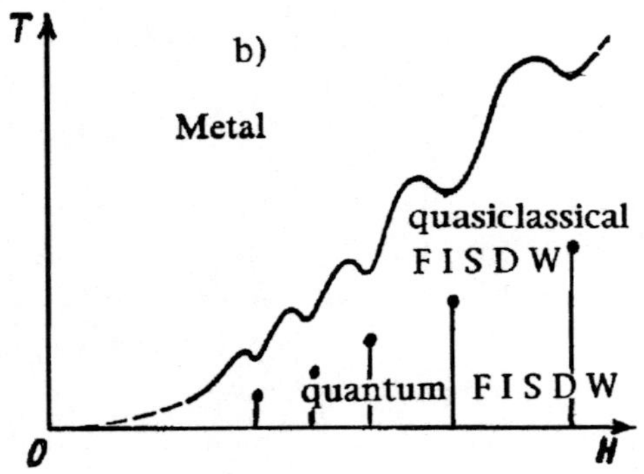

Fig. 3.10. The theoretical FISDW phase diagram [14], for small enough values of the parameter h in (3.42), consists of two regions: quantum FISDW and quasiclassical FISDW (see the text)

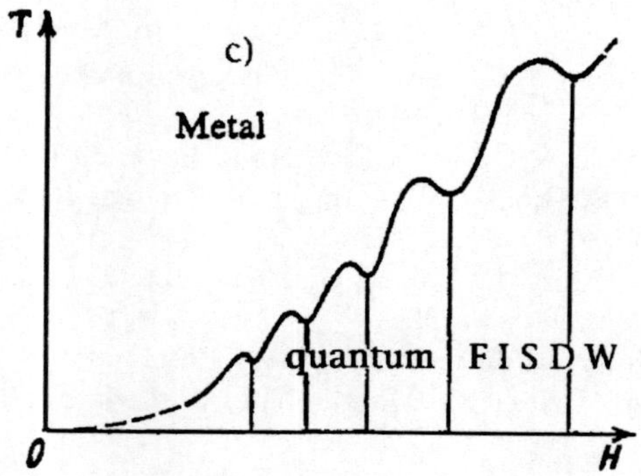

Fig. 3.11. The FISDW phase diagram, for large enough values of the parameter h (3.42), corresponds to noninteger discontinuous jumps of the vector nesting (3.32) between neighboring FISDW sub-phases

We note clear indications in favor of the theory [14] were observed in early experiments (e.g., [10, 11]), performed on $(TMTSF)_2PF_6$ compound by Chaikin's and Jerome's groups. Nevertheless, they were not appropriately interpreted in the absence of the theory [14] (see the discussions in [14]).

The main prediction of the above-discussed theory – the existence of two distinct regions in the FISDW phase diagram in the case of small values of the parameter h – has now been confirmed by Kornilov et al. [15]. The authors of [15] have directly observed a hysteresis and the first-order phase transitions at low temperatures, which disappear at high enough temperatures within the FISDW phase diagram in $(TMTSF)_2PF_6$.

Acknowledgments

The author is thankful to N.N. Bagmet, P.M. Chaikin, L.P. Gorkov, M. Heritier, A. Kornilov, M. Naughton, and V. Pudalov for fruitful discussions. This work was partially supported by the NSF grant DMR-0705986.

References

1. P.M. Chaikin, Mu-Yong Choi, J.F. Kwak, J.S. Brooks, K.P. Martin, M.J. Naughton, E.M. Engler, R.L. Greene, Phys. Rev. Lett. **51**, 2333 (1983)
2. M. Ribault, D. Jerome, J. Tuchendler, C. Weyl, K. Bechgaard, J. Phys. (Paris) Lett. **44**, L-953 (1983)
3. L.P. Gor'kov, A.G. Lebed, J. Phys. (Paris) Lett. **45**, L-433 (1984)
4. M. Heritier, G. Montambaux, P. Lederer, J. Phys. (Paris) Lett. **45**, L-943 (1984); G. Montambaux, M. Heritier, P. Lederer, Phys. Rev. Lett. **55**, 2078 (1985)
5. P.M. Chaikin, Phys. Rev. B **31**, 4770 (1985)
6. A.G. Lebed, Zh. Eksp. Teor. Fiz. **89**, 1034 (1985) [Sov. Phys. JETP, **62**, 595 (1985)]

7. L. Keldysh, Yu. Kopaev, Fiz. Tverd. Tela **6**, 2791 (1964)
8. T. Ishiguro, K. Yamaji, G. Saito, *Organic Superconductors*, 2nd edn. (Springer, Berlin Heidelberg New York, 1998)
9. A. Virosztek, L. Chen, K. Maki, Phys. Rev. B **34**, 3371 (1986)
10. S.T. Hannahs, J.S. Brooks, W. Kang, L.Y. Chiang, P.M. Chaikin, Phys. Rev. Lett. **63**, 1988 (1989)
11. J.R. Cooper, W. Kang, P. Auban, G. Montambaux, D. Jerome, K. Bechgaard, Phys. Rev. Lett. **63**, 1984 (1989)
12. V.M. Yakovenko, Phys. Rev. B **43**, 11353 (1991)
13. D. Poilblanc, M. Heritier, G. Montambaux, P. Lederer, J. Phys. C Solid State Phys. **19**, L321 (1986)
14. A.G. Lebed, Phys. Rev. Lett. **88**, 177001 (2002)
15. A.V. Kornilov, V.M. Pudalov, Y. Kitaoka, K. Ishida, T. Mito, J.S. Brooks, J.S. Qualls, J.A.A.J. Perenboom, N. Tateiwa, T.C. Kobayashi, Phys. Rev. B **65**, 060404R (2002)

4

Cascade of FISDW Phases: Wave Vector Quantization and its Consequences

M. Héritier

4.1 Introduction

The Bechgaard salts, $(TMTSF)_2X$ (where $X = PF_6, ClO_4, ReO_4$, etc.), exhibit an extraordinary rich variety of different physical phenomena, which are mainly due to their quasi-one-dimensional character [1]. They are also considered as strongly correlated fermions systems, often exhibiting in certain conditions non-Fermi liquid behaviors. However, one of the most original phenomena that they display is certainly the cascade of Field Induced Spin Density Wave (FISDW) phase transitions, which has been successfully discussed in a Fermi liquid picture, at temperature well below the 1D-3D crossover, i.e., much lower than the coupling energy between one-dimensional chains.

The first experimental observations of these FISDW phases were transport measurements of the resistivity [2], and Hall effect [3, 4] in $(TMTSF)_2\,ClO_4$ exhibiting a phase transition induced by an applied field of a few Teslas, from the normal metallic phase to an insulating phase. It was then rapidly shown [1] that, in increasing field, a whole cascade of phase transitions between SDW subphases was induced, and that the Hall effect exhibited a plateau behavior within each subphase. Our purpose, here, is an historical survey about some points of what appeared to us as a fascinating scientific adventure (see, also, the chapters of A.G. Lebed and V.M. Yakovenko in this volume).

The crucial point for understanding this phenomenon has been done, first, by Gork'ov and Lebed [5], within a Fermi liquid picture and a weak coupling approach. They considered a quasi-one-dimensional electron gas, described by a strongly anisotropic tight binding mean field Hamiltonian, with a hierarchy of the nearest neighbor electron transfer integrals, $t_{\mathrm{a}}\,(\sim 3000\,\mathrm{K}) \gg t_{\mathrm{b}}\,(\sim 300\,\mathrm{K}) \gg t_{\mathrm{c}}\,(\sim 10\,\mathrm{K})$. They pointed out the crucial appearance, under an applied magnetic field H, perpendicular to the most conducting planes, of a logarithmic divergence in the spin staggered static susceptibility, $\chi_0\,(2k_{\mathrm{F}}, T, H)$, because of the open quasi-nested Fermi surface. A very simple semi-classical argument [6] indicates that the electron orbits become one-dimensional under magnetic field, because the electron wave packet motion in the field oscillates in real space with a finite amplitude limited by the field. Gork'ov and Lebed were the first to prove, by their quantum calculation, that this orbital effect of the field restores the 1D logarithmic divergence of the noninteracting susceptibility. This was the starting point of the understanding of the FISDW phenomenon. At that time, Gork'ov and Lebed disregarded an interpretation of the Hall plateaux in terms of the Quantum Hall Effect because, then, there were no sign of any significant decrease of the longitudinal resistivity coinciding with the Hall plateaux. They pointed out the thermodynamic nature of the phenomenon, which they described as a phase transition between the metallic phase, stable in zero field, and a unique FISDW, with periodic reentrance of the normal metallic phase [5].

4.2 FISDW Wave Vector Quantization

However, shortly after the work by Gork'ov and Lebed [5] and Héritier et al. [7] suggested that, in fact, the step-like Hall voltage was indeed a new form of Quantized Hall Effect, intimately connected with the cascade mechanism. Their argument was based on the discovery that the most divergent loop, in the presence of the magnetic field, is obtained for a quantized field dependent longitudinal component of the wave vector :

$$\mathbf{Q} = \left(Q_x = 2k_{\mathrm{F}} - n\frac{2\pi}{x_0}, Q_y \simeq \frac{\pi}{b}, Q_z \simeq \frac{\pi}{c} \right) . \tag{4.1}$$

Here $x_0 = h/ebH$ is the magnetic length, which appears if one considers the area bx_0 threaded by one quantum flux, $\phi_0 = h/e$, between two neighboring chains separated by b under a magnetic field H. The energy scale associated with this magnetic length is $\hbar\omega_{\mathrm{c}} = \hbar v_{\mathrm{F}}\pi/x_0 = ev_{\mathrm{F}}bh/2$, where ω_{c} is the semi-classical frequency of the electron wave packet motion, traveling the Brillouin zone from $-\pi/b$ to $+\pi/b$. The behavior of the noninteracting electron spin susceptibility $\chi_0\,(2k_{\mathrm{F}})$ as a function of H and T is given in Fig. 4.1 [8]. Analogous results were obtained by Yamaji [9], at zero temperature.

4.3 Quantum Cascade of Phase Transitions

According to Héritier et al. [7], the index n appearing in the wave vector x-component of the SDW instability labels each SDW subphase and decreases by one unit from sub-phase to sub-phase as H increases. At the same time, this index is the number of exactly filled Landau bands of unpaired particles left by the SDW condensation in a situation of imperfect nesting. When H varies, a competition develops between the condensation energy of the SDW order and the diamagnetic energy: the former is lowered if electrons and holes condense

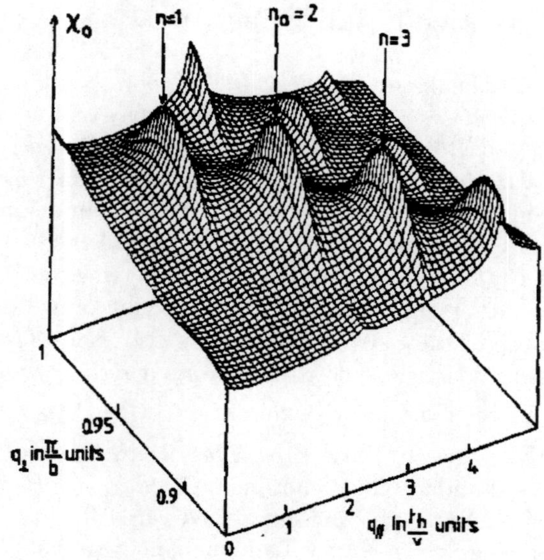

Fig. 4.1. Staggered spin susceptibility in the normal phase in the presence of a magnetic field, at fixed $Q_z = \pi/c$ (from [7]). A series of peaks appear under field parallel to the c axis. The peaks have a quantized component along the a axis. As the field varies, the peak intensities vary, and the absolute maximum shifts from one peak to an other

and increase the amplitude of the order parameter; the latter is lowered if Landau bands are exactly filled and the Fermi level sits in the middle of the gap between two Landau bands; accordingly, the SDW wave vector varies at fixed n so that the pockets of unpaired particles have exactly the right number for an integral number of filled Landau bands below the Fermi level. The FISDW wave vector varies smoothly with the increasing field, until it becomes energetically favorable to jump to the quantum number $(n-1)$. This picture was later confirmed by the analytical mean field and weak coupling theory of the condensed phase [10,11]. First, Lebed [10] showed that the free energy might be mapped by the BCS energy with effective coupling constant, and cut-off energy being magnetic field and pressure dependent nontrivial functions. He made a conclusion that the transition lines between sub-phases are vertical and of the weak first order, which was consistent with experiments. Later on, the same conclusions were published by Maki et al. [12] and Poilblanc et al. [11], who also included the role of the secondary gaps [11]. It is possible to interpret the quantization condition as a nesting quantization: the area between one sheet of the normal state Fermi surface and the other sheet translated by Q is quantized in terms of the area quantum $eH/\hbar$. Hence the name "Quantized Nesting Model" given to this theoretical model [7].

4.4 Novel Quantized Hall Effect

There is now overwhelming evidence for the thermodynamic nature of the cascade of FISDW phases [13] and for the occurrence of a novel type of Quantized Hall Effect [14]. Both aspects are intimately connected. Very well defined Hall plateaux with the ratios 1:2:3:4:5 are observed in $(TMTSF)_2PF_6$, for example. The quantization of the Hall effect is an immediate consequence of the "Quantized Nesting Model," which interprets the cascade of FISDW. In the ordered FISDW phase, the order parameter $\Delta(x)$ acts as an effective potential which couples electronic states, not only at wave vectors k and $k+Q$ because of the SDW ordering, but also at $k+Q-N\frac{2\pi}{x_0}(k_x/|k_x|)$. Therefore, the quasiparticle spectrum exhibits a series of gaps open at $k = \frac{1}{2}\left(Q_x - N\frac{2\pi}{x_0}\right)$. The free energy is minimized when the Fermi level lies in the middle of the largest of these gaps. This occurs when $Q_x = 2k_F + N\frac{2\pi}{x_0}$. We have separate Landau bands containing $1/2\pi b x_0 = eH/h$ states per unit surface. At zero temperature, each quantized FISDW phase has either completely filled or completely empty Landau bands. If the FISDW is pinned by some mechanism, for example by impurities, only single particle excitations contribute to the conductivity. Since perfect nesting in the c direction makes the problem effectively two-dimensional, Laughlin's gauge invariance arguments [15] tell us that the single atomic layer Hall conductivity is exactly quantized at zero temperature in units of e^2/h. We have

$$\sigma_{xy} = ne^2/h. \tag{4.2}$$

The value of n is precisely the value of N, which labels the FISDW subphase, where $Q_x = 2k_\mathrm{F} + N\frac{2\pi}{x_0}$. The proof was given by Poilblanc et al. [16], using an approach due to Streda [17]. This result was rederived by Yakovenko [18], using the Kubo formula. He used the topological properties of the wave functions in reciprocal space, which result from the variation of their phase factor upon transporting them along closed contours.

Obviously, this mechanism for the quantization of the Hall effect results from cooperative phenomena fixing the FISDW wave vector so as to maintain the Fermi level in a gap of itinerant states. It is therefore completely different from the usual QHE observed in the electron gas of the semiconductor heterojunctions.

Acknowledgments

It is a great pleasure to have the opportunity to acknowledge helpful collaborators G. Montambaux, P. Lederer, D. Poilblanc, S. Kaddour, S. Haddad. I also thank D. Jérome, M. Ribault, C. Pasquier, P. Garoche, F. Pesty, L.P. Gork'ov, A. Lebed, V. Yakovenko, K. Maki, K. Yamaji, P. Chaikin for many helpful, stimulating, and friendly discussions, and many other colleagues that it is impossible to quote all of them.

References

1. For a review, see T. Ishiguro and K. Yamaji, *Organic Superconductors*, (Springer-Verlag, Berlin, 1990) and references therein; J. Phys. IV **10** (2000) and references therein.
2. J.F. Kwak, J.E. Shirber, R.L. Greene et al., Phys. Rev. Lett. **46**, 1296 (1981)
3. M. Ribault et al., J. Phys. Lett. **44**, 1268 (1983)
4. P.M. Chaikin et al., Phys. Rev. Lett. **51**, 953 (1983)
5. L.P. Gor'kov, A.G. Lebed, J. Phys. Lett. **45** L433 (1984)
6. C. Kittel, *Quantum Theory of Solids*, (Wiley, New York, 1987)
7. M. Heritier, G. Montambaux, P. Lederer, J. Phys. Lett. **45**, L943 (1984)
8. G. Montambaux, M. Heritier, P. Lederer, Phys. Rev. Lett. **55**, 2078 (1985)
9. K. Yamaji, J. Phys. Soc. Jpn. **54**, 1034 (1985)
10. A.G. Lebed, Sov. Phys. JETP **62**, 595 (1985)
11. D. Poilblanc, M. Heritier, G. Montambaux, P. Lederer, J. Phys. C Solid State Phys. **19**, L321 (1986)
12. K. Maki, Phys. Rev. B **33**, 4826 (1986); L. Chen, K. Maki, V. Virostek, Phys. B **143**, 444 (1986)
13. P. Garoche, R. Brusetti, K. Bechgaard, J. Phys. Lett. **43**, L417 (1982); T. Takahashi, D. Jérome, K. Bechgaard, J. Phys. Lett. **43**, L565 (1983); P. Garoche, R. Brusetti, K. Bechgaard, Phys. Rev. Lett. **49**, 1346 (1982); M.J. Naughton et al., Phys. Rev. Lett. **55**, 969 (1985); F. Pesty, P. Garoche, K. Bechgaard, Phys. Rev. Lett. **55**, 2495 (1985)
14. J.R. Cooper, W. Kang, P. Auban, et al., Phys. Rev. Lett. **63**, 1984 (1989); S.T. Hannahs et al., Phys. Rev. Lett. **60**, 1189 (1988); M.J. Naughton et al., Phys. Rev. Lett. **61**, 621 (1988); L. Balicas, Kriza, F.I.B. Williams, Phys. Rev. Lett. **75**, 2000 (1995)
15. R.B. Laughlin, Phys. Rev. B **23**, 563 (1981)
16. D. Poilblanc, G. Montambaux, M. Héritier et al., Phys. Rev. Lett. **58**, 270 (1987)
17. P. Streda, J. Phys. C **15**, L1299 (1982)
18. V.M. Yakovenko, Phys. Rev. B **43**, 11353 (1991)

Part II

General Reviews

5

La Tour des Sels de Bechgaard

S.E. Brown, P.M. Chaikin, and M.J. Naughton

The interplay between electronic correlations and dimensionality effects in the Bechgaard salts $(TMTSF)_2X$ are manifested in a diversity of ground states, from spin–density wave (SDW) insulator to superconductivity. Of particular interest in this class of materials, known for quasi-one-dimensional (q1d) behavior, are the nontrivial effects of magnetic fields. While pressure destabilizes the SDW in $(TMTSF)_2PF_6$ in favor of a superconducting ground state, a magnetic field restores it via a cascade of intermediate field-induced SDW phases. Below the threshold for onset of the field-induced phases, angular magnetoresistance oscillations are used to map the q1d Fermi surface. However, Lebed's magic angle effect, which associates a set of observations with specific orientations of the magnetic field, remains a conundrum within the framework of standard semiclassical transport and Fermi liquid theory. Finally, the superconducting state remains topical more than a quarter century after the discovery of organic superconductivity in $(TMTSF)_2PF_6$, because of its robustness against applied magnetic fields. Indeed, H_{c2} exceeds the paramagnetic limiting field by a significant factor, prompting a number of investigations into the symmetry of the order parameter, with the goal of discerning how the conventional suppression of superconductivity by magnetic fields is avoided.

5.1 Introduction to the Bechgaard Salts

This chapter provides an introduction to the present understanding of the physics of the $(TMTSF)_2X$ molecular organic conductors, which is frequently claimed by some of us to possess most of the interesting condensed phases and exhibit all of the conduction mechanisms known to man. On any given day, one can arrange to make a specimen a superconductor, a metal, a semimetal, a semiconductor, or a correlated insulator, exhibiting single particle, superconducting, quantum Hall, and sliding density wave transport. Even this is not the most interesting aspect of $(TMTSF)_2X$. All of this rich physics follows from an interplay between the quantum essence of the electron, its spatial, q1d confinement, and the resulting enhancement of correlation effects. As such, it is of essentially no consequence that the organic conductors are organic or carbon based. Rather, it is the low-dimensional nature that matters, and all of the new physics revealed in the q1d organic conductors will be present in any other system, organic or otherwise, that properly mimics the q1d correlated electron nature of $(TMTSF)_2X$. This statement has been proven true already for several other q1d organic systems.

The series of organic charge transfer compounds $(TMTSF)_2X$ have become known as the Bechgaard salts [1], and the range of properties and phases alluded to above are explored by tuning the temperature, the applied magnetic field, and the pressure through mechanical or chemical means [2, 3]. $(TMTSF)_2PF_6$ was the original organic superconductor, discovered in Orsay in 1979 [4]. For that system, high pressure is necessary to suppress a SDW phase before superconductivity is observable below $T \sim 1$ K. Replacing the PF_6 counterion for ClO_4 eliminates the need for high pressure: $(TMTSF)_2ClO_4$ is superconducting below $T_c = 1.4$ K even without it. Still, quench-cooling the system to suppress ordering of the ClO_4 counterions restabilizes the SDW ground state. In both superconductors, new phases are observed when a magnetic field is applied perpendicular to the most conducting layers [5, 6]. This is the cascade of field-induced SDW transitions; they are an emergent phenomenon stabilized by a magnetic field that acts to restore the low-dimensional nature of the orbital electronic properties which, in the case of the PF_6 salt, were destroyed by the high pressure. In what follows, we provide a window into these and many other properties that make the Bechgaard salts remarkable and unique. We begin with the fundamentals: crystal structure, electronic energy spectrum, and transport properties. These are followed by sections describing the properties and phases more particular to the Bechgaard salts: the effects of magnetic field, including the stabilization of the field-induced SDW phases and the novel magnetoresistance oscillations observed under rotation of the field relative to the crystal axes. Finally, we describe the unusual features of the superconducting state, a story which seems far from the end.

5.1.1 Crystal Structure and Electronic Band Structure

It is natural to expect a stacking arrangement for the planar TMTSF molecules when crystallized. In the charge-transfer salts formed with singly charged counterions such as PF_6 or ClO_4, the stacks are arranged to form the three-dimensional, anisotropic structure shown for two views of the ClO_4 salt in Fig. 5.1 [7, 8]. The crystal space group is triclinic, with lattice constants $a = 7.297$, $b = 7.711$, and $c = 13.522$ Å, and angles $\alpha = 83.39°$, $\beta = 86.27°$, and $\gamma = 71.01°$. Referring to the Bechgaard salts as q1d acknowledges the underlying stacked nature and resulting conductivity anisotropy. The donor molecules are separated along the stack direction by nearly twice the selenium

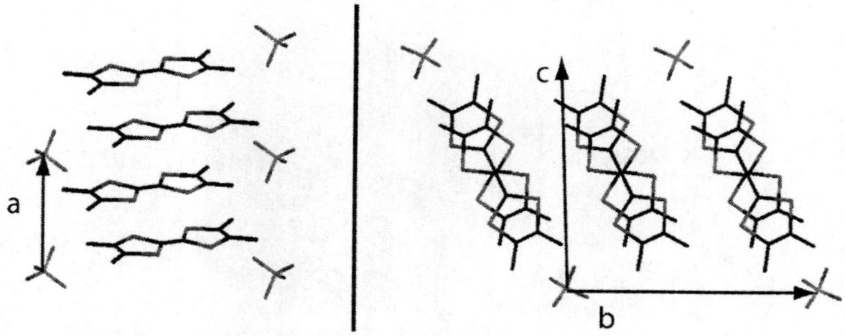

Fig. 5.1. Crystal structure for $(TMTSF)_2ClO_4$in the anion-ordered state, projected along the **b**-axis (*left*), and the stacking, **a**-axis

atom Van der Waals radius, ~3.8 Å, and the Se–Se interaction is strongest in this direction. There is significant interaction between stacks as well; the donor molecules and anion sublattices are segregated in layers is typical of many conducting organic charge-transfer salts. This aspect is reminiscent of "modulation doping," and is also common to many layered inorganic compounds such as the cuprate high-temperature superconductors [9]. But unlike the inorganic materials, these organic compounds have fixed 2:1 stoichiometry, so the charge density is restricted to a commensurate value: transfer of a single electronic charge from every two donor molecules to the counterion leads to a practically inert counterion sublattice and relatively small conductivities in the interlayer direction. For many physical properties described below, such as analyzing semiclassical orbital trajectories in a magnetic field, it is useful to define an orthogonal set of lattice vectors, $(\mathbf{a}, \mathbf{b}', \mathbf{c}^*)$, with $\mathbf{c}^* \perp (\mathbf{a}, \mathbf{b})$, and $\mathbf{b}' \perp (\mathbf{a}, \mathbf{c}^*)$.

The electronic energies are calculated in a tight-binding approximation [10,11],

$$E(\mathbf{k}) = -2t_a \cos(k_a a) - 2t_b \cos(k_b b) - 2t_c \cos(k_c c). \tag{5.1}$$

This assumes that the close Se–Se contact is the dominant overlap, and that the molecular orbital approximation applies. As the highest occupied molecular orbital consists of out-of-plane molecular π-orbitals, the largest overlap is intrastack and it is antibonding ($t_a < 0$). In the standard analysis, there is a partial cancellation for hopping integrals in the **b** direction for the TMTSF salts; thus, an enhancement of phenomena associated with one-dimensional physics is expected. The same applies to the family of isostructural salts made with tetramethyltetrathiafulvalene (TMTTF) donors. The overlaps in the third direction are smaller still: molecular separation is greater because of the interpenetrating counterion layers. In the approximation of [11], the integral ratios are $(4t_a : 4t_b) \approx (0.5\,\mathrm{eV} : 0.05\,\mathrm{eV})$ for the full width of the 1/4-filled band. A small dimerization of the intermolecular distances along the stack folds the band into an equivalent 1/2-filled system. The resulting Fermi surface for hole carriers for an orthorhombic cell and a small **c**-axis

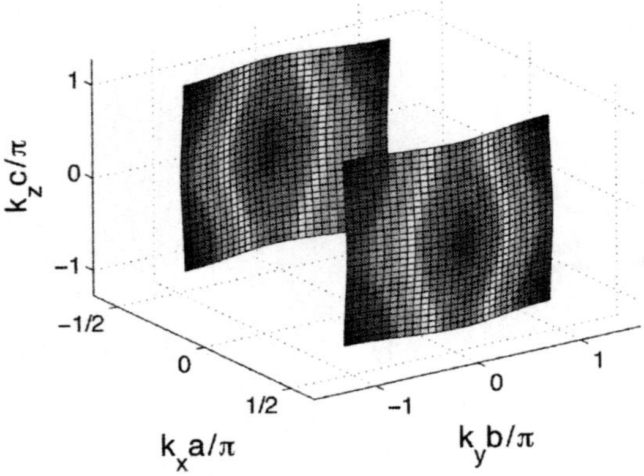

Fig. 5.2. The two sheets of the open Fermi surface of TMTSF salts, in the simplified case of orthorhombic crystal structure. The color coding is included to emphasize the warping and refers to $|k_x|$

overlap appears in Fig. 5.2. Here, the quasi one-dimensional (q1d) character is evident from the separate sheets, even though the temperature equivalent to the transverse bandwidth is greater than room temperature. The larger transfer integrals $(t_a : t_b) \approx (0.25\,\mathrm{eV} : 0.025\,\mathrm{eV})$ were obtained from measurements of the plasma frequency [12], and these increase slightly on cooling due to lattice contraction. Later, we describe the bandwidths inferred from magnetoresistance oscillations [13], which give $t_b \sim 25 - 30\,\mathrm{meV}$ for the PF_6 salt and $t_b \sim 12\,\mathrm{meV}$ in the anion-ordered state of the ClO_4 salt. These estimates agree roughly with the conductivity anisotropy according to the expectation $\sigma_a/\sigma_b \sim (t_a a)^2/(t_b b)^2$ [14–16].

At sufficiently high temperatures, where the conductivity is small enough that the scattering rate is less than the transverse hopping integral, there is only diffusive interstack transport. The crossover from one-dimensional to higher-dimensional coherence is expected in a temperature range of order room temperature when correlations are neglected. Experimentally, where the crossover occurs is a point of controversy. Based on NMR relaxation rate measurements, it was suggested that correlations drive the actual crossover to a lower temperature, ~30 K [17, 18]. Later optical [19] and low-frequency transport experiments [16,20] were interpreted to place the crossover at much higher temperatures.

5.1.2 The Ambient-Pressure Spin–Density Wave State in $(TMTSF)_2PF_6$, and Effect of Pressure

A good starting point for appreciating the different phases of the Bechgaard salts is the ambient-pressure SDW of $(TMTSF)_2PF_6$, observed at temperatures below $T_{SDW} = 12.1\,\mathrm{K}$. For $T > T_{SDW}$, the resistivity is anisotropic, but decreasing for all directions. Below the transition temperature, the Fermi surface is destroyed by the instability and the transport is activated. The magnetic nature of the ground state was confirmed by many methods, including magnetic susceptibility [21] and antiferromagnetic resonance [22]. With the application of approximately 0.6 GPa hydrostatic pressure, the transition temperature $T_{SDW} \to 0$, and the ground state is superconducting. Insofar as the transition temperatures are much smaller than the Fermi temperature, weak-coupling theories should apply to the ground states and phase transitions, and in fact, the semiconducting gap of the SDW state inferred from transport measurements is close to $\Delta = 1.7 k_B T_{SDW}$.

The SDW transition is one of the expected ground states for q1d electronic systems. In one dimension, the Lindhardt electronic susceptibility increases on lowering the temperature as $\chi_0(Q = 2k_F) = 2\mu_B^2 N(E_F) \ln(W/k_B T)$. In the presence of repulsive interactions U, the high-symmetry normal state is unstable when $U\chi_0(\mathbf{q})/2\mu_B^2 > 1$, with W a cutoff energy of the order of the bandwidth. A gap opens on the Fermi surface, leading to a net lowering of the total energy. There are important implications from including the interstack hoppings of the q1d systems. First, fluctuations are reduced so as to produce a phase transition at finite temperature. Increasing the interstack hopping leads to worsened nesting conditions, and the energy gain of the SDW phase relative to the normal phase diminishes. Generically, a first-order transition is expected into the metallic state at $T = 0$ as the interstack interaction is increased [23]. The nesting is critical when $t_b^2/t_a = O(\Delta)$, and $(TMTSF)_2PF_6$ is a classic example of this behavior [2, 24] (Fig. 5.3).

5.1.3 A Broader Context for Correlation Effects: the TMTTF Salts

If the Se atoms of the TMTSF molecule are replaced with S, we have the TMTTF molecule, and its 2:1 salts are isostructural to the Bechgaard series. At ambient pressure, they are insulating in the sense that from temperatures in the range of 200–300 K and below, the resistivity increases upon cooling,

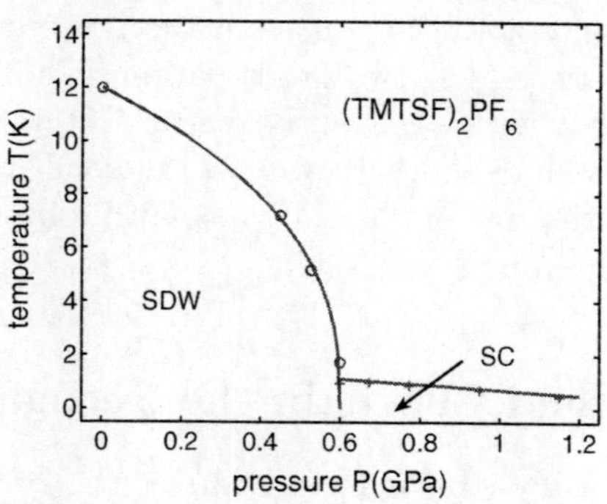

Fig. 5.3. Temperature/pressure phase diagram of $(TMTSF)_2PF_6$, showing spin–density wave (SDW) and superconducting (SC) phases

and they also exhibit transport anomalies at intermediate temperature that were called by Coulon et al., the "structureless" transition [25]. With applied pressure, these materials become good conductors and even superconductors, just like the TMTSF salts [26], suggesting that quantitative differences in dimensionality and correlations determine the physical properties, and furthermore, the relative importance of the different energy scales is implicitly controlled by the lattice constants through applied pressure, or by exchanging counterions or donor molecules.

There are two dominant routes to the insulating state in these compounds. The dimerization of the intermolecular distances along the stack turns on two-electron Umklapp scattering [27] with a coupling constant $g \sim DU/E_F$, with D the dimerization potential and U the repulsive on-site interaction [27, 28]. This route has long been taken as the principle reason for insulating, paramagnetic behavior in the TMTTF salts. However, with one hole for every pair of donors and strong anisotropy, including repulsive interactions from both nearest neighbor (U) and next-nearest neighbors (V) tends to stabilize an insulating, charge-ordered state [29–31]. Experiments on a number of TMTTF salts confirm that the charge-ordered (CO) state is ubiquitous to the TMTTF salts at ambient pressure [32,33]; indeed, the onset of the CO phase is Coulon's structureless transition.

If either of these two routes is dominant, there is a charge gap, but the system remains paramagnetic at elevated temperatures. The CO state lowers the symmetry, so it develops continuously below T_{CO}. The magnetic entropy is removed at a lower temperature, where either a spin gap opens, or long range magnetic order develops.

In fact, the CO amplitude ultimately controls the ground state symmetry; high pressure studies of $(TMTTF)_2AsF_6$ offer a clear example [34]. At ambient pressure, ^{13}C NMR spectroscopy shows this compound undergoes the CO transition at $T = 103\,K$, and a spin-Peierls transition to a nonmagnetic ground state at $T = 13\,K$. With applied pressure, T_{CO} decreases, whereas $dT_{SP}/dP > 0$. The part of the phase diagram accessed by experiment is shown in Fig. 5.4. The two second-order phase transition lines meet at a tetracritical point. The suppression of the SP phase in the presence of the CO indicates that the coupling between the two-order parameters is repulsive. The phase diagram indicates that with sufficiently strong CO amplitude, the SP phase would not form [35], and another ground state will take its place. Indeed, replacing SbF_6 counterions for the AsF_6 leads to a larger CO amplitude and an antiferromagnetic ground state [36].

5.2 Magnetic Field Effects in the Bechgaard Salts

5.2.1 A Little History and a Few Equations

During the 1960s and 1970s, there was an intense search for organic materials which exhibited metallic conductivity. These efforts were a reaction to the

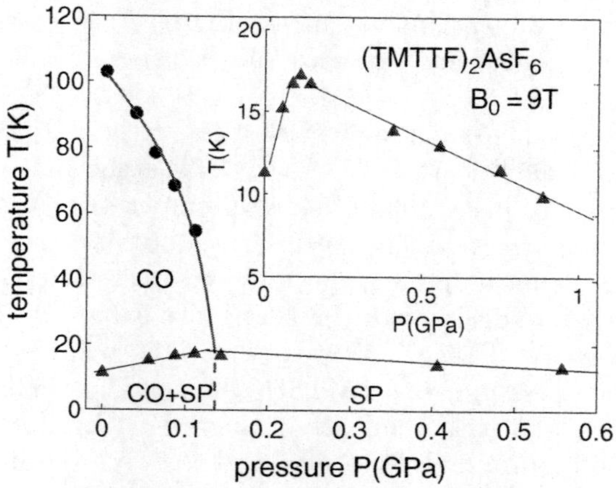

Fig. 5.4. Pressure–temperature phase diagram of $(TMTTF)_2AsF_6$, showing the demonstrating the repulsive interaction between CO and SP order parameters

newly discovered theory of superconductivity and the calculations and speculation of Little [37] that very high superconducting transition temperatures might result from electron–exciton interactions in organics and particularly in polymers. Previously, polymers, organic crystals, and organic charge transfer salts were insulators and semiconductors and were viewed as potentially useful for low conductivity applications such as xerography. Since the usual precondition for superconductivity was a metallic state, the hunt began for organic metals. The results from transport studies took on an increasingly important role, since high conductivity was the goal, but also because useful information about the characteristic energies and carrier densities were readily attainable.

The basic transport coefficients and their simplest material dependences are conductivity $\sigma = ne^2\tau/m = ne\mu$, thermopower S = entropy/carrier ($\sim E_g/2k_BT$ for semiconductors and $\sim(k/e)(k_BT/E_F)$ for metals), Hall coefficient, $R_H = 1/ne$, and magnetoresistance, $\Delta\rho(H)/\rho = (\omega_o\tau)^2 = (\mu H)^2$, where n is the carrier density, e is the electron charge, τ is the scattering time, m is the effective (band) mass, μ is the mobility, E_g is the energy gap, and $\omega_o = eH/mc$ is the cyclotron frequency. A simple explanation for single and two carrier Drude models and for Boltzmann transport is found in the recent paper of Wu et al. [38]. While the conductivity was, and remains, the most widely and easily measured property, it provides information on the product of carrier density and mobility. To separate the two effects, one needs the Hall coefficient, which uniquely measures carrier density, or the magnetoresistance, which provides information on mobility. The characteristic electronic energy of the carriers is obtained from a thermopower measurement. Before the mid-seventies, most of the organic conductors, typified by the q1d TCNQ charge transfer salts [39], remained metallic only slightly below room temperature, exhibiting metal–insulator transitions due to a combination of disorder, correlation and lattice distortion effects. The materials, therefore, had low mobilities, reduced dimensionality and strong electron correlations, making magnetotransport coefficients small and all transport coefficients difficult to interpret.

The situation changed markedly with the synthesis by Wudl [40] of the TTF molecule in the mid-70s. TTF–TCNQ was the first organic material which was metallic to below liquid nitrogen temperature [41]. At $T \sim 60$ and 40 K, there were sharp metal–insulator charge density wave transitions to a low temperature semiconducting phase. A few years later, the first organic superconductors were discovered, the Bechgaard salts, based on a derivative of the TTF molecule, TMTSF. With the metallic phase stable at low temperatures (under pressure for $(TMTSF)_2PF_6$ and at ambient pressure for $(TMTSF)_2ClO_4$), and with conductivity increasing by three-orders of magnitude from the room temperature value, these were the first high mobility organic metals as well as the first organic superconductors. However, as we have seen above, the band structure calculations as well as the proximity to a Peierls instability were consistent with a Fermi surface without any closed orbits. This would suggest an uninteresting landscape for magnetotransport and other magnetic field effects. Along the q1d x direction, the Brillouin zone is not filled, the carriers have a well-defined effective mass and the magnetotransport should be as for a single carrier Drude model (conventional Hall effect and negligible magnetoresistance). In the k_y and k_z directions, the Fermi surface touches the zone edges and the effective mass is ill-defined, having both electron- and hole-like regions. The simplest treatment of this situation would be a two-carrier Drude model [38], which has a nonsaturating quadratic magnetoresistance $\Delta\rho(H)/\rho = (\omega_o\tau)^2$. Of course, the absence of closed orbits eliminates the possibility of magnetic quantum oscillations such as the Schubnikov–de Haas (SdH) and de Haas–van Alphen (dHvA) effects. It was, therefore, a great surprise when a very large magnetoresistance (an order of magnitude resistance change in a field of $B = 7\,\mathrm{T}$) was found in the Bechgaard salts [5], accompanied by structure that greatly resembled SdH oscillations, in the early 1980s.

To theoretically understand the unusual behavior in a magnetic field, the first attempt was to see what happens to Landau quantization in an open orbit system. For closed orbits, one considers solutions to Schrödinger's equation in the plane perpendicular to the magnetic field. Without a field, the solution is a two-dimensional free electron band

$$E(\mathbf{k}) = (\hbar^2/2m)(k_x^2 + k_y^2) \tag{5.2}$$

with dispersion in the k_x and k_y directions. In the presence of a field the continuous spectrum is replaced by a set of discrete Landau levels

$$E_n = (n + 1/2)\hbar\omega_c \tag{5.3}$$

with no dispersion. In the two-dimensional open orbit case originally treated [42], the dispersion relation is

$$E(k_x, k_y) = (\hbar^2 k_x^2/2m) - 2t_b \cos(k_y b). \tag{5.4}$$

After a Landau–Peierls substitution and the choice of a Landau gauge $\mathbf{k} \to \mathrm{i}\nabla - e\mathbf{A}/\hbar c$, $\mathbf{A} = (0, Hx)$, the equation to be solved is

$$\frac{-\hbar^2}{2m}\frac{\partial^2}{\partial x^2}\psi(x) - 2t_b \cos\frac{2\pi}{\lambda}x\psi(x) = \varepsilon(k_x)\psi(x). \tag{5.5}$$

Although energy eigenvalues and eigenfunctions are not trivial to write down (they are Mathieu functions), what is clear is that the dispersion now consists of a set of one-dimensional bands in a reduced Brillouin zone defined by a new periodicity given by the reciprocal lattice vector of length $G = 2\pi/\lambda = eHb/hc$.

It is worth making the comparison between closed and open orbital quantization: for closed orbits the dispersion goes from two-dimensional to no dispersion, 0D, the magnetic length $l_H = (\hbar c/eH)^{1/2} = (\Phi_0/\pi H)^{1/2}$ defines the area of an orbit containing one flux quantum, $\Phi_0 = hc/e$, and the electrons circulate around this orbit with frequency ω_c. For open orbits, the 2D dispersion goes to 1D, the magnetic length $\lambda = 2\pi/G = \Phi_0/bH$ defines the area containing a flux quantum between adjacent chains and the electrons traverse along the Fermi surface from one edge of the Brillouin zone to the other with frequency $\omega_b = ev_{\mathrm{F}}bB/\hbar$. In real space, the semiclassical motion from open orbits corresponds to an electron performing oscillatory trajectory along y while moving in one direction in x. The total excursion along y is $b(t_b/\hbar\omega_b)$, so the orbit shrinks along y as the magnetic field increases. From both the nature of the orbits and the dispersion relation, the 2D open orbit system is one dimensionalized by application of a perpendicular magnetic field. However, the absence of sizable gaps in the 1D spectrum made it unlikely that the one dimensionalization alone could cause the SdH-like oscillations that were seen.

5.2.2 Field-Induced Spin–Density Waves

The presence of extremely large and unexplained magnetic phenomena in the Bechgaard salts led to more detailed and higher field studies by two groups. Independently and simultaneously, they presented Hall effect and other data which strongly suggested that the number of carriers was reduced in a series of transitions as the magnetic field was increased and that this was responsible for the "magneto-oscillations" [43, 44].

Since we know that 1D electronic systems are unstable to the formation of (spin or charge) density waves, such systems should have a Fermi surface instability. The idea of a field-induced SDW instability caused by one-dimensionalization was proposed by Gor'kov and Lebed [45]. (A less sophisticated model of the FISDW was independently proposed in [46].) The Peierls-like instability usually involves nesting the Fermi surface with a wavevector of $2k_{\mathrm{F}}$. However, in the presence of the magnetic field with the new magnetic reciprocal lattice vector G, the nesting can occur at $2k_F \pm nG$, with n an integer. The cascade of FISDW transitions occurs as the quantum number n takes on high values for small fields and decreases to the conventional SDW with $n = 0$ at high fields. Building on the Gor'kov–Lebed model, the nesting wavevectors and geometry for k_x and k_y were worked out by the Orsay group [47]. Also building on Gor'kov and Lebed's work, the FISDW state was further explored by Yamaji [87], Maki [49] and Yakovenko [50].

The original Hall measurements on the ClO_4 salt suggested the Hall resistance was quantized, as in the 2D semiconductor heterostructures exhibiting the quantum Hall effect [51]. But $(TMTSF)_2ClO_4$ also exhibited unquantized, opposite sign Hall steps (Ribault anomalies) [52]. It therefore took some time to experimentally and theoretically establish that these bulk crystals should in fact show a quantum Hall effect, and under what circumstances. The result of the quantized nesting model is that we expect a cascade of FISDW's as the magnetic field is increased. If we had a 2D system, then the instability would onset in arbitrarily weak field at $T = 0$. Since we have a three-dimensional system with dispersion along the k_z-axis, there is a threshold field even at $T = 0$, such that the FISDW gap which opens exceeds $4t_c$ (the bandwidth in the least conducting direction). Thus, at zero temperature, there is a second-order transition to an FISDW state with N-filled Landau bands and exhibiting QHE with $\rho_{xy} = h/Ne^2$ followed by a cascade of first-order phase transitions.

$(TMTSF)_2ClO_4$ has an anion ordering transition at $T_{AO} \approx 24\,\mathrm{K}$, which results in a halving of the Brilluoin zone in the **b** direction and the presence of a more complex band structure. This can roughly be associated with two distinct chains with two Fermi wavevectors k_{F1} and k_{F2} and two Fermi surfaces which may nest independently or in a coupled fashion. The phase diagram is more complicated than in the simplest quantized nesting model [53, 54]. We will return to the case of $(TMTSF)_2ClO_4$ later. Because of the complications in $(TMTSF)_2ClO_4$, attention switched in the late 1980s to the simpler $(TMTSF)_2PF_6$ salt which, however, is more difficult experimentally, since it is only metallic, superconducting and an FISDW system when pressure above $P_c \sim 0.6\,\mathrm{GPa}$ is applied. Experiments which established well-defined quantum Hall plateaus in the expected cascade were shortly thereafter observed independently by two groups [55, 56]. The original data are not as clean as subsequent data, the best of which are shown in Fig. 5.5 [57]. A complete phase diagram in temperature–pressure, magnetic field space followed in the early 1990s and is shown in Fig. 5.6. Here we see one of the most remarkable aspects of the Bechgaard salts. In a single crystal we have metal, SDW insulator, superconductor, and FISDW-QHE phases as T, P, and H are varied.

At pressures sufficient to suppress the anion ordering transition, the salt $(TMTSF)_2ClO_4$ behaves much like $(TMTSF)_2PF_6$. However, at ambient pressure with anion ordering present, the final FISDW transition from the semimetallic $N = 1$ QHE to the $N = 0$ SDW insulator is suppressed from $\sim 8T$ to at least 26 T, and unusual behavior persists to the highest fields studied. In particular, there are striking oscillations in the Hall resistance, which remains

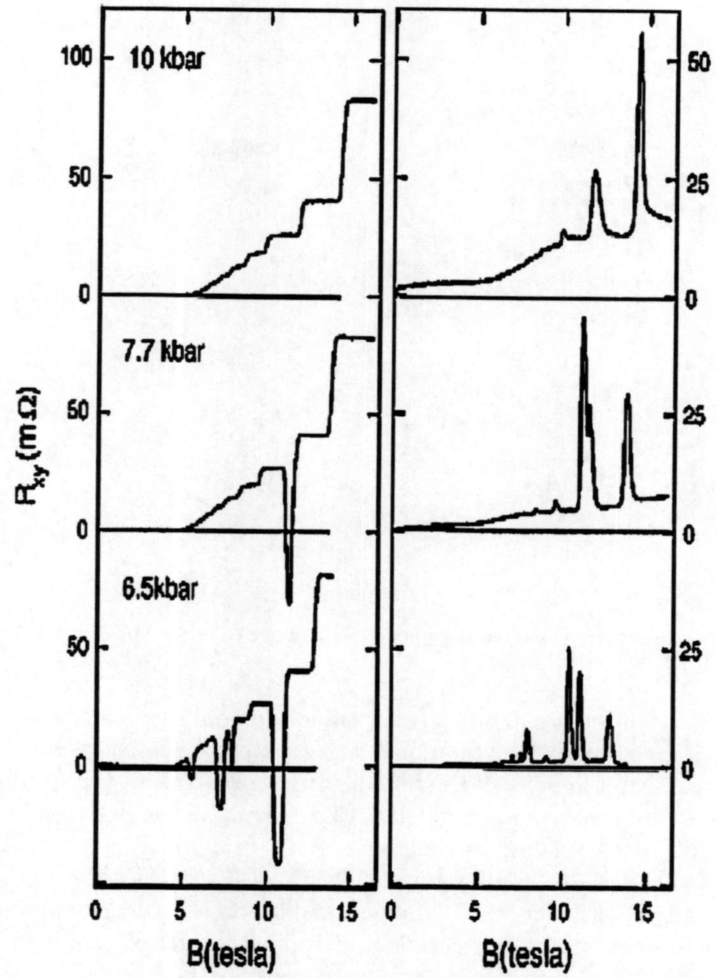

Fig. 5.5. Evolution of magnetoresistance and Hall effect of $(TMTSF)_2PF_6$ under pressure at 300 mK

unquantized. A primitive description of the ground state of the system below 26 T (SDW I) has two FISDW's and two nesting vectors which open gaps on both Fermi surfaces FS_1 and FS_2. Above 26 T (SDW II) there would be a different ground state. A single FISDW with a single nesting vector opens a gap directly on one Fermi surface, e.g., FS_1. However, the unnested Fermi surface is slave to the FISDW in two ways. Satellite gaps from FS_1 at $2k_{F1} + mG$ are swept through FS_2 periodically in $1/B$ as the field is increased, e.g., there is an induced gap at FS_2 whenever $2k_{F1} + mG = 2k_{F2}$ or $m = \hbar(2k_{F2} - 2k_{F1})/ebB$. This results in oscillations in the Hall coefficient [58], the magnetorestance and thermodynamic measurements, but basically leaves the system semimetallic. However, since $2k_{F1} + 2k_{F2} = 2\pi/a$ (the charge transfer is stoichiometric

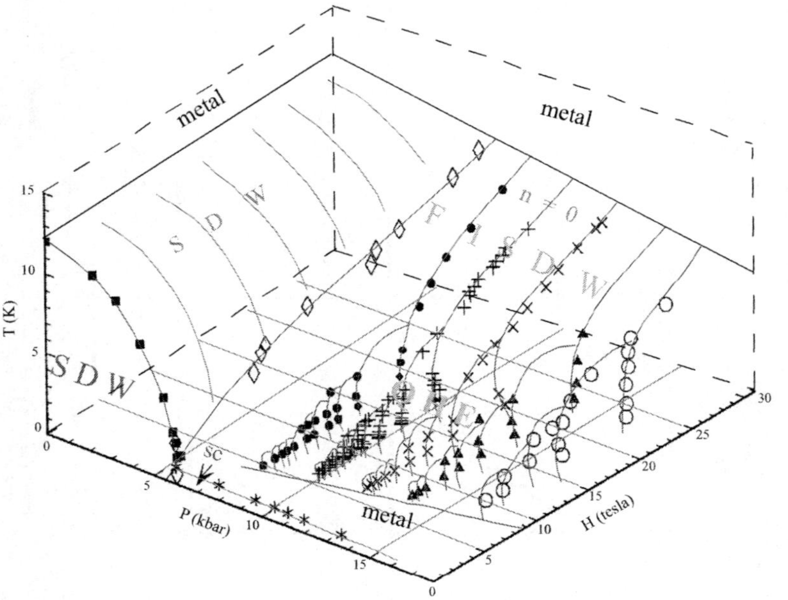

Fig. 5.6. Temperature, pressure, magnetic field phase diagram for $TMTSF_2PF_6$ [6]

at one quarter filling), a gap on FS_1 will induce an Umklapp gap on FS_2. This weak gap may eventually lead to an insulating state at high field. This description captures the periodicity of the high field Hall oscillations, but awaits both experimental and theoretical verification. The nature of the high-field state of $(TMTSF)_2ClO_4$ remains unresolved.

The finite threshold field for the FISDW as $T \to 0$ was, from the beginning, associated with the three dimensionality of the dispersion relation. A two-dimensional system is unstable and gapped in any perpendicular field and the FISDW gap grows as field is increased. For finite bandwidth along the **c**-axis, the FISDW gap must be larger than $4t_c$ for the transition to occur.

5.2.3 Angular Magnetoresistance Oscillations in Quasi-One-Dimensional Conductors

Soon after the initial Gor'kov–Lebed paper, Lebed [59] suggested that the effects of $4t_c$ could be eliminated by simply rotating the magnetic field away from the **c**-axis to have a Lorentz force on electrons moving in the **c** direction. In fact, he suggested that at certain magic angles, where the tilted field would produce commensurate electron motion along the **b** and **c** directions, the FISDW threshold field might be reduced to zero. Commensurate motion occurs when the frequency to cross the Brillouin zone in the **b** direction $\omega_b = ebB_z v_F/\hbar$ and the frequency to cross in the **c** direction $\omega_c = ecB_y v_F/\hbar c_o$ were rationally related (here, c_o is the speed of light), $\omega_b/\omega_c = p/q$ where p and

q are integers. Lebed argued that the dimensionality of the electronic spectrum would change when the field was rotated to these magic angles. This idea was soon followed by the suggestion of Lebed and Bak [14] that the dimensionality change on field rotation should also be seen in magnetoresistance. They predicted sharp resistance peaks at the magic angles.

These Lebed papers were the first in which it was suggested that a rich field of angular dependent magnetotransport phenomena was to be found in highly anisotropic organic conductors. The first experimental tests of these magic angle effects were the work on $(TMTSF)_2ClO_4$ by Osada et al. [61] and slightly later by Naughton et al. [62]. They found resistance dips at the magic angles, rather than the predicted peaks. It is also worth noting that a magic-angle resistance dip was first observed and published in [63], but that neither the authors nor subsequent readers recognized the significance of the dips until after Lebed's calculations and several experimental confirmations. Again, $(TMTSF)_2ClO_4$ is complicated by the anion ordering transition, and cleaner, more pronounced dips were later found in $(TMTSF)_2PF_6$. The fact that the anomalies in resistance are dips, not peaks, means that the theoretical explanation had to be revisited. There have been many subsequent theories, models and experiments, some of which will be discussed below. To date, there is not a successful model which describes even semiquantitatively the angular dependent magnetotransport in the Bechgaard salts for $\mathbf{b}-\mathbf{c}^*$ rotations.

To test some of the models for the Lebed oscillations, which are basically a $\mathbf{b}-\mathbf{c}$ plane field rotation, Danner et al. [13], tried a magnetic field rotation in the $\mathbf{a}-\mathbf{c}$ plane (we refer to the orthogonal set of lattice vectors as simply $\mathbf{a}$, $\mathbf{b}$, and $\mathbf{c}$). They found equally striking angular dependences, but on reflection realized that this phenomena had a straightforward explanation based on the band structure. In fact, on the Lebed oscillations which give information about the geometry of the Brillouin zone and the geometry of the crystal structure, the $\mathbf{a}-\mathbf{c}$ rotations provide a convenient measurement of the Fermi surface, and hence the band structure of the materials. The effect is similar to the Yamaji oscillations [64], which are found in quasi-two-dimensional systems with closed Fermi surfaces, and also used to find their Fermi surface parameters.

The directions for angular dependent magnetoresistance phenomena were completed with the measurement of and initial theory for the "third angular effect" by Yoshino [65] and Osada [66]. Like the $\mathbf{a}-\mathbf{c}$-axis rotation, this effect is directly given by a Boltzmann transport treatment of the q1d Fermi surface, and likewise can be inverted to provide information about the band structure. In a number of recent papers, it has become clear that even these three rotations, Lebed ($\mathbf{b}-\mathbf{c}$), Danner–Kang–Chaikin ($\mathbf{a}-\mathbf{c}$), and Yoshino–Osada ($\mathbf{b}-\mathbf{c}$), are not the whole story. The presence of a field along $\mathbf{a}$ strongly effects the magnetotransport for the original $\mathbf{b}-\mathbf{c}$ rotations (Lee–Naughton effect [67]).

Since simple Boltzmann transport does a good job for the $\mathbf{a}-\mathbf{c}$ and $\mathbf{a}-\mathbf{b}$ rotations, and since it is something that can be calculated or computed in a straightforward manner, it is worth trying to understand what the simplest theory can explain and what it cannot (a recent full quantum treatment and references to other treatments can be found in [68]). In some cases, we can look at the orbits and intuit what we expect; in others, it is appropriate to take what is known of the band structure and simply integrate the equation of motion to determine the orbits and then calculate the transport coefficients. In a periodic solid in a magnetic field, the electrons move on a constant energy surface, with their crystal momentum changing with time due to the Lorentz force:

$$\hbar\left(\frac{\partial \mathbf{k}}{\partial t}\right) = e\frac{\mathbf{v}}{c} \times \mathbf{B}. \tag{5.6}$$

Here, $\mathbf{v}$ is the group velocity $\mathbf{v} = \frac{1}{\hbar}\nabla_{\mathbf{k}}E(\mathbf{k})$ and $E(\mathbf{k})$ is the dispersion relation given in (5.1). The conductivity is related to the velocity–velocity correlation function, which has a simple form in Boltzmann transport,

$$\sigma_{ij} = e^2 \int \frac{\mathrm{d}\mathbf{k}}{4\pi^3}\tau v_i(\mathbf{k})\langle v_j(\mathbf{k})\rangle \left(-\frac{\partial f}{\partial E}\right), \tag{5.7}$$

where f is the Fermi function, τ is the scattering time. The velocity average over orbits weighted by the scattering time is given by:

$$\langle v_j(\mathbf{k})\rangle = \int \frac{\mathrm{d}t}{\tau}\mathrm{e}^{-t/\tau} v_j(\mathbf{k}(t)). \tag{5.8}$$

Given a band structure and scattering time, the equations can be readily integrated numerically. The programming and computations are laborious since the sampling grid in k space must be rather small to capture the important wiggles of the trajectories. On the other hand, we can gain a lot of insight by just looking at the orbits and seeing how they contribute to the velocity average in (5.8). If the velocity $\langle v_{\mathbf{k}}\rangle$ averages to zero, then the conductivity contribution from this orbit is zero. If all of the quasiclassical orbits for a particular field direction give $\langle v_{\mathbf{k}}\rangle = 0$, then a full quantum treatment will show the absence of dispersion in the direction of $\mathbf{k}$. This is the same mathematics and physics that we have seen in the one dimensionalization of the 2D open orbit in (5.8). The absence of dispersion in a certain direction means that the conductivity in that direction goes to zero. For a real system with finite scattering time, the resistance in such a direction will increase as $\frac{\Delta\rho}{\rho} = (\omega_{\mathrm{o}}\tau)^2$, without saturation.

Let us see what the orbits look like on the repeated zone Fermi surface at $k_x = +k_{\mathrm{F}}$. We will be concerned here with the conductivity along the **c**-axis, σ_{zz}. For simplicity (and without losing any physics), we continue to consider that the axes are orthorhombic. Four zones in the **b**–**c** plane are cartooned in Fig. 5.7. In Fig. 5.7a, we show the trajectories for magnetic field applied along the **b**, **c**, and **b**+**c** directions, which are relevant for the original Lebed prediction. For field along the **c** direction, we expect no σ_{zz} magnetoconductance classically, since the current and field direction are parallel and

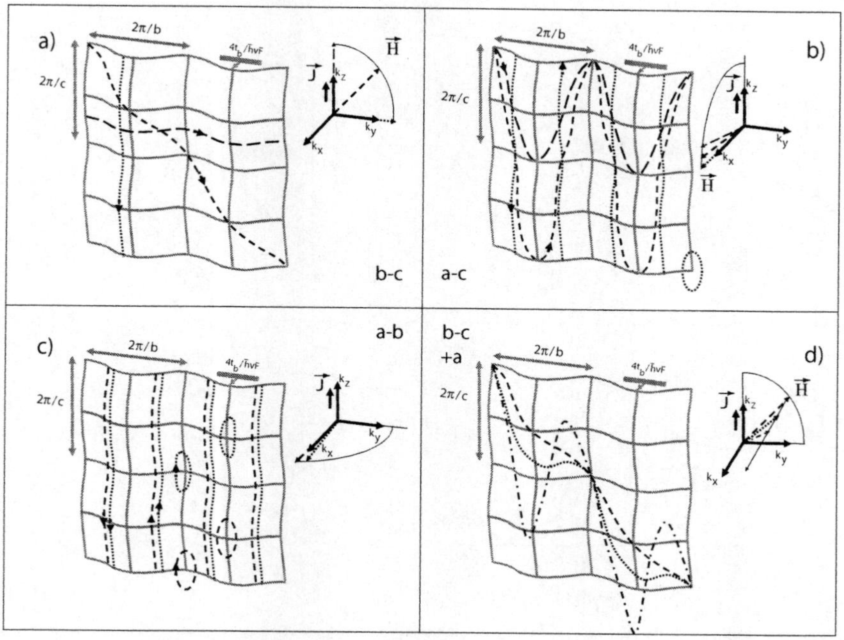

Fig. 5.7. Orbits over four Brillouin zones on the $+k_{Fx}$ Fermi surface for the Bechgaard salts showing trajectories corresponding to different field rotations. The dispersion relation is assumed in the simplest form with only t_a, t_b, and t_c contributions and the structure is taken as orthorhombic. We are interested here in σ_{zz} and hence in the velocity averaging along z. (**a**) For the original Lebed magic angle rotation, **b** to **c** the quasiclassical velocity along **c**, $\langle v_z \rangle$ averages to zero for all directions but $H//\mathbf{c}$, but the averaging is faster $\sim\cos\theta$ as $\boldsymbol{H} \to \mathbf{b}$. For this bandstructure in quasiclassics the only magic angle is **c**. (**b**) The geometry and trajectories for **a** to **c** (Danner–Kang–Chaikin) rotations. Here the magic angles from quasiclassics are directly related to the band parameters. The fastest averaging of $\langle v_z \rangle$ is when an orbit sweeps across the Fermi surface crossing an integral number of $2\pi/c$'s for each $2\pi/b$. (**c**) For the third angular dependence, **a** to **b** (Yoshino–Osada) rotation, the special angles involve points of inflection and also measure bandstructure properties. (**d**) The original Lebed magic angles do show up prominently in **b** to **c** rotation in the presence of a field along **a** (Lee–Naughton–Lebed oscillations). The orbits shown here are for **b**+**c**. The trajectory for no field along **a** as before averages $\langle v_z \rangle$ to zero quickly. However, for a field along **a** there is an asymmetry for the right and left halves of the Fermi surface and $\langle v_z \rangle \neq 0$. Note that for non-Lebed magic angles this orbit does not repeat and eventually every point on the Fermi surface is covered returning the average $\langle v_z \rangle$ to zero

the Lorentz force vanishes. The orbits for this field direction are all orthogonal to **c**, the velocities are not averaged along **c**, and hence there is no magnetoresistance. For an applied field along **b** and current along **c**, the Lorentz force is maximum and we expect classically the largest magnetoresistance for ρ_{zz}. In terms of velocity averaging, the trajectory averages uniformly over

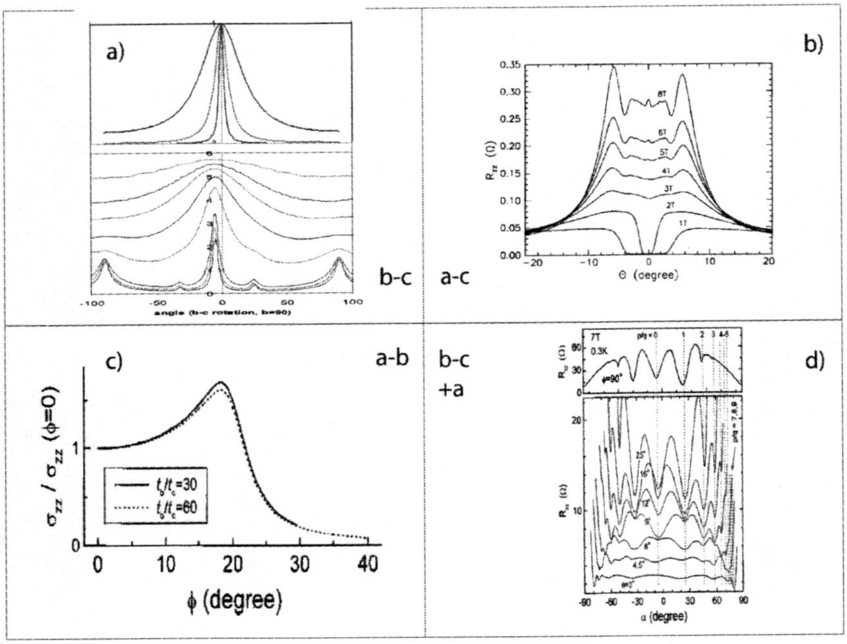

Fig. 5.8. Magnetoresistance corresponding to the trajectories shown in the previous figure. (**a**) The upper part of the figure shows the expected behavior of the conductance σ_{zz} for **b** to **c** rotations for $\omega_o \tau$ values of 3, 10, 30. The only magic angle is for field along **c**. The lower trace is data for the same rotation but from the experimental data of [70]. Unlike the quasiclassical Boltzmann result we see magic angles at $\mathbf{c}, \pm\mathbf{b}+\mathbf{c}, \pm 2\mathbf{b}+\mathbf{c}$ and at **b**. The last is not expected in any Boltzmann treatment. (**b**) Magnetoresistance data for the **a** to **c** rotation in $(TMTSF)_2ClO_4$ [13]. This corresponds quantitatively to the Boltzmann transport results. Note the small peak at zero which measures the smallest bandwidth t_c and is now known as a "squit." (**c**) Calculated magneto-conductance for the third angular dependence, **a** to **b**. The data closely follow the Boltzmann prediction. (**d**) Data for **b** to **c** rotations with various tilts to include a field component along **a** [67]. The upper curve with no **a** component corresponds to the inverse of the conductance data shown in (**a**). The Lebed magic angles here are fit by no classical model. However, the lower data with a large component along **a** is well fit by Boltzmann transport with the trajectories illustrated in the previous figure

the sinusoidally varying $v_z = 2ct_c cos(k_c c)$ to give $\langle v_z \rangle = 0$ and we therefore expect $\Delta\rho(\mathbf{H} \parallel \mathbf{b})/\rho = (\omega_b \tau)^2$. For any angle other than 0 ($\parallel$ **c**), including the **b**+**c** direction illustrated, $\langle v_z \rangle = 0$. But the averaging is not as fast, so we will obtain, in general, $\frac{\Delta\rho(\theta,H)}{\rho} = (\sin(\theta)\omega_o \tau)^2$. This angular dependence is shown in Fig. 5.8a. There is a simple $(\sin\theta)^2$ dependence and the dependence is smooth through the Lebed magic angle at **b**+**c**. The only magic angle here is $\theta = 0$ where the magnetoresistance vanishes.

Where, then, do the Lebed effects come from in a simple Boltzmann model? Osada [69] introduced a more complex band structure in which there is dispersion along the **b**+**c** direction (e.g., a term in $E(\mathbf{k})$ such as $-2t_{bc}\cos(\mathbf{k}_{bc}\cdot(\mathbf{b}+\mathbf{c}))$). A field along **b**+**c** does not give $\langle v_{\mathbf{b}+\mathbf{c}}\rangle = 0$, there is a field independent conductivity along **b**+**c** and, since there is a component of this along c, there is a contribution to σ_{zz} which does not depend on field. Assuming there are such contributions at all Lebed magic angles might explain the phenomena. However, there is no evidence of such large, next-nearest neighbor transfer integrals in experiments or in band structure calculations.

On the other hand, the Danner oscillations from an $\mathbf{a}-\mathbf{c}$ field rotation are straightforward to explain. They are shown schematically in Fig. 5.8b. To understand electron trajectories on the Fermi surface, it is important to recall that $\hbar\mathbf{v} = \nabla_{\mathbf{k}}(E(\mathbf{k}))$, and that since the Fermi surface is a surface of constant energy, the velocities are always directed perpendicular to the Fermi surface at each point. For field aligned along the **a** direction, there is no Lorentz force from the dominant velocity v_x, only from the weaker velocities in the **b** and **c** directions. The velocities are only finite away from the extrema of the Fermi surface and the largest v_y, for example, is found along the lines at $k_y = \pm\pi/2b$. There are orbits traversing up and down k_z at, or near, these lines, as suggested by the dotted paths in Fig. 5.7b. There are also some closed orbits around the Fermi surface extrema in this geometry, but they do not lead to quantum oscillations since they do not go from inside to outside of a closed Fermi surface as the magnetic field is increased. The orbits which traverse the Fermi surface in the **c** direction average v_z to zero and hence contribute to a nonsaturating magnetoresistance for σ_{zz}. However, the closed orbits and the orbits at, e.g., $k_y = 0$ do not average to zero.

When the field is tilted, so that there is a component along the **c** direction, the orbit trajectories move along both y and z directions as schematically shown by the dashed curves in Fig. 5.7b. There is averaging of v_z and in particular if an orbit sweeps periodically over an integral number of reciprocal lattices along c ($\delta k_z = 2\pi n/c$), then $\langle v_z\rangle = 0$ for all orbits. This condition for maxima in the magnetoresistance along z corresponds to tilt angles where the Bessel function $J_0(\gamma) = 0$, $\gamma = \frac{2t_b c}{\hbar v_{\mathrm{F}}}\frac{B_x}{B_z}$. Thus, the warping of the Fermi surface is directly measured by the magnitude of t_b. The magnetoresistance corresponding to this rotation is shown in Fig. 5.8b. The dominant "ear" at $\sim\pm 6^0$ is the first zero of the Bessel function. Note also the dimple at $\theta = 0$ which results from and is a measurement of the bandwidth in the least conducting **c** direction. This dimple has been redubbed a "squit" [71] in more recent literature, where its explanation remains unchanged, but where it is taken as evidence of "coherent" transport in the third direction.

Trajectories for the third angular effect (Yoshino–Osada) are shown in Fig. 5.7c. Here, aside from the closed orbits which "squash" and vanish as field is rotated from **a** toward **b**, there is little difference to be seen in the trajectories. The closed orbits do not play an important role. But as we have seen above, the best averaging of $\langle v_z \rangle$ to zero comes from the orbits at the inflection points near $k_y = \pi/2b$ where the v_y are maximum. These "effective" orbits, introduced by Lebed and Bagmet [72], give a maximum magnetoresistance for the field orientation which provides a good measure of the ratio t_b/t_c. The data for this rotation are shown in Fig. 5.8c.

Thus the **a**–**c** and **a**–**b** angular dependent magnetoresistance is readily and simply explained by conventional transport theory. Only the original Lebed **b**–**c** rotations are not so easily explained. An answer may come from recent calculations by Lebed, Bagmet, Ha, and Naughton [73–75], and by Osada. They look at the **b**–**c** rotations in the presence of a small field along **a**. The cartoon in Fig. 5.7d shows the trajectories which result from a field at the magic angle **b**+**c** as the field along **a** in increased. The Lorentz force from $v_y \times B_x$ deflects the orbit down on half of the periodic orbit and back up on other. It still goes to the same end points, but it no longer averages over the values of v_z as before. If the orbit were it would eventually hit all points on the Fermi surface and yield $\langle v_z \rangle = 0$, but for the magic angles $\langle v_z \rangle \neq 0$. This would suggest that Lebed magic angle effects vanish in the absence of a field along **a** and that most previous studies failed to carefully note this component while doing **b**–**c** sweeps. Recent experiments by Osada and Kang (see this volume) lend support to this idea.

While the presence of a field along **a** would allow an explanation of the magic angles in a **b**–**c** rotation, and might explain the overall behavior observed in the ClO_4 salt, the behavior of the PF_6 salt still remains unexplained. In Fig. 5.8, we see that aside from the sharp magic angle dips, the overall shape of the curve is different from the dependence expected from Boltzmann theory. In particular, the background on which the magic angles peaks ride is minimum for current perpendicular to the applied field (σ_{zz} has current going along z) and increased as field is rotated toward the current direction. The PF_6 data led Anderson to suggest a very different explanation of the Lebed magic angle effects [76]. His idea is that the q1d chains or the two-dimensional **a**–**b** planes are strongly correlated electron systems and only become Fermi liquids in the presence of the weak coupling in the interplane **c** direction. The magic angle effects then relate to decoupling the planes or chains and the "background" effects have to do with orbital motion in the decoupled planes.

What is important to emphasize is that there is a real space way of looking at the magic angle effects which is hidden in the **k**-space picture described so far. The Lebed magic angles correspond to magnetic fields directed along the real space lattice vectors in the **b**–**c** plane, in the direction from one chain to another chain. The essence of Anderson's argument is that at a magic angle, an electron can hop between chains (nearest, next nearest or further neighbors) without crossing field lines and hence with no Lorentz force or vector potential-induced phase shift.

One experiment which seems to indicate that electrons only move coherently in the planes defined by the chain axis and the interchain direction

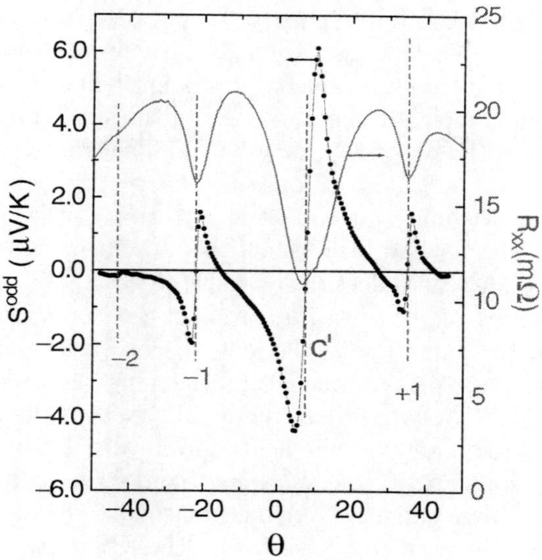

Fig. 5.9. Angle dependence of S_{ac}, a Nernst voltage, and the **a**-axis resistance at various magnetic fields, 0, 3.0, 5.0, and 7.5 T and temperature at 4.2 K. The resonant like structures at the Lebed magic angles suggest that the carrier transport is only coherent when the applied magnetic field is close to lattice planes and that the Lorentz force causes the carrier to deflect in one direction then the other as the field is rotated from above to below these planes

selected by the applied field is the Nernst effect measurement by Wu et al. [77]. The Nernst signal from a sample of $(TMTSF)_2PF_6$ is shown in Fig. 5.9. There are giant variations, reminiscent of resonances, as the field is rotated through the magic angles. A rather elaborate Boltzmann treatment of this problem fails to show any substantial Nernst effect at the magic angles. But a simple way of understanding the change of sign of the signal as one rotates through the magic angles is that the only velocities (a response to the temperature gradient) are in planes and that when the field is slightly above the plane the carriers are deflected (and produce a voltage) in one direction and when the field is slightly below the plane the carriers are deflected in the opposite direction. What is really unusual, however, is that the plane where the currents flow is not fixed in the crystal. The plane exhibiting coherent transport is the one close to the field direction when the field is near a magic angle.

5.3 Superconductivity in the Bechgaard Salts

5.3.1 Early Investigations of the Superconducting State

Superconductivity in the $(TMTSF)_2X$ system was discovered first by Jerome et al. [4] in $(TMTSF)_2PF_6$ in 1979 and then by Bechgaard et al. [78] in

$(TMTSF)_2ClO_4$ in 1981. Subsequently, X = AsF_6 [79], SbF_6 [80, 81], TaF_6 [80,81], FSO_3 [82], ReO_4 [83], and NbF_6 [84] have been found to superconduct. The initial measurements were magnetic susceptibility and resistivity versus temperature, which typically show an onset of the superconducting transition in the range 1.2–1.4 K, depending on anion X. The transition widths are typically rather broad, $\Delta T/T_c = \{T(\text{onset}) - T(R = 0)\}/T_c \sim$ 20–30%, in spite of the notion that the materials are clean with respect to extrinsic impurities. Bulk superconductivity had been established via magnetic measurements of, for example, the Meissner effect in X = ClO_4 [85].

From the series of experiments probing the superconducting state in $(TMTSF)_2X$ in the early 1980s, the general consensus was that the superconductivity was conventional in nature, meaning the order parameter was *s*-wave singlet. The evidence for conventionality was mainly from upper critical magnetic field H_{c2} and specific heat. Specifically, H_{c2} was shown have an anisotropy consistent with the crystal and band structures and a magnitude consistent with known pair-breaking mechanisms [86–88], and $C_v(T)$ [89] was shown to reveal an energy gap consistent with BCS theory.

5.3.2 Early Evidence for Unconventional Superconductivity

Nonetheless, some early experiments suggested unconventionality. These included nonmagnetic defect studies, NMR relaxation rate, and critical field. Rapid suppressions of T_c were observed in the defect studies, and an unusual temperature dependence to the spin–lattice relaxation rate as well as an absence of a Hebel–Slichter coherence peak were observed. These data, in combination with the proximity of the superconducting state to an SDW, led Abrikosov [90] in 1983 to suggest that the $(TMTSF)_2X$ materials were perhaps unconventional superconductors, such as *p*-wave triplet. Gor'kov and Jerome [91] had a similar hypothesis in 1985 upon analysis of the critical field data available at that time. Also, while most early critical magnetic field results were conventional, there were some indications of unusual behavior in $(TMTSF)_2X$. For example, Brusetti et al. [92] published a phase diagram for X = AsF_6 that seemed to show an anomalously large H_{c2} for field close to the **b**-axis.

Effect of Impurities on T_c

Conventional superconductors are only mildly perturbed by impurities that are not magnetic, or that do not induce magnetism. On the other hand, scattering by even nonmagnetic impurities is pair-breaking whenever the order parameter changes sign over the Fermi surface (i.e., in unconventional superconductors). A number of studies on the effect of nonmagnetic impurities on the superconducting transition in $(TMTSF)_2X$ were made in the 1980s. These studies reported significant suppression of T_c under anion [93] or cation [94] doping or radiation damage [95]. Each of these impurities is considered nonmagnetic, although there was some concern that X-ray irradiation in q1d systems produces unpaired spins [96]. In the Choi et al. work [95], radiation doses introducing defect concentrations of only a few 100 ppm were found to completely suppress superconductivity. This suppression of T_c is at least as severe, if not moreso, as in another putative unconventional superconductor, Sr_2RuO_4 [97]. As a result of these $(TMTSF)_2X$ defect studies, strong suspicion was raised with regard to the superconducting order parameter.

Nuclear Spin–Lattice Relaxation

The defect experiments are consistent with anisotropic superconductivity, but do not distinguish further between possible order parameters. The NMR relaxation rate is more constraining, in that it would indicate zeroes of the gap function on the Fermi surface. In the limit of weak magnetic fields, the nuclear spin–lattice relaxation rate T_1^{-1}in singlet superconductors arises from thermally excited quasiparticles. The temperature dependence of their population depends on the gap structure, and in particular the relaxation rate for $T \ll T_c$ is activated for fully gapped systems and varies more weakly, as a power law in T, in systems with gap anisotropy. To take an example, $T_1^{-1} \sim T^3$ for d-wave superconductors like the high-T_c cuprates which have line nodes in the gap structure [98,99]. A similar form is seen for the κ-$(ET)_2X$ quasi-two-dimensional systems [100–102].

Takigawa, Yasuoka, and Saito [103] performed the first experiments of the spin–lattice relaxation in the superconducting state of $(TMTSF)_2ClO_4$, probing the methyl group protons. To obtain the low-field behavior, they used a field-cycling technique, whereby the nuclear spins are polarized in the magnetic field, but the field is cycled to zero for the evolution time. Their results compare well to a T^3 dependence below T_c, consistent with a zero-field superconducting state with nodal lines on the open Fermi surface, but they did not explore temperatures low enough to be definitive. In a weak-coupling analysis, the behavior is identical for either singlet and triplet channels with order parameters [104]

$$\begin{aligned} \text{singlet} : \Delta(\mathbf{k}) &= \Delta_s \cos(bk_y), \\ \text{triplet} : \Delta(\mathbf{k}) &= \Delta_t \sin(bk_y). \end{aligned}$$

These two states have lines of nodes on the Fermi surface, which translates into a much smaller enhancement (i.e., the Hebel–Slichter peak) of the relaxation rate below the ordering temperature than would be expected for a fully gapped superconductor. On this point, the results are similar to what is observed for high-T_c compounds [98,99] as well as other organic superconductors such as the quasi-two-dimensional κ-$(ET)_2X$ systems [100–102].

Unfortunately, a more direct determination of the superconducting order parameter had not been made for technical reasons. For example, tunneling experiments were devised to detect phase shifts resulting from the

d-wave order parameter in the high-T_c cuprates. But for the early days of the Bechgaard salts, not even a reliable tunnel junction was available, so phase-sensitive measurements were not possible. And not for lack of trying, as interest in identifying the superconducting state was generated after reports of the upper critical fields greatly exceeding the paramagnetic limit in $(TMTSF)_2PF_6$ [105–107].

5.3.3 Recent Investigations: Triplet Superconductivity

With the detection of the various angular magnetoresistance effects in organic conductors, including $(TMTSF)_2X$, in the late 1980s came a renewed appreciation for the extreme anisotropy of these materials, and the strong sensitivity of magnetophysical effects to direction of magnetic field. The discoveries of the AMRO effects were of course initially motivated by Lebed's 1986 paper predicting dimensional crossovers at magic angles [108]. In that same year, Lebed published another paper suggesting a different dimensional crossover effect, this one relating to the superconducting state [109]. As a result of field-induced decoupling of the highly conducting layers by an in-plane magnetic field, orbital suppression of superconductivity was predicted to be dramatically weakened, leading to a divergent upper critical field, and reentrant superconductivity at very large field. This idea was furthered by the work of Dupuis, Montambaux, and Se de Melo (DMS) [110]. So, the stage was set for a new look at superconductivity in $(TMTSF)_2X$, and in q1d in general, starting with revising the H_{c2} issue.

Upper Critical Field H_{c2}

The first such reinvestigation was on $(TMTSF)_2ClO_4$ by the Buffalo group. For magnetic field in the $\mathbf{b}'$ direction (that suggested by Lebed to provide for dimensional crossover), Lee et al. [105, 106] found that the temperature at which the resistivity began to decrease upon cooling in field, interpreted as an onset of superconductivity, did not follow a conventional curve in $B - T$ space. That is, T_{onset} decreased very slowly as field was increased, leading to obvious upward curvature in $H_{c2}(T)$. In fact, T_{onset} actually *increased* slightly with increasing field above 4 T, hinting at reentrant superconductivity [105]. If interpreted as an upper critical field, then the H_{c2} along the $\mathbf{b}'$-axis derived from this T_{onset} would be several times larger than the Pauli, paramagnetic limit H_P. This limit is the maximum magnetic field in which a spin singlet superconductor can survive, and results from the Zeeman-like energy difference between opposite spins within a Cooper pair becoming comparable to the pair condensation energy. Since $H_P = 1.84T_c$ for an isotropic s-wave superconductor (it is smaller for anisotropic s-wave), this limit is about 2.1 T for $(TMTSF)_2ClO_4$ ($T_c = 1.13$ K), whereas the resistive onsets appeared to persist to 7 T. Thus, while the evidence for reentrance was insufficient for the authors to make such a claim, with only very small resistance decreases

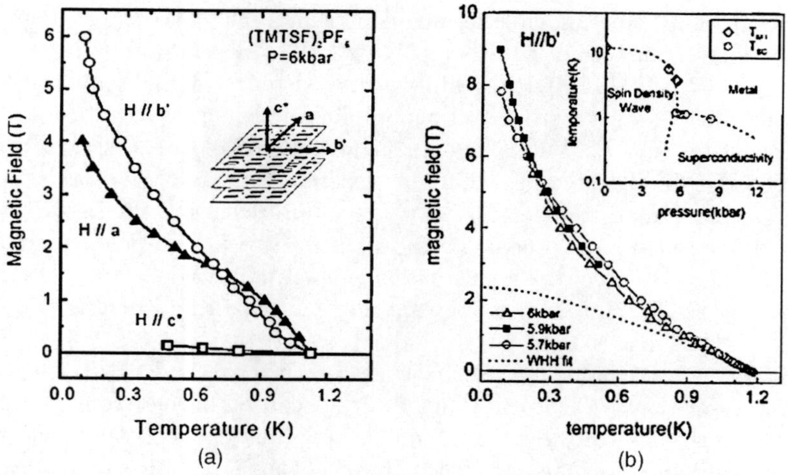

Fig. 5.10. (**a**) H_{c2} along the **a**, **b**$'$ and **c***-axes for $(TMTSF)_2PF_6$ at 0.6 GPa pressure, showing the lack of a Pauli limit and the anisotropy inversion near 0.7 K, 1.5 T. (**b**) H_{c2} for $\mathbf{B} \parallel \mathbf{a}$ at several pressures near the SDW phase

at the highest fields, it was becoming clear that $(TMTSF)_2ClO_4$ was not a conventional superconductor, and that the Lebed idea may have credence.

That work was followed by more extensive studies, this time on the PF_6 salt under pressure, again by Lee et al. [107, 111]. Here, the resistively determined H_{c2} was measured for magnetic field along the three principal, orthogonal axes **a**, **b**$'$, and **c***. Near T_c, all three $H_{c2}(T)$ curves appeared similar to those measured more than a decade prior, with $H^a > H^b > H^c$. However, above about 2 T, a crossover occurred, with H^b growing larger than H^a, and showing no signs of saturation; see Fig. 5.10a. Importantly, this behavior was independent of criterion used to determine the transition temperature (onset, midpoint, zero resistance, etc.) [107], and so was considered to accurately represent T_c and thus H_{c2}. As a result of this crossover, the upper critical field along the **b**$'$-axis was more than twice the Pauli limit at 0.1 K. For $(TMTSF)_2PF_6$, this limit is about 2.1 T, while the measured critical field exceeded 6 T. It was noted in [107] that, while strong spin–orbit scattering can lead to critical fields in excess of H_P even in s-wave superconductors, this can not explain the high critical fields for $(TMTSF)_2ClO_4$.

Data for the above work were taken at a pressure of 0.6 GPa, slightly above the critical pressure ($P_c \sim 0.55$ GPa) needed to suppress the SDW transition. Ever since Greene and Engler [112], however, it has been known that the competition between these two condensed phases can end in some sort of coexistence near this critical pressure. Lee et al. showed that this manifests itself in interesting ways in the critical field. In 2000, they showed that as pressure approaches P_c from above, $H_{c2}(T \sim 0)$ increases, reaching as high as 9 T at 0.1 K and 0.57 GPa, see Fig. 5.10b [113]. This is approximately four times the Pauli limit, now placing serious constraints on $(TMTSF)_2X$ being spin singlet superconductors. In other papers, Lee et al. [114, 115] showed again that near P_c, $(TMTSF)_2PF_6$ exhibited coexistence of SDW and superconducting phases, with resistivity upon cooling first increasing at T_{SDW} before falling to zero at T_c. Vuletic et al., in an extensive $P-T$-space study on $(TMTSF)_2PF_6$ [23], found a similar coexistence regime. Lee et al. [114] proposed a "slab" model, discussed below, as an alternative to the Lebed scenario to explain the rise in H_{c2} near P_c, and the subsequent high-field superconductivity (HSC). But even with this model, singlet pairing does not allow H_{c2} to exceed the Pauli limit, suggesting that nonsinglet pairing, such as spin triplet superconductivity, is at play.

Triplet superconductors, such as $L = 1$ p-wave or $L = 3$ f-wave, are not necessarily subject to a Pauli limit, because the electrons comprising the Cooper pairs can have equal spins, and thus do not experience a Zeeman energy split in magnetic field. However, they, like all superconductors, are still susceptible to orbital pair-breaking. The fact that the measured H_{c2} is so high in $(TMTSF)_2PF_6$ (and $(TMTSF)_2ClO_4$), at least for $\mathbf{B}$ in the apparently special $\mathbf{b}'$ direction, suggests that both spin and orbital pair-breaking are (independently) suppressed, due to triplet pairing and Lebed-DMS decoupling, respectively. This offers the startling if unlikely prospect of field-unlimited superconductivity.

At this point, all the evidence for large critical fields in $(TMTSF)_2X$ superconductors had been obtained from transport (resistivity) measurements, with the strongest case being made by the fact that even the zero resistance determination of T_c yielded H_{c2}'s above H_P. However, like any superconductor, diamagnetism is considered a more trustworthy and fundamental characteristic of the state, and so a magnetically determined $H_{c2}(T)$ phase line was desirable. Oh and Naughton [116] used a microcantilever magnetometer to record the magnetic torque emanating from a crystal of $(TMTSF)_2ClO_4$ in a magnetic field aligned along the $\mathbf{b}'$-axis. This device also allowed a simultaneous measurement of the resistivity. The resulting $H_{c2}(T)$ lines from the two measured quantities coincided, as shown in Fig. 5.11, with H_{c2} reaching 5.0 T at the lowest temperature of 25 mK. This showed that: (1) the previously employed method of determining H_{c2} from transport onset T_c's was reliable and (2) H_{c2} at low temperature was indeed anomalously large, exceeding the Pauli limit by at least 100%.

These second generation critical field studies thus provided compelling evidence for unconventional superconductivity, in particular spin triplet (p- or f-wave), with the added aspect of diminished orbital pair breaking for an in-plane magnetic field.

Thermal Conductivity

If the $(TMTSF)_2X$ materials are indeed triplet superconductors, evidence for this property would appear in other measurements. Especially important is

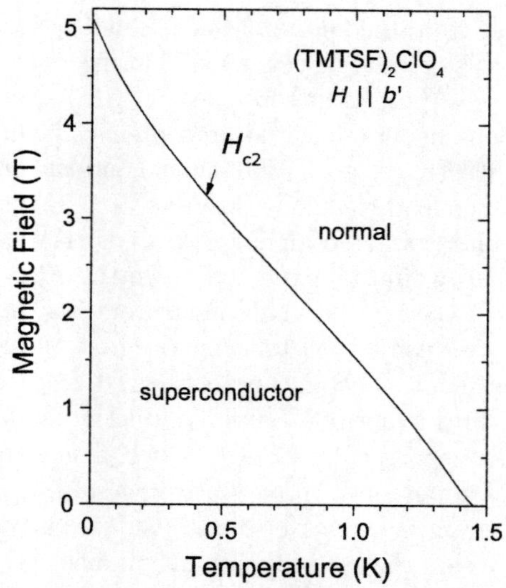

Fig. 5.11. H_{c2} along the $\mathbf{b}'$-axis for $(TMTSF)_2ClO_4$ as measured by both magnetization and resistance

order parameter phase-sensitive measurements such as thermal conductivity, spin–lattice relaxation rate, spin susceptibility, and tunneling. Belin and Behnia [117] measured $\kappa(T)$ in $(TMTSF)_2ClO_4$ deep into the superconducting state ($\sim T_c/10$) in zero and finite magnetic field, and found a T-dependence to the electronic component of κ that was incompatible with a Fermi surface energy gap containing nodes. They highlighted the contrast between their result (no nodes) and the Takigawa et al. [103] NMR result ($1/T_1 \sim T^3$) of a decade earlier, which suggested a Fermi surface with nodes. They also pointed out that the latter experiment was limited to $T > T_c/2$, not particularly close to absolute zero where conventional exponential T-dependence would be most evident. Finally, these authors remarked that their result is not necessarily incompatible with unconventional, even p-wave triplet pairing. As we will show later, there is indeed a p-wave state that is nodeless.

Tunneling

Tunneling, either quasiparticle or pair, is probably the most direct way to explore the symmetry of a superconducting order parameter [118], and thus to determine whether superconductivity is conventional or not. Nonetheless, only a very few tunneling experiments have been done on $(TMTSF)_2X$ materials. Perhaps this is due to inadequate (rough) surfaces, a puzzling fact that may also play a role in the ongoing dearth of high-resolution ARPES data on $(TMTSF)_2X$ crystals. A small number of papers were published in the 1980s related to tunneling in $(TMTSF)_2X$ [119–121], but these gave inconclusive or contrasting results. Moreover, each of these experiments involved tunnel junctions that contained either a normal metal (NIS) or a conventional superconductor (SNS' or SIS') electrode. That is, none involved tunneling between two TMTSF superconductors (SNS or SIS). Bando et al. [120] prepared planar junctions using deposited amorphous silicon as the barrier, and Pb as the counter electrode, such that their conductance signal was dominated by the gap of the lead electrode. More et al. [121] used GaSb (junction) and Sn (electrode) and extracted an energy gap of $2\Delta \sim 3.6\,\mathrm{meV}$, an order of magnitude larger than would be expected given the T_c.

More recently, Ha et al. [122] made measurements on a crossed bicrystal of $(\mathrm{TMTSF})_2\mathrm{ClO}_4$, ostensibly constituting the first all-organic superconductor tunneling experiment. The authors observed features they interpreted as consistent with unconventional superconductivity, including a substantial zero-bias conductance peak (ZBCP). Such a feature can not exist for s-wave conventional superconductors, since it results from the formation of midgap states at the junction barrier due to the inability of quasiparticles to cross, as a result of a sign change in the quantum mechanical superconducting order parameter. In s-wave symmetry, the order parameter has the same sign everywhere. They reported evidence for an energy gap of $2\Delta = 0.50\,\mathrm{meV}$, corresponding to $2\Delta/k_\mathrm{B}T_c \sim 4.2$, slightly larger than the BCS value of 3.6, suggesting slightly strong coupled, but nonetheless BCS, superconductivity. It is important to mention that all of these tunneling results continue to await independent confirmation.

NMR Knight Shifts and Spin–Lattice Relaxation Rates

Nuclear magnetic resonance has been a particularly important probe of magnetism and spin dynamics in organic conductors because conditions are good-to-excellent for it, and because neutron scattering, even elastic scattering, has been so far impractical. The single crystals available for experiments are often of very high quality, but they are historically small compared to the typical size used in neutron scattering experiments. To make magnetic signatures more difficult to detect, the moments are small and the unit cell is large making for small cross sections.

With respect to the superconducting state, different issues arise. As discussed above, the existence of upper critical fields in excess of the Pauli paramagnetic limit in the PF_6 and ClO_4 salts is inconsistent with homogeneous singlet pairing in the superconducting state. In principle, NMR is an excellent probe for testing the hypothesis of singlet vs. triplet pairing, because it is sensitive to the electron spin susceptibility through the electron–nucleus hyperfine interaction. Several nuclei are proven as good probes of electronic physics in the Bechgaard salts, including methyl protons, ^{77}Se, and ^{13}C, especially the double-bonded bridge carbons. The hyperfine coupling for all of these is anisotropic. For the ^{77}Se nuclei, the hyperfine coupling is dominated by molecular orbitals of predominantly Se p_z character, with the z-axis orthogonal to the plane of the molecule. Proton coupling is dipolar in nature and much weaker.

NMR in the Normal State

Of particular interest is the physical description of the normal state at low temperatures, close to other phase instabilities and in the regime where the magic angle effect is observed. As the NMR relaxation rates are dominated by the hyperfine interaction, it is a good spectroscopic probe of the density of magnetic excitations and a window into the physics of the normal state and the other broken-symmetry states. Unfortunately, neutron scattering is so far unsuccessful in probing the magnetic states of organic conductors, so NMR is the best available technique for information about spin dynamics and magnetic order parameters. The temperature dependence of the nuclear spin–lattice relaxation rates for either ^{13}C or ^{77}Se nuclei is determined by the hyperfine interaction and, in the case of the methyl group protons, by methyl group dynamics [18, 123].

As an example, we consider what NMR tells us about the normal state of the prototypical PF_6 salt. At temperatures greater than 30 K, $T_1^{-1} \sim \chi_s^2 T$, with χ_s the uniform spin susceptibility [18]. The form is similar to that for a Fermi liquid, though it has been interpreted to be the consequence of uniform ($\mathbf{q} \sim 0$) collective fluctuations of a Luttinger liquid for $T \gg 0$ [17]. In that case, the expectation is a crossover to Fermi liquid behavior as the interchain couplings become relevant, but before behavior of the various possible broken-symmetry states become apparent. The low temperature behavior is summarized in Fig. 5.12 for a range of pressures.

At low pressures, the approach to the SDW phase is associated with the steep increase in T_1^{-1}. With higher pressures, the SDW phase is suppressed, but it appears that an enhanced relaxation rate remains. At temperatures $T < \mathrm{O}\ (5\,\mathrm{K})$, there is a crossover to where $T_1^{-1} \sim T$ once again, but which a larger slope than what is seen at higher temperatures. Figure 5.12 also demonstrates the substantial reduction of the relaxation rate as pressure is increased.

In the case of a continuous transition, critical slowing down would lead to a divergence, and even where the transition is rounded or weakly discontinuous gives a clear signature of the transition. With a scaling analysis [17],

$$\begin{aligned} \frac{1}{T_1 T} &\sim \xi^{(2-\eta+z-d)}, \\ &\sim |T - T_{\mathrm{SDW}}|^{-\nu(2-\eta+z-d)}. \end{aligned}$$

Using mean-field values and the dynamical exponent $z = 2$ gives a divergence $|T - T_{\mathrm{SDW}}|^{-1/2}$; the experimental results are consistent with slowing down [18, 124]. Higher pressures reduce the ordering temperature $T_{\mathrm{SDW}} \to 0$ at P_c. Beyond that, the behavior is indistinguishable from a Fermi liquid (FL) state

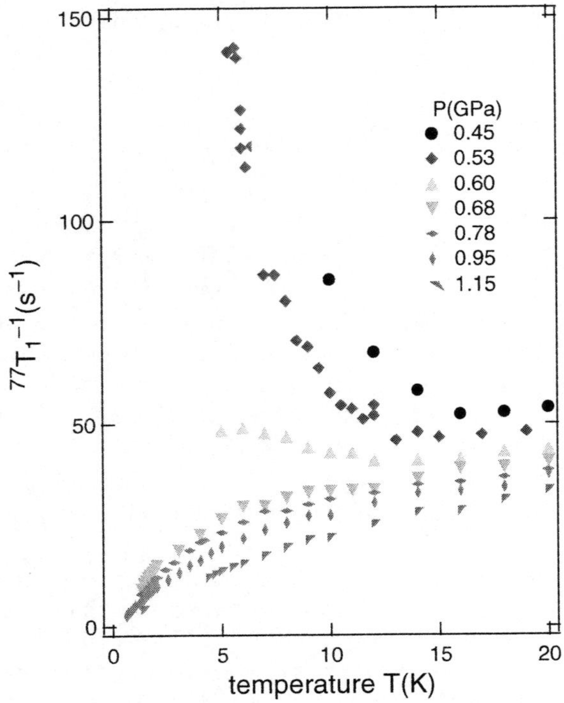

Fig. 5.12. Temperature dependence of ^{77}Se $T_1^{-1}$for $\mathbf{H} \parallel \mathbf{b}'$ over a range of pressures near the critical pressure P_c

where antiferromagnetic spin fluctuations (SF), with wavevector spanning the imperfectly nested Fermi surface are important. All of the data in Fig. 5.12 recorded at pressures greater than the critical pressure satisfies $T_1T \sim T + \Theta$ over a wide range of temperatures [125, 126], a form characteristic of itinerant systems with subcritical SF in two dimensions [127, 128]. In the inset, we show the pressure dependence of $1/T_1T$ vs. P for $(TMTSF)_2PF_6$. As pressure increases, the values approach that expected from a Korringa form for the relaxation.

NMR and the Spin Susceptibility of the Superconducting State

As described above, the NMR relaxation rate measurements by Takigawa and Saito were interpreted as evidence for an order parameter in $(TMTSF)_2ClO_4$ with gap nodes. With the observation of upper critical fields exceeding the paramagnetic limit for $(TMTSF)_2PF_6$ came the suggestion of triplet superconductivity. With singlet superconductivity in pure materials, the spin susceptibility is very small in the limit of small magnetic fields, so the hyperfine part of the Knight shift can distinguish between these two possibilities.

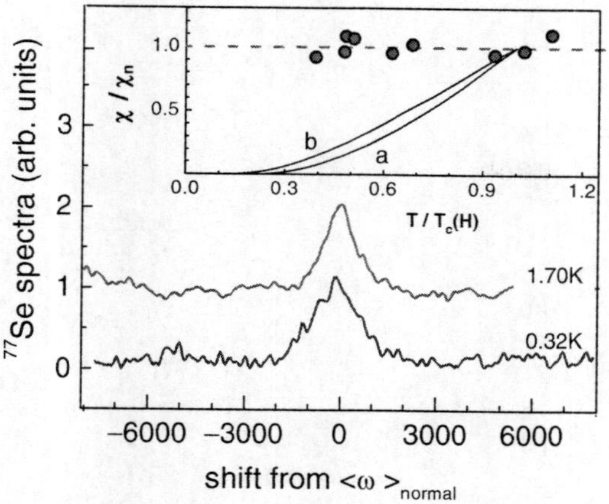

Fig. 5.13. The ^{77}Se spectrum in the normal ($T = 1.7$ K) and superconducting states. The inset shows the temperature dependence of the shift obtained, and indicates no detectable change below the transition temperature. The field is oriented $\mathbf{H} \parallel \mathbf{a}$

The temperature dependence of the Knight shift for $(TMTSF)_2PF_6$ at $P = 0.7$ GPa is shown in Fig. 5.13. To within experimental uncertainties, there is no detectable change in Knight shift over the entire temperature range. The ^{77}Se nucleus is just 8% abundant, so the sensitivity is weak and signal/noise is poor. To prevent heating from internal eddy currents due to the RF, very small NMR excitation pulses were used, and that leads to even smaller signals. While powder samples are normally introduced to avoid the heating problems, powder is impossible for this case because of the strict requirements for sample alignment relative to the magnetic field. The limits on the amplitude of RF pulses were determined by using the sample as its own bolometer: the sample temperature after RF pulses was inferred by making transport measurements in situ.

These experiments were interpreted as supporting the hypothesis for triplet pairing state in $(TMTSF)_2PF_6$. The equal-spin pairing state, in particular [129,130], would give rise to a spin susceptibility of this form, provided that the **d** vector is oriented in the direction of the magnetic field.

More recent experiments on the Knight shift in $(TMTSF)_2ClO_4$ open the possibility for a richer phase diagram within the superconducting state [126]. Knight shift measurements at applied fields less than $H_0 = 10$ kOe reveal a change in the ^{77}Se Knight shift consistent with a singlet superconducting state. Shown in Fig. 5.14 is the temperature dependence of the shift for $\mathbf{H} \parallel \mathbf{b}'$. For this direction, the hyperfine coupling is negative, so on entering the superconducting state the frequency increases. And because any diamagnetic shielding would lead to a frequency change in the opposite direction, the observations are unequivocally associated with the hyperfine field.

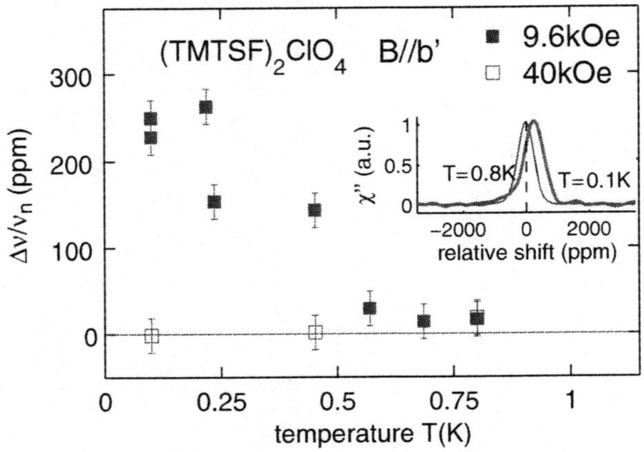

Fig. 5.14. Shift vs. temperature for magnetic field $H = 9.57\,\mathrm{kOe}$ parallel to $\mathbf{b}'$. The hyperfine coupling is negative for this direction, so the drop in spin susceptibility in the superconducting state leads to a positive change in the shift relative to that of the normal state. The inset shows spectra for the normal state ($T = 0.8\,\mathrm{K}$) and superconducting state ($T = 0.1\,\mathrm{K}$)

Also studied was the field dependence of the relaxation rate for magnetic fields $\mathbf{H} \parallel \mathbf{a}, \mathbf{b}'$ at $T = 0.1\,\mathrm{K}$. Observed was a sharp crossover, or phase transition for $H \sim 1.5 - 2\,\mathrm{T}$ for both directions, where the relaxation rate rapidly increased, approaching the normal state value. However, the in situ transport measurements indicate superconductivity survives to much greater field strengths, with onset $H_{c2} \geq 4\,\mathrm{T}$.

The two different regimes for the relaxation rate were described as "low-field superconductivity" (LSC) and "high-field superconductivity" [126], with the difference being a recovery of the spin susceptibility, presumably because of the presence of quasiparticles, in the HSC regime. The nature of the HSC remains unknown. It could be a triplet phase, or it could be in a Fulde–Ferrell–Larkin–Ovchinnikov (FFLO) state [131, 132], with a nonzero density of states at the Fermi limit so the usual paramagnetic limit does not apply. Prior to these measurements, there was no evidence for a phase transition within the superconducting state of either the PF_6- or the ClO_4-salt, (except for a solid–liquid vortex transition [116]) and consequently this possibility was not seriously considered. Further experiments will be needed to verify that the sharp change of the relaxation rate is symptomatic of a phase transition, not just of the vortex lattice, and to elucidate the nature of the high-field phase.

Recent Nonmagnetic Defect Studies

Recently, a thorough study of the effect of ReO_4 substitution for ClO_4 was done by the Orsay group [133]. In these studies, the scattering rate was tuned by adjusting the cooling rate on passing through the anion-ordering transition, or by introduction of up to 6% ReO_4 counterions. For both cases, the nonmagnetic nature of the impurities was verified by ESR techniques. T_c decreased linearly with the residual resistance, with $T_c \to 0$ when the mean free path was of the order of the coherence length. The results provided additional quantitative support for an interpretation in terms of unconventional superconductivity.

5.4 Phases and Properties Near the SDW-Superconductor Boundary

The superconducting properties of $(TMTSF)_2PF_6$ change rapidly in the range of pressures around P_c. In particular, a dramatic increase in upper critical fields is observed for fields aligned along $\mathbf{c}^*$ [114], along with a decrease in critical currents [23]. The latter observation was interpreted as evidence for phase segregation into SDW and superconducting domains, and an analysis of the temperature dependence of the critical fields indicate the interplay between these regions is influenced by the magnetic field.

The phase segregation scenario follows from a simple Landau theory [23, 115], which leads to a line of transitions dividing SDW insulator from the normal state that is of first order when $\mathrm{d}T_{\mathrm{SDW}}/\mathrm{d}P$ is sufficiently steep. The suppression of the SDW phase by pressure is commonly attributed to imperfect nesting, where the ground state energy of the SDW phase is increasing linearly with pressure (or inverse cell parameter $\mathbf{b}$). Depending on the experimental conditions, controlled pressure (stresses), or controlled volume (strains), there is simply a first order transition between the SDW and normal phases, or phase segregation is observed over a range total sample volumes.

A Landau thermodynamic potential for the difference between SDW and normal states $\delta f = f_{\mathrm{S}} - f_{\mathrm{N}}$ is written

$$\delta f = a(T - T_c)\phi^2 + u_4\phi^4 + u_6\phi^6. \tag{5.9}$$

In the usual way, $u_4 > 0$ gives a continuous transition, whereas u_4 negative causes it to be discontinuous. Thus, u_4 changing sign with the variation of another external parameter determines the tricritical point where $u_4 = 0$.

Consider the SDW transition temperature with no strain to be T_{c0}, and varying with strain x as $T_c(x) = T_{c0} - xT_c'(x)$. Adding a strain energy density $E_s = (1/2)Kx^2$ to the SDW free energy density and minimizing with respect to x recasts the fourth-order term as

$$[u_4 - \frac{a^2T_c'^2}{2K}]\phi^4. \tag{5.10}$$

The transport experiments reveal more complex behavior. As first seen by Greene and Engler [112], Vuletic et al. [23], and later Lee et al. [134] reported upon cooling first a resistive upturn characteristic of the SDW phase, and then a reentrant superconducting transition over a range of pressures $P < P_c$. In the case of Vuletic et al. [23], the segregation regime is almost 0.1 GPa. The phase segregation scenario was based on an analysis of the relative strength of the resistive upturn and the superconducting critical currents.

5.4.1 NMR Evidence for Phase Segregation for $P \approx P_c$

The evidence for phase segregation has been detected by NMR spectroscopy on methyl protons [115, 135]. At ambient pressure, the spectral width of the proton NMR lines is approximately $\Delta H \approx 50$ Oe at temperatures well below the transition, whereas the linewidths are just 10 Oe in the normal state. In the pressure regime where phase segregation occurs, the line develops features characteristic of both normal and SDW phases.

The progression is shown in Fig. 5.15. The evidence for the separated environment comes from the unusual shape: the center of the spectrum remains sharp as it does in the normal state, yet a significant fraction of the intensity moves to the broadened wings as SDW formation in part of the sample develops. Notably, the distribution of fields observed, as inferred from the formation of the wings, is nearly identical to what it is at ambient pressure. The sample was detected to be superconducting with measurements of the reflected RF power.

The experimental observations are consistent with the Landau theory provided that pressure is not a good control variable. Consider the simpler case of an isotropic system. Then, the steep variation of $\mathrm{d}T_P/\mathrm{d}P$ and the first order character of the phase transition, phase segregation occurs naturally if total volume is a good control parameter. In the real pressure cells the temperature is very low compared to the freezing temperatures of the fluid used to transmit the hydrostatic conditions. Hydrostatic pressure is strictly controlled for

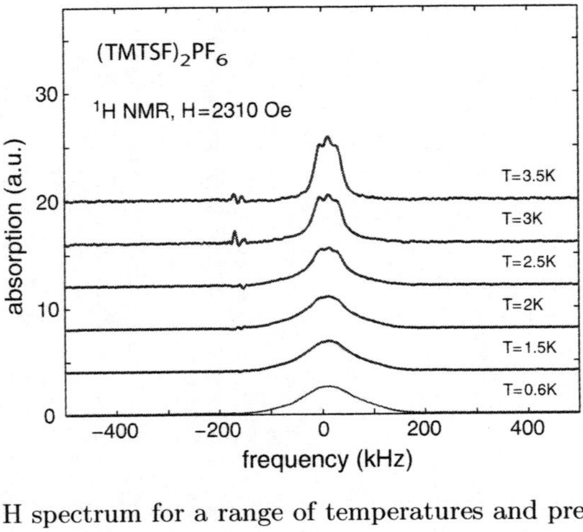

Fig. 5.15. The ^{1}H spectrum for a range of temperatures and pressure slightly less than $P_c \approx 0.6$ GPa

a small sample when the transmitting medium has no shear modulus and the cell body is relatively incompressible. Thus, at the low temperatures, deviations from hydrostatic conditions can occur, and the system under study can phase segregate even if pressure inhomogeneities are insignificant.

Grigoriev and Gor'kov have made an interesting proposal in which the SDW and normal states are either separated into mesoscopic domains (essentially a soliton-like lattice, and the normal state regions are antiphase domain walls), or are homogeneously coexisting over a range of pressure $P_{c1} < P < P_c$ [136,137]. The soliton lattice has lower energy than either homogeneous phase (SDW or normal state) for some choices of imperfect nesting. In the case of the soliton lattice, the normal state regions have thickness of order of the SDW coherence length. This has implications for the NMR lineshapes. The NMR lineshapes are consistent with the general picture in the sense that they are not described by a simple superposition of lineshapes from the normal state and SDW state. However, transport experiments [23] do not exhibit a significant decrease in superconducting transition temperature in the phase-segregation regime. Further experimentation is necessary before the situation can be clarified.

5.4.2 Critical Field Enhancement Close to the Superconductor-SDW Phase Boundary

In Fig. 5.16 we show the temperature dependence of the upper critical field for a range of pressures close to P_c. The field direction is normal to the layer of counterions, along which H_{c2} is smallest because of the large normal state cores. The steep increase in H_{c2} on lowering the pressure indicates that the effectiveness of the cores in destroying superconductivity is rapidly diminished. One very natural explanation is to assume that the orbital suppression of superconductivity is ineffective because the normal cores are not created in the same way as for higher pressures, which is possible if the field penetrates nonsuperconducting regions interpenetrating the superconducting ones. The characteristic length scale for the superconducting domains has to be smaller than the penetration depth to suppress the orbital currents.

Analysis of a slab model with the microscopic phase segregation shown in the right panel of Fig. 5.16 accounts well for the critical-field enhancement. As the field is increased from zero, there is good evidence for modification of the domain structure. Slabs with dimensions d perpendicular to the field smaller than the penetration depth have enhanced upper critical fields $H_{c2} \sim (\lambda/d)H_c$. The effect is due to a reduction of diamagnetic energy cost due to smaller screening currents. A bulk sample into which a superconductor–insulator boundary is introduced will have an energy associated with the boundary $\gamma/2$ and the screening currents contribute a kinetic energy cost, $K \sim (Hd/\lambda)^2$. The critical temperature is suppressed by an amount $\delta T_c \sim K + \gamma/d$. Minimizing with respect to the slab period d gives $H \propto (\lambda/\gamma)\delta T^{2/3}$. The enhancement described fits well to the results for pressures close to P_c [114].

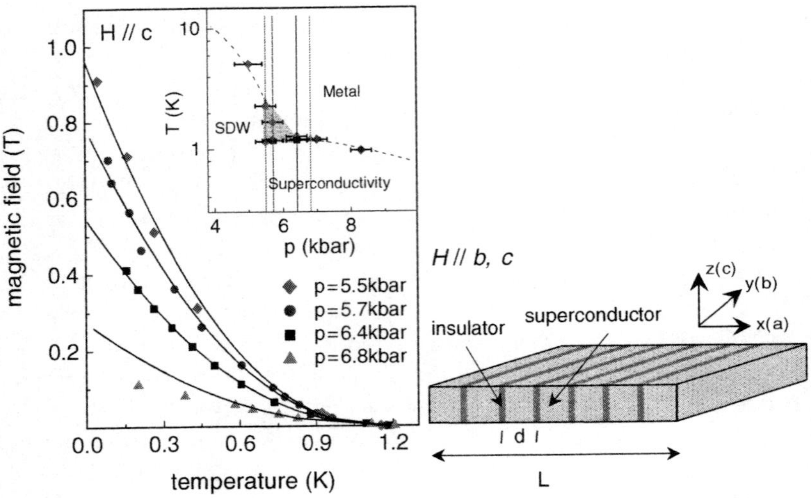

Fig. 5.16. *Left*: H_{c2} for $\mathbf{H} \parallel \mathbf{c}*$, over a range of pressures $P \sim P_c$. The inset shows the $P - T$-phase diagram. *Right*: Model analyzed for enhancement of the upper critical fields. The magnetic field is oriented in the $\mathbf{b} - \mathbf{c}$ plane, and the system alternates between SDW and superconductor along the high-conductivity **a**-axis with period d

An additional energy, describing the cost for introducing an SDW region of finite thickness sandwiched between superconducting material, acts as an additional contribution to the surface energy. As the difference in free energy between superconducting and SDW phases is decreased as $P \to P_c^+$, the cost is reduced and thus, the conditions are more favorable to critical field enhancement.

5.5 Conclusions and Conundra

Major remaining issues regarding superconductivity in $(TMTSF)_2X$ are those asked from the beginning: what is the order-parameter symmetry, and what is the pair-binding mechanism? While many of the experiments are consistent with triplet spin pairing, there are several triplet states available. Some are depicted in Fig. 5.17. Note that p_x is the only unconventional (non-s-wave) state without a Fermi surface node or nodes. The recent discovery that the low-field regime for $(TMTSF)_2ClO_4$ is a singlet superconductor adds new information as well as intrigue, since the nature of the high field phase is undetermined. In any case, its existence is linked to the large spin susceptibility, and the possibility for a transition between low-field and high-field regimes indicates that it is worthwhile to reexamine the possibility for a FFLO state.

A recurring theme of the $(TMTSF)_2X$ system appears to be phase inhomogeneity over different length scales, with slabs, solitons, phase segregation,

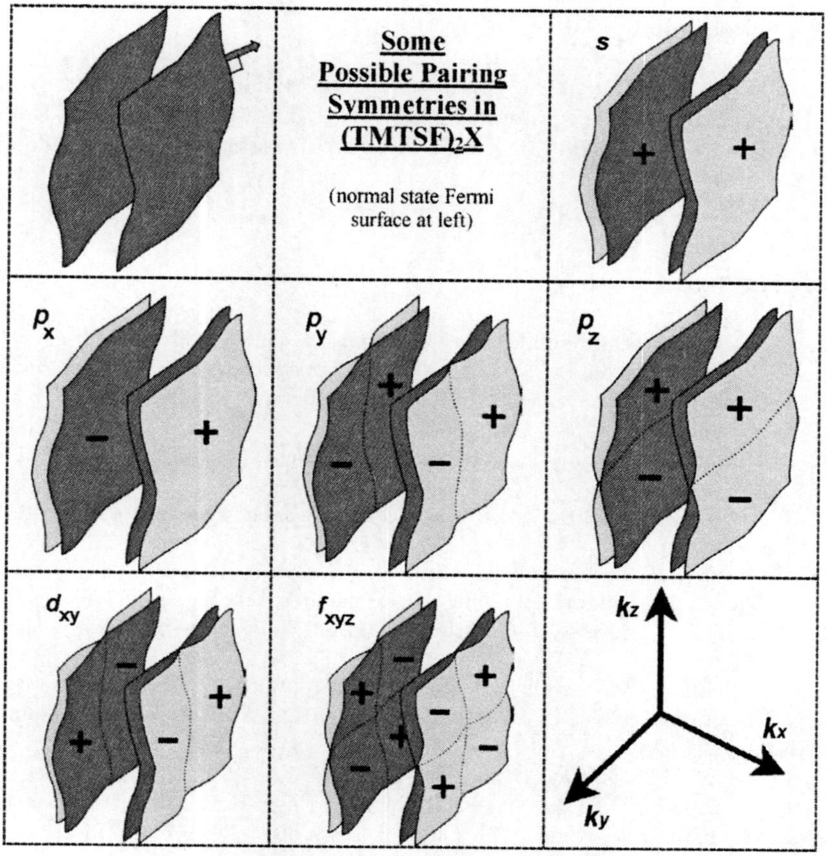

Fig. 5.17. Some of the various order parameter pairing symmetries possible in the $(TMTSF)_2X$ superconductors. The upper left shows the normal state Fermi surface, while the other figures sketch the Fermi surface gap function in k-space. Note that all states contain lines of gap nodes except s-wave and p_x-wave

and FFLO all involving spatial inhomogeneity. What is needed are real-space microscopic probes of the different and sometimes competing, ordered phases.

Many further questions are worth asking: Why is the fractional quantum Hall effect not observed? What is the very high field state of $(TMTSF)_2ClO_4$? Are there really several AMRO effects, or just different manifestations of one effect? Will clean tunneling (in the superconducting and SDW states) experiments ever be possible? Can these materials be doped with a gate electrode or by changing stoichiometry? Can one find a way to map the FIDSW gap structure and periodicity in magnetic field? The following chapters in this book will offer the reader a detailed tour of the complex tower of phenomena studied in the nearly three decades since the foundation of the field was laid by the initial discovery of organic superconductivity [4].

Acknowledgment

We wish to acknowledge support for research on organic conductors from the U.S. National Science Foundation, Division of Materials Research, through grant numbers DMR-050552 (SEB), DMR-0647327 (PMC), and DMR-0605339 (MJN).

References

1. K. Bechgaard et al., Solid State Commun. **33**, 1119 (1980)
2. T. Ishiguro, K. Yamaji, G. Saito, *Organic Superconductors*, *Springer Series in Solid State Sciences* (Springer, Berlin Heidelberg New York, 1998)
3. D. Jérome: Chem. Rev. **104**, 5565 (2004)
4. D. Jérome, A. Mazaud, M. Ribault, K. Bechgaard, J. Phys. Lett. **41**, L95 (1980)
5. J.F. Kwak, J.E. Schirber, R.L. Greene, E.M. Engler, Phys. Rev. Lett. **46**, 1296 (1981)
6. W. Kang, S.T. Hannahs, P.M. Chaikin, Phys. Rev. Lett. **70**, 3091 (1993)
7. N. Thorup, G. Rindorf, H. Soling, K. Bechgaard, Acta Cryst. B **37**, 1236 (1981)
8. B. Gallois, A. Meresse, J. Gaultier, R. Moret, Mol. Cryst. Liq. Cryst. **131**, 147 (1985)
9. J.G. Bednorz, K.A. Müller, Z. Phys. B **64**, 189 (1986)
10. M.H. Whangbo, W.M. Walsh, R.C. Haddon, F. Wudl, Solid State Commun. **43**, 637 (1982)
11. P.M. Grant, Phys. Rev. B **26**, 6888 (1982)
12. C.S. Jacobsen, D.B. Tanner, K. Bechgaard, Phys. Rev. B **28**, 7019 (1983)
13. G.M. Danner, W. Kang, P.M. Chaikin, Phys. Rev. Lett. **72**, 3714 (1994)
14. J. Moser et al., Phys. Rev. Lett. **84**, 2674 (2000)
15. G. Mihály, I. Kézsmárki, F. Zámborszky, L. Forró, Phys. Rev. Lett. **84**, 2670 (2000)
16. M. Dressel et al., Phys. Rev. B **71**, 075104 (2005)
17. C. Bourbonnais, J. Phys. I **3**, 143 (1993)
18. P. Wzietek et al., J. Phys. I **3**, 171 (1993)
19. V. Vescoli et al., Science **281**, 1181 (2001)
20. J. Moser et al., Eurphys. J. B **1**, 39 (1997)
21. J.C. Scott, H.J. Pedersen, K. Bechgaard, Phys. Rev. Lett. **45**, 2125 (1980)
22. J.B. Torrance, H.J. Pedersen, K. Bechgaard, Phys. Rev. Lett. **49**, 881 (1982)
23. T. Vuletic et al., Eur. Phys. J. B **25**, 319 (2002)
24. G.M. Danner, P.M. Chaikin, S.T. Hannahs, Phys. Rev. B **53**, 2727 (1996)
25. C. Coulon, S.S.P. Parkin, R. Laversanne, Phys. Rev. B **31**, 3583 (1985)
26. H. Wilhelm et al., Euro. Phys. J. B **21**, 175 (2001)
27. V.J. Emery, R. Bruinsma, S. Barisic, Phys. Rev. Lett. **48**, 1039 (1982)
28. D. Jerome, H.J. Shulz, Adv. Phys. **31**, 299 (1982)
29. H. Seo, H. Fukuyama, J. Phys. Soc. Jpn. **66**, 1249 (1997)
30. S. Mazumdar, R.T. Clay, D.K. Campbell, Phys. Rev. B **62**, 13400 (2000)
31. R.T. Clay, S. Mazumdar, D.K. Campbell, Phys. Rev. B **67**, 115121 (2003)
32. D.S. Chow et al., Phys. Rev. Lett. **85**, 1698 (2000)
33. P. Monceau, F.Y. Nad, S. Brazovskii, Phys. Rev. Lett. **86**, 4080 (2001)

34. F. Zamborszky et al., Phys. Rev. B **66**, 081103 (2002)
35. M. Kuwabara, H. Seo1, M. Ogata, J. Phys. Soc. Jpn. **72**, 225 (2003)
36. W. Yu et al., Phys. Rev. B **70**, 121101 (2004)
37. W.A. Little, Phys. Rev. **134**, A1416 (1964)
38. W. Wu, N. Ong, P. Chaikin, Phys. Rev. B **72**, 235116 (2005)
39. I.F. Schegolev, Phys. Status Solidi **12**, 9 (1972)
40. F. Wudl, G.M. Smith, E.J. Hufnagel, Chem. Commun. 1453 (1970)
41. M.J. Cohen, L.B. Coleman, A.F. Garito, A.J. Heeger, Phys. Rev. B **10**, 1298 (1974)
42. P.M. Chaikin, T. Holstein, A.M. Ya, Phil. Mag. B **48**, 457 (1983)
43. P.M. Chaikin et al., Phys. Rev. Lett. **51**, 2333 (1983)
44. M. Ribault et al., J. Phys. Lett. **44**, L953 (1983)
45. L.P. Gor'kov, A.G. Lebed, J. Phys. Lett. **45**, L433 (1984)
46. P.M. Chaikin, Phys. Rev. B **31**, 4770 (1985)
47. M. Herritier, G. Montambaux, P. Lederer, J. Phys. (Paris), Lett. **45**, L943 (1984)
48. K. Yamaji, J. Phys. Soc. Jpn. **54**, 1034 (1985)
49. A. Virosztek, L. Chen, K. Maki, Phys. Rev. B **34**, 3371 (1986)
50. V.M. Yakovenko, Europhys. Lett **3**, 1041 (1987)
51. K.V. Klitzing, G. Dorda, M. Pepper, Phys. Rev. Lett. **45**, 494 (1980)
52. M. Ribault et al., J. Phys. Lett. **45**, L935 (1984)
53. S.K. McKernan et al., Phys. Rev. Lett. **75**, 1630 (1995)
54. S. Haddad, S. Charfi-Kaddour, M. Heritier, R. Bennaceur, Phys. Rev. B (Condens. Matter Mater. Phys.) **72**, 085104 (2005)
55. S.T. Hannahs et al., Phys. Rev. Lett. **63**, 1988 (1989)
56. J.R. Cooper et al., Phys. Rev. Lett. **63**, 1984 (1989)
57. H. Cho, W. Kang, Phys. Rev. B **59**, 9814 (1999)
58. S. Uji et al., Phys. Rev. Lett. **94**, 077206 (2005)
59. A.G. Lebed, Pis'ma Zh. Eksp. Teor. Fiz. **43**, 174 (1986)
60. A.G. Lebed, P. Bak, Phys. Rev. Lett. **63**, 1315 (1989)
61. T. Osada et al., Phys. Rev. Lett. **66**, 1525 (1991)
62. M.J. Naughton et al., Phys. Rev. Lett. **67**, 3712 (1991)
63. J.F. Kwak, J.E. Schirber, R.L. Greene, E.M. Engler, Mol. Cryst. Liq. Cryst. **79**, 111 (1982)
64. K. Yamaji, J. Phys. Soc. Jpn. **58**, 1520 (1989)
65. H. Yoshino et al., J. Phys. Soc. Jpn. **64**, 2307 (1995)
66. T. Osada, S. Kagoshima, N. Miura, Phys. Rev. Lett. **77**, 5261 (1996)
67. I.J. Lee, M.J. Naughton, Phys. Rev. B **57**, 7423 (1998)
68. B.K. Cooper, V.M. Yakovenko, Phys. Rev. Lett. **96**, 037001 (2006)
69. T. Osada, S. Kagoshima, N. Miura, Phys. Rev. B **46**, 1812 (1992)
70. U.M. Scheven, E.I. Chashechkina, E. Lee, P.M. Chaikin, Phys. Rev. B **52**, 3484 (1995)
71. J. Singleton et al., Phys. Rev. Lett. **88**, 037001 (2002)
72. A.G. Lebed, N.N. Bagmet, Phys. Rev. B **55**, R8654 (1997)
73. A.G. Lebed, M.J. Naughton, Phys. Rev. Lett. **91**, 187003 (2003)
74. A.G. Lebed, N.N. Bagmet, M.J. Naughton, Phys. Rev. Lett. **93**, 157006 (2004)
75. H.I. Ha, A.G. Lebed, M.J. Naughton, Phys. Rev. B (Condens. Matter Mater. Phys.) **73**, 033107 (2006)
76. S.P. Strong, D.G. Clarke, P.W. Anderson, Phys. Rev. Lett. **73**, 1007 (1994)

77. W. Wu, I.J. Lee, P.M. Chaikin, Phys. Rev. Lett. **91**, 056601 (2003)
78. K. Bechgaard et al., Phys. Rev. Lett. **46**, 852 (1981)
79. M. Ribault, J. Pouget, D. Jerome, K. Bechgaard, C.R. Acad. Sci. Paris **291**, B145 (1980)
80. S.S.P. Parkin, M. Ribault, D. Jerome, K. Bechgaard, J. Phys. C **14**, 5305 (1981)
81. S.S.P. Parkin, M. Ribault, D. Jerome, K. Bechgaard, J. Phys. C **14**, L445 (1981)
82. R.C. Lacoe et al., Phys. Rev. B **27**, 1947 (1983)
83. S.S.P. Parkin, D. Jerome, K. Bechgaard, Mol. Cryst. Liq. Cryst. **79**, 213 (1981)
84. S. Sakata et al., Mol. Cryst. Liq. Cryst. **452**, 103 (2006)
85. K. Andres et al., Phys. Rev. Lett. **45**, 1449 (1980)
86. R.L. Greene et al., Mol. Cryst. Liq. Cryst. **79**, 183 (1982)
87. K. Murata et al., Mol. Cryst. Liq. Cryst. **79**, 283 (1982)
88. P.M. Chaikin, M.Y. Choi, R.L. Greene, J. Magn. Magn. Mater. **31–34**, 1268 (1983)
89. P. Garoche, R. Brusetti, D. Jérome, K. Bechgaard, J. Phys. Lett. **43**, L147 (1982)
90. A.A. Abrikosov, J. Low Temp. Phys. **53**, 359 (1983)
91. L.P. Gorkov, D. Jerome, J. Phys. Lett. **46**, L643 (1985)
92. R. Brusetti, M. Ribault, D. Jérome, K. Bechgaard, J. Phys. (Paris) **43**, 801 (1982)
93. S. Tomic et al., J. Phys. (Paris) Colloq. **44**, C3 (1983)
94. C. Coulon et al., J. Phys. (Paris) **43**, 1721 (1982)
95. M.Y. Choi et al., Phys. Rev. B **25**, 6208 (1982)
96. S. Bouffard et al., J. Phys. C: Solid State Phys. **15**, 2951 (1982)
97. A.P. Mackenzie et al., Phys. Rev. Lett. **80**, 161 (1998)
98. Y. Kitaoka et al., Phys. C **153–155**, 83 (1988)
99. T. Imai et al., J. Phys. Soc. Jpn. **57**, 2280 (1988)
100. S.M. DeSoto et al., Phys. Rev. B **52**, 10364 (1995)
101. H. Mayaffre et al., Phys. Rev. Lett. **75**, 4122 (1995)
102. K. Kanoda, K. Miyagawa, A. Kawamoto, Y. Nakazawa, Phys. Rev. B **54**, 76 (1996)
103. M. Takigawa, H. Yasuoka, G. Saito, J. Phys. Soc. Jpn. **56**, 873 (1987)
104. Y. Hasegawa, H. Fukuyama, J. Phys. Soc. Jpn. **56**, 877 (1987)
105. I.J. Lee, A.P. Hope, M.J. Leone, M.J. Naughton, Appl. Supercond. **2**, 753 (1994)
106. I.J. Lee, A.P. Hope, M.J. Leone, M.J. Naughton, Syn. Metals **70**, 747 (1995)
107. I.J. Lee, M.J. Naughton, G.M. Danner, P.M. Chaikin, Phys. Rev. Lett. **78**, 3555 (1997)
108. A.G. Lebed, JETP Lett. **43**, 174 (1986)
109. A.G. Lebed, JETP Lett. **44**, 114 (1986)
110. N. Dupuis, G. Montambaux, C.A.R. Sáde Melo, Phys. Rev. Lett. **70**, 2613 (1993)
111. M.J. Naughton, I.J. Lee, P.M. Chaikin, Syn. Metals **85**, 1481 (1997)
112. R.L. Greene, E.M. Engler, Phys. Rev. Lett. **45**, 1587 (1980)
113. I.J. Lee, P.M. Chaikin, M.J. Naughton, Phys. Rev. B **62**, R14669 (2000)
114. I.J. Lee, P.M. Chaikin, M.J. Naughton, Phys. Rev. Lett. **88**, 207002 (2002)
115. I.J. Lee et al., Phys. Rev. Lett. **94**, 197001 (2005)
116. J.I. Oh, M.J. Naughton, Phys. Rev. Lett. **92**, 67001 (2004)

117. S. Belin, K. Behnia, Phys. Rev. Lett. **79**, 2125 (1997)
118. I. Giaever, K. Megerle, Phys. Rev. **122**, 1101 (1961)
119. A. Fournel, Mol. Cryst. Liq. Cryst. **79**, 261 (1982)
120. H. Bando, K. Kajimura, T. Ishiguro, G. Saito, Mol. Cryst. Liq. Cryst. **119**, 131 (1985)
121. C. More et al., J. Phys. (France) Lett. **42**, L313 (1981)
122. H.I. Ha, J.I. Oh, J. Moser, M. Naughton, Syn. Metals **137**, 1215 (2003)
123. W. Clark et al., Syn. Metals **103**, 2166 (1999)
124. W.G. Clark et al., Int. J. Mod. Phys. B **16**, 3252 (2002)
125. W. Wu et al., Phys. Rev. Lett. **94**, 097004 (2005)
126. J. Shinagawa et al., Phys. Rev. Lett. (in press)
127. T. Moriya, Adv. Phys. **49**, 555 (2000)
128. A.J. Millis, H. Monien, D. Pines, Phys. Rev. B **42**, 167 (1990)
129. A.G. Lebed, K. Yamaji, Phys. Rev. Lett. **80**, 2697 (1998)
130. A.G. Lebed, K. Machida, M. Ozaki, Phys. Rev. B **62**, R795 (2000)
131. P. Fulde, R.A. Ferrell, Phys. Rev. **135**, A550 (1964)
132. A.I. Larkin, Y.N. Ovchinnikov, JETP **20**, 762 (1965)
133. N. Joo et al., Europhys. Lett. **72**, 645 (2005)
134. I.J. Lee et al., Phys. Rev. Lett. **88**, 170004 (2002)
135. W. Yu et al., Int. J. Mod. Phys. B **16**, 3090 (2002)
136. L.P. Gor'kov, P. Grigoriev, Europhys. Lett. **71**, 425 (2005)
137. L.P. Gor'kov, P.D. Grigoriev, Phys. Rev. B **75**, 020507 (2007)

6

Physical Properties of Quasi-Two-Dimensional Organic Conductors in Strong Magnetic Fields

S. Uji and J.S. Brooks

6.1 Introduction

Low-dimensional electronic systems have attracted much interest because of their exotic physical properties, such as superconductivity, charge- or spin-density wave formation, quantum Hall effect, and charge order. Among various low-dimensional systems, organic conductors have great advantages in the studies of the electronic states since they have a variety of crystal structures, leading to a broad range of physical properties, in spite of simple Fermi surface structure [1–3]. The high quality crystals that distinguish these materials enable us to obtain detailed information of the Fermi surface through the observation of quantum oscillation and associated magnetotransport phenomena as shown herein.

Most of physical properties in conducting materials are mainly determined by the electrons with the highest energy (Fermi energy), i.e., the electronic states on the Fermi surface. Therefore, investigation of the Fermi surface is essential to clarify the origins of their physical properties. In the last decade, experimental studies to gauge Fermi surface have been well developed and made a great contribution to the understanding of the electronic states of various organic conductors. Most studies have involved measurements of quantum oscillations in the magnetoresistance, Shubnikov-de Haas (SdH) effect, and in the magnetization, de Haas-van Alphen (dHvA) effect [4]. A few other measurements, quantum oscillations in the heat capacity and magnetothermal effect, and the angular dependent magnetoresistance oscillations are also known as powerful tools to investigate the Fermi surface.

First, we begin with a brief summary of the crystal structure and Fermi surface of a typical quasi-two-dimensional (Q2D) organic conductor, and then present the formalism of Landau quantization and quantum oscillations. We emphasize mainly on information we can obtain from the quantum oscillations and on the physics involved in the oscillations. For this purpose, we present representative, but simple examples to avoid confusing the readers by complicated details. In subsequent sections, we present some more specialized topics involving angular dependent magnetoresistance measurements, which complement and expand our understanding of the electronic structure of these materials.

6.2 Crystal Structure

Many organic conductors have layered structures composed of donor organic molecule layers and anion layers. A schematic crystal picture of a Q2D organic superconductor κ-(BEDT-TTF)$_2$Cu(NCS)$_2$ [5,6], where BEDT-TTF (ET) is bis(ethylenedithio)tetrathiofulvalene, is depicted in Fig. 6.1a. Various packing arrangements of the molecules are denoted by a Greek letter (κ in this case) because several different arrangements are generally achieved even with a same anion composition. The energy bands are derived from molecular orbitals of the ET molecules. Because of close stacking of the molecules along

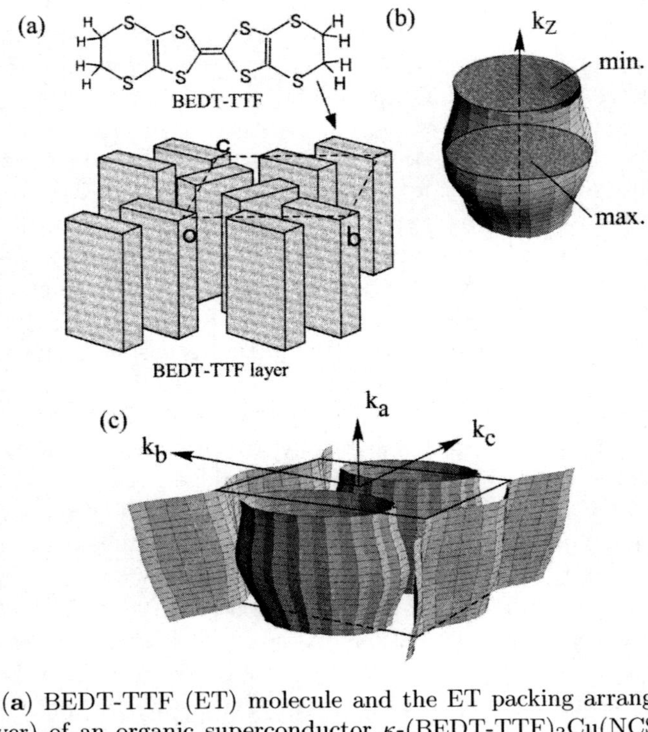

Fig. 6.1. (**a**) BEDT-TTF (ET) molecule and the ET packing arrangement (conducting layer) of an organic superconductor κ-(BEDT-TTF)$_2$Cu(NCS)$_2$. The ET molecule (conducting) and Cu(NCS)$_2$ anion (insulating) layers stack alternately along the a axis. (**b**) A quasi-two-dimensional Fermi surface. When the field is applied along the cylindrical (k_z) axis, two extremal (minimum and maximum) cross sections are well defined, indicated by shaded areas. (**c**) The calculated Fermi surface of κ-(BEDT-TTF)$_2$Cu(NCS)$_2$. The solid lines show the first Brillouin zone boundaries. There exist two open sheets of 1D Fermi surface and a closed 2D Fermi surface

the crystal b and c axes, the arrangement allows substantial overlap of the molecular orbitals, leading to finite transfer integrals between the adjacent molecules. The two ET molecules donate one electron to the $Cu(NCS)_2^{-1}$ anion. Therefore, the energy bands are partially filled, resulting in a metallic state. The ET molecule and anion (insulating) layers stack alternately along the a axis. The well-separated ET layers make the transfer integrals much smaller in this direction. Therefore, this salt has strong anisotropic electronic state, i.e., Q2D Fermi surface. Such strong anisotropy of the electronic state, resulting from a layered structure, is typical of most organic conductors.

The Q2D energy bands are often adequately described by a simple energy band written as

$$\epsilon(k) = \epsilon(k_x, k_y) - 2t_\perp \cos(k_z c), \tag{6.1}$$

where k_x and k_y are the components of the wavevector parallel to the conducting planes and k_z is that perpendicular to the planes. The quantity $t_\perp$ is the interlayer transfer integral. For Q2D systems, the condition $\epsilon(k_x, k_y) \gg t_\perp$ is well satisfied. The small value of $t_\perp$ results in a slight warping of the cylindrical Fermi surface along the k_z direction (Fig. 6.1b).

A schematic picture of the Fermi surface of κ-(BEDT-TTF)$_2$Cu(NCS)$_2$ calculated within a tight binding approximation is shown in Fig. 6.1c [7]. The solid lines show the first Brillouin zone (BZ) boundaries. Since this salt has two dimerized ET molecules (four molecules) in the unit cell (there are two holes in the energy bands), the band filling requires that the whole volume of the Fermi surface is equal to the first BZ. The Fermi surface intersects the BZ boundaries, and the low symmetry of the crystal structure causes band gaps at the boundaries. As a result, the Fermi surface splits into two parts: two open sheets of 1D Fermi surface and a closed 2D Fermi surface.

6.3 Landau Quantization and Quantum Oscillations

In the presence of a magnetic field H, the motion of a Bloch electron is described by the Lorentz force equation

$$\hbar\dot{\mathbf{k}} = -\mathbf{e}(\mathbf{v} \times \mathbf{H}), \tag{6.2}$$

where $\mathbf{k}$ is the wave vector of the electronic state and the velocity $\mathbf{v}$ of the electron motion is given by

$$\mathbf{v} = \frac{\mathbf{1}}{\hbar}\frac{\partial\varepsilon(\mathbf{k})}{\partial\mathbf{k}}, \tag{6.3}$$

where $\epsilon(k)$ is the energy dispersion of the electronic state. The electrons move on the constant energy surface in k space perpendicular to the magnetic field because the magnetic field does no work on the electrons. When the

Fermi surface is simply connected, closed trajectories are permitted, where the angular frequency of the cyclotron motion (cyclotron frequency) is

$$\omega_\mathrm{c} = \frac{eH}{m_\mathrm{eff}}, \tag{6.4}$$

where the m_eff is the cyclotron effective mass, defined by

$$m_\mathrm{eff} = \frac{\hbar^2}{2\pi}\left(\frac{\partial A_k}{\partial \varepsilon}\right). \tag{6.5}$$

This mass is generally different from the free electron mass m_0, as discussed later. Here A_k is the cross section of the Fermi surface perpendicular to the magnetic field direction. The cyclotron motion in k-space corresponds to real space motion, where the relation is

$$\hbar(\mathbf{k} - \mathbf{k_0}) = -\mathbf{e}(\mathbf{r} - \mathbf{r_0}) \times \mathbf{H}. \tag{6.6}$$

Here k_0 and r_0 are the initial positions in k- and real space, respectively. This equation is obtained by integrating (6.2). When the projected real space motion is closed, the motion is quantized by the Bohr–Sommerfeld rule,

$$\oint \mathbf{p} \cdot \mathbf{dr} = \oint (\hbar\mathbf{k} - \mathbf{eA}) \cdot \mathbf{dr} = 2\pi(n + \gamma)\hbar, \tag{6.7}$$

where A is the vector potential, n is the integer, and γ is a phase factor (which is $1/2$ for a parabolic band). This quantum condition is written in terms of total magnetic flux Φ through the real space orbit:

$$\Phi = \frac{2\pi(n + \gamma)\hbar}{e}. \tag{6.8}$$

This expresses that only an integral multiple of the fundamental flux $\Phi_0 = h/e$ (apart from γ) is allowed in the real space areas A_r^n of the closed orbits:

$$A_r^n H = \Phi_0(n + \gamma). \tag{6.9}$$

The allowed k-space areas corresponding to the real space areas are

$$A_k^n = \frac{(n + \gamma)H}{\Phi_0}. \tag{6.10}$$

This is known as the Onsager relation. The energy levels of these orbits are given by

$$\varepsilon_n = \hbar\omega_c(n + \gamma) + f(k_z), \tag{6.11}$$

where

$$f(k_z) = \hbar^2 k_z^2 / 2m_z, \tag{6.12}$$

for a parabolic band. These levels with energy of ϵ_n are called Landau levels.

The area A_k^n with quantum number n increases with increasing field, whereas the projected real space area decreases to conserve the number of the magnetic flux penetrating it, according to Bohr–Sommerfeld rule. As the field increases, the Landau level approaches an extremal cross section of the Fermi surface A_F and then the free energy increases to a maximum. If the field increases further, the highest Landau level with n becomes depleted (the nth level becomes empty, but the $(n-1)$th level is fully occupied), causing a sudden decrease of the free energy. The free energy then increases again until the next maximum is reached. The maxima occur whenever the area given by (6.10) is equal to the extremal area A_F, which is equally spaced with intervals periodic in $1/H$,

$$\Delta\left(\frac{1}{H}\right) = \frac{2\pi e}{\hbar}\frac{1}{A_\mathrm{F}} = \frac{1}{A_\mathrm{F}\Phi_0}. \tag{6.13}$$

This expression is directly related to the frequency of the quantum oscillations such as SdH and dHvA effects, as shown later.

6.4 Lifshitz and Kosevich (L-K) Formula

The treatment of quantum oscillation phenomena was first formulated by Lifshitz and Kosevich (L-K) [4, 8]. In the presence of the magnetic field, the thermodynamic potential Ω is given by

$$\Omega = -k_\mathrm{B}T\sum_i \ln\left[1-\exp\{(\zeta-\varepsilon_i)/k_\mathrm{B}T\}\right], \tag{6.14}$$

where we sum over all possible energy states. After some tedious calculations, we obtain the oscillatory part of the thermodynamic potential

$$\begin{aligned}\tilde{\Omega} = &\left(\frac{e}{2\pi\hbar}\right)^{3/2}\frac{e\hbar}{m_\mathrm{eff}}\frac{VH^{5/2}}{\pi^2 A_\mathrm{F}''^{1/2}}\\ &\times\sum_{p=1}^{\infty}\frac{1}{p^{5/2}}R_\mathrm{T}R_\mathrm{D}R_\mathrm{S}\cos\left[2\pi p\left(\frac{F}{H}-\frac{1}{2}\right)+\Psi\right],\end{aligned} \tag{6.15}$$

where A_F is the extremal area of the Fermi surface, i.e., maximum or minimum of the cross-section with respect to k parallel to the field, and

$$A_\mathrm{F}'' = \frac{\partial^2 A_\mathrm{F}}{\partial k_{//}{}^2} \tag{6.16}$$

gives the curvature of the Fermi surface (curvature factor). R_T, R_D, and R_S are reduction factors, to be defined below.

The last term

$$\cos\left[2\pi p\left(\frac{F}{H}-\frac{1}{2}\right)+\Psi\right] \tag{6.17}$$

shows periodic oscillations as a function of $1/H$ with the fundamental frequency F for $p = 1$. The terms for $p > 1$ correspond to harmonics of the oscillations with the frequencies pF. The phase factor Ψ is $-\pi/4(+\pi/4)$ for the maximum (minimum) cross section of the Fermi surface. The frequency F is given by

$$F = \frac{\Delta n}{\Delta(1/H)} = \frac{\hbar}{2\pi e} A_{\mathrm{F}}, \tag{6.18}$$

$$F[T] = 1.05 \times 10^4 \, A_{\mathrm{F}} \, \mathrm{\AA}^{-2}. \tag{6.19}$$

Generally, the Fermi surface is not spherical, i.e., the extremal cross-sections A_{F} of the Fermi surface are anisotropic. Therefore, the observation of the frequency F as a function of the magnetic field direction provides direct information about the Fermi surface structure.

Since the oscillatory magnetization components parallel and perpendicular to the field are, respectively, given by

$$\widetilde{M}_{//} = -\left(\partial\widetilde{\Omega}/\partial H\right)_{\zeta}, \quad \widetilde{M}_{\perp} = -\frac{1}{H}\left(\partial\widetilde{\Omega}/\partial\theta\right)_{\zeta,H}, \tag{6.20}$$

we obtain

$$\widetilde{M}_{//} = -\left(\frac{e}{\hbar}\right)^{3/2} \frac{e\hbar}{m_{\mathrm{eff}}} \frac{FVH^{1/2}}{\sqrt{2}\pi^{5/2} A_{\mathrm{F}}''^{1/2}} \times \sum_{p=1}^{\infty} \frac{1}{p^{3/2}} R_{\mathrm{T}} R_{\mathrm{D}} R_{\mathrm{S}} \sin\left[2\pi p\left(\frac{F}{H} - \frac{1}{2}\right) + \Psi\right], \tag{6.21}$$

and

$$\widetilde{M}_{\perp} = -\frac{1}{F}\frac{\partial F}{\partial\theta}\widetilde{M}_{//}. \tag{6.22}$$

In the differentiations, the small contributions coming from the $H^{5/2}$ terms and the θ dependence of $1/m_{\mathrm{eff}} A_{\mathrm{F}}''$ are neglected assuming $F/H \gg 1$. A typical example of the dHvA oscillations measured by magnetic torque, corresponding to $\widetilde{M}_{\perp}$, and the Fourier transform spectrum for κ-(BEDT-TTF)$_2$Cu(NCS)$_2$ are shown in Fig. 6.2a, b, respectively.

The oscillatory part of the effective density of states is

$$\widetilde{D}(H) = \left(\frac{2eH}{\hbar}\right)^{1/2} \frac{m_{\mathrm{eff}} V}{2^{3/2}\pi^{5/2} A_{\mathrm{F}}''^{1/2}} \times \sum_{p=1}^{\infty} \frac{1}{p^{1/2}} R_{\mathrm{T}} R_{\mathrm{D}} R_{\mathrm{S}} \cos\left[2\pi p\left(\frac{F}{H} - \frac{1}{2}\right) + \Psi\right]. \tag{6.23}$$

The probability of the scattering of the electrons is proportional to the number of states into which the electrons can be scattered. Therefore, the probability, which determines the electronic relaxation time τ and the conductivity, will oscillate in accordance with the oscillation of the density of states. When there exist a number of Fermi surfaces, their contribution to the total conductivity generally has a complicated functional form of involving separate densities of states. However, it is reasonable to assume that the total oscillation signal is given by the simple sum of the oscillations arising from each Fermi surface. The detailed theory by Adams and Holstein [9], developed for a spherical Fermi surface, gives the oscillatory term of the conductivity

$$\widetilde{\sigma} = \sigma_0(H)\left(1 + \frac{\widetilde{D}(H)}{D_0}\right), \tag{6.24}$$

where a purely numerical factor is omitted. The quantities σ_0 and D_0 are the nonoscillatory backgrounds of the conductivity and the density of states, respectively. This expression is valid only when $\widetilde{D}(H)/D_0 \ll 1$, i.e., the oscillation of the conductivity is much smaller than the nonoscillatory background. If this is not satisfied, the second order term of $\widetilde{D}(H)/D_0$ is not negligible, which causes large harmonics with phases different from those of the first order term. This makes the wave shape of the oscillations very complicated. Some organic conductors show large SdH oscillations, whose amplitudes reach a half (or more) of the total conductivity. In such cases, great care should be taken in the analysis of the wave shape. There are no such difficulties in the analysis of thermodynamic quantities such as dHvA oscillations.

6.4.1 Temperature Reduction Factor

At finite temperatures, the electrons can be thermally excited to higher Landau levels, which causes thermal damping of the oscillations. This damping effect gives the temperature reduction factor

$$R_T = \frac{pK\mu T/H}{\sinh(pK\mu T/H)}, \quad \mu = m_{\text{eff}}/m_0, \tag{6.25}$$

where the constant $K(=2\pi^2 m_0 k_B/e\hbar)$ is 14.7 (T K^{-1}). As $T/H \to 0$, the factor reaches unity. Only this reduction factor involves the temperature, and so the effective mass ratio μ is directly obtained from the temperature dependence of the oscillation amplitude. Usually, the oscillation amplitude divided by temperature is plotted as a function of temperature to obtain the mass. This is called mass plot (Fig. 6.2c).

6.4.2 Dingle Reduction Factor

Since real conductors include a lot of scattering centers such as impurities or lattice defects, the electrons on the cyclotron orbits are scattered by these

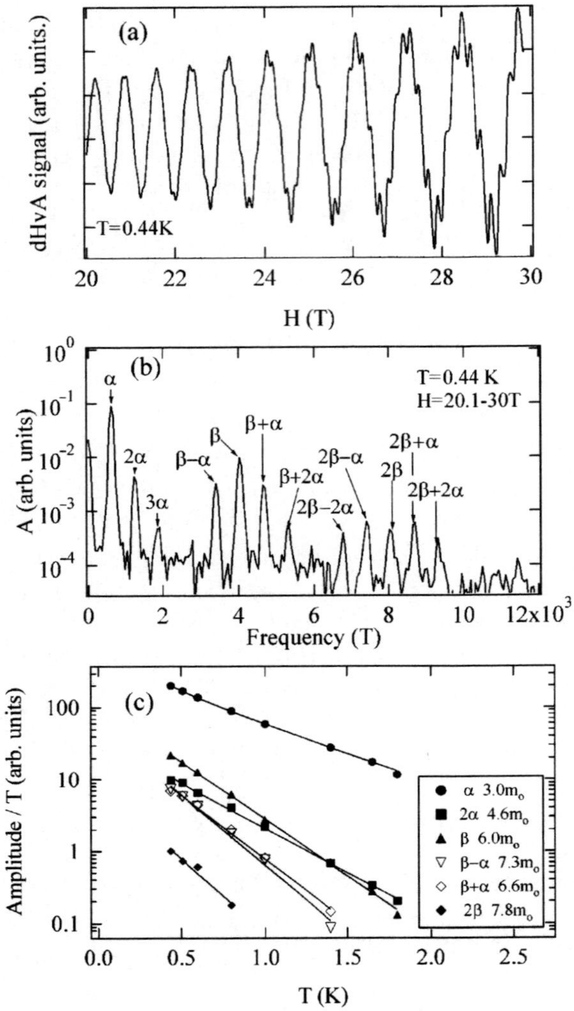

Fig. 6.2. (**a**) dHvA oscillations measured by magnetic torque in magnetic field perpendicular to the conducting layers for κ-(BEDT-TTF)$_2$Cu(NCS)$_2$ [10]. (**b**) Fourier transform spectrum of the oscillations between 20.1 and 30 T. Various combination frequencies of α and β are observed. (**c**) Mass plots for various oscillations. The oscillation amplitudes are obtained from Fourier transform spectra. The solid curves are the fitted results with (6.25). The determined masses corresponding to various orbits are also shown

centers and lose phase coherence. The scattering effect, which was first introduced by Dingle, effectively makes the Landau levels broaden according to the uncertainty principle. This gives the Dingle reduction factor

$$R_{\mathrm{D}} = \exp(-pKx\mu/H), \tag{6.26}$$

where x is called Dingle temperature, and is directly related to the scattering time τ,

$$x = \hbar / \, 2\pi k_{\mathrm{B}} \, \tau. \tag{6.27}$$

This factor can be also written as

$$R_{\mathrm{D}} = \exp(-p\pi/\omega_{\mathrm{c}}\tau), \tag{6.28}$$

i.e., less scattering (larger $\omega_{\mathrm{c}}\tau$) gives larger oscillations. Note that this factor also reaches unity as $x/H \rightarrow 0$. When the effective mass ratio is known, the Dingle temperature x is obtained from the field dependence of the oscillation amplitude.

6.4.3 Spin-Splitting Reduction Factor

In magnetic fields, the Landau levels are further split into a set of spin-up and spin-down levels by the Zeeman energy $g\mu_{\mathrm{B}}sH$, where g is the g-factor of the electrons and μ_{B} is the Bohr magneton. Each spin state ($s = 1/2$ or $-1/2$) leads to quantum oscillations with the same frequency F, but with different phases given by $\pi g\mu_{\mathrm{B}}H/\hbar\omega_{\mathrm{c}}$. The phase difference, which is independent of field, reduces the amplitude in addition to the other reduction factors. This reduction factor is called the spin-splitting reduction factor,

$$\begin{aligned} R_{\mathrm{S}} &= \cos(p\pi S), \\ S &= g\mu/2. \end{aligned} \tag{6.29}$$

In organic conductors, the g-factor is very close to the free electron value, 2.00. Therefore, when the effective mass ratio μ is equal to 3/2 for instance, the fundamental oscillation and the harmonics with odd p's disappear, and only the harmonics with even p's are observable. This is called the spin-splitting-zero condition, which is accidentally satisfied in a few organic conductors. A typical example [11] is shown in Fig. 6.3.

6.5 Other Oscillatory Effects

Since the thermodynamic potential oscillates as the field is varied, all the thermodynamic quantities also oscillate. Since the oscillations of the temperature and heat capacity C are small compared with the mean values, we obtain their oscillatory terms in the approximations [4, 12, 13],

$$\begin{aligned} \Delta\tilde{T} &= \frac{-T}{C}\int^{H}\frac{\partial\tilde{M}}{\partial T}\,\mathrm{d}H = \frac{1}{C}\frac{\partial\tilde{\Omega}}{\partial T} \\ &= \frac{T}{C}\left(\frac{e}{2\pi\hbar}\right)^{3/2}\frac{e\hbar}{m_{\mathrm{eff}}}\frac{VH^{5/2}}{\pi^2 {A''_{\mathrm{F}}}^{1/2}} \\ &\times\sum_{p=1}^{\infty}\frac{1}{p^{5/2}}R_{\mathrm{D}}R_{\mathrm{S}}\cos\left[2\pi p\left(\frac{F}{H}-\frac{1}{2}\right)+\Psi\right]\eta(Kp\mu T/H), \end{aligned} \tag{6.30}$$

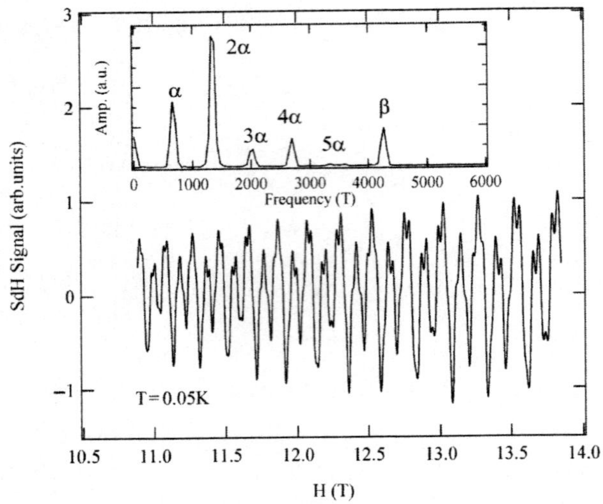

Fig. 6.3. SdH oscillations in magnetic field perpendicular to the conducting layers for a 2D organic conductor α-(ET)$_2$KHg(SCN)$_4$ [11]. *Inset*: Fourier transform spectrum of the oscillations. The Fermi surface topology (ignoring nesting and reconstruction) is similar to that of κ-(BEDT-TTF)$_2$Cu(NCS)$_2$ (Fig. 6.1c). The fundamental oscillations denoted by α and β, and the harmonics of α are observed. Note that the even harmonic multiples of α are relatively larger than the odd multiples. Since the effective mass is close to $1.5m_0$ and $g = 2$ for α, the spin-splitting-zero condition is almost satisfied

$$\begin{aligned}\Delta\widetilde{C} &= -T\frac{\partial}{\partial T}\left(\frac{\partial\widetilde{\Omega}}{\partial T}\right)\\ &= \left(\frac{e}{2\pi\hbar}\right)^{3/2}\frac{e\hbar}{m_{\text{eff}}}\frac{K\mu V H^{3/2}}{\pi^2 A_{\mathrm{F}}''^{1/2}}\\ &\quad\times\sum_{p=1}^{\infty}\frac{1}{p^{3/2}}R_{\mathrm{D}}R_{\mathrm{S}}\cos\left[2\pi p\left(\frac{F}{H}-\frac{1}{2}\right)+\Psi\right]\phi\left(Kp\mu T/H\right).\end{aligned} \tag{6.31}$$

In these formulae, the temperature reduction factors are, respectively, replaced with

$$\eta(y) = y\frac{1}{\sinh(y)} - y^2\frac{\cosh(y)}{\sinh(y)^2}, \tag{6.32}$$

$$\phi(y) = 2y\frac{\cosh(y)}{\sinh(y)^2} - y^2\frac{1+\cosh(y)^2}{\sinh(y)^3}. \tag{6.33}$$

Both factors vanish as $y(=Kp\mu T/H) \to 0$ (in a high field or low temperature limit). The factor η is always negative and has a minimum at $y = 2.5$. However, the factor ϕ changes the sign from positive to negative, passing zero

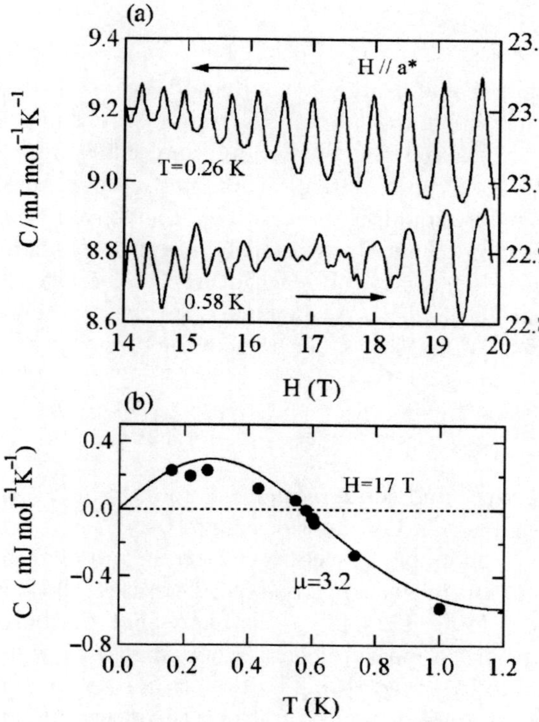

Fig. 6.4. (**a**) Quantum oscillations in the heat capacity of κ-(BEDT-TTF)$_2$Cu(NCS)$_2$. The node at 16.5 T for $T = 0.58$ K directly gives the effective mass of $3.2m_0$ [14]. (**b**) Temperature dependence of the heat capacity oscillation amplitude at 17 T. The negative values show that the phase of the oscillation is reversed. The solid curve is the fitted result with (6.33), which consistently gives the effective mass of $3.2m_0$

at $y = 1.6$ with increasing y. Therefore, if the oscillation amplitude of the heat capacity drops to zero as a function of field, we can directly obtain the effective mass only by one field sweep. The quantum oscillation of the heat capacity in κ-(BEDT-TTF)$_2$Cu(NCS)$_2$ is shown in Fig. 6.4a. At 0.58 K, the oscillation amplitude has a minimum at 16.5 T, but monotonically increases with field at 0.26 K. The node at 16.5 T corresponds to $y = 1.6$ in (6.33), giving the effective mass $3.2m_0$. The temperature dependence of the amplitude also has a minimum at 0.56 K, which consistently gives $3.2m_0$ (Fig. 6.4b). The absolute value of the heat capacity oscillation is easily experimentally obtained. Therefore, the results can be directly fitted with (6.31). This advantage allows us to estimate the curvature factor A'' of the Fermi surface. In κ-(BEDT-TTF)$_2$Cu(NCS)$_2$, the curvature factor was first experimentally obtained by this technique, $(2\pi/A'')^{1/2} = 6$, which gives the anisotropy of the transfer integral $t_{\parallel}/t_{\perp} = 140$ [14].

6.6 Effective Mass

The energy bands in many organic conductors are well reproduced by simple band structure calculations within a tight binding approximation. Such calculations tacitly assume that the ions and molecules are rigidly fixed, making a perfectly periodic potential, and that the Coulomb interaction between electrons are negligibly small. However, when electrons move in molecular layers, the lattice around them is distorted by electron–phonon interaction and also electrons are kept away from each other by electron–electron Coulomb interaction. Such effects lead to quasiparticles (interacting electrons) with renormalized effective masses. When the interactions are not too strong, the effective mass is expressed as [4]

$$m_{\mathrm{eff}} = m_{\mathrm{b}}(1 + \lambda_{\mathrm{e-p}})(1 + \lambda_{\mathrm{e-e}}), \tag{6.34}$$

where m_{b} is the bare band (unrenormalized) mass, which corresponds to that obtained from simple band calculations, and $\lambda_{\mathrm{e-p}}$ ($\lambda_{\mathrm{e-e}}$) is an enhancement factor due to electron–phonon (electron–electron) interaction.

The mass is renormalized in the electronic states whose energy levels are close to the Fermi level. Therefore, it is likely that the bare band mass m_{b} is determined by optical measurements, where high energy excitation spectra of the electrons are observed [15]. The cyclotron resonance measurements is believed to give the mass enhanced only by the electron–phonon interaction $m_{\mathrm{b}}(1+\lambda_{\mathrm{e-p}})$ [16–19] because all the electrons undergo cyclotron motion with the same phase at the resonance. (The relative distance between the electrons does not change at the resonance). In the quantum oscillation measurements, the mass is renormalized by both many body effects [4,20]. Extensive efforts to obtain the mass renormalization have been pursued in organic conductors. In some organic conductors, the effective mass obtained by quantum oscillation measurements can be several times the bare band mass [10,15,20,21].

6.7 Magnetic Breakdown

The Fermi surfaces of κ-(BEDT-TTF)$_2$Cu(NCS)$_2$ have a simple structure (Fig. 6.5a): a closed 2D orbit and two open 1D Fermi surface sheets. In the presence of a weak magnetic field perpendicular to the $k_{\mathrm{b}} - k_{\mathrm{c}}$ plane, only electrons on the 2D Fermi surface undergo cyclotron motion (closed orbit denoted by α) and those on the 1D Fermi surface move only in the k_c direction (open orbit)(Fig. 6.5b). However, at high fields, electrons have sufficiently high velocity due to the Lorentz force to allow tunneling from one part of the Fermi surface to another without any scattering. This coherent process (quantum tunneling) is called magnetic breakdown (MB). In Fig. 6.5a, the MB takes place at the zone boundaries between the 1D and 2D Fermi surfaces (denoted by arrows). The process by which electrons pass perpendicularly

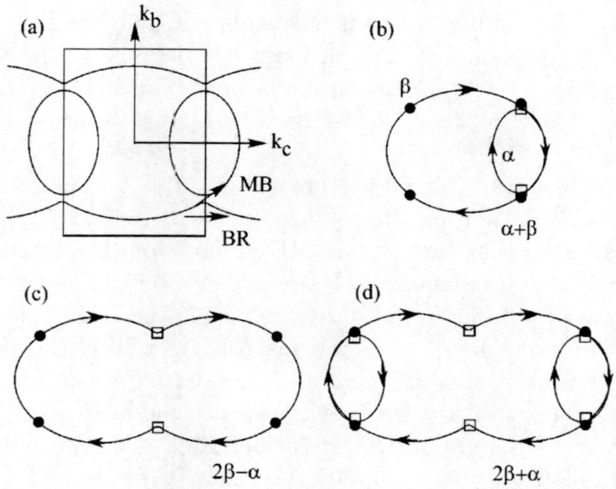

Fig. 6.5. (**a**) The Fermi surface in κ-(BEDT-TTF)$_2$Cu(NCS)$_2$. (**b**), (**c**), and (**d**) Various possible closed orbits by the MB and BR processes. The MB and BR points are denoted by closed circles and open squares, respectively. Note that the frequencies $\beta-\alpha$ and $2\beta-2\alpha$ experimentally observed (Fig. 6.2b) are not explained by this semiclassical MB treatment

through the zone boundaries on the same Fermi surface is Bragg reflection (BR). A semiclassical treatment gives the probability of the MB [4, 22],

$$P = \exp\left(-\frac{{\varepsilon_g}^2}{\hbar\omega_c\varepsilon_g}\right) = \exp\left(-\tfrac{H_0}{H}\right),$$
$$H_0 = \frac{m_{\text{eff}}}{\hbar e}\left(\frac{{\varepsilon_g}^2}{\varepsilon_F}\right), \tag{6.35}$$

where ϵ_g and ϵ_F are the energy gap at the zone boundary between the 1D and 2D Fermi surfaces, and the Fermi energy, respectively. The field H_0 is called MB field. The probability of BR is $Q = 1 - P$. In a high field limit $H \gg H_0$, P reaches unity, so only the MB process takes place. It means that all the electrons undergo cyclotron motion on the large area in k space (denoted by β). In an intermediate region ($P \simeq Q$), electrons can complete various closed orbits composed of α and β. Some examples are shown in Fig. 6.5. This semiclassical model well explains many of the combination frequencies observed in Fig. 6.2. However, the frequencies, $\beta-\alpha$ and $2\beta-2\alpha$, for instance, are not explained by this model.

The L-K formulae are obtained at fixed chemical potential. In conventional 3D metals, there exist a number of Fermi surfaces, which enable the electrons to move to lower energy Landau levels at any field. Therefore, the chemical potential is almost fixed (the chemical potential oscillation is negligibly small). However, in highly 2D systems where only one closed Fermi surface is present, the chemical potential oscillation is not negligible, and the L-K formula (and the wave shape) is modified in high magnetic fields ($\omega_c\tau \gg 1$) at very low temperatures ($\hbar\omega_c \gg kT$). When there coexist 1D and 2D Fermi surfaces and the MB takes place between the energy bands, the situation becomes more complicated.

Many studies have treated this subject [23–31]. A number of fully quantum mechanical calculations have been made to clarify the effect of the chemical potential oscillation on the dHvA effect and the origin of the semi-classically forbidden frequencies. At present, it is believed that the forbidden frequencies are of a purely quantum-mechanical origin and that the chemical potential oscillation plays an important role. However, the detailed interpretation still appears controversial. In addition, numerical calculations also show that the temperature dependence is also modified from the L-K formula. Since it gives only small corrections in a conventional temperature range, it is reasonable to estimate the effective mass based on the L-K formula.

6.8 Quantum Interference

When MB takes place between two open orbits along which the electrons travel in the same direction, an oscillatory behavior in the resistance, similar to SdH oscillations, will be observable. Such possibility was first studied by Stark and his coworkers in Mg [32]. Therefore, this effect is often called Stark quantum interference (QI). The geometry of this process is shown schematically in Fig. 6.6. The QI effect appears when an electron travels along two alternative paths labeled by 1 and 2 from the point A to B when the magnetic field is applied perpendicular to the plane. The probability amplitude along the path 1 from A to B is

$$q\,\mathrm{e}^{\mathrm{i}\phi_1}q, \tag{6.36}$$

where q is the BR probability amplitude ($Q = q^2$). Similarly, that along the path 2 is

$$ip\,\mathrm{e}^{\mathrm{i}\phi_2}ip \tag{6.37}$$

and $P = p^2$. The factor i comes from the condition that the transmitted and reflected wave functions at each junction must be orthogonal. The phase factor on each path is

$$\phi_j = \int_j \frac{e}{\hbar} A\,\mathrm{d}r, \tag{6.38}$$

where A is the vector potential. The total probability is given by

$$\begin{aligned} T &= \left|q\,\mathrm{e}^{\mathrm{i}\phi_1}q + ip\,\mathrm{e}^{\mathrm{i}\phi_2}ip\right|^2 = q^4 + p^4 - 2q^2p^2\cos(\phi_1 - \phi_2) \\ &= Q^2 + P^2 - 2QP\cos(\phi_1 - \phi_2). \end{aligned} \tag{6.39}$$

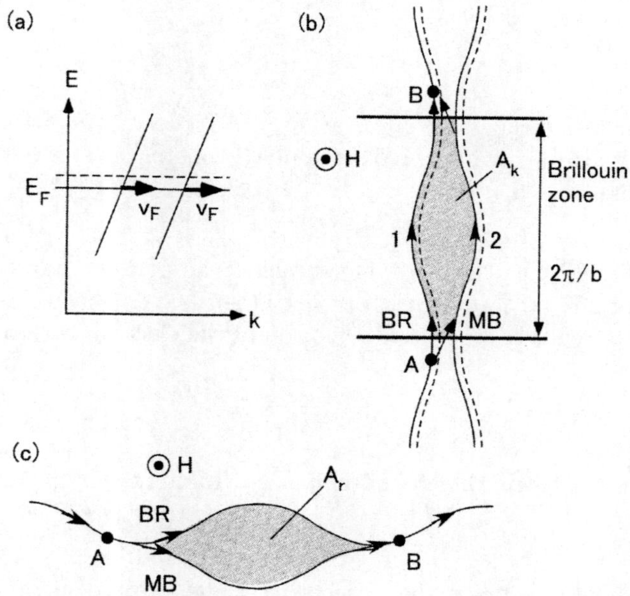

Fig. 6.6. (**a**) Two energy bands, (**b**) two sheets of 1D Fermi surface, causing the QI effect, and (**c**) real space trajectories corresponding to (**b**). The thick lines in (**b**) show the BZ boundaries. When an electron travels along two alternative paths labeled by 1 (two BR processes) and 2 (two MB processes) from the point A to B in magnetic field perpendicular to the plane, the QI effect will happen. The dotted lines show schematic trajectories with a small increase of energy, which corresponds to the energy level denoted by the dotted line in (**a**). Note that the area between the two trajectories (A_k) is not sensitive to the energy

The last term corresponds to the QI, whose phase is written as

$$\begin{aligned}\phi_1 - \phi_2 &= \int_1 \frac{e}{\hbar}\mathbf{A}\cdot \mathrm{d}\mathbf{r} - \int_2 \frac{e}{\hbar}\mathbf{A}\cdot \mathrm{d}\mathbf{r} = \oint \frac{e}{\hbar}\mathbf{A}\cdot \mathrm{d}\mathbf{r} \\ &= \int_S \frac{e}{\hbar}\nabla\times\mathbf{A}\cdot \mathrm{d}\mathbf{S} = \int_S \frac{e}{\hbar}\mathbf{H}\cdot \mathrm{d}\mathbf{S} = \frac{eH}{\hbar}A_r\,. \end{aligned} \tag{6.40}$$

Since the area in the real space between the two trajectories

$$\int_S \mathrm{d}S = A_r \tag{6.41}$$

corresponds to the area in k space

$$A_k = \left(\frac{eH}{\hbar}\right)^2 A_r, \tag{6.42}$$

we obtain

$$\cos(\phi_1 - \phi_2) = \cos\left(\frac{2\pi F}{H}\right), \quad F = \frac{\hbar}{2\pi e} A_k. \tag{6.43}$$

This term has the same form as the SdH effect. The QI effect is suppressed by scattering, which breaks the phase coherence. This reduction factor is written as

$$R_{\text{scatter}} = \exp\left[-(t_1 + t_2)/\tau\right], \tag{6.44}$$

where t_j is time that the electron completes the periodic motion in the BZ. When there are two open sheets of Fermi surface as shown in Fig. 6.6, the Lorentz force leads to a sinusoidal motion of the electron with a period

$$t_j = \frac{h}{bev_{\mathrm{F}j}H}, \tag{6.45}$$

where $v_{\mathrm{F}j}$ is the Fermi velocity of the path j. Taking the frequency $\omega_{\mathrm{c}} = 2\pi/t$, we obtain

$$R_{\text{scatter}} = \exp(-4\pi/\omega_{\mathrm{c}}\tau). \tag{6.46}$$

Here, we assumed that the trajectories 1 and 2 are identical. It should be noted that this factor is similar in form to the Dingle reduction factor R_D.

The QI effect can be distinguished from conventional SdH oscillations, because no cyclotron motion of electrons (no Landau quantization) is required. Therefore, the QI effect is observable only in the resistance, but not in any thermodynamical quantities. Another important feature is that the QI effect should be insensitive to temperature in a case as shown in Fig. 6.6. Even if the electron energy changes a little, the area between the two trajectories does not change. This corresponds to the zero mass formulation for the SdH oscillation (i.e., no thermal damping).

Oscillatory behavior in the resistance for a quasi 1D organic conductor $(TMTSF)_2ClO_4$ may be caused by the QI effect [33]. At temperatures above 5.5 K, this salt has Fermi surface as shown in Fig. 6.6. (No closed orbits are permitted.) In field perpendicular to the plane, the inplane resistance shows an oscillation in high fields (Fig. 6.7). We note that the oscillation amplitude decreases with increasing temperature. This is simply caused by the increase of the electron–phonon scattering (decrease of scattering time). No oscillations have yet been observed in thermodynamic quantities.[1]

[1] Another possibility for the oscillation above 5.5 K has been proposed by A.G. Lebed, Phys. Rev. Lett. **24**, 4903 (1995). Quantum mechanical considerations show that the distance in k-space between the two open sheets of the Fermi surface changes as a function of the inverse field, and consequently the electron–electron scattering time, i.e., the resistance oscillates in magnetic fields. This theory predicts that the amplitude shows $1/T^2$ dependence for $T > T^*$, where T^* is a characteristic temperature depending on the magnetic field. However, the experimental results do not show the $1/T^2$ dependence.

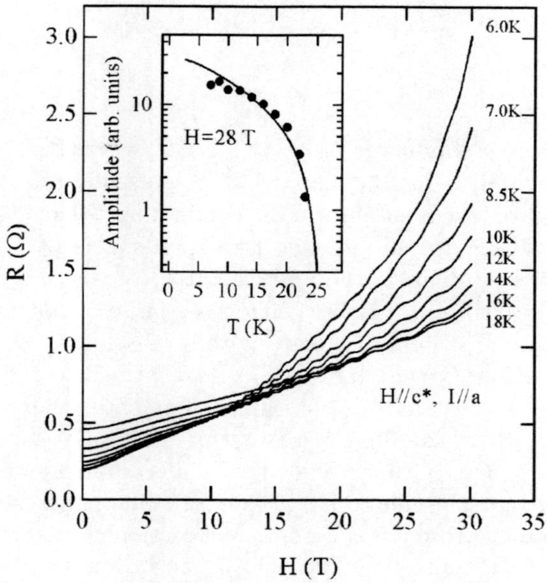

Fig. 6.7. Resistance of a quasi 1D organic conductor $(TMTSF)_2ClO_4$. The oscillatory behavior in high fields is believed to be the QI effect. *Inset*: Temperature dependence of the oscillation amplitude for $H = 28\,\mathrm{T}$. The solid line shows the calculated result, where the temperature dependent scattering time is obtained from the resistivity measurement

6.9 Internal Field

When an internal field H_{int} exists, which may be created by the exchange interaction with localized magnetic moments, the conduction electron spins are polarized by the internal field in addition to the external field. Since the energy bands are shifted by the Zeeman energy due to the internal field, the up and down spins have different Fermi surfaces (different cross sections). The oscillatory term for the fundamental frequency ($p = 1$) is modified as follows

$$\widetilde{D}(H) \propto Z_{\mathrm{down}} \cos\left[2\pi\left(\frac{F+\Delta F}{H} - \frac{1}{2}\right) + \Phi + \pi S\right] + Z_{\mathrm{up}} \cos\left[2\pi\left(\frac{F-\Delta F}{H} - \frac{1}{2}\right) + \Phi - \pi S\right], \tag{6.47}$$

where the prefactors Z_i include the temperature and Dingle reduction factors, which may be spin dependent. Note that the oscillatory term leads to

$$R_{\mathrm{s}} \cos\left[2\pi\left(\frac{F}{H} - \frac{1}{2}\right) + \Phi\right] \tag{6.48}$$

for $\Delta F = 0$. The shift ΔF is directly related to $H_{\rm int}$,

$$\Delta F = \frac{1}{4} g\mu H_{\rm int}. \tag{6.49}$$

In the 2D organic conductor λ-(BETS)$_2$FeCl$_4$, where BETS is bis(ethylenedithio)tetraselenafulvalene, the planar BETS molecules are stacked along the a and c axes, and consequently form 2D conducting layers [34]. This material has one closed 2D Fermi surface and two open sheets of 1D Fermi surface, whose topology is similar to that in Fig. 6.1c (see also Fig. 6.10 below). In the FeCl$_4$ anion (insulating) layers, there exist large $3d$ magnetic moments in the FeCl$_4$ anions in a high spin state with $s = 5/2$ [35]. A strong negative exchange interaction J (π-d interaction) between the π (conduction) electrons on the BETS molecules and the $3d$ localized moments creates an internal field ($H_{\rm int} \simeq J\langle s\rangle/g\mu_{\rm B}$), whose direction is antiparallel to H. At sufficiently high external field, $H_{\rm int}$ is saturated. When the field is applied perpendicular to the conducting layers, two oscillations with slightly different frequencies (Fig. 6.8) are observed, arising from two different cross sections of the 2D Fermi surfaces of the up and down spins [36]. The analysis according to (6.49) gives a large internal field of 32 T. For a nonmagnetic salt, λ-(BETS)$_2$GaCl$_4$, only a single frequency (no internal field) is seen as shown in Fig. 6.8.

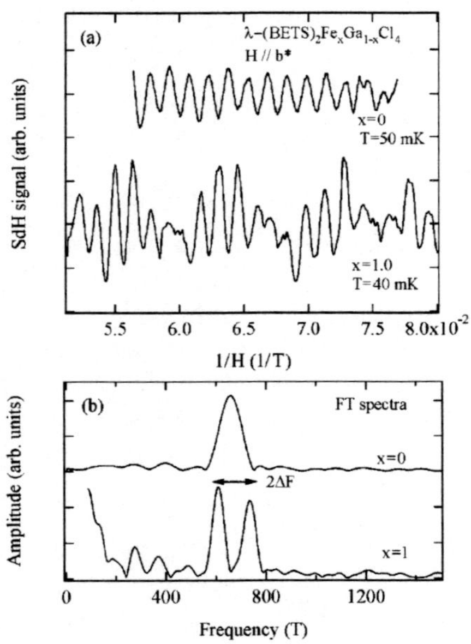

Fig. 6.8. (**a**) SdH oscillations of λ-(BETS)$_2$FeCl$_4$ ($x = 1$) and λ-(BETS)$_2$GaCl$_4$ ($x = 0$) in field perpendicular to the conducting layers. (**b**) FT spectra of the oscillations. Note that two peaks are evident only for the Fe salt ($x = 1.0$), where there exists large internal field $H_{\rm int}$. We obtain $H_{\rm int} = 32$ T, using $\mu = 4$ and $g = 2$

6.10 Special and Related Topics

In this section we present some specialized topics including field induced superconductivity and Fermi surface reconstruction where high magnetic fields have been crucial in their understanding.

6.10.1 Field Induced Superconductivity

Among intensive studies of superconductivity, the interplay between superconductivity and magnetism has been a subject of great interests for many years. In conventional superconductors, the superconductivity is destroyed by the application of a sufficiently high magnetic field. However, in some magnetic compounds, the superconductivity can be induced by high magnetic fields. λ-$(BETS)_2FeCl_4$, whose SdH oscillations are presented in the previous section, is known as the first material whose superconducting state is stabilized only under magnetic field (Fig. 6.9) [37–41].

At zero magnetic field, λ-$(BETS)_2FeCl_4$ shows a metal–insulator transition around 8 K. This is associated with the antiferromagnetic order of the Fe $3d$ moments [34]. The field of the order of 10 T breaks the antiferromagnetic insulating (AFI) phase and stabilizes a paramagnetic metallic (PM) phase. In this transition, the strong on-site Coulomb repulsion between the π electrons on the BETS molecules, and the π-d exchange interaction play an essential role [35, 42, 43].

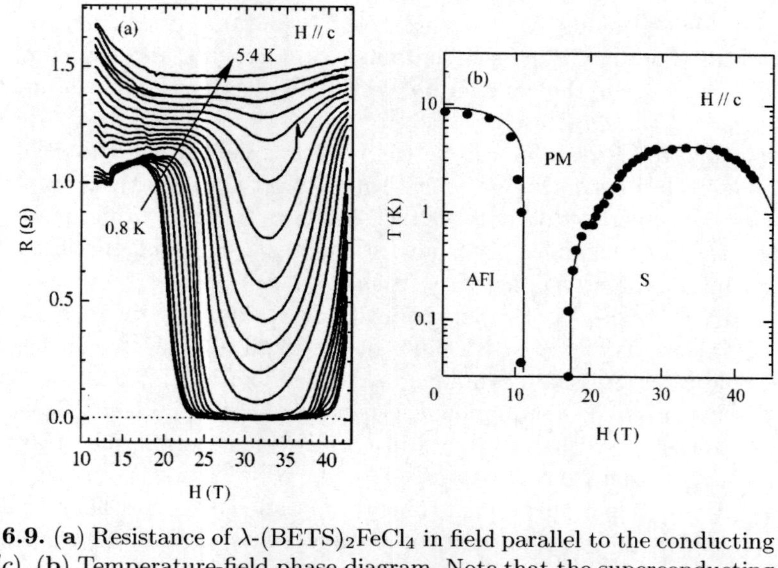

Fig. 6.9. (**a**) Resistance of λ-$(BETS)_2FeCl_4$ in field parallel to the conducting layers ($H//c$). (**b**) Temperature-field phase diagram. Note that the superconducting phase is stabilized only in high fields. PM, paramagnetic metal; AFI, antiferromagnetic insulator; S, superconductor

When the magnetic field is applied exactly parallel to the conducting *ac* layers, the resistance for $T = 0.8\,\mathrm{K}$ suddenly decreases at 20 T, and then superconductivity (S) appears. This superconductivity is broken at about 42 T. In general, superconductivity is broken under high magnetic field. However, for λ-$(BETS)_2FeCl_4$ and their alloys, superconductivity is induced by high magnetic fields.

The magnetic field has two mechanisms that destroy superconductivity. The first is the Zeeman effect, which breaks Cooper pairs if they are in a spin-singlet (but not a spin-triplet) state. The second is the so-called orbital effect, whereby the vortices penetrate into the superconductor and decrease the energy gained due to the Bose condensation (formation of Cooper pairs). For the case of 2D superconductors, the orbital effect is suppressed when the applied magnetic field is exactly parallel to the conducting layers because vortices are not formed. (The superconducting layers are weakly Josephson-coupled in real materials, and so the orbital effect is not suppressed completely, and Josephson vortices are formed in the insulating layers). This implies that the 2D superconducting state can survive up to a very high field as long as the field is parallel to the conducting layers, but does not explain why the superconductivity is induced by high fields.

Field induced superconductivity was first reported in the Chevrel compounds $ReMo_6S_8$, where the rare earth ion (Re) contributes a magnetic moment and the Mo_6S_8 structure provides an interpenetrating conduction electron matrix [44–46]. This superconductivity has been explained in terms of the Jaccarino–Peter (J-P) compensation effect [47, 48]. When magnetic field is applied to this material, the localized magnetic moments (with spin s) are easily aligned along the external field H. Because of a strong negative exchange interaction J between the magnetic moments and the conduction electrons, the spins of the conduction electrons feel a strong internal magnetic field ($H_{int} \simeq J\langle s\rangle/g\mu_\mathrm{B}$) created by the magnetic moments, whose direction is antiparallel to H. For $H = H_{int}$, the resultant field which the conduction electron spins feel becomes zero (the Zeeman effect is suppressed completely), and thus the superconductivity can be induced at fields around H_{int}. For small H, the superconductivity is also stable because the magnetic moments turn in random directions, i.e., H_{int} is small.

For λ-$(BETS)_2FeCl_4$, the Fe moments are paramagnetic in high fields, so the J-P effect is expected. Because of the highly 2D electronic state, the orbital field is suppressed for inplane fields. The fact that T_c has a maximum at 32 T shows that the conduction electrons see an internal field $H_{int} = 32\,\mathrm{T}$, which is consistent with the results in the previous section. The theoretical analysis suggest that there exists a S phase for $H < 2T$ and $T < 5.5K$ [39,41]. However, the AFI is more stable in this region, and so the low field S phase is not observable. The field induced superconductivity is also found in another magnetic BETS system, κ-$(BETS)_2FeBr_4$. This material undergoes an AF order of the Fe 3*d* moments at 2.5 K and then shows superconductivity at 1.4 K [49]. When the field is applied in the conducting layers, the S phase is broken at 3 T for $T = 36\,\text{mK}$, and then the S phase appears again between 10 and 16 T [50, 51]. This reentrant S phase is also caused by the J-P effect.

Although overall features of the field induced superconductivity in the BETS systems are explained by the J-P effect, there are still some important problems to be solved; the mechanism of the Cooper pairs (electron–phonon interaction, magnetic fluctuation, etc.), the symmetry of the energy gap (s-wave, p-wave, or d-wave), and the possibility of an FFLO state at lower temperatures near the two high and low external field critical boundaries as in Fig. 6.9b.

6.10.2 Angular Dependent Magnetoresistance and Fermi Surface Topologies

The phenomenon of field induced superconductivity mentioned in the last section also occurs in the alloy sequence [39] $\lambda - (BETS)_2Fe_xGa_{1-x}Cl_4$, where as the Fe concentration is decreased, H_{int} shows a corresponding decrease. This can be determined from the splitting of the SdH oscillation frequency as discussed in Sect. 6.9. In this section we discuss the angular dependent magnetoresistance oscillation (AMRO) properties for two alloys, $x = 1$ and $x = 0.8$. The AMRO method involves the measurement of the resistance of a single crystal, while it is rotated in a constant magnetic field. Recalling the warped cylindrical Fermi surface topology for a 2D layered conductor in Fig. 6.1, an AMRO experiment is carried out for a set of polar angle rotations (typically) $-\pi/2 < \theta < \pi/2$ over a set of azimuthal planes $0 < \phi < \pi$. Here θ is the angle between the magnetic field and the (z) axis of a cylindrical Fermi surface, and ϕ is the $(x\text{-}y)$ in-plane azimuthal angle. It is well established that for a warped closed orbit Fermi surface as depicted in Fig. 6.1, peaks in the interplane resistance will occur periodically with $\tan(\theta)$ [52–54]. In the case where the Fermi surface is closed, but elliptical and not circular in the x-y plane, the period will change with azimuthal angle ϕ, following the relationship

$$\delta(\tan(\theta)) = \pi/dk_{//}(\phi), \tag{6.50}$$

where d is the interlayer spacing and $k_{//}(\phi)$ is related to the in-plane Fermi momentum extrema by [55]

$$k_{//}(\phi)^2 = [k_{\text{Fmin}}\cos(\phi - \xi)]^2 + [k_{\text{Fmax}}\sin(\phi - \xi)]^2. \tag{6.51}$$

Here k_{Fmin} and k_{Fmax} are the extremal values of $k_{//}(\phi)$, and ξ is the azimuthal offset of the elliptical orbit with respect to a principal axis direction. As shown below, the in-plane Fermi momentum $k_{\text{F}}(\phi)$ can be obtained from the experimental (or fitted) $k_{//}(\phi)$ by a geometrical construction [56].

The Fermi surface for λ-(BETS)$_2$FeCl$_4$ computed from band structure [34] is shown in Fig. 6.10, where the closed hole orbit is predicted to be elliptical, and tilted away from the principal axes in the a-c plane. To experimentally determine this topology [57], the sample is rotated in a magnetic field as shown

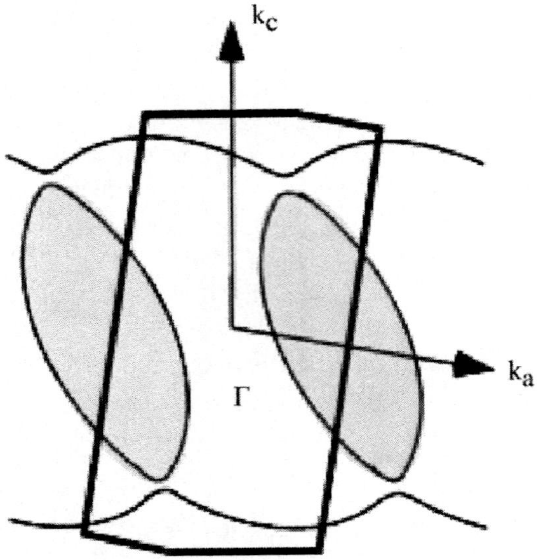

Fig. 6.10. Fermi surface of $\lambda - (BETS)_2FeCl_4$ after [34]

in Fig. 6.11a where we note that b^* is the interplane direction, perpendicular to the conducting planes. The resulting magnetoresistance oscillations (AMRO), plotted against tan(θ), show a periodic (δ) behavior (Fig. 6.11b). From the period, $k_{//}(\phi)$ can be computed from (6.50) and fitted to (6.51). $k_F(\phi)$ can be obtained by constructing lines perpendicular to the $k_{//}(\phi)$ lines, where the intersections of the constructed lines map out the Fermi surface (FS) perimeter, as shown in Fig. 6.11c. The general features of the experimental $k_F(\phi)$ are very similar to the elliptical FS in Fig. 6.10, but the cross-sectional area is about 40% of the first Brillouin zone, which is about twice that obtained from SdH oscillations (see Fig. 6.8). Additional corrugation due to the lower symmetry of the crystal structure has been suggested [57] as a possible origin for this difference.

A more detailed investigation of AMRO in these materials has been carried out on the alloy λ-$(BETS)_2Fe_{0.8}Ga_{0.2}Cl_4$ using a continuous rotation method [58], where the azimuthal angle ϕ advances about 7° for every 360° rotation of the polar angle θ. For this alloy concentration ($x = 0.8$), the field induced superconducting phase is centered around 25 T, less than the 32 T value for the pure iron system ($x = 1.0$) shown in Fig. 6.9. In the data presented herein, the sample was rotated through successive azimuthal angle ranges of 180° for 14 T and a subsequent 180° for 25 T. The full data set is shown in Fig. 6.12, where for the 25 T data, the sample enters the field induced superconducting state when ever the field is in-plane. The value $\phi = 0°$ degrees corresponds to a field rotation in the a′-b* plane as shown in Fig. 6.11c. We note that the superconductivity is partial due to the relatively high temperature of the experiment (1.8 K) with respect to T_c (3.7 K) for this alloy concentration [39].

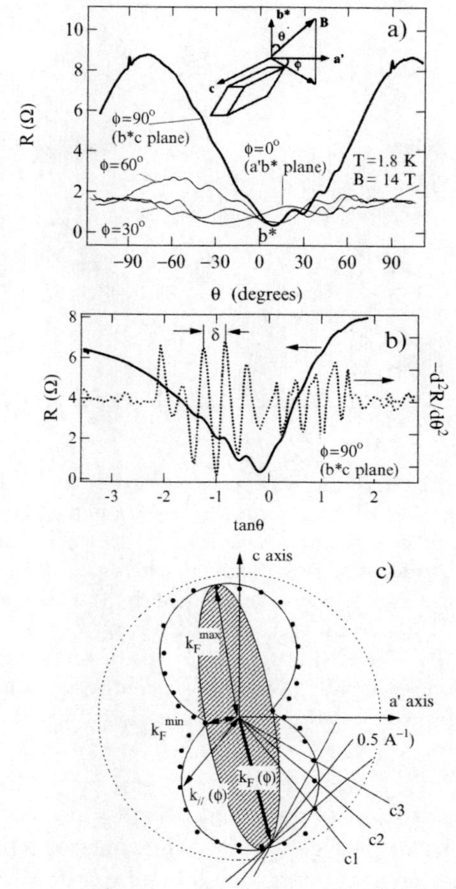

Fig. 6.11. AMRO study for λ-$(BETS)_2FeCl_4$ at 14 T and 1.8 K. (replotted after [57]). (**a**) AMRO for different values of ϕ. *Inset*: crystal axes and field direction in θ and ϕ. (**b**) Second derivative of AMRO signal for $\phi = 90°$ indicating $\tan(\theta)$ period δ. (**c**) Azimuthal plot of $k_{//}(\phi)$ from (6.50). Parameters and fit for (6.51) are also shown. To obtain $k_F(\phi)$, the intersection of lines perpendicular to radial lines along $k_{//}(\phi)$ values (i.e., c1, c2, c3,...) are constructed as shown, where for a complete 2π cycle, their intersection maps out $k_F(\phi)$ and the perimeter of the closed Fermi surface (shaded area)

However, at 14 T, the sample is in the normal state, and the coherence peak (see also Fig. 6.11a) is observed for in-plane fields. The coherence peak is believed to arise when the magnetic field is perpendicular to the axis of the warped cylindrical Fermi surface, where closed orbits can arise over a narrow range of angle [59]:

$$\Delta\theta_c \sim k_F t_z d / E_F, \qquad (6.52)$$

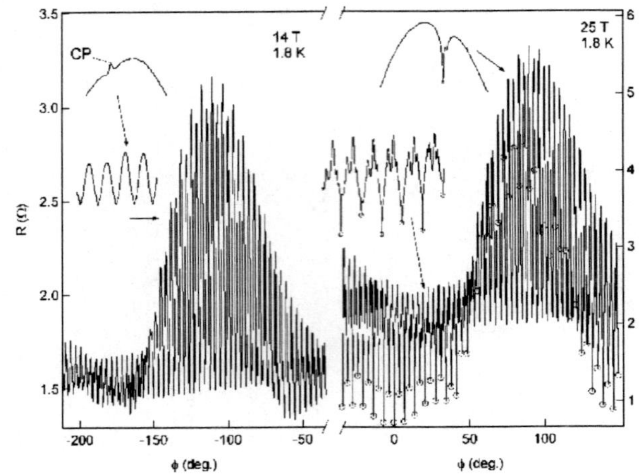

Fig. 6.12. AMRO data for λ-$(BETS)_2Fe_{0.8}Ga_{0.2}Cl_4$ at 1.8 K from a continuous rotation measurement for a 180° azimuthal rotation at 14 T in the normal state and a subsequent 180° azimuthal rotation at 25 T. Each full period (two cycles) corresponds to a 360° rotation in polar angle θ, where the azimuthal angle ϕ advances by ∼6.9°. The AMRO oscillations are evident in the fine structure of the 180° period waveforms. In the normal state at 14 T for the $B//a$-c plane orientation the "coherence peak" (CP) is evident (*upper left trace*), and at 25 T the field-induced superconducting phase is partially accessed (*sharp dips and open circles*) for $B//a$-c plane. The coherence peak and the dips are nearly coincident at 25 T as shown in the upper left inset

where $\Delta\theta_c$ is the width of the peak and t_z and d are the interplane transfer energy and thickness, respectively. Hence the coherence peak is considered as evidence for a finite, coherent interplane bandwidth, although there is some controversy about this interpretation. For λ-$(BETS)_2FeCl_4$ [57], t_z/E_F is estimated to be between 0.004 and 0.05, where we note that the peaks are most dominant for particular ranges of azimuthal angle. Hence the actual warped cylindrical structure may have higher degrees of corrugation.

A summary of the full 360° scan of azimuthal angle for the 14 and 25 T data is presented as a contour plot in Fig. 6.13, where the second derivative data is used to bring out the AMRO features. Here the more complex nature of the AMRO, which is asymmetric in the azimuthal plane, is apparent. Moreover, a crossing of AMRO oscillation patterns is evident when the period reaches a minimum. Although sample twining is always possible, which would lead to overlapping periods, the actual origin of the crossings are not at present clear. (The overlap would correspond to twins positioned at ∼65°.)

In Fig. 6.14 $k_{//}(\phi)$ for the $x = 0.8$ alloy system at 25 T is shown, compared with the $x = 1.0$ data in Fig. 6.11 above. Here fast Fourier transform methods were used to determine the $\tan(\theta)$ periods for each cycle of θ, which corresponds to an average value of ϕ, and $k(\phi)$ was computed from (6.51). Also shown is the corresponding loci of the value of the resistivity minimum due to

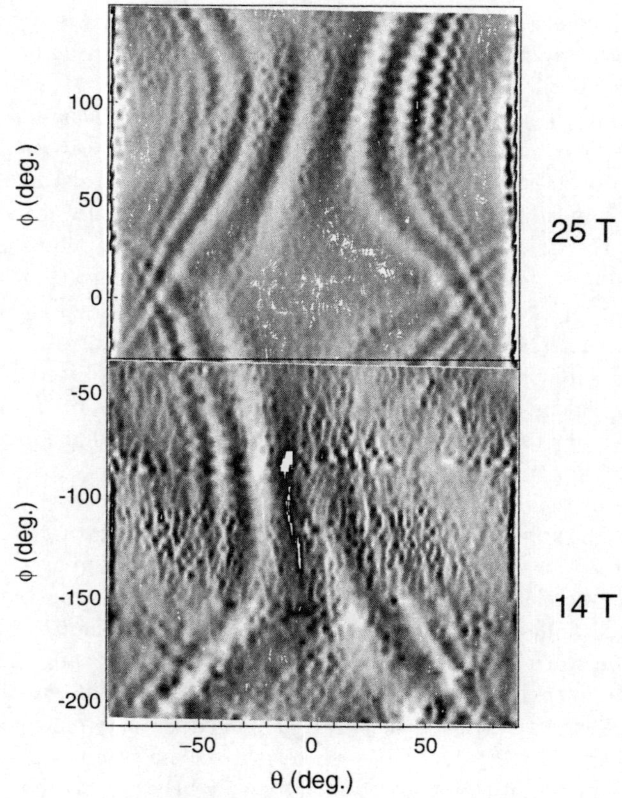

Fig. 6.13. Contour plot of second derivative of the AMRO data in Fig. 6.12 for the complete 2π azimuthal sample rotation for λ-$(BETS)_2Fe_{0.8}Ga_{0.2}Cl_4$ at 1.8 K. *Lower panel* – data taken at 14 T; *upper panel* – data taken at 25 T

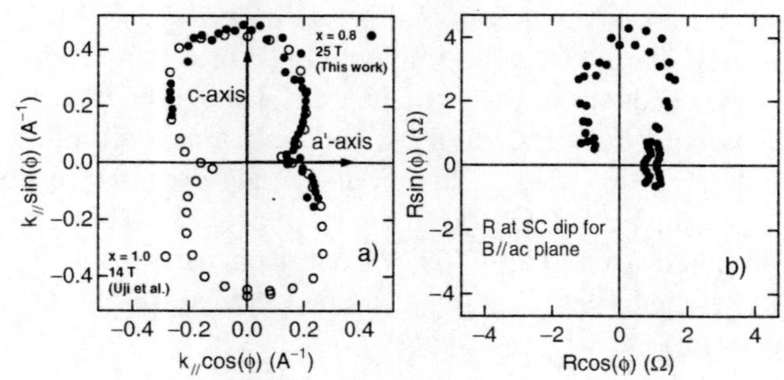

Fig. 6.14. Azimuthal angle-dependent features in λ-$(BETS)_2Fe_xGa_{1-x}Cl_4$. (**a**) $k//\phi$ for $x = 1.0$ at 14 T (from [57]) and $x = 0.8$ at 25 T from this work. (**b**) resistance minimum for $B//ac$-plane in the field induced superconducting phase

the onset of superconductivity for purely in-plane field, where we note there is a systematic asymmetry in the values for each 180° cycle of θ (the field is entering the sample from opposite sides for each cycle).

Our general observation is that for the two cases discussed, $x = 1.0$ and $x = 0.8$, the overall features of the Fermi surface topology are nearly identical. Common to both is the very large oscillatory magnetoresistance background that appears in addition to the AMRO signals, especially for ϕ values near $\pm 90°$ when the field is rotating in the approximate b-c plane. It is interesting to speculate whether it is possible to observe the effects of the exchange field splitting of the Fermi surface observed in the SdH effect (discussed in Sect. 6.9) in the AMRO signal as well. Although the multiple patterns in the AMRO data shown in Fig. 6.13 are suggestive of two shifted FS structures, as is the oscillating SC resistance minimum seen in Figs. 6.12 and 6.14, more work is necessary to determine the precise effects of the exchange field on the corrugated cylindrical geometry.

Further investigation of the AMRO was carried out in the antiferromagnetic insulating phase as shown in Fig. 6.15. After data in paramagnetic metallic state was taken (lower trace), the field was decreased into the AFI state and set near 9 T where the resistance was still measurable without significant phase shifts in the ac signal. The sample was then rotated as shown in the middle trace over a range of nearly $\pm 3 \times 360°$ in θ and $\pm 10°$ in ϕ. The striking differences between the metallic and AFI signals are the π and the 2π periods, respectively, i.e., in the AFI phase, the sample must be rotated through a complete 2π cycle in θ to return to the original resistance state. (Subsequent peaks in the resistance change slightly due to the variation in ϕ as well.) The 2π periodicity in the AFI phase is represented in the form of a hysteresis cycle in Fig. 6.16, where other features in the cycle such as resistance jumps with some degree of reproducibility appear. These jumps, which occur vs. magnetic field, seem to be a characteristic of the AFI state in both resistance [60] and in microwave cavity experiments [61]. A simple model for this behavior is one where AF domains are present, and where the conductivity is strongly coupled to the magnetism, which is the case for strongly coupled π-d materials as in the present case. Hence for changes in the field, and/or in-plane and out-of plane field, the resistance will follow shifts in AF domain structure. Moreover, in this model, if the domains respond relatively easily to changes in field direction, it seems reasonable that the sample must be rotated through a complete 360° orientation to recover the initial state. The AFI state alone has many interesting properties, and deserves future more detailed investigation.

6.10.3 High Field Aspects of the α-(BEDT-TTF)$_2$MHg(SCN)$_4$ Salts

The low temperature and high magnetic field behavior of the special class of quasi-two-dimensional conductors α-(BEDT-TTF)$_2$MHg(SCN)$_4$, first

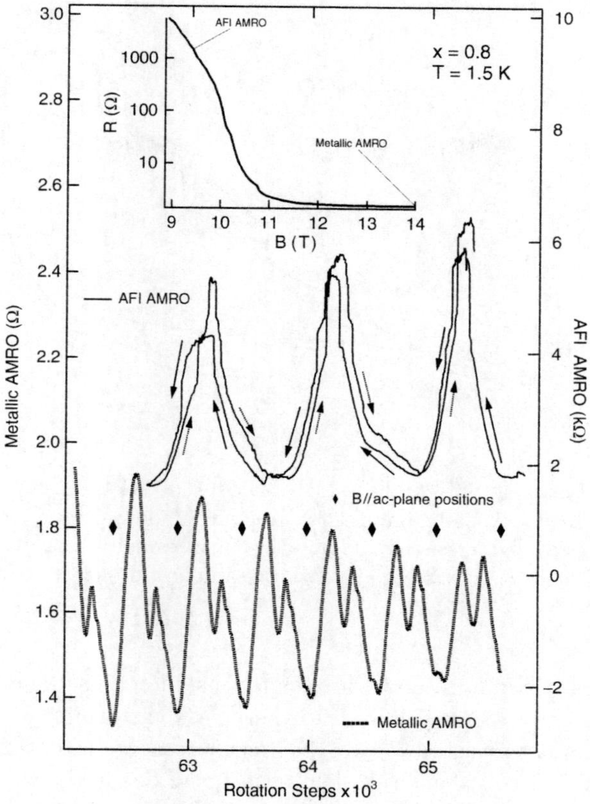

Fig. 6.15. Comparative AMRO measurements in the antiferromagnetic insulating (AFI) and metallic states. The diamonds represent the $B//ac$-plane positions. *Inset*: resistance vs. field at 1.5 K indicating regions of the measurements

synthesized by Oshima et al. [62], has provided a rich resource for the study of low-dimensional phenomena for over 15 years. Here M = K, Rb, NH_4, or Tl. Two highly motivating aspects of these materials were the discovery of superconductivity for $M = NH_4$ ($T_c \sim 1.2$ K) [63] and the appearance of a magnetic field induced phase transition above about 25 T at low temperatures by Osada et al. [64]. Subsequently, filamentary superconductivity was observed for M = K by Ito et al. [65] and bulk superconductivity in this same compound was first induced by uniaxial stress methods [66] and later with both uniaxial strain [67] and hydrostatic pressure [68, 69].

Superconductivity notwithstanding the magnetic field dependent properties in the normal state of the M = K, Rb, and Tl (here after the "density wave" compounds) are probably the most remarkable aspects of these systems. The crystal structure and Fermi surface [70] are shown in Fig. 6.17 and Fig. 6.18a, respectively. An essential feature is that below a transition temperature $T_p \approx 8$ to 12 K, the electronic structure changes from one with open

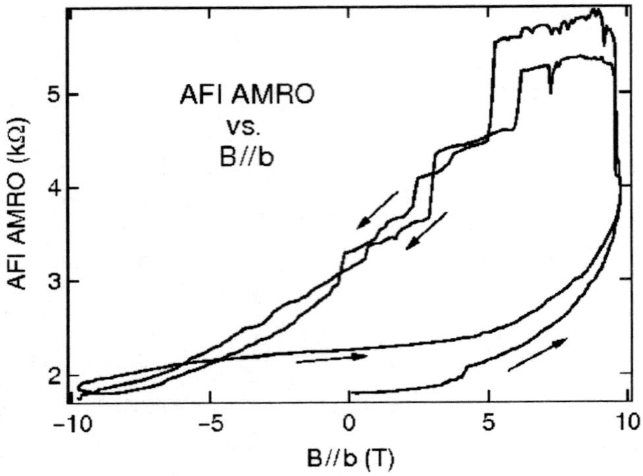

Fig. 6.16. Hysteresis effects based on AMRO in AFI region. $B//b$ refers to the magnetic field perpendicular to the ac-plane under rotation. Unlike standard "butterfly-shaped" hysteresis behavior, where a mirror image appears for negative field, the data require a complete cycle to the same initial field to return to the same resistance value

and closed orbit bands to one which is reconstructed as shown in Fig. 6.18b. This key discovery was made by the Chernogolovka group [71] using angular dependent magnetoresistance (AMRO) measurements as described below.

The nature of this transition has been hotly debated, but at present the most likely ground state is believed to be one with charge density wave (CDW) character [72–75]. An example of the magnetoresistance (MR) of α-(BEDT-TTF)$_2$KHg(SCN)$_4$ is shown in Fig. 6.19. As is the case for the other density wave compounds M = Tl and Rb, the MR rises rapidly, but above a peak value around 10 T, the MR decreases and quantum oscillations with a strong second harmonic appear. (A general survey of the magnetic field dependence of the α-(BEDT-TTF)$_2$MHg(SCN)$_4$ materials at ambient and high pressure is given by Brooks et al. [76]). Above 20 T a "kink field" B_k feature appears where the MR drops dramatically, and a new, complex set of oscillations are present. The upper inset shows that the temperature dependence of the oscillation amplitudes above B_k is nonmonotonic, indicating yet another phase boundary in the vicinity of 2.5 K. Hence at low temperatures magnetic fields above B_k do not completely restore the metallic ground state, and a Fröhlich-like ground state has been suggested for the high field, low temperature phase [77]. In the lower inset of Fig. 6.19 a schematic of the phase diagram for the density wave materials showing the metallic, DW1, and DW2 phases, which, as discussed below, are expected to be charge density wave in character.

Below T_p and B_k AMRO investigations can be used to reveal the details of the reconstructed Fermi surface structure. As might be expected by inspection of the difference between Fig. 6.18a,b, in the case of the α-BEDT-TTF

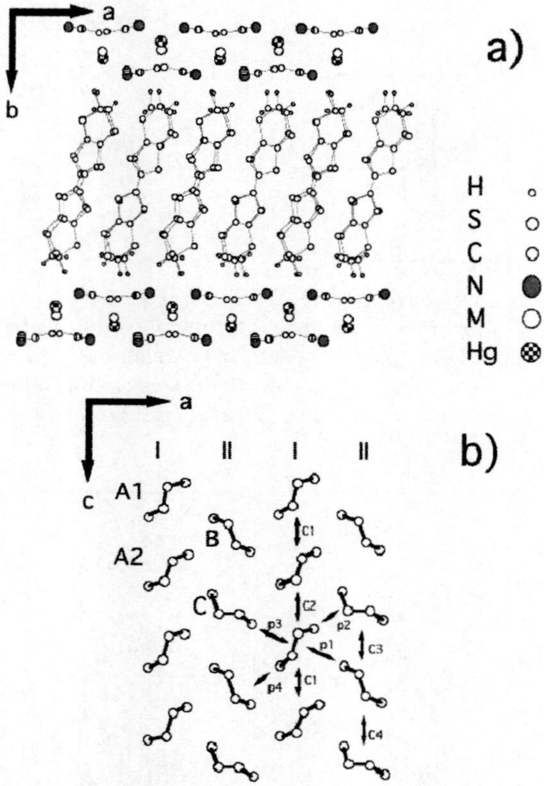

Fig. 6.17. Crystal structure of α-(BEDT-TTF)$_2$MHg(SCN)$_4$ where M = K, Rb, Tl, or NH$_4$ (redrawn after [70]). (**a**) Side view of the a-c plane. The conductivity occurs along the BEDT-TTF stacks in the a-c plane, which are separated along the b-axis by the insulating MHg(SCN)$_4$ anion layer. (**b**) Top view of BEDT-TTF α stacking motif. A1, A2, B, and C are inequivalent sites and comprise the unit cell: ($a = 10.082$ Å, $b = 20.565$ Å, $c = 9.933$ Å for M = KHg [62]). The transverse overlap integrals $p1$-$p4$ are significantly larger than the longitudinal integrals $c1$-$c4$, and give rise to a 2D energy band [70]

density wave compounds, the AMRO in the reconstructed ground state shows a fundamentally different character from that observed for a standard warped-cylindrical FS.[2] As an example, we present here the case for α-(BEDT-TTF)$_2$RbHg(SCN)$_4$ where we have used the same continuous rotation apparatus described in the previous section to map out the AMRO patterns in the DW1 phase. This is shown in Fig. 6.20 for a nearly 2π range

[2] For a more complete discussion of this subject in the α-BEDT-TTF systems, please see for instance the review article by Kartsovnik and Laukhin [78] for AMRO studies, Hanasaki et al. [79] for AMRO on α-(BEDT-TTF)$_2$KHg(SCN)$_4$ where the DW1 state is systematically removed with pressure, and Uji et al. [80] for a detailed study of SdH oscillations in the reconstructed FS DW1 phase.

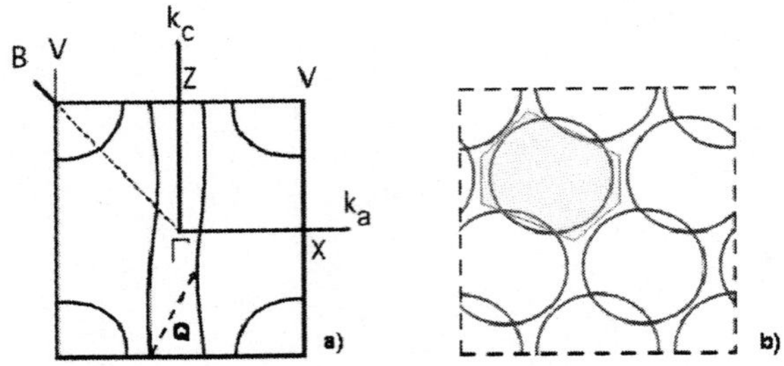

Fig. 6.18. (**a**) Fermi surface of α-(BEDT-TTF)$_2$MHg(SCN)$_4$ based on the structure shown in Fig. 6.17 (after [70]). Both open orbit and closed orbit bands are present in the metallic state, but below T_p the open orbit bands nest with wave vector **Q**. (**b**) As first proposed by Kartsovnik et al. [71], the nesting can produce a reconstructed Fermi surface that contains small closed pockets and a new set of open orbit bands along the **Q** direction

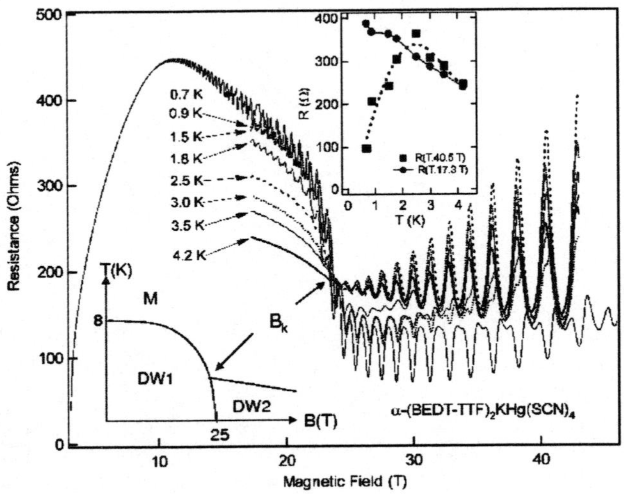

Fig. 6.19. High field magnetoresistance of α-(BEDT-TTF)$_2$KHg(SCN)$_4$ for different temperatures below $T_p = 8\,\mathrm{K}$. The data were taken in the Hybrid magnet at the NHMFL where the superconducting outsert is swept up to about 15 T, and the resistive insert is then swept between the outsert field and 45 T. *Upper insert*: temperature dependence of resistance below B_K (*circles*) and above B_K (*squares*). *Lower insert*: schematic phase diagram of the density wave compounds

of azimuthal angle for 17.5 T and 1.5 K. For $B//b$, the resistance shows broad peaks of essentially equal value, but for $B//ac$-plane, the value of the resistance (open circles) varies with respect to the in-plane field direction (ϕ). In some regions, the MR is actually greater near $B//ac$ than for $B//b$. Representative wave forms are shown for different ϕ regions vs. one 2π period of θ. Unlike the peaks that distinguish the AMRO oscillations in a warped cylindrical FS, here the AMRO periods are manifested in the dips, which have a

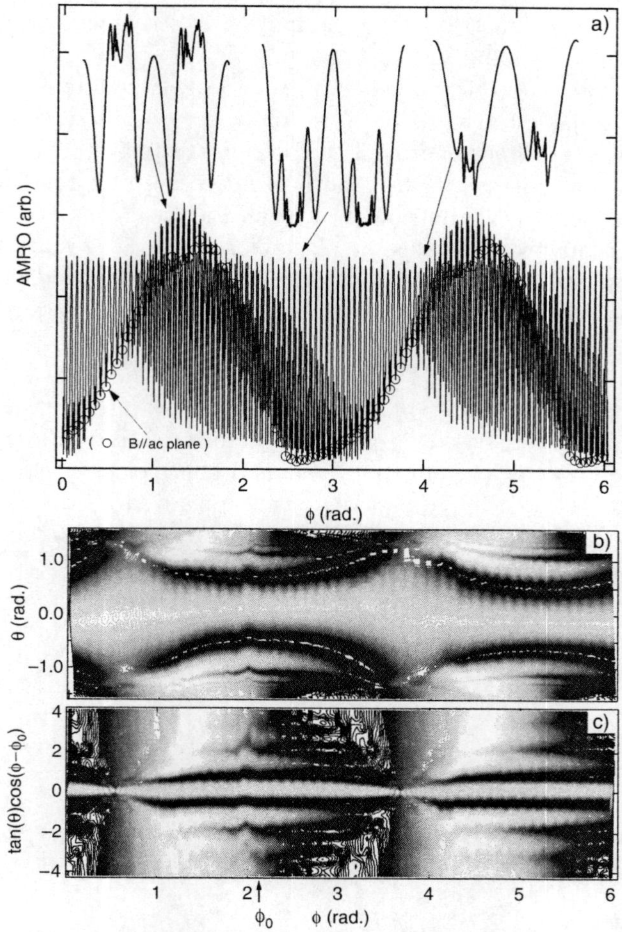

Fig. 6.20. (**a**) AMRO data for α-(BEDT-TTF)$_2$RbHg(SCN)$_4$ at 1.5 K and 17.5 T for a full 2π azimuthal rotation. Open circles represent the orientation where $B//ac$-plane. Upper traces are expanded regions of the AMRO for 2π changes in θ at the locations shown. Note that the $B//ac$-plane resistance sometimes exceeds the $B//b$ resistance, the latter remaining essentially constant. (**b**) Contour image of AMRO data for θ vs. ϕ. (**c**) Contour image of AMRO data for $\tan(\theta)\cos(\phi-\phi_0)$ vs. ϕ where $\phi_0 = 2.11$ rad. Note that the ϕ axis is the same for panels (**a**), (**b**), and (**c**)

minimum period when the field is directed in a plane defined by ϕ_0. A general relationship that describes the periodicity is [79, 81]

$$\tan(\theta)\cos(\phi-\phi_0) = [c_{\text{eff}}/b_{\text{eff}}](n-\delta), \tag{6.53}$$

where c_{eff} and b_{eff} are the effective in-plane and interplane lattice parameters that describe the new periodicity of the reconstructed Fermi surface, and δ

accounts for nonsymmetric interplane molecular stacking (i.e., the dips may not be symmetric around $\theta = 0$).

In Fig. 6.20b a contour representation of the AMRO data is given vs. θ and ϕ, and in Fig. 6.20c the same data are plotted vs. $\tan(\theta)\cos(\phi - \phi_0)$, where $\phi_0 = 2.11$ was adjusted to provide the best fit to (6.53). The period, represented by $c_{\mathrm{eff}}/b_{\mathrm{eff}}$, is $\approx$1.35, in good agreement with a previous study [82]. However, unlike [82], we do not observe an identical, but orthogonal set of dips, and perhaps twinning was the reason for the earlier results. We do, however, notice that near the positions $\phi_0 \pm \pi/2$, the dip period decreases, i.e., the dips appear closer to $\theta = 0$ for this field direction. We return to this observation later.

In Fig. 6.21 a mnemonic plot of the data is shown. Since we were unable to determine the crystallographic directions, they cannot be exactly specified in the plot. However, based on previous studies [82] the direction ϕ_0 lies at approximately 24° away from the c-axis, and hence clockwise from ϕ_0, either the c-axis at 24°, or the a-axis at 66° will appear. Note that the directions in 6.21 are essentially mirror reflections (i.e., $k_a \leftrightarrow -k_a$) of the directions in Fig. 6.18b.

The behavior of the MR for $B//ac$-plane is shown vs. azimuthal direction in Fig. 6.22. The ϕ dependence shows a complicated, but repeatable dependence (between $0 > \phi > \pi$ and $\pi > \phi > 2\pi$).

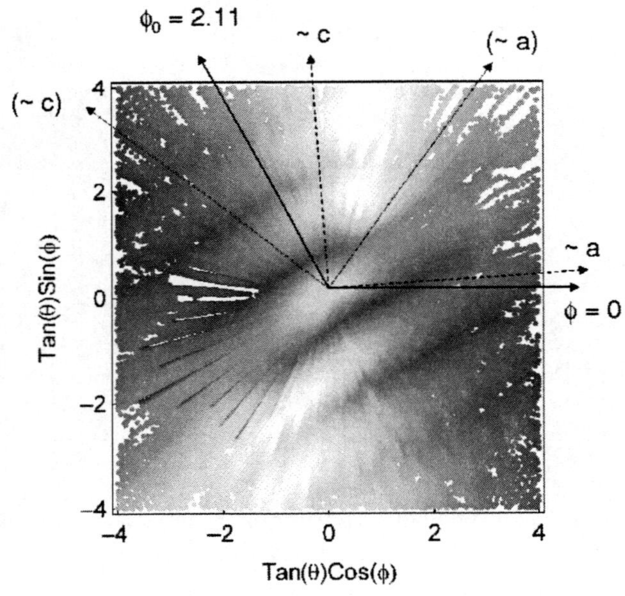

Fig. 6.21. Mnemonic plot of the AMRO data for α-(BEDT-TTF)$_2$RbHg(SCN)$_4$ at 1.5 K and 17.5 T for a full 2π azimuthal rotation, showing the dip locations along the ϕ_0 direction in the crystal with period 1.35. Based on previous measurements for the angle between ϕ_0 and the c-axis [82], the approximate $c(a)$-axis and $a(c)$-axis directions are shown (see text)

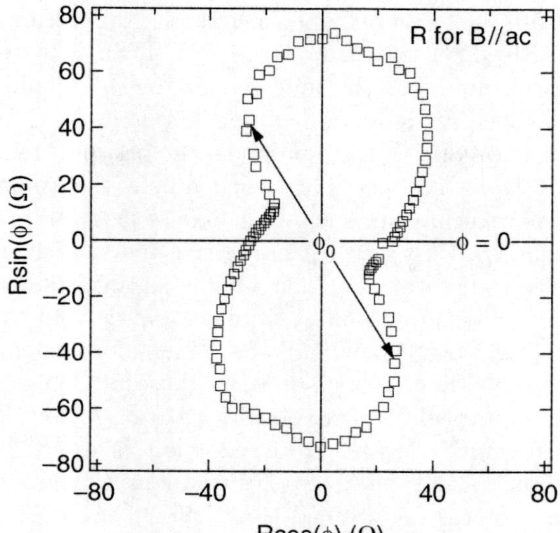

Fig. 6.22. Resistance of α-(BEDT-TTF)$_2$RbHg(SCN)$_4$ at 1.5 K and 17.5 T for the field orientation $B//ac$-plane. Within the uncertainties of the crystal direction, the maximum and minimum in-plane MR occur when the field is directed near the principal axes of the crystal

6.10.4 Discussion

The systematic AMRO data shown for α-(BEDT-TTF)$_2$RbHg(SCN)$_4$ are consistent with previous results for all of the density wave compounds, in that the AMRO dips reveal a "1D" periodicity in the electronic structure for $T < T_p$ and $B < B_k$ in a direction approximately 20–30° from the c-axis. The period is determined from the ratio $c_{\text{eff}}/b_{\text{eff}}$, which in the present case is ≈1.35. However, unlike the Bechgaard salts [83–87], the period does not correspond to a simple ratio of the in-plane to interplane lattice constants. In the present case (where $a \simeq 10, b \simeq 20, c \simeq 10$ Å), a period of c/b or a/b would yield ~0.5. Since b_{eff} must be directly related to the interplane spacing (i.e., $b_{\text{eff}} = b$), the in-plane effective reconstructed superlattice period c_{eff} must be much larger (e.g. 27 Å to yield a period of (6.53)).

Sasaki and Toyota [81] have given a nice geometric argument for the reconstructed super-lattice periods based on the real-space crystal structure. Following [81], and in reference to Fig. 6.17b,the band structure indicates that the strongest π orbital overlaps are along alternating A1-B-A2-C and C-A2-B-A1 molecular chain directions along stacking directions at $\approx \pm\, 30°$ either to the right or left of the c-axis direction. Considering that these chains are stacked in the a-c plane, and due to the further low symmetry of the crystal structure, six such chains are proposed for the superlattice periodicity (27 Å). This leads to the required $c_{\text{eff}}/b_{\text{eff}} = 1.35$, and the direction normal to the chains, based on the crystal structure, is ≈30°. In contrast, for the

1D Bechgaard salts, only one chain is necessary to describe the AMRO dip period in $(TMTSF)_2PF_6$ [87]. However, in $(TMTSF)_2ClO_4$ the tetrahedral ClO_4 anion orders, doubling the unit cell in the interchain direction. This leads to a corresponding reduction of the period by 1/2 since two chains make the superlattice [86]. Correspondingly, the proposal by Sasaki and Toyota is that *six chains* are necessary to compete the superlattice, and as they discuss, some mechanism such as a Peierls instability transition along the nesting vector Q (see Fig. 6.18) may drive the system into the stacked-chain superlattice configuration. Their model suggests that a nesting vector $\mathbf{Q} = (k_a/6, k_b/2, k_c/3)$ is involved. A temperature dependent X-ray study of α-$(BEDT\text{-}TTF)_2KHg(SCN)_4$ by Foury-Leylekian et al. [88] has shown several important results: (1) there is a superlattice structure which appears even at room temperature, and which increases dramatically below T_p; (2) the correlation length (half width of the satellite reflection) is about three BEDT-TTF units at room temperature, and increases to about six units below 100 K; (3) the primary wave vector determined is $\mathbf{Q} = (0.13, 0.1, 0.42)$, which can also nest the open orbit Fermi surface in Fig. 6.18. That a charge density wave-type instability involved is further supported by detailed ^{13}C NMR studies of α-$(BEDT\text{-}TTF)_2RbHg(SCN)_4$ by Miyagawa et al. [89] who find no evidence for local moments or spin fluctuations near T_P, as would be found in a spin density wave ground state.

6.11 Summary

In this article we have shown that the magnetic field plays two crucial roles in the advancement of our understanding of low dimensional materials. The first is that the magnetic field helps us to reveal the details of physical, electronic, and magnetic structure in low-dimensional materials, and theoretical models and experimental capabilities associated with high magnetic fields continue to improve and multiply. Newly synthesized materials, and even "old" materials find new physics when these advances are made. Second, high magnetic fields induced new phases and states of matter in low-dimensional materials. Hence these advanced theoretical and experimental tools can be applied yet again at high fields in newly discovered field-induced phases. Because of great technological advances in the last decade, these exciting activities are now routine in dc fields up to almost 50 T. We trust that in the future, this vibrant area of research involving the science of low-dimensional materials will allow systematic explorations in fields approaching 100 T at timescales where more sophisticated measurements will be viable. Unlike other scientific areas such as high energy physics where orders of magnitude are necessary to reveal new science, in high field condensed matter, incremental increases in available fields can produce entirely new physical states.

Acknowledgments

JSB wishes to acknowledge support from NSF-DMR 0203532 and 0602859, and also the help of his students and post-doctorals over many years. A portion of this work was performed at the National High Magnetic Field Laboratory (NHFML), which is supported by NSF Cooperative Agreement No. DMR-0084173, by the State of Florida and by the DOE. The authors are grateful for the assistance of the NHMFL technical staff and Resistive and Hybrid Magnet Groups without which much of this work could not have been possible. We also thank V. Williams for designing and constructing the continuous rotation probe, and the groups of A. Kobayashi, H. Kobayashi, and M. Tokumoto for their collaborations and for providing high quality samples.

References

1. T. Ishiguro, K. Yamaji, G. Saito, *Organic Superconductors*, 2nd edn. (Springer, Heidelberg, 1998)
2. J. Wosnitza, *Fermi Surfaces of Low-Dimensional Organic Metals and Superconductors* (Springer-Verlag, Berlin Heidelberg New York, 1996)
3. J. Singleton, Rep. Prog. Phys. **63**, 1111 (2000)
4. D. Shoenberg, *Magnetic Oscillations in Metals* (Cambridge University Press, Cambridge, 1984)
5. H. Urayama, H. Yamochi, G. Saito, K. Nozawa, T. Sugano, M. Kinoshita, S. Sato, K. Oshima, A. Kawamoto, J. Tanaka, Chem. Lett. **17**, 55 (1988)
6. H. Urayama, H. Yamochi, G. Saito, S. Sato, A. Kawamoto, J. Tanaka, T. Mori, Y. Maruyama, H. Inokuchi, Chem. Lett. **173**, 463 (1988)
7. K. Oshima, T. Mori, H. Inokuchi, H. Urayama, H. Yamochi, G. Saito, Phys. Rev. B **38**, 938 (1988)
8. I.M. Lifshitz, A.M. Kosevich, Sov. Phys. JETP **2**, 636 (1956)
9. E.N. Adams, T.D. Holstein, J. Phys. Chem. Sol. **10**, 2546 (1959)
10. S. Uji, M. Chaparala, S. Hill, P.S. Sandhu, J. Qualls, L. Seger, J.S. Brooks, Synth. Met. **86**, 1573 (1997)
11. S. Uji, H. Aoki, J.S. Brooks, A.S. Perel, G.J. Athas, S.J. Klepper, C.C. Agosta, D.A. Howe, M. Tokumoto, N. Kinoshita, Y. Tanaka, H. Anzai, Solid State Commun. **88**, 683 (1993)
12. B. McCombe, G. Seidel, Phys. Rev. **155**, 633 (1967)
13. P.F. Sullivan, G. Seidel, Phys. Rev. **173**, 679 (1968)
14. V.A. Bondarenko, S. Uji, T. Terashima, C. Terakura, S. Tanaka, S. Maki, J. Yamada, S. Nakatsuji, Synth. Met. **120**, 1039 (2001)
15. J.M. Caulfield, W. Lubczynski, F.L. Pratt, J. Singleton, D.Y.K. Ko, W. Hayes, M. Kurmoo, P. Day, J. Phys. Condens. Matter **6**, 2911 (1994)
16. K.F. Quader, K.S. Bedell, G.E. Brown, Phys. Rev. Lett. **36**, 156 (1987)
17. J. Singleton, F.L. Pratt, M. Doporto, T.J.B.M. Janssen, M. Kurmoo, J.A.A.J. Perenboom, W. Hayes, P. Day, Phys. Rev. Lett. **86**, 2500 (1992)
18. H. Ohta, Y. Yamamoto, K. Akioka, M. Motokawa, T. Sasaki, T. Fukase, Synth. Met. **86**, 2011 (1997)

19. S. Hill, J.S. Brooks, J.S. Qualls, T. Burgin, B. Fravel, L.K. Montgomery, J. Sarrao, Z. Fisk, Physica B **246–247**, 110 (1998)
20. J. Wosnitza, G.W. Crabtree, H.H. Wang, U. Geiser, J.M. Williams, K.D. Carlson, Phys. Rev. **45**, 3018 (1992)
21. S. Uji, T. Terashima, H. Aoki, J.S. Brooks, R. Kato, H. Sawa, S. Aonuma, M. Tamura, M. Kinoshita, Phys. Rev. B **50**, 15597 (1994)
22. R.W. Stark, L.M. Falicov, *Progress in Low Temperature Physics V*, ed. by C.J. Gorter, (North-Holland, Amsterdam, 1967)
23. N. Harrison, J. Caulfield, J. Singleton, P.H.P. Reinders, F. Herlach, W. Hayes, M. Kurmoo, P. Day, J. Phys. Condens. Matter **8**, 5415 (1996)
24. M. Nakano, J. Phys. Soc. Jpn, **66**, 19 (1997)
25. P.S. Sandhu, J.H. Kim, J.S. Brooks, Phys. Rev. B **56**, 11566 (1997)
26. J.Y. Fortin, T. Ziman, **80**, 3117 (1998)
27. J.H. Kim, S.Y. Han, J.S. Brooks, Phys. Rev. B **60**, 3213 (1999)
28. K. Kishigi, Y. Hasegawa, Phys. Rev. B **65**, 205405 (2002)
29. V.M. Gvozdikov, A.G.M. Jansen, D.A. Pesin, I.D. Vagner, P. Wyder, Phys. Rev. B **68**, 155107 (2003)
30. K. Kishigi, Y. Hasegawa, Phys. Rev. B **72**, 045410 (2005)
31. J.Y. Fortin, E. Perez, A. Auouard, Phys. Rev. B **71**, 155101 (2005)
32. R.W. Stark, C.B. Friedberg, J. Low Temp. Phys. **14**, 111 (1974)
33. S. Uji, J.S. Brooks, S. Takasaki, J. Yamada, H. Anzai, Solid State Commun. **103**, 387 (1997)
34. H. Kobayashi, H. Tomita, T. Naito, A. Kobayashi, F. Sakai, T. Watanabe, P. Cassoux, J. Am. Chem. Soc. **118**, 368 (1996)
35. L. Brossard, R. Clerac, C. Coulon, M. Tokumoto, T. Ziman, D.K. Petrov, V.N. Laukhin, M.J. Naughton, A. Audouard, F. Goze, A. Kobayashi, H. Kobayashi, P. Cassoux, Eur. Phys. J. B **1**, 439 (1998)
36. S. Uji, C. Terakura, T. Terashima, T. Yakabe, Y. Terai, M. Tokumoto, A. Kobayashi, F. Sakai, H. Tanaka, H. Kobayashi, Phys. Rev. B **65** 113101 (2002)
37. S. Uji, H. Shinagawa, T. Terashima, T. Yakabe, Y. Terai, M. Tokumoto, A. Kobayashi, H. Tanaka, H. Kobayashi, Nature **410**, 908 (2001)
38. L. Balicas, J.S. Brooks, K. Storr, S. Uji, M. Tokumoto, H. Kobayashi, A. Kobayashi, V. Barzykin , L.P. Gor'kov, Phys. Rev. Lett. **87**, 067002 (2001)
39. S. Uji, T. Terashima, C. Terakura, T. Yakabe, Y. Terai, S. Yasuzuka, Y. Imanaka, M. Tokumoto, A. Kobayashi, F. Sakai, H. Tanaka, H. Kobayashi, L. Balicas, J.S. Brooks, J. Phys. Soc. Jpn. **72**, 369 (2003)
40. S. Uji, S. Yasuzuka, M. Tokumoto, H. Tanaka, A. Kobayashi, B. Zhang, H. Kobayashi, E.S. Choi, D. Graf, J.S. Brooks, Phys. Rev. B **72**, 184505 (2005)
41. S. Uji, J.S. Brooks, J. Phys. Soc. Jpn. **75**, 051014 (2006)
42. O. Cepas, R.H. McKenzie, J. Merino, Phys. Rev. B **65** 100502 (2002)
43. C. Hotta, H. Fukuyama, J. Phys. Soc. Jpn. **69** 2577 (2000)
44. M. Ishino, N. Kobayashi, Y. Muto, *Ternary Superconductors*, ed. by G.K. Shenoy et al. (North Holland, Amsterdam, 1981) p. 95
45. H.W. Meul, C. Rossel, M. Decroux, O. Fischer, G. Remenyl, A. Briggs, Phys. Rev. Lett. **53**, 497 (1984)
46. C.M. Giroud, J.-L. Genicon, R. Tournier, C. Geantet, O. Pena, R. Horyn, M. Sergent, J. Low Temp. Phys. **69**, 419 (1987)
47. V. Jaccarino, M. Peter, Phys. Rev. Lett. **9**, 290 (1962)

48. O. Fischer, Helv. Phys. Acta **45**, 331 (1972)
49. H. Fujiwara, H. Kobayashi, Bull. Chem. Soc. Jpn. **78**, 1181 (2005)
50. T. Konoike, S. Uji, T. Terashima, M. Nishimura, S. Yasuzuka, K. Enomoto, H. Fujiwara, B. Zhang, H. Kobayashi, Phys. Rev. B **70**, 094514 (2004)
51. T. Konoike, S. Uji, T. Terashima, M. Nishimura, S. Yasuzuka, K. Enomoto, H. Fujiwara, E. Fujiwara, B. Zhang, H. Kobayashi, Phys. Rev. B **72**, 094517 (2005)
52. K. Yamaji, J. Phys. Soc. Jpn. **58**, 1520 (1989)
53. K. Kajita, Y. Nishino, T. Takahashi, R. Kato, H. Kobayashi, W. Sasaki, A. Kobayashi, Y. Iye, Solid State Commun. **70**, 1189 (1989)
54. M.V. Kartsovnik, V.N. Laukhin, S.I. Pesotskii, I.F. Schegolev, V.M. Yakovenko, J. Phys. I **2**, 89 (1992)
55. A.A. House, H. Harrison, S.J. Blundell, I. Deckers, J.P. Day, Phys. Rev. B **53**, 9127 (1996)
56. V.G. Peschansky, J.A.R. Lopez, T.G. Yao, J. Phys. I France **1**, 1469 (1991)
57. S. Uji, H. Shinagawa, C. Terakura, T. Terashima, T. Yakabe, Y. Terai, M. Tokumoto, A. Kobayashi, H. Tanaka, H. Kobayashi, Phys. Rev. B **64**, 024531 (2001)
58. J.S. Brooks, V. Williams, E. Choi, D. Graf, M. Tokumoto, S. Uji, F. Zuo, J. Wosnitza, J.A. Schlueter, H. Davis, R.W. Winter, G.L. Gard, K. Storr, New J. Phys., in press
59. N. Hanasaki, S. Kagoshima, T. Hasegawa, T. Osada, N. Miura, Phys. Rev. B **57**, 1336 (1998)
60. J.S. Brooks, L. Balicas, K.A. Storr, H. Kobayashi, H. Tanaka, A. Kobayashi, S. Uji, M. Tokumoto, Synth. Met. **133–134**, 485 (2003)
61. I.B. Rutel, S. Okubo, J.S. Brooks, E. Jobiliong, H. Kobayashi, A. Kobayashi, H. Tanaka, Phys. Rev. B **68**, 144435 (2003)
62. M. Oshima, H. Mori, G. Saito, K. Oshima, Chem. Lett. **1989**, 1159 (1989)
63. H.H. Wang, K.D. Carlson, U. Geiser, W.K. Kwok, M.D. Vashon, J.E. Thompson, N.F. Larsen, G.D. McCabe, R.S. Hulscher, J.M. Williams, Phys. C Supercond. **166**, 57 (1990)
64. T. Osada, R. Yagi, A. Kawasumi, S. Kagoshima, N. Miura, M. Oshima, G. Saito, Phys. Rev. B **41**, 5428 (1990)
65. H. Ito, H. Kaneko, T. Ishiguro, H. Ishimoto, K. Kono, S. Horiuchi, T. Komatsu, G. Saito, Solid State Commun **85**, 1005 (1993)
66. C.E. Campos, J.S. Brooks, P.J.M.v. Bentum, J.A.A.J. Perenboom, S.J. Klepper, P. Sandhu, S. Valfells, Y. Tanaka, T. Kinoshita, N. Kinoshita, M. Tokumoto, H. Anzai, Phys. Rev. B **52**, R7014 (1995)
67. S. Kagoshima, R. Kondo, Chem. Rev. **104**, 5593 (2004)
68. D. Andres, M.V. Kartsovnik, W. Biberacher, K. Neumaier, H. Muller, J. Phys. IV France **12**, 87 (2002)
69. D. Andres, M.V. Kartsovnik, W. Biberacher, K. Neumaier, E. Schuberth, H. Mller, Phys. Rev. B 72, 174513/7 (2005)
70. H. Mori, S. Tanaka, M. Oshima, G. Saito, T. Mori, Y. Maruyama, H. Inokuchi, Bull. Chem. Soc. Jpn. **63**, 2183 (1990)
71. M.V. Kartsovnik, A.E. Kovalev, N.D. Kushch, J. Phys. I France **3**, 1187 (1993)
72. P.L. Kuhns, J.S. Brooks, T. Caldwell, W.G. Moulton, A.P. Reyes, N. Biskup, A.M. Kini, J.A. Schlueter, H.H. Wang, U. Geiser, J.M. Williams, Solid State Commun **109**, 637 (1999)

73. M.V. Kartsovnik, D. Andres, W. Biberacher, P.D. Grigoriev, E.A. Shuberth, H. Muller, J. Phys. IV France **114**, 191 (2004)
74. D. Andres, M.V. Kartsovnik, P.D. Grigoriev, W. Biberacher, H. Muller, Phys. Rev. B **68**, 291191(R) (2003)
75. N. Harrison, C.H. Mielke, A.D. Christianson, N. Harrison, J.S. Brooks, M. Tokumoto, Phys. Rev. Lett. **86**, 1586 (2001)
76. J.S. Brooks, X. Chen, S.J. Klepper, S. Valfells, G.J. Athas, Y. Tanaka, T. Kinoshita, N. Kinoshita, M. Tokumoto, H. Anzai, C.C. Agosta, Phys. Rev. B **52**, 14457 (1995)
77. N. Harrison, C.H. Mielke, J. Singleton, J.S. Brooks, M. Tokumoto, J. Phys. Condens. Matter **13**, L389 (2001)
78. M.V. Kartsovnik, V.N. Laukhin, J. Phys. I France **6**, 1753 (1996)
79. N. Hanasaki, S. Kagoshima, N. Miura, G. Saito, J. Phys. Soc. Jpn. **65**, 1010 (1996)
80. S. Uji, T. Terashima, H. Aoki, J.S. Brooks, M. Tokumoto, N. Kinoshita, Y. Tanaka, H. Anzai, Phys. Rev. B **54**, 9332 (1996)
81. T. Sasaki, N. Toyota, Phys. Rev. B **49**, 10120 (1994)
82. M.V. Kartsovnik, H. Ito, T. Ishiguro, H. Mori, T. Mori, G. Saito, S. Tanaka, J. Phys. Condens. Matter **6**, L479 (1994)
83. A.G. Lebed, JETP Lett. **43**, 174 (1986)
84. A.G. Lebed, P. Bak, Phys. Rev. Lett. **63**, 1315 (1989)
85. T. Osada, A. Kawasumi, S. Kagoshima, N. Miura, G. Saito, Phys. Rev. Lett. **66**, 1525 (1991)
86. M.J. Naughton, O.H. Chung, M. Chaparala, X. Bu, P. Coppens, Phys. Rev. Lett. **67**, 3712 (1991)
87. W. Kang, S.T. Hannahs, P.M. Chaikin, Phys. Rev. Lett. **69**, 2827 (1992)
88. P. Foury-Leylekian, S. Ravy, J.-P. Pouget, H. Mller, Synth. Met. **137**, 1271 (2003)
89. K. Miyagawa, A. Kawamoto, K. Kanoda, Phys. Rev. B **56**, R8487 (1997)

7

Magnetic Properties of Organic Conductors and Superconductors as Dimensional Crossovers

A.G. Lebed and S. Wu

We suggest a unified theory to describe unconventional magnetic properties of metallic, superconducting, and field-induced spin(charge)–density wave [FIS(C)DW] phases in organic conductors with open sheets of Fermi surface. It is based on ideas about 3D → 2D, 3D → 1D, and 1D → 2D crossovers for electron spectra and wave functions, which occur due to quantum interference effects when electrons move in a magnetic field in the extended Brillouin zone.

Our theoretical results are applicable to quasi-one-dimensional (Q1D) compounds in an arbitrary inclined magnetic field and to quasi-two-dimensional (Q2D) compounds in a parallel magnetic field. Using the above-mentioned approach, we describe Lebed Magic Angle and Interference Commensurate oscillations in a metallic phase, FIS(C)DW phase diagrams, the Reentrant Superconductivity phenomenon, and some other non-trivial magnetic effects.

For historical surveys about theoretical discovery of the first 3D → 2D crossover, see the chapters by Lebed and by Heritier. For experimental properties of Q1D and Q2D conductors and superconductors, see the chapters by Brown, Chaikin, and Naughton, by Uji and Brooks, by Kartsovnik, by Singleton et al., by Harrison et al., by Kanoda, and by Hill. See also the chapter by Jerome about discovery of superconductivity in Q1D $(TMTSF)_2X$ materials and the chapters by Lebed and Si Wu and by Cherng, Zhang, and Sa de Melo about triplet versus singlet superconductivity scenarios in these materials.

7.1 Introduction

It is well known that most magnetic properties of traditional metals and superconductors are described by the standard theories [1, 2]. Meanwhile, a number of highly anisotropic layered quasi-one-dimensional (Q1D) and quasi-two-dimensional (Q2D) organic conductors and superconductors [3–5] and Q2D high-T_c superconductors [6–9] were synthesized during the last two

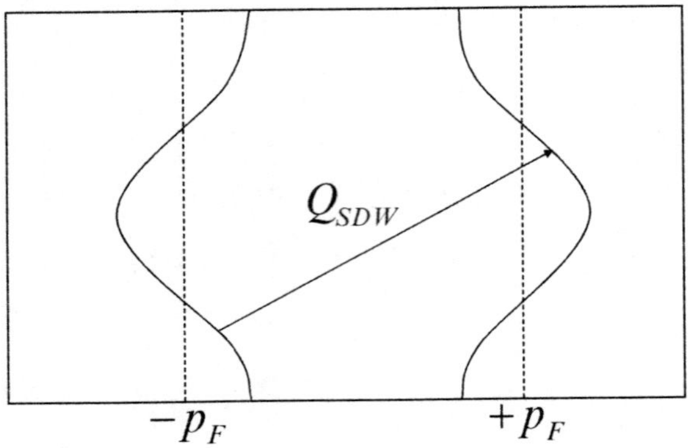

Fig. 7.1. Typical Q1D electron spectrum of $(TMTSF)_2X$ organic superconductors [see (7.1)], where $\mathbf{Q}_{SDW}$ is a wave vector of spin-density-wave (SDW) phase at H=0 [see Sect. 7.7]

decades. The remarkable high-T_c superconductors have enriched solid state physics by their unconventional d-wave superconducting pairing [6], unusual vortex phase diagrams in a magnetic field [7], non-Fermi-liquid effects in non-superconducting phases [8], and some other novel phenomena [9].

Low-dimensional organic superconductors (see Fig. 7.1) share some of these properties with the high-T_c compounds. Nevertheless, the major and most original contributions of the organic conductors and superconductors to solid state physics are quite different.

Indeed, in layered Q1D conductors, Fermi surface (FS) consists of two open sheets:

$$\begin{aligned}\varepsilon^{\pm}(\mathbf{p}) = \pm v_{\mathrm{F}}(p_x \mp p_{\mathrm{F}}) - 2t_b \cos(p_y b^*) - 2t_b' \cos(2p_y b^*) \\ - 2t_c \cos(p_z c^*) - 2t_c' \cos(2p_z c^*)\end{aligned} \quad (7.1)$$

[$p_{\mathrm{F}} v_{\mathrm{F}} \gg 2t_b \gg 2t_c \simeq 2t_b' \gg 2t_c'$; $+(-)$ stands for right(left) sheet of FS (see Fig. 7.1)]. For open FS (7.1), where Landau level quantization [1] is not possible, other quantum effects – Bragg reflections of electrons moving in the extended Brillouin zone (see Fig. 7.2) – determine unusual magnetic properties [10, 12, 13, 15–29]. In Q2D conductors and superconductors, the above-mentioned Bragg reflections determine unusual magnetic properties in a parallel magnetic field [10, 14].

As a result, Q1D and Q2D organic compounds demonstrate the following unique experimental properties, which are not observed in other materials: (1) a cascade of the magnetic field-induced spin-density-wave (FISDW) phase transitions [30–40] [see Sect. 7.7]; (2) three-dimensional quantum Hall effect (3D QHE) associated with these FISDW phases [32–40] (see also the chapter by Yakovenko); (3) very rich Fermi-liquid (FL) and non-Fermi-liquid (n-FL) angular-dependent magnetic oscillations such as Lebed magic angle (LMA)

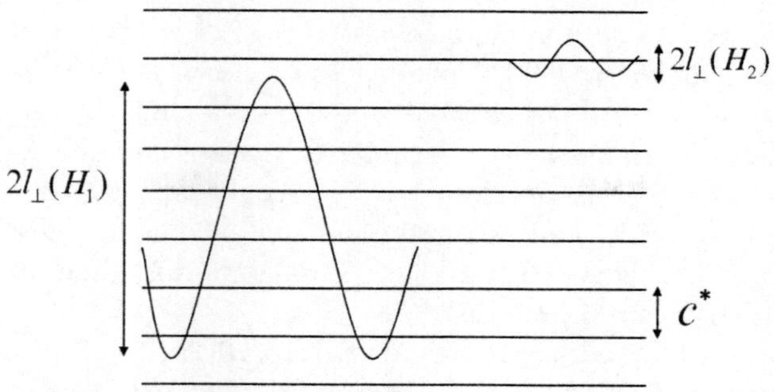

Fig. 7.2. Electron trajectories in a real space in Q1D (see Sect. 7.3.2 and [10, 11]) and Q2D (see Sect. 7.3.2 and [12–14]) conductors become finite and periodic along direction, perpendicular to the conducting layers. This results in quasi-classical (QC) $3D \rightarrow 2D$ crossovers [11, 15] if $\ell_\perp(H_1) \gg c^*$. If $\ell_\perp(H_2) \leq c^*$, quantum limit (QL) $3D \rightarrow 2D$ crossovers occur [10, 12, 14] which correspond to small probabilities of electron jumping from one conducting layer to another [14]. In the case of superconducting pairing, magnetic field decreases the "sizes" of Cooper pairs which cannot be larger than $\ell_\perp(H) \sim 1/H$ (see Sect. 7.8)

phenomena [41–57] (see Sects. 7.3.5 and 7.4.4), Yamaji–Kartsovnik oscillations [3], Danner–Kang–Chaikin (DKC) oscillations [58], the third angular effect (TAE) [59–64], and the interference commensurate (IC) oscillations [59,61–64] (see Sects. 7.3.6 and 7.4.5; observed in metallic phases of Q1D and Q2D conductors); (4) the LMA phenomenon in FISDW phases [53] (see Sect. 7.7); (5) novel type of cyclotron resonance (CR) related to open electron orbits [65–69] (see Sect. 7.5); (6) magnetic field-induced charge-density-wave (FICDW) phase diagrams [15, 70–76] (see Sect. 7.7).

In addition, there exist strong experimental evidences for: (7) n-FL effects in a parallel magnetic field in a metallic phase of $(TMTSF)_2PF_6$ conductor [55,77,78]; (8) unusual stabilization of superconductivity in magnetic fields much higher than both the upper critical field, H_{c2} [1, 2], and the Clogston paramagnetic limiting field, H_p [1, 2], in $(TMTSF)_2X$ [79–81] and (BETS)-based [82–85] superconductors in the forms of reentrant superconducting (RS) [10, 12, 20, 21, 86] and Larkin–Ovchinnikov–Fulde–Ferrell (LOFF) [1, 84] phases (see Sect. 7.8); (9) unconventional superconductivity in $(TMTSF)_2PF_6$ compound [20, 21, 87, 88] (see another chapter by Lebed and Wu, the chapter by Brown, Chaikin, and Naughton and the chapter by Cherng, Zhang, and Sa de Melo); (10) nano-scale [quantum limit (QL)] magnetic properties [89–91], where typical "sizes" of quasi-classical (QC) electron orbits in a parallel magnetic field become smaller or comparable to the interlayer distances [10, 12, 14] [see Sect. 7.5].

An important point is that only a few of the above-mentioned features can be explained by straightforward extensions of the traditional theories of metallic and superconducting phases [1, 2] to Q1D and Q2D anisotropic electron spectra. As a result of intensive studies of an electron motion in a magnetic field and magnetic field dependences of electron–electron interactions, a concept of a change in effective space dimensionality for some one-body and many-body phenomena was formulated by Gor'kov and Lebed [15] and by Chaikin [11]. It was developed by Lebed et al. [10, 12–14, 17–29, 92], Heritier, Montambaux, Lederer, and Poilblanc [93–95], Maki and Virosztek [96], Yamaji [97], Yakovenko et al. [36, 98–100], Dupuis, Montambaux and Sa de Melo [102–106] and others [107, 108].

This concept allowed one to explain [11, 13, 20, 21, 25–29, 92–100] and predict [10, 12, 17, 65, 70, 71] some of the above-described experimental properties. The simplest 3D → 2D crossover in a magnetic field, parallel to conducting planes of a layered Q1D conductor (7.1), was first suggested by Gorkov and Lebed [15] and by Chaikin [11]. It is illustrated in Fig. 7.2, where electron motion in a real space becomes periodic and finite perpendicular to conducting $(\mathbf{a}, \mathbf{b})$ planes. As shown in [10, 12, 14–16, 92, 93], this results in "two-dimensionalization" of electron–hole and electron–electron interactions leading to FISDW [11, 15, 16, 92, 93] and FICDW [70, 71] instabilities and to RS phase [10, 12], surviving at $H \gg H_{c2}$. This approach allowed Yakovenko [36, 98] and Poilblanc, Montambaux, and Heritier to explain the 3D QHE associated with FISDW subphases.

More complex 3D → $1D$ crossover was suggested by Lebed and Bak [17] and by Osada et al. [51] to explain LMA phenomena, experimentally observed in metallic phases of Q1D $(TMTSF)_2X$ and some other conductors [41–57]. Recently, Lebed, Bagmet, Ha, and Naughton [25–29] have suggested two new types of 1D → 2D crossovers in a magnetic field, which are due to interferences effects occurring when electrons move along open orbits in the extended Brillouin zone (see Fig. 7.2). These crossovers allow to explain rich angular magnetoresistance LMA and IC oscillations [41–57, 61–64], experimentally observed in $(TMTSF)_2X$, $(DMET\text{-}TSeF)_2X$, and κ-$(ET)_2Cu(NCS)_2$ materials.

Note that the above-mentioned dimensional crossovers happen in situations, where typical "sizes" of QC electron orbits perpendicular to the conducting planes' direction are much larger than the interlayer distances, $l_\perp(H) \gg c^*$ (see Fig. 7.2). We call such dimensional crossovers QC [14]. As anticipated by Lebed [10, 12] (see also [102–105] by Sa de Melo et al. and by Dupuis and Montambaux), in experimental magnetic fields, $H \simeq 10$–$50\,\mathrm{T}$, there may occur different dimensional crossovers – the so-called QL ones, where the typical "sizes" of QC orbits $l_\perp(H) \leq c^*$ (see Fig. 7.2). Nevertheless, QC approach used in [10–13, 15–29, 92, 92–108] was not allowed to study these particular QL properties of layered Q1D and Q2D conductors. A fully quantum mechanical theory of QL 3D → 2D crossovers in Q2D conductors has been recently suggested by Lebed in [14], where related momentum quantization law is introduced [see Sect. 7.5].

7.2 Our Goals

In this chapter, we theoretically demonstrate that most unconventional magnetic properties of organic conductors and superconductors can be explained in terms of different dimensional crossovers in a magnetic field. These include properties of metallic, FISDW, FICDW, and superconducting phases in Q1D conductors in an inclined magnetic field as well as properties of Q2D conductors and superconductors in a parallel magnetic field. Later, we also develop adequate physical picture and mathematical language to describe QL dimensional crossovers in a parallel magnetic field, where electrons are characterized by small probabilities to jump from one conducting layer to another [14].

The above-described program is closely related to the most important experimental discoveries in the area of organic superconductors and conductors. Among them, are such well-known phenomena as FISDW phase diagrams and related 3D QHE, discovered by Chaikin's and Ribault–Jerome's groups [30–32], and LMA phenomena, discovered by Naughton's [49, 52], Boebinger's [50], Osada's [51], and Chaikin's [53] groups. Our approach successfully explains the recent experimental discoveries of IC angular magnetic oscillations by Naughton's group [59,61,62] and by Yoshino et al. [63], recent discoveries of FICDW phase diagrams by Brooks' [72] and Kartsovnik's [74] groups, recent discoveries of a novel high magnetic field superconducting phase by Uji's and Brooks' groups [40], and some others. We also hope the spectacular RS phenomenon suggested by us [10, 12] is discovered in full in the nearest future. The existing experimental data by Naughton's and Chaikin's groups [79–81] and by Uji's and Brooks' groups [40, 82–85], gave a hint of its possible existence in $(TMTSF)_2X$ and (BETS)-based superconductors.

The plan of our review is as follows. In Sect. 7.3, we describe different types of 3D $\rightarrow$ 2D, 3D $\rightarrow$ 1D, and 1D $\rightarrow$ 2D crossovers, using QC language of an electron motion along open FS in Q1D and Q2D conductors. We calculate an average velocity of electrons, perpendicular to conducting chains, along their trajectories and show that it is zero for "incommensurate" directions of a magnetic field and is nonzero for some special "commensurate" directions. The latter correspond to LMA [17,26,27,41–57] and IC [24,25,28,29,59,61–64] angular magnetic oscillations.

In Sect. 7.4, we apply quantum mechanical Peierls substitution method for open orbits [10,12,13,15–29] to 3D $\rightarrow$ 2D, 3D $\rightarrow$ 1D, and 1D $\rightarrow$ 2D crossovers. We introduce momentum quantization law in a magnetic field, which characterizes open electron orbits, in contrast to textbook Landau energy levels quantization law [1]. We demonstrate that the above-mentioned dimensional crossovers correspond to localizations and de-localizations of electron wave-functions in an inclined magnetic field [24–29]. De-localization of electron wave functions is shown to happen at "commensurate" (i.e., LMA and IC) directions of a magnetic field.

In Sect. 7.5, we solve a fully quantum mechanical problem for 3D $\rightarrow$ 2D crossovers in a Q2D conductor in a parallel magnetic field. We derive momentum quantization law both in QC case, where "sizes" of electron orbits are larger than interplane distances, and QL case, where these "sizes" are less than interplane distances (see Fig. 7.2).

In Sect. 7.6, we apply results, obtained in the earlier sections, to calculate resistivity component, perpendicular to conducting layers of a Q1D conductor (7.1), and compare our results [24,25,25–29] to the existing experimental data on LMA and IC angular oscillations. In particular, we show that experimentally observed IC magnetoresistance oscillations are in excellent qualitative and quantitative agreements with the suggested theories [25, 26, 28]. We also discuss experiment [29], which directly proves 3D → 1D → 2D crossovers at IC directions of a magnetic field.

In Sect. 7.7, we apply the results of Sect. 7.4 to describe FISDW [11, 15, 16, 30–40, 92–108] and FICDW [15, 71–76] phase diagrams, experimentally observed in a number of Q1D conductors. These include theoretical description of a so-called quantized nesting model [92–100, 108] and more recent developments, including quantum and QC FISDW phases [22].

In Sect. 7.8, we apply the results of Sect. 7.4 to describe RS phenomenon [10, 12] – superconductivity stable at extremely high magnetic fields [79–85]. This spectacular phenomenon was first suggested by Lebed [10, 12] and developed by Dupuis, Montambaux, and Sa de Melo [102].

7.3 Dimensional Crossovers in a Magnetic Field

In this section, we consider different types of dimensional crossovers in a magnetic field by means of QC equations of motion.

7.3.1 Open Fermi Surfaces

Let us consider typical types of open electron trajectories in a magnetic field. The most generic case is provided by Q1D materials from chemical family $(TMTSF)_2X$, where $X = PF_6$, ClO_4, AsF_6, ReO_4, and some others. They are also often called Bechgaard salts [3]. The most well-known Q1D conductors, $(TMTSF)_2PF_6$ and $(TMTSF)_2AsF_6$, are characterized by a simple anisotropic tight-binding electron spectrum

$$\varepsilon(\mathbf{p}) = -2t_a \cos(p_x a^*/2) + 2t_b \cos(p_y b^*) + 2t_c \cos(p_z c^*) \tag{7.2}$$

($t_a \gg t_b \gg t_c$), which can be linearized near right and left sheets of FS

$$\varepsilon(\mathbf{p}) = \pm v_{\mathrm{F}}(p_x \mp p_{\mathrm{F}}) + 2t_b \cos(p_y b^*) + 2t_b' \cos(2p_y b^*) + 2t_c \cos(p_z c^*) \tag{7.3}$$

(see Fig. 7.3).

Another important for application conductor $(TMTSF)_2ClO_4$, which is a superconductor at ambient presser, is characterized by the so-called "anion

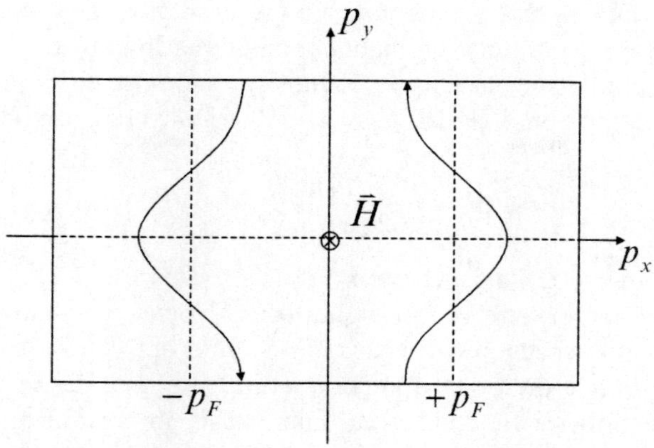

Fig. 7.3. Typical Q1D electron spectrum of $(TMTSF)_2X$ organic conductors [see (7.3)]. In a magnetic field, perpendicular to the conducting chains, $\mathbf{H} \parallel \mathbf{c}$, electrons move along two open sheets of FS

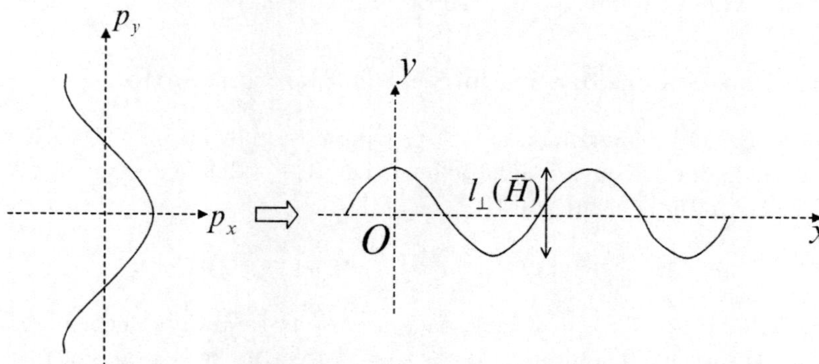

Fig. 7.4. Electron trajectories in a magnetic field for a Q1D metal (7.3) in momentum and real spaces

gap", Δ, in its electron spectrum [3]. In this case, Q1D electron spectrum correspond to four sheets of FS:

$$\varepsilon^i(\mathbf{p}) = \pm v_F(p_x \mp p_F) + (-1)^i \sqrt{4t_b^2 \cos^2(p_y b^*) + \Delta^2} + 2t_c \cos(p_z c^*), \quad (7.4)$$

where $i = 1, 2$.

A common property of all types of Q1D electron spectra [e.g. (7.3) and (7.4)] is that electrons move along open FS in a magnetic field perpendicular to conducting chains, $\mathbf{H} \parallel \mathbf{c}(\mathbf{z})$ (see Fig. 7.3). As shown later, when electrons move in the extended Brillouin zone along such trajectories, Bragg reflections result in the appearance of novel type of quantum effect-momentum quantization law [14,27–29] (see Fig. 7.4). [Note that most Q1D FS in organic conductors possess also Q2D anisotropy since the corresponding materials are layered and, thus, $t_b \gg t_c$. Therefore, in the case of Q1D FS (7.3) and (7.4), we use both terms: conducting chains and conducting layers.]

It is important that such type of motion (along open trajectories in the extended Brillouin zone) is also possible in Q2D layered conductors

$$\varepsilon(\mathbf{p}) = \frac{p_x^2 + p_y^2}{2m} + 2t_c \cos(p_z c^*) \tag{7.5}$$

in a parallel magnetic field, where $p_\mathrm{F}^2/2m \gg t_c$ (see Fig. 7.2). As it is known, Q2D superconductors are the most important for experimental applications and the most numerous low-dimensional conductors. For instance, (BEDT)- and (BETS)-based organic superconductors, all high-T_c superconductors, triplet superconductor Sr_2RuO_4, and many other important materials belong to Q2D materials.

And, finally, there exist important chemical families of low-dimensional conductors, where electrons move both along open and closed FS in a magnetic field. The most well-known example is the so-called α-phases of (BEDT)-based materials, where FICDW phase diagrams are observed.

7.3.2 "Two-Dimensionalization" of an Electron Motion

Let us consider a phenomenon of "two-dimensionalization" of an electron motion along open FS in a magnetic field (i.e., 3D → 2D crossover [10,12–28]), using QC equations of motion [1],

$$\mathrm{d}\mathbf{p}/\mathrm{d}t = (e/c)[\mathbf{v}(\mathbf{p}) \times \mathbf{H}], \quad \mathbf{v}(\mathbf{p}) = \partial\varepsilon(\mathbf{p})/\partial\mathbf{p}. \tag{7.6}$$

As it follows from (7.6), in a real space, vector $\mathbf{dr}$ is always perpendicular to vectors $\mathbf{dp}$ and $\mathbf{H}$. Therefore, an electron trajectory in a real space can be obtained from an electron trajectory in a reciprocal space by its rotating into $\pi/2$ (see Fig. 7.4).

Here, we analytically calculate electron trajectories in a real space for simplest Q1D electron spectrum (7.3) with $t_b' = 0$. In this case, from equations of motion (7.6), we obtain:

$$p_y b^* = \omega_b t + p_y^0 b^*, \quad \omega_b = eHv_\mathrm{F} b^*/c, \quad v_y(t) = -2t_b b^* \sin(p_y^0 b^* + \omega_b t). \tag{7.7}$$

Electron trajectories in a real space are given by the straightforward integration:

$$y(t) = \int v_y(t)\mathrm{d}t = (2t_b b^*/\omega_b)\cos(p_y^0 b^* + \omega_b t) = l_\perp(H)\cos(p_y^0 b^* + \omega_b t),$$
$$l_\perp(H) = 2b^*(t_b/\omega_b) \sim 1/H \tag{7.8}$$

(see Fig. 7.4).

Note that the trajectories (7.8) directly demonstrate 3D → 2D crossover in a magnetic field for a Q1D electron spectrum. Indeed, electron motion in a magnetic field becomes periodic and restricted along $\mathbf{b}(\mathbf{y})$-axis. It is easy to show that, in this case, an average velocity component, v_y, is zero along the trajectory

$$\langle v_y \rangle_t = \lim_{T\to\infty} \frac{1}{T} \int_0^T (-v_y^0) \sin(p_y^0 b^* + \omega_b t) = 0, \tag{7.9}$$

which means that electrons do not move along $\mathbf{b}(\mathbf{y})$-axis effectively. As shown in Sect. 7.4.2, quantum mechanical energy spectrum of such "two-dimensionalized" Q1D electrons is indeed pure 2D and that corresponding electron wave functions are localized within $(\mathbf{a},\mathbf{b})$-plane. It is important that the above described 3D $\to$ 2D crossover is responsible for such experimentally observed unconventional magnetic properties as FISDW [30–40], FICDW [70–76], and 3D QHE [32–40].

Note that 3D $\to$ 2D crossovers happen even in a weak magnetic field, where the "sizes" of electron trajectories (7.8) are lager than the interplane distances, $l_\perp(H) \geq c^*$. Later, we call such crossovers QC. According to [10], in experimental magnetic fields of the orders of $H = 5$–$50\,\mathrm{T}$, QL dimensional crossovers can happen, where "sizes" of the trajectories become less than the interplane distances, $l_\perp(H) \leq c^*$. For theory of QL crossovers, see Sect. 7.5 and [14].

Here, we consider 3D $\to$ 2D crossover for a Q2D conductor with electron spectrum (7.5) in a parallel magnetic field (see [12,14]). In Q2D case, Lorentz force in the equations of motion (7.6),

$$\frac{\mathrm{d}p_z}{\mathrm{d}t} = \left(\frac{e}{c}\right) v_\mathrm{F} H \sin\varphi, \tag{7.10}$$

depends on an electron position on FS [i.e., on angle ϕ (see Fig. 7.5)]. Therefore, the "sizes"

$$l_\perp(H,\varphi) = 2c^* t_c/\omega_c \sin\varphi \tag{7.11}$$

of the electron trajectories

$$z(t,\varphi) = l_\perp(H,\varphi)\cos(p_z^0 c^* + \omega_c \sin\varphi\ t), \quad \omega_c = e v_\mathrm{F} H c^*/c \tag{7.12}$$

also depend on electron positions on the Q2D FS (see Fig. 7.2). In particular, it is possible that a situation where some electrons are under conditions of QL dimensional crossover whereas the others are under conditions of QC one. Note that QL 3D $\to$ 2D crossovers in Q2D conductors are responsible for the RS phenomenon [10, 12, 79–86] – superconducting phase stable in ultra-high magnetic fields.

7.3.3 Periodic and Quasi-Periodic Trajectories

Let us discuss electron motion in a Q1D conductor in a magnetic field, perpendicular to its conducting chains

$$\mathbf{H} = (0, H\sin\alpha, H\cos\alpha) \tag{7.13}$$

(see Fig. 7.6).

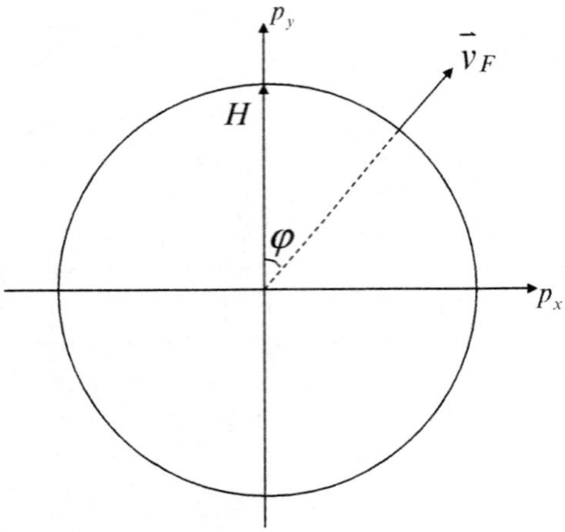

Fig. 7.5. Q2D electrons with spectrum (7.5) in a magnetic field

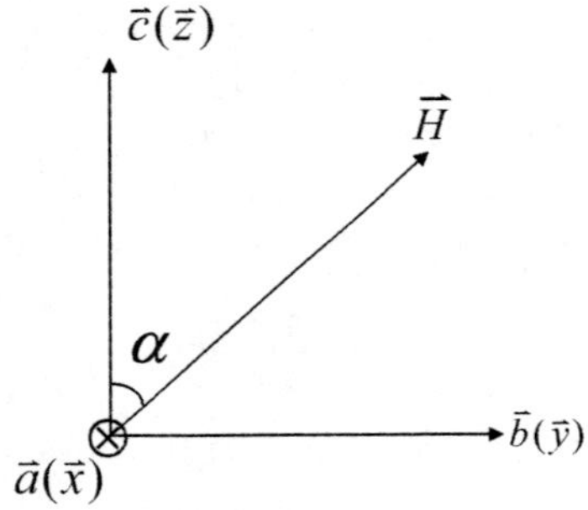

Fig. 7.6. Magnetic field is perpendicular to the conducting chains of a Q1D conductor and makes an angle α with **c**(**z**)-axis, which is perpendicular to the conducting planes

In this section, we consider the most general linearized Q1D electron spectrum

$$\varepsilon(\mathbf{p}) = v_x(p_y)[p_x - p_x(p_y)] + 2t_b \cos(p_y b^*) + 2t_c \cos(p_z c^*), \qquad (7.14)$$

and, thus, take into account of the dependence of the velocity component $v_x(p_y)$ on a position on FS (i.e., on a momentum component p_y). For Q1D electrons (7.14) in a magnetic field (7.13), equations of motion (7.6) can be written as

$$\mathrm{d}p_y/\mathrm{d}t = -(e/c)v_x(p_y)H\cos\alpha, \quad \mathrm{d}p_z/\mathrm{d}t = (e/c)v_x(p_y)H\sin\alpha. \qquad (7.15)$$

Below, it is convenient to use dimensionless variables, $\tilde{y} = p_y b^*$ and $\tilde{z} = p_z c^*$, and to introduce two cyclotron frequencies:

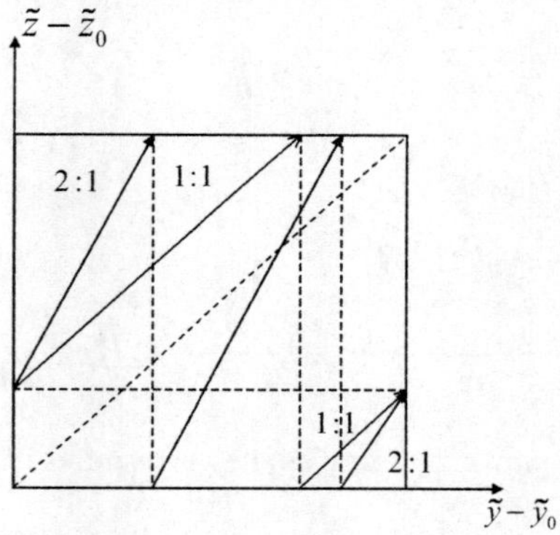

Fig. 7.7. Periodic ("commensuarate") trajectories in Brillouin zone, corresponding to $n/m = 1$ and $n/m = 2$ in (7.20)

$$\omega_b(p_y) = v_x(p_y)\omega_b/v_{\mathrm{F}}, \quad \omega_b = ev_{\mathrm{F}}b^*H/c,$$
$$\omega_c(p_y) = v_x(p_y)\omega_c/v_{\mathrm{F}}, \quad \omega_c = ev_{\mathrm{F}}c^*H/c. \tag{7.16}$$

In this case, we can represent equations of motion (7.15) in a more convenient form,

$$\mathrm{d}\tilde{y}/\mathrm{d}t = -\omega_b(p_y)\cos\alpha, \quad \mathrm{d}\tilde{z}/\mathrm{d}t = \omega_c(p_y)\sin\alpha, \tag{7.17}$$

and solve them,

$$\tilde{y} = \tilde{y}_0 - \omega_b(p_y)\cos\alpha\cdot t, \quad \tilde{z} = \tilde{z}_0 + \omega_c(p_y)\sin\alpha\cdot t. \tag{7.18}$$

It is important that electron trajectories become periodic

$$\tilde{z} - \tilde{z}_0 = \frac{n}{m}(\tilde{y} - \tilde{y}_0) \tag{7.19}$$

at some "commensurate" directions of a magnetic field,

$$\tan\alpha = (n/m)(b^*/c^*), \tag{7.20}$$

which are called LMA directions [17] (see Fig. 7.7). In Sects. 7.3.5 and 7.4.4, we show that 1D → 2D crossovers happen at LMA directions of a magnetic field (7.20), which are responsible for experimentally observed numerous LMA phenomena [41–57].

7.3.4 "One-Dimensionalization" of Electron Motion

Let us consider electron motion in a magnetic field (7.13), perpendicular to the conducting chains, in a real space for the simplified Q1D spectrum

$$\varepsilon(\mathbf{p}) = v_{\mathrm{F}}(p_x - p_{\mathrm{F}}) + 2t_b\cos(p_y b^*) + 2t_c\cos(p_z c^*), \tag{7.21}$$

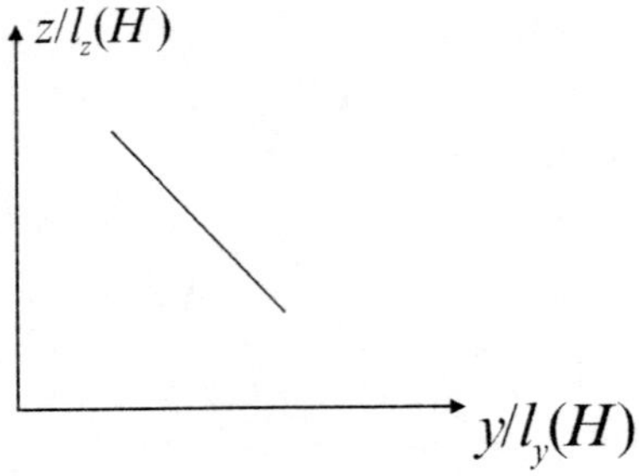

Fig. 7.8. Electron trajectory in a real space corresponding to the LMA direction of the field (7.20) with $m = n = 1$

where v_F does not depend on a momentum component p_y. This problem is solved in [17], where 3D → 1D crossover is suggested. For simplified electron spectrum (7.21) and (7.17) can be rewritten as:

$$\begin{aligned} \mathrm{d}\tilde{y}/\mathrm{d}t &= -\omega_b(p_y)\cos\alpha, \quad & \mathrm{d}\tilde{z}/\mathrm{d}t &= \omega_c(p_y)\sin\alpha, \\ v_y &= 2t_b b^* \sin\tilde{y}, & v_z &= 2t_c c^* \sin\tilde{z}. \end{aligned} \tag{7.22}$$

This system of equations can be analytically solved:

$$\begin{aligned} y &= l_y(H)\cos(\omega_b\cos\alpha\cdot t), \quad & z &= -l_z(H)\cos(\omega_c\sin\alpha\cdot t), \\ l_y(H) &= 2t_b b^*/\omega_b\cos\alpha, & l_z(H) &= 2t_c c^*/\omega_c\sin\alpha. \end{aligned} \tag{7.23}$$

Equation (7.23) directly demonstrates 3D → 1D crossover in a magnetic field (7.13). Indeed, according to (7.23), electron motions along both **y** and **z** axes are periodic and restricted. This means that electrons can move freely only along conducting chains and, thus, they become effectively 1D. To illustrate this 1D nature of an electron motion in a magnetic field, we calculate its average velocity along **y** and **z** axes and show that both of them are zero:

$$\begin{aligned} \langle v_z\rangle &= -\lim_{T\to\infty}\left(\frac{1}{T}\right)\frac{2t_c c^*}{\omega_c\sin\alpha}\Big[\cos(\omega_c\sin\alpha T + \tilde{z}_0) - \cos\tilde{z}_0\Big] = 0, \\ \langle v_y\rangle &= \lim_{T\to\infty}\left(\frac{1}{T}\right)\frac{2t_b b^*}{\omega_b\cos\alpha}\Big[\cos(\omega_b\cos\alpha T - \tilde{y}_0) - \cos\tilde{y}_0\Big] = 0. \end{aligned} \tag{7.24}$$

The simplest trajectory in a real space for magnetic field (7.13) with $m = n = 1$ in (7.20) is shown in Fig. 7.8. The corresponding localized wave functions and 1D electron spectrum are determined in Sect. 7.4.4.

7.3.5 Lebed Magic Angles as 1D → 2D Crossovers

In this section, we consider more realistic Q1D spectrum (7.14), where we take into account p_y-dependence of the velocity component $v_x(p_y)$ (see Fig. 7.9). This problem is considered in [27] and it reveals a nontrivial physical meaning

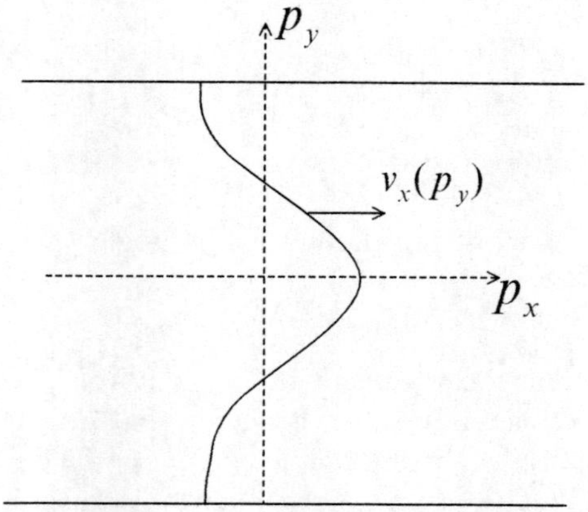

Fig. 7.9. Right sheet of the most general Q1D electron spectrum (7.14), where velocity component $v_x(p_y)$ depends on the position on FS

of LMA phenomena as 1D → 2D crossovers. Related effects are experimentally observed in a number of organic conductors in their metallic phases [41–57].

Let us rewrite (7.22) in the following way:

$$\mathrm{d}\tilde{y}/\mathrm{d}t = -\omega_b(\tilde{y})\cos\alpha, \quad \mathrm{d}t = -\mathrm{d}\tilde{y}/\omega_b(\tilde{y})\cos\alpha. \tag{7.25}$$

Then, an average over time velocity along **z**-axis can be expressed as:

$$\begin{aligned}\langle v_z\rangle_t &= \lim_{T\to\infty}\frac{1}{T}\int_0^T 2t_c c^*\sin(p_z c^*)\mathrm{d}t = -\lim_{T\to\infty}\frac{1}{T}\int_0^T \frac{2t_c c^*}{\omega_b(\tilde{y})\cos\alpha}\sin(p_z c^*)\mathrm{d}\tilde{y}\\ &= -\lim_{T\to\infty}\frac{1}{T}\frac{2t_c c^*}{\cos\alpha}\int_0^T \frac{\sin(\tilde{z}_0 + \tilde{y}\frac{c^*}{b^*}\tan\alpha)}{\omega_b(\tilde{y})}\mathrm{d}\tilde{y},\end{aligned} \tag{7.26}$$

where $\omega_b(\tilde{y}) = [v_x(\tilde{y})/v_{\mathrm{F}}]\omega_b$.

It is important that integration over $\tilde{y}$ in (7.26) corresponds to interference effects between density of states, $1/v_x(\ldots)$, and velocity component, $v_z(\ldots)$, perpendicular to conducting layers. As shown in [27], these interference effects occur when electron move in the extended Brillouin zone in a magnetic field and are due to Bragg's reflections of electron waves. From (7.26), it directly follows that, at LMA ("commensurate") directions of a magnetic field

$$\tan\alpha = N(b^*/c^*) \tag{7.27}$$

(where N is an integer), the interference effects become constructive, which results in nonzero value of the average velocity

$$\langle v_z\rangle = 2t_c c^*\sin\tilde{z}_0\langle\cos N\tilde{y}\cdot\sum_M A_M\cos M\tilde{y}\rangle = t_c c^* A_N\sin\tilde{z}_0 \neq 0, \tag{7.28}$$

where

$$f(\tilde{y}) = \sum_M A_M \cos(M\tilde{y}), \quad A_M = \frac{1}{\pi}\int_{-\pi}^{\pi} f(\tilde{y})\cos(M\tilde{y})\mathrm{d}\tilde{y},$$
$$v_\mathrm{F}/v_x(\tilde{y}) = 1 + f(\tilde{y}). \tag{7.29}$$

It is easy to show that, at non-Lebed magic angle (n-LMA) directions of a magnetic field

$$\tan\alpha \neq N(b^*/c^*) \tag{7.30}$$

destructive interference effects result in $\langle v_z \rangle = 0$ in (7.26).

On the basis of this analysis, we conclude that dimensionality of electron motion increases at LMA directions of a magnetic field (7.27). In Sect. 7.4.4, we show that this QC picture corresponds to quantum mechanical 1D $\rightarrow$ 2D crossovers for electron spectra and wave functions. For calculation of a one-particle conductivity under LMA 1D $\rightarrow$ 2D crossovers, see Sect. 7.6.1.

7.3.6 Interference Commensurate Oscillations as 1D $\rightarrow$ 2D Crossovers

Let us study more complicated electron trajectories in a magnetic field, which correspond to IC oscillations, experimentally observed in metallic phases of a number of organic conductors [25,28,29,59–64]. For this purpose, we consider motion of Q1D electrons described by the simplified electron spectrum (7.21), in an inclined magnetic field

$$\mathbf{H} = (\cos\theta\cos\varphi, \cos\theta\sin\varphi, \sin\theta)H \tag{7.31}$$

(see Fig. 7.10) with

$$v_y = 2t_b b^* \sin(p_y b^*), \qquad v_z = 2t_c c^* \sin(p_z c^*). \tag{7.32}$$

For the experimental case of $(\mathrm{TMTSF})_2\mathrm{X}$ conductors, $t_c \ll t_b$, equations of motion (7.6) can be written as:

$$\mathrm{d}p_z/\mathrm{d}t = -\left(\frac{e}{c}\right)H\left[2t_b b^* \sin(p_y b^*)\cos\theta\cos\varphi + v_\mathrm{F}\cos\theta\sin\varphi\right],$$
$$\mathrm{d}p_y/\mathrm{d}t = -\left(\frac{e}{c}\right)v_\mathrm{F} H\sin\theta. \tag{7.33}$$

As in Sect. 7.3.3, it is convenient to rewrite equations of motion (7.33) using dimensionless variables, $\tilde{y} = p_y b^*$ and $\tilde{z} = p_z c^*$

$$\mathrm{d}\tilde{y}/\mathrm{d}t = -\omega_b(\theta), \quad \mathrm{d}\tilde{z}/\mathrm{d}t = \omega_c(\theta,\varphi) - \omega_c^*(\theta,\varphi)\sin\tilde{y}, \tag{7.34}$$

where

$$\omega_b(\theta) = \omega_b\sin\theta, \quad \omega_b = ev_\mathrm{F}Hb^*/c, \omega_c(\theta,\varphi) = \omega_c\cos\theta\sin\varphi, \quad \omega_c = ev_\mathrm{F}Hc^*/c,$$
$$\omega_c^*(\theta,\varphi) = \omega_c^*\cos\theta\cos\varphi, \ \omega_c^* = (v_y^0/v_\mathrm{F})(eHv_\mathrm{F}c^*/c). \tag{7.35}$$

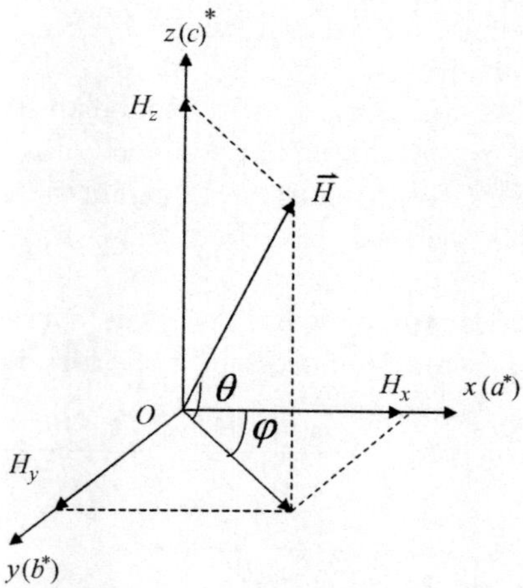

Fig. 7.10. Experimental geometry for observation of the interference commensurate (IC) oscillations [25,28,29]

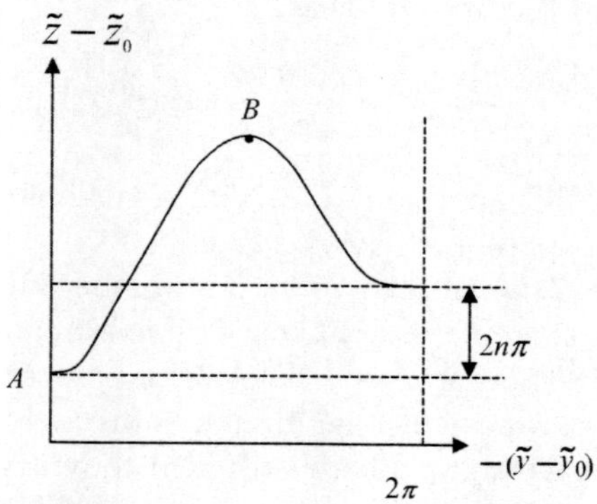

Fig. 7.11. The simplest IC trajectory (7.34) corresponding to $N = 1$ in (7.37)

Note that (7.34) can be solved analytically. As a result, we obtain:

$$\tilde{y} - \tilde{y}_0 = -\omega_b(\theta)t,$$
$$\tilde{z} - \tilde{z}_0 = -\frac{\omega_c(\theta,\varphi)}{\omega_b(\theta)}(\tilde{y} - \tilde{y}_0) - \frac{\omega_c^*(\theta,\varphi)}{\omega_b(\theta)}(\cos\tilde{y} - \cos\tilde{y}_0). \quad (7.36)$$

It is important that IC trajectories (7.36) (see Fig. 7.11), which are much more complicated than LMA ones (7.18), can be also subdivided into "commensurate" trajectories, which appear at "commensurate" directions of a magnetic field,

$$\sin\varphi = N(b^*/c^*)\tan\theta, \quad (7.37)$$

and "incommensurate" ones, which occur at arbitrary directions of a magnetic field. Note that "commensurate" trajectories correspond to periodic electron motion in a magnetic field in Brillouin zone (see Fig. 7.11), whereas electron motion in real space is never periodic.

Periodicity of electrons' motion at "commensurate" directions (7.37) of a magnetic field (7.31) results in the appearance of constructive interference effects which, as shown later (see also [28, 29]), increase the dimensionality of wave functions. In this section, we discuss these 1D → 2D crossovers using QC language.

Since IC trajectories are known (7.36), an average over time velocity component along **z**-axis can be expressed in the form:

$$\langle v_z \rangle_t = v_z^0 \sin\left[p_z^0 c^* + \frac{\omega_c(\theta,\varphi)}{\omega_b(\theta)} p_y^0 b^* + \frac{\omega_c^*(\theta,\varphi)}{\omega_b(\theta)} \cos(p_y^0 b^*)\right]$$
$$\left\langle \cos\left[N p_y b^* - \frac{\omega_c^*(\theta,\varphi)}{\omega_b(\theta)} \cos(p_y b^*)\right]\right\rangle_{p_y b^*}, \qquad (7.38)$$

where the average in (7.38) is taken over momentum component $p_y b^* \sim t$ [see (7.36)]. Integrating (7.38), it is possible to show that $\langle v_z(p_y)\rangle = 0$ for any "incommensurate" trajectory and that

$$\langle v_z \rangle_t = \sin\left[p_z^0 c^* + \frac{\omega_c(\theta,\varphi)}{\omega_b(\theta)} p_y^0 b^* + \frac{\omega_c^*(\theta,\varphi)}{\omega_b(\theta)} \cos(p_y^0 b^*)\right] J_N\left(\frac{2t_b c^* \cos\varphi}{v_F \tan\theta}\right) \qquad (7.39)$$

for "commensurate" IC directions (7.37) of a magnetic field. (For electron trajectories in a real space, see Figs. 7.12 and 7.13.)

As shown in Sect. 7.4.5 and [28, 29], quantum mechanical meaning of these 1D → 2D crossovers is an increase in space dimensionality of electron spectra and wave functions. Experimentally, 1D → 2D crossovers at IC directions of a magnetic field (7.37) are seen as deep minima of resistivity, which is lower for 2D electrons than for 1D ones. There exist also convincing experimental data [29] on a direct proof of 1D → 2D crossovers in $(TMTSF)_2ClO_4$ conductor.

From (7.39), it also follows that there are extra zeros of an average velocity, which correspond to zeros of Bessel functions

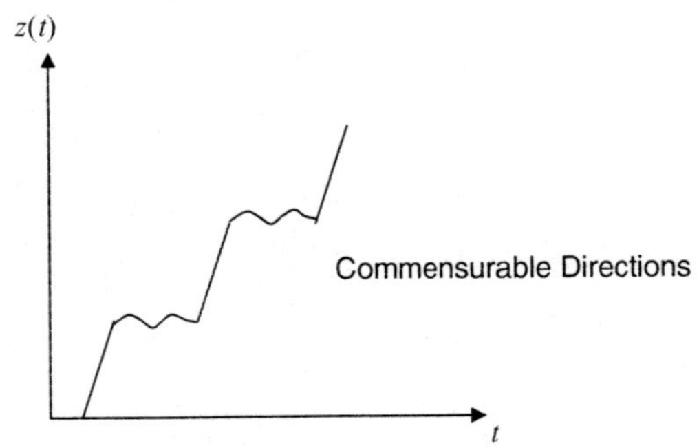

Fig. 7.12. Typical electron trajectory for "commensurate" (IC) direction of a magnetic field (7.37)

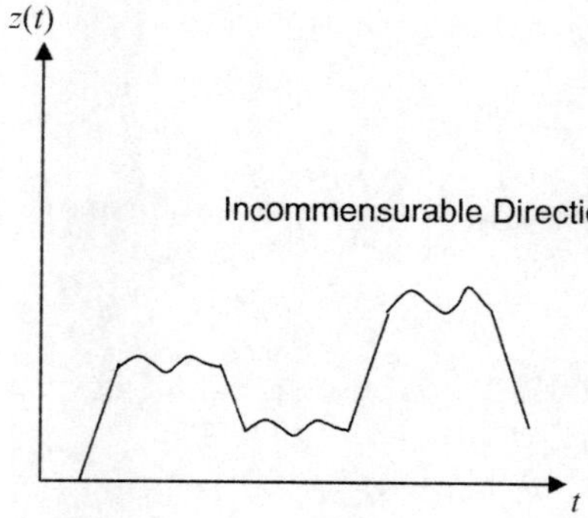

Fig. 7.13. Typical electron trajectory for "incommensurate" (non-IC) direction of a magnetic field

$$J_N\left(\frac{2t_b c^* \cos\varphi}{v_\mathrm{F}\tan\theta}\right) = 0. \tag{7.40}$$

Note that the latter zeros are not of a topological origin and due to special cos-like shape of FS (7.21) [25,28,29]. We relate them to experimentally observed [25, 29, 59–64] extended DKC oscillations. For calculation of one-particle conductivity under IC 1D → 2D crossovers, see Sect. 7.6.2.

7.4 Quantum Mechanics of Dimensional Crossovers

In this section, we quantum mechanically describe different types of dimensional crossovers, considered in Sect. 7.3 by means of QC equations of motion. In particular, we show that, in the absence of Landau levels quantization in a magnetic field for open electron orbits, other quantum effects – Bragg reflections – play an important role.

7.4.1 Momentum Quantization Law

In this section, we demonstrate that Bragg reflections, which occur when electrons move along open orbits in the extended Brillouin zone, result in momentum quantization law [14, 25–29], instead of textbook Landau levels quantization [1]. The above-mentioned momentum quantization law is shown to be a consequence of 3D → 2D crossovers phenomenon, which is quantum mechanically described later.

To consider quantum mechanical problem, let us make use of FL description of quasi-particles in a magnetic field [15]

$$\mathbf{H} = (0, 0, H), \quad \mathbf{A} = (0, Hx, 0) \tag{7.41}$$

by means of Peierls substitutions [10,12,13,15–29]

$$p_x - p_\mathrm{F} \to -\mathrm{i}\,(\mathrm{d}/\mathrm{d}x), \quad p_y b^* \to p_y b^* - (e/c)A_y b^* = p_y b^* - (\omega_b/v_\mathrm{F})x. \quad (7.42)$$

If we represent electron wave functions in the form

$$\phi_\varepsilon^\pm(x,y) = \mathrm{e}^{\pm i p_\mathrm{F} x} \cdot \psi_\varepsilon^\pm(x,y), \qquad (7.43)$$

then we obtain the following Schrödinger equation:

$$\hat{\varepsilon}^+(\hat{p}) \cdot \psi_\varepsilon^+(p_y,x) = \delta\varepsilon \cdot \psi_\varepsilon^+(p_y,x), \quad \delta\varepsilon = \varepsilon - \varepsilon_\mathrm{F}, \qquad (7.44)$$

where

$$\psi_\varepsilon^+(x,y) = \int_0^{2\pi/b^*} \psi_\varepsilon^+(p_y,x)\mathrm{e}^{\mathrm{i}p_y y}\, d(p_y b^*)/2\pi, \qquad (7.45)$$

also see Fig. 7.14.

As a result of Peierls substitution (7.42), Schrödinger equation (7.44) for wave functions in a mixed space–momentum representation, $\psi_\varepsilon^+(p_y,x)$, can be expressed as:

$$\left(\mp \mathrm{i} v_\mathrm{F} \frac{\mathrm{d}}{\mathrm{d}x} + 2t_b \cos\left[p_y b^* - \frac{\omega_b x}{v_\mathrm{F}}\right]\right)\psi_\varepsilon^\pm(p_y,x) = \delta\varepsilon \cdot \psi_\varepsilon^\pm(p_y,x). \qquad (7.46)$$

It is important that (7.46) can be solved analytically,

$$\psi_\varepsilon^\pm(p_y,x) = \exp\left(\pm\mathrm{i}\frac{\delta\varepsilon}{v_\mathrm{F}}x\right) \exp\left[\pm\frac{2\mathrm{i}t_b}{\omega_b}\left(\sin\left[p_y b^* - \frac{\omega_b x}{v_\mathrm{F}}\right] - \sin[p_y b^*]\right)\right], \quad (7.47)$$

where

$$\phi_\varepsilon^\pm(p_y,x) = \mathrm{e}^{\pm\mathrm{i}p_\mathrm{F} x} \cdot \psi_\varepsilon^\pm(p_y,x). \qquad (7.48)$$

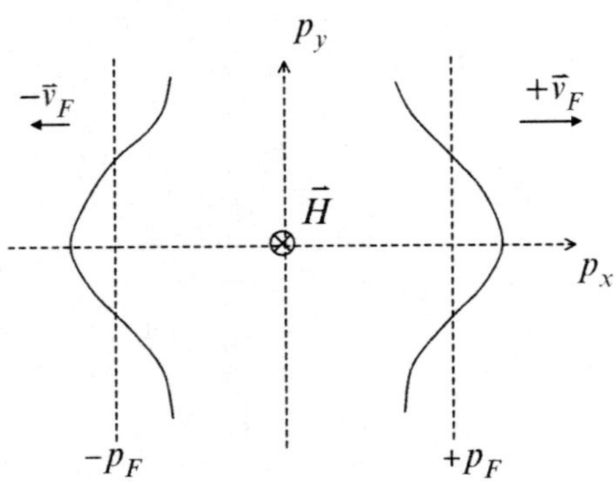

Fig. 7.14. Typical Q1D FS in a perpendicular to the conducting chains magnetic field, **H** ∥ **c**, corresponds to an infinite set of 1D FS (see Fig. 7.15)

Let us show that (7.47) and (7.48) directly demonstrate 3D $\rightarrow$ 2D crossovers in a magnetic field. For this purpose, we calculate Fourier component of wave functions (7.47) with respect to variable x

$$\psi_{\varepsilon}^{+}(p_y, x) = \mathrm{e}^{\mathrm{i}\frac{\delta\varepsilon}{v_\mathrm{F}}x} \sum_{m=-\infty}^{\infty} A_m(p_y)\mathrm{e}^{\mathrm{i}m\frac{\omega_b}{v_\mathrm{F}}x}, \tag{7.49}$$

where

$$A_m(p_y) = \frac{\omega_b}{2\pi v_\mathrm{F}} \int_{-\pi v_\mathrm{F}/\omega_b}^{\pi v_\mathrm{F}/\omega_b} \mathrm{e}^{-i\frac{\delta\varepsilon}{v_\mathrm{F}}x} \psi_{\varepsilon}^{+}(p_y, x)\mathrm{e}^{-im\frac{\omega_b}{v_\mathrm{F}}x}\mathrm{d}x. \tag{7.50}$$

As seen from (7.49), electron energy, $\delta\varepsilon$, does not depend on momentum component p_y, which is a consequence of 3D $\rightarrow$ 2D crossover. Moreover, from (7.50) it follows the following momentum quantization law: if electron has a definite energy, $\delta\varepsilon$, then its momentum component along $\mathbf{x}(\mathbf{a})$-axis is quantized:

$$p_x = \delta\varepsilon/v_\mathrm{F} + m(\omega_b/v_\mathrm{F}), \tag{7.51}$$

where ω_b/v_F is a momentum quantum (see Fig. 7.15).

It is possible to calculate Fourier coefficients $A_m(p_y)$ in (7.50) analytically

$$\begin{aligned} A_m(p_y) &= \frac{\omega_b}{2\pi v_\mathrm{F}} \int_{-\pi v_\mathrm{F}/\omega_b}^{\pi v_\mathrm{F}/\omega_b} \mathrm{e}^{-im\frac{\omega_b}{v_\mathrm{F}}x} \mathrm{e}^{\frac{2it_b}{v_\mathrm{F}}\left(\sin[p_y b^* - \frac{\omega_b x}{v_\mathrm{F}}] - \sin[p_y b^*]\right)} \mathrm{d}x \\ &= \mathrm{e}^{-\frac{2it_b}{v_\mathrm{F}}\sin[p_y b^*]} \cdot \mathrm{e}^{-imp_y b^*} \cdot \mathrm{e}^{-im\frac{\pi}{2}+i\phi} \int_0^{2\pi} \frac{\mathrm{d}\varphi}{2\pi} \mathrm{e}^{-im\varphi} \cdot \mathrm{e}^{-\frac{2it_b}{\omega_c}\cos\varphi} \\ &= \mathrm{e}^{\mathrm{i}\tilde{\phi}} \cdot J_{-m}\left(\frac{2t_b}{\omega_c}\right), \end{aligned} \tag{7.52}$$

where

$$|A_m(p_y)|^2 = J_m^2\left(\frac{2t_b}{\omega_b}\right) \tag{7.53}$$

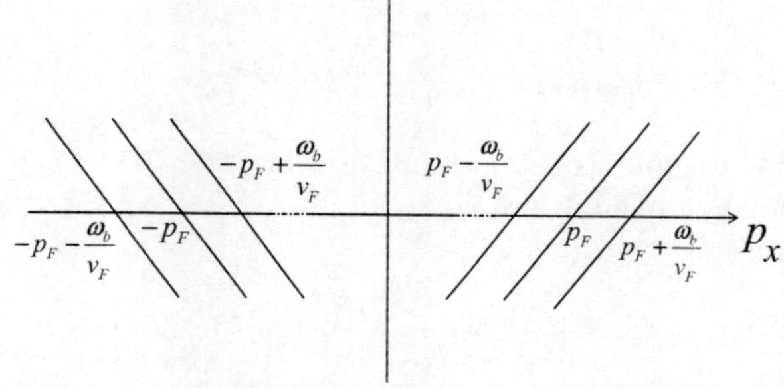

Fig. 7.15. The momentum quantization law [see (7.51)]

is a probability to have quantized momentum $\delta\varepsilon/v_{\mathrm{F}} + m(\omega_b/v_{\mathrm{F}})$ for an electron with definite energy $\delta\varepsilon$. (Note that the results of this section can be extended to a Q2D electron spectrum of an arbitrary shape [12].)

The above-described 3D → 2D crossover and momentum quantization law (7.51) play a central role in theories of FISDW, FICDW, and RS phases [see Sects. 7.7, and 7.8]. In Sect. 7.5, we solve fully quantum mechanical problem and show that momentum quantization law (7.51) is very general phenomenon and not restricted by Peierls substitution method. It survives in the cases of Q1D and Q2D FS of arbitrary shapes both under QC and QL dimensional crossover.

7.4.2 3D → 2D Crossovers

In this section, we show that electron wave functions become 2D (i.e., localized on conducting planes) in a magnetic field (7.41). To demonstrate "two-dimensionalization" of wave functions (7.47), it is convenient to make their Fourier transformations with respect to p_y and, thus, to calculate them in a real space.

As a result, we obtain

$$\psi_\varepsilon^+(y = nb^*, x) = \int_0^{2\pi} \frac{\mathrm{d}p_y}{2\pi} \psi_\varepsilon^+(p_y, x) \cdot \mathrm{e}^{\mathrm{i}p_y nb^*}, \tag{7.54}$$

where

$$|\psi_\varepsilon^+(y = nb^*, x)|^2 = |\int_0^{2\pi} \frac{\mathrm{d}\varphi}{2\pi} \mathrm{e}^{\mathrm{i}n\varphi} \cdot \mathrm{e}^{\frac{2it_b}{\omega_b} \cos\varphi}|^2 = J_n^2\left(\frac{2t_b}{\omega_b}\right). \tag{7.55}$$

Note that in (7.54), wave functions in a real space are calculated at $y = nb^*$ (i.e., on conducting planes). In this case, (7.55) gives a probability that wave function, centered at $y = 0$, is extended to conducting plane with $y = nb^*$ (see Fig. 7.2). As seen from (7.55), this probability quickly goes to zero at $n > 2t_b/\omega_b$ and, therefore, electron wave functions (7.54) are localized. Note that the results of this section can be extended to Q2D electron spectrum of an arbitrary shape [see Sect. 7.5].

7.4.3 3D → 1D Crossovers

In this section, we quantum mechanically describe 3D → 1D crossovers in an inclined magnetic field [see Sect. 7.3.4]. Let us consider electrons with simplified Q1D spectrum

$$\varepsilon^\pm(\mathbf{p}) = \pm v_{\mathrm{F}}(p_x \mp p_{\mathrm{F}}) + 2t_b \cos(p_y b^*) + 2t_c \cos(p_z c^*) \tag{7.56}$$

in an inclined magnetic field (7.13), which is perpendicular to conducting chains

$$\mathbf{H} = (0, H \sin\alpha, H \cos\alpha), \quad \mathbf{A} = (0, \cos\alpha, -\sin\alpha)Hx. \tag{7.57}$$

The above-mentioned problem is considered in [17]. In this case, we have to make Peierls substitutions for both p_y and p_z projections of electron momentum:

$$p_y \to p_y - (e/c)\, Hx \cos\alpha, \quad p_z \to p_z + (e/c)\, Hx \sin\alpha, \tag{7.58}$$

and to use QC expression for momentum operator projection along the conducting chains

$$p_x \mp p_{\mathrm{F}} \to -\mathrm{i}(\mathrm{d}/\mathrm{d}x). \tag{7.59}$$

As a result, electron Hamiltonian can be written as:

$$\begin{aligned}\hat{\varepsilon}^{\pm}(p_y, p_z, x) = \mp \mathrm{i} v_{\mathrm{F}} \frac{\mathrm{d}}{\mathrm{d}x} &+ 2t_b \cos\left[p_y b^* - \frac{\omega_b(\alpha)}{v_{\mathrm{F}}} x\right] \\ &+ 2t_c \cos\left[p_z c^* + \frac{\omega_c(\alpha)}{v_{\mathrm{F}}} x\right]\end{aligned} \tag{7.60}$$

with

$$\omega_b(\alpha) = (eHv_{\mathrm{F}} b^*/c) \cos\alpha, \quad \omega_c(\alpha) = (eHv_{\mathrm{F}} c^*/c) \sin\alpha. \tag{7.61}$$

As usual, we represent wave functions in the form (7.43) and count energy from Fermi level, $\delta\varepsilon = \varepsilon - \varepsilon_{\mathrm{F}}$. In this case, Schrödinger equation,

$$\hat{\varepsilon}^{+}(p_y, p_z, x)\psi_{\varepsilon}^{+}(p_y, p_z, x) = \delta\varepsilon \cdot \psi_{\varepsilon}^{+}(p_y, p_z, x), \tag{7.62}$$

can be rewritten as,

$$\begin{aligned}\left\{-\mathrm{i} v_{\mathrm{F}} \frac{\mathrm{d}}{\mathrm{d}x} + 2t_b \cos\left[p_y b^* - \frac{\omega_b(\alpha)}{v_{\mathrm{F}}} x\right] + 2t_c \cos\left[p_z c^* + \frac{\omega_c(\alpha)}{v_{\mathrm{F}}} x\right]\right\} \psi_{\varepsilon}^{+}(p_y, p_z, x) \\ = \delta\varepsilon \cdot \psi_{\varepsilon}^{+}(p_y, p_z, x)\end{aligned} \tag{7.63}$$

and analytically solved

$$\begin{aligned}\psi_{\varepsilon}^{+}(p_y, p_z, x) = \exp\left(\mathrm{i}\frac{\delta\varepsilon}{v_{\mathrm{F}}} x\right) \exp\left\{+\frac{2\mathrm{i} t_b}{\omega_b(\alpha)}\left(\sin\left[p_y b^* - \frac{\omega_b(\alpha)}{v_{\mathrm{F}}} x\right] - \sin[p_y b^*]\right)\right\} \\ \times \exp\left\{-\frac{2\mathrm{i} t_c}{\omega_c(\alpha)}\left(\sin\left[p_z c^* - \frac{\omega_c(\alpha)}{v_{\mathrm{F}}} x\right] - \sin[p_z c^*]\right)\right\}.\end{aligned} \tag{7.64}$$

As it directly follows from (7.64), electron energy levels in a magnetic field are 1D and do not depend on two momentum components, p_y and p_z.

It is convenient to rewrite wave functions (7.64) as the corresponding Fourier series:

$$\psi_\varepsilon^+(p_y,p_z,x) = \exp\left(\mathrm{i}\frac{\delta\varepsilon}{v_\mathrm{F}}x\right)\sum_{m_1}\sum_{m_2} J_{m_1}(2t_b/\omega_b)J_{m_2}(2t_c/\omega_c)$$
$$\exp\left[\mathrm{i}m_1\frac{\omega_b(\alpha)}{v_\mathrm{F}}x - \mathrm{i}m_2\frac{\omega_c(\alpha)}{v_\mathrm{F}}x\right]. \qquad (7.65)$$

As seen from (7.65), 1D electron spectrum possesses very peculiar momentum quantization law with two different momentum quanta, $\omega_b(\alpha)$ and $\omega_c(\alpha)$. Indeed, electron with definite energy, $\delta\varepsilon$, is characterized in (7.65) by the following quantized momentum component along the conducting chains:

$$p_x = p_\mathrm{F} + \delta\varepsilon/v_\mathrm{F} + m_1\omega_b(\alpha)/v_\mathrm{F} + m_2\omega_c(\alpha)/v_\mathrm{F}. \qquad (7.66)$$

Note that, at LMA directions (7.20) of a magnetic field (7.13), momentum quanta (7.66) become commensurate. As shown in [17], this results in increase of electron–hole pairing, which is experimentally observed as minima in FISDW threshold field [33].

7.4.4 Lebed Magic Angles as 1D → 2D Crossovers

In Sect. 7.3.5, we showed, using QC equations of electron motion that LMA directions (7.27) of a magnetic field correspond to 1D → 2D crossovers. In this section, we develop quantum mechanical theory of such dimensional crossovers for electron wave functions and spectrum. As shown above for simplified Q1D electron spectrum (7.56), dimensionality of electron spectrum and wave functions become 1D in a magnetic field perpendicular to the conducting chains.

Here, we consider realistic Q1D electron spectrum (7.14) and show that p_y-dependence of electron velocity, $v_x(p_y)$ (see Fig. 7.16), results in 1D → 2D

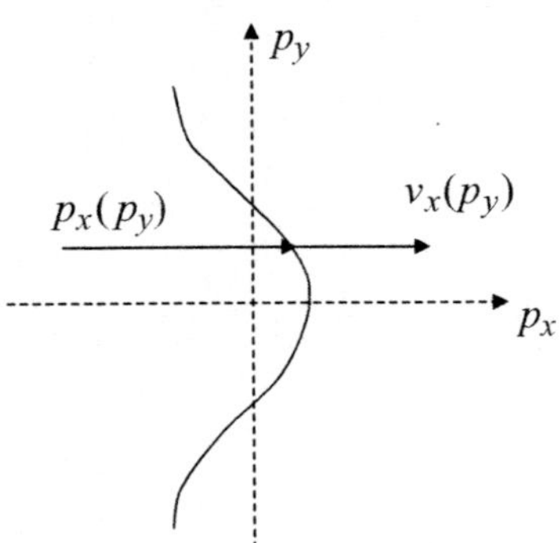

Fig. 7.16. Electron velocity component along the conducting chains and Fermi momentum depend on position on FS, $v_x(p_y)$ and $p_x(p_y)$

de-localization of electron wave functions at LMA (7.27) directions of the field. We use Peierls substitution method for a Q1D electron spectrum [27]

$$p_x \to -\mathrm{i}(\partial/\partial x), \quad p_{y,z} \to p_{y,z} - (e/c)A_{y,z}. \tag{7.67}$$

As a result, we obtain the following Hamiltonian:

$$\begin{aligned}\hat{H} = v_x\Big[p_y b^* - \frac{\omega_b(\alpha)}{v_\mathrm{F}}x\Big]\Big\{-\mathrm{i}\frac{\mathrm{d}}{\mathrm{d}x} - p_x\Big[p_y b^* - \frac{\omega_b(\alpha)}{v_\mathrm{F}}x\Big]\Big\} \\ + 2t_b \cos\Big[p_y b^* - \frac{\omega_b(\alpha)}{v_\mathrm{F}}x\Big] + 2t_c \cos\Big[p_z c^* - \frac{\omega_c(\alpha)}{v_\mathrm{F}}x\Big],\end{aligned} \tag{7.68}$$

where

$$\omega_b(\alpha) = eHb^* v_\mathrm{F}\cos\alpha/c, \quad \omega_c(\alpha) = eHc^* v_\mathrm{F}\sin\alpha/c. \tag{7.69}$$

[Note that, although operators $-\mathrm{i}(\partial/\partial x)$ and $v_x\big[p_y b^* - \frac{\omega_b(\alpha)}{v_\mathrm{F}}x\big]$ do not commute in Hamiltonian (7.68), it is not important in QC case, where $\omega_b(\alpha) \ll t_c, t_b \ll \varepsilon_\mathrm{F}$].

In QC limit, Schrödinger equation for Hamiltonian (7.68) can be solved:

$$\begin{aligned}\psi_{\delta\varepsilon}(x,p_y) = \frac{1}{\sqrt{v_x\big[p_y - \omega_b(\alpha)x/v_\mathrm{F}\big]}} \exp\Big\{\mathrm{i}\int_0^x p_x\big[p_y - \omega_b(\alpha)u/v_\mathrm{F}\big]\mathrm{d}u\Big\} \\ \times \exp\Big\{\mathrm{i}\int_0^x \frac{\delta\varepsilon \mathrm{d}u}{v_x\big[p_y - \omega_b(\alpha)u/v_\mathrm{F}\big]}\Big\}.\end{aligned} \tag{7.70}$$

This solution is valid at

$$\omega_b(\alpha) \ll t_c, t_b, \tag{7.71}$$

which is equivalent to condition that the "sizes" of electron trajectories in a magnetic field are larger than the interplane distances, $l_y(\alpha), l_z(\alpha) \gg c^*$. Note that the physical meaning of QC solution (7.70) is that probability to find electron at point x is inversely proportional to its velocity along the chains

$$|\psi_\varepsilon(x,p_y)|^2 \sim 1/v_x\big[p_y - \omega_b(\alpha)x/v_\mathrm{F}\big], \tag{7.72}$$

and, thus, inversely proportional to the density of states.

Below, we define operator of velocity component along **z**-axis:

$$\hat{v}_z(p_z) = v_z^0 \sin\Big(p_z c^* - \big[\omega_c(\alpha)/v_\mathrm{F}\big]x\Big) \tag{7.73}$$

and calculate the expectation value of this velocity component

$$\begin{aligned}\langle\hat{v}_z(p_z)\rangle &= \lim_{L\to\infty}\frac{1}{L}\int_0^L \frac{1}{\langle 1/v_x\big[p_y b^* - \omega_b(\alpha)u/v_\mathrm{F}\big]\rangle_x}\,\frac{\sin\big[p_z c^* - \omega_c(\alpha)x/v_\mathrm{F}\big]}{v_x\big[p_y b^* - \omega_b(\alpha)u/v_\mathrm{F}\big]}\mathrm{d}x \\ &= \begin{cases} 0, & \text{if } \omega_c \neq N\omega_b \\ \neq 0, & \text{if } \omega_c = N\omega_b, \end{cases}\end{aligned} \tag{7.74}$$

which is nonzero only at LMA directions (7.27) of a magnetic field.

It is possible to simplify the integral (7.74)

$$\langle \hat{v}_z(p_z) \rangle = v_z \sin(p_z c^*), \quad \tilde{v}_z = \frac{v_z^0}{\langle 1/v_x \rangle} \int_0^{2\pi} \frac{\cos(N\tilde{x})}{v_x(\tilde{x})} \mathrm{d}\tilde{x}. \tag{7.75}$$

If we define function, $1 + f(\tilde{y}) = 1/v_x(\tilde{y})$, and expand it into Fourier series

$$f(\tilde{y}) = \sum_{m=-\infty}^{+\infty} A_m \cos(m\tilde{y}), \quad A_N = \frac{1}{\pi} \int_{-\pi}^{\pi} f(\tilde{y}) \cos(N\tilde{y}) \mathrm{d}\tilde{y}, \tag{7.76}$$

the expectation value of velocity component can be written as:

$$\langle \hat{v}_z(p_z) \rangle = \begin{cases} 0, & \text{if } \omega_c \neq N\omega_b \\ \frac{v_z^0 A_N}{2}, & \text{if } \omega_c = N\omega_b \end{cases} \tag{7.77}$$

Using the similar procedure, we can define expectation value of kinetic energy along **z**-axis. As a result, we obtain:

$$\langle \hat{\varepsilon}(p_z) \rangle = \begin{cases} 0, & \text{if } \omega_c \neq N\omega_b \\ 2t_c \frac{A_N}{2} \cos(p_z c^*), & \text{if } \omega_c = N\omega_b. \end{cases} \tag{7.78}$$

Note that (7.77) and (7.78) directly demonstrate 1D $\rightarrow$ 2D crossovers for electron spectrum, which happen at LMA directions (7.27) of a magnetic field. Indeed, as it follows from (7.77) and (7.78), both expectations values of kinetic energy (7.78) and velocity (7.77) increase their dimensionality at LMA directions.

Let us show that electron wave functions are localized at arbitrary directions of a magnetic field and de-localized at LMA directions (7.27). For this purpose, we calculate z-dependence of wave functions by means of their Fourier transformations with respect to p_z component of electron momentum

$$\begin{aligned} \Psi(z = Nc^*) &= \int_0^{2\pi} \frac{\mathrm{d}p_z}{2\pi} \mathrm{e}^{\mathrm{i}p_z Nc^*} \exp\left\{2\mathrm{i}t_c \sqrt{I_1^2 + I_2^2} \cos(p_z c^* + \varphi)\right\} \\ &= \mathrm{e}^{-\mathrm{i}N\varphi} J_N\left(2t_c \sqrt{I_1^2 + I_2^2}\right), \end{aligned} \tag{7.79}$$

where integrals I_1 and I_2 are equal

$$I_{1(2)} = \int_0^x \frac{\cos(\sin)\left[p_z c^* + \frac{\omega_c(\alpha)}{v_\mathrm{F}} u\right]}{v_x\left[p_y - \omega_b(\alpha) u / v_\mathrm{F}\right]} \mathrm{d}u \tag{7.80}$$

with $J_N(\ldots)$ being Bessel function of Nth order.

To study the behavior of wave functions (7.79) and (7.80) at large values of coordinate z, the following property of Bessel functions is crucial: $J_N(y)$ is extended function of N if $N < |y|$ and exponentially decays with N at $N > |y|$. Using this property, we can conclude that wave functions (7.79) are extended if $\sqrt{I_1^2 + I_2^2} \to \infty$ as $x \to \infty$ and are localized if $\sqrt{I_1^2 + I_2^2} < M$ at any x. If we take into account (7.80), we can conclude that these localization – de-localization crossovers for wave functions happen at LMA directions (7.27) of a magnetic field.

Note that the described above 2D $\to$ 1D crossovers are responsible for experimentally observed LMA phenomena in a magnetic field. Although there is still no theory which can explain quantitatively all complex features of experimentally observed LMA phenomena [41–57], we believe that the above-described 2D $\to$ 1D crossovers give adequate qualitative descriptions. For calculation of resistivity component, perpendicular to conducting layers, ρ_{zz}, and comparison of the theory with the corresponding experimental data, see Sect. 7.6.1.

7.4.5 Interference Commensurate Oscillations as 2D $\to$ 1D Crossovers

In this section, we develop quantum mechanical theory of 1D $\to$ 2D crossovers responsible for experimentally observed IC resistive oscillations [59–64]. The corresponding QC trajectories are considered in Sect. 7.3.6. For quantum mechanical calculations of resistivity component ρ_{zz} and its comparison with the existing experimental data, see Sect. 7.6.2.

Let us consider the simplified Q1D electron spectrum

$$\hat{\varepsilon}^{\pm}(\mathbf{p}) = \pm v_{\mathrm{F}}(p_x \mp p_{\mathrm{F}}) + 2t_b f(p_y b^*) + 2t_c \cos(p_z c^*), \tag{7.81}$$

in an inclined magnetic field

$$\begin{aligned} \mathbf{H} &= (\cos\theta\cos\phi, -\cos\theta\sin\phi, \sin\theta)\ H, \\ \mathbf{A} &= (0, x\sin\theta, x\cos\theta\sin\phi - y\cos\theta\cos\phi)H \end{aligned} \tag{7.82}$$

(see Fig. 7.10).

We use Peierls substitution method, elaborated in [25–29],

$$p_x - p_{\mathrm{F}} \to -\mathrm{i}(\partial/\partial x), \quad p_{y,z} \to p_{y,z} - (e/c)A_{y,z}, \quad y \to +\mathrm{i}(\partial/\partial p_y). \tag{7.83}$$

In this case, electron Hamiltonian can be written as:

$$\begin{aligned} \hat{\varepsilon}^{+}(\mathbf{p}) = &-\mathrm{i}v_{\mathrm{F}}\frac{\partial}{\partial x} + 2t_b f\left[p_y b^* - \frac{\omega_b(\theta)x}{v_{\mathrm{F}}}\right] \\ &+ 2t_c \cos\left[p_z c^* + \frac{\omega_c(\theta,\varphi)x}{v_{\mathrm{F}}} - \mathrm{i}\frac{\tilde{\omega}_c(\theta,\varphi)}{v_{\mathrm{F}}}\frac{\partial}{\partial p_y}\right], \end{aligned} \tag{7.84}$$

where cyclotron frequencies in (7.84) are defined by (7.35).

The corresponding Schrödinger equation can be analytically solved and the wave functions can be written as:

$$\psi^{+}_{\delta\varepsilon}(x,p_y,p_z) = \exp\left(\frac{\mathrm{i}\delta\varepsilon x}{v_{\mathrm{F}}}\right)\exp\left(-\frac{2\mathrm{i}t_b}{v_{\mathrm{F}}}\int_0^x f\left[p_y b^* - \frac{\omega_b(\theta)u}{v_{\mathrm{F}}}\right]\mathrm{d}u\right)$$
$$\times \exp\left(-\frac{2\mathrm{i}t_c}{v_{\mathrm{F}}}\int_0^x \cos\left\{p_z c^* + \frac{\omega_c(\theta,\varphi)}{v_{\mathrm{F}}}u\right.\right.$$
$$\left.\left. + A\left(f\left[p_y b^* - \frac{\omega_b(\theta)u}{v_{\mathrm{F}}}\right] - f[p_y b^*]\right)\right\}\right), \tag{7.85}$$

where $A = (2t_b b^*/v_{\mathrm{F}})[\tilde{\omega}_c(\theta,\varphi)/\omega_b(\theta)]$.

Here, we demonstrate that wave functions (7.85) exhibit 1D $\rightarrow$ 2D crossovers at the commensurate IC directions (7.37) of a magnetic field, corresponding to LMA directions (7.27) of (**b**,**c**)-component of a magnetic field. We define operator of velocity component along **z**-axis in the presence of a magnetic field using the results of [25–28]

$$\hat{v}_z = \frac{\partial\hat{\varepsilon}^{+}(\mathbf{p})}{\partial p_z} = v_z^0 \sin\left[p_z c^* + \frac{\omega_c(\theta,\varphi)x}{v_{\mathrm{F}}} - \mathrm{i}\frac{\tilde{\omega}_c(\theta,\varphi)}{v_{\mathrm{F}}}\frac{\partial}{\partial p_y}\right]. \tag{7.86}$$

Using wave functions (7.85), we can write expectation value of the velocity in the following form:

$$\langle\hat{v}_z\rangle = v_z^0 \lim_{L\to\infty}\frac{1}{2L}\int_{-L}^{L}\mathrm{d}x \sin\left\{p_z c^* + \frac{\omega_c(\theta,\varphi)x}{v_{\mathrm{F}}}\right.$$
$$\left. + A\left[f\left[p_y b^* - \frac{\omega_b(\theta)x}{v_{\mathrm{F}}}\right] - f[p_y b^*]\right]\right\}. \tag{7.87}$$

It is important that the expectation value of velocity operator (7.87) can be rewritten as a summation of infinite number of waves:

$$\langle\hat{v}_z\rangle = v_z^0 \lim_{N\to\infty}\frac{1}{2N}\sum_{n=-N}^{N}\int_{-\pi v_{\mathrm{F}}/\omega_b(\theta)}^{\pi v_{\mathrm{F}}/\omega_b(\theta)} \sin\left\{p_z c^* - \frac{\omega_c(\theta,\varphi)x}{v_{\mathrm{F}}} - 2\pi n\frac{\omega_c(\theta,\varphi)}{\omega_b(\theta)}\right.$$
$$\left. + A\left(f\left[p_y b^* - \frac{\omega_b(\theta)x}{v_{\mathrm{F}}}\right] - f[p_y b^*]\right)\right\}, \tag{7.88}$$

which reveals its quantum interference nature.

Note that, for "incommensurate" directions of a magnetic field (7.82), $\omega_c(\theta,\varphi) \neq N\omega_b(\theta)$ and, thus, the interference effects in integral (7.88) are destructive. Therefore, in this case, $\langle\hat{v}_z\rangle = 0$. In contrast, "commensurate" directions of the field $\omega_c(\theta,\varphi) = N\omega_b(\theta)$ and, thus,

$$\sin\varphi = N\left(\frac{b^*}{c^*}\right)\tan\theta \tag{7.89}$$

[which corresponds to LMA projections of a magnetic field (7.82) on $(\mathbf{b},\mathbf{c})$-plane]. In this case, the expectation value of velocity is not zero and can be calculated analytically for a typical case, where $f(p_y b^*) = \cos(p_y b^*)$.

As a result, we obtain,

$$\langle v_z \rangle = v_z^0 \sin\Big[p_z c^* - \frac{\omega_c(\theta,\varphi)}{\omega_b(\theta)} p_y b^* - A\cos(p_y b^*)\Big] J_N(A). \tag{7.90}$$

In the same way, we can calculate expectation value of kinetic energy, corresponding to electron motion along **z**-axis and to demonstrate that it is zero for all "incommensurate" directions and nonzero for "commensurate" IC directions (7.37). To summarize, we come to a conclusion that dimensionality of electron spectrum become higher (2D) at commensurate IC directions of a magnetic field, which is a consequence of 1D $\rightarrow$ 2D crossovers.

Here, we show that the above mentioned 1D $\rightarrow$ 2D crossovers directly reveal themselves as localization – de-localization crossovers for electron wave functions. We calculate z-dependence of wave functions by taking Fourier components of (7.85) with respect to variable p_z,

$$\begin{aligned} \Phi^+(x, p_y, z = Nc^*) = &\int_0^{2\pi} \frac{\mathrm{d}p_z}{2\pi} \exp(\mathrm{i}p_z Nc^*) \\ &\times \exp\Big(-\frac{2\mathrm{i}t_\perp}{v_\mathrm{F}} \int_0^x \mathrm{d}u \cos\Big\{p_z c^* + \frac{\omega_c u}{v_\mathrm{F}} \\ &+ A\Big(\cos\Big[p_y b^* - \frac{\omega_b(\theta) u}{v_\mathrm{F}}\Big] - \cos[p_y b^*]\Big)\Big\}\Big). \end{aligned} \tag{7.91}$$

After rather complicated but straightforward integrations, we obtain,

$$\Phi^+(x, p_y, z = Nc^*) = \mathrm{e}^{-\mathrm{i}N\gamma} J_N\Big(\frac{2t_\perp}{v_\mathrm{F}} \sqrt{I_1^2 + I_2^2}\Big), \tag{7.92}$$

where

$$I_{1(2)} = \int_0^x \mathrm{d}u \cos(\sin)\Big[\frac{\omega_c u}{v_\mathrm{F}} + A\Big(f\Big[p_y b^* - \frac{\omega_b(\theta) u}{v_\mathrm{F}}\Big] - f[p_y b^*]\Big)\Big]. \tag{7.93}$$

Note that analysis of wave functions (7.92) can be done in the same manner as in Sect. 7.4.4. As a result, we come to conclusion that wave functions (7.92) become de-localized if the integrals $I_{1,2}$ are divergent at $x \rightarrow \infty$. It is important that integrals (7.93) can be written as summations of electron waves in the extended Brillouin zone

$$\begin{aligned} I_1(x = 2\pi M_0 v_\mathrm{F}/\omega_b) = &\sum_{M=0}^{M_0} \int_0^{2\pi v_\mathrm{F}/\omega_b} du \cos\Big\{\frac{\omega_c u}{v_\mathrm{F}} + 2\pi M \frac{\omega_c(\theta,\varphi)}{\omega_b(\theta)} \\ &+ A\Big(f\Big[p_y b^* - \frac{\omega_b(\theta) u}{v_\mathrm{F}}\Big] - f[p_y b^*]\Big)\Big\}, \end{aligned} \tag{7.94}$$

which reveal interference nature of the above described 1D $\rightarrow$ 2D crossovers.

As we discussed in the earlier sections, the above-described interference effects occur when electron move along open FS in a magnetic field in the extended Brillouin zone (see Fig. 7.11). At arbitrary directions of a magnetic field (7.82), the interference effects in (7.94) are destructive which lead to $1D$ electron spectrum. At "commensurate" directions (7.37) of the field, the interference effects in (7.94) become constructive, which increases electron spectrum and wave functions dimensionality up to $2D$.

7.4.6 Q2D Case: 3D → 2D Crossover and a Momentum Quantization Law

In this section, we apply quantum mechanical formalism to describe 3D → 2D crossovers in an important for applications case – a Q2D conductor with electron spectrum [12]

$$\varepsilon(\mathbf{p}) = \frac{1}{2m}(p_x^2 + p_y^2) - 2t_\perp \cos(p_z c^*), \quad t_\perp \ll \varepsilon_F, \tag{7.95}$$

in a parallel magnetic field, $\mathbf{H} = (0, H, 0)$ (see Fig. 7.5). As we show later, in this case, it is also possible to introduce momentum quantization law related to "two-dimensionalization" of the wave functions in a magnetic field.

Here, we choose a convenient gauge, $\mathbf{A} = (0, 0, -Hx)$, where electron wave functions have a simple form:

$$\Psi_\varepsilon(x, y, z) = \exp(\mathrm{i}p_y y)\exp(\mathrm{i}p_z z)\psi_\varepsilon(x, p_y, p_z). \tag{7.96}$$

The corresponding Schrödinger equation for wave function, $\psi_\varepsilon(x, p_y, p_z)$, can be obtained by means of the Peierls substitution, $p_z \to p_z - (e/c)Hx$,

$$\left[\frac{1}{2m}\left(-\frac{\mathrm{d}^2}{\mathrm{d}x^2} + p_y^2\right) - 2t_\perp \cos\left(p_z c^* - \frac{\omega_c x}{v_F}\right)\right]\psi_\varepsilon(x, p_y, p_z) = \varepsilon\, \psi(x, p_y, p_z), \tag{7.97}$$

where $\omega_c = ev_F c^* H/c$, v_F is the Fermi velocity, $\varepsilon_F = mv_F^2/2$.

Taking account of $t_\perp \ll \varepsilon_F$, we can ignore the existence of small closed orbits and can represent the solutions of (7.97) in the following form:

$$\psi_\varepsilon^\pm(x, p_y, p_z) = \frac{\exp\{\pm \mathrm{i}p_x^0 x \pm \mathrm{i}m\delta\varepsilon x/p_x^0 \mp \mathrm{i}(\frac{\lambda p_F}{2p_x^0})[\sin(p_z c^* - \frac{\omega_c x}{v_F}) - \sin(p_z c^*)]\}}{\sqrt{p_x^0}}, \tag{7.98}$$

where $\delta\varepsilon = \varepsilon - \varepsilon_F$, $p_F = mv_F$, $p_x^0 = \sqrt{p_F^2 - p_y^2}$, $\lambda = 4t_\perp/\omega_c$; sign $+(-)$ stands for $p_x > 0$ $(p_x < 0)$.

Wave functions (7.98) directly demonstrate 3D → 2D crossover in a parallel magnetic filed since electron energy does not depend on a momentum component p_z. Note that wave functions (7.98) are Bloch-like ones and correspond to continuous energy spectrum,

$$\delta\varepsilon^\pm(\mathbf{p}) = \pm p_x^0(p_x \mp p_x^0)/m + N\omega_c(p_x^0/p_F), \tag{7.99}$$

where p_x is a quasi-momentum limited to the magnetic Brillouin zone, $|p_x| < \omega_c/2v_{\mathrm{F}}$, N is an integer.

It is important that 2D electron spectrum (7.99) can be rewritten in a familiar momentum quantization law form:

$$p_x \mp p_x^0 = \delta\varepsilon^{\pm}/v_{\mathrm{F}} + N[\omega_c(p_x^0/p_{\mathrm{F}})/v_{\mathrm{F}}]. \qquad (7.100)$$

As shown in Sect. 7.8, the above-described 3D $\rightarrow$ 2D crossover is responsible for the appearance of RS phase in Q2D conductors in a parallel magnetic field. Some experimental evidences in a favor of the existence of RS phase in layered Q1D and Q2D superconductors are found in [79–85].

7.5 Q2D Conductor: A Fully Quantum Mechanical Problem

Note that, in the earlier sections, wave functions and electron spectra were found by means of the Peierls substitution method, which is justified for QC dimensional crossovers, where "sizes" of electron trajectories in a magnetic field are bigger than the interplane distance, $l_\perp(p_x, p_y, H) \geq c^*$. In this section, we present a solution of a fully quantum mechanical problem for energy levels and wave functions of electrons with Q2D spectrum (7.95) in a parallel magnetic field, which is also valid in the case of QL dimensional crossovers with $l_\perp(p_x, p_y, H) \leq c^*$. The above-mentioned problem is solved in [14].

To determine electron wave functions in Q2D conductor (7.95) in parallel magnetic field, we make use of QC description of electron motion within conducting $(\mathbf{x}, \mathbf{y})$-planes and solve fully quantum mechanical problem for electron motion between the planes. After Peierls substitutions for in-plane momenta, $p_x \rightarrow p_x - (e/c)A_x$, $p_y \rightarrow -\mathrm{i}(\mathrm{d}/\mathrm{d}y)$, $p_z \rightarrow -\mathrm{i}(\mathrm{d}/\mathrm{d}z)$ [14], we can represent electron Hamiltonian in the form:

$$\hat{H} = \varepsilon_{\|}\left(-\mathrm{i}\frac{\mathrm{d}}{\mathrm{d}x} + \frac{eHz}{c}, -\mathrm{i}\frac{d}{dy}\right) + \frac{1}{2m}\left(-\mathrm{i}\frac{\mathrm{d}}{\mathrm{d}z}\right)^2 - \frac{V}{m}\sum_{n=-\infty}^{\infty}\delta(z - c^* n), \qquad (7.101)$$

where the last term introduces potential energy of crystalline lattice along $\mathbf{z}$-axis, $V > 0$, $\delta(\ldots)$ is Dirac delta-function.

Note that Hamiltonian (7.101) is exact one for an isotropic Q2D case. As it follows from general theory [1], the suggested method disregards only corrections of the order of $\omega_c^2(p_x, p_y, H)/\varepsilon_{\mathrm{F}}$ to electron energy for arbitrary function $\varepsilon_{\|}(p_x, p_y)$. Note that QL conditions correspond to $t_\perp \sim \omega_c(p_x, p_y, H)$, and, thus, the above-mentioned corrections are of the order of $t_\perp^2/\varepsilon_F \ll \omega_c(p_x, p_y, H)$ under QL conditions, where $\omega_c(p_x, p_y, H)$ is a characteristic energy scale in a magnetic field. Therefore, Hamiltonian (7.101) allows to study both QC and QL dimensional crossovers.

Arbitrary solution of Schrödinger equation for Hamiltonian (7.101) can be written as

$$\Psi_\varepsilon(x,y,z) = \exp(\mathrm{i}p_x x)\exp(\mathrm{i}p_y y)\;\Psi_\varepsilon(p_x,p_y;z) \tag{7.102}$$

which corresponds to free electron motion within $(\mathbf{x},\mathbf{y})$-planes. After substitution of (7.102) into Hamiltonian (7.101), it can be rewritten as follows:

$$\hat{H} = \varepsilon_\parallel\left(p_x + \frac{eHz}{c}, p_y\right) - \left(\frac{1}{2m}\right)\frac{\mathrm{d}^2}{\mathrm{d}z^2} - \frac{V}{m}\sum_{n=-\infty}^{\infty}\delta(z - c^* n). \tag{7.103}$$

By expanding in-plane energy in powers of H, it is easy to make sure that Schrödinger equation for Hamiltonian (7.103) with the same accuracy can be expressed as:

$$\left[-\left(\frac{1}{2m}\right)\frac{d^2}{dz^2} + \omega_c(p_x,p_y,H)\left(\frac{z}{c^*}\right) - \frac{V}{m}\sum_{n=-\infty}^{\infty}\delta(z-c^*n)\right]\Psi_\varepsilon(p_x,p_y;z)$$
$$= [\varepsilon - \varepsilon_\parallel(p_x,p_y)]\Psi_\varepsilon(p_x,p_y;z), \quad \omega_c(p_x,p_y,H) = ev_x(p_x,p_y)c^*H/c. \tag{7.104}$$

It is possible to prove [14] that, if we use tight binding approximation for solutions of (7.104)

$$\Psi_{\varepsilon_N}(p_x,p_y;z) = \sum_{m=-\infty}^{\infty} A_{m-N}(p_x,p_y)\;\Phi_{\varepsilon_0}(z - c^* m) \tag{7.105}$$

[where $\Phi_{\varepsilon_0}(z - c^*m)$ is wave function of individual mth layer at $H = 0$, corresponding to energy $\varepsilon_0 < 0$, $|\varepsilon_0| \sim \varepsilon_\mathrm{F}$], then we disregard only corrections of the order of $\omega_c^2(p_x,p_y,H)/[\varepsilon_\parallel(p_x,p_y),\varepsilon_0] \sim t_\perp^2/\varepsilon_\mathrm{F}$ to electron energy.

Therefore, equation,

$$[\varepsilon - \varepsilon_0 - \varepsilon_\parallel(p_x,p_y) - m\omega_c(p_x,p_y,H)]A_m(p_x,p_y)$$
$$= -\,A_{m+1}(p_x,p_y)t_\perp - A_{m-1}(p_x,p_y)t_\perp \tag{7.106}$$

[where $t_\perp = (V^2/m)\exp(-Vc^*)$], which can be derived after substitution of wave-functions (7.105) into Hamiltonian (7.104) by means of tight binding approximation for the nearest neighbors, has the same accuracy as Hamiltonian (7.101). Therefore it can be used to describe 3D $\rightarrow$ 2D QL dimensional crossovers.

At given in-plane momenta, p_x and p_y, (7.106) is mathematically equivalent to the so-called Stark–Wannier ladder equation in electric field. Using [14], we can express wave functions and energy levels in the following way:

$$\Psi_{\varepsilon_N}(p_x,p_y;z) = \sum_{m=-\infty}^{\infty} J_{m-N}[2t_\perp/\omega_c(p_x,p_y,H)]\;\Phi_{\varepsilon_0}(z - c^*m),$$
$$\varepsilon_N(p_x,p_y) = \varepsilon_0 + \varepsilon_\parallel(p_x,p_y) + N\omega_c(p_x,p_y,H), \tag{7.107}$$

where $J_N[\ldots]$ is Bessel function of Nth order.

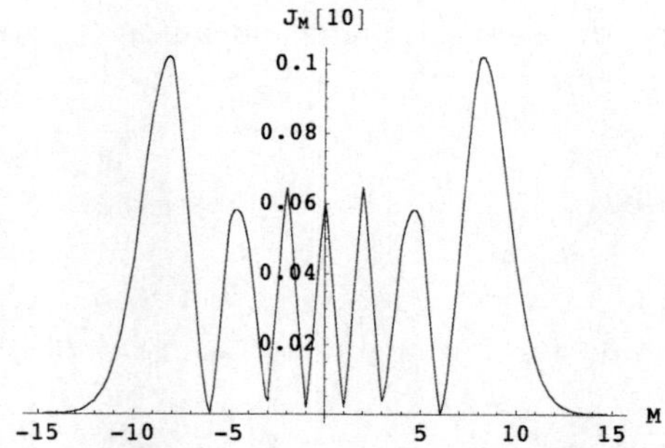

Fig. 7.17. Wave function (7.107) in QC case, where it occupies about 20 conducting layers

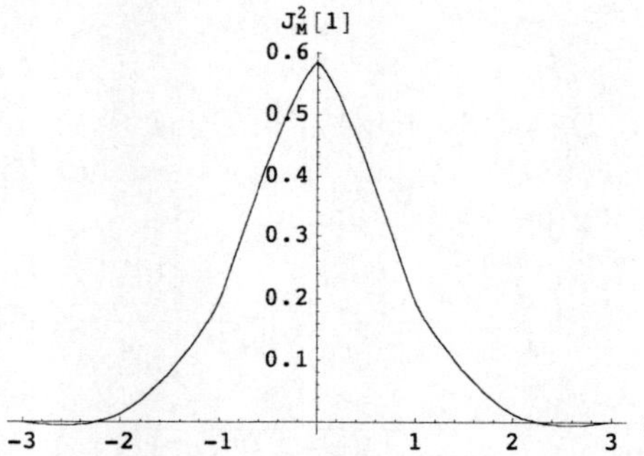

Fig. 7.18. Wave function (7.107) in QL case, where it is localized on one layer

Equation (7.107) represents the main result of this section. In contrast to textbook extended Bloch waves with complex envelope, $\exp(ikz)$ [1], the envelope functions in (7.107) are real functions localized on the Nth conducting layer (see Figs. 7.17 and 7.18). Therefore, we conclude that, in a parallel magnetic field, all wave functions are localized on layers with energy gap between two neighboring wave functions being $\omega_c(p_x, p_y, H)$. Equation (7.107) is valid both in QC and QL cases. As seen from Fig. 7.18, in QL case, wave function (7.107) is essentially localized on a single conducting layer with probability to jump on the neighboring layers being small quantity. In contrast, in QC case, the localization length of wave function (7.107) is approximately equal to corresponding size of QC electron orbit (see Fig. 7.17).

Later, we show that quantization law (7.107) leads to unusual ac infrared properties and a method to investigate Q2D FS is suggested. For these purposes, we calculate ac conductivity component, perpendicular to conducting layers, $\sigma_\perp(H, \omega)$, using known wave functions and energy spectrum (7.107). Let us find matrix elements of momentum operator, $\hat{p_z} = -\mathrm{i}(\mathrm{d}/\mathrm{d}z)$, responsible for interactions between electrons and electric field, $\mathbf{E} \parallel \mathbf{z}$. It is possible to make sure that the matrix elements are nonzero only for wave functions with the same in-plane momenta, p_x and p_y, and energies $\varepsilon_1 - \varepsilon_2 = \pm\omega_c(p_x, p_y, H)$,

$$p_z^{N,N+1} = \int \Psi_N^*(z)\left(-\mathrm{i}\frac{\mathrm{d}}{\mathrm{d}z}\right)\Psi_{N+1}(z)\mathrm{d}z = -p_z^{N+1,N} = -imc^* t_\perp . \quad (7.108)$$

[In other words, only optical transitions between electrons, localized on neighboring conducting layers and having the same in-plane momenta, are allowed].

To calculate $\sigma_\perp(H,\omega)$, we make use of the following extension of Kubo formalism:

$$\sigma_\perp(H,\omega) = -\mathrm{i}\frac{2e^2}{m^2 V}\sum_{N_1,N_2}\frac{|p_z^{N_1,N_2}|^2}{(E_{N_1}-E_{N_2})}\frac{[n(E_{N_2})-n(E_{N_1})]}{(E_{N_2}-E_{N_1}-\omega-i\nu)}, \quad \nu \to 0, \quad (7.109)$$

where $n(E)$ is Fermi–Dirac distribution function, V is a volume. After substituting matrix elements (7.108) and energy spectrum (7.107) in (7.109) and straightforward calculations, we obtain:

$$\sigma_\perp(H,\omega) \sim \mathrm{i}\int \frac{\mathrm{d}p}{|\mathbf{v}_\mathrm{F}(p_x,p_y)|}\left[\frac{1}{\omega_c(p_x,p_y,H)-\omega-\mathrm{i}\nu} + \frac{1}{-\omega_c(p_x,p_y,H)-\omega-\mathrm{i}\nu}\right], \quad \nu\to 0. \quad (7.110)$$

[Note that in exact (7.110) we omit only some factor, which is not significant for further consideration. Integration in (7.110) is made along 2D contour $\varepsilon_\parallel(p_x,p_y) = \varepsilon_\mathrm{F}$; $\mathbf{v}_\mathrm{F}(p_x,p_y) = \mathrm{d}\varepsilon_\parallel(p_x,p_y)/\mathrm{d}\mathbf{p}$; we use the approximation $n(E_{N_2}) - n(E_{N_1}) = (E_{N_2} - E_{N_1})\,[\mathrm{d}n(E)/\mathrm{d}E]$ since $|E_{N_2} - E_{N_1}| = \omega_c(p_x,p_y,H) \ll \varepsilon_\mathrm{F}$].

It is convenient to write explicitly the real and imaginary parts of conductivity (7.110):

$$\Re\,\sigma_\perp(H,\omega) \sim \int \frac{\mathrm{d}p}{|\mathbf{v}_\mathrm{F}(p_x,p_y)|}\,\delta[\omega_c(p_x,p_y,H)-\omega] = \begin{cases} \neq 0, \omega < \omega_c^{\max}(H) \\ 0, \omega > \omega_c^{\max}(H) \end{cases}, \quad (7.111)$$

$$\Im\sigma_\perp(H,\omega) \sim \int \frac{dp}{|\mathbf{v}_\mathrm{F}(p_x,p_y)|}\left[\frac{1}{\omega_c(p_x,p_y,H)-\omega} - \frac{1}{\omega_c(p_x,p_y,H)+\omega}\right], \quad (7.112)$$

where $\omega_c^{\max}(H)$ is the maximum value of energy gap, $\omega_c(p_x,p_y,H)$, corresponding to maximum value of velocity, $v_x^{\max} = max|v_x(p_x,p_y)|$, on the contour of integration [see Fig. 7.19 and (7.104)]; integral in (7.112) is determined as its principle value.

The main difference between (7.111) and (7.112) and the results of [12,13] is that (7.111) and (7.112) are valid both in QC and QL cases, whereas

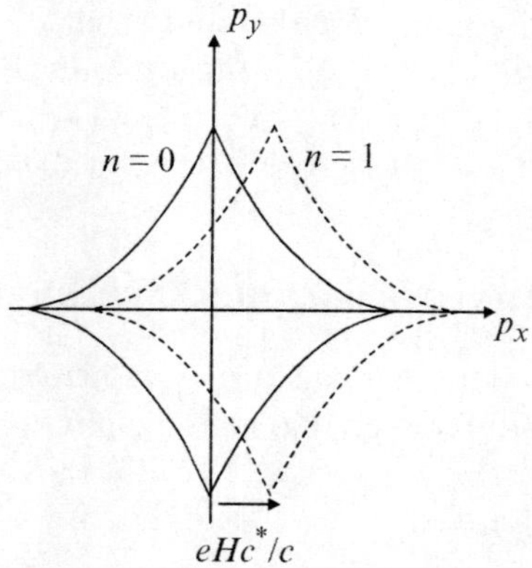

Fig. 7.19. Two FS, corresponding $N = 0$ and $N = 1$ in (7.107)

the results [12, 13] are essentially QC. Another difference is that (7.111) and (7.112) describe optical conductivity (i.e., conductivity in the absence of impurities), in contrast to kinetic equation results [13]. From (7.111) and (7.112), it follows that ac properties in a parallel magnetic field are unusual. Indeed, integration of δ-function in (7.111) results in nonzero value of real part of conductivity for ac frequencies at $0 < \omega < \omega_c^{\max}(H)$ (see Fig. 7.19). Therefore, electrons absorb electromagnetic waves at $0 < \omega < \omega_c^{\max}(H)$ (in the absence of impurities), in contrast to text book properties of metals.

Let us demonstrate that real part of conductivity (7.111) diverges at resonant frequency:

$$\omega = \omega_c^{\max}(H) = ev^{\max} Hc^*/c. \tag{7.113}$$

Indeed, in the vicinity of its maximum, $\omega_c(p_x, p_y, H) \simeq \omega_c^{\max}(H) - A(H)|\mathbf{p}|^2$, with $\mathbf{p}$ being momentum component perpendicular to $\mathbf{v}_F(p_x, p_y)$ at point, where $|v_x(p_x, p_y)|$ takes its maximum. In this case, integral (7.111) can be estimated as

$$\Re\sigma_\perp(H, \omega) \sim \frac{1}{\sqrt{\omega_c^{\max}(H) - \omega}}, \quad \omega_c^{\max}(H) - \omega \ll \omega_c^{\max}(H). \tag{7.114}$$

Therefore, by measuring $\omega_c^{\max}(H)$ at different directions of the field it is possible to determine angular dependence of $v_x^{\max}$ [68]. We stress, however, that the physical meaning of resonant frequency (7.113) at high magnetic fields, where electrons are almost completely localized on conducting layers (see Fig. 7.18), is completely different from its kinetic equation interpretation of Kovalev et al. [68] method.

To summarize, in this section, wave functions and electron spectrum of a Q2D conductor in a parallel magnetic field are determined. A method to test FL picture in Q2D organic and high-T$_c$ materials is suggested. We hope that this method is a useful experimental tool to study FL versus n-FL behavior in low-dimensional compounds. We also think that 3D $\to$ 2D QL dimensional crossover and quantization law (7.107), suggested in this section, will be useful for studies of RS phase [10, 12] and for explanations of unusual phenomena, observed in high parallel magnetic fields (see, for example, [89–91]).

7.6 Angular Magnetoresistance Oscillations

The goal of this section is a quantitative description of angular magnetic oscillations experimentally observed in metallic phases of low-dimensional conductors with open sheets of FS such as $(TMTSF)_2X$, $(DMET\text{-}TSeF)_2X$, and $\kappa\text{-}(ET)_2Cu(NCS)_2$ materials.

7.6.1 Lebed Magic Angles in a Metallic Phase

In this section, we quantitatively describe LMA angular magnetic oscillations, experimentally observed in $(TMTSF)_2X$ ($X = PF_6$ and ClO_4), $(DMET\text{-}TSeF)_2X$, and $\kappa\text{-}(ET)_2Cu(NCS)_2$ materials. For this purpose, we calculate conductivity of a layered Q1D metal (7.14) in an inclined magnetic field (7.13) (see Fig. 7.6), using known electron spectrum and wave functions (7.70) (see [27]).

To calculate the conductivity, $\sigma_{zz}(H,\alpha)$, within FL approach for noninteracting quasi-particles, let us introduce a QC operator of velocity component v_z in a magnetic field [27]

$$\hat{v}_z(p_z,x) = -v_z \sin[p_z c^* + \omega_c(\alpha)x/v_F], \quad v_z = 2t_c c^* . \tag{7.115}$$

Since wave functions (7.70) and the velocity operator (7.115) are known, we can calculate $\sigma_{zz}(H,\alpha)$ by means of Kubo formalism. As a result, we obtain,

$$\sigma_\perp(H,\alpha) \sim \left\langle \frac{1}{v_x(p_y)} \int_{-\infty}^{0} d(b^*u)\, \frac{\cos[n(\alpha)b^*u]}{\omega_b(p_y+u,\alpha)} \exp\left[-\int_u^0 \frac{d(b^*u_1)}{\tau\omega_b(p_y+u_1,\alpha)} \right] \right\rangle_{p_y}, \tag{7.116}$$

where

$$\omega_b(p_y,\alpha) = \omega_b(\alpha)\ [v_x(p_y)/v_F]\,, \quad \omega_c(p_y,\alpha) = \omega_c(\alpha)\ [v_x(p_y)/v_F],$$
$$n(\alpha) = \omega_c(\alpha)/\omega_b(\alpha) \tag{7.117}$$

and $< \cdots >_{p_y}$ stands for averaging over p_y.

After straightforward but rather complicated integrations, (7.116) can be rewritten as,

$$\frac{\sigma_\perp(H,\alpha)}{\sigma_\perp(0)} = \frac{1}{1+h_c^2(H)} - h_c^2(H) \int_{-\infty}^{0} du\ \exp(u)\cos[h_c(H)u] \times \left\langle \exp\left[\int_0^u f[y+u_1 h_b(H)]du_1 \right] - 1 \right\rangle_y ,$$
$$f(y) = v_F/v_x(y) - 1, \qquad h_b(H) = \omega_b(\alpha)\tau, \quad h_c(H) = \omega_c(\alpha)\tau. \tag{7.118}$$

Since, in Q1D case, $\rho_{zz}(H,\alpha) \simeq 1/\sigma_{zz}(H,\alpha)$, (7.118) solves the problem of defining $\rho_{zz}(H,\alpha)$ for electrons with open orbits spectrum (7.14) in an inclined magnetic field (7.13) [27].

To make our results more intuitive, we consider an important limiting case of (7.118) – a so-called clean limit, where $\omega_c(\alpha)\,\tau \gg 1$. In this case, (7.118) can be significantly simplified:

$$\frac{\sigma_\perp(H,\alpha)}{\sigma_\perp(0)} = \frac{1}{1+[\omega_c(\alpha)\tau]^2} - \tan^2\alpha\left(\frac{c^*}{2b^*}\right)^2 \sum_{n=1}^{\infty} \frac{A_n^2}{n^2}\left(\frac{2}{1+[\omega_c(\alpha)\tau]^2}\right.$$
$$\left. - \frac{1}{1+[\omega_c(\alpha)-n\omega_b(\alpha)]^2\tau^2} - \frac{1}{1+[\omega_c(\alpha)+n\omega_b(\alpha)]^2\tau^2}\right), \qquad (7.119)$$

where A_n are the Fourier coefficients of function $f(y) = v_\mathrm{F}/v_x(y) - 1$

$$A_N = \frac{1}{\pi}\int_{-\pi}^{+\pi} f(y)\cos(Ny)\,\mathrm{d}y. \qquad (7.120)$$

Equation (7.118) [and its clean limit (7.119)] is the main result of this section, and is distinct from all other models of conduction in such anisotropic systems. Equation (7.119) directly demonstrates that $\sigma_{zz}(H,\alpha)$ maxima [i.e., $\rho_{zz}(H,\alpha)$ minima] are related to minima in the denominators. They occur at LMA directions, defined by $\omega_c(\alpha) = n\omega_b(\alpha)$ [see (7.27)]. In Fig. 7.20, we present numerical simulations of (7.118) and (7.119) for three qualitatively different variants of Q1D spectrum (7.14), corresponding to $(TMTSF)_2PF_6$, κ-$(ET)_2Cu(NCS)_2$, and $(TMTSF)_2ClO_4$ conductors. As seen, $(TMTSF)_2PF_6$ exhibits only one LMA minimum, while the others exhibit several minima with large indices N in (7.27). We stress that all these qualitative features, as well as a doubling of period of LMA minima in the case of $(TMTSF)_2ClO_4$, are in good agreement with existing experimental data [41–57] and can be related to peculiarities of Q1D electron spectra in the above compounds [27].

We point out that the existing alternative model to describe LMA and AMRO effects in $\rho_\perp(H,\alpha)$, due to Osada et al. [51], while important from

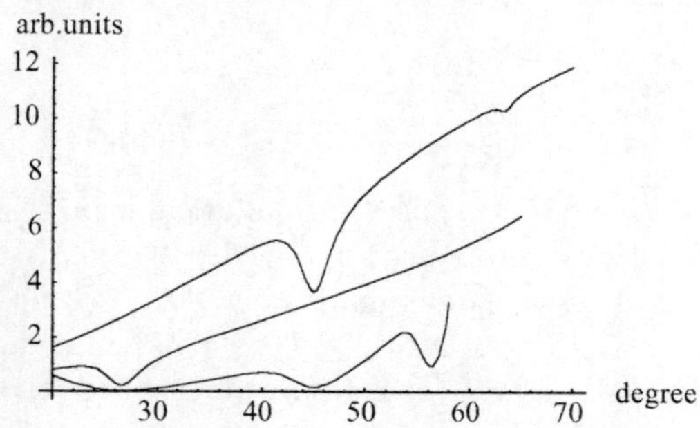

Fig. 7.20. LMA effects in a resistivity, calculated by means of formulas (7.118) and (7.119) for three different Q1D conductors. *Upper curve*: $(TMTSF)_2ClO_4$, middle curve: $(TMTSF)_2PF_6$, *lower curve*: $(ET)_2Cu(NCS)_2$

methodological and historical points of view, lacks direct physical meaning. The reason is that the transfer integrals, $t_{n,m}$, in that model are exponentially small in the framework of a realistic tight-binding model [3] of the Q1D electron spectrum. The 1D $\rightarrow$ 2D crossovers and interference effects, suggested in this section, have a real physical origin, related to the p_y-dependence of the density of states, $1/v_x(p_y)$, in (7.116) and (7.118).

From a classical point of view, in our case, the interference effects occur because of the fact that electrons spend a disproportionate amount of time on those parts of FS (7.14), where the density of states is larger. Such effects are not possible within the Osada model [51] and all its variants, which use a linearized electron spectrum with a *constant* density of states. For example, the weighting factors in (7.119) depend on the magnetic field orientation (i.e., on $\tan\alpha$) and, thus, their physical meaning is completely different from the angle-independent effective transfer integrals, postulated in [51].

We suggest that the 1D$\rightarrow$2D crossovers, introduced in this section, can form a basis for further FL and n-FL theories of more complex LMA phenomena.

7.6.2 Interference Commensurate Oscillations

In this section, we suggest a quantitative theory of IC oscillations, experimentally observed in a metallic phase of different Q1D conductors [see Sects. 7.3.6 and 7.4.5]. Below, we make use of the results of Sect. 7.4.5, where electron wave functions and spectrum are calculated and where IC oscillations are interpreted in terms of 1D $\rightarrow$ 2D crossovers.

Let us calculate one-particle conductivity, $\sigma_{zz}(H, \theta, \phi)$, perpendicular to conducting layers. For this purpose, we apply Kubo formalism and use the expressions for matrix elements and electron spectrum from Sect. 7.4.5. As a result we obtain (see [25]):

$$\sigma_{zz}(H,\theta,\phi) = \sigma_{zz}(0) \sum_{N=-\infty}^{+\infty} \left(\frac{J_N^2[\omega_c^*(\theta,\phi)/\omega_b(\theta)]}{1+\tau^2[\omega_c(\theta,\phi) - N\omega_b(\theta)]^2} \right), \tag{7.121}$$

where for a Q1D conductor (7.81)

$$\rho_{zz}(H,\theta,\phi) \simeq 1/\sigma_{zz}(H,\theta,\phi). \tag{7.122}$$

Note that (7.121) and (7.122) provide analytical expressions for experimentally measured resistivity component $\rho_{zz}(H,\theta,\phi)$ [25,59,61–63]. As seen from (7.122), $\sigma_{zz}(H,\theta,\phi)$ possesses maxima [i.e., $\rho_{zz}(H,\theta,\phi)$ possesses minima] at $\omega_c(\theta,\phi) = N\omega_b(\theta)$ if $\omega_c(\theta,\phi), \omega_b(\theta) \geq 1/\tau$. This coincides with the "commensurability" condition (7.89). It is important to note that this theory [25], based on (7.121) and (7.122), predicts no angular oscillations at LMA directions of a magnetic field (7.27) since $\omega_c^* = 0$ in (7.121). Therefore, the earlier interpretations [60–64] of the IC oscillations as a simple combination of LMA effects and DKC oscillations is too oversimplified.

We also stress that integration of the quasi-periodic function in (7.87) corresponds to interference effects in matrix elements of velocity operator (7.86) and, thus, directly demonstrates the interference nature of IC oscillations. Indeed, two different periods in (7.87) become "commensurate" at $\omega_c(\theta,\phi) = N\omega_b(\theta,\phi)$ [i.e., for "commensurate" trajectories (7.37)] and, therefore, the integral (7.87) is increased due to these interference effects.

Let us consider (7.121) at small enough angles, $\phi \ll \pi/2$ and $\theta \ll 2t_b c^*/v_F$. In this case, one can make use of an asymptotic expression for the Bessel functions, $J_N(2t_b c^*/v_F \tan\theta) \simeq \cos(2t_b c^*/v_F \tan\theta - \pi/4 - \pi N/2)$. Therefore, depending on the value of parameter $2t_b c^*/v_F \tan\theta$, the Bessel functions of even orders are bigger than those of odd orders or vise versa. At $\omega_c\tau, \omega_b\tau \simeq 1$, this results in the appearance of minima of $\rho_{zz}(\phi,\theta)$ in (7.121) and (7.122) only for all even or only for all odd values of the integer N. We call this phenomenon even–odd angular resonance (see Fig. 7.21). It is important that (7.121) demonstrates also another kind of angular oscillations (i.e., the so-called extended DKC oscillations) related to zeros of the Bessel functions. By analyzing (7.121) and (7.122), it is possible to show that $\rho_{zz}(H,\theta,\phi)$ is characterized by an unusual linear behavior for "non-commensurate" directions of a magnetic field and small $\theta \ll 2t_b b^*/v_F$,

$$\rho_{zz}(H,\theta,\phi) \sim |H|, \tag{7.123}$$

whereas, for "commensurate" directions (7.89), $\rho_{zz}(H,\theta,\phi)$ saturates with increasing magnetic field. The latter effect is different from the linear magnetoresistance predicted in [13].

In Figs. 7.21 and 7.22, we compare the experimental data [61,62], obtained on $(TMTSF)_2PF_6$ compound, with (7.121) and (7.122) using the same values of parameters, $t_a/t_b = 8.5$ and $\omega_c(\theta = 0, \phi = \pi/2, H = 1\ T)\ \tau = 15$, for two theoretical curves. These curves not only demonstrate qualitative but quantitative agreement between theory [25] and experiment in a broad region of magnetic field orientations, $\phi \leq 20°$. Note that, for $\theta = 3°$, $\rho_{zz}(H,\theta,\phi)$

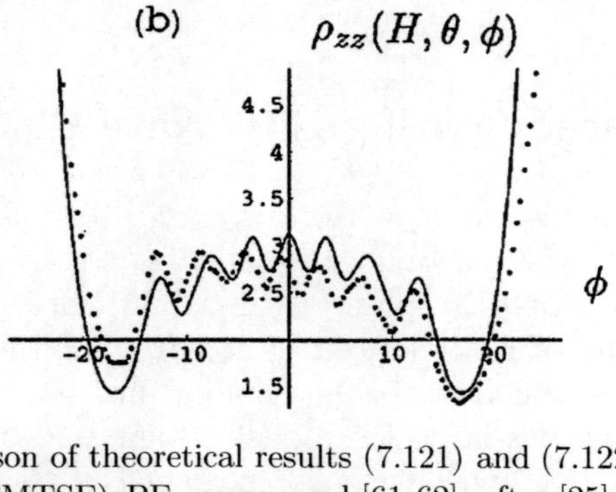

Fig. 7.21. Comparison of theoretical results (7.121) and (7.122) with experimental data, obtained on $(TMTSF)_2PF_6$ compound [61,62], after [25]. [Note that only odd angular resonances (7.89) appear]

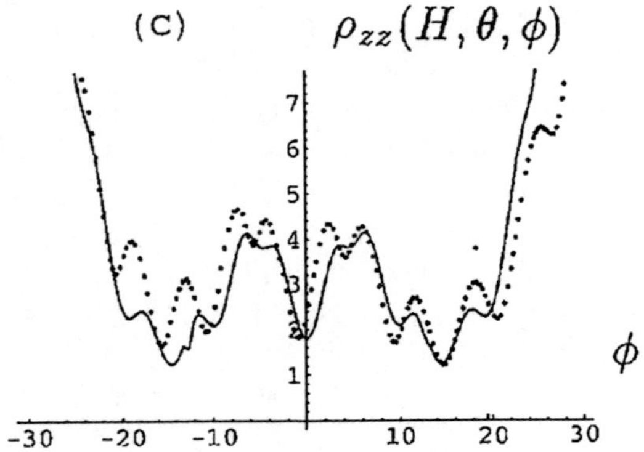

Fig. 7.22. Comparison of theoretical results (7.121) and (7.122) with experimental data [61,62], obtained on $(TMTSF)_2PF_6$ compound, after [25]. [Note that both even and odd angular resonances (7.29) appear]

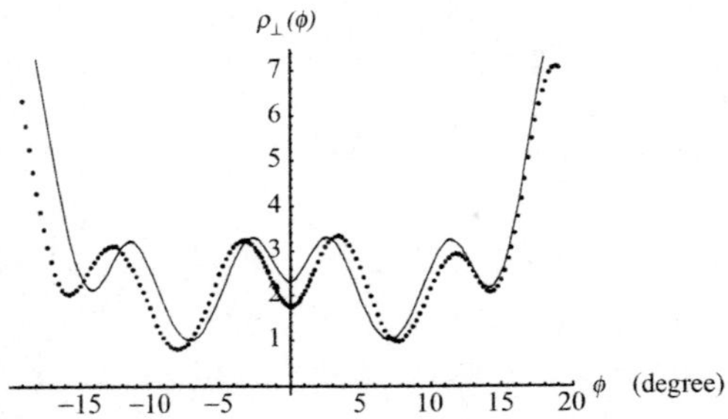

Fig. 7.23. Comparison of theoretical results [28] with experimental data [29], obtained on $(TMTSF)_2ClO_4$ compound, after [28]

minima appear both theoretically and experimentally only for odd integers N in (1.89) (see Fig. 7.21) which is an agreement with the odd angular resonance effect discussed above. In Fig. 7.23, we compare experimental data [29], obtained on $(TMTSF)_2ClO_4$ compound, with the corresponding theory [28].

7.7 Field-Induced Spin-Density-Wave Phases

In this section, we discuss how the discovery of a $2D \rightarrow 1D$ crossover in a magnetic field by Gorkov and Lebed [15] and by Chaikin [11] allows to explain an experimentally observed instability of a metallic phase with respect to formation of field-induced spin(charge)-density-wave [FIS(C)DW] phases. We consider in details the most important theoretical concept in this area – the quantized nesting (QN) model – suggested by Heritier, Montambaux, and Lederer [93] and developed by Lebed [92] and by Virosztek and Maki [96].

For the first time, we present an expression for a free energy of the FISDW subphases, which is valid not only in the vicinity of the phase transition line but also at arbitrary low temperatures. We also discuss some important features of FISDW phase diagrams, which are beyond QN model.

7.7.1 Peierls Spin(Charge)-Density-Wave Instability

It is well known that a pure 1D electron gas is unstable with respect to the spin(charge)-density-wave [S(C)DW] formation since its electron spectrum

$$\varepsilon^{\pm}(\mathbf{p}) = \pm v_{\mathrm{F}}(p_x \mp p_{\mathrm{F}}), \tag{7.124}$$

possesses the so-called nesting property [3],

$$\varepsilon(p) + \varepsilon(p + 2p_{\mathrm{F}}) = 0. \tag{7.125}$$

[Here $+(-)$ stands for right (left) sheet of 1D FS, p_{F} and v_{F} are the Fermi momentum and Fermi velocity, correspondingly].

Using qualitative language, we can say that, in 1D case, electrons and holes move along the conducting chains with the opposite velocities (see Fig. 7.24). The corresponding 1D quantum mechanical problem is always characterized by bound states for an attractive potential in the Schrödinger equation. Therefore, in the case of attractive electron–hole interactions, electrons and holes create the Peierls electron–hole pairs, which correspond to a semiconducting S(C)DW phase with the order parameter,

$$\Delta_{\mathrm{S(C)DW}}(x) = \Delta_{\mathrm{S(C)DW}} \cos(2p_{\mathrm{F}}x). \tag{7.126}$$

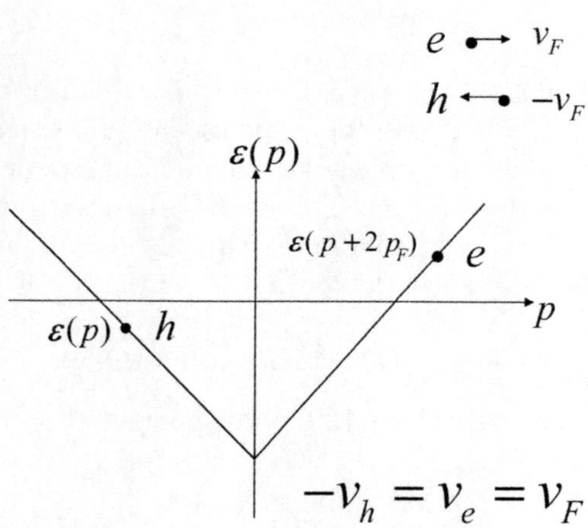

Fig. 7.24. For pure 1D electron spectrum, electrons and holes move with the opposite velocities, $v_{\mathrm{e}} = -v_{\mathrm{h}} = v_{\mathrm{F}}$, along the conducting chains. This results in the Peierls instability of a metallic phase (*see the text*)

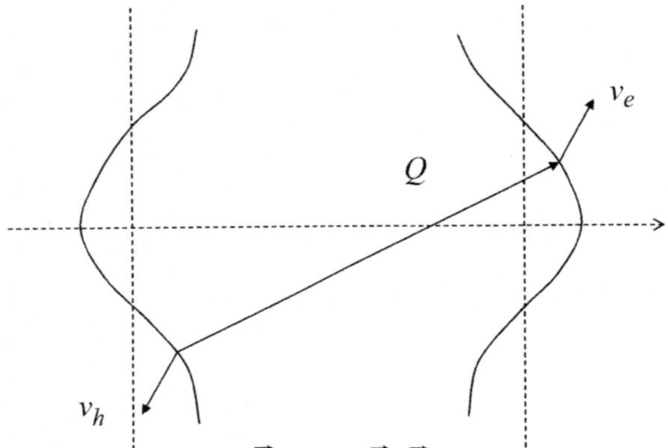

Fig. 7.25. For the simplest variant of the tight-binding spectrum (7.127), electron and hole, which organize the Peierls electron–hole pair, move with the opposite velocities, $\mathbf{v}_h(\mathbf{p}) = -\mathbf{v}_e(\mathbf{p}+\mathbf{Q})$

The finite degrees of "quasi-one-dimensionality" in real electron spectra can destroy the above-mentioned physical picture and can make a metallic phase to be stable with respect to the S(C)DW phase formation. Nevertheless, it is known that a Q1D spectrum, corresponding to the simplest variant of tight-binding model [3, 15]

$$\varepsilon^{\pm}(\mathbf{p}) = \pm v_F(p_x \mp p_F) + 2t_b \cos(p_y b^*) + 2t_c \cos(p_z c^*), \tag{7.127}$$

preserves the nesting condition

$$\varepsilon(\mathbf{p}) + \varepsilon(\mathbf{p}+\mathbf{Q_0}) = 0. \tag{7.128}$$

As seen from Fig. 7.25, in this case, electron and hole move with the opposite velocities, $\mathbf{v}_e(\mathbf{p}) = -\mathbf{v}_h(\mathbf{p}+\mathbf{Q_0})$, along some direction. It is important that this property is valid for any electron momentum, $\mathbf{p}$, defining the electron position on the Q1D FS (7.127). This makes the electron–hole interactions to be effectively 1D, and, therefore, in the case of attractive interactions, a ground state of the Q1D conductor (1.127) is the S(C)DW phase with the order parameter

$$\Delta_{S(C)DW}(\mathbf{r}) = \Delta_{S(C)DW} \cos(\mathbf{Q_0 r}). \tag{7.129}$$

Note that the order parameter (7.129) corresponds to the so-called ideal nesting vector,

$$\mathbf{Q_0} = (2p_F, \pi/b^*, \pi/c^*). \tag{7.130}$$

Electron spectrum (7.127), corresponding to an electron jumping between the nearest neighboring molecular sites, naturally appears in a number of organic metals, where distances between the 1D conducting chains are large enough. This allowed Yamaji [3] to suggest that the SDW phase, experimentally observed in $(TMTSF)_2PF_6$ conductor at ambient pressure, is due to the ideal nesting properties (7.128) of its electron spectrum.

He also pointed out that the SDW phase in a sister compound $(TMTSF)_2$ ClO_4 is unstable due to the existence of some small (but important) corrections to its electron spectrum

$$\begin{aligned}\varepsilon^{\pm}(\mathbf{p}) = & \pm v_F(p_x \mp p_F) + 2t_b \cos(p_y b^*) + 2t_c \cos(p_z c^*) \\ & + 2t'_b \cos(2p_y b^*) + 2t'_c \cos(2p_z c^*), \end{aligned} \tag{7.131}$$

where $t'_b \ll t_b$ and $t'_c \ll t_c$ are the so-called anti-nesting terms [3, 15].

Note that, if the main anti-nesting term in (7.131) is large enough, $t'_b \geq T_{SDW}$, then electrons and holes move with quite different velocities. In this case, the corresponding quantum mechanical problem becomes 3D and the Peierls electron–hole pairs do not appear [3, 15] even in the case of the attractive electron–hole interactions.

7.7.2 Field-Induced Spin-Density-Wave Phases and 3D → 2D Crossovers

Nevertheless, in 1983, Chaikin's [30] and Ribault–Jerome's [3] experimental groups found that some phase transitions, which are roughly periodic in $1/H$, appear in $(TMTSF)_2ClO_4$ conductor in a magnetic field. These transitions were interpreted by Gor'kov and Lebed [15], by Heritier, Montambaux, and Lederer [93], by Chaikin [11], and by Lebed [92] as the phase transitions between different FISDW subphases.

In this section, we describe how 2D → 1D crossovers in a magnetic field, discovered by Gor'kov and Lebed [15], explain instability of a Q1D metallic phase (7.131) with respect to the FISDW formation. Let us consider a realistic Q1D electron spectrum (7.131) in a perpendicular to the conducting chains magnetic field

$$\mathbf{H} = (0, 0, H), \quad \mathbf{A} = (0, Hx, 0). \tag{7.132}$$

It is important that the nesting terms, t_b and t_c, preserve the nesting condition (7.128), whereas the anti-nesting terms, t'_b and t'_c, in (7.131) destroy it in the absence of a magnetic field. In fact, in the layered Q1D organic conductors, the parameter t'_c is very small, $t'_c \simeq 0.1$ K, and, therefore, we neglect it below. On the other hand, the parameter t'_b is typically of the order of 10 K and plays, as we show below, a crucial role. Therefore, instead of the Q1D spectrum (7.131), we can consider the following 2D electron spectrum with one anti-nesting term:

$$\varepsilon^{\pm}(\mathbf{p}) = \pm v_F(p_x \mp p_F) + 2t_b \cos(p_y b^*) + 2t'_b \cos(2p_y b^*). \tag{7.133}$$

[Note that we also set $t_c = 0$ in (7.131), since t_c preserves the nesting condition (7.128) and, thus, is not important for the further consideration].

At first, we consider an instability of a metallic phase with respect to the FISDW formation, using qualitative arguments related to the QC equations of motion for electrons,

$$\frac{\mathrm{d}\mathbf{p}}{\mathrm{d}t} = \frac{e}{c}[\mathbf{v}(\mathbf{p}) \times \mathbf{H}], \quad \mathbf{v}(\mathbf{p}) = \frac{\mathrm{d}\varepsilon(\mathbf{p})}{\mathrm{d}\mathbf{p}}, \tag{7.134}$$

and for holes,

$$\frac{\mathrm{d}\mathbf{p}}{\mathrm{d}t} = -\frac{e}{c}[\mathbf{v}(\mathbf{p}) \times \mathbf{H}], \quad \mathbf{v}(\mathbf{p}) = \frac{\mathrm{d}\varepsilon(\mathbf{p})}{\mathrm{d}\mathbf{p}}. \tag{7.135}$$

From (7.134) and (7.135), it follows that $\mathbf{dp} \perp \mathbf{dr}$, and, thus, a quasi-particle trajectory in a real space can be obtained from that in a reciprocal space by rotating it into $\pi/2$ (see Fig. 7.26). Note that a main feature of the electron spectrum (7.133) is that it is periodic in the extended Brillouin zone. Therefore, the quasi-particles trajectories, in a perpendicular to the conducting chains magnetic field (7.132), become periodic and restricted along **y**-axis in a real space (see Figs. 7.26 and 7.27).

As it follows from Fig. 7.27, in a magnetic field, electrons and holes can move freely only along the conducting chains (i.e., **x**-axis), whereas their

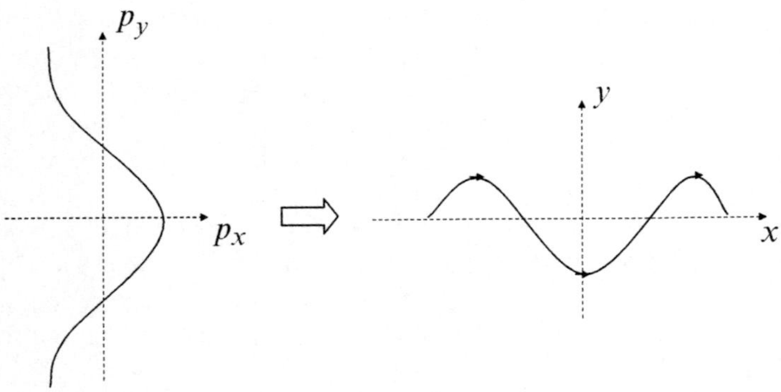

Fig. 7.26. Electron trajectory in a real (x, y)-space can be obtained from that in a reciprocal (p_x, p_y)-space by a rotation operation [see (7.134) and (7.135) and the text]

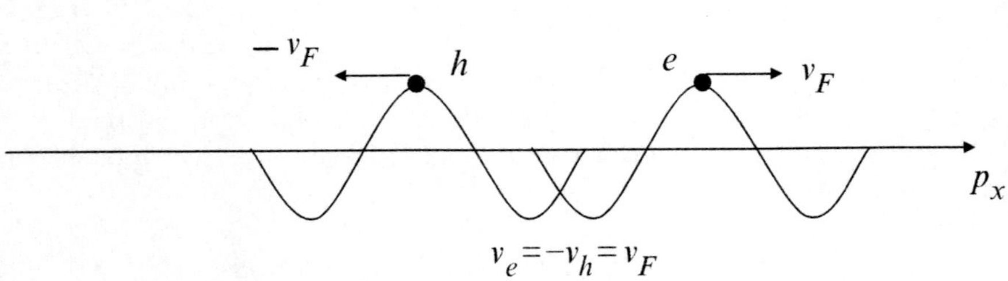

Fig. 7.27. Electron–hole interactions are "one-dimensionalized" in a magnetic field (7.132) [15] (*see the text*)

trajectories along **y**-axis are localized. This means that electrons and holes cannot go to infinity along **y**-axis and, thus, the electron–hole pairing in a magnetic field is "one-dimensionalized" [15]. 1D metal is known [3, 15] to be unstable against the S(C)DW phase formation, which is a reason why the Q1D metal (7.133) is unstable with respect to the FISDW phase formation in an arbitrary weak magnetic field [15].

Let us consider the suggested in [15] 2D → 1D crossover in some more details. The QC equations of motion (7.134) for an electron, located on a right sheet of the Q1D spectrum (7.133), can be written in a magnetic field (7.132) as:

$$\frac{\mathrm{d}p_y}{\mathrm{d}t} = \frac{e}{c} v_\mathrm{F} H, \quad p_y b^* = p_y^1 b^* + \omega_c(H)t, \quad \omega_c(H) = \frac{e}{c} v_\mathrm{F} H b^*. \tag{7.136}$$

Therefore, the electron velocity component along **y**-axis is

$$v_y^e(t, p_y^1) = -2t_b b^* \sin[p_y^1 b^* + \omega_c(H)t] - 4t_b' b^* \sin[2p_y^1 b^* + 2\omega_c(H)t] \tag{7.137}$$

and, thus, the electron trajectory can be expressed as:

$$y_e(t, p_y^1) = \frac{2t_b b^*}{\omega_c(H)} \cos[p_y^1 b^* + \omega_c(H)t] + \frac{2t_b' b^*}{\omega_c(H)} \cos[2p_y^1 b^* + 2\omega_c(H)t], \tag{7.138}$$

where p_y^1 is an initial electron momentum at $t = 0$.

On the other hand, the QC equations of motion (7.135) for a hole, located on a left sheet of the Q1D spectrum (7.133), in a magnetic field (7.132), can be written as:

$$\frac{\mathrm{d}p_y}{\mathrm{d}t} = -\frac{e}{c}(-v_\mathrm{F})H = \frac{e}{c} v_\mathrm{F} H, \quad p_y b^* = p_y^2 b^* + \omega_c(H)t. \tag{7.139}$$

Therefore, the hole velocity component along **y**-axis is

$$v_y^h(t, p_y^2) = -2t_b b^* \sin[p_y^2 b^* + \omega_c(H)t] - 4t_b' b^* \sin[2p_y^2 b^* + 2\omega_c(H)t] \tag{7.140}$$

and, thus, the hole trajectory can be expressed as:

$$y_h(t, p_y^2) = \frac{2t_b b^*}{\omega_c(H)} \cos[p_y^2 b^* + \omega_c(H)t] + \frac{2t_b' b^*}{\omega_c(H)} \cos[2p_y^2 b^* + 2\omega_c(H)t], \tag{7.141}$$

where p_y^2 is an initial hole momentum at $t = 0$.

Let us consider an electron–hole pair which satisfies the nesting condition (7.128), where $p_y^2 b^* = p_y^1 b^* + \pi$. As it follows from (7.137) and (7.140), in the absence of the anti-nesting term (i.e. at $t_b' = 0$), the electron and hole velocity components along **y**-axis are opposite at any moment of time. Therefore, only the anti-nesting term, t_b', prevents to have the ideal electron–hole FISDW pairs with the ideal nesting vector (7.130).

Nevertheless, as is seen from (7.138) and (7.141), the quasi-particles trajectories are localized along **y**-direction in a magnetic field (7.132) even in the presence of the anti-nesting term, t_b' (see Fig. 7.27). Therefore, electrons and holes always organize some nonideal FISDW pairs in an arbitrary weak magnetic field, whereas, in high magnetic fields, where the anti-nesting term, t_b', is not important, they organize the ideal FISDW pairs.

This point about the ideal and nonideal FISDW phases is clarified in Sect. 7.7.3. Here, we estimate a value of a magnetic field, above which the FISDW paring become ideal. As seen from (7.138) and (7.141), the ideal FISDW electron–hole pairing is achieved at $2t_b'/\omega_c(H) \leq 1$, where the contributions of the anti-nesting terms to electron and hole trajectories are less than an interchain distance, b^*,

$$\delta y \simeq \frac{2t_b' b^*}{\omega_c(H)} \leq b^*. \tag{7.142}$$

Equation (7.142) corresponds to some crossover magnetic field,

$$H_1 = \frac{4t_b' c}{ev_F b^*}, \tag{7.143}$$

which completely "one-dimensionalizes" the electron–hole interactions. (Note that, although the FISDW phase is stable in an arbitrary weak magnetic field [15], at magnetic fields lower than H_1, it is characterized by the nonideal nesting vector and, thus, by lower transition temperature [11, 15, 92, 93]).

The problem about the FISDW phase formation, which is discussed above using qualitative arguments, was first considered in [15]. In particular, in [15], the above mentioned problem is rigorously solved by means of the Green functions method. It is shown that a generalized susceptibility of the Q1D electrons (7.133) diverges at low temperatures (see Fig. 7.28). This indicates about the FISDW instability and defines the FISDW transition temperature (see Fig. 7.29). (Note that, in [15], the FISDW problem is solved for the ideal nesting vector (7.130), which corresponds to minimum of a free energy at high enough magnetic fields, $H \geq H_1$ [see (7.143) and Fig. 7.29].

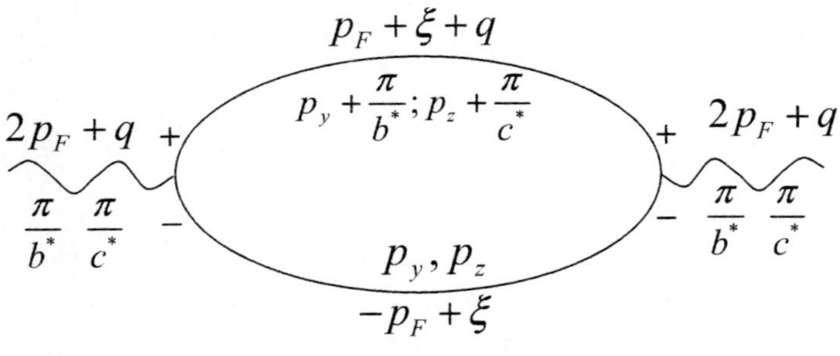

Fig. 7.28. A typical contribution to the mean field theory of a generalized electron susceptibility with respect to the FISDW formation [15]

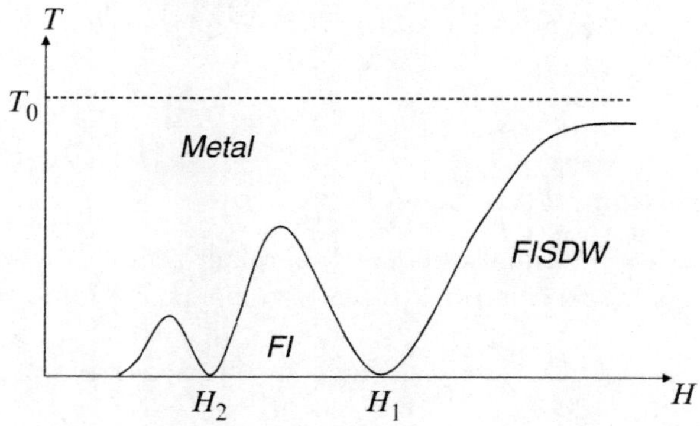

Fig. 7.29. *Solid line*: a solution of (7.144), which defines a transition temperature to the FISDW phase with the ideal nesting vector [i.e. for $q = 0$ in (7.145)]

The main mathematical result of [15] is that a phase boundary between the metallic phase (7.133) and the FISDW phase (7.130) can be described as:

$$\frac{1}{g} = \int_d^\infty J_0\left(\frac{4t_b'}{\omega_c(H)} \sin\left[\frac{\omega_c(H)x}{v_F}\right]\right) \frac{2\pi T \, dx}{v_F \sinh\left(\frac{2\pi T x}{v_F}\right)}, \tag{7.144}$$

where $g \ll 1$ is a dimensionless coupling constant, d is a cut-off distance.

Note that the Bessel function, $J_0(\ldots)$, is a periodic function in integral (7.144) and, thus, the integral logarithmically diverges at low temperatures at any value of a magnetic field. This means that (7.144) always has a solution and, therefore, the metallic phase is absolutely unstable with respect to the FISDW phase formation at low enough temperatures. This is the main physical result of [15], which is illustrated in Fig. 7.29, where (7.144) is graphically solved.

We stress again that (7.144) is written for the ideal nesting vector (7.130), which minimizes a free energy of the FISDW phase at high enough magnetic fields, $H \geq H_1$ (7.143). Although, as anticipated in [15], some other quantized nesting (QN) vectors may correspond to minima of a free energy at lower magnetic fields, $H \leq H_1$, nevertheless the FISDW problem for arbitrary magnetic fields is fully solved only in [92, 93].

7.7.3 Quantized Nesting Model

In this section, we discuss the so-called QN model, introduced by Heritier, Montambaux, and Lederer [93] and developed by Lebed [92] and by Virosztek and Maki [96]. From the results of the previous section, it follows that only the anti-nesting term, t_b', is responsible for possible deviations of the FISDW wave vector from its ideal nesting value (7.130). Let us suppose that $H \leq H_1$

and consider a phase transition to the FISDW phase, characterized by more general nesting vector

$$\mathbf{Q_0} = (2p_\mathrm{F} + q, \pi/b^*, \pi/c^*). \tag{7.145}$$

Momentum Quantization Law

In this section, we introduce a momentum quantization law, which allows to consider the QN model using qualitative arguments. For more details, see also Sect. 7.4.1.

Let us consider Q1D electrons, which are characterized only by the major anti-nesting term

$$\varepsilon^{\pm}(\mathbf{p}) = \pm v_\mathrm{F}(p_x \mp p_\mathrm{F}) + 2t'_b \cos(2p_y b^*), \tag{7.146}$$

since only the anti-nesting term in (7.133) can be responsible for $q \neq 0$ in (7.145).

We introduce electron wave functions near right and left sheets of Q1D FS (7.133) in the form:

$$\Psi^{\pm}_{\varepsilon}(x,y) = \exp(\pm i p_\mathrm{F} x)\psi^{\pm}_{\varepsilon}(x,y), \tag{7.147}$$

and expend the amplitudes $\psi^{\pm}_{\varepsilon}(x,y)$ into Fourier series with respect to a coordinate y,

$$\psi^{\pm}_{\varepsilon}(x,y) = \int_0^{2\pi} \psi^{\pm}_{\varepsilon}(x,p_y)\exp(\mathrm{i}p_y y)\frac{\mathrm{d}p_y}{2\pi}. \tag{7.148}$$

In a magnetic field (7.132) perpendicular to the conducting chains, the Schrödinger equation for the electron wave functions (7.148),

$$\left(\mp \mathrm{i}v_\mathrm{F}\frac{\mathrm{d}}{\mathrm{d}x} + 2t'_b \cos\left[2p_y b^* - \frac{2\omega_c(H)x}{v_\mathrm{F}}\right]\right)\psi^{\pm}_{\varepsilon}(x,p_y) = \delta\varepsilon\ \psi^{\pm}_{\varepsilon}(x,p_y), \tag{7.149}$$

can be obtained from (7.133) by using the Peierls substitution method [15]

$$p_x - p_\mathrm{F} \to -\mathrm{i}\left(\frac{\mathrm{d}}{\mathrm{d}x}\right), \quad p_y b^* \to p_y b^* - \left(\frac{e}{c}\right) A_y b^*, \tag{7.150}$$

where $\delta\varepsilon = \varepsilon - \varepsilon_\mathrm{F}$.

It is important that (7.149) can be solved analytically

$$\psi^{\pm}_{\varepsilon}(x,p_y) = \exp\left(\pm\mathrm{i}\frac{\delta\varepsilon}{v_\mathrm{F}}x\right)\exp\left[\pm\frac{\mathrm{i}t_b}{\omega_c}\left(\sin\left[2p_y b^* - \frac{2\omega_c x}{v_\mathrm{F}}\right] - \sin[2p_y b^*]\right)\right]. \tag{7.151}$$

Let us show that (7.151) directly demonstrates the 3D → 2D crossover in a magnetic field, described in Sect. 7.7.2, and a momentum quantization law. For this purpose, we calculate the Fourier component of the wave function (7.151) with respect to variable x,

$$\psi^{\pm}_{\varepsilon}(x,p_y) = \exp\left(\pm\mathrm{i}\frac{\delta\varepsilon}{v_\mathrm{F}}x\right)\sum_{m=-\infty}^{\infty} A_m(p_y)\exp\left(\mathrm{i}\frac{2\omega_c}{v_\mathrm{F}}mx\right), \tag{7.152}$$

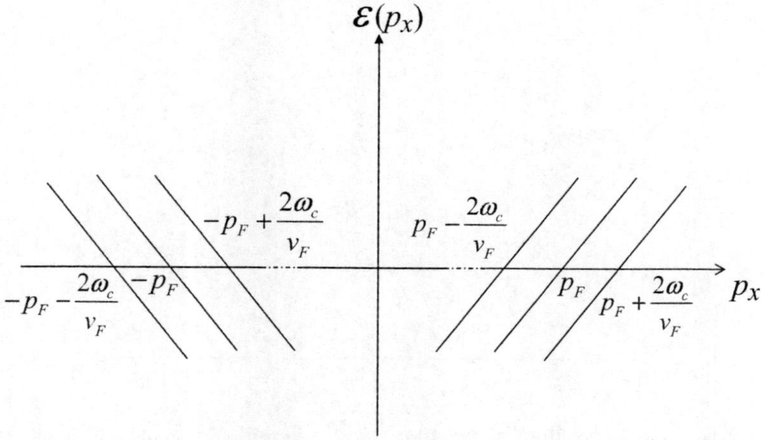

Fig. 7.30. Energy levels in the vicinities of $\pm p_F$ [see (7.154)]

where

$$A_m(p_y) = \frac{\omega_c(H)}{\pi v_F} \int_{-\pi v_F/2\omega_c}^{\pi v_F/2\omega_c} \exp\left(\mp \mathrm{i}\frac{\delta\varepsilon}{v_F}x\right) \exp\left(-\mathrm{i}\frac{2\omega_c}{v_F}mx\right) \psi_\varepsilon^{\pm}(x, p_y)\mathrm{d}x. \tag{7.153}$$

As seen from (7.151), electron energy, $\delta\varepsilon$, does not depend on a momentum component, p_y, which is an indication of the 2D $\to$ 1D crossover. Moreover, (7.152) introduces the following momentum quantization law: if an electron has a definite energy, $\delta\varepsilon$, then its momentum component along **x**-axis is quantized

$$p_x = \pm p_F \pm \delta\varepsilon/v_F + [2\omega_c(H)/v_F]n, \tag{7.154}$$

where $2\omega_c(H)/v_F$ is a momentum quantum (see Fig. 7.30).

The introduced above-momentum quantization law (7.154) plays a central role in the QN model for the FISDW subphases transitions. Indeed, as it follows from (7.154), the nesting condition (7.128) is obeyed in a magnetic field (7.132) for a number of wave vectors (7.145) with the quantized value of q

$$q = n[2\omega_c(H)/v_F], \tag{7.155}$$

where n is an integer.

Metal-FISDW Phase Transitions Line

Let us introduce the QN model, which was first suggested by Heritier, Montambaux, and Lederer [93, 94] and developed by Lebed [92] and by Virosztek and Maki [96]. To do this, we calculate a general susceptibility of the metallic phase (7.133) with respect to the FISDW pairing, corresponding to the nesting vector (7.145).

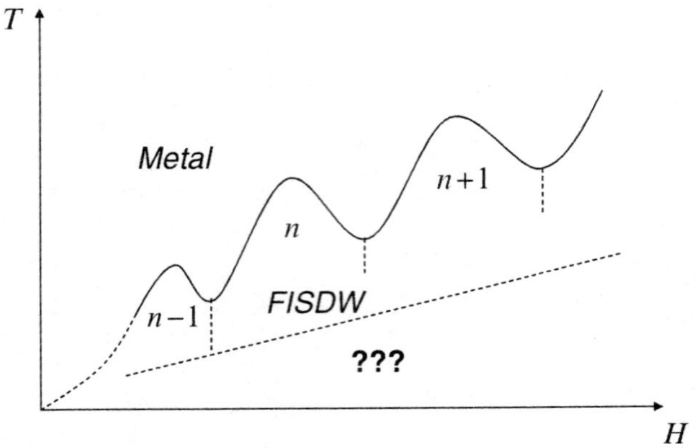

Fig. 7.31. Transition line between the metallic phase (7.133) and the FISDW subphases (7.155)

As a result, we obtain the following equation [93]:

$$\frac{1}{g} = \int_d^\infty J_0\left(\frac{4t_b'}{\omega_c(H)}\sin\left[\frac{\omega_c(H)x}{v_\mathrm{F}}\right]\right)\cos(qx)\frac{2\pi T\ dx}{v_\mathrm{F}\sinh\left(\frac{2\pi T x}{v_\mathrm{F}}\right)}, \qquad (7.156)$$

where there is an extra term, $\cos(qx)$, in comparison with (7.144).

From (7.156), it directly follows that the integral (7.156) is now divergent as $T \to 0$ at a number of quantized nesting vectors (7.155). This means that (7.156), which defines a transition line between the metallic and the FISDW phases, has solutions for a number of quantized nesting vectors (7.155) at low enough temperatures.

To obtain the phase transition line, we need to maximize the transition temperature (7.156) with respect to a quantum number n in (7.155). As a result, the following phase transition line is obtained by Heritier, Montambaux, and Lederer [93] (see Fig. 7.31).

We stress that (7.156) is valid only in the Ginzburg–Landau (GL) region and, thus, defines only transition line between the metallic phase and the FISDW subphases. To obtain the whole phase diagram, we need to calculate a free energy of the FISDW subphases out of the GL area, which was first performed by Lebed [92] and by Vitosztek and Maki [96].

Phase Transitions Between FISDW Subphases

Let us discuss the main results of [92], where a free energy of the Q1D electrons (7.133) are calculated in the FISDW phase and the phase transitions between different FISDW subphases are described (see Fig. 7.32).

We stress that the above-mentioned problem cannot be analytically solved for an arbitrary relationship between the FISDW order parameter, Δ_{FISDW},

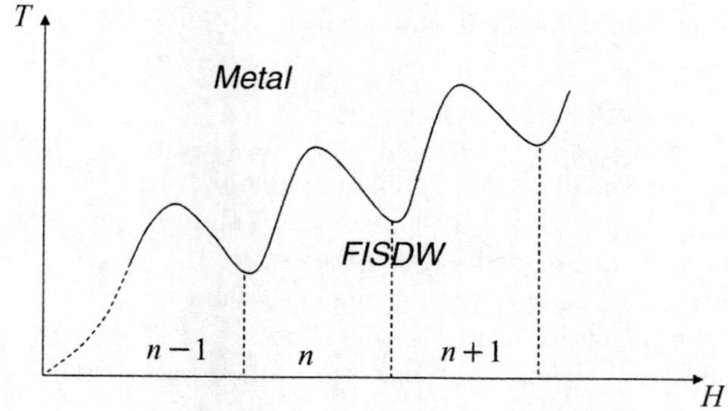

Fig. 7.32. The FISDW phase diagram within the QN model [92,93], where *vertical dashed lines* correspond to the first-order phase transitions between different FISDW subphases, characterized by quantized nesting vectors (7.155)

and the magnetic field dependent cyclotron frequency, $\omega_c(H)$. Nevertheless, as pointed in [92], in many cases

$$\omega_c(H)/\Delta_n \geq 1, \tag{7.157}$$

where

$$\Delta_{\text{FISDW}}(x,y,z) = \Delta_n \exp\left(\mathrm{i}\frac{\pi y}{b^*}\right)\exp\left(\mathrm{i}\frac{\pi z}{c^*}\right)\exp\left(\mathrm{i}\left[2p_{\text{F}} + 2n\frac{\omega_c(H)}{v_{\text{F}}}\right]x\right) \tag{7.158}$$

corresponds to the FISDW phase with the quantized nesting vector (7.155). [Note that the condition (7.157) corresponds to the so-called magnetic breakdown phenomenon, which occurs through the FISDW gap, Δ_n. It is important that the QN model, introduced in Sect. 7.7.2, is also based on the condition (7.157)].

As shown in [92], a free energy of the FISDW phase can be calculated at any temperature under the breakdown condition (7.157). In this case, it is possible to show that an arbitrary term of the free energy expansion in the presence of a magnetic field can be obtained from the expression of the BCS free energy by the replacement [92]

$$\Delta^2 \to \Delta_n^2 \; J_n^2[2t_b'/\omega_c(H)], \tag{7.159}$$

where Δ is the BCS gap, $J_n[\ldots]$ is the Bessel function of the nth order.

The terms of the second order in Δ_n are an exception. They, however, can be evaluated separately and correspond to magnetic field and pressure-dependent effective coupling constant [92],

$$\frac{1}{g} \to \frac{\ln(t_b'/t_b^0)}{J_n^2[2t_b'/\omega_c(H)]}, \tag{7.160}$$

and a magnetic field dependent cutoff constant [92]

$$\Omega = \gamma \omega_c(H)/\pi^2, \tag{7.161}$$

where t_b^0 is a critical value of the anti-nesting parameter, t_b', which destroys the SDW phase in the absence of a magnetic field, γ is the Euler constant. (In other words, as shown in [92], it is possible to summarize an infinite number of the Feynman diagrams defining the free energy).

In this form the free energy in a magnetic field has local minima for subphases with the quantized wave vectors q from (7.155) and the entire sum of Feynman diagrams is transformed into the functional

$$\delta F(n,H) = \Delta_n^2 \ln\left(\frac{t_b'}{t_b^0}\right) - 4T \int_0^{\Omega} d\xi \ln\left(\frac{\cosh\left[(\xi^2 + J_n^2[2t_b'/\omega_c(H)]\Delta_n^2)^{1/2}/2T\right]}{\cosh(\xi/2T)}\right). \tag{7.162}$$

From here we must determine the phase diagram. If we use the condition that the variation of functional (7.162) with respect to Δ_n vanish determines the equilibrium value of the gap for the nth subphase, Δ_n, then we can rewrite (7.162) as:

$$\delta F(n,H) = \int_0^{\infty} \mathrm{d}\xi \left[\frac{J_n^2(2t_b'/\omega_c(H))\Delta_n^2}{(\xi^2 + J_n^2[2t_b'/\omega_c(H)]\Delta_n^2)^{1/2}} \tanh \frac{[\xi^2 + J_n^2(2t_b'/\omega_c(H))\Delta_n^2]^{1/2}}{2T} - 4T \ln \frac{\cosh\left[(\xi^2 + J_n^2[2t_b'/\omega_c(H)]\Delta_n^2)^{1/2}/2T\right]}{\cosh(\xi/2T)}\right]. \tag{7.163}$$

[In the convergent integral (7.163) the upper limit is set to infinity].

Since a free energy (7.163) of each FISDW subphase now depends explicitly only on combination $\Delta_n^2\ J_n^2[4t_b'/\omega_c(H)]$, it follows that the optimum values of the wave vector q in (7.155) is reached at the maximum of the Bessel function $J_n^2[4t_b'/\omega_c(H)]$. Thus, the phase diagram of a layered Q1D conductors consists of the FISDW subphases determined by the wave vectors q from (7.155) for which $J_n^2[4t_b'/\omega_c(H)]$ is maximum.

The phase diagram of a Q1D conductor (7.133), obtained in [92], is shown schematically in Fig. 7.32, with the transition temperatures from metallic to the FISDW nth subphase, $T_n(H)$, being related to the energy gap, $\Delta_n(H)$, in the same way as in the BCS theory:

$$T_n(H) = \left(\frac{\gamma}{\pi}\right)\Delta_n(H). \tag{7.164}$$

[Note that each FISDW subphase is accompanied by a so-called 3D quantum Hall effect (QHE) [31,32]. For theories of the 3D QHE, see the chapter and [98] by Yakovenko and [95] by Poilblanc et al.]

7.7.4 Beyond Quantum Nesting Model

We point out that the magnetic breakdown condition (7.157), which may be rewritten as:

$$h = \omega_c(H)/\pi T_n(H) \geq 1, \tag{7.165}$$

is well fulfilled in $(TMTSF)_2ClO_4$ conductor [22]. On the other hand, as stressed in [22], it is not fulfilled in a sister compound $(TMTSF)_2PF_6$ under pressure and in some other materials.

Below, we discussed the results of [22] by Lebed, where the metal-FISDW phase boundary is theoretically analyzed for an arbitrary value of the parameter h. We also discuss in brief the experimental work [101] by Kornilov and Pudalov, performed on the $(TMTSF)_2PF_6$ conductor, where the main results of [22] is experimentally confirmed. In [22], it is shown that an account of a finite transition temperature, $T_n(H) \neq 0$, changes the main qualitative consequences of the QN model, described in Sect. 7.7.3.

In contrast to the QN model, it is shown [22] the following: (1) The longitudinal wave vector of the FISDW subphases (7.155) is not strictly quantized [i.e., n is not an integer in (1.155) unless $n = 0$]. (2) For small enough values of the parameter $h < h_c \simeq 1$, the FISDW phase diagram consists of two regions (a) a low-temperature region ("quantum FISDW") where there exist discontinuous (but noninteger) jumps of the FISDW wave vector (i.e., the first order transitions between different FISDW phases) (see Fig. 7.33); (b) a high temperature region ("quasi-classical FISDW") where the jumps and the first-order transitions disappear, but the FISDW wave vector (7.155) is still a nontrivial oscillating function of a magnetic field. (3) For large enough values of the parameter $h > h_c \simeq 1$, the FISDW phase diagram consists of a cascade of the first-order phase transitions between the different FISDW subphases. They are characterized by discontinuous (but noninteger) jumps of the FISDW wave vector (see Fig. 7.34).

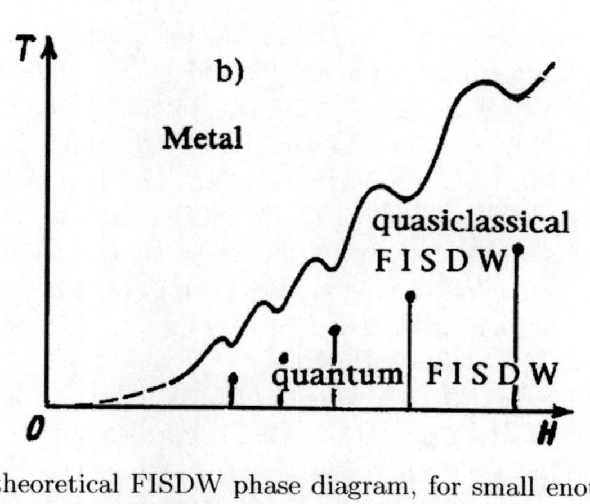

Fig. 7.33. The theoretical FISDW phase diagram, for small enough values of the parameter h (7.165), consists of two parts (*see the text*)

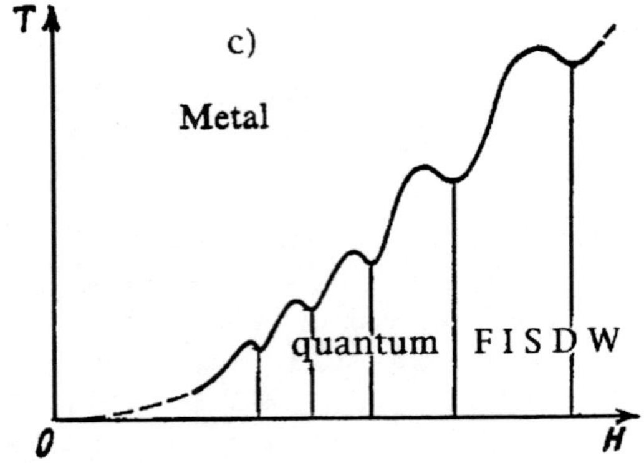

Fig. 7.34. The FISDW phase diagram, for large enough values of the parameter h (7.165), corresponds to noninteger discontinuous jumps of the vector nesting (7.155) between neighboring FISDW subphases

We note that the clear indications in favor of the theory [22] were observed in early experiments (e.g., in [31,32]), performed on $(TMTSF)_2PF_6$ compound by Chaikin's and Jerome's groups. Nevertheless, they were not appropriately interpreted in the absence of the comprehensive theory [22].

Nowadays, the main peculiarity of the above-discussed theory – the existence of two distinct regions in the FISDW phase diagram in the case of small values of the parameter h – was confirmed by Kornilov and Pudalov [101]. They have directly observed a hysteresis and the first-order phase transitions at low temperatures, which disappear at high enough temperatures within the FISDW phase diagram in $(TMTSF)_2PF_6$.

7.8 Reentrant Superconductivity Phenomenon

In this section, we consider the most spectacular phenomenon [related to the absence of the paramagnetic limit in $(TMTSF)_2PF_6$ superconductor for $\mathbf{H} \parallel \mathbf{b}$] – the so-called RS phase. This phase was first suggested by Lebed [10,12] and developed by Dupuis, Montambaux, and Sa de Melo [102]. The RS phase is shown [10,12] to be stable in magnetic fields, $\mathbf{H} \parallel \mathbf{b}$, which are much higher than both the upper critical field, $H_{c2}^{\perp}$, and the Clogston paramagnetic limit, H_{p}. The most unusual feature of the RS phase is that, at high enough magnetic fields, $H \geq H_{c2}^{\perp}$, superconducting transition temperature may increase with increasing magnetic field, $\mathrm{d}T_c/\mathrm{d}H > 0$ (see Fig. 7.35). This is true for both Q1D [10,102] and Q2D [12] conductors.

Let us first discuss RS phenomenon using qualitative arguments. Experimental geometry of its observation [79–81] corresponds to a magnetic field directed along $\mathbf{b}'$-axis within $(\mathbf{a},\mathbf{b})$-plane. As shown in Sect. 7.7.2, in such magnetic fields, electron motion becomes 2D due to 3D $\rightarrow$ 2D crossover. In

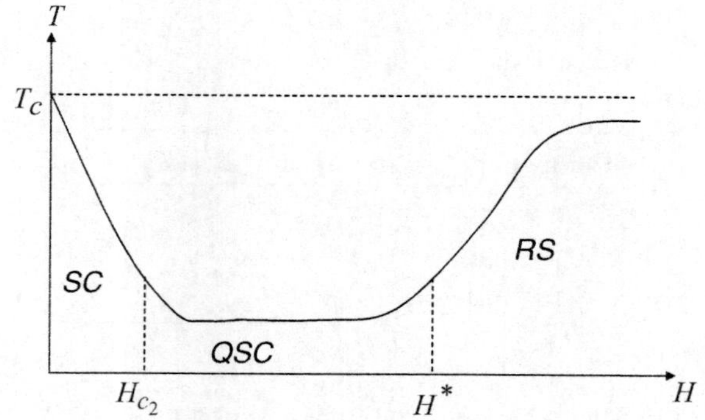

Fig. 7.35. Standard theory: superconductivity is destroyed by the upper critical field, H_{c2}. Refined theory, suggested in [10]: superconductivity is stable above the upper critical field, with transition temperature being a rising function of a magnetic field at magnetic field higher than some crossover field, $H > H^*$

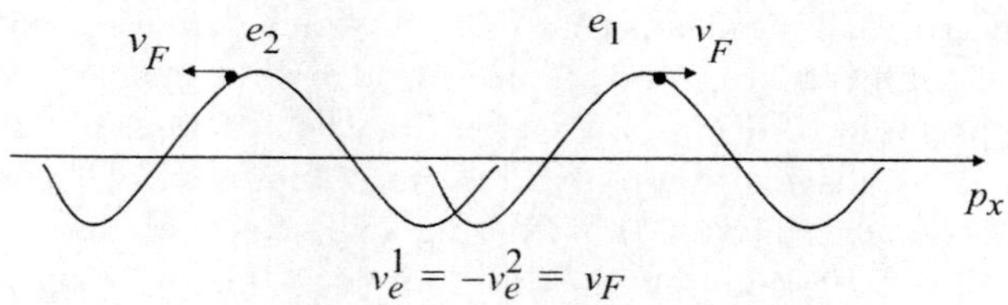

Fig. 7.36. Electron trajectories in a magnetic field, $\mathbf{H} \parallel \mathbf{b}$, become $2D$, since electron motions along **z**-axis become periodic and restricted

other words, electron motion becomes periodic and restricted along **z**-axis and, thus, electrons cannot go to infinity along **z**-axis (see Fig. 7.36).

Such "two-dimensionalization" of electron motion results in a stability of RS phase in an arbitrary magnetic field [10, 12, 13, 102] since 2D superconducting phase is stable in a parallel magnetic field. At high enough magnetic fields, $H > H^* = 2t_c/ev_{\mathrm{F}}c^* \simeq 15\ T$, transition temperature is expected to grow with increasing magnetic field (see Fig. 7.35).

The RS phenomenon is considered in [10,12,13,102] using Green function method. For superconducting order parameter, $\Delta(x)$, the following integral equation is obtained [10, 12], which defines superconducting transition temperature in a magnetic field,

$$\Delta(x) = \frac{g}{2} \int_{|x-x_1|>a} \mathrm{d}x_1 \, \frac{2\pi T}{v_{\mathrm{F}} \sinh\left(\frac{2\pi T|x-x_1|}{v_{\mathrm{F}}}\right)} \times J_0\left(2\lambda \sin\left[\frac{\omega_c(x-x_1)}{2v_{\mathrm{F}}}\right] \sin\left[\frac{\omega_c(x+x_1)}{2v_{\mathrm{F}}}\right]\right) \Delta(x_1), \quad (7.166)$$

where $\lambda = 4t_c/\omega_c$, $\omega_c = eHv_{\mathrm{F}}c^*/c$, g is a dimensionless coupling constant. [Note that (7.166) is valid for a triplet superconducting phase, where

Table 7.1. Experimental values of H^b_{c2} [79–81] are compared with the predictions of the standard Ginzburg–Landau–Abrikosov–Gorkov theory, H_{c2}, which disregards the RS phenomenon

Compound	Experimental $H^b_{c2}(T)$	H_{c2}
$X = PF_6$ [79]	6	2
$X = PF_6$ [80]	6	2.5
$X = PF_6$ [81]	9	2.5

paramagnetic destructive effects against superconductivity are absent. For more details, see another chapter by Lebed and Wu and Chap. 24].

From the integral equation (7.166), it directly follows that the triplet superconductivity is stable in an arbitrary magnetic field. Indeed, a choice of a periodic solution for superconducting order parameter in (7.166), $\Delta(x_1 + 2\pi v_F/\omega_c) = \Delta(x_1)$, make the integral (7.166) to be logarithmically divergent at low temperatures. This means that (7.166), which defines the superconducting transition temperature, always has periodic solutions. At high magnetic fields, these periodic solutions correspond to the RS phase, which is always stable and is qualitatively different from the BCS and LOFF phases.

In conclusion, we discuss the existing experimental data, obtained on $(TMTSF)_2PF_6$ [79–81] compound. Note that the expected values of the upper critical fields for this superconductor in a perpendicular to the conducting chains magnetic field, $\mathbf{H} \parallel \mathbf{b}$, in simplified theory, which does not take into account the existence of RS phase, are of the order of 2–3 T. From Table 7.1, it follows that the experimental upper critical fields are 3–4 times higher, which supports the RS scenario in $(TMTSF)_2PF_6$ compound.

On the other hand, the RS phase with $\mathrm{d}T_c/\mathrm{d}H > 0$ has not been detected yet. It is expected to occur in high magnetic fields, $H \geq 15\ T$ [10], in clean samples which demonstrate in a metallic phase a large quadratic magnetoresistance, $\Delta\rho/\rho \sim H^2 \gg 1$. To our surprise, magnetoresistance in the existing samples of $(TMTSF)_2PF_6$ and $(TMTSF)_2ClO_4$ conductors saturates in much lower magnetic fields, $H \simeq 5$–$10\,\mathrm{T}$, which may reflect intrinsic samples imperfections.

Acknowledgments

The author is thankful to N.N. Bagmet (Lebed), S.E. Brown, J.S. Brooks, P.M. Chaikin, L.P. Gorkov, M. Heritier, D. Jerome, M.V. Kartsovnik, W. Kang, A. Kornilov, M.J. Naughton, T. Osada, V.M. Pudalov, and J. Singleton for fruitful discussions. This work was partially supported by the NSF grant DMR-0705986.

References

1. A.A. Abrikosov, *Fundamentals of the Theory of Metals* (Elsevier, Amsterdam 1988)
2. M. Tinkham, *Introduction to Superconductivity*, 2nd edn. (McGraw-Hill, New York, 1996).
3. T. Ishiguro, K. Yamaji, G. Saito, *Organic Superconductors*, 2nd edn. (Springer, Heidelberg, 1998)
4. D. Jerome, Nature (London) **387**, 235 (1997)
5. D.G. Clark, S.P. Strong, P.M. Chaikin, E.I. Chashechkina, Science **279**, 2071 (1998)
6. C.C. Tsuei, J.R. Kirtley, Rev. Mod. Phys. **72**, 969 (2000)
7. G. Blatter, M.V. Feigel'man, V.B. Geshkenbein, A.I. Larkin, V.M. Vinokur, Rev. Mod. Phys. **66**, 1125 (1996)
8. C. Honerkamp, P.A. Lee, Phys. Rev. Lett. **92**, 177002 (2004)
9. O. Vafek, Z. Tesanovic, Phys. Rev. Lett. **91**, 237001 (2003)
10. A.G. Lebed, JETP Lett. **44**, 114 (1986)
11. P.M. Chaikin, Phys. Rev. B **31**, 4770 (1985)
12. A.G. Lebed, K. Yamaji, Phys. Rev. Lett. **80**, 2697 (1998)
13. A.G. Lebed, N.N. Bagmet, Phys. Rev. B **55**, R8654 (1997)
14. A.G. Lebed, Phys. Rev. Lett. **95**, 247003 (2005)
15. L.P. Gor'kov, A.G. Lebed, J. Phys. (Paris) Lett. **45** L-433 (1984)
16. A.G. Lebed, Sov. Phys. JETP **62**, 595 (1985)
17. A.G. Lebed, JETP Lett. **43**, 174 (1986); A.G. Lebed, P. Bak, Phys. Rev. Lett. **63**, 1315 (1989)
18. A.G. Lebed, Phys. Rev. Lett. **74**, 4903 (1995)
19. A.G. Lebed, J. Phys. (Paris) I **6**, 1819 (1996)
20. A.G. Lebed, Phys. Rev. B **59**, R721 (1999)
21. A.G. Lebed, K. Machida, M. Ozaki, Phys. Rev. B **62**, R795 (2000)
22. A.G. Lebed, Phys. Rev. Lett. **88**, 177001 (2002)
23. A.G. Lebed, JETP Lett. **78**, 138 (2003)
24. A.G. Lebed, M.J. Naughton, J. Phys. (Paris) IV **12**, 369 (2002)
25. A.G. Lebed, M.J. Naughton, Phys. Rev. Lett. **91**, 187003 (2003)
26. A.G. Lebed, N.N. Bagmet, M.J. Naughton, J. Phys. (Paris) IV **114**, 77 (2003)
27. A.G. Lebed, N.N. Bagmet, M.J. Naughton, Phys. Rev. Lett. **93**, 157006 (2004)
28. A.G. Lebed, H.-I. Ha, M.J. Naughton, Phys. Rev. B **71**, 132504 (2005)
29. H.-I. Ha, A.G. Lebed, M.J. Naughton, Phys. Rev. B **73**, 033107 (2005)
30. P.M. Chaikin, M.-Y. Choi, J.F. Kwak, J.S. Brooks, K.P. Martin, M.J. Naughton, E.M. Engler, R.L. Greene, Phys. Rev. Lett. **51**, 2333 (1983)
31. S.T. Hannahs, J.S. Brooks, W. Kang, L.Y. Chiang, P.M. Chaikin, Phys. Rev. Lett. **63**, 1988 (1989)
32. J.R. Cooper, W. Kang, P. Auban, G. Montambaux, D. Jerome, K. Bechgaard, Phys. Rev. Lett. **63**, 1984 (1989)
33. W. Kang, S.T. Hannahs, P.M. Chaikin, Phys. Rev. Lett. **70**, 3091 (1993)
34. S. Uji, C. Terakura, M. Takashita, T. Terashima, H. Aoki, J.S. Brooks, S. Tanaka, S. Maki, J. Yamada, S. Nakatsuji, Phys. Rev. B **60**, 1650 (1999)
35. N. Biskup, J.S. Brooks, R. Kato, K. Oshima, Phys. Rev. B **60**, R15005 (1999)
36. V.M. Yakovenko, H.-S. Goan, J. Phys. (Paris) I **6**, 1917 (1996)
37. K. Sengupta, N. Dupuis, Phys. Rev. B **68**, 094431 (2003)

38. S. Haddad, S. Charfi-Kaddour, M. Heritier, R. Bennaceur, Phys. Rev. B **72**, 085104 (2005)
39. D. Vignolles, A. Audouard, M. Nardone, L. Brossard, S. Bouguessa, J.-M. Fabre, Phys. Rev. B **71**, 020404 (R) (2005)
40. S. Uji, S. Yasuzuka, T. Konoike, K. Enomoto, J. Yamada, E.S. Choi, D. Graf, J.S. Brooks, Phys. Rev. Lett. **94**, 077206 (2005)
41. W. Wu, N.P. Ong, P.M. Chaikin, Phys. Rev. B **72**, 235116 (2005)
42. H. Ito, D. Suzuki, Y. Yokochi, S. Kuroda, M. Umemiya, H. Miyasaka, K-I. Sugiura, M. Yamashita, H. Tajima, Phys. Rev. B **71**, 212503 (2005)
43. S. Takahashi, S. Hill, S. Takasaki, J. Yamada, H. Anzai, Phys. Rev. B **72**, 024540 (2005)
44. W. Wu, I.J. Lee, P.M. Chaikin, Phys. Rev. Lett. **91**, 056601 (2003)
45. E.S. Choi, J.S. Brooks, H. Kang, Y.J. Jo, W. Kang, Phys. Rev. Lett. **95**, 187001 (2005)
46. H. Kang, Y.J. Jo, W. Kang, Phys. Rev. B **69**, 033103 (2004)
47. H. Kang, Y.J. Jo, S. Uji, W. Kang, Phys. Rev. B **68**, 132508 (2003)
48. E.I. Chashechkina, P.M. Chaikin, Phys. Rev. B **65**, 012405 (2002)
49. M.J. Naughton, O.H. Chung, L.Y. Chiang, J.S. Brooks, Mater. Res. Soc. Symp. Proc. **173**, 257 (1990)
50. G.S. Boebinger, G. Montambaux, M.L. Kaplan, R.C. Haddon, S.V. Chichester, L.Y. Chiang, Phys. Rev. Lett. **64**, 591 (1990)
51. T. Osada, A. Kawasumi, S. Kagoshima, N. Miura, G. Saito, Phys. Rev. Lett. **66**, 1525 (1991); T. Osada, S. Kogashima, N. Miura, Phys. Rev. B **46**, 1812 (1992)
52. M.J. Naughton, O.H. Chung, M. Chaparala, X. Bu, P. Coppens, Phys. Rev. Lett. **67**, 3712 (1991)
53. W. Kang, S.T. Hannahs, P.M. Chaikin, Phys. Rev. Lett. **69**, 2827 (1992)
54. K. Benhia, M. Ribault, C. Lenior, Europhys. Lett. **25**, 285 (1994)
55. E.I. Chashechkina, P.M. Chaikin, Phys. Rev. Lett. **80**, 2181 (1998)
56. T. Osada, H. Nose, M. Kuraguchi, Physica B **294–295**, 402 (2001)
57. P.M. Chaikin, J. Phys. (Paris) IV **12**, 189 (2002)
58. G.M. Danner, W. Kang, P.M. Chaikin, Phys. Rev. Lett. **72**, 3714 (1994)
59. M.J. Naughton, I.J. Lee, P.M. Chaikin, G.M. Danner, Synth. Met. **85**, 1481 (1997)
60. T. Osada, S. Kagoshima, N. Miura, Phys. Rev. Lett. **77**, 5261 (1996)
61. I.J. Lee, M.J. Naughton, Phys. Rev. B **57**, 7423 (1998)
62. I.J. Lee, M.J. Naughton, Phys. Rev. B **58**, R13343 (1998)
63. H. Yoshino, K. Saito, H. Nishikawa, K. Kikuchi, K. Kobayashi, I. Ikemoto, J. Phys. Soc. Jpn. **66**, 2248 (1997)
64. T. Osada, N. Kami, R. Kondo, S. Kagoshima, Synth. Met. **103**, 22024 (1999)
65. L.P. Gor'kov, A.G. Lebed, Phys. Rev. Lett. **71**, 3874 (1993)
66. A. Ardavan, J.M. Schrama, S.J. Blundell, J. Singleton, W. Hayes, M. Kurmoo, P. Day, P. Goy, Phys. Rev. Lett. **81**, 713 (1998)
67. A. Ardavan, S.J. Blundell, J. Singleton, Phys. Rev. B **60**, 15500 (1999)
68. A.E. Kovalev, S. Hill, K. Kawano, M. Tamura, T. Naito, H. Kobayashi, Phys. Rev. Lett. **91**, 216402 (2003)
69. S. Takahashi, S. Hill, S. Takasaki, J. Yamada, H. Anzai, Phys. Rev. B **72**, 024540 (2005)
70. D. Zanchi, A. Bjelis, G. Montambaux, Phys. Rev. B **53**, 1240 (1996)

71. A.G. Lebed, JETP Lett. **78**, 138 (2003)
72. J.S. Qualls, L. Balicas, J.S. Brooks, N. Harrison, L.K. Montgomery, M. Tokumoto, Phys. Rev. B **62**, 10008 (2000)
73. D. Graf, E.S. Choi, J.S. Brooks, M. Matos, R.T. Henrique, M. Almeida, Phys. Rev. Lett. **93**, 076406 (2004)
74. D. Andres, M.V. Kartsovnik, P.D. Grigoriev, W. Biberacher, H. Muller, Phys. Rev. B **68**, 201101(R) (2003)
75. R.D. McDonald, N. Harrison, L. Balicas, K.N. Kim, J. Singleton, X. Chi, Phys. Rev. Lett. **93**, 076405 (2004)
76. D. Graf, E.S. Choi, J.S. Brooks, M. Matos, R.T. Henrique, M. Almeida, Phys. Rev. Lett. **93**, 076406 (2004)
77. G.M. Danner, P.M. Chaikin, Phys. Rev. Lett. **75**, 4690 (1995)
78. S.P. Strong, D.G. Clarke, P.W. Anderson, Phys. Rev. Lett. **73**, 1007 (1994)
79. I.J. Lee, M.J. Naughton, G.M. Danner, P.M. Chaikin, Phys. Rev. Lett. **78**, 3555 (1997)
80. I.J. Lee, P.M. Chaikin, M.J. Naughton, Phys. Rev. Lett. **88**, 207002 (2002)
81. I.J. Lee, P.M. Chaikin, M.J. Naughton, Phys. Rev. B **65**, 180502(R) (2002)
82. S. Uji, H. Shinagawa, T. Terashima, T. Yakabe, Y. Terai, M. Tokumoto, A. Kobayashi, H. Tanaka, H. Kobayashi, Nature (London) **410**, 908 (2001)
83. L. Balicas, J.S. Brooks, K. Storr, S. Uji, M. Tokumoto, H. Tanaka, H. Kobayashi, A. Kobayashi, V. Barzykin, L.P. Gor'kov, Phys. Rev. Lett. **87**, 067002 (2001)
84. J.S. Brooks, S. Uji, E.S. Choi, I.B. Rutel, H. Kobayashi, A. Kobayashi, H. Tanaka, M. Tokumoto, Braz. J. Phys. **33**, 278 (2003)
85. T. Konoike, S. Uji, T. Terashima, M. Nishimura, S. Yasuzuka, K. Enomoto, H. Fujiwara, B. Zhang, H. Kobayashi, Phys. Rev. B **70**, 094514 (2004)
86. C.D. Vaccarella, C.A.R. Sa de Melo, Phys. Rev. B **63**, 180505 (2001)
87. I.J. Lee, S.E. Brown, W.G. Clark, M.J. Strouse, M.J. Naughton, W. Kang, P.M. Chaikin, Phys. Rev. Lett. **88**, 017004 (2002)
88. I.J. Lee, D.S. Chow, W.G. Clark, M.J. Strouse, M.J. Naughton, P.M. Chaikin, S.E. Brown, Phys. Rev. B **68**, 092510 (2003)
89. N.E. Hussey, M.N. McBrien, L. Balicas, J.S. Brooks, S. Horii, H. Ikuta, Phys. Rev. Lett. **89**, 086601 (2002)
90. N.E. Hussey, M. Kibune, H. Nakagawa, N. Miura, Y. Iye, H. Takagi, S. Adachi, K. Tanabe, Phys. Rev. Lett. **80**, 2909 (1998)
91. K. Waku, N.E. Hussey, F. Sakai, M. Nohara, H. Takagi, H. Nakagawa, N. Miura, S. Adachi, J. Phys. Soc. Jpn. Lett. **69**, 21 (2000)
92. A.G. Lebed, Sov. Phys. JETP **62**, 595 (1985)
93. M. Heritier, G. Montambaux, P. Ledered, J. Phys. (Paris) Lett. **46**, L-831 (1985)
94. G. Montambaux, M. Heritier, P. Lederer, Phys. Rev. Lett. **55**, 2078 (1985)
95. D. Poilblanc, G. Montambaux, M. Heritier, P. Lederer, Phys. Rev. Lett. **58**, 270 (1987)
96. A. Virosztek, K. Maki, Phys. Rev. B **34**, 3371 (1986)
97. K. Yamaji, Synth. Met. **13**, 29 (1986)
98. V.M. Yakovenko, Phys. Rev. B **43**, 11353 (1991)
99. N. Dupuis, V.M. Yakovenko, Phys. Rev. Lett. **80**, 3618 (1998)
100. K. Sengupta, H.-J. Kwon, V.M. Yakovenko, Phys. Rev. Lett. **86**, 1094 (2001)

101. A.V. Kornilov, V.M. Pudalov, Y. Kitaoka, K. Ishida, T. Mito, J.S. Brooks, J.S. Qualls, J.A.A.J. Perenboom, N. Tateiwa, T.C. Kobayashi, Phys. Rev. B **65**, 060404R (2002)
102. N. Dupuis, G. Montambaux, C.A.R. Sa de Melo, Phys. Rev. Lett. **70**, 2613 (1993)
103. R.D. Duncan, C.D. Vaccarella, C.A.R. Sa de Melo, Phys. Rev. B **64**, 172503 (2001)
104. C.D. Vaccarella, C.A.R. Sa de Melo, Phys. Rev. B **64**, 212504 (2001)
105. C.D. Vaccarella, C.A.R. Sa de Melo, Phys. Rev. B **63**, 180505(R) (2001)
106. K. Sengupta, N. Dupuis, Phys. Rev. B **68**, 094431 (2003)
107. K. Sengupta, I. Zutic, H.-J. Kwon, V.M. Yakovenko, S. Das Sarma, Phys. Rev. B **63**, 144531 (2001)
108. K. Sengupta, H.-J. Kwon, V.M. Yakovenko, Phys. Rev. Lett. **86**, 1094 (2001)

8

Layered Organic Conductors in Strong Magnetic Fields

M.V. Kartsovnik

Organic conductors, being characterized by an extremely high electronic anisotropy, rather simple Fermi surfaces, and high crystal quality, are excellent objects for studying general properties of quasi-two-dimensional metallic systems. In particular, their behavior in strong magnetic fields is found to be qualitatively different from that of usual three-dimensional metals or purely two-dimensional systems. We describe various novel high-field phenomena observed in layered organic conductors and illustrate how they can be used for investigation of their conducting system. It turns out that magnetic fields can be used not only as a probe for studying the electronic state but also as a control parameter for switching between different states. The family α-(BEDT-TTF)$_2$MHg(SCN)$_4$ with M = K, Tl, and Rb, as an impressing example of a strong influence of a magnetic field on electronic properties, will be discussed in detail. A brief overview is given on a few other layered compounds, showing interesting high-field properties and field-induced transitions.

8.1 Introduction

Strong magnetic fields have been known since many years as one of the most powerful tools for exploring the Fermi surfaces (FSs) of conventional metals [1–3]. First observations, in 1988, of the magnetic quantum oscillations in the salts κ-(BEDT-TTF)$_2$Cu(NCS)$_2$ [4] and β-(BEDT-TTF)$_2$IBr$_2$ [5] have shown this tool to be applicable to layered organic conductors and triggered tremendous activities in this direction, see [6] for a comprehensive survey of studies performed till the middle of 1990s. Nowadays, magnetic field experiments have become a necessary ingredient in complex characterization of newly synthesized compounds. They are also extensively used for gaining a deeper insight into electronic properties of already known materials.

Of course, one of prerequisites for successful application of magnetic field techniques to studying FS properties is a high crystal quality. Fortunately, this

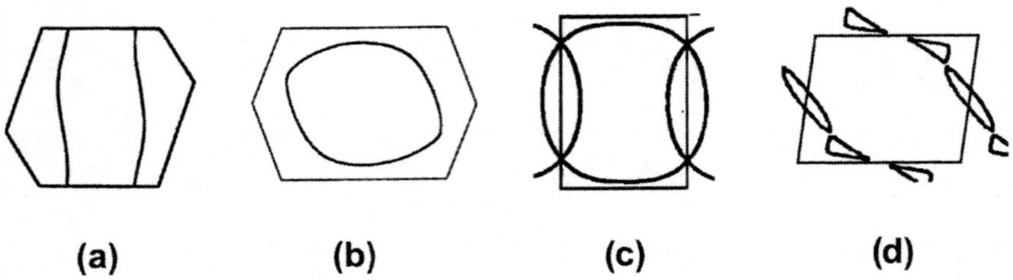

Fig. 8.1. 2D representation of typical FS geometries of organic conductors: (**a**) q1D salt $(TMTSF)_2AsF_6$ [7], and q2D compounds (**b**) β-$(BEDT\text{-}TTF)_2IBr_2$ [8], (**c**) κ-$(BEDT\text{-}TTF)_2Cu(NCS)_2$ [4], and (**d**) $(BEDO\text{-}TTF)_2ReO_4{\cdot}H_2O$ [9]

requirement can be fulfilled for many organic conductors, so that routinely available laboratory magnetic fields, $B \sim 10\,\text{T}$, are sufficient for creating the strong field regime, $\omega_c\tau \gg 1$, where ω_c is a characteristic cyclotron frequency and τ is the transport scattering time.

On the other hand, in addition to traditional methods of experimental Fermiology, such as magnetic quantum oscillations or cyclotron resonance, qualitatively new phenomena, associated with extremely high electronic anisotropy, have been found in organic conductors. In particular, the semiclassical magnetoresistance of these materials exhibits a number of novel features which can efficiently be used for quantitative description of various FS properties. As a matter of fact, rather simple FS topologies (see Fig. 8.1) make organic conductors perfect model objects for investigating general effects of a magnetic field on a quasi-one-dimensional (q1D) or quasi-two-dimensional (q2D) electronic system which can further be utilized for studying other low-dimensional compounds. Such general phenomena, characteristic of layered metallic systems, will be reviewed in the first part of this chapter.

In Sects. 8.2 and 8.3, the behavior of the semiclassical interlayer magnetoresistance in the strong field regime is addressed with the focus on the influence of the orientation of a magnetic field. We analyze qualitatively the physical origin of various angular effects in the low-frequency (basically, d.c.) magnetoresistance in terms of semiclassical electron orbits on a three-dimensional (3D), though highly anisotropic, FS and illustrate how they can be used for a detailed determination of the FS geometry. More rigorous theoretical derivations of the angular effects are presented in this book by Lebed and Wu (Chap. 7); high-frequency (microwave) magnetotransport properties are reviewed by Hill (Chap. 15).

Section 8.4 is devoted to the problem of an incoherent interlayer magnetotransport. In some compounds, the charge transfer between the layers is so weak that the FS can only be defined in two dimensions. Nevertheless, the interlayer conductivity may preserve the metallic character and exhibit many magnetic field properties common with those of the coherent magnetotransport. However, recent experiments have revealed some prominent deviations from the coherent 3D behavior, inconsistent with the present theoretical predictions.

It is well known that the Landau quantization of the electron motion on closed orbits in a strong magnetic field gives rise to oscillations of various field-dependent properties [2,10,11]. These magnetic quantum oscillations are particularly pronounced in layered organic conductors, due to their high two dimensionality. Section 8.5 provides a brief account of the Shubnikov–de Haas (SdH) and de Haas–van Alphen (dHvA) effects in organic conductors, focusing on deviations from the standard 3D behavior. More detailed reviews of the topic can be found in [6,12].

Due to the reduced dimensionality and relatively low carrier concentrations, many organic conductors exhibit strong electron correlations and, consequently, numerous instabilities of the normal metallic state, see, e.g., [13] and chapters in this book (Chaps. 3, 5, 7, 12, and 27). The energy scales of such instabilities are often so low that the ground states can be changed by applying a reasonably strong magnetic field. The second part of this chapter presents several examples of layered compounds, in which magnetic fields can effectively be used not only for monitoring the electronic state but also for switching between different states. As one of the most impressing examples, the salts α-(BEDT-TTF)$_2$MHg(SCN)$_4$ are considered in detail in Sect. 8.6. Numerous anomalies displayed by these salts in magnetic fields are discussed in view of the low-temperature charge-density-wave (CDW) state and associated changes in the electronic spectrum. Special emphasis is put on the rich "magnetic field–pressure–temperature" phase diagram, including, besides the normal metallic and low-field CDW states, different kinds of field-induced CDW (FICDW) states (see also the review by Bjeliš and Zanchi (Chap. 20) in this book) and superconductivity coexisting with the CDW [14]. In Sect. 8.7, several other examples of materials showing interesting high-field properties and field-induced phase transitions are presented.

8.2 Angle-Dependent Magnetoresistance Oscillations

In this section, we consider periodic oscillations of the interlayer magnetoresistance emerging when a constant magnetic field is turned from the direction normal to conducting layers toward the plane of the layers. Experimental observations are interpreted in terms of semiclassical electron motion in the 3D momentum space in the presence of the Lorentz force:

$$\mathbf{F}_\mathrm{L} = \mathrm{d}\boldsymbol{p}/\mathrm{d}t = e\boldsymbol{v} \times \boldsymbol{B}, \tag{8.1}$$

where $\boldsymbol{p}$, $\boldsymbol{v}$, and e are, respectively, the electron's momentum, velocity, and charge. In this model, electrons follow closed or open orbits on a 3D FS in the plane perpendicular to the field direction: $p_{\boldsymbol{B}} = \mathrm{const}$. The cyclotron frequency for a closed orbit is

$$\omega_\mathrm{c} = eB/m_\mathrm{c}, \tag{8.2}$$

where m_c is the cyclotron mass determined by the derivative of the orbit area S with respect to energy

$$m_c = (2\pi)^{-1}(\partial S/\partial \varepsilon)_{p_B=\text{const}}. \qquad (8.3)$$

In the case of an open orbit, the characteristic frequency of crossing one Brillouin zone in the direction perpendicular to $\boldsymbol{B}$ can also be expressed by (8.2) with the mass $m_c = \hbar/(v_F a)$, where v_F is the Fermi velocity and a is the relevant unit cell parameter of the crystal lattice. The electrical conductivity in the strong field limit is determined by the corresponding velocity component averaged over the cyclotron period and over all states on the FS. Therefore, the behavior of the magnetoresistance can be understood from a qualitative analysis of the average velocity.

In Sect. 8.2.1, angle-dependent magnetoresistance oscillations (AMRO) associated with a cylindrical FS are considered. As a model system for the observation of this kind of AMRO, the q2D metal β-$(BEDT\text{-}TTF)_2IBr_2$ is taken. This salt is characterized by a single cylindrical FS weakly warped along the axis of the cylinder. Taking into account a high quality of available single crystals, this compound is particularly suited for studying general strong field phenomena in q2D metals (see, e.g., [15, 16]).

Since many of layered organic conductors contain, in addition to a cylindrical FS, a pair of open sheets, as shown in Fig. 8.1c, they can also exhibit AMRO effects characteristic of purely q1D metals, such as the Bechgaard salts $(TMTSF)_2X$. These 1D AMRO phenomena will be presented in Sect. 8.2.2.

Since the crystal orientation in a magnetic field is of primary importance in most of the phenomena discussed in this chapter, we introduce notations with which the crystal vs. field orientation will be defined hereinafter. As shown in Fig. 8.2, we introduce the coordinate system with the xy-plane parallel to the plane of molecular layers and the z-axis normal to this plane. In cases of a strongly anisotropic in-plane conductivity, the x-axis will coincide with the highest conduction direction. The field orientation is defined by a polar angle θ, between the field direction and the z-axis, and an azimuthal angle φ which is formed by the projection of the field on the xy-plane and some characteristic direction in this plane, e.g., the x-axis. The transport current $\boldsymbol{I}$

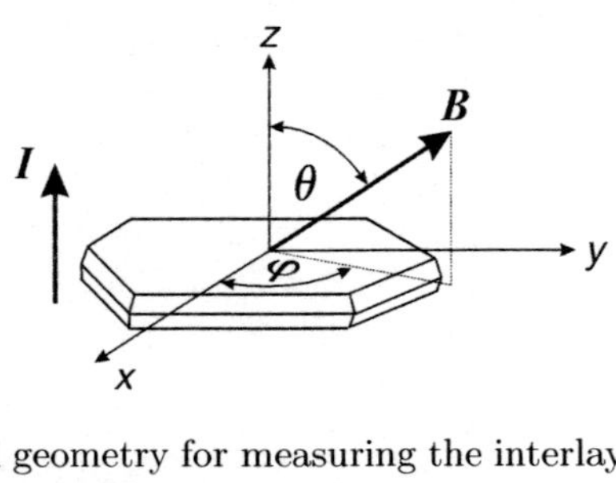

Fig. 8.2. Experimental geometry for measuring the interlayer magnetoresistance as a function of the magnetic field orientation

is always applied perpendicular to the layers, corresponding to the interlayer resistivity geometry.

8.2.1 Closed Orbits

Figure 8.3a shows an example of AMRO in the interlayer resistance of the layered superconductor β-(BEDT-TTF)$_2$IBr$_2$ recorded as a function of polar angle θ. The most prominent feature here is a regular series of sharp peaks, repeating with an increasing rate as θ approaches 90°. Soon after the first observation of AMRO in β-(BEDT-TTF)$_2$IBr$_2$ [17], similar oscillations were found in another salt, θ-(BEDT-TTF)$_2$I$_3$ [18], implying that they are a general property of q2D metals. The basic features of this oscillatory behavior can be summarized as follows [17–19]:

1. The peaks in the magnetoresistance repeat periodically with $\tan\theta$ and their positions are independent of temperature or magnetic field strength.
2. The field dependence of the interlayer resistivity changes dramatically upon changing the field orientation from a minimum to a peak in the AMRO. As shown in Fig. 8.3b, the sublinear, tending to saturation field dependence typical of the AMRO minimum (point 2 in Fig. 8.3a) transforms to a nonsaturating, approximately quadratic behavior in the AMRO maximum (point 1).
3. SdH oscillations are typically enhanced in the AMRO peaks.

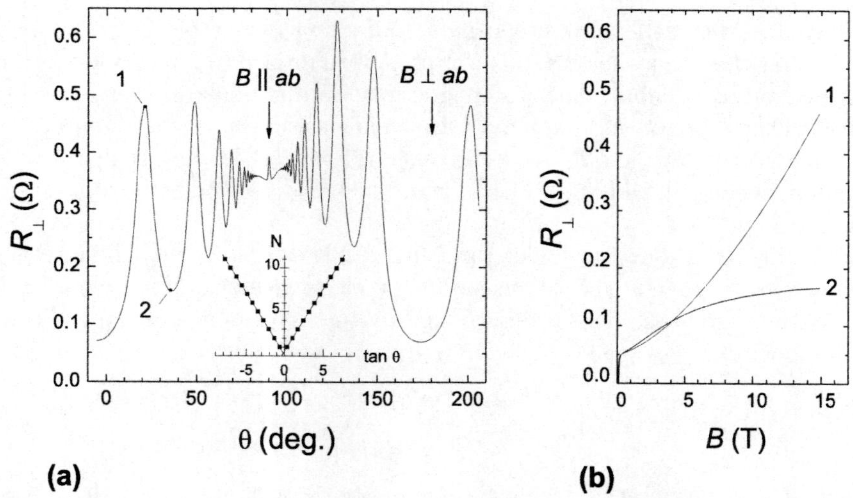

Fig. 8.3. Interlayer magnetoresistance of β-(BEDT-TTF)$_2$IBr$_2$: (**a**) angular dependence at $B = 15\,\mathrm{T}$ and $T = 1.4\,\mathrm{K}$, showing peaks periodically repeating in the $\tan\theta$ scale, as illustrated in the *inset* and (**b**) field dependence at the orientations corresponding to the first peak (*curve 1*) and to the successive minimum (*curve 2*) in the angular dependence

4. The amplitude of AMRO is much stronger for the interlayer magnetoresistance than for the intralayer one.

Feature (1) suggests that the AMRO must be associated with a specific geometry of electron orbits on a FS which is open in the direction perpendicular to the layers. Indeed, it turns out that, for a slightly warped cylindrical FS, the width of distribution of cyclotron orbit areas oscillates with changing the field orientation and almost vanishes at certain angles. In the case of the simplest q2D dispersion relation,

$$\varepsilon(\boldsymbol{p}) = \frac{p_{\parallel}^2}{2m} - 2t_z \cos(p_z a_z/\hbar), \quad t_z \ll \varepsilon_{\mathrm{F}}, \tag{8.4}$$

where $p_{\parallel} = (p_x^2 + p_y^2)^{1/2}$, m is the electron's effective mass, a_z is the interlayer spacing, $\varepsilon_{\mathrm{F}} = p_{\mathrm{F}}^2/2m$ is the Fermi energy, and t_z is the effective interlayer transfer integral, the difference between the maximum and minimum orbit areas is [20]

$$\Delta S \approx \frac{8\pi m t_z J_0(p_{\mathrm{F}} a_z |\tan\theta|/\hbar)}{\cos\theta} \tag{8.5}$$

and vanishes at the angles corresponding to zeros of the zeroth-order Bessel function $J_0(p_{\mathrm{F}} a_z |\tan\theta|/\hbar)$. For arguments of J_0 bigger than unity, this condition can be approximated as

$$|\tan\theta_N| = \frac{\pi\hbar}{p_{\mathrm{F}} a_z}(N - 1/4), \quad N = 1, 2, \ldots \tag{8.6}$$

Obviously, at θ_N, the system becomes similar to a perfectly 2D metal, in the sense that the energy spectrum is completely quantized in Landau levels. This should lead to an enhancement of magnetic quantum oscillations, in agreement with feature (3) above.[1] Moreover, the semiclassical interlayer conductivity σ_{zz} has been shown [22,23] to decrease sharply at θ_N, leading to peaks in the angular dependence of the resistivity $\rho_{zz} \approx 1/\sigma_{zz}$, periodically repeating in the $\tan\theta$ scale.

Qualitatively, one can understand this result as follows [8]. The interlayer velocity v_z averaged over the period of the electron motion on a closed orbit in a strong magnetic field is determined by the dependence of the orbit area on its position in $\boldsymbol{p}$-space:

$$\overline{v_z} = \overline{\partial\varepsilon/\partial p_z} = -\frac{\partial S(P_z)/\partial P_z}{(\partial S/\partial\varepsilon)_{P_z}} = -\frac{\partial S(P_z)/\partial P_z}{2\pi m_{\mathrm{c}}}, \tag{8.7}$$

where P_z is the point at which the plane of the orbit intersects the p_z-axis, and the cyclotron mass, $m_{\mathrm{c}} = m/\cos\theta$, monotonically changes with θ. The

[1] In some extremely anisotropic compounds, the interlayer bandwidth $4t_z$ is always smaller than the Landau level spacing $\hbar\omega_{\mathrm{c}}$ at fields necessary for the observation of quantum oscillations. In that case, the amplitude of the quantum oscillations is not affected by the AMRO, see, e.g., [21].

derivative $\partial S(P_z)/\partial P_z$ is, generally, finite, so that the interlayer velocity has a finite value for all P_z. Therefore, the conductivity σ_{zz}, which is determined by $\overline{v_z}(P_z)$, saturates with increasing the field. However, at the angles θ_N satisfying condition (8.6), $\partial S(P_z)/\partial P_z \approx 0$. This leads to the vanishing averaged velocity $\overline{v_z}$ and, hence, vanishing σ_{zz}. One can show [22, 23] that, as long as $\omega_c\tau \ll \varepsilon_F/t_z$, the interlayer conductivity decreases with magnetic field as $1/B^2$ at these angles. In organic metals, the ratio ε_F/t_z is typically $\sim 10^2$–10^3, whereas the value $\omega_c\tau = 10^2$ can hardly be achieved even for the cleanest samples at $B < 100\,\mathrm{T}$. Therefore, the resistivity ρ_{zz} is expected to increase proportional to B^2 at θ_N at all accessible, nondestructive magnetic fields. This conclusion is qualitatively consistent with the experimentally observed behavior illustrated in Fig. 8.3b.

The above consideration allows us to interpret the AMRO phenomenon as a periodically (in $\tan\theta$) repeating *dimensional crossover* between a strongly anisotropic, yet still 3D, transport and an almost ideally 2D case with a vanishingly small effective interlayer transfer. Indeed, quantum mechanical calculations [24, 25] have shown that, under a strong magnetic field, the effective interlayer tunneling amplitude oscillates as a function of θ and almost vanishes at the angles satisfying (8.6).

Equation (8.6) shows that the AMRO period is directly related to the Fermi momentum. However, this formula has been derived for an ideal cylindrical FS with a circular transverse cross section and the simplest cosine warping along p_z. In a more realistic case of an anisotropic in-plane dispersion $\varepsilon(p_x, p_y)$, the AMRO period depends on the orientation of the field component parallel to the layers, $\boldsymbol{B}_\parallel$. Then p_F on the right-hand side of (8.6) should be replaced by the maximum projection $p_B^{\max}$ of the Fermi momentum on the direction $\boldsymbol{B}_\parallel$ [23]. The importance of this correction is seen from Fig. 8.4 which illustrates how $p_B^{\max}$ is defined and related to the momentum on the FS. Now, by measuring AMRO periods for different azimuthal angles φ, it is possible to determine $p_B^{\max}(\varphi)$ and graphically deduce the shape and size of the FS in the p_xp_y-plane [8]. One can, of course, derive the in-plane Fermi momentum analytically, from the known dependence $p_B^{\max}(\varphi)$, in the way similar to that suggested by Kovalev et al. [26]:

$$\left|\boldsymbol{p}_\parallel^{\max}\right| = \sqrt{(p_B^{\max})^2 + (\mathrm{d}p_B^{\max}/\mathrm{d}\varphi)^2}, \tag{8.8}$$

$$\psi = \varphi + \arctan\left(\frac{\mathrm{d}p_B^{\max}/\mathrm{d}\varphi}{p_B^{\max}}\right), \tag{8.9}$$

with notations $\boldsymbol{p}_\parallel^{\max}$ and ψ explained in Fig. 8.4.

For approximate estimations, one can assume the FS cross section to have an elliptic form. Then the principal semiaxes of the ellipse, $\boldsymbol{p}_1$ and $\boldsymbol{p}_2$, are related to $p_B^{\max}$ as [27]

$$\left[p_B^{\max}(\varphi)\right]^2 = (p_1\cos\varphi)^2 + (p_2\sin\varphi)^2, \tag{8.10}$$

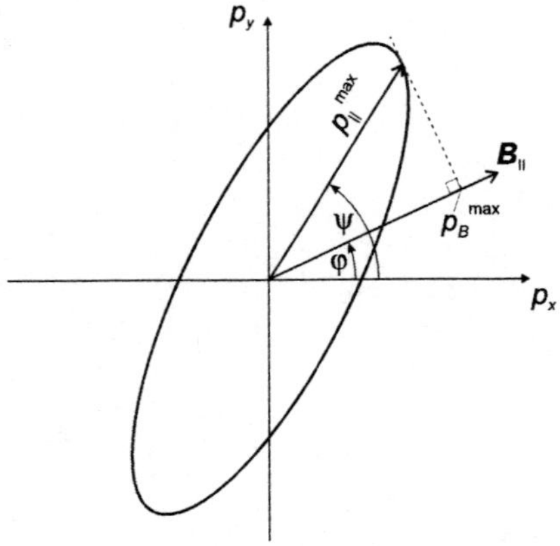

Fig. 8.4. Transverse cross section of a cylindrical FS. $\boldsymbol{B}_{\parallel}$ is the field component parallel to xy-plane; $\boldsymbol{p}_{\parallel}^{\max}$ is the in-plane component of the Fermi momentum whose projection on the direction $\boldsymbol{B}_{\parallel}$ attains the maximum value $p_B^{\max}$

where φ is counted from the direction of $\boldsymbol{p}_1$. However, one should keep in mind that organic conductors have generally rather low-symmetric crystal structures and irregular FSs. For example, in β-(BEDT-TTF)$_2$IBr$_2$, the shape of the FS cross section clearly deviates from an ellipse [8, 15]. Therefore, for exact determination of the FS, it is preferable to use the direct procedure described above.

It is possible to further generalize the expression for θ_N by taking into account an oblique warping of the Fermi cylinder, i.e., when the interlayer dispersion, $-2t_z\cos(p_z a_z/\hbar + \alpha)$, includes a phase offset depending on the in-plane momentum $\boldsymbol{p}_{\parallel}$: $\alpha(p_x, p_y) = -\alpha(-p_x, -p_y)$. If this phase can be represented by a linear combination, $\boldsymbol{u}\cdot\boldsymbol{p}_{\parallel} \equiv u_x p_x + u_y p_y$, the AMRO position is expressed in the form [8]:

$$|\tan\theta_N| = \frac{\pi\hbar(N-1/4) \pm \boldsymbol{u}\cdot\boldsymbol{p}_{\parallel}^{\max}}{p_B^{\max} a_z}, \tag{8.11}$$

where sign + (−) is chosen depending on the sign of $\tan\theta_N$. Thus, the warping direction can be estimated from the data on the asymmetry of the AMRO positions with respect to $\theta = 0°$.

Finally, we note that higher harmonics in the interlayer dispersion, besides damping the amplitude of AMRO, should lead to a shift of the AMRO position [28]. Usually, such a shift is quite small and difficult to distinguish from the experimental error. One can, however, judge about the presence of the higher harmonics by inspecting the angular dependence of both, the semiclassical magnetoresistance and the SdH oscillations. The presence of an appreciable second harmonic in the interlayer dispersion has been predicted to shift the position of the maximum of the SdH amplitude from the AMRO peak [28]. For example, the data on β-(BEDT-TTF)$_2$IBr$_2$ shown in Fig. 8.5

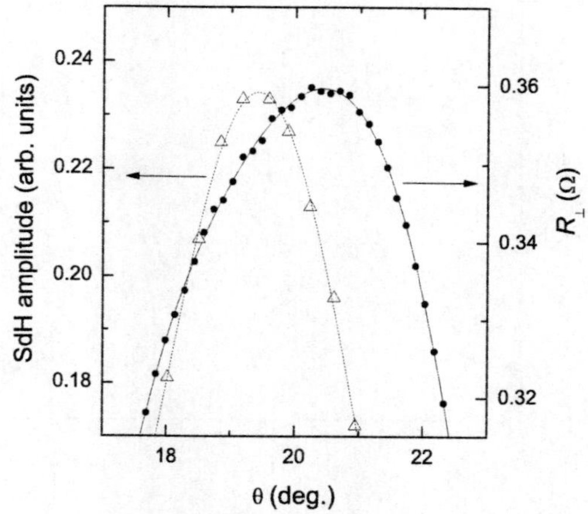

Fig. 8.5. Amplitude of the SdH oscillations (*open triangles*) and interlayer resistance (*solid circles*) of β-(BEDT-TTF)$_2$IBr$_2$ as a function of angle θ in the vicinity of the first AMRO peak [29]. The maximum in the SdH amplitude is shifted by $\approx 1^\circ$ from that of the semiclassical resistance, suggesting the presence of a higher harmonic contribution to the interlayer dispersion

reveal a difference of $\approx 1^\circ$ in the positions of the first AMRO peak and SdH maximum that suggests a considerable anharmonicity of the FS warping in this compound.

8.2.2 Open Orbits

FSs of many layered compounds, e.g., α-, κ-, and λ-type salts of BEDT-TTF and its derivatives, include, in addition to a cylindrical part, a pair of corrugated sheets which are open in the plane of the layers (see, e.g., Fig. 8.1c). This may lead to new features in magnetoresistance in comparison to those considered above. In this section, we make a brief overview of angular oscillations which are specific to open Fermi sheets and can, therefore, be called 1D AMRO, by contrast to the 2D AMRO associated with a cylindrical FS.

Consider a metallic system with a FS consisting of sheets extended perpendicular to the x-axis and warped along the y-axis, much stronger than along the z-axis. The most prominent examples of such q1D metals with strongly anisotropic coupling between the 1D chains are the Bechgaard salts (TMTSF)$_2$X [13]. For such compounds, the electron dispersion relation can be written in the lowest order tight-binding approximation as

$$\varepsilon(\boldsymbol{p}) = -2t_x \cos(p_x a_x/\hbar) - 2t_y \cos(p_y a_y/\hbar) - 2t_z \cos(p_z a_z/\hbar) - \varepsilon_{\mathrm{F}}. \quad (8.12)$$

Here x- and z-axes are associated with the most and the least conducting directions, respectively, so that the transfer integrals are $t_x \gg t_y \gg t_z$; a_x, a_y, and a_z are the corresponding lattice constants. The dispersion along the x-axis is often approximated to be linear near the Fermi level, so that

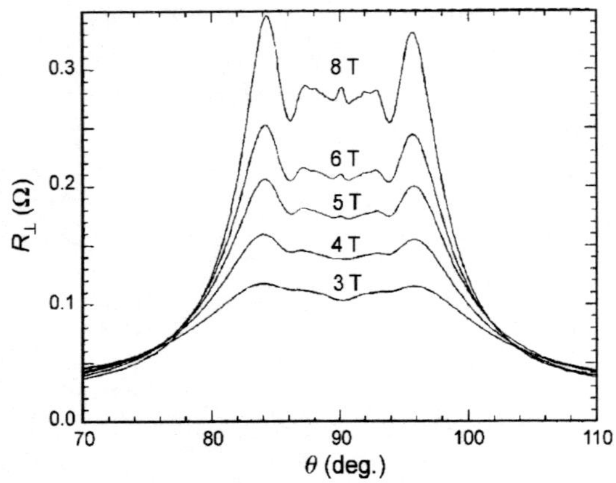

Fig. 8.6. DKC oscillations in the metallic state of $(TMTSF)_2ClO_4$ [30]: the field is rotated in the xz-plane; the resistance is measured along the z-direction

$$\varepsilon(\boldsymbol{p}) \approx v_F(|p_x| - p_F) - 2t_y \cos(p_y a_y/\hbar) - 2t_z \cos(p_z a_z/\hbar), \tag{8.13}$$

where v_F and p_F are, respectively, the Fermi velocity and Fermi momentum at $t_y = t_z = 0$.

By contrast to the case of a cylindrical FS, where a change in the azimuthal orientation of the field rotation plane may only cause a gradual change in the AMRO position, in the present case, qualitatively different phenomena are observed depending on whether the field is rotated in the xz-plane or the yz-plane.

Starting with the xz rotation, Fig. 8.6 shows the oscillations of the interlayer magnetoresistance of $(TMTSF)_2ClO_4$ observed by Danner, Kang, and Chaikin (DKC) [30] as the field was tilted from the direction of the lowest conductivity toward the chain direction. Despite the apparent similarity to the 2D AMRO, the origin of these DKC oscillations is quite different [30]. Assume the linearized dispersion (8.13) and consider the electron motion under the Lorentz force (8.1) in a magnetic field $\boldsymbol{B} = (B\sin\theta, 0, B\cos\theta)$. At a finite θ (not too close to $\pm\pi/2$: $|\theta| - \pi/2 \gg t_z/t_x$), the trajectories of electrons are extended along the p_y-axis; the frequency of crossing one Brillouin zone along p_y is

$$\omega_y = \frac{a_y}{\hbar}\left|\frac{\mathrm{d}p_y}{\mathrm{d}t}\right| = \frac{ev_F a_y}{\hbar} B_z. \tag{8.14}$$

At the same time, the interlayer momentum p_z oscillates

$$p_z(t) = p_z(0) + \frac{2t_y}{v_F}\tan\theta \sin(\omega_y t). \tag{8.15}$$

At small θ, the amplitude of the oscillation, $4t_y \tan\theta/v_F$, is smaller than the size of the Brillouin zone in the p_z-direction; the interlayer velocity v_z takes only a part of allowed values and its average over the periodic motion, $\overline{v_z}$, is, in general, nonzero. However, when, with increasing θ, the amplitude becomes equal to $2\pi\hbar/a_z$, the average $\overline{v_z}$ becomes composed of equal positive and negative contributions over the path and, therefore, vanishes. The resistivity is maximum at this point. The next zero of $\overline{v_z}$ and peak in ρ_{zz} are expected when the orbit covers exactly two Brillouin zones in the p_z-direction, then three, and so on. Thus, the period of the magnetoresistance oscillations can be expressed as

$$\Delta(\tan\theta) = \frac{2\pi\hbar/a_z}{4t_y/v_\mathrm{F}} = \frac{\pi\hbar v_\mathrm{F}}{2t_y a_z}. \tag{8.16}$$

Taking into account that the motion along p_z is a sinusoidal rather than a saw-tooth function of time, a more exact condition for $\overline{v_z} = 0$ is determined by zeros of the Bessel function $J_0[2t_y a_z \tan\theta/(\hbar v_\mathrm{F})]$ [30]. However, since the argument of the Bessel function is bigger than unity in the range of interest, (8.16) is a good expression for estimating the oscillation period.

The DKC oscillations can effectively be used for determination of FS warping along p_y: provided the interlayer distance a_z and the Fermi velocity along the chains are known, one can determine the transfer integral t_y from the oscillation period (8.16). For $(\mathrm{TMTSF})_2\mathrm{ClO}_4$ in the anion-ordered state, the value $t_y = 12\,\mathrm{meV}$ was obtained [30]. Above the ordering temperature, t_y should be doubled, yielding $t_y = 24\,\mathrm{meV}$. This value is comparable to $t_y \approx 32\,\mathrm{meV}$ found for $(\mathrm{TMTSF})_2\mathrm{PF}_6$ under a pressure of $\approx$10 kbar [31].

Another kind of 1D AMRO emerges at rotating a magnetic field in the yz-plane, i.e., in the plane of the open FS: $\boldsymbol{B} = (0, B\sin\theta, B\cos\theta)$. In this case, electron orbits are extended both in the p_y- and p_z-directions. The frequencies of crossing one $\boldsymbol{p}$-space unit cell along p_z,

$$\omega_z = \frac{e v_\mathrm{F} a_z}{\hbar} B_y, \tag{8.17}$$

and along p_y (8.14) are different and the electron motion is, generally, aperiodic. However, as was first noted by Lebed [32], when the field is directed parallel to a translation vector of the crystal lattice, $\boldsymbol{R}_{mn} = (0, ma_y, na_z)$, i.e., when

$$\tan\theta = \frac{m}{n}\frac{a_y}{a_z}, \tag{8.18}$$

where m and n are integers, the frequencies are commensurate (one can consider this as a kind of resonance) and the electron motion in $\boldsymbol{p}$-space becomes periodic. One can expect this periodicity to affect the transport properties. Indeed, Naughton et al. [33] and Osada et al. [34] have observed an anomalous decrease of magnetoresistance in the metallic state of $(\mathrm{TMTSF})_2\mathrm{ClO}_4$ at the *Lebed magic angles* (LMAs) satisfying condition (8.18). Later, the same effect has been found in other q1D compounds: $(\mathrm{TMTSF})_2\mathrm{X}$ ($\mathrm{X} = \mathrm{PF}_6$, ReO_4) [35–38], $(\mathrm{DMET\text{-}TSeF})_2\mathrm{X}$ ($\mathrm{X} = \mathrm{AuCl}_2$, AuI_2) [39, 40], $(\mathrm{DMET})_2\mathrm{CuCl}_2$ [41], and (BEDT-TTF)(TCNQ) [42]. Examples of the LMA oscillations in the high-pressure metallic state of the Bechgaard salts $(\mathrm{TMTSF})_2\mathrm{X}$ ($\mathrm{X} = \mathrm{PF}_6$, ReO_4) are shown in Fig. 8.7. Note that sharp dips of the magnetoresistance

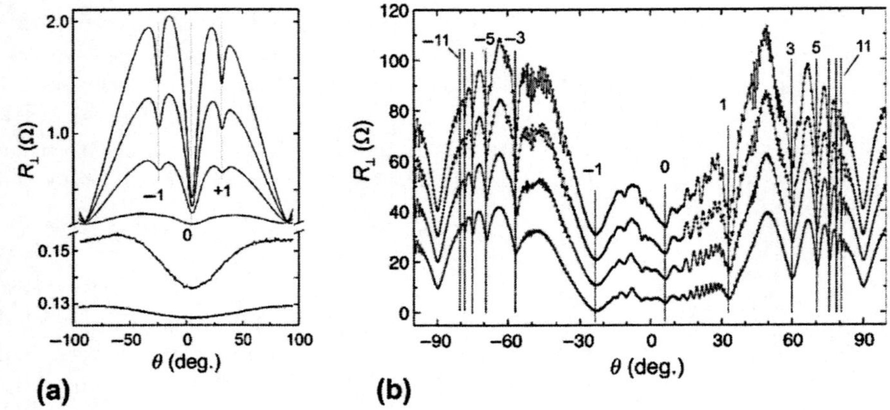

Fig. 8.7. Interlayer magnetoresistance of the high-pressure metallic state of the Bechgaard salts $(TMTSF)_2X$ at a magnetic field rotating in the yz-plane. Different curves correspond to different field intensities: (**a**) $X = PF_6$, $P = 10\,\text{kbar}$, $T = 1.3\,\text{K}$, $B = 0.1$, 0.3, 1, 3, 5, and 7 T (*bottom to top*) [35] and (**b**) $X = ReO_4$, $P = 12.2\,\text{kbar}$, $T = 1.5\,\text{K}$, $B = 5$, 6, 7, and 8 T (*bottom to top*) [36]. The magnetoresistance exhibits sharp dips at angles (8.18) with integer values of the ratio m/n

are observed at the angles satisfying condition (8.18) with integer values of m/n. For the ReO_4 salt, the ratio m/n takes only odd integer values due to a superstructure associated with anion ordering (see [43] for details).

To explain the LMA oscillations in q1D conductors, a number of theoretical models considering the field-induced density-wave instability [32, 44], many-body effects [45] and violations of the Fermi liquid behavior [31, 46] have been proposed. However, we consider here only one-particle Fermi liquid models as they appear to provide a satisfactory explanation for the case of normal metallic q2D compounds.

One of such models, proposed by Osada et al. [47], is based on including higher-order interchain transfer terms into the electronic dispersion (8.13):

$$\varepsilon(p_y, p_z) = -\sum_{k,l} t_{kl} \cos[(k a_y p_y + l a_z p_z)/\hbar]. \tag{8.19}$$

In this case, the solution of the semiclassical kinetic equation yields peaks in the conductivity at LMAs (8.18). This result has a simple physical meaning. At a general field orientation, all the (k, l)-th contributions to the velocity perpendicular to the chains, being proportional to $t_{kl} \sin\{[k a_y p_y(t) + l a_z p_z(t)]/\hbar\}$, oscillate around zero and their time averages vanish at strong fields. However, at the angles satisfying (8.18), the contribution from the hopping term t_{mn} is exactly parallel to $\boldsymbol{B}$. Therefore, it is not affected by the Lorentz force, remaining constant at an increasing field. As a result, the total average velocity saturates at a finite value, causing a peak in the conductivity. The magnitude of the effect depends on the significance of the (m, n)-th term in dispersion (8.19). Since the transfer integrals $t_{k,l}$ with $l > 1$ are usually negligibly small in the layered compounds, the conductivity peaks are mainly pronounced at LMA with integer indices, $m/n = N$, in agreement with experiment.

The above model is likely applicable to q2D metals, in which the open Fermi sheets are strongly warped and the electron mean free path extends over many unit cells in the y-direction. However, the presence of high-index features in some TMTSF salts (see, e.g., Fig. 8.7b) can hardly be attributed to high-order interchain transfer terms in these q1D compounds.

An alternative explanation of the LMA oscillations [48, 49] does not imply a long-range interchain transfer, instead, taking into account the nonlinearity of the electron spectrum (8.12) in the x-direction. The latter leads to a variation of the intrachain velocity v_x on the Fermi level with the interchain momentum p_y. As a result, the characteristic frequencies described by (8.14) and (8.17) with $v_x(p_y)$ substituted for v_F become dependent on p_y. In particular, there are "effective" parts on each orbit where the frequencies are the lowest and which, therefore, give the main contribution to the time-averaged velocity $\overline{v_z}$. Since ω_y and ω_z are generally incommensurate, the velocity $v_z \propto \sin[a_z p_z(t)/\hbar]$ at the "effective" parts takes all possible values and averages to zero. However, at commensurate field directions (8.18), the $\boldsymbol{p}$-space orbits become periodic and v_z is the same at different "effective" parts. This leads to a finite $\overline{v_z}$, hence, to an enhanced interlayer conductivity σ_{zz}. An explicit quantum mechanical analysis of a q1D system with a momentum-dependent intrachain velocity $v_x(p_y)$ was performed by Lebed and Naughton [49]. The conductivity σ_{zz} has indeed been found to exhibit peaks at LMA. This result was, however, associated with quantum interference effects rather than with "effective" parts on semiclassical electron orbits.

While the above consideration of the LMA effect assumes an orthorhombic crystal lattice, most of organic metals are triclinic. The generalization is straightforward: Consider a triclinic system with the unit cell vectors $\boldsymbol{a}$, $\boldsymbol{b}$, and $\boldsymbol{c}$, such that $\boldsymbol{a}$ is parallel to the most conducting direction (chains) and $\boldsymbol{ab}$ is the most conducting plane (layers). Then in an orthogonal coordinate system with $\hat{\boldsymbol{x}} \parallel \boldsymbol{a}$ and the xy-plane coplanar with $\boldsymbol{ab}$, the LMA condition is written as

$$\tan\theta = \frac{m}{n}\frac{b_y}{c_z} - \frac{c_y}{c_z} \tag{8.20}$$

(remind that for a strong layer-type anisotropy $m/n = N$), i.e., the field must lie along the component of a lattice vector in the plane normal to the chains.

So far, we discussed the magnetoresistance at a field rotating in either the xz- or yz-planes. Now let us turn to a more general orientation, $\boldsymbol{B} = B(\sin\theta\cos\varphi, \sin\theta\sin\varphi, \cos\theta)$. Figure 8.8 shows the angle-dependent interlayer magnetoresistance of the metallic state of $(TMTSF)_2PF_6$ reported by Lee and Naughton [50]. The lower panel shows the data at a field strongly inclined toward the xy-plane and rotating around the z-axis; the upper panel

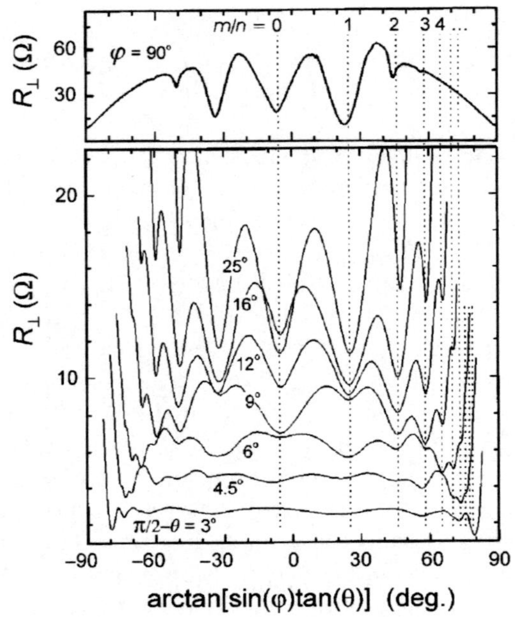

Fig. 8.8. Angular dependence of the interlayer magnetoresistance of $(TMTSF)_2PF_6$ at $P = 8.3\,$kbar, $B = 7\,$T, $T = 0.32\,$K [50]: (*upper panel*) field rotation in the yz-plane yields dips at LMAs with integer indices $|N| \leq 4$; (*lower panel*) field rotation around the z-axis; φ is the angle between the in-plane field component and the chain direction; different curves correspond to different tilt angles θ with respect to the z-axis; dips of the resistance are observed when the field orientation satisfies (8.21) with $|N| \leq 9$

represents a pure yz rotation. Vertical dashed lines are drawn through the dips of the resistance and correspond to the orientations:

$$\sin\varphi\,\tan\theta = N\frac{b_y}{c_z} - \frac{c_y}{c_z}, \tag{8.21}$$

i.e., when the field component normal to the chains satisfies (8.20) with integer values of m/n. Thus, (8.21) can be regarded as a generalized expression for LMAs. Note, however, that the resistance dips are seen up to much higher $|N|$ in the lower panel than in the upper one, suggesting a new mechanism of the oscillations to turn on when the field has a finite component along the x-axis in addition to those along the y- and z-axes. This was indeed confirmed by a number of theoretical works [51–54] (see also Chap. 7 for a review). It was shown that, under a magnetic field having a significant component along x, the interlayer conductivity of a system described by the dispersion relation (8.13) can be expressed as

$$\sigma_{zz}(B,\theta,\varphi) = \sigma_{zz}(0) \sum_{N=-\infty}^{\infty} \frac{J_N^2\left(\frac{2t_y a_z}{\hbar v_F}\tan\theta\cos\varphi\right)}{1+(\omega_y\tau)^2\left(N-\frac{a_z}{a_y}\tan\theta\sin\varphi\right)^2}, \tag{8.22}$$

where $\omega_y \propto B_z$ is the characteristic frequency (8.14) of electron motion in $\boldsymbol{p}$-space. When the field is directed at an LMA, the denominator of the corresponding Nth term in this expression reduces, producing a peak in the conductivity. The LMA oscillations are modulated by the generalized DKC effect represented by the Nth-order Bessel function J_N. Thus, (8.22) describes a superposition of both types of 1D AMRO.

This result can be explained in terms of semiclassical orbits in $\boldsymbol{p}$-space as proposed by Lebed and Naughton [53]. As shown by the upper curve in Fig. 8.9, the electron trajectories in the $p_y p_z$-plane are no longer straight lines, as it was in the case of a pure yz rotation, but oscillate with the amplitude $\Delta p_z = 4t_y \tan\theta\cos\varphi/v_F$. At an LMA field direction, an electron, following the trajectory in Fig. 8.9, is displaced along p_z by an integer number of unit cells, $N(2\pi\hbar/a_z)$, during one oscillation period, $T_y = 2\pi/\omega_y$. The interlayer velocity is then a periodic function of time (middle curve in Fig. 8.9); its average, contributed mainly by the trajectory parts near extremal values of $p_z(t)$, is generally nonzero. Thus, the electron acquires a finite shift along the z-axis as shown by the lower curve in Fig. 8.9. As a result, the conductivity σ_{zz} exhibits a peak. As noted above, the LMA peaks are modulated due to the DKC effect, which is directly related to the commensurability between amplitude of the oscillation Δp_z and the $\boldsymbol{p}$-space unit cell $2\pi\hbar/a_z$.

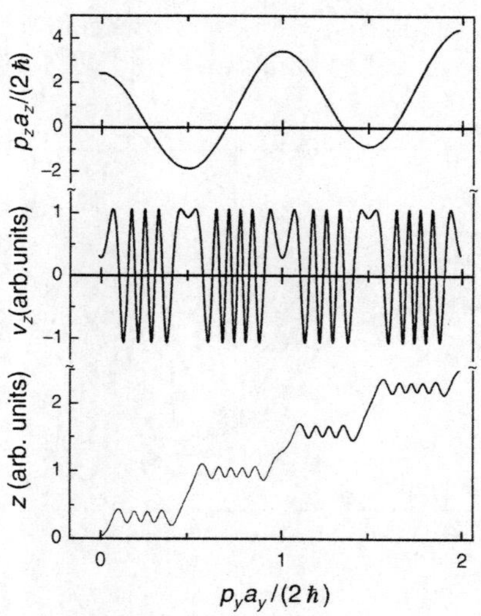

Fig. 8.9. Electron trajectory, interlayer velocity $v_z(t)$, and coordinate $z(t)$ under a magnetic field directed at the LMA with $N = 1$, see text (after [53])

From the quantum mechanical point of view, these generalized LMA–DKC oscillations are a result of interference of electron waves due to multiple Bragg reflections from the Brillouin zone boundaries [52, 53]. Recently, Cooper and Yakovenko [55] have proposed an elegant interpretation in terms of Aharonov–Bohm interference of electrons in the interlayer tunneling process.

An excellent quantitative agreement between the theory and experimental data on $(TMTSF)_2ClO_4$ has been obtained for $|\varphi| \leq 20°$ [56, 57]. Note, however, that at $|\varphi| \to 90°$ all the Bessel functions except J_0 vanish in (8.22), so that only the $N = 0$ peak remains in σ_{zz}. Therefore, one of the introduced above models [47, 49] should be invoked to account for the LMA oscillations at a pure yz rotation.

8.3 Other Effects of the Field Orientation on the Semiclassical Magnetoresistance

While the AMRO effects introduced above require a significant out-of-plane field component, the interlayer magnetoresistance also turns out to be very sensitive to the direction of the field aligned exactly or nearly parallel to conducting layers.

8.3.1 In-Plane Field Orientation

We start with the case of open Fermi sheets in which the azimuthal field orientation should obviously play an important role. Figure 8.10 shows the magnetoresistance of $(TMTSF)_2ClO_4$ as a function of the angle between the field lying in the plane and the chain direction [58]. It is maximum at the field

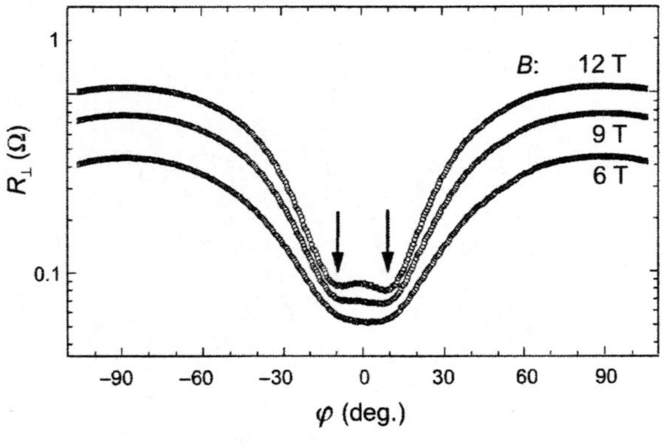

Fig. 8.10. Interlayer resistance of $(TMTSF)_2ClO_4$ under a magnetic field $\boldsymbol{B}$ rotating in the plane of the layers [58]; $T = 1.7\,\mathrm{K}$, φ is the angle between the field and the chain direction. *Arrows* point to minima of the magnetoresistance manifesting the "third angular effect" (TAE)

perpendicular to the chains. At this orientation, the charge carriers move in $\boldsymbol{p}$-space along p_z and the interlayer velocities oscillate around zero with the maximum frequency $\omega_z^{\max} = ev_\mathrm{F}a_zB/\hbar$ (see (8.17)). As the field is turned toward the chain direction, the frequency $\omega_z \propto B\sin\varphi$ decreases, leading to a decrease of the magnetoresistance. However, instead of coming smoothly to a minimum at $\varphi = 0°$, the $R_\perp(\varphi)$ curve shows a sudden change of the slope at $|\varphi| \approx 10°$ which develops into two clear minima at high fields.

This, so-called third angular effect, TAE (after the LMA and DKC oscillations as the first two angular effects in q1D conductors) was originally observed by Yoshino et al. [59] on $(\mathrm{DMET})_2\mathrm{I}_3$. Osada et al. [58] have recognized the effect as an inherent feature of the interlayer magnetoresistance of q1D conductors and associated it with a topological change of electron orbits on a q1D FS determined by the dispersion relation (8.13). Their numerical simulations have revealed a minimum in the magnetoresistance near the angle:

$$\varphi_\mathrm{c} = \arctan\left(\frac{2t_ya_y}{\hbar v_\mathrm{F}}\right), \tag{8.23}$$

where the same notations as in (8.13) are used.

The physical origin of the TAE was initially attributed to the appearance of small closed orbits on a warped (in both p_y- and p_z-directions) Fermi sheet at $|\varphi| \leq \varphi_\mathrm{c}$ [58]. However, as pointed out by Lebed and Bagmet [60], the effect can be explained without involving the closed orbits. Qualitatively, under a magnetic field parallel to the layers, electrons, generally, move on the FS along p_z under the Lorentz force. Their interlayer velocity, $v_z \propto \sin[a_zp_z(t)/\hbar]$, rapidly oscillates around zero and, hence, their contribution to the conductivity σ_{zz} diminishes with increasing the field. However, at $|\varphi| \leq \varphi_\mathrm{c}$, the in-plane velocity of some electrons on the FS is parallel or nearly parallel to $\boldsymbol{B}$. For such electrons, the Lorentz force is vanishingly small, therefore, they are the most effective in the interlayer charge transport. When the field is exactly directed at the critical angle φ_c, as schematically shown in Fig. 8.11, the number of these *effective* electrons is maximum (hatched part of the FS centered at the inflection point P_I in Fig. 8.11). This gives rise to a peak in σ_{zz} or, equivalently, a minimum in the interlayer resistivity. This explanation of the TAE has been confirmed by a detailed numerical analysis of contributions of different orbits to the interlayer conductivity [50, 61]. The calculations show that the conductivity in vicinity of φ_c is dominated by the electrons situating near the inflection line of the 3D FS, no matter whether the corresponding orbits are closed or open: even electrons sitting on the closed orbit do not complete the whole circulation during the scattering time, since the Lorentz force acting on them is vanishingly small.

The condition for the magnetoresistance minimum (8.23) can be used for evaluating the warping of the FS in the p_xp_y-plane. For example, it was applied to estimate the ratio t_y/t_x and its dependence on pressure in the $(\mathrm{DMET})_2\mathrm{X}$ salts [62, 63] in which the DKC oscillations, giving a similar information, are not observed.

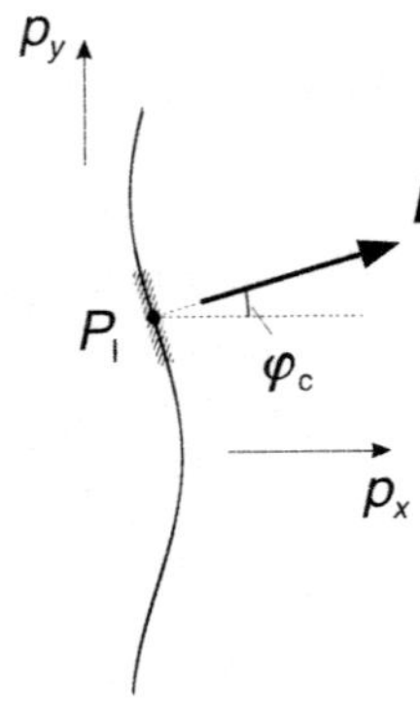

Fig. 8.11. The electrons on the FS, whose velocity is almost parallel to the magnetic field applied in the xy-plane, experience a vanishingly small Lorentz force; their interlayer velocity component does not change in time and, thus, they dominate in the interlayer conductivity σ_{zz}. When the field is directed perpendicular to the inflection point P_I of the open FS, the number of such *effective* electrons (corresponding to the hatched area around P_I) is maximum, providing a maximum of $\sigma_{zz}(\varphi)$ (after [60])

The concept of effective electrons [60] can also be applied to describe the effect of the in-plane field rotation in the case of a cylindrical FS. Indeed, independently of the FS topology within the plane, the relative number of the effective electrons dominating the interlayer conductivity is determined by the in-plane curvature of the FS, κ_FS, at the point $\boldsymbol{P}_B$ at which the in-plane velocity component $\boldsymbol{v}_\parallel(\boldsymbol{P}_B)$ is exactly parallel to $\boldsymbol{B}$. At a general field orientation, such that $\kappa_\mathrm{FS}(\boldsymbol{P}_B) \neq 0$, the interlayer resistivity was shown to be [60]

$$\rho_{zz} \approx \frac{1}{\sigma_{zz}} \propto v_\parallel^2(\boldsymbol{P}_B)\kappa_\mathrm{FS}(\boldsymbol{P}_B)B, \tag{8.24}$$

at moderately strong fields, $1 \leq \omega_{z0}\tau \leq (\varepsilon_\mathrm{F}/t_z)^\alpha$, where $\omega_{z0} = ea_z v_\parallel B/\hbar$ and $\alpha = 2/3$ for a q1D system (8.19) and 1/2 for a q2D system (8.4). The latter condition is normally fulfilled for clean samples of organic metals in fields $\sim$10 T.

The present model explains, at least, qualitatively, the very strong φ dependence of the magnetoresistance in β-$(BEDT\text{-}TTF)_2IBr_2$, pointing out a considerable nonellipticity of its FS cross section [15, 64]. On the other hand, if the shape of the cylindrical FS is known from an independent experiment, for example, from AMRO, the φ-dependent magnetoresistance (8.24) can be used to map the anisotropic in-plane Fermi velocity $v_\parallel(\varphi)$.

8.3.2 Coherence Peak

Turning back to the out-of-plane rotations, we note a common feature in the θ-dependent magnetoresistance of a q2D system (Fig. 8.3a) and, for the xy

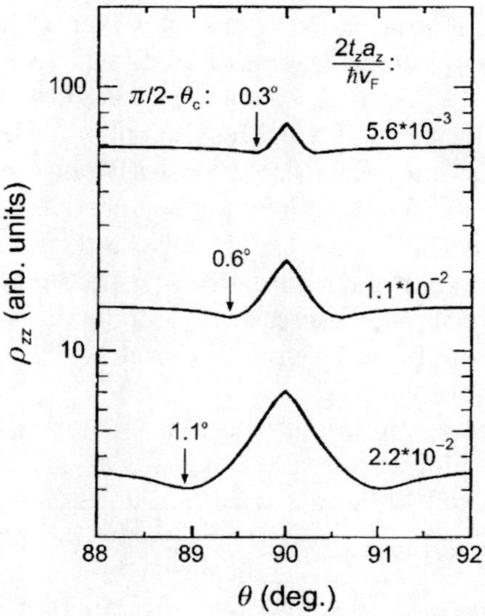

Fig. 8.12. The 90° peak feature in the angle-dependent interlayer resistivity of a q2D metal under a strong magnetic field, $\omega_c\tau(\theta = 0°) = 18$, calculated on the basis of the semiclassical kinetic equation [65]. One can see a clear correlation between the width of the peak and the anisotropy parameter $2t_z a_z/(\hbar v_F)$

rotation, of a q1D system (Fig. 8.6): in both cases, a narrow peak around $\theta = 90°$ is observed [8, 17, 30, 38]. A detailed study of the peak performed on a q2D compound β_H-(BEDT-TTF)$_2$I$_3$ [65] has shown that its width is independent of the field strength. This suggests that the effect is caused by a change in the geometry of cyclotron orbits at a certain orientation.

The peak has been reproduced by numerical simulations based on the semiclassical kinetic equation for both the q1D [30] and q2D [65] cases. The simulations have revealed a correlation between the peak width and the warping of the FS along p_z. Figure 8.12 shows the calculated resistivity of a q2D system (8.4) in the vicinity of $\theta = \pi/2$ [65]. One can see that, as θ approaches $\pi/2$, ρ_{zz} exhibits a local minimum followed by a sharp increase. The angle θ_c corresponding to the minimum is given by

$$\tan\left(\frac{\pi}{2} - \theta_c\right) \approx \frac{2t_z a_z}{\hbar v_F} = \frac{t_z}{\varepsilon_F}\frac{p_F a_z}{\hbar}. \qquad (8.25)$$

This expression is obviously analogous to that given above for the TAE critical angle (8.23) (note that the angle θ is counted from the *least* conducting direction, whereas in the TAE description φ is counted from the *most* conducting direction). This suggests the present effect to have a similar geometrical origin. Indeed, one can illustrate it by a scheme like that shown in Fig. 8.11 but with p_y replaced by p_z and with electron orbits extended in the $p_x p_z$-plane and tilted by a small angle $(\pi/2 - \theta)$ from the p_z-axis. Of course, the orbits are different in the q1D and q2D systems; however, their parts on the side of the FS facing the magnetic field are essentially the same. Again, similarly to the TAE case, the critical angle, θ_c, corresponds to the field direction normal to the inflection point of the FS in the p_xp_z-plane. The conductivity σ_{zz} at this field direction is determined by the effective electrons situated near the inflection point [66]. At $|\theta| < \theta_c$, there are no effective electrons: the velocities of all electrons on the FS have a significant component perpendicular to the field. At $\theta_c < |\theta| \leq \pi/2$, the conductivity is still dominated by effective electrons; however, their number decreases, as compared to that at θ_c, and their interlayer velocity rapidly vanishes as θ approaches $\pi/2$ [66]. This results in a sharp increase of the resistance.

From the above consideration, it is clear that the peak feature crucially depends on the warping of the FS along p_z, in other words, on the interlayer dispersion. As it will be discussed in Sect. 8.4, it is not always obvious that the latter exists at all. In some extremely anisotropic compounds, the interlayer hopping time, $\sim\hbar/t_z$, significantly exceeds the intralayer scattering time. Under this condition, the electron wave function is essentially localized within a layer and the interlayer dispersion is absent. As a result, the FS is only defined in two dimensions and the interlayer transport is incoherent. The problem of discriminating between the coherent and incoherent interlayer transport regimes has recently received much attention. In particular, the peak feature introduced here, indicating the presence of a well-defined 3D FS, can be used as an evidence of the coherent transport [67] and is, therefore, called a *coherence* peak.

8.4 Breakdown of the Interlayer Coherence as Seen from the Magnetotransport

The semiclassical description of the interlayer magnetoresistance presented in Sects. 8.2 and 8.3 presumes a small, however, finite dispersion for charge carriers in the interlayer direction and, therefore, a 3D (weakly warped along p_z) FS. This obviously implies that the mean free path across the layers significantly exceeds one unit cell or, equivalently, the scattering rate is much lower than the interlayer hopping rate:

$$1/\tau \ll t_z/\hbar. \tag{8.26}$$

If this condition is not fulfilled, the momentum p_z is undefined and the semiclassical 3D model is inappropriate. Generally speaking, two different regimes of the interlayer charge transport are possible in this situation.

In the *strongly incoherent* regime, the electron hopping between adjacent layers is entirely caused by interaction (say, with phonons or impurities), i.e., scattering processes. The electron states before and after hopping are uncorrelated; the interlayer transport is nonmetallic in this case. A well-known example of such a system is the high-T_c cuprate Bi-2201 (see, e.g., [68]): the interlayer resistivity of this compound $\rho_\perp(T)$ increases rapidly with cooling, above the superconducting transition, whereas the in-plane resistivity $\rho_\parallel(T)$ is metallic (though probably non-Fermi liquid [69]).

As to organic conductors, the author is aware of only one report on the layered compound θ-$(BETS)_4HgBr_4(C_6H_5Cl)$ [70] where metallic $\rho_\parallel(T)$ and nonmetallic $\rho_\perp(T)$ have been observed simultaneously at low temperatures. This, however, does not necessarily mean that all the other known organic conductors are fully coherent. In fact, one can consider a possibility of a weak overlap of the electron wave functions on adjacent layers, so that the interlayer transport is mostly caused by one-particle tunneling, while successive tunneling events are uncorrelated due to strong intralayer scattering: $\tau < \hbar/t_z$. In this *weakly incoherent* regime, the interlayer momentum is still undefined and the FS is purely 2D. However, the resistivity $\rho_\perp(T)$ has been predicted [71] to preserve the metallic temperature dependence determined by the intralayer scattering: $\rho_\perp(T) \sim (2t_z/\varepsilon_F)^2\rho_\parallel(T)$. Moreover, all the AMRO phenomena presented in Sect. 8.2 also persist in the weakly incoherent case [54, 67]. While the semiclassical description is no longer appropriate, they can be explained in terms of interference of the wave functions on adjacent layers in a tilted magnetic field. That the AMRO effects do not imply a 3D FS and can be theoretically obtained by considering just a two-layer system has recently been very clearly demonstrated by Yakovenko and Cooper [25, 55].

Figure 8.13 shows the angle-dependent magnetoresistance of a q2D metal calculated for the weakly incoherent regime [67]. Both the AMRO and nonoscillating background (away from $\theta = 90°$) closely resemble those obtained within the coherent 3D model [22, 65]. Likewise, the simulation of the xy rotation of a q1D metal [67] yields the same result as in the coherent case [30] at $|\theta| \leq 80°$.

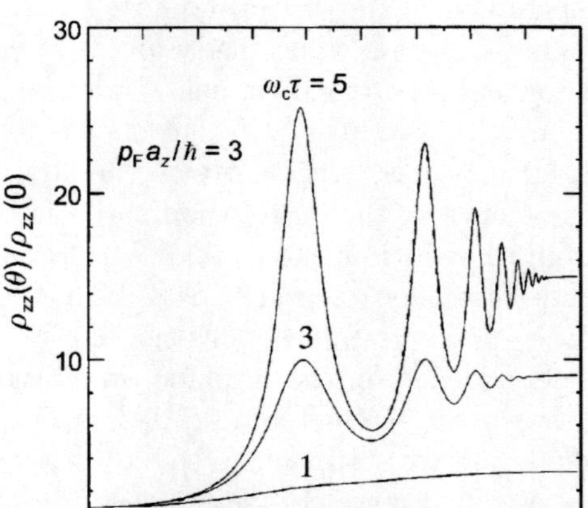

Fig. 8.13. Angle-dependent interlayer magnetoresistance of a q2D metal calculated within the weakly incoherent interlayer transport model [67]

Given the similarity of the major features of the zero- and high-field interlayer resistance in the coherent and weakly incoherent transport regimes, a question arises: is there a substantial physical difference between these regimes? McKenzie and Moses [67] proposed to use two fine features of the magnetoresistance at a field exactly or nearly parallel to the layers to distinguish between the coherent and weakly incoherent cases:

1. Field dependence $\rho_\perp(B)$ under a field parallel to the layers. It is predicted to be different in the two regimes at very high fields, such that $\omega_c\tau \gg (\varepsilon_F/t_z)^{1/2}$. However, one needs extremely strong fields, $\sim 10^2$ T, to use this criterion in real conditions. Moreover, experiments [19, 72, 73] show that, even in the well-studied coherent case, the exact field dependence can considerably deviate from the theoretical predictions [60, 74, 75].
2. Out-of-plane rotation near $\theta = 90°$. As described in Sect. 8.4, the coherence peak around 90° is associated with the geometry of cyclotron orbits on a 3D FS slightly warped along p_z and can only exist in the coherent regime. Its absence in the weakly incoherent transport model [67] is a natural consequence of the assumed strictly 2D FS.

The observation of the coherence peak has been used to prove the interlayer coherence and evaluate the interlayer transfer energy in a number of highly anisotropic compounds: κ-(BEDT-TTF)$_2$I$_3$ [73], α-(BEDT-TTF)$_2$NH$_4$Hg(SCN)$_4$ [76], κ-(BEDT-TTF)$_2$Cu(NCS)$_2$ [77], λ-(BETS)$_2$MCl$_4$ [78, 79], and κ-(BETS)$_4$FeBr$_4$ [80]. On the other hand, the absence of the peak in β″-(BEDT-TTF)$_2$SF$_5$CH$_2$CF$_2$SO$_3$ [73] suggests an incoherent interlayer transport in this salt.

Here, we will discuss in more detail a special case of α-(BEDT-TTF)$_2$KHg(SCN)$_4$. This compound has recently been found to allow a direct comparison between the coherent and weakly incoherent magnetotransport regimes [81, 82]: depending on the crystal quality, either of the two regimes can be obtained in this material, while the strong field criterion, $\omega_c\tau > 1$, is always achieved in conventional magnetic fields of a few tesla.

In the highest quality samples, the interlayer magnetoresistance of α-(BEDT-TTF)$_2$KHg(SCN)$_4$ is consistent with the 3D model: it displays a rather strong φ dependence at the xy rotation and a narrow coherence peak in θ sweeps [81]. From the width of the peak, $\Delta\theta \simeq 0.12$–$0.35°$, depending on φ, a very low interlayer transfer integral $t_z \simeq 15$–$45\,\mu$eV is obtained [81, 82].

By contrast, lower quality samples show no sign of the coherence peak and essentially no dependence on the in-plane orientation of the field. The basic differences between the "clean" and "dirty" samples hold for both the ambient-pressure density-wave state (see Sect. 8.6) and for the high-pressure normal metallic state [82]. Figure 8.14 illustrates these differences for the high-pressure state. The absence of the peak in the "dirty" sample suggests an incoherent interlayer transport. On the other hand, the prominent AMRO along with the metallic temperature dependence $R_\perp(T)$ point to the weak rather than strong incoherence.

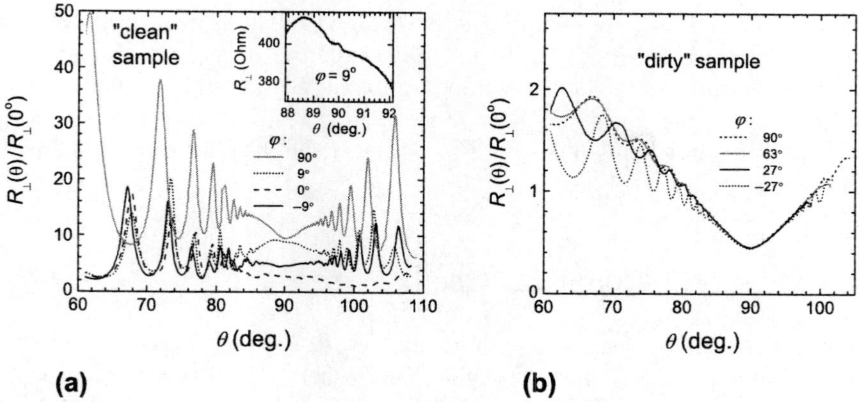

Fig. 8.14. Angle-dependent magnetoresistance of α-(BEDT-TTF)$_2$KHg(SCN)$_4$ under pressure, $P = 6.2\,\text{kbar}$, at $B = 20\,\text{T}$, $T = 1.4\,\text{K}$ [82]: (**a**) "clean" sample displays strong 2D AMRO and nonoscillating background, both depending on θ and φ, in agreement with the 3D Fermi liquid theory; *inset* shows the coherence peak; (**b**) "dirty" sample exhibits weaker AMRO, no coherence peak, and no φ dependence of the nonoscillating background; note that $R_\perp(\theta = 90°)/R_\perp(0°) < 1$

As one can see from Fig. 8.14b, the nonoscillating magnetoresistance background R_b is nearly independent of the azimuthal angle φ. Moreover, it decreases, becoming lower than $R(0°)$, as θ approaches 90°. This suggests that the interlayer transport is insensitive to the field component parallel to the layers. Indeed, it was shown [82] that $R_\mathrm{b}(\theta)$ of the "dirty" sample measured at different field strengths can be scaled to a single function of the out-of-plane field component, $B_\perp = B\cos\theta$. This result obviously contradicts not only the predictions of the 3D theory [60,74,75] but also the predictions of the weakly incoherent magnetotransport model [54,67]: similarly to the coherent case, the latter predicts the magnetoresistance to grow with increasing θ, saturating at $\theta \to 90°$ at a value significantly exceeding $R(0°)$, see Fig. 8.13.

Remarkably, the anomalous behavior of the "dirty" sample is observed in both the zero- and high-pressure states of α-(BEDT-TTF)$_2$KHg(SCN)$_4$ characterized by different FS geometries [83]. Moreover, a similar broad dip centered at $\theta = 90°$ was found in the angle-dependent magnetoresistance of other highly anisotropic materials: the purely q1D (TMTSF)$_2$PF$_6$ [31,35,84], purely q2D artificial GaAs/AlGaAs superlattice [85], and the above-mentioned β″-(BEDT-TTF)$_2$SF$_5$CH$_2$CF$_2$SO$_3$, combining open and cylindrical FSs [73].

Clarke and coworkers [86] have proposed a non-Fermi liquid model to explain the anomalous angular dependence in (TMTSF)$_2$PF$_6$. However, their model suggests incoherence to be a property of *clean* weakly coupled low-dimensional non-Fermi liquids arising due to strong electron correlations rather than due to disorder. By contrast, the above data on α-(BEDT-TTF)$_2$KHg(SCN)$_4$ reveal the coherence–incoherence transition (or crossover) at increasing disorder in the system. Moreover, in the coherent regime, both the AMRO behavior (especially the 2D AMRO in the pressurized state) and the background MR appear to be fully consistent with the Fermi liquid model.

Thus, while the demonstrated anomalous angular dependence of magnetoresistance seems to be a general feature of weakly incoherent layered systems, independent of the FS topology, the mechanism responsible for it remains unclear.

8.5 Magnetic Quantum Oscillations

Thus far we considered semiclassical magnetoresistance, neglecting quantization effects. However, in a strong magnetic field, the Landau quantization of the electron spectrum becomes important for compounds with closed FSs, giving rise to oscillations of various field-dependent properties. Quantum oscillations of resistivity and magnetization, i.e., the SdH and dHvA effects, were historically the first experimental methods of probing the FS and have been widely used in studies of conventional metals [1]. A brilliant comprehensive review of the standard 3D theory, and its applications to various materials, has been done by Shoenberg [2].

The SdH and dHvA oscillations have also been observed in numerous layered organic conductors, providing important information on their FS properties [6,12,87,88]. It was, however, realized that the extremely high electronic anisotropy of these compounds leads to significant violations of the usual 3D behavior. In this section, we briefly present main results of the theory and discuss its applicability to organic conductors. A more detailed survey of the recent status of the problem can be found in [12].

8.5.1 Lifshitz–Kosevich Formula for the de Haas–van Alphen Effect

The general formula for magnetization oscillations in a 3D metal has been obtained by Lifshitz and Kosevich (LK) [89]. In the case of a single extremal cross section of the FS, the magnetization component $\tilde{M}$ along the field can be written as follows

$$\tilde{M} = -\left(\frac{e}{2\pi\hbar}\right)^{3/2} \frac{S_{\text{ext}} B^{1/2}}{\pi^2 m_{\text{c}} |S''|_{\text{ext}}^{1/2}} \sum_{r=1}^{\infty} \frac{1}{r^{3/2}} \sin\left[2\pi r\left(\frac{F}{B} - \frac{1}{2}\right) \pm \frac{\pi}{4}\right] R_{\text{T}} R_{\text{D}} R_{\text{S}}, \tag{8.27}$$

where S_{ext} is the area of the extremal cross section in $\boldsymbol{p}$-space,

$$F = \frac{S_{\text{ext}}}{2\pi e\hbar} \tag{8.28}$$

is the fundamental frequency of the oscillations, m_{c} is the effective cyclotron mass (8.3), $(S'')_{\text{ext}} = (\partial^2 S/\partial p_B^2)_{\text{ext}}$ is the FS curvature along $\boldsymbol{B}$ at the extremal cross section, and the sign "+" or "−" is chosen depending on whether the cross section is minimal or maximal, respectively. The factors R_T, R_D, and R_S describe the damping of the oscillation amplitude due to the temperature smearing of the Fermi edge, broadening of Landau levels caused by a finite electron lifetime, and the spin-splitting effect, respectively:

$$R_T = \frac{Krm^*T/B}{\sinh(Krm^*T/B)}, \tag{8.29}$$

$$R_D = \exp(-Krm^*T_D/B), \tag{8.30}$$

and

$$R_S = \cos\left(\frac{\pi}{2}rgm^*\right), \tag{8.31}$$

where $m^* = m_c/m_e$ is the cyclotron mass in units of the free electron mass, $K \equiv 2\pi^2 k_B m_e/(\hbar e) \approx 14.7\,\mathrm{T/K}$ (k_B is the Boltzmann constant), $T_D = \hbar/(2\pi k_B \tau)$ is the so-called Dingle temperature determined by the crystal quality [90], and g is the Landé factor. If the FS contains multiple extremal cross sections, they additively contribute to the oscillations.

The damping factors R_T and R_D exponentially decrease with increasing the harmonic index r; further, the harmonic amplitudes are reduced by the prefactor $r^{-3/2}$. Therefore, the dHvA oscillations in 3D materials have in most cases a sinusoidal shape and it is sufficient to consider the behavior of the fundamental harmonic.

8.5.2 Shubnikov–de Haas Oscillations

The theory of the SdH oscillations is more complex than that of the dHvA effect since the former are entirely caused by deviations from the τ approximation and one should, in principle, consider a detailed problem of scattering processes modified by quantizing magnetic field [91]. Fortunately, it is usually possible to obtain a satisfactory description by following Pippard's idea [3] that the scattering probability and hence the resistivity are proportional to the density of states around the Fermi level, $D(\varepsilon_F)$. The oscillatory part of $D(\varepsilon_F)$ is easier to derive (see, e.g., [2]) and it can be expressed through the field derivative of the dHvA oscillations:

$$\tilde{D}(\varepsilon_F) \propto \left(\frac{m_c B}{S_{ext}}\right)^2 \frac{\partial \tilde{M}}{\partial B}. \tag{8.32}$$

As a result, the oscillations of the conductivity can be written as

$$\tilde{\sigma}/\sigma_0 \sim \tilde{D}(\varepsilon_F)/D(\varepsilon_F) = \sum_{r=1}^{\infty} \frac{1}{r^{1/2}} a_r \cos\left[2\pi\left(\frac{F}{B} - \frac{1}{2}\right) \pm \frac{\pi}{4}\right], \tag{8.33}$$

where

$$a_r \propto \frac{m_c B^{1/2}}{(S'')_{ext}^{1/2}} R_T R_D R_S \tag{8.34}$$

and σ_0 is the background conductivity. Thus, one can extract from the SdH oscillations basically the same information as from the thermodynamic dHvA oscillations.

8.5.3 Quantum Oscillations in Organic Metals

The 3D description presented in Sects. 8.5.1 and 8.5.2 assumes that the FS is intersected by many Landau tubes at a given field, i.e., that the Landau level spacing $\hbar\omega_c$ is much smaller than the electron bandwidth in the direction along the magnetic field. At this condition, only a small fraction of electrons, situating at or near an extremal cross section of the FS, contributes to quantum oscillations. In particular, this leads to a very low relative amplitude of the SdH signal. Even without taking into account the damping factors (8.29)–(8.31), it is of the order of $(B/F)^{1/2}$ [2]. Taking a typical value $F \sim 10^4$ T, one obtains for reasonable fields of ~10 T the amplitude of ~3% of the background conductivity. This estimation is further decreased due to the damping factors, so that typical amplitudes are as low as $10^{-4}\sigma_0$. By contrast, in highly anisotropic layered metals, almost all electrons on the FS may contribute to the oscillations. For example, the interlayer bandwidth, $W_\perp \simeq 4t_z$, in such compounds as α-(BEDT-TTF)$_2$MHg(XCN)$_4$, β″-(BEDT-TTF)$_2$SF$_5$CH$_2$CF$_2$SO$_3$, and κ-(BEDT-TTF)$_2$X with X = I$_3$, Cu(NCS)$_2$ (see Sects. 8.6, 8.7.1, and 8.7.2) is ≤ 0.02 meV, and the basic assumption of the 3D model is violated already at fields of ~1–3 T directed perpendicular to the layers. In less anisotropic materials, the 2D limit may be reached by increasing the field to 20–30 T or by choosing the field direction close to a peak in the 2D AMRO (8.11). In both cases, *all* electrons on the FS cylinder contribute to quantum oscillations that generally leads to a dramatic enhancement of their amplitude and a considerable anharmonicity of their waveform.

Figure 8.15 shows an example of the interlayer magnetoresistance measured on α-(BEDT-TTF)$_2$TlHg(SeCN)$_4$ [87]. The amplitude of the SdH oscillations amounts to ~100 times the background at the field of 50 T. The analysis of the temperature dependence of the oscillation amplitude using the standard formulas (8.29) and (8.33) yields an effective cyclotron mass which surprisingly changes with the field, as shown in the inset in Fig. 8.15. A similar field-dependent mass was reported for the isostructural salt α-(BEDT-TTF)$_2$NH$_4$Hg(SCN)$_4$ [93]. In both salts, an anomalously high amplitude of higher harmonics of the fundamental frequency has been noted [92, 93].

The dHvA oscillations have also been found to violate the 3D behavior in a number of experiments, exhibiting an anomalous harmonic content [94–96] as well as deviations from the temperature and field dependence predicted by the LK theory, see, e.g., [97–100].

Thus, it is obvious that a modification of the standard theory, taking into account the high two dimensionality of the electronic spectrum, is necessary for a correct description of magnetic quantum oscillations in organic conductors. In the following, we will separately consider the moderately anisotropic

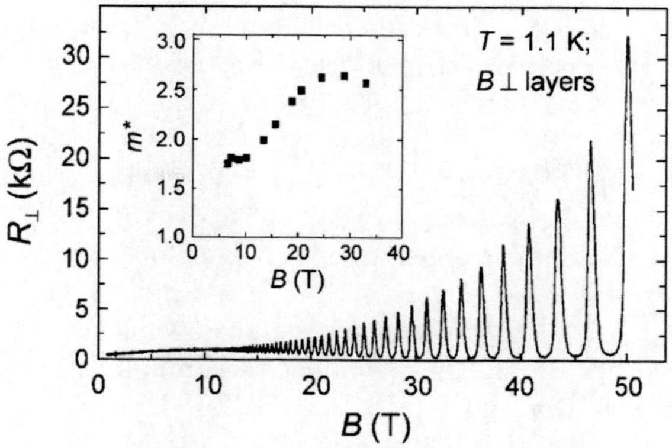

Fig. 8.15. Giant SdH effect in α-(BEDT-TTF)$_2$TlHg(SeCN)$_4$ [87]. *Inset*: the apparent cyclotron mass determined according to the standard 3D approach [92]

case, when the Landau level spacing is smaller (though not much smaller) than the interlayer bandwidth, and the highly 2D limit, when the entire conducting band becomes quantized.

Moderately Anisotropic Case: $\hbar\omega_c \leq 4t_z$

In layered organic conductors, the magnetic quantum oscillations are caused by closed orbits on cylindrical FSs. If the electronic anisotropy is not too high (see Sect. 8.4), the FS is defined in three dimensions and two extremal cross sections of the cylinder contribute to the oscillations: the maximum S_{max} and minimum S_{min}. As long as the Landau level spacing $\hbar\omega_c$ does not exceed the interlayer bandwidth, dHvA oscillations are quite well described by the 3D LK formula (8.27).

Since the difference $\Delta S = S_{max} - S_{min}$ is small, the corresponding electron parameters entering (8.27) are approximately the same. Then, the first harmonic of the dHvA signal exhibits beats:

$$\tilde{M}_1 \propto \sin\left[2\pi\left(\frac{F}{B} - \frac{1}{2}\right)\right]\cos\left(2\pi\frac{F_{beat}}{B} - \frac{\pi}{4}\right), \tag{8.35}$$

where F is the average of the frequencies F_{max} and F_{min} corresponding to the maximum and minimum cross sections, respectively, and

$$F_{beat} = (F_{max} - F_{min})/2 = \Delta S/(4\pi e\hbar). \tag{8.36}$$

Thus, the beat frequency provides an evaluation of the FS warping. In the case of the q2D dispersion (8.4), the following simple relationships hold: $F/B = \varepsilon_F/(\hbar\omega_c)$, $F_{beat}/B = 2t_z/(\hbar\omega_c)$, and, therefore,

$$\frac{2t_z}{\varepsilon_F} = \frac{F_{beat}}{F}. \tag{8.37}$$

As discussed in Sect. 8.2.1, the difference between the maximum and minimum areas of cyclotron orbits on a cylindrical FS oscillates as a function of the tilt angle θ, vanishing at the angles (8.11) corresponding to peaks in the semiclassical AMRO.

Keeping in mind that typical oscillation frequencies in organic conductors are $\sim 10^3$ T and the fields necessary for detecting the oscillations are of the order of a few Tesla, one can expect the beating behavior to be observed on the compounds with the ratio $2t_z/\varepsilon_F \sim 10^{-3}$–$10^{-2}$. Indeed, beating dHvA oscillations with $F_{\text{beat}}/F \sim 10^{-2}$ and the angular dependence $F_{\text{beat}}(\theta)$ consistent with (8.5) have been found in a number of compounds, e.g., in β-(BEDT-TTF)$_2$X with X = IBr_2 and I_3 [101], κ-(BEDT-TTF)$_2$Cu[N(CN)$_2$]Br [102], and (BEDT-TTF)$_4$[Ni(dto)$_2$] [103].

Of course, similar beats have also been found in the quantum oscillations of the magnetoresistance of the same compounds [17,102–105]. It has, however, been noted that the positions of the nodes in the SdH signal *do not coincide* with those in the dHvA signal [102–104], by contrast to what is expected from (8.32) and (8.33). This behavior illustrated in Fig. 8.16 indicates that the SdH effect violates the standard 3D behavior already when the Landau level spacing becomes comparable with the interlayer transfer integral: $\hbar\omega_c \leq t_z$. In the conventional case (Sect. 8.5.2), the oscillations of σ_{zz} are entirely caused by the oscillating density of states, whereas in the 2D limit oscillations of the interlayer velocity are dominant [98]. In the intermediate regime, $\hbar\omega_c \sim t_z$, both contributions are important. The interference of the two oscillating

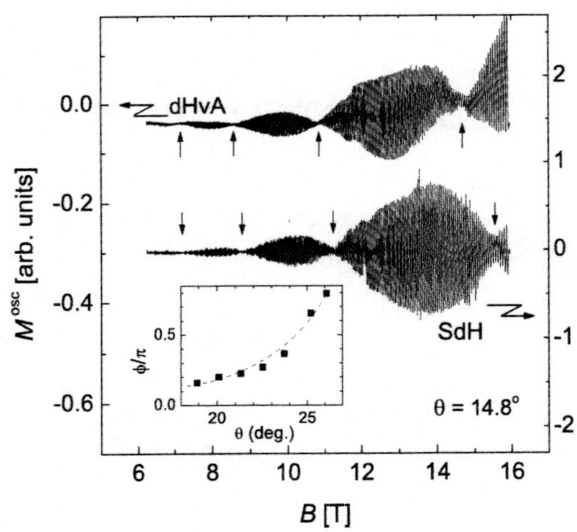

Fig. 8.16. Beating dHvA and SdH oscillations in β-(BEDT-TTF)$_2$IBr$_2$ [104]. The node positions of the SdH signal (*downward arrows*) are clearly shifted from those of the dHvA signal (*upward arrows*). When the field is tilted from the normal to the layers toward the AMRO peak at $\theta = 32°$, the beat frequency decreases and the phase shift of the SdH oscillations, shown in the *inset*, increases, exceeding the theoretical limit of $\pi/2$ [64]

quantities gives rise to a field-dependent phase shift [104, 106]:

$$\phi = \arctan\eta, \quad \eta \sim \frac{\hbar\omega_c}{2\pi t_z}, \tag{8.38}$$

in the factor responsible for the modulation of the SdH oscillations, analogous to the cosine factor in the right-hand side of (8.35). The theoretically predicted phase shift [106] is able to describe the experimental data [104] at $\hbar\omega_c \leq t_z$. However, the experiment shows that at $\eta \geq 1$ the phase shift increases faster than predicted and can even exceed the theoretical limit of $\pi/2$, as shown in the inset in Fig. 8.16 [64].

Another remarkable consequence of the interfering oscillations of the density of states and the interlayer velocity is slow oscillations having the frequency equal to $2F_{\text{beat}}$ [106, 107]. By contrast to the fundamental SdH signal, these oscillations are virtually insensitive to temperature and sample inhomogeneities. They can, therefore, be used for evaluations of the FS warping and scattering time even at relatively high temperatures and low fields at which the fundamental SdH oscillations are not yet resolved [107].

Extremely Anisotropic Case: $\hbar\omega_c \gg 4t_z$

A modification of the LK formula (8.27) for the dHvA effect in an ideally 2D was proposed by Shoenberg [108] and confirmed later by a more rigorous theoretical analysis [109]. The *Lifshitz–Kosevich–Shoenberg* (LKS) formula (also known as the 2D LK formula) represents the magnetization oscillations in the harmonic expansion form similar to (8.27):

$$\tilde{M} = -\frac{e}{2\pi\hbar}\frac{S}{\pi^2 m_c a_z}\sum_{r=1}^{\infty}\frac{1}{r}\sin\left[2\pi r\left(\frac{F}{B}-\frac{1}{2}\right)\right] R_T R_D R_S, \tag{8.39}$$

where F is defined by (8.28) with S_{ext} being simply the area of the FS cross-section S, which is constant in the 2D case.

The LKS formula successfully describes dHvA oscillations in some cases; for example, it accounts for the "inverse saw-tooth" shape of the oscillations in β″-(BEDT-TTF)$_2$SF$_5$CH$_2$CF$_2$SO$_3$ shown in Fig. 8.24a (see Sect. 8.7.1). However, there are a few restrictions as to its general applicability [12]. The strongest restriction is that it assumes a field-independent chemical potential:

$$\mu(B) = \text{const}. \tag{8.40}$$

This assumption may turn inappropriate in highly 2D materials. Indeed, considering, for simplicity, the case $T = 0$, the condition (8.40) implies that the lowest unoccupied state remains always at ε_F, independent of B. When the Landau levels are sharp ($k_B T_D \ll \hbar\omega_c$), this condition can be fulfilled only if there are additional field-independent states (e.g., impurity states or another conduction band with a sufficiently high density of states) or the total carrier concentration is allowed to change (high magnetostriction). Otherwise the chemical potential must be, at least partially, pinned to the nearest Landau level and hence oscillate with changing B. This problem has been extensively studied in the recent years (see [12] and references therein).

In the most general way [110,111], the oscillations of μ are incorporated in the dHvA effect by adding an additional term $2\pi r\tilde{\mu}/(\hbar\omega_c)$ in the sine function in (8.39), where the oscillatory part of the chemical potential $\tilde{\mu}$ is defined by the transcendental equation:

$$\tilde{\mu}(B) = \frac{\hbar\omega_c}{\pi(1+n_R)} \sum_{r=1}^{\infty} \frac{1}{r} \sin\left[2\pi r\left(\frac{F}{B} + \frac{\tilde{\mu}(B)}{\hbar\omega_c} - \frac{1}{2}\right)\right] R_T R_D R_S, \qquad (8.41)$$

and $n_R = D^R(\varepsilon_F)/D^{2D}(\varepsilon_F)$ is the ratio between the densities of states of the field-independent (reservoir) conducting band and of the 2D band at zero field. If $n_R \gg 1$, the amplitude of $\tilde{\mu}(B)$ is small and the LKS formula (8.39) can be applied to describe the dHvA effect. Since in many organic conductors a cylindrical FS coexists with open Fermi sheets, the latter act as the electron reservoir stabilizing the chemical potential. However, the density of states of this reservoir is, generally, of the same order as that of the 2D band, so that $n_R \sim 1$. Therefore, the oscillations of the chemical potential are only reduced by the damping factors (8.29)–(8.31) and may play an important role at low temperatures and high fields (typically, for clean samples, $T < 1\,\text{K}$, $B > 20\,\text{T}$, see, e.g., [98,99]).

If more than one 2D band contributes to the dHvA effect, the equations for the oscillatory magnetization and chemical potential should be generalized to sum up the corresponding contributions. Then the presence of more than one fundamental frequencies leads to frequency mixing effects in the resulting magnetization [112]. An illustrative example is κ-(BEDT-TTF)$_2$Cu(NCS)$_2$ which displays two fundamental frequencies, F_α and F_β, caused by quantization of the semiclassical and magnetic breakdown orbits (see Sect. 8.7.2). The oscillations $\tilde{\mu}(B)$ give rise to combination frequencies, $F_\beta - F_\alpha$, $F_\beta - 2F_\alpha$, etc., which do not correspond to any single quantized orbit [94,95]. On a qualitative level, such frequencies are a result of a communication between two quantized 2D bands via the chemical potential: if, for example, the $\mu(B)$ is pinned to a Landau level of the α-band and moves up with the latter (at increasing field), the frequency F' with which it is crossed by the Landau levels of the β-band is shifted from F_β by an amount close to the α-frequency, i.e., $F' \simeq F_\beta - F_\alpha$.

Unfortunately, by contrast to the dHvA effect, a comprehensive theory of the SdH effect in q2D metals is still lacking. The main problem is that the quantum oscillations of magnetoresistance crucially depend on details of the charge transfer and scattering processes whose nature in the organic conductors is far from being well understood. Despite the considerable amount of experimental data on the quantum oscillations of the interlayer magnetoresistance in the highly 2D limit, $\hbar\omega_c \gg 4t_z$, only few attempts of their theoretical description have been done [98,113,114]. In particular, it has been shown [113] that, in addition to the semiclassical (Boltzmann) contribution, there is a purely quantum term in the expression for the oscillating conductivity $\tilde{\sigma}_{zz}$. The quantum correction increases by increasing the ratio $\hbar\omega_c/t_z$, becoming of the same order of magnitude as the semiclassical term in the clean ($\omega_c\tau \gg 1$), highly 2D limit. The combination of the two terms gives rise to a "pseudogap" in the spectral conductivity $\sigma_{zz}(\varepsilon)$ at integer filling numbers, $\mu/(\hbar\omega_c) = n$, leading to an activation temperature dependence $\sigma_{zz}(T)$ [113]. Such a temperature dependence has indeed been observed on β″-(BEDT-TTF)$_2$SF$_5$CH$_2$CF$_2$SO$_3$ in fields above 20 T [115]. It has, however, been noted in [113] that the assumed self-consistent Born approximation may no longer be valid in the highly 2D case at strong fields, causing certain disagreements between the theory and experimental data [115,116]. Thus, a further development of the theory relies on a rigorous consideration of the scattering processes in these extremely anisotropic compounds.

8.6 High-Field Studies of the Low-Temperature Electronic State in α-(BEDT-TTF)$_4$MHg(SCN)$_4$

The family α-(BEDT-TTF)$_4$MHg(XCN)$_4$ includes a number of isostructural salts with M = K, Tl, Rb, and NH$_4$ and X = S, Se, and S$_x$Se$_{1-x}$ [117–122] and has a few BETS analogs [123]. All these compounds exhibit a very pronounced layer-type electronic anisotropy. In particular, some of them exhibit giant quantum oscillations (see, e.g., [87, 92, 93]), revealing a high degree of two dimensionality. However, due to the specific character of the intermolecular interactions within the conducting layer illustrated in Fig. 8.17, a part of carriers can be considered as q1D with a preferable conduction along the crystallographic ***a***-axis [120]. Thus, the FSs of these salts [120, 124] contain both a cylindrical part and a pair of weakly warped open sheets. The q1D and q2D bands are separated by a substantial gap near the Fermi level. Pulsed field dHvA studies of α-(BEDT-TTF)$_4$KHg(SCN)$_4$ [96] reveal a MB gap between the bands, $\Delta\varepsilon = 23 \pm 2$ meV. This value is considerably smaller than predicted theoretically (see, e.g., [125, 126]), however, it is still of the same order of magnitude as the Fermi energy, $\varepsilon_F \sim 40$ meV, estimated from magnetic quantum oscillations.

8.6.1 Magnetotransport Properties and the Fermi Surface Reconstruction in the Salts with M = K, Tl, and Rb

In the following, we will focus on the low-temperature properties of three members of this family, with X = S and M = K, Tl, and Rb, which are characterized by very similar crystallographic parameters [117, 120]: $a = 10.06 \pm 0.02$ Å,

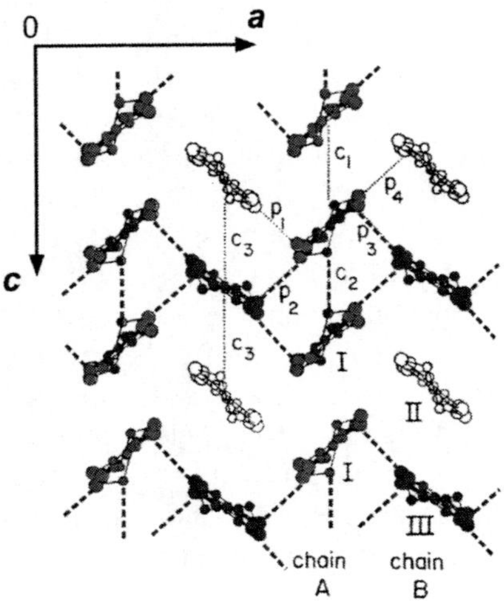

Fig. 8.17. Structure of cation radical layers of α-(BEDT-TTF)$_4$MHg(XCN)$_4$ (M = Tl, K, NH$_4$; X = S, Se) in the crystallographic ***ac***-plane with three nonequivalent positions of BEDT-TTF (*roman characters*) [120]. The electrical conductivity is governed by the transfer integrals c_i and p_j. Chains of dimers coupled by c_2 and donors III give rise to q1D electron-like carriers with a high conductivity along ***a***. The hole-like carriers originate from coupling of the chains through donors II and, feeling all the interactions, form a q2D band

$b = 20.56 \pm 0.01$ Å, $c = 9.95 \pm 0.02$ Å, $\alpha \approx 103.6°$, $\beta \approx 90.5°$, and $\gamma \approx 93.2°$ (remind that the conducting plane is ***ac***), and have similar physical properties. All the three compounds (hereafter referred to as the K, Tl, and Rb salts, respectively) undergo a phase transition at a temperature $T_\mathrm{p} \simeq 8$–10 K as evidenced by resistive [118, 127, 128], magnetic [129, 130], ESR [128, 131], specific heat [132], and thermal expansion [133] experiments.

These salts have been of particular interest since the first magnetotransport experiments revealing a bunch of striking anomalies. Some of the anomalies are illustrated in Fig. 8.18. In the field perpendicular to the layers, the interlayer resistance has been found to rapidly increase below T_p [118, 127, 134–136], reaching at $B_\mathrm{max} \simeq 10$–12 T a maximum which can amount, in clean samples, $\sim 10^2$ times the zero-field value at $T \leq 1$ K. At further increasing the field, the magnetoresistance exhibits a strong downturn, ending often with a sharp *kink* structure at the field $B_\mathrm{k} \simeq 23.5$, 27, and 32 T, for the K [134], Tl [135], and Rb salt [137], respectively. In the interval $B_\mathrm{max} \leq B \leq B_\mathrm{k}$, the dependence $R(B)$ shows a hysteresis which is especially pronounced at the kink [135, 138–140].

The SdH oscillations arising at ~ 8 T, i.e., shortly before B_max, are characterized by a rather complicated spectrum. The fundamental frequency – $F_\alpha = 670$–675 T for the K [134, 141] and Tl [118, 142] salts and 654 T for the Rb one [136] – corresponds to a cylindrical FS occupying ≈16% of the Brillouin zone, in reasonable agreement with the band structure calculations. However,

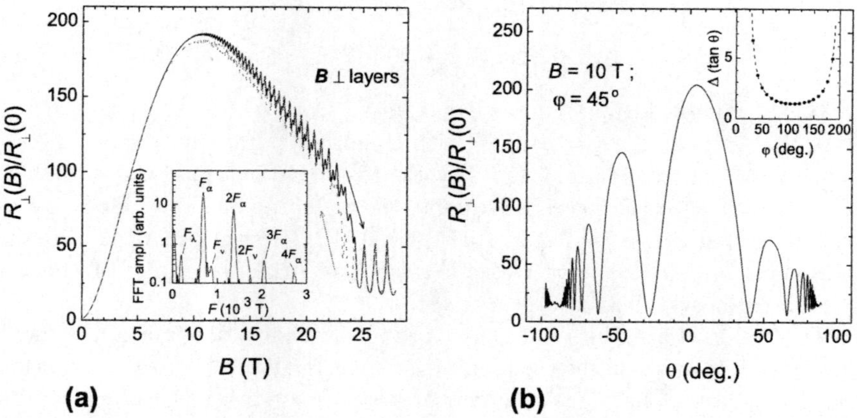

Fig. 8.18. Magnetoresistance of the K salt at $T = 1.4\,\mathrm{K}$: (**a**) up- (*black solid line*) and downward (*gray dashed line*) field sweeps at $\boldsymbol{B} \parallel$ layers; a hysteresis is most pronounced at the kink transition, $B \approx 23\,\mathrm{T}$. *Inset*: the fast Fourier transformation of the SdH oscillations in the field range 12–22.5 T; (**b**) example of the angular dependence at $B = 10\,\mathrm{T}$, showing strong 1D AMRO. The *inset* shows the dependence of the AMRO period on the azimuthal angle φ which is fitted by the formula: $\Delta(\tan\theta) = 1.26/\sin(\varphi - 20.5°)$ (*dashed line*); φ is counted from the crystallographic axis $\boldsymbol{a}$

at temperatures below 2 K, an anomalously strong second harmonic, $2F_\alpha$, emerges, leading to splitting of the fundamental oscillations [136,139,141–145]. Additionally, some crystals exhibit a low frequency, $F_\lambda \approx 180\,\mathrm{T}$, and its combinations with F_α [139,143–145]. Moreover, yet another frequency, $F_\mu = 775\,\mathrm{T}$ [145] as well as very slow oscillations at $\sim$10 T with an anomalous field and temperature dependence and a high-frequency $F_\beta \simeq 4{,}200\,\mathrm{T}$, corresponding to 100% of the Brillouin zone [146], have been reported for temperatures below 1 K . By contrast, above B_k, the oscillation spectrum consists of only F_α and its higher harmonics [96,135,139], with a weak contribution of the MB frequency F_β detected in the 60 T pulsed field experiment [96].

Turning to the angle-dependent semiclassical magnetoresistance, all three salts display strong oscillations [134,147,148] which have initially been interpreted as 2D AMRO (see Sect. 8.2.1) due to the cylindrical FS [134]. However, more detailed studies [147] have revealed a behavior more typical of the LMA oscillations. Indeed, as shown in Fig. 8.18b, the $R(T)$ dependence exhibits sharp dips, by contrast to peaks expected in the case of the 2D AMRO, and the dip positions obey the condition [143,147]:

$$\tan\theta_N \sin(\varphi - \varphi_0) = N\Delta_0 + \tan\theta_0, \tag{8.42}$$

having the same form as the generalized LMA condition (8.21).

The parameters entering (8.42) have been determined as

$$\Delta_0 = 1.35 \pm 0.03, \quad \varphi_0 = 24 \pm 2°, \quad \theta_0 = 27 \pm 1° \tag{8.43}$$

for the Tl and Rb salts [147–149] and

$$\Delta_0 = 1.26 \pm 0.02, \quad \varphi_0 = 20 \pm 2^\circ, \quad \theta_0 = 27 \pm 0.5^\circ \tag{8.44}$$

for the K salt [144, 150], where φ is counted from $\boldsymbol{a}$. For the latter salt, Iye et al. [151] reported $\varphi_0 = 20 \pm 2^\circ$ with the other parameters being the same as in (8.44), whereas Sasaki and Toyota [152] have obtained the same values as given in (8.43). Despite this quantitative discrepancy, the main message from these data is, as noted in [147]: the AMRO reveal LMA-like directions in a plane which does not coincide with any principal plane of the room temperature crystal structure.

To explain this result, it was proposed [143] that a new periodicity with a wave vector $\boldsymbol{Q}$ arises in the electronic system at the phase transition mentioned above. Assuming that $\boldsymbol{Q}$ is commensurate with the original reciprocal lattice and reconstructs the Brillouin zone, giving rise to a new characteristic plane, revealed by the AMRO, this wave vector was evaluated from the AMRO parameters (8.43) and (8.44):

$$\boldsymbol{Q}^{\mathrm{Tl,Rb}} = \zeta \boldsymbol{K}_a/6 + \boldsymbol{K}_c/3 + (\eta - 1/2)\boldsymbol{K}_b/3 \tag{8.45}$$

for the Tl and Rb salts [143, 148] and

$$\boldsymbol{Q}^{\mathrm{K}} = \zeta \boldsymbol{K}_a/8 + 3\boldsymbol{K}_c/8 + 3(\eta - 1/2)\boldsymbol{K}_b/8 \tag{8.46}$$

for the K salt [150]. Here $\boldsymbol{K}_a$, $\boldsymbol{K}_b$, and $\boldsymbol{K}_c$ are the reciprocal lattice vectors and $\zeta, \eta = +1$ or -1. The latter uncertainty comes from the fact that the real triclinic structure is very close to a more symmetrical orthorhombic side-centered one and the difference could not be resolved within the experimental accuracy.

Noting that the component of $\boldsymbol{Q}$ along k_x is close to the distance, $2k_{\mathrm{F}}$, between the open Fermi sheets predicted by the band structure calculations, it was suggested [143] that $\boldsymbol{Q}$ is the wave vector of a density wave leading to the nesting of the open FS. The remaining cylindrical FS was proposed to be reconstructed due to opening small gaps at the new Brillouin zone boundary, producing strongly warped open sheets in the $\boldsymbol{K}_b\boldsymbol{Q}$-plane.

Taking into account the rather complex commensurability factors in (8.45) and (8.46), one can suggest that these expressions are approximations for, in fact, an incommensurate superstructure. In this case, one could also expect folding of the FS and opening of small gaps at the points of the cylindrical FS satisfying the condition $\varepsilon(\boldsymbol{k}) = \varepsilon(\boldsymbol{k} + n\boldsymbol{Q})$ (see, e.g., [153]). The magnitude of the gaps, being determined by the amplitude of the nth harmonics of the superstructure potential, rapidly decreases with increasing n. Therefore, for a weak potential, only the gaps at the points where $\varepsilon(\boldsymbol{k}) = \varepsilon(\boldsymbol{k} \pm \boldsymbol{Q})$ are significant. Thus, even in an incommensurate case, a construction shown in Fig. 8.19, analogous to that proposed in [143], can be made.

The above model accounts, at least qualitatively, for many of the observed magnetotransport anomalies (see, e.g., [135, 137, 149]). The low-temperature

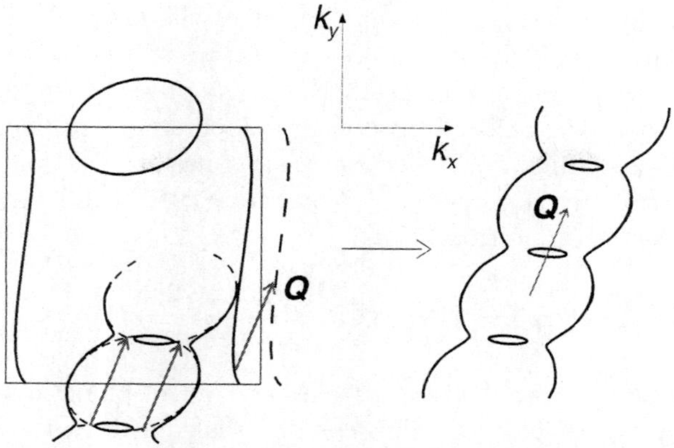

Fig. 8.19. FS reconstruction in α-(BEDT-TTF)$_4$MHg(SCN)$_4$ caused by the periodic potential (8.45) and (8.46). The q1D carriers are completely gapped due to the nesting of the open sheets. In addition, small gaps appear at the points of the cylindrical FS satisfying the condition $\varepsilon(\boldsymbol{k}) = \varepsilon(\boldsymbol{k} \pm \boldsymbol{Q})$; this gives rise to a pair of strongly warped open sheets and narrow cylinders

magnetoresistance below 10–15 T is determined mostly by the orbits on the new open Fermi sheets. The AMRO can be explained by Osada's model [47] (the strong warping of the sheets along $\boldsymbol{Q}$ implies the existence of high-order transfer terms t_{k1} in (8.19)) or by other LMA mechanisms discussed in Sect. 8.2.2. The small closed portions of the FS in Fig. 8.19 cause weak oscillations at $F_\lambda \approx 180\,$T. At increasing B, the MB through the gaps between the open and closed orbits occurs, leading to a gradual decrease of the semiclassical magnetoresistance [137] as well as to the SdH oscillations at F_α and combinations of F_α and F_λ. At $B > B_{\rm max}$, the angle-dependent magnetoresistance $R(\theta)$ reveals a gradual development of the 2D AMRO consistent with the original cylindrical FS. The MB scenario in the low-temperature state is corroborated by high-pressure experiments [154, 155]: with applying pressure, the MB gap is suppressed that leads to a drastic decrease, from 6–8 T to $\sim$2 T, of a threshold field at which the α-frequency oscillations emerge. The present simple model does not explicitly explain the anomalously strong second harmonic of the α-oscillations. While the origin of $2F_\alpha$ is still controversial [156, 157], the existing models do not contradict the proposed FS reconstruction.

There is no clear explanation at present for the other SdH frequency, F_μ [145], as well as for F_β and extremely low frequencies reported for $B < 10\,$T [146]. One should note, however, that these frequencies have not been reproduced by other authors so far. To explain the multiple frequencies, a much more severe reconstruction of the FS has been proposed [158]. This model, however, requires an unphysically strong commensurate superstructure potential, comparable in the magnitude with the crystal lattice potential.

On the other hand, it implies a major part of the original q1D band to be unnested which is unlikely if the low-temperature state is a density wave.

Another explanation of the 1D AMRO in the present compounds has been suggested by Yoshioka [159]. Introducing a superstructure potential into the Hamiltonian describing a q2D electron gas, he has argued that the interlayer tunneling is strongly suppressed in a magnetic field at all directions except those satisfying the condition:

$$Q_{xy} \tan\theta \cos\varphi + Q_z = N\frac{2\pi}{a_z}, \tag{8.47}$$

where Q_{xy} and Q_z are, respectively, the in- and out-of-plane components of the superstructure wave vector $\boldsymbol{Q}$. This condition is equivalent to (8.42), yielding the same $\boldsymbol{Q}$ for the parameter sets (8.45) and (8.46) as given above. At θ and φ satisfying (8.47), the magnetoresistance was predicted [159] to be the same as in the normal q2D system, without the superstructure. In particular, it must be enhanced at the angles corresponding to peaks in the 2D AMRO. However, no correlation has been found between the resistance at the angles (8.42) and the 2D AMRO structure, observed above the critical temperature or above the critical pressure. On the contrary, the resistance at these angles may be significantly lower than the value expected for the normal 2D AMRO (see, e.g., data in [149]).

An alternative model, associating the observed 1D AMRO with an unconventional symmetry of the density-wave order parameter, is reviewed by Dora et al. [160]. It should be noted that this model implies thermodynamic properties to show features correlating with the AMRO. However, no such correlation has been found in the experiment so far.

8.6.2 Magnetic Field–Temperature Phase Diagram: Evidence of a CDW Ground State

Taking into account the anisotropic change of the magnetic susceptibility at $T_{\rm p}$ [129], the above-discussed low-temperature superstructure was initially attributed to a spin-density wave (SDW). However, later experiments [130] have shown the magnetization to be an easy-plane type rather than easy-axis commonly observed on SDW organic conductors [13]. Furthermore, ESR [161] and NMR [162] experiments have failed to find a resolvable change in the corresponding line shapes, thus setting the upper limit of the static magnetic moment as low as $10^{-4}\mu_B$ per molecule [162]. A key to clarifying the problem was given by detailed resistive [87,163–166] and magnetization [163,167,168] studies of the B–T phase diagram. A representative example of the ambient-pressure B–T diagram obtained from magnetic torque measurements in a field almost perpendicular to the layers [167] is shown in Fig. 8.20a. The experimental phase lines closely resemble the shape of the B–T diagram predicted for a CDW system [169–172] (see also Chap. 20 by Bjeliš and Zanchi) which is

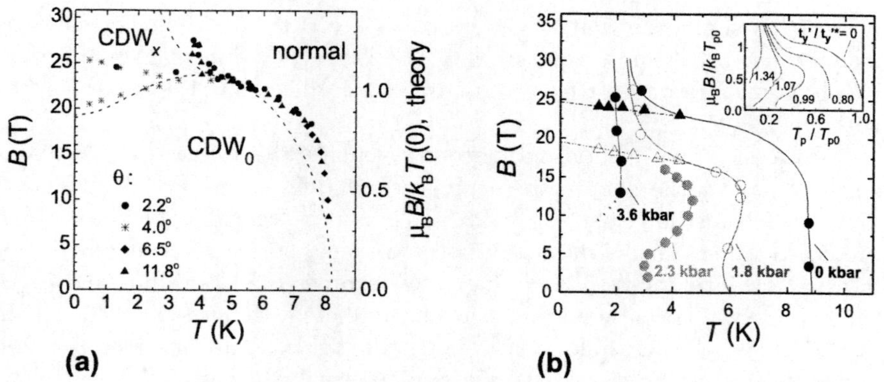

Fig. 8.20. B–T phase diagram of the K salt for fields almost perpendicular to the layers: (**a**) the ambient-pressure diagram obtained from torque measurements [167] (*symbols; scale on the left*) is nicely described by the theoretically predicted phase diagram of a perfectly nested CDW system [169] (*dashed line; scale on the right*); the deviation at $B < 20\,\mathrm{T}$ is ascribed to an imperfect nesting; (**b**) phase diagrams obtained from the resistive measurements at different pressures [166]. The shift and variation of the shape of the phase lines are qualitatively consistent with the theoretical predictions [169] shown in the *inset* (see text)

shown by the dashed lines in Fig. 8.20a. The temperature of the phase transition shifts, in a magnetic field normal to the layers, to lower values with an increasing rate until about $0.5T_\mathrm{p}(B=0)$. This is caused by the Pauli paramagnetic effect analogous to the paramagnetic pair breaking in superconductors. Below $\sim 0.5T_\mathrm{p}(B=0)$, a strong field applied at a constant temperature leads to a first-order transition [169, 172] from the CDW$_0$ state, characterized by a constant wave vector $\boldsymbol{Q}_0 = \boldsymbol{Q}(B=0)$, to the CDW$_x$ state with a wave vector component along the x-axis depending on field, $Q_x^{\mathrm{Pauli}}(B)$ (where the superscript emphasizes the Pauli paramagnetic origin of the effect). The temperature of the transition from the normal metallic state to CDW$_x$ saturates, so that the latter state survives in high fields.

The theoretically proposed CDW phase diagram is very similar to those of a clean superconductor (in the absence of the orbital pair breaking) [173] and of a spin-Peierls system [174]. Since both the superconducting and spin-Peierls orderings can be ruled out for the present compounds, the data in Fig. 8.20a are a strong argument for the CDW nature of the low-temperature state [164, 167, 175]. Recent low-temperature X-ray studies [176] support this conclusion although some problems with the sample dependence and thermal cycling have been encountered.

The CDW scenario is consistent with the above-discussed magnetotransport data. The anomalous magnetoresistance, in particular, the 1D AMRO, is only observed below T_p [150]. At low temperatures and fields above B_k, which corresponds to the CDW$_x$ state, the CDW gap is considerably reduced and, given, additionally, a strong MB at fields $\geq 20\,\mathrm{T}$, the magnetoresistance behavior is similar to that in the normal state [135, 149, 152, 177]. However, magnetoresistance and torque studies [163, 168, 178–180] reveal some features which are clearly distinct from the normal metallic properties but consistent with the CDW scenario.

While the experimental and theoretical phase diagrams in Fig. 8.20a are very similar in the shape, there are certain differences. For example, the first-order transition field, $B_{\mathrm{k}} \approx 23.5\,\mathrm{T}$, is about two times higher than the predicted value. This is, however, not surprising, since the actual CDW gap is usually considerably higher than the mean field BCS value used in the theory. Anyway, even being higher than predicted, the critical field remains in the range of accessible steady magnetic fields. This provides a unique opportunity to study the CDW state in the entire region of the B–T phase diagram.

Another obvious difference between the theory and experiment in Fig. 8.20a is that the latter reveals a weaker $T_{\mathrm{p}}(B)$ dependence at B below $\sim 0.7B_{\mathrm{k}}$. This difference has been proposed [167] to be due to an imperfect nesting of the q1D FS, which is not included in the theoretical line. If the nesting is imperfect, the Pauli effect is superposed by the orbital effect acting in the opposite direction [169]. The latter has exactly the same nature as in the well-studied SDW case (see, e.g., [13] for a review). Since the orbital effect is determined by the out-of-plane component of magnetic field, it should vanish at $\boldsymbol{B} \parallel$ layers. Indeed, specific heat [132] and magnetization [167] experiments confirm the expected difference between $T_{\mathrm{p}}(B)$ dependence in the parallel and perpendicular magnetic fields.

The orbital effect on a CDW becomes more evident under a moderately high pressure [166]. As might be expected, pressure worsens the nesting conditions for the q1D band, leading to a gradual suppression of the CDW state in these compounds [83, 166, 181–184]. Figure 8.20b shows the phase diagrams for the K salt at different pressures. At $P \sim 2\,\mathrm{kbar}$, which is close to the critical value $P_{\mathrm{c}} \approx 2.5\,\mathrm{kbar}$, the orbital effect is clearly manifested by an *increase* of the transition temperature at intermediate fields, $\sim$5–12 T. This result is fully consistent with the predictions for an imperfectly nested CDW system [169].

The inset in Fig. 8.20b shows the field dependence of the transition temperature between the normal and CDW states calculated [169] for different values of the second-order interchain transfer integral t'_y which characterizes the nesting conditions. At $t'_y = 0$, the nesting is perfect and the phase line is identical to that shown in Fig. 8.20a. When t'_y becomes equal to a critical value $t'^{*}_y \cong k_{\mathrm{B}}T_{\mathrm{p0}}$ (where T_{p0} is the zero-field transition temperature of a perfectly nested system), the CDW is no longer stable at zero field. However, it can be induced by a finite field applied normal to the layers. In agreement with the experiment [166], the orbital effect leads to a positive slope $\mathrm{d}T_{\mathrm{p}}/\mathrm{d}B$ on a part of the phase line when t'_y/t'^{*}_y is close to unity.

8.6.3 Field-Induced CDW Transitions

At $t'_y/t'^*_y \geq 1$, the theory [169, 185, 186] predicts a cascade of field-induced transitions between CDW subphases with quantized values of the wave vector $\boldsymbol{Q}$. The likely occurrence of this phenomenon in α-(BEDT-TTF)$_2$KHg(SCN)$_4$ has been suggested in [166]. Recent, more detailed studies [154, 155] have provided further arguments for the FICDW transitions in this compound.

Figure 8.21a shows the low-temperature ($T = 0.1\,$K) magnetoresistance of the K salt recorded at different pressures. At zero field, the CDW is completely suppressed by the critical pressure $P_c \approx 2.5\,$kbar [14, 166]. However, as evidenced by a detailed analysis of the field- and angle-dependent magnetoresistance, the density-wave state is reentered in a field $B \geq 2$–$3\,$T perpendicular to the layers. Besides the strong rapid SdH oscillations dominated by the α-frequency, very slow oscillations, approximately periodic in inverse field (frequency $\simeq 20\,$T), emerge at $P \geq P_c$. This is exactly the pressure region at which the FICDW transitions are expected [169, 186]. The shift of the oscillation positions with temperature [155] and a hysteresis structure correlating with the oscillations [154] indicate that these features are due to field-induced transitions rather than the usual SdH oscillations.

Of course, the FICDW transitions have the same origin as the well-known FISDW phenomenon. However, in the present case, the orbital effect competes with the suppressing Pauli effect. This leads to a significantly lower effective coupling constant $g_{\text{eff}}^{\text{CDW}}(B)$ than in the SDW case [186]. Therefore, the FICDW transitions are restricted to much lower temperatures and a relatively narrow pressure range. One can see, for example from Fig. 8.21a, that the features associated with the transitions are strongest at 3 and 3.5 kbar and become weaker already at 4 kbar.

The interplay between the orbital and Pauli effects has been predicted to lead to an enhancement of the effective coupling constant at the field directions at which the spin splitting becomes commensurate with the orbital quantization [171, 186]:

$$|\cos\theta_M| = \frac{1}{M}\frac{2\mu_B}{ev_F a_z}, \quad M = 1, 2, \ldots, \tag{8.48}$$

where μ_B is the Bohr magneton. The data shown in Fig. 8.21b support this prediction of a *commensurate splitting* (CS) effect. As expected, at tilting the field, the positions of the FICDW features scale with the perpendicular component, $B_z = B\cos\theta$. However, at certain angles [154], ≈42°, 65°, and 73.5°, the phase of the oscillations is inverted. This resembles the spin-zero effect in magnetic quantum oscillations [2] which is often observed in organic conductors [6]. However, there is an important difference: in our case, a nonmonotonic hysteresis, which vanishes at the "spin-zero" angles, reappears and is strongly enhanced at the angles θ_M situated in the middle between the "spin-zero" positions in the $1/\cos\theta$ scale [154]. This is illustrated in Fig. 8.21b by the up- and downward sweeps of the magnetoresistance at CS angle $\theta_M = 57.7°$. Note

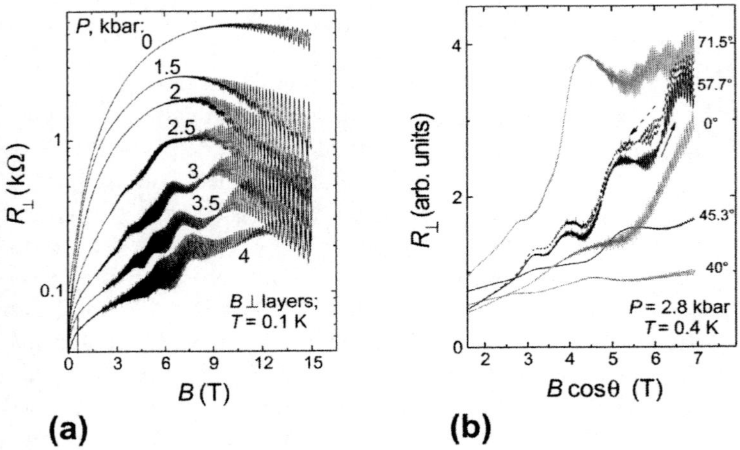

Fig. 8.21. Field-dependent magnetoresistance of the K salt at low temperatures: (**a**) in the perpendicular field, at different pressures between 0 and 4 kbar [155]; note slow oscillations emerging at $P \geq 2.5$ kbar which are associated with FICDW transitions (see text); (**b**) magnetoresistance at 2.8 kbar vs. the field component perpendicular to the layers, at different tilt angles θ [154]

that the FICDW features are quite strong at this angle at the temperature of 0.4 K, whereas in the perpendicular field they are already rather weak due to the temperature smearing effect.

The FICDW transitions presented above are the consequence of the quantizing orbital effect of the magnetic field. In this respect, they are very much alike the FISDW phenomenon. The necessary condition for both effects is the suppression of the zero-field density-wave ground state. The latter can be achieved in the present compounds by applying sufficiently high pressure. Now, we will consider another kind of FICDW transitions which can occur even at ambient pressure. These transitions originate from a *superposition* of the orbital quantization and the Pauli effect and, thus, are a unique property of the CDW state.

Figure 8.22a shows the interlayer magnetoresistance and torque of the K salt measured at zero pressure in a magnetic field strongly tilted toward the conducting layers [179]. The kink transition, B_{k}, is shifted to ≈ 12.5 T at this field orientation. Additionally, both the magnetoresistance and torque display a complex structure above B_{k}. As a matter of fact, this structure, found in magnetic torque [178], was the first indication that the low-temperature state of these salts is not normal even at the fields above the kink transition. Figure 8.22b summarizes the results of detailed studies of the new features at high tilt angles: they emerge at the highest fields at $\theta \geq 65°$ and rapidly shift down toward B_{k} as θ approaches 90°.

The hysteretic character of the anomalies suggests that they are associated with multiple first-order phase transitions. The increasingly high sensitivity of

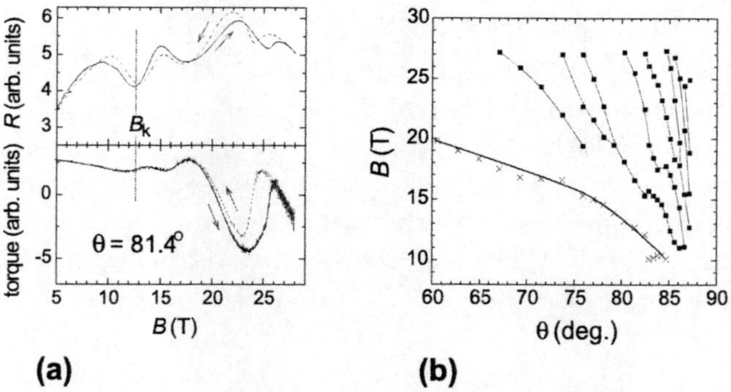

Fig. 8.22. Field-induced transitions in the K salt at ambient pressure, at high tilt angles [179]: (**a**) up- (*solid line*) and downward (*dashed line*) field sweeps of the magnetoresistance (*upper panel*) and torque (*lower panel*) at $\theta = 81.4°$, $T = 0.4\,\mathrm{K}$; (**b**) B–θ phase diagram showing the kink transition (*crosses*) and the new FICDW transitions (*squares*) detected by torque measurements

the transitions to changes in the tilt angle near 90° suggests that the orbital effect determined by the field component, $B_z = B\cos\theta$, plays a crucial role. On the other hand, it is important that they occur at $B > B_\mathrm{k}$, i.e., in the fields producing a very strong Pauli effect. It is, therefore, natural to suggest that the new transitions originate from an interplay between the Pauli and orbital effects on the high-field CDW_x state.

To explain these transitions, a simple qualitative model was proposed [179], taking into account that the CDW wave vector $\boldsymbol{Q}$ becomes field dependent in the CDW_x state. It was assumed that $\boldsymbol{Q}$ shifts to keep one spin subband fully gapped while the other subband becomes partially metallic. Then, the situation for the second subband is analogous to that in the "conventional" FICDW or FISDW case. One can expect a FICDW instability to be enhanced at a quantized set of the wave vectors:

$$Q_{xN}^{\mathrm{orbit}}(\boldsymbol{B}) = Q_{x0}(B) + N\frac{2ea_y B_z}{\hbar}, \tag{8.49}$$

where $Q_{x0}(B) = 2k_\mathrm{F} - 2\mu_\mathrm{B}B/(\hbar v_\mathrm{F})$ corresponds to the optimal nesting condition of this subband. As a result, a series of discontinuous jumps of $Q_x(B)$ is expected between successive levels $Q_{xN}^{\mathrm{orbit}}(\boldsymbol{B})$ in the vicinity of the continuous field-dependent $Q_x^{\mathrm{Pauli}}(B)$, predicted by Zanchi et al. [169].

While this model qualitatively explains the major features of the new FICDW transitions [179], one certainly needs a more rigorous theoretical description (see Chap. 20). In particular, one should take into account that $\boldsymbol{Q}(B)$ most likely splits in the CDW_x state [172], so that both spin subbands are maintained fully gapped. Nevertheless, one can still expect the CDW response function to be enhanced at the pointed above discrete set of Q_x [187].

8.7 Other Organic Conductors: Probing and Controlling Electronic Properties by Strong Magnetic Fields

In this section, several recent examples of layered organic conductors studied in strong magnetic fields are described. It is shown that not only FSs of these materials can be probed, but also the electronic ground state, itself, can be modified by a magnetic field.

8.7.1 β and β″ Salts

Layered β- and β″-type salts of BEDT-TTF and its derivatives are characterized by simple structures of the donor layers: the donor molecules are packed in homogeneous stacks parallel to each other. The difference between the two types of packing consists in the molecule inclination in a stack and, therefore, in intermolecular packing modes (see, e.g., [188]).

Salts of the β-type have the simplest structure with two molecules per unit cell. Given the formal charge of 1/2 hole per molecule, band structure calculations yield a single cylindrical FS occupying one half of the Brillouin zone [188]. Thus, provided the crystal quality is sufficiently high, these materials are perfectly suited for studying the q2D magnetotransport. For example, the first two ambient-pressure q2D superconductors, β-$(BEDT\text{-}TTF)_2X$ with $X = I_3$ and IBr_2 [189,190], have been used extensively for exploring both semiclassical and quantum magnetoresistance phenomena characteristic of q2D metals, see, e.g., [12, 16]. More recent experiments have revealed a similar behavior in a number of other β salts of BEDT-TTF [191, 192] and its nonsymmetric derivative BDA-TTP [193,194]. Both the SdH oscillations and 2D AMRO generally confirm theoretical predictions for the FSs in these materials. The only exclusion is the superconductor β-$(BDA\text{-}TTP)_2SbF_6$ whose FS derived from AMRO considerably deviates in the shape from the predicted one [193]. It should be noted that the effective cyclotron mass in this salt, $m_c = (12.4 \pm 1)m_e$ (where m_e is the free electron mass), is the largest yet found in an organic conductor, indicating the importance of many-body effects.

In β″ salts, intermolecular interactions are more complex due to an inclination of the molecule planes with respect to the stacking directions. The FS topology is very sensitive to subtle details of the molecule packing [188]. Furthermore, some of these salts undergo phase transitions with changing temperature, pressure, or magnetic field. Therefore, magnetoresistance measurements have been widely used for experimental characterization of the FS and its dependence on external conditions, see, e.g., [27, 195–200].

As concerns magnetic field phenomena, perhaps, the most interesting material is the fully organic superconductor, β″-$(BEDT\text{-}TTF)_2SF_5CH_2CF_2SO_3$. Figure 8.23 shows the angle-dependent magnetoresistance of this

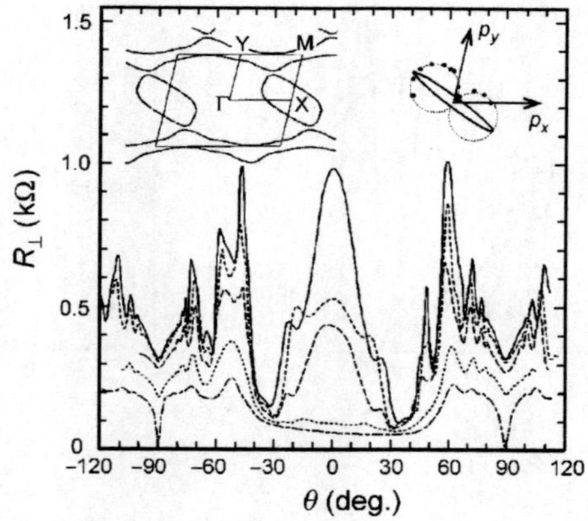

Fig. 8.23. Angle-dependent magnetoresistance of β″-(BEDT-TTF)$_2$SF$_5$CH$_2$CF$_2$SO$_3$ recorded at $T = 0.5$ K, at the fields (*from bottom to top*): 10, 15, 23, 28, and 33 T [73]. The 2D AMRO are clearly seen in the 15 T curve, being superposed by strong SdH oscillations at the higher fields. The *insets* show the theoretical FS (*top left*) [196] and the cylindrical part of the FS derived from the AMRO (*top right*) [201]

salt recorded at different fields [73]. At the fields below 20 T, it is dominated by the 2D AMRO, which reveal a cylindrical FS [73,201]. The cross section of the cylinder determined from the AMRO period, as described in Sect. 8.2.1, is shown in the top right inset in Fig. 8.23. Comparing with the theoretical prediction [196] (top left in Fig. 8.23), it is much more elliptic and its area is smaller by a factor 3. At the fields above 20 T, the AMRO are superposed by very large SdH oscillations.

Examples of dHvA and SdH oscillations in the field perpendicular to the layers are shown in Fig. 8.24. Due to the extremely high two dimensionality of the electronic system, the oscillations strongly violate the conventional 3D behavior [115,202]. On the other hand, they have successfully been described by the 2D theory, assuming a field-independent chemical potential [113,202]. In particular, the "inverse saw-tooth" shape of the dHvA oscillations shown in Fig. 8.24a is perfectly fitted by the LKS formula (8.39). It should be noted, however, that the reason for a perfect pinning of the chemical potential in such a strongly 2D system is unclear at present (see [12] for a discussion).

Interestingly, both the dHvA and SdH oscillations have been observed in β″-(BEDT-TTF)$_2$SF$_5$CH$_2$CF$_2$SO$_3$ not only in the normal metallic but also in the superconducting state [203,204]. Moreover, the dHvA oscillations were found to be even *enhanced* upon entering the superconducting region [204]. This highly unusual behavior has been attributed to a crossover from the incoherent interlayer transport regime in the normal state (see Sect. 8.4) to a 3D coherent regime established via Josephson coupling between the layers in the superconducting state [205].

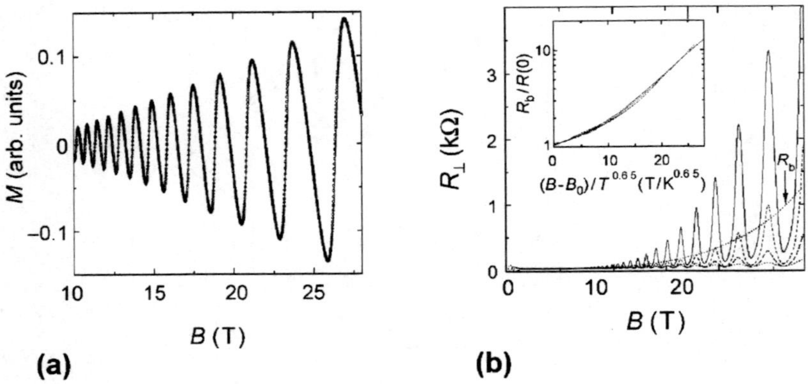

Fig. 8.24. Magnetic quantum oscillations in β″-(BEDT-TTF)$_2$SF$_5$CH$_2$CF$_2$SO$_3$: (**a**) dHvA oscillations (*symbols*) at $T = 0.44\,\mathrm{K}$ [202] have a sharp "inverse sawtooth" shape which is perfectly described by the LKS formula (8.39) (*solid line*); (**b**) magnetoresistance at the temperatures (*bottom to top*): 2.4, 1.4, 0.9, and 0.44 K, and its nonoscillating background $R_\mathrm{b}(B)$ (*dotted line*) extracted for $T = 0.44\,\mathrm{K}$; the *inset* demonstrates the dynamical scaling behavior of $R_\mathrm{b}(B,T)$ [116]

Apart from the quantum oscillations, the monotonic magnetoresistance component also shows an unusual behavior. It was found [116, 206] to rapidly grow at increasing the field or at lowering the temperature when both the field and the current were applied normal to the layers. The field dependence of the nonoscillating background R_b analyzed in [116] strongly violates Kohler's rule but can be described by a universal dynamical scaling relation characteristic of a quantum phase transition. While the correct procedure of extracting the background from a strongly oscillating magnetoresistance has been debated [113, 115], one can see from Fig. 8.24b that even *in the dips* of the SdH oscillations the resistance rapidly grows with cooling at $B \geq 20\,\mathrm{T}$. This is definitely not a normal metallic behavior and may indeed be a sign of a field-induced metal–insulator transition, as proposed in [116].

8.7.2 κ-(BEDT-TTF)$_2$X Salts

The κ-type salts of BEDT-TTF with the anions $\mathrm{X}^- = \mathrm{Cu(NCS)}_2^-$, Cu[N(CN)$_2$]Br$^-$, and Cu[N(CN)$_2$]Cl$^-$ (further on abbreviated as NCS, Br, and Cl salts, respectively) have constantly been of high interest due to the highest among organic superconductors critical temperatures $T_\mathrm{c} > 10\,\mathrm{K}$ [13] and an intricate competition between the normal metallic, superconducting, and Mott-insulating electronic states, see [207, 208] and references therein. Comprehensive reviews of numerous high-field studies of these compounds, providing important information on their FSs and other characteristics of the electronic system, can be found in [6, 88]. Here, we just briefly mention a few recent results illustrating (1) the general magnetotransport properties, (2) a

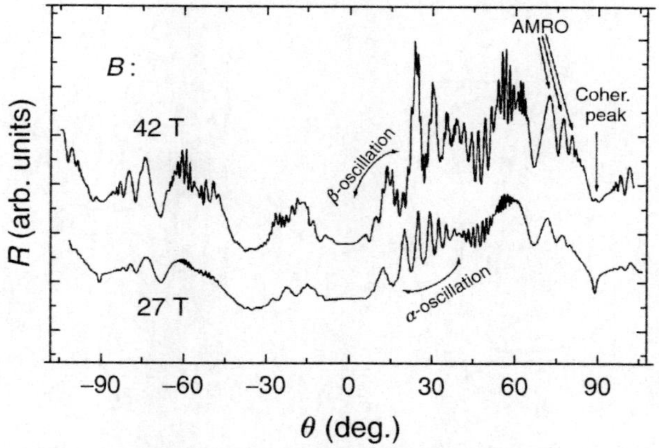

Fig. 8.25. Interlayer magnetoresistance of κ-(BEDT-TTF)$_2$(NCS)$_2$ measured as a function of the tilt angle θ at $T = 0.49\,\mathrm{K}$, at high magnetic fields [209]

FS reconstruction detected by magnetic oscillation experiments, and (3) a field-induced electronic phase transition.

Figure 8.25 shows the angular dependence of the magnetoresistance $R_\perp(B,\theta)$ of the best studied, NCS salt [209]. It possesses features characteristic of a system with coexisting cylindrical and open FSs (see Fig. 8.1c). One can clearly see AMRO associated with semiclassical α-orbits on the cylindrical FS as well as SdH oscillations due to semiclassical α- and MB β-orbits. The analysis in [209] reveals both the 2D and 1D AMRO in the $R_\perp(\theta)$ sweeps, in agreement with previous works [210, 211]. Near the parallel orientation, $\theta = 90°$, the 42 T curve displays a small coherence peak yielding an estimation of the interlayer transfer integral, $t_z \simeq 0.04\,\mathrm{meV}$ [77]. In the 27 T curve, a sharp dip around $\theta = 90°$ corresponds to the onset of the superconducting transition. The quantum oscillations, besides the fundamental frequencies $F_\alpha \approx 600/\cos\theta\,\mathrm{T}$ and $F_\beta \approx 3{,}900/\cos\theta\,\mathrm{T}$, contain a number of combination harmonics, $mF_\beta \pm nF_\alpha$, which point to the significance of chemical potential oscillations in this highly 2D material (see Sect. 8.5.3).

Turning to the other two salts, their FSs are predicted [212, 213] to be very similar to that of the NCS salt. The only significant difference consists in the absence of the gap Δ_0 between the cylindrical part and open sheets at the Brillouin zone boundary due to the centrosymmetric crystal structure of the Br and Cl salts. Indeed, SdH oscillations and AMRO found in the Cl salt under pressure (at $P = 1\,\mathrm{bar}$, the compound is a Mott insulator [207]) have shown a dominant contribution from the large β-orbit enveloping the whole FS [214]. However, the α-oscillations with the frequency $F_\alpha \sim 600\,\mathrm{T}$ (at $\boldsymbol{B} \perp$ layers) were also observed, indicating that Δ_0 is nonzero, although much smaller than in the NCS salt. Similarly, β-oscillations [102, 215] and very weak α-oscillations have been found in the Br salt at ambient pressure. However, high-pressure measurements have revealed no F_α but, instead, clear

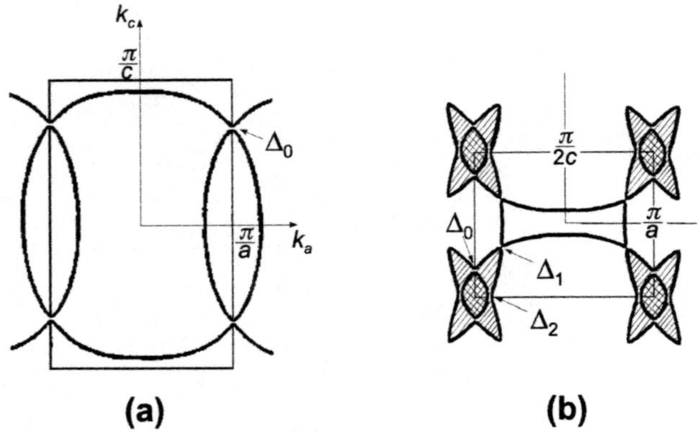

Fig. 8.26. FS of κ-(BEDT-TTF)$_2$Cu[N(CN)$_2$]Br: (**a**) without and (**b**) with taking into account a superstructure arising below 200 K [218]. The slow oscillations observed under pressure can be explained by additional small MB gaps Δ_1 and Δ_2 formed by the superstructure potential [217]

contributions of lower frequencies: $F_\alpha/4$ and $F_\alpha/2$ [216,217]. This change in the SdH spectrum has been attributed [217] to folding of the predicted FS along the k_c-direction and formation of additional gaps Δ_1 and Δ_2, as shown in Fig. 8.26. Such a reconstruction is consistent with a superstructure transition below 200 K detected at ambient pressure by X-rays [218]. The fact that the low-frequency oscillations become only visible under a high pressure suggests the superstructure to be enhanced by pressure. It would be interesting to perform a high-pressure X-ray study to check this suggestion.

As mentioned above, the Cl salt undergoes at ambient pressure a Mott transition into an insulating ground state. However, under a pressure exceeding 0.2–0.3 kbar (depending on temperature), the compound becomes metallic and superconducting with $T_c \approx 13$ K. The position with respect to the metal–insulator transition line on the phase diagram can be precisely tuned by changing temperature and pressure. Moreover, it was recently found to be sensitive also to a magnetic field [219]. Figure 8.27 shows an example of the field-dependent resistance of the Cl salt under pressure. At $B = 0$, the compound is metallic although very close to the phase boundary. With increasing the field, the resistance exhibits a discontinuous increase by an order of magnitude at $B \approx 6$ T, signaling a transition into the insulating state. In agreement with the theoretical prediction [220], this is a first-order phase transition, as evidenced by a strong hysteresis with respect to the field sweep direction. Kagawa et al. [219] have succeeded, by tuning the temperature and pressure, to obtain the field-induced transition into the insulating state directly from the superconducting state. Similar results have also been obtained on the Br salt which was driven close to the metal–insulator boundary by deuterium substitution and cooling rate control [221].

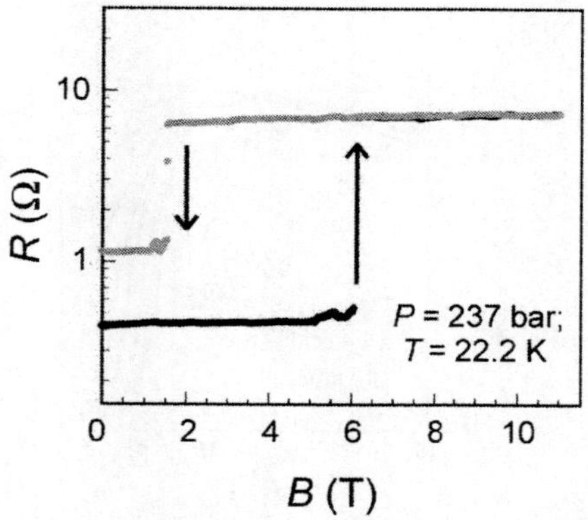

Fig. 8.27. Field-induced metal–insulator transition in κ-(BEDT-TTF)$_2$Cu[N(CN)$_2$] Cl [219]

8.7.3 λ and κ Salts of (BETS) with Magnetic Anions

Among the large number of conductors based on the BETS (or BEDT-TSF) donor molecule, the salts with Fe-containing anions have attracted particular attention due to an interplay between magnetic ordering in the anion layers and conducting properties of the molecular layers, see [222] for a recent review. The most remarkable consequence of this interplay is the *field-induced superconductivity* (FISC) which was first discovered in λ-(BETS)$_2$FeCl$_4$ [223]. At zero field, the ground state of this compound is an antiferromagnetic insulator [224]. However, at a field above 10 T, it becomes metallic [225,226] and, ultimately, the superconducting state is stabilized in the field between 17 and 42 T directed parallel to the conducting planes [223, 227]. This fascinating behavior, reflected in the field-dependent resistance, as shown in Fig. 8.28, is caused by the interaction between the conducting π electrons of the BETS layers and d electrons of Fe^{3+} ions in the anion layers. In particular, the FISC is consistently explained by the Jaccarino–Peter compensation effect [228,229]. Due to the exchange interaction with the magnetic moments localized on Fe^{3+}, the conduction electrons experience an exchange field $\boldsymbol{B}_\mathrm{J}$ directed against the external field $\boldsymbol{B}$. The material becomes superconducting when $B_\mathrm{eff} = |B - B_\mathrm{J}|$ is smaller than the paramagnetic Chandrasekhar–Clogston limit [230], provided the orbital pair-breaking mechanism is suppressed. The latter condition is fulfilled due to the intrinsic high two dimensionality of the conducting system and, additionally, due to the magnetically aligned $FeCl_4^-$ tetrahedra inhibiting the interlayer tunneling of singlet Cooper pairs [227].

The suppression of the orbital effect of magnetic field on the superconductivity makes λ-(BETS)$_2$FeCl$_4$ a promising candidate for the observation of the LOFF (Larkin–Ovchinnikov–Fulde–Ferrel) state. Indeed, the superconducting dome on the B–T diagram exceeds the estimated paramagnetic limit

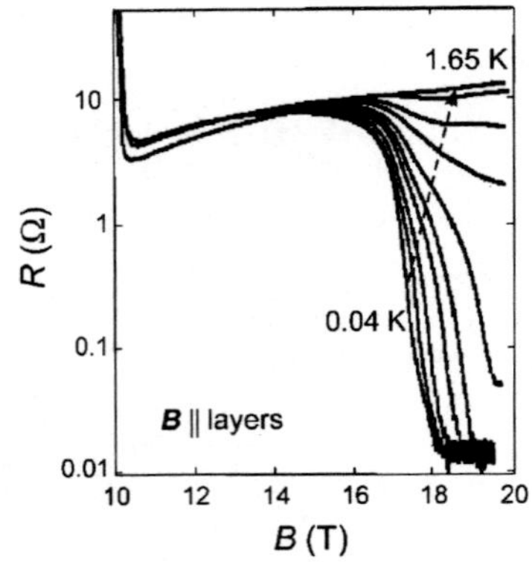

Fig. 8.28. Interlayer resistance of λ-$(BETS)_2FeCl_4$ as a function of magnetic field aligned parallel to the layers, measured at different temperatures between 40 mK and 1.65 K [223]. The transition from the antiferromagnetic insulating state to the metallic state at $B \approx 10\,T$ is followed by the superconducting transition at $B \simeq 17\,T$

and the shape of its phase boundary seems to be consistent with the existence of the LOFF state [231].

Besides the pure λ-$(BETS)_2FeCl_4$ salt, the FISC has been found in the salts with mixed anions, $Fe_xGa_{1-x}Cl_4^-$ [232] and $FeCl_{4-x}Br_x$ [233]. In the former alloy, the antiferromagnetic insulating state is gradually suppressed at decreasing x while the onset of superconductivity is being shifted to lower fields. At $x = 0.45$, the phase boundaries of the insulating and superconducting states merge, so that a direct field-induced transition between them, without an intermediate normal metallic state, is possible [232]. In alloys with the FeCl/Br anions, an increase of the Br content leads to an expansion of the insulating region of the B–T diagram but does not affect significantly the superconductivity. As a result, the normal metallic state shrinks and, again, a direct insulator–superconductor transition is observed in fields ~20–30 T at $x = 0.5$ and 0.7 [233].

Another example of FISC is given by κ-$(BETS)_2FeBr_4$. At zero field, this compound is characterized by a coexistence of an antiferromagnetic order in the anion layers (Néel temperature, $T_N = 2.5\,K$) with superconductivity ($T_c = 1.1\,K$) [234,235]. Under a magnetic field oriented parallel to the layers, the superconducting and antiferromagnetic states are successively suppressed; however, starting from 10 T a reentrant transition into the superconducting state occurs [236,237] (see Fig. 8.29).

The exchange interaction between the localized d electrons of Fe^{3+} and the itinerant π electrons of BETS layers is also reflected in the behavior of magnetic quantum oscillations observed in the present compounds. Figure 8.30 shows the beating SdH oscillations in λ-$(BETS)_2FeCl_4$ recorded in the magnetic field perpendicular to the layers [78]. Two corresponding frequencies, $F_1 = 608\,T$ and $F_2 = 738\,T$, reveal extremal FS cross sections with the areas

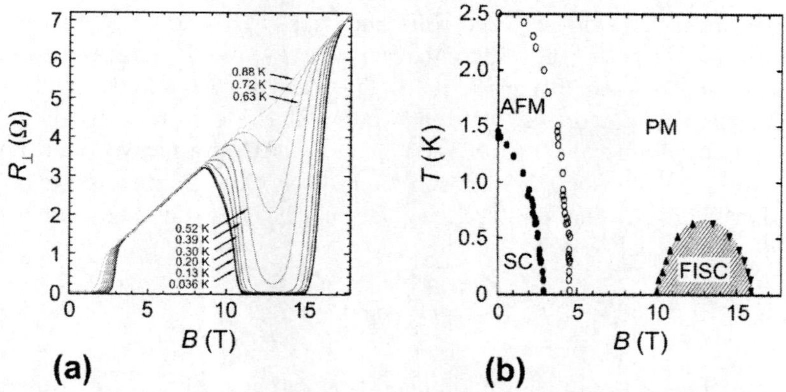

Fig. 8.29. Antiferromagnetic superconductor κ-(BETS)$_2$FeBr$_4$ in a magnetic field parallel to the c-axis (crystallographic short axis in the layer plane) [236]: (**a**) field-dependent interlayer resistance at different temperatures; (**b**) B–T diagram comprising the antiferromagnetic metallic (AFM), low-field superconducting (SC), paramagnetic metallic (PM), and FISC states

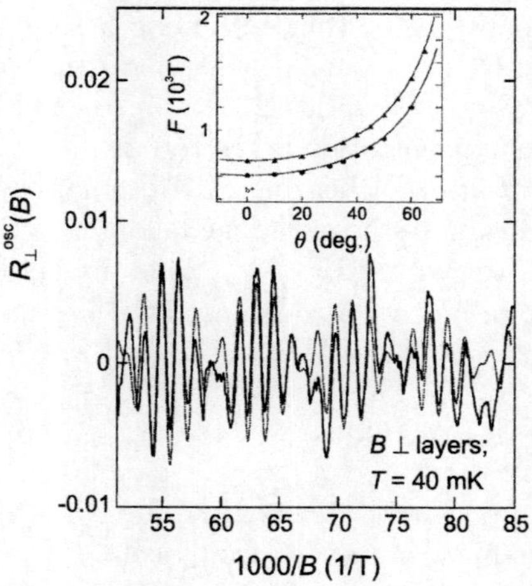

Fig. 8.30. Beating SdH oscillations in λ-(BETS)$_2$FeCl$_4$ revealing two fundamental frequencies, $F_1 = 608\,$T and $F_2 = 738\,$T [78]. At a tilted field, both frequencies precisely follow the $1/\cos\theta$ law (*inset*) and so does the frequency difference $\Delta F = F_2 - F_1$

equal to 14 and 17% of the Brillouin zone area, respectively. This result agrees with the band structure calculations yielding a FS consisting of a pair of open sheets and a cylinder occupying 18% of the Brillouin zone [226]. The beating is, at first glance, a consequence of a relatively strong warping of the cylindrical FS along the z-direction (see Sect. 8.5.3). The ratio $t_z/\varepsilon_F \approx 0.05$ evaluated from the beat frequency, according to (8.37), is close to the upper estimate obtained from the width of the coherence peak [78]. However, as seen from the inset in Fig. 8.30, the difference between F_1 and F_2 changes monotonically, $\propto 1/\cos\theta$, with tilting the field. This clearly contradicts the oscillating angular dependence of the difference between the extremal cyclotron orbit areas for a cylindrical FS expressed by (8.5). An alternative interpretation of the beating invokes the spin-splitting effect [238]. In the presence of an exchange field B_J, the spin-splitting factor (8.31) is modified and becomes field dependent [2]:

$$R_\mathrm{S} = \cos\left[\frac{\pi}{2}gm^*\left(1 - \frac{B_\mathrm{J}}{B}\right)\right] \tag{8.50}$$

(here we consider only the fundamental harmonic). Now R_S modulates the oscillation amplitude periodically in $1/B$, similarly to the effect of the FS warping (8.35). However, the angular dependence of the beat frequency, $F_\mathrm{beat} = gm^*B_\mathrm{J}/4$, being determined by that of the cyclotron mass m^*, is simply proportional to $1/\cos\theta$, in agreement with the experimental observation [78]. The exchange field, $B_\mathrm{J} = 32\,\mathrm{T}$, evaluated from F_beat is in excellent agreement with the value expected from the phase diagram, assuming the Jaccarino–Peter scenario [227], thus providing a solid argument in favor of the latter. Similar estimations have been performed on a few compounds of the λ-(BETS)$_2$Fe$_x$Ga$_{1-x}$Cl$_4$ series [239] and reasonable correlation between the internal field and concentration, x, of the magnetic ions Fe^{3+} was found.

The same interpretation of beating SdH oscillations has been applied to κ-(BETS)$_2$FeBr$_4$ to evaluate the exchange field and *predict* the FISC state in this salt in the field range 11–13 T [238]. This prediction has been perfectly confirmed by later experiments as demonstrated in Fig. 8.29.

8.8 Concluding Remarks

Numerous new magnetic field effects described in this review clearly distinguish layered organic conductors from usual 3D metals. On the other hand, many of these effects should, in principle, be observable in other types of q2D conductors. Indeed, the 2D AMRO (Sect. 8.2.1) and coherence peak (Sect. 8.3.2) have already been observed and used for characterization of the electronic system in a high-T_c cuprate [240], in the unconventional superconductor Sr_2RuO_4 [241], and in the intercalated graphites [242].

Magnetic quantum oscillations are strongly enhanced in organic conductors due to their high two dimensionality, providing important information about their FS properties additionally to that obtained from the semiclassical magnetoresistance. However, despite numerous recent studies of dHvA and SdH effects in q2D systems (see Sect. 8.5.3), more work has to be done to gain quantitative agreement between theory and experiment. Already in the relatively simple case of purely 2D dHvA oscillations with a constant chemical potential, observed in β''-(BEDT-TTF)$_2$SF$_5$CH$_2$CF$_2$SO$_3$ (Sect. 8.7.1), a problem of the cyclotron mass being dependent on field and on the harmonic index was encountered [100]. Even more difficult is the situation with the q2D SdH effect. The latter crucially depends on details of the charge transfer and scattering whose nature, especially in the most anisotropic compounds, is far from being clear, as demonstrated in Sect. 8.4.

The examples presented in Sects. 8.6 and 8.7 show that strong magnetic fields not only are an extremely powerful tool for exploring FS properties, but also can be used for manipulating the electronic states of these materials. Of course, there remain a plenty of questions for future work.

While the phase diagram of the α-(BEDT-TTF)$_2$MHg(SCN)$_4$ salts can be explained, on the whole, in terms of the field effect on an imperfectly nested CDW system (Sects. 8.6.2 and 8.6.3), the electronic properties of the high-field states are not well understood. Already at ambient pressure, the CDW$_x$ state exhibits anomalous features which were successively attributed to a quantum Hall effect [88], a field-induced Fröhlich superconductivity [243], and a novel unconventional quantum fluid [180]. Although these anomalies are most likely related to an interplay between the strong Landau quantization of the metallic q2D band and the CDW state, as proposed, e.g., in [180], their exact origin is still to be established. The properties of recently discovered two kinds of FICDW states are even less studied. More experimental and theoretical work is needed for a better understanding of the role of the Pauli effect superposed on the quantizing orbital effect.

The κ-(BEDT-TTF)$_2$X salts, exhibiting an intricate phase diagram with different competing instabilities of the normal metallic state, have been a subject of intense studies in the last decade [207, 208]. However, only very recently, it was found that the metal–insulator transition in these compounds can be driven by magnetic field (Sect. 8.7.2). This phenomenon is definitely worth further investigation.

Acknowledgments

I am very grateful to my coworkers and colleagues D. Andres, W. Biberacher, J. Brooks, E. Canadell, P. Chaikin, P. Grigoriev, A. Kovalev, A. Lebed, R. McKenzie, T. Osada, V. Peschansky, S. Pesotskii, R. Shibaeva, S. Uji, J. Wosnitza, and V. Yakovenko for their valuable contributions and numerous illuminating discussions and comments on the topics presented in this review.

References

1. A.P. Cracknell, K.C. Wong, *The Fermi Surface* (Oxford University Press, London, 1973)
2. D. Shoenberg, *Magnetic Oscillations in Metals* (Cambridge University Press, Cambridge, 1984)
3. A.B. Pippard, *Magnetoresistance in Metals* (Cambridge University Press, Cambridge, 1989)

4. K. Oshima, T. Mori, H. Inokuchi, H. Urayama, H. Yamochi, G. Saito, Phys. Rev. B **38**, 938 (1988)
5. M.V. Kartsovnik, V.N. Laukhin, V.I. Nizhankovskii, A.A. Ignat'ev, JETP Lett. **47**, 363 (1988)
6. J. Wosnitza, *Fermi Surfaces of Low-Dimensional Organic Metals and Superconductors* (Springer, Berlin Heidelberg New York, 1996)
7. P.M. Grant, J. Phys. (Paris) **44**, C3 (1983)
8. M.V. Kartsovnik, V.N. Laukhin, S.I. Pesotskii, I.F. Schegolev, V. Yakovenko, J. Phys. I **2**, 89 (1992)
9. S.S. Khasanov, B.Z. Narymbetov, L.V.Z.L.P. Rozenberg, R.P. Shibaeva, N.D. Kushch, E.B. Yagubskii, R. Rousseau, E. Canadell, Eur. Phys. J. B **1**, 419 (1998)
10. A.A. Abrikosov, *Fundamentals of the Theory of Metals* (Elsevier Science Publishers, Amsterdam, 1988)
11. N.W. Ashcroft, N.D. Mermin, *Solid State Physics* (Saunders College, Philadelphia, 1976)
12. M.V. Kartsovnik, Chem. Rev. **104**, 5737 (2004)
13. T. Ishiguro, K. Yamaji, G. Saito, *Organic Superconductors*, 2nd edn. (Springer, Berlin Heidelberg New York, 1998)
14. D. Andres, M.V. Kartsovnik, W. Biberacher, K. Neumaier, E. Schuberth, H. Müller, Phys. Rev. B **72**, 174513 (2005)
15. M.V. Kartsovnik, in *Molecular Low-Dimensional and Nanostructured Materials for Advanced Applications*, ed. by A. Graja, B.R. Bulka, F. Kajzar (Kluwer, Dordrecht, 2002), pp. 159–168
16. M.V. Kartsovnik, V.G. Peschansky, Low Temp. Phys. **31**, 185 (2005)
17. M.V. Kartsovnik, P.A. Kononovich, V.N. Laukhin, I.F. Schegolev, JETP Lett. **48**, 541 (1988)
18. K. Kajita, Y. Nishio, T. Takahashi, W. Sasaki, R. Kato, H. Kobayashi, A. Kobayashi, Y. Iye, Solid State Commun. **70**, 1189 (1989)
19. M.V. Kartsovnik, P.A. Kononovich, V.N. Laukhin, S.I. Pesotskii, I.F. Schegolev, Sov. Phys. JETP **70**, 735 (1990)
20. K. Yamaji, J. Phys. Soc. Jpn. **58**, 1520 (1989)
21. A.E. Kovalev, M.V. Kartsovnik, N.D. Kushch, Solid State Commun. **87**, 705 (1993)
22. R. Yagi, Y. Iye, T. Osada, S. Kagoshima, J. Phys. Soc. Jpn. **59**, 3069 (1990)
23. V.G. Peschansky, J.A.R. Lopes, T.G. Yao, J. Phys. I **1**, 1469 (1991)
24. Y. Kurihara, J. Phys. Soc. Jpn. **61**, 975 (1992)
25. V.M. Yakovenko, B.K. Cooper, cond-mat/0507120 (unpublished)
26. A.E. Kovalev, S. Hill, K. Kawano, M. Tamura, T. Naito, H. Kobayashi, Phys. Rev. Lett. **91**, 216402 (2003)
27. A.A. House, N. Harrison, S.J. Blundell, I. Deckers, J. Singleton, F. Herlach, W. Hayes, J.A.A.J. Perenboom, M. Kurmoo, P. Day, Phys. Rev. B **53**, 9127 (1996)
28. V.G. Peschansky, JETP **94**, 1035 (2002); H.R. Atalla, Low Temp. Phys. **29**, 593 (2003)
29. M.V. Kartsovnik, S.I. Pesotskii, W. Biberacher (unpublished)
30. G.M. Danner, W. Kang, P.M. Chaikin, Phys. Rev. Lett. **72**, 3714 (1994)
31. G.M. Danner, P.M. Chaikin, Phys. Rev. Lett. **75**, 4690 (1995)
32. A.G. Lebed, JETP Lett. **43**, 174 (1986)

33. M.J. Naughton, O.H. Chung, L.Y. Chiang, J.S. Brooks, Mater. Res. Soc. Symp. Proc. **173**, 257 (1990); M.J. Naughton, O.H. Chung, M. Chaparala, X. Bu, P. Coppens, Phys. Rev. Lett. **67**, 3712 (1991)
34. T. Osada, A. Kawasumi, S. Kagoshima, N. Miura, G. Saito, Phys. Rev. Lett. **66**, 1525 (1991)
35. E.I. Chashechkina, P.M. Chaikin, Phys. Rev. Lett. **80**, 2181 (1998)
36. H. Kang, Y.J. Jo, S. Uji, W. Kang, Phys. Rev. B **68**, 132508 (2003)
37. W. Kang, S.T. Hannahs, P.M. Chaikin, Phys. Rev. Lett. **69**, 2827 (1992)
38. I.J. Lee, M.J. Naughton, Phys. Rev. B **58**, R13343 (1998)
39. N. Biškup, J.S. Brooks, R. Kato, K. Oshima, Phys. Rev. B **62**, 21 (2000)
40. Y. Shimojo, M.A. Tanatar, T. Ishiguro, R. Kato, J. Phys. Soc. Jpn. **71**, 393 (2002)
41. H. Ito, Y. Yokochi, D. Suzuki, H. Tanaka, S. Kuroda, K. Enomoto, S. Uji, M. Umemiya, H. Miyasaka, K.i. Sugiura, M. Yamashita, Synth. Met. **156**, 162 (2005)
42. S. Yasuzuka, C. Terakura, T. Yakabe, Y. Terai, H.M. Yamamoto, R. Kato, S. Uji, Synth. Met. **135–136**, 647 (2003)
43. W. Kang, Unusual Magic Angles effects in Bechgaard salts. In *This volume*, Chap. 13
44. G. Montambaux, P.B. Littlewood, Phys. Rev. Lett. **62**, 953 (1989); L. Chen, K. Maki, Synth. Met. **29**, F493 (1989)
45. P.M. Chaikin, Phys. Rev. Lett. **69**, 2831 (1992); A.G. Lebed, P. Bak, Phys. Rev. Lett. **63**, 1315 (1989); V.M. Yakovenko, Phys. Rev. Lett. **68**, 3607 (1992); A.G. Lebed, J. Phys. I **4**, 351 (1994); J. Phys. I **6**, 1819 (1996); A.T. Zheleznyak, V.M. Yakovenko, Synth. Met. **70**, 1005 (1995); U. Lundin, R.H. McKenzie, Phys. Rev. B **70**, 235122 (2004)
46. S.P. Strong, D.G. Clarke, P.W. Anderson, Phys. Rev. Lett. **73**, 1007 (1994); D.G. Clarke, S.P. Strong, P.M. Chaikin, E.I. Chashechkina, Science **279**, 2071 (1998)
47. T. Osada, S. Kagoshima, N. Miura, Phys. Rev. B **46**, 1812 (1992)
48. K. Maki, Phys. Rev. B **45**, 5111 (1992)
49. A.G. Lebed, N.N. Bagmet, M.J. Naughton, Phys. Rev. Lett. **93**, 157006 (2004); J. Phys. IV **114**, 77 (2004)
50. I.J. Lee, M.J. Naughton, Phys. Rev. B **57**, 7423 (1998)
51. R.H. McKenzie, P. Moses, Phys. Rev. B **60**, R11241 (1999)
52. A.G. Lebed, M.J. Naughton, Phys. Rev. Lett. **91**, 187003 (2003); J. Phys. IV **12**, 369 (2002)
53. A.G. Lebed, M.J. Naughton, cond-mat/0304591 (unpublished)
54. T. Osada, Physica E **12**, 272 (2002); T. Osada, M. Kuraguchi, Synth. Met. **133–134**, 75 (2003); T. Osada, K. Kobayashi, E. Ohmichi, Synth. Met. **135–136**, 653 (2003)
55. B.K. Cooper, V.M. Yakovenko, Phys. Rev. Lett. **96**, 037001 (2006)
56. A.G. Lebed, H.-I. Ha, M.J. Naughton, Phys. Rev. B **71**, 132504 (2005)
57. H.-I. Ha, A.G. Lebed, M.J. Naughton, Phys. Rev. B **73**, 033107 (2006)
58. T. Osada, S. Kagoshima, N. Miura, Phys. Rev. Lett. **77**, 5261 (1996)
59. H. Yoshino, K. Saito, K. Kikuchi, H. Nishikawa, K. Kobayashi, I. Ikemoto, J. Phys. Soc. Jpn. **64**, 2307 (1995)
60. A.G. Lebed, N.N. Bagmet, Phys. Rev. B **55**, R8654 (1997)
61. H. Yoshino, K. Murata, J. Phys. Soc. Jpn. **68**, 3027 (1999)

62. H. Yoshino, K. Murata, K. Saito, H. Nishikawa, K. Kikuchi, I. Ikemoto, Phys. Rev. B **67**, 035111 (2003)
63. H. Yoshino, A. Oda, K. Murata, H. Nishikawa, K. Kikuchi, I. Ikemoto, Synth. Met. **120**, 885 (2001)
64. M. Kartsovnik, P. Grigoriev, W. Biberacher, A. Gröger, D. Andres, S. Pesotskii, N. Kushch, Synth. Met. **133–134**, 111 (2003)
65. N. Hanasaki, S. Kagoshima, T. Hasegawa, T. Osada, N. Miura, Phys. Rev. B **57**, 1336 (1998)
66. V.G. Peschansky, M.V. Kartsovnik, Phys. Rev. B **60**, 11207 (1999); J. Low Temp. Phys. **117**, 1717 (1999)
67. R.H. McKenzie, P. Moses, Phys. Rev. Lett. **81**, 4492 (1998); P. Moses, R.H. McKenzie, Phys. Rev. B **60**, 7998 (1999)
68. Y. Ando, G.S. Boebinger, A. Passner, N.L. Wang, C. Geibel, F. Steglich, Phys. Rev. Lett. **77**, 2065 (1996); S. Martin, A.T. Fiory, R.M. Fleming, L.F. Schneemeyer, J.V. Waszczak, Phys. Rev. B **41**, 846 (1990)
69. P.W. Anderson, Science **256**, 1526 (1992)
70. R.B. Lyubovskii, S.I. Pesotskii, S.V. Konovalikhin, G.V. Shilov, A. Kobayashi, H. Kobayashi, V.I. Nizhankovskii, J.A.A.J. Perenboom, O.A. Bogdanova, E.I. Zhilyaeva, R.N. Lyubovskaya, Synth. Met. **123**, 149 (2001)
71. N. Kumar, A.M. Jayannavar, Phys. Rev. B **45**, 5001 (1992); G. Soda, D. Jérome, M. Weger, J. Alizon, J. Gallice, H. Robert, J.M. Fabre, L. Giral, J. Phys. (Paris) **38**, 931 (1977); S. Shitkovsky, M. Weger, H. Gutfreund, J. Phys. (Paris) **39**, 711 (1978)
72. E. Ohmichi, Y. Maeno, S. Nagai, Z.Q. Mao, M.A. Tanatar, T. Ishiguro, Phys. Rev. B **61**, 7101 (2000)
73. J. Wosnitza, J. Hagel, J.S. Qualls, J.S. Brooks, E. Balthes, D. Schweitzer, J.A. Schlueter, U. Geiser, J. Mohtasham, R.W. Winter, G.L. Gard, Phys. Rev. B **65**, 180506(R) (2002)
74. V.G. Peschansky, JETP **85**, 337 (1997)
75. A.J. Schofield, J.R. Cooper, Phys. Rev. B **62**, 10779 (2000)
76. N. Hanasaki, S. Kagoshima, N. Miura, G. Saito, Phys. Rev. B **63**, 245116 (2001)
77. J. Singleton, P.A. Goddard, A. Ardavan, N. Harrison, S.J. Blundell, J.A. Schlueter, A.M. Kini, Phys. Rev. Lett. **88**, 037001 (2002)
78. S. Uji, H. Shinagawa, C. Terakura, T. Terashima, T. Yakabe, Y. Terai, M. Tokumoto, A. Kobayashi, H. Tanaka, H. Kobayashi, Phys. Rev. B **64**, 024531 (2001)
79. C. Mielke, J. Singleton, M.-S. Nam, N. Harrison, C.C. Agosta, B. Fravel, L.K. Montgomery, J. Phys.: Condens. Matter **13**, 8325 (2001)
80. T. Konoike, S. Uji, T. Terashima, M. Nishimura, S. Yasuzuka, K. Enomoto, H. Fujiwara, E. Fujiwara, B. Zhang, H. Kobayashi, Phys. Rev. B **72**, 094517 (2005)
81. M. Kartsovnik, S. Simonov, D. Andres, W. Biberacher, N. Kushch, H. Müller, J. Phys. IV **131**, 265 (2005)
82. M.V. Kartsovnik, D. Andres, S.V. Simonov, W. Biberacher, I. Sheikin, N.D. Kushch, H. Müller, Phys. Rev. Lett. **96**, 166601 (2006)
83. N. Hanasaki, S. Kagoshima, N. Miura, G. Saito, J. Phys. Soc. Jpn. **65**, 1010 (1996)
84. H. Kang, Y.J. Jo, W. Kang, Phys. Rev. B **69**, 033103 (2004)

85. M. Kuraguchi, E. Ohmichi, T. Osada, Y. Shiraki, Synth. Met. **133–134**, 113 (2003)
86. D.G. Clarke, S.P. Strong, Adv. Phys. **46**, 545 (1997); D.G. Clarke, S.P. Strong, P.M. Chaikin, E.I. Chashechkina, Science **279**, 2071 (1998)
87. M.V. Kartsovnik, V.N. Laukhin, J. Phys. I (France) **6**, 1753 (1996)
88. J. Singleton, Rep. Prog. Phys. **63**, 1111 (2000)
89. I.M. Lifshitz, A.M. Kosevich, Sov. Phys. JETP **2**, 636 (1956)
90. R.B. Dingle, Proc. R. Soc. A **211**, 517 (1952); Yu.A. Bychkov, Sov. Phys. JETP **12**, 977 (1961)
91. E.N. Adams, T.D. Holstein, J. Phys. Chem. Solids **10**, 254 (1959)
92. V.N. Laukhin, A. Audouard, H. Rakoto, J. Broto, F. Goze, G. Coffe, L. Brossard, J. Redoules, M. Kartsovnik, N. Kushch, L. Buravov, A. Khomenko, E. Yagubskii, S. Askenazy, P. Pari, Physica B **211**, 282 (1995)
93. P.S. Sandhu, G.J. Athas, J.S. Brooks, E.G. Haanappel, J.D. Goettee, D.W. Rickel, M. Tokumoto, N. Kinoshita, Y. Tanaka, Surf. Sci. **361/362**, 913 (1996)
94. F.A. Meyer, E. Steep, W. Biberacher, P. Christ, A. Lerf, A.G.M. Jansen, W. Joss, P. Wyder, K. Andres, Europhys. Lett. **32**, 681 (1995)
95. S. Uji, M. Chaparala, S. Hill, P.S. Sandhu, J. Qualls, L. Seger, J.S. Brooks, Synth. Met. **85**, 1573 (1997)
96. M.M. Honold, N. Harrison, M.-S. Nam, J. Singleton, C.H. Mielke, M. Kurmoo, P. Day, Phys. Rev. B **58**, 7560 (1998)
97. M. Tokumoto, A.G. Swanson, J.S. Brooks, C.C. Agosta, S.T. Hannahs, N. Kinoshita, H. Anzai, M. Tamura, H. Tajima, H. Kuroda, A. Ugawa, K. Yakushi, Physica B **184**, 508 (1993)
98. N. Harrison, R. Bogaerts, P.H.P. Reinders, J. Singleton, S.J. Blundell, F. Herlach, Phys. Rev. B **54**, 9977 (1996)
99. N. Harrison, C.H. Mielke, D.G. Rickel, J. Wosnitza, J.S. Qualls, J.S. Brooks, E. Balthes, D. Schweitzer, I. Heinen, W. Strunz, Phys. Rev. B **58**, 10248 (1998)
100. J. Wosnitza, S. Wanka, J. Hagel, E. Balthes, N. Harrison, J.A. Schlueter, A.M. Kini, U. Geiser, J. Mohtasham, R.W. Winter, G.L. Gard, Phys. Rev. B **61**, 7383 (2000)
101. J. Wosnitza, G. Goll, D. Beckmann, S. Wanka, D. Schweitzer, W. Strunz, J. Phys. I **6**, 1597 (1996)
102. H. Weiss, M.V. Kartsovnik, W. Biberacher, E. Balthes, A.G.M. Jansen, N.D. Kushch, Phys. Rev. B **60**, 16259(R) (1999)
103. M. Schiller, W. Schmidt, E. Balthes, D. Schweitzer, H. Koo, M.H. Whangbo, I. Heinen, T. Klausa, P. Kircher, W. Strunz, Europhys. Lett. **51**, 82 (2000)
104. P.D. Grigoriev, M.V. Kartsovnik, W. Biberacher, N.D. Kushch, P. Wyder, Phys. Rev. B **65**, 060403(R) (2002)
105. W. Kang, G. Montambaux, J.R. Cooper, D. Jérome, P. Batail, C. Lenoir, Phys. Rev. Lett. **62**, 2559 (1989)
106. P.D. Grigoriev, Phys. Rev. B **67**, 144401 (2003)
107. M.V. Kartsovnik, P.D. Grigoriev, W. Biberacher, N.D. Kushch, P. Wyder, Phys. Rev. Lett. **89**, 126802 (2002)
108. D. Shoenberg, J. Low Temp. Phys. **56**, 417 (1984)
109. P.D. Grigoriev, I.D. Vagner, cond-mat/0009409 (unpublished); T. Champel, V.P. Mineev, Philos. Mag. B **81**, 55 (2001)
110. P.D. Grigoriev, JETP **92**, 1090 (2001)

111. T. Champel, Phys. Rev. B **64**, 054407 (2001)
112. A.S. Alexandrov, A.M. Bratkovsky, Phys. Rev. Lett. **76**, 1308 (1996); J.-Y. Fortin, T. Ziman, Phys. Rev. Lett. **80**, 3117 (1998); T. Champel, Phys. Rev. B **65**, 153403 (2002); K. Kishigi, Y. Hasegawa, Phys. Rev. B **65**, 205405 (2002); M.A. Itskovsky, Phys. Rev. B **68**, 054423 (2003); J.-Y. Fortin, E. Perez, A. Audouard, Phys. Rev. B **71**, 155101 (2005)
113. T. Champel, V.P. Mineev, Phys. Rev. B **66**, 195111 (2002); cond-mat/0509026 (unpublished)
114. V.M. Gvozdikov, Phys. Rev. B **70**, 085113 (2004)
115. M.-S. Nam, A. Ardavan, J.A. Symington, J.S.N. Harrison, C.H. Mielke, J.A. Schlueter, R.W. Winter, G.L. Gard, Phys. Rev. Lett. **87**, 117001 (2001)
116. J. Wosnitza, S. Wanka, J. Hagel, H.V. Löhneysen, J.S. Qualls, J.S. Brooks, E. Balthes, J.A. Schlueter, U. Geiser, J. Mohtasham, R.W. Winter, G.L. Gard, Phys. Rev. Lett. **86**, 508 (2001)
117. H. Mori, S. Tanaka, K. Oshima, M. Oshima, G. Saito, T. Mori, Y. Maruyama, H. Inokuchi, Solid State Commun. **74**, 1261 (1990); M. Oshima, H. Mori, G. Saito, K. Oshima, Chem. Lett. 1159 (1989)
118. N.D. Kushch, L.I. Buravov, M.V. Kartsovnik, V.N. Laukhin, S.I. Pesotskii, R.P. Shibaeva, L.P. Rozenberg, E.B. Yagubskii, A.V. Zvarikina, Synth. Met. **46**, 271 (1992)
119. L.I. Buravov, N.D. Kushch, V.N. Laukhin, A.G. Khomenko, E.B. Yagubskii, M.V. Kartsovnik, A.E. Kovalev, L.P. Rozenberg, R.P. Shibaeva, M.A. Tanatar, V.S. Yefanov, V.V. Dyakin, V.A. Bondarenko, J. Phys. I **4**, 441 (1994)
120. R. Rousseau, M.-L. Doublet, E. Canadell, R.P. Shibaeva, S.S. Khasanov, L.P. Rozenberg, N.D. Kushch, E.B. Yagubskii, J. Phys. I (France) **6**, 1527 (1996)
121. S.I. Pesotskii, R.B. Lyubovskii, V.I. Nizhankovskii, W. Biberacher, M.V. Kartsovnik, K. Andres, J.A.A.J. Perenboom, N.D. Kuchsh, E.B. Yagubskii, H. Kobayashi, JETP **90**, 527 (2000)
122. T. Sasaki, T. Ozawa, H. Mori, S. Tanaka, T. Fukase, N. Toyota, J. Phys. Soc. Jpn. **65**, 213 (1996)
123. L.K. Montgomery, B.W. Fravel, J.C. Huffman, C.C. Agosta, S.A. Ivanov, Synth. Met. **85**, 1521 (1997); N.D. Kushch, L.I. Buravov, S.I. Pesotskii, R.B. Lyubovskii, E.B. Yagubskii, M.G. Kaplunov, E.V. Golubev, B.Z. Narymbetov, S.S. Khasanov, L.V. Zorina, L.P. Rozenberg, R.P. Shibaeva, A. Kobayashi, H. Kobayashi, J. Mater. Chem. **8**, 897 (1998)
124. H. Mori, S. Tanaka, M. Oshima, G. Saito, T. Mori, Y. Maruyama, H. Inokuchi, Bull. Chem. Soc. Jpn. **63**, 2183 (1990); L. Ducasse, A. Fritsch, Solid State Commun. **91**, 201 (1994); D.-K. Seo, M.-H. Whangbo, B. Fravel, L.K. Montgomery, Solid State Commun. **100**, 191 (1996)
125. J.-Y. Fortin, J. Bellissard, M. Gusmão, T. Ziman, Phys. Rev. B **57**, 1484 (1998)
126. J.H. Kim, S.Y. Han, J.S. Brooks, Phys. Rev. B **60**, 3213 (1999)
127. T. Sasaki, N. Toyota, M. Tokumoto, N. Kinoshita, H. Anzai, Solid State Commun. **75**, 93 (1990)
128. N. Kinoshita, M. Tokumoto, H. Anzai, J. Phys. Soc. Jpn. **60**, 2131 (1991)
129. T. Sasaki, H. Sato, N. Toyota, Synth. Met. **42**, 2211 (1991); H. Mori, I. Hirabayashi, S. Tanaka, T. Mori, H. Inokuchi, K. Oshima, G. Saito, Synth. Met. **56**, 2443 (1993)
130. P. Christ, W. Biberacher, W. Bensch, H. Müller, K. Andres, Synth. Met. **86**, 2057 (1997)

131. N. Kinoshita, M. Tokumoto, H. Anzai, J. Phys. Soc. Jpn. **59**, 3410 (1991)
132. A. Kovalev, H. Müller, Synth. Met. **86**, 1997 (1997); A.E. Kovalev, H. Müller, M.V. Kartsovnik, Sov. Phys. JETP **86**, 578 (1998)
133. M. Köppen, F. Kromer, M. Lang, F. Steglich, T. Sasaki, N. Toyota, H. Tajima, Synth. Met. **86**, 2059 (1997)
134. T. Osada, R. Yagi, A. Kawasumi, S. Kagoshima, N. Miura, M. Oshima, G. Saito, Phys. Rev. B **41**, 5428 (1990)
135. M.V. Kartsovnik, D.V. Mashovets, D.V. Smirnov, V.N. Laukhin, A. Gilewski, N.D. Kushch, J. Phys. I **4**, 159 (1994)
136. J.S. Brooks, S.J. Klepper, C.C. Agosta, M. Tokumoto, N. Kinoshita, Y. Tanaka, S. Uji, H. Aoki, G.J. Athas, X. Chen, H. Anzai, Synth. Met. **56**, 1791 (1993)
137. R.H. McKenzie, G.J. Athas, J.S. Brooks, R.G. Clark, A.S. Dzurak, R. Newbury, R.P. Starret, A. Skougarevsky, M. Tokumoto, N. Kinoshita, T. Kinoshita, Y. Tanaka, Phys. Rev. B **54**, R8289 (1996)
138. T. Sasaki, N. Toyota, Solid State Commun. **82**, 447 (1992)
139. F.L. Pratt, J. Singleton, M. Doporto, A.J. Fisher, T.J.B.M. Janssen, J.A.A.J. Perenboom, M. Kurmoo, W. Hayes, P. Day, Phys. Rev. B **45**, 13904 (1992)
140. J.S. Brooks, C.C. Agosta, S.J. Klepper, M. Tokumoto, N. Kinoshita, H. Anzai, S. Uji, H. Aoki, A.S. Perel, G.J. Athas, D.A. Howe, Phys. Rev. Lett. **69**, 156 (1992)
141. T. Sasaki, N. Toyota, M. Tokumoto, N. Kinoshita, H. Anzai, Solid State Commun. **75**, 97 (1990); M. Tokumoto, A.G. Swanson, J.S. Brooks, C.C. Agosta, S.T. Hannahs, N. Kinoshita, H. Anzai, J.R. Anderson, J. Phys. Soc. Jpn. **59**, 2324 (1990)
142. M.V. Kartsovnik, A.E. Kovalev, V.N. Laukhin, S.I. Pesotskii, N.D. Kushch, JETP Lett. **55**, 339 (1992)
143. M.V. Kartsovnik, A.E. Kovalev, N.D. Kushch, J. Phys. I (France) **3**, 1187 (1993)
144. J. Caulfield, S.J. Blundell, M.S.L. du Croo de Jongh, P.T.J. Hendriks, J. Singleton, M. Doporto, F.L. Pratt, A. House, J.A.A.J. Perenboom, W. Hayes, M. Kurmoo, P. Day, Phys. Rev. B **51**, 8325 (1995)
145. A. House, C.J. Hawoth, J.M. Caulfield, S. Blundell, M.M. Honold, J. Singleton, W. Hayes, S.M. Hayden, P. Meeson, M. Springford, M. Kurmoo, P. Day, J. Phys.: Condens. Matter **8**, 10361 (1996)
146. S. Uji, H. Aoki, J.S. Brooks, A.S. Perel, S.J. Klepper, C.C. Agosta, D.A. Howe, M. Tokumoto, N. Kinoshita, Y. Tanaka, H. Anzai, Solid State Commun. **88**, 683 (1993); S. Uji, H. Aoki, M. Tokumoto, T. Kinoshita, N. Kinoshita, Y. Tanaka, H. Anzai, Phys. Rev. B **49**, 732 (1994); S. Uji, T. Terashima, H. Aoki, J.S. Brooks, M. Tokumoto, N. Kinoshita, T. Kinoshita, Y. Tanaka, H. Anzai, Phys. Rev. B **54**, 9332 (1996)
147. M.V. Kartsovnik, A.E. Kovalev, V.N. Laukhin, S.I. Pesotskii, J. Phys. I (France) **2**, 223 (1992)
148. M.V. Kartsovnik, H. Ito, T. Ishiguro, H. Mori, T. Mori, G. Saito, S. Tanaka, J. Phys.: Condens. Matter **6**, L479 (1994)
149. M.V. Kartsovnik, A.E. Kovalev, V.N. Laukhin, I.F. Schegolev, H. Ito, T. Ishiguro, N.D. Kushch, H. Mori, G. Saito, Synth. Met. **70**, 811 (1995)

150. A.E. Kovalev, M.V. Kartsovnik, R.P. Shibaeva, L.P. Rozenberg, I.F. Schegolev, N.D. Kushch, Solid State Commun. **89**, 575 (1994); M.V. Kartsovnik, A.E. Kovalev, R.P. Shibaeva, L.P. Rozenberg, N.D. Kushch, Physica B **201**, 459 (1994)
151. Y. Iye, R. Yagi, N. Hanasaki, S. Kagoshima, H. Mori, H. Hujimoto, G. Saito, J. Phys. Soc. Jpn. **63**, 674 (1994)
152. T. Sasaki, N. Toyota, Phys. Rev. B **49**, 10120 (1994)
153. M.Ya. Azbel, Sov. Phys. JETP **19**, 634 (1964); M.Ya. Azbel, Phys. Rev. Lett. **43**, 1954 (1979)
154. D. Andres, Effects of high magnetic fields on a charge-density-wave system. Ph.D. Thesis, Technical University of Munich, Munich (2005); M.V. Kartsovnik, D. Andres, W. Biberacher, C.R. Chimie **10**, 6 (2007); D. Andres, M.V. Kartsovnik, W. Biberacher, K. Neumaier, I. Sheikin, H. Müller, preprint: arXiv: 0801.2696
155. M. Kartsovnik, D. Andres, P. Grigoriev, W. Biberacher, H. Müller, Physica B **346–347**, 368 (2004)
156. T. Sasaki, N. Toyota, Phys. Rev. B **48**, 11457 (1993); T. Sasaki, T. Fukase, Phys. Rev. B **59**, 13872 (1999)
157. N. Harrison, Phys. Rev. Lett. **83**, 1395 (1999); N. Harrison, N. Biskup, J.S. Brooks, L. Balicas, M. Tokumoto, Phys. Rev. B **63**, 195102 (2001)
158. N. Harrison, E. Rzepniewski, J. Singleton, P.J. Gee, M.M. Honold, P. Day, M. Kurmoo, J. Phys.: Condens. Matter **11**, 7227 (1999)
159. D. Yoshioka, J. Phys. Soc. Jpn. **64**, 3168 (1995)
160. K. Maki, B. Dóra, A. Virosztek, Unconventional density waves in organic conductors and in super-conductors. In: *This volume*, Chap. 19
161. R. Tsuchiya, T. Nakamura, T. Takahashi, T. Sasaki, N. Toyota, Synth. Met. **70**, 965 (1995)
162. K. Miyagawa, A. Kawamoto, K. Kanoda, Phys. Rev. B **56**, R8487 (1997)
163. M.V. Kartsovnik, W. Biberacher, E. Steep, P. Christ, K. Andres, A.G.M. Jansen, H. Müller, Synth. Met. **86**, 1933 (1997)
164. N. Biskup, J.A.A.J. Perenboom, J.S. Qualls, J.S. Brooks, Solid State Commun. **107**, 503 (1998)
165. C. Proust, A. Audouard, A. Kovalev, D. Vignolles, M. Kartsovnik, L. Brossard, N. Kushch, Phys. Rev. B **62**, 2388 (2000)
166. D. Andres, M.V. Kartsovnik, W. Biberacher, H. Weiss, E. Balthes, H. Müller, N. Kushch, Phys. Rev. B **64**, 161104(R) (2001)
167. P. Christ, W. Biberacher, M.V. Kartsovnik, E. Steep, E. Balthes, H. Weiss, H. Müller, JETP Lett. **71**, 303 (2000)
168. N. Harrison, L. Balicas, J.S. Brooks, M. Tokumoto, Phys. Rev. B **62**, 14212 (2000)
169. D. Zanchi, A. Bjeliš, G. Montambaux, Phys. Rev. B **53**, 1240 (1996)
170. A.I. Buzdin, V.V. Tugushev, JETP **85**, 428 (1983)
171. D. Zanchi, A. Bjeliš, G. Montambaux, J. Phys. IV **9**, 203 (1999)
172. P.D. Grigoriev, D.S. Lyubshin, Phys. Rev. B **72**, 195106 (2005)
173. A.I. Larkin, Yu.N. Ovchinnikov, Sov. Phys. JETP **20**, 762 (1964); P. Fulde, R.A. Ferrel, Phys. Rev. A **135**, 550 (1964)
174. L.N. Bulaevskii, A.I. Buzdin, D.I. Khomskii, Solid State Commun. **27**, 5 (1978); R.A.T. de Lima, C. Tsallis, Phys. Rev. B **27**, 6896 (1983); J.C. Bonner, J.A. Nothby, I.S. Jacobs, L.V. Interrante, Phys. Rev. B **35**, 1791 (1987)

175. R.H. McKenzie, cond-mat/9706235 (unpublished)
176. P. Foury-Leylekian, S. Ravy, J.P. Pouget, H. Müller, Synth. Met. **137**, 1271 (2003)
177. A. House, W. Lubczynski, S. Blundell, J. Singleton, W. Hayes, M. Kurmoo, P. Day, J. Phys.: Condens. Matter **8**, 10377 (1996)
178. P. Christ, W. Biberacher, H. Müller, K. Andres, E. Steep, A.G.M. Jansen, Physica B **204**, 153 (1995); P. Christ, W. Biberacher, H. Müller, K. Andres, E. Steep, A.G.M. Jansen, Synth. Met. **70**, 823 (1995); P. Christ, W. Biberacher, A.G.M. Jansen, M.V. Kartsovnik, A.E. Kovalev, N.D. Kushch, Surf. Sci. **361/362**, 909 (1996)
179. D. Andres, M.V. Kartsovnik, P.D. Grigoriev, W. Biberacher, H. Müller, Phys. Rev. B **68**, 201101(R) (2003)
180. N. Harrison, J. Singleton, A. Bangura, A. Ardavan, P.A. Goddard, R.D. McDonald, L.K. Montgomery, Phys. Rev. B **69**, 165103 (2004)
181. A. Schegolev, V. Laukhin, A. Khomenko, M. Kartsovnik, R. Shibaeva, L. Rozenberg, A. Kovalev, J. Phys. I (France) **2**, 2123 (1992)
182. T. Kouno, T. Osada, M. Hasumi, S. Kagoshima, N. Miura, M. Oshima, H. Mori, T. Nakamura, G. Saito, Synth. Met. **56**, 2425 (1993)
183. J. Brooks, K. Chen, S. Klepper, S. Valfells, G. Athas, Y. Tanaka, T. Kinoshita, M. Tokomoto, H. Anzai, C.C. Agosta, Phys. Rev. B **52**, 14457 (1995)
184. M.V. Kartsovnik, D. Andres, W. Biberacher, P. Christ, E. Steep, E. Balthes, H. Weiss, H. Müller, N.D. Kushch, Synth. Met. **120**, 687 (2001)
185. L.P. Gor'kov, A.G. Lebed, J. Phys. Lett. **45**, 433 (1984)
186. A.G. Lebed, JETP Lett. **78**, 138 (2003)
187. P.D. Grigoriev (unpublished)
188. T. Mori, Bull. Chem. Soc. Jpn. **71**, 2509 (1998)
189. E.B. Yagubskii, I.F. Schegolev, V.N. Laukhin, P.A. Kononovich, M.V. Kartsovnik, A.V. Zvarykina, L.I. Buravov, JETP Lett. **39**, 12 (1984)
190. J.M. Williams, H.H. Wang, M.A. Beno, T.J. Emge, L.M. Sowa, P.T. Copps, F. Behroozi, L.N. Hall, K.D. Carlson, G.W. Crabtree, Inorg. Chem. **23**, 3839 (1984)
191. S. Uji, T. Terashima, S. Yasuzuka, J. Yamaura, H.M. Yamamoto, R. Kato, Phys. Rev. B **68**, 064420 (2003)
192. S. Yasuzuka, S. Uji, H.M. Yamamoto, J.-I. Yamaura, C. Terakura, T. Terashima, T. Yakabe, Y. Terai, R. Maeda, R. Kato, J. Phys. Soc. Jpn. **74**, 679 (2005)
193. E.S. Choi, E. Jobilong, A. Wade, E. Goetz, J.S. Brooks, J. Yamada, T. Mizutani, T. Kinoshita, M. Tokumoto, Phys. Rev. B **67**, 174511 (2003)
194. E.S. Choi, D. Graf, J.S. Brooks, J. Yamada, H. Akutsu, H. Kikuchi, M. Tokumoto, Phys. Rev. B **70**, 024517 (2004)
195. A. Audouard, P. Auban-Senzier, V.N. Laukhin, L. Brossard, D. Jérome, N.D. Kushch, Europhys. Lett. **34**, 599 (1996)
196. D. Beckmann, S. Wanka, J. Wosnitza, J.A. Schlueter, J.M. Williams, P.G. Nixon, R.W. Winter, G.L. Gard, J. Ren, M.-H. Whangbo, Eur. Phys. J. B **1**, 295 (1998)
197. J. Hagel, J. Wosnitza, C. Pfleiderer, J.A. Schlueter, J. Mohtasham, G.L. Gard, Phys. Rev. B **68**, 104504 (2003)
198. A. Audouard, V.N. Laukhin, L. Brossard, T.G. Prokhorova, E.B. Yagubskii, E. Canadell, Phys. Rev. B **69**, 144523 (2004)

199. A.I. Coldea, A.F. Bangura, J. Singleton, A. Ardavan, A. Akutsu-Sato, H. Akutsu, S.S. Turner, P. Day, Phys. Rev. B **69**, 085112 (2004)
200. R.B. Lyubovskii, S.I. Pesotskii, M. Gener, R. Rousseau, E. Canadell, J.A.A.J. Perenboom, V.I. Nizhankovskii, E.I. Zhilyaeva, O.A. Bogdanova, R.N. Lyubovskaya, J. Mater. Chem. **12**, 483 (2002)
201. J. Wosnitza, S. Wanka, J.S. Qualls, J.S. Brooks, C.H. Mielke, N. Harrison, J.A. Schlueter, J.M. Williams, P.G. Nixon, R.W. Winter, G.L. Gard, Synth. Met. **103**, 2000 (1999)
202. J. Wosnitza, S. Wanka, J. Hagel, E. Balthes, N. Harrison, J.A. Schlueter, A.M. Kini, U. Geiser, J. Mohtasham, R.W. Winter, G.L. Gard, Phys. Rev. B **61**, 7383 (2000); J. Hagel, S. Wanka, J. Wosnitza, E. Balthes, J.A. Schlueter, A.M. Kini, U. Geiser, J. Mohtasham, R.W. Winter, G.L. Gard, Synth. Met. **120**, 813 (2001)
203. J. Wosnitza, S. Wanka, J. Hagel, R. Häussler, H.V. Löhneysen, J.A. Schlueter, U. Geiser, P.G. Nixon, R.W. Winter, G.L. Gard, Phys. Rev. B **62**, R11973 (2000)
204. J. Wosnitza, J. Hagel, P.J. Meeson, D. Bintley, J.A. Schlueter, J. Mohtasham, R.W. Winter, G.L. Gard, Phys. Rev. B **67**, 060504(R) (2003)
205. V.M. Gvozdikov, J. Wosnitza, Low Temp. Phys. **32**, 109 (2006); V.M. Gvozdikov, J. Wosnitza, cond-mat/0502462 (unpublished)
206. F. Zuo, X. Su, P. Zhang, J.S. Brooks, J. Wosnitza, J.A. Schlueter, J.M. Williams, P.G. Nixon, R.W. Winter, G.L. Gard, Phys. Rev. B **60**, 6296 (1999)
207. K. Kanoda, Hyperfine Interact. **104**, 235 (1997)
208. S. Lefebvre, P. Wzietek, S. Brown, C. Bourbonnais, D. Jérome, C. Mézière, M. Fourmigué, P. Batail, Phys. Rev. Lett. **85**, 5420 (2000); P. Limelette, P. Wzietek, S. Florens, A. Georges, T.A. Costi, C. Pasquier, D. Jérome, C. Mézière, P. Batail, Phys. Rev. Lett. **91**, 016401 (2003); F. Kagawa, K. Miyagawa, K. Kanoda, Nature **436**, 534 (2005)
209. P.A. Goddard, S.J. Blundell, J. Singleton, R.D. McDonald, A. Ardavan, A. Narduzzo, J.A. Schlueter, A.M. Kini, T. Sasaki, Phys. Rev. B **69**, 174509 (2004)
210. M.V. Kartsovnik, A.E. Kovalev, V.N. Laukhin, H. Ito, T. Ishiguro, N. Kushch, H. Anzai, G. Saito, Synth. Met. **70**, 819 (1995)
211. M.-S. Nam, M.M. Honold, C. Proust, N. Harrison, C.H. Mielke, S.J. Blundell, J. Singleton, W. Hayes, M. Kurmoo, P. Day, Synth. Met. **103**, 1905 (1999)
212. A.M. Kini, U. Geiser, H.H. Wang, K.D. Carlson, J.M. Williams, W.K. Kwok, K.G. Vandervoort, J.E. Thompson, D.L. Stupka, D. Jung, M.-H. Whangbo, Inorg. Chem. **29**, 2555 (1990)
213. U. Geiser, A.J. Schultz, H.H. Wang, D.M. Watkins, D.L. Stupka, J.M. Williams, J.E. Schirber, D.L. Overmyer, D. Jung, J.J. Nova, M.-H. Whangbo, Physica C **174**, 475 (1991)
214. M.V. Kartsovnik, W. Biberacher, K. Andres, N.D. Kushch, JETP Lett. **62**, 905 (1995); Y. Yamauchi, M.V. Kartsovnik, T. Ishiguro, M. Kubota, G. Saito, J. Phys. Soc. Jpn. **65**, 354 (1996)
215. H. Weiss, M.V. Kartsovnik, W. Biberacher, E. Steep, A.G.M. Jansen, N.D. Kushch, JETP Lett. **66**, 202 (1997); C.H. Mielke, N. Harrison, D.G. Rickel, A.H. Lacerda, R.M. Vestal, L.K. Montgomery, Phys. Rev. B **56**, 4309 (1997)

216. M.V. Kartsovnik, G.Y. Logvenov, H. Ito, T. Ishiguro, G. Saito, Phys. Rev. B **52**, R15715 (1995)
217. H. Weiss, M.V. Kartsovnik, W. Biberacher, E. Steep, E. Balthes, A. Jansen, K. Andres, N. Kushch, Phys. Rev. B **59**, 12370 (1999)
218. Y. Nogami, J.-P. Pouget, H. Ito, T. Ishiguro, G. Saito, Solid State Commun. **89**, 113 (1994)
219. F. Kagawa, T. Itou, K. Miyagawa, K. Kanoda, Phys. Rev. Lett. **93**, 127001 (2004)
220. L. Laloux, A. Georges, W. Krauth, Phys. Rev. B **50**, 3092 (1994)
221. H. Taniguchi, K. Kanoda, A. Kawamoto, Phys. Rev. B **67**, 014510 (2003)
222. H. Kobayashi, H. Cui, A. Kobayashi, Chem. Rev. **104**, 5265 (2004)
223. S. Uji, H. Shinagawa, T. Terashima, T. Yakabe, Y. Terai, M. Tokumoto, A. Kobayashi, H. Tanaka, H. Kobayashi, Nature **410**, 908 (2001)
224. A. Kobayashi, T. Udagawa, H. Tomita, T. Naito, H. Kobayashi, Chem. Lett. 2179 (1993)
225. F. Goze, V.N. Laukhin, L. Brossard, A. Audouard, J.P. Ulmet, S. Askenazy, T. Naito, H. Kobayashi, A. Kobayashi, M. Tokumoto, P. Cassoux, Europhys. Lett. **28**, 427 (1994)
226. L. Brossard, R. Clerac, C. Coulon, M. Tokumoto, T. Ziman, D.K. Petrov, V.N. Laukhin, M.J. Naughton, A. Audouard, F. Goze, A. Kobayashi, H. Kobayashi, P. Cassoux, Eur. Phys. J. B **1**, 439 (1998)
227. L. Balicas, J.S. Brooks, K. Storr, S. Uji, M. Tokumoto, H. Tanaka, H. Kobayashi, A. Kobayashi, V. Barzykin, L.P. Gor'kov, Phys. Rev. Lett. **87**, 067002 (2001)
228. V. Jaccarino, M. Peter, Phys. Rev. Lett. **9**, 290 (1962)
229. O. Fisher, Helv. Phys. Acta **45**, 331 (1972)
230. B.S. Chandrasekhar, Appl. Phys. Lett. **1**, 7 (1962); A.M. Clogston, Phys. Rev. Lett. **9**, 266 (1962)
231. M. Houzet, A. Buzdin, L. Bulaevskii, M. Maley, Phys. Rev. Lett. **88**, 227001 (2002); H. Shimahara, J. Phys. Soc. Jpn. **71**, 1644 (2002)
232. S. Uji, T. Terashima, C. Terakura, T. Yakabe, Y. Terai, S. Yasuzuka, Y. Imanaka, M. Tokumoto, A. Kobayashi, F. Sakai, H. Tanaka, H. Kobayashi, L. Balicas, J.S. Brooks, J. Phys. Soc. Jpn. **72**, 369 (2003)
233. S. Uji, S. Yasuzuka, M. Tokumoto, H. Tanaka, A. Kobayashi, B. Zhang, H. Kobayashi, E.S. Choi, D. Graf, J.S. Brooks, Phys. Rev. B **72**, 184505 (2005); S. Uji, S. Yasuzuka, H. Tanaka, M. Tokumoto, B. Zhang, H. Kobayashi, E.S. Choi, D. Graf, J.S. Brooks, J. Phys. IV (France) **114**, 391 (2005)
234. E. Ojima, H. Fujiwara, K. Kato, H. Kobayashi, H. Tanaka, A. Kobayashi, M. Tokumoto, P. Cassoux, J. Am. Chem. Soc. **121**, 5581 (1999); H. Fujiwara, E. Fujiwara, Y. Nakazawa, B.Zh. Narymbetov, K. Kato, H. Kobayashi, A. Kobayashi, M. Tokumoto, P. Cassoux, J. Am. Chem. Soc. **123**, 306 (2001)
235. H. Kobayashi, A. Kobayashi, P. Cassoux, Chem. Soc. Rev. **29**, 325 (2000)
236. T. Konoike, S. Uji, T. Terashima, M. Nishimura, S. Yasuzuka, K. Enomoto, H. Fujiwara, B. Zhang, H. Kobayashi, Phys. Rev. B **70**, 094514 (2004); T. Konoike, H. Fujiwara, B. Zhang, H. Kobayashi, M. Nishimura, S. Yasuzuka, K. Enomoto, S. Uji, J. Phys. IV (France) **114**, 223 (2004)
237. H. Fujiwara, H. Kobayashi, E. Fujiwara, A. Kobayashi, J. Am. Chem. Soc. **124**, 6816 (2002)
238. O. Cépas, R.H. McKenzie, J. Merino, Phys. Rev. B **65**, 100502 (2002)

239. S. Uji, T. Terakura, T. Terashima, T. Yakabe, Y. Terai, M. Tokumoto, A. Kobayashi, F. Sakai, H. Tanaka, H. Kobayashi, Phys. Rev. B **65**, 113101 (2002)
240. N.E. Hussey, M. Abdel-Jawad, A. Carrington, A.P. Mackenzie, L. Balicas, Nature **425**, 814 (2003)
241. E. Ohmichi, H. Adachi, Y. Mori, Y. Maeno, T. Ishiguro, T. Oguchi, Phys. Rev. B **59**, 7263 (1999)
242. K. Enomoto, T. Yamaguchi, T. Konoike, M. Nishimura, S. Uji, T. Enoki, M. Suzuki, I.S. Suzuki, Synth. Met. **154**, 289 (2005)
243. N. Harrison, C.H. Mielke, J. Singleton, J.S. Brooks, M. Tokumoto, J. Phys.: Condens. Matter **13**, L389 (2001)

9

High-field Magnetoresistive Effects in Reduced-Dimensionality Organic Metals and Superconductors

J. Singleton, R.D. McDonald, and N. Harrison

The large charge-transfer anisotropy of quasi-one- and quasi-two-dimensional crystalline organic metals means that magnetoresistance is one of the most powerful tools for probing their bandstructure and interesting phase diagrams. Here we review various magnetoresistance phenomena that are of interest in the investigation of metallic, superconducting, and charge-density-wave organic systems.

9.1 Introduction

Quasi-two-dimensional crystalline organic metals and superconductors are very flexible systems in the study of many-body effects and unusual mechanisms for superconductivity [1–7]. Their "soft" lattices enable one to use relatively low pressures to tune the same material through a variety of low-temperature groundstates, for example, from Mott insulator via intermingled antiferromagnetic and superconducting states to unusual superconductor [4, 6, 7]. Pressure also provides a sensitive means of varying the electron–phonon and electron–electron interactions, allowing their influence on the superconducting ground state to be mapped [3,4,8]. The self-organizing tendencies of organic molecules mean that organic metals and superconductors are often rather clean and well-ordered systems; as we shall see later, this enables the Fermi-surface topology to be measured in a very great detail using modest magnetic fields [3,9]. Such information can then be used as input parameters for theoretical models [3]. And yet the same organic molecules can adopt a variety of configurations, leading to "glassy" structural transitions and mixed phases in otherwise very pure systems [4, 10, 11]; these states may be important precursors to the superconductivity in such cases [11].

Intriguingly, there seem to be at least two (or possibly three) distinct mechanisms for superconductivity [3, 12–14] in the quasi-two-dimensional organic conductors. The first applies to half-filled-band layered charge-transfer salts, such as the κ-, β-, and β'- packing arrangements of salts of the form (BEDT-TTF)$_2$X, where X is an anion molecule; the superconductivity appears to be mediated by electron correlations/antiferromagnetic fluctuations [3–5]. The second mechanism applies to, e.g., the β'' phase BEDT-TTF salts [4]; it appears to depend on the proximity of a metallic phase to charge order [11–13]. Finally, there may be some instances of BCS-like phonon-mediated superconductivity [14].

The main purpose of this chapter is to discuss the role that high magnetic fields and magnetoresistance measurements can play in unraveling the above-mentioned properties of quasi-one- and quasi-two-dimensional organic metals and superconductors. Hence, we shall spend some time discussing the high-field magnetotransport experiments that have helped to measure the Fermi surfaces of charge-transfer salts of molecules such as BEDT-TTF and BETS. In addition to their invaluable role in mapping the bandstructure, high magnetic fields allow one to tune some of the organic conductors into some new and intriguing phases; magnetoresistance phenomena can then be used to delineate the phase boundary of the new state. Examples include field-induced superconductivity [15] and exotic states such as the Fulde–Ferrell–Larkin–Ovchinnikov (FFLO) phase [16]. The phase diagram of the latter state in κ-(BEDT-TTF)$_2$Cu(NCS)$_2$ is shown in Fig. 9.1; its derivation is a good illustration of the general utility of high fields and magnetoresistive phenomena. First, a conventional ($\sim$30 Hz) measurement of the magnetoresistance was used to very precisely orient the sample in the magnetic field and to measure the superconducting-to-resistive transition [16]. Subsequently, high-frequency (MHz) magnetoresistance measurements that are sensitive to changes in dissipation *within* the zero-resistance state allowed the FFLO to type II superconductivity boundary to be measured [16]. In view of recent doubts about the proposed FFLO state in CeCoIn$_5$ [17], organic conductors such as κ-(BEDT-TTF)$_2$Cu(NCS)$_2$ [16] and λ-(BETS)$_2$GaCl$_4$ [18] are perhaps as yet the only systems in which the FFLO has been truly observed. Later in this paper, we shall describe other recent observations of field-induced phases in crystalline organic metals, which are related to the FFLO but which result in insulating states.

The remainder of this chapter is organized as follows. Section 9.2 describes the Fermi-surface topologies of some typical organic charge-transfer salts, concentrating on the features that influence the magnetoresistance in high fields; a brief mention is also made of the deficiencies of bandstructure calculations. Magnetoresistive phenomena are discussed in Sect. 9.3, including measurements of the effective dimensionality, angle-dependent magnetoresistance oscillations (AMROs), the magnetoresistivity tensor, and Fermi-surface-traversal resonances. The Shubnikov-de Haas effect is treated in Sect. 9.4, with a focus on the effects of reduced dimensionality and the extraction of quantities such as the quasiparticle scattering rate and effective mass. Sections 9.5 and 9.6 discuss some of the phenomena associated with charge-density waves above the Pauli paramagnetic limit. Finally, some thoughts about future prospects are given in Sect. 9.7.

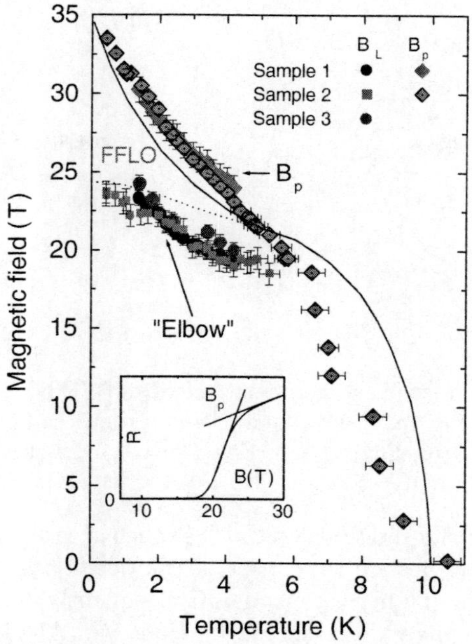

Fig. 9.1. Observation of the Fulde–Ferrell–Larkin–Ovchinnikov (FFLO) phase in κ-(BEDT-TTF)$_2$Cu(NCS)$_2$ [16]. The points labeled $B_{\rm P}$ denote the resistive upper critical field for two different samples (see inset for an illustrative example). The $B_{\rm L}$ points denoting the phase boundary between the mixed phase and FFLO state were deduced using simultaneous megahertz differential susceptibility measurements; the change in vortex stiffness that accompanies entry into the FFLO state causes an "elbow" in the field-dependent susceptibility. "Sample 3" is a measurement on a third sample under different conditions of electric field; consistency of the phase boundaries for the three samples shows that the effect is not due to artifacts of vortex pinning. The curves are a theoretical model due to Shimahara for the upper critical field and FFLO (see [16] for details)

9.2 Intralayer Fermi-Surface Topologies

The defining property of a metal is that it possesses a *Fermi surface*, that is, a constant-energy surface in k-space which separates the filled electron states from empty electron states at absolute zero ($T = 0$). The shape of the Fermi surface is determined by the dispersion relationships (energy vs. $\mathbf{k}$ relationships) $E = E(\mathbf{k})$ of each partially filled band and the number of quasiparticles to be accommodated. As is mentioned elsewhere in this book, the crystal structures of the organic metals and superconductors that feature in this article are mostly layered (or chain-like), with planes of anions (and perhaps other space-filling molecules [11,19]) alternating with layers of the cation molecules whose overlapping molecular orbitals provide the electronic bands. The main consequence of this structural anisotropy is that the intralayer

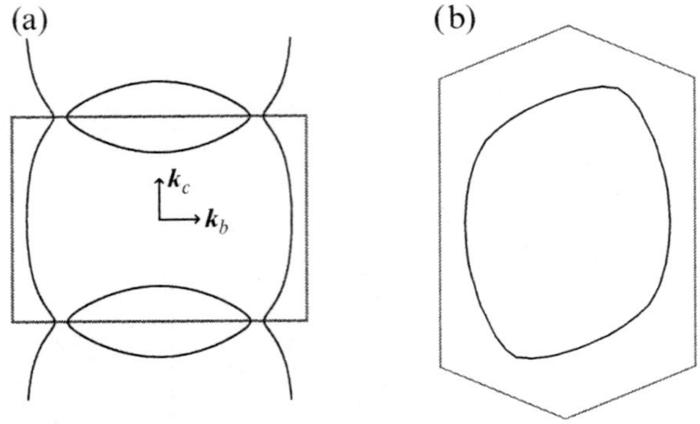

Fig. 9.2. (**a**) Brillouin zone and Fermi-surface of κ-(BEDT-TTF)$_2$Cu(NCS)$_2$, showing the open, quasi-one-dimensional sections, and the closed, quasi-two-dimensional pocket [20–22]. (**b**) Brillouin zone and Fermi-surface of β-(BEDT-TTF)$_2$IBr$_2$ [28]

(or interchain) transfer integrals tend to be much greater by a factor $\sim 10^2$–10^3 than the interlayer ones, so that most of the quasiparticle dispersion occurs parallel to the cation planes (or chains). Consequently, for many experiments, including the Shubnikov-de Haas and de Haas-van Alphen effects, the properties of the Fermi surface appear almost exactly two dimensional. We shall return to the consequences of this fact in later sections.

Figure 9.2 shows sections (parallel to the highly conducting planes) through the first Brillouin zone and Fermi surfaces of two typical BEDT-TTF salts. The example in Fig. 9.2a is κ-(BEDT-TTF)$_2$Cu(NCS)$_2$ [20–22]; κ-phase BEDT-TTF salts have a dimerized arrangement of BEDT-TTF molecules such that there are four per unit cell, each pair (or dimer) jointly donating one hole. The overall Fermi-surface cross-section is therefore the same as that of the Brillouin zone. However, the Fermi surface intersects the Brillouin zone boundaries in the **c** direction, such that band gaps open up (see, e.g., Chap. 2 of [23]). The Fermi surface thus splits into open (electron-like) sections (often known as *Fermi sheets*) running down two of the Brillouin-zone edges and a closed hole pocket (referred to as the "α pocket") straddling the other; it is customary to label such sections "quasi-one-dimensional" and "quasi-two-dimensional," respectively. The names arise because the group velocity **v** of the electrons is given by [23, 24]

$$\hbar \mathbf{v} = \nabla_{\mathbf{k}} E(\mathbf{k}). \tag{9.1}$$

The Fermi surface is a surface of constant energy; (9.1) shows that the velocities of electrons at the Fermi surface will be directed perpendicular to it. Therefore, referring to Fig. 9.2, electrons on the closed Fermi-surface pocket can possess velocities that point in any direction in the (k_b, k_c) plane; they have freedom of movement in two dimensions and are said to be *quasi-two-dimensional.* By contrast, electrons on the open sections have velocities predominently directed parallel to k_b and are *quasi-one-dimensional.*

κ-phase BEDT-TTF superconductors κ-(BEDT-TTF)$_2$X can be made with a variety of other anion molecules, including X = Cu[N(CN)$_2$]Br (11.8 K), Cu[N(CN)$_2$]Cl (12.8 K (under pressure)), and I$_3$ (4 K); here the number in parentheses represents T_c. In all of these salts, the Fermi-surface topology is very similar to that in Fig. 9.2a; small differences in the symmetry of the anion layer lead to variations in the gap between the quasi-one-dimensional and quasi-two-dimensional Fermi-surface sections [14, 25, 26]. A summary of the detailed differences and effective is given in Sect. 3.2 of [14] (see also [25]). We shall see later (Sect. 9.3.3) that magnetic breakdown [14], in which the field-induced motion of quasiparticles causes them to tunnel across the gaps between the Fermi-surface sections, leading to new magnetic quantum oscillation frequencies, is a common phenomenon in these salts [27].

Figure 9.2b shows the Fermi-surface topology and Brillouin zone of β-(BEDT-TTF)$_2$IBr$_2$ [28]. In this case (see Fig. 9.2b) there is one hole per unit cell, so that the Fermi surface cross-sectional area is half that of the Brillouin zone; only a quasi-two-dimensional pocket is present.

As mentioned above, the bandstructure of a charge-transfer salt is chiefly determined by the packing arrangement of the cation molecules. The β, κ, β'', λ, and α phases have tended to be the most commonly studied. The latter four phases all have predictable Fermi surfaces consisting of a quasi-two-dimensional pocket plus a pair of quasi-one-dimensional Fermi sheets (the pocket arrangement differs from phase to phase) [14, 25, 29]; the β-phase is alone in possessing a Fermi surface consisting of a single quasi-two-dimensional pocket [26].

As much of the rest of this article will be about using magnetoresistance to measure Fermi surfaces, it is worth including a brief note on the deficiencies of bandstructure calculations habitually applied to the organics. The bandstructures of crystalline organic metals have usually been calculated using the extended Hückel (tight-binding)[1] approach, which employs the highest occupied molecular orbitals (HOMOs) of the cation molecule [25]. Section 5.1.3 of [26] discusses this approach and cites some of the most relevant papers. Whilst this method is usually quite successful in predicting the main features of the Fermi surface (e.g., the fact that there are quasi-one-dimensional and quasi-two-dimensional Fermi-surface sections), the details of the Fermi-surface topology are sometimes inadequately described (see, e.g., [30]). This can be important when, for example, the detailed corrugations of a Fermi sheet govern the interactions that determine its low-temperature ground state [20, 30]. A possible way around this difficulty is to make slight adjustments of the transfer integrals so that the predicted Fermi surface is in good agreement with experimental measurements [20–22, 30]. In the β'' and λ phases the predicted bandstructure seems very sensitive to the choice of basis set, and the

[1] Simple introductions to the tight-binding model of bandstructure are given in [23, 24].

disagreement between calculation and measurement is often most severe (see, e.g., [31–34]).

More sophisticated Hubbard-unrestricted Hartree–Fock band calculations have been carried out for κ-(BEDT-TTF)$_2$Cu(NCS)$_2$ [35]. These calculations attempt to take into account many-body effects, and are successful in reproducing a number of experimental properties. They also indicate the importance of both antiferromagnetic fluctuations and electron–phonon interactions in κ-(BEDT-TTF)$_2$Cu(NCS)$_2$, a fact important in the proposed mechanisms for superconductivity [4, 14]. More recently, techniques such as dynamical mean field theory (DMFT) have been applied to quasi-two-dimensional organic superconductors [4, 36, 37], predicting some aspects of the complex phase diagram of the κ-phase BEDT-TTF salts.

9.3 High-Field Magnetotransport Effects

9.3.1 Measurements of the Effective Fermi-Surface Dimensionality via the SQUIT Peak

We remarked above that the electronic properties of quasi-two-dimensional organic metals are very anisotropic. Many band-structure-measuring techniques chiefly give information about the intralayer topology of the Fermi surface. However, it is important to ask whether the Fermi surface is exactly two-dimensional or whether it extends in the interlayer direction, i.e., is three-dimensional.

This question is of quite general interest, as there are many correlated electron systems that have very anisotropic electronic bandstructure. In addition to the organic superconductors [14, 38], examples include the "high-T_c" cuprates [39], and layered ruthenates [40] and manganites [41]. Such systems may be described by a tight-binding Hamiltonian in which the ratio of the interlayer transfer integral $t_\perp$ to the average intralayer transfer integral $t_{||}$ is $\ll 1$ [14, 38, 42]. The inequality $\hbar/\tau > t_\perp$ [43], where τ^{-1} is the quasiparticle scattering rate [38, 39, 42], frequently applies to such systems, suggesting that the quasiparticles scatter more frequently than they tunnel between layers. Similarly, under standard laboratory conditions, the inequality $k_B T > t_\perp$ often holds, hinting that thermal smearing will "wipe out" details of the interlayer periodicity [45].

The question has thus arisen as to whether the interlayer charge transfer is coherent or incoherent in these materials, i.e., whether or not the Fermi surface is a three-dimensional entity extending in the interlayer direction [14, 38, 42]. Incoherent interlayer transport is used as a justification for a number of theories that are thought to be pivotal in the understanding of reduced-dimensionality materials (see e.g. [38, 45]). Moreover, models for unconventional superconductivity in κ-phase BEDT-TTF salts invoke the nesting properties of the Fermi surface [20, 46, 47]; the degree of nesting might depend on whether the Fermi surface is two dimensional or three dimensional (see [14], Sect. 3.5).

In this context, the experimental situation is at first sight complicated because many apparently solid experimental tests for coherence in organic metals and superconductors have been deemed to be inconclusive [42]; e.g., semiclassical models can reproduce AMRO [48] and FTR data [49,50] equally well when the interlayer transport is coherent or "weakly coherent" [42]. However, it turns out that magnetoresistance *can* yield an unambiguous measurement of interlayer coherence, by way of a feature in the interlayer magnetoresistivity ρ_{zz} known as the *SQUIT* (Suppression of QUasiparticle Interlayer Transport) or *coherence peak*, observed for exactly in-plane fields (Figs. 9.3a, b).

To see how this comes about, consider a simple tight-binding expression for the interlayer (z-direction) dispersion [14,51]; $E(k_z) = -2t_\perp \cos(k_z a)$. Here $t_\perp$ is the interlayer transfer integral and a is the unit-cell height in the z direction. The introduction of such an interlayer dispersion, paired with an in-plane two-dimensional dispersion relationship, will result in a three-dimensional Fermi surface with a sinusoidally modulated Fermi-surface cross-section (see [23], Chap. 8). A more realistic version of the same idea is shown schematically for κ-(BEDT-TTF)$_2$Cu(NCS)$_2$ in Fig. 9.3c [53, 55, 56] (compare Fig. 9.2a). If the Fermi surface is extended in the interlayer direction, a magnetic field applied exactly in the intralayer plane can cause a variety of orbits on the sides of the Fermi surface (shown schematically in Figs. 9.3d, e) via the Lorentz force

$$\hbar(\mathrm{d}\mathbf{k}/\mathrm{d}t) = -e\mathbf{v} \times \mathbf{B}, \tag{9.2}$$

where $\mathbf{v}$ is given by (9.1); this results in orbits on the Fermi surface, in a plane perpendicular to $\mathbf{B}$ [14, 24]. Numerical modeling using a Chambers equation approach and a realistic parameterization of the Fermi surface [51] shows that the closed orbits about the belly of the Fermi surface are very effective in averaging $v_\perp$, the interlayer component of the velocity. Therefore, the presence of such orbits will lead to an increase in the resistivity component ρ_{zz} [51, 57, 58]. On tilting $\mathbf{B}$ away from the intralayer direction, the closed orbits cease to be possible when $\mathbf{B}$ has turned through an angle Δ, where [51]

$$\Delta(\text{in radians}) \approx v_\perp / v_{||}. \tag{9.3}$$

Here $v_\perp$ is the maximum of the interlayer component of the quasiparticle velocity and $v_{||}$ is the intralayer component of the quasiparticle velocity in the plane of rotation of $\mathbf{B}$. Therefore, on tilting $\mathbf{B}$ through the in-plane orientation, one expects to see a peak in ρ_{zz}, of angular width 2Δ, if (and only if [42]) the Fermi surface is extended in the z direction. It is this peak that is referred to as the "coherence peak" or the "SQUIT" peak (Figs. 9.3a, b). By using (9.1) and (9.3) and measured details of the intralayer Fermi-surface topology, it is possible to use Δ to deduce $t_\perp$ [51] to considerable accuracy.

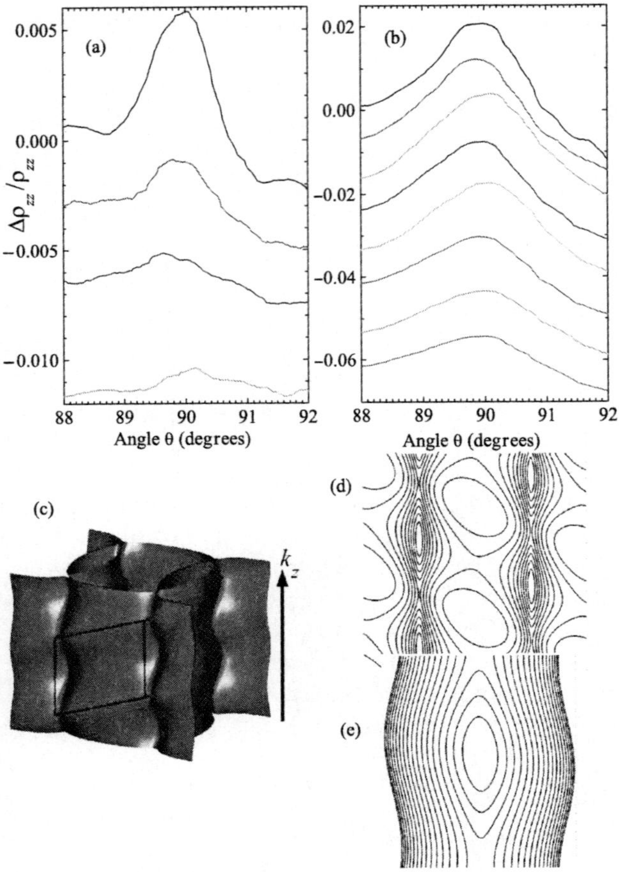

Fig. 9.3. Illustration of the "SQUIT" or "coherence peak" for in-plane fields in κ-(BEDT-TTF)$_2$Cu(NCS)$_2$ (after [52]). (**a**) The 45 T magnetoresistance of κ-(BEDT-TTF)$_2$Cu(NCS)$_2$ close to a tilt angle $\theta = 90°$ plotted as $\Delta\rho_{zz}/\rho_{zz\mathrm{BG}}$, the fractional change in ρ_{zz} from the more slowly varying background. Data for temperatures $T = 5.3$ (highest), 7.6, 9.6, and 13.1 K (lowest) are shown, offset for visibility. In this plane of rotation, closed orbits on the quasi-one-dimensional Fermi-surface sections (see (**d**) below) are responsible for the SQUIT, observed as a peak at $\theta = 90°$. (**b**) Similar data for a plane of rotation in which the SQUIT is caused by closed orbits on the quasi-two-dimensional Fermi-surface sections (see (**e**) below); the traces are for $T = 5.3$ (highest), 7.6, 8.6, 9.6, 10.6, 12.1, 13.1, and 14.6 K (lowest). Each trace has been offset for clarity. (**c**) Three-dimensional representation of the Fermi surface of κ-(BEDT-TTF)$_2$Cu(NCS)$_2$(after [51]); the finite interlayer transfer integral gives the corrugations (shown greatly exaggerated) on the sides of the FS. Quasi-one-dimensional and quasi-two-dimensional Fermi-surface sections are shown in red and blue, respectively. (**d**) Consequent field-induced closed orbits on the side of the quasi-one-dimensional Fermi-surface sections when $\theta = 90°$ and the field $\mathbf{B}$ is parallel to $\mathbf{k_b}$. (**e**) Similar closed orbits on the quasi-two-dimensional section when $\theta = 90°$ and $\mathbf{B}$ is parallel to $\mathbf{k_c}$ (see Fig. 9.2a). Orbits such as those in (**d**) and (**e**) give rise to the SQUIT peak in ρ_{zz} (see (**a**) and (**b**))

Figures 9.3a, b show typical data for κ-(BEDT-TTF)$_2$Cu(NCS)$_2$. The observation of a peak in ρ_{zz} close to $\theta = 90°$ allows the interlayer transfer integral to be estimated to be $t_\perp \approx 0.065$ meV [51]. This may be compared with intralayer transfer integrals ~150 meV [21].

Such data are of great interest because they illustrate that the criteria frequently used to delineate interlayer incoherence are rather a poor guide to reality. For example, a temperature of $T \approx 15$ K ($k_B T \approx 30 t_\perp$) leads one to expect incoherent interlayer transport via the criteron $k_B T > t_\perp$ proposed by Anderson [45], yet the peak in ρ_{zz} shown in Fig. 9.3b unambiguously demonstrates a three-dimensional Fermi-surface topology [52].

Demonstrations of interlayer coherence have been carried out on the quasi-two-dimensional organic conductors β-(BEDT-TTF)$_2$IBr$_2$ [28], κ-(BEDT-TTF)$_2$Cu$_2$(CN)$_3$ [59] (under pressure), α-(BEDT-TTF)$_2$NH$_4$Hg(SCN)$_4$ [57] (under pressure), β-(BEDT-TTF)$_2$I$_3$ [57], κ-(BEDT-TTF)$_2$Cu(NCS)$_2$ [22], λ-(BETS)$_2$GaCl$_4$ [34], and β''-(BEDT-TTF)$_2$SF$_5$CH$_2$CF$_2$SO$_3$ [60, 61]. In the latter example, no peak was observed, suggesting incoherent interlayer transport or no warping. In all of the other instances, the ρ_{zz} data demonstrate a Fermi surface, which is extended in the interlayer direction.

Inspired by this success, the technique has recently been extended to systems such as cuprate superconductors [62].

9.3.2 Mechanisms for Angle-Dependent Magnetoresistance Oscillations in Quasi-Two-Dimensional Organic Metals

Whilst they give very accurate information about the cross-sectional areas of the Fermi-surface sections, magnetic quantum oscillations do not provide any details of their *shape* (see Sect. 9.4). Such information is usually derived from angle-dependent magnetoresistance oscillations (AMROs) [14, 29, 48, 51, 63]. AMROs are measured by rotating a sample in a fixed magnetic field whilst monitoring its resistance; the coordinate used to denote the position of AMROs is the polar angle θ between the normal to the sample's quasi-two-dimensional planes and the magnetic field [48, 63]. It is also very informative to vary the plane of rotation of the sample in the field; this is described by the azimuthal angle ϕ [48, 51, 52, 63]).

As has been mentioned in the previous section, many quasi-two-dimensional charge-transfer salts of molecules such as BEDT-TTF and BETS exhibit the SQUIT or coherence peak, showing that they possess a well-defined three-dimensional Fermi surface, even at quite elevated temperatures [51]. In such cases, the AMROs can be modeled using a Boltzmann-transport approach, which treats the time evolution of quasiparticle velocities across a three-dimensional Fermi surface.[2]

In such a picture, AMROs result from the averaging effect that the semi-classical orbits on the Fermi surface have on the quasiparticle velocity. Both

[2] As mentioned elsewhere in this book, the picture for TMTSF salts can be rather different.

quasi-one-dimensional and quasi-two-dimensional Fermi-surface sections can give rise to AMROs; in the former case, the AMROs are sharp dips in the resistivity, periodic in $\tan\theta$; in the latter case, one expects *peaks*, also periodic in $\tan\theta$ [48, 63]. To distinguish between these two cases, it is necessary to carry out the experiment at several different ϕ (some fine cautionary hints are given in [51]). The ϕ-dependence of the AMROs can be related directly to the shape of a quasi-two-dimensional Fermi-surface section; in the case of a quasi-one-dimensional sheet, the AMROs yield precise information about the sheet's orientation [14, 29, 48, 51, 63]. Typical data are shown in Fig. 9.4. Numerical modeling of such data (using a Boltzmann transport approach) allows a detailed three-dimensional picture of the Fermi surface to be built up (see Fig. 9.5a).

A third type of AMRO has been observed in the angle-dependent magnetoresistance of κ-(BEDT-TTF)$_2$Cu(NCS)$_2$subjected to high pressures. The experiments employ a miniature diamond-anvil cell, attached to a cryogenic goniometer, providing full two axis rotation at ^{3}He temperatures [64]. The apparatus is placed in a 33 T Bitter magnet. A plethora of AMROs is observed at each pressure (Fig. 9.5b), caused by field-induced quasiparticle orbits across the Fermi surface. Raising pressure suppresses the gap between the quasi-two-dimensional pocket and quasi-one-dimensional sheets of the Fermi surface (see Fig. 9.2a), increasing the probability of magnetic breakdown (see Sect. 9.3.3 and [27] for a more detailed explanation of magnetic breakdown). This permits AMROs due to breakdown orbits about the complete Fermi surface.

Finally, note that increasing the temperature gradually suppresses the AMROs (see Figs. 9.6a, b). Modeling of this phenomenon shows that it can be described by a temperature-dependent scattering rate, $\tau^{-1} = \zeta + \chi T^2$, where ζ and χ are constants [52]. The exponent suggests that the suppression of AMROs is due to electron–electron scattering. Another interesting feature of this suppresion is that the same scattering rate appears to apply to orbits on the quasi-one-dimensional and quasi-two-dimensional parts of the Fermi surface. This suggests that mechanisms for superconductivity in organic metals that invoke a large variation in scattering rate over the FS (e.g., "FLEX" methods [20]) may be inappropriate for κ-(BEDT-TTF)$_2$Cu(NCS)$_2$

9.3.3 Further Clues about Dimensionality in the Resistivity Tensor Components

In the past there has been some confusion as to the origin of AMROs in quasi-two-dimensional charge-transfer salts of BEDT-TTF; and in particular, the component of the resistivity tensor in which the oscillations occur. There are few reliable measurements of the in-plane conductivity or resistivity $\rho_{||}$ of quasi-two-dimensional crystalline organic metals [3, 14]; experiments involving conventional edge contacts are problematic [3, 14]. Because of the very large resistivity anisotropy, such data are almost often dominated by the much larger interplane resistivity component, ρ_{zz} [14]. To circumvent this problem, a

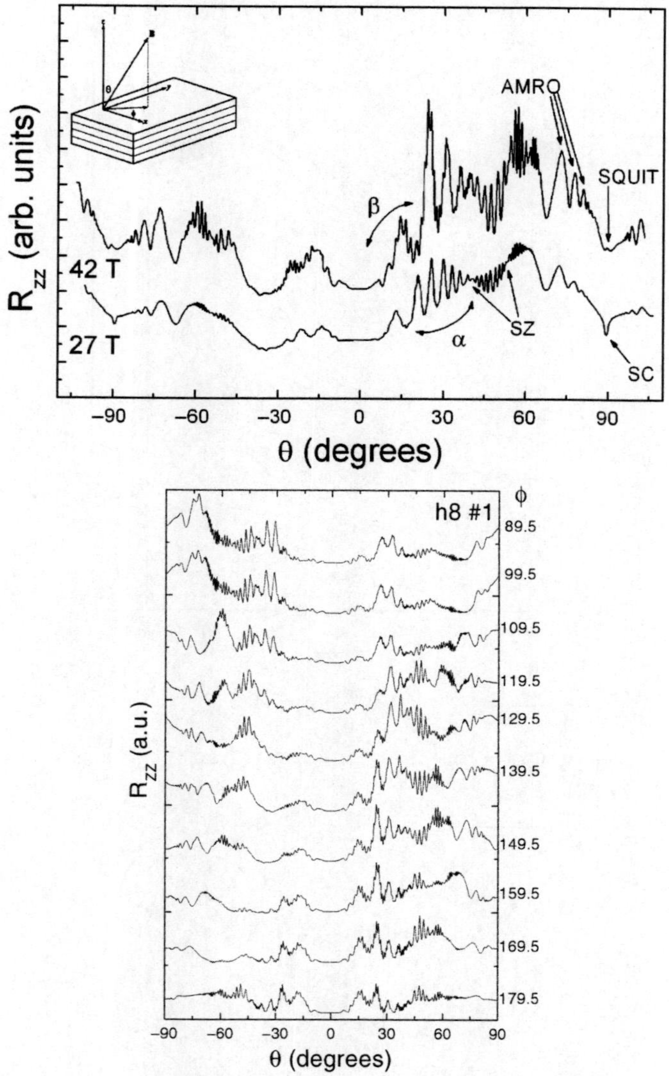

Fig. 9.4. (**a**) Typical θ dependence of the magnetoresistance of κ-(BEDT-TTF)$_2$Cu(NCS)$_2$ [51]. The data shown are for a hydrogenated sample at 490 mK, $\phi = 149°$ (where ϕ is the azimuthal angle), and fields of 27 T (*lower*) and 42 T (*upper*). The data have been offset for clarity. Some representative features are indicated; Shubnikov-de Haas (SdH) oscillations due to the Q2D pockets (α) and the breakdown orbit (β); spin zeros in the SdH amplitudes (SZ); the onset of the superconducting transition (SC); angle-dependent magnetoresistance oscillations (AMRO), whose positions are field independent; and the resistive SQUIT peak in the presence of an exactly in-plane magnetic field (in-plane Peak). The inset diagram is included to illustrate the measurement geometry. (**b**) The angle-dependent interlayer magnetoresistance of the same sample at various values of the azimuthal angle ϕ; $T = 500$ mK and $B = 42$ T

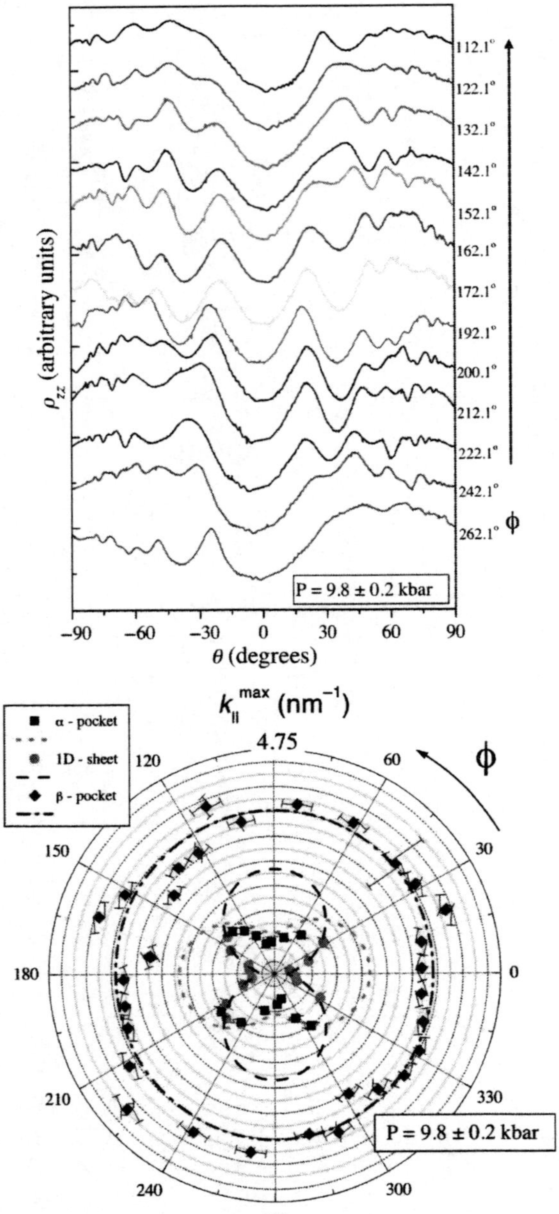

Fig. 9.5. (*Top*) Typical angle-dependent magnetoresistance oscillations (AMROs) at a magnetic field of 30 T and a temperature of 1.5 K; the pressure is 9.8 kbar. Here, θ denotes the angle between the normal to the sample's quasi-two-dimensional planes and the field; ϕ describes the plane of rotation. (*Bottom*) Polar plot of the periodicities (in units of $\tan\theta$) of the various AMRO series. The inset key gives the mechanism for the features, with the blue diamonds representing the oscillations due to magnetic breakdown [64]

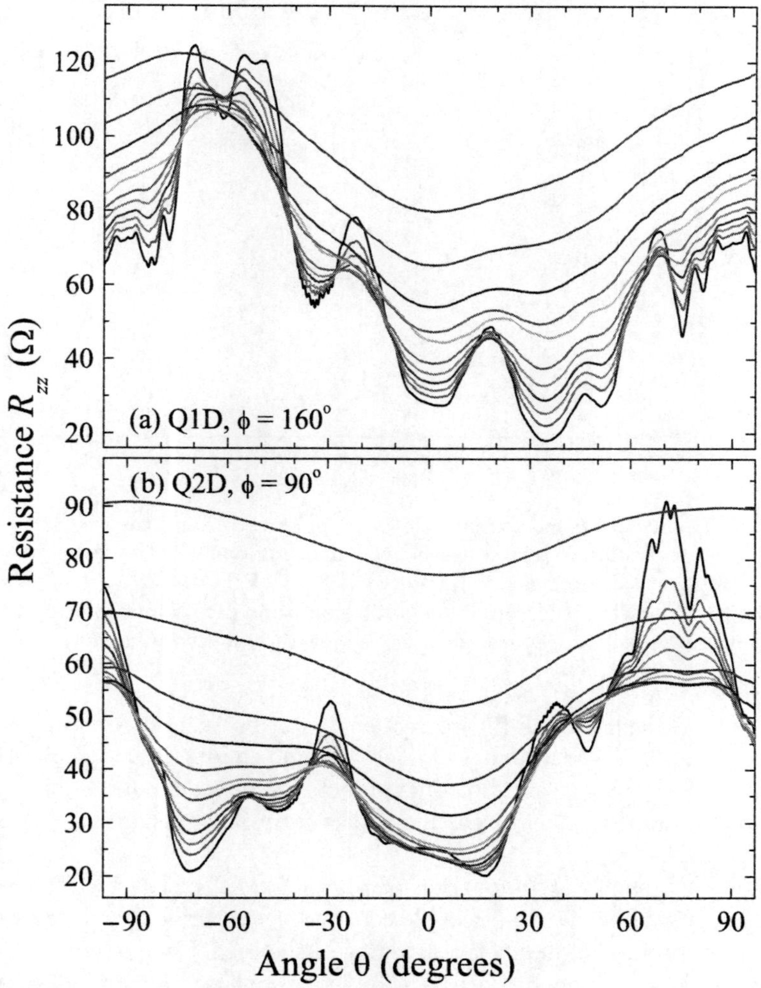

Fig. 9.6. Suppression of AMROs by increasing temperature shown as interlayer resistance $R_{zz}(\propto \rho_{zz})$ of a κ-(BEDT-TTF)$_2$Cu(NCS)$_2$ crystal vs. tilt angle θ for various constant T ($B = 45\,\mathrm{T}$). (**a**) Data for $\phi = 160°$, a plane of rotation at which ρ_{zz} is determined by phenomena on the quasi-one-dimensional Fermi-surface sections. In order of increasing R_{zz} at $\theta = 35°$, the curves are for $T = 5.3$, 6.5, 7.6, 8.6, 9.6, 10.6, 12.1, 13.1, 14.6, 17.1, 19.6, and 29.3 K respectively. (**b**) Similar data for $\phi = 90°$; here ρ_{zz} features are associated with the quasi-two-dimensional Fermi-surface. In order of increasing R_{zz} at $\theta = -70°$, the curves are for $T = 5.3$, 7.6, 8.6, 9.6, 10.6, 12.1, 13.1, 14.6, 17.1, 19.6, 24.5, and 29.3 K, respectively

number of authors have turned to a MHz skin depth technique to measure the field dependence of the in-plane resistivity [3,34,65,66]; this technique is very suitable for pulsed magnetic fields [3,34,66]. Samples for such experiments are mounted within a small coil, which forms part of the tank circuit of a

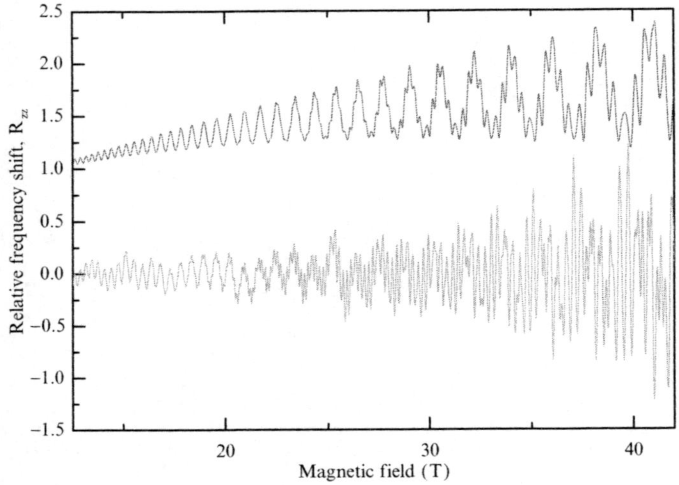

Fig. 9.7. Simultaneous measurement of ρ_{zz} (*upper trace*) and the frequency shift of a tunnel diode oscillator (related to the in-plane component of the resistivity) (*lower trace*). The sample is a single crystal of κ-(BEDT-TTF)$_2$Cu(NCS)$_2$; the temperature is 470 mK. Note that the rapid magnetic quantum oscillations due to magnetic breakdown are much more prominent in the lower data set (after [50])

tunnel-diode oscillator; shifts in frequency can be related to changes in the skin depth and hence to changes in the in-plane conductivity [34, 65]. In some experiments [3, 65], top and bottom contacts are also mounted on the sample, so that simultaneous measurements of the interplane resistivity component, ρ_{zz}, can be made.

Figure 9.7 shows a comparison of ρ_{zz} and the frequency shift measured simultaneously. The most noticeable contrast between the two data sets is the much more prominent magnetic breakdown (the higher frequency) oscillations in the (in-plane) frequency data. Although a quantitative model of such oscillations poses some theoretical difficulties, it is easy to see qualitatively why magnetic breakdown will influence the in-plane conductivity much more than it does the interplane conductivity. Magnetic breakdown represents the tunneling of quasiparticles from the one Fermi-surface section to another [27]. This will affect the way in which a quasiparticle's velocity evolves with time, and hence the conductivity. As virtually all of the dispersion of the electronic bands is in-plane, a magnetic breakdown event will have a relatively large effect on the time dependence of the in-plane component of a quasiparticle's velocity. By contrast, the warpings in the interplane direction of both sections of the the Fermi surface of κ-(BEDT-TTF)$_2$Cu(NCS)$_2$ seem to be rather similar [22]; hence magnetic breakdown will have comparatively little effect on the time evolution of the interplane component of the quasiparticle velocity.

Figure 9.8 contrasts the angle dependence of the two components of the resistivity at fixed magnetic field. While ρ_{zz} exhibits strong AMROs, the

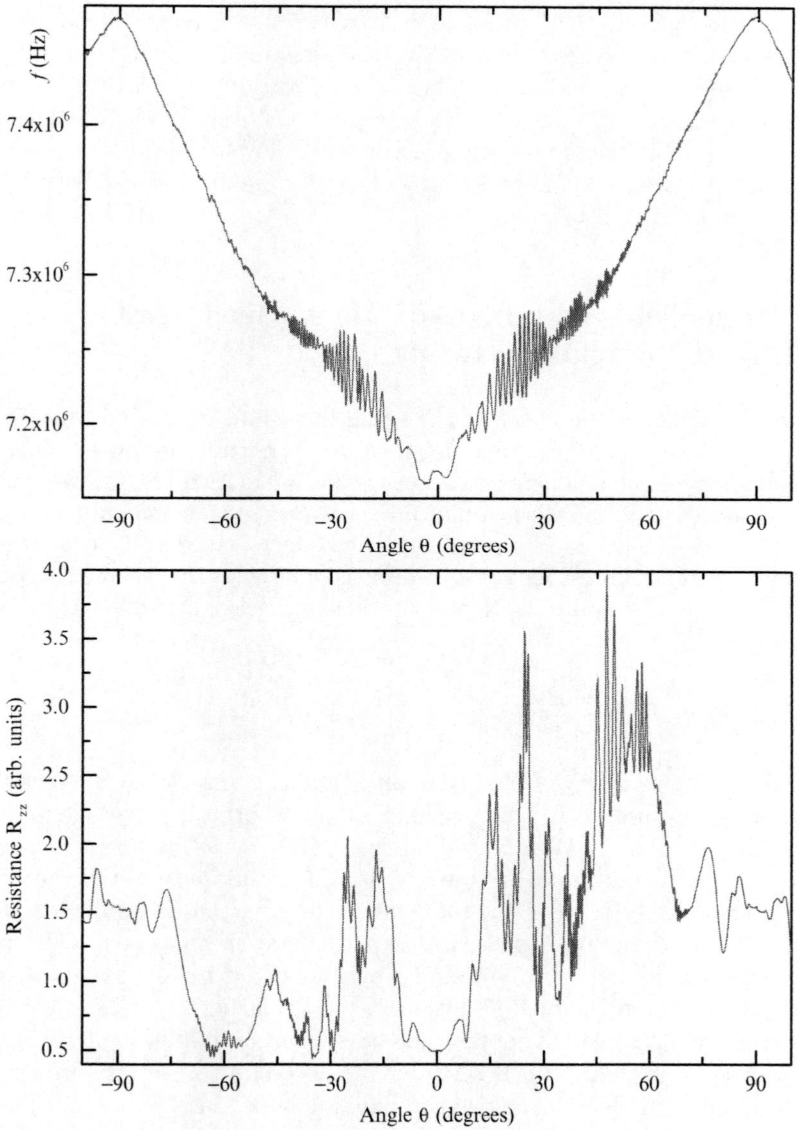

Fig. 9.8. Comparison of the magnetic-field orientation dependence of the interlayer resistance R_{zz} (*lower trace* – proportional to ρ_{zz}) and that of the frequency shift of a tunnel diode oscillator (*upper trace* – related to $\rho_{||}$) measured simultaneously. $\theta = 0$ corresponds to the field being normal to the quasi-two-dimensional planes of κ-(BEDT-TTF)$_2$Cu(NCS)$_2$. The temperature is 470 mK and the static magnetic field is 42 T. The rapid oscillations close to $\theta = 0$ in both figures are Shubnikov-de Haas oscillations. The slower oscillations, periodic in $\tan\theta$ and only seen in the lower trace, are AMROs. The peak denoting interlayer coherence is visible at $\theta = 90°$ in the lower figure (after [50])

frequency (depending the in-plane resistivity) shows none. This is entirely consistent with the expectations of semiclassical models of AMROs [48, 51].

It should be noted that a high-frequency variant of AMROs, known as the *Fermi-surface-traversal resonance* (FTR), has been developed. This high-frequency (GHz) magneto-optical technique allows additional information about the topology and corrugations of quasi-one-dimensional Fermi sheets to be deduced [14, 49, 67].

9.4 High-Field Shubnikov-de Haas Measurements and Quasiparticle Scattering

Above we saw that the effect of the small but finite interlayer transfer integral on the Fermi-surface topology is very important in phenomena such as angle-dependent magnetoresistance oscillations (AMROs). However, from the perspective of magnetic quantum oscillatory phenomena such as the Shubnikov-de Haas and de Haas-van Alphen effects, the Fermi surface properties behave in an almost ideally two-dimensional way. To see how this occurs, consider the Landau quantization of the quasiparticle states due to a magnetic field **B** [23, 24]:

$$E(\mathbf{B}, k_z, l) = \frac{\hbar e|\mathbf{B}|}{m^*}(l + \frac{1}{2}) + E(k_z) \equiv \hbar\omega_c(l + \frac{1}{2}) + E(k_z). \qquad (9.4)$$

Here $E(k_z)$ is the energy of the (unmodified) motion parallel to $\mathbf{B}$, l is the Landau quantum number $(0, 1, 2, \dots)$, and m^* is an orbitally averaged effective mass; the angular frequency $\omega_c = eB/m^*$ (the *cyclotron frequency*) corresponds to the semiclassical frequency at which the quasiparticles orbit the Fermi surface [23, 24]. (For the moment we neglect the Zeeman term due to spin [14].) In virtually all practical experiments, the magnetic field is perpendicular to the highly conducting planes or tilted by angles less than 60° from this direction. In such situations, $E(k_z)$ will be $\sim t_\perp$; for most of the quasi-two-dimensional charge-transfer salts, the Landau-level energy spacing will eclipse $t_\perp$ in fields of order 1-5 T. The necessity to use a rigorously two-dimensional approach to analyze the Shubnikov-de Haas (resistivity) and de Haas-van Alphen (magnetization) oscillations [14, 29, 68] in these cases cannot be overemphasized; see [27, 69] for a thorough discussion of this point.

9.4.1 The Deduction of Quasiparticle Scattering Rates

Recently, it has been proposed that the dependence of superconducting properties on the quasiparticle scattering rate is an excellent way of identifying the mechanism for superconductivity in quasi-two-dimensional organics [1]. Unfortunately, this is not as straightforward as it might seem. A measure of the scattering rate in metallic systems is often derived from the rate at which magnetic quantum oscillations (such as de Haas-van Alphen oscillations) grow in amplitude with increasing magnetic field; the dominant term in the Lifshitz–Kosevich formula (see, e.g., [68,69]) describing this phenomenon is $\exp[-14.7m^*_{\rm CR}T_{\rm D}/B]$ (SI units). The constant describing the phase-smearing of the oscillations due to Landau-level broadening is the so-called *Dingle temperature*, $T_{\rm D}$.

If one were to assume that $T_{\rm D}$ is due *solely* to scattering (i.e., the energy width of the Landau levels detected by the de Haas-van Alphen effect is associated only with their finite lifetime due to scattering), then $T_{\rm D}$ would be related to the scattering rate $\tau^{-1}_{\rm dHvA}$ by the expression

$$T_{\rm D} = \frac{\hbar}{2\pi k_{\rm B}\tau_{\rm dHvA}}. \tag{9.5}$$

In many experimental works, it is assumed that the $\tau_{\rm dHvA}$ deduced from $T_{\rm D}$ in this way is a true measure of scattering rate; however, in quasi-two-dimensional organic metals this is almost certainly not the case. The problem becomes apparent if such scattering times are compared with the $\tau_{\rm CR}$ deduced from cyclotron resonance experiments [50]. In such cases the following inequality is found:

$$\tau_{\rm CR} > \tau_{\rm dHvA}. \tag{9.6}$$

For some layered metals (see, e.g., the work of Hill [70]), the $\tau_{\rm CR}$ measured in cyclotron resonance has been found to be four to ten times larger than $\tau_{\rm dHvA}$. An example of this is shown in Fig. 11 of [50], where the insertion of the scattering rate inferred from Shubnikov-de Haas oscillations into a model for the cyclotron resonance produces linewidths that are too much broad. A more realistic linewidth is obtained with a longer scattering time. As we shall now describe, spatial inhomogeneities are the likely culprit for the difference of scattering times.

Screening is less effective in systems containing low densities of quasiparticles (such as organic metals), compared to that in elemental metals; hence variations in the potential experienced by the quasiparticles can lead to a spatial variation of the Landau-level energies (see Fig. 9.9). Even in the (hypothetical) complete absence of scattering, Harrison [71] has shown that this spatial variation would give the Landau level a finite energy width (see Fig. 9.9) and therefore lead to an apparent Dingle temperature

$$T_{\rm D} = \frac{\bar{x}[1-\bar{x}]F'(\bar{x})^2 a}{\pi k_{\rm B} m^*}\sqrt{\frac{\hbar e^3}{2F}}. \tag{9.7}$$

Here F is the magnetic-quantum-oscillation frequency and $F' = {\rm d}F/{\rm d}x$; x represents the (local) fractional variation of the quasiparticle density due to the potential fluctuations and $\bar{x}$ is its mean.

The Dingle temperature measured in experiments will therefore usually represent a combination of the effects described by (9.5) and (9.7). Hence the

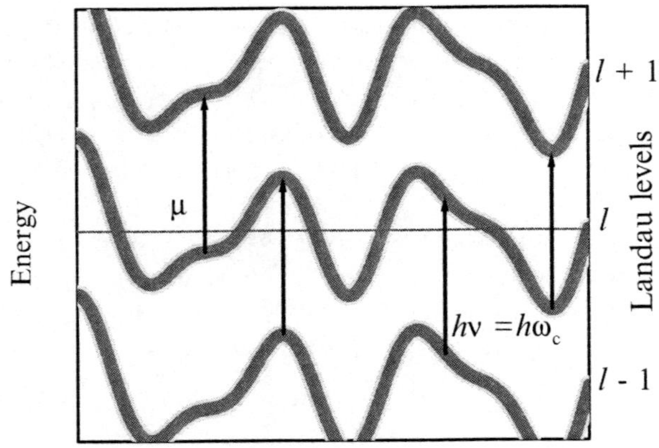

Fig. 9.9. Cartoon of the effect of spatial inhomogeneities on Landau-level widths and energies. The variations in the potential experienced by the electrons make the Landau levels (shaded curves) move up and down in energy as one moves through the sample. As the field is swept, the levels will move up through the chemical potential μ and depopulate, resulting in the de Haas-van Alphen and Shubnikov-de Haas effects. The dingle temperature essentially parameterizes the movement of the *total energy width of a Landau level* through μ; hence it will measure a width that includes the energy variation due to inhomogeneities. By contrast, cyclotron resonance (shown by vertical arrows) is a "vertical" transition; it will measure just the true width of the Landau levels due to scattering (represented by shading)

simple-minded use of (9.5) to yield τ_{dHvA} from T_{D} will tend to result in a parameter that is an overestimate of the true scattering rate (see Fig. 9.9). By contrast, cyclotron resonance (shown by vertical arrows in Fig. 9.9) is a "vertical" transition (due to the very low momentum of the photon); it will measure just the true width of the Landau levels due to scattering (represented by shading) [50].

Providing a thorough treatment of the sample's high-frequency magneto-conductivity is made [50], then the measured scattering rate deduced from a cyclotron resonance experiment is a good measure of the energy width of the Landau levels due to their finite lifetime. Once this has been realized, cyclotron resonance can be used to give quantitative details of the quasiparticle scattering mechanisms. By contrast, the *apparent* scattering rate deduced from de Haas-van Alphen and Shubnikov-de Haas oscillations can contain very significant contributions from spatial inhomogeneities [71].

Finally, one should add that the scattering rate τ_{σ}^{-1} measured in a zero-field *in-plane* conductivity experiment can be very significantly different from τ_{CR}^{-1} because the two measurements are sensitive to different types of scattering process. One of us has discussed this issue in detail in two recent papers [72, 73]; in particular, in the case of the κ-phase BEDT-TTF salts, there is an (as-yet) unexplained quatitative discrepancy between the size of the in-plane conductivity and other measures of the quasiparticle properties [73]. Further work to resolve this question is necessary.

Finally, "jitter" in the Brillouin zone boundary may be yet another source of scattering. This is predicted to give rise to "hot spots" where the Fermi surface intersects the Brillouin zone boundary [74].

9.5 Charge-Density Waves at Fields above the Pauli Paramagnetic Limit

Intense magnetic fields (B) impose severe constraints on spin-singlet paired electron states. Superconductivity is one example of a ground state where this is true, although orbital diamagnetic effects usually destroy superconductivity at lower magnetic fields than does the Zeeman effect [16, 75]. Charge-density wave (CDW) systems, by comparison, are mostly free from orbital effects [76], and so can only be destroyed by coupling B directly to the electron spin. While most CDW systems have gaps that are too large to be destroyed in laboratory-accessible fields [76], several new compounds have been identified within the last decade that have gaps ($2\Psi_0$) as low as a few meV, bringing them within range of state-of-the-art static magnetic fields.

As we shall discuss in Sect. 9.6, α-(BEDT-TTF)$_2M$Hg(SCN)$_4$(where M = K, Tl, or Rb; $2\Psi_0 \sim 4\,\text{meV}$) is one example that has been extensively studied [77]. However, it has a complicated phase diagram in a magnetic field owing to the imperfect nature of the nesting [78]; closed orbits exist after the Fermi-surface reconstruction, which become subject to Landau quantization in a magnetic field [79], potentially modifying the ground state. By contrast, (Per)$_2M$(mnt)$_2$ (where M = Pt and Au) appears to be fully gapped [80]. However, the existence of spin $\frac{1}{2}$ moments on the Pt sites makes the M = Au system a pristine example of a small-gap, fully dielectric CDW material.

Measurements of the CDW transition temperature T_P (where the subscript "P" stands for "Peierls") in (Per)$_2$Au(mnt)$_2$ ($T_\text{P} = 11\,\text{K}$ at $B = 0$) as a function of B indicate that it is suppressed in a predictable fashion [81], allowing a Pauli paramagnetic limit of $B_\text{P} \approx 37\,\text{T}$ to be inferred. However, the closure of the CDW gap with field is in fact considerably more subtle [82]; a finite transfer integral t_a perpendicular to the nesting vector produces a situation analogous to that in an indirect-gap semiconductor, where the minimum energy of the empty states above the chemical potential μ is displaced in k-space from the maximum-energy occupied states below μ [82]. Consequently, Landau quantization of the states above and below μ is possible, leading to a thermodynamic energy gap $E_\text{g}(B, T)$ of the form [82]

$$E_\text{g}(B, T) = 2\Psi(T) - 4t_a - g\mu_\text{B}B + \gamma\hbar\omega_\text{c}. \tag{9.8}$$

Here, $\Psi(T)$ is the temperature-dependent CDW order parameter ($\Psi(T) \to \Psi_0$ as $T \to 0$), ω_c is a characteristic cyclotron frequency in the limit $B \to 0$, and γ is a nonparabolicity factor; $g \approx 2$ is the Landé g-factor and μ_{B} is the Bohr magneton. Note that the Landau quantization *competes* with Zeeman splitting; however, at sufficiently high B it becomes impossible to sustain closed orbits, leading to a straightforward dominance of the Zeeman term [82].

Another subtlety in assessing the field-dependent thermodynamic gap (and hence the Pauli paramagnetic limit) in $(\mathrm{Per})_2\mathrm{Au}(\mathrm{mnt})_2$ stems from the complicated nature of the low-temperature electrical conductivity, which contains contributions from both the sliding collective mode of the CDW and thermal excitation across the gap (see Fig. 9.10a) [82, 83]. This leads to a measured resistivity $\rho_{yy} \approx (\sigma_{\mathrm{T}} + j_y/E_{\mathrm{t}})^{-1}$, where σ_{T} is the conductivity due to thermal excitation across the gap and j_y/E_{t} is the contribution from the collective mode, j_y being the current density, and E_{t} is the threshold electric field to depin the CDW. Now (9.8) contains Ψ, which is T-dependent; moreover, E_{t} may also depend on T. Thus, Arrhenius plots are in general curved (see Fig. 9.10b), with a slope

$$\frac{\partial \ln \rho_{yy}}{\partial(1/T)} \approx \frac{\frac{1}{2k_{\mathrm{B}}}(E_{\mathrm{g}} - T\frac{\partial E_{\mathrm{g}}}{\partial T}) - \frac{j_y T^2}{\sigma_{\mathrm{T}} E_{\mathrm{t}}^2}\frac{\partial E_{\mathrm{t}}}{\partial T}}{1 + \frac{j_y}{\sigma_{\mathrm{T}} E_{\mathrm{t}}}}. \tag{9.9}$$

With appropriate choice of temperature and bias regimes, it is possible to make a reliable estimate of E_{g} from plots such as those in Fig. 9.10b. On the other hand, it can be shown [84] that poorly chosen experimental conditions can easily lead to errors in the size of the derived gap.

Once accurate values of $E_{\mathrm{g}}(B)$ have been obtained, the method of [82] can be used to fit (9.8) by adjusting the parameters $2\Psi_0 + 4t_a$, $4t_a$, and v_{F}, where v_{F} is the Fermi velocity in the metallic state; in the CDW state it is used to parameterize the quasiparticle dispersion. A good fit is obtained using $t_a = 0.20 \pm 0.01$ meV, $v_{\mathrm{F}} = (1.70 \pm 0.05) \times 10^5\,\mathrm{m\,s^{-1}}$, and $2\Psi_0 = 4.02 \pm 0.04$ meV [82]. These parameters correspond to $E_{\mathrm{g}} = 3.21 \pm 0.07$ meV at $B = 0$, $T = 0$; the derived transfer integrals t_a and t_b($\approx$188 meV) are in good agreement with theory [85] and thermopower data [80]. (Note that these band parameters exclude the possibility of field-induced CDW (FICDW) states of the kind proposed in [86] in $(\mathrm{Per})_2M(\mathrm{mnt})_2$ salts.)

Armed with the band parameters, one obtains a reliable estimate of the Pauli paramagnetic limit; $B_{\mathrm{P}} = (\Delta_0 + 2t_a)/\sqrt{2}gs\mu_{\mathrm{B}} \approx 30$ T. This corresponds to a sharp drop in measured resistance R ($\propto \rho_{yy}$), as shown in Fig. 9.11 (left side), which displays data recorded at $T = 25$ mK. At such temperatures, there are very few thermally activated quasiparticles indeed, leaving only the CDW collective mode to conduct; this gives rise to a R in Fig. 9.11 that is strongly dependent on current. On passing through $B_{\mathrm{P}} \approx 30$ T, R drops very sharply, and there is also hysteresis between up- and down-sweeps of the field. The latter effect could be the consequence of a first-order phase transition on reaching B_{P}, compounded by CDW pinning effects.

However, the most interesting observation about Fig. 9.11 is the fact that the strongly nonlinear I-V characteristics persist at fields well above B_{P}, as

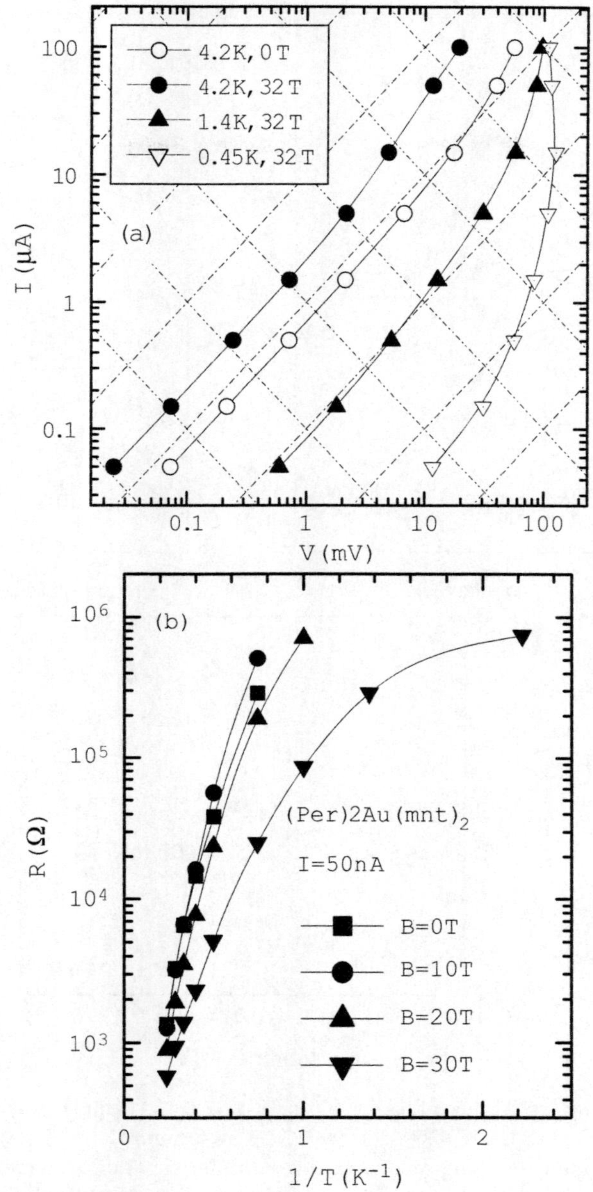

Fig. 9.10. (**a**) Nonlinear current-vs.-voltage characteristics of $(\mathrm{Per})_2\mathrm{Au(mnt)}_2$ plotted on a logarithmic scale for various temperatures and fields (see inset key). The negative-slope diagonal lines are contours of constant power and the positive-slope diagonal lines are contours of constant resistance, providing a guide as to when the sample's behavior is dominated by ohmic, thermally activated conduction rather than sliding. (**b**) Arrhenius plots of resistance R ($\propto \rho_{yy}$, logarithmic scale) vs. $1/T$ with $I = 50\,\mathrm{nA}$ for $(\mathrm{Per})_2\mathrm{Au(mnt)}_2$ at several different B (after [82])

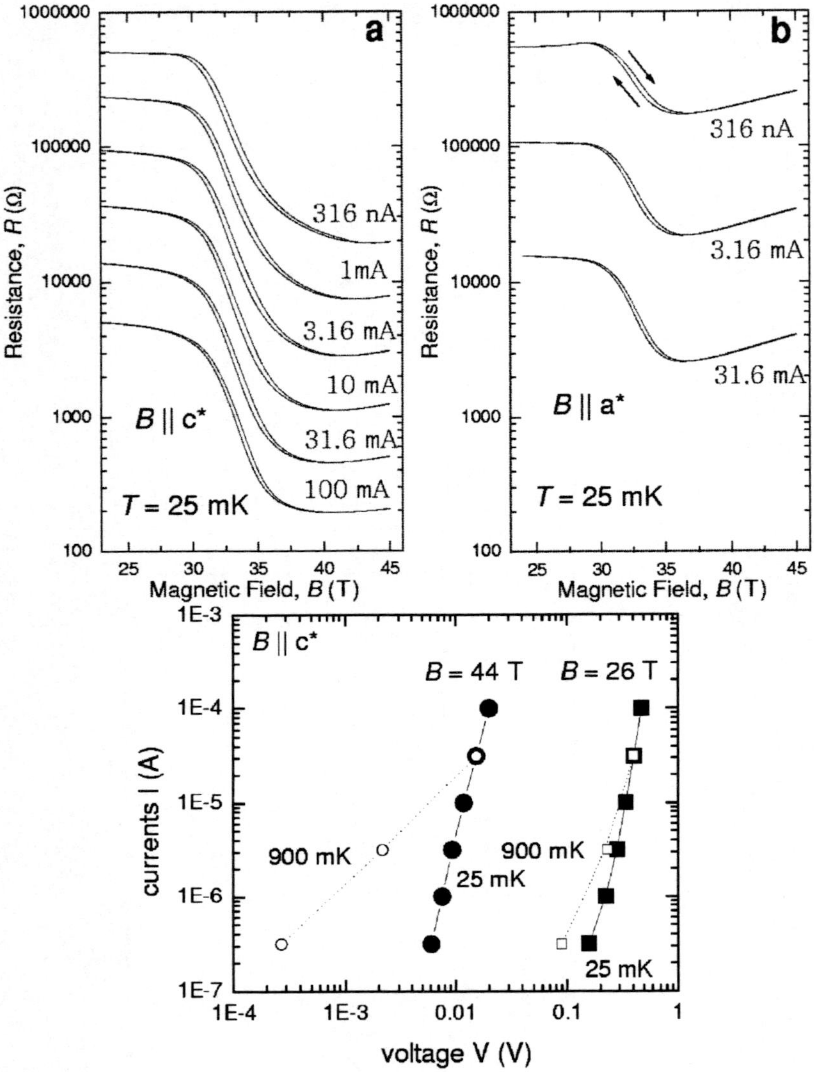

Fig. 9.11. (*Top*) resistance of a single crystal of $(Per)_2Au(mnt)_2$ measured at 25 mK for fields between 23 and 45 T, for two different orientations c^* (**a**) and a^* (**b**) of B perpendicular to its long axis b, at several different applied currents. The lowest resistance for a given current occurs for B parallel to c^*, which is perpendicular to a^*. The dependence of the resistance on current signals nonohmic behavior. (*Bottom*) nonlinear current-vs.-voltage characteristic of $(Per)_2Au(mnt)_2$ plotted on a log–log scale for magnetic fields (26 and 44 T) above (*circles*) and below (*squares*) B_P. Filled symbols connected by solid lines represent data taken at 25 mK while open symbols connected by dotted lines represent data taken at 900 mK (after [83])

can be seen from both R data and I-V plots (right-hand side of Fig. 9.11). One can conclude from these data that it is the CDW depinning voltage that changes at B_{P}. This is probably a consequence of the CDW becoming incommensurate or of the order parameter of the charge modulation becoming considerably weakened [76,83]. Evidence for the latter is obtained by repeating the I-V measurements at slightly higher temperatures of 900 mK (Fig. 9.11, right side). This temperature is sufficient to restore Ohmic behavior for $B > B_{\mathrm{P}}$, suggesting that a reduced gap for $B > B_{\mathrm{P}}$ allows quasiparticles to be more easily excited.

There have been some interesting proposals for FICDW phases (e.g., [86]). These mechanisms require the system not to be completely gapped; instead possessing a closed orbit for one of the spins that would then have a Landau gap at the chemical potential. Such a situation typically leads to the quantum Hall effect and a metallic behavior of longitudinal resistivity [87]. Current would then be able to flow without the CDW having to be depinned. These effects nevertheless appear to be absent from the data of Fig. 9.11; the continuation of the nonlinear CDW electrodynamics for $B > B_{\mathrm{P}}$ suggests that metallic behavior is not regained. Instead, both spin components of the Fermi surface are most likely to be gapped independently with differing nesting vectors, leading to an exotic CDW phase that has some analogies with the FFLO state of superconductors. The presence of two distinct, spin-polarized CDWs with different periodicities will furnish separate spin and charge modulations that could in principle be detected using a diffraction experiment [83].

It has been suggested [88] that FICDWs occur in charge-density wave (CDW) systems in strong magnetic fields when orbital quantization facilitates nesting of quasi-one-dimensional Zeeman-split bands. The free energy is minimized at low integral Landau subband filling factors ν by the formation of a Landau gap at the Fermi energy [88]. Hence, as is the case in field-induced spin-density wave states (FISDW) [87], orbital quantization is implicit in FICDW formation, yielding a Hall conductivity $\sigma_{xy} \approx 2\nu e^2/ah$ (where $a \sim 20\,\mathrm{\AA}$ is the layer spacing) and a longitudinal conductivity $\sigma_{xx} \sim \sigma_0 \exp[-\Delta/k_{\mathrm{B}}T]$ that is very small and thermally activated ($\sigma_{xx} \ll \sigma_{xy}$). Inversion of the conductivity tensor yields

$$\rho_{xx} \approx (ah/2\nu e^2)^2 \sigma_0 \mathrm{e}^{-\frac{\Delta}{k_{\mathrm{B}}T}} \ll \rho_{xy} \approx ah/2\nu e \; \{\nu \geq 1\} \tag{9.10}$$

$$\rho_{xx} \approx (1/\sigma_0)\mathrm{e}^{+\frac{\Delta}{k_{\mathrm{B}}T}} \gg ah/2e^2 \gg \rho_{xy} \approx 0 \quad \{\nu = 0\}. \tag{9.11}$$

Samples of $(\mathrm{Per})_2\mathrm{Pt}(\mathrm{mnt})_2$ have the ideal topology ($1{,}000 \times 50 \times 25\,\mu\mathrm{m}$) for deriving reliable estimates of ρ_{xx} from the R_{xx} data of [86], yielding $130 \lesssim \rho_{xx} \lesssim 400{,}000\,\mu\Omega\,\mathrm{cm}$, greatly exceeding the maximum metallic resistivity $ah/2e^2 \approx 26\,\mu\Omega\,\mathrm{cm}$ throughout all of the proposed FICDW phases by as much as a factor of 20,000. The data of [86] are therefore consistent with $\nu = 0$ throughout. For the quantized nesting model to apply, each FICDW state would have different values of ν, only one of which can be 0.

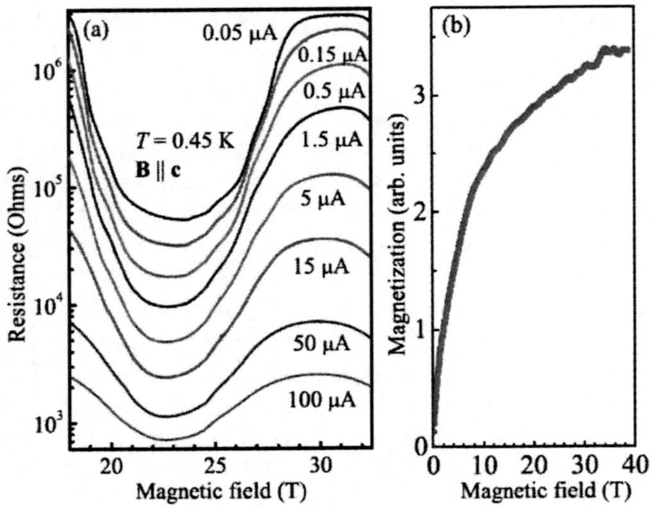

Fig. 9.12. (**a**) Resistance vs. field for $(\text{Per})_2\text{Pt}(\text{mnt})_2$ at various currents measured by the present authors. (**b**) Magnetization M of many randomly oriented $(\text{Per})_2\text{Pt}(\text{mnt})_2$ crystals at $T = 0.50\,\text{K}$; M does not saturate by $B \approx 40\,\text{T}$ [89]

Varying the current, I, causes R_{xx} and thus ρ_{xx} vary considerably, reaching values in Fig. 9.1a that exceed $ah/2e^2$ by a factor of $\approx 10^7$. Its strong dependence on the current density j is consistent with a sliding collective mode contribution to σ_{xx} (for all fields $B < 33\,\text{T}$ and $\nu = 0$), yielding

$$\rho_{xx} \approx [\sigma_0 \exp -(\Delta/k_\text{B}T) + j/E_\text{t}]^{-1}, \tag{9.12}$$

where E_t is a threshold CDW depinning electric field [89], which may itself depend on T. Note that the near-linear I vs. voltage V-plots for $IV < 2\,\mu\text{W}$ (for different values of B, T, and I in Fig. 2 of [89]) suggests that heating is not significant factor for $I < 5\,\mu\text{A}$ in Fig. 9.12a.

Hence, the behavior of the fully gapped $(\text{Per})_2M(\text{mnt})_2$ system does not fit the usual definition of "magnetoresistance" but is the consequence of magnetic field-induced changes in the electric field E_t required to depin the CDW from the lattice, where $\nu = 0$ throughout. Such behavior is inconsistent with the quantized nesting model, which requires different values of ν for each subphase [88]; thus the steps in ρ_{xx} probably correspond to field-induced changes in E_t.

The cooperative dimerization of the Pt spins in $(\text{Per})_2\text{Pt}(\text{mnt})_2$ can easily provide a mechanism for additional phase transitions or changes in E_t compared to $(\text{Per})_2\text{Au}(\text{mnt})_2$. The Pt spins couple strongly to both the CDW, via distortions of the crystal lattice, and B, as shown by the fact that they dominate the total magnetic susceptibility (Fig. 9.12b). Their effect on the phase diagram is likely to be significant until all spins are fully aligned by a field $B \gtrsim 40$ T (Fig. 9.1b).

9.6 New Quantum Fluid in Strong Magnetic Fields with Orbital Flux Quantization

α-(BEDT-TTF)$_2$KHg(SCN)$_4$ is undoubtedly one of the most intriguing of BEDT-TTF-based charge-transfer salts [77–79]. Like many other such materials, it possesses both two-dimensional (2D) and one-dimensional (1D) Fermi-surface (FS) sections. However, the 1D sheets are unstable at low temperatures, causing a structural phase transformation below $T_P = 8$ K into a CDW state [87]. Imperfect nesting combined with the continued existence of the 2D hole FS pocket gives rise to complicated magnetoresistance and unusual quantum-oscillation spectra at low magnetic fields and low temperatures [14]. At high fields, the CDW undergoes a number of transformations into new phases, many of which have been suggested to be field-induced CDW phases [78].

Undoubtedly the most exotic aspect of this material is its transformation into an unusual CDW state above a characteristic field $B_k = 23$ T (known as the "kink" transition); B_k is now known to correspond to the CDW Pauli paramagnetic limit [14, 79], like (Per)$_2M$(mnt)$_2$. Such a regime is reached in α-(BEDT-TTF)$_2$KHg(SCN)$_4$ owing to the unusually low value of T_P [79]. At fields higher than B_k, Zeeman splitting of the energy bands makes a conventional CDW ground state energetically unfavourable [90], possibly yielding a novel modulated CDW state like that proposed for (Per)$_2$Au(mnt)$_2$ (see Sect. 2 and [82, 83]). In α-(BEDT-TTF)$_2$KHg(SCN)$_4$ this state is especially unusual due to the existence of the 2D pocket, which appears to be ungapped by CDW formation [79]. The CDW and 2D hole pocket screen each other, with pinning of the CDW then enabling a nonequilibrium distribution of orbital magnetization throughout the bulk [79]. Consequently, such a state exhibits a critical state analogous to that in type II superconductors. Some of us have discussed this in more detail in another Chapter of this book.

9.7 Summary

We have tried to illustrate why some members of the high-field community are fond of charge-transfer salts. This enthusiasm is likely to continue for some time, as the charge-transfer salts offer great versatility as a plaything for studying the formation of bandstructure. Using the known self-organizational properties of small organic molecules, one can really indulge in "molecular architecture," in which the structure of a charge-transfer salt is adjusted to optimize a desired property [25]. The most imaginative essays in this field involve the use of molecules that introduce a further property that modifies the electronic behavior, such as chirality or the presence of magnetic ions [91]; in the latter case, one of the aims is to manufacture an organic Kondo system. Some recent experimental data from three such compounds are presented in Fig. 9.13; the salts in question have the generic formula

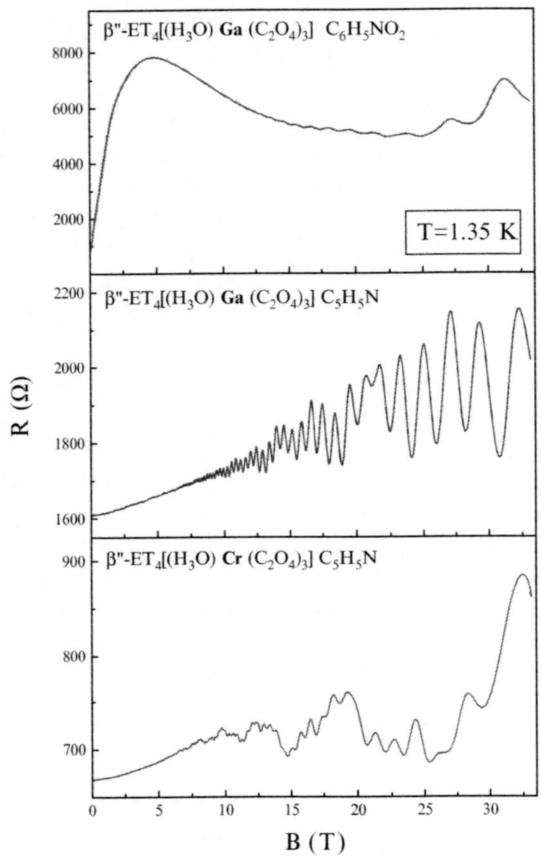

Fig. 9.13. Interlayer resistance of three different charge-transfer salts of the form β''-(BEDT-TTF)$_2$[H$_3$OM(C$_2$O$_4$)$_3$] (Sol) (ET is an abbreviation of BEDT-TTF). The magnetic field is perpendicular to the highly conducting planes (after [11, 19])

β''-(BEDT-TTF)$_2$[H$_3$OM(C$_2$O$_4$)$_3$] (Sol), where M is a transition ion and Sol is an incorporated solvent molecule. The M ion allows one to introduce magnetic moments in a controllable way, whereas changing the solvent molecule allows fine details of the unit cell structure to be altered [91]. As Fig. 9.13 shows, such adjustments cause very distinct changes to the Fermi-surface topology, reflected in the magnetic quantum oscillation spectra [11, 19]. Some of these salts appear to be superconductors.

There are also many reasons for continuing to study charge-transfer salts at high magnetic fields. A particular goal is the ultraquantum limit, in which only one quantized Landau-level is occupied; phenomena such as yet more varieties of field-induced superconductivity have been predicted to occur once such a condition is attained [3, 14, 19]. Furthermore, there are many open questions as to the role of chiral Fermi liquids in such fields [14]. Another area of considerable interest is the observation of magnetic breakdown. At fields above 50 T, the magnetic energy of the holes in the organic superconductors is starting to become a substantial fraction of their total energy, and one gradually starts to approach the famous Hofstadter "butterfly" limit [14].

Acknowledgement

This work is funded by US Department of Energy (DoE) grant LDRD 20040326ER. Work at NHMFL is performed under the auspices of the National Science Foundation, DoE, and the State of Florida.

References

1. B.J. Powell, R.H. McKenzie, J. Phys. Condens. Matter **16**, L367 (2004)
2. M.T. Dressel, N. Drichko, Chem. Rev. **104**, 5689 (2004)
3. J. Singleton, C.H. Mielke, Contemp. Phys. **43**, 63 (2002)
4. B.J. Powell, R.H. McKenzie, Phys. Rev. Lett. **94**, 047004 (2005)
5. B.J. Powell, R.H. McKenzie, Phys. Rev. B **69**, 024519 (2004)
6. F. Kagawa, T. Itou, K. Miyagawa, K. Kanoda, Phys. Rev. Lett. **93**, 127001 (2004)
7. S. Lefebvre, P. Wzietek1, S. Brown, C. Bourbonnais, D. Jrome, C. Mzire, M. Fourmigu, P. Batail, Phys. Rev. Lett. **85**, 5420 (2000)
8. R.D. McDonald, A.K. Klehe, J. Singleton, W. Hayes, J. Phys. Condens. Matter **15**, 5315 (2003)
9. P.A. Goddard, S.J. Blundell, J. Singleton, R.D. McDonald, A. Ardavan, A. Narduzzo, J.A. Schlueter, A.M. Kini, T. Sasaki, Phys. Rev. B **69**, 174509 (2004)
10. J. Singleton, C.H. Mielke, W. Hayes, J.A. Schlueter, J. Phys. Condens. Matter **15**, L203 (2003)
11. A.I. Coldea, A.F. Bangura, J. Singleton, A. Ardavan, A. Akutsu-Sato, H. Akutsu, S.S. Turner, S.S.P. Day, Phys. Rev. B **69**, 085112 (2004)
12. J. Merino, R.H. McKenzie, Phys. Rev. Lett. **87**, 237002 (2001)
13. P.A. Goddard, S.W. Tozer, J. Singleton, A. Ardavan, A. Abate, M. Kurmoo, J. Phys. Condens. Matter **14**, 7345 (2002)
14. J. Singleton, Rep. Prog. Phys. **63**, 1111 (2000)
15. S. Uji, H. Shinagawa, T. Terashima, T. Yakabe, Y. Terai, M. Tokumoto, A. Kobayashi, H. Tanaka, H. Kobayashi, Nature **410**, 908 (2001)
16. J. Singleton, J.A. Symington, M.S. Nam, A. Ardavan, M. Kurmoo, P. Day, J. Phys. Condens. Matter **12**, L641 (2000)
17. V.F. Mitrovic, M. Horvatic, C. Bethier, G. Knebel, G. Lapertot, J. Flouquet, Phys. Rev. Lett., (in press)
18. M.A. Tanatar, T. Ishiguro, H. Tanaka, H. Kobayashi, Phys. Rev. B **66**, 134503 (2002)
19. A.F. Bangura, A.I. Coldea, J. Singleton, A. Ardavan, A. Akutsu-Sato, T. Akutsu, S.S. Turner, P. Day, T. Yamamoto, K. Yakushi, Phys. Rev. B **72**, 014543 (2005)
20. J. Schmalian, Phys. Rev. Lett. **81**, 4232 (1998)

21. J.M. Caulfield, W. Lubczynski, F.L. Pratt, J. Singleton, D.Y.K. Ko, W. Hayes, M. Kurmoo, P. Day, J. Phys. Condens. Matter **6**, 2911 (1994)
22. J. Singleton, P.A. Goddard, A. Ardavan, N. Harrison, S.J. Blundell, J.A. Schlueter, A.M. Kini, Phys. Rev. Lett. **88**, 037001 (2002)
23. J. Singleton, *Band Theory and Electronic Properties of Solids*, (Oxford University Press, Oxford, 2001)
24. N.W. Ashcroft, N.D. Mermin, *Solid State Physics*, (Saunders, Philadelphia, 1976)
25. T. Mori, Bull. Chem. Soc. Jpn. **71**, 2509 (1998); T. Mori, H. Mori, S. Tanaka, Bull. Chem. Soc. Jpn. **72**, 179 (1999); T. Mori, Bull. Chem. Soc. Jpn. **72**, 2011 (1999)
26. T. Ishiguo, K. Yamaji, G. Saito, *Organic Superconductors*, (Springer, Berlin Heidelberg New York, 1998)
27. N. Harrison, J. Caulfield, J. Singleton, P.H.P. Reinders, F. Herlach, W. Hayes, M. Kurmoo, P. Day, J. Phys. Condens. Matter **8**, 5415 (1996)
28. M.V. Kartsovnik, V.N. Laukhin, S.I. Pesotskii, I.F. Schegolev, V.M. Yakovenko, J. Phys. I (Paris) **2**, 89 (1990)
29. J. Wosnitza, *Fermi Surfaces of Low-Dimensional Organic Metals and Superconductors*, (Springer, Berlin Heidelberg New York, 1996)
30. N. Harrison, E. Rzepniewski, J. Singleton, P.J. Gee, M.M. Honold, P. Day, M. Kurmoo, J. Phys. Condens. Matter **11**, 7227 (1999)
31. D. Beckmann, S. Wanka, J. Wosnitza, J.A. Schlueter, J.M. Williams, P.G. Nixon, R.W. Winter, G.L. Gard, J. Ren, M.H. Whangbo, Eur. Phys. J. B **1**, 295 (1998); J. Wosnitza, S. Wanka, J.S. Qualls, J.S. Brooks, C.H. Mielke, N. Harrison, J.A. Schleuter, J.M. Williams, P.G. Nixon, R.W. Winter, G.L. Gard, Synth. Met. **103**, 2000 (1999); F. Zuo, X. Su, P. Zhang, J.S. Brooks, J. Wosnitza, J.A. Schlueter, J.M. Williams, P.G. Nixon, R.W. Winter, G.L. Gard, Phys. Rev. B **60**, 6296 (1999); J. Wosnitza, Physica B **246–247**, 104 (1998)
32. M. Doporto, J. Singleton, F.L. Pratt, J. Caulfield, W. Hayes, J.A.A.J. Perenboom, I. Deckers, G. Pitsi, M. Kurmoo, P. Day, Phys. Rev. B **49**, 3934 (1994)
33. A.A. House, N. Harrison, S.J. Blundell, I. Deckers, J. Singleton, F. Herlach, W. Hayes, J.A.A.J. Perenboom, M. Kurmoo, P. Day, Phys. Rev. B **53**, 9127 (1996)
34. C.H. Mielke, J. Singleton, M.-S. Nam, N. Harrison, C.C. Agosta, B. Fravel, L.K. Montgomery, J. Phys. Condens. Matter **13**, 8325 (2001)
35. E. Demiralp, W.A. Goddard III, Phys. Rev. B **56**, 11907 (1997)
36. J. Liu, J. Schmalian, N. Trivedi, Phys. Rev. Lett. **94**, 127003 (2005)
37. H.O. Jeschke, G. Kotliar, Phys. Rev. B. **71**, 085103 (2005)
38. D.G. Clarke, S.P. Strong, Adv. Phys. **46**, 545 (1997); J. Phys. Condens. Matter **8**, 5415 (1996)
39. L.B. Ioffe, A.J. Millis, Science **285**, 1241 (2000)
40. C. Bergemann, S.R. Julian, A.P. Mackenzie, S. NishiZaki, Y. Maeno, Phys. Rev. Lett. **84**, 2662 (2000)
41. C.N.R. Rao, J. Mater. Chem. **9**, 1 (1999)
42. R.H. McKenzie, P. Moses, Phys. Rev. Lett. **81**, 4492 (1998); Phys. Rev. B **60**, 11241 (1999)
43. This is roughly equivalent to the Mott-Ioffe-Regel criterion [44]; see e.g. [42] or Section 7.2 of [38]

44. N.F. Mott, E.H. Davies, *Electronic Properties of Non-Crystalline Materials*, (Taylor and Francis, London, 1975); A.F. Ioffe, A.R. Regel, Prog. Semicond. **4**, 237 (1960)
45. P.W. Anderson, *The Theory of Superconductivity in the High T_c Cuprates*, (Princeton University Press, Princeton, 1997), p. 50
46. K. Kuroki, H. Aoki, Phys. Rev. B **60**, 3060 (1999)
47. R. Louati, S. Charfi-Kaddour, A. Ben Ali, R. Bennaceau, M. Heritier, Synth. Met. **103**, 1857 (1999)
48. M.S. Nam, S.J. Blundell, A. Ardavan, J.A. Symington, J. Singleton, J. Phys. Condens. Matter **13**, 2271 (2001)
49. J.M. Schrama, J. Singleton, R.S. Edwards, A. Ardavan, E. Rzepniewski, R. Harris, P. Goy, M. Gross, J. Schlueter, M. Kurmoo, P. Day, J. Phys. Condens. Matter **13**, 2235 (2001)
50. J. Singleton, R.S. Edwards in *High Magnetic Fields, Science and Technology*, ed. by F. Herlach, N. Miura. Theory and Experiment, vol 2 (World Scientific, Singapore, 2003), p. 85
51. P.A. Goddard, S.J. Blundell, J. Singleton, R.D. McDonald, A. Ardavan, A. Narduzzo, J.A. Schlueter, A.M. Kini, T. Sasaki, Phys. Rev. B **69**, 174509 (2004)
52. J. Singleton, P.A. Goddard, A. Ardavan, S.J. Blundell, A. Coldea, J. Schlueter, Phys. Rev. Lett. submitted, `arXiv:cond-mat/0610318`
53. The existence of such a modulation might suggest that a beating between "neck and belly" frequencies would be observed in the dHvA effect [14]; however, in careful low-field studies [54], no such beating has been observed. We shall see that this is because the typical cyclotron energy is rather greater than $t_\perp$, even at the lowest fields used.
54. T. Sasaki, W. Biberacher, K. Neumaier, W. Hehn, K. Andres, T. Fukase, Phys. Rev. B **57**, 10889 (1998)
55. When viewed in the extended zone scheme, the quasi-two-dimensional section of this type of Fermi surface (blue in Fig. 9.3(c)) is often known as a "coke-bottle" Fermi surface (USA) or a "Cumberland sausage" Fermi surface (Europe). (The latter is a hand-made sausage with weak but approximately regular corrugations.)
56. This simple discussion deals with an interlayertransfer integral is directed along the normal to the layers. However, most real charge-transfer salts have rather lower symmetry, and it is imperative that the correct direction of the interlayer transfer integral be taken into account; see Ref. [51]
57. T. Osada, S. Kagoshima, N. Miura, Phys. Rev. Lett. **77**, 5261 (1996); N. Hanasaki, S. Kagoshima, T. Hasegawa, T. Osada, N. Miura, Phys. Rev. B **57**, 1336 (1998); *ibid.* **60**, 11210 (1999)
58. V.G. Peschansky, M.V. Kartsovnik, Phys. Rev. B **60**, 11207 (1999); I.J. Lee, M.J. Naughton, Phys. Rev. B **57**, 7423 (1998)
59. E. Ohmichi, H. Ito, T. Ishiguro, T. Komatsu, G. Saito, J. Phys. Soc. Jpn. **66**, 310 (1997)
60. V.M. Gvozdikov, J. Wosnitza, Low Temp. Phys. **32**, 109 (2006)
61. M.S. Nam, S.J. Blundell, A.Ardavan, J. Symington, J. Singleton, J. Phys. Condens. Matter **13**, 2271 (2001)
62. N.E. Hussey, J. Phys. Chem. Solids **67**, 227 (2006)
63. S.J. Blundell, J. Singleton, J. Phys. I **6**, 1837 (1996)

64. A. Bangura, A. Ardavan, S.J. Blundell, J. Singleton, P.A. Goddard, J. Schlueter, Phys. Rev. Lett., submitted (2006).
65. N. Harrison, C.H. Mielke, J. Singleton, J.S. Brooks, M. Tokumoto, J. Phys. Condens. Matter **13**, L389 (2001)
66. C.C. Agosta, T. Coffey, Z. Bayindir, I. Mihut, C. Martin, M. Tokumoto, Int. J. Mod. Phys. B **30**, 3227
67. A.E. Kovalev, T. Ishiguro, T. Kondo, G. Saito, Phys. Rev. B **62**, 103 (2000)
68. D. Shoenberg, *Magnetic Oscillations in Metals*, (Cambridge University Press, Cambridge, 1984)
69. N. Harrison, R. Bogaerts, P. Reinders, J. Singleton, S.J. Blundell, F. Herlach, Phys. Rev. B **54**, 9977 (1996)
70. S. Hill, Phys. Rev. B **55**, 4931 (1997); ibid. **62**, 8699 (2000)
71. N. Harrison, J. Singleton, J. Phys. Condens. Matter **13**, L463 (2001)
72. J. Singleton, in *Encyclopedia of Condensed Matter Physics*, vol. 1, ed. by F. Bassani, G.L. Liedl, P. Wyder, (Elsevier, Oxford, 2005), p. 343
73. J. Singleton, C.H. Mielke, W. Hayes, J.A. Schlueter, J. Phys. Condens. Matter **15**(12), L203 (2003)
74. I. Mihut, C.C. Agosta, C. Martin, C.H. Mielke, T. Coffey, M. Tokumoto, M. Kurmoo, J.A. Schlueter, P.A. Goddard, N. Harrison, Phys. Rev. B **73**, 125118 (2006)
75. M. Tinkham, *Introduction to Superconductivity* (McGraw-Hill, New York, 1994)
76. G. Grüner, *Density Waves in Solids, Frontiers in Physics 89*, (Addison-Wesley, Reading, 1996)
77. N. Harrison, L. Balicas, J.S. Brooks, M. Tokumoto, Phys. Rev. B **62**, 14212 (2000)
78. D. Andres, M.V. Kartsovnik, P.D. Grigoriev, W. Biberacher, H. Muller, Phys. Rev. B **68**, 201101 (2003)
79. N. Harrison, J. Singleton, A. Bangura, A. Ardavan, P.A. Goddard, R.D. McDonald, L.K. Montgomery, Phys. Rev. B **69**, 165103 (2004)
80. M. Matos, G. Bonfait, R.T. Henriques, M. Almeida, Phys. Rev. B **54**, 15307 (1996)
81. D. Graf, J.S. Brooks, E.S. Choi, S. Uji, J.C. Dias, M. Almeida, M. Matos, Phys. Rev. B **69**, 125113 (2004)
82. R. McDonald, N. Harrison, J. Singleton, A. Bangura, P.A. Goddard, A.P. Ramirez, X. Chi, Phys. Rev. Lett. **94**, 106404 (2005)
83. R. McDonald, N. Harrison, L. Balicas, K.H. Kim, J. Singleton, X. Chi, Phys. Rev. Lett. **93**, 076405 (2004)
84. R.D. McDonald, N. Harrison, P. Goddard, J. Singleton, X. Chi, cond-mat 0408408 (2004)
85. R.T. Henriques, J. Phys. C **17**, 5197 (1984)
86. D. Graf, E.S. Choi, J.S. Brooks, M. Matos, R.T. Henriques, M. Almeida, Phys. Rev. Lett. **93**, 076406 (2004)
87. P.M. Chaikin, J. Phys. I (France) **6**, 1875 (1996)
88. A.G. Lebed, JETP Lett. **78**, 138 (2003)
89. R.D. McDonald, N. Harrison, preprint (2006)
90. R.H. McKenzie, cond-mat/970635
91. P. Day, Phil. Trans. R. Soc. London **357**, 3163 (1999)

10

Energy and Dielectric Relaxations in Bechgaard–Fabre Salts

P. Monceau, J.-C. Lasjaunias, K. Biljaković, and F. Nad

We present an overview of energy and dielectric relaxation measurements on Bechgaard–Fabre salts, which enlarge the already very rich phase diagram. It is shown that a new generic property of the whole family of $(TMTTF)_2X$ Fabre salts is charge ordering, which occurs in the temperature range of 40–200 K. This new electronic charge superstructure at $4k_F$, a generalization of the classical Wigner crystal, manifests itself by a ferroelectric character. Specific heat measurements in different controlled kinetic conditions in the vicinity of the sub-phase at 3.5 K in the spin–density wave state of $(TMTSF)_2PF_6$ show features characteristic of freezing in supercooled liquids, indicating a glass transition. Thermodynamical measurements below 1 K on both $(TMTTF)_2X$ and $(TMTSF)_2X$ salts reveal extremely long dynamics of energy relaxation and nonequilibrium phenomena (aging). For incommensurate Bechgaard salts, the relaxation time distribution is Gaussian while for commensurate ground states in Fabre salts, it shows a bimodal character.

10.1 Introduction

Low-dimensional electronic systems serve as a workshop on problems of strong correlations. The richest opportunities are found in the family of charge transfer salts M_2X formed of stacks of organic molecules tetramethyltetrathiofulvalene (M = TMTTF) or tetramethyltetraselenafulvalene (M = TMTSF) with anions PF_6, AsF_6, SbF_6, SCN, etc. as counterions. The ions occupy loose cavities delimited by the methyl groups of the organic molecules. These materials show almost all known ground states at low temperature: a metal, a paramagnetic insulator, spin/charge density waves (SDW/CDW), a spin-Peierls state, and finally a superconducting state [1]. In parallel, there is a set of several different structural types due to anion ordering (AO), associated with slight arrangements in X chains [2]. Their transition temperatures $T_a \approx$ 100–200 K are much higher than that corresponding to magnetic transitions occurring in the range of $T_c \approx$ 1–20 K. The electron transfer integral along the chain direction for both sulfur and selenium compounds is typically $t_a \approx 3{,}000\,\mathrm{K}$, to be compared with the values in transverse directions: $t_b \approx 300\,\mathrm{K}$ and $t_c \approx 10\,\mathrm{K}$. So, at temperatures higher than t_b, the system is in the confinement limit [3, 4]. A cross-over towards higher dimensionality is expected to occur at T^* when T becomes smaller than t_b.

Structurally the Bechgaard–Fabre salts exhibit a dimerization of the intermolecular distance; consequently, the conduction band is split into a filled lower band separated from a half-filled upper band by a dimerization gap, Δ_ϱ, the 3/4 filling resulting from the 2:1 stoichiometry of $(TMTTF)_2X$ and $(TMTSF)_2X$ salts.

When globally considered, a generic, well accepted earlier, pressure vs. temperature phase diagram accounts for the general properties of $(TMTTF)_2X$ and $(TMTSF)_2X$ salts [5]. The emerging picture from this phase diagram is the existence of a, so called one-dimensional Mott charge localization between room temperature and the magnetic ordering at low T around 1–20 K. This insulating state exists for all the sulphur compounds with a charge gap having an activation energy of the order of Δ_ϱ. Under pressure this correlation gap decreases, the band width broadens, and the sulfur compounds behave similar to the selenium ones. At low T, the increase of pressure induces a cascade of transitions with the ground state changing from spin-Peierls (SP) to antiferromagnetic (AF), then incommensurate SDW, and finally superconductivity [6].

However, this unified phase diagram does not include all the Bechgaard–Fabre salts. It neglects the anion ordering inducing subtle changes in the crystal symmetry, which modify strongly the electronic properties [7], and above all, it does not take into account the newly discovered phase transition into a charge ordered (CO) state [8–10] generic for all the $(TMTTF)_2X$ compounds [11]. This phase transition is a 3D transition that occurs in the temperature range where the insulating state was described earlier as a Mott–Hubbard insulating state. Such a charge ordered state, or a $4k_F$ charge density wave, is essentially the generalization of the classical Wigner lattice. At ambient pressure, the ground state of $(TMTTF)_2X$ is formed of a combination of the CO and SP or AF order parameters with competitive interaction [12].

For $(TMTSF)_2PF_6$ and $(TMTSF)_2AsF_6$ Bechgaard salts, the ground state is a spin density wave. The magnitude of the SDW transition anomaly in the specific heat is too large to be explained by the electron spin contribution alone, which implies that the lattice is also involved in the transition [13]. Recently, X-ray diffuse scattering experiments have revealed the coexistence of a $2k_F$ and $4k_F$ charge density wave in the SDW state of $(TMTSF)_2PF_6$ in a narrow temperature (12–10.7 K) range below the SDW transition temperature at $T_{SDW} = 12\,\mathrm{K}$ [14]. These CDWs were considered to be purely electronic without involving lattice displacements. These experiments have been extended down to 2 K and performed also on $(TMTSF)_2AsF_6$, $(TMTSF)_2Br$, and $(TMTSF)_2ClO_4$ [15]. The coexistence in the ground state of $(TMTSF)_2PF_6$ of a $2k_F$ CDW and of a $2k_F$ SDW modulation has been explained theoretically [16, 17] in the framework of extended 1D Hubbard models coupled to the lattice or in terms of the next-nearest-neighbor Coulomb interactions between electrons causing the $4k_F$ CDW. The amplitudes of the $2k_F$ and $4k_F$ CDWs decrease with decreasing temperature below 3–4 K. It is in this temperature range where specific heat measurements in different controlled kinetic conditions in $(TMTSF)_2PF_6$ exhibit a jump showing all the characteristics of freezing in supercooled liquids, indicating a glass transition in the SDW ground state [18]. This discontinuity in the lattice contribution as well as another around 2 K occur at the same temperatures where nuclear spin relaxation shows an anomalous behavior, ascribed to sub-SDW phases [19].

Resulting from interaction with randomly distributed impurities, the ground state of Bechgaard–Fabre salts exists with many metastable states. The disordered nature of the ground state is the best manifested by thermodynamical measurements at very low temperature. Additional low energy excitations (LEE) to regular phonons contribute below 1 K to the specific heat, C_P. In the same temperature range, the nonexponential enthalpy relaxation with "aging" effects prove the nonlinearity of the specific heat and reveal the broad distribution of relaxation times for recovering the thermodynamical equilibrium [20]. These features are reminiscent of the low-T thermodynamical properties of structural glasses or orientationally disordered crystals, i.e., "orientational glasses." In the case of a commensurate ground state as in $(TMTTF)_2PF_6$ and $(TMTTF)_2Br$, the relaxation time distribution shows a bimodal character [21].

The organization of the present review is as follows: Sect. 10.2 is devoted to the estimation of Coulomb interactions in Bechgaard–Fabre salts. Charge ordering with the associated ferroelectric transition is discussed in Sect. 10.3. Comparison of the low temperature specific heat of $(TMTSF)_2X$ and $(TMTTF)_2X$ salts in the temperature range 40–0.08 K is presented in Sect. 10.4.

10.2 Coulomb Interactions

As known in the extensive theoretical and experimental works since the last three decades, the ground state realized in quasi-1D systems results from the complexity of their crystal structure and from many types of interactions inherent to these crytals: electron–electron, electron–phonon interactions, magnetic interactions, and for molecular crystals interplay between intrastack and interstack, interaction of molecular chains with anion chains, etc. For the $(TMTTF)_2X$ salts, electron–electron correlation phenomena play the leading role.

While the Bechgaard salts $(TMTSF)_2X$ display a metallic behavior down to low temperature where a transition in a SDW state occurs below ~12 K, $(TMTTF)_2X$ salts exhibit a charge localization in the temperature range 100–200 K, with a maximum in the conductivity at T_ϱ and a thermally activated variation below T_ϱ, revealing strong Coulomb interaction effects in these sulfur salts [22].

Quasi-1D $(TMTTF)_2X$ conductors consist of molecular chains along the highest conductivity axis with two electrons per four molecules, which corresponds to 1/4-filling in terms of holes. These molecular chains are slighly dimerized due to intermolecular interaction. As a result, with decreasing temperature a dimerized gap Δ_ϱ opens with an effective 1/2-filling of the upper conduction band. Two intermolecular transfer integrals along molecular stacks, t_1 and t_2, have to be considered [23]. Quantum chemistry calculations show that in $(TMTTF)_2PF_6$ the dimerization decreases with decreasing temperature and that at low T, the t_1/t_2 ratio is about 1.1–1.2 [23].

In this context two types of theoretical models have been essentially developed. In the frame of the so-called g-ology models [24–26], the electron–electron correlations are considered as a perturbation to the one-electron approach. These models have been used for describing the low-energy properties of these salts that exhibit the features of a Luttinger liquid rather than those of a Fermi liquid [3, 4].

The second group of models includes the various versions of the Hubbard model [27–34]. The extended Hubbard model takes into account the interaction between charge carriers on the site of the host lattice (on-site interaction) with the characteristic energy U as well as the interaction between charge carriers on the neighboring sites (near-neighbor interaction) with the characteristic energy V. In the case of $(TMTTF)_2X$ compounds, at temperatures above the transition in spin ordered states, the electron–electron interaction is determined by long-range Coulomb interaction and it is stronger than spin interaction. This is one of the reasons of spin–charge separation observed in such 1D conductors [1]. In the frame of the extended Hubbard model, it was shown that the dimerized energy gap is strengthened taking into account on-site and near-neighbor interactions [27]. At the same time, the spectrum of spin excitations remains gapless, which also corresponds to spin–charge separation [4].

One important result of the extended Hubbard model approach concerns the formation of a $4k_F$ CDW superstructure of Wigner crystal type in such 1D compounds with decreasing temperature. Using Monte Carlo techniques, it was shown that, for large enough U and V magnitudes, strictly one-site interaction results only in a weak $4k_F$ CDW, while long-range near-neighbor Coulomb interaction can produce a marked singularity at $4k_F$ [27]. Analogously, using mean-field approximation [30, 33], it was recently shown that for a one-dimensional molecular chain with or without dimerization, the form of the developed superstructure depends considerably on the magnitude of the near-neighbor interaction V; at V above some critical value V_c a $4k_F$ CDW superstructure occurs with charge disproportionation depending on V. Estimation of V/U and V/t magnitudes in $(TMTTF)_2PF_6$, obtained from quantum chemistry calculations and from optical conductivity [23, 34], yields values for V/U in the range 0.4–0.5 and for V/t in the range of 2–3, manifesting the essential role played by the long-range Coulomb interaction in this compound.

10.3 Charge Ordering and Ferroelectric Transition

The first direct experimental evidence for the existence of a charge-ordered (CO) state as theoretically predicted [30] was obtained by means of NMR studies in the quasi-one-dimensional $(\text{DI-DCNDI})_2\text{Ag}$ (in short, DI-Ag) compound (where DI-DCNQI is 2,5-diiodo-N,N′-dicyanoquinonediimine) [35]. The planar DCNQI molecules are uniformly stacked along one-dimensional columns without visible dimerization at room temperature, which corresponds to a 1/4-filled band. The particular aspect of the structure recently reported [36] is that the interchain configuration has a spiral symmetry with a translation of seven molecules along the chain axis by one turn. This specific odd spiral structure is not compatible with the doubling of the unit cell in the CO state, which may induce frustration in the charge ordering [37] (see below). It was shown that, with decreasing temperature below 220 K, the ^{13}C-NMR spectra are splitted, pointing out the appearance of nonequivalent differently charged molecules along the chain axis [35]. This charge disproportionation saturates to ∼3:1 at temperatures below 130 K. These results were confirmed by X-ray studies. X-ray diffraction patterns have revealed at 30 K the existence of $4k_\text{F}$ satellite reflections, which corresponds well to the CO detected by NMR [38]. Charge distribution has also been investigated by Raman vibrational spectroscopy [39].

It is known that the study of the dielectric permittivity is one of the most direct techniques for identifying a phase transition and the frequency dependence of relaxation. We then have measured the complex conductance $G(T,\omega)$ of $(\text{TMTSF})_2\text{PF}_6$ and $(\text{TMTTF})_2\text{X}$ salts in a wide temperature range between 2 and 300 K and for frequencies between 10^2 and 10^7 Hz. The magnitude of the dielectric constant was calculated by the standard equation $\varepsilon' = \text{Im}G/\omega$. All the data were measured with a small ac amplitude within the linear response.

10.3.1 Low Frequency Permittivity in the SDW State of $(\text{TMTSF})_2\text{PF}_6$ at Low Temperature

Below the SDW phase transition and the formation of the SDW energy gap, the electrical conductivity $\sigma(T)$ of $(\text{TMTSF})_2\text{PF}_6$ is governed by single electrons thermoactivated across the gap, at least in the temperature range $T_\text{SDW} > T > T_\text{SDW}/2$. The concentration of those electrons is enough to screen the various SDW inhomogeneous excitations (mainly phase SDW excitations). At $T_\text{SDW}/2 \geq T$, the reduced electron concentration is approximately the same as the defect concentration (mainly dislocations and solitons). Then, at low T, the conductivity is essentially determined by these SDW collective excitations. Below 2 K, $\sigma(T)$ deviates from a simple Arrhenius form to a hopping type conductivity [40]. The temperature dependence of the real part of the dielectric constant ε', as shown in Fig. 10.1, exhibits a maximum which shifts to lower temperatures with decreasing frequency [40]. The interpretation of the low frequency dielectric permittivity in $(\text{TMTSF})_2\text{PF}_6$ follows step by

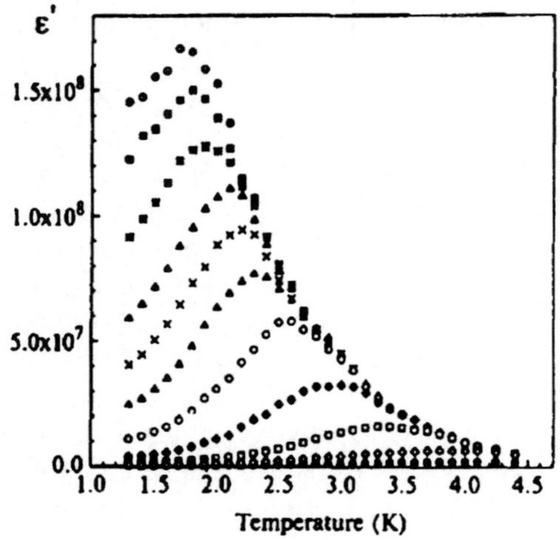

Fig. 10.1. Temperature dependence of the real part ε' of the dielectric permittivity $(TMTSF)_2PF_6$ at fixed frequencies from the top to the bottom at 0.111, 0.5, 1.1, 2, 3, 5, 11, 30, 100, 300, and 1000 (in kHz) (from [40])

step that proposed for CDWs [41]. In the vicinity of T_{SDW} down to $T_{SDW}/2$, the interaction of combined SDW-CDW with impurities has essentially a weak pinning collective character. At lower T, because the reduction of the screening and the resulting hardening of the superstructure, the pinning becomes essentially local induced by strong pinning centers [42,43]. Local deformations in the superstructure at these pinning centers are nucleated and dominate the kinetic properties of the system. The dielectric permittivity results from the summation of polarization effects and dipole interactions between these randomly distributed solitons. The main cause of the ε' growth is due to the increase of the superstructure rigidity with decreasing screening of its defects by free carriers. The more rigid SDW tries to be more homogeneous. As a result, the SDW coherence length and the dielectric constant will increase (till a huge value of 10^7–10^8) [40]. However, at lower T, dynamic retarding effects occur and the SDW has not enough time to respond to the ac perturbation, and its response therefore (i.e., the ε' magnitude) decreases.

Analysis of the temperature dependence of the $\varepsilon'(\omega)$ and $\varepsilon''(\omega)$ curves shows that the relaxation time has two branches: with decreasing temperature the long-time branch diverges near some temperature (α-type relaxation), while the short-time branch increases monotonically (β-type relaxation). These results provide evidence for a transition of the SDW state of $(TMTSF)_2PF_6$ into a glassy-like state at low temperature below around 2 K [18]. The time dependent specific heat that is observed below 1 K [20] has been ascribed to the relaxation between metastable states that originate at strong pinning centers [44].

10.3.2 First Experiments on $(TMTTF)_2X$ Indicating Charge Ordering

Diffuse X-ray scattering on $(TMTTF)_2PF_6$ have also revealed a $2k_F$ displacive lattice instability of spin-Peierls type. These fluctuations develop below 100 K and condense at the SP phase transition $T_{SP} = 19\,K$ [2, 45] into satellite reflections, recently observed by neutron scattering [46]. In the case of $(TMTTF)_2Br$, 1D-SP fluctuations developed below 70 K were seen to vanish at the AF transition at $T_N = 13\,K$, indicating the competition between the AF and the SP orders [45].

Thus after our measurements on $(TMTSF)_2PF_6$, it was quite natural for us to study the possible polarization response of these fluctuating superstructures in $(TMTTF)_2PF_6$ and in $(TMTTF)_2Br$. The great surprise we had was the observation of a huge peak (10^5 at 1 MHz) of the real part of the dielectric constant ε' in $(TMTTF)_2Br$ [47] and in $(TMTTF)_2PF_6$ [8] occuring at a temperature T_{CO} below the temperature T_ϱ for charge localization but above the temperature of magnetic ordering. The interpretation of this huge peak we gave was a possible evidence for a charge-induced-correlated state [8], the first step opening the field of charge ordering in $(TMTTF)_2X$ salts.

NMR studies [10] on $(TMTTF)_2PF_6$ and $(TMTTF)_2AsF_6$ have then shown that ^{13}C-NMR lines split at the same temperature at which the dielectric permittivity exhibits divergence. This splitting, proportional to the charge disproportionation, occurs on the initially homogeneous charged chains. An important factor for the stabilization of the CO state can be the interaction of electrons on molecular chains with the periodic anion potential of surrounded anion chains. In this context, it was proposed [48, 49] that the charge disproportionation, which develops as a result of electron–electron interactions, can be stabilized by a shift of the anion chains as a whole (transition with a wave-vector $\boldsymbol{q} = 0$). This shift breaks the symmetry center of the unit cell, which yields in its turn a spontaneous dipole moment associated with the charge imbalance on the two molecules in the unit cell. Consequently, this provides ferroelectric properties characteristic for the CO state [48, 49]. On the base of the extended Hubbard model by method of exact diagonalization of small clusters, it was shown that, indeed, charge ordering can be accompanied by a small uniform displacement of anions from their symmetric positions within the unit cell [50].

Previous reports on $(TMTTF)_2X$ salts had suggested the occurrence of a phase transition in the same temperature range than T_{CO}. It showed itself in changes of the resistance [51] and of the thermopower [52] in weak features in high frequency (~GHz) dielectric permittivity [53]. However, X-ray investigations brought no observation of superlattice reflections and then, they were called "structureless" transitions [45,51]. But none of these reports have either identified the nature of the phase transition, the charge ordering, and its ferroelectric properties, nor measured the charge gap associated to it.

10.3.3 Conductivity

Figure 10.2 shows the variation of the real part of the conductance G, normalized by its room temperature value G_0, as a function of inverse temperature for $(TMTTF)_2X$ samples with centro-symmetric anions (CSA), X = PF_6, AsF_6, and SbF_6. The analogous dependence for non-centrosymmetric anions (NCSA) are shown in Fig. 10.3, X = ReO_4 and SCN.

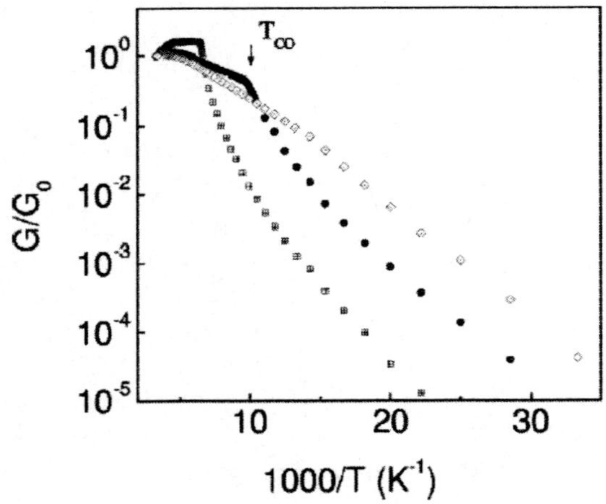

Fig. 10.2. Real part of the conductance G normalized to its room temperature value G_0 as a function of inverse temperature at frequency 1 kHz for $(TMTTF)_2X$ salts with centrosymmetrical anions X = SbF_6 (*squares*), AsF_6 (*circles*), and PF_6 (*diamonds*) (from [11])

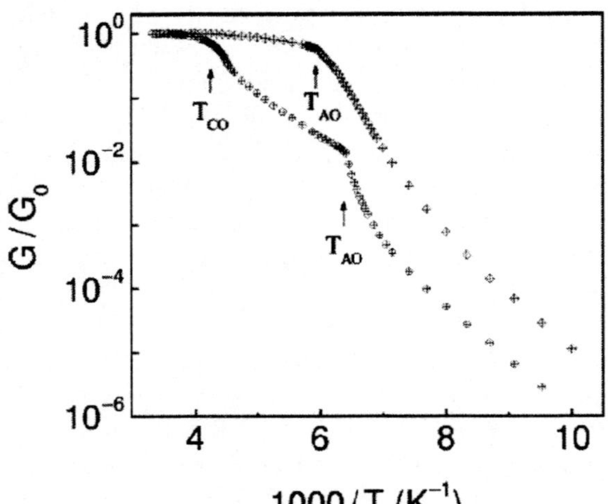

Fig. 10.3. Real part of the conductance G normalized to its room temperature value G_0 as a function of inverse temperature at frequency 1 kHz for $(TMTTF)_2X$ salts with non-centrosymmetrical anions X = ReO_4 (*circles*), SCN (*diamonds*) (from [54])

Table 10.1. Parameters of $(TMTTF)_2X$ conductors

Anion X	T_ϱ (K)	Δ_1 (K)	T_{CO} (K)	Δ_2 (K)	T_{AO} (K)	Δ_3 (K)	Low temp. state
Br	≈200	75	28	–	–	–	AFM
PF_6	250	300	70	370	–	–	SP
AsF_6	230	175	100.6	360	–	–	SP
SbF_6	≈170	–	154	500	–	–	AFM
BF_4	220	580	83	750	39	–	–
ReO_4	290	800	227.5	1,400	154	2,000	–
SCN	265	500	169	–	169	2,000	AFM

For the majority of $(TMTTF)_2X$ the temperature dependence of the conductance has common features: by cooling from room temperature the conductance is growing first, reaches a maximum at some temperature T_ϱ and below that it decreases approximately linearly in the $\log G/G_0(1/T)$ scale (Fig. 10.2). The magnitude of T_ϱ and the slope of the $G/G_0(1/T)$ dependence, i.e., the activation energy Δ_1 in that temperature range, are listed in Table 10.1. One has to note that $(TMTTF)_2SbF_6$ shows a somewhat different behavior: the initial growth of conductance is followed by a wide plateau without any strongly pronounced maximum (Fig. 10.2), or if any in the extreme vicinity of the CO transition.

In all these $(TMTTF)_2X$ compounds a bend is observed below T_ϱ in the $G/G_0(1/T)$ dependencies near some temperature T_{CO}. The position of this bend and the corresponding magnitude of T_{CO} can be determined more exactly in the temperature dependence of the logarithmic derivative $\mathrm{d}\log G/\mathrm{d}(1/T)$. The appropriate values of T_{CO} are shown in Table 10.1. As can be seen from Figs. 10.2 and 10.3, after some transitional temperature interval at $T < T_{CO}$, the $G/G_0(1/T)$ dependence again takes the form of a close linear variation with a new activation energy Δ_2, the magnitude of which is also indicated in Table 10.1.

At lower temperatures the behavior of the CSA conductors differs from that of NCSA ones. In the case of CSA, the thermally activated decrease of conductance continues down to the transition temperature in the magnetic ordered state T_{MO}. This corresponds to a transition to the spin-Peierls state for the PF_6 anion at $T_{SP} = 19\,\mathrm{K}$ [55], and into the antiferromagnetic state at $T_N = 8\,\mathrm{K}$ for the SbF_6 anion [47]. Δ_2 is larger than Δ_1, a clear manifestation of the opening of a new charge gap below T_{CO}. In Br salt, the values of $T_{CO} = 28\,\mathrm{K}$ and $T_N = 15\,\mathrm{K}$ are relatively close to each other. In the temperature range below T_{CO}, a different behavior is observed for $(TMTTF)_2X$ conductors with NCSA tetrahedral anions BF_4 and ReO_4. In these compounds, after some activated variation of $G(T)$ below T_{CO}, an anion ordering occurs at T_{AO} with formation of a superstructure with a wave-vector $\boldsymbol{q} = (1/2, 1/2, 1/2)$ for both salts [2]. In the case of SCN anion, there is no noticeable anomaly in the $G(T)$ dependence between T_ϱ and the anion transition temperature at $T_{AO} = 169\,\mathrm{K}$ with $\boldsymbol{q} = (0, 1/2, 1/2)$ [56]. As can be seen from Fig. 10.3, ReO_4

and SCN show below T_{AO} a thermoactivated decrease of G with a similar large activation energy, $\Delta_3 = 2{,}000\,K$ (Table 10.1) [58]. In the case of BF_4 salt, due to the lower magnitude of $T_{AO} = 39\,K$, the $G(T)$ decrease is close to our experimental resolution, which prevents the evaluation of Δ_3 in this salt [11]. At lower temperature an AF transition occurs in the case of the SCN anion [56].

10.3.4 Dielectric Permittivity

Figure 10.4 shows the temperature dependence of $\varepsilon'(T)$ at frequency 100 kHz for CSA:PF_6 [8,9], AsF_6 [48] and SbF_6 [57], and NCSA:BF_4 [11] and ReO_4 [57] anions. As an be seen the $\varepsilon'(T)$ dependencies do not show any visible anomalies in the temperature range near the maximum of $G(T)$ at $T \approx T_\varrho$. Narrow peaks in the $\varepsilon'(T)$ dependencies are observed at T_{CO}, where an abrupt drop of conductance occurs (Figs. 10.2 and 10.3). In the case of AsF_6, SbF_6, BF_4, and ReO_4, the $\varepsilon'(T)$ magnitude shows a divergence at T_{CO}. Indeed, the maximum of ε' reaches the huge value of the order of 10^6.

The $\varepsilon'(T)$ dependence in the range of small ε' magnitudes is shown in Fig. 10.5 in a semi-logarithmic scale for NCSA $(TMTTF)_2ReO_4$ and $(TMTTF)_2$ SCN.

$(TMTTF)_2$ ReO_4 (as well as $(TMTTF)_2$ BF_4) presents the unique opportunity to study in the same compound the CO phase transition at $T_{CO} = 227.5\,K$ and the anion ordering [2] at T_{AO} with a distortion wave-vector $\boldsymbol{q} = (1/2, 1/2, 1/2)$. The difference between these two transitions is obvious as seen in Fig. 10.5: a divergence of ε' at T_{CO} and a jump-wise drop of ε' at T_{AO}. The case of $(TMTTF)_2$ SCN is more particular [57]. Anion ordering [2] with $\boldsymbol{q} = (0, 1/2, 1/2)$ occurs at $T_{AO} = 169\,K$. The magnitude of ε' shows also

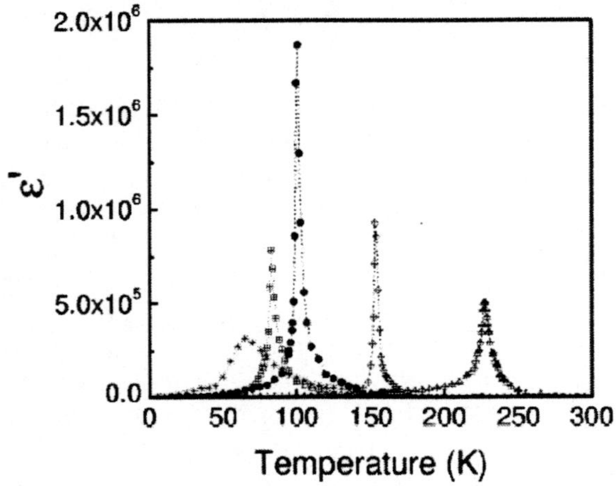

Fig. 10.4. Temperature dependence of the real part of the dielectric permittivity ε' at 100 kHz for $(TMTTF)_2X$ with X = PF_6 (*stars*), BF_4 (*squares*), AsF_6 (*circles*), SbF_6 (*diamonds*), ReO_4 (*triangles*) (from [11])

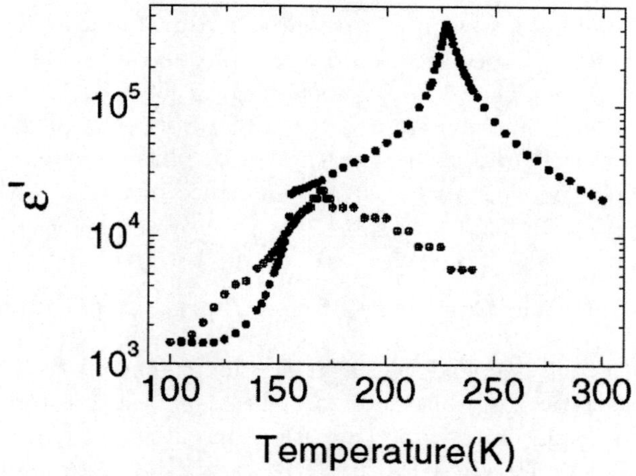

Fig. 10.5. Temperature dependence of the real part of the dielectric permittivity ε' on a semi-logarithmic scale at frequency of 1 MHz: $(TMTTF)_2ReO_4$ (*closed circle*), $(TMTTF)_2SCN$ (*encircled plus*) (from [58])

a maximum at the same temperature. However, the temperature evolution of ε' of $(TMTTF)_2$ SCN is very different from that of other CO transitions in $(TMTTF)_2$ X salts: ε' begins to grow below 240 K, its maximum is only 2×10^4 and the decrease of ε' below T_{AO} occurs more slowly. In fact, this unique transition in $(TMTTF)_2$ SCN has probably the both characteristics: CO and AO. Parameters of $(TMTTF)_2$ X conductors, including T_ϱ, T_{CO}, T_{AO}, and the different gaps are gathered in Table 10.1.

It is well known that application of pressure is a powerful technique for modifying electronic spectra in quasi-1D conductors such as $(TMTTF)_2$ X salts formed from weakly coupled one-dimensional chains. Using NMR spectroscopy, it was shown that T_{CO} in $(TMTTF)_2$ AsF_6 is rapidly suppressed with P above 0.15 GPa, while T_{SP} increases, demonstrating thus the competition between the two condensed phases [12]. In the case of $(TMTTF)_2$ SbF_6, the maximum of $G(T)$ associated with the Mott–Hubbard localization disappears under pressure [54]. With increasing pressure the peak in $\varepsilon'(T)$ corresponding to the CO transition shifts to lower temperature and broadens. The frequency dependence of the dielectric permittivity also changes considerably. For $P > 0.24$ GPa, the $\varepsilon'(T)$ curves become frequency dependent [54]. The major effect of pressure is to increase the band width relative to the near-neighbor Coulomb interaction, V, and thus to destabilize the CO phase. ^{13}C-NMR spectroscopy studies have shown that the temperature dependence of the NMR splitting is not monotonous [59]: it first increases, passes through a maximum before decreasing at lower T. This decreasing coincides with the freezing of counterion motion, detected by ^{19}F-NMR spectrocopy. With pressure, the AF transition is suppressed and is no more detected above 0.5 GPa, but the ^{13}C-NMR spectrum is consistent with a singlet (SP) ground state [59]. Therefore, it was proposed that a strong CO instability as observed in $(TMTTF)_2$ SbF_6 impedes the SP phase (as found in $(TMTTF)_2AsF_6$ and $(TMTTF)_2PF_6$) but yields an AF ground state [59, 60]. The SP instability would be recovered under pressure. $(TMTTF)_2SbF_6$ has then to be located on the extreme low pressure axis in the (P, T) phase diagram [59, 60]. This result is a clear evidence of the strong influence of charge ordering on the phase diagram of $(TMTTF)_2$ X salts.

10.3.5 Ferroelectric Character of the Charge Ordered State

The form of the $\varepsilon'(T)$ dependences near T_{CO} indicates that the CO transitions have a ferroelectric (FE) character and in many aspects these transitions are similar to displacement-type transitions in classical FE [61]. Indeed, the divergence of the dielectric permittivity ε' near T_{CO} corresponds practically exactly to a FE second order transition usually described by a Curie law: $\varepsilon'(T) = A/|T - T_{CO}|$. Figure 10.6 shows the $1/\varepsilon'(T)$ dependence for CSA (X = PF_6, AsF_6, and SbF_6) and NCSA (X = BF_4 and ReO_4). As can be seen, these dependencies are really close to be linear in a relatively wide temperature range near T_{CO}. In the case of the most sharp transitions (AsF_6 and SbF_6), the slope of the $1/\varepsilon'(T)$ branch (i.e., the magnitude of the Curie constant A) at temperatures below T_{CO} is twice that at $T > T_{CO}$, in agreement with the theory of a second order FE transition. In the case of ReO_4 salt this ratio is equal to 1.5, probably because of the influence of the orientational disorder of the semi-symmetric (tetrahedral) anions on the CO transition. For PF_6 anion, the CO transition is much broader, although the qualitative features of the $1/\varepsilon'(T)$ dependence are similar to those in the CSA salts.

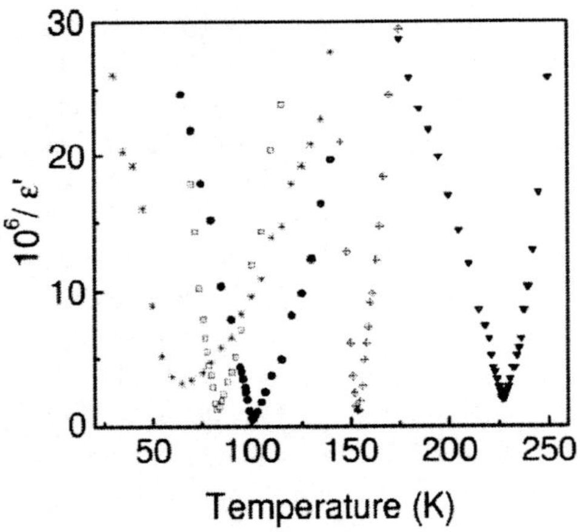

Fig. 10.6. Inverse of the real part of the dielectric permittivity ε' as a function of temperature at frequency 100 kHz for $(TMTTF)_2$ X with X = PF_6 (*stars*), BF_4 (*squares*), AsF_6 (*circles*), SbF_6 (*diamonds*), ReO_4 (*triangles*) (from [11])

Deviation of the $1/\varepsilon'(T)$ dependence from the Curie law and the Landau theory in the extreme vicinity of T_{CO} is observed for all anions under investigation. The origin of such a disagreement between theory and experiment is currently associated with fluctuations and with defects in the crystal lattice. In the present viewpoint about second order phase transitions, the Landau theory does not take into account the strong strength of fluctuations of the order parameter in the immediate proximity to T_{CO}, which can lead to a deviation from the Curie law. Although the samples under investigation are of a high quality, however, they include a lot of impurities, structural defects, and dislocations. More pure the $(\mathrm{TMTTF})_2$ X samples with a more perfect crystal structure, more sharp anomalies are observed in the $G(T)$ dependence and more strong is the ε' divergence near T_{CO}.

Divergence of the relaxation time, τ, as well as slower relaxation processes are typical features of classical FE [61,67]. The divergence of τ was observed in many compounds belonging to classical FE and it is one of the distinctive features of the displacive transition into the FE state. Such a sharp maximum in the temperature range of the relaxation time has been observed from measurements of the frequency dependence of the imaginary part of dielectric losses, i.e., the imaginary part of the dielectric permittivity of $(\mathrm{TMTTF})_2$ AsF_6 [62]. Moreover, in a small temperature range below T_{CO}, the $\varepsilon''(f)$ curves show the development of a low frequency shoulder that involves some slower relaxation processes. In the present case, these processes may correspond to the motion of domain walls within the domain structure developed at $T < T_{\mathrm{CO}}$. In spite of high homogeneity, the $(\mathrm{TMTTF})_2$ X samples contain still impurities and defects. As well known [61], even with a small disorder, i.e., with an arbitrary weak random field, the sample with an homogeneous order parameter prefers to break into domains. The most probably, such a domain structure develops at $T < T_{\mathrm{CO}}$. The polydomain FE sample is divided into many parts with different polarization. By analogy with classical FE, the polarization of $(\mathrm{TMTTF})_2$ X crystals under the electric field, associated with motion of domain walls, yields the main contribution to the magnitude of their dielectric permittivity. However, in spite of the mentioned similarity between the FE transitions in these 1D conductors and 3D classical FE, differences between them have to be noted. In $(\mathrm{TMTTF})_2$ X conductors the anisotropy of electric properties is large. The conductivity along molecular chains has a coherent character while in transverse directions it is not coherent [1]. Therefore, the FE state in $(\mathrm{TMTTF})_2$ X conductors cannot be considered as fully equivalent to that in classical three-dimensional FE. The question about the physical nature of the CO state below T_{CO} requires additional studies. In particular, in the attempt to measure polarization–depolarization phenomena under an increasing electric field, the main difficulty is Ohmic heating effects, resulting from the relatively high conductivity of $(\mathrm{TMTTF})_2$ X compounds in comparison with the conductivity of typical classical FE [61, 67]. Nonlinear optical measurements with the search of second harmonic generation would be very interesting with this respect.

10.3.6 Deuteration

In addition to usual effects on a given crystal such as temperature, pressure, and others, the exchange of hydrogen by deuterium in the methyl group of the TMTTF molecule has been used in the study of $(TMTTF)_2$ X compounds. By measurements of the spin susceptibility, some indications on the increase of T_{CO} in AsF_6 and PF_6 salts were obtained [63].

Figure 10.7 shows the temperature dependences of the real part of the dielectric permittivity ε' for deuterated and non-deuterated $(TMTTF)_2$ AsF_6 and $(TMTTF)_2$ SbF_6 samples [64]. All these curves exhibit a pronounced peak at the temperature for the CO phase transition, where the dielectric constant shows a divergence (see Fig. 10.4).

As can be seen from Fig. 10.7, the effect of deuteration leads to a considerable shift of the peaks $\varepsilon'(T)$ curves to higher temperatures by 16 K for the AsF_6 salt and by 13 K for the SbF_6 salt. In addition, the magnitudes of ε' maxima decrease slightly. In other respects the forms of $\varepsilon'(T)$ dependences near T_{CO} for deuterated samples of AsF_6 and PF_6 salts are well described by the Curie law $1/\varepsilon'(T) \sim |T - T_{CO}|$, analogous to hydrogenated samples [57]. The frequency dependences of $\varepsilon'(T)$ curves do not appreciably change as a result of deuteration for these both salts.

The temperature dependence of the dielectric permittivity $\varepsilon'(T)$ for deuterated and hydrogenated $(TMTTF)_2ReO_4$ are shown in Fig. 10.8 on a semi-logarithmic scale for a better visibility of the low temperature part of the $\varepsilon'(T)$ curve.

We observed a shift of the $\varepsilon'(T)$ peak, corresponding to an increase of the CO transition (Fig. 10.5) by 7 K. The form of the $\varepsilon'(T)$ dependence near

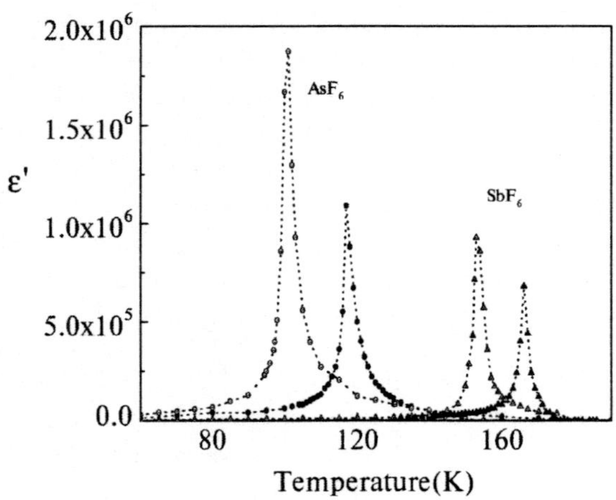

Fig. 10.7. Temperature dependences of the real part of the dielectric permittivity ε' at frequency 100 kHz for deuterated (*full circles*) and hydrogenated (*circles with cental dot*) $(TMTTF)_2$ AsF_6 and analogous dependencies for deuterated (*full triangle*) and hydrogenated (*triangle with central dot*) for $(TMTTF)_2$ SbF_6 (from [64])

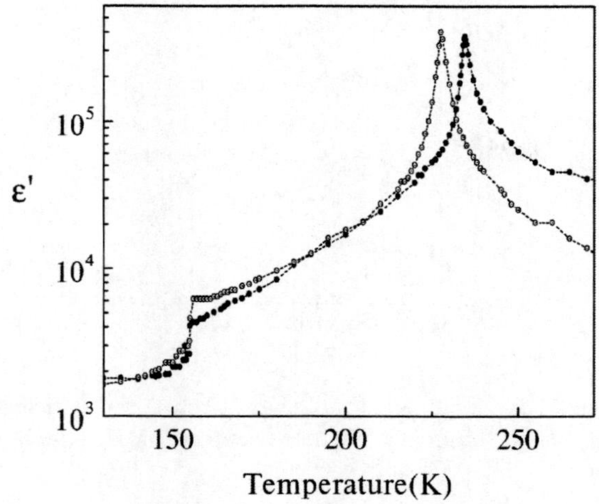

Fig. 10.8. Temperature dependences of the real part of the dielectric permittivity ε' at frequency 1 MHz for deuterated (*full circles*) and hydrogenated (*circles with cental dot*) $(TMTTF)_2$ ReO_4 (from [64])

T_{CO} for the deuterated ReO_4 salt is similar to the hydrogenated one with, however, a smaller amplitude of the ε' peak. The second anomaly observed near the temperature of the anion ordering at $T_{AO} \approx 155$ K (Fig. 10.7) is not affected by deuteration. While under pressure the conductivity at room temperature increases and T_{CO} decreases, deuteration yields the opposite effect: σ_o decreases and T_{CO} increases [64]. Estimations of the ratio V/t in deuterated samples, taking into account variations of the parameters of the unit cell, are consistent with the growth of this ratio [63, 64]. As shown in theoretical works [30, 33] the magnitude of V/t governs the formation of the CO state. Neutron scattering has revealed very weak intensity spin-Peierls superlattice peaks in $(TMTTF)_2$ PF_6 [46, 63] and in $(TMTTF)_2$ AsF_6 [65]. Deuterated $(TMTTF)_2$ PF_6 exhibits a SP temperature at 12.9 K with respect to 16.4 K for hydrogenated one [63, 65]. Thus, deuteration acts as a negative pressure. Effects of CH_3/CD_3 methyl groups were also considered. Effect of molecular motion was analyzed using ^{19}F-NMR spectroscopy. ^{19}F spin relaxation rate data show two maxima at around 135 K and 210 K [63] occurring at the same temperatures in hydrogenated and deuterated samples. These peaks were assigned to rotation modes for the SbF_6 anions with different activation energies.

10.3.7 Discussion

Thus dielectric permittivity has been measured in the combined SDW-CDW ground state of the Bechgaard $(TMTSF)_2PF_6$, in the temperature range of the CO phase transition in the Fabre $(TMTTF)_2PF_6$ with centrosymmetric

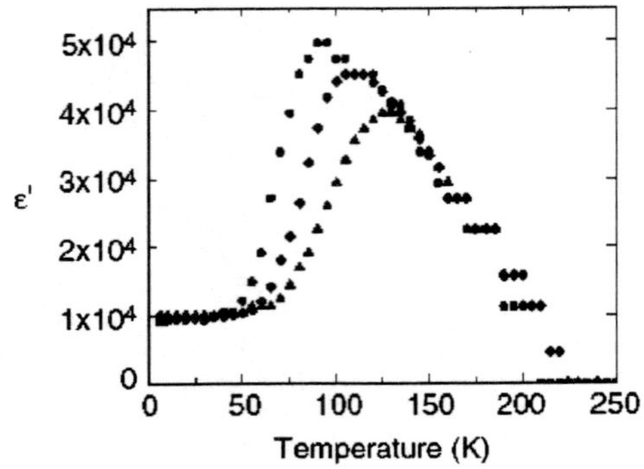

Fig. 10.9. Temperature dependencies of the real part of the dielectric permittivity ε' of $(DI\text{-}DCNQI)_2Ag$ at frequency 100 kHz (*circles*), 1 MHz (*diamonds*), and 5 MHz (*triangles*) (from [66])

anions $(TMTTF)_2Br$, $(TMTTF)_2PF_6$, $(TMTTF)_2AsF_6$, and $(TMTTF)_2SbF_6$ and with nonsymmetric anions $(TMTTF)_2ReO_4$, $(TMTTF)_2BF_4$, and $(TMTTF)_2SCN$ and finally in $(DI\text{-}DCNQI)_2Ag$. The temperature dependence of ε' for this last compound [66] is shown in Fig. 10.9.

The growth of ε' above the resolution of our measurements becomes noticeable below 220 K temperature at which ^{13}C-NMR spectra start to split, indicating the occurrence of charge disproportionation on chains [35]. With decreasing temperature, the magnitude of ε' continues to increase and is independent of frequency down to 140 K. At lower T, ε' reaches a maximum at a temperature depending on frequency: with decreasing frequency the maximum of ε' increases and its position in the temperature scale is shifted to lower temperature. Such variation is typical of the slowing-down behavior of relaxation phenomena. These $\varepsilon'(\omega, T)$ dependencies are at variance with those of $(TMTTF)_2AsF_6$, which show no dispersive response near T_{CO} below 1 MHz. They resemble those (except the magnitude of ε') of $(TMTTF)_2PF_6$, characteristic of a diffuse phase transition in ferroelectrics with compositional heterogeneities, so-called relaxator ferroelectrics [11, 62]. Sources of this heterogeneity can be less well crystallized samples, cracks, or defects. Careful analysis of the ^{13}C-NMR spectra has revealed [37] a small line around the zero shift in addition to the splitted lines caused by the charge ordering. It was argued [37] that these features are manifestations of the frustration in the charge ordering creating kink and anti-kink in the $(DI\text{-}DCNDI)_2Ag$ with a spiral topology. Moreover, from the temperature dependence of infrared spectra, it was concluded [68] that a $4k_F$ dimerization along molecules in the stacks begins from room temperature and that, below 150–200 K, there are signs characteristic of a $2k_F$ tetramerized periodicity.

In the SDW state of $(TMTSF)_2PF_6$ (Fig. 10.1) the maximum of ε' at 100 kHz reaches $1 \sim 2 \times 10^8$; at the same frequency $\varepsilon'_{\rm max}$ is $\sim 2 \times 10^6$ for $(TMTTF)_2AsF_6$ and other CSA salts. This value is only $\sim 2 \times 10^4$ for $(TMTTF)_2SCN$ (at 1 MHz) and of the same order $\sim 5 \times 10^4$ at 100 kHz for $(DI\text{-}DCNQI)_2Ag$. The small amplitude of ε' for $(TMTTF)_2SCN$ can be understood [49] by the transverse components of the combined CO-AO phase transition ($\mathbf{q} = 0, 1/2, 1/2$) which can be interpreted as yielding an antiferroelectric character to the phase transition. With this respect, the phasing of the charge ordering on adjacent chains of $(DI\text{-}DCNQI)_2Ag$ has to be clarified.

10.4 Thermodynamical Properties

Specific heat measurements allow to investigate structural, electronic, and magnetic transitions and to determine the electron and phonon contributions. They can also give very important informations on all excitations in the system and, moreover, the disorder character of the system under study can be manifested in the thermodynamical investigations. Thus, the lattice contribution, the anomaly at the low-temperature magnetic phase transition (spin density wave (SDW), antiferromagnetic (AF), spin-Peierls(SP), etc.), the low-energy excitations (LEE) have been intensively studied in the Bechgaard–Fabre salts.

The main technique we used is a transient heat-pulse method. In this method, the sample is loosely connected to the thermal bath by a thermal link. In normal conditions, C_P is calculated from the increment of the temperature $\Delta_T = T - T_0$ caused by a heat pulse, by using the exponential decay following the initial T increase: $T(t) - T_0 = (T_P - T_0)\exp(-t/\tau)$, with the relaxation time $\tau = CR_l$, where C is the total heat capacity and R_l is the thermal resistivity of the thermal link to the regulated cold sink. In the experiments, $\Delta T/T_0$ is always kept between 3 and 10%, depending on the heat capacity $C(T)$ variation, T_0 being the reference temperature before the heating pulse, stabilized within 2×10^{-4} during several hours.

However, in quasi-1D organic as well as inorganic systems, we have observed a progressive deviation from an exponential decay of the transient for temperatures $T < 0.5\,K$, indicating a time-dependent, or non-equilibrium, specific heat [69]. By convention, we define C_P on a "short time scale" by the amplitude of the initial decay $(T_P - T_0)$ of the transient, the value obtained in this way corresponding to the minimum value of C_P. In the nonequilibrium case, C_P was found to depend upon the time delivery of energy (or "waiting time"). With a long heat delivery, the power input $\dot{Q}_P$ should be regulated in order to maintain constant the temperature increment $(T_i - T_0)$ to a few percent above the reference T_0. After switching off the power, the transient $\Delta T(t)$ is recorded. The heat capacity is determined by the total heat release through the link R_l obtained by *integration* over time of the instantaneous release.

As known well, the physical properties of the Bechgaard salts are also very sensitive to the cooling rate, particularly in the case of perchlorate $(TMTSF)_2$ ClO_4 for which an anion ordering (AO) takes place at $T_{AO} = 24\,K$. When T is swept down very slowly, the AO transition takes place and the ground state is superconducting [1]. If the sample is quenched from $T > T_{AO}$, this ordering does not occur and the SDW ground state is achieved. All the following data are taken with a cooling rate of 15 h, between room temperature and liquid nitrogen temperature, then 10 h, from 77 to 4.2 K. Below 20 K, the cooling rate was typically $0.3\,\mathrm{K\,h^{-1}}$. For $(TMTSF)_2\,ClO_4$ the data correspond to a quench from 40 to 20 K through T_{AO} at a rate of $3.3\,\mathrm{K\,s^{-1}}$, which yields the SDW transition at 4.5 K.

Figure 10.10a, b shows the overall features of C_{P} of $(TMTSF)_2\,PF_6$ in the form of the T dependence of C_{P}/T^3 from 1.8 to 30 K in Fig. 10.10a [13], and of $C_{\mathrm{P}}(T)$ from 0.08 to 2 K in Fig. 10.10b [70].

The SDW transition appears as a shallow anomaly at $T_{\mathrm{SDW}} = 12.2\,\mathrm{K}$. The deviation from the usual Debye T^3 law, seen in C_{P}/T^3 as a bump with a maximum at 7 K, is ascribed to specific low-energy phonons. In addition, C_{P}/T^3 shows an abrupt drop at $T = 3.5\,\mathrm{K}$ depending on the experimental dynamic conditions: transient heat pulse or quasi-adiabatic methods [71]. It was explained as an evidence of a glass-like transition due to the freezing of internal degrees of freedom in the disordered SDW [18, 71]. This jump also occurs in the T-range where SDW sub-transitions were detected by NMR [19, 72] and nonlinear transport properties [73]. Below 2 K (see Fig. 10.10b), C_P deviates gradually from T^3 to finally reach a T^{-2} variation. This hyperfine contribution is the high-T tail of a Schottky anomaly resulting from metastable low-energy excitations (LEE) of the SDW ground state, closely separated in energy. When the contribution of LEE to the specific heat dominates over the other terms (phonons and addenda), nonequilibrium phenomena occur with aging in the heat response [20]. In the following, we will analyze in more details all these features of C_{P} of Bechgaard–Fabre salts, in particular the phonon contribution, SDW and CDW sub-phases, low-energy excitations, and nonequilibrium thermodynamics and aging phenomena.

10.4.1 Lattice Contribution

The best way to represent the lattice contribution is a C_{P}/T^3 plot vs. T. Comparison [74] between two selenides, $(TMTSF)_2PF_6$ and $(TMTSF)_2AsF_6$, and two sulfides, $(TMTTF)_2PF_6$ and $(TMTTF)_2Br$, is shown in Fig. 10.11.

The lattice contribution follows a T^3 law only in a small T-interval below 3 K. Quantitatively and qualitatively there is a strong similarity between both selenide salts on one hand and both sulfide ones on the other. Therefore, it can be concluded that the lattice term is mainly determined by the TMTSF or TMTTF molecules forming the stacks, the role of anions being minor. Analysis of C_{P} yields a common $\beta = C_{\mathrm{P}}/T^3 = 14.5\,\mathrm{mJ\,mol^{-1}\,K^{-4}}$ for $(TMTSF)_2\,PF_6$ and $(TMTSF)_2\,AsF_6$ corresponding to a Debye temperature of 200 K. In the case of $(TMTTF)_2\,Br$ and $(TMTTF)_2\,PF_6$, the lattice contribution β is estimated to be $7.8\,\mathrm{mJ\,mol^{-1}\,K^{-4}}$ [74].

In a larger T-range, Fig. 10.12 shows C_{P}/T^3 for $(TMTSF)_2\,PF_6$ (as in Fig. 10.10a), $(TMTSF)_2\,AsF_6$, $(TMTSF)_2\,ClO_4$ in the quenched SDW state, and $(TMTTF)_2Br$ [13]. One can see that C_{P}/T^3 exhibits a bump around 7 K, not only for $(TMTSF)_2\,PF_6$, but also for $(TMTSF)_2AsF_6$ and $(TMTTF)_2Br$.

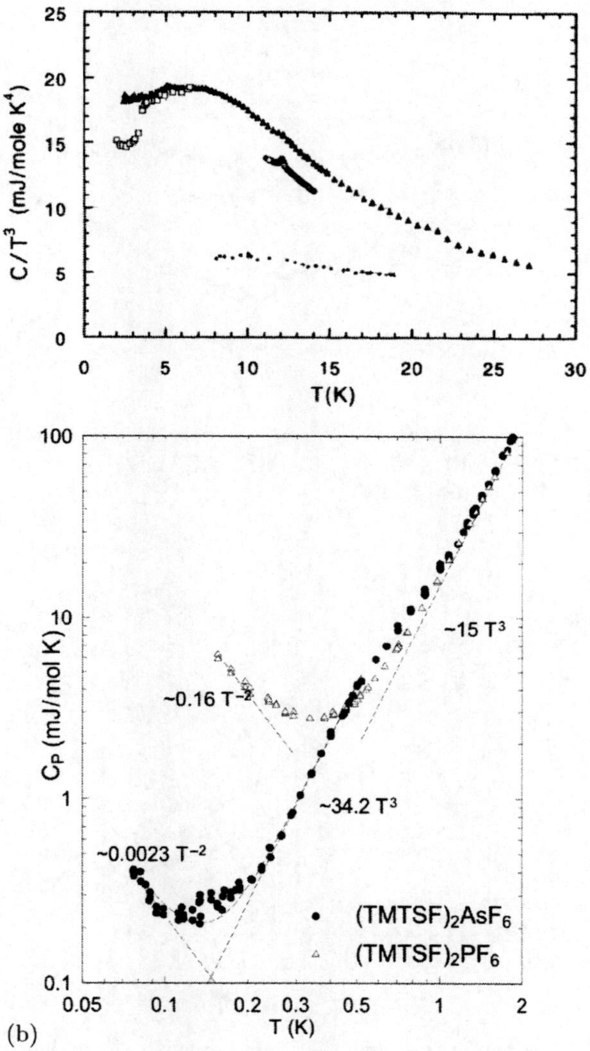

(a)

(b)

Fig. 10.10. (**a**) C_P/T^3 vs. temperature for $(TMTSF)_2$ PF_6 measured with quasi-adiabatic (*filled triangle*), transient-heat-pulse (*square* and *circle*), and ac technique (*filled circle*) (from [13]). (**b**) Specific heat of $(TMTSF)_2PF_6$ and $(TMTSF)_2AsF_6$, defined on short-time scale, below 2 K in a log–log plot. The low energy excitations (LEE) in $(TMTSF)_2AsF_6$ is reduced by a factor 70 compared to the similar term in $(TMTSF)_2PF_6$ as evidenced by the reduction of the hyperfine T^{-2} term (from [70])

For $(TMTSF)_2ClO_4$ in the quenched SDW state, this bump is located at lower T, around 4 K.

Such an excess specific heat to the usual Debye T^3 law is currently ascribed to low-energy modes. In CDWs they originate from low-lying transverse acoustic or optical modes related to the particularity of the 1D structure [13, 75].

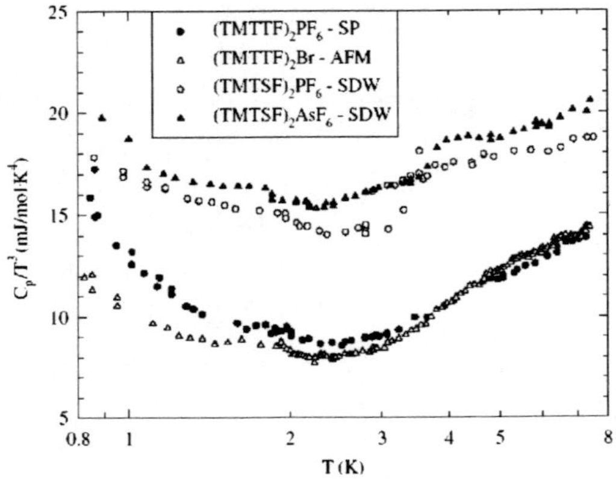

Fig. 10.11. Variation of the specific heat, C_P, defined on the short time scale divided by T^3 of $(\mathrm{TMTSF})_2\mathrm{PF}_6$, $(\mathrm{TMTSF})_2\mathrm{AsF}_6$, $(\mathrm{TMTTF})_2\mathrm{PF}_6$, and $(\mathrm{TMTTF})_2\mathrm{Br}$, between 0.8 and 8 K (from [74])

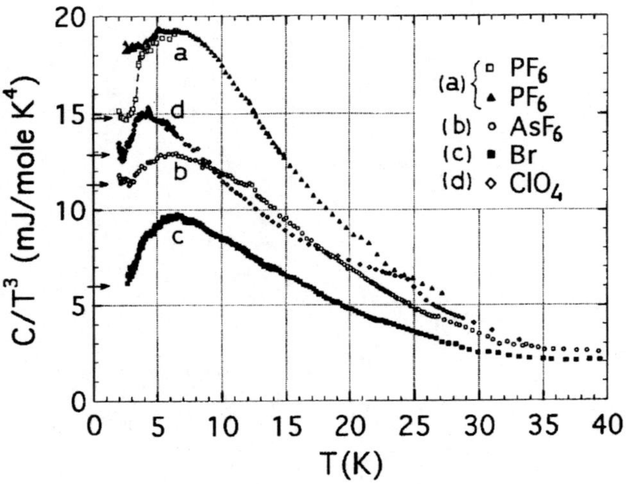

Fig. 10.12. Comparison of the temperature dependence of C_P/T^3 for $(\mathrm{TMTSF})_2\mathrm{PF}_6$ (the same as in Fig. 10.10a) with two different time scales, $(\mathrm{TMTSF})_2\mathrm{AsF}_6$, $(\mathrm{TMTTF})_2\mathrm{Br}$, and $(\mathrm{TMTSF})_2\mathrm{ClO}_4$ in the quenched SDW state for which the excess of specific heat at the anion ordering transition is visible at 24 K. The arrows indicate the low temperature cubic regime $\beta\ T^3$ below 3 K (from [13])

An alternative (or complementary) explanation lies on "phason" excitations [76] inherent to the incommensurate CDW ground states. In the case of organic compounds, the low energy "intra"-molecular phonon modes have been revealed by far infrared (IR) spectroscopy [77, 78], and modeled as Einstein oscillators for evaluating their contribution to C_P.

Independently of the contribution of these Einstein modes, the substantial decrease of C_P/T^3 in the high T range cannot be explained by the usual assumption of deviation from the T^3 law starting around $T \approx \vartheta_D/20$. This deviation can be taken into account by using a T-dependent phonon background T^α with $\alpha < 3$, characteristic of a system with anisotropic force constants. Good fits to C_P/T^3 for $(TMTSF)_2$ PF_6, $(TMTSF)_2$ AsF_6, and $(TMTTF)_2$ Br have been made based on the assumption of two main contributions: an acoustic background which changes from T^3 below a crossover temperature T^* towards T^α above T^*, and two Einstein modes based on far IR experiments [77,78] (see [13] for details). For $(TMTSF)_2PF_6$, the fit yields $\beta = 14.8\,\mathrm{mJ\,mol^{-1}\,K^{-4}}$ as previously determined, $T^* = 2.5\,\mathrm{K}$ and $\alpha = 2.4$. A sub-cubic regime for the lattice contribution can be interpreted within the model of Genensky and Newell [79] due to the strong anisotropy of the force constants along or between adjacent chains. In an intermediate T-range, the abnormal contribution varying as $T^{2.5}$ for the transverse vibrations and the regular T^3 contribution from longitudinal vibrations result in a total specific heat varying as T^α, with the power-law coefficient α in the range between 2.5 and 3. The temperature T^* of the crossover between the low-T Debye regime and the T^α regime is determined by the ratio $\Gamma/(4\kappa)^{1/2}$ [79], Γ being the force constant of adjacent chains and κ being the force constant for bending along the chains. The present analysis yields a T^* value of about 2.5 K, a temperature lower than in the case of polymeric chain compounds like hexagonal Se [80] or crystalline polyethylene [80,81]. The power law exponent α is 2.4 ± 0.1, close to the limit case for the Genensky–Newell model, implies the predominance of the contribution of the transverse modes in the lattice specific heat. A similar analysis [82] performed on slowly cooled and quenched $(TMTSF)_2$ ClO_4 yields the coefficient $\alpha = 2.7$ for both states, that means that the force constants anisotropy is less for the perchlorate salt than for the other salts.

10.4.2 SDW and Sub-SDW Phase Transitions

In Fig. 10.12, the SDW transition appears only as a shallow anomaly in the total specific heat around 12 K for $(TMTSF)_2$ PF_6, $(TMTSF)_2$ AsF_6, and $(TMTTF)_2$ Br and ≈4.5 K for the quenched SDW in $(TMTSF)_2$ ClO_4. On the other side, for this latter compound, the AO is clearly identified from the anomaly at 24 K.

The excess specific heat, dC_P, is obtained by the subtraction from the total C_P of a background determined by a polynomial fit to the data away from the transition [see [13]]. The excess specific heats are plotted in Fig. 10.13 as dC_P/T vs. T for $(TMTSF)_2$ PF_6, $(TMTSF)_2$ AsF_6, and $(TMTTF)_2$ Br, and directly as C_P/T^3 for the quenched $(TMTSF)_2$ ClO_4.

While the shape is perfectly symmetric and therefore reminiscent of that for a first-order transition, no sign of 1st order effects (latent heat, hysteresis) was observed. In Fig. 10.13, the solid lines represent a "mean-field" (MF) analysis usually used for a second-order phase transition, which defines an ideal jump ΔC_P at the transition temperature T_C, by equalizing the entropy for experimental and idealized curves (the equal-area analysis). With this

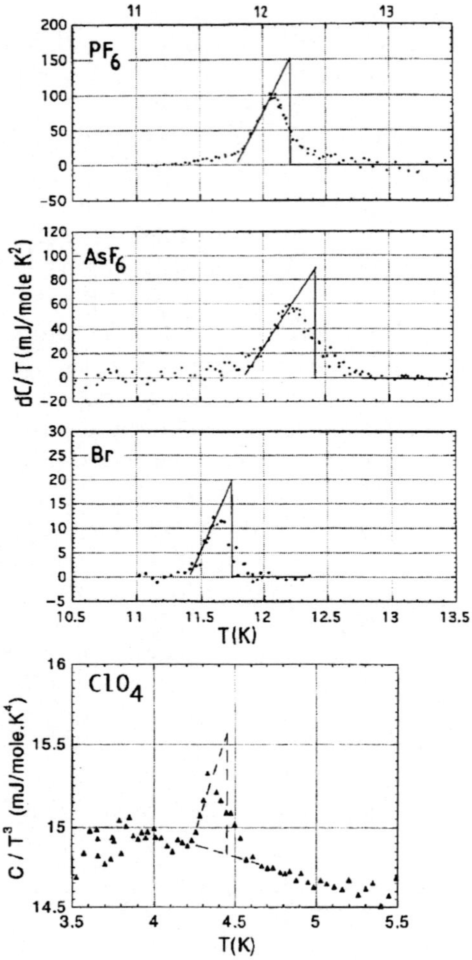

Fig. 10.13. Specific heat at the SDW transition of $(TMTSF)_2$ PF_6, $(TMTSF)_2$ AsF_6, $(TMTTF)_2Br$ as dC/T [13]. Specific heat at the SDW transition of $(TMTSF)_2$ ClO_4 in the SDW quenched state reported in a C_P/T^3 plot [82]

analysis, as shown in Fig. 10.13, the SDW transitions were found to be 12.2 K for $(TMTSF)_2$ PF_6, 12.4 K for $(TMTSF)_2$ AsF_6, and at 4.45 K for $(TMTSF)_2$ ClO_4 in the quenched state; the AF transition for $(TMTTF)_2$ Br being at 11.7 K. The magnitudes of the normalized specific heat jump $\Delta C_P/T_C$ of the SDW transitions are 150, 88, and 14 mJ mol^{-1} K^{-2} for the PF_6, AsF_6, and ClO_4 compounds, respectively, and 19 mJ mol^{-1} K^{-2} for the AF transition in the Br compound. That means that the contributions for the SDW transitions are, respectively, 4.5%, 2.5%, and 4.7% of the total C_P for the three first samples, and less than 1% for the Br compound, which is at the limit of the experimental detection.

Assuming that the jump in C_P is solely of electronic origin and using the usual BCS mean-field expression, $\Delta C = 1.4\ \gamma T_C$ for the electronic condensation, this yields γ values of 107, 63, and 10 $\mathrm{mJ\,mol^{-1}\,K^{-2}}$ for the PF_6, AsF_6, and ClO_4 compounds, respectively, and 13.6 $\mathrm{mJ\,mol^{-1}\,K^{-2}}$ for the Br compound. The γ value for quenched ClO_4 is similar to the value that we have obtained in the metallic phase of this salt, obtained by very slow cooling, with $\gamma = 11\,\mathrm{mJ\,mol^{-1}\,K^{-2}}$ [82]. The good agreement with the MF analysis indicates that free electrons are completely condensed due to the opening of the SDW gap, as expected for a contribution of itinerant electrons above T_C. However, γ values are considerably higher for the PF_6 and AsF_6 salts, for which large discrepancies with the MF analysis are reported (see a detailed discussion on this point in [13]). It was shown before that the lattice specific heat deviates less from the Debye law for $(TMTSF)_2\ ClO_4$ ($\alpha = 2.7$ rather than 2.5 for other salts). Then, the effective dimensionality is higher for $(TMTSF)_2\ ClO_4$, which may explain why the MF analysis is more correct for this compound than for $(TMTSF)_2\ PF_6$ and other salts.

In Fig. 10.10a, one can see the abrupt jump at 3.5 K in $(TMTSF)_2$ PF_6 observed under particular time scale conditions (transient heat pulses), whereas under quasi-adiabatic (QA) conditions [71] this jump is absent down to 2.5 K. In the QA technique, after cooling, the sample is isolated from the thermal bath and reheated adiabatically step by step, each heating process (by about 3% of T) being used to measure the heat capacity. Duration of heating was about 100 s and the total data acquisition was about 4–5 min per point. In addition, data obtained with the transient heat pulses but for various "annealing" conditions are reported in more details in Fig. 10.14. The first obvious result is that although an excellent agreement between the two techniques is obtained between 4 and 7 K, the characteristic jump, which occurs around 3–3.5 K (depending on thermal history) with the transient heat pulse technique, does not appear down to 2.5 K with the QA technique.

With respect to glass transition of supercooled liquids, the jump in specific heat represents the freezing-in of the configurational degrees of freedom over the experimental time scale. A characteristic property of the glass transition is its strong sensitivity on the thermal history of the system and to kinetics of the measurement. We have demonstrated [18] the sensitivity of the specific heat jump of $(TMTSF)_2PF_6$ to the thermal history and in particular to any isothermal annealing process performed in the vicinity of the glass transition (close to T_g). It is also known for glasses that the temperature of the jump is sensitive to the frequency of the measurement. Specific heat spectroscopy has shown that the anomaly occurs at lower temperature when the frequency decreases. The QA technique is per definition characterized by the absence of relaxation of the temperature of the sample after the energy input. By comparison, in the transient technique, first, the duration of energy input (heat pulse) is about 100 times shorter, and second, there is a relaxation of temperature towards its initial value after each heat pulse with a time constant of several tens seconds. In that sense the latter technique, compared to the d.c.

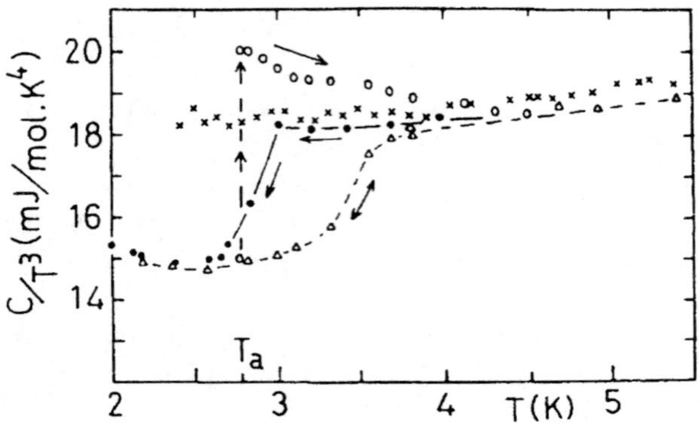

Fig. 10.14. Temperature dependence of the total specific heat of $(TMTSF)_2PF_6$ in a C_P/T^3 plot in the vicinity of the glass transition, obtained in very different kinetic conditions: Data (*cross*) are obtained with the quasi-adiabatic technique. Other data are obtained with the transient-heat-pulse technique: (*open triangle*) correspond to a regular heating or cooling at a rate of $0.5\,\mathrm{K\,h^{-1}}$ (reversible cycle) between 2.9 and 4.5 K for an unannealed sample. After a long annealing at $T_a = 2.77\,\mathrm{K}$ for 40 h, a large endothermic maximum appears on reheating (*open circle*), whereas on subsequent cooling (*closed circle*), C_P follows the equilibrium value down to 3 K (from [83])

adiabatic one, is more similar to an a.c. technique [84], despite its moderate frequency (0.1–0.001 Hz). This can explain the absence of C_p jump down to 2.5 K in the QA measurements.

Above T_g, all degrees of freedom – vibrational and configurational – can be excited during the experimental time span. This corresponds to the specific heat C_{PE} of the supercooled liquid in thermodynamical (or metastable) equilibrium. Here C_{PE} varies very closely to a cubic law up to 8 K ($C_{PE} = \beta T^3$, with $\beta = 18.5$–$19\,\mathrm{mJ\,mol^{-1}\,K^{-4}}$). Below T_g, there remains only vibrational contribution of the "glass," C_{PG}, which corresponds here to the lattice term following the Debye law: $C_{PG} = \alpha T^3$, with $\alpha = 15\,\mathrm{mJ\,mol^{-1}\,K^{-4}}$. We note that C_{PE} obeys a cubic law like the vibrational specific heat. In comparison to usual glass forming systems, where T_g occurs at several 100 K, here T_g appears at very low T, in the liquid helium T-range. In this region, C_P is dominated by the lattice specific heat ($\propto T^3$) and the transition appears as a jump in a C_P/T^3 diagram. Contrary to $(TMTSF)_2PF_6$, $(TMTSF)_2AsF_6$ does not show any annealing effects or hysteretic behavior. However, it exhibits important dynamical effects [70, 83]. It has to be noted that the CDW superlattice spots observed in X-ray measurements [15] disappear below 3–4 K in $(TMTSF)_2PF_6$ and that their amplitude in $(TMTSF)_2AsF_6$ is very small.

Finally, the sub-phase transition around 3–3.5 K in $(TMTSF)_2PF_6$ and $(TMTSF)_2AsF_6$ corresponds to a dynamic glassy transition. The jump in C_P around 3.5 K in $(TMTSF)_2PF_6$ has the characteristics of freezing in supercooled liquids. Concerning the second sub-phase transition detected at $T = 1.9\,\mathrm{K}$ by NMR [19], a discontinuity in specific heat is observed (see

Fig. 10.11) at the same temperature as well in the selenides $(TMTSF)_2PF_6$ and $(TMTSF)_2AsF_6$ as in sulfides $(TMTTF)_2PF_6$ and $(TMTTF)_2Br$. These results allow us to conclude that this discontinuity is not related to any (electronic) superstructure induced in these salts, but probably results from the stacks along the chains.

10.4.3 Low Energy Excitations

As seen in Figs. 10.10b and 10.11, below 2 K, C_P of $(TMTSF)_2PF_6$ deviates progressively from a T^3 law on decreasing T, due to the presence of the additional contribution of LEEs. This deviation is accentuated below 0.5 K, simultaneously with the occurrence of nonexponential kinetics. Finally, below 0.3 K, an upturn appears with a T^{-2} variation.

Short time C_P data are well described below 3 K to the lowest T by

$$C_\mathrm{P} = C_\mathrm{h} T^{-2} + \mathrm{A} T^{\nu} + \beta T^3 .$$

The two first terms are the "hyperfine" and the power law contributions of the LEE and the third one is the usual phonon contribution.

For $(TMTSF)_2PF_6$ one finds [20] $C_\mathrm{P} = 0.065\,T^{-2} + 5\,T^{1.2} + 14.5\,T^3$ mJ mol^{-1} K^{-1} to be compared to

$$C_\mathrm{P} = 0.42\,T^{-2} + 5\,T^{1.2} + 7.8\,T^3 \ \mathrm{mJ\,mol^{-1}\,K^{-1}}$$

for $(TMTTF)_2PF_6$. The temperature dependence of the C_LEE contribution in $(TMTSF)_2PF_6$, $(TMTTF)_2PF_6$, and also in $(TMTTF)_2Br$ showing a power law dependence is drawn in Fig. 10.15. It appears that the contributions of the LEE are nearly identical in the commensurate SP and in the incommensurate SDW ground states. The T dependence for $(TMTTF)_2Br$ is $T^{0.44}$, but the amplitude of the LEE contribution is comparable to the two other compounds.

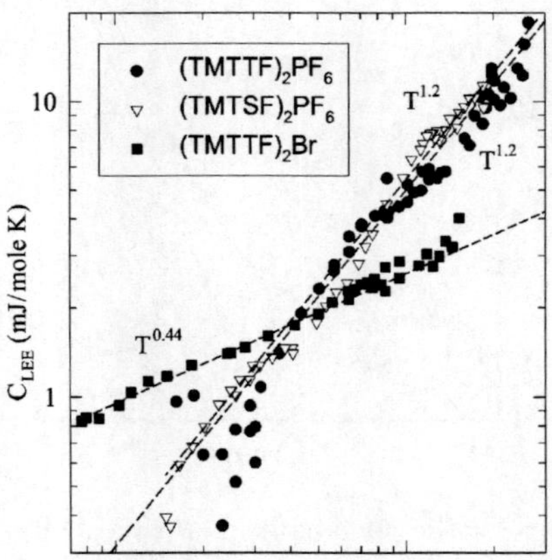

Fig. 10.15. Contribution of the low energy excitations (LEE) to the short time scale specific heat of $(TMTSF)_2PF_6$, $(TMTTF)_2Br$, and $(TMTTF)_2PF_6$

For $(TMTSF)_2AsF_6$, below 0.5 K, C_P is considerably smaller than that for $(TMTSF)_2PF_6$ (Fig. 10.10b) m which corresponds to a large reduction of the LEE contribution. This reduction reveals a new behavior as an intermediate T^3 regime below 0.25 and 0.5 K with an amplitude β^* twice as large as the regular phononic part. It was tentatively ascribed to the phason contribution of the incommensurate SDW modulation [70].

10.4.4 Nonequilibrium Phenomena

In the conditions where the contribution of the LEE to the specific heat dominates over the other terms (phonons and addenda), we have observed a highly nonexponential relaxation in almost all the (charge and spin) density wave systems. It was also shown that C_P depends upon the time delivery of energy (or "waiting" time t_w). Figure 10.16 shows the temperature dependence of C_P, determined by the integration technique as explained above, with different t_w for $(TMTSF)_2PF_6$ [20]. As soon as t_w increases, C_P shows a nonlinear effect: at a fixed temperature, C_P increases with t_w, revealing a saturation, i.e., there is no more evolution of $C_P(t)$ at $T = 0.2$ K when t_w increases from 12 to 24 h. We have interpreted this saturation by the fact that the SDW sub-system has reached its own thermodynamic equilibrium. In a time scale

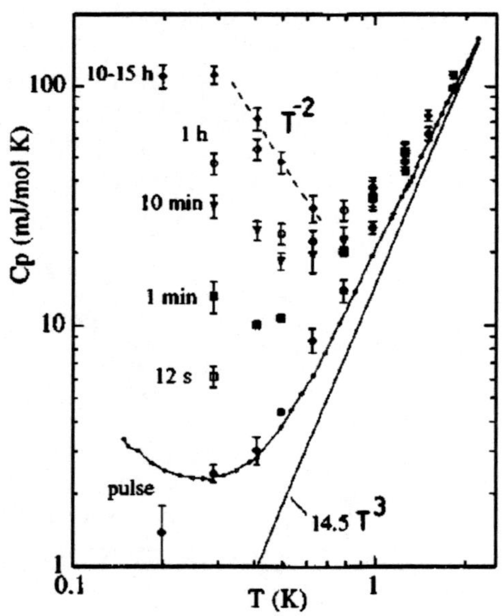

Fig. 10.16. Dependence of the specific heat of $(TMTSF)_2PF_6$ on the duration of energy delivery t_w : from a pulse of less than 1 s up to 10–15 h. The continuous curve represents data calculated from the initial T increment in response to heat pulses. Other data are obtained by integration of the total energy released. The amplitude of the Schottky anomaly tail in T^{-2} is largely increased for large t_w (from [20])

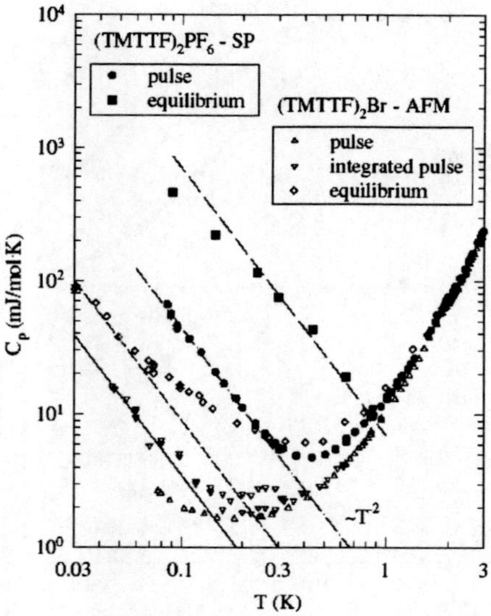

Fig. 10.17. Dependence of the specific heat of $(TMTTF)_2PF_6$ and $(TMTTF)_2Br$ (down to 30 mK) obtained from pulses and at the thermodynamical equilibrium (derived from integration of the energy release). The lines (continuous or dashed) represent the hyperfine T^{-2} contribution of the LEE (from [74])

of $\sim 10^4$ s, at $T = 0.2$ K, the time dependence effect on C_P in the case of $(TMTSF)_2PF_6$ is tremendously large. In Fig. 10.16 one can see that C_P for $t_w \sim 10$ h is 70 times larger than the value defined on a short-time scale.

Similar to $(TMTSF)_2PF_6$, the specific heat of $(TMTTF)_2PF_6$ and $(TMTTF)_2Br$ are strongly dependent on this time scale (Fig. 10.17). It is noted that C_P for $(TMTTF)_2PF_6$ is much larger than that for $(TMTTF)_2Br$ for similar dynamical conditions.

This nonlinear behavior is clearly related to the kinetics of the LEEs. The progressive deviation of C_P from the background T^3 law for increasing t_w is the manifestation of the decoupling of the LEEs from the phonons. The temperature dependence of C_P at equilibrium indicates the tail of a Schottky anomaly ($\propto T^{-2}$) with its maximum being at lower T (<30 mK for $(TMTTF)_2Br$).

Figure 10.18a shows the time dependence of the heat relaxation $\Delta T(t)$ with increasing t_w for $(TMTSF)_2PF_6$ [20,21]. This relaxation can be satisfactorily fitted with a phenomenological streched-exponential function [85].

$$\Delta T(t)/\,\Delta T_0 = \exp(-t/\tau)^{\beta},$$

with β the streching parameter. In the high T limit the relaxation recovers an exponential decay, so $\beta(T > 1\,\mathrm{K}) = 1$. However, $\beta(T)$ decreases when T is decreased, leading to a very slow relaxation.

A better characterization of a nonexponential relaxation can be obtained by the relaxation rate $S(t) = \mathrm{d}(\Delta T/\Delta T_0)/\mathrm{d}\ln\tau$, as shown in Fig. 10.18b.

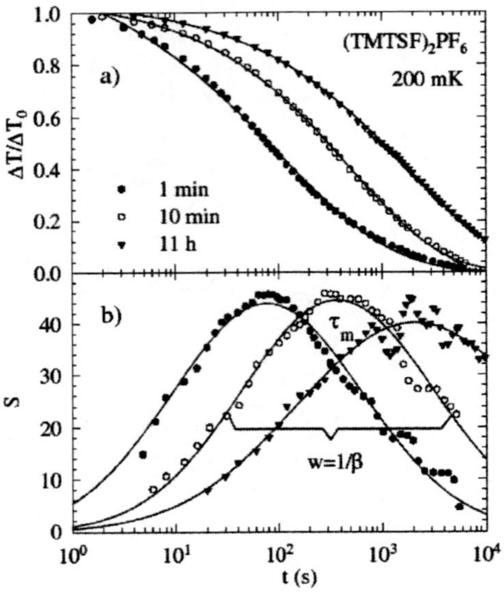

Fig. 10.18. Homogeneous broadening and shift of the heat relaxation with increasing duration t_w (aging) of the small heat perturbation (1 mn, 10 mn, 11 h.) for (TMTSF)$_2$PF$_6$ (**a**) relaxation $\Delta T/\Delta T_o$ vs. log$t$ with corresponding stretched exponential fit $\exp(-t/\tau)^{\beta}$, (**b**) $S(t) = d(\Delta T/\Delta T_o)/d\ln t$ fitted to a Gaussian function $G(\ln t)$ with a width w (from [21])

$\Delta T(t)/\Delta T_0$ can be written in terms of a distribution of relaxation times $P(\ln \tau)$ as

$$\Delta T(t)/\Delta T_0 = \int_{\ln \tau_0}^{\infty} P(\ln \tau) \exp(-t/\tau) \mathrm{d} \ln \tau,$$

where τ_0 is the microscopic time. Since $P(\ln \tau)$ varies slowly with τ, one can replace the exponential by a step function from which one deduce $S(t) = P(\ln t)$ [86]. As in spin glasses there is a "homogenous" broadening and shift of the relaxation peak with t_w. However, since our system eventually reaches thermodynamic equilibrium at $T_0 + \Delta T$ for sufficiently long t_w [69], aging is then interrupted [87].

On the other side, in the case of (TMTTF)$_2$PF$_6$, heat relaxation, while still nonexponential, exhibits quite different properties. Instead of the homogenous broadening reflecting a broad distribution of relaxation times and a shift of the spectra with t_w as in (TMTSF)$_2$PF$_6$ (Fig. 10.18b) as in real aging effects, (TMTTF)$_2$PF$_6$ demonstrates "discrete bands" of relaxation times. Their distribution are narrower and the relaxation times are smaller (saturation of aging is reached within a few tens of minutes as compared to days in (TMTSF)$_2$PF$_6$ at a similar T). Figure 10.19 shows a bimodal redistribution of the relaxation spectrum in the thermodynamical equilibrium for (TMTTF)$_2$PF$_6$ [88], indicating that heat perturbation modifies different parts

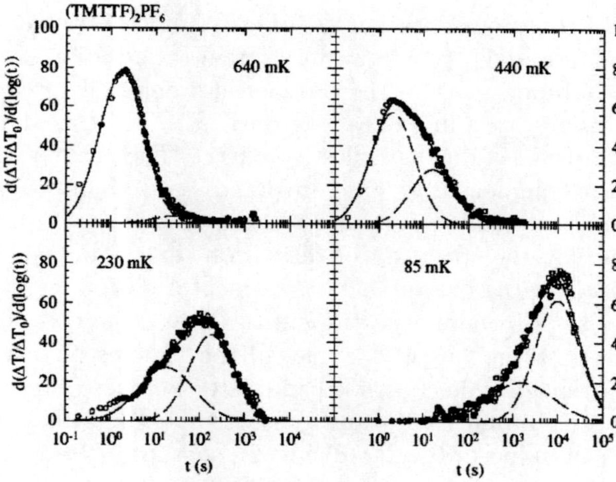

Fig. 10.19. Relaxation time spectra at the *thermodynamical* equilibrium for $(TMTTF)_2PF_6$ at given temperatures. A possible deconvolution corresponds to the sum of two Gaussian distributions corresponding to fast and slow LEE responses, the fast contribution being predominant at higher T and the slow one at lower t; (from [88])

of the spectrum in contrast to the homogenous shift of the entire spectrum as seen Fig. 10.18.

As shown in Fig. 10.19, the "discrete bands" in the relaxation rate at the thermodynamical equilibrium can be well deconvoluted into two Gaussian functions representing the slow and the fast LEE responses. Thus in the commensurate spin-Peierls ground state of $(TMTTF)_2PF_6$, two dynamically distinct entities appear below 1 K. At higher T, the fast contribution plays the dominant role in the dynamics. But, at the lowest T, the slow entity dominates both the dynamics and the thermodynamics. A similar deconvolution has also been carried out for the commensurate AF ground state of $(TMTTF)_2Br$, revealing also a bimodal dynamics [88].

The incommensurate systems as $(TMTSF)_2PF_6$, but also charge density wave systems as o-TaS_3 [69], show a homogenous shift of the relaxation rate distribution with t_w when T is decreased below 0.5 K, whereas commensurate systems such as spin-Peierls $(TMTTF)_2PF_6$ and antiferromagnetic $(TMTTF)_2Br$ show a bimodal dynamics [21]. Models giving explanation for these distinctly different dynamics have recently appeared [89, 90]. As proposed [42–44] the local deformations of a density wave in the strong pinning limit are bisolitons, obtained by minimizing the elastic pinning energies, and characterized by a ground state at energy E_0, separated from another state at energy $E_0 + \Delta E$ by a "bounce" state at energy $E_0 + \Delta V$, therefore, defining an effective two-level system. As a qualitative difference between commensurate and incommensurate systems [89, 90] one has $\Delta E = 0$ in the commensurate case and $\Delta E > 0$ in the incommensurate case, so that the populations of the effective two-level system does not couple to temperature variations in the commensurate case. It is suggested [89, 90] that, in addition, the degeneracy of the classical ground state in the commensurate case is lifted by quantum fluctuations that restore a finite heat response. The long time dynamics in the incommensurate case is due to collective effects. The issue of the long time dynamics in the commensurate case, involving both quantum and collective effects, is mainly an open theoretical question.

To explain why the spectrum of relaxation time is bimodal in the commensurate case, one can note that experimental data suggest that in the commensurate system there exist two entities that relax with approximately the same energy barrier but with values of the "microscopic time" differing by one order of magnitude. Possible candidates for these two entities would be (1) the dipole solitonic excitations generated by the strong pinning impurities and (2) self-induced disorder due to 2π solitons between microdomains separating the two spin-Peierls ground states. Alternatively, one may observe that $(TMTTF)_2PF_6$ has a charge gap $\Delta_c = 200\,K$ coexisting with a spin gap $\Delta_s = 20\,K$, deduced from the spin-Peierls transition temperature from the mean field approximation. The coherence length in the charge channel $\xi_c = \hbar v_F/\Delta_c \approx 10a_0$ is thus much smaller than the coherence length in the spin channel $\xi_s = \hbar v_F/\Delta_s \approx 100a_0$ (where a_0 is the lattice parameter and v_F the Fermi velocity). The two channels are weakly coupled so that the charge soliton, being smaller, relaxes faster than the spin soliton. In both cases – commensurate or incommensurate systems – values of internal relaxation time at equilibrium are thermally activated: $\tau_{eq} = \tau_0 \exp(\Delta/kT)$, with an activation energy Δ of the order of 1 K, which implies the very rapid increase of the slow dynamics at temperatures much lower than 1 K (see Fig. 10.19 and [88]) and τ_0 of the order of second.

10.4.5 Effect of Magnetic Field

Figure 10.20 shows the variation of the LEE contribution at short time scale of $(TMTSF)_2$ PF_6 for different magnetic fields applied between 0 and 2 T along two crystallographic axis at different temperatures below 0.5 K [91,92]. There is an increasing effect of the field at lowering temperature, in particular for low field (<0.5 T). A sharp maximum develops at $H = 0.2\,T$, precursor of a magnetic phase transition, probably for $T < 0.1\,K$. This anomaly is independent of the direction of the applied field.

As for the slow heat relaxation properties, the field effect is universal over all the CDW/SDW compounds we have investigated up to now. We can briefly resume the results [93]:

- Instead of a sharp maximum as shown in Fig. 10.20, a transition towards a new metastable branch, i.e., a new value of the LEE contribution, is induced by the field at very low temperature (always around 100 mK). This

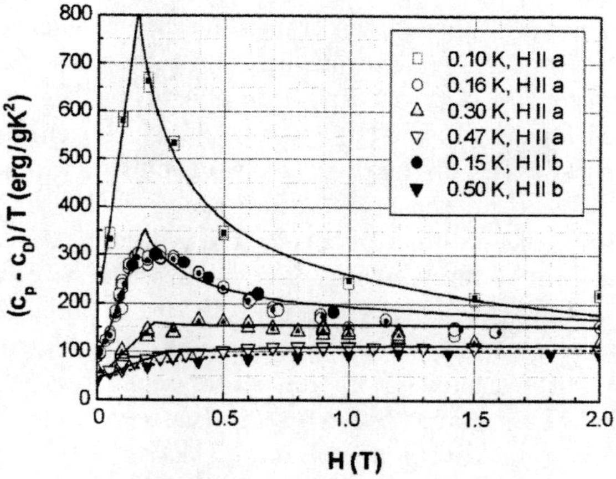

Fig. 10.20. Contribution of the low energy excitations C_P-C_D to the specific capacity of $(TMTSF)_2$ PF_6 divided by T as a function of a magnetic field applied parallel to a or b' axis at different temperatures. The data at $T = 0.5$ K show the contribution of the quasi-linear term T^α (with $\alpha = 1.2$, see Fig. 10.15). The lines are attempts for fitting the Schottky anomaly contribution, predominant below 0.3 K (from [91,92])

metastable state is very difficult to cure, either by high field application, or high temperature excursion (at least above the CDW/SDW transition). The switch to the new state corresponds to hysteretic effects of $C_P(H)$ during the first field application.

– At field higher than around 1 T, C_P increases again with the field, and can reach quadratic H^2 variations.

The dynamics of the transients following a heat pulse is also sensitive to H, and can show an evolution towards a pure exponential decay with longer time constants. That means that the relaxation time spectrum is strongly affected by the high field.

These different properties are presently under theoretical investigation, in the continuation of the previous model of strong pinning impurities distributed at random, at the origin of bisolitons, and the slow dynamics due to these collective excitations [89,90].

10.5 Conclusions

We have reviewed a corpus of experiments on energy and dielectric relaxations on Bechgaard and Fabre salts. Measuring on one hand the specific heat of $(TMTSF)_2PF_6$ in the SDW state under different kinetic conditions and on the other hand measuring the dielectric permittivity and the relaxation time in the same temperature range has allowed to demonstrate that around 3.5–4 K, a glass transition takes place. This transition corresponds to the sub-SDW phase determined by NMR measurements.

The importance of electron correlation effects resulting from long-range Coulomb interactions in $(TMTTF)_2X$ compounds has been revealed in dielectric measurements by the discovery of a new ground state – a charge ordered state – which appears in the temperature range above the temperature of magnetic ordering. The stability of this charge ordered state is expected to be provided by the shift of anion chains as a whole with the wave vector $q = 0$. This charge ordered state is characterized by the periodic redistribution of charge between molecules in the molecular chains (charge disproportionation) with a doubling of the electronic superstructure, which corresponds to an electronic $4k_F$ CDW. This charge ordered state has some characteristic features of a ferroelectric state: divergence of the dielectric permittivity and of the relaxation time near T_{CO} according to a Curie law and also development of a domain structure below T_{CO}. However, the loss of the center of symmetry and the corresponding change of the crystallographic space group has yet to be determined. For non-centrosymmetric anions, a phase transition in an anion ordered state, which occur below T_{CO}, is accompanied by the development of a lattice superstructure with the wave-vector $q = (1/2, 1/2, 1/2)$. The sequence of these phase transitions is completed by a magnetic ordering transition at still lower temperatures. $(TMTTF)_2SCN$ is specific in the sense that charge ordering and anion ordering occur at the same temperature. The wave-vector for the anion ordering being $q = (0, 1/2, 1/2)$ allows charge polarization along the chain axis with antiphase between adjacent chains; that may yield an antiferroelectric behavior.

The disordered state character of the ground state of Bechgaard–Fabre salts have been the best determined by thermodynamical measurements at very low temperature. It was shown that additional low energy excitations contribute to the specific heat in addition to the normal phonons. Enthalpy relaxation phenomena that strongly develop below 0.5 K have be shown to be nonexponential, exhibiting aging effects that reveals a broad distribution of relaxation times. The amplitude of the energy released, as well as its sensitivity to the power-input duration, is remarkably large among disordered systems, in particular in comparison to structural glasses. This metastability can be interpreted as originating from local deformations (bisolitons) of the SDW (but similar phenomena occur in CDW systems) at the site of strong pinning centers, characterized by two-level states in the energy landscape. Indeed, the enthalpy relaxation occurs in the temperature range where C_P is dominated by the T^{-2} tail of Schottky anomalies (resulting from 2-level states), the amplitude being very sensitive to the time span. In addition to their common character of broad distributions of relaxation times, one can differentiate between incommensurate and commensurate ground states, in particular by the bimodal (slow vs. fast entities) property of the relaxation for the commensurate case.

Acknowledgement

We are grateful to S. Brazovskii and R. Mélin for their help in many aspects of this review. We thank D. Staresinić for his participation to some parts of this work and in the preparation of the manuscript. This work was supported in part by RFBR (grant 05-02-17578) and partly in the frame of CNRS-RAS Associated European Laboratory between IRE-RAS and CNRS. Financial support from Euro-Net is also acknowledged.

References

1. D. Jérome, in *Organic Conductors*, ed. by J.P. Farges (Marcel Dekker, New York, 1994) p. 405
2. J.P. Pouget, S. Ravy, J. Phys. I (France) **6**, 1505 (1996)
3. C. Bourbonnais, D. Jérome, in *Advances in Synthetic Metals*, ed. by P. Bernier, S. Lefrant, G. Bidan (Elsevier, Amsterdam, 1999), p. 206
4. T. Giamarchi, Chem. Rev. **104**, 5037 (2004)
5. C. Bourbonnais, D. Jérome, Science **281**, 1156 (1998)
6. D. Jérome, A. Mazaud, M. Ribault, J. Phys. Lett. (France) **41**, 95 (1980)
7. S. Brazovskii, V. Yakovenko, Sov. Phys. JETP **62**, 1340 (1985); J. Phys. Lett. (France) **46**, 111 (1985)
8. F. Nad, P. Monceau, J.M. Fabre, J. Phys. IV (France) **9**, Pr10-361 (1999)
9. F. Nad, P. Monceau, C. Carcel, J.M. Fabre, Phys. Rev. B **62**, 1753 (2000)
10. D.S. Chow, F. Zamborsky, B. Alavi, D.J. Tantillo, A. Baur, C.A. Merlic, S.E. Brown, Phys. Lett. **85**, 1698 (2000)
11. F. Nad, P. Monceau, J. Phys. Soc. Jpn., **75**, 051005 (2006)
12. F. Zamborsky, W. Yu, W. Raas, S.E. Brown, B. Alavi, C.A. Merlic, A. Baur, Phys. Rev. B **66**, 081103 (2002)
13. H. Yang, J.C. Lasjaunias, P. Monceau, J. Phys. Condens. Matter **11**, 5083 (1999)
14. J.P. Pouget, S. Ravy, J. Phys. I (France) **6**, 1501 (1996)
15. S. Kagoshima, Y. Saso, M. Maesato, R. Kondo, T. Hasegawa, Solid State Commun. **110**, 479 (1999)
16. N. Kobayashi, M. Ogata, J. Phys. Soc. Jpn. **66**, 3356 (1997)
17. S. Mazumdar, R. Ramasesha, T. Clay, D.K. Campbell, Phys. Rev. Lett. **82**, 1522 (1999)
18. J.C. Lasjaunias, K. Biljaković, F. Nad, P. Monceau, K. Bechgaard, Phys. Rev. Lett. **72**, 1283 (1994)
19. T. Takahashi, T. Harada, Y. Kobayashi, K. Kanoda, K. Suzuki, K. Murarta, G. Saito, Synth. Met. **41–43**, 3985 (1991)
20. J.C. Lasjaunias, K. Biljaković, P. Monceau, Phys. Rev. B **53**, 7699 (1996)
21. J.C. Lasjaunias, R. Mélin, D. Staresinić, K. Biljaković, B. Souletie, Phys. Rev. Lett. **94**, 245701 (2005)
22. C. Coulon, P. Delhaes, S. Flandrois, R. Lagnier, E. Bonjour, J.M. Fabre, J. Phys. (France) **43**, 1059 (1982)
23. A. Fritsch, L. Ducasse, J. Phys. I (France) **1**, 855 (1991); ibid **6**, 583 (1996)
24. J. Solyom, Adv. Phys. **28**, 201 (1979)
25. S. Barisic, S. Brazovskii, in *Recent Developments in Condensed Matter Physics*, vol 1, ed. by J.T. Devreese (Plenum, New York, 1981), p. 327

26. V.J. Emery, R. Bruisma, S. Barisic, Phys. Rev. Lett. **48**, 1039 (1982); V.J. Emery, in *Highly Conducting One-Dimensional Solids*, ed. by J.T. Devreese, R.P. Evrard, V.E. van Doren (Plenum, New York, 1979), p. 247
27. J.E. Hirsch, D.J. Scalapino, Phys. Rev. B **27**, 7169 (1983); **29**, 5554 (1984)
28. F. Mila, X. Zotos, Europhys. Lett. **24**, 133 (1993)
29. K. Penc, F. Mila, Phys. Rev. B **49**, 9670 (1994); **50**, 11429 (1994)
30. H. Seo, H. Fukuyama, J. Phys. Soc. Jpn. **66**, 1249 (1997)
31. Y. Shibata, S. Nishimoto, Y. Ohta, Phys. Rev. B **64**, 235107 (2001)
32. M. Tsuchiizu, H. Yoshioka, Y. Suzumura, J. Phys. Soc. Jpn. **70**, 1460 (2001)
33. H. Seo, J. Phys. Soc. Jpn. **69**, 805 (2000)
34. F. Mila, Phys. Rev. B **52**, 4788 (1995)
35. K. Hiraki, K. Kanoda, Phys. Rev. Lett. **80**, 4737 (1998)
36. R. Kato, Bull. Chem. Soc. Jpn. **73**, 515 (2000)
37. K. Kanoda, T. Itou, K. Hiraki, J. Phys. IV (France) **131**, 21 (2005)
38. Y. Nogami, K. Oshima, K. Hiraki, K. Kanoda, J. Phys. IV (France) **9**, Pr10-357 (1999)
39. K. Yamamoto, K. Yakushi, K. Hiraki, T. Takahashi, M. Meneghetti, C. Pecile, Synth. Met. **135–136**, 563 (2003)
40. F. Nad, P. Monceau, K. Bechgaard, Solid State Commun. **95**, 655 (1995)
41. F. Nad, P. Monceau, Phys. Rev. B **51**, 2052 (1995)
42. A. Larkin, JETP **105**, 1793 (1994)
43. A. Larkin, S. Brazovskii, Solid State Commun. **93**, 275 (1995)
44. Yu. Ovchinnikov, K. Biljaković, J.C. Lasjaunias, P. Monceau, Europhys. Lett. **34**, 645 (1996)
45. J.P. Pouget, S. Ravy, Synth. Met. **85**, 1523 (1997)
46. P. Foury-Leylekian, D. Le Bollc'h, B. Hennion, S. Ravy, A. Moradpur, J.P. Pouget, Phys. Rev. B **70**, 180405 (2004)
47. F. Nad, P. Monceau, J.M. Fabre, Eur. Phys. J. B **3**, 301 (1998)
48. P. Monceau, F. Nad, S. Brazovskii, Phys. Rev. Lett. **86**, 4080 (2001)
49. S. Brasovskii, J. Phys. IV (France), Pr9-149 (2002); Synth. Met. **133**, 301 (2003); S. Brazovskii, J. Phys. IV (France) **114**, 9 (2004); S. Brazovskii, this issue.
50. J. Riera, D. Poilblanc, Phys. Rev. B **62**, 16243 (2000); **63**, 241102 (2001)
51. R. Laversanne, C. Coulon, B. Gallois, J.P. Pouget, R. Moret, J. Phys. Lett. (France) **45**, L393 (1984)
52. C. Coulon, S.S.P. Parkin, R. Laversanne, Phys. Rev. B **31**, 3583 (1985)
53. H.H.S. Javadi, R. Laversanne, A.J. Epstein, Phys. Rev. B **37**, 4280 (1988)
54. M. Nagasawa, F. Nad, P. Monceau, J.M. Fabre, Solid State Commun. **136**, 262 (2005)
55. P. Wzietek, F. Creuzet, C. Bourbonnais, D. Jérome, K. Bechgaard, P. Batail, J. Phys. (France) **3**, 171 (1993)
56. C. Coulon, A. Maaroufi, J. Amiell, E. Dupart, S. Flandrois, P. Delhaes, R. Moret, J.P. Pouget, J.P. Morand, Phys. Rev. B **26**, 6322 (1982)
57. F. Nad, P. Monceau, C. Carcel, J.M. Fabre, J. Phys. Condens. Matter **12**, L435 (2001)
58. F. Nad, P. Monceau, C. Carcel, J.M. Fabre, J. Phys. Condens. Matter **13**, L717 (2001)
59. W. Yu, F. Zhang, F. Zamborsky, B. Alavi, A. Baur, C.A. Merlic, S.E. Brown, Phys. Rev. B **70**, 121101 (2004)
60. F. Zhang, S.E. Brown, S. Lefebvre, P. Wzietek, J. Phys. IV (France) **131**, 9 (2005)

61. M.E. Lines, A.M. Glass, in *Principles and Applications of Ferroelectrics and Related Materials* (Clarendon Press, Oxford, 1977)
62. F. Nad, P. Monceau, L. Kaboud, J.M. Fabre, Europhys. Lett. **73**, 567 (2006)
63. T. Nakamura, private comunication; K. Furukawa, T. Hara, T. Nakamura, J. Phys. Soc. Jpn. **74**, 3288 (2005)
64. F. Nad, P. Monceau, T. Nakamura, K. Furukawa, J. Phys. Condens. Matter **17**, L399 (2005)
65. J.P. Pouget, P. Foury-Leylekian, D. Le Bolloc'h, B. Hennion, S. Ravy, C. Coulon, V. Cardoso and A. Moradpour, J. Low Temp. Phys., **142**, 147 (2006)
66. F. Nad, P. Monceau, H. Hiraki, T. Takahashi, J. Phys. Condens. Matter **16**, 7107 (2004)
67. L.E. Cross, Ferroelectrics **76**, 241 (1987)
68. M. Meneghetti, C. Pecile, K. Yakushi, K. Tamamoto, K. Kanoda, K. Hiraki, J. Solid State Chem. **168**, 632 (2002)
69. K. Biljaković, J.C. Lasjaunias, P. Monceau, F. Levy, Phys. Rev. Lett. **67**, 1902 (1991)
70. J.C. Lasjaunias, K. Biljaković, D. Staresinić, P. Monceau, S. Takasaki, J. Yamada, S.-I. Nakatsuji, H. Anzai, Eur. Phys. J. B **7**, 541 (1999)
71. J. Odin, J.C. Lasjaunias, K. Biljaković, P. Monceau, K. Bechgaard, Solid State Commun. **91**, 523 (1994)
72. K. Nomura, H. Hosokawa, N. Matsunaga, M. Nagasawa, T. Sambongi, H. Anzai, Synth. Met. **70**, 1295 (1995)
73. A. Hoshikawa, K. Nomura, S. Takasaki, J. Yamada, S. Nakatsuji, H. Anzai, M. Tokumoto, N. Kinoshita, J. Phys. Soc. Jpn. **69**, 1457 (2000)
74. J.C. Lasjaunias, J.P. Brison, P. Monceau, D. Steresinić, K. Biljaković, C. Carcel, J.M. Fabre, J. Phys. Condens. Matter **14**, 837 (2002)
75. H. Requardt, R. Currat, P. Monceau, J.E. Lorenzo, A.J. Dianoux, J.C. Lasjaunias, J. Marcus, J. Phys. Condens. Matter **9**, 8639 (1997)
76. K. Biljaković, J.C. Lasjaunias, F. Zougmoré, P. Monceau, F. Levy, L. Bernard, R. Currat, Phys. Rev. Lett. **57**, 1907 (1986)
77. J.E. Eldridge, G.S. Bates, Mol. Cryst. Liq. Cryst. **119**, 183 (1985)
78. H.K. Ng, T. Timusk, K. Bechgaard, J. Phys. Coll. (France) **44**, C3867 (1983)
79. S.M. Genensky, G.F. Newell, J. Chem. Phys. **26**, 486 (1957)
80. B. Wunderlich, H. Baur, Adv. Polym. Sci. **7**, 151 (1970)
81. W. Reese, J. Macromol. Sci. Chem. **3**, 1257 (1969)
82. H. Yang, J.C. Lasjaunias, P. Monceau, J. Phys. Condens. Matter **12**, 7183 (2000)
83. K. Biljaković, J.C. Lasjaunias, P. Monceau, J. Phys. IV (France) **9**, Pr10-27 (1999)
84. J. Coreneus, B. Alavi, S.E. Brown, Phys. Rev. Lett. **70**, 2322 (1993)
85. K. Biljaković, in *Phase Transitions and Relaxation in Systems with Competing Energy Scales*, ed. by T. Riste, D. Sherrington (Kluwer, Dordrecht, 1993), p. 339
86. L. Lundgren, P. Svedlindh, P. Nordblad, O. Beckman, Phys. Rev. Lett. **51**, 911 (1983)
87. J.P. Bouchaud, E. Vincent, J. Hamman, J. Phys. I (France) **4**, 139 (1994)
88. J.C. Lasjaunias, K. Biljaković, D. Staresinić, P. Monceau, J. Phys. IV (France) **12**, Pr9-23 (2002)
89. R. Mélin, K. Biljaković, J.C. Lasjaunias, P. Monceau, Eur. Phys. J. B **26**, 417 (2002)

90. R. Mélin, K. Biljaković, J.C. Lasjaunias, Eur. Phys. J. B **43**, 489 (2005)
91. J.C. Lasjaunias, S. Sahling, K. Biljaković, P. Monceau, J. Phys. IV (France) **12**, Pr9-27 (2002)
92. S. Sahling, J.C. Lasjaunias, K. Biljaković, P. Monceau, J. Low Temp. Phys. **133**, 273 (2003)
93. S. Sahling, J.C. Lasjaunias, R. Mélin, P. Monceau, G. Reményi, Eur. Phys. J. B **59**, 9 (2007)

11

Ferroelectricity and Charge Ordering in Quasi-1D Organic Conductors

S.A. Brazovskii

The family of molecular conductors TMTTF/TMTSF-X demonstrates almost all known electronic phases in parallel with a set of weak structural modifications of "anion ordering" and mysterious "structureless" transitions. Only in early 2000s their nature became elucidated by discoveries of a huge anomaly in the dielectric permittivity and by the NMR evidences for the charge ordering (disproportionation). These observations have been interpreted as the never expected ferroelectric transition. The phenomenon unifies a variety of different concepts and observations in quite unusual aspects or conjunctions: ferroelectricity of good conductors, structural instability toward the Mott–Hubbard state, Wigner crystallization in a dense electronic system, the ordered $4K_{\mathrm{F}}$ CDW, richness of physics of solitons, interplay of structural and electronic symmetries. The corresponding theory of the "combined Mott–Hubbard state" deals with orthogonal contributions to the Umklapp scattering of electrons coming from the two symmetry breaking effects: the built-in nonequivalence of bonds and the spontaneous nonequivalence of sites. The state gives rise to several types of solitons, all of them showing in experiments. On this basis, we can interpret the complex of existing experiments, and suggest future ones, such as optical absorption and photoconductivity, combined ferroelectric resonance and the phonon antiresonance, plasma frequency reduction.

11.1 Introduction: History and Events

The discovery of first organic superconductors (Bechgaard, Jerome, Ribault, et al. 1979/1980, see[1] [12, 13]) gave rise to a number of directions which none could foresee in advance. A great deal of experiments and theoretical speculations were done already within first years. Still a firework of

[1] We start the literature list with proceedings of most important scientific conferences and reviews collections [1–11].

phases was so amusingly rich that it feeds intensive research till our days. The 20th anniversary was marked by a new discovery of ferroelectricity and charge ordering/disproportionation whose consequences are the subject of this review.

We shall address only quasi-1D systems, and among them mostly the family of the Bechgaard–Fabre salts $(TMTCF)_2X$, with extensions to related materials. Typically, the selenium subfamily $(TMTSF)_2X$ was a basis for research on the metallic phase: particularly the superconductivity and diverse spectacular magnetic oscillations. The sulphur subfamily $(TMTTF)_2X$ was suitable for effects of electronic correlations, particularly the Mott–Hubbard dielectrization. The studies were mostly concentrated upon electronic phase transitions, which take place between 1 and 20 K, and on fashionable "non-Fermi liquid" aspects of the normal state at higher T. Actually this "normal" part of the phase diagram was filled, for most of compositions X, with tiny structural transitions of "anion orderings" observed for all noncentrosymmetric counterions X ([14, 15] for a review). Intentionally or not, these features were not appreciated either in experiments on electronic properties or in theories – except notes by Emery [16] and a systematic approach of Yakovenko and the author [17, 18].

The most powerful sleeping bomb was hidden in mysterious "structureless transitions" registered in sulphur subfamily for centrosymmetric anions (X = Br, AsF_6, etc.) at higher temperatures of about 100–250 K ([19–22] for a recent review). Unexplained, unattended by the community and finally forgotten, these transitions have been waiting for a revenge since mid-1980s. Only in 2000s their nature was identified, almost by chance, as something never expected: the ferroelectricity (FE) [23], or sometimes antiferroelectricity (AFE), see reviews [24,25]. Moreover, the ferroelectricity happened to be a consequence of a hidden charge ordering/disproportionation (CO/CD) [26].[2] The whole high temperature region of the phase diagram happened to be symmetry broken. More history excursions can be found in [22, 27, 28], see also Appendix 4.

The phenomenon is gaining ground, as it is being found in more and more compounds [22, 29, 30]. Also the charge ordering is registered now in layered organic materials, see [31, 32] for reviews, accompanied by the anomalously high dielectric permittivity [33], also signs of the ferroelectricity have been indicated [34]. The ferroelectricity was observed also in dielectric mixed-stacks organic compounds showing the neutral-ionic transition [35]. The energy scale (up to 300 K in terms of temperatures and ~1,000 K in terms of energy gaps) of the charge ordering effect is so high that it allows to anticipate its hidden

[2] The terms charge ordering, charge disproportionation, and charge localization have been utilized in the literature to identify the described phenomena. Here we shall typically use a more easily pronounced version of charge ordering, and reserve the term charge disproportionation for the same effect to underline nonequal molecular charges. The charge ordering transition temperature will be called T_{CO}; its other notations in the literature were typically T_0 or T_{CD}.

presence also in the metallic subfamily $(TMTSF)_2X$; the first direct sign has been brought by the case of $(TMTSF)_2FSO_3$ [32]. At least the effect must determine the properties of many materials where its presence has already been confirmed.

How could it happen, that such a dominant effect, which is so clear/obvious today, was missed, with so many misleading consequences? Actually there was more than enough of warnings through decades. In the prehistoric epoch,[3] it was the structural $4K_F$ anomaly, for that time this discovery could also be classified as a never expected one ([36–39] for a theory; [40] for a review). Its impact upon electronic properties has been shown via the strong effect of the $4K_F + 2K_F$ lock-in [41], unfortunately this spectacular observation was washed out from memory. The coexisting $4K_F$ and $2K_F$ charge density waves (CDWs) with the exact trimerization were observed in NMP-TCNQ [38]; remarkably, the T dependent conductivity resembles the one of $(TMTTF)_2X$. The sequence of transitions (spontaneous dimerization followed by the spin-Peierls (SP) tetramerization) in MEM-TCNQ [42] can be compared directly with today's cases of $(TMTTF)_2PF_6$ and $(TMTTF)_2ReO_4$, see Sect. 11.5.2.

Soon after the discovery of organic superconductors, a very rich information has been accumulated on subtle structural transitions of the anion orderings (see the review [15]), but these "gifts of the magi" were not exploited as much as they deserve. As an exemption, in mid-1980s the interplay of electronic and structural properties was emphasized in the earlier construction of the universal phase diagram [17, 18]; its milestones were just the cases of the intersite charge ordering in $(TMTTF)_2SCN$ and the interchain charge disproportionation in $(TMTSF)_2ClO_4$ (in contemporary language).

In retrospect of the structureless transitions, the events started to accelerate in 1997 when a theoretical work by Seo and Fukuyama [43] indicated that the AFM state may be accompanied by the spontaneous inequivalence of on-cite occupations – the charge disproportionation. Next year, the NMR experiments by Hiraki and Kanoda [44] have allowed to register the charge disproportionation in a new compound $(DI\text{-}DCNQI)_2Ag$, with understanding of observation of the Wigner crystallization or equivalently the condensation of the $4K_F$ anomaly.[4] The SDW ordering was present indeed, but at a temperature T_{SDW} which is more than one order of magnitude lower than the charge ordering one T_{CO}. One year later, in 1999, Nad and Monceau [46], pursuing other goals of physics of sliding incommensurate SDWs, have faced the structureless transitions. Their methodic has allowed to precisely pinpoint the transition temperatures and to show that the already pronounced anomaly

[3] A unique review connecting the science of organic metals throughout all decades is offered in [13].

[4] The charge ordering had been already found [45] in substituted perylene salts $(TMP)_2X$ ($X = PF_6, AsF_6$), which was left unattended, may be because of the accent upon effects of disorder. The charge ordering was identified via combined observations of the $4K_F$ CDW by the X-ray and of the doubling of sites by the C^{13} NMR.

in microwave response [21] becomes unsustainably high in the low frequency experiments of [46]. This strong anomaly and the precise localization of its temperature have attracted the attention of the Brown group, and in 2000 they registered by the NMR the spontaneous inequivalence of sites since its very onset at $T_{\rm CO}$. It took another year to finally resolve the mystery of the structureless transitions in 2001 [23]. The remarkable improvement in the experimental techniques has allowed to obtain anomalies as sharp as shown in Fig. 11.2. This, together with the suggested theory [23, 47], left no doubt that we are dealing with a least expected phase transition to the ferroelectric state. This second order transition manifests itself in the giant anomaly of the dielectric permittivity $\varepsilon(T)$, Fig. 11.2, which perfectly fits the Landau–Curie–Weiss law, Fig. 11.7 below.

The ferroelectricity is ultimately coupled with the charge ordering, which is followed by a fast formation, or a steep increase, of the conductivity gap Δ. There is no sign of a spin gap formation [48] or of a spin ordering down to the ten times lower scale of $T_{\rm e}$. Hence we are dealing with a surprising ferroelectric version of the Mott–Hubbard state which usually was associated rather with magnetic orderings. The anomalous diverging polarizability is coming from the electronic system, even if ions are very important to choose and stabilize the long range 3D order. The ferroelectric transition in $(TMTTF)_2X$ is a particular, bright manifestation of a more general phenomenon of charge ordering, which now becomes recognized as a common feature of organic and some other conductors [31, 32, 49].

This rich history tells us about the necessity for reconciliation of different branches of synthetic metals which have been almost split for two decades.[5] Indeed, the major success in finding the ferroelectric anomaly was due to precise low frequency methods for the dielectric permittivity ε [50]: designed for pinned CDWs in inorganic chain conductors [51], the methods were applied also to SDWs in organic conductors [52], and finally to the structureless transitions [46]. On the theory side, the author's approach of the Combined Mott–Hubbard state [23, 47, 53, 54] has been derived from a similar experience [55, 56] in a model of conducting polymers.

11.2 Hierarchy of Phases in Quasi-1D Organic Conductors

Low-dimensional electronic systems serve as a workshop on both particular and general problems of strong correlations, degenerate collective ground states, their symmetry and topological features. The richest opportunities have been opened by the family of first quasi-1D organic superconductors: the

[5] The conference ICSM-82 [1] was both the summit in science of synthetic metals and the last meeting where its all directions were present simultaneously and in mutual recognition.

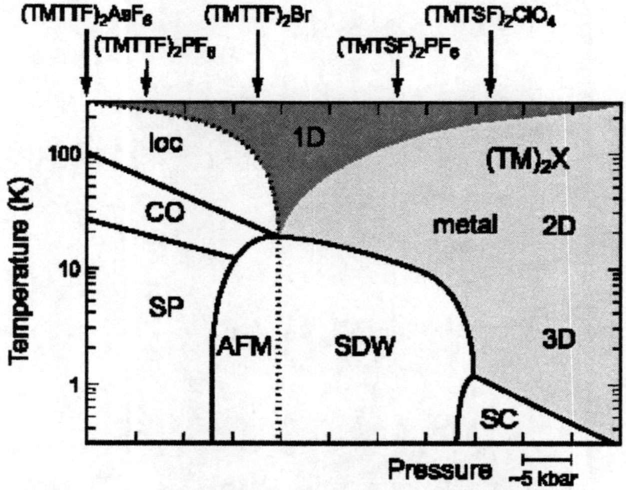

Fig. 11.1. Schematic phase diagram of the $(TMTCF)_2X$ family. SC – superconductivity; AFM – antiferromagnet – i.e., commensurate SDW; SP – spin-Peierls; CO – charge ordering, loc – charge localization (CO pretransitional effect); 1D, 2D, 3D – dimensional regimes. After Dressel, Dumm et al.

Bechgaard–Fabre salts $(TMTSF)_2X$, $(TMTTF)_2X$. These compounds demonstrate, at low temperatures T, transitions to almost all known electronic phases, see [12, 13] and Fig. 11.1. These are: a normal metal, a regime with the Mott–Hubbard charge localization (the (para)magnetic insulator – MI in classification of [17, 18]), a spin-Peierls (i.e., a CDW of bonds dimerization in a framework of strong electronic repulsion), a spin–density wave (SDW), an antiferromagnet (AFM) (i.e., the doubly commensurate SDW), a SDW induced by high magnetic fields (FISDW), and finally a superconductivity (SC) (of still a suggestive nature [57], see also Appendix 4). Temperatures of electronic transitions $T = T_e$ range within 1 K for the SC, 10–15 K for SDWs, 10–20 K for the spin-Peierls phase, see Fig. 11.1.

11.2.1 Structural Transitions of the Anion Ordering

There is a set of several weakly different structural types due to the anion ordering (AO) [15], which are fine arrangements of chains of singly charged counterions X. The temperatures T_{AO} of these weak structural transitions are usually about 100 K. The AOs are characterized, as any superstructure, by their wave numbers $\mathbf{q} = (q_{\parallel}, q_{\perp}) \neq 0$, where $q_{\parallel}$ and $q_{\perp}$ are the components in the stack and the interstack directions. The AOs were always observed for noncentrosymmetric ions (typically tetrahedral X = ClO_4, ReO_4, BF_4, etc., except for the special case of the linear ion X = SCN). The orientational ordering was supposed to be a leading mechanism, see [58] for a theory, with positional displacements (arrows in Fig. 11.3) being its consequences only. But

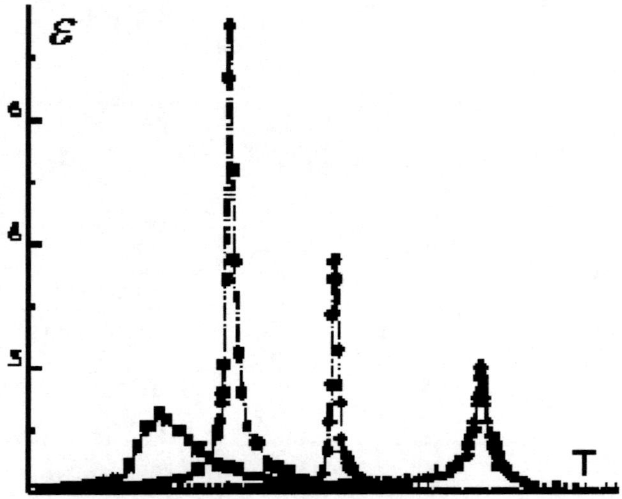

Fig. 11.2. Linear plots of the temperature dependence of the real part of the dielectric permittivity ε' at 100 kHz for X = PF_6 (*asterisks*), BF_4 (*squares*), AsF_6 (*circles*), SbF_6 (*diamonds*), ReO_4 (*triangles*). Reproduced from [25], see also figures in [24]

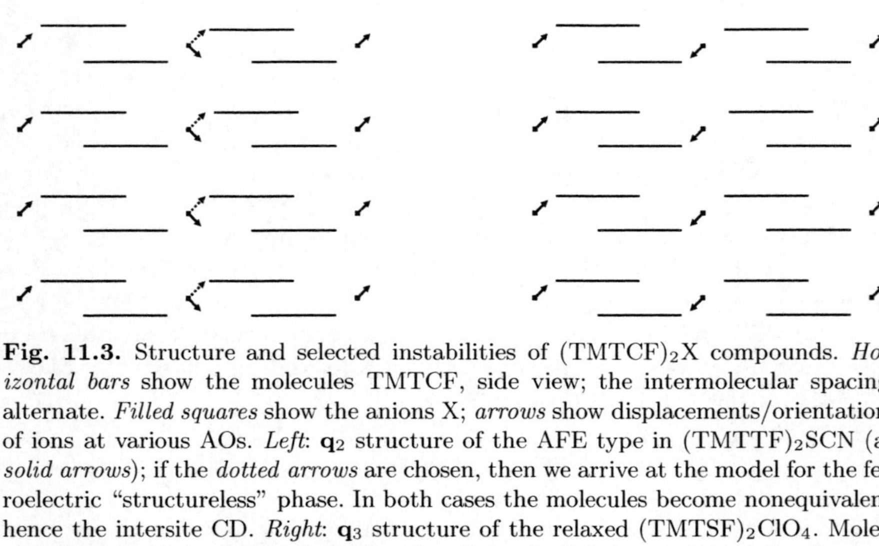

Fig. 11.3. Structure and selected instabilities of $(TMTCF)_2X$ compounds. *Horizontal bars* show the molecules TMTCF, side view; the intermolecular spacings alternate. *Filled squares* show the anions X; *arrows* show displacements/orientations of ions at various AOs. *Left*: $\mathbf{q}_2$ structure of the AFE type in $(TMTTF)_2SCN$ (all *solid arrows*); if the *dotted arrows* are chosen, then we arrive at the model for the ferroelectric "structureless" phase. In both cases the molecules become nonequivalent, hence the intersite CD. *Right*: $\mathbf{q}_3$ structure of the relaxed $(TMTSF)_2ClO_4$. Molecular sites are kept equivalent within the chain, but chains environment alternate, hence the interchain CD. Other tetramerized structures are shown in [15]

there is also a universal mechanism of the Earnshow instability [59], inherent to all structures of classical charged particles, which can trigger independently the displacive component in materials with both centrosymmetric and noncentrosymmetric anions. (See more in the Appendix 1.)

Already within the unperturbed crystal structure at $T > T_{\mathrm{CO}}$, the tiny dimerization of bonds by anions X in $(TMTCF)_2X$ provokes the dielectrization

[18,60,61]. Regular alternation of ions, positioned against every second intermolecular spacing, dimerizes the intermolecular distances. It doubles the on-stack unit cell, hence changes the mean electronic occupation from 1/2 per molecule to 1 per dimer. The bond alternation gives rise [60] to the relatively small Umklapp scattering U_b, which opens (according to Dzyaloshinskii and Larkin [62], Luther and Emery [63,64], see Sects. 11.4 and 11.8) the route to the Mott–Hubbard insulator. (See [53] and [65] for a history introduction and Appendix 4 for some quotations).

11.2.2 Charge Ordering Transitions

At similar or even higher range, $T = T_{\mathrm{CO}} \approx 100$–$200\,\mathrm{K}$, also other unidentified transitions were observed sometimes in the TMTTF subfamily. Their signatures were seen in conductivity [19], microwave permittivity [21], in thermopower [20]. But there was no trace of any lattice effects, hence the title structureless transitions was assigned. Recently their mysterious nature has been elucidated by discoveries of the huge anomaly in the dielectric permittivity ε at T_{CO} [23,24] and of the charge disproportionation seen by the NMR [26] at $T < T_{\mathrm{CO}}$. It is still an experimental challenge to refine the structure; a scheme for its reconstruction is outlined in this section.

The ferroelectric anomaly signifies that the term structureless is not quite correct: this is not an isomorphic modification. While the Brave lattice is based on the same unit cell indeed, the full space symmetry is changed by loosing the inversion center, which is the necessary condition for the ferroelectric state.

The NMR experiments [26] have clearly detected the appearance at the T_{CO} of the site nonequivalence as a sign of the CO/CD. Recently the charge disproportionation was also confirmed by means of the molecular spectroscopy [66], see Fig. 11.17 below. While bonds are usually dimerised already within the basic structure, see Fig. 11.3, the molecules stay equivalent above T_{CO}, the latter symmetry being lifted by the charge ordering transition. At presence of the inequivalence of bonds, the additional inequivalence of sites lifts the inversion symmetry, hence the allowance for the chain's polarization leading to the ferroelectricity.

The confidence for this identification comes from a good fortune, that the 3D pattern of the charge ordering appears in two, AFE and FE, forms. Both types AFE and FE are the same paramagnetic insulators (the MI phase of [17, 18]), which is shown by comparison of electric and magnetic measurements. The NMR brought the same observation [26] both for the structureless transitions and for the particular AFE phase in $(\mathrm{TMTTF})_2\mathrm{SCN}$; also the permittivities at high frequencies and temperatures are very similar. The structure of the $(\mathrm{TMTTF})_2\mathrm{SCN}$ has been already known as a very rare type of $\boldsymbol{q}_2 = (0, 1/2, 1/2)$. The site inequivalence was already identified by the structural studies, see [15] and Fig. 11.3. The structure was completely refined; the anions displacements are unilateral for each column while undulating among them.

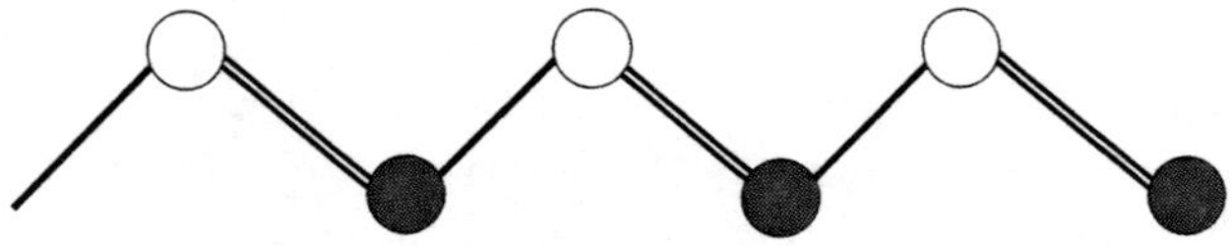

Fig. 11.4. The scheme of a conjugated polymer of the $(AB)_x$ type, realized recently as the modified polyacetylene [72]. *Dark and light circles* correspond to chemically inequivalent monomers (actually the carbon atoms with different ligands). Single and double connections show the expected spontaneous dimerization of bonds. Curiously, this model takes just the form of the hypothetical design by W. Little for a polymeric superconductor (D. Jerome, introduction to ISCOM-03 [3]), which suggestion gave birth to the whole science of synthetic metals

So, what is the difference between the $\boldsymbol{q}_2$ case and the structureless transitions? The $\boldsymbol{q}_2$ structure already showed the common polarization along a single stack, $q_{\|} = 0$, but alternating in perpendicular directions, $\boldsymbol{q}_{\perp} \neq 0$ (Fig. 11.3); it gives, as we can interpret today, the AFE ordering. For the structureless transition all displacements must be identical $\boldsymbol{q} = 0$ among the chains/stacks, thus leading fortunately to the ferroelectric state. We arrive at the quite sound conjecture that the ferroelectric "structureless" state corresponds to the scheme of Fig. 11.3 where we should choose the dotted arrows instead of solid ones. Without the advantage of having the structural information [15] on the $(TMTTF)_2SCN$, our understanding of the nature of the ferroelectric compounds would be more speculative. This case can be viewed as a corner stone, or Rosetta stone, in a sense that it belongs to two classes of anion ordering and charge ordering systems and transferrers a complementary information.

In any case, FE or AFE, the polar displacement gives rise to the joint effect of the built-in and the spontaneous contributions to the dimerization, due to alternations of both bonds and sites.[6] None of these two types of dimerization changes the unit cell of the zigzag stack which basically contains two molecules, hence $q_{\|} = 0$ for the charge ordering wave vector. But their sequence lifts the mirror (glide plane) and then the inversion symmetries which must lead to the on-stack electric polarization. This interference resembles the orthogonal mixing in the "combined Peierls state" [55] in conjugated polymers of the $(AB)_x$ type (Fig. 11.4). Here also the bond dimerization is the $q_{\|} = 0$ transition, just because of the backbone zigzag structure: what looks for electrons as a period doubling is structurally the lifting of the glide plane symmetry.

[6] Both contributions can be of the built-in type in the particular case of the $(TMTSF)_{0.5}(TMTTF)_{0.5}$ mixture [67], where the alternation of S/T-containing molecules provides the site dimerization. Oppositely, both coexisting dimerizations may appear spontaneously, but these two independent symmetry breakings require for two successive second order phase transitions or for the first order one. The later option may be relevant to observations [30, 33].

11.2.3 Overlapping and Coexistence of Phases

The structureless displacive instability and the usual orientational AOs can be independent. This is supported by finding of a sequence of the ferroelectric and the anion ordering transitions in $(TMTTF)_2ReO_4$ [24]. We will show in Sect. 11.5.2 that this true present from the Nature gives an access to compound solitons, as to intriguing events of the spin-charge reconfinement. At the same time, the case of $(TMTTF)_2SCN$, where $q_{\|} = 0$ while $q_{\perp} \neq 0$, presents the corner stone belonging to both types of AOs and charge ordering transitions.

Notice finally that the $q_{\|} = 0$ structure is not ultimately the on-chain charge ordering. The so-called $\boldsymbol{q}_3 = (0, 1/2, 0)$ type of the anion ordering [15], observed in the relaxed phase of the $(TMTSF)_2ClO_4$, shows the interchain redistribution of charges. This is a kind of the "incommensurability transition" (with an unusually reversed order in comparison with the conventional lock-in to the commensurable state) which may be considered [17,18] as a prerequisite of the superconductivity.

11.2.4 Electronic Mechanism of the Charge Ordering

To get an idea of the driving force behind the hidden structureless transition we shall follow the example of the "combined Peierls state" [68,69] developed for polymers like the modified polyacetylene. It describes a joint effect of built-in and spontaneous contributions to the dimerization, hence to the electronic gap. Since this concept was formulated [70], it has been widely used in studies of conducting polymers gaining a clear experimental support in diverse observations, especially in optics and ESR, see [71]. Particularly relevant is the experimentally recent [72] case of the "orthogonal mixing" relevant to polymers of the $(AB)_x$ type (Fig. 11.4). Here the built-in gap Δ_s comes from the site dimerization due to the AB alternation, while the spontaneous contribution to the total gap Δ comes from the dimerization of bonds Δ_b, like in the generic Peierls effect. Importantly, the two gaps add in quadrature: $\Delta = \sqrt{\Delta_\mathrm{in}^2 + \Delta_\mathrm{ex}^2}$ – the amplitudes of electron's scattering (Fig. 11.11 right) are shifted by $\pi/2$.

For the present case of the charge ordering, the Peierls effect is substituted by the Mott–Hubbard one. Also the built-in and the spontaneous effects are interchanged: the built-in one comes from inequivalence of bonds while the spontaneous one comes from nonequivalence of sites. The charge gap $\Delta = \Delta(U)$ appears as a consequence of both contributions to the Umklapp scattering, and it is a function of the total amplitude $U = \sqrt{U_\mathrm{s}^2 + U_b^2}$. (But the observable gap $\Delta(U)$ is not additive in quadratures any more.)

The electronic energy F_e depends only on the charge gap Δ, which is a function of only the total $U = \sqrt{U_\mathrm{s}^2 + U_b^2}$. The energy of lattice distortions F_l depends only on the spontaneous site component U_s: $F_l = (K/2)U_\mathrm{s}^2$, where

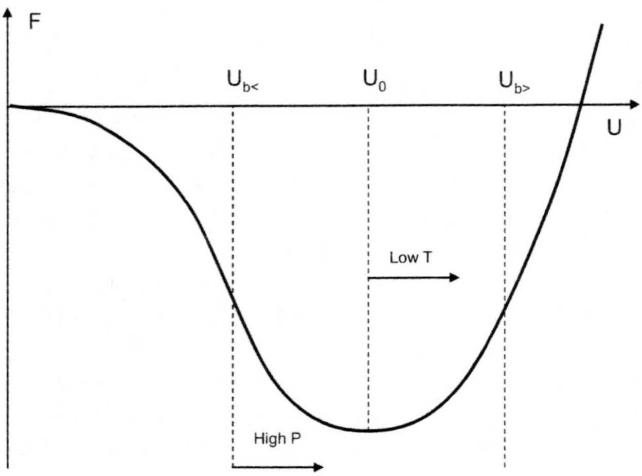

Fig. 11.5. Ground state energy $F(U)$ as a function of the total magnitude $U = \sqrt{U_s^2 + U_b^2}$ of the commensurability (Umklapp) potential. Its minimum at U_0 is reachable if $U_b < U_0$ corresponding to a position $U_{b<}$ at the U-axis of the figure; then the spontaneous U_s is created to augment U_b to U_0. It does not happen if $U_b > U_0$ corresponding to the position $U_{b>}$, then there is no transition. The *arrows* show that these two regimes can interchange by either changing of U_0 by temperature or of U_b by pressure

K is an elastic constant. Thus the total energy can be written in terms of the total U:

$$F(U) = F_e(U) + 1/2KU^2 - 1/2KU_b^2,\ U \geq U_b.$$

The ground state is determined by its minimum over U, but unusually at the constraint $U \geq U_b$. We can encounter three possibilities for the energy function $F(U)$ (Fig. 11.5):

(a) It has no minimum except $U = 0$: no charge ordering at any condition
(b) It has a minimum at some value $U = U_0 < U_b$: charge ordering is possible but not reached yet
(c) It has a minimum at some value $U = U_0 > U_b$, which now determines the ground state. Since the value U_0 increases with decreasing temperature, there will be a phase transition at $U_0(T) = U_b$ provided that $U_0(0) > U_b$

The phase transition in the regime (c) can be reversed if U_b is made to increase passing above U_0, then the charge ordering disappears. This is what seems to happen typically in experiments under pressure [73, 74].

How does this minimum at U_0 appear actually? In principle, the electron energy $F_e(U) < 0$ is always gained by opening the gap Δ which reduces the zero point fluctuations. But to overcome the energy loss $\sim U_s^2$ from lattice deformations and charge disproportionation, we need $F_e(U) \sim -U^{2-\zeta}$ with $\zeta \geq 0$. For conventional $2K_F$ CDWs it is always the case with $F_e(U) \sim -U^2 \ln E_F/U$ corresponding to the limit $\zeta \to 0$. For our case of the $4K_F$

modulations the condition is reached only at large enough interactions [39] with the marginal case $\zeta = 0$ corresponding to the so-called Luther–Emery line [63,64], see more in Sect. 11.4. We can view it also as a general criterium of the $4K_{\rm F}$ instability [39].

11.2.5 Electric Polarization and Ferroelectricity

In principle, there are three contributions to the electric polarizability:

1. *Intergap electronic polarizability* is regular at $T_{\rm CO}$: $\varepsilon_\Delta \sim \omega_{\rm p}^2/\Delta^2$, where $\omega_{\rm p}$ is the plasma frequency. At the transition, where $\Delta(T \approx T_{\rm CO})$ is still well below its low T value $\Delta(0)$, ε_Δ can be as large as $\sim 10^3$ which corresponds indeed to the background upon which the anomaly at $T_{\rm CO}$ is developed. The value of the gap well corresponds to the lower $\varepsilon_\Delta \approx 100$ as it was already estimated in [21].
2. *Ion displacements* could already lead to the macroscopic polarization, like in usual ferroelectrics, but taken alone they cannot explain the observed giant magnitude of the effect. Taking the typical parameters [15] of AOs (recall the SCN case as the corner stone), the ionic displacive contribution may be estimated (see Appendix 2.1) as $\varepsilon_i \sim 10^1 T_{\rm CO}/|T - T_{\rm CO}|$, which is by 10^{-3} below a typical experimental value $\varepsilon \approx 2.5 \times 10^4 T_{\rm CO}/(T - T_{\rm CO})$ [23].
3. *Collective electronic contribution* can be estimated roughly [23], see Appendix 2.1, as a product of the above two $\varepsilon_{\rm el} \sim \varepsilon_i \varepsilon_\Delta$, which provides both the correct T dependence and the right order of magnitude of the effect. The anomalous diverging polarizability is coming from the electron subsystem, even if the instability is triggered by the ions, whose role is to stabilize the long range 3D ferroelectric order, and to discriminate between FE, AFE, or more complex patterns [30], see more in the Appendix 2.2.

11.3 Electronic Properties

Here we shall give a short summary of experimental results on electronic properties of $(\mathrm{TMTTF})_2\mathrm{X}$ compounds.[7]

11.3.1 Permittivity

Typical plots of the temperature dependence of the dielectric permittivity ε[8] in Figs. 11.2 and 11.6 demonstrate very sharp (even in the log scale of Fig. 11.6) ferroelectric peaks. Most of the cases show the purely monodomain "initial" ferroelectric permittivity, ideally fitting the Curie law $\varepsilon \sim |T - T_{\rm CO}|^{-1}$, Fig. 11.7 – even with the right ratio = 2 of the Landau theory for inclinations

[7] Experimental plots of this chapter are the results by Monceau et al. see [24,25] for a broad collection.

[8] We imply everywhere, unless specified, the real part of ε: Re$\varepsilon = \varepsilon'$.

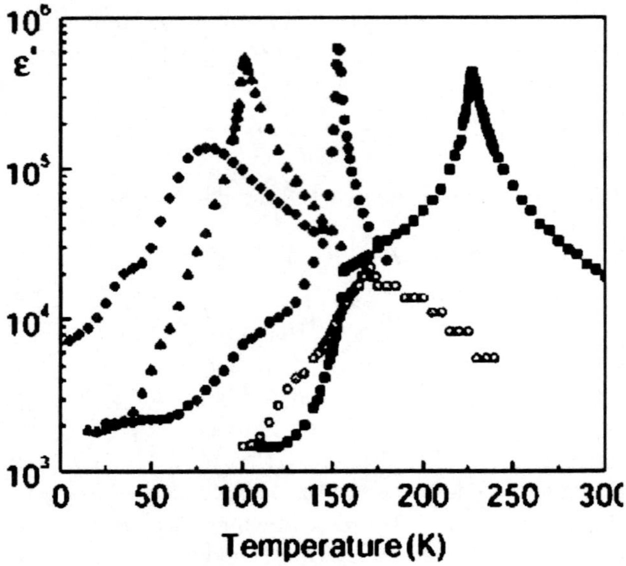

Fig. 11.6. $\log \varepsilon'$ vs. T at $f = 1\,\text{MHz}$ for FE cases $X = PF_6$ (*open diamond*), AsF_6 (*open triangle*), SbF_6 (*filled circle*), ReO_4 (*filled square*) and for the AFE X = SCN (*open circle with dot*). Reproduced from [24]

at $T \gtrless T_{CO}$ [23].[9] In general, we observe the frequency dependent depolarization of the monodomain ferroelectricity, instead of the more usual long time hysteresis of the repolarization. Still abundant normal carriers screen the ferroelectric polarization at the surface, which eliminates the need for the domain structure. Low-temperature studies are necessary to find the remnant polarization. Radiational damage or other disorder will also help to pin the domain walls and freeze the polarization, which should give rise to the conventional hysteresis curve.

The AFE case of X = SCN in Fig. 11.6 shows a smooth maximum of ε, rather than a divergent peak, as it should be. Nevertheless, the high T slopes of all curves for $\varepsilon(T)$ look very similar. It tells us that the ferroelectric state is gradually developing already within the high $T > T_{CO}$ 1D regime, before the 3D interactions discriminate between the in-phase FE and the out-of-phase AFE orderings. Recall that, contrary to these low frequencies, in the microwave range the X = SCN compound showed [21] the strongest response with respect to ferroelectric cases; hence the purely 1D regime of the ferroelectricity was recovered. This may also be the key to the AFE/FE choice.

[9] A relatively rounded anomaly in the PF_6 case of Fig. 11.7 correlates with its stronger frequency dispersion, Fig. 11.8: it can be a pinning of ferroelectric domain walls, in other words a hidden hysteresis. For an extended information on frequency dispersion, see [77, 78] and the review [25].

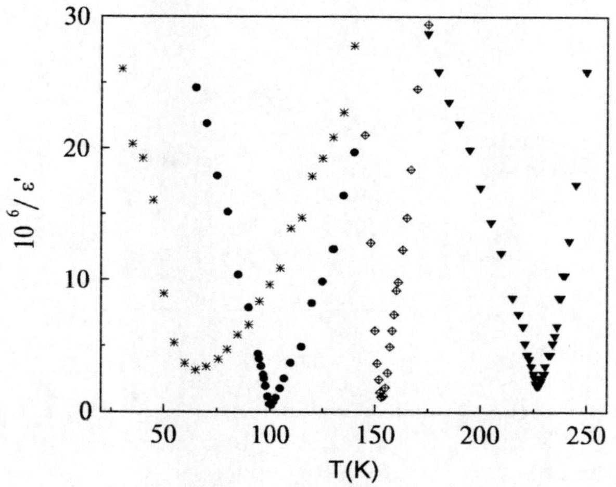

Fig. 11.7. Temperature dependence of the inverse of the real part of the dielectric permittivity ε' of $(TMTTF)_2X$ with X = PF_6, AsF_6, SbF_6, ReO_4 (in the order from left to the right) at the frequency of 100 Hz. Reproduced from [23, 24].

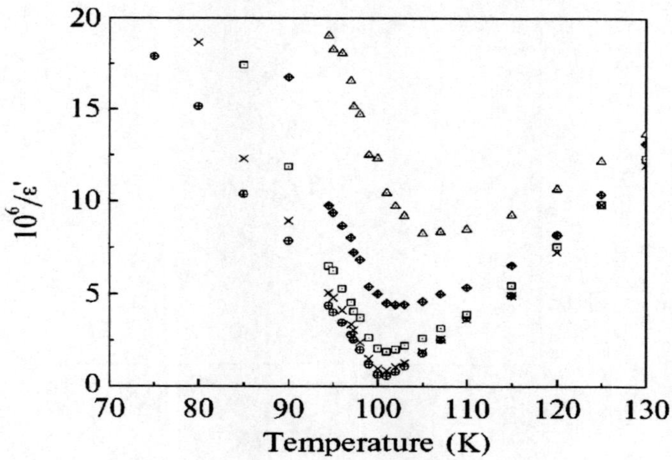

Fig. 11.8. Plots of $10^6/\varepsilon'(T)$ of $(TMTTF)_2AsF_6$ at 10,000, 3,000, 1,000, 300, 100 KHz. Nad et al. unpublished

Indeed, for highly polarizable units containing the already polar ion SCN, the dominating Coulomb forces will always lead to an AFE structure.

The X = ReO_4 case shows at 150 K a subsequent first order anion ordering transition of tetramerization. Here ε drops down (Fig. 11.6, see also [24,25] for details) which might be caused (via the factor ω_p^2/Δ^2) by the increase of the gap Δ as it is seen at the conductivity plot in Fig. 11.10. The case of X = PF_6 shows the spin-Peierls transition at $T = 19$ K. While being symmetrically equivalent to the anion ordering in X = ReO_4 case, it is of the second order, so that in $\varepsilon(T)$ it shows up only as a shoulder.

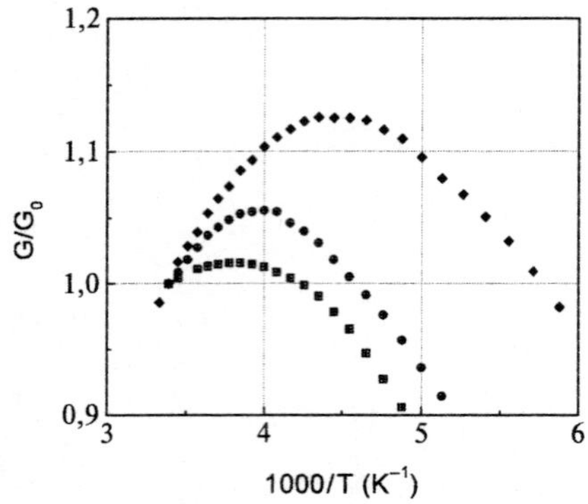

Fig. 11.9. Plots for the normalized conductivities G/G_0 (G_0 is the RT value) in the $T \approx T_\rho$ region. X = SbF_6 (*squares*), AsF_6 (*circles*), PF_6 (*diamonds*). The plots $G(T)$ show maxima at $T = T_\rho > T_{CO}$. Reproduced from [24, 25]

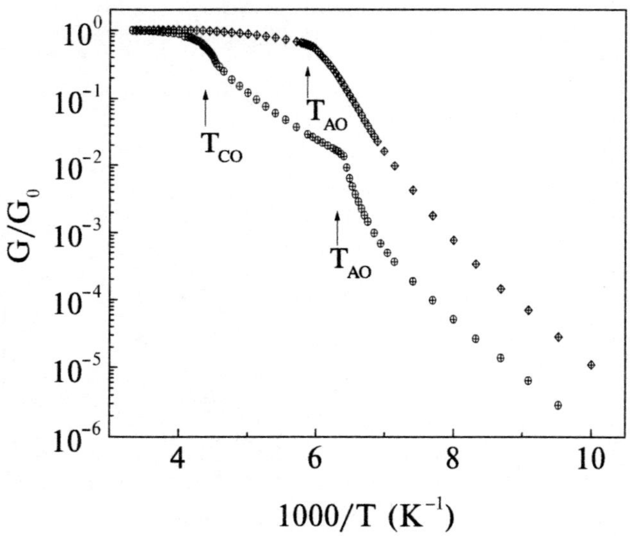

Fig. 11.10. Ahrenius plots for normalized conductivities G/G_0 in a wide T region. X = ReO_4 (*circles*), X = SCN (*diamond*). Reproduced from [47, 75]

11.3.2 Conductivity

Figure 11.9 gives several examples of the conductivity $G(T)$ within the high T region around T_ρ. Figure 11.10 shows, for selected examples, the conductivity $G(T)$ within a broad range of temperatures (see more cases in [24,25]). Typical plots correlate with old data [21],[10] but with more insight available today:

[10] In reviewed experiments by Nad et al., the conductivity was extracted via $\Im\varepsilon$, within the same experiment as measuring the $\mathrm{Re}\varepsilon$.

(1) Clear examples for conduction by charged spinless particles – holons = solitons: there are no gaps in spin susceptibility $\chi \approx cnst$ [48].
(2) There is no qualitative difference in $G(T)$ (as well as in NMR! [26]) between FE cases and the AFE one of the SCN: only the on-chain charge ordering is important, not the type of the interchain pattern!
(3) Coexistence of both types of transitions, CO and AO, in $(TMTSF)_2ReO_4$; the subsequent anion ordering increases the conduction gap and opens the spin gap [76].
(4) Gap contribution of the spontaneous site dimerization develops very fast, and soon it dominates over the bond dimerization gap. The last may not be seen at all – recall that the ferroelectric anomaly extends by at least 30 K above T_{CO}, Fig. 11.7, signifying the pretransitional CO, which is able to workout a pseudogap.

At $T < T_{\mathrm{CO}}$ the charge ordering adds more to the charge gap Δ, which is formed now by joint effects of alternations of bonds and sites. The conductivity $G(T)$ plots in Fig. 11.9 show this change by kinks at $T = T_{\mathrm{CO}}$, turning down to higher activation energies at low T. The steepness of G just below T_{CO} reflects the growth of the charge ordering contribution to the gap, which is expected to be $\delta\Delta(T) \sim \sqrt{T_{\mathrm{CO}} - T}$, see Sect. 11.4; it must correlate with $\varepsilon^{-1} \sim (T_{\mathrm{CO}} - T)$. Indeed, what may look as the enhanced gap (the tangent for the Ahrenius plot) near T_{CO}, actually can be its T dependence expected as $\Delta(T) = \sqrt{\Delta_b^2 + C\Delta_{\mathrm{s}}(0)(1 - T/T_{\mathrm{CO}})}$, $C \sim 1$. The differential plotting of $\Delta(T) = -\mathrm{d}\log G/\mathrm{d}(1/T)$ would be helpful.

So marginal effect of the built-in bond dimerization opens the route to compounds with equivalent bonds [22, 29, 44, 79].

$(DMtTTF)_2ClO_4$: Here, the ordered state has been already identified [30] as a complex (incommensurate in the transverse direction) AFE structure, which develops following a fascinating pseudo-first-order phase transition at $T = 150$ K. This material shows [22] the characteristic resistivity rise below quite typical $T_\rho \approx 120$–150 K, which does not affect, as expected, the susceptibility χ until the low $T_{\mathrm{SDW}} \approx 10$ K AFM ordering. The charge ordered phase is waiting to be tested for the NMR splitting and for the ε anomaly, or at least for traditional signatures of structureless transitions like the thermopower.

$(EDT\text{-}TTF\text{-}CONMe_2)_2X$: A similar situation is expected in this new material [22, 29, 79], which is also the Mott insulator with the conductivity gap $\approx$1,350 K. Unfortunately, there is a lack of information for this compound; vaguely it was declared [79] as a clear case of fourfold commensurability effects, following the "no CO/CD" scenario for $(TMTCF)_2X$ family, see Sect. 11.8.2.

$(DI\text{-}DCNQI)_2Ag$: The nondimerized compound, where the observation of the charge ordering [44] has started the modern trend. Here also the activation energy value $\approx$490 K is well above the $T_{\mathrm{CO}} \approx 220$ K.

Now we can close the circle to return to the $(TMTTF)_2X$ series and to guess that the observed conductivity downturn at $T_\rho > T_{\mathrm{CO}}$ may be mostly due to the fluctuational pseudogap coming from the charge ordering proximity.

There may be only a minor contribution from the bond dimerization specific to $(TMTTF)_2X$.

11.4 Ferroelectric Mott–Hubbard Ground State

11.4.1 Choosing the Theory Approach

The earliest theoretical prediction [43] applies literally to a situation expected in 3D, or at most 2D systems, where the charge ordering is set up simultaneously and in ultimate conjunction with the AFM/SDW order. In $(TMTTF)_2X$ type cases, the pronouncedly 1D electronic regime brings its particular character, as well as it allows for a specifically efficient treatment [23], which also happens to be particularly well suited to describe the ferroelectric transition.

We are using the bosonization procedure, see reviews [64, 80, 81], which is most adequate to describe low energy excitations and collective processes. It takes into account automatically the separation of spins and charges, which is a common feature of our systems. All information about basic interactions, whatever they are (on-site, neighboring sites, long range Coulomb, lattice contribution, see Sect. 11.8.1) is concentrated in a single parameter ($\gamma = K_\rho$ – in modern notations). This approach allows to efficiently use the symmetry arguments and classification. It allows to naturally interpret the solitonic spinless nature of elementary excitation – the thermally activated charge carriers. Most importantly for our goals, this approach provides a direct access to the dielectric permittivity. The procedure easily covers also the secondary anion ordering or spin-Peierls transitions at lower T. A rigorous way to describe a correlated 1D electronic system, bosonization also provides a physically transparent phenomenological interpretation in terms of fluctuating $4K_F$ density wave, i.e., a local Wigner crystal, subject to a weak commensurability potential. See more arguments in [18], and some quotations in Appendix 4.

Among other approaches, recall also the traditional line of RG theories in recent quasi-1D versions [82, 83]. The today's results do not pass the major test for the charge ordering (at least it was merely overlooked), hence they cannot be applied to $(TMTTF)_2X$ type cases, even if a work was deducted to this purpose [82].

Recall also numerical studies like exact diagonalization of small clusters with short range interactions or quantum chemistry methods [84, 85]. Usually they pass the test for the charge ordering, but experience difficulties to obtain its ferroelectric type as the most favorable one. Even in case of a success for the ground state, the access to polarization and to elementary excitations will not be sufficient, or even possible, particularly for more sophisticated cases of combined symmetry breakings, Sect. 11.5.2.

11.4.2 Ground State and Symmetry Breaking

We introduce the charge phase φ as for 2K$_F$ CDW or SDW modulations $\sim \cos(\varphi + (x - a/2)\pi/2a)$ – the origin is taken at the inversion center in-between the two molecules. Here a is the intermolecular distance; recall that $\pi/4a = K_F$ is the Fermi number in the parent metallic phase. Later on we shall need also the spin counting phase θ: φ and θ together define the CDW order parameters completely as $\mathcal{O}_{CDW} \sim \exp(i\varphi)\cos\theta$. Gradients of these phases φ'/π and θ'/π give local concentrations of the charge and the spin. The energy density (potential and kinetic) of charge polarizations is

$$\frac{1}{4\pi\gamma}\left\{(\partial_x\varphi)^2 v_\rho + (\partial_t\varphi)^2/v_\rho\right\}. \tag{11.1}$$

Here $v_\rho \sim v_F$ is the charge sound velocity and γ is the main control parameter, which depends on all interactions; its origin is discussed in Sect. 11.8.1. (γ is the same as γ_ρ of [17, 18] or K_ρ of our days.)

In addition to (11.1), there is also the commensurability energy H_U coming from the Umklapp scattering of electrons (Fig. 11.11a left). For our particular goals it is important to notice several interfering sources for the weak twofold commensurability, i.e., two contributions to the Umklapp interaction. Their forms can be derived from the symmetry alone [23], as we shall sketch now.

Consider the nondimerized system with 1/2 electrons per site. It possesses translational and inversion symmetries $x \to x + 1$ and $x \to -x$ which corresponds to phase transformations $\varphi \to \varphi + \pi/2$ and $\varphi \to -\varphi$. The lowest order invariant contribution to the Hamiltonian is $H_4 \sim U_4 \cos 4\varphi$. This is the fourfold commensurability energy which is usually very small, recall the conventional CDWs. The reason is not only a smallness of U_4 which is coming

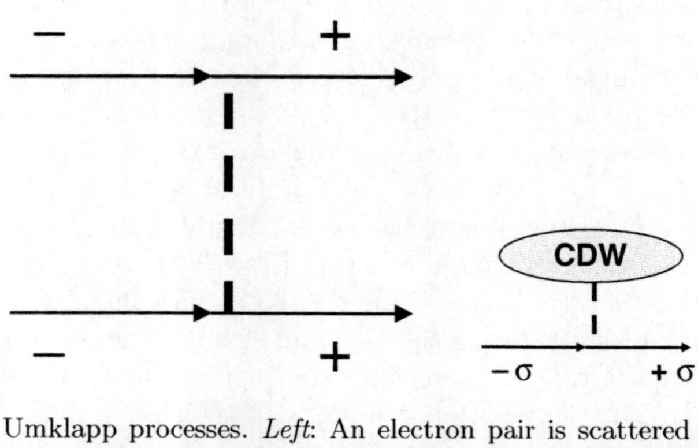

Fig. 11.11. Umklapp processes. *Left*: An electron pair is scattered from $-K_F$ to $+K_F$; the quasimomentum deficiency $4K_F$ is absorbed by the periodicity of the dimerised lattice. *Right*: At presence of the tetramerization providing the $2K_F$ CDW background, a single electron is scattered from $-K_F$ to $+K_F$ (conserving the spin σ); the deficiency $2K_F$ is absorbed by the CDW order corresponding to the tetramerization

from Umklapp interaction of eight particles, half of them staying high away from the Fermi energy. In addition, it is renormalized as $\sim \exp[-8\langle\varphi^2\rangle]$, that is it becomes small, being a product of two large numbers in the negative exponent, see more in Sect. 11.8.2.

Consider next the site dimerization which modulates alternatively the onsite energies. The preserved symmetries are $x \to x+2$ and $x \to 1-x$ (reflection with respect to the molecular cite center). Hence the invariance is required with respect to $\varphi \to \varphi + \pi$ and $\varphi \to \pi/2 - \varphi$ and we arrive at $H_U^s = -U_\mathrm{s} \sin 2\varphi$.

Consider finally the bond dimerization. The symmetry $x \to x+1$ is broken while the symmetries $x \to x+2$ and $x \to -x$ are preserved. The Hamiltonian must be invariant under the corresponding transformations of φ: $\varphi \to \varphi + \pi$ and $\varphi \to -\varphi$ and we arrive at $H_U^b = -U_b \cos 2\varphi$.

At presence of both types of dimerization, the nonlinear Hamiltonian, to be added to (11.1), becomes

$$\begin{aligned} H_U &= -U_b \cos 2\varphi - U_\mathrm{s} \sin 2\varphi = -U \cos(2\varphi - 2\alpha), \\ U &= (U_b^2 + U_\mathrm{s}^2)^{1/2}, \ \tan 2\alpha = U_\mathrm{s}/U_b. \end{aligned} \tag{11.2}$$

The $4k_\mathrm{F}$ charge modulation will be

$$\rho_{4K_\mathrm{F}} \sim \psi_-^2 \psi_+^{\dagger 2} + HC \sim \cos(\pi x/a + 2\alpha) = (-1)^{x/a} \cos(2\alpha),$$

where $\psi_\pm$ are one-electron operators near the points $\pm K_\mathrm{F}$. The $2K_\mathrm{F}$ bond and site CDWs are proportional to

$$\rho_{2K_\mathrm{F}}^{b,s} \sim \{\psi_- \psi_+^\dagger \mp h.c.\} \sim \cos\theta \{\cos, \sin\}(\pi x/2a + \alpha).$$

Their mean values are zero, because of the spin factor which vanishes in average $\langle \cos\theta \rangle = 0$ until the spin gap is established below the spin-Peierls temperature, see Sect. 11.5.2.

The U_s comes from the electronic charge ordering and ionic displacements coupled by long range 3D Coulomb and structural interaction, which are well described by the mean field approach. But the electronic degrees of freedom must be treated exactly at a given U_s. The renormalization, due to quantum fluctuations of phases, leaves the angle α invariant, but it reduces U in (11.2) down to $U^* \sim \Delta^2/\hbar v_\mathrm{F}$ ($U^* \neq 0$ only at $\gamma < 1$); it determines the gap $\Delta \sim U^{1/(2-2\gamma)}$. (In scaling relations, we imply units E_F for Δ and E_F/a for U.) The spontaneous charge ordering $U_\mathrm{s} \neq 0$ requires that $\gamma < 1/2$, far enough from $\gamma = 1$ for noninteracting electrons. The magnitude $|U_\mathrm{s}|$ is determined by competition between the electronic gain of energy and its loss $\sim U_\mathrm{s}^2$ from the lattice deformation and charge redistribution (recall Sect. 11.2.4 and Appendix 2.2). Three-dimensional ordering of signs $U_\mathrm{s} = \pm|U_\mathrm{s}|$ discriminates the FE and AFE states.

A conceptual conclusion is that the Mott–Hubbard state can be energetically favorable, then the system will mobilize an available auxiliary parameter to reach it. In our case, this parameter is the site disproportionation, which very fortunately is registered as the ferroelectricity.

11.5 Elementary Excitations

11.5.1 Solitons

Solitons Seen in Most Common Cases

For a given U_s, the ground state is doubly degenerate between $\varphi = \alpha$ and $\varphi = \alpha \pm \pi$, which allows for phase $\pm\pi$ solitons [86] with the energy $E_\pi = \Delta$; they are the charge $\pm e$ spinless particles – the (anti)holons. Figure 11.12 illustrates the origins of the potential $U_s \sin(2\varphi)$ by showing the sequence of zeros and extrema ± 1 of the electron wave function Ψ taken at E_F. The upper row corresponds to one of two correct ground states, energy minima at $\varphi = 0$, π. The degeneracy "$\Psi = \pm 1$ at good sites" gives rise to π-solitons as kinks between these two signs of the wave function (Fig. 11.13 left).

The π solitons always determine the conductivity at $T < T_\rho$. Thus in $(TMTCF)_2X$ they are observed in conductivity at both $T \gtrless T_{CO}$; in compounds without the built-in dimerization – at $T < T_{CO}$. An expected characteristic feature of solitonic conductivity is its strong reduction and enhanced activation energy for the conductivity in the interchain direction as it has been observed in $(TMTTF)_2X$ [87]. Dynamics of solitons have been accessed in recent tunneling experiments [88] on incommensurate CDWs. There are also other cases of CDWs where conducting chains are modulated by counterion columns. One is the very first known CDW compound, the Pt chain KCP – in the stoichiometric version $K_{1.75}[Pt(CN)_4]1.5H_2O$. It was shown theoretically [89], with an interpretation of available experimental data, that the "weak twofold commensurability" induced by columns of counterions gives

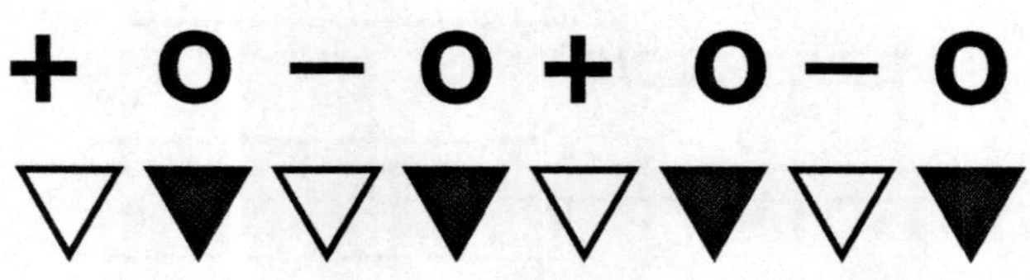

Fig. 11.12. Degeneracy scheme for electrons on a lattice with the site dimerization. *Lower row*: *white/black triangles* show the good/bad sites. *Upper row*: signs $\pm$ and zeros of the electronic wave function at E_F. Shown configuration: $\varphi = 0$ or π: $\Psi = \pm 1$ or $\Psi = \mp 1$ at good sites and $\Psi = 0$ at bad sites

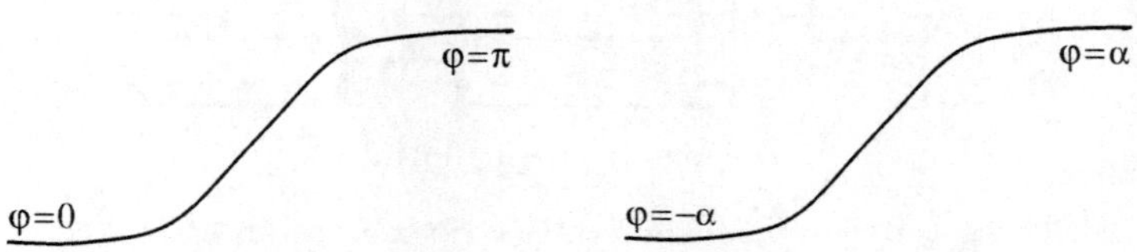

Fig. 11.13. Phase profiles $\varphi(x)$ of charged solitons. *Left*: π solitons = (anti)holons; spinless charge $\pm e$ particles seen in conductivity. *Right*: α solitons = domain walls of the ferroelectric polarization, seen in dispersion of the permittivity

rise to the physics of phase π solitons. The mechanism of spinless conduction by the e-charged π-solitons might be universal to all systems with the charge ordering, irrespective to the ordering pattern. Actually the π-soliton is a 1D vision of a vacancy or an addatom in the Wigner crystal, which is the most universal view of the charge ordering. We shall return to this topics in the Sect. 11.6.2.

Ferroelectric Solitons

In the charge ordered phase, $U_s \neq 0$ gives rise to the ferroelectric ground state, if the same α is chosen for all stacks. The state is the two-sublattice AFE if α signs alternate as in the $(TMTTF)_2SCN$, and more complex patterns of α are possible as it has been found in new compounds [30].

Spontaneous U_s itself can change the sign between different domains of ferroelectric displacements. Then the electronic system must also adjust the mean value of φ in the ground state from α to $-\alpha$ or to $\pi - \alpha$. Hence the ferroelectric domain boundary $U_s \Leftrightarrow -U_s$ requires for the phase α-solitons, Fig. 11.13, of the increment $\delta\varphi = -2\alpha$ or $\pi - 2\alpha$, whichever is smaller; it will concentrate the noninteger charge q: $q/e = -2\alpha/\pi$ or $q/e = 1 - 2\alpha/\pi$ per chain.

Above the 3D ordering transition $T > T_{CO}$, the α-solitons can be seen as individual particles, charge carriers. Such a possibility requires for the fluctuational 1D regime of growing charge ordering. It seems to be feasible in view of a strong increase of ε at $T > T_{CO}$ even for the AFE case of the X = SCN in Fig. 11.6, which signifies the growing single chain polarizability before the 3D interactions enter the game (Fig. 11.14).

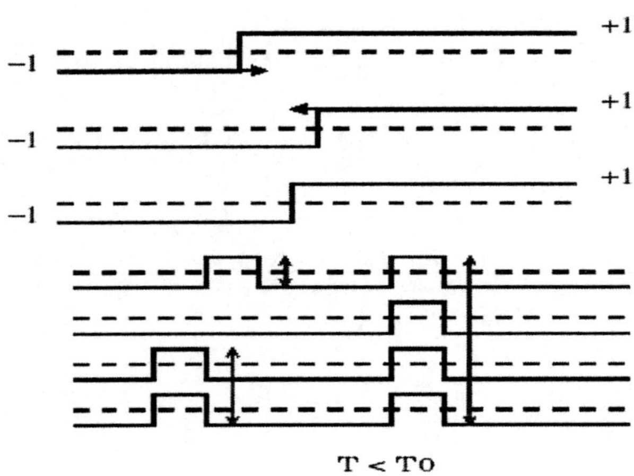

Fig. 11.14. Crossover from on-chain ferroelectric solitons into ferroelectric domain walls at $T < T_{CO}$. *Upper three lines*: aggregation of solitons into the wall. *Lower three lines*: binding of solitons into pairs with subsequent aggregation into the bubble – the nucleus of the opposite polarization. Reproduced from [90]

Below T_{CO}, the α-solitons must be aggregated into domain walls [90], separating domains of opposite ferroelectric polarization. The noticeable asymmetry in the frequency dependence of ε' above and below T_{CO} (Fig. 11.8) might be just due to this aggregation. The nonsymmetry shows up much stronger in the frequency dependence of the imaginary part of the permittivity ε'', see figures in [25, 77, 91]. The peak in $\varepsilon''(\omega)$ determines the relaxation time τ showing two temperature regimes:

(1) Near T_{CO}, $\tau(T)$ grows sharply corresponding to the expected slowing down of the critical collective mode.
(2) At low T, τ grows exponentially following the activated law with the energy similar to the one of the conductivity, Δ; hence the relaxation is external, via resistance of the electrically coupled gas of charge carriers – π-solitons.
(3) Below the main peak in $\varepsilon''(\omega)$, a long tail appears at $T < T_{CO}$, showing a weak secondary maximum. This tail may be well interpreted as the internal relaxation of the ferroelectric polarization through the motion of pinned domain walls – aggregated α-solitons.

11.5.2 Effects of Subsequent Transitions: Spin-Charge Reconfinement and Combined Solitons

Physics of solitons is particularly sensitive to a further symmetry lowering, and the subsequent anion ordering of the tetramerization in $(TMTTF)_2ReO_4$ [15, 24, 47] is a very fortunate example. This true present from the Nature demonstrates clearly the effect of the spin-charge reconfinement. Similar effects are expected for the spin-Peierls state, e.g., in $(TMTTF)_2PF_6$, but the clarity of the 1D regime in ReO_4 is uniquely suitable to keep the physics of solitons on the scene.

The conductivity plot for the ReO_4 case in Fig. 11.10 shows, at $T = T_{AO}$, the jump in Δ (actually even in $G(T)$, see details in [24, 25]) which is natural for the first-order transition. We argue that the new higher Δ at $T < T_{AO}$ comes from special topologically coupled solitons which explore both the charge and the spin sectors.

Now we must invoke also the phase θ describing the spin σ degree of freedom, such that θ'/π is the smooth spin–density. Its free distortions are described by the Hamiltonian

$$H_\theta = \frac{1}{4\pi}\left\{(\partial_x\theta)^2 v_\sigma + (\partial_t\theta)^2/v_\sigma\right\}, \tag{11.3}$$

where $v_\sigma \sim v_F$ is the spin sound velocity. (The absence of the factor γ^{-1} in (11.3) in comparison with (11.1) is not a typing mistake.)

The additional deformation of tetramerization (Fig. 11.15) acts upon electrons as a $2K_F$ CDW (Fig. 11.11 right panel) thus adding the energy

$$H_V = -V\cos(\varphi - \beta)\cos\theta. \tag{11.4}$$

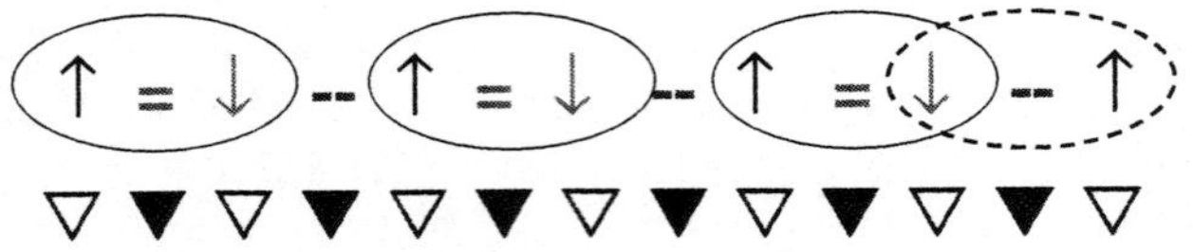

Fig. 11.15. Effect of the tetramerization leading to the energy (11.4). *White and black triangles* show the site inequivalence after the charge ordering. Tetramerization adds the inequivalence of spin-exchange bonds double and single. *Ellipses* show the preferable spin singlets. The last *dashed ellipse* shows an alternative spin configuration which was equivalent above the $T_{\rm AO}$; now, at $T < T_{\rm AO}$, it becomes the spin excitation

(Here the shift β, mixing of site and bond distortions, reflects the lack of the inversion symmetry, since $T_{\rm AO}$ is already below $T_{\rm CO}$.) Within the reduced symmetry of Fig. 11.15, the invariant Hamiltonian becomes

$$H_U + H_V = -U\cos(2\varphi - 2\alpha) - V\cos(\varphi - \beta)\cos\theta. \tag{11.5}$$

Figure 11.15 suggests a schematic illustration for the effect of the tetramerization. Inequivalence of bonds double and single between good sites endorses ordering of spin singlets. Also it prohibits translation by the two-site distance which was explored by the $\delta\varphi = \pi$ soliton. But its combination with the soliton ($\delta\theta = \pi$, which shifts the sequence of singlets) carrying the unpaired spin, is still allowed as the self-mapping.

Formation at $T < T_{\rm AO}$ of the new V-term (11.4) destroys the spin liquid, which existed at $T > T_{\rm AO}$ on top of the charge ordered state. The V term in (11.5) lifts the continuous θ-invariance, thus opening at $T < T_{\rm AO}$ the spin gap $\Delta_\sigma \sim V^{2/3}$, as it is known for the spin-Peierls transitions [92, 93]. Moreover, it lifts even the discrete invariance $\varphi \to \varphi + \pi$ of H_U, thus prohibiting the π solitons to exist alone; now they will be confined in pairs (either neutral or 2e – charged) tightened by spin strings. But the joint invariance

$$\varphi \to \varphi + \pi \ , \ \theta \to \theta + \pi$$

is still present in (11.4), giving rise to *compound topological solitons* [94] (cf. [93]).

The major effects of the tetramerization V-term are the following:

(a) To open the spin gap $2\Delta_\sigma$ corresponding to creation of new $\{\delta\theta = 2\pi, \delta\varphi = 0\}$ uncharged spin-triplet solitons
(b) To prohibit former $\delta\varphi = \pi$ charged solitons, the holons, now they are confined in pairs bound by spin strings, the activation $\gtrsim 2\Delta$ is required
(c) To allow for topologically bound spin-charge compound solitons $\{\delta\varphi = \pi\}$, $\delta\theta = \pi$, which leave the Hamiltonian (11.5) invariant, the activation $\gtrsim \Delta$ is required

For the last compound particle (c), the quantum numbers are as for a normal electron: the charge e and the spin 1/2, but their localization is very

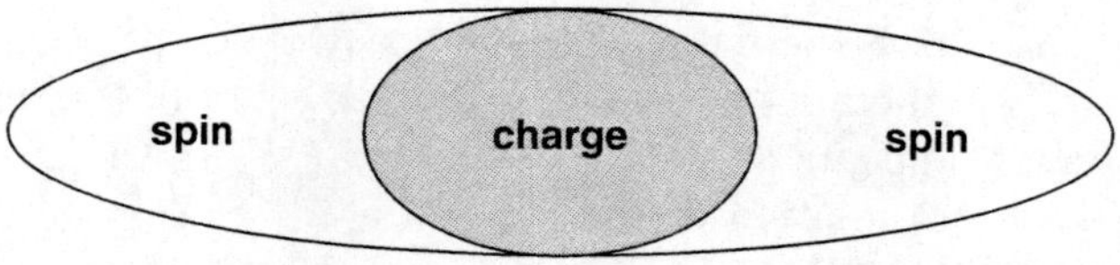

Fig. 11.16. Illustration of confinement below the tetramerization transition. Different scales of spin and charge distributions within the compound solitons: charge e, $\delta\varphi = \pi$ is concentrated sharply within $\xi_\rho \sim \hbar v/\Delta$; spin 1/2, $\delta\theta = \pi$ is concentrated loosely within $\xi_\sigma \sim \hbar v/\Delta_\sigma$

different, Fig. 11.16. The compound soliton is composed with the charge e core soliton (here $\delta\varphi = \pi$ within the shorter length $\xi_\rho \sim \hbar v_F/\Delta$) which is supplemented by spin 1/2 tails of the spin soliton (here $\delta\theta = \pi$ within the longer length $\xi_\sigma \sim \hbar v_F/\Delta_\sigma \gg \xi_\rho$). This complex of two topologically bound solitons gives the carriers seen at $T < T_{AO}$ at the conductivity plot for the $X = ReO_4$, Fig. 11.10 and [47].

Similar effects should take place below intrinsically electronic transitions, particularly relevant can be the spin-Peierls one for $X = PF_6$. But there the physics of solitons will be shadowed by 3D electronic correlations, which are not present yet for the high T_{AO} of $X = ReO_4$ case.

11.6 Optics

11.6.1 Optics: Collective and Mixed Modes

Optics: Phase Mode

Beyond the ferroelectric phase (e.g., the AFE state or the disordered one at $T > T_{CO}$) the optical absorbtion starts at the frequency $\omega_t < 2\Delta$, which is the bottom of the spectrum of phase excitations $\omega^2 = (v_\rho k)^2 + \omega_t^2$. In the quasi-classical limit of small $\gamma \ll 1$, ω_t can be interpreted as the frequency of phase oscillations of the $4K_F$ CDW around the minimum of the commensurability energy (11.2). The renormalized value U^* of U in (11.2) can be expressed through this observable parameter as

$$U^* = \frac{\omega_t^2}{8\pi\gamma v_\rho}, \quad \omega_t \approx \pi\gamma\Delta < \Delta \ (\gamma \ll 1). \tag{11.6}$$

As a "transverse" frequency of the optical response, ω_t gives the background dielectric susceptibility:

$$\varepsilon_\Delta(\omega) = \frac{\omega_p^{*2}}{(\omega_t^2 - \omega^2)}, \quad \omega_p^{*2} = \gamma\frac{v_\rho}{v_F}\omega_p^2, \quad \omega_p^2 = 8\frac{e^2 v_F}{\hbar s}, \tag{11.7}$$

where ω_{p}^{*} and ω_{p} are the actual plasma frequency of the parent metal and its bare value and s is the area per stack. In exact theory of the quantum sine-Gordon model, ω_{t} appears as the fist bound state E_1 of the pair of solitons, see [95–97]. Its exact relation to the gap 2Δ is $\omega_{\mathrm{t}} = E_1 = 2\Delta \sin(\pi\tilde{\gamma}/2)$ as explained below in (11.9).

Optical Permittivity $\varepsilon(\omega)$ Near the Ferroelectric Transition

We shall present, without derivation, the formula for the mixed electron–phonon contribution to the permittivity, valid at $T \geq T_{\mathrm{CO}}$ – above and approaching the ferroelectric transition:

$$\frac{\varepsilon(\omega)}{\varepsilon_\infty} = 1 + \frac{(\omega_{\mathrm{p}}^{*}/\omega_{\mathrm{t}})^2(1-(\omega/\omega_0)^2)}{(1-(\omega/\omega_0)^2)(1-(\omega/\omega_{\mathrm{t}})^2) - Z}, \quad Z = \left(\frac{\omega_{\mathrm{cr}}}{\omega_{\mathrm{t}}}\right)^{2-4\gamma} \leq 1. \tag{11.8}$$

Here ω_0 is a frequency of the molecular mode coupled to the charge ordering; $\omega_{\mathrm{cr}}(T)$ is the critical value of the collective mode frequency $\omega_{\mathrm{t}}(T)$, below which the spontaneous ferroelectric charge ordering takes place. Near the critical point $Z(T_{\mathrm{CO}}) = 1$ we see here:

Fano antiresonance	at ω_0, just as observed in [66] – Fig. 11.17
Combined electron–phonon resonance	at $\omega_{0\mathrm{t}}^2 \approx \omega_0^2 + \omega_{\mathrm{t}}^2$
FE soft mode	at $\omega_{\mathrm{fe}}^2 \approx (1-Z)(\omega_0^{-2} + \omega_{\mathrm{t}}^{-2})^{-1}$.

Near T_{CO} the ferroelectric mode might be overdamped; then it grows in frequency following the order parameter, that is as $\omega_{\mathrm{fe}} \sim \varepsilon^{-1/2}$ (which can

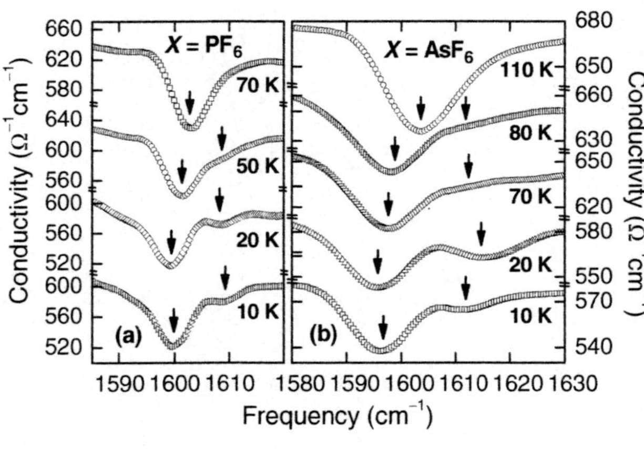

Fig. 11.17. CO effect upon the molecular mode. The figure demonstrates two effects: (1) Fano antiresonance at any T/T_{CO} which might be due the stack polarizability; (2) its splitting at $T < T_{\mathrm{CO}}$ due the CD – in accordance with the NMR. Reproduced from [66]

yield two orders of magnitude at low T) to become finally comparable with $\min\{\omega_0, \omega_t\}$. Suggesting smooth T dependencies $\omega_t(T)$ and $\omega_{cr}(T)$, we find the critical singularity at $\omega = 0$ as $\varepsilon(T) = A|T/T_{CO} - 1|^{-1}$. It develops upon the already big gapful contribution $A \sim (\omega_p^*/\omega_t)^2 \sim 10^3$ in agreement with experimental values $\varepsilon \sim 10^4 T_{CO}/(T - T_{CO})$. All that confirms that the ferroelectric polarization comes mainly from the electronic system, even if the corresponding displacements of ions are very important to choose and stabilize the long range 3D order.

11.6.2 Optics: Solitons

Thanks to detailed information on the sine-Gordon model, we can clearly formulate the expectations for optical properties related to physics of solitons (see details in [95] and [96] for a review; the contemporary stage and comparison of different approaches can be found in [97]).

The most general feature of the optical spectra, valid through the whole gapful regime $\gamma < 1$, is the two-particle gap $E_g = 2\Delta$ corresponding to the creation of a pair of $\pm\pi$ solitons. Contrary to the common sense intuition and the elementary theory, the absorption $I(\omega)$ has no singularity at the threshold E_g. The optical density of states law $DOS \sim (\omega - E_g)^{-1/2}$ is compensated by vanishing of the matrix element. As a result, the absorption starts smoothly as $I \sim (\omega - E_g)^{1/2}$. Realistically, the gap will be seen only as the nonsingular threshold for the photoconductivity, where other absorption mechanisms are excluded. For the built-in Mott state without the charge ordering, $1/2 < \gamma < 1$, there is no absorption below E_g. This case was studied in detail in [98].

But for the Mott state due to the spontaneous charge ordering, $\gamma < 1/2$, there might be also sharp peaks of the optical absorption even below the two-particle gap $E_g = 2\Delta$. Actually the spectral region $\omega < 2\Delta$ is filled by a sequence of quantum breathers, bound states of two solitons at energies

$$E_n = 2\Delta \sin(\frac{\pi}{2}\tilde{\gamma}n), \tilde{\gamma}^{-1} = \gamma^{-1} - 1. \qquad (11.9)$$

Here the substitution of γ for $\tilde{\gamma}$ takes into account quantum corrections to the control parameter itself [96]. The modes with odd numbers n are optically active. The primary, lowest bound state E_1 gives the corrected value for the collective mode ω_t. It reduces to the classical result of (11.6) in the limit of small $\gamma \ll 1$, where $E_1 = 2\Delta \sin(\pi/2\tilde{\gamma}) \approx \pi\gamma\Delta = \omega_t < 2\Delta$.

At a sequence of special values of $\gamma_n = 1/2n$, starting from $\gamma = 1/2$ downwards, the next bound state splits off from the gap. Only at these moments, the absorption edge singularity blows up from the smooth dependence $\sim (\omega - E_g)^{1/2}$ to the divergency $DOS \sim (\omega - E_g)^{-1/2}$. (There is a crossover near these values of γ [97], where a rounded edge singularity can be observed.) This property opens an amusing possibility to observe spectral anomalies at special values of the monitoring parameter, varying it, e.g., by applying pressure.

We see that the scheme changes qualitatively just at the borderline for the charge ordering instability $\gamma = 1/2$: from the essentially quantum regime $1/2 < \gamma < 1$ (with $E_g = 2\Delta$ and no separate ω_t at all) to the quasiclassical low γ regime with a sequence of peaks between 2Δ and ω_t. This fact was not quite recognized in existing interpretations, see [83], of intriguing optical data [99–101]. For example, the detailed theoretical work [98] was all performed for the case equivalent to $\gamma > 1/2$, in our notations, and cannot be applied to the $(TMTTF)_2X$ case as it was supposed. The studies of [95] and [97] are quite applicable, with the adjustment to the twofold commensurability, which we have implied above.

The available experiments, see Figs. 11.18 and 11.19, can be interpreted as observations of the collective mode – solitons' bound state. Photoconductivity experiments (distinguished from the bolometric effect!) are necessary to discriminate between two absorption mechanisms: at ω_t and above 2Δ.

Recall that quenching of the edge singularity is not just a peculiar property of the sin-Gordon model. It takes place also in theory [102] of generic 1D semiconductors such as conjugated polymers, as a consequence of the final state interaction via the long range Coulomb forces. Merging of local and Coulomb interaction is still an unattended issue in theory of solitons. A broader review on experimental manifestations of solitons in quasi-1D conductors can be found in [103].

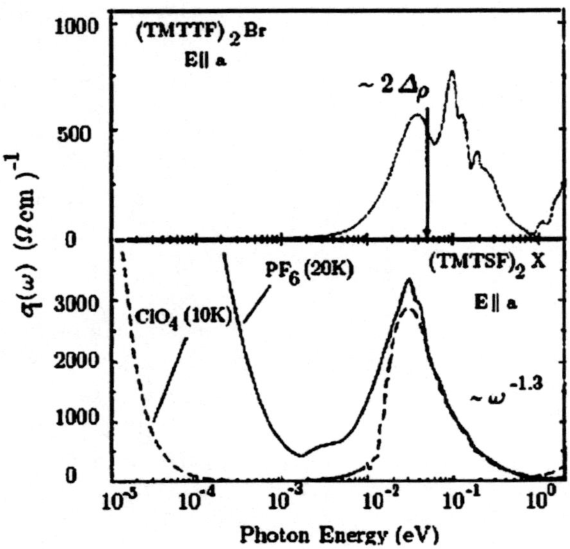

Fig. 11.18. Comparison of optical absorption of TMTTF and TMTSF compounds in a wide ω range. It seems that after subtracting molecular peaks at the high $\omega > 2\Delta_\rho$ slope (*upper panel*) the gaps shapes will be quite similar in both TMTTF and TMTSF cases. Is it a way to prove the hidden charge ordering in the metallic state of the Se subfamily? Reproduced from [99]

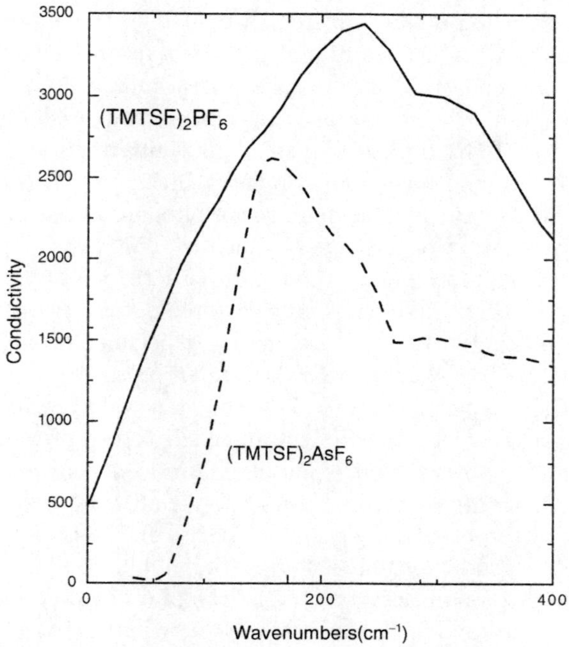

Fig. 11.19. Optical spectra of two $(TMTSF)_2X$ salts. The case of AsF6 shows a clear excitonic peak which may be the mode ω_t. The case of $X = PF_6$ shows a more smeared feature, probably due to the noticeably high SDW gap within the charge ordering one. The shoulders seen at frequencies above the peaks can be well interpreted as the two-particle gaps 2Δ. Then they will determine the thresholds in photoconductivity. M. Dressel, unpublished

11.6.3 Optics: Summary

Now we summarize shortly the main expected optical features [54]:

1. *In any case of the CO*, for both FE or AFE orders, we expect
   (1a) Strongest absorption feature comes from the phase mode, an analogy of the exciton as the bound kink-antikink pair at ω_t
   (1b) Two-particle gap 2Δ (e.g., the photoconductivity) lies higher than ω_t, it is given by free pairs of π-solitons, $\Delta = E_\pi$
   (1c) Spectral region $\omega_t < \omega < 2\Delta$ may support also quantum breathers – higher bound states of solitons
2. *In case of the ferroelectric order* we expect additionally or instead of 1
   (2a) Fano antiresonance at the bare phonon mode frequency ω_0 coupled to the CO
   (2b) Combined electron–phonon resonance at $\omega_{0t} > \omega_0, \omega_t$, which substitutes for 1a
   (2c) Ferroelectric soft mode at ω_{fe}, evolving from $\omega_{fe} = 0$ at $T < T_{CO}$

Unfortunately, the optical picture of $(TMTTF)_2X$ compounds is complicated by multiple phonon lines filling just the relevant region of the optical gap. But this complication may not be all in vain, if it is viewed also as another indication for the charge ordering. Indeed, surprisingly (kept noticed since early 1980s [104]) high intensity of molecular vibrations in TMTTF may be just due to the inversion symmetry lifting by the charge ordering, recall [105] for well-confirmed mechanisms of phonons activation by the CDW formation. Oppositely, weaker phonon lines in TMTSF [104] may speak in favor of a fluctuational regime of the charge ordering in accordance with the pseudogap, rather than a true gap, in electronic optical transitions.

The whole obstacle can be overcome by experiments on low gap charge ordered states like in $(TMTTF)_2Br$, or under pressure. There, as one can guess from other experiments [73, 74], the gap value will fall below the region of intensive molecular vibrations, which today prevent the observations. Figure 11.18 shows that it is happening indeed, and just for X = Br as expected: the electronic spectrum starts to split off from the vibrational one.

Recall finally the great experience of optics in another type of strongly correlated 1D electronic systems: conjugated polymers. They were studied by such a complex of optical techniques as photoconductivity, stimulated photoemission, photoinduced absorption, electro-absorption, time resolved measurements, see [106].

A popular interpretation (see [83] for a review) of optics [99–101] and sometimes conductivity [22] in TMTCF – type compounds neglects the dimerization (either generic from bonds or from sites via the charge ordering) and relies upon the weaker fourfold commensurability effects. They give rise to the energy $\sim U_4 \cos 4\varphi$ originated by higher order (eight particles collisions) Umklapp processes; its stabilization would require a stronger e–e repulsion (or slowing down, see Sect. 11.8.1) corresponding to $\gamma < 1/4$. While not excluded in principle, this mechanism does not work in TMTTF case, already because this scenario does not invoke any charge ordering instability. Moreover, the experiment shows, Fig. 11.9, that even small increments of the dimerization, just below the second order transition at T_{CO}, immediately transfer to the activation energy, hence the domination of the twofold commensurability.

11.7 Fate of the Metallic TMTSF Subfamily

By now, the revaluation definitely concerns mostly the TMTTF subfamily, whose members usually, by the temperature T_{CO}, already fall into the charge-gap regime starting at $T_\rho > T_{\mathrm{CO}}$, see Fig. 11.9. The TMTSF compounds are highly conductive, which does not allow for measurements of the low frequency permittivity ε. Also the NMR splitting [26, 107] will probably be either too small or broadened because of expected disordered or temporal nature of the charge ordering in the metallic phase (see [32] for a proved example of another system with the charge ordering). Nevertheless the transition may be there, just being hidden or existing in a fluctuation regime, as may be realized, e.g., in stripe phases of high-T_c cuprates [108]. If and when confirmed, the entire picture of intriguing abnormal metallic state [109] will have to be revised following the TMTTF case.

The signature of the ferroelectric charge ordered state may have been already seen in optical experiments [99–101], as we have discussed above. Indeed, the Drude-like low frequency peak appearing within the pseudogap (Fig. 11.18 lower panel) can be interpreted now as the optically active mode of the ferroelectric polarization; the joint lattice mass, see Sect. 11.8.1, will naturally explain its otherwise surprisingly low weight. Recall here the earlier conjectures on collective nature of the conductivity peak [12, 13, 110], derived from incompatibilities of IR optics, conductivity and NMR. Vice versa, the ferroelectric mode must exist in TMTTF compounds, whose identification is the ultimate goal. Even the optical pseudogap itself [99–101], being unexpectedly large for TMTSF compounds, Fig. 11.18 top, with their less pronounced dimerization of bonds, can be largely due to the hidden spontaneous dimerization of sites. Even the shapes of the gap and the pseudogap in TMTTF and TMTSF cases appear to look similar, if the first one is cleaned from molecular modes (Fig. 11.18 upper panel). Optical experiments will probably be elucidated when addressed to members of the $(TMTTF)_2X$ family, showing the charge ordering with a particularly reduced value of the associated gap (below typical molecular vibrations – down to the scale of the pseudogap in $(TMTSF)_2X$).

Hidden existence of the charge ordering and the local ferroelectricity, at least in fluctuating regime, is the fate of the TMTSF compounds and the major challenge is to detect it, as it was seen explicitly in TMTTF compounds. Key effects of anion orderings, particularly the opportunity to compare relaxed and quenched phases of $(TMTSF)_2X$, also are waiting for attention.

11.8 Origin and Range of Basic Parameters

11.8.1 Generic Origins of Basic Parameters: Interactions Among Electrons or with Phonons?

The combination of optical and conductivity data can provide a deeper insight into the nature of observed regimes. The value of Δ is already known as the conductivity activation and 2Δ can be found independently as the photoconductivity threshold; ω_t is measurable through the optical absorption. Then their ratio will provide the basic microscopic parameter γ.

The full quantitative implementation requires to resolve for divergence (2,3 or even more times!) [101, 104] in values of such a basic and usually robust parameter as the plasma frequency, if it is extracted from different parts of the spectrum. This discrepancy can signify the strong renormalization $\omega_p^* \ll \omega_p$, which can develop while the probe frequency decreases from the bare scale $\omega_p > 1\,\text{eV}$ to the scale $\omega_t, \omega_0 \sim 10^{-2}\,\text{eV}$.

Remind the full (kinetic $\sim C_{\text{kin}}$ and potential $\sim C_{\text{pot}}$) energy of elastic deformations for the charge phase φ:

$$\frac{\hbar v_{\text{F}}}{4\pi}\left\{(\partial_{\text{t}}\varphi)^2 C_{\text{kin}}/v_{\text{F}}^2 + (\partial_x\varphi)^2 C_{\text{pot}}\right\} \equiv \frac{1}{4\pi\gamma}\left\{(\partial_{\text{t}}\varphi)^2/v_\rho + (\partial_x\varphi)^2 v_\rho\right\},$$

where

$$\gamma = \frac{1}{C_{\text{pot}}C_{\text{kin}}}, \; \frac{v_\rho}{v_{\text{F}}} = \frac{C_{\text{pot}}}{C_{\text{kin}}}, \; \frac{\omega_{\text{p}}^*}{\omega_{\text{p}}} = \left(\gamma\frac{v_\rho}{v_{\text{F}}}\right)^{1/2} = \frac{1}{C_{\text{kin}}}.$$

The potential parameter C_{pot} is "1 + e–e repulsion contribution." The kinetic parameter C_{kin} is "1+lattice adjoined mass." The parameter γ contains a product of C's – not distinguishable separately; the velocity v_ρ contains a ratio of C's – not distinguishable separately. But ω_{p}^* contains only C_{kin}, which gives a direct access to the joint lattice dynamics. (Another factor for reduction of the parameter γ, the Coulomb hardening $C_{\text{pot}} > 1$, acts upon γ and velocity v_ρ,[11] but cancels in their product which gives ω_{p}^*.) The lowering of ω_{p}^* singles out the effect of the effective mass enhancement $C_{\text{kin}} > 1$, which is due to coupling of the charge ordering with $4K_{\text{F}}$ phonons [113].[12]

The final step is to notice that the mass enhancement will not be effective above the $4K_{\text{F}}$ phonon frequency ω_{ph}; actually C_{kin} is a function of ω: $C_{\text{kin}}(\omega) = C_{\text{kin}}(0)\omega_{\text{ph}}^2/(\omega_{\text{ph}}^2 + \omega^2)$. It explains the difference in values of ω_{p}^* extracted from high and from intermediate frequency ranges. If true, then the charge ordered state is a kind of a polaronic lattice (as it was already guessed for some CDWs [115]). It resembles another Wigner crystal: electrons on the liquid Helium surface, see [116], where self-trapped electrons gain the effective mass from surface deformations – the ripplons.

11.8.2 Where are We?

There are still several fascinating questions to understand:

Why the charge ordering is so common and appears at such a high energy/temperature scale?
Why do we see it instead of the abnormal metal (Luttinger Liquid regime)?
Why does it develop spontaneously before the 1/4 filling effects have a chance to be seen?

We may get an idea for the answers to these questions by analyzing the sequence of various regimes, as it appears by changing the control parameter γ.

Phenomenological Hamiltonian may have the following typical components (we restrict it here to terms containing only the charge phase φ).

---

11 In CDWs the Coulomb hardening, as well as the very common mass enhancement [111], are confirmed experimentally [112].

12 X-ray scattering gives a direct evidence for the coupling of the $4K_{\text{F}}$ electronic density with either lattice displacements [36,37,40] or intramolecular modes [114].

$H_\varphi \sim \gamma^{-1}[v_\rho(\partial_x\varphi)^2 + v_\rho^{-1}(\partial_t\varphi)^2]$ – from the electron liquid or fluctuating $4K_F$ CDW (`basic`)

$+U_1\cos(\varphi)$ – from tetramerization or spin-Peierls (`spontaneous, frequent`)

$+U_2\cos(2\varphi)$ – from dimerization (`built-in or spontaneous, typical`)

$+U_3\cos(3\varphi)$ – from trimerization: TTF-TCNQ under pressure, NMP-TCNQ (`special`)

$+U_4\cos(4\varphi)$ – built-in, from host lattice (`typical`).

The parameter $\gamma = K_\rho$ controls quantum fluctuations of the phase φ which will, or will not, destroy the above nonlinearities; it defines:

1. Survival, against renormalization to zero, of nonlinearities $\sim U_i$
2. Their spontaneous generation – known for U_1, U_2
3. Physics of solitons and the collective mode
4. Relevance of the interchain coupling: metal/insulator branching

We can quote the following regimes:

$\gamma < 1$: Renormalized $U_2 \neq 0$; the charge gap is originated in case of built-in dimerization (of either bond or site types). This is the generic Mott–Hubbard state, any repulsion is sufficient to stabilize it. Solitons = holons appear as free excitations, giving both the thermal activation energy in conductivity, and the optical absorption threshold. This regime is certainly valid for all $(TMTTF)_2X$. It is not applicable to nondimerized materials like DMtTTF and EDT-TTF-CONMe salts [29, 79]. They still would be metallic, unless falling into the next regime described below, which does happen actually.

$\gamma < 1/2 = 0.5$: $4K_F$ anomaly appears in X-ray scattering at high T, as registered in fractionally filled cases of TTF-TCNQ and its derivatives [37,38,40]. In 1/4 filled compounds the spontaneous site dimerization potential $\sim U_s$ is formed, hence no need for a bare commensurability/Umklapp potential $\sim U_b$. This regime is proved to be valid, by observations of the ferroelectric response and the NMR splitting, in most of $TMTTF_2X$, and by X-rays [30] in $(DMtTTF)_2ClO_4$.[13] All materials are waiting for determining the optical signatures, see Sect. 11.6.3.

$\gamma < 4/9 = 0.39$: Trimerization lock-in of $4K_F + 2K_F$ superlattices (confirmed in TMTTF-TCNQ under a special pressure [41] and in NMP-TCNQ [38]).

$\gamma < 1/4 = 0.25$: Effects of 1/4 filling may come to play (cf. the lock-in of sliding of CDWs under stress in MX_3 conductors [117,118]).

$\gamma < \sqrt{5} - 2 \approx 0.24$: Ultimate SDW instability, even for incommensurate cases. (The confinement index of the electron pair in the course of the

[13] The last compound, as well as $(EDT\text{-}TTF\text{-}CONMe)_2X$, is still waiting for the standard (ferroelectricity, NMR) characterization.

interchain hopping is $(1/\gamma - \gamma)/2 < 2$ [17,18]). Seemingly, it is not the case of TMTSF metals: they need the HMF support to form the FISDW.

$\gamma \approx 0.23$: A guess from high $\omega \gg \Delta$ tails ($\sim\omega^{-1/3}$, see Fig. 11.18) of optical absorption [100], see also [83].

$\gamma < 2/9 \approx 0.22$: Spontaneous trimerization which is not observed: TTF-TCNQ needs a precise pressure to pinpoint exactly 1/3.

$\gamma < 3 - \sqrt{8} \approx 0.17$: Last feature of electrons' Fermi surface disappears. (The electron Green function index is $(\gamma + 1/\gamma + 2)/4 < 2$.) This regime was guessed from ARPES experiments in $(TMTSF)_2X$ [119], but not seen in seemingly more correlated TTF-TCNQ [120].

$\gamma = 0.125$: Spontaneous 1/4 filling in totally incommensurate chains. But the usual CO, the $4K_F$ condensation, would has happened already well before.

Resume: Most of qualitative effects of electronic correlations would appear at $\gamma < 1/2$, where the system is already unstable with respect to the charge ordering. The existing experiments on most studied materials pose the following constraints. $(TMTTF)_2X$: $\gamma < 0.5$; $(TMTSF)_2X$: $\gamma > 0.24$; TTF-TCNQ: $0.22 < \gamma < 0.39$.

11.9 Conclusions and Perspectives

We have presented the key issues of related phenomena of the ferroelectricity and the charge ordering in organic metals. In $(TMTTF)_2X$ the dielectric permittivity ε demonstrates clear cases of the ferroelectric and antiferroelectric phase transitions. The combination of conductivity and magnetic susceptibility proves the spinless nature of charge carriers. The independence and occasional coexistence of structureless ferroelectric transitions and usual anionic ones brings the support of structural information. The sequence of symmetry breakings gives access to physics of three types of solitons: π-solitons (holons) are observed via the activation energy Δ in conductivity; noninteger α-solitons (ferroelectric domain walls) provide the low frequency dispersion; topologically coupled compound spin-charge solitons determine the conductivity below a subsequent structural transition of the tetramerization. The photoconductivity gap 2Δ will be given by creations of soliton–antisoliton pairs. The lower optical absorption comes from the collective electronic mode: in the ferroelectric case it becomes the mixed electron–phonon resonance coexisting with the Fano antiresonance. The ferroelectric soft mode evolves from the overdamped response at T_{CO}. The reduced plasma frequency signifies the slowing down of electrons' collective motion by the adjoint lattice mass; it recalls that the charge ordering has a $4K_F$ lattice counterpart in accordance with the X-ray experience [36].

We propose that a latent charge ordering in the form of ferroelectricity exists even in the Se subfamily $(TMTSF)_2X$, giving rise to the unexplained yet low frequency optical peak and the enhanced pseudogap. Another, interchain type of the charge disproportionation, known in the relaxed $(TMTSF)_2ClO_4$, is still waiting for attention; possibly it is present, perhaps also in a hidden form, in other superconducting cases. Discoveries of the ferroelectric anomaly and of the related charge ordering call for a revaluation of the phase diagram of the $(TMTTF)_2X$ and similar compounds and return the attention to the interplay of electronic and structural properties. In this respect, we point to the old theory [17, 18] for the synergetic phase diagram of these materials.

The ferroelectricity discovered in organic conductors, beyond its own virtues, is the high precision tool to diagnose the onset of the charge ordering and the development of its order. The wide range of the ferroelectric anomaly ($T_{CO}\pm 30$ K) suggests that the growing charge ordering dominates the whole region below and even above these already high temperatures. Even higher is the on-chain energy scale, from 500 up to 2,000 K, as given by the conduction gaps formed at lower T, and by optical features. Recall also the TTF-TCNQ with ever present, up to the RT, $4K_F$ fluctuations. All that appears at the *grand unification* scale, which knows no differences with respect to interchain couplings, anion orderings, ferro- and antiferroelectric types, between sulphur and selenium subfamilies, between old faithful incommensurate or weakly dimerized compounds, and the new quarter filled ones. Hence the formation of the electronic crystal (however we call it: *disproportionation, ordering, localization, or Wigner crystallization of charges*; $4K_F$ *density wave, etc.*) must be the starting point to consider lower phases and the frame for their properties.

On the theory side, the richness of symmetry-defined effects of the charge ordering, ferroelectricity, AFE, and various AOs (see [17,18] for earlier stages) allows for efficient qualitative assignments and interpretations.

Still there are standing questions on charge ordering and ferroelectricity:

Why is the charge ordering so common? Is it universal?
Why is the astonishing ferroelectricity so frequently encountered as a form of charge ordering?
Why do the AFE and more complex patterns appear on other occasions?
Is it a spontaneously created Mott–Hubbard state?
Is it a Wigner crystal, if yes then of what: electrons or polarons?
Is it a $4K_F$ CDW: driven by electrons and stabilized by the lattice?
Role of anions: is there a key to the FE/AFE choice?

Are there other challenges ignored since decades? Examples: the plasma frequency mystery, the special anion ordering structure for superconducting phase.

The last remark reminds us that the story of hidden surprises may not be over. Another, interchain, type of the charge disproportionation, known in the relaxed $(TMTSF)_2ClO_4$ (see more in [15]), is still waiting for attention, possibly being present in a latent form in other superconducting cases. This additional anion ordering can contribute to the controversial physics of inhomogeneous state at the SC/SDW boundary under pressure (see Brown, Chaikin and Naughton, and Jérome and Bourbonnais, in this volume). The coexistence of charge ordering and anion ordering transitions in $(TMTTF)_2ReO_4$ warns for this possibility in other cases as well. If it is not found to be the case, then we will be forced to accept two different types of superconductivity in $(TMTSF)_2ClO_4$ and in $(TMTSF)_2PF_4$ under pressure; the state of today's theories [82] still ignores the challenges from this anion ordering.

We conclude with the overview of challenges for future studies:

Hidden existence of charge ordering and ferroelectricity in the metallic *Se* subfamily
Optical identification of gaps and soft modes
Physics of solitons via conductivity, optics, NMR
Ferroelectric hysteresis, relaxation, domain walls
More exploration of anion ordering and other structural information

We finish to say that new events call for a substantial revision of the contemporary picture of the most intriguing family of organic metals and its neighbors, and for further efforts to integrate various approaches to their studies.

Acknowledgments

The author acknowledges collaboration with P. Monceau and F. Nad, discussions with S. Brown, H. Fukuyama, J.-P. Pouget, and S. Ravy, comments from R. Ramazashvili and S. Teber. This work was partly supported by the INTAS grant 7972 and by the ANR program (the project BLAN07-3-192276).

Appendix 1: Earnshow Instability: Ion in the Cage

Empirically we see a systematic difference between the usual $\boldsymbol{q} \neq 0$ anion ordering and the $\boldsymbol{q} = 0$ ferroelectric transitions. The first ones are always observed for noncentrosymmetric anions, so that the orientational ordering was supposed to be a principle mechanism [58], with positional displacements being its consequences. The accent on the orientational ordering was probably one of reasons, why the hidden $\boldsymbol{q} = 0$ transitions were not understood initially. Oppositely to $\mathbf{q} \neq \mathbf{0}$ AOs, the $\boldsymbol{q} = 0$ ones are mostly observed in systems with the noncentrosymmetric anions. Then we should think about a universal mechanism related only to the positional instability. This is apparently the case of the Earnshow theorem [59], which states that a classical crystal with only Coulomb interactions is never stable. Our compounds may be regarded as such, as long as the cavity for the ion is large enough and they are allowed to move to minimize the Coulomb energy of the charge transfer. The Earnshow instability will probably develop itself as a displacement $\mathbf{u}$ toward one of the two closest molecules along the diagonal connecting the nearest molecules on neighboring stacks (Fig. 11.3). Then in all cases, both centrosymmetric and noncentrosymmetric, there is a double well potential (or the potential is flattened [121] at least) for either displacements or orientations, or for both. It can also happen that for noncentrosymmetric ions, with orientation as another degree of freedom, new directions of $\boldsymbol{u}$ are enforced. Then the former ones may be abandoned, what happens probably in the relaxed $(TMTSF)_2ClO_4$.[14] Otherwise, the former displacive minima can be preserved which opens the possibility for a sequence of transitions of T_{CO} and T_{AO} types, as it was observed in case of $(TMTTF)_2ReO_4$.

We suggest an elementary illustration of the purely ionic instability. Consider a single ion in the cage formed by four oppositely charged molecules from two neighboring stacks. Let a is the intramolecular spacing (stack period) and h is the distance from the ion to the stack. If we allow for the ion displacements δa along the stack, then the energy of Coulomb interactions will change as

$$\delta W \approx \frac{(h^2 - a^2/2)}{(h^2 + a^2/4)^{5/2}} (\delta a)^2 .$$

The system is unstable with respect to the longitudinal displacement δa if $2h^2 < a^2$. Otherwise, if $2h^2 > a^2$, it is unstable with respect to the transverse displacement δh. These two cases correspond to instabilities observed in $(TMTTF)_2SCN$ and $(TMTSF)_2ClO_4$.

The quantitative criterium discriminating the types of instabilities may change if we improve calculations by taking into account the actual charge distribution over the large molecule, and their whole array. But the qualitative statement on the instability, the Earnshow theorem, can be violated only by quantum mechanical effects like the orientational energy of methyl end groups forming the cage. Altogether, the double well potential for the ion will be formed.

These expectations give a sound interpretation of the very strong isotope effect upon deuteration [25, 122]. The T_{CO} enhancement can be caused [121] by the shortening of the (C–D) bonds, in comparison with the (C–H) ones, in methyl groups, thus widening the ion's cage and reducing its stability.

Appendix 2: Permittivity Sources

Appendix 2.1: Estimations for the Ionic Contribution to ε

A purely ionic contribution near the $\boldsymbol{q} = 0$ instability is expected to be

$$\varepsilon_i = 4\pi \frac{e^2}{\Omega} \frac{\partial u}{\partial eE} \sim 4\pi \frac{e^2}{a} \left(\frac{u_0}{a}\right)^2 \frac{1}{T_{CO}} \frac{T_{CO}}{T - T_{CO}} \sim 10^1 \frac{T_{CO}}{T - T_{CO}} ,$$

[14] In this sense, the traditional interpretation of specifics of small anions as the "negative chemical pressure" should be revised. A small ion is allowed to realize larger varieties of its displacive instability [121], one of them fortunately gives rise to the $\boldsymbol{q}_3$ structure favorable to superconductivity.

where Ω is the unit cell volume, $u_0 \sim 0.1a$ is a guess for the equilibrium displacement at low T, and also we have estimated $\partial u/\partial eE \sim u_0^2/T_{\mathrm{CO}}$. We see that the anomaly may develop upon the background value of only 10^1 which is three orders of magnitude below the experimental scale of $\varepsilon \approx 2.5 \times 10^4 T_{\mathrm{CO}}/|T - T_{\mathrm{CO}}|$ [23].

Appendix 2.2: Phase Instability

Consider the interference of ionic displacements and the charge ordering. Suppose the stability of the anionic system is controlled by a parameter K such that $K = 0$ would correspond to the instability due to short range interactions of charges calculated without the contribution from electronic correlations. For homogeneous deformations and in presence of an external electric field E, the energy per stack reads

$$\frac{1}{\pi}E\varphi - U_b \cos 2\varphi - U_{\mathrm{s}} \sin 2\varphi + \frac{K}{2}U_{\mathrm{s}}^2. \tag{11.10}$$

For shortness we do not distinguish bare and renormalized values. Expanding in small φ, the minimization yields

$$U_{\mathrm{s}} = \frac{E/2\pi}{U_b K - 1}, \quad \varphi = -\frac{E/\pi}{U_b - K^{-1}}, \quad \varepsilon = \frac{4e\varphi}{sE} = \frac{4e^2/\pi s}{U_b - K^{-1}}.$$

We see that the transition of a joint electron-ion instability takes place at $K = U_b^{-1}$, well above the point $K = 0$ of a purely ionic one.

To keep nonlinearity at hand and to access the new ground state below the instability, we should exclude from (11.10) only U_{s} to arrive at the energy (for $E = 0$)

$$F(\varphi) = -U_b \cos 2\varphi + \frac{1}{2K}\cos^2 2\varphi + \text{const.}$$

At $U_b > K^{-1}$ there is only one minima at $\varphi = 0$ but at $U_b = K^{-1}$ this point becomes unstable which originates the anomaly in ε. At $U_b < K^{-1}$ the two new minima appear at $\varphi = \pm\alpha$ (first closely, with $\alpha \approx \sqrt{(1 - KU_b)/2}$) which makes the chain to be polarized.

Appendix 3: Competing Philosophies for Organic Conductors

Since long time, there are two opposing philosophies for interpretations of these magnificent materials:

A. *Generic picture* summarized in [109] implies that the sequence of electronic phases follows a smooth variation of basic parameters reducible to the effective pressure, see Fig. 5 in [109]. The majority of compounds with noncentrosymmetric anions were abandoned, presumably their AOs were thought to exhort ill defined or undesirable complications.

The advantages are appealing:

(a) Concentration on simplest examples avoiding structural effects
(b) Generality in a common frame of strongly correlated systems driven mostly by basic parameters – interactions vs. the bandwidth
(c) Extensive use of experiments under pressure.

But there is also another side of the coin:

(1) Concentration on only simplest examples avoiding the rich information [15] on correlation of electronic and structural properties
(2) Necessity to introduce the case of the noncentrosymmetric anion ClO_4 to demonstrate the appearance of the superconductivity without pressure.[15]
(3) Accent upon pressure as a universal parameter
(4) Ignoring the structureless transitions, which are typical just for these selected compounds with centrosymmetric anions; hence loosing the dominating effect of the CO/FE.

B. *Specific picture* developed in [17,18,60,123] suggested the synergy of structural and electronic phase transitions with the accent upon compounds with AOs. It extends naturally to new observations on charge ordering and ferroelectricity. Its main statements are the following (see [17] and Chap. 6 of [18] for applications):

(a) Displacive, rather than orientational, mechanisms are driving the AOs (the Earnshow instability of separated charges)
(b) Each fine structural change exerts a symmetrically defined effect which triggers a particular electronic state
(c) 1D "g-ological" phase diagram results in 2D, 3D phase transitions only when it is endorsed by appropriate symmetry lowering effects
(d) Main proof for the 1D physics of the Mott transitions is given by the $\boldsymbol{q}_2$ structure of the $(TMTTF)_2SCN$. (Today it is shown as the AFE case of the CO.)
(e) Superconductivity appears only if the system is drawn away from the half filling thus avoiding the Mott insulator state. It happens in the relaxed phase of the $(TMTSF)_2ClO_4$ thanks to the unique $\boldsymbol{q}_3$ type of the anion ordering leading to inequivalence (CD in today's terms) of chains.[16]

---

[15] The logic of the "effective pressure" demands to show for this compound only the quenched phase with the SDW state, rather than the relaxed phase where the superconductivity appears only after the particular structural transition of the AO.

[16] This is a purely defined case of what today is called the "internal doping." Estimations for its magnitude, i.e., the potential $\pm W$ of the interchain charge disproportionation, range from moderate $W \approx 50\,\mathrm{K}$ [124, 125] to high $W \approx 250\,\mathrm{K}$ [126] and even higher [123] values. The uncertainty comes from different interpretation of the fast oscillations.

But there are difficulties of this picture as well:

1. In applications to *Se* compounds there are common problems of any quasi-1D approach confronting the success of the band picture for the FISDWs (and vice versa!).
2. There are cases of the superconducting state without observation of the particular $\boldsymbol{q}_3$ type of the AO.[17]
3. Missing opportunity to view the anion ordering transition in $(TMTTF)_2$ SCN case as the AFE charge ordering.
4. Unawareness of the later discoverd structureless transitions.

Appendix 4: History Excursions

Correlation between electronic phases and fine structural effects of the AOs has been noticed [17, 18, 60] and exploited in details [17, 18] long time ago. Experiments generally prove this correlation but also demonstrated deviations, see [15], from the unique correspondence suggested at the earlier stage. The discrepancies are related to the third variable ingredient: the electronic dispersion in the interchain direction. The recent events call again for a unifying picture of electronic and structural effects which returns us to suggestions already made two decades ago. Below we quote from ill-known publications written in early mid-1980s, whose views become relevant nowadays.

Extracts from [17]. For further details and applications see Chap. 6 in [18].

...This theory permits us to suggest a general model for the phase diagram of the Bechgaard salts in a way that the variation of electronic states is mainly determined by the crystal symmetry changes.
...1D divergent susceptibilities give rise to observable phenomena only if the pair coherence is preserved in the course of the interchain tunneling. In the gapless regime it can be maintained by proper interchain electronic phase shifts which can appear due to some symmetry changes.
...Experimental data show us the following correlation between the anionic structure (characterized by the wave vector $\boldsymbol{q}$) and the state of the electronic system:[18]

1. Unperturbed structure. Bonds are dimerized. PD: M→MI→ SDW. The last two phases are clearly separated only in TMTTF subfamily

[17] Nevertheless, the recent views on independent AOs allow to suggest that the $\boldsymbol{q}_3$ structure is still there, at least in local or dynamic form without the long range order. Low T structural studies under pressure, of particularly $(TMTSF)_2PF_6$, are required.

[18] Additional notations here: PD – phase diagram, MI – magnetic/Mott insulator.

2. $\boldsymbol{q}_2 = (0, 1/2, 1/2)$. The molecules are not equivalent. PD: M→MI→SDW (or CDW = spin-Peierls). T_{MI} and T_{SDW} are well separated (X = SCN: $T_{\text{MI}} = 160\,\text{K}$, while $T_{\text{SDW}} = 7\,\text{K}$)
3. $\boldsymbol{q}_3 = (0, 1/2, 0)$. The neighboring stacks are not equivalent. PD: M→SC→ FISDW
4. $\boldsymbol{q}_4 = (1/2, 1/2, 1/2)$. The tetramerization. PD: M→I transition being driven externally by the AO

...The rare case 2 helps us to fix the model for the whole family: a strongly correlated 1D state with the separation of charge- and spin degrees of freedom. The typical case 1 qualitatively corresponds to the same model while the separation is less pronounced and interpretation may be controversial.[19]

The most important for appearance of the SC is the case 3: the alternating potentials lead to some redistribution of the charge between the two types of stacks, hence the system is driven from the twofold commensurability which removes the Umklapp scattering, destroys the Mott–Hubbard effect and stabilizes the conducting state down to lower temperatures where the SC can appear.[20]

Extracts from [60].

Here are some extracts from [60],[21] which itself was an extension of earlier observations on effects of counterions in charge transfer CDWs of the KCP type [89]. A better known later publication is [61].

...We propose an alternative explanation of $(\text{TMTSF})_2\text{PF}_6$, based on the fact that this material possesses a weak dimerization gap Δ. This gap is due to the environment of the given chain, which, unlike the chain itself, does not posses a screw symmetry along the chain axis. Without the effect of the environment the band is quarter filled. The environment (PF_6, etc.) opens a small gap Δ in the middle of this band, which therefore becomes half filled. Hence also small are the corresponding constant for the Umklapp scattering: $g_3 \sim g_1 \Delta / E_\text{F}$. The effect of g_3 appears only below sufficiently low temperature $T_3 \sim E_\text{F} g^{1/2} (g_3/g)^{1/g}$, $g = 2g_2 - g_1$.
...Assuming the pressure suppresses g_3 and with it T_3, the Josephson coupling J of superconducting fluctuation will finally overcome the Umklapp scattering. This interpretation explains the observations in $(\text{TMTSF})_2\text{PF}_6$ as a result of competition of the two small (off chain) parameters, g_3 and J, rather than as a result of the accidental cancelation of the large coupling constants $2g_2$ and g_1 (D. Jerome and H. Schulz, Adv. Phys. **31**, 299 (1982)).

[19] Fortunately it is unambiguous today, two decades later.
[20] This is a clear case of the "internal doping" in today's terminology.
[21] Reference [60] was probably the first theoretical work performed in response to the discovery of the organic superconductivity within first weeks. Hence it speaks only about the case of X = PF6: the zero pressure superconductor X = ClO4 was not discovered yet.

...In this way there appears a region in the phase diagram where the superconductivity exists in absence of g_3, but where the CDW is introduced by g_3.
...A closer examination of the model shows that it is the triplet[22] superconductivity.

References

1. *Proceedings of the ICSM 1982*, ed. by R. Comès, J. Phys. Fr. Colloq. C3, **44** (1983)
2. *Proceedings of the ISCOM 2001 (Yamada Conference)*, ed. by G. Saito, Synth. Met. **133–134** (2003)
3. *Proceedings of ISCOM 2003*, ed. by P. Batail, E. Canadell, N. Dupuis, et al., J. Phys. IV, **114** (2004)
4. *Proceedings of ISCOM 2005*, ed. by J. Brooks, J. Low Temp. Phys., **142**, 3–4 (2006)
5. *Proceedings of the ICSM 2002*, ed. by C.Q. Wu, Y. Cao, X. Sun, D.B. Zhu, Synth. Met. **135–137** (2003)
6. *Proceedings of the ICSM 2004*, ed. by L.A.P. Kane-Maguire, D.L. Officer, Synth. Met. **152** (2005)
7. *Proceedings of ECRYS 2002*, ed. by S. Brazovskii, N. Kirova, P. Monceau, J. Phys. Fr. IV, **12** (2002)
8. *Proceedings of ECRYS 2005*, ed. by S. Brazovskii, N. Kirova, P. Monceau, J. Phys. Fr. IV, **131** (2005)
9. J.P. Farges (ed.), *Organic Conductors* (Dekker, New York, 1994)
10. *Common Trends in Synthetic Metals and High-T_c Superconductors* (I.F. Schegolev memorial volume), ed by S. Brazovskii, J. Phys. I, **6** (1996)
11. P. Monceau, *Electronic Properties of Inorganic Quasi 1D Compounds* (Reidel, Dordrecht, 1985)
12. D. Jérome, in [9], p. 405
13. D. Jérome, Chem. Rev. **104**, 5565 (2004)
14. R. Moret, J.-P. Pouget, R. Comès, K. Bechgaard, in [1], p. 957
15. J.-P. Pouget, S. Ravy, in [10], p. 1501
16. V.J. Emery, reference therein [1], p. 977
17. S. Brazovskii, V. Yakovenko, J. Phys. Lett. **46**, 111 (1985)
18. S. Brazovskii, V. Yakovenko, Sov. Phys. JETP **62**, 1340 (1985)
19. R. Laversanne, C. Coulon, B. Gallois, et al., J. Phys. Lett. **45**, L393 (1984)
20. C. Coulon, S.S.P. Parkin, R. Laversanne, Phys. Rev. B **31**, 3583 (1985)
21. H.H.S. Javadi, R. Laversanne, A.J. Epstein, Phys. Rev. B **37**, 4280 (1988)
22. C. Coulon, in [3], p. 15
23. P. Monceau, F. Nad, S. Brazovskii, Phys. Rev. Lett. **86**, 4080 (2001)
24. F. Nad, P. Monceau, C. Carcel, et al., in [2], p. 265; F. Nad, in [7], p. 133
25. F. Nad, P. Monceau, J. Phys. Soc. Jpn. **75**, 051005 (2006)

[22] This proposal, apparent from already existing by that old time theories – see e.g. [80], is back to the agenda today, 25 years later.

26. D.S. Chow, F. Zamborszky, B. Alavi, et al., Phys. Rev. Lett. **85**, 1698 (2000); S. Brown, in [2,5,7]; F. Zamborszky, W. Yu, W. Raas, et al., Phys. Rev. B **66**, R081103 (2002)
27. W. Yu, F. Zamborszky, B. Alavi, et al., in [3], p. 35
28. H. Seo, M. Ogata, M. Kuwabara, in [3], p. 29
29. P. Batail, in [8], p. 307
30. S. Ravy, P. Foury-Leylekian, D. Le Bolloc'h, et al., in [3], p. 81
31. H. Fukuyama, in [2], p. 257, and [5], p. 673
32. T. Takahashi, K. Hiraki, S. Moroto, et al., in [8], p. 133
33. F. Nad, P. Monceau, at ECRYS 2005, in [8], unpublished
34. H. Matsui, H. Tsuchiya, T. Suzuki, et al., Phys. Rev. B **68**, 155105 (2003)
35. M. Buron, E. Collet, M.H. Lemèe-Cailleau, et al., in [7], p. 357; H. Okamoto, T. Koda, Y. Tokura, et al., Phys. Rev. B **43**, 8224 (1991); S. Horiuchi, R. Kumai, Y. Okimoto, et al., Phys. Rev. Lett. **85**, 5210 (2000)
36. J.-P. Pouget, S.K. Khanna, F. Denoyer, et al., Phys. Rev. Lett. **37**, 437 (1976)
37. S. Kagoshima, T. Ishiguro, H. Anzai, J. Phys. Soc. Jpn. **41**, 2061 (1976)
38. S.K. Khanna, J.-P. Pouget, R. Comès, et al., Phys. Rev. B **21**, 486 (1980)
39. V.J. Emery, Phys. Rev. Lett. **37**, 107 (1976)
40. J.-P. Pouget, in *Semiconductors, Semimetals*, vol. 27, ed. by E.M. Conwell (Academic Press, New York, 1988), p. 87
41. A. Andrieux, H.J. Schulz, D. Jérome, K. Bechgaard, Phys. Rev. Lett. **43**, 227 (1979)
42. S. Huizinga, J. Kommandeur, G.A. Sawatzky, et al., Phys. Rev. B **19**, 4723 (1979)
43. H. Seo, H. Fukuyama, J. Phys. Soc. Jpn. **66**, 124 (1997); H. Seo, in [7], p. 205
44. K. Hiraki, K. Kanoda, Phys. Rev. Lett. **80**, 4737 (1998)
45. V. Ilakovac, S. Ravy, A. Moradpour, et al., Phys. Rev. B **52**, 4108 (1995)
46. F. Nad, P. Monceau, J. Fabre, J. Phys. IV **9**, 361 (1999); F. Nad, P. Monceau C. Garcel, et al., Phys. Rev. B **62**, 1753 (2000); J. Phys. C **12**, L435 (2000)
47. S. Brazovskii, P. Monceau, F. Nad, in [5], p. 1331; cond-mat/0304483
48. S.S.P. Parkin, J.C. Scott, J. Thorance, E.M. Engler, in [1], p. 1111; J. Thorance, in [1], p. 799
49. H. Alloul, in [8], p. 27
50. P. Monceau, F. Nad, in [10], p. 2121
51. F. Nad, P. Monceau, J. Phys. IV C2, **3**, 343 (1993)
52. F. Nad, P. Monceau, K. Bechgaard, Solid State Commun. **95**, 655 (1995)
53. S. Brazovskii, in [2], p. 301; also cond-mat/0304076
54. S. Brazovskii, in [7], p. 149; also cond-mat/0306006
55. S. Brazovskii, N. Kirova, JETP Lett. **33**, 4 (1981); S. Brazovskii, N. Kirova, S.I. Matveenko, Sov. Phys. JETP **59**, 434 (1984)
56. M.J. Rice, E.J. Mele, Phys. Rev. Lett. **49**, 1455 (1982)
57. See reviews of this volume by D. Jérome, A. Lebed, C. Sa de Melo, D. Podolsky, et al.
58. R. Bruinsma, V.J. Emery, J. Phys. C3, **44**, 1155 (1983)
59. L.D. Landau, E.M. Lifshitz, *Electrodynamics of Continuous Media* (Pergamon, Oxford, 1960)
60. S. Barisic, S. Brazovskii, in *Recent Developments in Condensed Matter Physics*, vol. 1, ed. by D. Devreese (Plenum Press, New York, 1981), p. 142
61. V.J. Emery, R. Bruinsma, S. Barisic, Phys. Rev. Lett. **48**, 1039 (1982)

62. I.E. Dzyaloshinskii, A.I. Larkin, Sov. Phys. JETP, **34**, 422 (1972)
63. A. Luther, V.J. Emery, Phys. Rev. Lett. **33**, 589 (1974)
64. V.J. Emery, in *Highly Conducting One-Dimensional Solids* (Plenum, New York, 1979) p. 247
65. S. Brazovskii, Synth. Met. **152**, 309 (2005)
66. M. Dumm, M. Abaker, M. Dressel, in [8], p. 55
67. V. Ilakovac, S. Ravy, J.-P. Pouget, et al., Phys. Rev. B, **50**, 7136 (1994)
68. S. Brazovskii, N. Kirova, in *Sov. Sci. Rev. Sect. A*, ed. by I.M. Khalatnikov, vol. 5 (Harwood Acadmic, London, 1984), p. 99
69. S. Brazovskii, N. Kirova, S.I. Matveenko, Sov. Phys. JETP **59**, 434 (1984)
70. S. Brazovskii, N. Kirova, JETP Lett. **33**, 4 (1981)
71. L. Lauchlan, S. Etemad, T.-C. Chung, et al., Phys. Rev. B **24**, 3701 (1981)
72. I. Gontia, S.V. Frolov, M. Liess, et al., Phys. Rev. Lett. **82**, 4058 (1999)
73. M. Nagasawa, F. Nad, P. Monceau, et al., in [3], p. 119
74. W. Yu, F. Zhang, F. Zamborszky, et al., Phys. Rev. B **70**, 121101 (2005)
75. F. Nad, P. Monceau, C. Carcel, J.M. Fabre, J. Phys. C, **13**, L717 (2001)
76. T. Nakamura, J. Phys. Soc. Jpn. **72**, 213 (2003)
77. F. Nad, P. Monceau, L. Kaboub, J.M. Fabre, Europhys. Lett. **73**, 567 (2006)
78. F. Nad, P. Monceau, M. Nagasawa, et al., in [8], p. 15
79. K. Heuzé, M. Fourmigué, P. Batail, et al., Adv. Mater. **15**, 1251 (2003)
80. J. Sólyom, Adv. Phys. **28**, 201 (1979)
81. A.O. Gogolin, A.A. Nersesyan, A.M. Tsvelik, *Bosonization Approach to Strongly Correlated Systems* (Cambridge University Press, Cambridge, 1999)
82. C. Bourbonnais, R. Duprat, in [3], p. 3
83. T. Giamarchi, this volume, p. 719, and in [3], p. 23
84. S. Mazumdar, in [5], p. 1317; R.T. Clay, S. Mazumdar, D.K. Campbell, Phys. Rev. B **67**, 115121 (2003)
85. J. Riera, D. Poilblanc, Phys. Rev. B **63** 241102 (2001)
86. M.J. Rice, A.R. Bishop, J.A. Krumhansl et al., Phys. Rev. Lett. **36**, 432 (1976)
87. P. Auban-Senzier, D. Jérome, C. Carcel et al., in [3], p. 41; B. Korin-Hamzić, E. Tafra, M. Basletić et al., Phys. Rev. B **73** 115102 (2006)
88. Yu.I. Latyshev, P. Monceau, S. Brazovski et al., in [8], p. 197; and Phys. Rev. Lett. **95**, 266402 (2005); ibid., **96**, 116402 (2006); S. Brazovskii, Yu.I. Latyshev, S.I. Matveenko, et al., in [8], p. 77
89. S. Brazovskii, S. Gordyunin, JETP Lett. **31**, 371 (1980)
90. S. Teber et al., in [7]; J. Phys. C, **13**, 4015 (2001); J. Phys. C, **14**, 7811 (2002)
91. P. Monceau, H.-C. Lasjaunias, K. Biljakovic, and F. Nad, this volume, p. 277
92. V.Ya. Krivnov, A.A. Ovchinnikov, Sov. Phys. JETP **63**, 414 (1986)
93. H. Fukuyama, H. Takayama, in [11], p. 41
94. S. Brazovskii, cond-mat/0204147,0006355
95. H. Kleinert, K. Maki, Phys. Rev. B **19**, 6238 (1979)
96. K. Maki, in [11], p. 125
97. F.H.L. Essler, F. Gebhard, E. Jeckelmann, Phys. Rev. B **64**, 125119 (2001)
98. D. Controzzi, F.H.L. Essler, A.M. Tsvelik, Phys. Rev. Lett. **86**, 680 (2001)
99. V. Vescoli, L. Degiorgi, W. Henderson et al., Science **281**, 1181 (1998)
100. V. Vescoli, L. Degiorgi, K.P. Starkey, and L.K. Montgomery, Solid State Commun. **111**, 507 (1999)
101. A. Schwartz, M. Dressel, G. Grüner et al., Phys. Rev. B **58**, 1261 (1998); M. Dressel, A. Schwartz, G. Grüner, Phys. Rev. Lett. **77**, 398 (1996)

102. N. Kirova, S. Brazovskii, Curr. Appl. Phys. **4**, 473 (2004); Thin Solid Films **403**, 419 (2002); N. Kirova, S. Brazovskii, A.R. Bishop, Synth. Met. **100**, 29 (1999)
103. S. Brazovskii, Proceeding of ISCOM 2007, Solid State Sciences (2008), to be publ., arXiv:0801.3202v1
104. C. Jacobsen, D.B. Tanner, K. Bechgaard, Phys. Rev. B **28**, 7019 (1983); D. Pedron, R. Bozio, M. Meneghetti, et al., Phys. Rev. B **49**, 10893 (1994)
105. M.J. Rice, Phys. Rev. Lett. **37**, 36 (1976)
106. N. Kirova, S. Brazovskii, Synth. Met. **41**, 139 (2004)
107. S. Fujiyama, T. Nakamura, Phys. Rev. B **70**, 045102 (2004)
108. J. Tranquada, in [7], p. 239 and reference therein
109. C. Bourbonnais, D. Jérome, in *Advances in Synthetic Metals*, (Elsevier, New York, 1999); cond-mat/9903101
110. H.K. Ng, T. Timusk, K. Bechgaard, Phys. Rev. B **30**, 5842 (1984); H.K. Ng, T. Timusk, D. Jérome, et al., Phys. Rev. B **32**, 8041 (1985); N. Cao, T. Timusk, K. Bechgaard, in [10], p. 1719
111. See articles in [11].
112. B. Hennion, J.-P. Pouget, M. Sato, Phys. Rev. Lett. **68**, 2374 (1992) and ibid., **69**, 3266 (1992)
113. A. Finkelstein, S. Brazovskii, J. Phys. C **14**, 847 (1981); Solid State Commun. **38**, 745 (1981)
114. Y. Nogami, K. Oshima, K. Hiraki, et al., J. Phys. IV **9**, 357 (1999); Synth. Met. **102**, 1778 (1999)
115. L. Perfetti, H. Berger, A. Reginelli, et al., Phys. Rev. Lett. **87**, 216404 (2001); L. Perfetti, S. Mitrovic, G. Margaritondo, et al., Phys. Rev B **66**, 075107 (2002)
116. V. Shikin, Sov. Phys. Usp Fiz. Nauk **158**, 127 (1989)
117. V.B. Preobrazhensky, A.N. Taldenkov, J. Phys. IV C2, **3**, 57 (1993); V.B. Preobrazhensky, A.N. Taldenkov, V.V. Frolov, Synth. Met. **55-57**, 2617 (1993)
118. G.X. Tessema, M.J. Skove, Y.T. Tseng, J. Phys. IV C2, **3**, 53 (1993); J. Kuh, Y.T. Tseng, K. Wagner, et al., Phys. Rev. B **57**, 14576 (1998)
119. F. Zwick, S. Brown, G. Margritondo, et al., Phys. Rev. Lett. **79**, 3982 (1997)
120. F. Zwick, D. Jérome, G. Margaritondo, et al., Phys. Rev. Lett., **81**, 2974 (1998)
121. J.-P. Pouget, P. Foury-Leylekian, D. Le Bolloc'h, et al., in [4], p. 147
122. F. Nad, P Monceau, T. Nakamura, et al., J. Phys. C **17**, L399, (2005); K. Furukawa, T. Hara, T. Nakamura, J. Phys. Soc. Jpn. **74**, 3288 (2005)
123. S. Brazovskii, V. Yakovenko, Sov. Phys. JETP Lett. **43** (1986) 134
124. S. Haddad, S. Charfi-Kaddour, C. Nickel, M. Hèritier, and R. Bennaceur, Phys. Rev. Lett. **89**, 087001 (2002); S. Haddad, M. Hèritier, and S. Charfi-Kaddour, this volume, p. 605
125. H.I. Ha, A.G. Lebed, M.J. Naughton, Phys. Rev. B **73**, 033107 (2006)
126. D. Radic, A. Bjelis, D. Zanchi, Phys. Rev. B **69**, 014411 (2004); and in [8], p. 281.

12

Interacting Electrons in Quasi-One-Dimensional Organic Superconductors

C. Bourbonnais and D. Jérome

This review highlights the main features of the temperature–pressure phase diagram of the Bechgaard and Fabre salts series of quasi-one-dimensional organic superconductors. We go over the various electronic and structural instabilities found experimentally in the normal state of the sulfur $(TMTTF)_2X$ series at relatively high temperature and show how these are strongly influenced by the one-dimensional character of electronic degrees of freedom. The problem of three-dimensional long-range order is then discussed for the Fabre series and the mechanisms responsible for the spin-Peierls and Néel phase transitions are depicted. The influence of pressure on the relative stability of these phases and the emergence of quasi-particles when the Fabre series evolves toward the Bechgaard $(TMTSF)_2X$ salts series are presented. Itinerant antiferromagnetism, density-wave, and uncoventional superconductivity are described and the microscopic origin of their interplay is discussed.

12.1 Introduction

Superconductivity in organic materials has emerged in 1979 from an important background of preexisting knowledge and experimental techniques. All previous studies undertaken since 1973, which had been mostly performed on the (TTF – TCNQ) series of charge transfer organic conductors, had failed to reveal superconductivity using chemistry and (or) pressure to suppress the density-wave or the so-called Peierls instability inherent to one-dimensional conductors. A breakthrough, which contributed to the discovery of organic superconductivity, has been the synthesis of the molecule TMTSF by Bechgaard and coworkers [1].

Actually, in the early 1970s, leading ideas governing the search for new materials likely to exhibit good metallicity and possibly superconductivity were driven by the possibility to minimize the role of electron–electron repulsions and at the same time to increase the electron–phonon interaction, while keeping the overlap between conducting stacks as large as possible. This led to the synthesis of the new electron donating molecule TMTSF, presenting much analogy with the previously known fulvalene donors in which the redox potential $(\Delta E)_{1/2}$ can be minimized [2,3], by utilizing selenium instead of sulfur as heteroatoms [4]. The next step was quite encouraging since the use of a high pressure has allowed to remove the instability due to the divergence of the Peierls channel down to the lowest temperatures in the two-chain conductor TMTSF – DMTCNQ [5]. A lucky situation has also been the synthesis of a series of 1D organic salts based on the radical cationic molecule TMTTF (the sulfur analog of the TMTSF molecule) and on a variety of inorganic monoanions such as ClO_4, BF_4, or SCN [6,7]. A conducting character could thus be anticipated from the intermolecular overlap of partly filled highest molecular orbitals (HOMO) of individual molecules. The compounds, $(TMTTF)_2X$, were all insulating at ambient pressure but their crystal structure is the prototype of the $(TMTSF)_2X$ series in which superconductivity has been subsequently discovered. In the rest of this article when we mention a $(TM)_2X$ compound, this means that the organic molecule can be either TMTSF or TMTTF. The structure exhibits a face to face packing of flat molecular units along the a direction and the formation of molecular layers in the a-b planes separated by anions stacks (Fig. 12.1). The overall symmetry is triclinic, not too far from orthorhombic, often taken as the approximate structure by theoreticians. In addition, the structure reveals an important peculiarity namely, the anions are located in centrosymmetrical cavities lying slightly above or below each molecular plane with a zigzag stacking of the molecules along the a direction. This structure leads in turn to a weak alternation of the interplanar distance (dimerization and a concomitant splitting) of the HOMO conduction band into a filled lower sub-band separated from a half-filled upper (hole-like) band by a gap Δ_D at $\pm 2k_F$, called the dimerization gap in the new Brillouin zone. However, on account of the finite transverse dispersion, this dimerization gap does not lead to a genuine gap in the middle of the density of states as given from the extended-Hückel band calculation. The only claim that can be made is that these conductors show commensurate band filling (three-quarter filled with electrons or one-quarter empty with holes). This originates from the 2:1 stoichiometry. Consequently, according to a noninteracting particle band calculation, all compounds in the $(TMTTF)_2X$ series should be found conducting.

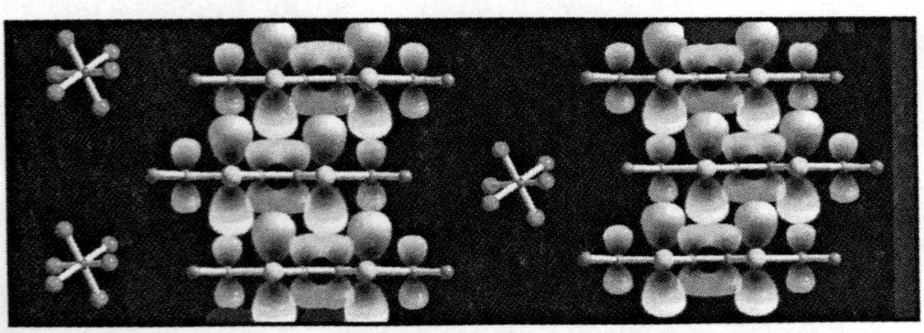

Fig. 12.1. $(TM)_2X$, view of the cationic and anionic stacking perpendicular to the stacking axis, *courtesy* of J. Ch. Ricquier, IMN, Nantes

Table 12.1. Calculated band parameters for three representative members of the $(TM)_2X$ series according to the room temperature crystallographic data in [8]

	$(TMTTF)_2PF_6$	$(TMTSF)_2PF_6$	$(TMTSF)_2ClO_4$
t_1	137	252	258
t_2	93	209	221
$\bar{t}$	115	230	239
$\frac{\Delta t}{\bar{t}}$	0.38	0.187	0.155
$\bar{t}_{\perp b}$	13	58	44

The average intra and interstack interactions are given in lines 3 and 5, respectively. The bond dimerization is shown in line 4. All energies are in meV

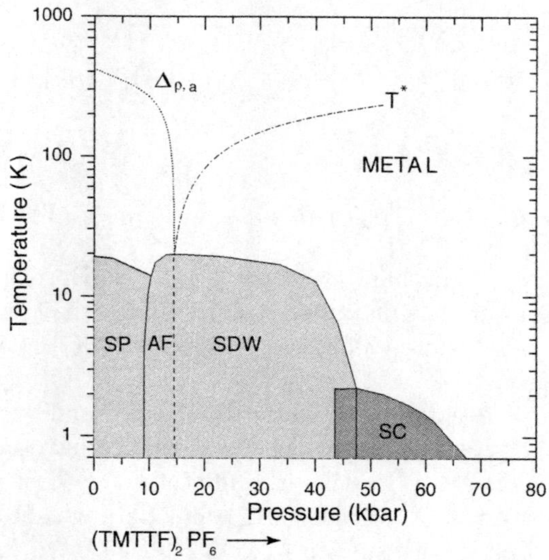

Fig. 12.2. Phase diagram of $(TMTTF)_2PF_6$, as determined from transport measurements. After [11,12]

In Table (12.1) we report the band parameters of different members of the $(TMTSF)_2X$ series, band parameter calculation as computed from crystallographic data [8]. The sulfur compounds exhibit bands that are significantly narrower and their crystallographic structure is more dimerized than those of the selenide compounds. $(TMTTF)_2Br$ (not listed in Table 12.1) is, however, an exception among the sulfur compounds with a dimerization of 0.13, which is smaller than the value calculated for $(TMTSF)_2ClO_4$. This might be due to the calculation of electronic bands based on rather old and less accurate crystallographic data than those used for the other compounds [9]. All $(TM)_2X$ compounds with diversified anions can be gathered on a generic phase diagram displaying a wealth of different physical properties [10]. The gross features of the $(TM)_2X$ phase diagram are shown in Figs. 12.2 and 12.3. Compounds on the left hand side of the phase diagram, such as $(TMTTF)_2PF_6$, are insulators below room temperature with a broad metal-to-insulator transition,

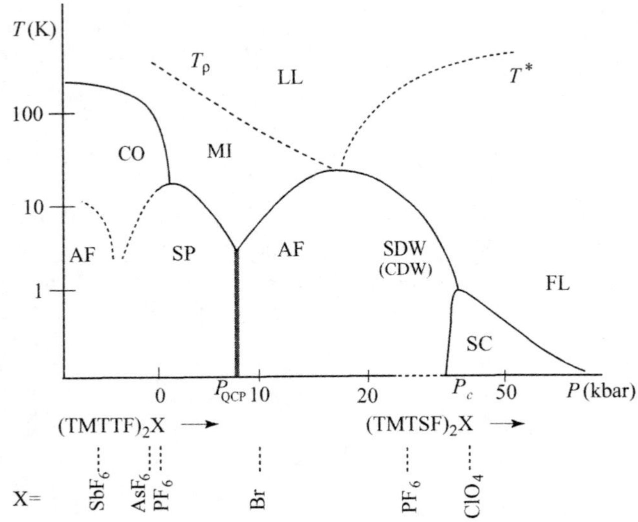

Fig. 12.3. The generic phase diagram of $(TM)_2X$

while those on the right hand side of $(TMTTF)_2Br$ exhibit an extended temperature regime with a metallic behavior and a sharp transition towards an insulating ground state. Therefore, the cause of the insulating nature of some members of the $(TMTTF)_2X$ series should be determined in relation to the role of e–e repulsion and low dimensionality as we shall show later. Although the most extensive pressure studies have been performed on $(TMTSF)_2PF_6$ and $(TMTTF)_2PF_6$, recent studies of other compounds of the $(TMTTF)_2X$ series with X = ReO_4, BF_4 and Br have shown that the main features observed under pressure in $(TMTTF)_2PF_6$ or in $(TMTSF)_2PF_6$ are also observed in other systems [11]. At this stage we may emphasize that the band filling of these materials is commensurate and in addition the existence of a dimerization in the crystal structure of the $(TM)_2X$ series raises quite a challenging problem for the physics of one-dimensional conductors, since with the axial dimerization the conduction band becomes half-filled while it is originally quarter-filled from stoichiometry considerations. The commensurate band filling opens a new scattering channel for the carriers between both sides of the Fermi surface as then the total momentum transfer for two (four) electrons from one side of the 1D Fermi surface to the other is equal to a reciprocal lattice vector (Umklapp scattering for half (quarter)-filled bands).

12.2 Elements of Theory for Interacting Electrons in Low Dimension

In this section we shall depict some of the main results of the theory of quasi-one-dimensional metals. Given the pronounced one-dimensional anisotropy of the compounds, it is natural to first consider the 1D limit. To this end, the study of susceptibilities of noninteracting electrons is particularly revealing of the natural instabilities that can take place in one dimension. Take for example the bare *Peierls* susceptibility of the system, which is the response to the formation of $2k_\mathrm{F}$ electron–hole pairs, the constituents of charge-density-wave (CDW) and spin-density-wave (SDW) correlations (here k_F being the Fermi wave vector). In one dimension, the energies of an electron (hole) state at k and a hole (electron) state at $k-2k_\mathrm{F}$ are connected through the *nesting* relation $\epsilon(k)=-\epsilon(k-2k_\mathrm{F})$ of the electron spectrum $\epsilon(k)$ close to the Fermi level. The summation over a macroscopic number of intermediate electron–hole states linked by this relation leads to an infrared logarithmic singularity of the form $\chi^0_\mathrm{P}(2k_\mathrm{F},T)\sim(\pi v_\mathrm{F})^{-1}\ln(E_\mathrm{F}/T)$. Similarly, the *Cooper* susceptibility χ^0_C, which probes the formation of pairs (two holes or two electrons) of particles of total momentum zero that are connected through the inversion property of the spectrum $\epsilon(k)=\epsilon(-k)$, also gives rise to a logarithmic divergence $\chi^0_\mathrm{C}(T)\sim(\pi v_\mathrm{F})^{-1}\ln(E_\mathrm{F}/T)$ – a singularity that is actually found in any dimension.

What thus really makes one dimension so peculiar resides in the fact that both singularities refer to the same set of electronic states and will then interfere one another [13]. In the presence of interactions, the interference is found to all order of perturbation theory for the scattering amplitudes of electrons with opposite Fermi velocities and it modifies the nature of the electron system in an essential way. In the framework of the 1D electron gas model, the selected emphasis put by these infrared singularities on electronic states close to the Fermi level allows us to define interactions with respect to the Fermi points $\pm k_\mathrm{F}$ [14,15]. Thus for a rotationally invariant system of length L, the Hamiltonian of the electron gas model can be written in the form

$$
\begin{aligned}
H = &\sum_{k,p,\sigma} \epsilon_p(k) a^\dagger_{p,k,\sigma} a_{p,k,\sigma} \\
&+ \frac{1}{L} \sum_{\{k,q,\sigma\}} g_1\, a^\dagger_{+,k_1+2k_\mathrm{F}+q,\sigma} a^\dagger_{-,k_2-2k_\mathrm{F}-q,\sigma'} a_{+,k_2,\sigma'} a_{-,k_1,\sigma} \\
&+ \frac{1}{L} \sum_{\{k,q,\sigma\}} g_2\, a^\dagger_{+,k_1+q,\sigma} a^\dagger_{-,k_2-q,\sigma'} a_{-,k_2,\sigma'} a_{+,k_1,\sigma} \\
&+ \frac{1}{2L} \sum_{\{p,k,q,\sigma\}} g_3\, a^\dagger_{p,k_1+p2k_\mathrm{F}+q,\sigma} a^\dagger_{p,k_2-p2k_\mathrm{F}-q+pG,\sigma'} a_{-p,k_2,\sigma'} a_{-p,k_1,\sigma} \\
&+ \frac{1}{L} \sum_{\{p,k,q,\sigma\}} g_4\, a^\dagger_{p,k_1+q,\sigma} a^\dagger_{p,k_2-q,-\sigma} a_{p,k_2,-\sigma} a_{-p,k_1,\sigma}
\end{aligned}
\qquad (12.1)
$$

where $\epsilon_p(k)=v_\mathrm{F}(pk-k_\mathrm{F})$ is the electron spectrum energy for right ($p=+$) and left ($p=-$) going electrons; g_1 and g_2 are the back and forward scattering amplitudes, respectively; whereas g_3 corresponds to the Umklapp scattering, a process made possible at half-filling, where the reciprocal lattice vector $G=4k_\mathrm{F}=2\pi/a$ enters in the momentum conservation law; finally, one has the coupling g_4 by which two electrons near k_{F} (resp. $-k_{\mathrm{F}}$) experience a small momentum transfer that keeps them on the same branch [15].

In the one-loop perturbation theory, the electron scattering amplitudes are corrected by the aforementioned Cooper and Peierls logarithmic singularities. These logarithms are scale invariant quantities as a function of energy or temperature, which allow us to write down scaling or flow equations for the various scattering amplitudes. After all cancelations because of Cooper–Peierls interference, we have [14, 15],

$$\begin{aligned} \tilde{g}_1' &= -\tilde{g}_1^2 + \dots \\ (2\tilde{g}_2' - \tilde{g}_1') &= \tilde{g}_3^2 + \dots \\ \tilde{g}_3' &= \tilde{g}_3(2\tilde{g}_2 - \tilde{g}_1) + \dots, \end{aligned} \tag{12.2}$$

where $\tilde{g}_i' = (\mathrm{d}/\mathrm{d}\ell)\tilde{g}_i$ and $\ell = \ln E_{\mathrm{F}}/T$ is the logarithmic – loop – variable. The long wavelength spin excitations are governed by the $\tilde{g}_1 \equiv g_1/\pi v_\sigma$ ($v_\sigma = v_{\mathrm{F}}(1 - g_4/2\pi v_{\mathrm{F}})$) coupling, which is decoupled from both $\tilde{g}_3 \equiv g_3/\pi v_\rho$ ($v_\rho = v_{\mathrm{F}}(1 + g_4/2\pi v_{\mathrm{F}})$) and the combination $2\tilde{g}_2 - \tilde{g}_1 \equiv (2g_2 - g_1)/\pi v_\rho$ connected to charge excitations. In the physically relevant repulsive sector for systems like $(\mathrm{TM})_2\mathrm{X}$ where $g_{1,2} > 0$, and owing to the existence of a small dimerization gap $\Delta_{\mathrm{D}} \ll E_{\mathrm{F}}$ of organic stacks (Table 12.1), weak half-filled Umklapp scattering $g_3 \approx g_1 \Delta_{\mathrm{D}}/E_{\mathrm{F}}$ is present [16, 17]. Thus for $g_1 - 2g_2 < | g_3 |$, both $2g_2 - g_1$ and g_3 are relevant variables for the charge and scale to the strong coupling sector, where a charge gap Δ_ρ is found below the temperature scale $T_\rho(\sim\Delta_\rho/2)$. This can be seen as a $4k_{\mathrm{F}}$ charge localization responsible for a Mott insulating (MI) state. On the other hand, the solution $\tilde{g}_1(T) = \tilde{g}_1/(1 + \tilde{g}_1 \ln E_{\mathrm{F}}/T)$ for the g_1 coupling, which follows from (12.2), is marginally irrelevant and scales to zero, leaving the spins degrees of freedom gapless as shown by the calculation of the uniform spin susceptibility [14, 18],

$$\chi_\sigma(T) = \frac{2\mu_{\mathrm{B}}^2(\pi v_\sigma)^{-1}}{1 - \frac{1}{2}\tilde{g}_1(T)}. \tag{12.3}$$

The spin susceptibility decreases monotonically as a function of temperature and is unaffected by the occurrence of a charge gap. The electron system develops singularities, however, for staggered density-wave response. Thus the $2k_{\mathrm{F}}$ SDW or antiferromagnetic response, which is governed by the combination of couplings $\tilde{g}_2(\ell) + \tilde{g}_3(\ell)$ that flows to strong coupling, develops a power law singularity of the form

$$\chi_{\mathrm{AF}}(2k_{\mathrm{F}}, T) \propto (\pi v_{\mathrm{F}})^{-1}(T/\Delta_\rho)^{-\gamma}, \tag{12.4}$$

where the power law exponent $\gamma = \tilde{g}_2(T_\rho) + \tilde{g}_3(T_\rho) \sim 1$. The response for the $2k_{\mathrm{F}}$ charge-density-wave "on bonds," called the bond-order-wave (BOW) response, which is governed by the combination of couplings $\tilde{g}_2(\ell) + \tilde{g}_3(\ell) - 2\tilde{g}_1(\ell)$, also develops a power law singularity in temperature

$$\chi_{\mathrm{BOW}}(2k_{\mathrm{F}}, T) \propto (\pi v_{\mathrm{F}})^{-1}(T/\Delta_\rho)^{-\gamma_{\mathrm{BOW}}}. \tag{12.5}$$

Here the exponent $\gamma_{\mathrm{BOW}} \sim 1$ is essentially the same as the one for AF response – the amplitude of the latter being larger, however [19]. When $2k_{\mathrm{F}}$ phonons are included, their coupling to singular BOW correlations yields a lattice instability of the spin–Peierls (SP) type. It is worthwhile to note that all the above properties of the 1D electron gas model find some echo in the phase diagram of $(\mathrm{TM})_2\mathrm{X}$ (Fig. 12.3).

12.2.1 Some Results of the Bosonization Picture [20]

We now turn to the description of the one-dimensional electron gas using the bosonization method. A major property of interacting electrons in one dimension is that long wavelength charge or spin-density-wave oscillations constructed by the combination of electron–hole pair excitations at low energy form extremely stable excitations [20, 21]. Quasi-particles excitations, like those taking place in a Fermi liquid (FL) for example, are absent at low energy and are replaced by these collective acoustic excitations for both spin (σ) and charge (ρ) degrees and freedom, thus allowing the construction of a phase representation of the electron gas Hamiltonian. The Fermi field

$$\psi_{p,\sigma}(x) = L^{-\frac{1}{2}} \sum_k a_{p,k,\sigma}\, \mathrm{e}^{\mathrm{i}kx}$$

$$\sim \lim_{\alpha_0 \to 0} \frac{\mathrm{e}^{\mathrm{i}pk_{\mathrm{F}}x}}{\sqrt{2\pi\alpha_0}} \exp\Big(-\frac{\mathrm{i}}{\sqrt{2}}[p(\phi_\rho + \sigma\phi_\sigma) + (\theta_\rho + \sigma\theta_\sigma)]\Big), \tag{12.6}$$

can be expressed in terms of the spin and charge phase fields $\phi_{\nu=\rho,\sigma}$ [20, 22] (α_0 is a short distance cut-off). These satisfy the commutation relations

$$[\Pi_{\nu'}(x'), \phi_\nu(x)] = -\mathrm{i}\delta_{\nu\nu'}\delta(x - x'), \tag{12.7}$$

where $\Pi_\nu(x)$ is the momentum conjugate to $\phi_\nu(x)$ and is defined by $\theta_\nu(x) = \pi \int \Pi_\nu(x')\, \mathrm{d}x'$. In this phase variable representation the full electron gas Hamiltonian takes the form

$$H = \sum_{\nu=\rho,\sigma} \frac{1}{2} \int \left[\pi u_\nu K_\nu \Pi_\nu^2 + u_\nu (\pi K_\nu)^{-1} \left(\frac{\partial \phi_\nu}{\partial x}\right)^2 \right] \mathrm{d}x$$
$$+ \frac{2g_1}{(2\pi\alpha_0)^2} \int \cos(\sqrt{8}\phi_\sigma)\, \mathrm{d}x + \frac{2g_3}{(2\pi\alpha_0)^2} \int \cos(\sqrt{8}\phi_\rho)\, \mathrm{d}x. \tag{12.8}$$

The harmonic part of the phase Hamiltonian on the first line corresponds to the Tomonaga–Luttinger model, which is exactly solvable. The spectrum shows no quasi-particles but only collective excitations and all the properties of the model then become entirely governed by the velocity u_ν and the "stiffness constant" K_ν of acoustic excitations, which depend on interactions. This corresponds to the physics of the so-called Luttinger (LL) or Tomonaga–Luttinger liquid. In the Tomonaga–Luttinger limit the power law singularity of the AF spin response $\chi_{\mathrm{AF}}(2k_{\mathrm{F}},T) \sim T^{-\gamma}$ is confirmed and the exponent

$$\gamma = 1 - K_\rho \tag{12.9}$$

is expressed in terms of the charge stiffness constant K_ρ. The absence of quasi-particle excitations is captured by the power law decay of the density of states at the Fermi level

$$N(0) \sim (\pi v_{\mathrm{F}})^{-1}\Big(\frac{T}{E_{\mathrm{F}}}\Big)^{\alpha}, \tag{12.10}$$

with the exponent

$$\alpha = \frac{1}{4}(K_\rho + 1/K_\rho - 2). \tag{12.11}$$

The quasi-particle weight at the Fermi level $z(T) \sim T^{\alpha}$ follows a similar power law decrease.

In the presence of the sine Gordon terms due to the backscattering and Umklapp couplings in the phase Hamiltonian (12.8), an exact solution cannot be found in the general case. However, a perturbative scaling procedure can be used for the various parameters that define the Hamiltonian [20]. For rotationally invariant repulsive couplings, g_1 is marginally irrelevant as found in the many-body description (Sect. 12.2), and only the flow in the charge sector essentially matters. In low order, one can write the following flow equations

$$\begin{aligned}\frac{\mathrm{d}K_\rho}{\mathrm{d}l} &= -\frac{1}{2}K_\rho^2\tilde{g}_3^2,\\ \frac{\mathrm{d}\tilde{g}_3}{\mathrm{d}l} &= \tilde{g}_3(2-2K_\rho).\end{aligned} \tag{12.12}$$

For repulsive couplings, the bare $K_\rho = K_\rho(g_4, 2g_2 - g_1) < 1$, g_3 is relevant and scales to strong coupling, as found in the previous fermion scaling description in (12.2), while $K_\rho^*(l \gg 1) \to 0$. An expression for the charge gap can be found

$$\Delta_\rho \sim W\tilde{g}_U^{1/[2(1-n^2K_\rho)]}, \tag{12.13}$$

where for half-filling Umklapp $n = 1$ and $\tilde{g}_U = \tilde{g}_3$ [20].

A charge gap is not limited to half-filling but may be present for other commensurabilities too [23]. At quarter-filling, for example, the transfer of four particles from one side of the Fermi surface to the other leads to the Umklapp coupling

$$H_{1/4} \simeq \frac{2g_{1/4}}{(2\pi\alpha_0)^2}\int \mathrm{d}x \cos(2\sqrt{8}\phi_\rho). \tag{12.14}$$

The phase argument of this term differs and leads to a distinct flow equation

$$\frac{\mathrm{d}\tilde{g}_{1/4}}{\mathrm{d}l} = (2 - 8K_\rho)\tilde{g}_{1/4}, \tag{12.15}$$

which goes to strong coupling if $K_\rho < 1/4$, namely for sizable long-range Coulomb interaction [24]. The value of the insulating gap is given by (12.13) by taking $n = 2$ and $\tilde{g}_U = \tilde{g}_{1/4}$ at quarter-filling ($\tilde{g}_{1/4} \sim (U/W)^3$ in the Hubbard limit) [24,25]. It is worth noting that in the special situation where the quarter-filled chains are weakly dimerized, both half-filling and quarter-filling Umklapp couplings are present in practice and should interfere one another [26]. In effect for materials like $(TM)_2X$, stoichiometry imposes half a carrier (hole) per TM molecule, a concentration that cannot be modified by applying pressure. Consequently, uniformly spaced molecules along the stacking axis should lead to a situation where a unit cell contains 1/2 carriers, i.e., the conduction band is quarter-filled. However, the nonuniformity of molecular packing has been observed in early structural studies of the $(TMTTF)_2X$ crystals [8]. The dimerization of the overlap between molecules occurs along the stacks, a situation that is pronounced in the sulfur series, although it is also encountered in some members of the $(TMTSF)_2X$ series (Table 12.1). The impact of such a dimerization on the electronic structure is generally quantified by a modulation of the intra-stack overlap integral, because both longitudinal and transverse molecular displacements can contribute to the intermolecular overlap and could make them half-filled band compounds.

12.2.2 The Role of Interchain Coupling

Electronic materials like $(TM)_2X$ can only be considered as close realizations of 1D interacting fermion systems so that interchain coupling, though small, must be taken into account in their description. A nonzero intermolecular overlap perpendicular to the chains yields finite interchain hopping integrals $t_{\perp b}$ and $t_{\perp c}$ along the b and c directions, respectively. These play an essential role either in the restoration of Fermi liquid (FL) quasi-particles or in establishing long-range order. Considering a square lattice of $N_\perp$ chains, the electron spectrum takes the form

$$E_p(\mathbf{k}) = \epsilon_p(k) - 2t_{\perp b}\cos k_{\perp b} - 2t_{\perp c}\cos k_{\perp c}, \tag{12.16}$$

where $\mathbf{k} = (k, k_{\perp b}, k_{\perp c})$ and $t_{\perp c} \ll t_{\perp b} \ll E_{\mathrm{F}}$. Owing to the strong anisotropy in the transfer integrals, a one-electron coherent motion in the transverse direction is not present at all temperatures. In the noninteracting case for example, the temperature scale below which thermal fluctuations no longer blur the transverse quantum mechanical coherence for the electron is simply $T_{x^1} \sim t_{\perp b}$, which can be seen as a one-particle dimensionality crossover. In the presence of interactions, however, the quasi-particle weight $z(T) \sim (T/E_{\mathrm{F}})^\alpha$ being reduced by the LL behavior, the condition becomes $T_{x^1} \sim z(T_{x^1})t_{\perp b}$ [27], namely

$$T_{x^1} \sim t_{\perp b}\Big(\frac{t_{\perp b}}{E_{\mathrm{F}}}\Big)^{\alpha/1-\alpha}. \tag{12.17}$$

The one-particle crossover scale T_{x^1} then decreases in the presence of interactions. When a Mott insulating phase takes place in the 1D domain, α reaches unity, T_{x^1} vanishes, and the transverse band motion is not possible and the single-particle coherence becomes spatially confined along the stacks.

Transverse coherence is nevertheless possible but it is achieved through two-particle pair hopping, a mechanism for long-range order that is not present in the Hamiltonian at the start but which emerges when interactions along the stacks combine with $t_{\perp,b,c}$ in the one-dimensional region [28–31]. For repulsive interactions and commensurate fillings, the most important pair hopping processes that gradually emerges as a function of energy is an interchain antiferromagnetic exchange,

$$\delta H_{\perp} = \sum_{i,j} \int \mathrm{d}x \, J_{\perp i,j} \, \mathbf{S}_i(x) \cdot \mathbf{S}_j(x).$$

In the one-dimensional regime, it is governed at the one-loop level by the flow equation in the Fourier space

$$\tilde{J}'_{\perp}(\mathbf{q}_0) = \tilde{f}(\ell) + \tilde{J}_{\perp}(\mathbf{q}_0)\gamma(\ell) - \frac{1}{2}(\tilde{J}_{\perp}(\mathbf{q}_0))^2, \tag{12.18}$$

where $\tilde{f}(\ell) \simeq -2\sum_{i=b,c}[(\tilde{g}_2(\ell) + \tilde{g}_3(\ell))t_{\perp,i}/W]^2 \mathrm{e}^{(2-2\alpha(\ell))\ell}$ and $\mathbf{q}_0 = (2k_{\mathrm{F}}, \pi, \pi)$ is the modulation wave vector of AF order. Here $\alpha(\ell)$ and $\gamma(\ell)$ are scale dependent power law exponents of the quasi-particle weight (12.11) and antiferromagnetic susceptibility (12.4), respectively. Depending on the sign of $2 - 2\alpha(\ell) - \gamma(\ell)$, different cases can be considered. Thus when a Mott gap is formed at $\ell_\rho = \ln E_{\mathrm{F}}/T_\rho$, $2 - 2\alpha(\ell) - \gamma(\ell)$ becomes negative above ℓ_ρ and the solution of (12.18) can be put in the following simple Stoner form at temperature T

$$\tilde{J}(\mathbf{q}_0) \approx \frac{\tilde{J}_{\perp b} + \tilde{J}_{\perp c}}{1 - (J_{\perp b} + J_{\perp c})\chi_{\mathrm{AF}}(2k_{\mathrm{F}}, T)}, \tag{12.19}$$

where the low temperature 1D AF susceptibility $\chi_{\mathrm{AF}}(2k_{\mathrm{F}}, T)$ is given by (12.4), for $\gamma = 1$ ($K_\rho^* = 0$), and

$$J_{\perp b,c} \sim \pi v_{\mathrm{F}} \frac{t^{*2}_{\perp b,c}}{\Delta_\rho^2} \tag{12.20}$$

is the effective exchange in the b and c directions at the scale T_ρ. From (12.19), a singularity will then occur at

$$T_{\mathrm{c}} \sim \frac{(t^{*2}_{\perp b} + t^{*2}_{\perp c})}{\Delta_\rho}, \tag{12.21}$$

signaling a transition to a Néel ordered state. This result indicates that T_{c} – essentially dominated by the exchange in the b direction – increases as the Mott temperature T_ρ decreases (Fig. 12.4). When the commensurability effects are decreasing and T_ρ eventually merges with the critical behavior of the transition, antiferromagnetism becomes itinerant and corresponds to a

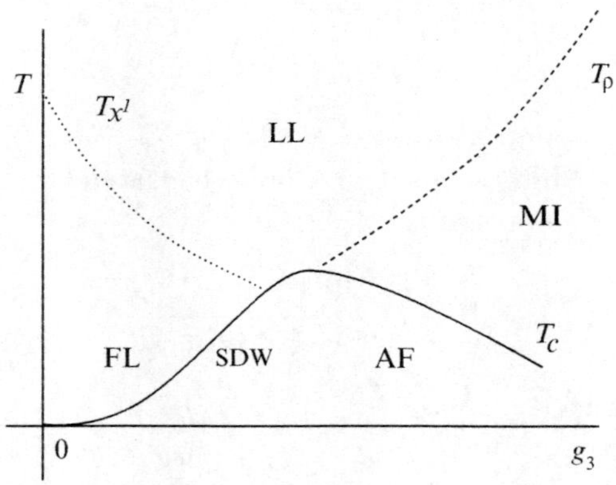

Fig. 12.4. Schematic phase diagram of a quasi-one-dimensional electron system at repulsive coupling calculated from the scaling theory. Here $g_3 \propto g_{1,2}$ is the running Umklapp parameter for the amplitude of interaction

SDW state. The interchain exchange enters in the weak coupling domain where $2 - 2\alpha(\ell) - \gamma(\ell)$ is small but positive. This modifies the critical temperature, which reads

$$T_c \sim \tilde{g}^{*2} t^*_{\perp b}, \tag{12.22}$$

where $\tilde{g}^* = \tilde{g}^*_2 + \tilde{g}^*_3$ and $t^*_{\perp b} = z t_{\perp b}$ are, respectively, the effective amplitude of electron–electron coupling and transverse hopping close to the transition. The calculations show that T_c starts to decrease in this domain giving rise to a maximum in T_c [30, 32, 33]. When the strength of 1D correlations becomes weaker, one quickly arrives at the regime where T_{x^1} becomes larger than the above T_c so that the one-electron motion is no longer confined along the stacks and interchain coherence develops before the onset of criticality linked to the transition.

The temperature T_{x^1} then becomes the scale below which the nesting of the whole Fermi surface becomes coherent. By looking at the effective spectrum (12.16) in which, $t_{\perp b,c} \to t^*_{\perp b,c} = z(T_{x^1}) t_{\perp b,c}$, perfect nesting occurs at $\mathbf{q}_0 = (2k_F, \pi, \pi)$, where the electron–hole symmetry $E^*_-(\mathbf{k}) = -E^*_+(\mathbf{k} + \mathbf{q}_0)$ holds. It follows that electron–hole excitations within the energy shell $\sim T_{x^1}$ above and below the coherent – warped – Fermi surface lead to a logarithmically singular response $\chi^0(\mathbf{q}_0, T) \sim (\pi v_F)^{-1} \ln T_{x^1}/T$ in the Peierls channel. This singularity is also found in the perturbation theory of the scattering amplitudes and for repulsive interactions, and it yields an instability of the normal state towards SDW long-range order. When perfect nesting prevails, a not too bad approximation consists of neglecting the interference between the Cooper and Peierls channels (we shall revert to the problem of interference below T_{x^1} later in Sect. 12.4.6). This corresponds to the ladder diagrammatic summation. In the scaling theory language, the ladder approximation corresponds to the flow equation

$$\frac{\mathrm{d}\tilde{J}}{\mathrm{d}\ell} = \frac{1}{2}\tilde{J}^2 + ..., \tag{12.23}$$

for an effective coupling constant $\tilde{J} = \tilde{g}_2 + \tilde{g}_3 - \tilde{J}_\perp$ that defines the net attraction between an electron and a hole separated by $\mathbf{q}_0$. The integration leads to the Stoner expression

$$\tilde{J}(T) = \frac{\tilde{J}^*}{1 - \frac{1}{2}J^*\chi^0(\mathbf{q}_0, T)}, \tag{12.24}$$

where $\tilde{J}^*$ is the effective SDW coupling obtained from (12.18) and (12.2) at $\ell_{x^1} = \ln E_\mathrm{F}/T_{x^1}$, and $\chi^0(\mathbf{q}_0, T) = (\pi v_\mathrm{F})^{-1} \ln T_{x^1}/T$. The above expression leads to a BCS singularity at the SDW critical temperature

$$T_\mathrm{c} = T_{x^1}\,\mathrm{e}^{-2/J^*}, \tag{12.25}$$

which decreases as the interactions decrease. Nesting frustration is required to suppress the transition [34, 35]. When nesting deviations are sufficiently strong, however, the FL remains unstable. Actually, when the partial but finite interference between the Peierls and the Cooper channels is restored, the system turns out to develop a superconducting instability. We shall return to this in Sect. 12.4.6.

12.3 The Fabre Salts Series

12.3.1 The Generic $(TM)_2X$ Phase Diagram

Although the $(TM)_2X$ generic phase diagram can be established by measuring the transport properties of various compounds, it has been most rewarding to use a single compound, namely $(TMTTF)_2PF_6$, and the help of a high pressure to span the generic diagram in Figs. 12.2 and 12.3. The study of this strongly insulating system $(TMTTF)_2PF_6$ under high pressure has been very useful not only because it has led to the stabilization of superconductivity in a strongly insulating sulfur compound [12], but also because its location at the left end of the phase diagram has allowed several key properties of quasi 1D conductors to be carefully monitored under pressure, for instance, the longitudinal or transverse transports activation and the one-dimensional deconfinement arising under pressure [11], which will be discussed later in Sect. 12.3.3. The phase diagram in Fig. 12.2 where $(TMTTF)_2PF_6$ is the reference compound at ambient pressure has been obtained from the temperature dependences of $\rho_c(T)$ and $\rho_a(T)$ to be discussed below. In the low pressure region (see Fig. 12.5, $P <10$ kbar) $\rho_c(T)$ and $\rho_a(T)$ are both activated, with $\Delta_{\rho,c}$ being about 30% larger than $\Delta_{\rho,a}$. At higher pressures, the activation of ρ_c follows a gentle decrease, while $\Delta_{\rho,a}$ collapses abruptly at a pressure of about 14 kbar, which also marks the onset of a nonmonotonous temperature

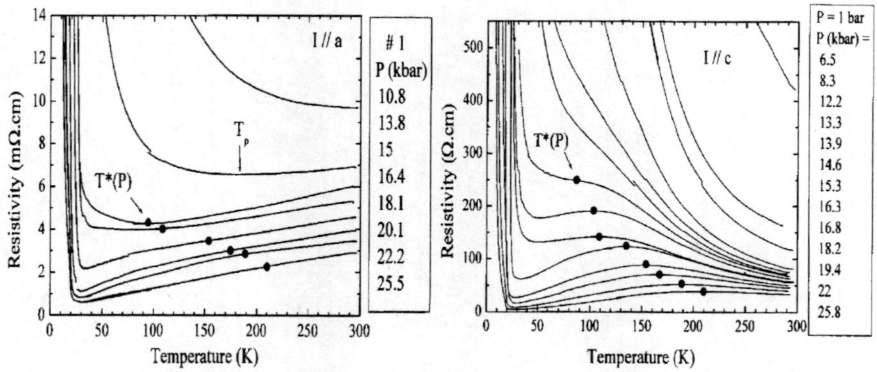

Fig. 12.5. $(TMTTF)_2PF_6$, longitudinal (*left*) and transverse (*right*) resistances vs. temperature at different pressures. After [11]

dependence of ρ_c with a maximum arising at the temperature denoted $T^\star$. Above 14 kbar, $\rho_a(T)$ displays a metallic temperature dependence down to the sharp metal–insulator transition below 20 K, while $\rho_c(T)$ remains indicative of a weakly insulating state above the temperature $T^\star$, which increases under pressure. The different pressure dependences observed for longitudinal and transverse transport will be considered in the context of the pressure-induced deconfinement in Sect. 12.3.3.

12.3.2 Longitudinal Transport

A major problem encountered with the $(TM)_2X$ materials (as well as in most organic conductors) is the very strong pressure (or volume) dependence of their electronic properties, particularly in transport measurements [36–38]. This strong volume dependence is going along a particularly large thermal expansion. Hence, the only temperature dependence that can be compared with the prediction of the theory is the one measured at a constant volume. As all temperature dependences are obtained experimentally under constant pressure, a transformation to the constant volume T-dependence must be performed [37,38], in order to make a significant comparison with the theory. Fortunately, this correction is not that relevant for the case of $(TMTTF)_2PF_6$ under high pressure since the compressibility of these materials is known to decrease under pressure [39]. This is no longer the case when the metallic phase is already stable at ambient pressure as we shall see later for the case of $(TMTSF)_2PF_6$ (Sect. 12.4.1).

The insulating character of these materials with partly filled conduction bands is not expected in the framework of noninteracting electrons. The reason for this must be the existence of strong repulsive interactions between 1D carriers. One explanation for the different values and pressure dependences of the activation energies can be taken as an evidence for in-chain conduction

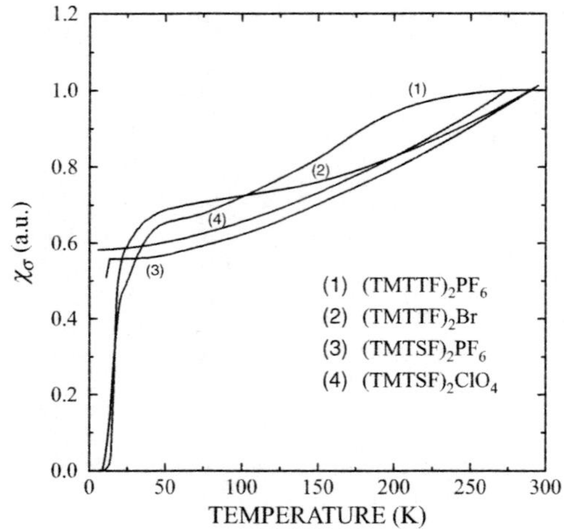

Fig. 12.6. Temperature dependence of the uniform spin susceptibility χ_σ of $(TM)_2X$ compounds at ambient pressure. After [41] ($(TMTTF)_2PF_6$); [42] ($(TMTTF)_2Br$); [43] ($(TMTSF)_2PF_6$); [44] ($(TMTSF)_2ClO_4$)

made possible by thermally excited 1D objects similar to solitons in conducting polymers [40], whereas transverse transport requires the excitation of real quasi-particles (QPs) through a Mott–Hubbard gap larger than the soliton gap as we shall see. Contrasting with the strong pressure dependence of the transport properties, the susceptibility is hardly sensitive to the location of a specific compound in the generic phase diagram. The spin susceptibility is dropping by about 40% between room and low temperature for all materials, although the actual magnitude and the low temperature behavior depend on the compound (see Fig. 12.6). According to the 1D result 12.3 in the high temperature domain, a value of $\tilde{g}_1 \sim 0.5$ can reasonably account for the temperature dependence of the magnetic susceptibility.

Although the insulating behavior of $(TMTTF)_2X$ salts can be clearly ascribed to their commensurate band filling, a closer examination is needed to determine which of the Umklapp scattering channels 1/2- or 1/4-filled is the most active. In the presence of such a gap, the transport is activated at low temperature $\rho_a \propto \mathrm{e}^{\Delta_{\rho,a}/T}$ but is expected to vary according to the power law

$$\rho_a \propto T^{4n^2 K_\rho - 3} \tag{12.26}$$

in the high temperature regime, i.e., $T > \Delta_{\rho,a}$ [23]. The material resembles a metal at high temperature along the longitudinal direction. In the high T 1D regime ($T > t_{\perp b}$), the picture of noncoupled chains is approached. Therefore, the density of quasiparticle states should resemble the situation that prevails in a Luttinger liquid namely, $N(\omega) \sim |\,\omega\,|^\alpha$, where α is related to a bare K_ρ through (12.11), forgetting about the influence of the Mott gap (supposedly smaller than the temperature).

12.3.3 Transverse Transport and Deconfinement

Under a pressure higher than 14 kbar the behavior of the $(TMTTF)_2PF_6$ resistance along the direction of the weakest coupling, i.e., along the c-axis, displays an insulating character with a maximum around 80–100 K and becomes metallic at lower temperatures, although remaining several orders of magnitude above the Mott–Ioffe critical value, which is considered as the limit between metal and insulating-like transport [45]. Figure 12.7 displays the temperature dependence of ρ_c in $(TMTTF)_2PF_6$ and also for other members of the $(TM)_2X$ family with different anions such as $(TMTTF)_2BF_4$ and $(TMTTF)_2Br$.

The insulating character of the transverse transport has been interpreted as the signature of a non-Fermi liquid behavior for carriers within planes (chains) [47]. When transverse transport along the c-direction is incoherent, transverse conductivity probes the physics of the a-b planes, and conductivity in terms of the transverse coupling $t_{\perp c}$ is expressed in the tunneling approximation as

$$\sigma_c(\omega, T) \propto t_{\perp c}^2 \iint \mathrm{d}x \mathrm{d}\omega' A(x, \omega') A(x, \omega + \omega') \frac{f(\omega') - f(\omega' + \omega)}{\omega}, \qquad (12.27)$$

where $A_{1D}(x, \omega)$ is the one-electron spectral function of a single chain and $f(\omega)$ is the Fermi–Dirac function. For a-b planes made of an array of weakly

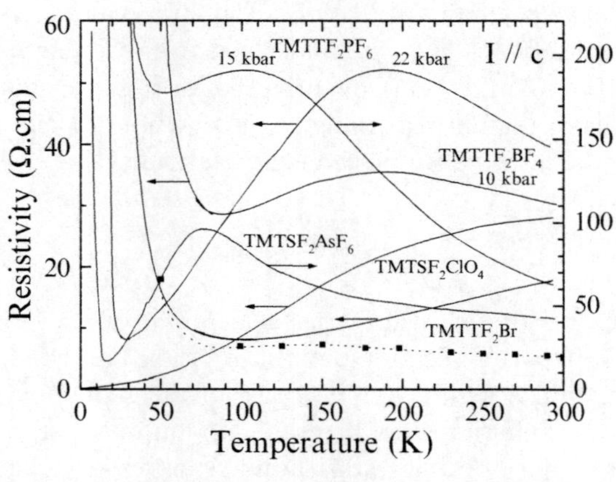

Fig. 12.7. Temperature dependence of the transverse transport along the c-axis in several compounds belonging to the $(TM)_2X$ series studied under pressure. The temperature of the maximum of resistivity shows the location of the dimensional crossover $T^\star$, which is strongly pressure dependent for each compound. The resistivity upturn at low temperatures represents the stabilization of the insulating SDW phase. In case of the selenium compound $(TMTSF)_2ClO_4$, the crossover lies around room temperature, whereas for the sulfur compound $(TMTTF)_2Br$, a resistivity maximum is seen only after the constant volume correction is taken into account. After [46]

interacting Luttinger chains, (12.27) leads to a power law temperature dependence for the c-axis conduction. The temperature at which the c-axis transport switches from an insulating to a metallic temperature dependence corresponds to a crossover between two regimes: a high temperature regime with no QP weight at the Fermi energy (possibly a TL liquid in the 1D case) and another regime in which the QP weight increases with decreasing temperature. This interpretation does not imply that the transport along the c-direction must also become coherent below the cross-over. The c-axis transport may well remain incoherent with a FL being established in the a-b plane at temperatures below $T^\star$. The temperature dependence of the resistivity along the least conducting direction is thus expressed as [48]

$$\rho_c(T) \propto T^{1-2\alpha}. \tag{12.28}$$

Consequently, the temperature dependence of transport properties along the a and c-axes above $T^\star$ should possibly lead to a consistent determination of K_ρ.

Now, regarding $(TMTTF)_2PF_6$, we are facing a very interesting system, since the evolution from a Mott insulator to a metal can be carefully studied under pressure in a single sample and a decrease in compressibility under pressure makes constant volume correction less significant for temperature dependences measured at high pressures. Turning to the evaluation of the correlation coefficient from the temperature dependence of ρ_c, we end up fitting the data for $(TMTTF)_2PF_6$ in the pressure domain around 12 kbar, Fig. 12.5, with a very small value of K_ρ (or large values of α) which is not compatible with the value $K_\rho = 0.23$ derived from the far infrared (FIR, see below) and NMR data [42]. Consequently, the Mott gap seems to be important in this temperature regime governing the excitation for the motion of single particles along c. Tentatively, one can expect a transverse resistivity behaving according to [48],

$$\rho_c(T) \propto T^{1-2\alpha} \mathrm{e}^{\Delta_{\rho,c}/T}. \tag{12.29}$$

Since the Mott–Hubbard gap varies as a power of K_ρ, even a small variation in the ratio between the Coulomb interaction and the bandwidth under pressure can explain a significant decrease of all gaps moving from the left to the right in the generic phase diagram. To summarize, it is interesting to have a look at the data of longitudinal and transverse transports obtained in $(TMTTF)_2PF_6$ under pressure displayed on Fig. 12.8. The transverse transport along c is due to the hopping of quasi-particle and therefore requires an activation through a gap, which is the remanence of the Mott–Hubbard gap. It survives the onset of the dimensional cross-over. Thus we make the important identification

$$T^\star \equiv T_{x^1} \tag{12.30}$$

between the temperature at which the metallic behavior in ρ_c is restored and the single-particle dimensionality crossover (12.17). From Fig. 12.8, we

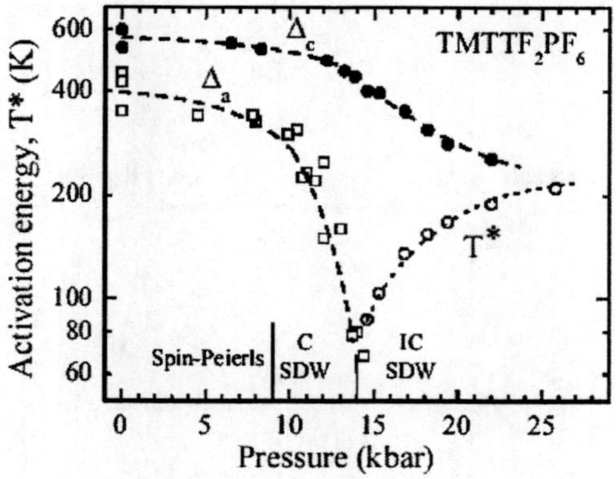

Fig. 12.8. Pressure dependence of the transport activation in $(TMTTF)_2PF_6$. The activation for the c axis transport ($\Delta_c \equiv \Delta_{\rho,c}$, in the text) although decreasing under pressure survives up to high pressures, while the longitudinal transport ($\Delta_a \equiv \Delta_{\rho,a}$, in the text) is no longer activated above 14 kbar when the dimensional cross-over arises at the finite temperature $T^\star$. After [11]

see that $T^\star$ goes to zero when the insulating behavior in the longitudinal transport is restored, which is at 14 kbar for a system like $(TMTTF)_2PF_6$, namely at the onset of electron confinement where the renormalization of $t^*_{\perp b,c}$ is strong (12.17). As for the longitudinal transport, it proceeds via the thermal excitation of 1D objects similar to the solitons in conducting polymers through a gap smaller than the quasi-particle gap [40]. They loose their one-dimensional character and thus acquire a metallic power law temperature dependence when the transverse coupling becomes pertinent under pressure (i.e. above 14 kbar).

12.3.4 Far Infrared Response in the $(TM)_2X$ Series

An other signature of the Mott–Hubbard gap has been given by the frequency-dependent conductivity $\sigma(\omega)$ measured in various salts of the $(TM)_2X$ series exhibiting very different values of the conductivity at room temperature [49] (see Fig. 12.9). The peak of the conductivity at a frequency ω_0 correlates with the magnitude of the room temperature conductivity, namely both sulfur salts $(TMTTF)_2PF_6$ and $(TMTTF)_2Br$, which are insulating, exhibit a conductivity peak around $\omega_0 \approx 1{,}000\,\text{cm}^{-1}$. In $(TMTSF)_2PF_6$ the peak occurs around $200\,\text{cm}^{-1}$, that is, very close to the zero frequency. Both optical and transport data give $2\Delta_{\rho,a} = 800$–$1{,}000$ K in $(TMTTF)_2PF_6$ at ambient pressure [11]. The difference between K_ρ for selenium and sulfur compounds ($K_\rho = 0.18$ for the latter material) is a result of the difference between their bare bandwidths, since on-site repulsion, being a molecular property, is likely to be less sensitive to pressure than the intermolecular overlap along the stacking axis.

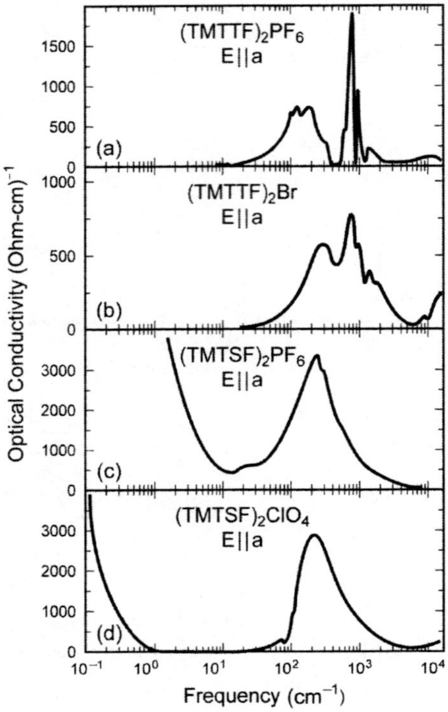

Fig. 12.9. Frequency dependence of the optical conductivity of several $(TM)_2X$ compounds for the light polarized along the a axis at ambient pressure and $T \approx 20\,K$. After [50]

Nuclear Magnetic Resonance

The measurement of the temperature dependent nuclear spin-lattice relaxation rate in NMR denoted by T_1^{-1} is another tool that has played a quite important role in the description of low energy electron spin correlations in $(TM)_2X$ [42]. The connection between nuclear relaxation and the electron spin dynamics is given by the Moriya T_1^{-1} expression

$$T_1^{-1} = |\, A\, |^2\, T \int \frac{\chi''(\boldsymbol{q},\omega)}{\omega}\, \mathrm{d}^D q, \tag{12.31}$$

which is taken in the zero Larmor frequency limit ($\omega \to 0$) and where A is proportional to the hyperfine matrix element. The relaxation of nuclear spins gives relevant information about the static, dynamics, and dimensionality D of electronic spin correlations. In $D = 1$, this expression gives a relatively easy access to the interaction parameter K_ρ that enters in most power laws expressions in one dimension [18, 27, 51]. According to (12.3) and (12.4), the enhancement of the imaginary part of the spin susceptibility χ'' occurs at

$q \sim 0$ and $q \sim 2k_F$, which yields

$$T_1^{-1} \simeq C_0 T \chi_\sigma^2(T) + C_1 T^{K_\rho}. \tag{12.32}$$

Here C_0 and C_1 contains weak logarithmic corrections in temperature. As a function of temperature, two different behaviors can be singled out. At high temperature, where uniform spin correlations dominate and those at $2k_F$ are small, the relaxation rate is governed by the $T\chi_\sigma^2(T)$ term. In the low temperature domain, however, $2k_F$ spin correlations are singularly enhanced, while uniform correlations remain finite so that $T_1^{-1} \sim T^{K_\rho}$.

Consider, for example, the insulating compounds $(TMTTF)_2X$, we have seen in Sect. 12.2.1 that the renormalized charge stiffness $K_\rho^* \to 0$ essentially vanishes below the Mott scale. The resulting behavior for the relaxation rate becomes

$$T_1^{-1} \sim C_1 + C_0 T \chi_\sigma^2. \tag{12.33}$$

As shown in Fig. 12.10, this behavior indeed emerges for $(TMTTF)_2PF_6$ salt when the relaxation rate is combined to the spin susceptibility data ($T\chi_\sigma^2$) in the MI phase above three-dimensional ordering [51, 52]. A similar linear behavior of T_1^{-1} vs. $T\chi_\sigma^2$, with a finite intercept confirming the $K_\rho^* = 0$ value of the charge stiffness, is invariably found in all insulating materials down to the low temperature domain that surrounds the three-dimensional magnetic or lattice distorted long-range order [37, 42, 53].

As one moves to the right-hand-side in the phase diagram of Fig. 12.3, we see that the T_1^{-1} vs. $T\chi_\sigma^2$ law is relatively well obeyed over a large temperature domain in the normal phase. If one takes $(TMTSF)_2PF_6$ for example, deviations are seen only below 150 K or so, indicating that K_ρ would be finite over the whole temperature range [42]. For $(TMTSF)_2ClO_4$, T_1^{-1} enhancement coming from antiferromagnetic correlations emerges at even lower temperature (∼30 K) [27].

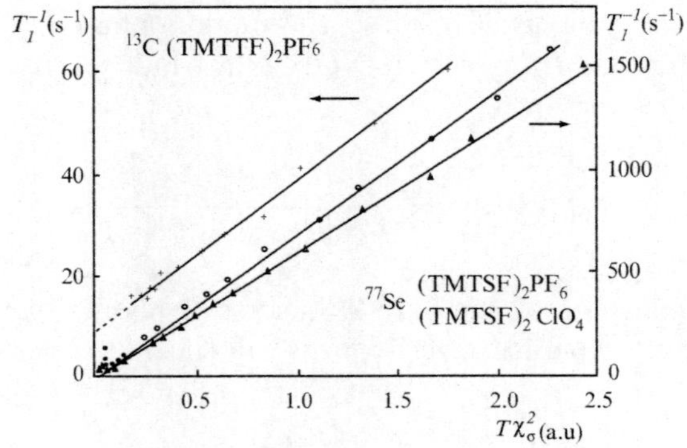

Fig. 12.10. Temperature dependence of NMR nuclear relaxation rate plotted as T_1^{-1} vs. $T\chi_\sigma^2$ for ^{13}C $(TMTTF)_2PF_6$ (*crosses*), ^{77}Se $(TMTSF)_2PF_6$ (*open circles*), and ^{77}Se $(TMTSF)_2ClO_4$ (*open triangles*). After [51]

12.3.5 The Ordered States at Low Temperature

The Spin-Peierls Instability of the Sulfur Compounds

The two compounds $(TMTTF)_2PF_6$ and $(TMTTF)_2AsF_6$ of the Fabre series develop an instability of the MI phase that involves both spins and lattice degrees of freedom. X-ray diffuse scattering measurements of Pouget et al. [54] on both compounds have soon revealed the existence of diffuse scattering lines at the wave vector $2k_F$, showing the onset of 1D lattice softening below the temperature scale $T^0_{SP} \approx 60\,K$ for $(TMTTF)_2PF_6$ and 45 K for $(TMTTF)_2AsF_6$. These lines condense into satellite reflections at the transition temperature $T_{SP} \approx 19\,K$ for $(TMTTF)_2PF_6$ salt and 15 K for $(TMTTF)_2AsF_6$, where a static lattice distortion takes place at the wave vector $\mathbf{q}_0 = (2k_F, \pi, \pi)$. Recent elastic neutron scattering experiments did confirm the existence of such a static distortion at $\mathbf{q}_0$ for $(TMTTF)_2PF_6$ below T_{SP} [55]. Since both the softening and the transition occurs in the Mott insulating state where only spin excitations are gapless, one thus deals with a spin-Peierls (SP) transition with a pronounced quasi-1D character. The expected nonmagnetic nature of both SP fluctuations and long-range order has been confirmed by the temperature dependence of the spin susceptibility (Fig. 12.6) and nuclear spin-lattice relaxation rate [42,56]; these quantities are reduced in the fluctuation regime and show thermal activation below T_{SP}.

Following the example of the Peierls transition, the SP instability proceeds from the coupling of singular 1D bond-order-wave (BOW) electronic correlations to acoustic phonons at $2k_F$ [57–59]. Unlike the Peierls case, however, the coupling becomes singular in the MI state instead of the metallic phase. The microscopic theory predicts a power law singularity in the 1D electronic BOW response below the Mott scale (12.5), where for weakly dimerized chains systems like $(TMTTF)_2X$, the power law exponent $\gamma_{BOW} = 1\,(K^*_\rho = 0)$, as also found for the antiferromagnetic spin response (12.9). The enhancement of the electron–phonon interaction at $2k_F$ by BOW correlations can be worked out by perturbation theory [57,58]. In the random phase approximation, the static temperature electron–electron vertex part induced by the exchange of $2k_F$ phonons takes the form

$$\Gamma_{ph}(2k_F, T) = \frac{g^0_{ph}}{1 - g^0_{ph}\chi_{BOW}(2k_F, T)}, \tag{12.34}$$

where g^0_{ph} is the square of the bare $2k_F$ electron–phonon matrix element. A singularity of the temperature vertex part will then develop at the mean field temperature

$$T^0_{SP} = c\,\tilde{g}^0_{ph} T_\rho, \tag{12.35}$$

where $c \gtrsim 1$. Since there is no phase transition in one dimension, T^0_{SP} is not a true transition temperature but a temperature scale of lattice fluctuations that can be identified with the softening temperature seen in X-ray experiments.

Although higher order fluctuations corrections will bring back the transition at $T = 0$, T^0_{SP} remains the right temperature scale for the onset of short-range fluctuations [57].

As for long-range SP order, it is driven by an interchain interaction, which we denoted $V_\perp$, whose contributions combine Coulomb, interchain hopping and three-dimensional phonons [32, 58]. These coupling favor staggered bond order transversally to the chains at the wave vector $\mathbf{q}_0$. A molecular field treatment of interchain coupling, which takes into account one-dimensional fluctuations rigorously, leads to the mean-field expression for the spin-Peierls susceptibility

$$\chi_{\mathrm{SP}}(\mathbf{q}_0, T) = \frac{\chi_{\mathrm{1D,SP}}(2k_F, T)}{1 - V_\perp \chi_{\mathrm{1D,SP}}(2k_F, T)}, \tag{12.36}$$

where $\chi_{\mathrm{1D,SP}}$ is the 1D spin-Peierls fluctuations susceptibility. The singularity in χ_{SP} occurs at

$$T_{\mathrm{SP}} = T^0_{\mathrm{SP}} f(V_\perp / T^0_{\mathrm{SP}}), \tag{12.37}$$

where from the singular behavior of $\chi_{\mathrm{1D,SP}}$ at low temperature [57], it is found that $f(V_\perp/T^0_{\mathrm{SP}}) \sim 1/3$ for $V_\perp/T^0_{\mathrm{SP}} \ll 1$ [58]. The resulting estimation $T_{\mathrm{SP}} \sim T^0_{\mathrm{SP}}/3$ apparently holds for most electronically driven quasi-1D structural transition in the adiabatic limit [60]. The observed values of the ratio $T_{\mathrm{SP}}/T^0_{\mathrm{SP}}$ in $(\mathrm{TMTTF})_2\mathrm{PF}_6$ and $(\mathrm{TMTTF})_2\mathrm{AsF}_6$ are compatible with this estimation.

The respective T_{SP} of $(\mathrm{TMTTF})_2\mathrm{AsF}_6$ and $(\mathrm{TMTTF})_2\mathrm{PF}_6$ evolve differently under low pressure. For $(\mathrm{TMTTF})_2\mathrm{AsF}_6$, T_{SP} first increases under pressure and reaches a maximum when the charge ordering temperature T_{CO} (see below) merges with T_{SP} at 1.5 kbar, and finally decrease at higher pressure [62]. Charge order breaks the inversion symmetry in the unit cell and acts as a $4k_{\mathrm{F}}$ charge-density-wave potential on molecular sites. The potential reduces the strength of BOW correlations [63], and in turn the amplitude of both T^0_{SP} and T_{SP}. For $(\mathrm{TMTTF})_2\mathrm{PF}_6$, however, T_{CO} occurs at much lower temperature [64], and T_{SP} shows essentially a constant decrease under pressure. It finally goes down rapidly, extrapolating to zero at $P_{\mathrm{QCP}} \approx 9$ kbar, where it competes with a Néel state [52, 61, 65]. This critical pressure corresponding to the zero temperature extrapolation of the critical lines can be put in the category of a quantum critical point (Fig. 12.11) [61].

The decrease of T^0_{SP} and T_{SP} under pressure can be qualitatively understood if one considers that the drop in T_ρ weakens electronic BOW correlations, which according to (12.35), reduces T^0_{SP} (Figs. 12.2 and 12.11). As T^0_{SP} carries on decreasing with pressure, it will reach values that become small compared to the typical energy ω_{D} of $2k_{\mathrm{F}}$ phonons ($\omega_{\mathrm{D}} \sim 100$ K in these materials [66]), where quantum effects enter into play [67–69]. In effect, the above expression for T^0_{SP} has been obtained for "static" phonons, in the so-called adiabatic approximation where the molecules are supposed to have an infinite mass. Adiabaticity is a reasonable assumption provided that

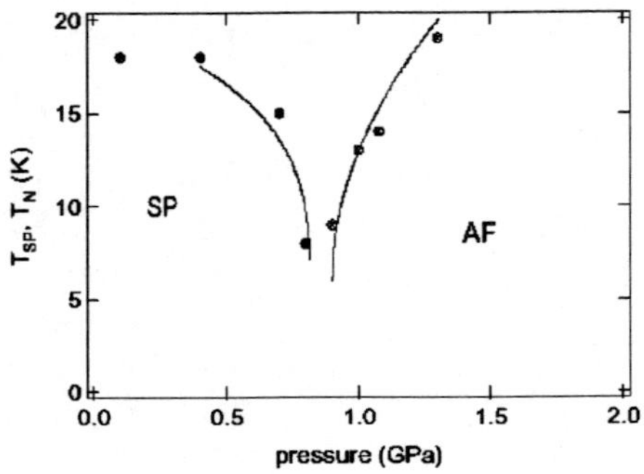

Fig. 12.11. The phase diagram of $(TMTTF)_2PF_6$ near the quantum critical point at P_{QCP}. After [61]

$\omega_D/2\pi T^0_{SP} \ll 1$ [67]. As T^0_{SP} decreases and ω_D increases under pressure, however, the adiabatic condition will not be satisfied any more and the lattice softening will be reduced by the zero point motion of the molecular lattice. It has been shown that a quantum–classical crossover is expected at $\omega_D/\pi T^0_{SP}(P) \approx 1$, where quantum corrections completely suppress T^0_{SP} and in turn T_{SP} [67,68]. For a system like $(TMTTF)_2PF_6$, this would take place after a reduction of T_ρ by a factor of two or so, which according to Fig. 12.3, corresponds to a pressure of the order of P_{QCP}.

Charge Ordering [70]

The MI state of most members of the Fabre series is characterized by another temperature scale connected to a different type of long-range order. The study of the temperature dependence of electrical permittivity for $(TMTTF)_2PF_6$ and $(TMTTF)_2AsF_6$ has indeed revealed the existence of a singularity in the dielectric constant at 70 and 100 K, respectively [71,72]. This singularity, not seen in the magnetic susceptibility (Fig. 12.6), is associated with an instability in the charge sector. The nature of this state was clarified at the same time by Chow et al. [64], who showed from NMR that the instability is actually a continuous phase transition towards a charge disproportionation in the unit cell. NMR does not tell, however, at which wave vector this charge ordered (CO) state takes place. In this regard, Monceau et al. [73] suggested that the anion lattice may undergo a uniform displacement when coupled to the $4k_F$ electron charge instability along the stacks. In this picture, the CO instability in $(TMTTF)_2X$ would be akin in most cases to a ferroelectric phase transition with a divergent dielectric constant.

The experimental identification of a charge-ordered state in $(TMTTF)_2X$ salts lifted a sizable part of the veil surrounding the nature of the so-called "structureless" phase transition that was detected much earlier from transport measurements for several members of the $(TMTTF)_2X$ series [74,75]. It also puts an additional scale in the generic phase diagram of Fig. 12.3. Moreover, for a compound like $(TMTTF)_2SbF_6$ with a larger centrosymmetrical anion, it was shown from NMR under pressure that the existence of both a CO transition at $T_{\mathrm{CO}} \simeq 150\,\mathrm{K}$ and a Néel state at $T_c \simeq 7\,\mathrm{K}$ forces to extend the pressure scale of Fig. 12.3 further on the left [76], where a Néel rather than a spin-Peierls state is stable. This Néel state is suppressed under pressure and replaced by a nonmagnetic phase, presumably of the spin-Peierls type such as those found in $(\mathrm{TMTTF})_2\mathrm{PF}_6$ and $(\mathrm{TMTTF})_2\mathrm{AsF}_6$ [76], and which has been discussed earlier. At ambient pressure, $(\mathrm{TMTTF})_2\mathrm{SbF}_6$ would then be found at the *left* of $(\mathrm{TMTTF})_2\mathrm{PF}_6$, which defines the origin on the pressure scale of Fig. 12.3. Other compounds like the 7 K antiferromagnet $(\mathrm{TMTTF})_2\mathrm{SCN}$ would also be located in the same region of this extended phase diagram.

It is not clear, however, how far on the right-hand-side of Fig. 12.3 CO ordering is found. As mentioned above, it is known to be rapidly suppressed under pressure for compounds like $(\mathrm{TMTTF})_2\mathrm{AsF}_6$ and $(\mathrm{TMTTF})_2\mathrm{PF}_6$. However, it has been claimed to be present in compound like $(\mathrm{TMTTF})_2\mathrm{Br}$ at ambient pressure [71].

The possibility for a CO order state, as a $4k_{\mathrm{F}}$ charge instability, is predicted to take place in purely quarter-filled one-dimensional system for small charge stiffness $K_\rho < 1/4$ (12.15), namely for sizable long-range Coulomb interaction where it coincides with an insulating state [24, 77]. It has been found also for models of interacting electrons in weakly dimerized chains with and without anions displacements [26, 78], in the framework of mean-field theory [79] and numerical calculations [80].

The Néel Order

Sufficiently above P_{QCP} in Fig. 12.3, antiferromagnetic correlations within the MI state are much less affected by lattice SP fluctuations, which are sizably weaker in this pressure range. This is shown in the phase diagram by the absence of a spin pseudo gap from the temperature dependent nuclear relaxation rate in $(\mathrm{TMTTF})_2\mathrm{PF}_6$ at 13 kbar and in $(\mathrm{TMTTF})_2\mathrm{Br}$ at 1 bar [42]. This is also confirmed in the case of $(\mathrm{TMTTF})_2\mathrm{Br}$ by X-ray diffuse scattering experiments at ambient pressure [81]. The observation in these conditions of a temperature independent nuclear relaxation rate for both materials indicates that the power law exponent of the singularity in the AF response in (12.4) below the Mott scale T_ρ is $\gamma = 1$ ($K_\rho^* = 0$) [42]. We have seen in Sect. 12.2.2 that in the presence of a Mott gap, $\Delta_\rho > t_{\perp b}$ (here $\Delta_\rho \equiv \Delta_{\rho,a}$), electron–hole bound pairs are formed and a coherent electron band motion in the transverse directions cannot take place. The propagation of order in the transverse directions leading to a Néel ordered state is provided by the antiferromagnetic interchain exchange $J_{\perp b,c}$ given by (12.20). We have seen that the temperature scale for the Néel ordering is determined by the singularity of the exchange coupling at the one-loop level (12.18), which leads to $T_{\mathrm{c}} \propto 1/\Delta_\rho$ (12.21).

This result indicates that T_{c} – essentially dominated by the exchange in the b direction – increases as the Mott gap Δ_ρ decreases, a feature commonly observed in the Fabre series (Figs. 12.2 and 12.3) [82, 83]. It

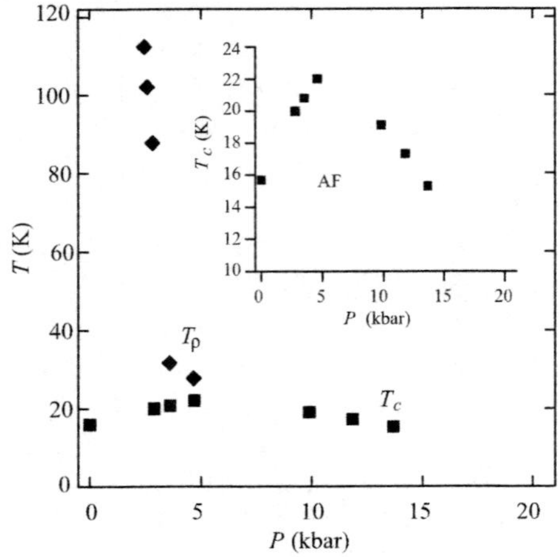

Fig. 12.12. Variation of the critical AF critical temperature as a function of hydrostatic pressure in $(TMTTF)_2Br$. *Inset*: a zoom of the maximum of T_c. After [82]

is worth noting that in the quarter-filled Mott insulator compound $(EDT\text{-}TTF\text{-}CONMe_2)_2AsF_6$, the interchain exchange is also the driving force of antiferromagnetic long-range order. In these systems too, T_c is found to increase as Δ_ρ is decreasing, a behavior that proved to be independent of the commensurability of Umklapp scattering processes behind the insulating gap (Fig. 12.4).

When the pressure is further increased, the Mott insulating and Néel critical scales meet and then the spins order themselves directly from the metallic state. Antiferromagnetism becomes itinerant in character and corresponds to a SDW state. The interchain exchange enters in the weak coupling sector and continues to be active, albeit on a relatively small pressure range with T_c given by (12.22). The calculations show that T_c starts to decrease in this restricted pressure domain giving rise to a maximum in T_c seen in experiments for $(TMTTF)_2Br$ (Fig. 12.12) [82, 83], $(TMTTF)_2PF_6$ [12] (Fig. 12.2), and mixed selenium–sulfur compounds $(TMDMTSF)_2PF_6$ [84, 85]. This weak domain coupling quickly evolves to a regime where $t^*_{\perp b}$ and then the single electron transverse coherence length along the b direction is increasing rapidly under pressure, signaling the beginning of a coherent band motion perpendicular to the chains. This yields the onset of electronic deconfinement and coherent nesting of the Fermi surface at $T^\star$ [86]. The value of T_c in this latter domain also decreases when the couplings decrease under pressure (Fig. 12.4).

12.4 The Bechgaard Salts

12.4.1 The Metallic Phase

The strongly metallic character of the $(TMTSF)_2PF_6$ salt has been one of the highlights in the search for organic conductors [87]. The temperature

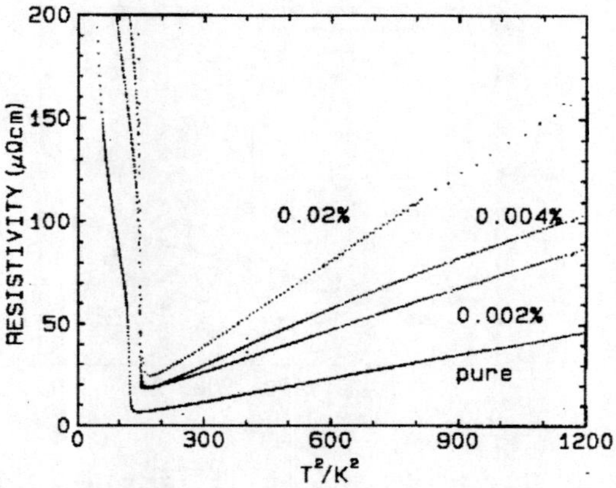

Fig. 12.13. Temperature dependence of the $(TMTSF)_2PF_6$ longitudinal resistivity plotted vs. T^2 for pure and irradiated samples. After [88]

dependence of the longitudinal resistivity follows a power law $T^{1.4}$ from 300 K down to about 100 K. Below 35 K, the resistivity of $(TMTSF)_2PF_6$ or $(TMTSF)_2AsF_6$ is quadratic in temperature $\rho_a(T) = \rho_0 + AT^2$, which is valid down to the metal–insulator transition due to the onset of an itinerant antiferromagnetic state at 12 K (Fig. 12.13).

For high quality samples the resistance ratio $\rho_a(300\,K)/\rho_0$ can reach values as large as 800 [88] (Fig. 12.13). Furthermore, an interesting behavior encountered in $(TMTSF)_2PF_6$ materials (and also in most organic conductors) is the very strong pressure (or volume) dependence of their electronic properties, particularly the transport property [36–38]. In addition, the thermal expansion of these materials is particularly large. Hence, the only temperature dependence that can be compared with the prediction of the theory is the one measured at a constant volume. As all temperature dependences are obtained under constant pressure, a constant volume transformation must be performed. An example is given in Fig. 12.14 by the longitudinal transport of $(TMTSF)_2PF_6$ behaving at high temperature similar to T^2 under ambient pressure but varying sublinearly ($\sim T^{0.93}$) from 300 to 150 K, once the volume correction is taken into account [38]. The experimental power law of longitudinal resistivity leads in turn to $n^2 K_\rho = 0.98$ according to the theory of resistivity [23]. Note that a similar power law for the temperature dependence of longitudinal transport ($\sim T^{0.93}$) can also be observed for two sulfur-compounds, BF_4 and PF_6, under high enough pressure when the correction to constant volume becomes negligible.

In the early days of the research on $(TMTSF)_2X$ compounds, the lattice dimerization was believed to govern entirely the amplitude of the Mott–Hubbard gap [16, 17], when the half-filled scenario is privileged, namely ($n = 1$). Hence, $n^2 K_\rho = 0.98$ leads to a bare value of K_ρ close to unity implying a weakly coupled electron gas. This situation of a very weak coupling is difficult to reconcile with an enhancement of the spin susceptibility

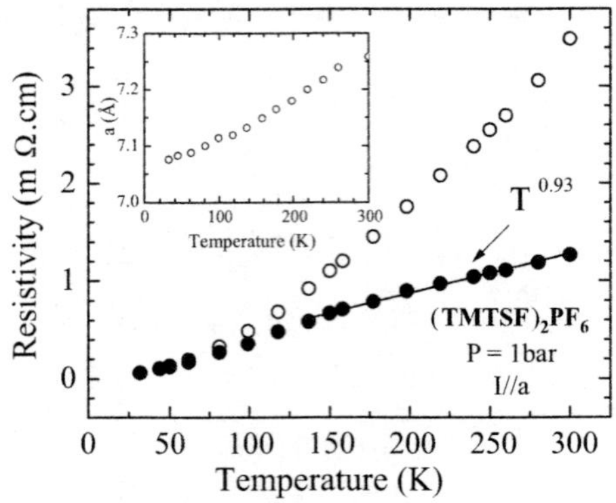

Fig. 12.14. Temperature dependence of the $(TMTSF)_2PF_6$ longitudinal resistivity at constant volume showing quasi linear T-dependence, with the thermal dependence of the lattice parameter a displayed in the inset. After [38]

and the characteristic enhancement of the nuclear spin relaxation rate [42,89], but an additional argument against this weak coupling is made possible by the unusual behavior of transverse transport (20.6) [48]. The weak coupling value for $K_\rho \approx 1$ derived from the temperature dependence of longitudinal transport and the optical data (see below) would imply $\alpha \approx 0$ and consequently a metal-like temperature dependence for $\rho_c(T)$, which is *at variance* with the data.

More recently, an alternative interpretation based on new experimental results has been proposed assuming that the 1/4-filled scattering could justify the existence of the Mott gap in the entire $(TM)_2X$ series [23]. With such a hypothesis ($n = 2$), the fit of the experimental data would thus lead to $K_\rho = 0.23$ and $\alpha = 0.64$ (see below, discussion on optical response) [90]. This bare value for K_ρ agrees fairly well in the 1/4-filled scenario at the very least at high temperature where the influence of half-filling Umklapp should be weak [26, 89]. This value that would imply $U/W = 0.7$ for the Hubbard parameter is compatible with plasma edge measurements and the enhancement in the spin susceptibility [42, 89]. Such a strong coupling value for the bare K_ρ implies that a system such as $(TMTSF)_2PF_6$ lies at the border between a 1D Mott insulator and a Luttinger liquid, though slightly on the insulating side. $(TMTTF)_2Br$ is another particularly interesting system, in which the pressure coefficient of the resistivity is very large. Hence, it is the correction to constant volume, which makes the maximum in ρ_c to appear around 150 K, while this maximum is absent in the constant pressure runs (see Fig. 12.7).

The other approach to the correlation coefficient is given by the far infrared optical studies of $(TMTSF)_2PF_6$, which have been very helpful for the determination of K_ρ since the FIR gap of about $\Delta_{\rho,a} = 200\ \mathrm{cm}^{-1}$ in $(TMTSF)_2PF_6$ has been attributed to the signature of the Mott–Hubbard gap [90]. Consequently, the frequency dependence of the conductivity above the Mott gap is closely linked to the dynamics of the excited carriers in the 1D regime. The theory predicts a power law dependence for the optical conductivity at frequencies larger than the Mott gap [23] namely,

$$\sigma_{1,a}(\omega) \sim \omega^{4n^2K_\rho - 5}, \tag{12.38}$$

at $\omega > 2\Delta_{\rho,a}$, see Fig. (12.15).

According to the optical experimental data (see Fig. 12.16 [90]), $\sigma_{1,a}(\omega) \propto \omega^{-1.3}$ at high frequency leading to $n^2K_\rho = 0.93$. This value for the correlation coefficient is fairly close to the one derived above from parallel transport data but none of these experiments allow by themselves to discriminate between half or quarter-filled Umklapp scattering.

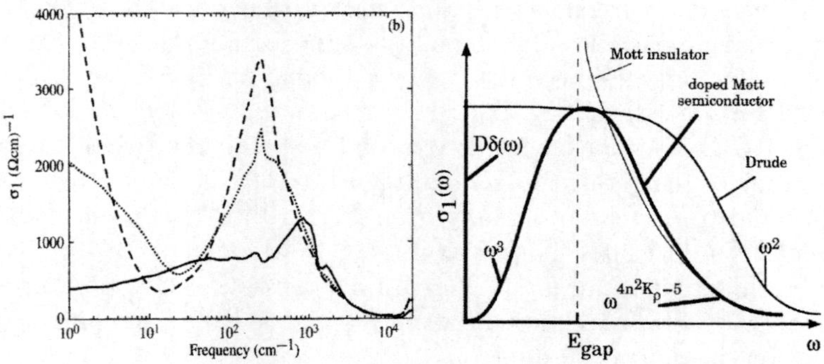

Fig. 12.15. Far infrared optical conductivity (**E**$\|a$) data of $(TMTSF)_2PF_6$, experiment [90] for $T = 300, 100, 20$ K (*left*) and theory for a doped Mott insulator [23] (*right*)

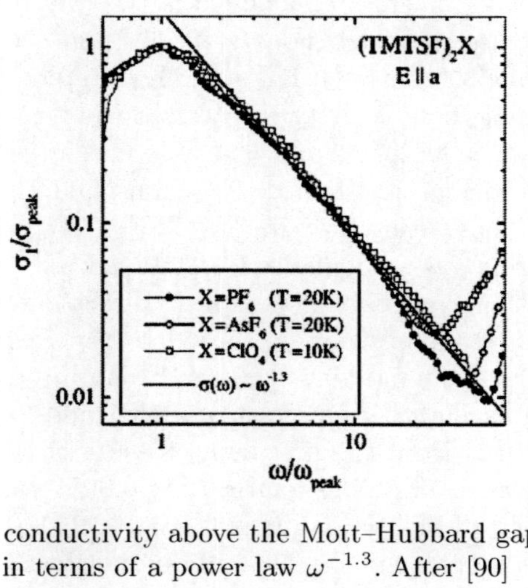

Fig. 12.16. Optical conductivity above the Mott–Hubbard gap in several selenide conductors analyzed in terms of a power law $\omega^{-1.3}$. After [90]

12.4.2 Pseudogap and Zero Frequency Mode in the Metallic Phase of $(TMTSF)_2X$

It is also most illuminating to have a look at the conductivity in the far infrared regime. A large gap of order 1,000 K is observed in the frequency dependence of the FIR conductivity of sulfur compounds [91]. This is in line with the activation energy of the DC conductivity in those compounds. However, the surprise arose for selenium compounds, which behave apparently like normal metals as far as DC transport is concerned, in spite of the marked gap observed in the FIR regime at low temperature. The apparent normal behavior of the resistivity varying quadratically in temperature for $(TMTSF)_2ClO_4$ or $(TMTSF)_2PF_6$ above the SDW transition could lead to the misleading conclusion of a 2 or 3D Fermi gas in which the temperature dependence of the transport is governed by e–e scattering. However, the analysis of the conductivity in terms of the frequency reveals quite a striking breakdown of the Drude theory for single-particles. The inability of the Drude theory to describe the optical conductivity has been noticed by a number of experimentalists working on $(TMTTF)_2X$ with X = ClO_4, PF_6, or SbF_6 [92]. When the reflectance of $(TMTSF)_2ClO_4$ in the near-infrared is analyzed with the Drude model in the whole range of temperatures from 300 down to 30 K, the electron scattering rate is found to decrease gradually from $2.5 \times 10^{14}\,\mathrm{s}^{-1}$ at room temperature to $1.3 \times 10^{14}\,\mathrm{s}^{-1}$ at 30 K [93]. Even if the RT value is not far from the value for DC conductivity, a drastic difference emerges at low temperature as σ_{DC} increases by a factor about 100 between RT and 30 K [94], as compared to the factor 2 for the optical lifetime.

Another striking feature of the optical conductivity has been noticed when the Kramers–Krönig transformation of the reflectance is performed in a broad frequency domain for $(TMTSF)_2ClO_4$ as well for all conducting materials at low temperature. Given the usual Drude relation $\sigma_{\mathrm{DC}} = \omega_{\mathrm{p}}^2\tau/4\pi$ between transport lifetime and plasma frequency data (the plasma frequency has been found nearly temperature independent [49, 93]) and the measured resistance ratio for ρ_a of about 800 between RT and 2 K obtained in good quality measurements, the Drude conductivity in the frequency range $\approx 40\,\mathrm{cm}^{-1}$ should amount to at least $4{,}000\,\Omega\mathrm{cm}^{-1}$ [95, 96]. The measured optical conductivity is at most of the order of $500\,\Omega\mathrm{cm}^{-1}$ [95]. Consequently, the rise in the conductivity as $\omega \to 0$ has been taken in $(TMTSF)_2ClO_4$ as well as in the other salts with PF_6 or SbF_6 as an evidence for a hidden zero frequency mode. This mode is actually so narrow that it escapes a direct determination from K–K analysis of the reflectance, which is limited to the frequency domain above $10\,\mathrm{cm}^{-1}$. Estimates of the mode width have been obtained using the DC conductivity and the oscillator strength Ω_{p}^2 of the mode with $\sigma_{\mathrm{DC}} = \Omega_{\mathrm{p}}^2\tau/4\pi$, where Ω_{p} is measured from the first zero crossing of the dielectric constant. This procedure gives a damping factor $\Gamma = 0.005$ and $0.09\,\mathrm{cm}^{-1}$ at 2 and 25 K respectively, in $(TMTSF)_2ClO_4$ [96] (Fig. 12.17). The confirmation of a very long scattering time for the DC conduction has also been brought by

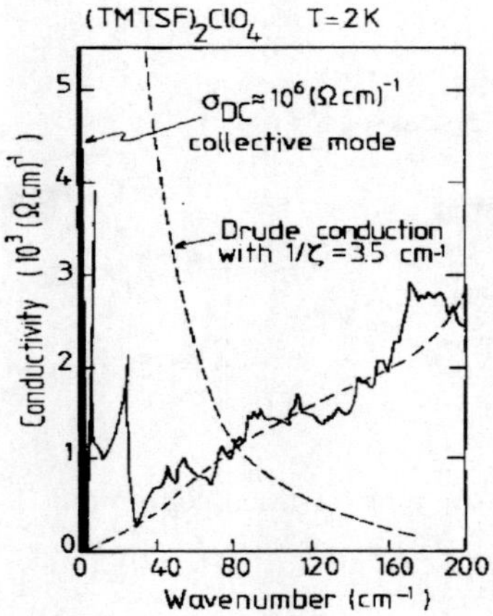

Fig. 12.17. Far infrared data of $(TMTSF)_2ClO_4$. The dashed line is the Drude behavior with $1/\tau = 3.5\,\mathrm{cm}^{-1}$ and $\omega_p = 10^4\,\mathrm{cm}^{-1}$. After [96]

the rapid suppression of T_c by nonmagnetic defects in the nonconventional superconductor $(TMTSF)_2ClO_4$ leading to $\Gamma_c = 0.56\ \mathrm{cm}^{-1}$ at low temperature [97], meaning that the electron lifetime at low temperature is actually much longer than the value inferred from a Drude description.

There is now a wealth of experimental evidences showing the development of a narrow frequency mode in the Mott gap of $(TMTSF)_2ClO_4$ and related conducting compounds. From FIR data in $(TMTSF)_2PF_6$, it has also been shown that the narrow mode carries only a small fraction (a few percent) of the total spectral weight [49, 90], but it is this mode that explains the very large value of the DC conduction observed at low temperature.

12.4.3 Quarter-Filled Compounds

Even if the debate between 1/2- and 1/4-fillings may be relevant for $(TM)_2X$, this is no longer the case for new synthesized compounds in a family whose general structure precludes any dimerization. The structural peculiarity of the salt $(EDT-TTF-CONMe_2)_2AsF_6$ is the absence of inversion center between adjacent molecules in stacks and instead the presence of a glide symmetry plane [98] (see Fig. 12.18). The analysis of the transport data of $(EDT-TTF-CONMe_2)_2AsF_6$ has shown that in spite of the existence of a glide symmetry plane, the carriers are localized, and even more localized than in the most insulating salts of the Fabre series known at present. Since the localization in this compound cannot be ascribed to a 1/2-Umklapp scattering or to the Anderson localization, 1/4-Umklapp scattering seems to be the only channel left to explain carrier localization in this commensurate 1D conductor. Under ambient pressure, given the total bandwidth deduced from quantum chemistry [$W = W_{\mathrm{tot}}(P = 1\,\mathrm{bar}) = 0.350\,\mathrm{eV}$ (3,850 K)] and the experimental Mott gap ($2\Delta_{\rho,a} = 2{,}700\,\mathrm{K}$), the theory [23] gives, in the case of quarter

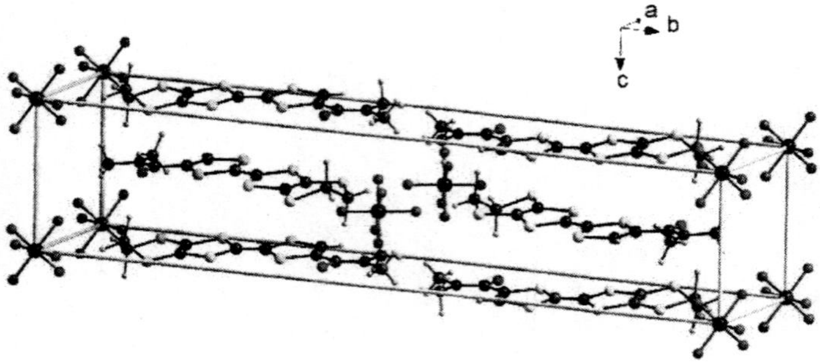

Fig. 12.18. Structure of the quarter-filled compound (EDT-TTF-CONMe$_2$)$_2$AsF$_6$. After [98]

filling *stricto sensu*, $2\Delta_{\rho,a} = 2W(U/W)^{3/2(1-4K_\rho)}$. This leads to the bare value $K_\rho = 0.1$, with a reasonable $U/W = 0.7$.

The Mott gap of (EDT – TTF – CONMe$_2$)$_2$AsF$_6$ is much larger than the expected value of the bare interstack overlap $t_\perp$, which makes according to (12.17) the single particle hopping between neighboring stacks nonpertinent in the pressure regime less than 20 kbar, since the transverse hopping is renormalized to zero on account of a strong intrachain electron–hole interaction.

The interaction over bandwidth ratio $U/W \approx 0.7$ is also in fair agreement with the result of a crude analysis of the spin susceptibility of S-salts [42, 89] and indicates that these compounds lie in the strong coupling sector. With increasing pressure, the gap decreases steadily up to the pressure of 20 kbar above which it disappears sharply due to the competing transverse coupling. Below 15 kbar, the gap of (EDT – TTF – CONMe$_2$)$_2$AsF$_6$ (1,350 K) is about equal to the gap of (TMTTF)$_2$X measured under ambient pressure [47]. Since the logarithmic pressure dependences of the gap seem to be similar for both compounds, we may say that (EDT – TTF – CONMe$_2$)$_2$AsF$_6$ can also be considered as part of the generic Fabre–Bechgaard salt diagram, provided the origin of the pressure axis is shifted to the left by 15 kbar. Hence, it is reasonable to expect that the TL parameter K_ρ increases from left to right in the (TM)$_2$X diagram, since both optical and transport data suggest $K_\rho = 0.23$ in the Se compounds, whilst it is only of the order of 0.1 in sulfur compounds. Let us add that the uniformly stacked 1D conductor (DI – DCNQI)$_2$Ag, another quarter-filled compound, reveals localization properties quite similar to those observed in the (TM)$_2$X series [99]. The normal state of this system is insulating at low pressure probably due to strong electron correlations, but at pressures exceeding 15 kbar, the longitudinal resistivity is metallic above 100 K with a quasi linear temperature dependence leading to $K_\rho = 0.25$ in the quarter-filled band hypothesis. This result shows once more that the conductor lies at the border with the quarter-filled Mott localized insulator.

12.4.4 A Robust 1D Compound: $(TTDM\text{-}TTF)_2Au(mnt)_2$

For the sake of completeness we shall mention the behavior of an interesting organic salt, which unlike all $(TM)_2X$ compounds has failed to reveal the usual suppression of the insulating phase under pressure [100]. The $(TTDM\text{-}TTF)_2Au(mnt)_2$ system, in spite of a strong structural analogy with the $(TM)_2X$ materials exhibiting stacks of donors arranged in layers with short interchain contacts, shows a unique extreme 1D character together with a strong bond dimerization. According to extended Huckel calculations, the dimerization gap at the middle of the HOMO band amounts to 0.027 eV, i.e 13% of the upper band dispersion and the interchain coupling within the layers is practically zero. As the smallness of this contact is due to terminal sulfur atoms in intermolecular contacts not participating to the HOMO of the molecule, we can expect a survival of the 1D character under pressure and no pressure-induced dimensionality crossover. Consequently, contrasting with the members of the Fabre–Bechgaard series, the Mott–Hubbard insulating nature persists up to a pressure of 25 kbar [100], which is usually large enough to severely decrease (if not suppress) the localization in the latter family as shown in the present article.

12.4.5 The Spin-Density-Wave Phase

At the core of the unity shown by the phase diagram of Fig. 12.3 is the shift on the pressure scale when selenium is substituted for the sulfur atom and yields the Bechgaard salts series $(TMTSF)_2X$ [87, 101]. The selenium series was at the start considered more promising compared to previous organic compounds, mainly because the metal–insulator transition at ambient pressure occurs only below 20 K after high metallic conductivity have been attained [87].

In the two compounds of the series $(TMTSF)_2PF_6$ and $(TMTSF)_2AsF_6$, the transition occurs at $T_c \approx 12\,K$ at 1 bar [87, 101, 102]. The transition early showed all the characteristics of SDW long-range ordering [103, 104]. Similar SDW is also found in $(TMTTF)_2X$ but at much higher values on the pressure scale. In $(TMTTF)_2PF_6$, for example, about 40 kbar of pressure is needed to reach a $T_c \sim 10\,K$ (Fig. 12.2) [12, 105], whereas for $(TMTTF)_2Br$ [82, 106] and $(TMTTF)_2BF_4$ [107] about 10 and 27 kbar must be applied, respectively.

The gradual emergence of a plasma edge in the b direction below 100 K [108, 109] and the recovery of transverse metallic and longitudinal Fermi liquid ($\sim T^2$) resistivity below $T^\star$ for $(TMTSF)_2PF_6$ indicate that the transverse electron band motion has developed some coherence at the onset of the SDW transition. The mechanism of the instability will then naturally follow from the property of nesting of the Fermi surface based on the property $E^*_{-p}(\mathbf{k}) = -E^*_p(\mathbf{k}+\mathbf{q}_0)$ of the spectrum (12.16) for a special wave vector $\mathbf{q}_0$, called the nesting vector. The relevance of the Fermi surface for the transition has been confirmed by the determination of the modulation wave vector of SDW by

NMR [110], which coincides with the best nesting vector obtained by band calculations [8]. In the simplified model spectrum (12.16) for an approximate orthorhombic lattice, perfect nesting is found at $\mathbf{q}_0 = (2k_F, \pi, \pi)$. Deviations with respect to this ideal situation, however, are likely to be found in practice. This amounts to use the spectra

$$E^*_p(\mathbf{k}) = v(pk - k_F) - 2t^*_{\perp b} \cos k_{\perp b} - 2t^*_{\perp c} \cos k_{\perp c} - 2t'_{\perp b} \cos 2k_{\perp b}, \quad (12.39)$$

which contains small next-to-nearest-neighbor hopping $t'_{\perp b}$ along the b direction ($t'_{\perp b} \ll t_{\perp b}$). This leads to the modified nesting condition

$$E^*_p(\mathbf{k}) = -E^*_{-p}(\mathbf{k} + \mathbf{q}_0) + 4t'_{\perp b} \cos 2k_{\perp b}, \quad (12.40)$$

with $k_{\perp b}$ dependent nesting frustration.

The determination of the temperature scale for the SDW instability follows the analysis given in Sect. 12.2.2, where the ladder result (12.24) becomes

$$\tilde{J}(T) = \frac{\tilde{J}^*}{1 - \frac{1}{2}\tilde{J}^* \chi^0(\mathbf{q}_0, T)}, \quad (12.41)$$

where

$$\pi v_F \chi^0(\mathbf{q}_0, T) = \ln \frac{T_{x^1}}{T} + \psi\Big(\frac{1}{2}\Big) - \Re e \Big\langle \psi\Big(\frac{1}{2} + \mathrm{i}\frac{t'_{\perp b} \cos 2k_{\perp b}}{\pi T}\Big)\Big\rangle_{k_{\perp b}} \quad (12.42)$$

is the bare susceptibility in the presence of nesting frustration $t'_{\perp b}$ ($\psi(x)$ is the Digamma function and $\langle \ldots \rangle_{k_{\perp b}}$ is an average over $k_{\perp b}$) [111]. As mentioned previously, the above expressions differ slightly from previous mean-field approaches [34,35,111,112], in that the contribution of intermediate electron–hole excitations to χ^0 has been here restricted to an energy shell $\pm T_{x^1}$ (instead of $\pm E_F$) around the Fermi level. More energetic excitations extending up to the Fermi energy are one-dimensional in character and are governed by (12.2) [17, 28, 30, 113]. The above Stoner form also neglects the finite coupling between electron–hole and electron–electron pairings, an interference that persists even below T_{x^1}. This approximate weak coupling description of SDW remains qualitatively correct, however, as long as $t'_{\perp b}$ does not reach too large values, that is where the interference between density-wave and superconductivity can change the nature of the ground state (see Sec. 12.4.6).

The Stoner form (12.41) develops a singularity at the critical temperature T_c that depends on $t'_{\perp b}$, the main parameter that is standardly used to mimic the actual effect of pressure on the SDW state [34, 35, 111, 114]. A finite $t'_{\perp b}$ will then reduce T_c with respect to the BCS limiting value $T^0_c = T_{x^1}\, \mathrm{e}^{-2/\tilde{J}^*}$ at perfect nesting (12.25), a feature of the model that was soon linked with experiments done under pressure [101, 102, 115, 116] (inset of Fig. 12.24). Besides the monotonic increase of T_{x^1} and the decrease of $\tilde{J}^*$ under pressure, the detrimental influence of $t'_{\perp b}$ on T_c remains the most dominant effect.

The expression (12.41) predicts that at the approach of the critical value $t'^{\mathrm{cr}}_{\perp b} \sim 0.7T^0_{\mathrm{c}}$, T_{c} rapidly goes down to zero – though the possibility of SDW at incommensurate $\mathbf{q}_0$ and very low temperature may introduce a change in the slope of T_{c} near P_{c} [117]. To $t'^{\mathrm{cr}}_{\perp b}$ will then correspond a critical pressure P_{c} for the suppression of SDW; this variation of T_{c} agrees with the characteristic pressure profile generally observed for $(\mathrm{TMTSF})_2\mathrm{X}$ at moderate pressure (Fig. 12.24) and for $(\mathrm{TMTTF})_2\mathrm{X}$ at higher pressure (Figs. 12.2 and 12.3). The fitness of the model to describe the T_{c} of the SDW state in $(\mathrm{TMTSF})_2\mathrm{X}$ can be further assessed if one considers the influence of a transverse magnetic field on T_{c}. A perpendicular magnetic field $H \| \mathrm{c}^*$ tends to confine the motion of electrons along the chain direction and gradually restores better nesting conditions. This in turn increases T_{c} with H [111], consistently with early field dependent measurements of T_{c} in $(\mathrm{TMTSF})_2\mathrm{PF}_6$ [118].

Intrusion of Charge-Density-Wave Order

The reexamination of diffuse scattering X-ray patterns of $(\mathrm{TMTSF})_2\mathrm{PF}_6$ by Pouget and Ravy [81,119] revealed the emergence, besides SDW, of a charge-density-wave superstructure (CDW) at T_{c}; both having the same modulation vector $\mathbf{q}_0$. These results were subsequently confirmed by Kagoshima et al. [120], who also found a similar superstructure in $(\mathrm{TMTSF})_2\mathrm{AsF}_6$, but with a weaker amplitude. These results came as a surprise since at variance with ordinary Peierls phenomena, the CDW order is not preceded by any lattice softening in the normal state (the $2k_{\mathrm{F}}$ diffuse scattering lines do exist at high temperature but their amplitudes become vanishingly small in the vicinity of T_{c} in the normal state [54,119]). The CDW superstructure would then be entirely electronic in character with no lattice displacement involved.

In connection with these X-ray results, it is worth mentioning the earlier optical conductivity measurements of Ng et al. [121] on the isostructural member of the series $(\mathrm{TMTSF})_2\mathrm{SbF}_6$. The results show the growth in the infrared of new phonon lines at T_{c}, precisely those usually expected for the excitations of a CDW superstructure; their temperature dependent intensity follows roughly the one of the SDW order parameter below T_{c}. CDW phonon lines in the far infrared conductivity have also been found in the metallic phase of $(\mathrm{TMTSF})_2\mathrm{ClO}_4$ at low temperature [122], indicating that $2k_{\mathrm{F}}$-CDW and SDW correlations apparently coexist in the normal phase [27,42].

On theoretical grounds the possibility for SDW and CDW to coexist has been analyzed recently by considering the redistribution of charge and spin in the unit cell, a possibility that emerges when its internal – two-molecules – structure and the long-range Coulomb interaction are taken into account. In the framework of extended Hubbard model, numerical and mean-field approaches show that charge and spin can be so rearranged in the unit cell that SDW, BOW, and CDW can coexist [59,79,123,124]. Moreover, it was shown recently that when interchain Coulomb interaction is included, this can favor – even for small amplitude – BOW and CDW correlations while not affecting SDW [125].

12.4.6 Some Features of the Superconducting State

The Superconducting Transition

For all cases of superconductivity in the $(TM)_2X$ series, the first evidence has been provided by a drop of the resistivity below the critical temperature and the suppression of this drop under magnetic field. We shall focus the presentation on the two members of the $(TMTSF)_2X$ series, which have attracted most attention:

- $(TMTSF)_2PF_6$, because this has been the first organic superconductor to be found by transport measurements [101] and subsequently confirmed by magnetic shielding [126, 127], and also because the electronic properties of the 1D electron gas on the organic stacks are only weakly (if at all) affected by the centrosymmetrical anions PF_6. The finding of a very small and still nonsaturating resistivity under ambient pressure reaching the value of $10^{-5}\,\Omega^{-1}\,\mathrm{cm}^{-1}$ at 12 K triggered further pressure studies at a pressure of 9 kbar in a dilution refrigerator, which led to the discovery of a zero resistance state below 1 K (see Fig. 12.19). The nonsaturation of the resistivity had been taken as a signature of superconducting precursor effects. We shall come again to this important question later.
- $(TMTSF)_2ClO_4$, because it is the only member of the $(TM)_2X$ series displaying superconductivity at ambient pressure (Fig. 12.20). However, the study of the superconducting state in $(TMTSF)_2ClO_4$ is meeting the problem of the ClO_4 anions ordering at 24K, doubling the periodicity along the *b*-axis [60]. Consequently, great care must be taken to cool the sample slowly enough in order to reach a well anion-ordered state (R-state) at low

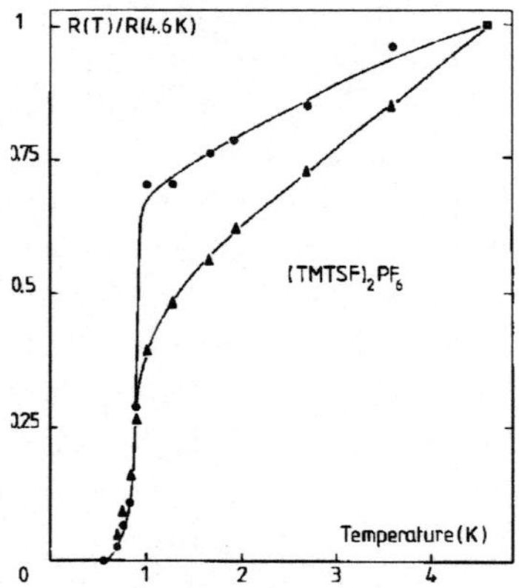

Fig. 12.19. $(TMTSF)_2PF_6$, first observation of organic superconductivity under pressure. After [101]

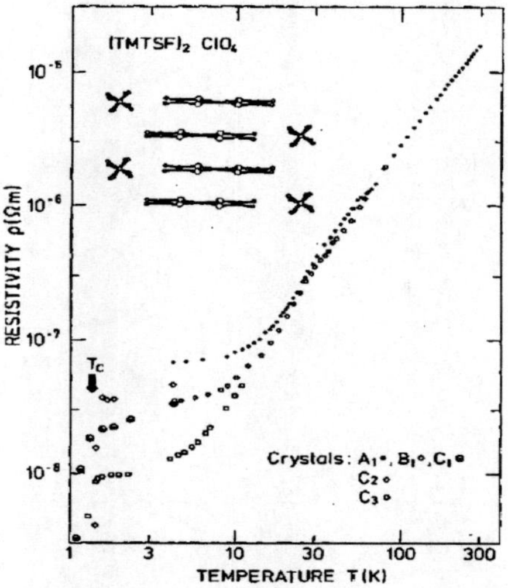

Fig. 12.20. $(TMTSF)_2ClO_4$, first observation of organic superconductivity at ambient pressure. After [94]

temperature; otherwise superconductivity is faced to its great sensitivity to disorder, a very important feature for organic superconductors that will be discussed more extensively later.

Since $(TMTSF)_2ClO_4$ is an ambient pressure superconductor, the thermodynamic evidence of the phase transition has been obtained from specific heat on single crystals [128]. The electronic contribution to the specific heat of $(TMTSF)_2ClO_4$ in a C_e/T vs. T plot (Fig. 12.21) displays a very large anomaly around 1.2 K [128]. Above 1.22 K, the total specific heat obeys the classical relation in metals $C/T = \gamma + \beta T^2$, where the Sommerfeld constant for electrons $\gamma = 10.5\,\mathrm{mJ\,mol^{-1}\,K^{-2}}$, corresponding to a density of states at the Fermi level $N(E_F) = 2.1\,\mathrm{states\,eV^{-1}\,mol^{-1}}$ for the two spin directions [128]. The specific heat jump at the transition amounts then to $\Delta C_e/\gamma T_c = 1.67$, i.e., only slightly larger than the BCS ratio for an s-wave superconductor. The behavior of $C_e(T)$ in the superconducting state leads to the determination of the thermodynamical critical field $H_c = 44 \pm 2$ Oe and the quasi-particle gap $2\Delta = 4\,\mathrm{K}$. T_c is depressed at a rate of $1.1\,\mathrm{mK\,Oe^{-1}}$, when a magnetic field is applied along the c^* axis [129]. Comparing the value of the density of states derived from the specific heat and the value of the Pauli susceptibility [43] lends support to a weak coupling Fermi liquid picture (at least in the low temperature range) [89].

Another confirmation of organic superconductivity has been provided by the measurement of thermal conduction in $(TMTSF)_2ClO_4$ [130]. From the difference between thermal conductivity in magnetic field (larger than the critical field) and in zero applied field, the authors of [130] have been able to extract the electronic contribution of the thermal conductivity below T_c down

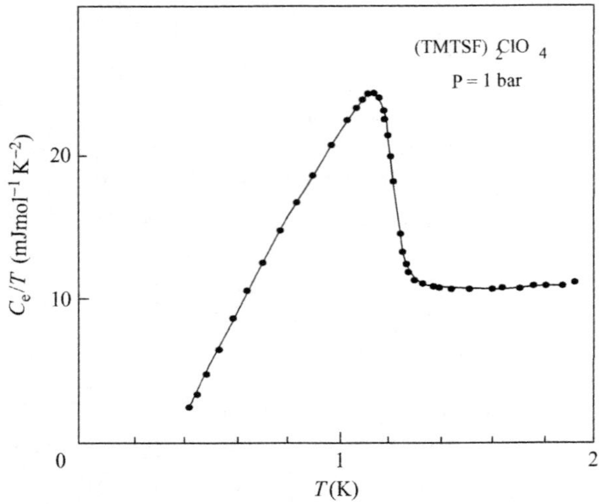

Fig. 12.21. Electronic contribution to the specific heat of $(TMTSF)_2ClO_4$, plotted as C_e/T vs. T. After [128]

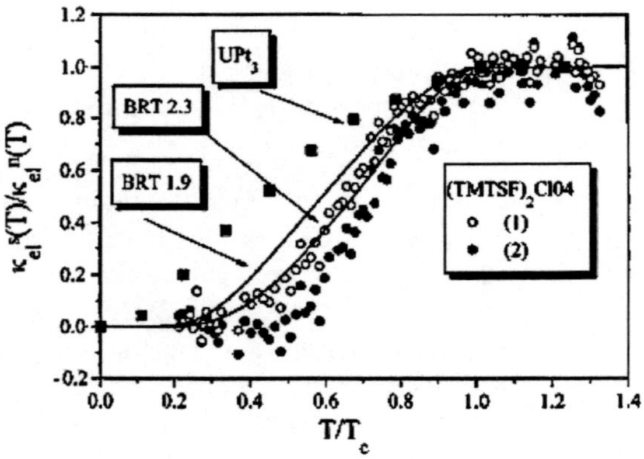

Fig. 12.22. $(TMTSF)_2ClO_4$, normalized electronic thermal conductivity compared to the BRT theory and the data in the unconventional heavy fermion superconductor. After [130]

to about $T_c/5$ (Fig. 12.22). These data lead to a ratio $\Delta(0)/k_B T_c = 2$ within the Bardeen Rickaysen and Teword (BRT) theory of the thermal conduction in the superconducting state [131]. Such a ratio is in fair agreement with the specific heat jump data mentioned above. The saturation of the electronic thermal conduction observed at low temperature in Fig. 12.22 is in favor of a well defined gap in the quasi-particle spectrum (possibly due to the interplay with the anion gap of this particular compound), but does not necessarily imply s-pairing for the orbital symmetry of the superconducting wave function (see the discussion in Sect. 12.4.6). The results of thermal conductivity contrast with those NMR spin-lattice relaxation rate measurements obtained earlier by Takigawa et al. on $(TMTSF)_2ClO_4$ [132], which show the absence of an Hebel Slichter peak at T_c and a power law dependence $T_1^{-1} \propto T^3$ for protons NMR (^{1}H) – these two features being compatible with the existence of nodes for the superconducting gap [133]. A similar algebraic dependence on temperature for T_1^{-1} has been found for ^{77}Se nuclei in $(TMTSF)_2PF_6$ above the critical pressure P_c [134].

The onset of superconducting order has also been detected by a shift of the muon precession frequency entering the superconducting state below T_c under a field of 180 G [135]. However, the superconducting state is not accompanied by any enhancement of the muon relaxation rate according to the data of [135] and the recent zero field data of [136]. Such a behavior is *at variance* with the behavior of Sr_2Ru0_4 [137], in which the increase of the relaxation rate in a zero applied field suggests the development of spontaneous magnetic fields and is taken as a possible (but not unique) evidence for time-reversal symmetry breaking and triplet superconductivity in this oxide material.

Critical Fields

The anisotropic character of the electronic structure already known from the anisotropy of the optical data in the normal phase is reflected in a severe anisotropy of the critical fields H_{c2} measured along the three principal directions in $(TMTSF)_2ClO_4$ [138–141]. Early data in $(TMTSF)_2ClO_4$ [141] are not in contradiction with the picture of singlet pairing but no data were given below 0.5 K, the temperature domain where it would be most rewarding to see how H_{c2} compares with the Pauli limit when H is perfectly aligned along the a and b' axes. This study has been revisited quite recently in perfectly aligned magnetic fields down to 0.2 K [142].

The linearity of the critical fields with temperature in the vicinity of the T_c suggests an orbital limitation in the Ginzburg–Landau formalism for the critical field and rules out a Pauli limitation, which would favor a $(1-T/T_c)^{1/2}$ dependence [139, 143].

Furthermore, in support to an early suggestion [143], the band structure parameters of $(TMTSF)_2ClO_4$ can explain the values and the anisotropies of the critical fields assuming the existence of nodes of the superconducting order parameter [142]. These results imply that critical fields values calculated without any contribution from the spin–orbit coupling can overcome the Pauli limit at low temperature by factors of two or more [144, 145]. The situation for $(TMTSF)_2PF_6$ may, however, be quite different as the possibility of a nonhomogenous superconducting phase in the vicinity of the critical pressure opens another possibility for an enhancement of the superconducting critical fields, vide infra (Fig. 12.23).

Superconductivity and Pressure

The pressure dependence of T_c is admittedly a remarkable feature for the $(TMTSF)_2X$ compounds, since it is the pressure parameter that enabled

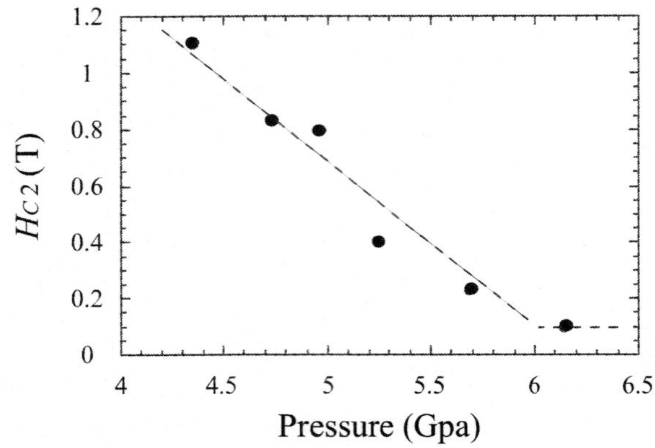

Fig. 12.23. $(TMTTF)_2PF_6$, critical fields along $c^\star$ under pressure in the SDW/SC coexistence regime. The lines are guides to the eyes. After [146]

organic superconductivity to be discovered. However, pressure also suppresses the superconducting phase very quickly. As far as $(TMTSF)_2PF_6$ is concerned, in the vicinity of the critical pressure the pressure coefficient amounts to $\delta \ln T_c/\delta P = 11\%\,\mathrm{kbar}^{-1}$ leading in turn to a Gruneisen constant for superconductivity $\delta \ln T_c/\delta \ln V = 18$ at 9 kbar [147], with the compressibility data measured under pressure (at 16 kbar $\delta \ln V/\delta P = 0.7\%\,\mathrm{kbar}^{-1}$) [39] in the same compound. This value is indeed sizably larger than 7, the value which is obtained in tin, the elemental superconductor exhibiting the strongest sensitivity to pressure [148]. A look at the $(TMTSF)_2PF_6$ phase diagram shows that the strong pressure dependence of T_c is, however, restricted to the close vicinity of the border with the SDW phase.

The pressure coefficient of superconductivity in $(TMTSF)_2ClO_4$ is even more dramatic since then $\delta \ln T_c/\delta \ln V = 36$ [149] using the compressibility of $1\%\,\mathrm{kbar}^{-1}$ (this is the value measured for $(TMTSF)_2PF_6$ at ambient pressure [39], since to the best of our knowledge compressibility data for $(TMTSF)_2ClO_4$ are still missing). However, this remarkable sensitivity of T_c in $(TMTSF)_2ClO_4$ might actually be related to the very specific problem of anion ordering in this compound as it has been suggested from the recent study on the sensitivity of T_c against the presence of nonmagnetic disorder [97]. Anion ordering reveals an uprise of the ordering temperature under pressure [149–151], which can be derived from the pressure dependence of a small kink in the temperature dependence of the resistivity, the signature of the ordering, moving from 24 up to 26.5 K under 1.5 kbar and corroborated by studies at even higher pressures [152]. Together with this uprise, there exists a slowing down in the dynamics of the anions needed for the ordering. Hence, high pressure studies require a special attention to the cooling rate, which must be kept low enough to allow anion ordering at low temperature. This may be the explanation for the discrepancy between high pressure data showing the signature of anion ordering up to 8 kbar [151], and the absence of ordering claimed from the interpretation of magneto-angular oscillations [153].

SDW-SC Coexistence

A situation of nonhomogenous superconductivity have been clearly identified near the border between SDW and superconductivity in the $(TMTSF)_2X$ phase diagram [116]. At high pressure ($P > 9.4$ kbar), the superconducting phase emerges from a metallic state and can be reasonably thought of as homogeneous with a critical current density along the a-axis $J_c = 200\,\mathrm{A\,cm^{-2}}$. Below this critical pressure, there exists a pressure domain for $(TMTSF)_2PF_6$ (≈1 kbar wide) in which a superconducting signature is observed at a nearly pressure independent temperature below the onset of a SDW instability, where the critical current density is greatly reduced, $J_c \approx 10\,\mathrm{A\,cm^{-2}}$. This feature points in favor of a coexistence of SDW and SC macroscopic domains consisting of metallic (SC) tubes parallel to the a axis (Fig. 12.24). The existence of coexisting macroscopic regions of SDW and SC order is also supported by a recent NMR investigation performed at a pressure slightly lower than the critical pressure for the establishment of the homogenous state [154]. This latter study has enabled a quantification of the relative volume fractions in the SDW-metallic regime using the proton NMR linewidth as the local probe. A related consequence of the existence of macroscopic insulating domains in the superconducting phase allowing a channeling of the lines

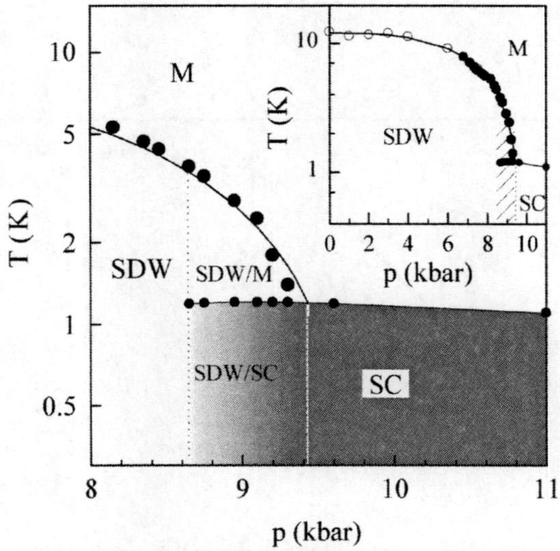

Fig. 12.24. Coexistence between SDW and superconductivity in $(TMTTF)_2PF_6$ in the vicinity of the critical pressure for suppression of the SDW ground state. After [116]

of force in the material is the large increase of the upper critical field H_{c2} [134], which had already been reported long time ago in $(TMTSF)_2PF_6$ with the second confirmation of organic superconductivity [155], and also later in $(TMTSF)_2AsF_6$ [102].

It must be kept in mind that the slab formation in the vicinity of the SDW state with the formation of insulating domains is not related to the penetration of the magnetic field in a type II superconductor but to the result from a competition between insulating and conducting phases. The claim for the existence of tubular domains $\|a$ is based on an analysis of the resistance in the vicinity of the critical pressure at low temperature above T_c and of the critical currents in the superconducting state [116]. The theoretical approach relies on a variational model leading to an inhomogeneous phase with an energy lower than the energy of the homogenous states (Metallic or Insulating SDW). Since it is the transverse b parameter that is expected to govern the respective stability of the SDW and metallic phases at a pressure lower than the critical one but close enough to the homogenous critical line, the formation of macroscopic metallic domains with a smaller b is energetically favored in-between b expanded SDW domains [156]. Similarly, at pressures larger than the critical pressure, insulating SDW domains should be present between metallic regions (see Fig. 12.25). This interpretation is at variance with a model based on similar experimental data for the critical fields in the coexistence regime in which the formation of thin superconducting slabs perpendicular to a sandwiched between SDW insulating domains is the result of a self organization process

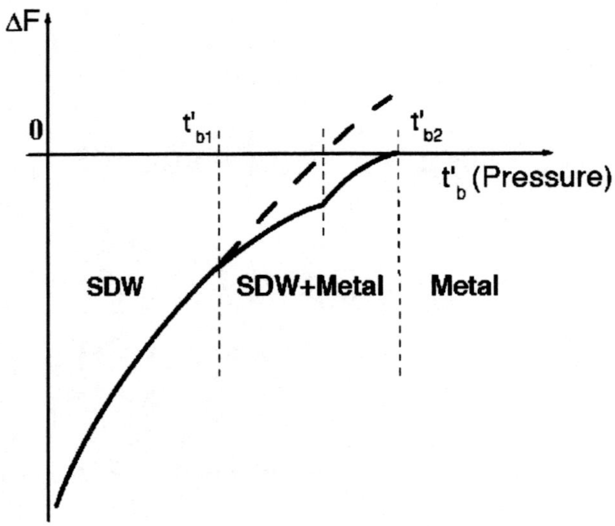

Fig. 12.25. Sketch of the theory showing from the free energy vs. the parameter t'_b, the region of coexistence between SDW and superconductivity in the vicinity of the critical pressure [116]. $t'_b(\equiv t'_{\perp b}$ in the text) parameterizes the deviation from the perfect nesting of the Q-1D Fermi surface. After [116]

taking advantage of the largest field penetration length perpendicular to the direction of the field [134].

Another approach has been taken by Gorkov and Grigoriev for the interpretation of the SDW/SC coexistence regime in $(TMTSF)_2PF_6$ [157]. This model predicts the existence of soliton domain walls in the SDW phase close to the critical pressure between the uniformly gapped SDW phase and the SC phase. As these domain walls should be perpendicular to the molecular stacking axis, a strong anomaly of the transport anisotropy could be anticipated related to the formation of conducting slabs perpendicular to the a axis in the coexistence pressure regime. However, no such anomaly has been detected experimentally [158].

Recent data obtained with $(TMTSF)_2PF_6$ under very high pressure, where the coexistence regime is much broader than that for $(TMTSF)_2PF_6$, have clearly shown that the critical field for $H\|c^*$ is enhanced by a factor 10 at the border with the SDW phase. In this pressure domain, it can reach 1 T, (Fig. 12.23), while it amounts to about 0.1 T at very high pressure when the superconducting state is homogenous [146], a value similar to the observation in the R-state of $(TMTSF)_2ClO_4$.

Finally, an experimental study performed in $(TMTSF)_2ReO_4$ under pressure has revealed the existence of conducting filaments parallel to a in the pressure domain close to 10 kbar when metallic and insulating domains coexist at low temperature as a consequence of two coexisting anion orders in this material [159].

Superconductivity and Nonmagnetic Defects

It is the remarkable sensitivity of organic superconductivity to irradiation [160, 161] that led Abrikosov to suggest the possibility of triplet pairing in these materials [162]. Although irradiation was recognized to be an excellent method for the introduction of defects in a controlled way [163], defects thus created can be magnetic [164], and the suppression of superconductivity by irradiation induced defects as a signature of nonconventional pairing must be taken with "a grain of salt" since local magnetic moments can also act as strong pair-breakers on s-wave superconductors. Several routes have been followed to introduce an intrinsically nonmagnetic perturbation modulating the potential seen by the carriers on the organic stacks. Nonmagnetic disorder has been achieved substituting TMTTF for TMTSF on the cationic stacks of $(TM)_2X$ salts with PF_6 [7] and ClO_4 salts [165]. However, in both situations cationic alloying induces drastic modifications of the normal state electronic properties, since the SDW transition of $(TMTSF)_2PF_6$ is quickly broadened and pushed towards higher temperature upon alloying [166].

Leaving the cation stack uniform, scattering centers can also be created on the anion stacks with the solid solution $(TMTSF)_2ClO_{4(1-x)}ReO_{4x}$, where Tomić et al. first mentioned the suppression of superconductivity upon alloying with a very small concentration of ReO_4 anions [167, 168]. In the case of a solid solution with tetrahedral anions such as ClO_4 or ReO_4, one is confronted to two potential sources of nonmagnetic disorder that act additively on the elastic electronic lifetime according to the Mathiessen's law. First the modulation due to the different chemical natures of the anions and second a disorder due to a progressive loss of long-range ordering at T_{AO} in the $(TM)_2X$ solid solution, although X-ray investigations have revealed that long-range order is preserved up to 3% ReO_4 with a correlation length $\xi_a > 200$ Å [169]. Studies of superconductivity in $(TMTSF)_2ClO_4$ performed under extremely slow cooling conditions have shown that T_c is a fast decreasing function of the nonmagnetic disorder [97], where the residual resistivity along the c^* axis has been used for the measure of the disorder in the alloys with different concentrations (Fig. 12.26). It must be emphasized that the residual resistivity is derived from a fit of the temperature dependence of the normal state resistivity according to a Fermi liquid model below the anion ordering temperature of 24 K, namely $\rho(T) = \rho_0^- + AT^2$. This treatment of the resistivity below 10 K or so allows us to remove the influence of superconducting precursor effects above the ordering temperature.

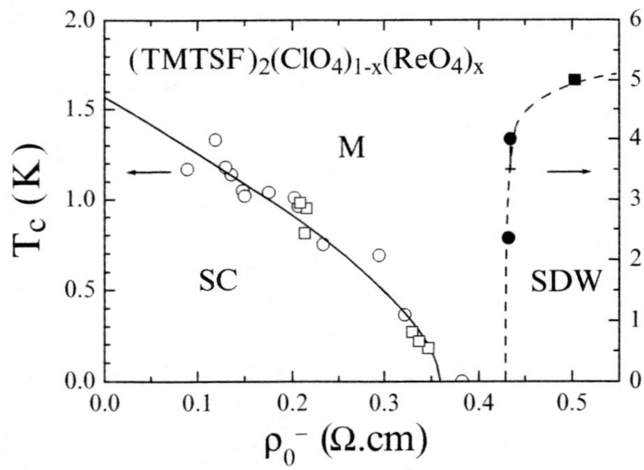

Fig. 12.26. Phase diagram of $(TM)_2X$ governed by nonmagnetic disorder. All open circles refer to the very slowly cooled samples in the R-state with different ReO_4^- contents. Open squares are data from the same samples corresponding to slightly larger cooling rates although keeping a metallic behavior above T_c. A 10% sample with $\rho_0^- \approx 0.32\,\Omega$ cm has provided four different T_c depending on the cooling rate. One sample (8%) did not reveal any ordering down to 0.1 K. These data show that the residual resistivity is a better characterization for the disorder than the nominal ReO_4^- concentration. Full dots (15 and 17%) are relaxed samples exhibiting a SDW ground state. The vertical bar is the error bar for a sample in which a maximum of the logarithmic derivative could not be clearly identified and therefore the actual SDW temperature should lie below 4 K, the temperature of minimum resistivity. The full square is the Q-state of a 6% sample. The continuous line at the Metal-SC transition is the best fit of the data with the diGamma function model providing $T_c^0 = 1.57$ K. The dashed line at the Metal-SDW transition is only a guide for the eye. After [170]

The suppression of T_c is clearly related to the enhancement of the scattering rate in the solid solution. Since the additional scattering cannot be ascribed to magnetic scattering according to the EPR checks showing no additional traces of localized spins in the solid solution, the data in Fig. 12.26 cannot be reconciled with the picture of a superconducting gap keeping a constant sign over the whole ($\pm k_F$) Fermi surface. They require a picture of pair breaking in a superconductor with an unconventional gap symmetry. The conventional pair breaking theory for magnetic impurities in usual superconductors has been generalized to the case of nonmagnetic impurities in unconventional materials and the correction to T_c obeys the following relation [171, 172],

$$\ln\left(\frac{T_c^0}{T_c}\right) = \psi\left(\frac{1}{2} + \frac{\alpha T_c^0}{2\pi T_c}\right) - \psi\left(\frac{1}{2}\right), \qquad (12.43)$$

with $\psi(x)$ being the Digamma function, $\alpha = \hbar/2\tau k_B T_c^0$ the depairing parameter, τ the elastic scattering time, and T_c^0 the limit of T_c in the absence of any scattering. From the data in Fig. 12.26, the best fit leads to $T_c^0 = 1.57\,\mathrm{K}$ and a critical scattering for the suppression of superconductivity of $1/\tau_{cr} = 1.85\,\mathrm{cm}^{-1}$. Accordingly, $1/\tau$ amounts to $0.56\,\mathrm{cm}^{-1}$ in the pristine $(TMTSF)_2ClO_4$ sample. Such a value for the inverse carrier life time is admittedly significantly smaller than the predicted width at half height, namely $1/\tau \approx 2\,\mathrm{cm}^{-1}$ assuming a classical Drude behavior involving the temperature dependence of the DC conductivity and the longitudinal plasma frequency [49]. The present derivation of the electron life time compares fairly well with far infrared optical measurements leading to a zero frequency conductivity peak with a width less than 2–4 cm^{-1} [92, 95]. Our results support the existence of a very narrow zero frequency peak carrying a minor fraction of the total spectral weight, which is probably the signature of a correlated low-dimensional Fermionic gas.

The sensitivity of T_c to nonmagnetic disorder cannot be reconciled with a model of conventional superconductors. The gap must show regions of positive and negative signs on the Fermi surface, which can be averaged out by a finite electron lifetime due to elastic scattering. As these defects are local, the scattering momentum of order $2k_F$ can mix $+k_F$ and $-k_F$ states and therefore the sensitivity to nonmagnetic scattering is still unable to tell the difference between p_x and d orbital symmetries for the superconducting wave function. A noticeable progress could be achieved paying attention to the spin part of the wave function. In the close vicinity of T_c, orbital limitation for the critical field is expected to prevail and therefore the analysis of the critical fields close to T_c does not imply a triplet pairing [143]. When the magnetic field is oriented along the intermediate b-axis, violations of the Pauli limitation have been claimed in $(TMTSF)_2PF_6$ [144] and recently in $(TMTSF)_2ClO_4$ superconductors [145]. However, it must be kept in mind that in all these experiments under transverse magnetic field along the b' axis, the electronic structure is profoundly affected by the application of the field, which tends to localize the electrons as shown by the normal state crossing over from a metallic to an insulating state when investigated with a current along c^* [142]. Furthermore it is still unclear whether the superconducting phase remains uniform under very strong transverse field [173, 174].

The nature of the superconducting coupling in $(TM)_2X$ conductors is still intensively studied and debated. The absence of temperature dependence of the ^{13}C Knight shift through the critical temperature at a pressure where $(TMTSF)_2PF_6$ is superconducting implies a triplet pairing [134]. However, the sample thermalization during the time of the NMR experiment has been questioned [175] and this result will have to be reconfirmed. The nature of the coupling in $(TM)_2X$ superconductors has not yet reached a consensus. This is due in part to the lack of unambiguous experimental data for samples exhibiting superconductivity in the very low temperature region. This is *at variance* with the singlet coupling found in 2D organic superconductors with a T_c in the 10 K range as clearly indicated by the Knight shifts measurements in the superconducting state [176, 177]. It can be noticed that in spite of the established singlet coupling, the critical fields H_{c2} of 2D superconductors can also greatly exceed the paramagnetic limit in the parallel geometry [178–180].

An Approach to the Mechanism of Superconductivity

Given the experimental uncertainty about the nature of superconductivity in the Bechgaard and Fabre salts, it is therefore still premature to privilege the triplet scenario for pairing over the singlet one. In any case, however, superconductivity certainly differs from what is commonly seen in ordinary metals and this raises the question of the possible causes of unconventional pairing in these materials [181]. This problem perplexed almost every one in the field from the start, since the requirements for a traditional phonon-mediated mechanism for pairing are apparently not met [182]. Antiferromagnetism that completely surrounds superconductivity in the phase diagram represents the main obstacle for an effective attraction mediated by phonons to take place. Superconductivity is indeed invariably replaced by an SDW instability whether one moves backward on the pressure scale or whether at fixed pressure $P > P_c$, one moves along the magnetic field axis $\mathbf{H}(\| c^*)$, where a cascade of field-induced SDW states is found [183]. $(TMTSF)_2ClO_4$ is another example that illustrates how close $(TM)_2X$ are to the threshold of a SDW instability at $P > P_c$. This compound presents an anion ordering on slow cooling and is already a superconductor at ambient pressure ($P_c < 1$ bar, Fig. 12.20); it develops an SDW instability either beyond a critical alloying [167, 170] (Fig. 12.26) or by just cranking up the cooling rate [167, 184, 185]. Additional experimental weight supporting underlying coupling conditions for SDW comes from the properties of the normal state at $P > P_c$. NMR experiments show indeed a strong enhancement of the spin-lattice relaxation rate T_1^{-1} as a function of temperature, revealing the existence of strong antiferromagnetic spin fluctuations in a very broad temperature domain of the metallic state above the superconducting T_c (Fig. 12.27) [27, 42, 52, 186]. Therefore, all this goes to show that even

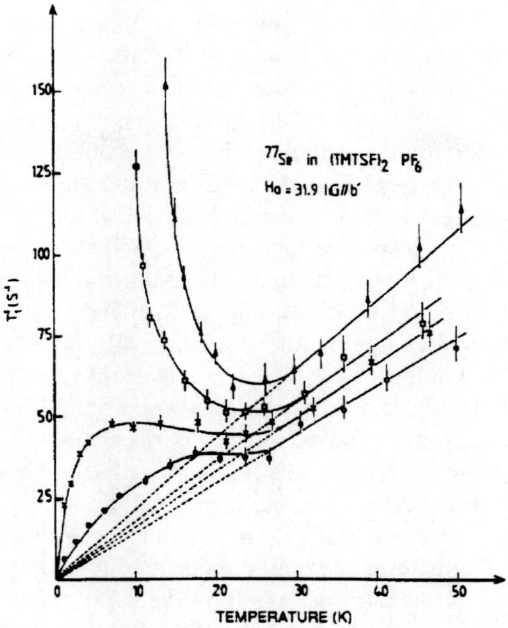

Fig. 12.27. Temperature dependence of the nuclear spin relaxation rate T_1^{-1} of $(TMTSF)_2PF_6$ at $P = 1$ bar (*triangles*), 5.5 kbar (*squares*), 8 kbar (*crosses*), and 10 kbar (*full circles*). Deviations with respect to a linear temperature behavior result from antiferromagnetic fluctuations. After [52]

in the presence of superconductivity, the interactions in $(TM)_2X$ at $P > P_c$ remain repulsive and favorable to a SDW state, only nesting conditions are apparently changing in each case.

According to the model described in Sect. 12.4.5, these fluctuations are made at the microscopic level of electron–hole pairs at $(\mathbf{k}, \mathbf{k} \pm \mathbf{q}_0)$. These will then coexist with electron–electron (and hole–hole) pairing at $(\mathbf{k}, -\mathbf{k})$, namely those responsible for the superconducting instability of the normal state at $P > P_c$. Since these two different pairings refer to the same electronic excitations around the Fermi surface, there will be some intrinsic dynamics or interference between them. We already came up against the problem of interfering pairing instabilities in the one-dimensional case (Sect. 12.2). For repulsive interactions and perfect nesting at $2k_F$, we have seen that interference between electron–hole and electron–electron pairing is maximum for a 1D – two points – Fermi surface and enters as a key ingredient in the formation of either a Luttinger liquid or a Mott insulating phase at commensurate filling [13–15]. At finite $t_{\perp b}$ and for temperature well below $T^{\star}$, however, the outcome differs and may provide a logical link between SDW and superconductivity.

The connection between superconductivity and density-wave correlations in isotropic systems goes back to the work of Kohn and Luttinger in the mid sixties [187]. They showed that the coupling between electron-hole (density-wave) and electron–electron correlations, albeit very small, is still present for a spherical Fermi surface. In this isotropic limit, $2k_{\rm F}$ Friedel (charge) fluctuations act as a oscillating pairing potential for electrons giving rise to a purely electronic mechanism for superconductivity at large angular momentum. Emery suggested that this nonphonon mechanism should be working in the spin sector as well, being boosted by the proximity of a SDW state in the quasi-1D geometry in order to yield experimentally reachable $T_{\rm c}$ [188] – an effect that was early on confirmed in the framework of renormalized mean-field theory [31, 189, 190] and various RPA approaches [191–193]. However, these approaches amount to extract an effective superconducting coupling from short-range density-wave correlations, which in turn serves as the input of a ladder diagrammatic summation. It turns out that the ladder theory, as a single channel approximation, neglects the quantum interference between the different kinds of pairings, and as such it cannot capture the dynamical emergence of superconductivity.

Because of the finite value of $t_{\perp b}$, interference becomes nonuniform along the Fermi surface. This introduces a momentum dependence in the scattering amplitudes, which can be parameterized by the set of transverse wave vectors for in going $(k_{\perp 1}k_{\perp 2})$ and outgoing $(k'_{\perp 1}k'_{\perp 2})$ electrons (here $k_\perp \equiv k_{\perp b}$). The generalization of the 1D scaling equations (12.2) to now $k_\perp$-dependent interactions $g_{i=1,2,3}(k'_{\perp 1}k'_{\perp 2}k_{\perp 2}k_{\perp 1})$ in the quasi-1D case, where both $t_{\perp b}$ and $t'_{\perp b}$ are present, has been worked out recently [114, 125, 194, 195]. The results can be put in the following schematic form:

$$\begin{aligned}\partial_\ell g_i(k'_{\perp 1}k'_{\perp 2}k_{\perp 2}k_{\perp 1}) = \sum_{k_\perp}\sum_{n,n'=1}^{3} \Big\{ & \epsilon^{nn'}_{{\rm C},i}\, g_n(\{k_\perp\})\, g_{n'}(\{k_\perp\})\mathcal{L}'_{\rm C}(k_\perp, q_{{\rm C}\perp}) \\ & + \epsilon^{nn'}_{{\rm P},i}\, g_n(\{k_\perp\})\, g_{n'}(\{k_\perp\})\mathcal{L}'_{\rm P}(k_\perp, q_{{\rm P}\perp}, t'_{\perp b})\Big\}. \end{aligned} \tag{12.44}$$

Here $\mathcal{L}'_{\rm C,P} = \partial_\ell \mathcal{L}_{\rm C,P}$ where $\mathcal{L}_{\rm C,P}$ are the Cooper (electron–electron) and Peierls (electron–hole) loops, with $q_{{\rm C,P}\perp}$ as their respective $\{k_\perp\}$-dependent transverse momentum variables, and $\epsilon^{nn'}_{{\rm C,P},i} = \pm 1$ or 0. By integrating these flow equations, the singularities shown by interactions signal instabilities of the normal state at a critical temperature $T_{\rm c}$. The nature of ordering is determined by the profile of interactions in $\{k_\perp\}$ space, which in turn corresponds to a divergence of a given order parameter susceptibility χ_μ. Feeding these equations with a realistic set of bare parameters for the repulsive intrachain interactions g_i and the band parameters $t_{\perp b}$ and $E_{\rm F}$ in $({\rm TM})_2{\rm X}$, it is possible to follow the instabilities of the normal state as a function of nesting deviations parameter $t'_{\perp b}$, which simulates the main influence of pressure in the model [114, 125, 194–196].

Thus at perfect nesting, when $t'_{\perp b} = 0$, the normal state develops a SDW instability at $T^0_{\rm c} \sim 20\,{\rm K}$, which for small Umklapp scattering corresponds to

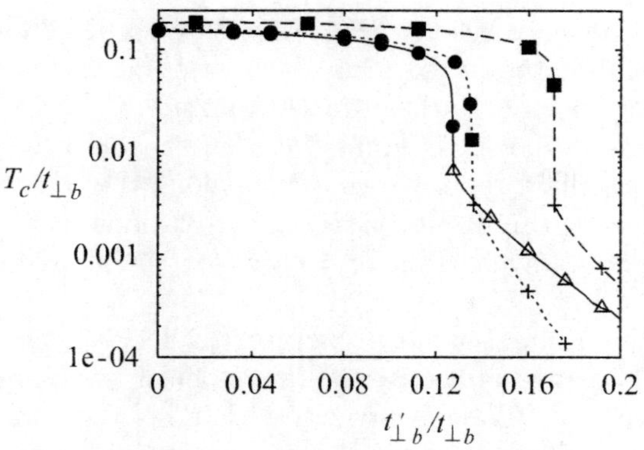

Fig. 12.28. Calculated phase transition temperature of the quasi-one-dimensional electron gas model as a function of the nesting deviation parameter $t'_{\perp b}$ for repulsive intrachain interactions $g_{1,2,3}$. Continuous line ($\tilde{g}_1^{\perp} = 0$), dotted line ($\tilde{g}_1^{\perp} = 0.11$), and dashed line ($\tilde{g}_1^{\perp} = 0.14$). After Nickel et al., [125]

the range of T_c expected in most of $(TMTSF)_2X$ at ambient pressure and $(TMTTF)_2X$ at relatively high pressure. The range of T_c roughly squares with the one obtained in the single channel approximation with no interference below T_{x^1} (§12.4.5). As $t'_{\perp b}$ increases, T_c is gradually decreasing until the critical range $t'^{\rm cr}_{\perp b} \simeq 0.8T_c^0$ is reached where the SDW is suppressed (full circles, Fig. 12.28). The metallic phase remains unstable at finite temperature, however, but the instability now takes place in the superconducting channel (open triangles, Fig. 12.28). The order parameter is of the form $\Delta(k_\perp) = \Delta \cos k_\perp$ and has nodes at $k_\perp \pm \pi/2$; it corresponds to a interstack singlet or $d_{x^2-y^2}$-wave pairing. Therefore, for repulsive intrachain interactions, an attraction between electrons can be dynamically generated from the interference between Cooper and Peierls scattering channels. The attraction between carriers on neighboring chains can be seen as being mediated by spin fluctuations. The fact that SCd and SDW instability lines meet at the maximum of the superconducting $T_c \sim 1\,$K and that the ratio $T_c^0(\text{SC}d)/T_c^0(\text{SDW}) \sim 1/20$, together with their respective $t'_{\perp b}$ dependence, are worth noticing features in regard to the experimental phase diagram (Fig. 12.3).

Regarding the possible symmetries of the superconducting order parameter, an analysis of the momentum dependence of the scattering amplitude $g_i(\{k_\perp\})$ reveals that for the electron gas model defined with only *intra*chain repulsive interactions, the strongest superconducting instability is invariably found in the singlet SCd-wave channel [114,196]. Triplet superconductivity in the p_x channel, which has been proposed on phenomenological grounds as a possible candidate to describe superconductivity in the Bechgaard salts [197], is strongly suppressed. In effect, the triplet SCp_x superconductivity, which has a gap order parameter $\Delta_r = r\Delta$ with $r = \text{sign}\, k_x$, is an intrachain pairing that is subjected to the strongest repulsive part of the oscillating potential produced by SDW correlations [189]. More favorable conditions for triplet pairing do exist but they take place at higher angular momentum, in the interchain *f*-wave channel with an order parameter $\Delta_r(k_\perp) = r\Delta \cos k_\perp$, a possibility that was shown to come out from the mean-field analysis [198]. However, for intrachain repulsive couplings alone the renormalization group analysis show that the amplitude of triplet correlations are always subordinate to those of the SC*d*-wave channel, which yields the highest superconducting T_c [125,196].

Following the Kohn–Luttinger picture, triplet pairing at high angular momentum is actually connected to charge-density-wave fluctuations [187]. The presence of a CDW superstructure that coexists with a SDW state in the Bechgaard salts (see Sect. 12.4.5) has in this respect stressed their importance in these salts close to P_c. Following the example of most quasi-one-dimensional systems in which a CDW superstructure is found [199,200], *inter*chain Coulomb interaction is a physically relevant coupling that must be taken into account in the presence of charge correlations. By including, besides the g_i, *inter*chain backward ($g_1^\perp$), forward ($g_2^\perp$), and Umklapp ($g_3^\perp$) scattering amplitudes, one defines the so-called extended electron gas model [201,202]. For realistic repulsive couplings, this model allows us to expand the range of possibilities of both superconducting and density-wave long-range orders. The RG solution of (12.44) in the $T - t'_{\perp b}$ phase diagram shows that $g_1^\perp$ plays a key role on the one hand, in the stability of SDW and SC*d* orders, and on the other hand in the emergence of triplet superconductivity and CDW order [125,194]. The RG results depicted in Fig. 12.29 indeed show that beyond a relatively small threshold in $g_1^\perp$, SC*d* long-range order is no longer stable and the instability of the normal state is rather found in the interchain triplet *f*-wave channel above a critical $t'_{\perp b}$. The SC*f* T_c are comparable to SC*d* but show stronger "pressure" coefficient along the $t'_{\perp b}$ axis (Fig. 12.28). The results also show that the strength of SDW correlations remain essentially unaffected and

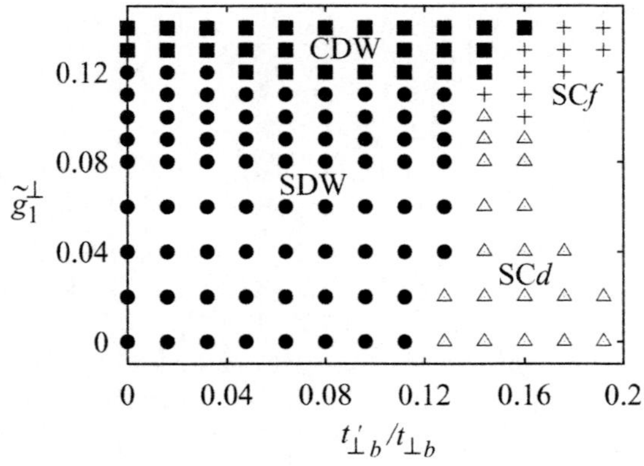

Fig. 12.29. Calculated phase diagram of the extended quasi-one-dimensional electron gas model for repulsive couplings. After Nickel et al. [125]

can still dominate the normal phase in addition to CDW correlations, whose amplitude grows with the strength interchain coupling. Although SDW order can precede the triplet SCf instability along the $t'_{\perp b}$ "pressure" scale near the interchain coupling threshold, it is very close in stability with a CDW superstructure (Figs. 12.28 and 12.29).

Given the observation of a close proximity between SDW and CDW in the phase diagram of the Bechgaard salts, it follows that not only singlet SCd but also SCf are serious candidates for the description of superconductivity in these materials. Since many parameters of the model are likely to change under pressure besides $t_{\perp b}$, this puts some haziness about how one actually moves in the phase diagram of Fig. 12.29, and in turn on the most stable type of superconductivity in these compounds.

12.5 Conclusion and Outlook

In this brief account we went through the main physical properties of the Bechgaard and Fabre salts series of organic superconductors. The global phase diagram that was gradually built over the years around these two series of compounds under either hydrostatic or chemical pressure stands out as a model of unity for the physics of low-dimensional correlated systems. Much effort went to explain the multifaceted phase diagram of $(TM)_2X$ as a whole, an attempt that also proved to be an active quest of unity for the theory.

In this respect, while the theoretical description of the spin-Peierls and antiferromagnetic instabilities in the phase diagram of $(TM)_2X$ does not meet any serious conceptual difficulty, their strong competition at the putative quantum critical point on the pressure scale is to our knowledge without precedent in the field of low-dimensional compounds. Although the quantum criticality that is behind this competition would certainly gain to be further clarified on experimental ground, its comprehension clearly challenges the traditional framework of critical phenomena and fosters some new conceptual focus in unifying antiferromagnetism and a lattice distorted spin liquid phase. In the last few years, the phenomenon of charge ordering has also played an important part in improving and even expanding the structure of the phase diagram of $(TM)_2X$. Though its observation has been so far restricted to members of the Fabre salts series, this phenomenon raises important questions about the influence of charge disproportionation in the relative stability of the spin-Peierls and Néel states in this series of compounds.

The interplay of different types of commensurability in weakly dimerized quarter-filled compounds like the $(TM)_2X$ is another issue that is at the heart of a better understanding of strong electronic correlations that characterize the properties normal phase in $(TM)_2X$. This problem is linked to the persistent issue of the dimensionality crossover or about how the restoration of a Fermi liquid is achieved in quasi-one-dimensional conductors like the $(TM)_2X$.

As the starting point of the study of the Bechgaard salts more than 25 years ago, superconductivity is certainly one of the hardest part of the phase diagram to both explain and characterize. In spite of the recent experimental advances, which again confirm the nonconventional nature of superconductivity in $(TM)_2X$, the problem of the symmetry of the superconducting order parameter, as well as the issue of the presence and location of nodes for the gap, are all enduring questions for which a consensus of views has yet to be reached. Given the experimental constraints and difficulties tied to the use of extremely low temperature and high pressure conditions in $(TM)_2X$, these questions will certainly continue to consume major experimental efforts in the next few years.

The mechanism of organic superconductivity in quasi-one-dimensional molecular crystals is a related key issue in want of a satisfactory explanation. The extensive experimental evidence in favor of the systematic emergence of superconductivity in $(TM)_2X$ just below their stability threshold for antiferromagnetism has shown the need for a unified description of electronic excitations at the core of both density-wave and superconducting correlations. In this matter, the recent progress achieved by the renormalization group method have resulted in definite predictions about the possible symmetries of the superconducting order parameter when a purely electronic mechanism is involved – predictions that often differ from phenomenologically based approaches to superconductivity. The results for the electron gas model, albeit appealing when confronted to existing data, remain only indicative, however, of what may be the actual origin of superconductivity in these complex materials. In this respect, the future progress on the experimental side will be certainly decisive for the theory.

Acknowledgments

A large part of the results presented in this review are based on a long term activity in the domain of low-dimensional organic superconductors with a large number of contributors at Orsay and Sherbrooke. We acknowledge in particular the recent contributions of P. Auban-Senzier, J.C. Nickel, R. Duprat, and N. Dupuis. This extended activity would not have been possible without the fruitful cooperation of our colleagues in Chemistry, J.M. Fabre (Montpellier), K. Bechgaard (Copenhagen), and P. Batail (Angers).

References

1. K. Bechgaard, D.O. Cowan, A.N. Bloch, J. Chem. Soc. Chem. Commun. **22**, 937 (1974)
2. A. Garito, A. Heeger, Acc. Chem. Res. **7**, 232 (1974)
3. E. Engler et al., J. Am. Chem. Soc. **99**, 5909 (1977)
4. J. Andersen et al., Acta Cryst. B **34**, 1901 (1978)

5. A. Andrieux et al., J. Phys. Paris **40**, 1199 (1979)
6. G. Brun et al., C.R. Acad. Sci. (Paris) **284C**, 211 (1977)
7. C. Coulon et al., J. Phys. (Paris) **43**, 1059 (1982)
8. L. Ducasse et al., J. Phys. C **19**, 3805 (1986)
9. K. Behnia et al., Phys. Rev. Lett. **74**, 5272 (1995)
10. D. Jérome, Science **252**, 1509 (1991)
11. P. Auban, D. Jérome, C. Carcel, J. Fabre, J. Phys. IV (France) **114**, 41 (2004); (to be published)
12. D. Jaccard et al., J. Phys. Condens. Matter **13**, 89 (2001)
13. Y.A. Bychkov, L.P. Gorkov, I. Dzyaloshinskii, Sov. Phys. JETP **23**, 489 (1966)
14. I.E. Dzyaloshinskii, A.I. Larkin, Sov. Phys. JETP **34**, 422 (1972)
15. J. Solyom, Adv. Phys. **28**, 201 (1979)
16. S. Barisic, S. Brazovskii, in *Recent Developments in Condensed Matter Physics*, vol. 1, ed. by J.T. Devreese (Plenum, New York, 1981), p. 327
17. V.J. Emery, R. Bruisma, S. Barisic, Phys. Rev. Lett. **48**, 1039 (1982)
18. C. Bourbonnais, J. Phys. I (France) **3**, 143 (1993); H. Nélisse, C. Bourbonnais, H. Touchette, Y.M. Vilk, A.M.S. Tremblay, Eur. Phys. J. B **12**, 351 (1999); Y. Fuseya, M. Tsuchiizu, Y. Suzumura, C. Bourbonnais, J. Phys. Soc. Jpn. Lett. **74**, 3159 (2005)
19. M. Kimura, Prog. Theor. Phys. **63**, 955 (1975)
20. T. Giamarchi, (this book); T. Giamarchi, *Quantum Physics in One Dimension* (Oxford University Press, Oxford, 2004)
21. J. Voit, Rep. Prog. Phys. **58**, 977 (1995)
22. H.J. Schulz, G. Cuniberti and P. Pieri, in *Field Theory for Low-Dimensional Condensed Matter Systems*, G. Morandi and P. Sodano and A. Tagliacozzo and V. Tognetti, Editors, Springer Verlag (New York), 9 (2000)
23. T. Giamarchi, Phys. B **230–232**, 975 (1997)
24. F. Mila, X. Zotos, Europhys. Lett. **24**, 133 (1993)
25. H.J. Schulz, in *Strongly Correlated Electronic Materials: The Los Alamos Symposium 1993*, ed. by K.S. Bedell et al. (Addison-Wesley, Reading, MA, 1994), p. 187
26. M. Tsuchiizu, H. Yoshioka, Y. Suzumura, J. Phys. Soc. Jpn. **70**, 1460 (2001)
27. C. Bourbonnais et al., J. Phys. (Paris) Lett. **45**, L755 (1984)
28. C. Bourbonnais, L.G. Caron, Phys. B **143**, 450 (1986)
29. S. Brazovskii, Y. Yakovenko, Sov. Phys. JETP **62**, 1340 (1985)
30. C. Bourbonnais, L.G. Caron, Int. J. Mod. Phys. B **5**, 1033 (1991)
31. C. Bourbonnais, L.G. Caron, Europhys. Lett. **5**, 209 (1988)
32. C. Bourbonnais, in *Les Houches, Session LVI (1991), Strongly interacting fermions and high-T_c superconductivity*, ed. by B. Doucot, J. Zinn-Justin (Elsevier Science, Amsterdam, 1995), p. 307
33. C. Bourbonnais, in *Trends in High Magnetic Fields Science*, ed. by L.P.L.C. Berthier, P. Boebinger, G. Martinez (Springer, Heidelberg, 2002); arXiv. cond-mat/0204345
34. B. Horovitz, H. Gutfreund, M. Weger, Mol. Cryst. Liq. Cryst. **79**, 591 (1982)
35. K. Yamaji, J. Phys. Soc. Jpn. **51**, 2787 (1982)
36. J.R. Cooper, Phys. Rev. B **19**, 2404 (1979)
37. D. Jérome, in *Organic Conductors: Fundamentals and Applications*, ed. by J.-P. Farges (Dekker, New York, 1994), pp. 405–494

38. P. Auban-Senzier, D. Jérome, J. Moser, in *Physical Phenomena at High Magnetic Fields*, ed. by R.S.Z. Fisk, L.P. Gorkov (World Scientific, Singapore, 1999), p. 211
39. B. Gallois, Ph.D. Thesis, University of Bordeaux I, 1987
40. S. Brazovskii, S.A. Gordyunin, JETP Lett. **31**, 371 (1980)
41. F. Creuzet et al., Synth. Met. **19**, 277 (1987)
42. P. Wzietek et al., J. Phys. I **3**, 171 (1993)
43. N. Miljak, J.R. Cooper, K. Bechgaard, J. Phys. Coll. **44**, 893 (1983)
44. N. Miljak, J.R. Cooper, Mol. Cryst. Liq. Cryst. **119**, 141 (1985)
45. N. Mott, *Metal–Insulator Transitions* (Taylor and Francis, London, 1974)
46. P. Auban-Senzier, Orsay work, to be published
47. J. Moser et al., Eur. Phys. J. B **1**, 39 (1998)
48. A. Georges, T. Giamarchi, N. Sandler, Phys. Rev. B **61**, 16393 (2000)
49. C. Jacobsen, D. Tanner, K. Bechgaard, Phys. Rev. B **28**, 7019 (1983)
50. M. Dressel, Naturwissenschaften **90**, 337 (2003)
51. C. Bourbonnais et al., Phys. Rev. Lett. **62**, 1532 (1989)
52. F. Creuzet et al., Synth. Met. **19**, 289 (1987)
53. B. Gotschy et al., J. Phys. I (France) **2**, 677 (1992)
54. J. Pouget et al., Mol. Cryst. Liq. Cryst. **79**, 129 (1982)
55. P. Foury-Leylekian et al., Phys. Rev. B **70**, 180405(R) (2004)
56. M. Dumm et al., Phys. Rev. B **61**, 511 (2000)
57. C. Bourbonnais, B. Dumoulin, J. Phys. I (France) **6**, 1727 (1996)
58. L.G. Caron, C. Bourbonnais, F. Creuzet, D. Jerome, Synth. Met. **19**, 69 (1987)
59. J. Riera, D. Poilblanc, Phys. Rev. B **62**, R16243 (2000)
60. J.P. Pouget, in *Low Dimensional Conductors and Superconductors*, ed. by D. Jérome, L.G. Caron (Plenum Press, New York, 1987), p. 17
61. D. Chow et al., Phys. Rev. Lett. **81**, 3984 (1998)
62. F. Zamborszky et al., Phys. Rev. B **66**, 081103(R) (2002)
63. V. Ilakovac et al., Phys. Rev. B **50**, 7136 (1994)
64. D.S. Chow et al., Phys. Rev. Lett. **85**, 1698 (2000)
65. L.G. Caron et al., Synth. Met. **27B**, 123 (1988)
66. J.P. Pouget et al., Phys. Rev. Lett. **37**, 437 (1976)
67. L.G. Caron, C. Bourbonnais, Phys. Rev. B **29**, 4230 (1984)
68. L.G. Caron, S. Moukouri, Phys. Rev. Lett **76**, 4050 (1996)
69. H. Bakrim, C. Bourbonnais, Phys. Rev. B **76**, 195115 (2007)
70. For more details about the charge ordered state in the $(TMTTF)_2X$ series, see the contributions of P. Monceau et al.; S.B. Brown and P.M. Chaikin; and S. Brazovskii
71. F. Nad, P. Monceau, C. Carcel, J.M. Fabre, Phys. Rev. B **62**, 1753 (2000)
72. F. Nad, P. Monceau, C. Carcel, J.M. Fabre, J. Phys. Condens. Matter **12**, L435 (2000)
73. P. Monceau, F. Nad, S. Brazovskii, Phys. Rev. Lett **86**, 4080 (2001)
74. C. Coulon, S.S.P. Parkin, R. Laversanne, Phys. Rev. B **31**, 3583 (1985)
75. R. Laversanne et al., J. Phys. (Paris) Lett. **45**, L393 (1984)
76. W. Yu et al., Phys. Rev. B **70**, R121101 (2004)
77. H. Yoshioka, M. Tsuchiizu, Y. Suzumura, J. Phys. Soc. Jpn. **70**, 762 (2001)
78. J. Riera, D. Poilblanc, Phys. Rev. B **63**, R241102 (2001)
79. H. Seo, H. Fukuyama, J. Phys. Soc. Jpn. **66**, 1249 (1997)
80. R.T. Clay, S. Mazumdar, D.K. Campbell, Phys. Rev. B **67**, 115121 (2003)

81. J.P. Pouget, S. Ravy, Synth. Met. **85**, 1523 (1997)
82. B.J. Klemme et al., Phys. Rev. Lett. **75**, 2408 (1995)
83. S.E. Brown et al., Synth. Met. **86**, 1937 (1997)
84. P. Auban et al., J. Phys. (France) **50**, 2727 (1989)
85. P. Auban, D. Jérome, I. Johannsen, K. Bechgaard, Synth. Met. **41**, 2285 (1991)
86. C. Bourbonnais, D. Jérome, Science **281**, 1156 (1998)
87. K. Bechgaard et al., Solid State Commun. **33**, 1119 (1980)
88. S. Tomic et al., J. Phys. I (France) **1**, 1603 (1991)
89. C. Bourbonnais, D. Jérome, in *Advances in Synthetic Metals, Twenty Years of Progress in Science and Technology*, ed. by P. Bernier, S. Lefrant, G. Bidan (Elsevier, New York, 1999), pp. 206–261, arXiv:cond-mat/9903101
90. A. Schwartz et al., Phys. Rev. B **58**, 1261 (1998)
91. C. Jacobsen, in *Low Dimensional Conductors and Superconductors*, ed. by D. Jérome, L. Caron (Plenum Press, New York, 1987), p. 253
92. T. Timusk, in *Low Dimensional Conductors and Superconductors*, ed. by D. Jérome, L. Caron (Plenum Press, New York, 1987), p. 275
93. K. Kikuchi et al., Solid State Commun. **42**, 433 (1982)
94. K. Bechgaard, M. Carneiro, M. Olsen, F. Rasmussen, Phys. Rev. Lett. **46**, 852 (1981)
95. H. Ng, T. Timusk, D. Jérome, K. Bechgaard, Phys. Rev. B **32**, 8041 (1985)
96. H. Ng, T. Timusk, K. Bechgaard, J. Phys. **44**, C3 (1983)
97. N. Joo et al., Eur. Phys. J. B **40**, 43 (2004)
98. K. Heuzé et al., Adv. Mater. **15**, 1251 (2003)
99. T. Itou et al., Phys. Rev. Lett. **93**, 216408 (2004)
100. E.B. Lopes et al., Eur. Phys. J. B **29**, 27 (2002)
101. D. Jérome, A. Mazaud, M. Ribault, K. Bechgaard, J. Phys. (Paris) Lett. **41**, L95 (1980)
102. R. Brusetti, M. Ribault, D. Jerome, K. Bechgaard, J. Phys. (Paris) **43**, 801 (1982)
103. K. Mortensen, Y. Tomkiewiez, T.D. Schultz, E.M. Engler, Phys. Rev. Lett. **46**, 1234 (1982)
104. A. Andrieux, D. Jérome, K. Bechgaard, J. Phys. (Paris) Lett. **42**, L87 (1981)
105. T. Adachi et al., J. Am. Chem. Soc. **122**, 3238 (2000)
106. L. Balicas et al., J. Phys. I (France) **4**, 1539 (1994)
107. P. Auban-Senzier et al., Synth. Met. **133–134**, 11 (2003)
108. C.S. Jacobsen, D. Tanner, K. Bechgaard, Phys. Rev. Lett. **46**, 1142 (1981)
109. V. Vescoli et al., Science **281**, 1181 (1998)
110. T. Takahashi, Y. Maniwa, H. Kawamura, G. Saito, Phys. B **143**, 417 (1986)
111. G. Montambaux, Phys. Rev. B **38**, 4788 (1988)
112. L. Gor'kov, J. Phys. I (France) **6**, 1697 (1996)
113. C. Seidel, V.N. Prigodin, J. Phys. (Paris) Lett. **44**, L403 (1983)
114. R. Duprat, C. Bourbonnais, Eur. Phys. J. B **21**, 219 (2001)
115. D. Jérome, H. Schulz, Adv. Phys. **31**, 299 (1982)
116. T. Vuletic et al., Eur. Phys. J. B **25**, 319 (2002)
117. Y. Hasegawa, H. Fukuyama, J. Phys. Soc. Jpn. **55**, 3978 (1986)
118. J.F. Kwak et al., Phys. Rev. Lett. **56**, 972 (1986)
119. J.P. Pouget, S. Ravy, J. Phys. I (France) **6**, 1501 (1996)
120. S. Kagoshima et al., Solid State Commun. **110**, 479 (1999)
121. H.K. Ng, T. Timusk, K. Bechgaard, Phys. Rev. B **30**, 5842 (1984)

122. N. Cao, T. Timusk, K. Bechgaard, J. Phys. I (France) **6**, 1719 (1996)
123. S. Mazumdar, R. Ramasesha, R.T. Clay, D.K. Campbell, Phys. Rev. Lett. **82**, 1522 (1999)
124. M. Kobayashi, M. Ogata, K. Yonemitsu, J. Phys. Soc. Jpn. **67**, 1098 (1998)
125. J.C. Nickel, R. Duprat, C. Bourbonnais, N. Dupuis, Phys. Rev. Lett. **95**, 247001 (2005)
126. M. Ribault, G. Benedek, D. Jérome, K. Bechgaard, J. Phys. Lett. Paris **41**, L (1980)
127. K. Andres et al., Phys. Rev. Lett. **45**, 1449 (1980)
128. P. Garoche, R. Brusetti, K. Bechgaard, Phys. Rev. Lett. **49**, 1346 (1982)
129. R. Brusetti, P. Garoche, K. Bechgaard, J. Phys. C **16**, 2495 (1983)
130. S. Belin, K. Behnia, Phys. Rev. Lett. **79**, 2125 (1997)
131. J. Bardeen, G. Rickaysen, L. Teword, Phys. Rev. **113**, 982 (1959)
132. M. Takigawa, H. Yasuoka, G. Saito, J. Phys. Soc. Jpn. **56**, 873 (1987)
133. Y. Hasegawa, H. Fukuyama, J. Phys. Soc. Jpn. **56**, 877 (1987)
134. I.J. Lee et al., Phys. Rev. Lett. **88**, 17004 (2002)
135. L.P. Le et al., Phys. Rev. B **48**, 7284 (1993)
136. G.M. Luke et al., Phys. B **326**, 378 (2003)
137. G.M. Luke et al., Nature **394**, 558 (1998)
138. D.U. Gubser et al., Mol. Cryst. Liq. Cryst. **79**, 225 (1982)
139. R.L. Greene et al., Mol. Cryst. Liq. Cryst. **79**, 183 (1982)
140. K. Murata et al., Mol. Cryst. Liq. Cryst. **79**, 283 (1982)
141. K. Murata et al., Jpn. J. Appl. Phys. **26**, 1367 (1987)
142. N. Joo, unpublished results
143. L.P. Gorkov, D. Jérome, J. Phys. Lett. **46**, L643 (1985)
144. I.J. Lee, P.M. Chaikin, M.J. Naughton, Phys. Rev. B **62**, R14669 (2000)
145. J.I. Oh, M.J. Naughton, Phys. Rev. Lett. **92**, 067001 (2004)
146. H. Wilhelm (2004), private communication
147. H.J. Schulz et al., J. Phys. Lett. **42**, L (1981)
148. L. Jennings, C. Swenson, Phys. Rev **112**, 31 (1958)
149. D. Mailly, Ph.D. Thesis, Thesis Université d'Orsay, 1983
150. F. Creuzet, D. Jérome, A. Moradpour, Mol. Cryst. Liq. Cryst. **119**, 297 (1985)
151. K. Murata et al., Mol. Cryst. Liq. Cryst. **119**, 245 (1985)
152. F. Guo et al., J. Phys. Soc. Jpn. **67**, 3000 (1998)
153. W. Kang, S. Hannahs, P. Chaikin, Phys. Rev. Lett. **70**, 3091 (1993)
154. I.J. Lee et al., Phys. Rev. Lett. **94**, 197001 (2005)
155. R.L. Greene, E.M. Engler, Phys. Rev. Lett. **45**, 1587 (1980)
156. M. Héritier, G. Montambaux, P. Lederer, J. Phys. (Paris) Lett. **45**, L943 (1984)
157. L.P. Gorkov, P.D. Grigoriev, Europhys. Lett. **71**, 425 (2005)
158. P. Auban-Senzier, to be published (2008)
159. C. Colin, P. Auban-Senzier, C.R. Pasquier, K. Bechgaard, Eur. Phys. Lett. **75**, 301 (2006)
160. S. Bouffard et al., J. Phys. C **15**, 295 (1982)
161. M.-Y. Choi et al., Phys. Rev. B **25**, 6208 (1982)
162. A.A. Abrikosov, J. Low Temp. Phys. **53**, 359 (1983)
163. L. Zuppirolli, in *Low Dimensional Conductors and Superconductors*, ed. by D. Jérome, L.G. Caron (Plenum Press, New York, 1987), p. 307
164. M. Sanquer, S. Bouffard, Mol. Cryst. Liq. Cryst. **119**, 147 (1985)
165. I. Johannsen et al., Mol. Cryst. Liq. Cryst. **119**, 277 (1985)

166. K. Mortensen, E.M. Engler, Phys. Rev. B **29**, 842 (1984)
167. S. Tomic, D. Jérome, P. Monod, K. Bechgaard, J. Phys. (Paris) Lett. **43**, L839 (1982)
168. B. Korin-Hamzić, L. Forró, J.R. Cooper, K. Bechgaard, Phys. Rev. B **38**, 11177 (1988)
169. S. Ravy, R. Moret, J.P. Pouget, R. Comès, Phys. B **143**, 542 (1986)
170. N. Joo et al., Europhys. Lett. **72**, 645 (2005)
171. K. Maki, H. Won, S. Haas, Phys. Rev. B **69**, 012502 (2004)
172. A.I. Larkin, Sov. Phys. JETP **21**, 153 (1965)
173. A.G. Lebed, JETP Lett. **44**, 114 (1986)
174. N. Dupuis, G. Montambaux, C.A.R.S. de Melo, Phys. Rev. Lett. **70**, 2613 (1993)
175. The claim for triplet superconducting pairing in $(TM)_2X$ is based essentially on NMR investigations (^{77}Se Knight shift and spin-lattice relaxation time T_1) performed on $(TMTSF)_2PF_6$ under pressure [134]. A major difficulty regarding this experiment is the actual control of the temperature of the electron spins during the time of the NMR pulse and acquisition. It is not unlikely that the electron spins can reach a high temperature state for the experiment leading to the determination of the Knight shift, see also recent data obtained by Shinagawa et al. on $(TMTSF)_2ClO_4$ (Proceedings of ISCOM 05) and Phys. Rev. Lett. **98**, 147002 (2007). An other possibility could be the existence of a field-induced transition within the superconducting state between a low field singlet phase with d-wave symmetry and a high field phase exhibiting either a triplet symmetry or a Fulde-Ferrell-Larkin-Ovchinikov modulated superconducting gap.
176. H. Mayaffre et al., Phys. Rev. Lett **75**, 4951 (1995)
177. K. Kanoda, K. Miyagawa, A. Kawamoto, Y. Nakazawa, Phys. Rev. B **54**, 76 (1996)
178. S. Kamiya et al., Phys. Rev. B **65**, 134510 (2002)
179. F. Zuo et al., Phys. Rev. B **61**, 750 (2000)
180. D. Jérome, C.R. Pasquier, in *One-Dimensional Organic Superconductors*, ed. by A.V. Narlikar (Springer Verlag, Berlin, 2005)
181. N. Dupuis, C. Bourbonnais, J.C. Nickel, Fiz. Nizk. Temp. **32**, 505 (2006), arXiv:cond-mat/0510544
182. V.J. Emery, J. Phys. (Paris) Coll. **44**, 977 (1983)
183. W. Kang, S.T. Hannahs, P.M. Chaikin, Phys. Rev. Lett. **70**, 3091 (1993)
184. T. Takahashi, D. Jérome, K. Bechgaard, J. Phys. (Paris) Lett. **43**, L565 (1982)
185. H. Schwenk, K. Andres, F. Wudl, Phys. Rev. B **29**, 500 (1984)
186. W. Wu et al., Phys. Rev. Lett. **94**, 097004 (2005)
187. W. Kohn, J.M. Luttinger, Phys. Rev. Lett. **15**, 524 (1965)
188. V.J. Emery, Synth. Met. **13**, 21 (1986)
189. M.T. Béal-Monod, C. Bourbonnais, V.J. Emery, Phys. Rev. B **34**, 7716 (1986)
190. L.G. Caron, C. Bourbonnais, Phys. B **143**, 453 (1986)
191. D.J. Scalapino, E. Loh, J.E. Hirsch, Phys. Rev. B **34**, R8190 (1986)
192. H. Shimahara, J. Phys. Soc. Jpn. **58**, 1735 (1989)
193. H. Kino, H. Kontani, J. Phys. Soc. Jpn. **68**, 1481 (1999)
194. J.C. Nickel, R. Duprat, C. Bourbonnais, N. Dupuis, Phys. Rev. B **73**, 094616 (2006)
195. C. Bourbonnais, R. Duprat, J. Phys. IV **114**, 3 (2004)

196. Y. Fuseya, Y. Suzumura, J. Phys. Soc. Jpn. **74**, 1263 (2005)
197. A.G. Lebed, K. Machida, M. Ozaki, Phys. Rev. B **62**, R795 (2000)
198. K. Kuroki, R. Arita, H. Aoki, Phys. Rev. B **63**, 094509 (2001)
199. J.P. Pouget, R. Comes, in *Charge Density Waves in Solids*, ed. by L.P. Gor'kov, G. Gruner (Elsevier Science, Amsterdam, 1989), p. 85
200. S. Barisic, A. Bjelis, in *Theoretical Aspects of Band Structures and Electronic Properties of Pseudo-One-Dimensional Solids*, ed. by H. Kaminura (D. Reidel, Dordrecht, 1985), p. 49
201. L.P. Gor'kov, I.E. Dzyaloshinskii, Sov. Phys. JETP **40**, 198 (1974)
202. P.A. Lee, T.M. Rice, R.A. Klemm, Phys. Rev. B **15**, 2984 (1977)

Part III

Unusual Properties of a Metallic Phase

13

Unusual Magic Angles Effects in Bechgaard Salts

W. Kang

We reviewed nontrivial angular magnetoresistance oscillations of Bechgaard salts, the representative quasi-one-dimensional conductors. The most frequently studied three compounds, $(TMTSF)_2ClO_4$, $(TMTSF)_2PF_6$, and $(TMTSF)_2ReO_4$, share the same background magnetoresistance and overall resonance structure regardless of external pressure and do not show dramatic change of metallic properties from one to another. However, weak periodic potential induced by anion ordering plays much more important role than it has been expected. It is responsible for the rich variety of Lebed resonances we have observed in various Bechgaard salts. The mechanism for the formation of a two-dimensional Fermi surface in $(TMTSF)_2FSO_3$ is unique and deserves further investigation.

13.1 Introduction

Extensive experimental and theoretical studies have been made on the angular magnetoresistance oscillations in various organic conductors with low-dimensional electronic structures. Several distinct resonance effects have been reported depending on the rotating plane of magnetic field and on the sample dimensionality [1–9]. They often gave a novel way to determine directly some band parameters.

Before starting this paper, it is useful to review briefly the two representative resonance phenomena of the low-dimensional electrons. Lebed and Bak predicted that quasi-one-dimensional metals would exhibit magnetoresistance peaks at special orientations of the magnetic field in the *bc*-plane where the periods of electron motions along the k_b and k_c directions on an open Fermi surface sheet are commensurate [1, 10]. The so-called magic angles are given for the triclinic lattice by

$$\tan\theta = \frac{p}{q}\frac{b\sin\gamma}{c\sin\beta\sin\alpha^*} - \cot\alpha^*, \tag{13.1}$$

in which θ denotes the angle that the magnetic field makes from the c^* direction, p and q are small integers, b, c, β, γ, and α^* are lattice parameters [5,6]. From its nature, such a behavior (called Lebed resonance hereafter) is present only in the systems with quasi-one-dimensional Fermi surfaces, such as $(TMTSF)_2X$ and α-(BEDT-TTF)KHg$(SCN)_4$.

On the other hand, in the case of quasi-two-dimensional metals with the cylindrical Fermi surface with weak modulation along the least conducting k_c direction, there are special angles at which the distribution of the Fermi surface cross section disappears completely. The electron eigenenergies near the Fermi surface are then fully quantized into Landau levels and the magnetoresistance is very much enhanced [3]. More realistic analysis with a consideration of the general triclinic structure was developed by Kartsovnik et al. [4]. The angles where the magnetoresistance has maxima are given in the simplest form when a magnetic field rotates in a plane containing the symmetry axis, as [3,4]:

$$ck_{\mathrm{F}}\,|\tan\theta_N| = \pi\left(|N| - \frac{1}{4}\right) \pm C(\phi), \tag{13.2}$$

where c is the distance between adjacent conducting planes, k_{F} is the average Fermi wavevector parallel to the major axis, N is the index of resonance, θ_N is the angle for the Nth peak measured from the axis of the cylindrical Fermi surface, and $C(\phi)$ is a function of azimuthal angle ϕ and disappears for the tetragonal system. The Fermi wavevector k_{F} can be readily determined quantitatively from the $\tan\theta_N$ versus N plot in (13.2). However, the nature of angular magnetoresistance oscillations in quasi-two-dimensional systems are more likely semiclassical without involving the Landau quantization, because they are observed at relatively high temperature or in dirty samples. Both numerical and analytic calculation of interlayer conductivity reproduced the same angular magnetoresistance as those observed experimentally [11,12].

Most of the early results on the Lebed resonance in $(TMTSF)_2ClO_4$ [5,6] and then in $(TMTSF)_2PF_6$ [13] were understood with (13.1), at least qualitatively. However, the more experimental results have been filed, the more unusual properties continue to be discovered in various compounds and in various conditions. We will discuss in this paper those unusual properties in the angular magnetoresistance of Bechgaard salts.

13.2 Fractional Magic Angle Effects in $(TMTSF)_2ReO_4$

At ambient pressure, $(TMTSF)_2ReO_4$ undergoes a metal–insulator transition at 180 K driven by an anion ordering with a characteristic wavevector (1/2, 1/2, 1/2). External pressure suppresses the (1/2, 1/2, 1/2) anion ordering and encourages formation of the anion ordering with another wavevector (0, 1/2, 1/2). According to the refined P–T phase diagrams, the metallic state remains stable down to low temperature at pressure above 10.5 kbar and the superconducting transition occurs at $T_c \sim 1.6$ K [9]. It is interesting to note

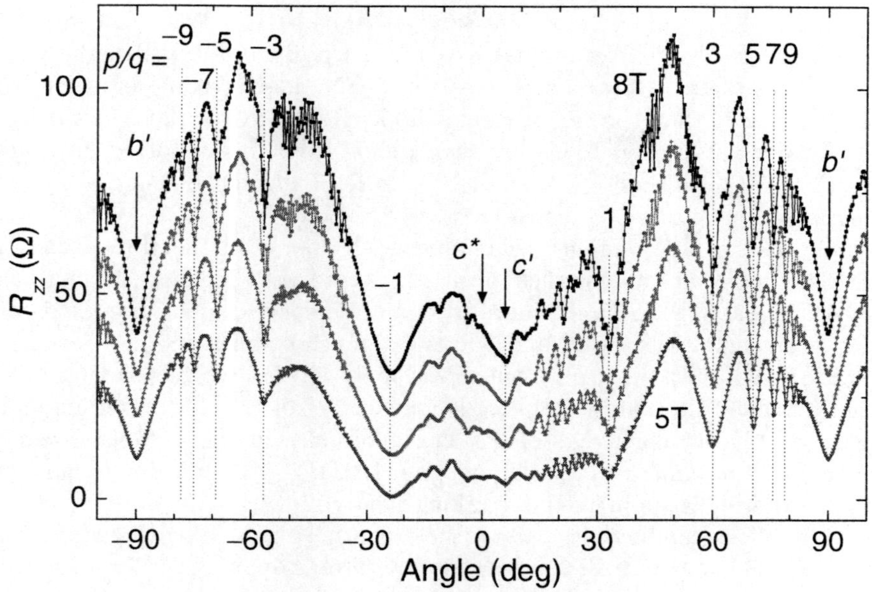

Fig. 13.1. Angle dependence of R_{zz} of $(TMTSF)_2ReO_4$ when the magnetic field rotates in the $b'c^*$-plane. The field was increased from 5 to 8 T in increments of 1 T. Successive curves are shifted by 10 Ω for clarity. Applied pressure was 12.2 kbar and temperature was 1.5 K. Both rapid oscillations and resonances are well developed. The resonances occur exclusively where $p/q =$ odd integers, as indicated in the figure

that the effect of superlattice formation simply cancels in (13.1) for the (0, 1/2, 1/2) superlattice, because both the lattice parameters along the b- and c-axes are doubled.

Figure 13.1 shows the angular magnetoresistance of $(TMTSF)_2ReO_4$ at 12.2 kbar with the magnetic field rotating in the $b'c^*$-plane [14]. The shape of the angular magnetoresistance curves appears very complicated between $-60°$ and $+60°$ due to the Shubnikov–de Haas like oscillations (also called rapid oscillations). Although the rapid oscillations are observed in most superconducting $(TMTSF)_2X$ salts in some conditions, $(TMTSF)_2ReO_4$ is special in that they are clearly observed in both the angular magnetoresistance and in the conventional magnetoresistance. The frequency of the rapid oscillations can be obtained by tracing the resistance as a function of $H_{c^*} = H \cos\theta$. For the curves in Fig. 13.1, the frequencies of the positive(16°–58°) and negative (−56° to −35°) angles are 312 ± 2 T and agrees with the value obtained from conventional magnetoresistance measurements [15].

In addition to the rapid oscillations, the angular magnetoresistance curves for $(TMTSF)_2ReO_4$ show clear resonance dips. The largest dip appears at $-24°$, corresponding to $p/q = -1$, and a similarly large dip appears at $+33°$, corresponding to $p/q = +1$. Ignoring the rapid oscillations, the angular magnetoresistance between $\pm 40°$ resembles that previously reported for

$(TMTSF)_2PF_6$ under low pressure (<9 kbar) [16, 17]. However, the $(TMTSF)_2ReO_4$ curve continues to show sharp dips up to very large values of θ (more than 11 dips in each direction). The most striking feature in this set of data is that the magic angles appear only where the p/q values in (13.1) are odd integers. This behavior differs from that reported previously for $(TMTSF)_2ClO_4$ and $(TMTSF)_2PF_6$, both of which show magic angles at any integer values of p/q in (13.1).

When the ReO_4 anions order alternately along all the three axes, the TMTSF molecules adjust their positions accordingly, making the molecular sites unequal. TMTSF molecules are mainly displaced along the c-axis and displacements along the a- and b-axes are much smaller; moreover, pairs of TMTSF molecules in a unit cell move apart or closer together laterally in the $b'c^*$-plane alternately [18]. In the case of the (0, 1/2, 1/2) ordering under pressure, the arrangements of TMTSF molecules in the $b'c^*$-plane can be regarded to be similar to those under the (1/2, 1/2, 1/2) ordering because the anion ordering is suppressed only along the a-axis.

The c-axis conductivity at zero field can be regarded as the summation of contributions along all the possible interlayer hoppings [19, 20]. If a finite magnetic field rotates in the $b'c^*$-plane, electron motion along the field direction dominates, with electron motion along other directions being suppressed by the component of the field perpendicular to those directions. Such an argument can be tested for $(TMTSF)_2ReO_4$ from the angular dependence of the conductance ($\sigma_{zz} \sim 1/R_{zz}$), Fig. 13.2. It exhibits peaks at magic angles and is negligibly small elsewhere, in good agreement with previous results for $(TMTSF)_2PF_6$ [19, 20]. Results for both of the compounds provide support for the above-mentioned theory, namely that the c-axis conductivity at zero field can be regarded as the summation of contributions along all the possible interlayer hoppings.

Now, let us try to understand the absence of the Lebed resonances for $p/q =$ even integers. The interchain electron motion along some representative directions depicted again in Fig. 13.3a on a simplified scheme of the arrangement of TMTSF molecules in $(TMTSF)_2ReO_4$ projected in the $b'c^*$-plane. Arrows represent a pair of TMTSF molecules and their directions indicate two different molecular rearrangements. When the magnetic field is applied in the direction of line $p/q = 1$, the electrons move along the TMTSF molecule pairs with the same lateral displacement, forming a uniform molecular chain. In the case of line $p/q = 2$, however, the electrons must move along TMTSF molecule pairs that are alternately displaced. TMTSF molecule pairs along this line form a dimerized molecular chain, which is vulnerable to the formation of a kind of Peierls gap. The formation of such a gap means that electrons can barely move along the line $p/q = 2$ at low temperature, despite it being a major crystallographic orientation. As a result, the conductivity is significant only when the field is parallel to one of the lattice directions along which all the TMTSF molecular pairs are equivalent (solid lines in Fig. 13.3a). Using the general convention that we have used for p and q in (13.1), the solid lines

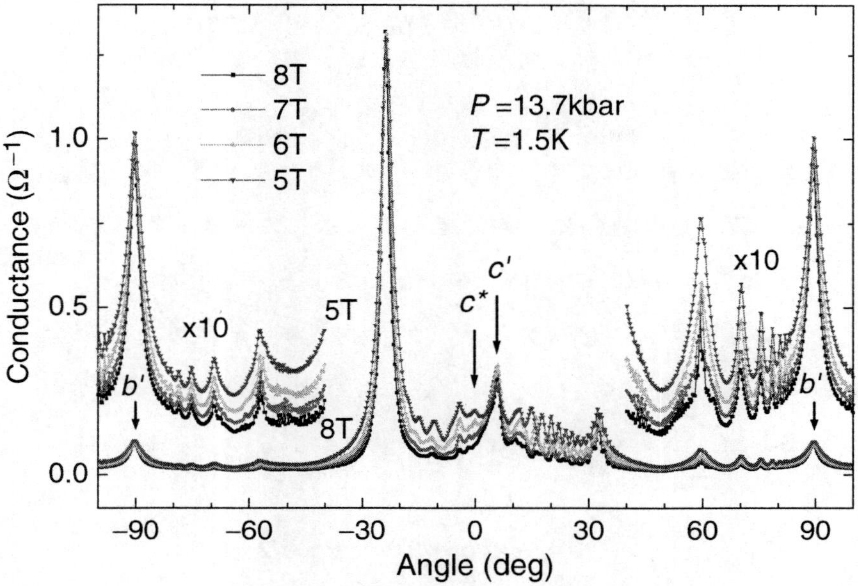

Fig. 13.2. The angle-dependent magnetoconductance obtained by inversion of R_{zz} in Fig. 13.1. The conductance data at large angles are also shown magnified by a factor of ten

satisfy the condition that both p and q are odd integers. This simple analysis explains successfully the absence of the Lebed resonance in $(TMTSF)_2ReO_4$ when either p or q is an even integer.

The validity of our model can also be tested for $(TMTSF)_2ClO_4$ with anion ordering only along the b-axis, shown schematically in Fig. 13.3b. Again, dashed lines represent the directions in which electron motion suffers from a dimerization gap and therefore the resonance effect is expected to be negligible. A local minimum in R_{zz} is anticipated when the magnetic field is aligned parallel to one of the solid lines. The q values for the solid lines are always odd integers, but not the dotted lines. When $q = 1$, the resonance effect can occur for all the integer values because both odd and even integers are possible for p, which agrees with previous experimental results [5, 6]. Although $(TMTSF)_2ClO_4$ appears similar to $(TMTSF)_2PF_6$ on replacing lattice parameter b with $2b$, the difference between the two samples is exposed when $p/q =$ nonintegers. For instance, $p/q = 1/3$ is allowed in $(TMTSF)_2ClO_4$ but $p/q = 1/2$ is not, whereas both are allowed in $(TMTSF)_2PF_6$. Experimentally, it is difficult to discern the sub-resonances with $q > 1$ because the conductivity of $(TMTSF)_2X$ along the c-axis is about 1,000 times smaller than that along the b-axis [6]. A substantially large field is necessary to observe resonances for $|p/q| < 1$. However, $(TMTSF)_2ClO_4$ enters in the FISDW state above 6.1 T at 1.5 K and the magic angle effect with $|p/q| < 1$ becomes again difficult to discern [5].

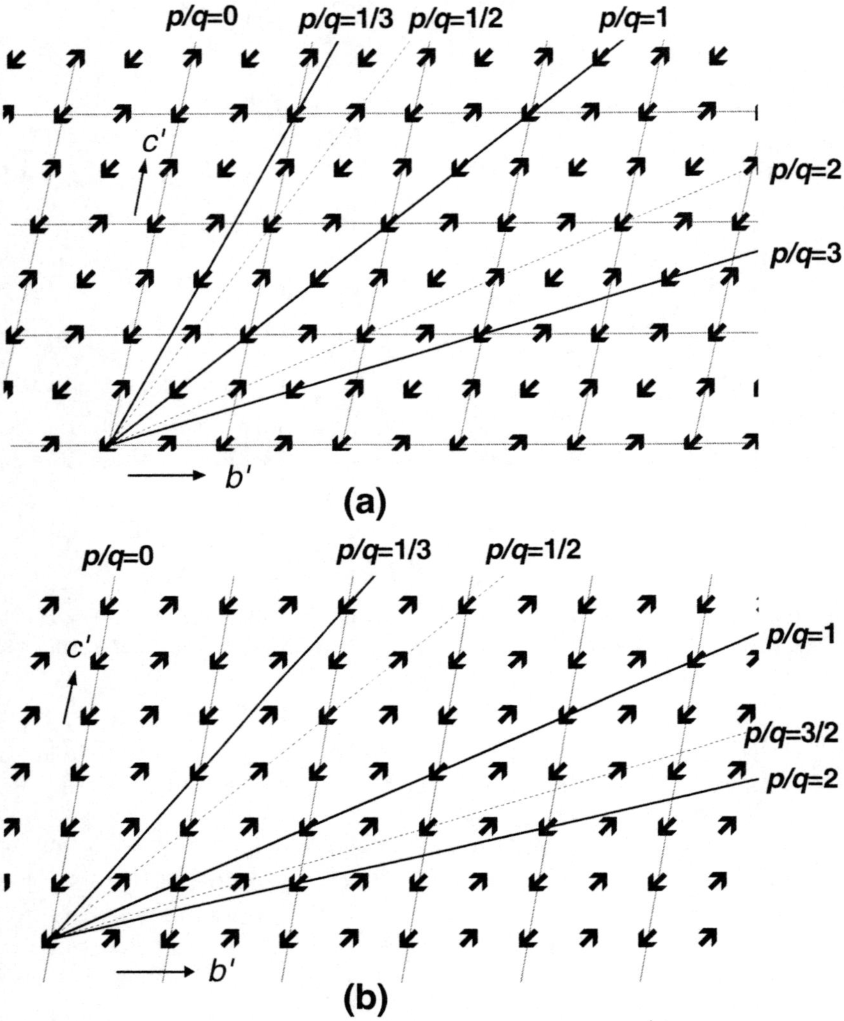

Fig. 13.3. Simplified schemes of molecular sites projected to the $b'c^*$-plane for (**a**) $(TMTSF)_2ReO_4$ and (**b**) $(TMTSF)_2ClO_4$ after (0, 1/2, 1/2) and (0, 1/2, 0) anion ordering, respectively. Each *arrow* indicates a pair of TMTSF molecules and its orientation represents the molecular displacement. There are two *arrow* types: (↙) represents a pair of TMTSF molecules that are closer to each other and (↗) represents molecules that have moved apart laterally in the $b'c^*$-plane. *Solid lines* join molecules with the same displacements and *dashed lines* with alternating displacements

Contrary to the behavior of $(TMTSF)_2ClO_4$ and $(TMTSF)_2PF_6$, the resonance effects for $p/q =$ nonintegers are readily detected in $(TMTSF)_2ReO_4$. Although the peaks due to sub-resonance effects are small and overwhelmed by the rapid oscillations, those peaks that are out of phase with the rapid

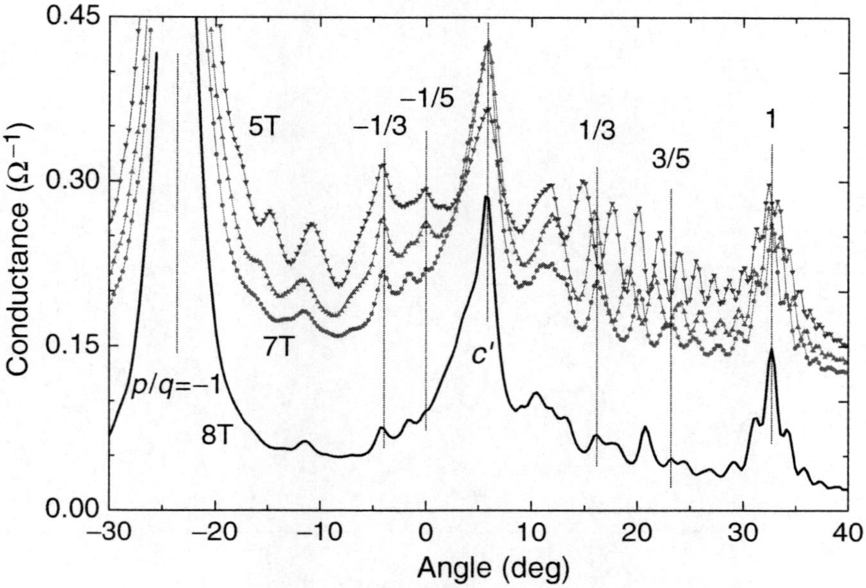

Fig. 13.4. Plot of σ_{zz} between $-30°$ and $+40°$. Sub-resonances are distinguished from rapid oscillations by their fixed positions. A small offset is applied to the upper three curves to make the lowest curve clear

oscillations, and whose positions do not move with increasing field, can be easily distinguished from the rapid oscillations in the conductance plot (Fig. 13.4). An increasing number of sub-resonances can be distinguished as the magnetic field is increased. At 13.5 T, sub-resonances with $|p/q| > 1$ as well as those with $|p/q| < 1$ were observed, even in the $R_{zz}(\theta)$ plot (Fig. 13.5). In Fig. 13.5, the sub-resonances are marked with arrows together with the corresponding p/q values. These sub-resonances provide further confirmation that the magic angle peaks appear only when both p and q are odd integers for the (0, 1/2, 1/2) anion ordering superlattice.

To understand fully the properties of $(TMTSF)_2ReO_4$, a mechanism that turns on the conducting states at the Fermi level whenever the magnetic field is aligned along a resonance direction and that switches them off if the field is not appropriately aligned is required [19, 20].

In summary, the Lebed resonances observed in the angular magnetoresistance of $(TMTSF)_2ReO_4$ showed obvious differences from those previously reported for $(TMTSF)_2ClO_4$ and $(TMTSF)_2PF_6$. The angular magnetoresistance of $(TMTSF)_2ReO_4$ showed a remarkably large number of resonances and sub-resonances, enabling us to make a detailed study of these features. Unlike other organic conductors of the form $(TMTSF)_2X$, $(TMTSF)_2ReO_4$ exhibits Lebed magic angles only when both p and q are odd integers. This effect can be accounted for by considering the very small but finite molecular displacements inside a unit cell due to anion ordering. When electrons move

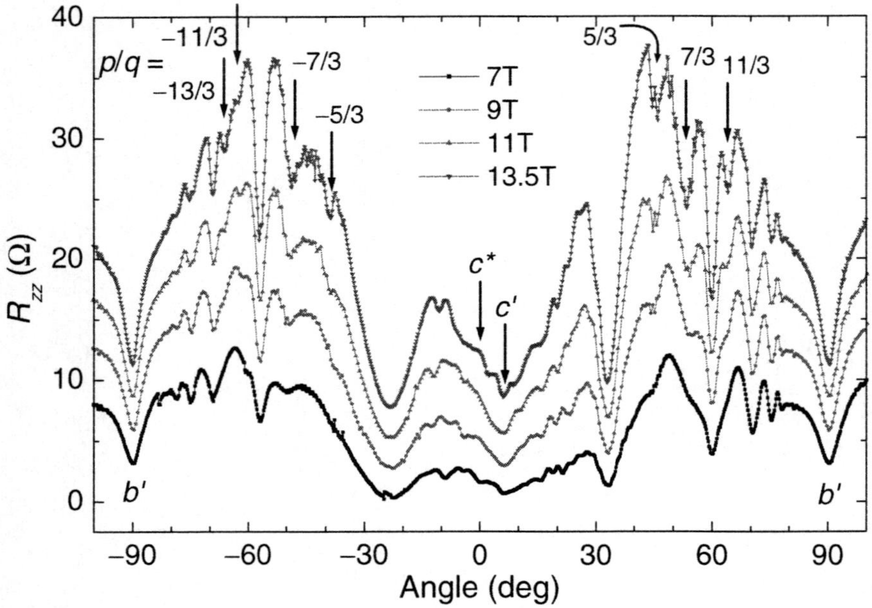

Fig. 13.5. Angular magnetoresistance of a $(TMTSF)_2ReO_4$ sample above 7 T under 12.0 kbar and at 1.5 K. Successive curves are shifted by 2.5 Ω for clarity. Only some sub-resonance dips for $|p/q| > 1$ are indicated, where both p and q are again odd integers

along the direction of alternating TMTSF molecular displacements, a type of dimerization gap prevents coherent motion. As a result, coherent interlayer electron motion survives only in those directions for which both p and q are odd integers, and hence the molecular displacements are identical. Manifestation of higher order warping of the Fermi surface due to weak anion ordering potential along the second and the third conduction direction and the observation of well-developed resonances at angles as large as 83° ($p/q = 13$) indicates that the Bechgaard salts are electronically extremely clean systems and that the electron–electron correlation is very important in these systems.

13.3 Two Kinds of Angular Magnetoresistance Resonances of $(TMTSF)_2PF_6$: Pressure Dependence or Sample Dependence

$(TMTSF)_2PF_6$ has long been considered as ideal because it has the simplest band structure among Bechgaard salts. Angular magnetoresistance resonances are of course readily observed in this compound. However, the background of R_{zz} in $(TMTSF)_2PF_6$ showed an anomalous pressure dependence. Considering only the background, R_{zz} at relatively low pressure has a minimum around c^* direction, increases as the field moves toward $\pm b'$ direction, and develops a sharp minimum around the $\pm b'$ directions [16, 19]. (We will call such a behavior type I behavior.) On the other hand, at high pressure, angular magnetoresistance has a $(\cos)^{\alpha}$-like ($1 < \alpha < 2$) background superposed by resonance dips around c' direction [19,21], very similar to that of the intraplane resistance R_{xx} (type II behavior). A crossover from type I to type II has been reported to take place somewhere between 6.0 and 8.3 kbar [16, 21] or between 8.5 and 9.2 kbar [19].

Chashechkina and Chaikin suggested that the zero-field interlayer conductivity could be described as the sum of contributions from all hopping directions between neighboring molecular chains which is independently removed by a perpendicular field [19]. This model is strongly supported by the observation of only odd-indexed Lebed resonance in $(TMTSF)_2ReO_4$ [14] discussed in Sect. 13.2. With an additional anomalous in-plane contribution of the order of t'_b, the unnested bandwidth, which is destroyed by a field perpendicular to the b-axis, they could not only successfully fit the angular dependence at fixed fields but also explain the crossover from type I to type II background magnetoresistance [19].

However, the type II background of R_{zz}, being at the center of the FL-NFL transition theory [22, 23] does hardly conciliate any semi-classical theory [24–26]. In the simplest form, the Lorentz force is always zero for a field parallel to a current and maximum for a field perpendicular to the current, which favors the type I behavior for R_{zz}. Such an effect can be ignored only when the coherence in the c direction is wiped out. Therefore, the crossover of background magnetoresistance must imply a drastic change of electron transport mechanism even at a moderate pressure.

A systematic study of angular dependent interlayer resistance R_{zz} under various pressures as well as magnetic fields must be performed to elucidate the mechanism of the crossover. Special attention should be paid to perform as many different pressure experiments as possible with the same sample to exclude any ambiguity which might be caused by sample dependence. The experimental result shows that sample dependence rather than pressure dependence is more plausible for the observed crossover.

Figure 13.6 shows the temperature dependence of resistance of three samples, one for R_{xx} and two for R_{zz} [27]. Measurements were performed in the order of 11.3, 7.5, 6.7, and 11.8 kbar. R_{xx} decreases monotonically with temperature regardless of pressure while R_{zz} of the sample #3 develops a pronounced maximum at an intermediate temperature. As pressure increases, the amount of resistance increase lessens and the temperature of maximum resistance ($T_{\max}$) shifts toward higher temperature with a good agreement with the results in [28] and [29]. Although resistances of both R_{zz} samples rise with cooling, the ratios of resistance at maximum to that at room temperature are very much different from each other, by 1.02 and 1.30, respectively, at 7.5 kbar. Electron motions along two conducting chains are incoherent above $T_{\max}$ and reduction of the interchain electron tunneling along the c-axis is responsible

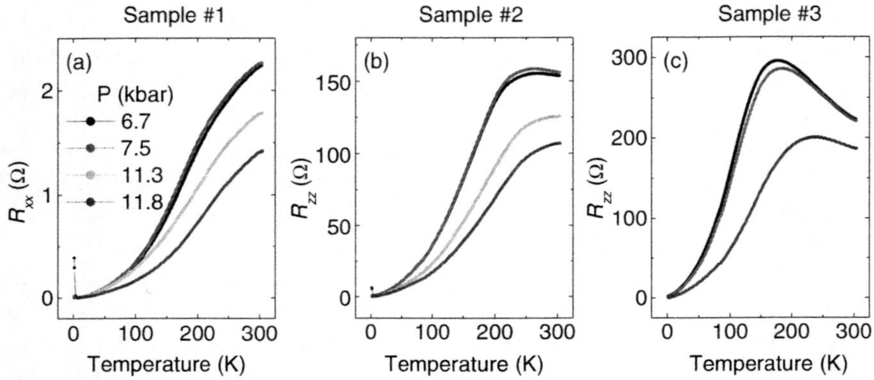

Fig. 13.6. Temperature dependence of resistance for the first three samples under four different pressures. $R_{xx}(T)$ decreases monotonically with temperature while $R_{zz}(T)$ is characterized with a maximum at intermediate temperature. Two graphs of $R_{zz}(T)$ are contrasted by the amount of resistance increase at maximum. $R_{\mathrm{max}}/R_{\mathrm{RT}} = 1.02$ and 1.30, respectively, for (**b**) and (**c**) under 7.5 kbar. Although the curves at 6.7 and at 7.5 kbar are quite similar, the latter could be distinguished by the absence of the field-induced spin–density waves until 8 T

for the resistance increase on cooling [29]. Therefore, it is bizarre that R_{zz} of sample #2 shows only a little change with lowering temperature.

Presented in Fig. 13.7 is angular magnetoresistance of the three samples measured with a field rotating in the $b'c^*$-plane at four different pressure values. All the data except the first row were obtained at a fixed field of 8 T. At 6.7 kbar, the pressure is slightly so low that a field-induced spin–density wave state is beginning to develop already at 8 T and the angular magnetoresistance under 6 T is presented instead. Difference of angular magnetoresistance among samples is obvious. While the type II background (superposed by Lebed resonance dips) is dominant for the first two samples, resistance for the sample #3 is dominated by the type I background. At this point, we need to remind ourselves that all the measurements were carried out with the same set of samples over the broadest pressure range ever studied, yet there is no pressure induced crossover in the background of R_{zz}. Instead, two characteristic backgrounds are independently observed in two different samples for all over the pressure range; the sample #2 conserves the type II background whereas the sample #3 the type I background until 11.8 kbar. This is a clear contrast to the previous reports [16, 19, 21] where a pressure-induced crossover from type I to type II background was mentioned.

Now, it is necessary to explain the behavior of sample #2. First, although its electrical wires were arranged to measure R_{zz}, the angular magnetoresistance background is akin to that of R_{xx}. We also notice that the temperature dependence of R_{zz} of sample #2 lies between that of R_{xx} (sample #1) and that of R_{zz} (sample #3). Increase of resistance down to T_{max} is merely 2%

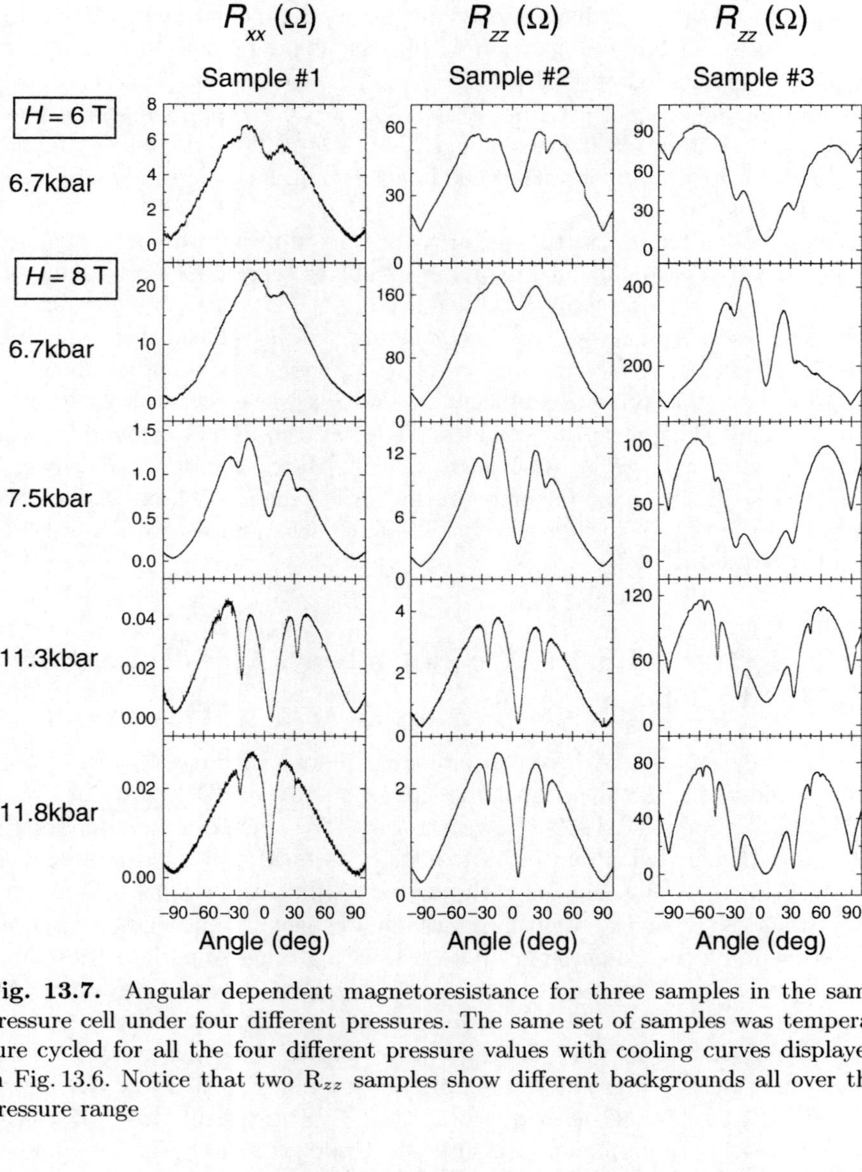

Fig. 13.7. Angular dependent magnetoresistance for three samples in the same pressure cell under four different pressures. The same set of samples was temperature cycled for all the four different pressure values with cooling curves displayed in Fig. 13.6. Notice that two R_{zz} samples show different backgrounds all over the pressure range

even at the lowest pressure of 6.7 kbar. Finally, overall resistance of sample #2 in various experiments is about one-tenth of that of sample #3, again in-between R_{xx} and R_{zz} while all the samples studied here have the similar size. So, resistance of the sample #2 is suspected to measure a kind of mixture of R_{xx} and R_{zz}. In fact, the current path in the sample hardly well-defined and R_{zz} is inevitably contaminated with R_{xx}. Another possibility is that the crystal is internally twinned. As the current always flows along the lowest possible resistance path, it is natural that both $R_{zz}(T)$ and $R_{zz}(H)$ of this sample show similar behaviors to those of R_{xx}. Calculating $\sigma_{zz}(H,\theta)$ from semiclassical Boltzmann equation within the single-relaxation-time approximation, it is anticipated that $\rho_{zz}(H,\theta) \propto \rho^0_{zz}[1+(\omega_c\tau)^2\sin^2\theta]$ where θ is the angle of magnetic field with the c^* axis and $\omega_c = eHv_{\mathrm{F}}c/h$ is the frequency with which an electron traverses the Brillouin zone along the c-axis in a magnetic field.[1] It supports also the type I background for $R_{zz}(\theta)$, giving a simple and clean explanation.

However, after failing to find any pressure induced crossover of angular magnetoresistance behavior in spite of extensive search, we may exclude the possibility of crossover from type I to type II of R_{zz} when the external pressure increases. Absence of crossover was also verified in another compound, $(TMTSF)_2ReO_4$. Although some of the R_{zz} configured samples showed an R_{xx}-like behavior, pressure induced crossover has never been observed in any samples [30]. Therefore, it is evident that the type I background of angular magnetoresistance is intrinsic to the interlayer resistance R_{zz} even in $(TMTSF)_2PF_6$. As a result, many of the earlier analyses based on the observation of the type II background in $(TMTSF)_2PF_6$ and its crossover need to be revisited [19, 22, 23].

13.4 Bechgaard Salts Are Not Always One-Dimensional: $(TMTSF)_2FSO_3$

The Bechgaard salts with centrosymmetric anions such as $X = PF_6$, AsF_6, SbF_6, undergo a metal to insulator transition to the SDW state at around 12 K during cooling [31, 32]. However, the SDW transition is suppressed at pressures higher than about 6 kbar, and the SC state appears at temperatures below $T_c = 1.2$ K [33]. In salts with noncentrosymmetric anions such as ClO_4, ReO_4, and NO_3, on the other hand, the anions usually undergo orientational ordering and form a superlattice potential with a range of periodicities. When the ClO_4 salt is slowly cooled at ambient pressure, the anion ordering (AO) with wavevector $\mathbf{q}_1 = (0, 1/2, 0)$ occurs below 24 K [34] and the SC state stabilizes below 1.4 K [35]. Thermal quenching of the same system leads to the insulating SDW ground state with disordered anions below 6 K [36]. In the ReO_4 salt, the AO with $\mathbf{q}_2 = (1/2, 1/2, 1/2)$ [18] leads to an insulating state below 180 K at ambient pressure [37]. Under pressures higher than about 8 kbar, another type of AO with $\mathbf{q}_3 = (0, 1/2, 1/2)$ dominates [38] and the SC state emerges below 1.3 K [39]. In the NO_3 salt, the planar NO_3 anions order with $\mathbf{q}_4 = (1/2, 0, 0)$ at around 45 K [40]; the metallic nature of this salt remains intact through this transition, and on further cooling a transition to a SDW state starts at around 10 K [41]. The metallic state is recovered above

1 This equation is the simplified form of (11) of [25] by putting $t_b/t_a \sim 0$.

8 kbar without any sign of superconductivity for pressures up to 24 kbar and temperatures down to 50 mK [42].

Additional effects due to permanent electric dipole moments residing on the anions have been studied for the salt $(TMTSF)_2FSO_3$ [43–45]. The tetrahedral structure of the FSO_3 anion, in which the sulfur atom lies at the center, resembles the structures of other tetrahedral anions such as ClO_4 and ReO_4. However, the asymmetric distribution of two types of atoms with different electronegativity at the four apices in FSO_3 causes this anion to have a permanent electric dipole moment. At ambient pressure, $(TMTSF)_2FSO_3$ undergoes a metal to insulator transition at around 89 K with formation of the anion superstructure with $\mathbf{q}_2 = (1/2, 1/2, 1/2)$ [46]. However, the structural ordering of the FSO_3 anions can only be considered complete when their electric dipole moments are ordered. No information is currently available on the ordering of dipole moments in $(TMTSF)_2FSO_3$.

$X = FSO_3$ anions are unique in that they carry permanent electric dipole moments [43,44]. The first studies showed only a simple pressure–temperature phase diagram in which the physical ground states were often not well defined [44,47]. However, our recent reinvestigation showed that $(TMTSF)_2FSO_3$ has a very reproducible and very complex phase diagram [48], which is presumably related with the additional degree of freedom given by the electric dipole moments.

Angular magnetoresistance of $(TMTSF)_2FSO_3$ under 6.2 kbar and at 1.8 K for five different magnetic fields is shown in Fig. 13.8a [49]. Oscillations are clearly seen superposed on slowly varying background. Strikingly, their overall feature is quite different from the Lebed oscillations observed in other Bechgaard salts [5, 6, 13, 30]. Neither the angular positions of dips nor those of peaks correspond to the angles predicted for the Lebed resonance (indicated with broken arrows on the 13.5 T curve).

Surprisingly, the angular positions of resistance peaks in Fig. 13.8a are well fitted by (13.2), as shown in Fig. 13.8b. In fact, the overall angular magnetoresistance is very similar to those observed in the textbook examples of quasi-two-dimensional electron systems such as β-$(BEDT\text{-}TTF)_2X$, $X = IBr_2$ [4] and I_3 [50]. Figure 13.8b reveals some interesting facts. First, the left side can be fitted with $\tan\theta_N = 1.712(N - 0.0713)$ and the right side with $|\tan\theta_N| = 1.712(|N| - 0.495)$. The slopes of two sides agree with each other remarkably well and give 0.139 Å^{-1} for the Fermi wavevector. Second, the phases or the intercepts with x-axis of the linear fittings are far different from the standard Yamaji formula and strongly asymmetric. This reflects the triclinic structure of this compound for which the corrugation is not symmetric.

Another important result suggesting the quasi-two-dimensional Fermi surface in $(TMTSF)_2FSO_3$ is the narrow peak of width 4.0° observed when the magnetic field nearly parallel to the b'-axis, shown enlarged in the inset of Fig. 13.8b. The existence of a small belly due to the third directional warping of otherwise two-dimensional cylindrical Fermi surface, that is, the coherent

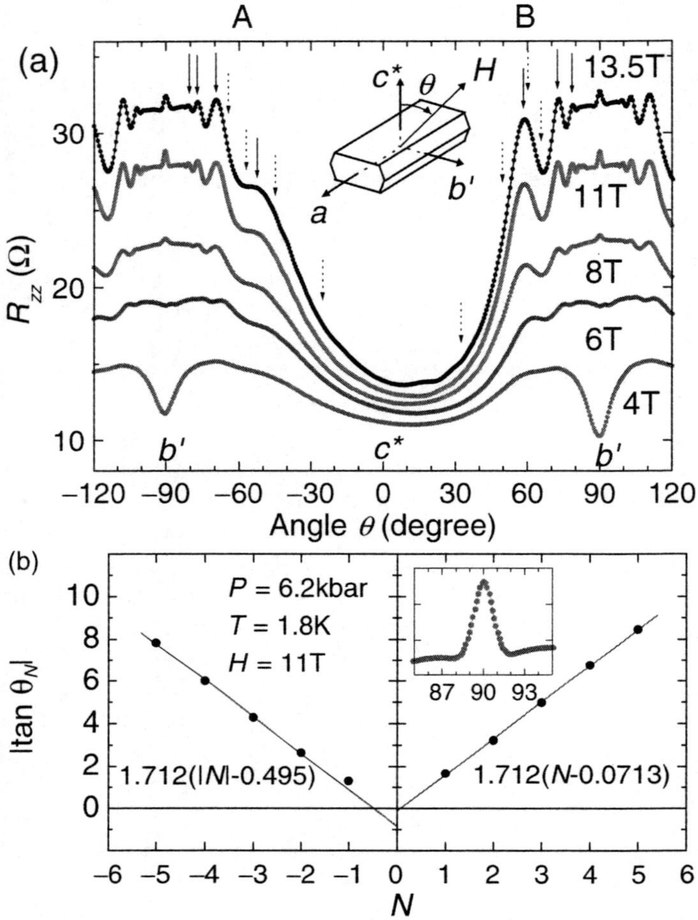

Fig. 13.8. (**a**) Angular magnetoresistance of $(TMTSF)_2FSO_3$ under 6.2 kbar and at 1.8 K. The angles predicted by (13.1) are indicated with *broken arrows*. The peak of the 11 T curve around b' direction is enlarged in (**b**). (**b**) Plot of $|\tan\theta_N|$ versus N, where θ_N are angles for the peaks in angular magnetoresistance (*solid arrows* in (**a**)). Parts A and B correspond to the sides for negative and positive N, respectively. Lines are the least square fittings. The point for $N = -1$ is excluded because its angular position is difficult to be determined precisely

interlayer transport is the origin of this peak [50–52]. The half width of the peak is related to the transfer integral along the third direction such as

$$\frac{\pi}{2} - \theta_{\mathrm{c}} = \frac{2m^* t_c c}{\hbar^2 k_{\mathrm{F}}}, \tag{13.3}$$

where θ_{c} is the critical angle above which the resistance increases rapidly [50]. Using the values of m^* and k_{F} obtained by the temperature dependence of

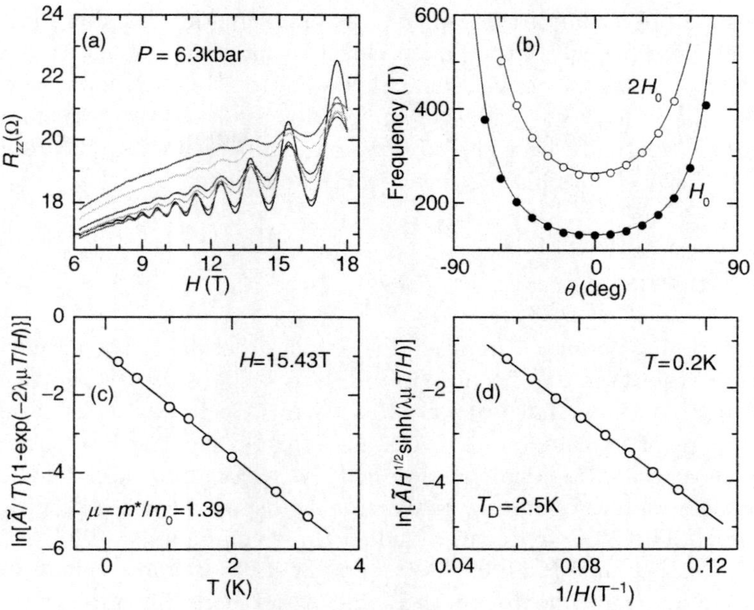

Fig. 13.9. (**a**) Magnetoresistance, $R_{zz}(H)$ at temperatures 0.2, 1.0, 1.6, 2.0, 2.7, and 3.5 K from below. (**b**) Plot of two lowest harmonic frequencies versus the angle between the magnetic field and the c^*-axis. *Solid lines* are fittings to $nH_0/\cos\theta$. (**c**) Fitting to the Lifshitz–Kosevich formula with $\mu = 1.39$ as a function of temperature. $\tilde{A} = \Delta R/R$ is the oscillating amplitude of resistance normalized by background resistance at 15.43 T. (**d**) Fitting to the Lifshitz–Kosevich formula as a function of $1/H$ at 0.2 K

amplitudes of Shubnikov–de Haas oscillations (see below) and the fitting to Eq. (13.3), respectively, we have $t_c \simeq 1.01 \times 10^{-3}$ eV, which can be compared with values reported by Danner et al. [7] from the fitting of ρ_{zz} of $(TMTSF)_2ClO_4$ in the *ac*-rotation.

The third argument that supports the idea of two-dimensional Fermi surface in $(TMTSF)_2FSO_3$ is the temperature and field dependence of Shubnikov–de Haas oscillations (Fig. 13.9). Oscillations arising from conventional closed orbits are in general well interpreted with the Lifshitz–Kosevich formula [53]. In a simplified form in which only the first harmonic is considered and spin-splitting effect is ignored, the Lifshitz–Kosevich formula states that the oscillation amplitude $\Delta R/R$ can be represented as

$$\Delta R/R \propto T \exp(-\lambda\mu T_{\mathrm{D}}/H)/\sqrt{H} \sinh(\lambda\mu T/H), \tag{13.4}$$

where $\lambda = 2\pi^2 k_{\mathrm{B}} m_0/e\hbar = 14.7\,\mathrm{T\,K^{-1}}$, μ is the effective cyclotron mass in relative units of the free electron mass m_0, and T_{D} is the Dingle temperature. Although Shubnikov–de Haas-like oscillations are frequently observed in magnetoresistance of other $(TMTSF)_2X$, they do not arise from closed electron orbits but from a complex interplay between magnetic breakdown and Bragg reflection between the warped open Fermi surfaces [54]. While they are also periodic in $1/H$ as the conventional Shubnikov–de Haas oscillations, the temperature dependence of oscillation amplitude has one or two maxima and the oscillation eventually vanishes at very low temperature. Application of the Lifshitz–Kosevich formula has never worked for them.

13.5 Summary

Weak periodic potential induced by anion ordering plays a much more important role than has been expected. It is responsible for the richness of Lebed resonances we have observed in various Bechgaard salts. Systematic investigation of the compounds discussed in this review especially as a function of hydrostatic and uniaxial pressure will be particularly useful for the further understanding of interlayer electron transport in a metal with strong anisotropy. The most frequently studied three compounds, $(TMTSF)_2ClO_4$, $(TMTSF)_2PF_6$, and $(TMTSF)_2ReO_4$, do not show dramatic change of metallic properties from one to another, and they share the same background magnetoresistance and overall resonance structure. Then, the rich variety of resonant features in detail must be attributed to the different anion ordering potential in each compound. The mechanism for the formation of a two-dimensional Fermi surface in $(TMTSF)_2FSO_3$ is unique and deserves further investigation.

Acknowledgments

Most of the work presented here was performed in collaboration with H.Y. Kang, Y.J. Jo. Work on $(TMTSF)_2FSO_3$ and work on $(TMTSF)_2ReO_4$ were performed in collaboration with O.H. Chung and S. Uji, respectively. Over the years, facilities at various places such as Ewha Womans University (Seoul, Korea), Korea Basic Science Institute (Daeduk, Korea), National Institute for Materials Science (Tsukuba, Japan), and National High Magnetic Field Laboratory (Tallahassee, USA) have been used. Research grants came from the Korea Science and Engineering Foundation and the Korea Research Foundation.

References

1. A.G. Lebed, Pis'ma Zh. Eksp. Teor. Fiz. **43** 137 (1986) [JETP Lett. **43**, 174 (1986)]
2. A.G. Lebed, N.N. Bagmet, M.J. Naughton, Phys. Rev. Lett. **93**, 157006 (2004)
3. K. Yamaji, J. Phys. Soc. Jpn. **58**, 1520 (1989)

4. M.V. Kartsovnik, V.N. Laukhin, S.I. Pesotskii, I.F. Schegolev, V.M. Yakovenko, J. Phys. I **2**, 89 (1991)
5. T. Osada, A. Kawasumi, S. Kagoshima, N. Miura, G. Saito, Phys. Rev. Lett. **66**, 1525 (1991)
6. M.J. Naughton, O.H. Chung, M. Chaparala, X. Bu, P. Coppens, Phys. Rev. Lett. **67**, 3712 (1991)
7. G.M. Danner, W. Kang, P.M. Chaikin, Phys. Rev. Lett. **72**, 3714 (1994)
8. T. Osada, S. Kagoshima, N. Miura, Phys. Rev. Lett. **77**, 5261 (1996)
9. Y.J. Jo, H. Kang, W. Kang, Synth. Met. **120**, 1043 (2001)
10. A.G. Lebed, P. Bak, Phys. Rev. Lett. **63**, 1315 (1989)
11. R. Yagi, Y. Iye, T. Osada, S. Kagoshima, J. Phys. Soc. Jpn. **59**, 3069 (1990)
12. V.G. Peschansky, J.A.R. Lopes, T.G. Yao, J. Phys. I **1**, 1469 (1991)
13. W. Kang, S.T. Hannahs, P.M. Chaikin, Phys. Rev. Lett. **69**, 2827 (1992)
14. H. Kang, Y.J. Jo, S. Uji, W. Kang, Phys. Rev. B **68**, 132508 (2003)
15. W. Kang, J.R. Cooper, D. Jérome, Phys. Rev. B **43**, 11467 (1991)
16. I.J. Lee, M.J. Naughton, Phys. Rev. B **58**, R13343 (1998)
17. E.I. Chashechkina, P.M. Chaikin, Phys. Rev. Lett. **80**, 2181 (1998)
18. R. Moret, J.P. Pouget, R. Comes, K. Bechgaard, Phys. Rev. Lett. **49**, 1008 (1982)
19. E.I. Chashechkina, P.M. Chaikin, Phys. Rev. B **65**, 012405 (2001)
20. E.I. Chashechkina, I.J. Lee, S.E. Brown, D.S. Chow, M.J. Naughton, P.M. Chaikin, Synth. Met. **119**, 13 (2001)
21. I.J. Lee, M.J. Naughton, Phys. Rev. B **57**, 7423 (1998)
22. S.P. Strong, D.G. Clarke, P.W. Anderson, Phys. Rev. Lett. **73**, 1007 (1994)
23. D.G. Clarke, S.P. Strong, P.M. Chaikin, E.I. Chashechkina, Science **279**, 2071 (1998)
24. T. Osada, S. Kagoshima, N. Miura, Phys. Rev. B **46**, 1812 (1992)
25. K. Maki, Phys. Rev. B **45**, 5111 (1992)
26. P. Moses, R.H. McKenzie, Phys. Rev. B **63**, 024414 (2000)
27. H. Kang, Y.J. Jo, W. Kang, Phys. Rev. B **69**, 033103 (2004)
28. J.R. Cooper, L. Forró, B. Korin-Hamzić, K. Bechgaard, A. Moradpour, Phys. Rev. B **33**, 6810 (1986)
29. J. Moser, M. Gabay, P. Auban-Senzier, D. Jérome, K. Bechgaard, J.M. Fabre, Eur. Phys. J. B **1**, 39 (1998)
30. H. Kang, Y.J. Jo, H.C. Kim, H.C. Ri, W. Kang, Synth. Metals **120**, 1051 (2001)
31. J.C. Scott, H.J. Pedersen, K. Bechgaard, Phys. Rev. Lett. **45**, 2125 (1980)
32. K. Mortensen, Y. Tomkiewicz, T.D. Schultz, E.M. Engler, Phys. Rev. Lett. **46**, 1234 (1981)
33. D. Jérome, A. Mazaud, M. Ribault, K. Bechgaard, J. Phys. (France) Lett. **41**, L-95 (1980)
34. J.P. Pouget, G. Shirane, K. Bechgaard, J.M. Fabre, Phys. Rev. B **27**, 5203 (1983)
35. K. Bechgaard, K. Carneiro, M. Olsen, F. Rasmussen, C.S. Jacobsen, Phys. Rev. Lett. **46**, 852 (1981)
36. T. Takahashi, D. Jérome, K. Bechgaard, J. Phys. (Paris) Lett. **43**, L565 (1982)
37. C.S. Jacobsen, H.J. Pedersen, K. Mortensen, G. Rindorf, N. Thorup, J.B. Torrance, K. Bechgaard, J. Phys. C **15**, 2651 (1982)
38. R. Moret, S. Ravy, J.P. Pouget, R. Comes, K. Bechaard, Phys. Rev. Lett. **57**, 1915 (1986)

39. S.S.P. Parkin, D. Jérome, K. Bechgaard, Mol. Cryst. Liq. Cryst. **79**, 213 (1981)
40. J.P. Pouget, R. Moret, R. Comes, K. Bechgaard, J. Phys. (Paris) Lett. **42**, L5203 (1981)
41. P. Baillargeon, C. Bourbonais, S. Tomić, P. Vaca, C. Coulon, Synth. Met. **27**, B83 (1989)
42. A. Mazaud, Ph.D. thesis, Université Paris-Sud, Orsay, 1981
43. F. Wudl, E. Aharon-Shalom, D. Nalewajek, J.V. Waszczak Jr., W.M. Walsh Jr., L.W. Rupp, P. Chaikin, R. Lacoe, M. Burns, T.O. Poehler, M.A. Beno, J.M. Williams, J. Chem. Phys. **76**, 5497 (1982)
44. R.C. Lacoe, S.A. Wolf, P.M. Chaikin, F. Wudl, E. Aharon-Shalom, Phys. Rev. B **27**, 1947 (1983)
45. R.C. Lacoe, P.M. Chaikin, F. Wudl, E. Aharon-Shalom, J. Phys. (Paris), Colloq. **44**, C3-767 (1983)
46. R. Moret, J.P. Pouget, R. Comes, K. Bechgaard, J. Phys. (Paris), Colloq. **44**, C3–957 (1983)
47. S. Tomić, Ph.D. thesis, Université Paris-Sud, Orsay, 1986
48. Y.J. Jo, E.S. Choi, H. Kang, I.S. Seo, O.H. Chung, W. Kang, Phys. Rev. B **67**, 014516 (2003)
49. W. Kang, O.H. Chung, Y.J. Jo, H. Kang, I.S. Seo, Phys. Rev. B **68**, 073101 (2003)
50. N. Hanasaki, S. Kagoshima, T. Hasegawa, T. Osada, N. Miura, Phys. Rev. B **57**, 1336 (1998)
51. V.G. Peschansky, M.V. Kartsovnik, Phys. Rev. B **60**, 11207 (1999)
52. N. Hanasaki, S. Kagoshima, T. Hasegawa, T. Osada, N. Miura, Phys. Rev. B **60**, 11210 (1999)
53. D. Shoenberg, *Magnetic Oscillations in Metals* (Cambridge University Press, Cambridge, 1984)
54. S. Uji, J.S. Brooks, M. Chaparala, S. Takasaki, J. Yamada, H. Anzai, Phys. Rev. B **55**, 12446 (1997)

14

Versatile Method to Estimate Dimensionality of Q1D Fermi Surface by Third Angular Effect

H. Yoshino and K. Murata

This chapter describes a method to determine shape of quasi-one-dimensional (Q1D) Fermi surfaces by measuring the field-orientation dependence of the magnetoresistance. One observes a pair of minima of the magnetoresistance when the magnetic field is rotated within the most conducting plane of Q1D metals at low temperature ($\leq$4.2 K) and high magnetic field ($\geq$5 T). This phenomenon is called the "third angular effect (TAE)" and angular width of the minima is related to the ratio of bandwidth t_y/t_x. We consider physical origin of the TAE and its application to Q1D TMTSF and DMET salts. Two pairs of Q1D Fermi surfaces and pressure dependence of t_y/t_x in $(TMTSF)_2ClO_4$ are carefully discussed as important examples.

14.1 Third Angular Effect

Third angular effect (TAE) is one of the angular-dependent magnetoresistance oscillations (AMROs) observed for quasi-one-dimensional (Q1D) metals such as $(TMTSF)_2ClO_4$ and $(DMET)_2I_3$, where TMTSF = tetramethyltetraselenafulvalene and DMET = dimethyl(ethylenedithio)diselenadithiafulvalene. Several kinds of AMROs have been discovered starting with the first report of Lebed resonance that is observed for the rotation of the magnetic field **B** around the x-axis [1,2] (Fig. 14.1). Here, we define the first, second, and third conducting axes as x, y', and z^* ones, respectively. When **B** is rotated around the z^*-axis or, in other words, within the most conducting xy-plane of a Q1D metal as in Fig. 14.2a with $\theta_1 = 0$ or Fig. 14.2b with $\theta_2 = 0$, two minima of magnetoresistance (MR) are observed at low temperature and high magnetic field. This phenomenon was observed for an organic superconductor $(DMET)_2I_3$ for the first time [3]. Figure 14.3 shows the x-axis and z^*-axis MR of $(DMET)_2I_3$ measured at 4.2 K up to 7 T. Two local minima develop at $\phi_c = -15°$ and 15° on increasing magnitude of **B**. This is called the TAE since there are also the second effect "Danner–Kang–Chaikin oscillations" for the field rotation around y'-axis [4].

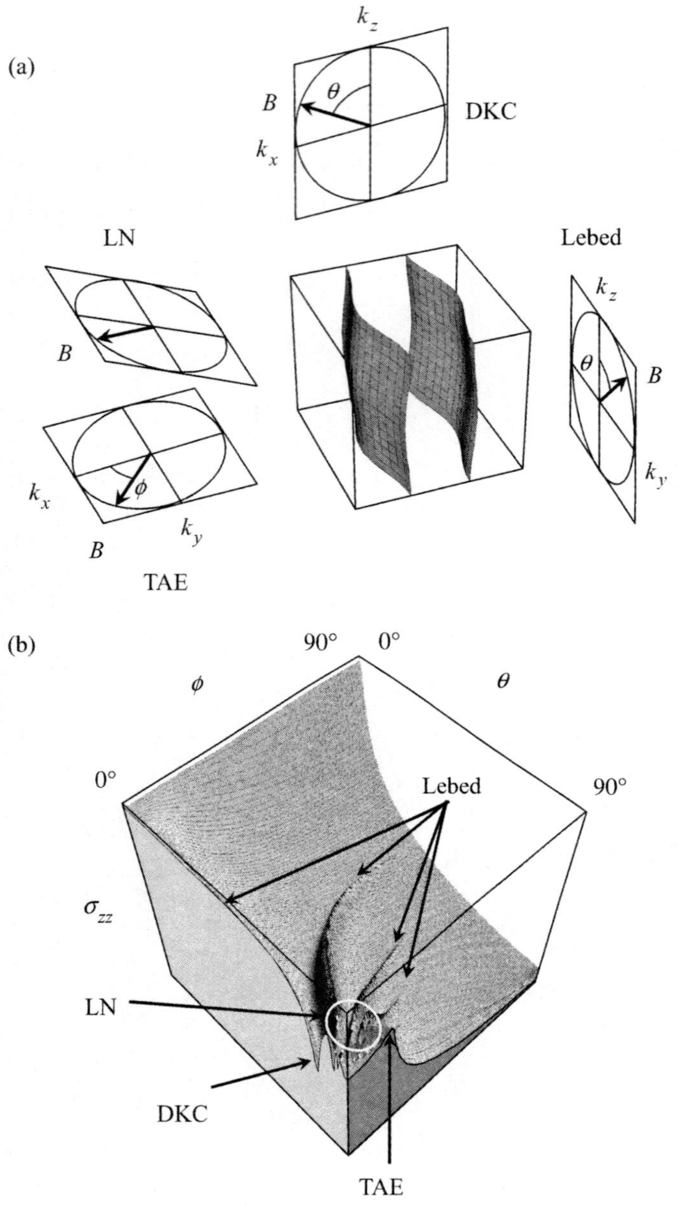

Fig. 14.1. (**a**) Typical planes of rotation of the magnetic field to observe Lebed resonance, Danner–Kang–Chaikin (DKC) oscillations, third angular effect (TAE), and Lee–Naughton (LN) oscillations in Q1D metals. (**b**) Calculated angular-dependent magnetoconductivity σ_{zz} for a model system of $(TMTSF)_2ClO_4$ using (14.3) and (14.23) with $t_x : t_y : t_z = 300 : 30 : 1$, $\tau = 10^{-11}$ s, and $\mathbf{B} = 10\,\mathrm{T}$. The anion-ordering transition at 24 K is not considered

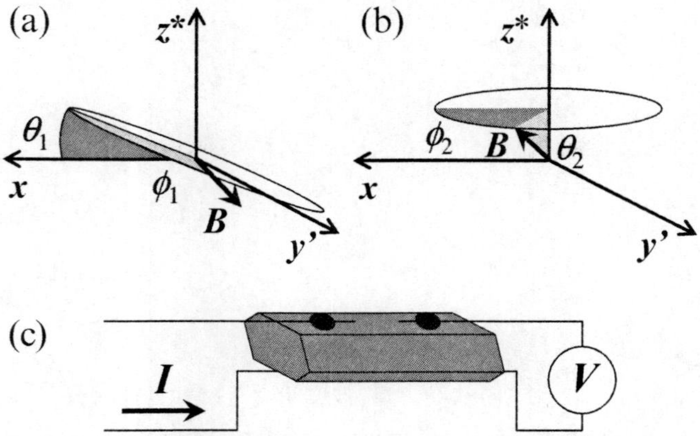

Fig. 14.2. Definitions of angles θ and ϕ relative to the crystal axes used mainly in (**a**) experimental and (**b**) theoretical works. The first, second, and third conducting axes are defined as x, y', and z^* ones, respectively. (**c**) Typical arrangement of electrodes on the crystal to measure the z^*-axis electrical resistance of Q1D conductors such as $(TMTSF)_2ClO_4$ and $(DMET)_2I_3$

It may be worth noting that the definition of the angles θ and ϕ is different from literature to literature even in the works related to the AMROs in the Q1D metals. The definition in Fig. 14.2a is preferred in experimental works, while that in Fig. 14.2b (or $\theta = \pi/2 - \theta_2$) is usually used in theoretical works. One needs a combination of two magnets to realize the rotation of **B** in Fig. 14.2b experimentally.

It has been revealed that the TAE is a general phenomenon commonly observed for many Q1D conductors ($(DMET)_2I_3$ [3,5], $(TMTSF)_2ClO_4$ [6–9], $(TMTSF)_2PF_6$ (0.85 GPa) [10,11], $(DMET)_2AuBr_2$ [12], $(DMET)_2AuCl_2$ [13], and $(DMET)_2CuCl_2$ [14]). Furthermore, the TAE anomalies of MR appear not at insulating (spin-density-wave, SDW) state of $(TMTSF)_2PF_6$ and $(DMET)_2AuI_2$ but at metallic state. This suggests that the origin of the TAE is closely related to Q1D Fermi surface as well as the other AMRO effects.

14.2 Origin of TAE

14.2.1 Requirement to Observe TAE

After the possible change in the electronic state caused by the magnetic field was pointed out [3], two kinds of semiclassical interpretations based on the Fermiology have been proposed by Osada et al. [6,7] and Lebed and Bagmet [15,16], respectively. Osada et al. pointed out the importance of closed orbital motion of the carriers on the Fermi surface affected by the Lorentz force (Fig. 14.4). They claimed that the z-component of the carrier velocity v_z is

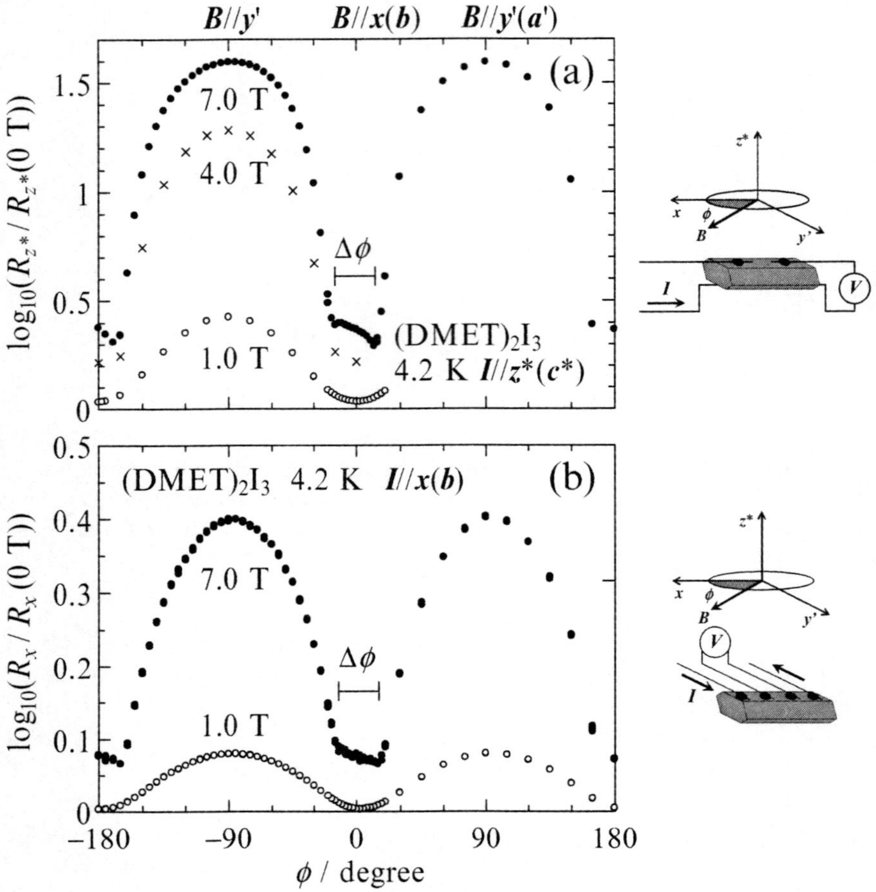

Fig. 14.3. Angular dependence of magnetoresistance of $(DMET)_2I_3$ for the electrical current along (**a**) $z^*(c^*)$- and (**b**) $x(b)$-axes. The magnetic field **B** was rotated within the most conducting $xy(ab)$-plane and the angle ϕ is zero for $\mathbf{B} \parallel x$ (from [3])

averaged to be zero due to the closed orbital motion on the Q1D Fermi surface. In other words, the carriers drawing the closed orbits do not contribute to the electrical conduction along the z-axis. Thus, the electrical resistivity along the z-axis ρ_z shows anomalous increase in the angle region where the closed orbital motion exists. On rotating the magnetic field off from the x-axis in the xy-plane, the closed orbits disappear at an angle that is close to ϕ_c where the MR shows a minimum.

The requirement of the Fermi surface to observe the TAE was confirmed by measuring the angular dependence of the MR of $(TMTSF)_2PF_6$ at ambient pressure [17] and $(TMTSF)_2ClO_4$ in the Q-state [12]. The details of the Q- and R-states of $(TMTSF)_2ClO_4$ are described in Sect. 14.4. $(TMTSF)_2PF_6$ and $(TMTSF)_2ClO_4$ (Q-state) undergo the SDW transition at $T_{\mathrm{SDW}} = 12$

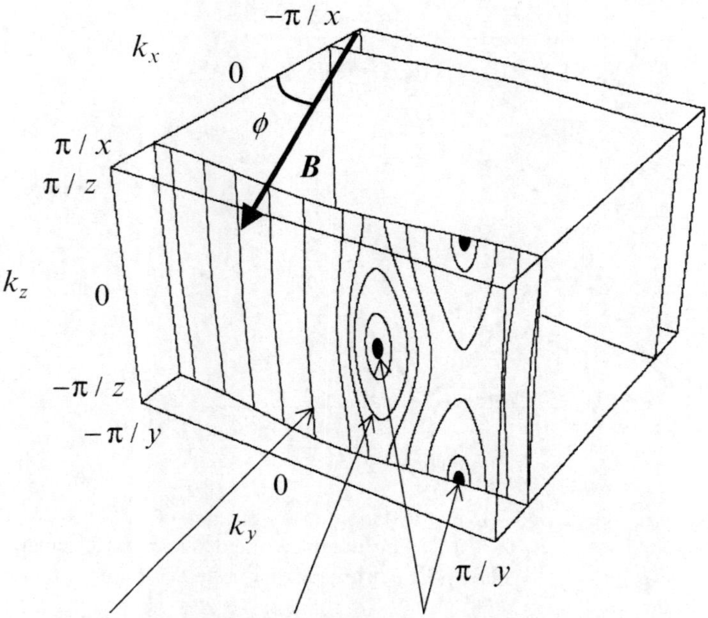

Fig. 14.4. Open and closed orbits caused by Lorentz force on a quasi-one-dimensional Fermi surface. *Effective carriers* have velocity parallel to the applied magnetic field **B**

and 7 K, respectively. The Q1D Fermi surface disappears below T_{SDW} and transport properties become semiconducting. Figure 14.5 shows the field-orientation dependence of the MR of $(\mathrm{TMTSF})_2\mathrm{PF}_6$ at 4.2 K and ambient pressure. No TAE anomaly is recognized around $\phi \sim 0$, while Lee and Naughton et al. [10,11] observed the anomaly in the metallic state at 0.85 GPa. No TAE anomaly was also confirmed for the Q-state of $(\mathrm{TMTSF})_2\mathrm{ClO}_4$ at ambient pressure as is described in Sect. 14.5.

14.2.2 Semiclassical Picture of TAE

According to the semiclassical magnetotransport theory, the electrical conductivity tensor in the magnetic field is calculated by integrating the time dependence of carrier velocity $v_i(t)$ ($i = x$, y, and z) that is described by the equation of motion:

$$\hbar\dot{\mathbf{k}} = -e\mathbf{v} \times \mathbf{B}, \tag{14.1}$$

$$\dot{\mathbf{v}} = \frac{1}{\hbar}\frac{\partial \mathcal{E}(\mathbf{k})}{\partial \mathbf{k}}. \tag{14.2}$$

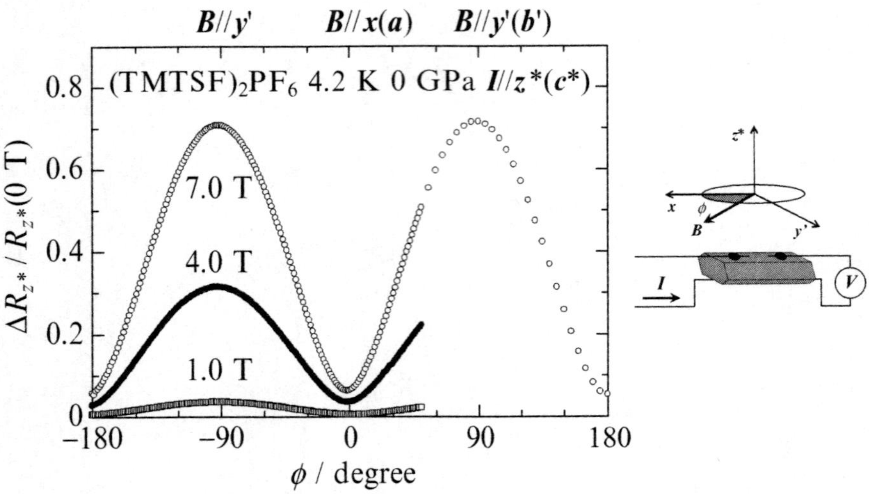

Fig. 14.5. Angular dependence of the magnetoresistance of $(TMTSF)_2PF_6$ at 4.2 K and ambient pressure up to 7.0 T. The magnetic field was rotated within the most conducting xy-plane. $(TMTSF)_2PF_6$ undergoes the SDW transition at 12 K and does not show the TAE

In the present case, it is reasonable to assume a Q1D tight-binding energy dispersion:

$$\mathcal{E}(\mathbf{k}) = -2t_x \cos(k_x x) - 2t_y \cos(k_y y) - 2t_z \cos(k_z z), \tag{14.3}$$

with $t_x : t_y : t_z = 300 : 30 : 1$ for Q1D conductors such as $(TMTSF)_2X$. In this chapter, all the transfer integrals t_x, t_y, and t_z are assumed to be positive. By solving the Boltzmann equation under the relaxation time approximation ($\tau = \text{const.}$), one gets the following expression to calculate the conductivity tensor elements σ_{ij} under the magnetic field. As long as one uses the dispersion relation (14.3), σ_{ij} is a function of t_x, t_y, t_z, τ, and $\mathbf{B}$:

$$\sigma_{ij}(t_x, t_y, t_z, \tau, \mathbf{B}) = \frac{2e^2}{V} \sum_k \left(-\frac{\mathrm{d}f}{\mathrm{d}\mathcal{E}}\right) v_i(\mathbf{k}, 0)\langle v_j \rangle, \tag{14.4}$$

$$\langle v_j \rangle = \int_{-\infty}^{0} v_j(\mathbf{k}, t) \mathrm{e}^{t/\tau} \mathrm{d}t, \tag{14.5}$$

where e is the electron charge, V is the volume of the system, f is the Fermi distribution function, t is the time, and $\langle v_j \rangle$ is the jth component of $\mathbf{v}$ averaged by time until the carrier is scattered, respectively. Osada et al. [6, 7] calculated the angular dependence of magnetoresistance and reproduced the TAE anomaly for an orthorhombic model system at 0 K.

On the other hand, Lebed and Bagmet [15, 16] proposed a theory based on the concept of effective carriers that are free from Lorentz force due to their carrier velocity parallel to the magnetic field. They claimed that the effective carriers govern the electrical conduction along the z-axis in the clean limit $\omega\tau \gg 1$, because v_z of noneffective carriers is averaged to be zero by their orbital motion on the Fermi surface. Since the density of the effective carriers diverges at the angle ϕ_{i}, where the normal to the Fermi surface at the inflection point matches with the magnetic field direction, the MR shows singular anomaly at ϕ_{i}.

Lebed and Bagmet obtained the following expressions for σ_{zz}:

$$\sigma_{zz}(H) = \frac{e^2 t_z^2 z\tau}{\pi^2} \oint \frac{\mathrm{d}l}{|\mathbf{v}_n(\alpha)|\,(1+\omega_{\mathrm{c}}^2(\alpha)\tau^2)}, \tag{14.6}$$

$$\omega_{\mathrm{c}}(\alpha) = \frac{e}{c}|\mathbf{v}_n(\alpha)|Hz\sin\alpha, \tag{14.7}$$

where H is the magnetic field, c is the light velocity, z is the lattice constant along the interplane direction, $\mathbf{v}_n$ is the in-plane velocity component, α is the angle between H and $\mathbf{v}_n$, and $\omega_{\mathrm{c}}(\alpha)$ is a characteristic frequency of an electron motion in the direction perpendicular to the conducting plane, respectively. This is a solution of the Boltzmann equation by assuming anisotropic two-dimensional (2D) Fermi surface. Then, a linearized form of the Q1D dispersion relation (14.3),

$$\mathcal{E}(\mathbf{k}) = \pm\hbar v_{\mathrm{F}}(k_x \mp k_{\mathrm{F}}) - 2t_y\cos(k_y y) - 2t_z\cos(k_z z), \tag{14.8}$$

is applied to calculate σ_{zz}. The structure of the TAE was also reproduced by this model.

Lee and Naughton [11] calculated angular dependence of the time average of v_z at typical points on the Q1D Fermi surface and claimed that ϕ_{c} is much smaller than ϕ_{i} for real Q1D metals. According to them, not the closed orbits but the effective carriers are responsible for the TAE. Their evaluation of ϕ_{c} from the maximum of $\langle v_z\rangle(\phi)$ at $(k_y, k_z) = (0, \pi/2z)$ on the Fermi surface (the point indicated by the arrow in Fig. 14.7a), however, neglects the closed orbits that still remain around the inflection point after the carrier at $(0, \pi/2z)$ starts to draw an open orbit. This can be clearly shown by drawing orbital motion of carriers on the Fermi surface at typical orientations of the magnetic field as follows.

Yoshino and Murata [18] showed detailed steps of calculating σ_{zz} using (14.3) and (14.4). The lattice constants $a/2 = 7.068/2$ Å, $b = 7.638$ Å, and $c = 13.123$ Å of $(\mathrm{TMTSF})_2\mathrm{ClO}_4$ [19] are chosen for x, y, and z, respectively, as realistic parameters. Small dimerization of TMTSF molecules along the x-axis and doubling of y due to anion ordering at 24 K are ignored. The angles α, β, and γ are assumed to be $\pi/2$ for simplicity. The integration in (14.5) was carried out numerically. The derivative, $\mathrm{d}f/\mathrm{d}\mathcal{E}$, becomes a delta function at 0 K. Then, the $\mathbf{k}$-space summation is carried out over k_y and k_z in the first Brillouin zone.

The TAE is commonly observed for ρ_{zz} or, in other words, for the out-of-plane MR. This is now calculated as one element of the inverse tensor of

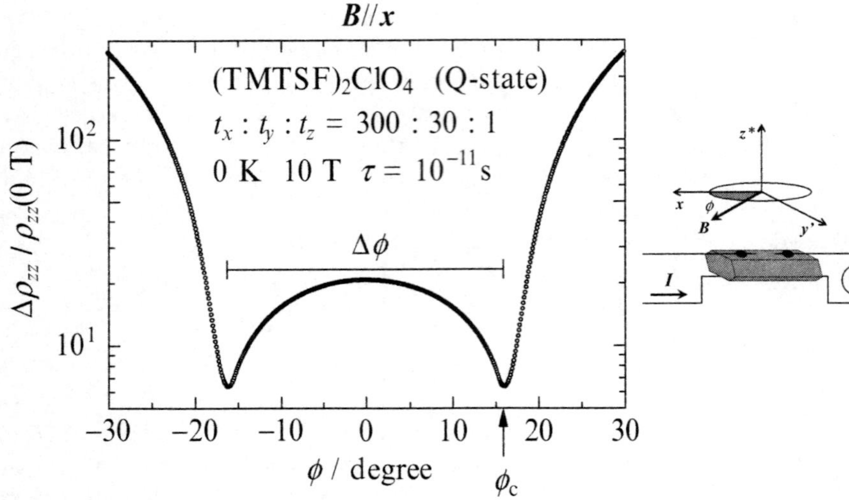

Fig. 14.6. Calculated angular dependence of the zz-component of magnetoresistivity tensor ρ_{zz} of $(TMTSF)_2ClO_4$ at 0 K and 10 T using (14.3) and (14.4) with $t_x = 300\,\text{meV}$, $t_x : t_y : t_z = 300 : 30 : 1$, and $\tau = 10^{-11}$ s. Slight dimerization of TMTSF molecules along the x-axis and anion-ordering transition at 24 K in the real system are ignored. Time step of the numerical integration is 10^{-14} s. The $k_y k_z$-plane is divided into 512×512 sites. The magnetic field is rotated within the xy-plane and ϕ is measured from the x-axis (from [18])

$\boldsymbol{\sigma}$. Figure 14.6 shows the angular dependence of σ_{zz} calculated by assuming $t_x = 300\,\text{meV}$ and $t_x : t_y : t_z = 300 : 30 : 1$. It is needed to divide the first Brillouin zone into 512×512 sites to get precise angular dependence of each $\boldsymbol{\sigma}$ tensor element, while 64×64 ones are sufficient just for ρ_{zx}, ρ_{zy}, and ρ_{zz}. As for time integration, $\tau = 10^{-11}$ s is assumed and the time step, $\Delta t = 10^{-14}$ s, is enough fine to get precise result for the angle $\phi \leq 30°$ and the magnitude of the magnetic field $B < 30\,\text{T}$, where ϕ is measured from the x-axis within the xy-plane.

Figure 14.7 shows an projection of a Q1D Fermi surface onto the $k_y k_z$-plane. Time average of v_z at each **k**-point calculated by (14.5) is shown as density plots for several orientations of the magnetic field between $\phi = 0°$ and 20°. Typical trajectories of carriers are also drawn by solid curves. The intensity of white and black shows the magnitude of $\langle v_z \rangle$ in the plus and minus directions, respectively. Thus, gray means $\langle v_z \rangle$ is almost zero. At $\phi = 0° (\mathbf{B} \parallel k_x)$, there exist the closed orbits of carriers on the convex ($k_y \sim 0$) and concave ($k_y \sim \pi/y$) parts of the Q1D Fermi surface (Fig. 14.7a). *One can find that $\langle v_z \rangle$ is nonzero in the area where the closed orbits are allowed, while it is almost zero on open orbits.* Since the carrier velocity **v** is almost parallel to the magnetic field on the closed orbits, the carriers can complete the orbits for few times at most, while the carriers can run across

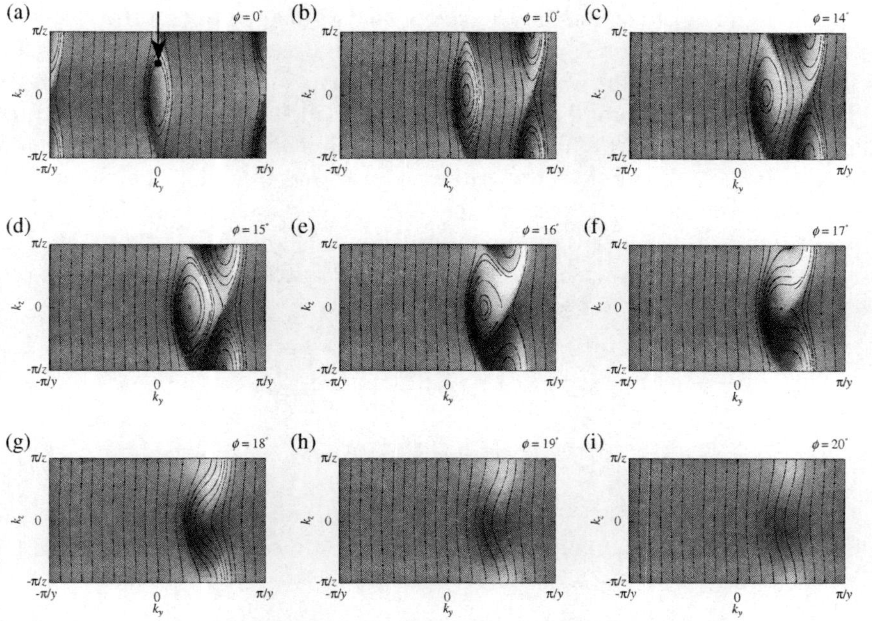

Fig. 14.7. Density plot of the time average of v_z ($\langle v_z \rangle$) of carriers on the Q1D Fermi surface of $(TMTSF)_2ClO_4$ with the same condition as in Fig. 14.6. The angle ϕ is measured from the x-axis within the xy-plane. The intensity of *white and black* scales the magnitude of $\langle v_z \rangle$ in plus and minus directions, respectively. *Solid curves* are typical traces of carriers caused by the Lorentz force. The **k**-point $(k_y, k_z) = (0, \pi/2z)$ indicated by the *arrow* in (**a**) is mentioned in the text (from [18])

the first Brillouin zone from three to more than ten times on the open orbits. Thus, the v_z is less averaged to zero in the area where the closed orbits are allowed. Namely, *the carriers on the closed orbits themselves are effective* for the electrical conduction along the z-axis.

On rotating the magnetic field from the k_x- (or x-) axis, the group of convex closed orbits and that of concave ones get closer and closer (see $\phi = 10°$ and 14° of Fig. 14.7b, c) and their rims merge with each other at 15°. Above 15° (Fig. 14.7d), meander Ω-shaped open orbits appear between the convex and concave closed orbits (16° and 17° in Fig. 14.7e, f). Around 16°, which is almost the same as the minimum angle $\phi_c = 16.1°$ of the TAE (Fig. 14.6), the carriers on most of the closed and Ω orbits are scattered before terminating the travel on each orbit. On these semiclosed orbits, the averaging of $\mathbf{v}$ is distinctly incomplete and, therefore, the intensity of white and black remains nonzero. This is the origin of the sharp anomalies of ρ_{zz} at ϕ_c in Fig. 14.6. One can still see a small closed orbit near the inflection point $(k_y, k_z) = (\pi/2y, 0)$ at 17° closed to ϕ_i where the magnetic field is parallel to the normal to the Fermi surface at the inflection point. Only the open orbits are allowed at $\phi > \phi_i = 17°$ (18°, 19°, and 20° in Fig. 14.7g–i). *Thus, ϕ_c is very close to ϕ_i.*

The number of carriers on the closed and Ω-shaped open orbits decreases when the system becomes more one dimensional, i.e., $t_x \gg t_y > t_z$. Then the closed and Ω orbits shrink along the k_y-axis and become very similar to the open orbits. The averaging of v_z to zero is still incomplete on these quasi-open orbits due to the small Lorentz force. This corresponds to the effective carriers in Lebed's model noted above and shows that not the shape of the orbits but the incomplete averaging of v_z is essential to the TAE. Indeed, Sugawara et al. [20] derived an analytical expression of ρ_{zz} of a Q1D system represented by (14.8) when the magnetic field is rotated within the xy-plane *only by considering the open orbits*:

$$\rho_{zz}(B,\phi) = \rho_{zz}(0)\sqrt{\frac{2(P^2+Q^2)}{P+\sqrt{P^2+Q^2}}}, \tag{14.9}$$

$$P = 1-(\omega\tau)^2\left(\sin^2(\phi)-\eta^2\cos^2(\phi)\right), \tag{14.10}$$

$$Q = 2\omega\tau\sin(\phi), \tag{14.11}$$

$$\rho_{zz}(0) = \frac{\pi\hbar yz}{e^2 v_{\mathrm{F}}\tau\zeta^2}, \tag{14.12}$$

$$\eta = \frac{2t_y y/\hbar}{v_{\mathrm{F}}}, \tag{14.13}$$

$$\zeta = \frac{2t_z z/\hbar}{v_{\mathrm{F}}}, \tag{14.14}$$

$$\omega = \frac{ev_{\mathrm{F}}Bz}{\hbar}, \tag{14.15}$$

where η and ζ are the anisotropy parameters of carrier velocity between the x- and y-axes, and between the x- and z-axes, respectively.

It should be noted that the TAE anomaly observed in σ_{xx} (Fig. 14.3b) is not reproduced by the numerical and analytical calculations described in this section, though sharp anomalies appear in the calculated σ_{xx} (and ρ_{xx}) at ϕ_{i} [18]. Recently, Kaneyasu et al. [21] studied the effect of electron correlation on the scattering time/quasiparticle lifetime τ as well as the carrier velocity within the semiclassical calculation. They claimed that the AMROs experimentally observed in angular dependence of σ_{xx} in the Q1D metals are reproduced by considering the momentum dependence of τ.

14.2.3 Quantum Mechanical Picture of TAE

The discussion up to now is based on the Fermi liquid picture assuming Q1D dispersion relations (14.3) or (14.8) and on the solution of the Boltzmann equation (14.4). Recently, Lebed et al. [22–24] have developed another Fermi liquid picture but based on the Kubo formalism to discuss the dimensional crossover from one to two dimension at magic angles of the Lee–Naughton (LN) oscillations. They proposed several equations to describe σ_{zz} at the general orientation of magnetic field. The simplest form for the linearized dispersion relation (14.8) is as follows

$$\frac{\sigma_{zz}(B,\theta,\phi)}{\sigma_{zz}(0)} = \sum_{n=-\infty}^{+\infty} \left(\frac{J_n^2\left(\omega_z^*(\theta,\phi)/\omega_y(\theta)\right)}{1+\tau^2\left(\omega_z(\theta,\phi)-n\omega_y(\theta)\right)^2} \right), \tag{14.16}$$

$$\omega_y(\theta) = \frac{e}{c} v_{\mathrm{F}} B y \sin(\theta), \tag{14.17}$$

$$\omega_z(\theta,\phi) = \frac{e}{c} v_{\mathrm{F}} B z \cos(\theta)\sin(\phi), \tag{14.18}$$

$$\omega_z^*(\theta,\phi) = \frac{v_y^0}{v_{\mathrm{F}}} \frac{e v_{\mathrm{F}} B z}{c} \cos(\theta)\cos(\phi), \tag{14.19}$$

$$v_y^0 = 2t_y y, \tag{14.20}$$

where θ and ϕ represent the direction of the inclined magnetic field

$$\mathbf{B} = B(\cos\theta\cos\phi, -\cos\theta\sin\phi, \sin\theta), \tag{14.21}$$

$J_n(\cdots)$ is the nth Bessel function, and x, y, and z are the lattice constants. The definition of the angles θ and ϕ is that in Fig. 14.2b. Equation (14.16) also reproduces the TAE structure of $(\mathrm{TMTSF})_2\mathrm{ClO}_4$ in the R-state as well as that of the LN oscillations. Lebed et al. [22–25] claimed that the origin of the complex LN oscillations is related to interference effects due to commensurate electron trajectories in a tilted magnetic field in the extended Brillouin zone.

Another interpretation of the AMROs in the Q1D metals by Osada et al. [26–28] does not require coherent coupling of the 1D chains along the third conducting z-direction any more. Starting from two Q1D layers,

$$\mathcal{E}^{(2\mathrm{D})}(\mathbf{k}) = \hbar v_{\mathrm{F}}(|k_x| - k_{\mathrm{F}}) - 2t_y\cos(k_y y), \tag{14.22}$$

weakly interacting with a hopping probability t_z, they derived the interlayer dc conductivity σ_{zz} by the Kubo formalism after calculating the matrix element of interlayer tunneling as the overlap integral between the layers in the tilted magnetic field $\mathbf{B} = (B_x, B_y, B_z)$:

$$\sigma_{zz} = \frac{4}{2\pi y z \hbar v_{\mathrm{F}}} \left(\frac{e t_z z}{\hbar}\right)^2 \sum_{n=-\infty}^{+\infty} J_n^2\left(\frac{4t_y}{v_{\mathrm{F}} y e B_z}\sin\left(\frac{e y z B_x}{2\hbar}\right)\right) \times \frac{\tau}{1+v_{\mathrm{F}}^2\left(n(yeB_z/\hbar) \pm (zeB_y/\hbar)\right)^2\tau^2}. \tag{14.23}$$

Not only the TAE but also the other AMROs in the Q1D metals are reproduced by (14.23). It should be noted that the angular dependence of σ_{zz} for the xy-rotation (the TAE configuration) is obtained in the limit $B_z \to 0$. In this picture, the AMROs are interpreted as the result of resonant tunneling of electrons between the 2D layers in the tilted magnetic field.

Recently, Cooper and Yakovenko [29] also proposed formulation of the ac interlayer conductivity $\sigma_{zz}(\mathbf{B},\omega)$ in terms of Aharonov–Bohm interference in interlayer electron tunneling in the two-layer system,

$$\frac{\sigma_{zz}(\mathbf{B},\omega)}{\sigma_{zz}(0,0)} = \sum_{n=-\infty}^{+\infty} \frac{J_n^2(B'_x)}{1+(\omega_c\tau)^2(n-B'_y \mp \omega/\omega_c)^2}, \tag{14.24}$$

$$\omega_c = \frac{eyv_F B_z}{\hbar c}, \tag{14.25}$$

$$B'_x = \frac{B_x}{B_z}\frac{2t_y d}{\hbar v_F}, \tag{14.26}$$

$$B'_y = \frac{B_y}{B_z}\frac{d}{y}, \tag{14.27}$$

and in the multilayer system,

$$\frac{\sigma_{zz}(\mathbf{B},\omega)}{\sigma_{zz}(0,0)} = \sum_m \sum_{n=-\infty}^{+\infty} \frac{t'^2_m J^2_{n-m}(B'_x)}{1+(\omega_c\tau)^2(n-B'_y \mp \omega/\omega_c)^2}, \tag{14.28}$$

$$t'^2_m = t^2_m / \sum_l t_l^2, \tag{14.29}$$

where a linearized 2D dispersion relation similar to (14.22) is assumed for each layer with d being the interlayer distance and t_m being the interlayer tunneling amplitudes between the chains shifted by m units in the y-direction.

Equations (14.16), (14.23), and (14.24) with $\omega = 0$ are similar to one another, namely σ_{zz} is expressed as a sum of squared Bessel functions. The anomalies in AMROs in the Q1D metals are understood as a result of resonance between the electrons in different layers/chains at magic angles. The magic angles are the same as that predicted by the semiclassical theory discussed above since they are related to the shape of Q1D Fermi surface.

14.3 Estimation of Dimensionality t_y/t_x by TAE

The origin of the TAE is closely related to the corrugation of the Q1D Fermi surface within the xy-plane and its critical angle ϕ_c, where ρ_{zz} = min is very close to the inflection angle ϕ_i determined by the curvature of the Q1D Fermi surface as described in Sect. 14.2. This means, in reverse, that we can determine the curvature of a Q1D Fermi surface by utilizing the TAE. In this section, we show the detailed procedure that is applicable even for triclinic systems.

Suppose a model Q1D metal (Fig. 14.8) with lattice parameters, x, y, and γ, which correspond to, for example, $a/2$, b, and γ of $(TMTSF)_2ClO_4$ and $(TMTSF)_2PF_6$; and $b/2$, a, and γ of $(DMET)_2I_3$, respectively. This is an approximation to ignore the weak dimerization of TMTSF or DMET

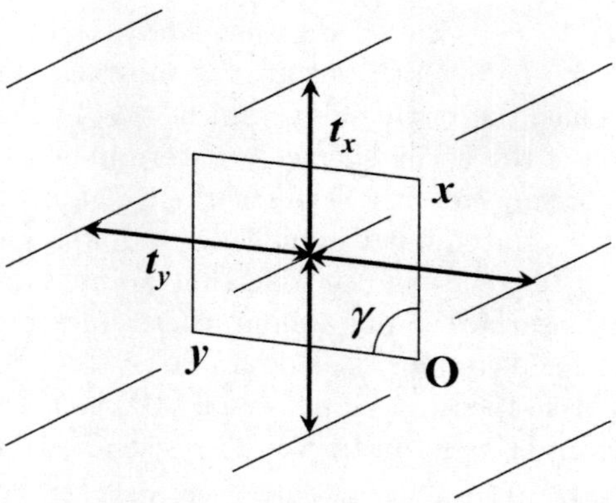

Fig. 14.8. A model Q1D system with lattice constants x, y, and γ

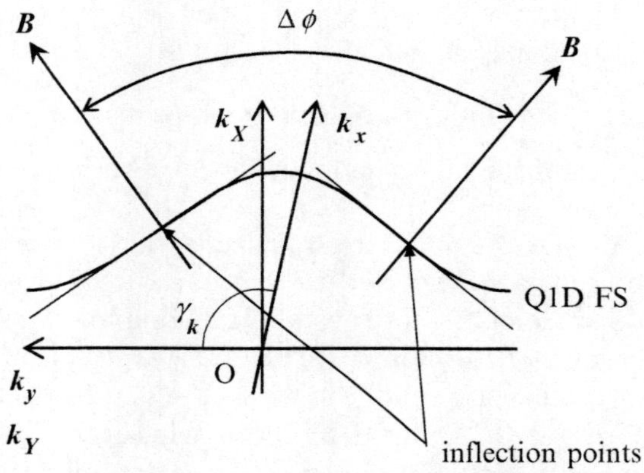

Fig. 14.9. Relation between inflection points and the TAE anomaly with the angular width $\Delta\phi$

molecules in the 1D column. The neighboring sites are interacting with transfer integrals t_x and t_y along x- and y-axes, respectively. The dispersion relation becomes as follows by the tight-binding approximation:

$$\mathcal{E}(\mathbf{k}) = -2t_x \cos(k_x x) - 2t_y \cos(k_y y). \qquad (14.30)$$

In the $k_x k_y$ space, a pair of Q1D Fermi surfaces/sheets spreads along the k_y-axis as in Fig. 14.9.

Since $\phi_\mathrm{c} \simeq \phi_\mathrm{i}$ depends only on the shape of the Fermi surface, a rewritten form of (14.30) is sufficient to calculate ϕ_i:

$$\varepsilon(\mathbf{k}) = -\cos(k_x x) - \xi \cos(k_y y), \qquad (14.31)$$

$$\varepsilon(\mathbf{k}) = \mathcal{E}(\mathbf{k})/2t_x, \qquad (14.32)$$

$$\xi = t_y/t_x. \qquad (14.33)$$

The ratio (ξ) of t_y to t_x is called *dimensionality* since this corresponds to the anisotropy of the bandwidth within the most conducting plane of the Q1D conductors. Once the band filling, which is 1/4 of holes in this case, is assumed, the shape of the Fermi surface is determined.

The slope of tangent to the Fermi surface against the k_y-axis is the largest in absolute magnitude at the inflection points. Some lines with slope in between those of tangents at the inflection points cross, at least, at two points on the Fermi surface. Some among these lines cross with the Fermi surface with depth smaller than the amplitude of the undulating Fermi surface along the k_z-axis in the three-dimensional (3D) picture. The cross sections between these lines and the Fermi surface correspond to the closed orbits. The closed orbits converge to the inflection point at inflection angles ϕ_{i} as seen in Fig. 14.7e. There are two kinds of inflection angles with opposite signs because of the triclinic symmetry of the crystal, i.e., $\gamma \neq 90°$. The inflection angles are expressed by the largest and the smallest slopes, $s_{\max}$ and $s_{\min}$, of the Fermi surface against the k_y-axis as follows:

$$\phi_{\mathrm{i1}} = \tan^{-1}(s_{\max}), \quad \phi_{\mathrm{i2}} = \tan^{-1}(s_{\min}). \tag{14.34}$$

The resulting width of the hump anomaly in the TAE is given by $\Delta\phi$ [rad] $= |\phi_{\mathrm{i1}}| + |\phi_{\mathrm{i2}}|$ [12].

Yoshino et al. [12, 13] numerically calculated ξ dependence of $\Delta\phi$ by using the lattice constants reported in the literature and estimated ξ from the experimental value of $\Delta\phi$ for several Q1D conductors. Gallois et al. [19] reported the crystal structure of $(TMTSF)_2PF_6$ at 1.7 K and 0.70 GPa that are close to the experimental condition where Lee and Naughton et al. [10,11] observed the TAE (at 0.3 K and 0.85 GPa), while the crystal structure of $(DMET)_2I_3$ has been determined only at room temperature and ambient pressure. Figure 14.10 is the result of the calculation by using the lattice parameters: x, y, and γ of 7.752/2 Å, 6.703 Å, and 78.25° of $(DMET)_2I_3$ [30,31] and 6.980/2 Å, 7.581 Å, and 70.30° of $(TMTSF)_2PF_6$ [19]. Above $\xi = 1/3$, the Fermi surface closes also along the k_y-axis and becomes quasi-2D. The difference in $\Delta\phi$ between $(DMET)_2I_3$ and $(TMTSF)_2PF_6$ at the same ξ is from the difference in the lattice parameters, especially y. From Fig. 14.10 and the observed $\Delta\phi$, ξ is estimated as 1/9.7 and 1/8.6 for $(DMET)_2I_3$ at ambient pressure and $(TMTSF)_2PF_6$ under pressure, respectively. It is worth noting that the lattice parameters of $(TMTSF)_2PF_6$ at room temperature and ambient pressure [32] give $\xi = 1/8.3$. Thus, the error in the estimated ξ by using the room temperature lattice parameters is considered to be very small for the DMET salts. The estimated values are in good agreement with that of 1/10 as is often referred. Figure 14.11 is the first Brillouin zone and Q1D Fermi surface of $(TMTSF)_2PF_6$ and $(DMET)_2I_3$ determined by the TAE.

The present method also gives $\xi = 1/10$, 1/9.8, and 1/10 for $(DMET)_2AuBr_2$ [12], $(DMET)_2AuCl_2$ [13], and $(DMET)_2CuCl_2$ [14], respectively. The four Q1D DMET salts have almost the same ξ irrespective of their different ground states from one another. For $(TMTSF)_2ClO_4$ at ambient pressure,

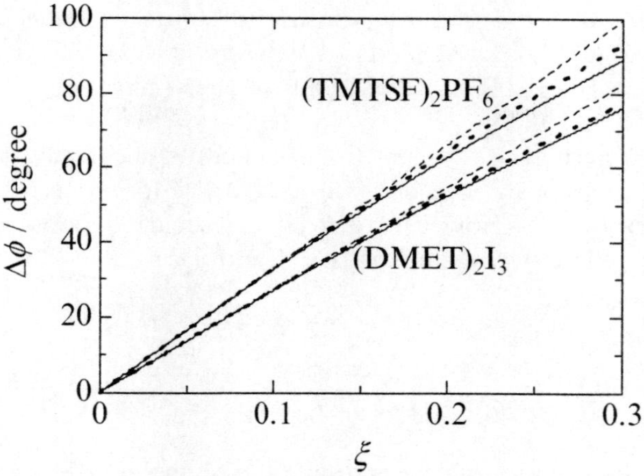

Fig. 14.10. The width of the TAE anomaly $\Delta\phi$ as a function of the dimensionality $\xi = t_y/t_x$ for $(TMTSF)_2PF_6$ and $(DMET)_2I_3$. *Solid curves* are obtained numerically by using (14.34) from [12], while *broken and dotted curves* are from approximated expressions of analytical inflection angle (14.48) and (14.50), respectively

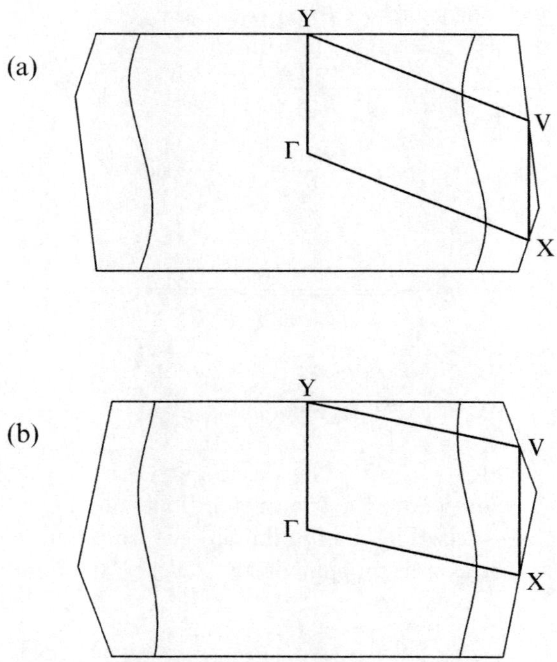

Fig. 14.11. The first Brillouin zone and Fermi surface of (**a**) $(TMTSF)_2PF_6$ at 0.85 GPa and (**b**) $(DMET)_2I_3$ at ambient pressure with the dimensionality $\xi = t_y/t_x$ [$\xi = 1/8.6$ for $(TMTSF)_2PF_6$ and $\xi = 1/9.7$ for $(DMET)_2I_3$] determined by the TAE

$\xi = 1/12$ was reported by measuring the pressure dependence of $\Delta\phi$ as is discussed in Sect. 14.4. Less one dimensionality of $(TMTSF)_2PF_6$ ($\xi = 1/8.6$) as compared with $(TMTSF)_2ClO_4$ is due to the application of the pressure of 0.85 GPa.

The TAE method to estimate ξ is very simple but it may be more convenient to give analytical relation between $\Delta\phi$ and ξ for the application to other Q1D conductors. We start with the following equation modified from (14.31) by normalizing lattice constants as $x = 1$ and y/x:

$$\varepsilon \equiv \mathcal{E}/(2t_x) = -\cos(k_x \cdot 1) - \xi\cos(k_y r), \tag{14.35}$$

$$\xi = t_y/t_x, \tag{14.36}$$

$$r = y/x. \tag{14.37}$$

Trace of the Q1D Fermi surface is drawn by solving (14.35) when $\varepsilon = \varepsilon_\mathrm{F}$. The solutions are

$$k_{x,\mathrm{F}} = \pm\cos^{-1}\left(-\varepsilon_\mathrm{F} - \xi\cos k_y r\right). \tag{14.38}$$

The inflection angle ϕ_i is the one between the normal to the Fermi surface at the inflection point and the most conducting x-axis within the xy-plane. The slope of the Fermi surface $\partial k_{X,\mathrm{F}}/\partial k_Y$ is calculated as

$$\begin{aligned}\frac{\partial k_{X,\mathrm{F}}}{\partial k_Y} &= \frac{\partial k_{X,\mathrm{F}}}{\partial k_{x,\mathrm{F}}}\frac{\partial k_{x,\mathrm{F}}}{\partial k_y}\frac{\partial k_y}{\partial k_Y} \\ &= \sin\gamma_k \cdot \frac{\partial k_{x,\mathrm{F}}}{\partial k_y} \cdot \left(\frac{\partial k_{x,\mathrm{F}}}{\partial k_y}\cos\gamma_k + 1\right)^{-1},\end{aligned} \tag{14.39}$$

$$\frac{\partial k_{x,\mathrm{F}}}{\partial k_y} = \frac{r\xi\sin k_y r}{\sqrt{1 - \left(-\varepsilon_\mathrm{F} - \xi\cos k_y r\right)^2}}, \tag{14.40}$$

$$k_{X,\mathrm{F}} = k_{x,\mathrm{F}}\sin\gamma_k, \tag{14.41}$$

$$k_Y = k_{x,\mathrm{F}}\cos\gamma_k + k_y, \tag{14.42}$$

$$\gamma_k = \pi - \gamma, \tag{14.43}$$

where X- and Y-axes denote Cartesian coordinates and the latter is taken to be parallel to the y-axis. The maximum and the minimum of $\partial k_{X,\mathrm{F}}/\partial k_Y$ are calculated under the condition that $\partial^2 k_{X,\mathrm{F}}/\partial k_Y^2 = 0$. Then, ϕ_i is calculated as follows:

$$\phi_\mathrm{i} = \tan^{-1}\left(\left.\frac{\partial k_{X,\mathrm{F}}}{\partial k_Y}\right|_{\max,\min}\right) = \tan^{-1}\left(\frac{s\sin\gamma_k}{\pm 2 + s\cos\gamma_k}\right), \tag{14.44}$$

$$s = \sqrt{2}r\sqrt{1 - \varepsilon_\mathrm{F}^2 + \xi^2 - \sqrt{-4\varepsilon_\mathrm{F}^2\xi^2 + \left(-1 + \varepsilon_\mathrm{F}^2 + \xi^2\right)^2}}. \tag{14.45}$$

When ε_F is approximated to be $\varepsilon_{F,1D} = 1/\sqrt{2}$, i.e., the Fermi energy for the 3/4-filled band in the pure 1D case, (14.45) is simplified as shown below:

$$s = r\sqrt{1 + 2\xi^2 - \sqrt{1 - 12\xi^2 + 4\xi^4}}. \tag{14.46}$$

The lowest order of Taylor's series of (14.44) with (14.46) gives the simplest form of ϕ_i and $\Delta\phi$ for the triclinic system:

$$\phi_i = \pm\sqrt{2}r\xi \sin\gamma_k, \tag{14.47}$$

$$\Delta\phi \text{ [rad]} = 2\sqrt{2}r\xi \sin\gamma_k = 2\sqrt{2}r\xi \sin\gamma. \tag{14.48}$$

Equation (14.48) matches with $\Delta\phi = 2\sqrt{2}r\xi$ that is obtained for the orthorhombic system with $\gamma = \pi/2$ [6]. The broken lines in Fig. 14.10 are drawn by using (14.48) and this equation is usually sufficient to estimate ξ since the difference between each solid curve and broken line is small especially below $\xi \sim 1/10$ for $(TMTSF)_2PF_6$ and $(DMET)_2I_3$, respectively.

One needs the second term of Taylor's series of (14.44) when one wants to calculate the position of asymmetric ϕ_i of the triclinic system or the dimensionality of the system is rather high ($\xi > 1$), while $\Delta\phi$ is the same as (14.48) due to the cancellation of the second term in (14.49):

$$\phi_i = \pm\sqrt{2}r\xi \sin\gamma_k - 2r^2\xi^2 \cos\gamma_k \sin\gamma_k. \tag{14.49}$$

Finally, one needs the third term of the Taylor's series in the region where the deviation of the broken lines from the solid curves in Fig. 14.9 is rather large, namely in case of $\xi > 1/7$:

$$\Delta\phi \text{ [rad]} = 2\sqrt{2}r\xi \sin\gamma_k \left(1 + \frac{\xi^2}{3}\left(3 + 2r^2\left(1 + 2\cos(2\gamma_k)\right)\right)\right). \tag{14.50}$$

14.4 Case of Two Pairs of Q1D Fermi Surfaces

In the previous sections, we do not take into account the anion-ordering transition of $(TMTSF)_2ClO_4$ at $T_{AO} = 24$ K at ambient pressure. Two kinds of orientations are possible for the tetragonal ClO_4^- anion in $(TMTSF)_2ClO_4$ but they are disordered by thermal motion at room temperature. Below T_{AO}, however, the anions order with the wave number $\mathbf{Q} = (0, 1/2, 0)$. Thus, the Brillouin zone as well as the Q1D Fermi surface is folded as in Fig. 14.12. The disordered and ordered states are usually called Q (quenched)- and R (relaxed)-states, respectively. The Q1D Fermi surface in the Q-state splits into two pairs in the R-state due to the Coulomb potential of the ordered anions. The shape of Q1D Fermi surface is not sinusoidal in general. Thus, the two pairs of the Fermi surfaces in the R-state are not identical and their warping is different from each other. Many researchers have investigated $(TMTSF)_2ClO_4$

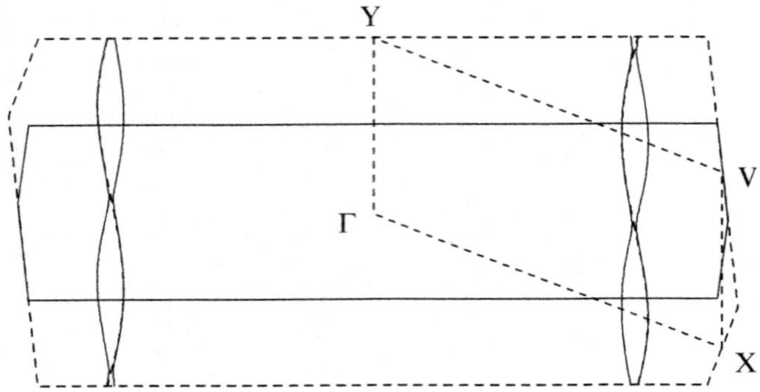

Fig. 14.12. The first Brillouin zones of the Q- (*broken lines*) and R-states (*solid lines*) of $(TMTSF)_2ClO_4$. Note that the slight dimerization of the TMTSF molecules along the x-axis is ignored, namely $x/2$, y, and z are used to draw this figure. The shape of the Fermi surfaces was determined by the TAE (see below) measured at ambient and hydrostatic pressures

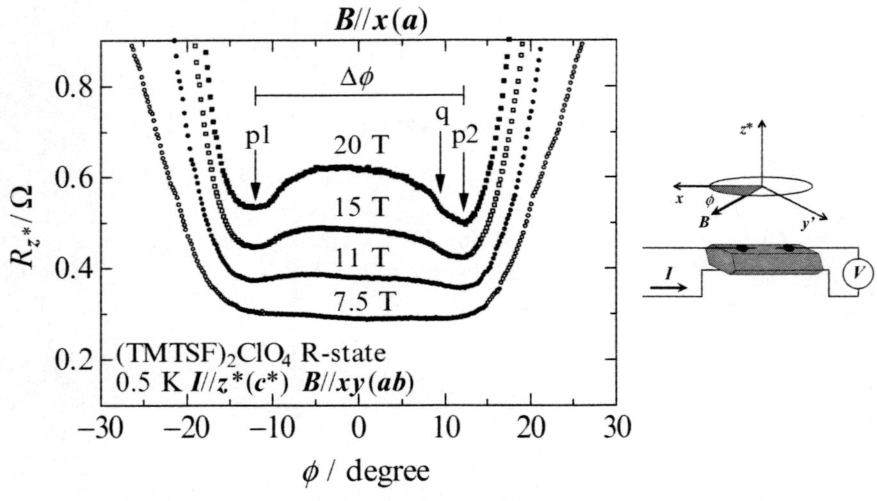

Fig. 14.13. Angular dependence of the z^*-axis magnetoresistance of $(TMTSF)_2ClO_4$ at 0.5 K in the R-state. The magnetic field was rotated within the xy-plane. The third angular effect anomaly with the width of $\Delta\phi$ is recognized (from [8])

in the R-state because superconductivity and other interesting phenomena such as field-induced SDW appear. The doubling of the Brillouin zone, however, makes the interpretation of the experimental result difficult. This is also the case with the TAE, i.e., two pairs of the TAE minima should be observed since the warping of each kind of Q1D Fermi surface is different from each other.

Figure 14.13 shows the angular dependence of the MR of $(TMTSF)_2ClO_4$ in the R-state at 0.5 K and up to 20 T. The TAE anomaly is appreciable above 7.5 T between $\phi_c = +12°$ and $-12°$. The definition of $\Delta\phi$ is shown in Fig. 14.13. A step anomaly is recognized at 9° as is indicated by the arrow q besides the TAE minima (p1 and p2) at $\phi_c = \pm 12°$.

The dispersion relations (14.51) from the perturbation theory are often used to describe the electronic state of the R-state, where the small dimerization of TMTSF molecules along the x-axis is ignored [33]:

$$\begin{aligned} \mathcal{E}(\mathbf{k}, E_g) = & -2t_x \cos(k_x x) \\ & \pm \sqrt{(2t_y \cos(k_y y))^2 + E_g^2} \\ & -2t_z \cos(k_z z). \end{aligned} \tag{14.51}$$

Here, E_g is the anion gap. Two kinds of Q1D Fermi surfaces of $(TMTSF)_2ClO_4$ give different $\Delta\phi$ from each other in general, namely the inner Fermi surface from the lower band gives narrower $\Delta\phi$. Thus, the anomalies p1 and p2 in Fig. 14.13 are probably originated from the outer Fermi surface and q from the inner one.

One can calculate the angular dependence of ρ_{zz} using (14.4) in combination with (14.51) by assuming that each Fermi surface contributes independently to the electrical conduction. Figure 14.14 shows the calculated ρ_{zz} of $(TMTSF)_2ClO_4$ in the Q- and R-states using common parameters

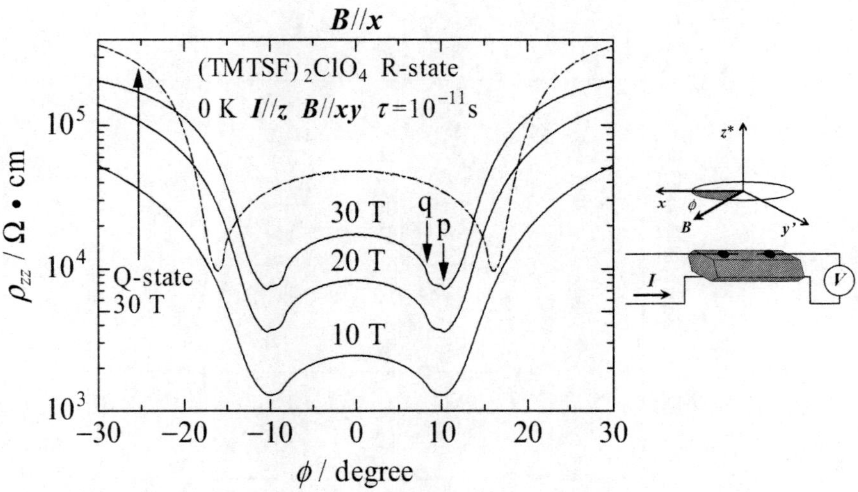

Fig. 14.14. Numerically calculated angular dependence of ρ_{zz} of $(TMTSF)_2ClO_4$ in the Q- (*dotted curve*) and R-states (*solid curves*) by the relaxation time approximation using (14.4) with (14.3) for the Q-state and (14.51) for the R-state, respectively. An orthorhombic lattice ($\alpha = \beta = \gamma = \pi/2$) was assumed and the lattice constants are from [19]. Other parameters needed for the numerical calculation are $t_x : t_y : t_z = 300 : 30 : 1$ and $\tau = 10^{-11}$ s for the both Q- and R-states and $E_g = 0.083 t_x$ for the R-state, respectively. The *arrows p and q* indicate the TAE anomalies from the outer and inner Fermi surfaces, respectively (from [8])

$t_x : t_y : t_z = 300 : 30 : 1$ and $\tau = 10^{-11}$ s, and for the R-state $E_g = 0.083 t_x$. As is clear from Fig. 14.12, the anion ordering makes the Fermi surface flatter. This is the reason why the calculated $\Delta\phi$ is wider in the Q-state than in the R-state. The arrows p and q in Fig. 14.14 indicate the TAE anomalies corresponding to the outer and inner Fermi surfaces. Overall structure of solid curves is very similar to the experimental result in Fig. 14.13. The lack of observing one of the TAE anomalies from the inner Fermi surface in Fig. 14.13 is because the real system is triclinic and the positions of the TAE anomalies from the inner and outer Fermi surfaces are closer in the left side than in the right one. The position of the four TAE anomalies is calculated as the inflection angle ϕ_i by using (14.51) with $t_z = 0$ and plotted as a function of E_g in Fig. 14.15. Here, the dimensionality $\xi = t_y/t_x$ is fixed to 1/12 from the TAE measurement under pressure, which is described in Sect. 14.5. Then, the experimental value of $\Delta\phi = 24°$ gives $E_g = 0.028 t_x = 0.17 t_y$. Several groups have independently estimated the magnitude of E_g by other methods. The comparison of the estimated E_g is described in [12] as well as in a recent report by Ha et al. [25]. Since $(TMTSF)_2ClO_4$ has the two pairs of Q1D Fermi surfaces with different nesting condition from each other, it is remarkably challenging to understand its ground states, namely phase transition from metallic to superconducting as well as complicated field-induced SDW states. Recent discussions on the magnitude of E_g or the anion potential relative to t_y are seen in [34–39].

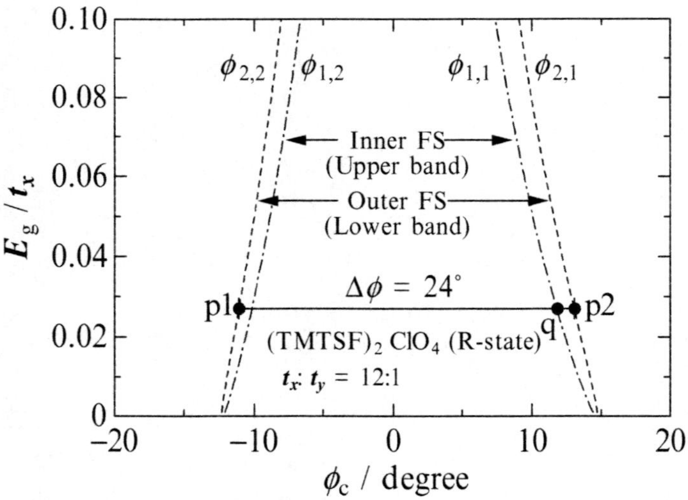

Fig. 14.15. The dependence of the angular positions ($\phi_{1,1}$, $\phi_{1,2}$, $\phi_{2,1}$, and $\phi_{2,2}$) of the third angular effect anomaly on the anion gap (E_g) determined from the inflection angle of the Fermi surface by using (14.51), assuming $\xi = t_y/t_x = 1/12$ and $t_z = 0$. The *broken and chain curves* are for the outer and inner Fermi surfaces, respectively

14.5 Pressure Dependence of the Dimensionality

The application of the TAE to investigate the pressure dependence of the dimensionality $\xi = t_y/t_x$ is described in this section. The sharp anomalies of the TAE give precise information of the shape of Q1D Fermi surface, for which conventional tools for Fermiology such as Shubnikov–de Haas effect are not applicable. Anisotropy of plasma frequency within the most conducting plane also gives some information on dimensionality. It is, however, usually difficult to carry out the optical measurement under pressure. *Thus, the TAE is very important as a tool to investigate the dimensionality of the Q1D conductors under pressure.*

The pressure dependence of the ground state of Q1D conductors is one of the most important problems that have attracted the attention of a lot of researchers. For TMTSF and TMTTF (=tetramethyltetrathiafulvalene) salts, spin-Peierls, SDW, superconducting, metallic states appear as hydrostatic and chemical pressures increase in a generalized temperature–pressure phase diagram [40]. The effect of pressure on the ground state is not so clear because it changes not only total bandwidth but also dimensionality. From the viewpoint of suppressing the SDW transition caused by Peierls instability of Q1D Fermi surface, pressure seems to increase the dimensionality of the system resulting in the worse nesting of the Fermi surface. There has been, however, no direct evidence for the dimensionality enhancement by pressure until recently the TAE is applied for this problem.

Yoshino and Murata et al. measured the TAE of $(TMTSF)_2PF_6$ [41], $(DMET)_2I_3$ [41], and $(TMTSF)_2ClO_4$ [42] under several pressures and found the increase in $\Delta\phi$ by pressure. Figure 14.16 shows the result for $(TMTSF)_2ClO_4$ up to 0.13 GPa [42]. In this measurement, the TAE was measured not only in the R-state (a) but also in the Q-state (b). Hydrostatic pressure was applied on the sample by using a Cu–Be clamped pressure cell with a mixture of olefin oils (Daphne 7373, Idemitsu CO, Ltd [43]) as pressure medium. The angular dependence of MR was measured at 1.2 K and 5 T by rotating the superconducting split magnet. The R-state is attained by the slow cool using a heater contacted on the cell. The cell was once cooled down to 4.2 K to realize the Q-state. The temperature around the sample was raised in short time above the anion-ordering temperature by another heater in the pressure medium, then rapid cooling to suppress the anion ordering was achieved simply by turning off the heater switch.

The definition of $\Delta\phi$ is shown in Fig. 14.16a for the result at 0.26 GPa in the R-state and in Fig. 14.16b for 1.28 GPa in the Q-state, respectively. The increase in $\Delta\phi$ is obvious for the result of the R-state. The increase in $\Delta\phi$ observed for $(TMTSF)_2PF_6$ [41] is due to the enhancement of dimensionality. In $(TMTSF)_2ClO_4$, however, the relative decrease in E_g as compared with the bandwidth ($\sim 4t_x$) should also be considered.

The TAE anomalies are appreciable only above 0.6 GPa in the Q-state as in Fig. 14.16b, while the density-wave transition seems to be suppressed above

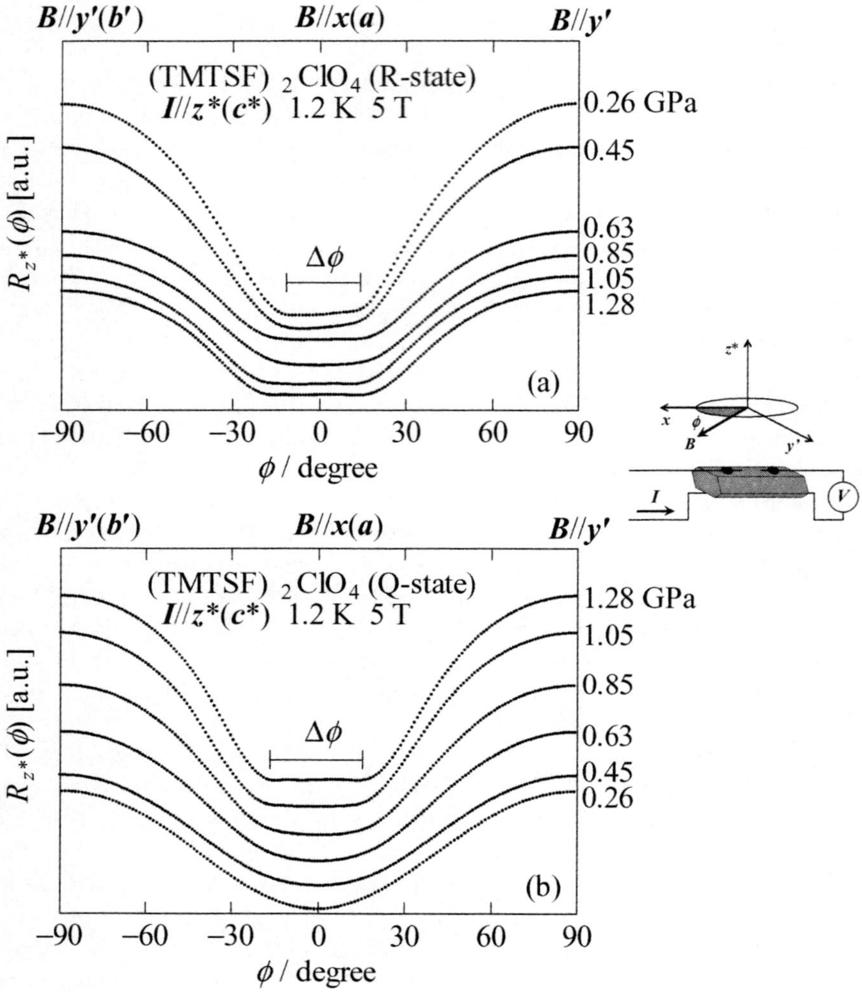

Fig. 14.16. Angular dependence of the $z^*(c^*)$-axis magnetoresistance of $(TMTSF)_2ClO_4$ in the (**a**) R- and (**b**) Q-states at 1.2 K and 5 T up to 1.28 GPa. The magnitude of the pressure is that at the lowest temperature (from [42])

0.27 GPa from the temperature dependence of the electrical resistivity. The random potential of the ClO_4^- anions probably suppresses the TAE anomaly at lower pressure in the Q-state.

Figure 14.17 shows the pressure dependence of $\Delta\phi$ in the R- and Q-states of $(TMTSF)_2ClO_4$. The extrapolation of the result for the R-state (open circles) from higher pressure to 0 GPa coincides with that reported previously (open square) [8]. On increasing pressure, $\Delta\phi$ becomes large, but the effect of pressure is larger below 0.6 GPa than above that, where $\Delta\phi$ is almost the same for the Q- and R-states.

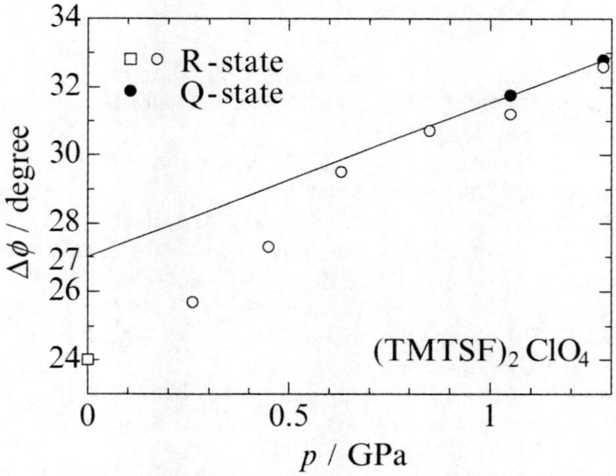

Fig. 14.17. Pressure dependence of the width ($\Delta\phi$) of the third angular effect anomaly of $(TMTSF)_2ClO_4$ in the R- (*open circles and square*) and Q-states (*closed squares*). The *solid line* shows extrapolation of closed circles to 0 GPa (from [42])

The solid line in Fig. 14.17 was drawn by assuming the linear pressure dependence of $\Delta\phi$ in the Q-state as is observed for other Q1D metals [41]. The extrapolated value of $\Delta\phi = 27°$ at 0 GPa in the Q-state, that will be observed if the anion-ordering and the density-wave transitions do not occur, gives $\xi = 1/12$ following the procedure described in Sect. 14.4. On the other hand, $\xi = 1/5.6$ is obtained for $\Delta\phi = 33°$ at 1.28 GPa in the Q-state by assuming the same lattice parameters as those at ambient pressure. Thus, the effect of the pressure on the dimensionality is rather high for $(TMTSF)_2ClO_4$. This is *one of the first direct evidences of the dimensionality enhancement of the Q1D conductors by hydrostatic pressure.*

References

1. A.G. Lebed, P. Bak, Phys. Rev. Lett. **63**, 1315 (1989)
2. T. Osada, S. Kagoshima, N. Miura, Phys. Rev. B **46**, 1812 (1992)
3. H. Yoshino, K. Saito, K. Kikuchi, H. Nishikawa, K. Kobayashi, I. Ikemoto, J. Phys. Soc. Jpn. **64**, 2307 (1995)
4. G.M. Danner, W. Kang, P.M. Chaikin, Phys. Rev. Lett. **72**, 3714 (1994)
5. H. Yoshino, K. Murata, T. Sasaki, K. Saito, H. Nishikawa, K. Kikuchi, K. Kobayashi, I. Ikemoto, J. Phys. Soc. Jpn. **66**, 2248 (1997)
6. T. Osada, S. Kagoshima, N. Miura, Phys. Rev. Lett. **77**, 5361 (1996)
7. T. Osada, Synth. Met. **86**, 2143 (1997)
8. H. Yoshino, A. Oda, T. Sasaki, T. Hanajiri, J. Yamada, S. Nakatsuji, H. Anzai, K. Murata, J. Phys. Soc. Jpn. **68**, 3142 (1999)
9. H. Yoshino, A. Oda, T. Sasaki, T. Hanajiri, J. Yamada, S. Nakatsuji, H. Anzai, K. Murata, Synth. Met. **103**, 2251 (1999)

10. M.J. Naughton, I.J. Lee, P.M. Chaikin, G.M. Danner, Synth. Met. **85**, 1481 (1997)
11. I.J. Lee, M.J. Naughton, Phys. Rev. B **57**, 7423 (1998)
12. H. Yoshino, K. Saito, H. Nishikawa, K. Kikuchi, K. Kobayashi, I. Ikemoto, J. Phys. Soc. Jpn. **66**, 2410 (1997)
13. H. Yoshino, K. Murata, K. Saito, H. Nishikawa, K. Kikuchi, I. Ikemoto, Phys. Rev. B **67**, 035111 (2003)
14. H. Ito, D. Suzuki, Y. Yokochi, S. Kuroda, M. Umemiya, H. Miyasaka, K.-I. Sugiura, M. Yamashita, H. Tajima, Phys. Rev. B **71**, 212503 (2005)
15. A.G. Lebed, N.N. Bagmet, Phys. Rev. B **55**, R8654 (1997)
16. A.G. Lebed, N.N. Bagmet, Synth. Met. **85**, 1493 (1997)
17. H. Yoshino, K. Murata, K. Saito, H. Nishikawa, K. Kikuchi, I. Ikemoto, J. Phys. Soc. Jpn. **65**, 3127 (1996)
18. H. Yoshino, K. Murata, J. Phys. Soc. Jpn. **68**, 3027 (1999)
19. B. Gallois, D. Chasseau, J. Gaultier, C. Hauw, A. Filhol, G. Bechgaard, J. Phys. (Paris) **44**, C3-1071 (1983)
20. S. Sugawara, T. Ueno, Y. Kawasugi, N. Tajima, Y. Nishio, K. Kajita, J. Phys. Soc. Jpn. **75**, 053704 (2006)
21. H. Kaneyasu, K. Kishigi, Y. Hasegawa, J. Phys. Soc. Jpn. **75**, 023709 (2006)
22. A.G. Lebed, M.J. Naughton, Phys. Rev. Lett. **91**, 187003 (2003)
23. A.G. Lebed, N.N. Bagmet, M.J. Naughton, Phys. Rev. Lett. **93**, 157006 (2004)
24. A.G. Lebed, H.I. Ha, M.J. Naughton, Phys. Rev. B **71**, 132504 (2005)
25. H.I. Ha, A.G. Lebed, M.J. Naughton, Phys. Rev. B **73**, 033107 (2006)
26. T. Osada, Physica E **12**, 272 (2002)
27. T. Osada, M. Kuraguchi, K. Kobayashi, E. Ohmichi, Physica E **18**, 200 (2003)
28. T. Osada, M. Kuraguchi, Synth. Met. **133–134**, 75 (2003)
29. B.K. Cooper, V.M. Yakovenko, Phys. Rev. Lett. **96**, 037001 (2006)
30. M.Z. Aldoshina, L.O. Atovmyan, L.M. Gol'denberg, O.N. Krasochka, R.N. Lyubovskaya, R.B. Lyubovskii, M.L. Khidekel', Dokl. Akad. Nauk SSSR **289**, 1140 (1986) (in Russian); Phys. Chem. **289**, 689 (1986) (English translation)
31. J.M. Williams, J.R. Ferraro, R.J. Thorn, K. Douglas Carlson, U. Geiser, H.H. Wang, A.M. Kini, M.-H. Wahangbo, *Organic Superconductors (Including Fullerenes)* (Prentice Hall, New Jersey, 1992)
32. N. Thorup, G. Rindorf, H. Soling, K. Bechgaard, Acta Crystallogr. B **37**, 1236 (1981)
33. P.M. Grant, J. Phys. (Paris) **44**, C3-847 (1983)
34. K. Kishigi, J. Phys. Soc. Jpn. **67**, 3825 (1998)
35. M. Miyazaki, K. Kishigi, Y. Hasegawa, J. Phys. Soc. Jpn. **68**, 313 (1999)
36. K. Sengupta, D. Dupuis, Phys. Rev. B **65**, 035108 (2001)
37. S. Haddad, S. Charfi-Kaddour, C. Nickel, M. Héritier, R. Bennaceur, Phys. Rev. Lett. **89**, 087001 (2002)
38. S. Uji, S. Yasuzuka, T. Konoike, K. Enomoto, J. Yamada, E.S. Choi, D. Graf, J.S. Brooks, Phys. Rev. Lett. **94**, 077206 (2005)
39. S. Haddad, S. Charfi-Kaddour, M. Héritier, R. Bennaceur, Phys. Rev. B **72**, 085101 (2005)
40. See for example, C. Bourbonnais, D. Jérome, Science **281**, 1155 (1998)
41. H. Yoshino, A. Oda, K. Murata, H. Nishikawa, K. Kikuchi, I. Ikemoto, Synth. Met. **120**, 885 (2001)
42. H. Yoshino, S. Shodai, K. Murata, Synth. Met. **133–134**, 55 (2003)
43. K. Murata, H. Yoshino, H.O. Yadav, Y. Honda, N. Shirakawa, Rev. Sci. Instrum. **68**, 2490 (1997)

15

Microwave Spectroscopy of Q1D and Q2D Organic Conductors

S. Hill and S. Takahashi

This chapter reviews recent experimental studies of a novel open-orbit magnetic resonance phenomenon. The technique involves measurement of angle-dependent microwave magneto-conductivity and is, thus, closely related to the cyclotron resonance and angle-dependent magnetoresistance techniques. Data for three contrasting materials are presented: $(TMTSF)_2ClO_4$, α-$(BEDT\text{-}TTF)_2KHg(SCN)_4$, and κ-$(BEDT\text{-}TTF)_2I_3$. These studies reveal important insights into the Fermiology of these novel materials, and provide access to important electronic parameters such as the in-plane Fermi velocity and quasiparticle scattering rate as well. It is argued that all three compounds exhibit coherent three-dimensional band transport at liquid helium temperatures, and that their low-energy magnetoelectrodynamic properties appear to be well explained on the basis of a conventional semiclassical Boltzmann approach. It is also suggested that this technique could be used to probe quasiparticles in nodal superconductors.

15.1 Introduction

Many of the novel broken symmetry states observed in organic conductors are driven by electronic instabilities associated with their low-dimensional Fermi surfaces (FSs), e.g., nesting instabilities [1–6]. Consequently, techniques which can probe the detailed topology of the FSs of organic charge-transfer salts have been widely employed by researchers in this field (for several recent reviews, see [7–10]). Examples include: the de Haas–van Alphen (dHvA) and Shubnikov–de Haas (SdH) effects [7–12]; angle-dependent magnetoresistance oscillations (AMRO) [7–10,13–27]; angle-resolved photoemission spectroscopy (ARPES) [28]; and cyclotron resonance (CR) [29–55]. With the exception of ARPES, all of the above techniques involve the use of strong magnetic fields and low temperatures. In the case of the SdH and dHvA effects, the magnetic field causes Landau quantization which leads to the magneto-oscillatory behavior of various thermodynamic and transport phenomena [9, 11]. The

CR and AMRO effects, meanwhile, are essentially semiclassical in origin, and are caused by the periodic motion of electrons induced by the Lorentz force [18,21,35,36]. Each of these techniques requires that the product $\omega_c\tau > 1$, where ω_c ($= eB/m^*$) is the cyclotron frequency and τ is the transport scattering time. Meanwhile, the SdH and dHvA effects additionally require that $\hbar\omega_c > \{k_BT, h/\tau_\varphi\}$, where τ_φ is the quantum lifetime [11]. These criteria are easily satisfied for many organic conductors due to their exceptional purity. However, instances of the application of such methods to inorganic oxides such as the cuprate and ruthenate superconductors are extremely rare [56–58]. Consequently, ARPES is probably the technique which has been most widely applied, and has contributed most significantly to the understanding of FS driven phenomena in unconventional (including high-T_c) superconductors [59].

In this chapter, we describe a new open-orbit magnetic resonance phenomenon, the so-called periodic orbit resonance (POR), which enables angle-resolved mapping of the in-plane Fermi velocity (v_F) for both quasi-one-dimensional (Q1D) and quasi-two-dimensional (Q2D) organic conductors [35,36,47–49]. As such, this technique is complimentary to ARPES, i.e., it can provide information concerning the in-plane momentum dependence of the density-of-states ($\propto 1/v_F$) and quasiparticle scattering rate (τ^{-1}). However, the POR phenomenon involves measurement of the bulk microwave conductivity [60]. Consequently, it is immune to surface effects which have been known to cause problems in ARPES measurements. Furthermore, the POR technique provides sub-millivolt energy resolution and is, thus, sensitive to extremely fine details of the FS topology [35, 36, 47, 49, 55]. We will illustrate the utility of this method for several organic conductors, including $(TMTSF)_2ClO_4$ and $(BEDT\text{-}TTF)_2X$ [$X = KHg(SCN)_4$ and I_3].

We begin by developing a theoretical framework which enables us to simulate the microwave magneto-conductance of a sample with a Q1D FS topology (see Fig. 15.1). We follow exactly the same semiclassical approach which has been adopted by various researchers to model AMRO data [18, 21]. Indeed, POR represents nothing more than an evolution of AMRO to high frequencies, such that $\omega \sim \omega_c$ and $\omega_c\tau > 1$, where ω is the measurement frequency and ω_c represents the characteristic frequency associated with any periodic motion of quasiparticles on the FS induced by the Lorentz force. AMRO occur due to the commensurate motion of electrons on the FS for certain applied field orientations (provided $\omega_c\tau > 1$). These commensurabilities affect the collective dynamical properties of the electronic system, giving rise to resistance minima ("resonances") when the field is aligned with real space lattice vectors. The idea of "magic angle" resonances goes back 20 years to Lebed [13], who first predicted the occurrence of angle-dependent magnetic field effects in organic conductors. Lebed's theory predicted dimensional crossovers (3D→2D→1D) arising from commensurate electronic motion [13, 14, 26]. Some authors have since suggested that this could lead to Fermi liquid/non-Fermi liquid (FL/NFL) crossovers [61], i.e., to fundamental differences in the thermodynamic ground state at, and away from, Lebed's "magic angles."

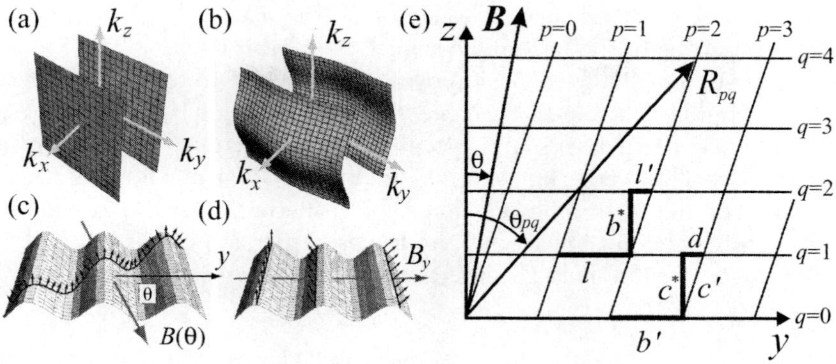

Fig. 15.1. Fermi surfaces corresponding to (15.1) with (**a**) $t_b = t_c = 0$ and (**b**) finite t_b and t_c, which causes a warping of the flat 1D FS in (**a**). The trajectories of quasiparticles on a warped FS (finite t_b, and $t_c = 0$) are shown in (**c**) and (**d**) for different field orientations; the *arrows* represent the quasiparticle velocities. In (**c**), the field results in an oscillatory v_y, whereas this is not the case in (**d**), where the field is applied along the FS warping direction. (**e**) The oblique real-space lattice appropriate to the tight-binding model represented by (15.5). The relevant real-space vectors $R_{pq} = (0, pb' + qd, qc^*)$ for $(TMTSF)_2ClO_4$, and $(0, pl + ql', qb^*)$ for α-$(BEDT\text{-}TTF)_2KHg(NCS)_4$ (Sect. 15.3.2)

When one moves to a rotating frame corresponding to incident microwave radiation of frequency $\omega \approx \omega_c$, one finds that the semiclassical commensurability effects discussed above occur at field orientations which depend on the frequency [27, 36, 47, 55]. In other words, there is nothing "magic" about the angles corresponding to the AMRO minima, as we will demonstrate from POR (AC AMRO) measurements presented in this chapter. Experiments also suggest that, at temperatures below 5 K and at moderate fields, the quasiparticle dynamics for all of the above-mentioned compounds appear to be coherent in all three directions [47, 49, 55]. Indeed, the presented results are consistent with a 3D Fermi-liquid state, and can be easily interpreted in terms of the semiclassical Boltzmann transport equation (or an equivalent quantum mechanical theory [27, 42]). These findings are somewhat at odds with recent thermal transport measurements [62–64] presented elsewhere in this book (Sect. 5.2.3), and do not support the idea of FL/NFL crossovers. Finally, we consider open-orbit POR in Q2D systems subjected to an in-plane magnetic field [49], and discuss the possibility of utilizing this technique to probe the normal quasiparticles in nodal superconductors [54].

15.2 The Periodic Orbit Resonance Phenomenon

We shall mainly limit discussion here to open-orbit POR; for detailed discussion of closed orbit POR observed in Q2D systems, refer to [35, 36]. As a starting point, we consider a Q1D system with a pair of corrugated FS sheets at $k_x = \pm k_F$ (see Fig. 15.1b). Such a FS is typical for the Q1D Bechgaard salts, where electrons delocalize easily along the TMTSF cation chains due to linear stacking of the partially occupied Se π-orbitals [2,4]. The FS corrugations arise from the weak orbital overlap transverse to the chain direction. Within a simple tight binding scheme, such an electronic band structure may be parameterized in terms of a set of highly anisotropic transfer integrals ($t_a : t_b : t_c \approx 200 : 20 : 1\,\mathrm{meV}$ for $(\mathrm{TMTSF})_2\mathrm{ClO}_4$ [4]). We begin a simple treatment of the POR phenomenon by considering an orthorhombic crystal structure, and by linearizing the x-axis dispersion about $\epsilon_\mathrm{F}(=0)$ and $k_x = \pm k_\mathrm{F}$ (we treat a more general case below). The energy dispersion is then expressed as

$$E(\boldsymbol{k}) = \hbar v_\mathrm{F}(|k_x| - k_\mathrm{F}) - 2t_b\cos(k_y b) - 2t_c\cos(k_z c), \tag{15.1}$$

where b and c are the y and z dimensions of the orthorhombic unit cell. The simplest case involves setting $t_b = t_c = 0$, in which case the FS consists of a pair of absolutely flat sheets at $k_x = \pm k_\mathrm{F}$ (Fig. 15.1a). A finite t_b results in a sinusoidal corrugation of the FS, with periodicity $2\pi/b$ directed along the b-(or y-)axis. This situation is illustrated in Fig. 15.1c, d.

We next consider the effect of the Lorentz force $[\hbar\dot{\boldsymbol{k}} = -e(v_\mathrm{F} \times B)]$ on a quasiparticle on the FS due to a magnetic field, B, applied within the yz-plane (i.e., parallel to the FS). The quasiparticle follows a trajectory over the corrugated FS which is perpendicular to B (Fig. 15.1c). For arbitrary orientation of the field within the yz-plane, this gives rise to motion which is periodic with a characteristic frequency

$$\omega_\mathrm{c} = 2\pi\frac{\dot{k}_y}{2\pi/b} = \frac{v_\mathrm{F}eBb}{\hbar}|\sin\theta|, \tag{15.2}$$

where θ is the angle between the magnetic field and the corrugation axis (b-axis in this case). This periodic k-space motion has important consequences for the conductivity which, in the Boltzmann picture, is governed by the time evolution of quasiparticle velocities averaged over the FS, i.e.,

$$\sigma_{ii}(\omega) \propto -\int\left(\frac{\partial f_{\boldsymbol{k}}}{\partial\varepsilon_{\boldsymbol{k}}}\right) v_i(\boldsymbol{k},0)\mathrm{d}^3\boldsymbol{k}\int_{-\infty}^{0} v_i(\boldsymbol{k},t)\mathrm{e}^{-\mathrm{i}\omega t}\mathrm{e}^{t/\tau}\mathrm{d}t, \tag{15.3}$$

where $i = x, y$, or z. Naturally, the time evolution of each velocity component is governed by the field strength and its orientation. For small corrugation ($t_b \ll t_a$), one can neglect variations in v_x to the first order of approximation. However, it is clear that v_y will acquire an oscillatory component with a frequency ω_c given by (15.2). Inserting this time dependence into (15.3) results in the following expression for the y-component of the conductivity tensor:

$$\mathrm{Re}\{\sigma_{yy}(\omega, B, \theta)\} = \frac{\sigma_\mathrm{o}}{2}\left[\frac{1}{1+(\omega+\omega_\mathrm{c})^2\tau^2} + \frac{1}{1+(\omega-\omega_\mathrm{c})^2\tau^2}\right], \tag{15.4}$$

where σ_o is the DC conductivity. This equation reduces to the simple Drude formula for the AC conductivity of a metal for $B = 0$ (i.e., $\omega_c = 0$). For finite magnetic field, (15.4) contains nonresonant and resonant terms where the ω in the denominator of the Drude formula is replaced by $(\omega + \omega_c)$ and $(\omega - \omega_c)$, respectively.

It is important to note that ω and ω_c essentially play identical roles in (15.4), hence the correspondence between the AMRO and POR phenomena. In a finite magnetic field, the DC conductivity $\sigma_{yy}(\omega = 0, \theta)$ exhibits a resonance (resistance minimum) when $\omega = \omega_c = 0$, i.e., when the magnetic field is directed along the b-direction corresponding to a "magic angle" ($\theta = 0$). From Fig. 15.1d, it is clear that when the applied field is directed along the corrugation axis (magic angle), it has no influence on either the y or z components of the quasiparticle velocities. Consequently, the usual Drude behavior is recovered, and both σ_{yy} and σ_{zz} exhibit a maximum at $\omega = 0$. As the field is rotated away from the magic angle, ω_c becomes finite and increases as $\sin\theta$. Consequently, $\sigma_{yy}(\omega = 0)$ decreases (ρ_{yy} increases); the finite ω_c essentially leads to a randomization of v_y between successive collisions and, therefore, to a suppression of σ_{yy}. This is the origin of the resistance minima observed in AMRO experiments (see further discussion below). For a finite ω_c ($\theta \neq 0$), meanwhile, the resonance can be recovered by setting $\omega = \omega_c$; in this situation, v_y remains static in a rotating frame corresponding to the driving frequency ω. This is the origin of the POR phenomenon [35, 36], which was originally discussed by Osada et al. [18]. What one sees, therefore, is that AMRO correspond to DC conductivity resonances observed by sweeping ω_c, while POR correspond to precisely the same resonances except that they are observed at finite ω, i.e., the Drude conductivity peak moves away from $\omega = 0$ to $\omega = \omega_c$. As we shall demonstrate, POR are also best observed by sweeping ω_c; this may be achieved either by varying B directly, or by rotating the field in exactly the same way as one would in an AMRO experiment. For the field rotation measurements (AC AMRO), we note that two resonances should be observed at $\theta = \pm\sin^{-1}\{\hbar\omega/v_F eBb\}$, not a single resonance at $\theta = 0$. Thus, from this semiclassical point of view, one clearly sees that the directions corresponding to AMRO minima are not really "magic."

The preceding discussion assumes the existence of an extended 3D FS, i.e., that the sample under investigation exhibits coherent 3D band transport. Given the extreme anisotropy of many organic conductors, it is natural to consider what would happen if the quasiparticle dynamics were incoherent in one or more directions. A number of authors have explored this limit in which the scattering rate exceeds the hopping frequency in a given direction within a crystal [65–67], e.g., $\tau^{-1} > t_c/\hbar$. In this case, it is meaningless to consider energy dispersion and FS warping in this direction; essentially, the lifetime broadening of quasiparticle states exceeds the bandwidth in that direction. McKenzie et al. have shown that, for layered materials, POR occur for both coherent and weakly incoherent interlayer transport [66]. Thus, the observation of POR does not necessarily imply the existence of a 3D FS. By weakly incoherent, it is implied that one cannot define an interlayer momentum ($\hbar k_z$), but that the in-plane momentum is conserved when a quasiparticle hops between layers [65, 66]. Therefore, it is likely that many of the conclusions drawn from the POR studies described in this chapter apply regardless of whether a truly 3D FS exists. Nevertheless, we will argue that all of the materials investigated in this chapter exhibit a coherent 3D Fermi liquid state at low temperatures (<5 K).

In principle, one could also observe POR by sweeping ω, just as one can use zero-field optical techniques to measure the Drude conductivity in a conventional metal [68]. However, the large effective masses ($\sim 2m_e$) and exceptionally long scattering times (tens of ps) found in many organic conductors at liquid helium temperatures make it very difficult to observe POR using broadband techniques, due to the lack of sensitivity and/or resolution at microwave frequencies [69], hence the need for high-sensitivity narrow-band cavity-based techniques [70–74]. As an aside, as noted above, frequency- and field-swept experiments are essentially equivalent in the Boltzmann theory. Consequently, one should be able to extract essentially the same information from a POR measurement as one would from a broadband optical measurement. We will make such comparisons in Sect. 15.5 of this chapter.

15.2.1 Modification of Theory for Realistic Crystal Structures

As shown in Sect. 15.2, a finite tight-binding transfer integral along the y-direction (perpendicular to the highly conducting x-axis) results in a single resonance in the y-component of the conductivity tensor (σ_{yy}) when $\omega = \omega_c$. In the preceding example, we set $t_c = 0$. Consequently, $v_z = 0$ for all momenta, and it is trivial to show that this leads to $\sigma_{zz} = 0$. If t_c were finite, the FS would acquire an additional corrugation along the orthogonal z-direction. For such an orthorhombic crystal, one can show that the Lorentz force leads to completely separable solutions for $v_y(t)$ and $v_z(t)$ (because the corrugation axes are orthogonal). In fact, $v_y(t)$ [$= 2bt_b \sin\{k_y(t)b\}/\hbar$] depends only on t_b and $v_z(t)$ [$= 2ct_c \sin\{k_z(t)c\}/\hbar$] depends only on t_c. Now that v_z also oscillates, one would expect to observe a DC AMRO minimum in σ_{zz} when the field is directed exactly along the c-axis. At finite frequencies, this resonance would evolve into two POR at angles $\theta' = \pm \sin^{-1}\{\hbar\omega/v_F eBc\}$ where, this time, θ' is the angle between the applied field and the c-(or z-)axis. The important point to note here is that only σ_{zz} is sensitive to t_c and only σ_{yy} is sensitive to t_b. Thus, one would need to independently measure the conductivities along orthogonal directions in such a crystal to extract information about the two transverse tight-binding transfer integrals. We note that an absolute measure of these parameters would require absolute measurements of the conductivity, which is extremely difficult using a narrow-band cavity perturbation technique [69–72]. However, as we shall now see, one can directly obtain information concerning the relative magnitudes of different transfer integrals from measurement of a single conductivity component for crystal structures with symmetry lower than orthorhombic [21, 47].

We now consider the quite general case of an oblique lattice [21, 47], as illustrated in Fig. 15.1e. We also consider the possibility of electron hopping not only between nearest-neighbor chains, but also to next-nearest-neighbors and so on. In the standard fashion, we label chains by a pair of indices p and q, as illustrated in Fig. 15.1e. For convenience, we treat the specific case of the $(TMTSF)_2ClO_4$ lattice for which the crystallographic a-axis corresponds to the chain direction, b' ($\perp a$, $\| ab$-plane) corresponds to the intermediate conducting direction, and c^* ($\perp ab$-plane) corresponds to the least conducting direction [4]. We then reference this lattice to a Cartesian coordinate system by aligning the crystallographic a and b'-directions with the x- and y-axes, respectively. The real-space vectors R_{pq} which define the locations of neighboring chains are then given by $(0, pb' + qd, qc^*)$, where p and q are integers and d is defined in Fig. 15.1e. One can then write down a linearized dispersion relation about ϵ_F $(= 0)$ and $k_x = \pm k_F$ in terms of a set of tight-binding transfer integrals, t_{pq}, between neighboring chains, i.e.,

$$\begin{aligned} E(\boldsymbol{k}) &= \hbar v_F(|k_x| - k_F) - \sum_{p,q} t_{pq} \cos(\boldsymbol{k} \cdot R_{pq}) \\ &= \hbar v_F(|k_x| - k_F) - \sum_{p,q} t_{pq} \cos[(pb' + qd)k_y + (qc^*)k_z]. \end{aligned} \tag{15.5}$$

Each transfer integral t_{pq} will produce a sinusoidal corrugation of the FS, with the corrugation axis defined by the real-space vector R_{pq}, and the period given by $2\pi/|R_{pq}|$; the amplitudes of the corrugations will scale with t_{pq}. More importantly, each t_{pq} will produce a sinusoidal modulation of the quasiparticle velocity components transverse to the chains, and to a corresponding resonance in the conductivity [21]. However, the dispersion relation given by (15.5) does not give rise to completely separable solutions for $v_y(t)$ and $v_z(t)$. In particular, hopping in the c-direction (finite q) will have a direct influence on $v_y(t)$. Furthermore, diagonal hopping terms involving finite p and q will affect both velocity components and, consequently, give rise to resonances in both σ_{yy} and σ_{zz}. Thus, in general, one may now expect to observe several POR harmonics (or AMRO minima) depending on the relative magnitudes of the higher order t_{pq} transfer integrals.

To parameterize this more general case, one can define a set of angles, θ_{pq}, corresponding to interchain directions (see Fig. 15.1e). Then,

$$\tan\theta_{pq} = \frac{pb'}{qc^*} + \frac{d}{c^*}, \tag{15.6}$$

where this angle is referenced to the z-direction [c^*-axis for $(TMTSF)_2ClO_4$] within the plane of the FS (yz-plane) [47]. These are the so-called Lebed magic angles [13]. One can then go through the same procedure as in the preceding section to consider the effect of an applied magnetic field. For field rotation in the plane of the FS, the characteristic POR frequencies are given by

$$\omega_{pq} = \frac{v_F e B R_{pq}}{\hbar} |\sin(\theta - \theta_{pq})| = \omega_{pq}^{\max} |\sin(\theta - \theta_{pq})|, \tag{15.7}$$

where θ is again measured relative to the z-axis, and $\omega_{pq}^{\max}$ ($= v_{\mathrm{F}}eBR_{pq}/\hbar$) is the maximum value of ω_{pq} occurring when $|\theta-\theta_{pq}| = 90°$ [47]. The transverse components of the conductivity tensor (σ_{yy} and σ_{zz}) may then be written as,

$$\mathrm{Re}\{\sigma_{ii}(\omega,\theta,B)\} \propto \sum_{p,q} A_{ii}^{pq}\left[\frac{1}{1+(\omega+\omega_{pq})^2\tau^2}+\frac{1}{1+(\omega-\omega_{pq})^2\tau^2}\right], \quad (15.8)$$

where $A_{yy}^{pq} = \{(pb'+qd)t_{pq}\}^2$ and $A_{zz}^{pq} = \{qc^*t_{pq}\}^2$. From these expressions, one can see that DC AMRO minima occur when $\theta = \theta_{pq}$ ($\omega_{pq} = 0$), provided the appropriate A_{yy}^{pq} is finite. However, in general, POR will be observed when $\omega = \omega_{pq}$, at angles given by (15.7), i.e.,

$$\theta_{\mathrm{POR}} = \theta_{pq} \pm \sin^{-1}(\omega/\omega_{pq}^{\max}). \quad (15.9)$$

Here, one again sees that the resonance condition reduces to the DC result, $\theta = \theta_{pq}$, when $\omega = 0$.

Each resonance, be it a DC AMRO minimum or a POR, corresponds to a given transfer integral, t_{pq}. As noted above, the amplitude of each POR is proportional to $A_{ii}^{pq} \propto t_{pq}^2$. Thus, the relative intensities of POR provide a direct measure of the fourier spectrum of the FS warping, since the t_{pq} represent the fourier amplitudes associated with each modulation vector R_{pq} [21]. We note that DC AMRO minima for a given p/q occur at the same angles, irrespective of the values of p and q. This is not the case for a finite frequency POR measurement, where the resonance depends on both θ_{pq} and R_{pq} through the prefactor in (15.7). Indeed, the POR condition (15.7) defines a set of curves in the 2D ω versus θ plane, whereas AMRO correspond simply to the 1D line of points corresponding to the $\omega = 0$ intercepts of the POR curves. The POR data presented in the following section have been performed at many different frequencies, both by sweeping the field (equivalent to sweeping ω) at a fixed orientation relative to the crystal, and by varying the orientation (sweeping θ) of a static field.

15.3 Experimental Observation of POR for Q1D Systems

In this section, we present experimental POR data obtained for two contrasting Q1D systems: $(\mathrm{TMTSF})_2\mathrm{ClO}_4$ and α-$(\mathrm{BEDT\text{-}TTF})_2\mathrm{KHg(SCN)}_4$. Microwave measurements were carried out using a millimeter-wave vector network analyzer and a high-sensitivity cavity perturbation technique; this instrumentation is described in detail elsewhere [73, 74]. To enable in situ rotation of the sample relative to the applied magnetic field, two different techniques were employed. The first involved a split-pair magnet with a 7 T horizontal field and a vertical access. Smooth rotation of the entire rigid microwave probe relative to the fixed field was achieved via a room temperature stepper motor (with 0.1° resolution). The second method involved in situ rotation of the end plate of a cylindrical cavity, mounted with its axis transverse to a 17 T superconducting solenoid. Details concerning this cavity, which provides an angle resolution of 0.18°, have been published elsewhere [74]; a combination of the two methods enables double-axis rotation capabilities.

15.3.1 $(TMTSF)_2ClO_4$

$(TMTSF)_2ClO_4$ undergoes a structural transition at $T_{AO} = 24$ K which involves an ordering of the ClO_4 anions (which lack a center of inversion) [4]. The anion order, which results in a doubling of the unit cell along the *b*-direction, is extremely sensitive to the cooling rate around T_{AO}. When cooled rapidly (quenched), at rates exceeding $50\,K\,min^{-1}$, the ClO_4 anions remain disordered and the system undergoes a transition to a spin–density wave (SDW) insulating ground state below 6 K. On the other hand, if the sample is cooled slowly (relaxed) through T_{AO} at a rate of less than $0.1\,K\,min^{-1}$, the anions order. Relaxed samples remain metallic and eventually become superconducting below about 1 K. Thus, great care was taken in all experiments on $(TMTSF)_2ClO_4$ to ensure reproducible cooling of the sample. Indeed, all of the data presented in this chapter were obtained for samples cooled at rates between 0.01 and $0.1\,K\,min^{-1}$ from 32 to 17 K. Even for these slow cooling rates, subtle differences are found in the data for different cooling rates.

Figure 15.2 shows changes in the microwave loss measured in a cylindrical TE011 ($f_{011} = 52.1$ GHz) cavity containing a single $(TMTSF)_2ClO_4$ crystal at a temperature of 2.0 K; the frequency is 52.1 GHz in (a) and 75.5 GHz in (b)

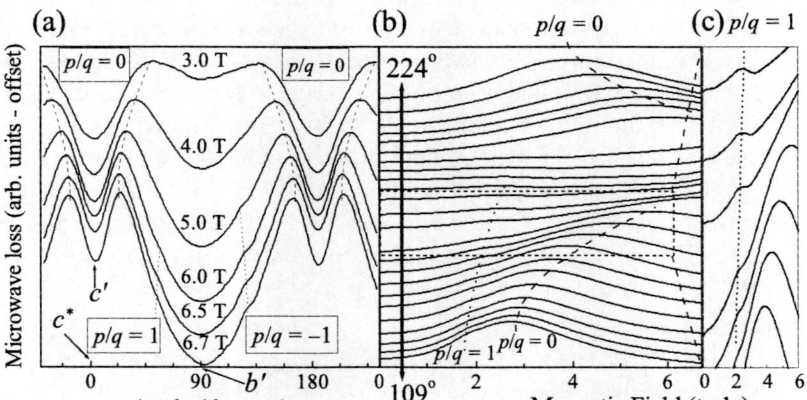

Fig. 15.2. (**a**) Angle- and (**b**, **c**) field-swept microwave absorption data ($\propto \sigma_{zz}$) for $(TMTSF)_2ClO_4$ (sample C) [55]. The peaks correspond to POR (offset for clarity), which have been labeled according to the scheme described in the main text. The field was rotated in the $b'c^*$-plane; the temperature was 2 K, and the frequency was 52.1 GHz in (**a**) and 75.5 GHz in (**b**, **c**). The angle step in (**b**) is 5° and the data in (**c**) correspond to an enlargement of the small resonances denoted as $p/q = 1$ in (**b**)

and (c). The sample was positioned so that the c^*-axis conductivity dominates the losses in the cavity arising from the perturbation due to the sample [60]. The c^*-axis conductivity is sufficiently low and the sample sufficiently thin so that microwave-induced c^*-axis currents penetrate more or less uniformly throughout the sample (depolarization regime). In this regime, the loss arising due to the sample is proportional to the conductivity [70–72]. Therefore, one can equivalently consider the vertical axis in Fig. 15.2 to represent AC conductivity. In Fig. 15.2a, data are plotted (offset) as a function of the orientation of the applied magnetic field within the $b'c^*$-plane, for several applied field strengths. In Fig. 15.2b, c, data are plotted (offset) as a function of the field strength, for many field orientations.

The first thing to note is the resemblance of the data in Fig. 15.2a to DC AMRO data, albeit inverted (conductivity rather than resistivity is plotted). Indeed, the quality of the data are comparable to the best AMRO measurements [25], even though the present technique is contactless and extremely sensitive to the mechanical stability of the instrumentation. Clear conductivity resonances (resistance minima) are seen either side of the positions close to 0° and 180° (labeled $p/q = 0$ and marked by dashed lines); the absolute conductivity is minimum (resistance maximum) close to 90°. Less obvious are weak conductivity resonances (labeled $p/q = \pm 1$ and marked by dotted lines) superimposed on the sloping background associated with $p/q = 0$ resonances. The most striking aspect of all of these resonances is the fact that their positions depend on the magnetic field strength, i.e., they do not occur at fixed field orientations (magic angles). This result is quite different from the DC limit, where AMRO minima are always observed at the same field orientations for any field strength. Nevertheless, as we will now explain, the trends observed here at high frequency are entirely consistent with (15.6)–(15.9).

To better understand the general form of the data in Fig. 15.2a, it is important to first calibrate the field orientation relative to the crystallographic axes. Separate measurements to higher magnetic fields (not shown, see [55]) reveal clear features associated with the field-induced spin–density wave (FISDW) transition. These features scale in field as $1/\cos\theta$ [75] and do not shift with frequency; θ is the angle between the applied field and the crystallographic c^* direction. In this way, one can identify the principal symmetry directions, and these are indicated in Fig. 15.2a.

Conductivity maxima (resonances). The c^*-axis conductivity (σ_{zz}) is dominated by hopping in the c direction, i.e., $p = 0$ and $q = 1$, or $p/q = 0$ (see Fig. 15.1). In low-field c^*-axis DC ($\omega = 0$) AMRO measurements, the deepest resistance minima (σ_{zz} maxima) occur when $\omega_{01} = 0$, i.e., when $\theta = \theta_{01}$ or, equivalently, when the field is directed along the appropriate FS warping direction (see Fig. 15.1e and [25]), which is c' (R_{01}) in this case (projection of c onto the FS ($b'c^*$-plane)). Here, we see in fact that σ_{zz} resonances *do not* occur when the field is directed along c'. This is due to the finite measurement frequency (denominator in (15.8)). One instead sees resonances when $\omega_{01} = \pm\omega$, i.e., either side of c' at angles given by (15.9). Nevertheless, in the high-field limit $\omega_{01}^{\max} \gg \omega$ ($\omega/\omega_{01}^{\max} \to 0$), the resonance condition reverts to the DC result. This is apparent in Fig. 15.2a where one sees that the two main conductivity peaks move together with increasing field such that they should eventually superimpose at $\theta = \theta_{01}(\equiv c')$ in the infinite field limit (see also Fig. 15.3 below). In addition to the $p/q = 0$ resonances, weaker $p/q = \pm 1$ resonances can be seen either side of 90°.

Conductivity minima. The c^*-axis conductivity should be minimum when $|\omega - \omega_{01}|$ is maximum (see (15.8)). For $\omega_{pq}^{\max} > \omega > 0$ there are two such directions: (1) when $\omega_{01} = 0$ ($\theta = \theta_{01}$), corresponding to the field applied along c' and (2) when $\omega_{01} = \omega_{01}^{\max}$ [$(\theta - \theta_{01}) = 90°$], corresponding to the field directed 90° away from c' ($\cong 5°$ away from b'). The data in Fig. 15.2a show precisely two such minima, with a conductivity resonance in between these directions. For $\omega > \omega_{pq}^{\max}$, there is no resonance: the conductivity nevertheless exhibits a nonresonant maximum when $\omega_{01} = \omega_{01}^{\max}$ and a minimum when $\omega_{01} = 0$, corresponding, respectively, to the field applied parallel and perpendicular to c'.

Further evidence that the POR *do not* occur at the magic angles can be seen in Fig. 15.2b, c, where the field strength is varied while its orientation is fixed. Again, one clearly observes a broad peak in σ_{zz} in Fig. 15.2b (labeled $p/q = 0$ and marked by dashed lines) whose position in field shifts upon varying the field orientation. As will be seen below (Fig. 15.3), this conductivity resonance corresponds to precisely the same $p/q = 0$ resonance observed in Fig. 15.2a. In fact, closer inspection also reveals clear evidence for the $p/q = \pm 1$ resonances – see dotted curve and expanded view in Fig. 15.2c. The very fact that one can observe these resonances without rotating the field indicates that the POR angles change upon varying the magnetic field strength, i.e., the POR (AC AMRO) angles *are not* "magic."

By combining all data obtained from the two methods, one can compile a 2D plot of all POR data (either resonance field versus angle or resonance angle versus field). In fact, the relevant parameters are the field orientation θ and the ratio ω/B or f/B, where f is the microwave frequency (see (15.7)). Figure 15.3 displays such 2D plots compiled from data obtained at several frequencies (45–76 GHz) for three different samples (A, B, and C); in every case, the field was nominally rotated within the $b'c^*$-plane. By plotting data in this way, one can see from (15.7) that each series of resonances should collapse onto a single sinusoidal arc given by $A_o \sin|\theta - \theta_{pq}|$, where the amplitude $A_o = ev_F R_{pq}/h$ gives a direct measure of the Fermi velocity (provided the R_{pq} are known). Earlier experiments on samples A and B (Fig. 15.3a) were conducted only in the fixed angle, swept field mode, and only $p/q = 0$ resonances were observed for these samples which were obtained from the same synthesis. As shown in Fig. 15.3a, to within the experimental error, all of the data for samples A and B collapse onto a single curve. Even data obtained for metallic samples cooled at different rates (A1 and A2) lie on the same curves, in spite of the fact that the POR linewidths differ significantly ($\tau \sim 1.5$–7 ps [76]). Experiments on sample A were conducted in a high-field magnet system, allowing for identification of

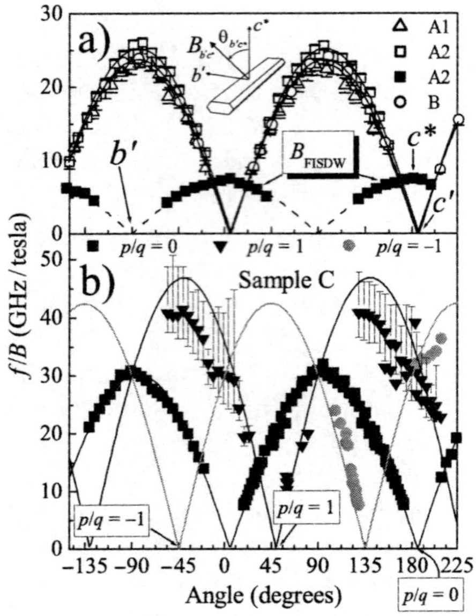

Fig. 15.3. Angle dependence of f/B for (**a**) samples A and B and (**b**) sample C, for field rotation in the $b'c^*$-plane; the principal crystal axes are indicated in (**a**) and the different POR have been labeled in both figures. A1 and A2 denote different cool downs for sample A. The *solid curves* are fits to (15.7) (see main text for detailed explanation). In (**a**), data are also included for the angle dependence of the FISDW transition (*solid squares*), and the inset depicts the experimental geometery

the FISDW transition (solid data points). The angle dependence of B_{FISDW} was used to confirm the orientation of the sample [55]. Using the accepted value, $R_{01} = 13.1\,\text{Å}$, and the value for A_o ($= 24 \pm 1\,\mathrm{GHz\,T^{-1}}$) deduced from Fig. 15.3a, a value of $v_F = 7.6 \pm 0.3 \times 10^4\,\mathrm{m\,s^{-1}}$ was obtained for samples A and B.

The data presented in Fig. 15.3b were obtained for a third sample (C) taken from a separate synthesis. When fully relaxed ($\tau \sim 6\,\mathrm{ps}$), this sample *does* show harmonic POR corresponding to $p/q = \pm 1$ (in addition to $p/q = 0$). Surprisingly, the obtained $A_o = 30 \pm 1\,\mathrm{GHz\,T^{-1}}$ is significantly higher for this sample, giving a value for the Fermi velocity of $v_F = 9.5 \pm 0.3 \times 10^4\,\mathrm{m\,s^{-1}}$. Very preliminary studies on yet another sample also yielded similarly high A_o values [48]. Therefore, it seems that sample quality has a direct influence on the electronic bandwidth. This could be related to the proximity of the SDW phase, as cleaner (more metallic) samples exhibit higher v_F values (larger bandwidth). There is also an apparent correlation between the higher v_F samples and the observation of higher harmonic POR. Whether this is due to the longer scattering times or due to differences in electronic structure is not clear. Interestingly, it has been demonstrated in a separate study that there is a direct correlation between the scattering time, τ, and the superconducting T_c, suggesting unconventional pairing in $(TMTSF)_2ClO_4$ [76].

The solid curves in Fig. 15.3b were generated using (15.7) in the following way: the value of A_o, and the b' and c' directions were determined on the basis of fits to the $p/q = 0$ POR data (solid squares); the lighter colored curves corresponding to the $p/q = \pm 1$ POR were then simulated simply by replacing R_{pq} and θ_{pq} with the appropriate (published [4]) values in (15.7). As can be seen, the agreement is rather good, especially at higher fields (smaller f/B). The error bars increase for the weaker $p/q = \pm 1$ POR at the highest f/B values because the absolute uncertainty in determining their location is relatively field independent (even increasing slightly with decreasing field). The discrepancy at low fields between the harmonic POR and the simulations could also indicate a possible field dependence of the electronic bandwidth (v_F). However, this discrepancy is barely outside of the accuracy of the measurement.

Each POR harmonic observed in Figs. 15.2 and 15.3b arises from a particular tight-binding transfer integral t_{pq}. The relative amplitudes of each of the POR harmonics are related directly to these integrals through the coefficients, A_{ii}^{pq} ($\propto t_{pq}^2$), in (15.8). The t_{pq} affect the nature of the FS warping (15.5) which, in turn, influences the tendency of the FS to nest. A higher harmonic content to the FS warping tends to suppress density-wave instabilities. Once again, the general trends observed in these experiments agree with other experimental observations, namely that the more metallic samples exhibit higher harmonic POR (higher harmonic FS warping). Nevertheless, it is quite clear from these measurements that interlayer hopping ($q = 1$) to the nearest-neighbor ($p = 0$) sites is considerably stronger than hopping to next-nearest-neighbor ($p = -1$) and third-nearest neighbor ($p = +1$) sites – see Figs. 15.1 and 15.2. In the following section, we present contrasting data for a system showing very high harmonic POR content. Due to the significantly higher conductivity in $(TMTSF)_2ClO_4$ along the b' direction, it has not been possible to observe $q = 0$ POR.

There are some important differences between the σ_{zz} POR data displayed in Fig. 15.2 and published DC AMRO data for $(TMTSF)_2ClO_4$ [25]. In the present work, POR are observed in the $\omega \rightarrow 0$ limit when the field is along c' ($p/q = 0$) and roughly $\pm 45°$ away from c' ($p/q = \mp 1$). DC measurements reveal very clear evidence for higher harmonic AMRO such as $p/q = \pm 2$ and ± 3. However, these are only seen in regions of parameter space that were inaccessible with the instrumentation used for the microwave measurements. Comparisons of interlayer ($q = 1$) POR and AMRO data obtained over comparable temperature, field and field orientation ranges reveal similar behaviors in terms of the relative strengths of the $p/q = 0$ and ± 1 resonances. However, a surprising aspect of the DC AMRO measurements is a strong, sharp $q = 0$ ($p/q = \infty$) resonance when the field is parallel to b' [25], though this dip is less obvious in separate earlier studies [22]. There is no evidence for such a resonance in the microwave measurements. The reason for this difference is not entirely clear. According to (15.8), σ_{zz} should not exhibit any $q = 0$ resonances, since $A_{zz}^{pq} = 0$ for $q = 0$. One possibility may be that the DC σ_{zz}

measurements are contaminated with σ_{yy}, which should display a dominant $p/q = \infty$ resonance. However, this may suggest additional (perhaps nonclassical) effects in the DC conductivity which do not influence the microwave conductivity [61].

15.3.2 α-(BEDT-TTF)$_2$KHg(SCN)$_4$

Like the Bechgaard salts, the Q2D α-(BEDT-TTF)$_2$MHg(SCN)$_4$ (M = NH_4, K, Tl, Rb) family of organic charge-transfer salts have a rich history in terms of studies of their Fermiology [4, 7, 8, 10, 21, 30–32, 34–36, 41, 42, 47]. Indeed, the first CR studies were performed on the M = K member of this family [30], which is also the focus of the present section. For this compound, the least conducting (z-) direction is parallel to the crystallographic b^*-axis [4,77]. Meanwhile, the conductivity within the ac-plane ($\perp b^*$) is rather more isotropic than the conductivity within the ab-plane for (TMTSF)$_2$ClO$_4$. Consequently, the room temperature FS of α-(BEDT-TTF)$_2$KHg(SCN)$_4$ has a more two-dimensional character [77, 78], comprising both open (Q1D) and closed (Q2D) pockets (see Fig. 1 in [47]). At low temperatures, matters are complicated by the fact that α-(BEDT-TTF)$_2$KHg(SCN)$_4$ undergoes a charge density wave (CDW) transition at 8 K [4, 79], which leads to a reconstruction of the room temperature FS [7, 8, 10, 19, 80]. Unlike the anion ordering in (TMTSF)$_2$ClO$_4$, the low temperature state in α-(BEDT-TTF)$_2$KHg(SCN)$_4$ is not sensitive to the cooling rate through this CDW transition. The precise nature of the reconstructed FS remains controversial. However, it is believed that the open sections nest, and that the closed pockets reconstruct in such a way as to give rise to new open FS sections together with smaller closed pockets [19,80]; one of the original proposals for the reconstructed FS is shown in Fig. 1 of [47]. This model serves to illustrate most aspects of the POR data presented in this section.

Due to the uncertainty associated with the low temperature FS, several early magnetooptical studies of α-(BEDT-TTF)$_2$KHg(SCN)$_4$ incorrectly attributed resonant absorptions to the conventional CR phenomenon [30, 31]. The first indications that they were in fact due to open-orbit POR came from angle-dependent cavity perturbation measurements by Ardavan et al. [41]. In these investigations, the applied magnetic field was rotated in two different planes perpendicular to the highly conducting ac-plane, which is easily identified from the plate-like shape of a typical single crystal. In the case of Q2D CR, the cyclotron frequency depends only on the magnitude of the field component perpendicular to the conducting layers and should, therefore, be insensitive to the particular plane of rotation [35]. The main clue that the resonances observed in α-(BEDT-TTF)$_2$KHg(SCN)$_4$ were due to Q1D POR came from the fact that the angle dependence varied strongly with the plane of rotation [36]. In this study, Ardavan et al. were able to fit data corresponding to two FS warping components to sinusoidal arcs of the form given by (15.7) [41].

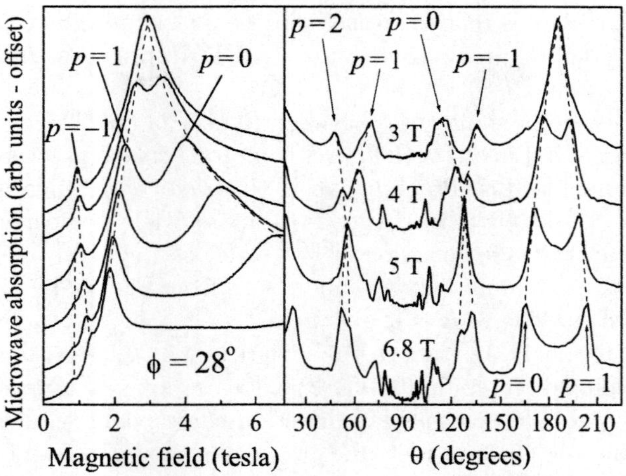

Fig. 15.4. Microwave absorption data ($\propto \sigma_{zz}$) for α-(BEDT-TTF)$_2$KHg(SCN)$_4$ obtained in (**a**) field- and (**b**) angle-swept modes (from [47]). All data were obtained at 2.2 K and 53.9 GHz for field rotation in a plane inclined at $\phi = 28°$ with respect to the Q1D FS, and the traces have been offset for clarity. The POR, corresponding to peaks in absorption, have been labeled according to the ratio of p/q

Improvements in spectrometer design (enhanced sensitivity and mechanical stability [73, 74]) have since enabled cavity perturbation measurements with improved signal-to-noise characteristics. Figure 15.4 displays microwave loss data obtained by Kovalev et al. [47] at 2.2 K and 53.9 GHz, both in field-swept (a) and angle-swept (b) mode. Similar to (TMTSF)$_2$ClO$_4$, the sample was positioned within the cavity so as to excite interlayer (b^*-axis) currents throughout the bulk of the sample [60]. Thus, the vertical scale in Fig. 15.4 is proportional to σ_{zz}. Unlike (TMTSF)$_2$ClO$_4$, however, the reduced in-plane anisotropy, together with the 8 K FS reconstruction, makes it impossible to visually identify the orientation of the open FS associated with the low-temperature state. Consequently, one does not initially know the relative angle between the field rotation plane and the Q1D FS (xz-plane). It is therefore necessary to extend the theory developed in Sect. 15.2.1 for arbitrary rotations [47], i.e., not just rotations within the plane of the FS. Before doing so, however, we present a detailed discussion of the data in Fig. 15.4.

The first point to note upon comparison with Fig. 15.2 is the significant increase in the harmonic content of the POR for α-(BEDT-TTF)$_2$KHg(SCN)$_4$. This is particularly apparent from the angle-swept data. Indeed, careful inspection of Fig. 15.4b reveals over 17 clear resonances within a 180° angle range, along with an apparent continuum of peaks as the field orientation approaches 90° ($\perp$ to b^*); several of these peaks have been labeled. This behavior is dramatically different from that observed for (TMTSF)$_2$ClO$_4$, and is indicative of a high harmonic content to the FS warping [21]. However, it is not unexpected, as DC AMRO data are equally rich [8]. As will be seen below, this behavior is related to the intrinsic quasi-two-dimensionality of this BEDT-TTF compound, along with the 8 K CDW transition and the resulting FS reconstruction [19, 80].

There are certain similarities between Fig. 15.2a and 15.4b, but also some key differences. First of all, both figures exhibit a broad conductivity minimum when the applied field is perpendicular to the least conducting direction (90°). Furthermore, the POR patterns for both materials exhibit symmetry about $\theta = 90°$. In contrast, the behavior around (or close to) $\theta = 0°$ and 180° is quite different for the two materials. As can be seen from Fig. 15.2a, the two $p/q = 0$ resonances either side of $\theta = 0$ diverge as $B \to 0$ ($\omega \gg \omega_{01}$ limit), whereas they appear to merge together in the high field ($\omega/\omega_{01} \to 0$) limit, suggesting that the warping direction in $(TMTSF)_2ClO_4$ is along c' (5° away from c^*); in other words, the relevant warping direction corresponds to nearest-neighbor interlayer hopping. By contrast, the two peaks shown in Fig. 15.4b merge at 3 T. In fact, these two peaks do not even correspond to the same POR harmonic. If one looks carefully at the labeling in Fig. 15.4b (*vide infra*), it is apparent that the $p/q = 0$ peaks will merge somewhere around $\theta = 150°$. This implies that the relevant R_{01} tight-binding hopping direction possesses a significant in-plane component. We note that none of the observed POR have a $\omega/\omega_{pq} \to 0$ intercept anywhere close to $\theta = 0$, implying that all relevant interlayer hopping matrix elements possess significant in-plane components. Even though α-$(BEDT\text{-}TTF)_2KHg(SCN)_4$ possesses a low-symmetry triclinic ($P\bar{1}$) structure, this degree of obliqueness of the R_{pq} vectors *cannot* be explained from high temperature (104 K) crystallographic data [77]. We shall discuss this point further below, along with the observation that the POR imply a very weak decay of the t_{pq} with increasing $|p|$.

As discussed above, before we can fit the POR peak positions to the model developed in Sect. 15.2.1, we must modify (15.7) slightly for an arbitrary plane of rotation of the applied field. We also adapt this equation and (15.6) so that they are appropriate for the α-$(BEDT\text{-}TTF)_2KHg(SCN)_4$ crystal structure (see Fig. 15.1e), for which b^* corresponds to the least conducting z-direction. For field rotation in a plane inclined at an angle ϕ with respect to the FS (yz-plane), the DC AMRO condition is given by

$$\tan \theta_{pq} = \frac{1}{\cos \phi}\left[\frac{pl}{qb^*} + \frac{l'}{b^*}\right], \tag{15.10}$$

where, as usual, the angles θ_{pq} are measured relative to the z-axis (see also Fig. 15.1e for a definition of the parameters l and l'). The modified resonance condition is then given by the following expression:

$$\frac{\omega_{pq}}{B \cos \theta} = \frac{v_F e b^*}{\hbar} \cos \phi |\tan \theta_{pq} - \tan \theta|. \tag{15.11}$$

From (15.11) it can be seen that plots of $f/B \cos \theta$ versus $\tan \theta \cos \phi$ should produce straight lines with slope $\pm eb^* v_F/h$, with offsets given by θ_{pq}. Such a

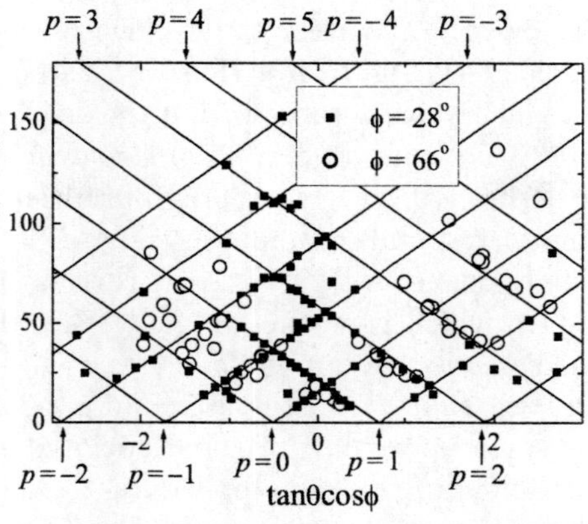

Fig. 15.5. A compilation of all data obtained for α-(BEDT-TTF)$_2$KHg(SCN)$_4$ scaled according to (15.11) (from [47]). Several branches have been labeled with the appropriate index p

plot is shown in Fig. 15.5 for data obtained for two planes of rotation ($\phi = 28°$ and $\phi = 66°$). As can clearly be seen, the data scale well, particularly in the lower central portion of the plot corresponding to field orientations well away from $\theta = 90°$. These results therefore confirm the Q1D nature of the POR, as previously reported by Ardavan et al. [41]. The solid lines represent the best fit to the data with v_F, l, and l' as the only free parameters. The ratios $l/b^* = 1.2$ and $l'/b^* = 0.5$ are in excellent agreement with DC AMRO measurements [81–83]. The obtained value of $v_F = 6.5 \times 10^4\,\mathrm{m\,s^{-1}}$, meanwhile, cannot be deduced from DC AMRO measurements. The deviation between the data and the fit in the peripheral regions of Fig. 15.5 may have several explanations. First of all, errors in the calibration of the field orientation will be amplified for small values of $\cos\theta$ and large values of $\tan\theta$. It is also likely that the above theory breaks down when the field is oriented close to the direction perpendicular to the plane of the Q1D FS (small $\cos\theta$, $\cos\phi$, and large $\tan\theta$) due to the Q2D nature of α-(BEDT-TTF)$_2$KHg(SCN)$_4$, i.e., one can no longer assume that v_x is a constant of the motion, resulting in nonseparability of all three velocity components. One may estimate an effective mass associated with the high-temperature Q2D FS from the obtained value of v_F. However, such a procedure involves making assumptions about the energy dispersion. Nevertheless, the obtained effective mass ($m^* = 1.6$–$2.4m_e$, depending on the assumption made about the energy dispersion) is in reasonable agreement with SdH and dHvA measurements [8] (see [47] for a more in depth discussion).

We end this section with a discussion of the very high harmonic content of the POR observed for α-(BEDT-TTF)$_2$KHg(SCN)$_4$. If one takes the view that each resonance corresponds to a given tight-binding transfer integral, then the present study (as well as DC AMRO measurements [8, 21]) would imply a very weak decay of the t_{pq} with increasing $|p|$ (for $q = 1$). Such a result would not justify the use of a tight-binding approximation, as it would imply significant orbital overlap to next-nearest and next–next-nearest, etc. sites in the lattice (see Fig. 15.1e). However, as pointed out by Blundell et al. [21], such a view is inappropriate for the low-temperature FS in α-(BEDT-TTF)$_2$KHg(SCN)$_4$ due to the 8 K reconstruction that results from the CDW superstructure [19, 80]. The more appropriate view to take is that each POR harmonic corresponds to a fourier component of the warping of the Q1D FS. Interestingly, inspection of Fig. 15.4 reveals similar amplitudes for the $p = 0$ and 1 POR, which have similar $|R_{pq}|$ (they differ in amplitude by less than 10%); the same is true for the $p = 2$ and -1 POR, for which $|R_{21}|$ and $|R_{-11}|$ differ by less than 7%. Thus, it would appear as though the POR intensities scale as some power law of the associated R_{pq} ($\propto b^*/|\cos\theta_{pq}|$). Meanwhile, the corresponding fourier amplitudes, t_{pq}, scale as the square root of the POR intensities, i.e., $A_{zz}^{pq} \propto t_{pq}^2$, see (15.8). We note that for the most extreme anharmonicity, the top hat (or square wave) function, the t_{pq} should scale as R_{pq}^{-1} and the POR intensities as R_{pq}^{-2}: R_{01} and R_{21} differ by roughly a factor of 2 [$(R_{01}/R_{21})^2 = 4$], while the corresponding $p/q = 0$ and $p/q = 2$ POR differ in intensity by about a factor of 6. Thus, the POR intensities imply an extremely high harmonic content to the FS warping, which can only be explained in terms of a reconstruction. It is also clear that the R_{pq} bear no simple relationship with the principal lattice vectors. Again, this is because the corrugations on the reconstructed FS are related to the CDW nesting vector, not the underlying lattice structure. Therefore, POR measurements are entirely consistent with proposed models for the low-temperature Fermi surface of α-(BEDT-TTF)$_2$KHg(SCN)$_4$ [19, 80].

15.4 Open-Orbit POR in a Q2D System

We conclude this experimental survey by describing a new open-orbit POR effect which was recently reported by Kovalev et al. in the Q2D organic conductor κ-(BEDT-TTF)$_2$I$_3$ [49]. The new effect is observed when the magnetic field is applied parallel to the highly conducting layers – the bc-plane for this particular compound. As before, the Lorentz force induces periodic quasiparticle trajectories on the warped FS such that $\dot{\boldsymbol{k}}$ is perpendicular to the applied field. However, in this case, the FS is a warped cylinder, as depicted in Fig. 15.6a. Nonetheless, because the FS is warped (finite dispersion along a), the Lorentz force leads to periodic modulations of the interlayer quasiparticle velocities, v_z, which in turn give rise to resonant contributions to σ_{zz} (see (15.3)). As can be seen from Fig. 15.6a, a range of different trajectories are induced: open-orbits, self-crossing orbits, and closed orbits [10, 84–86]. Furthermore, since $\dot{\boldsymbol{k}}$ is proportional to $(v_F \times B)$, it is apparent that the periodicities associated with each orbit will vary significantly over the FS, i.e., the POR frequencies associated with different states on the FS are not discrete, as was the case in the previous examples. The DC σ_{zz} (AMRO) is believed

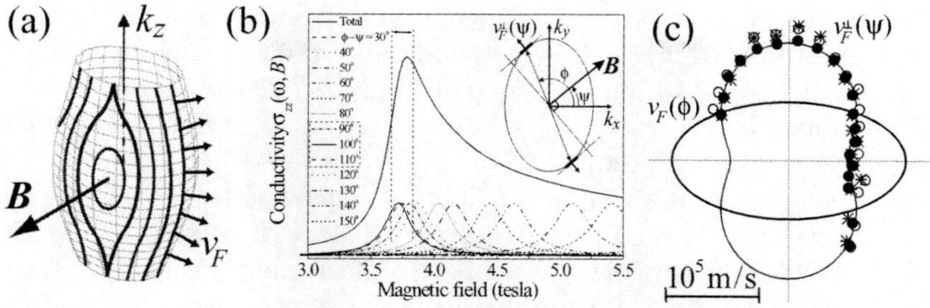

Fig. 15.6. (**a**) An illustration of the quasiparticle trajectories on a warped Q2D FS cylinder for a field oriented perpendicular to the cylinder axis. The resulting trajectories lead to v_z oscillations and to a resonance in σ_{zz} (see main text). The *solid curve* in (**b**) illustrates the conductivity resonance resulting from the electron trajectories in (**a**). Different parts of the FS contribute to different parts of the resonance (*dashed curves*). However, the density of resonances is highest for FS patches which are parallel to the applied field (see inset). Consequently, these FS regions dominate the POR. In (**c**), experimental data are plotted corresponding to $v_{\rm F}^{\perp}$ obtained by various different methods (see [49] for explanation); the *dashed curve* is a fit to the data and the *solid curve* represents the corresponding $v_{\rm F}(\phi)$

to be dominated by the self-crossing orbits, though this has been the subject of some debate [84, 85].

The situation at microwave frequencies is considerably simpler (provided $\omega\tau > 1$). As illustrated in Fig. 15.6b, $\sigma_{zz}(\omega)$ exhibits a peak (quasiresonance) which is dominated by contributions from quasiparticle states corresponding to vertical (symmetry equivalent) strips of the FS which are tangent to the applied field, as depicted in the inset to Fig. 15.6b. The reason for the dominance of these trajectories is twofold. First of all, every periodic trajectory has an associated frequency $\omega_{\rm c}(\boldsymbol{k})$. However, these frequencies vary from zero (smallest closed orbits) up to a maximum cut-off frequency, $\omega_{\rm c}^{\max}$, given by the maximum value of $|(v_{\rm F} \times B)|$. This cut-off gives rise to a singularity in $\sigma_{zz}(\omega)$. Note that $\omega_{\rm c}^{\max}$ corresponds to states with $v_{\rm F} \perp B$, i.e., states on FS sections which are tangent to B. Second, the density of momentum states diverges at the cut-off frequency, i.e., the number of states per unit frequency range diverges for the tangential patches. The resultant conductivity is obtained by integration over all states on the FS (or all frequencies from 0 to $\omega_{\rm c}^{\max}$). As illustrated in Fig. 15.6b, such an integration gives rise to a conductivity resonance. The actual value of $\omega_{\rm c}^{\max}$ ($= eBav_{\rm F}^{\perp}/\hbar$, where a is the interlayer spacing and $v_{\rm F}^{\perp}$ is the Fermi velocity associated with the tangential patches) corresponds to the point on the curve having the maximum slope, i.e., slightly to the low-field side of the peak in $\sigma_{zz}(B)$. Measurement of $\omega_{\rm c}^{\max}$ as a function of the field orientation ψ within the xy-plane yields a plot of $v_{\rm F}^{\perp}(\psi)$. The procedure for mapping $v_{\rm F}^{\perp}(\phi)$ is then identical to that of reconstructing the FS of a Q2D conductor from the measured periods of Yamaji oscillations [4, 87] (see [49] for more in-depth discussion).

Actual experimental data corresponding to $v_F^\perp(\psi)$, along with the deduced $v_F^\perp(\phi)$, are displayed in Fig. 15.6c. As with all of the previous POR investigations, a cavity perturbation technique was employed in such a way as to excite only interlayer currents [60, 73]. The interlayer (a) direction is easily identified due to the platelet shape of a typical κ-(BEDT-TTF)$_2$I$_3$ single crystal. Experiments were performed at 4.5 K (above the superconducting transition temperature of 3.5 K) and at a frequency of 53.9 GHz, corresponding to the TE011 mode of the employed cavity. Use of both the transmitted phase and amplitude enabled precise determination of $\omega_c^{max}(\phi)$ (see [49] for representative spectra) and the scattering time τ (= 5 ps). The FS of κ-(BEDT-TTF)$_2$I$_3$ consists of a network of overlapping weakly warped cylinders. The underlying lattice periodicity results in a removal of the degeneracies at the intersections of these cylinders, giving rise to the coexistence of smaller closed surfaces and open sheets. The deduced anisotropy in $v_F^\perp(\phi)$ is in good agreement with the known anisotropy associated with small Q2D FS for κ-(BEDT-TTF)$_2$I$_3$ [88], i.e., $v_F^x = 1.3 \times 10^5\,\mathrm{m\,s^{-1}}$ and $v_F^y = 0.6 \times 10^5\,\mathrm{m\,s^{-1}}$. Furthermore, one can estimate the effective mass associated with this Q2D pocket using the relation $m^* = \hbar(S_k/S_v)^{1/2}$, where S_k is the cross-sectional area of the FS in k-space and S_v the area enclosed by $v_F^\perp(\phi)$ in 2D velocity space. This procedure gives $m^* = 1.7m_e$ [49], while the experimental value deduced from SdH and dHvA measurements is $\sim 1.9m_e$ [89].

15.4.1 POR in Q2D Nodal Superconductors

In the case of the high-T_c superconductors and other candidate Q2D d-wave superconductors (including several organic conductors [90]), it is generally accepted that normal quasiparticles coexist with the superfluid along vertical line nodes on the original approximately cylindrical high-temperature FS [91] (see Fig. 15.7). These quasiparticles will dominate the low temperature low-

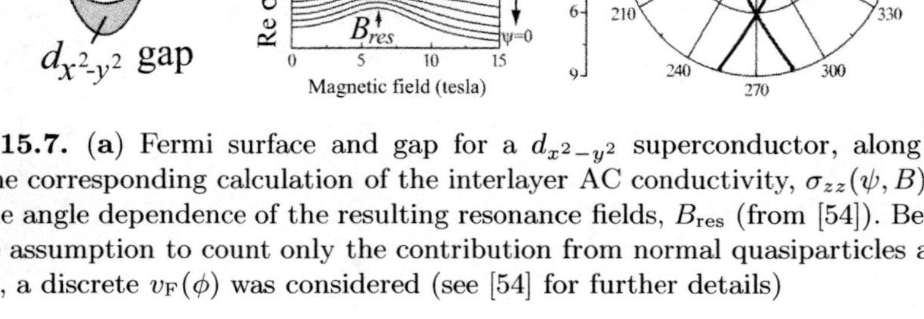

Fig. 15.7. (**a**) Fermi surface and gap for a $d_{x^2-y^2}$ superconductor, along with (**b**) the corresponding calculation of the interlayer AC conductivity, $\sigma_{zz}(\psi, B)$, and (**c**) the angle dependence of the resulting resonance fields, B_{res} (from [54]). Because of the assumption to count only the contribution from normal quasiparticles at the nodes, a discrete $v_F(\phi)$ was considered (see [54] for further details)

energy electrodynamics, including the microwave spectral range (all other single-particle excitations are gapped) [92–94]. A magnetic field applied parallel to the xy-plane $[B(\psi)]$ preserves in-plane momentum. Consequently, such a field will tend to drive quasiparticles along the vertical line nodes, thus preserving the open-orbit POR effect described above. What is more, the nodal quasiparticles will tend to be even longer lived than in the normal state due to the reduced phase space for scattering [92]. Therefore, this dominance of the nodal regions of the FS suggests that it may be possible to directly probe quasiparticles in nodal superconductors via open-orbit POR.

Figure 15.7a illustrates the situation discussed above for a system with a warped elliptical FS (only the cross section is shown), such as κ-(BEDT-TTF)$_2$X (X$=$I$_3$, Cu(NCS)$_2$, etc. [4]), with a $d_{x^2-y^2}$ superconducting gap (shaded area); the dashed curve represents the corresponding angle-dependent Fermi velocity, v_F. Within the superconducting state, the FS is gapped everywhere, apart from at the locations of the four line nodes. Figure 15.7b shows numerical simulations of the field dependent interlayer (z-axis) conductivity, $\sigma_{zz}(\psi, B)$, for different field orientations, due to the quasiparticles existing at the line nodes; refer to [54] for further details of these calculations. The peaks in conductivity correspond to the open-orbit POR described in Sect. 15.4. However, the angle dependence is now dominated by the line nodes, rather than the extremal regions of the FS. Thus, the conductivity exhibits two resonances corresponding to pairs of line nodes on opposite sides of the FS. The resonance field $B_\mathrm{res}(\psi)$ depends simply on the angle between the applied field and the lines joining these line node pairs – hence the two resonances. As shown in Fig. 15.7c, $B_\mathrm{res}(\psi)$ exhibits a fourfold symmetry characteristic of the $d_{x^2-y^2}$ gap, as opposed to the twofold symmetry characteristic of the original un-gapped elliptical FS (Fig. 15.6c).

In principle, one could apply this POR technique to any nodal superconductors which satisfy $\omega\tau > 1$. Using this method, it may be possible to measure both v_F (effective mass) and τ associated with nodal quasiparticles. Moreover, it may also be possible to confirm the symmetry of the superconducting gap from the angle dependence of the POR. The organic superconductors may be particularly attractive for such investigations due to their extreme purity. However, recent experiments suggest that it may also be possible to observe this effect in very pure high-T$_\mathrm{c}$ compounds such as $Y_2Ba_2Cu_3O_{6+x}$ [94] and $Tl_2Ba_2CuO_{6+\delta}$ [58].

15.5 Discussion and Comparisons with Other Experiments

The first four columns in Table 15.1 summarize optical constants deduced from the POR studies outlined in the previous sections, for: (1) (TMTSF)$_2$ClO$_4$, (2) α-(BEDT-TTF)$_2$KHg(SCN)$_4$, and (3) κ-(BEDT-TTF)$_2$I$_3$. The final four columns list various physical parameters obtained from the literature; unless

Table 15.1. Physical parameters deduced from these investigations and from the literature for (1) $(TMTSF)_2ClO_4$, (2) α-$(BEDT\text{-}TTF)_2KHg(SCN)_4$, and (3) κ-$(BEDT\text{-}TTF)_2I_3$

	τ (ps)	v_F (10^4 ms^{-1})	m^* (m_e)	$ne^2\tau/m^*$ (S cm^{-1})	τ_h (ps)	τ_φ (ps)	τ^{opt} (ps)	σ_o^{opt} (S cm^{-1})
1.	1–7*	7.6–9.5*	1.45–1.75*	$\sim 10^6$	0.7^a	–	>10	10^3–10^4
2.	15	6.5	1.6–2.4	3×10^5	0.75^b	0.8–2.5^d	1	$\sim 10^3$
3.	5	10	1.7–2.5	4×10^5	1^c	5^e	$\sim$0.1	$\sim$1,500

The transport scattering time (τ) and Fermi velocity (v_F) were deduced from POR measurements. The effective masses (m^*) for (2) and (3) were deduced from the obtained values of v_F, as described in [47] and [55], respectively; m^* for (1) was deduced on the basis of a tight binding model, as described in [47]. The quantity $ne^2\tau/m^*$ represents the DC conductivity deduced on the basis of the τ and m^* values obtained from the POR measurements, using literature values for the carrier concentration n [4, 77, 95]; we assume that a carrier density corresponding to only 20% of the Brillouin zone contributes to the conductivity in the low-temperature state of (2) [8]. τ_h is the interlayer hopping time deduced from: [a]the interlayer bandwidth [4]; [b]Ref. [47], [c]Ref. [96]. τ_φ is the quantum lifetime deduced from magneto oscillation data: [d]Ref. [8], [e]Ref. [96]. The optical scattering time (τ^{opt}) and DC conductivity (σ_o^{opt}) were taken from Refs. [97,98] for (1) at 10 K, from Ref. [99] for (2) at 6 K, and from Ref. [95] for (3) at 15 K
* Cooling rate dependent

otherwise indicated, these parameters represent low-temperature limiting values. The first point to note is that all three materials appear to display coherent 3D band transport at low temperatures. This can be seen by comparing the scattering times (either the transport time, τ, or the quantum lifetime, τ_φ) with the interlayer hopping times, τ_h, i.e., for all materials, $\tau_h < \{\tau, \tau_\varphi\}$. However, for all three materials, the difference is not so great (roughly an order of magnitude). Thus, one may expect a crossover to an incoherent regime at fairly low temperatures. For (3), the transport lifetime deduced from the POR measurements is in excellent agreement with the quantum lifetime deduced from dHvA studies [96], in spite of the fact that the samples were grown completely independently by different groups. For (2), meanwhile, there is a significant discrepancy between the transport and quantum lifetimes. Part of the reason for this discrepancy could be related to the complexities of the low-temperature state of α-$(BEDT\text{-}TTF)_2KHg(SCN)_4$ [8]. In particular, the reconstructed FS consists of multiply connected Q1D and Q2D sections, giving rise to an extremely rich pattern of quantum oscillations involving magnetic breakdown. It is also notable that the POR data presented in Sect. 15.3.2 clearly originate from quasiparticles belonging to an open FS, whereas quasiparticles responsible for quantum oscillatory phenomena necessarily involve closed orbits. Thus, a direct comparison between τ and τ_φ is probably inappropriate for (2). Nevertheless, a significant (order of magnitude) difference between these quantities has previously been noted for another member ($M = NH_4$) of the α-phase salts, which *does not* undergo a low-temperature FS reconstruction [35]. Such a difference can be explained in terms of an inhomogeneous broadening of Landau levels, which would show up in τ_φ, but not necessarily in the transport lifetime [11]. Due to the absence of closed pockets, a comparison between τ and τ_φ is not possible for (1).

As discussed in Sect. 15.2, ω and ω_c play essentially the same role in (15.4). It is, therefore, interesting to compare optical constants obtained from these measurements with published values deduced from more conventional optical methods. There is one caveat, however: most optical reflectivity measurements are limited to frequencies well above the scattering rates (τ^{-1}) deduced from these POR studies. For example, optical studies of (2) are limited to frequencies above 50 cm^{-1} (1.5 THz) [99], and those of (3) to above 500 cm^{-1} (15 THz) [95]. Consequently, most of the low-temperature Drude spectral weight does not even contribute to the measured reflectivity in these investigations. The only material for which low-frequency cavity perturbation data do exist (down to ∼5 cm^{-1}) is $(TMTSF)_2ClO_4$ [97, 98, 100, 101]. Nevertheless, one still has to rely on Kramers–Kronig and Hagen–Rubens extrapolation for $\omega \to 0$ to recover the Drude spectral weight [68]. Thus, such techniques ultimately rely on accurate measurements of the DC conductivity. It is perhaps not surprising, therefore, that the optical scattering rates, $\tau^{\rm opt}$, deduced for (2) and (3) are roughly half the minimum frequencies employed in the measurement, i.e., ∼30 and ∼300 cm^{-1}, respectively. In both cases, the deduced values of $\tau^{\rm opt}$ are nowhere near the scattering rates deduced from POR studies. In addition, there is at least a two orders of magnitude difference between the DC conductivities found in optical studies and those deduced from POR measurements which assume a 100% contribution of free carriers to the Drude peak.

For $(TMTSF)_2ClO_4$, the situation is only slightly better. The values for $\tau^{\rm opt}$ and $\sigma_0^{\rm opt}$ listed in Table 15.1 were taken from the more recent literature [97, 98]. While the value of $\tau^{\rm opt}$ is in reasonable agreement with the POR data (because of the lower frequencies employed), the value of $\sigma_0^{\rm opt}$ clearly is not; again, there is a factor of 100 difference between the upper range for $\sigma_0^{\rm opt}$ and the POR estimate. However, there exist earlier reports of considerably higher conductivities, including the original study of Bechgaard et al. [102], where low-temperature values in the 10^5–10^6 S cm^{-1} range were found. This agrees nicely with the POR value quoted in Table 15.1. It is notable that, when the larger conductivity is used to extrapolate optical data, an anomalously narrow low-energy Drude peak is obtained (width = 0.034 cm^{-1} or 1 GHz) [101]. This highlights the problems associated with characterization of the low-energy Drude spectral weight from conventional broadband optical techniques which are ordinarily limited to frequencies above about 10 cm^{-1}. However, several recent studies have suggested a general trend among many of the layered organic conductors: namely, that only a small fraction (as little as 1%) of the overall spectral weight is to be found in the low-energy free carrier (Drude) response. If this is indeed the case, then the DC conductivities deduced from

POR measurements should be revised downward by a factor of 100, bringing them into alignment with the σ_o^{opt} values.

The experimental situation presented above is far from ideal. While the POR measurements give reliable estimates of the quasiparticle scattering time from the resonance half-width, the spectrometer is not calibrated to measure the absolute loss due to the sample. Consequently, the absolute value of the conductivity is not obtained from these measurements. Furthermore, it is usually the interlayer conductivity that is measured, so one has to make the assumption that the measured scattering time is isotropic. Nevertheless, the obtained value of τ broadly agrees with other techniques such as SdH and dHvA. In the case of the optical studies, neither τ^{opt} or σ_o^{opt} are obtained directly for the three materials highlighted in this chapter, and the reported τ^{opt} values are inconsistent with many other observations, e.g., the dHvA and SdH effects. Nevertheless, it has been suggested that both the Bechgaard and α-phase BEDT-TTF salts exhibit dramatic deviations from a simple Drude response on the basis of these optical studies [97,99]. One of the main pieces of evidence is the missing spectral weight in the low-energy Drude peak. However, as already pointed out above, if one takes the POR scattering time and the DC conductivity reported by Bechgaard et al. [102], then one finds that the Drude peak in $(TMTSF)_2ClO_4$ accounts for 100% of the free-carrier spectral weight. Furthermore, the observation of magneto-oscillatory phenomenon in many of these materials is hard to reconcile with the notion of a Drude peak containing only 1% of the free-carrier spectral weight. Clearly, the only way to resolve these issues is through the development of techniques which accurately measure the low-energy electrodynamic properties (including DC) of these materials. This ought to be a relatively straightforward task, but available samples are often tiny and prone to twinning and microcracking, which can severely influence such measurements.

Finally, we note that very extensive low-energy electrodynamic measurements have been reported down to $0.1\,\mathrm{cm}^{-1}$ for the $(TMTSF)_2PF_6$ member of the Bechgaard family [97,98,103], as have detailed AMRO measurements [23]. Furthermore, the low-temperature properties of this material are not influenced by anion ordering. Therefore, it is highly desirable to make POR measurements on $(TMTSF)_2PF_6$. However, this requires the application of hydrostatic pressure (>7 kbar) to suppress a SDW phase that occurs below 12 K under ambient pressure conditions. We note that such techniques are currently under development (see also Sect. 16.2).

15.6 Summary and Conclusions

On the basis of detailed POR measurements, we present compelling evidence that the low-energy magnetoelectrodynamics of three contrasting organic conductors, $(TMTSF)_2ClO_4$, α-(BEDT-TTF)$_2$KHg(SCN)$_4$, and κ-(BEDT-TTF)$_2$I$_3$, can be explained on the basis of a conventional semiclassical

Boltzmann theory, and that all three materials exhibit coherent 3D band transport at liquid helium temperatures. We demonstrate that there is nothing "magic" about the angles at which DC resistance minima are observed in AMRO experiments, findings that do not support the notion of fundamentally different thermodynamic ground states at, and away from, the Lebed "magic angles." We also argue that the POR can account for 100% of the free-carrier spectral weight for $(TMTSF)_2ClO_4$. Again, this finding appears to conflict with claims of dramatic deviations from a simple Drude response on the basis of broadband optical measurements. Finally, we propose that the POR technique could be used to probe quasiparticles in nodal superconductors.

Acknowledgement

This work was supported by the National Science Foundation (DMR0239481) and by Research Corporation. The authors acknowledge useful discussions with D. Tanner, V. Yakovenko, M. Dressel and D. Maslov.

References

1. A.W. Overhauser, Phys. Rev. **128**, 1437 (1962)
2. S. Kagoshima, H. Nagasawa, T. Sambogni, *One-Dimensional Conductors* (Springer, Berlin Heidelberg New York, 1988)
3. G. Grüner, *Density Waves in Solids* (Addison Wesley, Reading, MA, 1994)
4. T. Ishiguro, K. Yamaji, G. Saito, *Organic Superconductors* (Springer, Berlin Heidelberg Germany, 1998)
5. R. Louati, S. Charfi-Kaddour, A.B. Ali, R. Bennaceur, M. Héritier, Phys. Rev. B **62**, 5957 (2000)
6. Y. Tanuma, K. Kuroki, Y. Tanaka, R. Arita, S. Kashiwaya, H. Aoki, Phys. Rev. B **66**, 094507 (2002)
7. J. Wosnitza, *Fermi Surfaces of Low-Dimensional Organic Metals and Superconductors, Springer Tracts in Modern Physics* (Springer, Berlin Heidelberg New York, 1996)
8. J. Singleton, Rep. Prog. Phys. **63**, 1111 (2000)
9. J. Singleton, *Band Theory and Electronic Properties of Solids* (Oxford University Press, Oxford, 2001)
10. M.V. Kartsovnik, Chem. Rev. **104**, 5737 (2004)
11. D. Shoenberg, *Magnetic Oscillations in Metals* (Cambridge University Press, Cambridge, 1984)
12. A.A. Abrikosov, *Fundamentals of the Theory of Metals* (Elsevier, New York, 1988)
13. A.G. Lebed, JETP Lett. **43**, 174 (1986)
14. A.G. Lebed, P. Bak, Phys. Rev. Lett. **63**, 1315 (1989)
15. G.S. Boebinger, G. Montambaux, M.L. Kaplan, R.C. Haddon, S.V. Chichester, L.Y. Chiang, Phys. Rev. Lett. **64**, 591 (1990)
16. T. Osada, A. Kawasumi, S. Kagoshima, N. Miura, G. Saito, Phys. Rev. Lett. **66**, 1525 (1991)

17. M.J. Naughton, O.H. Chung, M. Chaparala, X. Bu, P. Coppens, Phys. Rev. Lett. **67**, 3712 (1991)
18. T. Osada, S. Kagoshima, N. Miura, Phys. Rev. B **46**, 1812 (1992)
19. M.V. Kartsovnik, A.E. Kovalev, N.D. Kusch, J. Phys. I **3**, 1187 (1993)
20. G.M. Danner, W. Kang, P.M. Chaikin, Phys. Rev. Lett. **72**, 3714 (1994)
21. S.J. Blundell, J. Singleton, Phys. Rev. B **53**, 5609 (1996)
22. E.I. Chashechkina, P.M. Chaikin, Phys. Rev. B **56**, 13658 (1997)
23. E.I. Chashechkina, P.M. Chaikin, Phys. Rev. Lett. **80**, 2181 (1998)
24. I.J. Lee, M.J. Naughton, Phys. Rev. B **58**, R13343 (1998)
25. W. Kang, H. Kang, Y.J. Jo, S. Uji, Synth. Met. **133–134**, 15 (2003)
26. A.G. Lebed, N.N. Bagmet, M.J. Naughton, Phys. Rev. Lett. **93**, 157006 (2004)
27. B.K. Cooper, V.M. Yakovenko, Phys. Rev. Lett. **96**, 037001 (2006)
28. F. Zwick, S. Brown, G. Margaritondo, C. Merlic, M. Onellion, J. Voit, M. Grioni, Phys. Rev. Lett. **79**, 3982 (1997)
29. A.S. Perel, J.S. Brooks, C.J.G.N. Langerak, T.J.B.M. Janssen, J. Singleton, J.A.A.J. Perenboom, L.Y. Chiang, Phys. Rev. Lett. **67**, 2072 (1991)
30. J. Singleton, F.L. Pratt, M. Doporto, T.J.B.M. Janssen, M. Kurmoo, J.A.A.J. Perenboom, W. Hayes, P. Day, Phys. Rev. Lett. **68**, 2500 (1992)
31. S. Hill, J. Singleton, F.L. Pratt, M. Doporto, W. Hayes, T.J.B.M. Janssen, J.A.A.J. Perenboom, M. Kurmoo, P. Day, Synth. Met. **56**, 2566 (1993)
32. S. Hill, A. Wittlin, J. van Bentum, J. Singleton, W. Hayes, J.A.A.J. Perenboom, M. Kurmoo, P. Day, Synth. Met. **70**, 821 (1995)
33. S. Hill, P.S. Sandhu, M. Boonman, J.A.A.J. Perenboom, A. Wittlin, S. Uji, J.S. Brooks, R. Kato, H. Sawa, S. Aonuma, Phys. Rev. B **54**, 13536 (1996)
34. A. Polisski, J. Singleton, P. Goy, W. Hayes, M. Kurmoo, P. Day, J. Phys. Condens. Matter **8**, L195 (1996)
35. S. Hill, Phys. Rev. B **55**, 4931 (1997)
36. S.J. Blundell, A. Ardavan, J. Singleton, Phys. Rev. B **55**, R6129 (1997)
37. H. Ohta, Y. Yamamoto, K. Akioka, M. Motokawa, T. Sasaki, T. Fukase, Synth. Met. **86**, 2011 (1997)
38. K. Akioka, H. Ohta, Y. Yamamoto, M. Motokawa, K. Kanoda, Synth. Met. **86**, 2051 (1997)
39. H. Ohta, Y. Yamamoto, K. Akioka, M. Motokawa, K. Kanoda, Synth. Met. **86**, 1913 (1997)
40. S. Hill, J.A.A.J. Perenboom, T. Stalcup, N.S. Dalal, T. Hathaway, J.S. Brooks, Physica B **246–247**, 549 (1998)
41. A. Ardavan, J.M. Schrama, S.J. Blundell, J. Singleton, W. Hayes, M. Kurmoo, P. Day, P. Goy, Phys. Rev. Lett. **81**, 713 (1998)
42. A. Ardavan, S.J. Blundell, J. Singleton, Phys. Rev. B **60**, 15500 (1999)
43. C. Palassis, M. Mola, J. Tritz, S. Hill, S. Uji, K. Kawano, M. Tamura, T. Naito, H. Kobayashi, Synth. Met. **120**, 999 (2001)
44. Y. Oshima, N. Nakagawa, K. Akioka, H. Ohta, S. Okubo, M. Tamura, Y. Nishio, K. Kajita, Synth. Met. **103**, 1919 (1999)
45. H. Ohta, Y. Oshima, N. Nakagawa, K. Akioka, S. Okubo, K. Kanoda, Synth. Met. **103**, 1913 (1999)
46. Y. Oshima, H. Ohta, S. Okubo, K. Koyama, M. Motokawa, M. Tamura, Y. Nishio, K. Kajita, Synth. Met. **120**, 853 (2001)
47. A.E. Kovalev, S. Hill, J.S. Qualls, Phys. Rev. B **66**, 134513 (2002)
48. A. Kovalev, S. Hill, S. Takasaki, J. Yamada, H. Anzai, J. Appl. Phys. **93**, 8665 (2003)

49. A.E. Kovalev, S. Hill, K. Kawamo, M. Tamura, T. Naito, H. Kobayashi, Phys. Rev. Lett. **91**, 216402 (2003)
50. Y. Oshima, H. Ohta, K. Koyama, M. Motokawa, H.M. Yamamoto, R. Kato, J. Phys. Soc. Jpn. **71**, 1031 (2002)
51. Y. Oshima, H. Ohta, K. Koyama, M. Motokawa, H. Yamamoto, R. Kato, M. Tamura, Y. Nishino, K. Kajita, J. Phys. Soc. Jpn. **72**, 143 (2003)
52. Y. Oshima, M. Kimata, K. Kishigi, H. Ohta, K. Koyama, M. Motokawa, H. Nishikawa, Phys. Rev. B **68**, 54526 (2003)
53. S. Takahashi, A.E. Kovalev, S. Hill, S. Takasaki, J. Yamada, H. Anzai, J.S. Qualls, K. Kawano, M. Tamura, T. Naito, H. Kobayashi, Int. J. Mod. Phys. B **27–29**, 3499 (2004)
54. S. Takahashi, S. Hill, J. Appl. Phys. **97**, 10B106 (2005)
55. S. Takahashi, S. Hill, S. Takasaki, J. Yamada, H. Anzai, Phys. Rev. B **72**, 024540 (2005)
56. A.P. MacKenzie, Y. Maeno, Rev. Mod. Phys. **75**, 657 (2003)
57. S. Hill, J.S. Brooks, Z.Q. Mao, Y. Maeno, Phys. Rev. Lett. **84**, 3374 (2000)
58. N.E. Hussey, M. Abdel-Jawad, A. Carrington, A.P. Mackenzie, L. Balicas, Nature **425**, 814 (2003)
59. A. Damasceli, Z. Hussain, Z.-X. Shen, Rev. Mod. Phys. **75**, 437 (2003)
60. S. Hill, Phys. Rev. B **62**, 8699 (2000)
61. D.G. Clarke, S.P. Strong, P.M. Chaikin, E.I. Chashechkina, Science **279**, 2071 (1998)
62. W. Wu, I.J. Lee, P.M. Chaikin, Phys. Rev. Lett. **91**, 056601 (2003)
63. W. Wu, P.M. Chaikin, W. Kang, J. Shinagawa, W. Yu, S.E. Brown, Phys. Rev. Lett **94**, 097004 (2005)
64. E.S. Choi, J.S. Brooks, H. Kang, Y.J. Jo, W. Kang, Phys. Rev. Lett. **95**, 187001 (2005)
65. P. Moses, R.H. McKenzie, Phys. Rev. B **60**, 7998 (1999)
66. R.H. McKenzie, P. Moses, Phys. Rev. B **60**, 11241 (1999)
67. T. Osada, K. Kobayashi, E. Ohmichi, Synth. Met. **135–136**, 653 (2003)
68. M. Dressel, G. Grüner, *Electrodynamics of Solids: Optical Properties of Electrons in Matter* (Cambridge University Press, Cambridge, 2002)
69. M. Dressel, A. Schwartz, G. Grüner, L. Degiorgi, Phys. Rev. Lett. **77**, 398 (1996)
70. O. Klein, S. Donovan, M. Dressel, G. Grüner, Int. J. Infrared Millimeter Waves **1993**, 2423 (1993)
71. S. Donovan, O. Klein, M. Dressel, K. Holczer, G. Grüner, Int. J. Infrared Millimeter Waves **14**, 2459 (1993)
72. M. Dressel, O. Klein, S. Donovan, G. Grüner, Int. J. Infrared Millimeter Waves **14**, 2489 (1993)
73. M. Mola, S. Hill, P. Goy, M. Gross, Rev. Sci. Inst. **71**, 186 (2000)
74. S. Takahashi, S. Hill, Rev. Sci. Inst. **76**, 023114 (2005)
75. T. Osada, H. Nose, M. Karaguchi, Physica B **294–295**, 402 (2001)
76. S. Takahashi, S. Hill, S. Takasaki, J. Yamada, H. Anzai, AIP Conf. Proc. **850**, 619 (2006)
77. H. Mori, S. Tanaka, M. Oshima, G. Saito, T. Mori, Y. Maruyama, H. Inokuchi, Bull. Chem. Soc. Jpn. **63**, 2183 (1990)
78. C.E. Campos, P.S. Sandhu, J.S. Brooks, T. Ziman, Phys. Rev. B **53**, 12725 (1996)

79. N. Harrison, Phys. Rev. Lett. **83**, 1395 (1999)
80. N. Harrison, E. Rzepniewski, J. Singleton, P.J. Gee, M.M. Honold, P. Day, M. Kurmoo, J. Phys.: Condens. Matter **11**, 7227 (1999)
81. A. Kovalev, M. Kartsovnik, R. Shibaeva, L. Rozenberg, I. Schegolev, N. Kushch, Solid State Commun. **89**, 575 (1994)
82. Y. Iye, R. Yagi, N. Hanasaki, S. Kagoshima, H. Mori, H. Fujimoto, G. Saito, J. Phys. Soc. Jpn. **63**, 674 (1994)
83. J. Caulfield, S. Blundell, M. du Croo de Jongh, P. Hendriks, J. Singleton, M. Doporto, F. Pratt, A. House, J. Perenboom, W. Hayes, M. Kurmoo, P. Day, Phys. Rev. B **51**, 8325 (1995)
84. N. Hanasaki, S. Kagoshima, T. Hasegawa, T. Osada, N. Miura, Phys. Rev. B **57**, 1336 (1998)
85. V.G. Peschansky, M.V. Kartsovnik, Phys. Rev. B **60**, 11207 (1999)
86. J. Singleton, P.A. Goddard, A. Ardavan, N. Harrison, S.J. Blundell, J.A. Schlueter, A.M. Kini, Phys. Rev. Lett. **88**, 037001 (2002)
87. K. Yamaji, J. Phys. Soc. Jpn. **58**, 1520 (1989)
88. H. Kobayashi, K. Kawano, T. Naito, A. Kobayashi, J. Mater. Chem. **5**, 1681 (1995)
89. E. Balthes, D. Schweitzer, I. Heinen, H.J. Keller, W. Strunz, W. Biberacher, A.G.M. Jansen, E. Steep, Z. Phys. B **99**, 163 (1996)
90. R.H. McKenzie, Science **278**, 820 (1997)
91. C.C. Tsuei, J.R. Kirtley, Rev. Mod. Phys. **72**, 969 (2000)
92. A. Hosseini, R. Harris, S. Kamal, P. Dosanjh, J. Preston, R. Liang, W.N. Hardy, D.A. Bonn, Phys. Rev. B **60**, 1349 (1999)
93. J. Corson, J. Orenstein, S. Oh, J. O'Donnell, J.N. Eckstein, Phys. Rev. Lett. **85**, 2569 (2000)
94. P.J. Turner, R. Harris, S. Kamal, M.E. Hayden, D.M. Broun, D.C. Morgan, A. Hosseini, P. Dosanjh, G.K. Mullins, J.S. Preston, R. Liang, D.A. Bonn, W.N. Hardy, Phys. Rev. Lett. **90**, 237005 (2003)
95. M. Tamura, H. Tajima, K. Yakushi, H. Kuroda, A. Kobayashi, R. Kato, H. Kobayashi, J. Phys. Soc. Jpn. **60**, 3861 (1991)
96. J. Wosnitza, J. Hagel, J.S. Qualls, J.S. Brooks, E. Balthes, D. Schweitzer, J.A. Schlueter, U. Geiser, J. Mohtasham, R.W. Winter, G.L. Gard, Phys. Rev. B **65**, 180506 (2002)
97. A. Schwartz, M. Dressel, G. Gruner, V. Vescoli, L. Degiorgi, T. Giamarchi, Phys. Rev. B **58**, 1261 (1998)
98. M. Dressel, K. Petukhov, B. Salameh, P. Zornoza, T. Giamarchi, Phys. Rev. B **71**, 075104 (2005)
99. M. Dressel, N. Drichko, J. Schlueter, J. Merino, Phys. Rev. Lett. **90**, 167002 (2003)
100. H.K. Ng, T. Timusk, K. Bechgaard, J. Phys. Colloq. Fr. **44**, 867 (1983)
101. N. Cao, T. Timusk, K. Bechgaard, J. Phys. I Fr. **6**, 1719 (1996)
102. K. Bechgaard, K. Carneiro, M. Olson, F.B. Rasmussen, C.S. Jacobsen, Phys. Rev. Lett. **46**, 852 (1981)
103. S. Donovan, Y. Kim, L. Degiorgi, M. Dressel, G. Gruner, W. Wonneberger, Phys. Rev. B **49**, 3363 (1994)

Part IV

Field-Induced Spin(Charge)-Density Wave Phases

16

Magnetic Field-Induced Spin-Density Wave and Spin-Density Wave Phases in $(TMTSF)_2PF_6$

A.V. Kornilov and V.M. Pudalov

The electron system in $(TMTSF)_2PF_6$ is highly sensitive to the external parameters and, depending on pressure, magnetic field, temperature, etc., undergoes transitions to various phases such as the spin density wave (SDW) state, field-induced spin density wave (FISDW) state, etc. This chapter overviews experimental studies of the respective spin ordered states.

We describe here investigations of various effects which originate from the one-dimensional character of electron motion in $(TMTSF)_2PF_6$ in magnetic field; the latter gives birth to various spin-ordered phases, depending on pressure, magnetic field, and temperature. We describe here experimental studies of peculiarities of these phases and phase transitions. Observation of the open-orbit cyclotron resonance confirms theoretically predicted picture of the one-dimensionalization in magnetic field. The effective mass appears to be noticeably higher than anticipated; this makes a principle impact on the phase diagram of the FISDW state. We describe experiments in the FISDW regime, which revealed a fine structure of the phase diagram, with a low-temperature quantum domain and hysteresis type transitions and a high temperature semi-classical region with no hysteresis. We present experiments that clarify the origin of the so-called rapid oscillations (RO), a puzzling phenomenon intrinsic to the spin-ordered state. Finally, we describe experimental studies of the character of a phase in the close vicinity of the borders between the paramagnetic metal and antiferromagnetic insulator. The results provide a strong evidence for the phase separation: spontaneous formation of a macroscopically inhomogeneous state, which contains macroscopic inclusions of the minority phase, spatially separated from the majority phase.

16.1 Introduction

The $(TMTSF)_2PF_6$ compound has a quasi-one-dimensional (Q1D) electron system confined in a three-dimensional host lattice. The electron system is therefore highly sensitive to the external parameters and, depending on

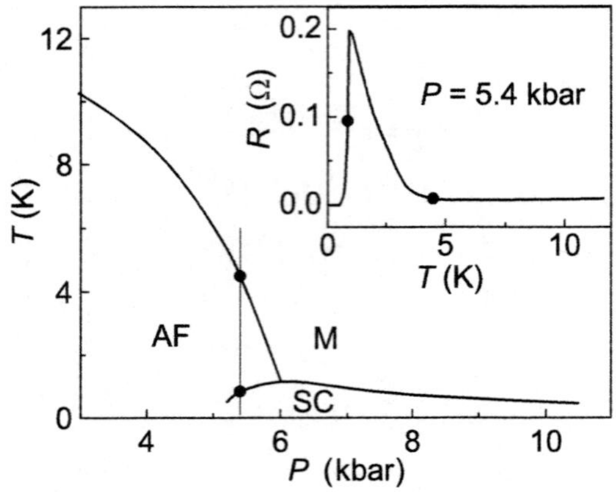

Fig. 16.1. *P–T* phase diagram for $(TMTSF)_2PF_6$ in zero magnetic field. M – metal, AF – antiferromagnetic insulator, SC – superconductor. *Vertical line* shows an isobaric trajectory $P = 5.4$ kbar, which crosses the M–AF and AF–SC phase borders. The inset shows the temperature dependence of the resistance that corresponds to the vertical trajectory at the main panel. The *bold dots* mark two corresponding phase transitions, from M to AF state, and from AF to SC state

pressure, magnetic field, temperature, etc., exhibits properties inherent of one-, two-, and three-dimensional systems; the examples such as the spin-density wave (SDW) state, field-induced spin-density wave (FISDW) state, quantum Hall effect, and superconductivity are considered in various chapters of this book.

The paramagnetic metallic state of the Q1D electronic system is unstable and at lowering temperatures undergoes a transition to the SDW state, which is an antiferromagnetic (AF) spin-ordered state (an insulator). Increasing pressure destroys the SDW state and makes the paramagnetic metallic state more favorable (see Fig. 16.1).

Further increase of the magnetic field causes one of the most remarkable phenomena – a cascade of transitions between different field induced SDW (FISDW) states, related with magnetic field driven changes of the nesting vector (for more details, see chapters by A.G. Lebed, M. Heritier, V.M. Yakovenko, S.E. Brown, P.M. Chaikin, and M.J. Naughton). The cascade of FISDW transitions terminates with the lowest order insulating state, which is suggested to be equivalent to the zero field SDW phase (at low pressures).

In the metallic state, there is an open Fermi surface (FS); the magnetic field (below the onset of the FISDW) applied perpendicular to the conducting plane causes one-dimensionalization of the electron motion. Nevertheless, the finite transfer integrals (perpendicular to the most conducting direction) lead to the periodic motion of the electrons in magnetic field. This motion was detected by observing cyclotron resonance in microwave range. The corresponding experiment is described in Sect. 22.2.

The measured effective mass appeared to be unexpectedly large. This result initiated theoretical and experimental revision of the so-called "standard model," which was developed to describe both the cascade of FISDW states and related quantum Hall effect. Detailed magnetoresistance measurements performed in high magnetic fields, in a wide range of temperatures and pressures (see in Sect. 25.1.2), have revealed new phases which subdivide the overall FISDW phase diagram into two regions. In the low-temperature (quantum) domain, the nesting vector is partially quantized by magnetic field, whereas in the high temperature (semiclassic) phase the nesting vector varies with field continuously.

One of the most interesting features of the SDW phase is the so-called "rapid oscillations" (RO), the resistivity oscillations periodic in the inverse field. Their existence is puzzling, because at first sight, for a Q1D system all electronic states are expected to be localized and, hence, quantum oscillations are not expected to occur. Another puzzle is that the RO have a nonmonotonic temperature dependence: as T decreases, their magnitude first grows, reaches a maximum and then diminishes. The detailed experimental studies (described in Sect. 22.4.2) have shown that RO exist only in the spin-ordered state and are caused by the formation in momentum space of closed two-dimensional orbits from the initial one-dimensional open FS. Temperature changes cause depopulation of the delocalized states (two-dimensional closed orbits) in favor of the localized ones, resulting in the disappearance of RO. These observations agree with the theory predicting the simultaneous existence of two spin-density waves.

The critical region and the character of transition between the metallic, AF-insulating state, and superconducting states are of a special interest. The experiments described in Sect. 22.5 revealed that an inhomogeneous state forms in the close vicinity of the phase boundaries; this state consists of inclusions of the minority phase into the majority phase.

16.2 Cyclotron Resonance on Open Orbits

The electron spectrum in the absence of magnetic field is effectively three-dimensional and corresponds to two warped sheets of the FS (see Fig. 16.2); in the tight-binding approximation it is of the form [1,2]:

$$\varepsilon_{\pm}(\mathbf{k}) = \pm v_{\mathrm{F}}(hk_x \mp hk_{\mathrm{F}}) + 2t_b \cos(k_b b) + 2t_c \cos(k_c c). \qquad (16.1)$$

The $\pm$ in (16.1) is for the right and left sheets of the FS, v_{F} is the Fermi velocity along the most conducting direction. a, b, and c are the principle lattice constants, x is along a; t_b and t_c are the transfer integrals along b and c, respectively.

When a magnetic field B is applied perpendicular to the a–b-plane, the motion of electrons *becomes effectively one-dimensional* [1–4]. The electron motion is limited along b and c, being extended along a direction, as

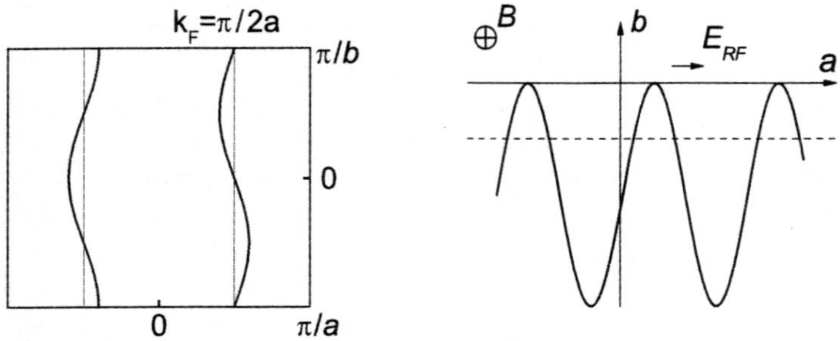

Fig. 16.2. A simplified view of the Fermi surface (*left panel*) and of the real space motion of electrons in the a–b-plane in the presence of magnetic field B along c-axes (*right panel*). a-axes coincides with the sample surface, *dashed line* depicts the skin layer, E_{RF} shows polarization of the electric component of the microwave field

schematically shown in Fig. 16.2. The amplitude of the periodic oscillations along b and c is $2t_b/\hbar\omega_c$ and $2t_c/\hbar\omega_c$, respectively. Here $\omega_c = eH/m^*$ is the cyclotron frequency, and m^* is the effective mass. The magnetic length $\lambda = h/(ebB)$ sets a period of motion along a, corresponding to the flux quantization $\Phi_0 \equiv h/e = \lambda bB$. The one-dimensionalization of electron motion in magnetic field is the key point in theoretical description of such fascinating effects in $(TMTSF)_2PF_6$ as, e.g., FISDW and RO (see Sects. 25.1.2 and 22.4.2, respectively). It is, therefore, highly desirable to test experimentally the above theoretical presumption by an unambiguous and direct experiment.

In 1993, Gorkov and Lebed [4] suggested to verify the one-dimensional sinusoidal type of electron motion by observation of the cyclotron resonance (CR). As Fig. 16.2 shows, when magnetic field is applied along c^*, electrons propagate along the a direction and periodically approach the crystal surface; therefore, they may be resonantly accelerated by the alternating electric field, polarized along a. A modification of such CR type experiment was performed by Ardavan et al. [5] for open orbits of the quasi-two-dimensional compound α-$(BEDT\text{-}TTF)_2KHg(SCN)_4$ at ambient pressure (for further details, see Chap. 15).

However, performing cyclotron resonance experiments on $(TMTSF)_2PF_6$ represent a hard task, because microwave measurements in a wide range of frequencies have to be done at high pressure, low temperatures, and high magnetic fields. This technical problem has been solved in [6], where a miniature nonmagnetic high pressure cell with a high aperture was developed. In this section, we describe an experiment [7] to observe the cyclotron resonance in $(TMTSF)_2PF_6$ under pressure. The results of the experiment (1) indeed, have demonstrated the one-dimensional character of electron motion in magnetic field and (2) enabled us to determine the Fermi velocity (effective mass) that is not affected by electron–electron interactions.

16.2.1 Experimental

We have fabricated a 18 mm diameter, nonmagnetic hydrostatic pressure cell for use up to 20 kbar, with sapphire pistons as window [6]. With its large aperture of 5 mm, this pressure cell allows spectroscopy from the mm-wave to ultraviolet range. Measurements were performed on a grid of crystals placed in the pressure medium against the sapphire window with their least conducting c^*-axis parallel to the magnetic field. One of the crystals was contacted to measure simultaneously the a-axis magnetoresistance. The waveguide, ended with the pressure cell, was placed into the ^{3}He cryostat, inside the 16.4 T superconducting magnet. The heterodyne detection technique [8] of the mm-wave radiation offers sensitivity to both amplitude and phase of the signal, so that both dispersion and absorption effects can be measured simultaneously. The setup was sufficiently sensitive to detect the weak signal reflected back from the sample inside the pressure cell.

16.2.2 Observations

Figure 16.3 shows some of the data with $(TMTSF)_2PF_6$ at a pressure of 10 kbar and at T = 600 mK. Resistance $R_{xx}(B)$ data in Fig. 16.3 show the onset of the cascade of the field-induced SDW, at field of ≈8 T (for more detail, see Sect. 25.1.2). The cascade reveals itself in step-like changes of the resistance which are periodic in $1/B$. This confirms that the pressure exceeds the critical value 6 kbar needed to suppress the SDW state (at $B = 0$). Consequently, at low fields $B <$ 8 T the sample is in the metallic state. Therefore, the electrons can freely propagate in the bulk (as shown schematically in Fig. 16.2), that is necessary for CR observation. The microwave signal also exhibits anomalies

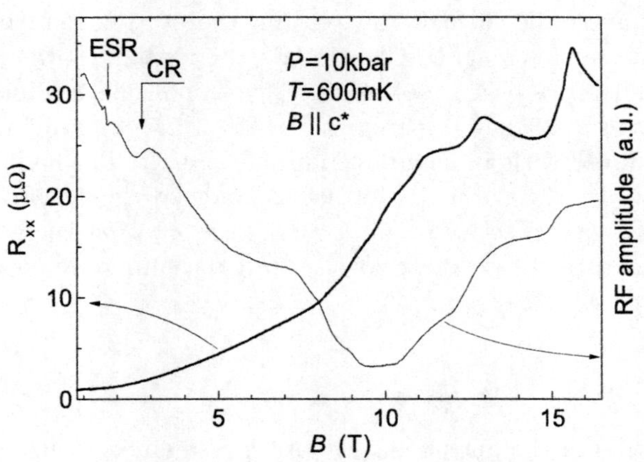

Fig. 16.3. Typical example of the microwave reflection amplitude at 50.1 GHz frequency (right axis) and resistance (left axis). P = 10 kbar, T = 600 mK

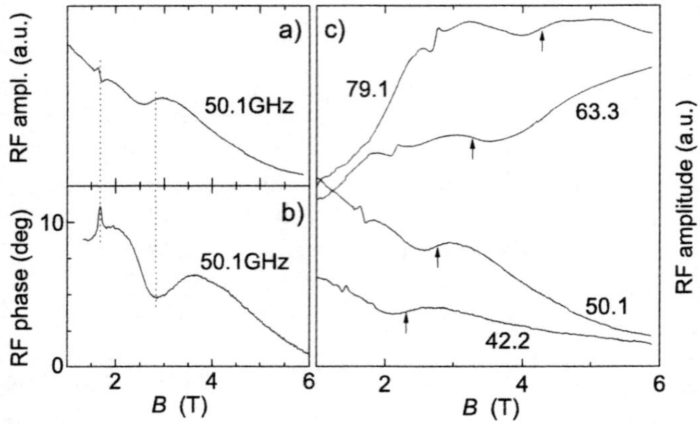

Fig. 16.4. Microwave reflection at 50.1 GHz frequency: (**a**) amplitude and (**b**) phase. (**c**) Microwave reflection amplitude for four frequencies. Frequency values (in GHz) are indicated next to each curve. *Arrows mark* the cyclotron resonance positions. $P = 10$ kbar and $T = 600$ mK for all panels

in amplitude of the reflected signal (see Fig. 16.3), which are correlated with the $R_{xx}(B)$ anomalies in high fields.

At low fields, below the threshold field of 8 T, there is a clear dip at ≈3 T; another, even sharper feature can be seen at somewhat lower field ≈2 T (see Fig. 16.3). These anomalies are missing in the $R_{xx}(B)$ curve. The corresponding features are rather weak and their analysis is not straightforward. To reveal true resonances on the top of the nonmonotonic background signal, we have detected simultaneously both, the amplitude and phase of the reflected microwave signal. As Fig. 16.4 shows, the low-field sharp features are present in both amplitude A and phase φ, and are well correlated with each other (see dotted lines in Fig. 16.4a, b). By plotting the $A(\varphi)$-dependence (for a given frequency and varying magnetic field) in polar coordinates, we unambiguously identify these features with resonances. The resonant field B for both features shifts upward as frequency F increases (see Fig. 16.4); Fig. 16.5 shows that this trend is well fitted with a linear dependence $F = \alpha B$. The higher field feature (marked with arrows in Fig. 16.4c) corresponds to the open orbit cyclotron resonance CR with effective mass $m = e/(2\pi\alpha) = 1.5m_e$, whereas the lower field sharp feature corresponds to the electron spin resonance (ESR) with effective g-factor ≈2.1.

16.2.3 Summary

The observation of the open-orbit cyclotron resonance confirms theoretically predicted picture of the one-dimensionalization in magnetic field where electrons propagate infinitely along a-axis and, simultaneously, perform oscillatory motion in perpendicular direction. The measured effective mass appears to be

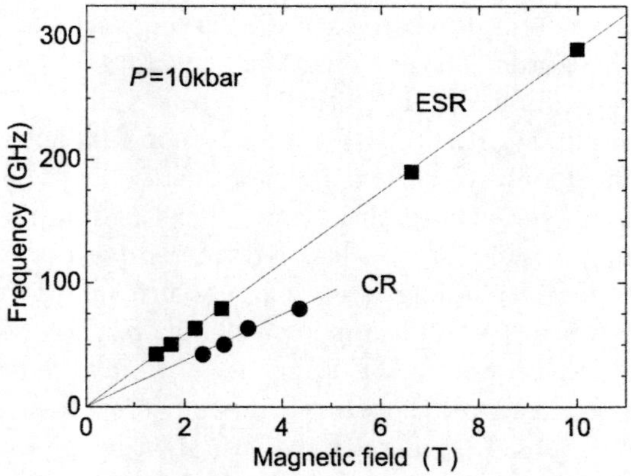

Fig. 16.5. Magnetic field dependences of the resonant frequency with $m^* \approx 1.5\, m_e$ for CR, and $g^* \approx 2.1$ for ESR

a factor of 1.5 higher than anticipated; this makes a principle impact on the FISDW phase diagram considered in Sect. 25.1.2.

16.3 Novel Phases in the Field-Induced Spin-Density Wave

Conduction in the materials of the $(TMTSF)_2X$ family, is highly anisotropic, with ratio of the components $\sigma_{xx} : \sigma_{yy} : \sigma_{zz} \sim 10^5 : 10^3 : 1$; this anisotropy leads to a rich phase diagram [1,9–11]). Magnetic field B applied in the least conducting direction z, first quenches the superconducting state and, further induces a cascade of phase transitions between FISDW states accompanied by the quantum Hall effect [12,13]. These phenomena have been experimentally discovered in many Q1D organic compounds of the (TMTSF)X family (X = PF_6, ClO_4, AsF_6) [12–17], and also in $(DMET\text{-}TSeF)_2Y$ (where Y = AuI_2 and $AuCl_2$) [18,19].

A so-called "standard" model was suggested in [1,3,10,14,17] to explain the metal–SDW transitions in magnetic field. Later it was developed into a "quantized nesting model" in [2,20–28] to describe a cascade of the first-order transitions between different FISDW sub-phases. According to this model, electrons condense in the SDW state whose period determines a nesting vector in the momentum space. Under the assumption of the electron–hole symmetry, the x-component of the nesting vector Q_x is quantized [2,11]:

$$Q_x = 2k_F - N\frac{eBb}{h}, \qquad (16.2)$$

where k_F is the Fermi wave vector, b is the size of the elementary cell in y-direction, and N is an integer. As magnetic field varies, N changes by an integer ΔN, causing step-like changes in the nesting vector, which result in the sequence of the first-order phase transitions.

According to the recent theoretical analysis [26, 29, 30], however, the electron–hole symmetry in the SDW state is not fulfilled unless $N = 0$ in (16.2) (for further detail, see review article by A. Lebed). As a result, (1) the nesting vector is not strictly quantized and (2) the step-like changes in the nesting vector may disappear above a certain temperature by transforming into oscillations. Correspondingly, as temperature increases, the first-order transitions with $\Delta N \approx 1$ may disappear, whereas FISDW state still persists. Thus, in contrast to the quantized nesting model which predicts the first-order phase transition to exist over the whole range of temperatures where FISDW develops, the novel model predicts that the first-order phase transitions may disappear above a certain temperature, T_0. The latter possibility depends on the parameter $\hbar\omega_c/(2\pi k_B T_0)$ [29], where ω_c is the cyclotron frequency. Our results of the ω_c measurements (see Sect. 22.2) demonstrate that the borderline $T_0(B)$ should fall into the FISDW domain of the phase diagram; this is in contrast to that for $(TMTSF)_2ClO_4$, where the cyclotron mass is a factor of two smaller.

In this section, we describe the experiments performed in [31] to verify the theoretical predictions of the two models above. We studied temperature dependence of the magnetoresistance in $(TMTSF)_2PF_6$ at various pressures. Specifically, we measured, at different pressures, a temperature evolution of the hysteresis intrinsic to magnetoresistance traces of $R(B)$. We observed that the hysteresis indeed disappears above a temperature T_0 whereas FISDW state still persists. We found such behavior to manifest itself over the whole explored range of the existence of the FISDW. According to our results, the total P–B–T phase diagram of the FISDW state can be subdivided into two domains, the "low-T" domain where the first-order phase transitions between FISDW subphases take place, and the "high-T" domain where the transitions between the FISDW states do not exhibit first-order behavior. This observation is in agreement with the novel model, where the low T-phase is treated as a "quantum FISDW" state with step-like changes in the nesting vector, whereas the high T-phase is treated as the "semiclassical FISDW" state where the nesting vector changes continuously and oscillates.

16.3.1 Experimental

Three samples (of a typical size $2 \times 0.8 \times 0.3\,\mathrm{mm}^3$) were grown by a conventional electrochemical technique. Measurements were made using either four ohmic contacts formed at the a–b-plane or eight contacts at two a–c-planes; in all cases 25 μm Pt wires were attached by a graphite paint to the sample along the most conducting direction a. The sample and a manganin pressure gauge were inserted into a Teflon cylinder placed inside a nonmagnetic 18 mm o.d. pressure cell [6] filled with Si-organic pressure transmitting liquid (polyethylsiloxane) [32]. The cell was mounted inside the liquid He^4, He^3, or He^3/He^4

chamber, in a bore of a 16 T superconducting magnet. For all measurements, the magnetic field was applied along the least conducting direction, c^*, of the crystal. Sample resistance was measured by four probe a.c. technique at 132 Hz frequency, with current 1–4 μA to avoid nonlinear effects. The out-of phase component of the measured voltage was found to be negligible in all measurements, indicating ohmic contacts to the sample.

The sample temperature was varied slowly, at a rate less than 0.25 K min^{-1} to avoid breaking the sample. The measured changes in the sample resistance were fully reproducible during the full run of measurements including temperature sweeps; this indicated that the sample quality did not change. The magnetoresistance was measured either at a constant T and varying magnetic field B, or at a constant B and varying T. Sample temperature was determined by RuO_2 resistance thermometer with a precalibrated magnetoresistance. Measurements were done in magnetic fields up to 16 T and for temperatures in the range from 30 to 1.4 K (mainly) and down to 0.12 K (partly). The most detailed results were obtained for pressures 7, 8, 10, and 14 kbar.

Figure 16.6 shows magnetoresistance traces measured (a) at $P = 10$ kbar in the temperature range 0.6–4.2 K and (b) at 8 kbar, 0.12 K. In agreement with earlier observations [33], when magnetic field exceeds the critical value (which is 0.16 T in our case), the superconductivity is quenched and the sample resistance starts gradually increasing. Further, this smooth dependence transforms into step-like changes in R. As temperature decreases, the step-like changes become steeper and appear at progressively lower fields. This behavior is also consistent with earlier observations [12–14, 17] and is interpreted as transitions between different sub-phases in FISDW [12, 13, 34]. This interpretation is further supported by the hysteresis between $R(B)$ traces for the field ramping up and down, which is clearly shown in Fig. 16.6. The hysteresis is also

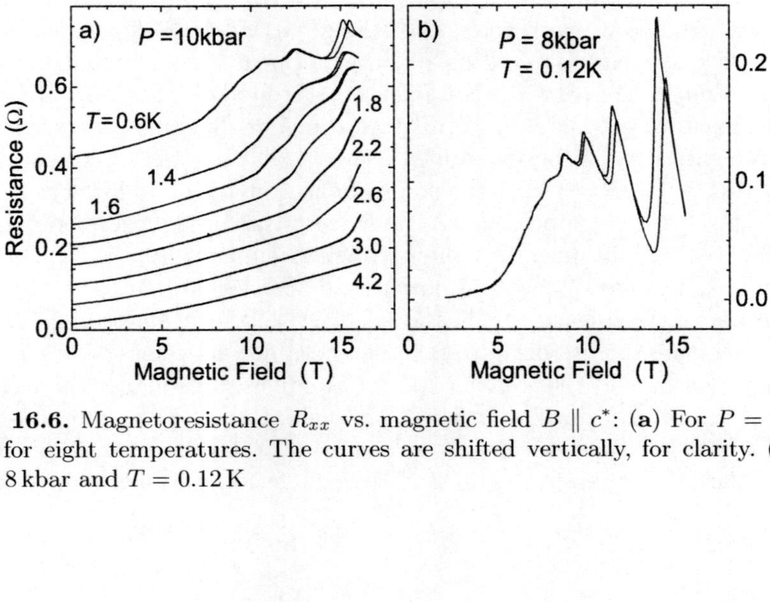

Fig. 16.6. Magnetoresistance R_{xx} vs. magnetic field $B \parallel c^*$: (**a**) For $P = 10$ kbar and for eight temperatures. The curves are shifted vertically, for clarity. (**b**) For $P = 8$ kbar and $T = 0.12$ K

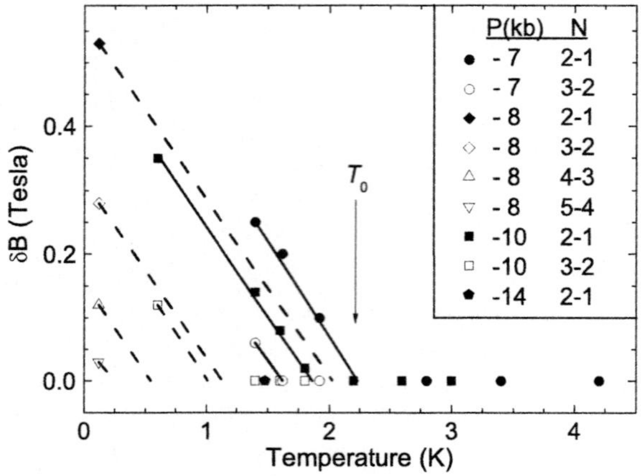

Fig. 16.7. Hysteresis width vs. temperature for several pressures. *Lines* are the guide to the eye. *Dashed lines* show an anticipated behavior in cases where only single data point was taken. The table shows pressure values P (in kbar) and the subphase numbers N between which the transition takes place. *Vertical arrow* depicts T_0 for one of the transitions, $N = 2 \longleftrightarrow 1$ at $P = 7\,\text{kbar}$

consistent with earlier observations [13] and signals the onset of the first-order phase transitions.

As temperature increases, the hysteresis weakens and tends to disappear as illustrated in Fig. 16.6a. Nevertheless, the steps in $R(B)$ persist to higher temperatures, being therefore non- or at least partly correlated with the hysteresis. To quantify the hysteresis strength, we calculated the maximal width of the hysteresis loop δB for each curve and plotted it in Fig. 16.7 as a function of temperature.

For three groups of the data in Fig. 16.7 the hysteresis width decreases linearly with temperature and vanishes at a certain temperature T_0; above $T = T_0$ it remains equal to zero. The falling part of these dependences were fitted with linear curves (solid lines), which appear to have the same slope. We plotted linear curves with the same slope through other single data points (dashed lines) to estimate T_0 for all transitions at different pressure values. Measurements at two other pressures, 7 and 14 kbar have shown qualitatively similar results. At $P = 7\,\text{kbar}$ the steps (transitions) shift to lower fields and persist up to higher temperatures. The hysteresis, δB, is bigger than that at 8 and 10 kbar and disappears at slightly higher temperature. At $P = 14\,\text{kbar}$, the trend is opposite: T_0 becomes lower than that for 10 kbar.

The above three features, (1) the existence of the hysteresis in $R(B)$ at low temperature, (2) its disappearance above a certain temperature T_0, and (3) the persistence of the steps in $R(B)$ to temperatures higher than T_0, are observed in our experiments for several transitions (see Figs. 16.6 and 16.7). It seems likely that these features are generic also to other transitions in the FISDW part of the phase diagram and that the hysteresis for higher N-values

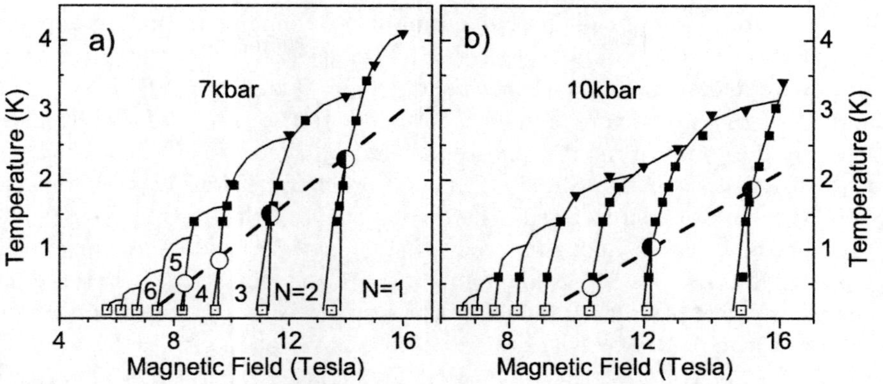

Fig. 16.8. B–T-phase diagram (**a**) for $P = 7\,\text{kbar}$, and (**b**) for $P = 10\,\text{kbar}$. *Solid squares* (*triangles*) correspond to the onset of the steps in $R(B)$ (and $R(T)$) curves measured directly at the specified pressures. *Open squares* are recalculated from the data taken at $P = 8\,\text{kbar}$. *Circles* denote T_0: *semi-closed symbols* are for the data measured directly at the specified pressure, *open ones* are recalculated from the data measured at $P = 8\,\text{kbar}$. *Dashed lines* correspond to the hysteresis width; other lines are guide to the eye

was not observed in our measurements just because T_0 for these transitions is lower than our lowest accessible temperature, 1.4 K (for the majority of measurements).

The B–T phase diagram in Fig. 16.8 summarizes the results of all measurements at $P = 7\,\text{kbar}$ and 10 kbar. The closed squares depict the onsets of the steps in $R(B)$ obtained from magnetic field sweeps at fixed temperatures and triangles are obtained from the temperature sweeps $R(T)$ at fixed field. In addition to the data taken directly, the open squares show the lower temperature data, $T = 0.12\,\text{K}$, taken at $P = 8\,\text{kbar}$ which has been recalculated to correspond to the data at $P = 7\,\text{kbar}$ and to 10 kbar. In this procedure, the data for 8 kbar were shifted in magnetic field according to the pressure coefficient $\mathrm{d}(B^{-1})/\mathrm{d}P = -0.015\,\mathrm{T}^{-1}\mathrm{kbar}^{-1}$ which we determined from the higher temperature data at $T = 1.4\,\text{K}$ for $P = 7$, 10, and 14 kbar. The hysteresis width (see Figs. 16.6 and 16.7) is depicted by the dashed lines.

In general, the P–B–T phase diagram in Fig. 16.8 is qualitatively similar to that previously reported [12, 13], but, in addition, displays the boundaries of hysteresis domains vs. temperature. The hysteresis domains for different transitions collapse at $T = T_0$; this was determined for seven transitions and the corresponding points are denoted with open circles. The separatrix points T_0 thus subdivide the phase boundaries into the two regions: the low temperature domain ($T < T_0$) of the hysteretic behavior and the high temperature domain ($T > T_0$) where the FISDW transitions develop without a hysteresis. The disappearance of the hysteresis with rising temperature at one fixed pressure was mentioned earlier [13] but to the best of our knowledge no studies of this effect followed.

The subdivison of the phase diagram in two domains is qualitatively consistent with the novel model for the FISDW suggested by Lebed [29,30], where the low-temperature domain corresponds to the "quantum FISDW" sub-phase and the high temperature domain corresponds to the "semiclassical FISDW." The subdivision into the two phases depends on the ratio $r = \hbar\omega_c/(2\pi k_B T_0)$; when it reaches a critical value r_c, the first-order transitions between different FISDW phases are suppressed. The x-component of the nesting vector (16.2) then becomes a continuous function of the magnetic field. According to the model [29,30], for $r < r_c$, at low-temperatures, the transitions between different phases take place with jumps in the nesting vector, are of the first order, and are accompanied by hysteresis of various physical quantities; however, the nesting vector is not strictly quantized. At higher temperatures $T > T_0$, the transitions between different FISDW phases take place without any jumps in the nesting vector, and are therefore not first-order phase transitions. The B–T phase diagram plotted in Fig. 16.8 for pressures $P = 7$ and 10 kbar looks qualitatively similar but for higher pressure, the new phase boundary is shifted to lower temperature by about (0.5–1) K.

Recently, the FISDW phase diagram was considered theoretically in [35]. The authors concluded that the deviations of Q_x from the quantized value may occur only for high order FISDW transitions, $N > 5$, and that at low temperatures the quantization of Q_x is always exact. These conclusions are in contrast with those of [29,30], and also seem to disagree with our experimental data. Thus, the theoretical predictions need in more detailed experimental verification; in particular, it should be verified whether or not the quantization of Q_x is exact, and whether it persists to high order transitions at very low temperatures. The corresponding experiments have not been performed yet.

16.3.2 Summary

From studies of the temperature and magnetic field dependences of the resistivity of the Q1D organic conductor, we found that its P–B–T phase diagram splits in two domains, where the transitions between different FISDW states take place (1) with a hysteresis as the first-order phase transitions (for low temperatures) and (2) without hysteresis (for high temperatures). This result is not expected within the quantized nesting model and is consistent with the recent suggestion by Lebed [29, 30] that the period of the spin structure in FISDW state can be either partially quantized or not quantized at all. Figure 16.8 shows the new phase boundary where such behavior occurs.

16.4 Rapid Oscillations

The FS of $(TMTSF)_2PF_6$ comprises two sheets, perpendicular to a, which are slightly corrugated due to small transfer integrals in b and c directions (see Fig. 16.2). The Q1D electron system undergoes a phase transition to the SDW state at low temperatures ($T < 12\,K$). One of the most interesting features of the SDW phase in this material is the oscillatory magnetoresistance periodic in the inverse magnetic field, $1/B$, the so-called RO [12, 13, 36–38]. Their existence, by itself, is puzzling, because at first sight, for a Q1D system all electronic states are expected to be localized and hence, quantum oscillations are not expected to occur. Another puzzle is that the RO have a nonmonotonic temperature dependence [12,13,36–38]: as T decreases, their magnitude first grows, reaches a maximum and then vanishes; these features are clearly in contrast with those of regular quantum oscillations (e.g., Shubnikov–de Haas oscillations).

Along with the disappearance of the RO, in the same temperature interval a number of other anomalies are observed in studies of NMR [39], microwave conductivity [40], spin susceptibility [41], specific heat [42], etc. This set of phenomena provides evidence for an interesting transformation in the SDW system with temperature; this is currently the focus of research interest and is far from being resolved.

The oscillations have been extensively studied since the early 1980s; they are thought to be related to the general properties of low-D physics, because RO are observed in almost all known Q1D organic compounds $(TMTSF)_2X$ family (X = PF_6, AsF_6, ClO_4, NO_3, etc.) [37, 38, 43–46] and in $(DMET\text{-}TSeF)_2AuCl_2$ [19]. Materials with noncentrosymmetric anions (e.g., ClO_4, NO_3, etc.) undergo a doubling of the lattice periodicity due to anion ordering. The corresponding energy gap in the electron spectrum is used in various models (either in combination with or without, a nesting vector) proposed for the explanation of RO. The models are based on the formation and quantization of either closed orbits or interference orbits [47–51]. These models are evidently inapplicable to materials with centrosymmetric anions (e.g., PF_6, AsF_6), where no anion ordering occurs; for this reason more sophisticated models have been proposed [37,38,52,53].

Almost all preceding studies of the RO have been performed at zero pressure. To verify the potential existence of the RO in the metallic phase, we applied a pressure $P > 6\,kbar$; this is known to suppress the SDW and to stabilize the metallic state (see Sect. 25.1 and [9]). Application of a sufficiently strong magnetic field along c^*-axis breaks down the metallic state and induces a cascade of FISDW phases, $N = i, i-1, \ldots$ (see Sect. 25.1.2 and [2, 9–11]), which terminates in the insulating $N = 0$ phase; the latter is believed to be the same SDW phase as that for zero pressure. In this section we describe the experiments [54–56], where the RO have been studied at high pressures both in metallic and insulating FISDW phases and at the same temperatures. To study the dependence of the magnetotransport on magnetic field orientation under pressure, a spherical pressure cell [57] containing the sample was rotated in situ in the bore of a superconducting magnet.

16.4.1 Experimental

Two samples (of typical size $2 \times 0.8 \times 0.3\,\mathrm{mm}^3$) were grown by conventional electrochemical techniques. 25 µm Pt wires were attached using graphite paint to the sample on the a–b-plane along the most conducting direction a. The sample and a manganin pressure gauge were inserted into a miniature spherical pressure cell with an outer diameter of 15 mm [57] filled with Si-organic pressure transmitting liquid [32]. The pressure was created and fixed at room temperature; the pressure values quoted throughout this chapter refer to the liquid helium temperatures [58]. The cell was mounted in a two-axes (θ, φ) rotation system placed in He^4 in a bore of either 17 or 21 T superconducting magnets. The rotation system enabled rotating the pressure cell (with a sample inside) around the main axis θ by 200° (with an accuracy ~0.1°) and around the auxiliary axis φ by 360° (with accuracy ~1°); this allowed us to set the sample at any desired orientation with respect to the magnetic field direction within 4π steradian. The sample resistance R_{xx} was measured using a four probe a.c. technique with an excitation current of 1–4 µA (to avoid nonlinear effects in the SDW state) at 16–132 Hz frequency. The out-of-phase component of the measured voltage was found to be negligible in all measurements, indicating good ohmic contacts to the sample.

16.4.2 On the Existence of Oscillations in Various Domains of the *P*–*B*–*T* Phase Space

Figure 16.9a shows the magnetoresistance measured at fixed temperature $T = 4.2\,\mathrm{K}$ for various pressures. At low pressure, ≈5 kbar, the sample is in the SDW state. This is illustrated by the temperature dependence of the resistance at zero magnetic field shown in the inset to Fig. 16.9. As temperature decreases below ≈6 K, the resistance sharply rises (see the upper curve in the inset), signaling the transition from a metallic to an insulating state; this is in accord with the known phase diagram [9]. In this insulating SDW state, at $T = 4.2\,\mathrm{K}$, as magnetic field increases, the resistance starts to oscillate above a field of approximately 10 T (see the upper curve in the panel (a)). The oscillations are periodic in $1/B$, as Fig. 16.9b shows. For higher pressures $P > 6$ kbar, the sample remains metallic at zero field, down to the lowest temperatures. This is demonstrated by the monotonic temperature dependence of the resistance with $\mathrm{d}R/\mathrm{d}T > 0$ for $P = 7.5$ kbar in the inset to the Fig. 16.9 (lower curve). For $P = 10$ and 15 kbar, the temperature dependences are similar to that for $P = 7.5$ kbar, and only slightly shifted to lower resistance values.

As the magnetic field rises, the sample experiences a transition from the $B = 0$ metallic state to the FISDW state. This is clearly shown on the curve for $P = 7.5$ kbar as a sharp factor of 50 increase in the resistance at a field ≈13 T (see the main panel of Fig. 16.9a). Immediately after the transition to the spin-ordered state, the magnetoresistance starts to oscillate; again the oscillations are periodic in $1/B$ (see Fig. 16.9c). As the pressure increases

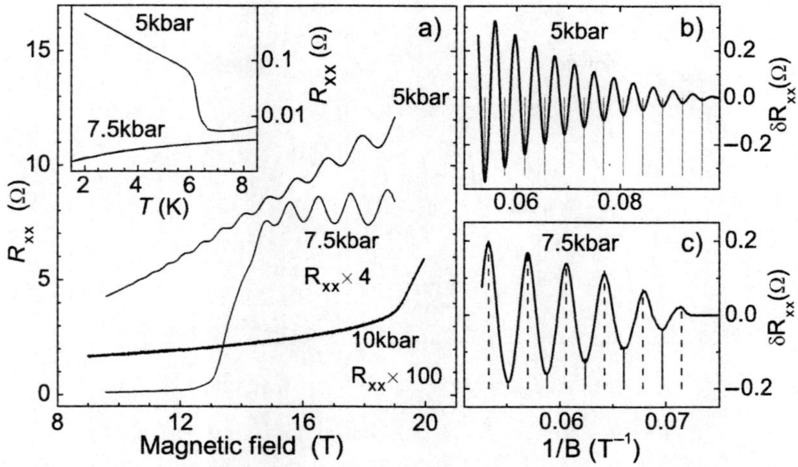

Fig. 16.9. (**a**) Magnetoresistance R_{xx} measured at $T = 4.2\,\mathrm{K}$ for $B \parallel c^*$ at different pressures. Scaling factors and pressure values are indicated next to each curve. The inset shows the temperature dependence of R_{xx} at $B = 0$ for $P = 5$ and 7.5 kbar. *Right panels*: oscillatory part of the magnetoresistance δR_{xx} vs. the inverse magnetic field at $T = 4.2\,\mathrm{K}$ for (**b**) $P = 5\,\mathrm{kbar}$ and (**c**) $P = 7.5\,\mathrm{kbar}$. *Vertical lines* are equidistant in $1/B$

further, the onset of the FISDW state shifts to progressively higher fields (see curve for $P = 10\,\mathrm{kbar}$). Importantly, there are no oscillations seen at the same temperature for $P = 10$ and 15 kbar (the latter is not shown in the figure) over the whole range of magnetic fields corresponding to the metallic state.

The magnetoresistance data at $P = 7.5\,\mathrm{kbar}$ are analyzed in more detail in Fig. 16.10a. The oscillations in the FISDW state are so large (~20% at 18 T) that can be easily extrapolated to lower fields. For this extrapolation we used the empirical field dependence of the RO amplitude measured at $P = 5\,\mathrm{kbar}$ (see Fig. 16.9) and scaled down (using one fitting parameter) to fit the data at $P = 7.5\,\mathrm{kbar}$ in the field range 14–19 T. The extrapolated oscillations are shown by dotted line in 16.10a. It appears however that no oscillations can be seen in the metallic state either in $R_{xx}(B)$ or in its derivative $\mathrm{d}R_{xx}/\mathrm{d}B(B)$ for $B < 13.5\,\mathrm{T}$ with accuracy $\delta R/R < 0.2\%$ (as the inset to Fig. 16.10a shows). This confirms that the oscillations (1) occur along with the onset of the FISDW state and (2) set in a step-like fashion, with a nonzero amplitude. It follows from the data presented in Figs. 16.9a and 16.10a that, at a given temperature and magnetic field, the existence of the RO depends on whether the system is in the spin-ordered or in the metallic state.

Figure 16.10b shows the temperature evolution of the magnetoresistance for a fixed pressure $P = 10\,\mathrm{kbar}$. At high temperature $T = 4.2\,\mathrm{K}$, the system is in the metallic state and the oscillations are missing. As temperature decreases, the system undergoes transition to the FISDW state (see the phase diagram in Fig. 16.8); this is concomitant with the appearance of

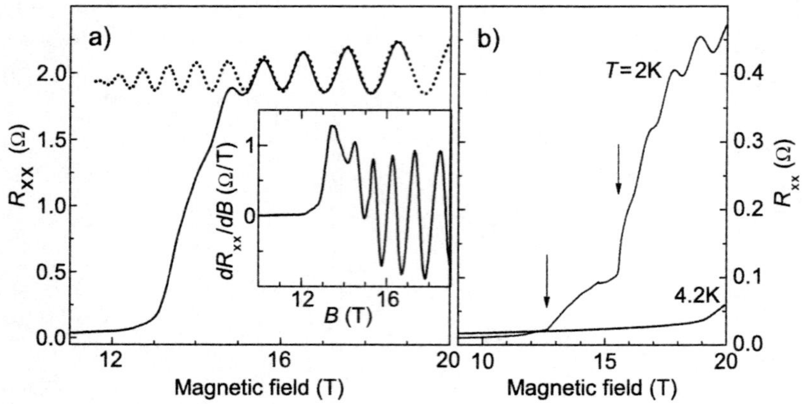

Fig. 16.10. Magnetoresistance $R_{xx}(B)$ measured for $B \parallel c^*$: (**a**) at $P = 7.5$ kbar and $T = 4.2$ K (*continuous line*). *Dotted curve* shows the extrapolation of the oscillatory magnetoresistance to lower fields. Inset shows the derivative $\mathrm{d}R/\mathrm{d}B$ that demonstrates a step-like onset of the oscillations at the FISDW transition ($B \approx 13$ T). (**b**) $R_{xx}(B)$ at $P = 10$ kbar for $T = 4.2$ and 2.0 K. *Vertical arrows* mark two FISDW phase transitions $N = 2 \Leftrightarrow 1$ and $N = 1 \Leftrightarrow 0$ [9, 11, 31]

the magnetoresistance oscillations. As above, we conclude that, for a given pressure and magnetic field, the existence of the RO depends on whether the system is in the spin-ordered or in the metallic state.

An interesting question is whether the RO exist only in the SDW and $N = 0$ FISDW phases or are a more general phenomenon, intrinsic also to other ($N \neq 0$) FISDW phases. This issue was investigated in [54, 56], where it was found that the oscillations exist not only in $N = 0$ but also in $N = 1$ phases, though their magnitude is much weaker in the latter case. The RO in the $N = 1$ phase are an extension of those in the $N = 0$ phase and, hence RO in both phases have a common origin.

The totality of our experimental data taken on two different samples in the ranges of temperature 1.4–8 K, magnetic field (up to 20 T), and pressures (0–15 kbar) proves that *the RO in* $(TMTSF)_2PF_6$ *are intrinsic only to the spin-ordered state and hence are caused by nesting.*

16.4.3 Existence of Delocalized States in the Spin-Ordered Phase

Figure 16.9b, c show two examples of the oscillatory part of the magnetoresistance $\delta R_{xx}/R_{xx}$ plotted vs. the inverse magnetic field for two pressure values. For these two examples, as well as for all other pressures and temperatures, studied in our experiment, the oscillations are periodic in $1/B$. It is therefore likely that the RO originate from Landau quantization. The latter is in contrast to the quantized nesting model [2, 9–11], wherein the spin-ordered state should be a totally gapped insulating state. The $1/B$ periodicity of the

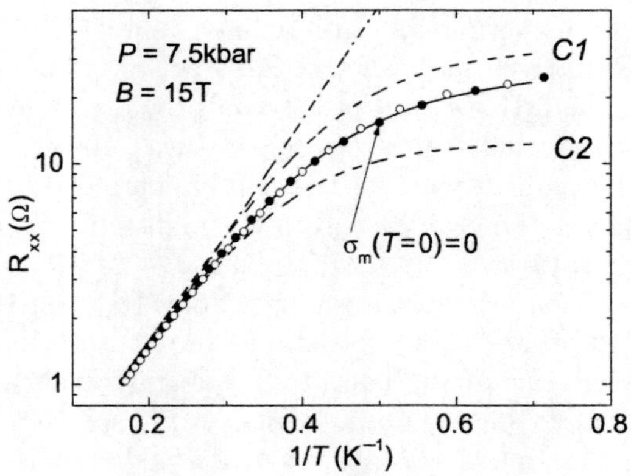

Fig. 16.11. Temperature dependence of the background magnetoresistance at $P = 7.5\,\mathrm{kbar}$ at a magnetic field of 15 T. The *bold line* shows experimental data; *dashed lines* $C1$ and $C2$ are examples of data fitting using a two-component conductivity model (16.3) with T-independent metallic term, $C1$ and $C2$, respectively ($C1 < C2$). The *dash-dotted line* shows the temperature-activated conduction of a single component insulator with 12 K energy gap. *Filled and empty circles* show examples of fitting using a two-component conductivity model (16.5) and (16.4) with T-dependent metallic term (described further in subsection 16.4.5)

oscillations thus strongly suggests the existence of delocalized carriers (and, correspondingly, closed orbits in the Brillouin zone).

The existence of the metallic closed orbits in the spin-ordered state may be also deduced from the data shown in Fig. 16.11 that demonstrates the monotonic dependence of the background magnetoresistance $R_{\mathrm{bg}} = R_{xx} - \delta R_{xx}$ on the inverse temperature. For a totally gapped insulator, the temperature dependence is expected to have an activating or hopping character, $R \propto \exp[(T_0/T)^p]$. The experimental data indeed have an approximately activated character. However, neither activated ($p = 1$) (dash-dotted line in Fig. 16.11), nor hopping ($p = 1/2, 1/3, 1/4$) temperature dependences can model the data over the whole temperature range $T = 1.4$–$8\,\mathrm{K}$ of the spin-ordered state. Taking the exponent p as a fitting parameter, one can achieve an approximate fitting only with p as low as $1/8$ or $1/7$, the values which have no apparent physical meaning. The lack of agreement is clearly caused by a "low-temperature" saturation of the data, which contrasts with hopping or activating dependences.

The saturation can be modeled by a straightforward addition of a "metallic conduction" component C_{m} to the temperature-activated (hopping) conduction of an insulator

$$\sigma_{xx} = \sigma_{\mathrm{ins}} + \sigma_{\mathrm{met}} = A\exp(-T_0/T) + C_{\mathrm{m}}. \tag{16.3}$$

The examples of fitting using (16.3) with two values $C_{\mathrm{m}} = C1$ and $C2$ are shown by dashed lines in Fig. 16.11. The fit with constant $C2$ provides good agreement with the experimental data at high temperatures, whereas the fit with $C1$ provides correct slope of the $R(T)$ at low temperatures; however, neither choice of the fitting constant C_{m} can provide quantitative agreement with the data in the whole range of temperatures. Nevertheless, this simple model reflects the main feature of the experimental $R(T)$ dependence and qualitatively explains the crossover from the high-temperature activated law to the low-temperature saturation of the data.

The two experimental results described above (1) the $1/B$-periodicity of the oscillatory part of the magnetoresistance and (2) the violation of activated temperature dependence of the background magnetoresistance, point at the *existence of delocalized states (closed orbits) in the spin-ordered phase. These are the states that are quantized in a magnetic field and give rise to the rapid oscillations.*

16.4.4 Magnetoresistance Oscillations in a Tilted Field

The existence of the delocalized states in the spin-ordered phase raises the question on the geometry and orientation (in momentum space) of the corresponding Fermi surface closed orbits responsible for the RO. To address this issue, we measured frequency of the magnetoresistance oscillations as a function of the magnetic field orientation. Examples of $R_{xx}(B)$-curves at $T = 4.2\,\mathrm{K}$ for a magnetic field tilted in the $c^* - b'$ plane are shown in Fig. 16.12a. When the field is tilted from the c^*-axis ($\theta = 0$), the frequency of oscillations (in inverse field) increases as $275/\cos\theta\,\mathrm{T}$ (see Fig. 16.12c). The same angular dependence has been obtained for tilting the field from c^*-axes in other planes (the corresponding results are not shown). This angular dependence is evidence that *the closed orbits are (1) two-dimensional and (2) lie in the a–b crystal plane.*

As the temperature is changed, neither the frequency nor its angle dependence varies (see Fig. 16.12a, b). This is illustrated further in Fig. 16.12c, where data for $T = 4.2$ and $2\,\mathrm{K}$ coincide with each other and follow the same angular dependence, $275/\cos\theta$. We conclude that *neither the size nor the orientation of the metallic closed orbits change with temperature.*

16.4.5 Temperature Dependence of the Oscillations

The normalized amplitude of oscillations, $\delta R/R$, exhibits a nonmonotonic temperature dependence that is illustrated in more detail in Fig. 16.13. The oscillation amplitude slowly rises as T decreases from high temperatures, reaching a maximum at $\approx 3\,\mathrm{K}$, and then sharply falls (see Fig. 16.13b). Since the size of the closed orbits in momentum space does not change with temperature (see Sect. 16.4.4), the oscillation amplitude reflects the population of the delocalized states (closed orbits). Therefore, the temperature dependence of the RO amplitude (Fig. 16.13b) signals a T-dependent redistribution of the carriers between the delocalized and localized states.

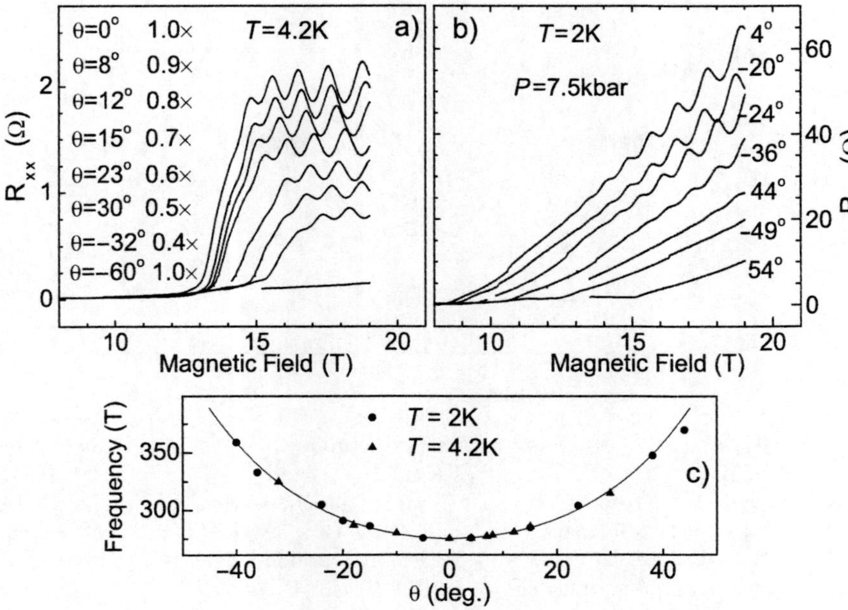

Fig. 16.12. Magnetoresistance $R_{xx}(B)$ for various magnetic field orientations in the $c^* - b'$ plane ($\theta = 0$ for $B||c^*$) at $P = 7.5\,\text{kbar}$ (**a**) for $T = 4.2\,\text{K}$ and (**b**) for $T = 2\,\text{K}$. For clarity, the curves on the panel (**a**) are scaled individually (the scaling factors are shown next to each curve). (**c**) Angular dependence of the oscillation frequency for $T = 4.2\,\text{K}$ (*triangles*) and $2\,\text{K}$ (*dots*). Continuous curve depicts the $275/\cos\theta$ dependence

We propose that the measured T-dependence of the oscillations amplitude (Fig. 16.13b) is indicative of the occupation of the "metallic" two-dimensional closed orbits n_{m}; the decrease in the occupation upon cooling causes the disappearance of the oscillations. Correspondingly, the metallic contribution σ_{met} to the two-component conduction model (16.3) should be taken to be temperature-dependent.

To estimate n_{m} versus temperature, we use an exponential function to describe the T-dependent carrier occupation.

$$n_{\text{m}}(T) = n_0 + n_2 \exp(-T_0/T). \tag{16.4}$$

This equation contains three fitting parameters: n_0 is the occupation number at $T = 0$, $n_0 + n_2$ is the occupation number at $T = \infty$, and T_0 is the crossover temperature. Although we do not think that the Lifshits–Kosevich formula for the two-dimensional case [59] may describe the RO amplitude, it certainly may have some physical relevance. We therefore attempted to fit the occupation number empirically by considering the nonmonotonic T-dependence of the RO amplitude (Fig. 16.13b) as a product of (1) the growth of the carrier

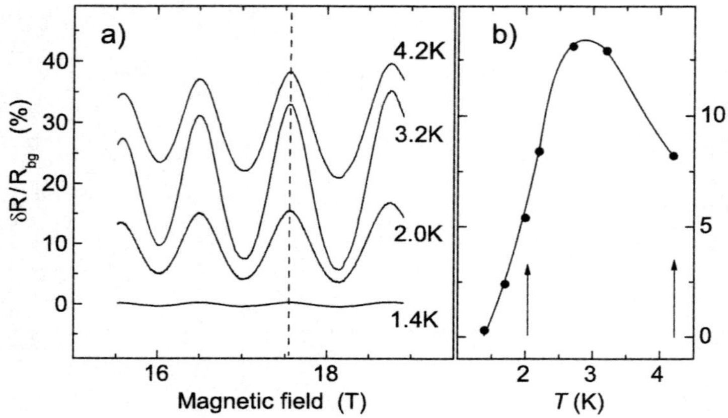

Fig. 16.13. (**a**) Examples of the oscillatory component of the magnetoresistance $(R - R_{\mathrm{bg}})/R_{\mathrm{bg}}$ for $B \parallel c^*$. The curves for 4.2, 3.2, and 2.0 K are offset vertically by 30, 20, and 10%, respectively. (**b**) Temperature dependence of the oscillation amplitude at $B \approx 17.5\,\mathrm{T}$ (marked on panel (**a**) with a *dashed line*). *Vertical arrows* show the temperature values at which the angular dependence of the oscillation frequency was measured (data shown in Fig. 16.12). Pressure $P = 7.5\,\mathrm{kbar}$ for all panels

occupation with temperature and (2) the decay of the amplitude of quantum oscillations with T.

$$n_{\mathrm{m}}(T) = n_0 + n_1 \left(\frac{\delta R}{R}\right)_{\mathrm{RO}} \left[\frac{2\pi^2 kT}{\hbar\omega_{\mathrm{c}} \sinh(2\pi^2 kT/\hbar\omega_{\mathrm{c}})}\right]^{-1} \qquad (16.5)$$

In (16.5), $(\delta R/R)_{\mathrm{RO}}$ is the measured RO amplitude, the term in square brackets is the anticipated temperature decay of the quantum oscillations in the two-dimensional "metal" with a fixed carrier density [59], the cyclotron mass is $1.5m_e$ [7], and n_0 and n_1 are the fitting parameters.

The filled and empty circles in Fig. 16.11 show the results of fitting with (16.4) and (16.5), respectively. As shown, both models fit the experimental data equally well. It also follows from the above fitting that the relative proportion of the delocalized states is of the order of a few percent. Moreover, using (16.4), a successful fit can be achieved by using only two variable parameters, fixing $n_0 = 0$. Based on our experimental data and the above models, we cannot conclude whether or not the amount of delocalized states vanishes to zero at zero temperature. However, the above comparison confirms our conjecture that *the occupation of the two-dimensional closed orbits which provide metallic conduction diminishes as temperature decreases, and, hence, causes the disappearance of the rapid oscillations.*

16.4.6 Discussion

Before comparing with theoretical models, we summarize our experimental results as follows:

(1) Rapid oscillations exist only in the spin-ordered state and are missing in the metallic state. This points at an intimate relation between the RO and the spin ordering.
(2) The $1/B$ periodicity of oscillations and the nonactivated (nonhopping) temperature dependence of the resistance suggests the existence (in the spin-ordered state) of the closed orbits in momentum space, which provide the metallic conduction at finite temperatures. These closed orbits are quantized in perpendicular field and give rise to the RO.
(3) We have found [54,56] that RO exist not only in the SDW and FISDW $N = 0$ phases but also in the higher order FISDW phase $N = 1$. The oscillation frequency does not change at the $N = 0 \Leftrightarrow 1$ transition; however, their amplitude essentially weakens and the phase possibly changes.
(4) The oscillation frequency varies as $1/\cos\theta$ as the magnetic field is tilted from the c^*-axes in any direction. This proves that the closed orbits are two-dimensional and lie in the a–b crystal plane.
(5) Decrease in temperature causes the amplitude of the RO to diminish; this directly shows that the occupation of the delocalized states decreases (and may even vanishes) as $T \to 0$.
(6) Despite drastic changes of the RO-amplitude, the frequency of the oscillations and its angular $1/\cos\theta$ dependence do not change with temperature. Therefore, the areas encircled by the two-dimensional closed orbits and their orientation do not vary with temperature.
(7) A decrease in temperature causes the amplitude of the RO to diminish; this directly shows that the occupation of the metallic closed orbits decreases as $T \to 0$. We interpret this fact as an evidence for redistribution of the carriers from a delocalized to a localized subsystem as temperature falls.

The experimental results described above fit best to a theory by Lebed [50, 53] that considers the influence of Umklapp processes on spin ordering [52]. Such processes can exist in a half-filled band with an energy spectrum consisting of two phase-shifted warped Fermi contours [60]. The Umklapp processes result in the appearance of an auxiliary nesting vector Q_1 beyond the primary nesting vector Q_0: $Q_1 = Q_0 - 2\pi/a = Q_0 - 4k_F$, as shown in Fig. 16.14. Correspondingly, in addition to the main gap Δ_0 caused by Q_0, the second gap Δ_1 appears in the energy spectrum. The calculations in [53] are performed for $T \geq \Delta_0 \gg \Delta_1$. This theory [53] presumes a commensurability of only the x-component of the nesting vector with the lattice, in contrast to other models [37, 38]. The main SDW with nesting vector Q_0 is responsible for the localization of the majority of the carriers, whereas the auxiliary spin-density wave with nesting vector Q_1 is responsible for the appearance of the delocalized carriers occupying closed orbits in momentum space.

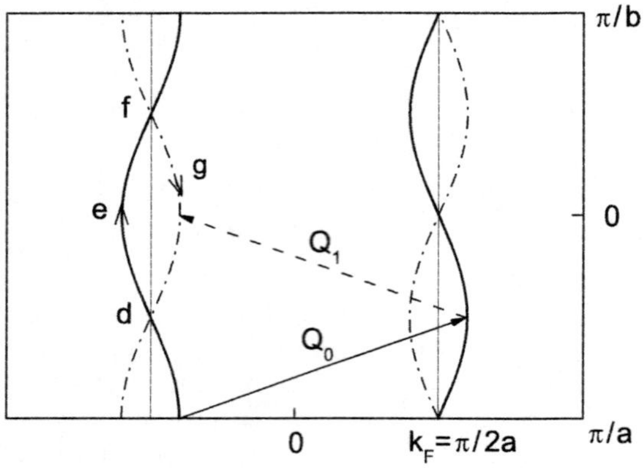

Fig. 16.14. Schematic view of the first Brillouin zone with a corrugated open Fermi surface (*two bold lines*). The actual FS-contours should be shifted upward by $\pi/4b$. The *thick arrow* shows the primary nesting vector Q_0, and the *dashed arrow* shows the secondary nesting vector, involving the Umklapp processes $Q_1 = Q_0 - 2\pi/a = Q_0 - 4k_F$ [50, 53]. *Short arrows* show direction of motion of carriers in a magnetic field perpendicular to the *a–b*-plane

Within the framework of this theory, the RO should arise only in the spin-ordered phase, as a result of the coexistence of the two spin-density waves with two nesting vectors, respectively. This prediction is in accord with our experimental data. The auxiliary SDW is caused by Umklapp processes and therefore has a small amplitude (that is expected to be proportional to the SDW gap). This agrees with our fitting (16.4) and (16.5), where the proportion of the delocalized states was estimated to amount to a few percent. The oscillation amplitude drastically weakens at the transition to higher order FISDW phase $N = 1$ [54, 56]. This is consistent with deterioration of nesting and weakening of the corresponding SDW amplitudes Q_0 and Q_1.

The model explains the existence of closed orbits (contour *d–e–f–g–d* in momentum space) as illustrated in Fig. 16.14 [61]. More over, according to the theory [53], no magnetic breakdown is required for the formation of the closed orbits and, hence, for the appearance of the oscillations in $(TMTSF)_2PF_6$. The closed orbits are two-dimensional and lie in the *a–b*-plane, which is also in agreement with our data. It is the magnetic field quantization of the closed orbits in momentum space that gives rise to the rapid oscillations. In theory, the size of the closed orbits in momentum space depends only on the warping of the FS (i.e., on the t_b transfer integral) and is independent of temperature and magnetic field. This is again in accord with our data.

The frequency of the oscillations calculated on the basis of the model, $4t_b/(\pi e b v_F)$, equals to 286 T, where e is the elementary charge, $b = 6.7$Å, $v_F = 1.11 \times 10^5 \mathrm{m\,s^{-1}}$ (as follows from the cyclotron resonance measurements [7]) and t_b is taken to be 200 K. The experimentally measured frequency at $P = 7.5$ kbar is 275 T (see the inset to Fig. 16.11), that is very close to the calculated value.

Thus, the theory explains qualitatively and, in part, quantitatively, almost all of our experimental results, except for the diminishing of the oscillation amplitude with decreasing temperature. To explain this experimental fact and complete the picture, the theory should be amended to account for the temperature dependence of the effectiveness of Umklapp processes.

16.4.7 Summary

From studies the magnetoresistance for $(TMTSF)_2PF_6$ in the spin ordered phase, we found that in this supposedly purely insulating phase, there exist delocalized states (at least, at finite temperature). These carriers occupy two-dimensional closed orbits in momentum space, lying in the a–b-plane. The closed orbits in momentum space are quantized in a perpendicular magnetic field giving rise to the rapid oscillations. Temperature decrease does not change the size or orientation of the closed orbits, but it does cause a redistribution of carriers from the delocalized to the localized states and hence, the disappearance of the RO. Our data agree qualitatively with a theory that considers two coexisting SDWs with two nesting vectors, respectively. In the theory, the second (auxiliary) SDW is formed due to Umklapp processes; to fit all our experimental data, its amplitude should weaken as temperature decreases. Our results thus shed light on the origin of the rapid oscillations in $(TMTSF)_2PF_6$.

16.5 Coexistence of the Antiferromagnetic and Metallic Phases in $(TMTSF)_2PF_6$

16.5.1 Introduction

The interplay (coexistence, segregation, or competition) of the magnetic spin ordering and the superconducting pairing of electrons is in the focus of the broad research interest. These effects are of key importance for understanding the rich physics of high T_c superconductors, heavy fermion compounds, and also organic conductors [9, 62–64]. Indeed, for these materials, having low-dimensional electron systems, the phase diagrams are surprisingly similar on the plane "pressure" P-temperature T (here by pressure we mean either externally applied pressure or internal "chemical pressure," i.e., dopant concentration) – see Fig. 16.1. The phase diagrams for these materials include a magnetically ordered phase, metallic, and superconducting phases [9, 62–64].

The origin of the superconducting (SC) phase in $(TMTSF)_2PF_6$ remains puzzling; there are experimental and theoretical results pointing at a triplet mechanism of electron pairing [74]. The SC phase neighbors with the two other interesting phases, the antiferromagnetic (AF) SDW state (which is an insulator) and metallic (M) state. Therefore, the issue of the character of the phase border and the origin of the transitions between magnetic, superconducting, and metallic phases (caused, e.g., by pressure changes) become of the fundamental importance [64, 66].

A microscopically mixed state has been suggested in [67] to occur in $(TMTSF)_2PF_6$ due to lack of the complete nesting over the entire Fermi surface. On the P–T phase diagram, the mixed state should occupy a narrow strip of pressures just below the critical P_{SDW} value of the transition to the insulating SDW state. At temperature $T = 1.4$ K, this region should be about 4% wide on the pressure scale. One should also admit an alternative possibility, where in the vicinity of the phase border, a macroscopically inhomogeneous state may arise. The inhomogeneous state incorporates inclusions of the minority phase embedded in the majority phase. As an example, it is well known that the two-dimensional Mott-type insulators tend to the formation of phase-inhomogeneous states [62, 68]. It is also known that martensitic transformations [69] concomitant with phase segregated states take place in such materials, where the free energy of electron system (including the magnetic energy of spin ordering) depends linearly on lattice deformation. If the inhomogeneous state with spatially separated phases emerged on the border of the magnetic and superconducting states, it would have demonstrated simultaneously magnetic and superconducting properties, similarly to those of the heterophase mixed state.

The mixed state as well as the state with spatial phase separation is expected to exhibit similar purely superconducting or purely magnetic properties far away from the phase boundary, so that their behaviors are indistinguishable. However, in the close vicinity of the phase boundary (T_0, P_0), properties of these two types of states are different. In principle, one may distinguish these two possibilities, if the system is forced to cross the phase border by varying pressure at constant temperature, along the horizontal trajectory in Fig. 16.1. In particular, for the inhomogeneous state with inclusions of the minority phase, one might expect such effects as prehistory and hysteresis: The properties of the system at a given point of the P–T phase diagram may depend on the pathway which the system has arrived at this point, due to a path-dependent concentration of segregated phases. In contrast, there is no reason to expect history effects for the mixed state.

Straightforward performing such experiment represents a hard technical task. However, some earlier measurements indicated the presence of an inhomogeneous state in $(TMTSF)_2PF_6$, in the vicinity of the critical pressure [70–72]. Specifically, in [70], from measurements of the Knight shift in ^{77}Se nuclear magnetic resonance, a quenching of the SDW state was found to be a slow function of pressure; this was interpreted as an indication of the coexistence of the SDW and metallic states. In [72], to explain an unusual anisotropy of the H_{c2} temperature dependence, the authors assumed a co-occurrence of the macroscopic domains of superconducting and insulating AF (i.e., SDW) states. Vuletic et al. [71] explored the narrow range of pressures around P_0 by measuring $R(T)$ at fixed pressure values (i.e., along the vertical trajectories in Fig. 16.1). The authors observed hysteresis effects within the insulating AF phase and fitted the set of the measured $R(T, P_i)$ curves using a simple percolation model (which modeled the inclusions of one phase into another one). It was concluded that the observed hysteresis in the $R(T)$-dependences reflects a macroscopically inhomogeneous state, i.e., a mixture of the two phases. We note, however, that the identification of the insulating and metallic phases is not trivial; at zero magnetic field the signature of the insulating SDW state is the onset of the $R(T)$ rise with cooling. Under circumstances when the system may contain a new phase with a priori unknown conduction, such procedure may be potentially ambiguous; therefore, the conclusions made in [71] require additional verification.

In this section we describe the experiments [73,74], where using a different experimental approach, we unambiguously detected the simultaneous presence of both, metallic an antiferromagnetic phases. Using the magnetic field dependence of the T–P phase border for this compound, we swept the magnetic field at a number of fixed pressure values in the vicinity of (T_0, P_0); the magnetic field caused changes of the phase boundary and the corresponding phase transitions between the AF and M states. Thus, the magnetic field was used in our experiment for both, driving the system through the phase transition (instead of pressure), and for reliably identifying the phase content. We observed strong prehistory effects in the resistivity (in the presence of the magnetic field). Besides, we found prehistory effects also in the character of the magnetic field dependence $R(B)$; these effects occur when the system crosses the phase boundary. These results evidence for the *macroscopically* inhomogeneous heterophase state in the vicinity of the AF–M border. Depending on the direction of the magnetic field sweeping, the minority phase extends across the phase border, deep into the majority phase. Observation of the hysteresis and prehistory effects is not consistent with the model of the microscopically mixed (coexisting) antiferromagnetic and metallic states. Thus, our results demonstrate that a macroscopically inhomogeneous state with spatially separated phases emerges at the transition from the antiferromagnetic to metallic phase; this is in agreement with the preceding data [70,71].

16.5.2 The Idea of the Experiment

For the Q1D compound $(TMTSF)_2PF_6$ at zero magnetic field, there is a narrow pressure range in the vicinity of $P \approx 6$ kbar, where the two electronic phase transitions take place as temperature decreases. First, there is a transition from the metallic phase [75] to the insulating AF phase (SDW), and further, from the AF state to the superconducting state. Figure 16.1 shows the corresponding phase diagram [9,71] which incorporates the domains of the AF, M, and SC phases. A vertical trajectory $P = 5.4$ kbar on the phase diagram corresponds to the temperature dependence of resistance $R(T)$, shown in the inset to Fig. 16.1. Crossings of the phase boundaries are marked with dots on the vertical trajectory (Fig. 16.1) and on the measured $R(T)$ dependence.

Traditional Approach: Varying P and T

To explore the character of the transition, one has to be able to unambiguously identify the phase character and component content in the vicinity of the phase boundary. The observation of the absolute resistivity solely at zero field can hardly provide the required information, because the resistance changes smoothly and insignificantly in the vicinity of the M–AF transition. To identify the pure homogeneous M and AF states, one could, in principle, make use of the temperature dependence of conduction that has a pseudoactivated character in the AF phase and diffusive character with the "metallic" sign $dR/dT > 0$ in the M phase. In practice, however, this would require $R(T)$ measurements over a broad temperature range, which is inaccessible in the AF phase. Indeed, for the most interesting regime in the vicinity of the contiguity of the three phases, $T_0 = 1.3\,\mathrm{K}$ and $P_0 = 6.1\,\mathrm{kbar}$, the temperature range of the existence of the AF phase is limited both from the high and low sides (see Fig. 16.1). Besides, direct studies of such transition by changing the pressure in situ (i.e., along horizontal trajectories in Fig. 16.1) at low temperatures represent a very hard technical task.

Alternative Approach: Varying P_0 and T_0

According to our measurements in magnetic field and the earlier results (see, e.g., [76]), the AF/M border $T_0(P_0)$ shifts to higher temperatures as magnetic field grows. Figure 16.15 shows schematically the changes of the border with magnetic field. Due to the smooth and monotonic dependence of T_0 on magnetic field, this dependence may be used for varying T_0. Thus, the system may

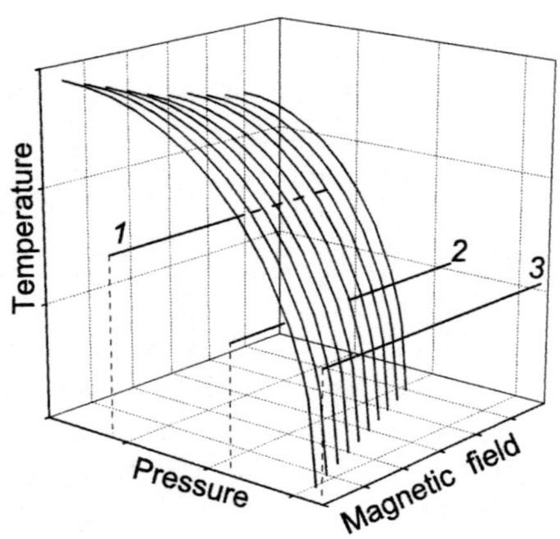

Fig. 16.15. Simplified P–B–T phase diagram for the nonsuperconducting phase space ($T \geq 1.14\,\mathrm{K}$). The details of the phase diagram at high pressures and in high fields are shown in the inset to Fig. 16.16b. For the entire phase of $(TMTSF)_2PF_6$, see [9]

be forced to cross the border by varying the magnetic field at fixed values of pressure and temperature. Figure 16.15 shows that when the initial P and T values (at zero field) are chosen in the vicinity but bigger than P_0, T_0, the phase trajectory of the system (trajectory 2) will cross the phase border as magnetic field grows. Thus, crossing the border occurs due to the changes in $T_0(B)$ and $P_0(B)$ at fixed T and P values. Besides, crossing the border in the presence of magnetic field causes qualitative changes in the behavior of magnetoresistance, which were used in [73, 74] for identifying the phase state and phase content of the system.

16.5.3 Experiment

The two studied $(TMTSF)_2PF_6$ single crystals had a bar shape with mirror-smooth flat surfaces. Their typical dimensions were $2 \times 0.8 \times 0.3\,\mathrm{mm}^3$ along the crystal directions a, b, and c, respectively. The two nominally equivalent $(TMTSF)_2PF_6$ single crystals showed qualitatively similar behavior and had slightly different resistivity values at low temperature. Four $25\,\mu\mathrm{m}$ Pt wires were attached to the contact pads, which were drawn by a graphite conductive paint on the sample surface. The current contacts completely covered the two (b–c) sample surfaces. The potential contact pads were made as parallel narrow strips on the (a–b) surface, across the whole sample width (i.e., along the b-axis). The current was directed strictly along the a-axis; the voltage drop between potential probes along the a-axis corresponded, in the first approximation, to the a-component of the resistivity ρ_a. In this geometry, the admixture of the ρ_b component to the measured ρ_a component was low, $\leq 10\%$. Due to the flat and mirror-smooth (a–b) surface of the samples, the admixture of the ρ_c component was negligibly low. The latter conclusion is confirmed by low residual sample resistance $\rho(T = 1.4\,\mathrm{K}) \approx 5 \times 10^{-6}\,\Omega\,\mathrm{cm}$.

Measurements were made by four-probe a.c. lock-in technique at 32 Hz frequency. The sample and a manganin pressure gauge were placed inside a nonmagnetic pressure cell [6] filled with Si-organic (polyethilenesiloxane) pressure transmitting liquid [32]; a required pressure was created at room temperature. The pressure values quoted in the paper are the low-temperature (4.2 K) values [58]. The ohmic character of the contacts to the sample was confirmed by the negligibly small out of phase component of the measured voltage drop between the contacts.

The pressure cell was mounted in a cryostat in the bore of a 16 T superconducting magnet. Measurements at temperatures $T \geq 1.4\,\mathrm{K}$ were done in the ^{4}He cryostat. For all measurements, the magnetic field was applied along the least conducting direction c^* of the crystal and the current was applied along a. Temperature of the pressure cell was measured using the RuO_2 resistance thermometer, and the heat contact of the sample to the liquid-helium bath was provided with Pt wires. To implement the idea of measurements with crossing the phase border due to magnetic field changes, the pressure value has to be set in the interval 6.2–6.4 kbar at $T \approx 1.4\,\mathrm{K}$.

16.5.4 Observations

The magnetic field dependence of the resistance is qualitatively different for three different trajectories (1, 2, 3) in the B–P–T phase space, depicted in Fig. 16.15b.

Trajectory 1

Figure 16.16a shows a typical example of the magnetic field dependence of the resistance, which corresponds to the trajectory 1 in Fig. 16.15b. For such trajectory, which entirely belongs to the AF domain, the resistance changes are not accompanied with a hysteresis, and R increases monotonically with magnetic field. In strong field, the so-called RO (see Sect. 22.4.2) appear on the background of the monotonic $R(B)$ growth [77]; such $R(B)$ dependence is typical for the AF phase [37]. The oscillations can be more clearly seen in the derivative $\mathrm{d}R(B)/\mathrm{d}B$, shown in the inset to Fig. 16.16a. As pressure increases (but still remains less than the critical $P_0(B)$ value), the resistance magnitude decreases, whereas the $R(B)$ dependence does not change qualitatively.

Trajectory 3

When the initial T, P values are chosen essentially greater than T_0, P_0, the trajectory 3 of the system (Fig. 16.15b) lies entirely in the M domain [75] over the whole range of magnetic field. Figure 16.16b shows that the magnetoresistance in this case has a character qualitatively different from that discussed above for the AF phase. As magnetic field increases, the smooth growth of $R(B)$ transforms into step-like changes, which are related to the developing cascade of transitions between the states with different nesting vector [9, 12, 13, 31]. In strong fields and at low temperatures, the transitions between the states with different nesting vector have a character of the first-order phase transitions [29–31]. Correspondingly, the step-like changes of $R(B)$ in strong fields and at low temperatures are accompanied with hysteresis in $R(B)$ [31]. Such hysteresis may be noticed in Fig. 16.16b in the vicinity of the step-like changes of $R(B)$ in strong fields.

The inset to Fig. 16.16b shows, on the B–T plane, the corresponding phase diagram, which includes different sub-phases of the FISDW [9, 12, 13, 31]. In Figure 16.16b, the $R(B)$ curve at $T = const.$ corresponds to the isothermal trajectory on the B–T phase diagram (shown in the inset to this figure) which sequentially crosses different sub-phases. The corresponding jumps in $R(B)$ are periodic in $1/B$ [12, 13] and well correspond to the phase boundaries on the known B–T phase diagram of the FISDW-regime (see inset to Fig. 16.16b) [31].

Each individual sub-phase has its own nesting vector and the quantized Hall resistance value [12, 13, 78]. Indices N for different sub-phases in Fig. 16.16b correspond to the number of filled Landau levels in the quantum

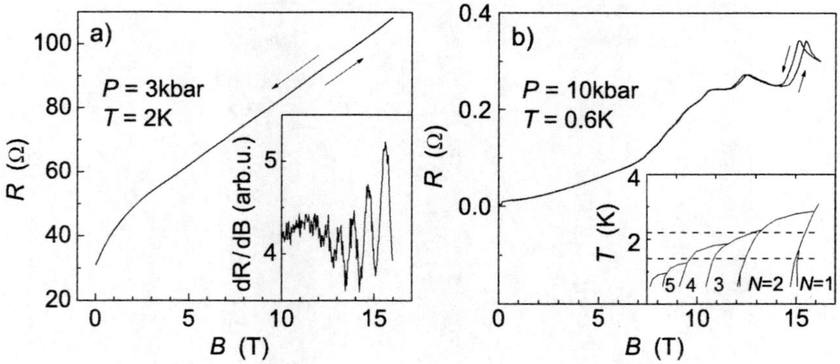

Fig. 16.16. (**a**) Variation of the resistance with magnetic field in the AF state, which corresponds to the motion along the trajectory 1 in Fig. 16.15b. In the inset, on the derivative $\mathrm{d}R/\mathrm{d}B$, one can see the RO, which are typical for the AF phase. (b) Variation of the resistance with magnetic field in the M state, (trajectory 3 in Fig. 16.15b). In high fields, one can see jumps in $R(B)$, which are typical for the M state and correspond to crossing the boundaries between the FISDW phases with different nesting vector. The inset shows the corresponding phase diagram for the FISDW region [31]

Hall effect, and, simultaneously, determine quantized changes of the nesting vector [29, 30]. When pressure P decreases (but still remains bigger than the critical P_0 value), the resistance magnitude smoothly increases whereas $R(B)$ dependence does not change qualitatively. The resistance jumps related to the phase transitions persist and monotonically shift to lower fields, thus indicating the shift of the phase boundaries [31].

Trajectory 2

When the initial P, T values are chosen in the vicinity but slightly bigger than P_0, T_0, the phase trajectory 2 is expected to cross the border with increasing magnetic field, as discussed above. The $R(T)$ dependence measured at $B = 0$ evidences for the true metallic initial state of the sample at $P = 6.4$ kbar (which is close to the critical value $P_0 \approx 6.1$ kbar). When magnetic field increases, the resistance changes insignificantly up to $B \approx 7$ T (see Fig. 16.17a). Upon further increase of B up to 16 T, the resistance sharply grows by three orders of magnitude. This growth indicates the transition from the metallic M to insulating AF phase. On the background of the growing monotonic component of the resistance, one can note the appearance of nonmonotonic periodic variations of resistance (starting from $B \approx 8$ T), which are absolutely atypical for the AF phase.

As magnetic field is swept down (from 16 to 7 T), a strong hysteresis (∼20%) is revealed in the resistance (Fig. 16.17a), whereas the nonmonotonic component of R practically disappears. The $R(B)$ hysteresis and the appearance and disappearance of the nonmonotonic component of resistance depend

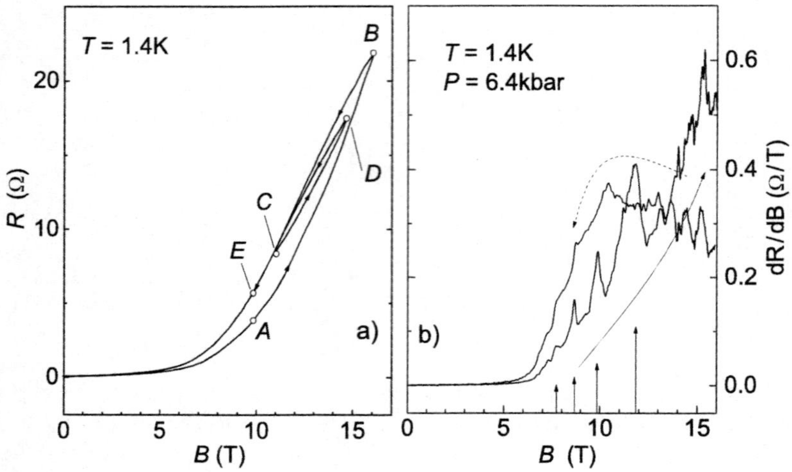

Fig. 16.17. (**a**) Variation of the resistance with magnetic field for the case of crossing the M–AF boundary (trajectory 2 in Fig. 16.15). The sharp growth of $R(B)$ at $B \approx 7\,\mathrm{T}$ corresponds to the M $\rightarrow$ AF transition. (**b**) Magnetic field dependence of the derivative $\mathrm{d}R/\mathrm{d}B$, which corresponds to the $R(B)$ curves in Fig. 16.17a upon increasing and decreasing field. *Vertical arrows* show the borders between the FISDW phases, depicted from the experimentally determined phase diagram (see [31] and the inset to Fig. 16.16b). The inset demonstrates periodicity of the $\mathrm{d}R/\mathrm{d}B$ peaks in $1/B$

only on the absolute magnetic field value and do not depend on its sign [73]. The magnitude of the hysteresis grows with magnetic field. Upon repeated magnetic field sweeps from 0 to 16 T, the above-described $R(B)$ dependence is completely reproduced. The nonmonotonic component of the resistance is more clearly seen in the derivative, $\mathrm{d}R/\mathrm{d}B$, which is shown in Fig. 16.17b. It is worthy to note that the nonmonotonic component is observed only when the magnetic field is increased and is practically invisible when the field is decreased from 16 T.

Vertical arrows in Fig. 16.17b depict the FISDW phase boundaries, which were experimentally determined in [31] from the jumps in $R(B)$ versus field in the FISDW area of the pure M phase. The location of peaks in $\mathrm{d}R/\mathrm{d}B$ in Fig. 16.17b coincides well with the arrows (i.e. with the anticipated borders of the FISDW phases), whereas the peaks are equidistant in $1/B$. For these two reasons, we may identify the observed peaks in $\mathrm{d}R/\mathrm{d}B$ with crossing the borders between the FISDW phases with $N = 6 \Longleftrightarrow 5$, $5 \Longleftrightarrow 4$, $4 \Longleftrightarrow 3$, and $3 \Longleftrightarrow 2$, correspondingly, on the FISDW phase diagram (Figs. 16.16b and 16.8) upon isothermal sweeping the field.

We wish to stress once more that the existence of the peaks and their periodicity in $1/B$ would be quite natural for the M state, but is absolutely unexpected for the AF state. On the $\mathrm{d}R/\mathrm{d}B$ dependence, the next peak ($N = 2 \Longleftrightarrow 1$) at $B \sim 14\,\mathrm{T}$ is not seen, despite the peak amplitude is known to strengthen with decreasing N [31]. The absence of this peak at $B = 14\,\mathrm{T}$

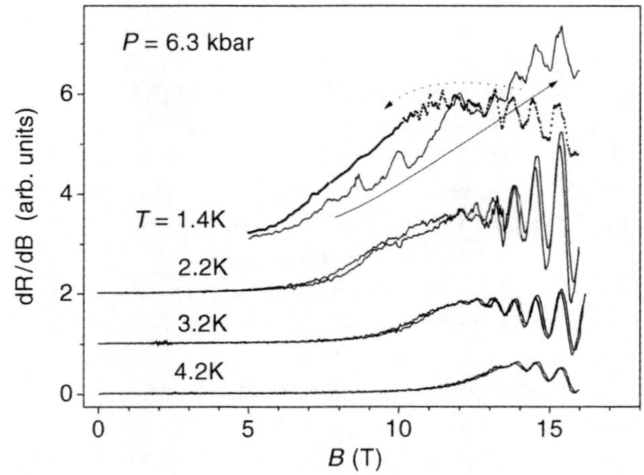

Fig. 16.18. Magnetic field dependence of the derivative $\mathrm{d}R/\mathrm{d}B$ for four temperatures. The *curves* show rapidly disappearing hysteresis as temperature grows, and a characteristic nonmonotonic temperature dependence of the rapid oscillations

indicates that the admixture of the M phase is much reduced and, hence, the system tends to the homogeneous AF state. The fully homogeneous AF state is not achieved yet, since the hysteresis in $R(B)$ curves (see Fig. 16.17a) does not disappear up to the highest field 16 T, though it is a factor of 2 weaker in the field interval (16–15) T than at fields (15–14) T. As an additional confirmation of this conclusion, we note that in stronger fields $B > 14\,\mathrm{T}$ rapid oscillations may be seen in Fig. 16.17b; the RO are characteristic for the AF phase in $(TMTSF)_2PF_6$ [77] (see Sect. 22.4.2).

Magnetic field dependences of $\mathrm{d}R/\mathrm{d}B$ for different temperatures are shown in Fig. 16.18. One can see that the hysteresis of $R(B)$ in fields sweeping up and down disappears as temperature increases. Note that at low temperature, the hysteresis reveals itself not only in the magnitude of $R(B)$ (and $\mathrm{d}R/\mathrm{d}B$) but also in the qualitatively different character of the $R(B)$ dependence. When the field is swept up, the $R(B)$ dependence exhibits jumps (marked by arrows in fields $B = 8$–$12\,\mathrm{T}$). The jumps are characteristic for the FISDW phase transitions in the M phase, whereas the system passed transformation to the AF phase starting from field $B \approx 6\,\mathrm{T}$; the strong growth of resistance in Fig. 16.17a and the appearance of RO evidence for this transformation. When the magnetic field is swept down, these anomalous jumps are almost invisible and one can see only the anticipated RO [77].

16.5.5 Discussion of the Results

The most essential results of our studies are as follows:

1. As expected, when the field increases (decreases) and drives the system along the trajectory 2 (see Fig. 16.15b), the system exhibits the phase

transition. The steep, a factor of 10^3, raise (fall) of the resistance at $B \approx 6\,\mathrm{T}$ evidences for this phase transformation.

2. At the transition from the M to AF phase, rather far away from the phase border, the magnetoresistance continues to exhibit residual signatures of the metallic (minority) phase. When the field grows, the signatures of the minority phase almost disappear and are not restored when the system again approaches the same phase border (i.e. as field decreases). In other words, at such transition, a strong hysteresis is observed in both, the magnitude of R and the character of the $R(B)$ dependence.
3. Upon return transformation from the AF to M phase (with decreasing field), a hysteresis in the magnitude of R is observed: at the transition, the resistance is noticeably higher than that for the pure M phase (or than the resistance value measured as the field grows from $B = 0$). The "true" value of R is restored only when the field is decreased to zero.

In view of the complicated character of the magnetoresistance behavior, which exhibits signatures of both phases, the experimental determination of the AF and M phases becomes of the principle importance. According to the existing theory [79], the SDW–M transition is expected to be either of the second, or weak first order. Experimental data are in agreement with this conclusion [80]. In the vicinity of the critical pressure, the SDW gap Δ, in general, might be small as compared to the antinesting parameter t'_b [79]. In this case, the pseudoactivated $R(T)$ dependence in the SDW phase would have a semimetallic character and would be indistinguishable from the "metallic" $R(T)$ dependence, thus making the identification of the two phases difficult. However, for the specific two-dimensional tight binding case in $(TMTSF)_2PF_6$, Δ does not depend on pressure in the vicinity of the critical pressure and is not small at the transition [79]. This agrees with experimental observations [81], where $R(T)$ was shown to have a pseudoactivated character with rather big gap in the vicinity of the critical pressure. We use therefore, the sharp growth of $R(T)$ (starting from $B \approx 7\,\mathrm{T}$ in Fig. 16.17a) as a firm indication of the onset of the insulating SDW phase.

16.5.6 Inhomogeneous State: Phase Separation or Phase Mixing?

Manifestly, our experimental results do not fit the behavior, anticipated for a microscopically mixed state made of the two coexisting phases. For such a state, the hysteresis effects and the dependence of the phase content on the prehistory should not occur. The behavior described above is also not typical for a homogeneously "overheated" or "overcooled" phases at the first-order phase transitions, because the minority phase disappears smoothly with no sharp jumps in R. Besides, for the second-order transition in a homogeneous system, neither hysteresis nor overheating/overcooling should take place. In the domain of the phase space, where only M or AF phase should exist, clear signatures of the opposite phase are observed beside the "correct" phase. Therefore, the appearance of the hysteresis and the distinct signatures of the presence of both phases in the same domain of the phase space, both evidence that the phase content of the system becomes inhomogeneous. From a theoretical viewpoint, the cascade of transitions could also exist in the AF phase (accompanied with the corresponding jumps in $R(B)$) [79]. However, such cascade has never been observed experimentally in the AF phase. Furthermore, even if the cascade of transitions occurs, as a homogeneous state, it would not give raise to prehistory effects such as observed in our experiment.

The main experimental results were observed on two different samples and found to be qualitatively similar (compare Figs. 16.17b and 16.18). The phase-inhomogeneous state is not a consequence of the inhomogeneity of the sample or of the external pressure. The experimental results which prove this are as follows:

1. The existence of the prehistory in the appearance of the phase-inhomogeneous state contradicts the assumption of the inhomogeneity of the sample or external pressure. Indeed, if such inhomogeneities exist, they would manifest always and cause broadening of the transition. However, in no case, the inhomogeneities may cause hysteresis effects in the character of the $R(B)$ changes, such as jumps in $\mathrm{d}R/\mathrm{d}B$ which manifest themselves when the field is swept up and disappear when the field is swept down.
2. Hysteresis in the character of $R(B)$ dependence arises only at pressure and temperature values in the vicinity of the phase border (T_0, P_0). No history effects are observed when the system is moved away from the phase boundary in either pressure or temperature axes. This may be seen, e.g., in Fig. 16.18, where in strong fields $B > 12\,\mathrm{T}$, rapid oscillations [77] have the same magnitude and phase for the field sweeping up and down.
3. The samples studied are of a high quality. This is confirmed by
   (a) Low residual resistance $\rho(1.4\mathrm{K}) \approx 5 \times 10^{-6}\,\Omega\,\mathrm{cm}$
   (b) High residual resistance ratio, $R(300\,\mathrm{K})/R(1.4\,\mathrm{K}) \sim 450$
4. Both, the homogeneity of the pressure over the sample volume and the high quality of the sample are demonstrated by the sharpness of the SDW transition at $T = 7\,\mathrm{K}$ in Fig. 16.20a.

One could attempt to interpret the $R(B)$ and $\mathrm{d}R/\mathrm{d}B$ data shown in Fig. 16.17 as a sequence of transitions M–FISDW–SDW between the corresponding homogeneous states; such sequence is traditional for the high pressure regime $P > P_0 \approx 6\,\mathrm{kbar}$. To interpret the data (Fig. 16.17) in such a way, one has to assume an unusual sequence of transitions $N = 2 \Rightarrow N = 0$ (in principle, it might be assumed to occur in small-sized droplets of the minority M phase). However, even with this assumption, the above interpretation would face a few problems and would not be able to explain our experimental data. First, as Fig. 16.17a shows, starting from $B \approx 6\,\mathrm{T}$, the resistance begins quickly growing (a factor of 1,000 at 16 T). This is typical for the transitions to the insulating SDW state, which has a large energy gap. In contrast, the known M–FISDW transitions are not accompanied with resistance changes larger than a factor of 2, due to much lower energy gaps. (see e.g., [12,13,31]) Therefore, the $R(B)$ growth at $B \approx 6\,\mathrm{T}$ can not be identified with M–FISDW transitions and we identified it with M–SDW transition. Second, if the resistance behavior with increasing field were treated in the above way, one would have to adopt that the FISDW phase arises as field increases and is almost lost as field decreases (see Fig. 16.17b). The hysteresis in the character of the $R(B)$ changes (i.e., the presence of the FISDW jumps on the way up and their absence on the way down) is an important evidence for the existence of the inhomogeneous state. Third, the hysteresis in $R(B)$ (see Fig. 16.17a) persists over the entire range of fields. This is not typical for the FISDW transitions, where the hysteresis can arise only in narrow regions of fields adjacent to the transitions between different FISDW states.

On the Temperature Dependence of the Observed Phenomena

As temperature increases, the hysteresis in both $R(B)$ and $\mathrm{d}R/\mathrm{d}B$ quickly vanishes (see Fig. 16.18). This might be caused by either complete disappearance of the minority M-phase or washing out the difference between resistivities in the insulating SDW phase and M-phase. Indeed, the T-dependences of resistivities in these phases are very different and the admixture of the minority phase will therefore produce stronger effect to the resistivity at lower temperatures. Based on the existing experimental data we can not distinguish between these two possibilities.

For the homogeneous FISDW phase in magnetic field, the number of the observed FISDW transitions decreases as T increases [12,13,31]. This is illustrated in the inset to Fig. 16.16b, which shows that the trajectory for $T = 2.2\,\mathrm{K}$ intersects the FISDW area at higher field than the trajectory for $T = 1.4\,\mathrm{K}$. For the studied inhomogeneous state, the evolution of $\mathrm{d}R/\mathrm{d}B$ with temperature (Fig. 16.18) qualitatively agrees with the above FISDW phase diagram: all four FISDW transitions observed at $T = 1.4\,\mathrm{K}$ are missing at $T \geq 2.2\,\mathrm{K}$ (as expected for the homogeneous state). This confirms once more that these peaks correspond to the FISDW transitions.

16.5.7 Prehistory Effects

The prehistory effect is the most unexpected among the results obtained, even more unexpected than the hysteresis. This phenomenon is illustrated in Fig. 16.19, where four different dependences $\mathrm{d}R(B)/\mathrm{d}B$ are compared; these dependences correspond to four trajectories (AB, BC, CD, and DE) shown in Fig. 16.17a. When the system crosses the M–AF phase border (at $B \approx 6\,\mathrm{T}$) and moves deep into the AF domain along the trajectory AB, the derivative $\mathrm{d}R(B)/\mathrm{d}B$ exhibits peaks (marked with vertical arrows). These peaks

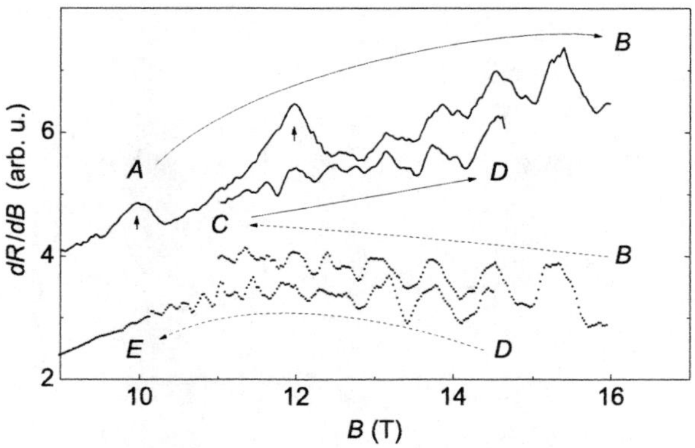

Fig. 16.19. History effects developing in dR/dB as magnetic field varies along the trajectories AB, BC, CD, and DE (see Fig. 16.17a). *Two arrows* mark the peaks in dR/dB corresponding to the FISDW transitions in the M phase; these peaks are missing as field decreases

correspond well to the resistance jumps observed in the M phase at crossing the borders between the FISDW sub-phases 4 $\Longleftrightarrow$ 3, and 3 $\Longleftrightarrow$ 2 [31]; correspondingly, the peaks have nothing in common with the AF phase in which the system is supposed to be for the given P, B, T values. The existence of these peaks evidences that, at least, a part of the sample has not transform into the insulating AF phase and remains in the M phase. In the field $B \approx 15\,$T, instead of the next anticipated peak (which would correspond to the FISDW transition 2 $\Longleftrightarrow$ 1), one can see only weak oscillations reminiscent of the RO in the AF phase. This points out that the amount of the M phase is substantially reduced (though it does not fully disappear yet, as discussed above.

Subsequent sweeps of the field down, up, and again down along the B→C, C→D, and D→E trajectories (which do not cross the AF–M border), reveal the two important features: (a) the hysteresis in $R(B)$ weakens by a factor of two (see Fig. 16.17a), (b) the FISDW transition (i.e., dR/dB peak at 12 T) does not show up anymore (see Fig. 16.19). These features confirm the above conclusion that the amount of the M phase is substantially reduced. However, we do not have a satisfactory explanation of an apparent disproportionality between a partial reduction in the hysteresis strength and complete disappearance of the FISDW peak at 12 T.

16.5.8 Phase Separation at Zero Magnetic Field

In the above-described experiments, the presence of the magnetic field was not of a principle importance. The role of the magnetic field was to produce a qualitative difference between the $R(B)$ dependences in the AF and M phases; this is necessary to identify crossing the border and to reveal the phase content of the inhomogeneous state. In our view, the phase-inhomogeneous state

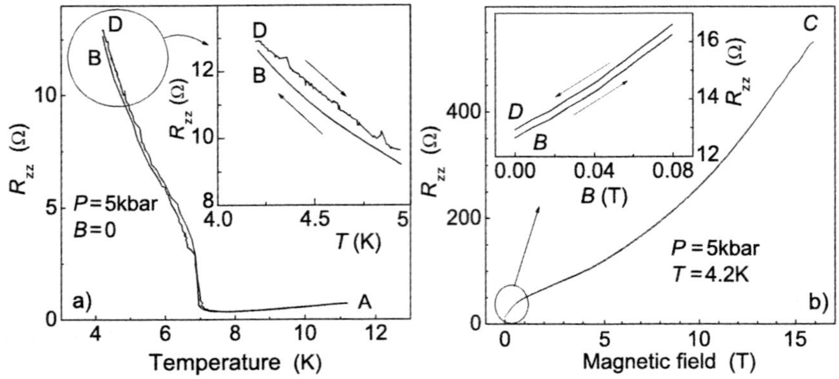

Fig. 16.20. (**a**) History effects in the temperature dependence $R(T)$ at zero field when the trajectory crosses the M–AF phase boundary; (**b**) Reestablishing the homogeneous state by sweeping the magnetic field to 16 T and back to zero. Pressure is $P = 5$ kbar

with inclusions of the minority phase imbedded into the majority phase must also arise in the transition from M to AF phase with decreasing temperature (see the phase diagram in Fig. 16.15a). In this case, however, the anticipated resistance changes are weak and of a quantitative rather than qualitative character. Such measurements have been already undertaken in [71], and our task was to test or confirm these results for the same samples in which we have explored the character of the transition in nonzero magnetic field.

For the experiment we have chosen the pressure $P = 5$ kbar, which is less than the critical value. The results are represented in Fig. 16.20. For this pressure and $B = 0$, the system transforms from the metallic to AF state as temperature decreases below $T = 7$ K [9, 71]. At the transition, the resistance sharply raises and further grows with decreasing the temperature; this behavior corresponds to the onset of the insulating state (SDW). The variations of the resistance with temperature along the trajectory AB are shown in Fig. 16.20a. The final resistance value at point B corresponds to the minimal temperature 4.2 K in this experiment.

According to the above assumption, at point B the system has a spatially inhomogeneous phase content: beyond the majority insulating AF phase, it also contains inclusions of the minority metallic M phase. The following experiment was done to check this assumption: the magnetic field was increased from 0 to 16 T; according to the above results, such strong field should substantially reduce the amount of the minority phase. Figure 16.20b shows the changes in the resistance with increase (trajectory BC) and subsequent decrease (trajectory CD) of the magnetic field. After magnetic field is decreased to zero, the system returns to a state (point D), which is similar to the initial one (point B). However, magnifying the data in the inset of Fig. 16.20b reveals a small (~0.5%) increase in the resistance at point D as compared to that at the initial point B. This minor difference evidences for decreasing the share of the well-conducting metallic phase. At both points B and D, the resistance does not change with time (over a few days), therefore, the observed difference is a stationary rather than nonstationary effect and is not related to temperature gradients. Since the observed hysteresis is weak, its interpretation could hardly be done without preliminary studies of much stronger hysteresis effects in magnetic field.

The minor difference between the resistances at points D and B is not caused by a magnetic field hysteresis (intrinsic to superconducting magnets): once the system arrived at point D, its resistance does not exhibit hysteresis anymore upon subsequent magnetic field sweeps up and down. In other words, as the field varies repeatedly from D to C and back, no irreversible changes in the resistance are observed. We conclude therefore that the inclusions of the metallic phase (if any) do not change anymore. Upon further increase of temperature at zero field, the resistance varies along the trajectory DA. If the sample is cooled down again, its resistance reproduces the trajectory AB and we arrive at point B within 0.1% accuracy, the result which evidences for restoring the phase-inhomogeneous state.

16.5.9 Summary

The experiments described above reveal hysteresis in the magnitude of the resistance and in the character of its variation with magnetic field, which develops at the transition from metallic to antiferromagnetic insulator state. Furthermore, we found that the behavior of the resistance with magnetic field becomes prehistory dependent. These results evidence unambiguously for the occurrence of the inhomogeneous state in the vicinity of the phase boundary between the M and AF phases; we conclude that this state consists of macroscopic inclusions of the minority phase imbedded into the majority phase. Our conclusions do not depend on any model assumptions about the spatial arrangement of the two-phase state, because for identifying the phase content we used qualitative difference in the magnetoresistance behavior in the AF and M states. Our results are in a good agreement with previous data [70, 71], obtained in different ways. Recently, similar coexistence of macroscopic regions of SDW and metal and strong hysteretic effects were observed in $(TMTSF)_2PF_6$ by NMR and transport type experiments in [82].

The observed phenomena of the phase separation are similar to those typically seen in martensitic transformations. If this analogy is not accidental, it suggests that the free energy of the electron system depends linearly on distortion of the charge or SDW system. This conjecture requires an experimental verification.

The minority phase inclusions can be eliminated almost completely as the system moves far away from the boundary, deep in the majority phase domain. The hysteresis in the magnitude and in the field dependence of the resistance does not depend on time; it is a stationary and well reproducible effect. The most striking evidence for the heterophase content was obtained from experiments in finite magnetic fields. However, similar phase-inhomogeneous state has also been observed in zero field, for the transition from metallic to antiferromagnetic insulator (SDW) state. These observations confirm that the magnetic field does not play an essential role in the occurrence of the heterophase state. Extending this analogy to the superconducting transition, we note an interesting possibility that the transition from the antiferromagnetic insulator to superconducting state might also occur via superconducting transition in inclusions of the minority metallic phase, rather than between the two homogeneous AF and SC states. This suggestion also requires an additional experimental verification.

16.6 Concluding Remark

The $(TMTSF)_2PF_6$ charge transfer salt was synthesized by K. Bechgaard in 1979 [11, 83] and since than was intensively studied worldwide. The experiments described in this chapter demonstrate that, despite more than a quarter of century of intensive research, this material, thanks to its rich phase diagram, continues to be an unique playground for exploring new exciting phenomena in low-dimensional physics.

Acknowledgements

The experiments described in this chapter were supported by INTAS, RFBR, Programs of the Russian Academy of sciences, and by the Presidential program "The state support of the leading scientific schools."

References

1. L.P. Gor'kov, A.G. Lebed, J. Phys. (Paris) Lett. **45**, L433 (1984)
2. M. Heritier, G. Montambaux, P. Lederer, J. Phys. (Paris) Lett. **45**, L943 (1984)
3. P.M. Chaikin, Phys. Rev. B **31**, 4770 (1985)
4. L.P. Gor'kov, A.G. Lebed, Phys. Rev. Lett. **71**, 3874 (1993)
5. A. Ardavan, J.M. Schrama, S.J. Blundell, et al., Phys. Rev. Lett. **81**, 713 (1998)
6. A.V. Kornilov, V.A. Sukhoparov, V.M. Pudalov, in *High Pressure Science and Technology*, ed. by W. Trzeciakowski (World Scientific, Singapore, 1996), pp. 63–65.
7. A.V. Kornilov, P.J.M. van Bentum, J.S. Brooks, et al., Synth. Met. **103**, 2246 (1999)
8. *Millimetre-Wave Vector Network Analyzer*, AB-millimetre, 52 Rhue Lhomond, F-75005 Paris, France
9. P.M. Chaikin, J. Phys. I (France) **6**, 1875 (1996)
10. For a review, see L.P. Gor'kov, Usp. Fiz. Nauk, **144**, 381 (1984) [Sov. Phys. Uspekhi **27**, 809 (1984)]

11. For a review, see T. Ishiguro, K. Yamaji, G. Saito, *Organic Superconductors*, 2nd edn. (Springer, Berlin Heidelberg New York, 1998)
12. S.T. Hannahs, J.S. Brooks, W. Kang L.Y. Chiang, P.M. Chaikin, Phys. Rev. Lett. **63**, 1988 (1989)
13. J.R. Cooper, W. Kang, P. Auban G. Montambaux, D. Jérome, K. Bechgaard, Phys. Rev. Lett. **63**, 1984 (1989)
14. M. Ribault, D. Jérome, D. Tuchender, et al., J. Phys. (France) Lett. **44**, L953 (1983)
15. W. Kang, S.T. Hannahs, P.M. Chaikin, Phys. Rev. Lett. **70**, 3091 (1993)
16. S. Uji, C. Terakura, M. Takashita, et al., Phys. Rev. B **60**, 1650 (1999)
17. P.M. Chaikin, M.-Y. Choi, J.F. Kwak, et al., Phys. Rev. Lett. **51**, 2333 (1983)
18. K. Oshima, et al., Synth. Met. **70**, 861 (1995)
19. N. Biskup, J.S. Brooks, R. Kato, K. Oshima, Phys. Rev. B **60**, R15001 (1999)
20. A.G. Lebed, Zh. Eksp. Teor. Fiz. **89**, 1034 (1985) [Sov. Phys. JETP, **62**, 595 (1985)]
21. K. Maki, Phys. Rev. B **33**, 4826 (1986)
22. K. Yamaji, Synth. Met. **13**, 29 (1986)
23. G. Montabaux, M. Heritier, P. Lederer, Phys. Rev. Lett. **55**, 2078 (1985)
24. N. Dupuis, V.M. Yakovenko, Phys. Rev. Lett. **80**, 3618 (1998)
25. D. Zanchi, G. Montabaux, Phys. Rev. Lett. **77**, 366 (1996)
26. D. Poilblanc, G. Montambaux, M. Heritier, P. Lederer, Phys. Rev. Lett. **58**, 270 (1987)
27. V.M. Yakovenko, Phys. Rev. B **43**, 11353 (1991)
28. K. Sengupta, H.-J. Kwon, V.M. Yakovenko, Phys. Rev. Lett. **86**, 1094 (2001)
29. A.G. Lebed, Pis'ma v ZhETF **72**, 205 (2000) [JETP Lett. **72**, 141 (2000)]
30. A.G. Lebed, Phys. Rev. Lett. **88**, 177001 (2002)
31. A.V. Kornilov, V.M. Pudalov, Y. Kitaoka, et al., Phys. Rev. B **65**, 060404 (2002)
32. A.S. Kirichenko, A.V. Kornilov, V.M. Pudalov, Instrum. Exp. Tech. **48**(6), 813 (2005)
33. R.L. Greene, P. Haen, S.Z. Huang, et al., Mol. Cryst. Liq. Cryst. **79**, 225 (1982)
34. J.F. Kwak, J.E. Schirber, R.L. Green, E.M. Engler, Phys. Rev. Lett. **46**, 1296 (1981)
35. K. Sengupta, N. Dupuis, Phys. Rev. B **68**, 094431 (2003)
36. J.P. Ulmet, P. Auban, A. Khmou, S. Askenazy, J. Phys. (Paris) Lett. **46**, L545 (1985)
37. S. Uji, J.S. Brooks, M. Chaparala, et al., Phys. Rev. B, **55**, 12446 (1997)
38. J.S. Brooks, J. O'Brien, R.P. Starrett, et al., Phys. Rev. B **59**, 2604 (1999)
39. W.G. Clark, M.E. Hanson, W.H. Wong, B. Alavi, J. Phys. (France) IV **3** 235 (1993)
40. J.L. Musfeldt, M. Poirier, P. Batail, S. Lenoir, Phys. Rev. B **51**, 8347 (1995)
41. T. Takahashi, Y. Maniwa, H. Kawamura, G. Saito, J. Phys. Soc. Jpn. **55**, 1364 (1986)
42. J.C. Lasjaunias, K. Biljacović, F. Nad', et al., Phys. Rev. Lett. **72**, 1283 (1994)
43. J.-P. Ulmet, A. Narjis, M.J. Naughton, J.M. Fabre, Phys. Rev. B **55**, 3024 (1997)
44. S. Uji, S. Yasuzuka, T. Konoike, K. Enomoto, et al., Phys. Rev. Lett. **94** 077206 (2005)
45. J.S. Qualls, C.H. Mielke, J.S. Brooks, et al., Phys. Rev. B **62**, 12680 (2000)
46. N. Biskup, L. Balicas, S. Tomić, et al., Phys. Rev. B **50**, 12721 (1994)
47. S.A. Brazovskii, V.M. Yakovenko, Pis'ma v ZhETF, **43**, 102 (1986) [JETP Lett. **43** 134 (1986)]

48. A.G. Lebed, P. Bak, Phys. Rev. B **40**, 11433 (1989)
49. V.M. Yakovenko, Phys. Rev. Lett. **68**, 3607 (1992)
50. A.G. Lebed, Phys. Rev. Lett. **74**, 4903 (1995)
51. K. Kishigi, K. Machida, Phys. Rev. B **53**, 5461 (1996)
52. K. Yamaji, J. Phys. Soc. Jpn. **56**, 1101 (1987)
53. A.G. Lebed, Phys. Scr. **T39**, 386 (1991)
54. A.V. Kornilov, V.M. Pudalov, A.-K. Klehe, A. Ardavan, J.S. Qualls, Pis'ma v ZhETF, **84**, 744 (2006) [JETP Lett. **84**, 628 (2006)]
55. A.V. Kornilov, V.M. Pudalov, A.-K. Klehe, A. Ardavan, J.S. Qualls, J. Singleton, J. Low Temp. Phys. **142**(3/4), 305 (2006)
56. A.V. Kornilov, V.M. Pudalov, A.-K. Klehe, A. Ardavan, J.S. Qualls, J. Singleton Phys. Rev. B **76**, 045109 (2007)
57. A.V. Kornilov, V.M. Pudalov, Instrum. Exp. Tech. **42**(1), 127 (1999)
58. The pressure reading of the manganen gauge at He temperatures was calibrated in a separate experiments, using the superconducting Sn gauge. The decrease in pressure with cooling from 290 K to helium temperatures was typically ≈0.3–0.4 GPa
59. A. Isihara, L. Smrčka, J. Phys. C: Solid State Phys. **19**, 6777 (1986)
60. For $(TMTSF)_2PF_6$, according to the NMR data [41], the two warped Fermi contours are shifted by $\pi/2b$ along b direction
61. Since Umklapp processes are elastic, the quasiparticles on dashed and thick lines in Fig. 16.14 have the same energy and the contour d–e–f–g–d is equipotential. This contour looks similar to the ordinary 2D "metallic pocket." However, this is not a real 2D "metallic pocket," because there is no 2D parabolic energy spectrum underneath the states shown by dashed lines in Fig. 16.14 (in contrast to the states shown by thick lines). These states can be considered similar to the known resonant states in semiconductors. The occupation number of these resonant states is determined by the intensity of the Umklapp processes; it is not related with the area of the d–e–f–g–d contour (as it would be in case of a real "metallic pocket"). The geometry of the dashed contour and, hence, the d–e–f–g–d area are firmly determined by the Q_1 vector.
62. J. Orenstein, A.J. Millis, Science, **288**, 468 (2000)
63. N.D. Mathur, F.M. Grosche, S.R. Julian, et al., Nature, **394**, 39 (1998)
64. S.S. Saxena, P. Agarwal, K. Ahilan, et al., Nature **406**, 587 (2000)
65. I.J. Lee, D.S. Chow, W.G. Clark, et al., Phys. Rev. B **68**, 092510 (2003); A.G. Lebed, K. Machida, M. Ozaki, Phys. Rev. **62**, R795 (2000); K. Kuroki, R. Arita, H. Aoki, Phys. Rev. B **63**, 094509 (2001)
66. S. Uji, H. Shinagawa, T. Terashima, et al., Nature, **410**, 908 (2001)
67. K. Yamaji, J. Phys. Soc. Jpn. **51**, 2787 (1982)
68. V.J. Emery, S.A. Kivelson, H.Q. Lin, Phys. Rev. Lett., **64**, 475 (1990)
69. Z. Nishiyama, M. Fine, M. Meshii, C. Wayman (eds.), *Martensitic Transformations* (Academic Press, New York, 1978); R. Gotthardt, J. van Humbeeck (eds.), *Proceedings of International Conference on Martensitic Transformations*, ICOMAT 95, Lausanne, Switzerland, 1995
70. L.J. Azevedo, J.E. Schirber, E.M. Engler., Phys. Rev. B, **27**, 5842 (1983)
71. T. Vuletić, P. Auban-Senzier, C. Pasquier, et al., Eur. Phys. J. B **25**, 319 (2002)
72. I.J. Lee, P.M. Chaikin, M.J. Naughton, Phys. Rev. Lett. **88**, 207002 (2002)
73. A.V. Kornilov, V.M. Pudalov, Y. Kitaoka, et al., Phys. Rev. B **69**, 224404 (2004)
74. A.V. Kornilov, V.M. Pudalov, Y. Kitaoka, et al., Pis'ma v ZhETF **78**, 26 (2003) [JETP Lett. **78**, 21 (2003)]

75. For simplicity, throughout the paper by "metallic" (M) we mean a state at $P > P_0(T)$. Strictly speaking, this state is metallic only in zero and low magnetic field. In quantizing magnetic fields, the FISDW state arises in this region of the phase space (see Sec. 25.1.2 and this state is considered to be semimetallic [12,13]
76. N. Matsunaga, K. Yamashita, H. Kotani, et al., Phys. Rev. B **64**, 052405 (2001)
77. On the origin of "rapid oscillations" (RO), see Sect. 22.4.2. In $(TMTSF)_2PF_6$, RO are observed in the antiferromagnetic SDW or $N = 0$ FISDW phase and are much weaker in the $N = 1$ FISDW phase [56]. For the purposes of the current chapter it is important only that for the one-dimensional Fermi surface, RO (1) have much shorter period than the FISDW anomalies, and (2) have a nonmonotonic temperature dependence of the amplitude. Due to these features, we used the RO as an additional tool to identify the SDW phase
78. V.M. Yakovenko, H.-S. Goan, Phys. Rev. B **58**, 10648 (1998)
79. S. Brazovski, L.P. Gor'kov, A.G. Lebed, Zh. Exp. Teor. Fiz **83**, 1198 (1982) [JETP **56**, 683 (1982)]
80. See., e.g., R. Brusetti, P. Garoche, K. Bechgaard, J. Phys. (Paris) Colloq. **44**, C3-805 (1983)
81. See, e.g., G.M. Danner, P.M. Chaikin, S.T. Hannahs, Phys. Rev. B **53**, 2727 (1996)
82. I.J. Lee, S.E. Brown, W. Yu, et al., Phys. Rev. Lett. **94**, 197001 (2005)
83. K. Bechgaard, et al., Solid State Comm. **33**, 1119 (1980)

17

Theory of the Quantum Hall Effect in Quasi-One-Dimensional Conductors

V.M. Yakovenko

This chapter reviews the theory of the quantum Hall effect (QHE) in quasi-one-dimensional (Q1D) conductors. It is primarily based on the author's papers (Phys. Rev. B 43:11353, 1991; J. Phys. I (France) 6:1917, 1996; Phys. Rev. B 58:10648, 1998; Phys. Rev. Lett. 86:1094, 2001). The QHE in Q1D conductors is closely related to the magnetic-field-induced spin–density wave (FISDW) observed in these materials. The theory of the FISDW is reviewed in this book by Lebed, by Héritier, and by Haddad et al. The FISDW experiment is reviewed by Brown, Chaikin and Naughton, and by Pudalov and Kornilov.

17.1 Introduction to Quasi-One-Dimensional Conductors

Organic metals of the $(TMTSF)_2X$ family, where TMTSF is tetramethyltetraselenafulvalene and X is an inorganic anion such as PF_6, are quasi-one-dimensional (Q1D) crystals consisting of parallel conducting chains formed by the organic molecules of TMTSF. The chain direction is denoted as $\boldsymbol{a}$ or x. The interchain coupling is much weaker in the $\boldsymbol{c}(z)$ direction than in the $\boldsymbol{b}(y)$ direction, so that the chains form weakly coupled two-dimensional (2D) layers. In a simple model, the electron dispersion $\varepsilon(\boldsymbol{k})$ in these materials is described by a tight-binding model representing tunneling between the TMTSF molecules

$$\varepsilon = 2t_a \cos(k_x a) + 2t_b \cos(k_y b) + 2t_a \cos(k_z d) + \dots , \tag{17.1}$$

where $\boldsymbol{k} = (k_x, k_y, k_c)$ is the electron wave vector. The intermolecule distances are $a = 0.73\,\mathrm{nm}$, $b = 0.77\,\mathrm{nm}$, and $d = 1.35\,\mathrm{nm}$ [1], and we approximate the triclinic crystal structure by the orthogonal one. The electron tunneling amplitudes $t_a \gg t_b \gg t_c$ are estimated as 250, 25, and 1.5 meV [1]. The band (17.1) is quarter-filled[1] by holes, because each anion X^- takes one electron.

[1] We ignore weak dimerization of the TMTSF molecules [1], which is not essential for our consideration.

The Fermi surface is open and consists of two disconnected sheets with k_x close to $\pm k_{\rm F}$, where $k_{\rm F} = \pi/4a$ is the Fermi momentum along the chains. In the vicinity of the Fermi surface, we can linearize the longitudinal electron dispersion $\varepsilon_{\|}(k_x)$. Measuring ε from the Fermi energy and neglecting t_c, we can approximate (17.1) as

$$\varepsilon_{\pm} = \varepsilon_{\|}(k_x) + \varepsilon_{\perp}(k_y b) = \pm\hbar v_{\rm F}(k_x \mp k_{\rm F}) + 2t_b \cos(k_y b) + 2t_b' \cos(2k_y b), \quad (17.2)$$

where the signs $\pm$ correspond to the two sheets of the Fermi surface, $v_{\rm F} = \partial\varepsilon_{\|}/\hbar\partial k_x \approx 10^5\,{\rm m\,s^{-1}}$ [2] is the Fermi velocity, and $\hbar$ is the Planck constant. In this chapter, we study the in-plane Hall conductivity σ_{xy} per one layer in a magnetic field $\boldsymbol{B}$ applied perpendicular to the $\boldsymbol{a}$–$\boldsymbol{b}$ layers. The interlayer coupling t_c is not essential for this consideration, and so most of the theory is presented for just one 2D layer, as in (17.2). The tunneling amplitude t_b' to the next-nearest chain in (17.2) is important for the FISDW theory.

17.2 Hall Effect in the Normal State

Let us briefly discuss the Hall effect in the normal state of $(\mathrm{TMTSF})_2\mathrm{X}$. The textbook formula says that the Hall coefficient is $R_{\rm H} = 1/nec$, where n is the carrier concentration per one layer, e is the electron charge, and c is the speed of light. However, the linearized model (17.2) gives $R_{\rm H} = 0$, because of the electron–hole symmetry. A non-zero result is obtained by taking into account the curvature β of the longitudinal electron dispersion [3–5]:

$$R_{\rm H}^{(\rm n)} = \frac{\beta}{nec}, \qquad \beta = \frac{k_{\rm F}}{v_{\rm F}}\frac{\partial^2\varepsilon_{\|}}{\partial k_x^2}. \quad (17.3)$$

For the quarter-filled tight-binding band (17.1) in $(\mathrm{TMTSF})_2\mathrm{X}$, $n = 1/2ab$ and $\beta = \pi/4$ [3–5], as opposed to $\beta = 1$ for the conventional parabolic dispersion. The experimentally measured $R_{\rm H}^{(\rm n)}$ [6, 7] is in overall agreement with (17.3), but it also exhibits some puzzling temperature dependence [5, 8].

The Hall resistivity $\rho_{xy} = R_{\rm H}^{(\rm n)} B = \beta(h/e^2)(2abB/\phi_0)$ is small and not quantized. This is because, for any realistic B, the magnetic flux through the area $2ab$ per one carrier is much smaller than the flux quantum $\phi_0 = hc/e$, and so the Landau filling factor is very high, of the order of 10^2–10^3.

17.3 Introduction to the Quantum Hall Effect in the FISDW State

When a strong magnetic field B is applied in the z direction perpendicular to the layers, the $(\mathrm{TMTSF})_2\mathrm{X}$ materials experience a cascade of phase transitions between the so-called magnetic-field-induced spin-density-wave states. These are true thermodynamic phase transitions, observed in specific heat [9, 10], magnetization [11], NMR [12], and virtually all other physical quantities [13, 14]. A detailed theory of the FISDW is presented in other chapters of this book, as well as in the book [1] and in the review volume [15]. According to the theory [16–19], the electron spin density in the FISDW state develops spontaneous modulation with the wave vector

$$\boldsymbol{Q} = (Q_x, Q_y, Q_z), \quad Q_x = 2k_{\mathrm{F}} - NG, \quad G = ebB/\hbar c, \quad E_B = \hbar v_{\mathrm{F}} G, \quad (17.4)$$

where G is the characteristic wave vector of the magnetic field, and N is an integer number (positive or negative). The magnetic length $l_x = 2\pi/G$ is defined so that the magnetic flux through the area bounded by l_x and by the interchain distance b is equal to the flux quantum: $Bl_x b = \phi_0$. Notice that $G \ll k_{\mathrm{F}}$, because $Bab \ll \phi_0$ for realistic magnetic fields. We also introduced the characteristic energy E_B of the magnetic field in (17.4).

The difference between Q_x and $2k_{\mathrm{F}}$ in (17.4) plays crucial role for the Hall effect. Let us calculate σ_{xy} in the FISDW state by naively counting the number of available carriers [16]. The FISDW with the wave vector $\boldsymbol{Q}$ (17.4) hybridizes the $\pm k_{\mathrm{F}}$ sheets of the Fermi surface and opens an energy gap. Some electron and hole pockets are formed above and below the energy gap because of imperfect nesting. The effective carrier concentration n is determined by their total area in the momentum space. This is the area between the two sheets of the Fermi surface (17.2) $k_x^{\pm}(k_y) = \pm k_{\mathrm{F}} \mp \varepsilon_{\perp}(k_y)/v_{\mathrm{F}}$, one of them shifted by the wave vector (Q_x, Q_y):

$$\begin{aligned} n &= \frac{2}{2\pi} \int_0^{2\pi/b} \frac{\mathrm{d}k_y}{2\pi} \{k_x^{+}(k_y) - [k_x^{-}(k_y - Q_y) + Q_x]\} \\ &= \frac{2}{2\pi} \int_0^{2\pi/b} \frac{\mathrm{d}k_y}{2\pi} \left(2k_{\mathrm{F}} - Q_x + \frac{\varepsilon_{\perp}(k_y - Q_y) - \varepsilon_{\perp}(k_y)}{v_{\mathrm{F}}}\right). \end{aligned} \quad (17.5)$$

Here the factor 2 accounts for two spin projections. The integral of the last term in (17.5) vanishes for any Q_y, because

$$\int_0^{2\pi/b} \varepsilon_{\perp}(k_y)\, \mathrm{d}k_y = 0, \quad (17.6)$$

whereas (17.4) and the first term in (17.5) give $n = 2NeB/hc$. Substituting this expression in the textbook formula $\sigma_{xy} = nec/B$, we find

$$\sigma_{xy} = 2Ne^2/h. \quad (17.7)$$

Formula (17.7) was also derived in [20] using the Středa formula.

Equation (17.7) represents the integer quantum Hall effect (QHE). The Hall effect becomes quantized, because the FISDW (17.4) eliminates most of the carriers, but leaves N fully occupied Landau levels in the remaining pockets [21]. With the increase of B, the system experiences a cascade of phase transitions between the FISDW states with different numbers N. This produces a series of the quantum Hall plateaus (17.7) separated by phase transitions, as indeed observed experimentally in $(\mathrm{TMTSF})_2\mathrm{X}$ [21–30]. In a simple case, N takes consecutive numbers $0, 1, 2, \dots$ with decreasing magnetic field B [21–24], but $N(B)$ may also show more complicated sequences, including sign changes [23, 25–30]. The bulk Hall conductivity is measured in experiment, and σ_{xy} per one layer is obtained by dividing by the number of conducting layers. The latter may depend on current distribution, so that the absolute values of σ_{xy} are not obtained very precisely [27], but the relative values of σ_{xy} do correspond to integer numbers. In the FISDW state, $\sigma_{xy} \gg \sigma_{xx}$, σ_{yy}, as expected for the QHE [23,25,26,29,30].

The derivation of (17.7) presented above is simple but not quite satisfactory. In the FISDW state, there are multiple gaps in the energy spectrum of electrons, including a gap at the Fermi level, and so the notion of metallic electron and hole pockets is not quite meaningful. Moreover, the magnetic energy E_B is typically greater than the FISDW gap Δ, so there is strong magnetic breakdown between the pockets, and the semiclassical Landau quantization using closed electron orbits is not well defined. After briefly summarizing the FISDW theory in Sect. 17.4, we present a more rigorous derivation of the QHE as a topological invariant [31] in Sect. 17.5. This formalism is also utilized for various generalizations presented in the subsequent sections.

17.4 Mathematical Theory of the FISDW

A magnetic field B applied perpendicular to the layers can be introduced in the electron dispersion (17.2) via the Peierls–Onsager substitution $k_y \to k_y - (e/\hbar c)A_y$ in the gauge $A_y = Bx$. As a result, the transverse dispersion $\varepsilon_\perp(k_y b - Gx)$ produces a periodic potential in the x direction with the wave vector G given by (17.4) [32,33]. Electrons also experience the periodic potential $\Delta_0 \cos(Q_x x + Q_y y)$ from the FISDW. To find the energy spectrum, let us decompose the electron wave function $\psi(x, k_y)$ into the components $\psi_+(x, k_y)$ and $\psi_-(x, k_y)$ with the longitudinal momenta close to $\pm k_\mathrm{F}$:

$$\psi(x, k_y) = \psi_+(x, k_y)\,\mathrm{e}^{+\mathrm{i}k_\mathrm{F}x} + \psi_-(x, k_y)\,\mathrm{e}^{-\mathrm{i}k_\mathrm{F}x}. \tag{17.8}$$

Introducing the two-component spinor $[\psi_+(x, k_y), \psi_-(x, k_y + Q_y)]$, we can write the electron Hamiltonian as a 2×2 matrix operating on this spinor

$$\hat{H} = \begin{pmatrix} -\mathrm{i}v_\mathrm{F}\partial_x + \varepsilon_\perp(k_y b - Gx) & \Delta_0 \mathrm{e}^{\mathrm{i}(Q_x - 2k_\mathrm{F})x} \\ \Delta_0 \mathrm{e}^{-\mathrm{i}(Q_x - 2k_\mathrm{F})x} & \mathrm{i}v_\mathrm{F}\partial_x + \varepsilon_\perp(k_y b + bQ_y - Gx) \end{pmatrix}. \tag{17.9}$$

Here the diagonal terms represent the electron dispersion (17.2) in the presence of a magnetic field. The off-diagonal terms represent the periodic potential of the FISDW (17.4). The wave vector $2k_\mathrm{F}$ is subtracted from Q_x because of the $\mathrm{e}^{\pm\mathrm{i}k_\mathrm{F}x}$ factors introduced in (17.8). We do not write the spin indices explicitly in (17.8) and (17.9), because they are not essential for our consideration. Equations (17.8) and (17.9) can be applied to pairing between ψ_+ and ψ_- with parallel or antiparallel spins, as discussed in more detail in [31].

Now let us make a phase transformation of the spinor components[2]

$$\begin{pmatrix} \psi_+ \\ \psi_- \end{pmatrix} = \begin{pmatrix} \psi'_+ \exp\left[(\mathrm{i}/E_B)\int^{k_y b - Gx} \varepsilon_\perp(\xi)\,\mathrm{d}\xi\right] \\ \psi'_- \exp\left[-(\mathrm{i}/E_B)\int^{k_y b - Gx} \varepsilon_\perp(Q_y b + \xi)\,\mathrm{d}\xi\right] \end{pmatrix}, \quad (17.10)$$

where E_B is the characteristic magnetic energy given in (17.4). Substituting (17.10) into (17.9), we obtain a new Hamiltonian acting on the spinor (ψ'_+, ψ'_-)

$$\hat{H}' = \begin{pmatrix} -\mathrm{i}v_\mathrm{F}\partial_x & \tilde{\Delta}(x) \\ \tilde{\Delta}^*(x) & \mathrm{i}v_\mathrm{F}\partial_x \end{pmatrix}. \quad (17.11)$$

As a result of the transformation (17.10), the $\varepsilon_\perp$ terms are removed from the diagonal in (17.11), but they re-appear in the off-diagonal terms

$$\tilde{\Delta}(x) = \Delta_0 \exp\left\{-\mathrm{i}NGx - \frac{\mathrm{i}}{E_B}\int^{k_y b - Gx} [\varepsilon_\perp(\xi) + \varepsilon_\perp(Q_y b + \xi)]\,\mathrm{d}\xi\right\}. \quad (17.12)$$

Since $\varepsilon_\perp(k_y)$ satisfies (17.6), the integral in (17.12) is a periodic function of $k_y b - Gx$, so $\tilde{\Delta}(x)$ can be expanded in a Fourier series with coefficients c_m

$$\tilde{\Delta}(x) = \Delta_0\,\mathrm{e}^{-\mathrm{i}NGx}\sum_m c_m\,\mathrm{e}^{\mathrm{i}m(k_y b - Gx)}. \quad (17.13)$$

Because the FISDW forms at a low transition temperature in high magnetic fields, it is appropriate to consider the limit $\Delta_0 \ll E_B$, where the FISDW potential is much weaker than the magnetic energy. In this case, when (17.13) is substituted into (17.11), each periodic potential in the sum (17.13) can be treated separately and opens a gap $2|\Delta_0 c_m|$ in the energy spectrum at the wave vectors $k_x = \pm[k_\mathrm{F} - (N+m)G/2]$ [34, 35]. The electron mini-bands separated by the energy gaps can be interpreted as the broadened Landau levels, with the number of states in each mini-band proportional to $G \propto B$.

The term with $m = -N$ in (17.13) opens a gap at the Fermi level. This is possible only if $2k_\mathrm{F} - Q_x$ is an integer multiple of G, i.e. there are N mini-bands between the gaps at $k_x = \pm Q_x/2$ and $k_x = \pm k_\mathrm{F}$. Let us focus on the gap at the Fermi level and omit the terms with $m \neq -N$ in the sum (17.13). In this single-gap approximation [34,35], the Hamiltonian (17.11) becomes the same as for a 1D density wave [36] with the effective amplitude $\Delta\mathrm{e}^{-\mathrm{i}\varphi}$:

$$\hat{H}' = \begin{pmatrix} -\mathrm{i}v_\mathrm{F}\partial_x & \Delta\,\mathrm{e}^{-\mathrm{i}\varphi(k_y)} \\ \Delta^*\mathrm{e}^{\mathrm{i}\varphi(k_y)} & \mathrm{i}v_\mathrm{F}\partial_x \end{pmatrix}, \quad \Delta = \Delta_0 c_{-N}, \quad \varphi = Nbk_y. \quad (17.14)$$

[2] This kind of transformation was first introduced in [32], which started development of the FISDW theory.

The Hamiltonian (17.14) has the gapped energy spectrum

$$E(p_x) = \sqrt{(v_{\rm F}p_x)^2 + |\Delta|^2}, \quad p_x = \hbar(k_x - k_{\rm F}). \tag{17.15}$$

Notice that the off-diagonal terms in (17.14) have the phase $\varphi(k_y)$, which does not matter for the energy spectrum (17.15), but plays crucial role in the QHE.[3]

The Fourier coefficients c_m (17.13) depend on Q_y and the ratio of the tunneling amplitudes t_b and t_b' (17.2) to the magnetic energy E_B. The values of N and Q_y in an FISDW state are selected in such a way as to maximize the coefficient $|c_{-N}(Q_y)|$ and, thus, the energy gap Δ (17.14) at the Fermi level for a given magnetic magnetic field.[4] As magnetic field changes, the optimal values of N and Q_y change. Since the parameter N must be integer in order to produce a gap at the Fermi level, it changes by discontinuous jumps, which produces a cascade of the FISDW transitions.

17.5 Quantum Hall Effect as a Topological Invariant

Suppose an electric field E_y is applied perpendicular to the chains. We can introduce it in the Hamiltonian (17.9) by the Peierls–Onsager substitution $k_y \to k_y - (e/\hbar c)A_y$ using the gauge $A_y = -E_y ct$, where t is time. Then, the periodic potential $\varepsilon_\perp[k_y b - G(x - vt)]$ starts to move with the velocity $v = cE_y/B$. This motion induces some current j_x along the chains, which constitutes the Hall effect. However, the FISDW periodic potential in the off-diagonal terms in (17.9) does not move, if the FISDW is pinned. The moving and non-moving periodic potentials are combined in the effective Hamiltonian (17.14), where the phase becomes time-dependent $\varphi = Nb(k_y + eE_y t/\hbar)$. The time-dependent phase means that the effective 1D density wave (17.14) slides along the chains, carrying the Fröhlich current [36]

$$j_x = \frac{2e}{2\pi b}\frac{\partial\varphi}{\partial t} = \frac{2Ne^2}{h}E_y = \sigma_{xy}E_y. \tag{17.16}$$

Equation (17.16) represents the quantum Hall effect and agrees with (17.7).[5]

3 The electron conductivity tensor for the FISDW state was calculated in [37], but quantized contribution to σ_{xy} was lost.

4 The single-gap approximation is not sufficient for a self-consistent calculation of thermodynamic quantities, such as the free energy and magnetization [38]. However, it is adequate for describing low-energy electron states relevant for the QHE.

5 A more rigorous treatment of two periodic potentials is presented in [39,40], using the methods of [41–43] and giving the same result (17.16).

The Hall conductivity at zero temperature can be also expressed in terms of a topological invariant called the Chern number [44, 45]:

$$\sigma_{xy} = -\mathrm{i}\frac{e^2}{\hbar}\sum_a \int \frac{\mathrm{d}k_x}{2\pi}\int \frac{\mathrm{d}k_y}{2\pi}\left(\frac{\partial\langle\psi_a|}{\partial k_x}\frac{\partial|\psi_a\rangle}{\partial k_y} - \frac{\partial\langle\psi_a|}{\partial k_y}\frac{\partial|\psi_a\rangle}{\partial k_x}\right)$$
$$= -\mathrm{i}\frac{e^2}{\hbar}\sum_a \int \frac{\mathrm{d}k_x}{2\pi}\int \frac{\mathrm{d}k_y}{2\pi}\left[\frac{\partial}{\partial k_x}\left(\langle\psi_a|\frac{\partial|\psi_a\rangle}{\partial k_y}\right) - \frac{\partial}{\partial k_y}\left(\langle\psi_a|\frac{\partial|\psi_a\rangle}{\partial k_x}\right)\right]. \quad (17.17)$$

Here $|\psi_a(k_x, k_y)\rangle$ are the normalized eigenvectors of the Hamiltonian. The integral is taken over the Brillouin zone, and the sum over a goes over all completely occupied bands, assuming there are no partially filled bands.

Let us apply (17.17) to the FISDW state [31], first setting $Q_y = 0$. The eigenfunctions of (17.14) are defined on the Brillouin zone torus $-k_\mathrm{F} \le k_x \le k_\mathrm{F}$ and $0 \le k_y \le 2\pi/b$ with the gap $\Delta\,\mathrm{e}^{-\mathrm{i}\varphi(k_y)}$ at $k_x = \pm k_\mathrm{F}$. Let us start with a wave function $|\psi_0\rangle$ at some point k_x^0 far away from $\pm k_\mathrm{F}$ and change k_x along a closed line encircling the torus at a fixed k_y. As we pass through $+k_\mathrm{F}$, the wave function transforms from ψ_+ to $\mathrm{e}^{\mathrm{i}\varphi}\psi_-$. The phase factor appears because the off-diagonal terms in (17.14) have the phase. When we return to the original point $k_x^0 + 2k_\mathrm{F}$ in the next Brillouin zone, the wave function becomes $\mathrm{e}^{\mathrm{i}\varphi}|\psi_0\rangle$. The first term in (17.17) is a full derivative in k_x, so it reduces to a difference taken between $k_x^0 + 2k_\mathrm{F}$ and k_x^0

$$-\mathrm{i}\frac{e^2}{h}\int_0^{2\pi/b}\frac{\mathrm{d}k_y}{2\pi}\left(\langle\psi_0|\mathrm{e}^{-\mathrm{i}\varphi(k_y)}\frac{\partial}{\partial k_y}\,\mathrm{e}^{\mathrm{i}\varphi(k_y)}|\psi_0\rangle - \langle\psi_0|\frac{\partial}{\partial k_y}\,|\psi_0\rangle\right)$$
$$= \frac{e^2}{h}\int_0^{2\pi/b}\frac{\mathrm{d}k_y}{2\pi}\frac{\partial\varphi(k_y)}{\partial k_y} = [\varphi(2\pi/b) - \varphi(0)]\,\frac{e^2}{h} = \frac{Ne^2}{h}. \quad (17.18)$$

Multiplied by the spin factor 2, (17.18) gives the same result as (17.7).

The second term in (17.17) gives zero. In this term, the expression under the integral can be rewritten as the difference $\langle\psi|\partial_{k_x}|\psi\rangle|_{k_y=0}^{k_y=2\pi/b}$. The Hamiltonian (17.14) at $k_y b = 2\pi$ is the same as at $k_y = 0$, and we can select the wave functions to be the same, thus the difference equals zero. It was shown in [31] that the results do not change when we take into account $Q_y \ne 0$, $Q_z \ne 0$, and the multiple gaps below the Fermi energy generated by the periodic potentials in (17.13).

17.6 Coexistence of Several Order Parameters

The formalism presented in Sect. 17.5 is particularly useful in the case where several FISDWs with different amplitudes Δ_j and numbers N_j coexist [31]. In this case, the off-diagonal terms in (17.14) become

$$\Delta(k_y) = \sum_j \Delta_j \exp(-\mathrm{i}bk_yN_j). \quad (17.19)$$

According to (17.18), the Hall conductivity is determined by the winding number of the complex function (17.19), i.e. by the number of times the phase of $\Delta(k_y)$ changes by 2π when k_y goes from 0 to $2\pi/b$. This integer number, taken with the opposite sign, must be substituted in (17.7) instead of N. So, when several FISDWs coexist, σ_{xy} is not a superposition of partial Hall conductivities, but is always given by the integer winding number of (17.19).

When two FISDWs coexist, σ_{xy} is given by the integer N_j whose partial gap $|\Delta_j|$ is bigger. To illustrate this, let us use a vector representation of complex numbers and a planetary analogy. Let us associate the first term in the sum (17.19) with a vector pointing from the Sun to the Earth, and the second term with a vector from the Earth to the Moon. As the parameter k_y increases, the Earth rotates around the Sun, and the Moon rotates around the Earth. The Hall conductivity is determined by the number of times the Moon rotates around the Sun. Clearly, this winding number is determined only by the bigger orbit of the Earth. If one partial gap $|\Delta_l|$ is bigger then the sum of all other partial gaps $|\Delta_l| > \sum_{j \neq l} |\Delta_j|$, then σ_{xy} is determined only by the biggest term in (17.19), i.e. $N = N_l$ in (17.7).

An FISDW state consisting of multiple periodic potentials was discussed in [46], and the QHE in this model was studied in [47]. Lebed pointed out in [48] that the Umklapp scattering requires coexistence of two FISDWs with N and $-N$. The QHE in this case was studied in [49] using the topological method, and it was found that σ_{xy} may take the values corresponding to N, $-N$, or zero. A more detailed study was presented in [50]. It was suggested that this effect may explain sign reversals of the QHE observed in $(TMTSF)_2PF_6$ [23, 29, 30]. An alternative theory of the QHE sign reversals was proposed in [51]. Sign changes of the QHE are also observed in $(TMTSF)_2ClO_4$ [25–27] and $(TMTSF)_2ReO_4$ [28], which have anion ordering. Coexistence of the FISDW pairings between different branches of the folded Fermi surface in this case was proposed in [13, 14].

17.7 Temperature Evolution of the Quantum Hall Effect

Equation (17.16) is a good starting point for the discussion of the temperature dependence of the QHE [39, 40, 52]. According to this equation, the Hall conductivity can be viewed as the Fröhlich conductivity of the effective 1D density wave (17.14). Thus, the temperature dependence of the QHE must be the same as the temperature dependence of the Fröhlich conductivity, which was studied in the theory of density waves [53, 54]. At $T \neq 0$, the electric current carried by the density-wave condensate is reduced with respect to the zero-temperature value by a factor $f(T)$, which also reduces σ_{xy}:

$$\sigma_{xy}(T) = f(T)\, 2Ne^2/h\text{TSeF}, \tag{17.20}$$

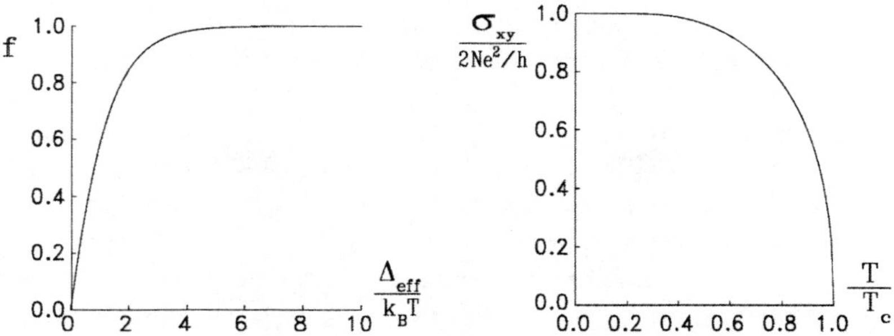

Fig. 17.1. (**a**) The temperature reduction factor $f(\Delta/T)$ of the Hall conductivity (17.20) given by (17.22). (**b**) The Hall conductivity (17.20) in the FISDW state as a function of temperature T normalized to the FISDW transition temperature T_c

$$f(T) = 1 - \int_{-\infty}^{\infty} \frac{\mathrm{d}p_x}{v_\mathrm{F}} \left(\frac{\partial E}{\partial p_x}\right)^2 \left[-\frac{\partial n_\mathrm{F}(E/T)}{\partial E}\right], \qquad (17.21)$$

where $E(p_x)$ is the electron dispersion (17.15) in the FISDW state, and $n_\mathrm{F}(\epsilon/T) = (\mathrm{e}^{\epsilon/T} + 1)^{-1}$ is the Fermi distribution function with $k_\mathrm{B} = 1$. Equations (17.20) and (17.21) have a two-fluid interpretation. The first term in (17.21) represents the FISDW condensate current responsible for the QHE. The second term represents the normal component originating from electron quasiparticles thermally excited above the energy gap. They equilibrate with the immobile crystal lattice and do not participate in the Fröhlich current, thus reducing the Hall coefficient. A simple derivation of (17.21) is given in [39, 40, 52].[6]

The function f (17.21) depends on Δ/T and can be written as [4, 54]

$$f\left(\frac{\Delta}{T}\right) = \int_0^{\infty} \mathrm{d}\zeta \, \tanh\left(\frac{\Delta}{2T}\cosh\zeta\right) / \cosh^2\zeta. \qquad (17.22)$$

The function f is plotted in Fig. 17.1a. It equals 1 at $T = 0$, where (17.20) reproduces the QHE, gradually decreases with increasing T, and vanishes at $T \gg \Delta$. Taking into account that the FISDW order parameter Δ itself depends on T and vanishes at the FISDW transition temperature T_c, it is clear that $f(T)$ and $\sigma_{xy}(T)$ vanish at $T \to T_\mathrm{c}$, where $\sigma_{xy}(T) \propto f(T) \propto \Delta(T) \propto \sqrt{T_\mathrm{c} - T}$. Assuming that the temperature dependence $\Delta(T)$ is given by the BCS theory [34, 35], we plot $\sigma_{xy}(T)$ in Fig. 17.1b. Strictly speaking $\sigma_{xy}(T)$ should not vanish at $T \to T_\mathrm{c}$, but approach the Hall conductivity of the metallic phase. However, as discussed in Sect. 17.2, the Hall effect in the normal state is very small, so this modification is not essential.

The function $f(T)$ (17.21) is qualitatively similar to the function $f_\mathrm{s}(T)$ that describes temperature dependence of the superconducting condensate density in the London case [55]. Both functions equal 1 at $T = 0$, but the

[6] Reference [4] discussed the Hall conductivity in the FISDW state at $T \neq 0$, but failed to reproduce the QHE at $T = 0$.

superconducting function behaves differently near T_c: $f_s(T) \propto \Delta^2(T) \propto T_c - T$. To understand the difference between the two functions, they should be considered at small, but finite frequency ω and wave vector q [40]. Equations (17.21) and (17.22) represent the dynamic limit, where $q/\omega = 0$. This is the relevant limit in our case, because the electric field is homogeneous in space ($q = 0$), but may be time-dependent ($\omega \neq 0$). The effective periodic potential (17.14) is also time-dependent in the presence of E_y, as shown in (17.16). On the other hand, for the Meissner effect in superconductors, where the magnetic field is stationary ($\omega = 0$), but varies in space ($q \neq 0$), the static limit $\omega/q = 0$ is relevant. The dynamic and static limits are discussed in more detail in [40].

In the derivation of (17.20), we assumed that the wave vector Q_x (17.4) is always quantized with an integer N, so that the energy gap is located at the Fermi level. While this is the case at $T = 0$ [56], Lebed pointed out in [57] that Q_x is not necessarily quantized at $T \neq 0$. Because of the thermally excited quasiparticles and multiple periodic potentials present in (17.13), the optimal value of N in (17.4) may be non-integer, which results in deviations from the QHE. Equation (17.20) was compared with the experimental temperature dependence of the Hall effect in the FISDW state using a limited data set in [58] and detailed measurements in [59]. A good quantitative agreement was found for small integer numbers $N \sim 1$, where the QHE is well defined. However, for the FISDW with bigger N and lower T_c at lower B, poor quantization of the Hall effect was found [60] at the experimentally accessible temperatures, in qualitative agreement with the theory of the non-quantized FISDW [56,57].

17.8 Influence of the FISDW Motion on the Quantum Hall Effect

In the derivation of (17.16), we assumed that the FISDW is pinned and produces only a static periodic potential. However, when the FISDW is subject to a strong or time-dependent electric field, it may move. It is interesting to study how this motion would affect the QHE [39, 40, 61].

Motion of the FISDW can be described by introducing a time-dependent phase[7] Θ of the FISDW amplitude Δ_0 in (17.9): $\Delta_0 \to \Delta_0 \mathrm{e}^{-\mathrm{i}\Theta}$. Then, this phase re-appears in the off-diagonal terms in (17.14) and contributes to the electric current (17.16) along the chains

$$j_x = \frac{2Ne^2}{h} E_y + \frac{2e}{2\pi b} \dot{\Theta}, \tag{17.23}$$

where the dot represents the time derivative. Equation (17.23) needs to be supplemented with an equation of motion for Θ. The latter can be obtained from the effective Lagrangian of the system derived in [39, 40, 61]

$$L = -\frac{Ne^2}{hc} \varepsilon_{ijk} A_i F_{jk} + \frac{\hbar}{4\pi b v_F} \dot{\Theta}^2 + \frac{e}{\pi b} \Theta E_x + \frac{eN}{2\pi v_F} \dot{\Theta} E_y. \tag{17.24}$$

[7] The phase Θ in this paper has the opposite sign to Θ in [39, 40].

Summation of over $(i, j, k) = (x, y, t)$ is implied in the first term, and F_{jk} is the electromagnetic field tensor . The first term in (17.24) is the Chern–Simons term responsible for the QHE [31, 62]. The second term is the kinetic energy of a moving FISDW. The third term, well known in the theory of density waves [36], represents potential energy in the electric field along the chains. The most important for us is the last term, which describes interaction between the FISDW motion and the electric field perpendicular to the chains [39, 40, 61]. This term is permitted by symmetry and has the structure of a mixed product $\boldsymbol{v}[\boldsymbol{E} \times \boldsymbol{B}]$. Here, $\boldsymbol{v}$ is the velocity of the FISDW, proportional to $\dot{\Theta}$ and directed along the chains, i.e. along the x-axis. The magnetic field $\boldsymbol{B}$ is directed along the z-axis; so, the electric field $\boldsymbol{E}$ enters through the component E_y. One should keep in mind that the magnetic field enters the last term in (17.24) implicitly, through the integer N, which depends on B and changes sign when B changes sign. Varying (17.24) with respect to Θ and phenomenologically adding the pinning and friction terms, we find the FISDW equation of motion

$$\ddot{\Theta} + \frac{1}{\tau}\dot{\Theta} + \omega_0^2 \Theta = \frac{2ev_{\mathrm{F}}}{\hbar} E_x - \frac{eNb}{\hbar}\dot{E}_y, \tag{17.25}$$

where τ is a relaxation time, and ω_0 is the pinning frequency.

Let us first consider the ideal case of a free FISDW without pinning and damping. If the electric field E_y is applied perpendicular to the chains, the last term in (17.25) induces such a motion of the FISDW that the second term in (17.23) exactly cancels to the first term, and the Hall effect vanishes. If the electric field E_x is parallel to the chains, we consider the perpendicular current j_y obtained by varying (17.24) with respect to A_y:

$$j_y = -\frac{2Ne^2}{h} E_x + \frac{eN}{2\pi v_{\mathrm{F}}} \ddot{\Theta}. \tag{17.26}$$

Using (17.25) for the ideal case, we see that the two terms in (17.26) cancel out, and the Hall effect vanishes. The cancellation of the QHE by the moving FISDW is in the spirit of Lenz's law, which says that a system responds to an external perturbation in such a way as to minimize its effect. Thus, the QHE exists only if the FISDW is pinned and does not move.

In a more realistic case with pinning and damping, we solve (17.25) by the Fourier transform from time t to frequency ω and substitute the result into (17.23) and (17.26) to obtain the ac Hall conductivity[8]

$$\sigma_{xy}(\omega) = \frac{2Ne^2}{h} \frac{\omega_0^2 - \mathrm{i}\omega/\tau}{\omega_0^2 - \omega^2 - \mathrm{i}\omega/\tau}. \tag{17.27}$$

The absolute value $|\sigma_{xy}|$ computed from (17.27) is plotted in Fig. 17.2 as a function of ω/ω_0 for $\omega_0\tau = 2$. It is quantized at $\omega = 0$ and has a resonance

[8] $\sigma_{xy}(\omega)$ for an FISDW was studied in [4], but failed to reproduce the QHE at $\omega = 0$.

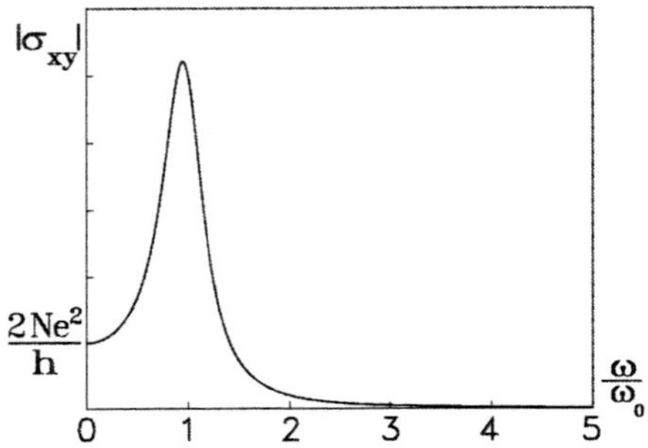

Fig. 17.2. The absolute value of the Hall conductivity $|\sigma_{xy}|$ computed from (17.27) for $\omega_0\tau = 2$ as a function of frequency ω normalized to the pinning frequency ω_0

at the pinning frequency. At higher frequencies, where pinning and damping can be neglected, and the FISDW behaves as a free inertial system, we find that $\sigma_{xy}(\omega) \to 0$.

Frequency dependence of the Hall conductivity in semiconducting QHE systems was measured using the technique of crossed wave guides in [63, 64]. No measurements of $\sigma_{xy}(\omega)$ have been done in $(TMTSF)_2X$ thus far, but they would be very interesting and can differentiate the QHE in the FISDW state from the conventional QHE in semiconductors. To give a crude estimate of the required frequency range, we quote the pinning frequency $\omega_0 \sim 3\,\mathrm{GHz}$ $\sim$0.1 K $\sim$10 cm for a regular SDW (not FISDW) in $(TMTSF)_2PF_6$ [65].

The FISDW can be also depinned by a strong dc electric field. In this case, the FISDW motion is controlled by dissipation, which is difficult to study theoretically on microscopic level. For a steady motion, the last terms in (17.25) and (17.26) drop out, so we expect no changes in σ_{xy}, but some increase in σ_{xx} due to the sliding FISDW [4].[9] If σ_{yy} also increases due to increased dissipation and excitation of quasiparticles, then we expect that $\rho_{xx} = \sigma_{yy}/(\sigma_{xy}^2 + \sigma_{xx}\sigma_{yy})$ and $\rho_{yy} = \sigma_{xx}/(\sigma_{xy}^2 + \sigma_{xx}\sigma_{yy})$ would increase when the FISDW starts to slide, whereas $\rho_{xy} = \sigma_{xy}/(\sigma_{xy}^2 + \sigma_{xx}\sigma_{yy})$ would decrease. The experimental measurements in $(TMTSF)_2PF_6$ are in qualitative agreement with these expectation [66], although earlier measurements in $(TMTSF)_2ClO_4$ [67] produced a different result.

The influence of steady motion of a regular charge-density wave (not FISDW) on the Hall conductivity was studied theoretically in [68] and experimentally in [69–71]. In this case, there is no QHE contribution from the condensate, and the effect is primary determined by the thermally excited normal carriers.

---

[9] In principle, due to the presence of $\boldsymbol{B}$, we can phenomenologically add a term proportional to E_y to (17.25) and a term proportional to $\dot{\Theta}$ to (17.26). These terms have dissipative origin and violate the time reversal symmetry, so they cannot be obtained from a Lagrangian. Deriving them from the Boltzmann equation is a difficult task.

17.9 Chiral Edge States

Thus far, we studied the QHE in the bulk. Generally, a system with the integer QHE characterized by the number N is expected to have N chiral edge states [72–74]. These electron states are localized at the boundaries of the sample and circulate along the boundary with some velocity v, as shown in Fig. 17.3. Excitations with the opposite sense of circulation are absent, so these states is chiral. Theory of the edge states in multilayered QHE systems was discussed phenomenologically in [75, 76].

The QHE can be equivalently formulated in terms of the chiral edge states [72, 73]. Suppose a small electric voltage V_y is applied across the sample. It produces a difference of chemical potentials between the opposite edges of the sample. The electron states in the bulk of the sample are gapped, so they would not respond to this perturbation. However, the edge modes are gapless, so the difference of chemical potentials produces imbalance $\delta n = 2eV_y/hv$ between the occupation numbers of the chiral modes at the opposite edges. Here, we utilized the 1D density of states $2/hv$, accounting for two spin projections. The chiral modes at the opposite edges propagate in the opposite directions, so the population imbalance between them generates the net edge current I_x in the x direction:

$$I_x = evN\,\delta n = evN\,\frac{2eV_y}{hv} = \frac{2Ne^2}{h}\,V_y. \tag{17.28}$$

Equation (17.28) represents the QHE, this time for the Hall conductance, rather than conductivity (17.7), which are the same in 2D. Notice that the velocity v of the chiral edge states cancels out in (17.28), and only their number N enters the final formula. The bulk and edge formulations of the QHE are equivalent. Whether the Hall current actually flows in the bulk or along the

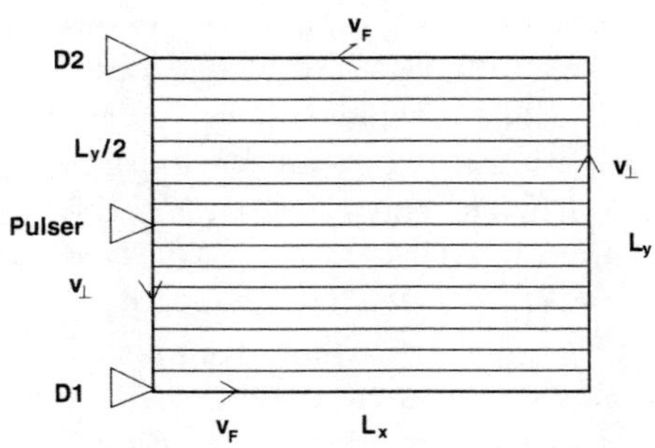

Fig. 17.3. Chiral edge modes circulate around the sample with the velocities $v_\perp$ and v_F in the direction of the *arrows*. The *thin parallel lines* represent conducting chains of a Q1D material. In the proposed time-of-flight experiment, a pulse from the pulser is detected at different times t and t' by the detectors D1 and D2

edge depends on where the voltage drops in the sample, but the overall Hall conductance does not depend on this.

Later, we discuss the structure of the edge states for the FISDW [39,77].[10]

17.9.1 Edges Perpendicular to the Chains

First, let us consider a 1D density wave occupying the positive semi-space $x > 0$ with an edge at $x = 0$. In this case, the wave function (17.8) must vanish at the edge: $\psi(x = 0) = 0$, so $\psi_+(x = 0) = -\psi_-(x = 0)$. With this boundary condition, the Hamiltonian (17.14) also admits a localized electron state, in addition to (17.15), with the energy $|E| < \Delta$ inside the gap:

$$E = -\Delta \cos\varphi, \quad \psi_\pm = \pm\frac{\mathrm{e}^{-\kappa x}}{\sqrt{\kappa}}, \quad \kappa = -\frac{\sin\varphi}{\xi}, \quad \xi = \frac{\hbar v_\mathrm{F}}{\Delta}. \tag{17.29}$$

The wave function (17.29) exponentially decays into the bulk at a length of the order of the coherence length ξ. Equation (17.29) is meaningful only when $\kappa > 0$, so the localized state exists on the left edge only for $\pi < \varphi < 2\pi$. However, a solution with $\kappa < 0$ is appropriate for the right edge at the opposite end of the sample. The edge state (17.29) is mathematically similar to the localized state at a kink soliton in a density wave [79].

Now, let us consider an FISDW occupying the positive semi-space $x > 0$ along the chains and extended in the y-direction. In this case, the phase $\varphi = Nbk_y$ in (17.14) depends on k_y. Substituting $\varphi(k_y)$ into (17.29), we find [77]

$$E(k_y) = -\Delta\cos(Nk_yb), \quad \kappa = -\frac{\sin(Nk_yb)}{\xi}, \quad \psi_\pm = \pm\frac{\mathrm{e}^{\mathrm{i}k_yy-\kappa x}}{\sqrt{\kappa}}. \tag{17.30}$$

The single bound state (17.29) is now replaced by the band (17.30) of the edge states labeled by the wave vector k_y perpendicular to the chains. These states (17.30) are localized along the chains and extended perpendicular to the chains. At the left edge of the sample, we require that $\kappa \propto -\sin(Nk_yb)$ is positive, which gives N branches of the edge states in the transverse Brillouin zone $0 < k_yb < 2\pi$. The complementary N branches of the edge states, determined by the condition $\kappa < 0$, exist at the right edge. The dashed and solid lines in Fig. 17.4 show the energy dispersion $E(k_y)$ (17.30) of the states localized at the left and right edges for $N = 2$. It is clear that the group velocities of the edge states $\partial E(k_y)/\hbar\partial k_y$ have opposite signs for the left and right edges. Thus, they carry a surface current around the sample, as indicated by the arrows in Fig. 17.3. The sense of circulation is determined by the sign of N, which is controlled by the sign of the magnetic field B.

[10] The edge states in the normal phase of Q1D conductors in a magnetic field were studied in [78].

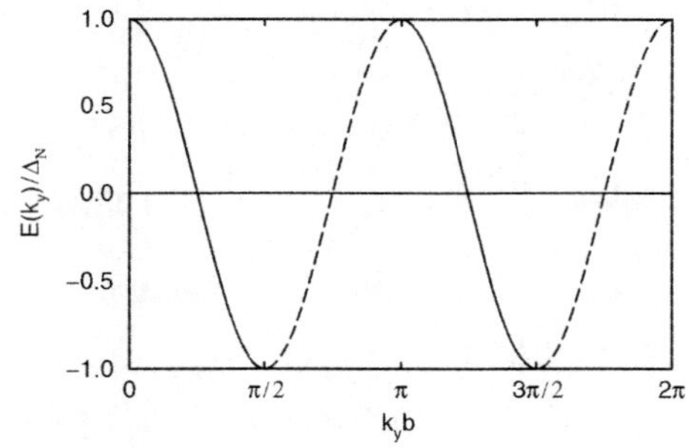

Fig. 17.4. Energy dispersion $E/(-\varDelta)$ (17.30) of the electron states localized at the right (*solid lines*) and left (*dashed lines*) edges of the sample as a function of the transverse momentum k_y for $N = 2$

The edge states bands are filled up to the Fermi level in the middle of the gap, and their group velocity at the Fermi level is

$$v_\perp = \frac{1}{\hbar}\frac{\partial E(k_y)}{\partial k_y}\bigg|_{E=0} = \frac{Nb\varDelta}{\hbar}, \tag{17.31}$$

The velocity $v_\perp$ is quite low, because it is proportional to the small FISDW gap $\varDelta$, so $v_\perp \ll v_F$.

17.9.2 Edges Parallel to the Chains

Now let us discuss the edges parallel to the chains. The effective Hamiltonian (17.14) operates on the electron wave functions $\psi'_\pm(k_x, k_y)$ labeled by the momentum k_y. The bulk energy spectrum (17.15) of (17.14) is degenerate in k_y. Thus, we can use the Wannier wave functions $\psi'_\pm(k_x, M) = \int e^{ik_y Mb}\psi'_\pm(k_x, k_y)\, dk_y/2\pi$ as a new basis [78, 80]. The wave function $\psi'_\pm(k_x, M)$ is localized across the chains around the chain with the number M. Introducing the destruction operators $\hat{a}_\pm(p_x, M)$ for this basis, we can rewrite the Hamiltonian (17.14) in the following form [80]:

$$\begin{aligned}\hat{H}' = \int \frac{dk_x}{2\pi}\sum_M & v_F p_x\,[\hat{a}^+_+(p_x, M)\,\hat{a}_+(p_x, M) - \hat{a}^+_-(p_x, M)\,\hat{a}_-(p_x, M)] \\ & + \varDelta\,[\hat{a}^+_+(p_x, M+N)\,\hat{a}_-(p_x, M) + \hat{a}^+_-(p_x, M)\,\hat{a}_+(p_x, M+N)]. \end{aligned} \tag{17.32}$$

As a consequence of the k_y-dependent phase in the off-diagonal terms in (17.14), the Hamiltonian (17.32) represents pairing between the $+k_F$ and $-k_F$ states localized at the different chains $M + N$ and M. For the chains in the bulk of the crystal, this pairing results in the gapped energy spectrum (17.15). However, the states at the edges are exceptional. The $+k_F$ states on the first N chains on one side of the crystal and the $-k_F$ states on the last N chains on the other side of the crystal do not have partners to couple with, so these states remain ungapped [39]. Thus, one side of the sample possesses N gapless chiral modes propagating along the edge with the velocity $+v_{\mathrm{F}}$, and the other side has N gapless chiral modes propagating in the opposite direction with the velocity $-v_{\mathrm{F}}$, as shown in Fig. 17.3.

17.9.3 Possibilities for Experimental Observation of the Chiral Edges States

Let us estimate parameters of the chiral edge states. The activation energy in the FISDW state with $N = 1$ was found to be $2\Delta = 6\,\mathrm{K}$ in $(\mathrm{TMTSF})_2\mathrm{ClO}_4$ at $B = 25\,\mathrm{T}$ [81]. Substituting $\Delta = 3\,\mathrm{K}$ and $N = 1$ into (17.31), we find $v_\perp = 300\,\mathrm{m\,s^{-1}}$, which is three orders of magnitude lower than $v_{\mathrm{F}} = 10^5\,\mathrm{m\,s^{-1}}$ [2]. Despite the big difference in velocities, the total currents of the parallel and perpendicular edge states are the same. Indeed, the slow perpendicular states with $v_\perp = Nb\Delta/\hbar$ have the large width $\xi = \hbar v_{\mathrm{F}}/\Delta$, whereas the fast parallel states with the velocity v_{F} have the narrow width Nb. The total edge current I carried by the perpendicular states (17.30) is

$$I = \frac{2Ne}{h}\int_0^{\pi/2bN} \frac{\partial E(k_y)}{\partial k_y}\,\mathrm{d}k_y = \frac{2Ne\Delta}{h} = \frac{ev_\perp}{\pi b} = 20\,\mathrm{nA}. \tag{17.33}$$

The same current is carried along the chains by the difference between the gapped and ungapped branches of the electron dispersion (17.15)

$$I = \frac{2Ne}{h}\int_{-\infty}^{0} \left(\frac{\partial E(p_x)}{\partial p_x} - v_{\mathrm{F}}\right) \mathrm{d}p_x = \frac{2Ne\Delta}{h}. \tag{17.34}$$

It is tempting to use (17.33) to calculate magnetization of the sample. However, there are additional contributions to the total magnetization coming from the edge states inside the energy gaps opened by the neglected terms in (17.13) below the Fermi level. Magnetization of the FISDW state was calculated in [38] using the bulk free energy and was measured in $(\mathrm{TMTSF})_2\mathrm{ClO}_4$ in [11].

The most convincing demonstration of the edge states would be the time-of-flight experiment [77] analogous to that performed in GaAs in [82,83]. The experimental setup is sketched in Fig. 17.3. An electric pulse is applied by the pulser at the center of the edge perpendicular to the chains. The pulse travels counterclockwise around the sample. For the typical sample dimensions $L_x = 2\,\mathrm{mm}$ and $L_y = 0.2\,\mathrm{mm}$, we find the times of flight $t_x = L_x/v_{\mathrm{F}} = 8\,\mathrm{ns}$ and $t_y = L_y/v_\perp = 0.67\,\mu\mathrm{s}$ using the values of $v_\perp$ and v_{F} quoted above for $N = 1$. Thus, the pulse will reach the detector D1 at the time $t = L_y/2v_\perp$ and the other detector D2 at the longer time $t' = 3t + 2L_x/v_{\mathrm{F}} \approx 3t$. The difference between the arrival times t and t' is a signature of the chiral edge states. The flight time t should exhibit discontinuities at the FISDW phase boundaries due to discontinuity of both N and Δ affecting $v_\perp$ in (17.31). The pulse must be shorter than $t = 0.33\,\mu s$ for clear resolution and longer than $\hbar/\Delta = 2.6\,ps$, so that only the low-energy excitations are probed.

Existence of the edge states was confirmed in a multilayered GaAs system by showing that the conductance G_{zz} perpendicular to the layers is proportional to the number of the edge states and the perimeter of the sample [84]. A similar experiment in $(TMTSF)_2AsF_6$ produced inconclusive results [85].

The specific heat per layer C_e of the gapless edge excitations is proportional to temperature at $T \ll \Delta$ [77]:

$$\frac{C_e}{T} = \frac{N\pi}{3\,\hbar}\left(\frac{2L_y}{v_\perp} + \frac{2L_x}{v_F}\right) \approx \frac{2\pi}{3\Delta}\frac{L_y}{b} \approx 3 \times 10^{-3}\,\frac{\mathrm{mJ}}{\mathrm{K^2\,mole}}, \qquad (17.35)$$

where the dominant contribution comes from the edges perpendicular to the chains. The ratio of C_e/T (17.35) to the bulk specific heat in the normal state $C_b^{(n)}/T = \pi L_x L_y / 3\hbar b v_F$ is roughly equal to the ratio of the volumes occupied by the bulk and edge states: $C_b^{(n)}/C_e = L_x/2\xi \approx 10^3$. The experimentally measured $C_b^{(n)}/T$ is $5\,\mathrm{mJ}\,(\mathrm{K^{-2}\,mol^{-1}})$ [13]. In the FISDW state, the bulk specific heat is exponentially suppressed because of the energy gap Δ and becomes smaller than C_e (17.35) at sufficiently low temperatures $T \leq \Delta/14$ [77]. This regime could have been possibly achieved in the specific heat measurements [13] performed at $T = 0.32\,\mathrm{K}$ and $B = 9\,\mathrm{T}$ in $(TMTSF)_2ClO_4$. According to (17.35), C_e/T must be discontinuous at the boundaries between the FISDW phases, where Δ changes discontinuously [38]. Notice that N cancels out in (17.35), in contrast to the phenomenological model [76].

Other possibilities for experimental observation of the edge states are discussed in [86]. We would like to mention that non-chiral midgap edge states are expected to exist in the superconducting p-wave state of $(TMTSF)_2X$ [87].

17.10 Generalization to the Three-Dimensional Quantum Hall Effect

Experiments show that the FISDW state in $(TMTSF)_2X$ depends only on the B_z component of a magnetic field. Thus, although the $(TMTSF)_2X$ crystals are three-dimensional (3D), they can be treated as a collection of essentially independent 2D layers, and the QHE can be studied within a 2D theory. Nevertheless, let us discuss a generalization of this theory to the 3D case.

Suppose a magnetic field has the B_z and B_y components perpendicular to the chains. As Lebed pointed out in [88], inserting them into (17.1) via the Peierls–Onsager substitution creates two periodic potentials in the x direction, $2t_b\cos(k_y b - G_1 x)$ and $2t_c\cos(k_z d + G_2 x)$, with the wave vectors G_1 and G_2. Thus, an FISDW may form with the wave vector [89]

$$Q_x = 2k_F - N_1 G_1 - N_2 G_2, \quad G_1 = ebB_z/\hbar c, \quad G_2 = edB_y/\hbar c. \qquad (17.36)$$

Repeating the derivation of Sect. 17.4 and introducing the electric field components E_x and E_y as in (17.16), we find the electric current per one chain [90, 91]

$$I_x = \frac{2e}{2\pi}\frac{\partial\varphi}{\partial t} = \frac{2e^2}{h}(N_1 b E_y - N_2 \mathrm{d} E_z). \tag{17.37}$$

Using the current density per unit area $j_x = I_x/bd$, we rewrite (17.37) as

$$\boldsymbol{j} = \frac{2e^2}{h}\boldsymbol{E}\times\boldsymbol{K}, \quad \boldsymbol{K} = \left(0, \frac{N_2}{b}, \frac{N_1}{d}\right). \tag{17.38}$$

Equation (17.38) with an integer vector $\boldsymbol{K}$ belonging to the reciprocal crystal lattice represents the general form of the QHE in a 3D system [92]. This formula was applied to the 3D FISDW (17.36) in [89, 91, 93] and to general lattices in [94–96]. The edge states picture, presented in Sect. 17.9, was generalized to the 3D FISDW in [97].

Although the 3D FISDW (17.36) does not realize in $(\mathrm{TMTSF})_2\mathrm{X}$, it may occur in other families of Q1D materials. It would be very interesting to investigate whether FISDW exists in the Q1D material $(\mathrm{DI\text{–}DCNQI})_2\mathrm{Ag}$, where $t_b = t_c$ by crystal symmetry, so there is no layered structure [98].

17.11 Conclusions and Open Questions

In summary, the QHE in the FISDW state is a direct consequence of the quantization and magnetic field dependence of the wave vector Q_x (17.4). As a result, N completely filled Landau bands are maintained between Q_x and $2k_\mathrm{F}$, whereas the Fermi sea plays the role of a reservoir. More rigorously, the QHE can be obtained as a topological invariant in terms of the winding number of a phase in the Brillouin zone. The topological approach is particularly useful when several FISDWs coexist. The Hall effect can be also viewed as the Fröhlich current of an effective density wave.[11] This allows us to derive its temperature dependence within a two-fluid picture. Motion of the FISDW cancels the Hall effect in the ideal case of a free FISDW and at high frequency. The QHE can be also formulated in terms of the chiral edge states circulating around the sample with low speed across the chains and high speed along the chains. The QHE can be also generalized for a three-dimensional FISDW. While the temperature dependence of the QHE was measured experimentally and found to be in agreement with the theory, the other theoretical results, such as the frequency dependence of σ_{xy} and the existence of the edge states, await experimental verification. It would be interesting to search for a 3D FISDW in $(\mathrm{DI\text{–}DCNQI})_2\mathrm{Ag}$.

[11] The theory presented in Sects. 17.4 and 17.16 was also applied to quantized adiabatic transport induced by surface acoustic waves in carbon nanotubes [99].

Later we list some open questions in the theory of the QHE and the FISDW. One of problems is that, in experiment, the dissipative components σ_{xx} and σ_{yy} tend to saturate at small, but finite values in the limit $T \to 0$ [23, 24]. The reason for this is not completely clear, but may be due to impurity scattering [100]. As a result, the Hall effect quantization is not as good as in semiconducting systems, especially at lower magnetic fields.

Predicting the sequence of N as a function of B is a task for the FISDW theory. For a simple model, the theory gives consecutive numbers [17, 38], as observed experimentally at higher pressures in $(TMTSF)_2PF_6$ [21–24]. However, at lower pressures, a complicated sequence of N with multiple sign reversals and a bifurcation in the B–T phase diagram is observed [29, 30]. This sequence is not fully understood, although theoretical attempts have been made [47–51].

Developing a detailed theory of the FISDW in $(TMTSF)_2ClO_4$ and $(TMTSF)_2ReO_4$ is even more difficult because of the period doubling in the $\boldsymbol{b}$ direction due to the anion ordering in these materials. Experiments show a complicated phase diagram for $(TMTSF)_2ClO_4$ [13,14]. Numerous theoretical scenarios for the FISDW in $(TMTSF)_2ClO_4$ are reviewed by Haddad et al. in this book and in [101]. Particularly puzzling is the behavior of $(TMTSF)_2ClO_4$ in very strong magnetic fields between 26 and 45 T, where it is supposed to be in the FISDW phase with $N = 0$, i.e. with zero Hall effect. Instead, the Hall coefficient and other quantities show giant oscillations as a function of the magnetic field [102]. This phenomenon is not fully understood, but may be related to the soliton theory [103].

References

1. T. Ishiguro, K. Yamaji, G. Saito, *Organic Superconductors*, 2nd edn. (Springer, Berlin Heidelberg New York, 1998)
2. S. Takahashi, S. Hill, S. Takasaki, J. Yamada, H. Anzai, Phys. Rev. B **72**, 024540 (2005)
3. J.R. Cooper, M. Miljak, G. Delplanque, D. Jérome, M. Weger, J.M. Fabre, L. Giral, J. Phys. (France) **38**, 1097 (1977); J.R. Cooper, B. Korin-Hamzić in *Organic Conductors*, ed. by J.-P. Farges (Dekker, New York, 1994) p. 359
4. A. Virosztek, K. Maki, Phys. Rev. B **39**, 616 (1989)
5. V.M. Yakovenko, A.T. Zheleznyak, Synth. Metals **103**, 2202 (1999)
6. G. Mihály, I. Kézsmárki, F. Zámborszky, L. Forró, Phys. Rev. Lett. **84**, 2670 (2000)
7. J. Moser, J.R. Cooper, D. Jérome, B. Alavi, S.E. Brown, K. Bechgaard, Phys. Rev. Lett. **84**, 2674 (2000)
8. V.M. Yakovenko, A.T. Zheleznyak, Synth. Metals **120**, 1083 (2001)
9. F. Pesty, P. Garoche, K. Bechgaard, Phys. Rev. Lett. **55**, 2495 (1985)
10. N.A. Fortune, J.S. Brooks, M.J. Graf, G. Montambaux, L.Y. Chiang, J.A.A.J. Perenboom, D. Althof, Phys. Rev. Lett. **64**, 2054 (1990)
11. M.J. Naughton, J.S. Brooks, L.Y. Chiang, R.V. Chamberlin, P.M. Chaikin, Phys. Rev. Lett. **55**, 969 (1985)

12. T. Takahashi, D. Jérome, K. Bechgaard, J. Phys. (France) **45**, 945 (1984)
13. U.M. Scheven, E.I. Chashechkina, E. Lee, P.M. Chaikin, Phys. Rev. B **52**, 3484 (1995)
14. S.K. McKernan, S.T. Hannahs, U.M. Scheven, G.M. Danner, P.M. Chaikin, Phys. Rev. Lett. **75**, 1630 (1995)
15. I.F. Schegolev Memorial Issue, J. Phys. I (France) **6**(12) (1996)
16. M. Héritier, G. Montambaux, P. Lederer, J. Phys. Lett. (France) **45**, L943 (1984)
17. G. Montambaux, M. Héritier, P. Lederer, Phys. Rev. Lett. **55**, 2078 (1985); Erratum **56**, 97 (1986)
18. L.P. Gor'kov, Sov. Phys. Usp. **27**, 809 (1985)
19. A.G. Lebed, Sov. Phys. JETP **62**, 595 (1985)
20. D. Poilblanc, G. Montambaux, M. Héritier, P. Lederer, Phys. Rev. Lett. **58**, 270 (1987)
21. P.M. Chaikin, W. Kang, S. Hannahs, R.C. Yu, Physica B **177**, 353 (1992)
22. S.T. Hannahs, J.S. Brooks, W. Kang, L.Y. Chiang, P.M. Chaikin, Phys. Rev. Lett. **63**, 1988 (1988)
23. J.R. Cooper, W. Kang, P. Auban. G. Montambaux, D. Jérome, K. Bechgaard, Phys. Rev. Lett. **63**, 1984 (1988)
24. S. Valfells, J.S. Brooks, Z. Wang, S. Takasaki, J. Yamada, H. Anzai, M. Tokumoto, Phys. Rev. B **54**, 16413 (1996)
25. R.V. Chamberlin, M.J. Naughton, X. Yan, L.Y. Chiang, S.-Y. Hsu, P.M. Chaikin, Phys. Rev. Lett. **60**, 1189 (1988)
26. M.J. Naughton, R.V. Chamberlin, X. Yan, S.-Y. Hsu, L.Y. Chiang, M.Ya. Azbel, P.M. Chaikin, Phys. Rev. Lett. **61**, 621 (1988)
27. W. Kang, D. Jérome, J. Phys. I (France) **1**, 449 (1991)
28. W. Kang, J.R. Cooper, D. Jérome, Phys. Rev. B **43**, 11467 (1991)
29. L. Balicas, G. Kriza, F.I.B. Williams, Phys. Rev. Lett. **75**, 2000 (1995)
30. H. Cho, W. Kang, Phys. Rev. B **59**, 9814 (1999)
31. V.M. Yakovenko, Phys. Rev. B **43**, 11353 (1991)
32. L.P. Gor'kov, A.G. Lebed, J. Phys. Lett. (France) **45**, L433 (1984)
33. P.M. Chaikin, Phys. Rev. B **31**, 4770 (1985)
34. D. Poilblanc, M. Héritier, G. Montambaux, P. Lederer, J. Phys. C **19**, L321 (1986)
35. A. Virosztek, L. Chen, K. Maki, Phys. Rev. B **34**, 3371 (1986)
36. G. Grüner, *Density Waves in Solids* (Addison-Wesley, New York, 1994)
37. S.-R. Chang, K. Maki, J. Low Temp. Phys. **66**, 357 (1987)
38. G. Montambaux, D. Poilblanc, Phys. Rev. B **37**, 1913 (1988); G. Montambaux, M.J. Naughton, R.V. Chamberlin, X. Yan, P.M. Chaikin, M.Ya. Azbel, Phys. Rev. B **39**, 885 (1989)
39. V.M. Yakovenko, H.-S. Goan, J. Phys. I (France) **6**, 1917 (1996)
40. V.M. Yakovenko, H.S. Goan, Phys. Rev. B **58**, 10648 (1998)
41. I. Dana, Y. Avron, J. Zak, J. Phys. C **18**, L679 (1985)
42. H. Kunz, Phys. Rev. Lett. **57**, 1095 (1986)
43. M. Kohmoto, J. Phys. Soc. Jpn. **62**, 659 (1993)
44. D.J. Thouless, in *The Quantum Hall Effect*, ed. by R.E. Prange, S.M. Girvin (Springer, Berlin Heidelberg New York, 1987) Chap. 4
45. M. Kohmoto, Phys. Rev. B **39**, 11943 (1989)

46. K. Machida, M. Nakano, J. Phys. Soc. Jpn. **59**, 4223 (1990); K. Machida, Y. Hori, M. Nakano, J. Phys. Soc. Jpn. **60**, 1730 (1991); Y. Hori, K. Machida, J. Phys. Soc. Jpn. **61**, 1246 (1992); K. Machida, Y. Hori, M. Nakano: Phys. Rev. Lett. **70**, 61 (1993)
47. K. Machida, Y. Hasegawa, M. Kohmoto, V.M. Yakovenko, Y. Hori, K. Kishigi, Phys. Rev. B **50**, 921 (1994); Y. Hasegawa, K. Machida, M. Kohmoto, V.M. Yakovenko, J. Supercond. **7**, 757 (1994)
48. A.G. Lebed, JETP Lett. **51**, 663 (1990); Physica B **169**, 368 (1991); Sov. Phys. JETP **72**, 1035 (1991), Phys. Scr. T **39**, 386 (1991)
49. N. Dupuis, V.M. Yakovenko, Phys. Rev. Lett. **80**, 3618 (1998)
50. N. Dupuis, V.M. Yakovenko, Phys. Rev. B **58**, 8773 (1998); Europhys. Lett. **45**, 361 (1999); Phys. Rev. B **61**, 12888 (2000).
51. D. Zanchi, G. Montambaux, Phys. Rev. Lett. **77**, 366 (1996).
52. H.-S. Goan, V.M. Yakovenko, Synth. Metals **85**, 1609 (1997)
53. P.A. Lee, T.M. Rice: Phys. Rev. B **19**, 3970 (1979); T.M. Rice, P.A. Lee, M.C. Cross, Phys. Rev. B **20**, 1345 (1979)
54. K. Maki, A. Virosztek, Phys. Rev. B **41**, 557 (1990); **42**, 655 (1990)
55. J.R. Schrieffer, *Theory of Superconductivity*, 2nd edn. (Perseus Books, Reading, MA, 1999) p. 218
56. K. Sengupta, N. Dupuis, Phys. Rev. B **68**, 094431 (2003)
57. A.G. Lebed, Phys. Rev. Lett. **88**, 177001 (2002); JETP Lett. **72**, 141 (2000)
58. V.M. Yakovenko, H.-S. Goan, J. Eom, W. Kang, J. Phys. IV (France) **9**, Pr10-195 (1999)
59. T. Vuletić, C. Pasquier, P. Auban-Senzier, S. Tomić, D. Jérome, K. Maki, K. Bechgaard, Eur. Phys. J. B **21**, 53 (2001)
60. A.V. Kornilov, V.M. Pudalov, Y. Kitaoka, K. Ishida, T. Mito, J.S. Brooks, J.S. Qualls, J.A.A.J. Perenboom, N. Tateiwa, T.C. Kobayashi, Phys. Rev. B **65**, 060404 (2002)
61. V.M. Yakovenko, J. Phys. IV (France), Colloque C2, **3**, 307 (1993); J. Supercond. **7**, 683 (1994)
62. V.M. Yakovenko, Phys. Rev. Lett. **65**, 251 (1990)
63. F. Kuchar, R. Meisels, G. Weimann, W. Schlapp, Phys. Rev. B **33**, 2965 (1986)
64. L.A. Galchenkov, I.M. Grodnenskii, M.V. Kostovetskii, O.R. Matov, JETP Lett. **46**, 542 (1987)
65. D. Quinlivan, Y. Kim, K. Holczer, G. Grüner, F. Wudl, Phys. Rev. Lett. **65**, 1816 (1990)
66. L. Balicas, Phys. Rev. Lett. **80**, 1960 (1998)
67. T. Osada, N. Miura, I. Oguro, G. Saito, Phys. Rev. Lett. **58**, 1563 (1987)
68. S.N. Artemenko, A.N. Kruglov, Sov. Phys. Solid State **26**, 1448 (1984); E.N. Dolgov, Sov. J. Low Temp. Phys. **10**, 747 (1984)
69. G.X. Tessema, N.P. Ong, Phys. Rev. B **23**, 5607 (1981)
70. S.N. Artemenko, E.N. Dolgov, A.N. Kruglov, Yu.I. Latshyev, Ya.S. Savitskaya, V.V. Frolov, JETP Lett. **46**, 891 (1984)
71. L. Forró, J.R. Cooper, A. Jánossy, K. Kamarás, Phys. Rev. B **34**, 9047 (1986); L. Forró, J.R. Cooper, A. Jánossy, M. Maki, Solid State Commun. **62**, 715 (1987); M. Petravić, L. Forró, J.R. Cooper, F. Levy, Phys. Rev. B **40**, 2885 (1989)
72. B.I. Halperin, Phys. Rev. B **25**, 2185 (1982)
73. C.L. Kane, M.P.A. Fisher, in *Perspectives in Quantum Hall Effects*, ed. by S. Das Sarma, A. Pinczuk (Wiley, New York, 1997) Chap. 4

74. Y. Hatsugai, Phys. Rev. Lett. **71**, 3697 (1993); Phys. Rev. B **48**, 11851 (1993); J. Phys. Cond. Mat. **9**, 2507 (1997)
75. J.T. Chalker, A. Dohmen, Phys. Rev. Lett. **75**, 4496 (1995)
76. L. Balents, M.P.A. Fisher, Phys. Rev. Lett. **76**, 2782 (1996)
77. K. Sengupta, H.-J. Kwon, V.M. Yakovenko, Phys. Rev. Lett. **86**, 1094 (2001)
78. V.M. Yakovenko, H.-S. Goan: Phys. Rev. B **58**, 8002 (1998)
79. S.A. Brazovskii, Sov. Phys. JETP **51**, 342 (1980)
80. V.M. Yakovenko, Europhys. Lett. **3**, 1041 (1987); Sov. Phys. JETP **66**, 355 (1987)
81. R.V. Chamberlin, M.J. Naughton, X. Yan, L.Y. Chiang, S.-Y. Hsu, P.M. Chaikin, Synth. Metals **27**, 41 (1988)
82. R.C. Ashoori, H.L. Stormer, L.N. Pfeiffer, K.W. Baldwin, K. West, Phys. Rev. B **45**, 3894 (1992)
83. G. Ernst, R.J. Haug, J. Kuhl, K. von Klitzing, K. Eberl, Phys. Rev. Lett. **79**, 3748 (1997)
84. D.P. Druist, P.J. Turley, K.D. Maranowski, E.G. Gwinn, A.C. Gossard, Phys. Rev. Lett. **80**, 365 (1998)
85. S. Uji, C. Terakura, M. Takashita, T. Terashima, H. Aoki, J.S. Brooks, S. Tanaka, S. Maki, J. Yamada, S. Nakatsuji, Phys. Rev. B **60**, 1650 (1999)
86. H.-J. Kwon, V.M. Yakovenko, K. Sengupta, Synth. Metals **133–134**, 27 (2003)
87. K. Sengupta, I. Zutić, H.-J. Kwon, V.M. Yakovenko, S. Das Sarma, Phys. Rev. B **63**, 144531 (2001)
88. A.G. Lebed, JETP Lett. **43**, 174 (1986)
89. G. Montambaux, P.B. Littlewood, Phys. Rev. Lett. **62**, 953 (1989); Erratum **63**, 818 (1989)
90. Y. Sun, K. Maki, Phys. Rev. B **49**, 15356 (1994)
91. Y. Hasegawa, Phys. Rev. B **51**, 4306 (1995)
92. B.I. Halperin, Jap. J. Appl. Phys. Suppl. **26-3**, 1913 (1987)
93. M. Koshino, H. Aoki, T. Osada, Phys. Rev. B **65**, 205311 (2002)
94. G. Montambaux, M. Kohmoto, Phys. Rev. B **41**, 11417 (1990)
95. M. Kohmoto, B.I. Halperin, Y.-S. Wu, Phys. Rev. B **45**, 13488 (1992)
96. M. Koshino, H. Aoki, K. Kuroki, S. Kagoshima, T. Osada, Phys. Rev. Lett. **86**, 1062 (2001); Phys. Rev. B **65**, 045310 (2002)
97. M. Koshino, H. Aoki, B.I. Halperin, Phys. Rev. B **66**, 081301 (2002)
98. T. Itou, K. Kanoda, K. Murata, T. Matsumoto, K. Hiraki, T. Takahashi, Phys. Rev. Lett. **93**, 216408 (2004); Phys. Rev. B **72**, 113109 (2005)
99. V.I. Talyanskii, D.S. Novikov, B.D. Simons, L.S. Levitov, Phys. Rev. Lett. **87**, 276802 (2001); P.J. Leek, M.R. Buitelaar, V.I. Talyanskii, C.G. Smith, D. Anderson, G.A.C. Jones, J. Wei, D.H. Cobden, Phys. Rev. Lett. **95**, 256802 (2005)
100. M.Ya. Azbel, P. Bak, P.M. Chaikin, Phys. Rev. Lett. **59**, 926 (1987)
101. S. Haddad, S. Charfi-Kaddour, M. Héritier, R. Bennaceur, Phys. Rev. B **72**, 085104 (2005)
102. S. Uji, S. Yasuzuka, T. Konoike, K. Enomoto, J. Yamada, E.S. Choi, D. Graf, J.S. Brooks, Phys. Rev. Lett. **94**, 077206 (2005)
103. A.G. Lebed, Phys. Rev. B **55**, 1299 (1997)

18

Orbitally Quantized Density-Wave States Perturbed from Equilibrium

N. Harrison, R. McDonald, and J. Singleton

We consider the effect that experimental changes in the magnetic induction **B** have in causing an orbitally quantized field-induced spin- or charge-density wave (FISDW or FICDW) state to depart from thermodynamic equilibrium. The competition between elastic forces of the density wave (DW) and pinning leads to the realization of a critical state that is in many ways analogous to that realized within the vortex state of type II superconductors. Such a critical state has been verified experimentally in charge-transfer salts of the composition α-(BEDT-TTF)$_2M$Hg(SCN)$_4$, but should be a generic property of all orbitally quantized DW phases. The metastable state consists of a balance between the DW pinning force and the Lorentz force on extended currents associated with drifting cyclotron orbits, resulting in the establishment of persistent currents throughout the bulk and to the possibility of a three-dimensional 'chiral metal' that extends deep into the interior of a crystal.

18.1 Introduction

There is a growing interest in broken-translational-symmetry phases that incorporate orbital quantization [1–10]. The interplay between orbital and periodic charge, spin or current degrees of freedom introduces additional constraints that can lead to new types of quantum order with radically different physical properties. Organic charge-transfer salts based on the TMTSF molecule provide us with a particularly vivid example. Orbital quantization constrains the allowed the values of the spin-density modulation vector $\mathbf{Q}_\nu$ for a given Landau level filling factor ν, leading to discrete field-induced spin-density wave (FISDW) phases with different quantized Hall conductances [3, 11, 12]. One consequence of orbital quantization of the energy spectrum into levels (or subbands), separated in energy by $\hbar\omega_\mathrm{c} = \hbar e|\mathbf{B}|/m$, that is common to all these systems [1–10], is the dependence of the equilibrium modulation vector $\bar{\mathbf{Q}}(\mathbf{B})$ (or the period $\Lambda = 2\pi/|\bar{\mathbf{Q}}(\mathbf{B})|$) on the magnetic induction **B**. Standard theoretical models treat FISDW or field-induced charge-density wave (FICDW) phases that are in thermal equilibrium [3,8,12], with $\bar{\mathbf{Q}}(\mathbf{B},\mathbf{r})$ being uniform over all spatial coordinates $\mathbf{r}$.

The present review considers the effect that density wave (DW) pinning has on changing the physical properties of orbitally quantized DW states [13]. Pinning inhibits maintenance of thermodynamic equilibrium as $\bar{\mathbf{Q}}(\mathbf{B},\mathbf{r})$ changes with $\mathbf{B}$, leading to the induction of non-equilibrium (metastable) states in response to a change in the magnetic field $\mathbf{H}$. This may simply be a direct change in the externally applied magnetic field $\mathbf{H} = \mathbf{B}/\mu_0 - \mathbf{M}$ or the field gradient associated with the application of an electrical transport current $\mathbf{j} = \nabla \times \mathbf{H}$. Competition between pinning and the elastic DW restoring force (due to its perturbation from equilibrium) results in a critical state analogous to that encountered in the vortex state of type II superconductors [14]. The critical region consists of a current associated with the drift of cyclotron orbits orthogonal to the pinning force that has an equivalent role to a pinned supercurrent. The pinning potential modifies the eigenvalues of the quasi-two dimensional electron (or hole) gas, giving rise to the possibility of a type of 'chiral metal' that extends deep into the bulk, rather than being restricted to the surface [15]. The finite spatial extent of such critical regions have the potential to radically transform the magnetotransport properties of the system as a whole. This can be an important factor in the realization of large Hall angles in bulk FISDW and FICDW systems, greatly enhanced conductances and persistent current phenomena [9,11,12,16]. We consider the specific cases of metastable FISDW and FICDW states in α-(BEDT-TTF)$_2M$Hg(SCN)$_4$ and (TMTSF)$_2X$.

18.2 Critical State

To understand the physics of pinning in the context of orbitally quantized DW states, it is instructive to refer to the simple model depicted in Fig. 18.1. We consider a linear relationship between $\bar{\mathbf{Q}}$ and $\mathbf{B}$, such that an increase in $\mathbf{B}$ causes the equilibrium period Λ of the spin- or charge-density wave modulation to shorten. Since real samples are of finite spatial extent, maintenance

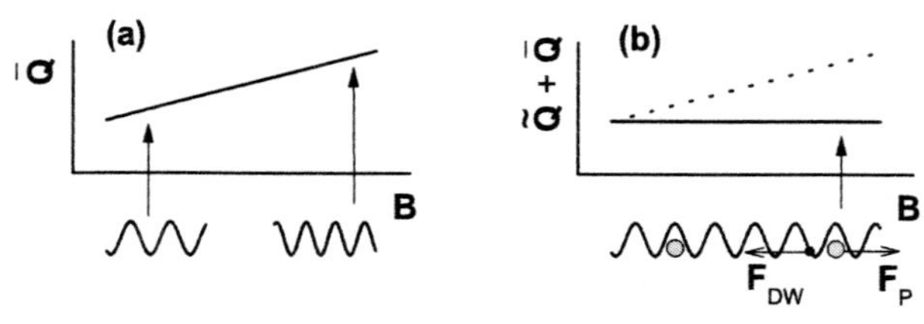

Fig. 18.1. (**a**) Example of an equilibrium modulation vector $\bar{\mathbf{Q}}$ that increases with $\mathbf{B}$, with an sketch showing how the period becomes shorter. (**b**) A plot showing how a pinning force $\mathbf{F}_\mathrm{P}$ prevents equilibrium (*dotted line*) from being achieved, leading to an additional non-equilibrium contribution $\tilde{\mathbf{Q}}$ (*solid line*)

of equilibrium requires the density modulation to undergo translation with respect to $\mathbf{r}$ in accordance with the continuity equation

$$\frac{\partial \rho_\Lambda}{\partial t} + \nabla \cdot \mathbf{j}_\Lambda = 0. \tag{18.1}$$

Here, ρ_Λ represents the physical quantity (e.g. charge, spin or current) affected by the modulation and $\mathbf{j}_\Lambda$ represents the current associated with its translational motion. Pinning inhibits maintenance of equilibrium, leading to non-equilibrium contribution $\tilde{\mathbf{Q}}$ such that the total $\mathbf{Q} = \tilde{\mathbf{Q}} + \bar{\mathbf{Q}}$ remains initially unchanged as depicted in Fig. 18.1b. The stored energy associated with the compression (or tension) of the density modulation is therefore analogous to that of a spring, with pinning sites then opposing the restoring force $\mathbf{F}_{\mathrm{DW}}$ by providing an equal and opposite pinning force $\mathbf{F}_{\mathrm{p}}$.

The physics of this exact situation has been extensively studied in the vortex state of type II superconductors [14]; in this case involving the spatial modulation of the supercurrent density of a flux-line lattice. The field-induced compression (or tension) of the modulation with respect to equilibrium in Fig. 18.1b leads to a build up of non-equilibrium stored energy

$$\tilde{\Phi} = \frac{\mu_0(\Delta \mathbf{H})^2}{2} \propto (\Delta \tilde{\mathbf{Q}})^2. \tag{18.2}$$

The surface of a sample then plays a pivotal role in determining how the system responds to a change $\Delta\mathbf{H}$ in the external magnetic field $\mathbf{H}$. Since pinning cannot support the infinite Lorentz force per vortex $\mathbf{F}_{\mathrm{DW}} \propto \nabla \partial \tilde{\Phi}/\partial \rho_\Lambda$ that would otherwise result were $\tilde{\Phi}$ to remain finite up to the sample surface, the vortex lattice undergoes translational motion so as to ensure that $\tilde{\Phi} = 0$ at the surface. This initiates the formation of a critical state, where upon the Lorentz force $\mathbf{F}_{\mathrm{DW}}$ is balanced by the maximum pinning force $\mathbf{F}_{\mathrm{p,max}}$ that the sample pinning centers can sustain. The critical region then propagates progressively further into the interior of the sample as the external magnetic field $\mathbf{H}$ is changed. The point $x = x_{\mathrm{c}}$ in Fig. 18.2 represents the furthest extent of the critical region into the sample for a given change $\Delta\mathbf{H}$ in $\mathbf{H}$. Within the 'critical' region (i.e. $0 < x < x_{\mathrm{c}}$) there exists a gradient in $\tilde{\mathbf{Q}}(\mathbf{r})$ and Λ, while beyond the critical region (i.e. $x > x_{\mathrm{c}}$) $\tilde{\mathbf{Q}}(\mathbf{r})$ remains approximately uniform.

Since the build up of non-equilibrium stored energy $\Phi(\tilde{\mathbf{Q}})$, orbital quantization and pinning are all common to flux-line lattices and field-induced

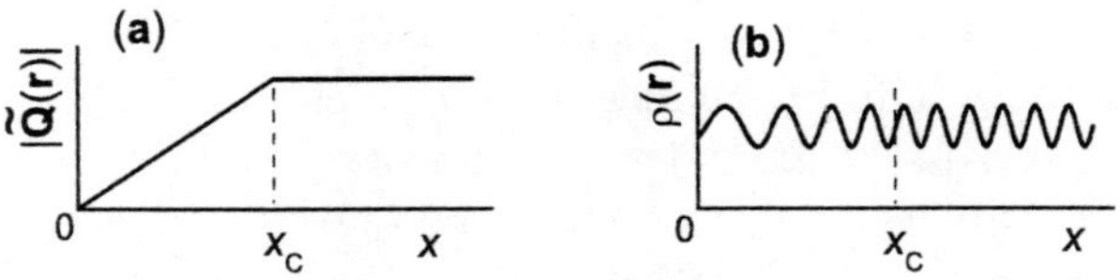

Fig. 18.2. A notional representation of a 'critical state' region between $\mathbf{r} = 0$ and $\mathbf{r} = [x_{\mathrm{c}}, 0, 0]$ depicting (**a**) the gradient in $\tilde{\mathbf{Q}}(\mathbf{r})$ and (**b**) the gradient in the modulation density

DW states, the response of both types of system to a change in magnetic field exhibits a high degree of similarity [9,17]. The primary difference in the case of DW systems is that the presence of states in the Landau gap (due to the quantum lifetime) causes the irreversible diamagnetic susceptibility $\tilde{\chi} = \partial M/\partial H$ to depart significantly from the ideal diamagnetic value of -1. A critical state nevertheless develops, yielding magnetic hysteresis that can be observed experimentally within orbitally quantized DW states. The existence of such a critical state has been directly demonstrated in charge-transfer salts of the composition α-(BEDT-TTF)$_2M$Hg(SCN)$_2$ (where M= K and Rb) [6,9,17,18]. Figure 18.3a–c shows examples in α-(BEDT-TTF)$_2$KHg(SCN)$_2$. A qualitatively similar magnetic hysteresis occurs deep within the $\nu = 1$ FISDW state of (TMTSF)$_2$ClO$_4$ [19] (shown in Fig. 18.3d), although the formation of a critical state remains to be studied.

18.3 Model for Non-equilibrium Field-Induced Density-Wave States

To model the critical state in orbitally quantized DW states, it is convenient to consider the surface of a three-dimensional crystal normal to the unit vector $\hat{\mathbf{x}}$, as represented graphically in Fig. 18.2. This enables us to consider a situation in which the components of the electric field $\mathbf{E} = [E_x(x), E_y(x), 0]$, current $\mathbf{j} = [0, j_y(x), 0]$ and magnetic flux density $\mathbf{B} = [0, 0, B_z(x)]$ vary only with respect to x in the continuity limit (i.e. $E_x/(\partial E_x/\partial x) \gg \Lambda$). We consider DW formation and Landau quantization to take place within layers orthogonal to the unit vector $\hat{\mathbf{z}}$. The total magnetic flux density $B_z = B_0 + \Delta B_z$ parallel to $\hat{\mathbf{z}}$ is composed of a uniform component $B_0 = \mu_0 H_0$ outside the sample, and a non-uniform perturbation $\Delta B_z(x)$ inside the sample. The gradient $-\partial \Delta B_z/\mu_0 \partial x$ can be considered to be the sum $j_y = j_{y,\mathrm{f}} + j_{y,\mathrm{m}}$ of free $j_{y,\mathrm{f}} = -\partial \Delta H_z/\partial x$ and pinned (or magnetic) $j_{y,\mathrm{m}} = -\partial \Delta M_z/\partial x$ currents, such that $\Delta B_z = \mu_0(\Delta H_z + \Delta M_z)$. Free currents $j_{y,\mathrm{f}}$ are those that contribute directly to the electrical transport in response to $\mathbf{E}$ while magnetic currents $j_{y,\mathrm{m}}$ are those resulting from the drift of extended states orthogonal to the pinning force $\mathbf{F}_\mathrm{p}$.

18.3.1 Materials of Interest

We consider two qualitatively different models of orbitally quantized DW states that apply to bulk crystalline materials, that have already been shown to be readily accessible to inductive experiments. The first Landau level spectrum depicted graphically in the left panel of Fig. 18.4 corresponds to the 'quantized nesting model' [12] that is generally representative of FISDW states in (TMTSF)$_2X$ salts (where X = PF$_6$, AsF$_6$ or ClO$_4$). A variant of this model may also apply to α-(BEDT-TTF)$_2M$Hg(SCN)$_4$ [8] (where M = K, Tl or Rb) for $\mathbf{B}$ applied within the layers (planes), or to (Per)$_2$Pt(mnt)$_2$ [10],

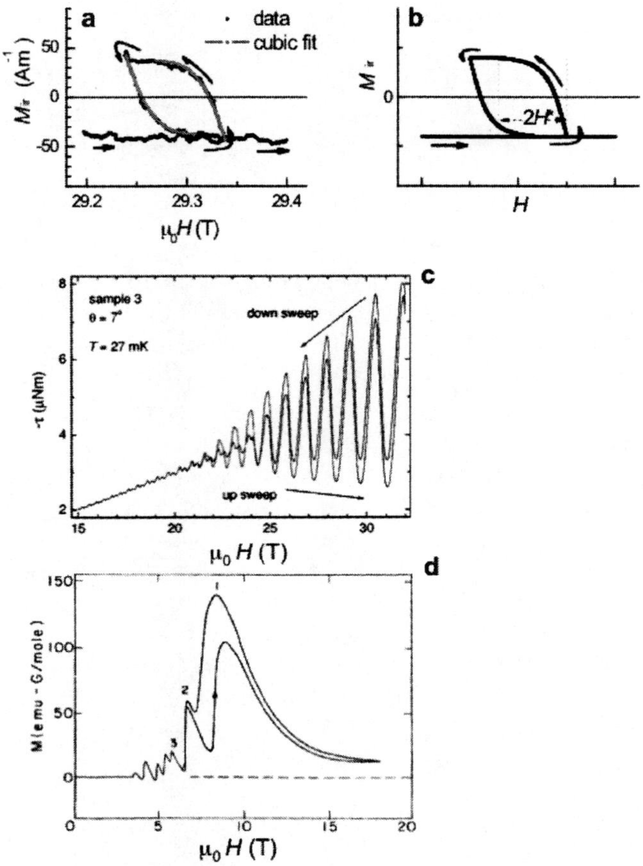

Fig. 18.3. Examples of the non-equilibrium magnetization of field-induced density-wave measured by way of magnetic torque. (**a**) A hysteresis loop obtained on cycling the magnetic field in α-(BEDT-TTF)$_2$KHg(SCN)$_2$ [9], together with a cubic fit (red). This plot shows only the irreversible contribution M_{irr} (the reversible de Haas–van Alphen part has been subtracted). (**b**) A simulation of a hysteresis loop for a hypothetical cylindrical sample, resulting in the cubic lineshape where H^* is the coercion field. Other geometries will yield a cubic lineshape to leading order. (**c**) The consequences of such hysteresis over an extended range of magnetic field leading to different up and down magnetic field sweeps, measured on a different sample [6]. (**d**) Qualitatively similar hysteresis over an extended range of magnetic field in (TMTSF)$_2$ClO$_4$ in which FISDW states are realized at ambient pressure [19]

provided there exists sufficient overlap of electronic orbitals between Perylene chains. The second energy level spectrum depicted graphically in Fig. 18.4 corresponds to a two band model in which the orbital quantization of a 2D hole pocket and DW formation occur on separate bands as in α-(BEDT-TTF)$_2M$Hg(SCN)$_4$ [9, 17] (where M = K, Tl or Rb) but that nevertheless share the same chemical potential μ. In both cases, we assume here that

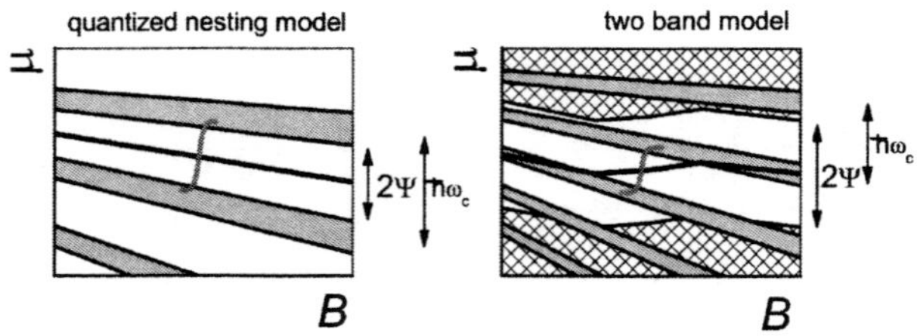

Fig. 18.4. Part of the energy level structure over a short range of μ and **B** for the 'quantized nesting' and 'two band' models. *Grey shading* represents Landau levels or Landau subbands while the *hatched area* represents the gapped density of states of a non-orbitally quantized SDW or CDW a second band in the two band model. The *thick black* and *red lines* represent loci of $\Delta\bar{\mu}(B_z)$ and $\Delta\bar{\mu}(B_z) + \Delta\tilde{\mu}(B_z)$, respectively

orbital quantization occurs only in layers orthogonal to $\hat{\mathbf{z}}$ and that the two spin states are degenerate. The latter occurs naturally as a consequence of SDW formation in strong magnetic fields when the spins lie orthogonal to **B** [12] or may also occur in a CDW system when the product mg (where g is the g-factor) is an integer number of free electron masses m_{e}, as is approximately the case in α-(BEDT-TTF)$_2$KHg(SCN)$_4$ [9, 17].

In spite of the differences in electronic structure, the non-equilibrium behaviors of these model systems are equivalent. **B** equally influences both systems through a change in the Landau-level degeneracy $D = m\omega_{\mathrm{c}}/\pi\hbar c$, where the cyclotron frequency is given by [9, 20]

$$\omega_{\mathrm{c}} = \frac{qB_z}{m}\left(1 + \frac{m\nabla\cdot\mathbf{E}}{2qB_z^2}\right)^{1/2}. \tag{18.3}$$

Here, q is the charge of the quasiparticles (positive for holes and negative for electrons) while 'c' is the interlayer spacing. The second term inside the bracket accounts for a possible divergence $\nabla\cdot\mathbf{E}$ in the electric field due to a small non-equilibrium charge density within the sample. This might occur where the diagonal components of the resistivity tensor ρ_{xx}, ρ_{yy} and ρ_{zz} to vanish, should the transport become ballistic under special conditions (see Sect. 4.3). The derivation of this term has been covered in detail in the literature [9, 20]; hence here we merely quote the result. The total degeneracy can be expressed as

$$D_0 + \Delta D = \frac{q}{\pi\hbar c}\left(B_0 + \Delta B_z + \frac{m}{2qB_0}\frac{\partial E_x}{\partial x}\right) \tag{18.4}$$

for the geometry considered in Fig. 18.2, where ΔD is then its perturbation from equilibrium. The mean density of states $\bar{g}_{\mathrm{2D}} = D/\hbar\omega_{\mathrm{c}} = m/\pi\hbar^2 c$ (averaging out the oscillatory component) remains independent of B_z.

The response of the system to ΔB_z depends on changes in the relative position of μ between the Landau levels in Fig. 18.4. This can be tracked by considering the fact that the total perturbed charge density of the 2D Landau levels ($\Delta\rho_{2D}$) and of the DW ($\Delta\rho_{DW}$) must obey Poisson's equation,

$$\Delta\rho_{2D} + \Delta\rho_{DW} = -\varepsilon\nabla\cdot\mathbf{E}, \tag{18.5}$$

hence preserving charge neutrality when $\nabla\cdot\mathbf{E} = 0$ (in the continuity limit). The more extreme situation occurs when the chemical potential μ resides between Landau levels, as depicted in Fig. 18.4. The perturbed charge density contribution due to the Landau levels in Fig. 18.4, for both DW models, is then given by

$$\Delta\rho_{2D} = \nu q\Delta D - \nu\beta q\Delta D - \beta q\bar{g}_{2D}\Delta\mu, \tag{18.6}$$

where the first term on the right-hand side represents the sum over all ν occupied Landau levels which shift in response to ΔB_z. The second term on the right-hand side is a correction (constant μ) for the existence of a finite density of states $\beta\bar{g}_{2D}$ between the Landau levels. If the Landau levels are broadened by a finite quantum lifetime τ, for example, then $\beta \approx 4/\pi\omega_c\tau \ll 1$. The third term on the right-hand side accounts for the change $\Delta\mu$ in μ. The latter is related to $\Delta\mathbf{Q}$ by

$$\Delta\mu = -\mathrm{sign}(q_{DW})\frac{\hbar}{2}\bar{\mathbf{v}}_F\cdot\Delta\mathbf{Q}, \tag{18.7}$$

where $\bar{\mathbf{v}}_F$ is the mean Fermi velocity of the band on which the DW forms and sign(q_{DW}) represents the sign of its carriers. We have $q_{DW} \equiv q$ in the case of the quantized nesting model in Fig. 18.4, whereas $q_{DW} = -q$ in the two band model. The equilibrium and non-equilibrium evolution of the DW is considerably simplified by using the variable $\Delta\mu$ instead of $\Delta\mathbf{Q}$. In a similar fashion to $\Delta\mathbf{Q}$, $\Delta\mu$ must consist of both equilibrium $\Delta\bar{\mu}$ and non-equilibrium $\Delta\tilde{\mu}$ components such that

$$\Delta\mu = \Delta\bar{\mu} + \Delta\tilde{\mu}. \tag{18.8}$$

18.3.2 Equilibrium Conditions

For the DW to achieve the equilibrium state described in standard theoretical models, the DW must adjust itself according to (18.1) so as to minimize the total free energy [9] (see later). This is accomplished when $\Delta\tilde{\mu} = 0$ in (18.8), in which case μ, represented by the thick black line in Fig. 18.4, attempts to lie in the middle of the DW gap. This occurs when

$$\Delta\rho_{DW} = -\nu q_{DW}\Delta D \quad \{\text{quantized nesting}\} \tag{18.9}$$

$$\Delta\rho_{DW} = q_{DW}\bar{g}_{DW}\Delta\bar{\mu} \quad \{\text{two band}\} \tag{18.10}$$

for the quantized nesting and the two band models, respectively. On inserting each expression for $\Delta\rho_{DW}$ [together with (18.6) and (18.4)] into (18.5), we obtain

$$\frac{m}{2qB_0}\frac{\partial E_x}{\partial x} + \Delta B_z = \alpha\frac{m}{\hbar q\nu}\Delta\bar{\mu}, \tag{18.11}$$

where $\alpha \equiv -1$ and $\alpha = (\eta + \beta)/(1 - \beta) \approx \eta$ for the for the quantized nesting and two band models, respectively. Note that the right-hand side of (18.5) is much smaller than the similar contribution $\propto \nabla \cdot \mathbf{E}$ originating from (18.3) and can therefore be neglected.

18.3.3 Pinning and Non-equilibrium Thermodynamics

As discussed in the introduction, pinning inhibits maintenance of an equilibrium DW state by preventing it from moving (or sliding) [13] so as to satisfy the continuity (18.1). The DW is no longer able to respond to B_z for $x > x_c$ in Fig. 18.2, instead remaining physically static such that $\Delta\rho_{\rm DW} = 0$. Under these circumstances $\Delta\tilde{\mu} \neq 0$, with (18.6) therefore yielding

$$\frac{m}{2qB_0}\frac{\partial E_x}{\partial x} + \Delta B_z = \frac{\beta m}{(1-\beta)\hbar q\nu}(\Delta\tilde{\mu} + \Delta\bar{\mu}). \tag{18.12}$$

Equation (18.12) describes the red line in Fig. 18.4, which yields precisely the same result for both the quantized nesting and two-band models. On substituting $\Delta\bar{\mu}$ from (18.11), this reduces to

$$\frac{m}{2qB_0}\frac{\partial E_x}{\partial x} + \Delta B_z = \gamma\frac{\beta m}{\hbar q\nu}\Delta\tilde{\mu}, \tag{18.13}$$

where $\gamma = 1$ and $\gamma = (\eta + \beta)/\eta(1 - \beta) \approx 1$ for the quantized nesting and two band models, respectively. Because $\beta \ll 1$ on the right-hand side of (18.13), the non-equilibrium state of the DW is *rapidly* established, even for a very small ΔB_z.

As with the critical state model of type II superconductors, any perturbation of the orbitally quantized DW state from equilibrium leads to the build up of stored energy. In the case of the DW, this energy is given to leading order by

$$\bar{\Phi}_{\rm DW} + \tilde{\Phi}_{\rm DW} \approx -g_{\rm DW}\left(\frac{\Psi^2}{2} - \left(\frac{\hbar v_{\rm F}}{2}\right)^2\left(\frac{\partial\phi}{\partial x_{\mathbf{Q}}}\right)^2\right) \tag{18.14}$$

where $\partial\phi/\partial x_{\mathbf{Q}}$ represents the gradient in the phase ϕ of the DW parallel to $\mathbf{Q}$ [13]. The first term on the right-hand side is the equilibrium part whereas the second term is the non-equilibrium part. Through (18.7), this second term can be expressed in terms of $\Delta\tilde{\mu}$ [21]. The magnetization change associated with 2D Landau level spectrum contributes as additional component to the free energy, which is given to leading order by

$$\tilde{\Phi}_{\rm 2D} \approx \beta g_{\rm 2D}\frac{\Delta\tilde{\mu}^2}{2}. \tag{18.15}$$

Were $\beta = 0$, as for a pristine system with an infinite quantum lifetime, pinning of the DW would cause μ to jump discontinuously between Landau levels, with no net change in the free energy $\tilde{\Phi}_{\rm 2D}$. This is a consequence of the fact that the free energy of a 2D electron gas in the canonical ensemble retains its minimum value irrespective of the location of μ within the gap. It is the existence of a finite density of states $\approx \beta g_{2D}$ within the gap that causes $\tilde{\Phi}_{2D}$ to become dependent on μ [22]; hence the β prefactor to (18.15).

On combining (18.14) and (18.15), we obtain

$$\tilde{\Phi} \approx (g_{DW} + \beta g_{2D})\Delta\tilde{\mu}^2 \tag{18.16}$$

for the total non-equilibrium stored energy. It is immediately apparent on inspection of (18.16) that the contribution $\propto \beta g_{2D}$ from the magnetization of the 2D Landau level spectrum is much smaller than that from the DW. This implies that the effective magnetic current originating from the gradient in the de Haas–van Alphen magnetization due to a gradient in $\Delta\tilde{\mu}$ caused by pinning is much smaller than that associated directly with the drift of the cyclotron orbits orthogonal to the pinning force. As will soon become apparent, it is this latter contribution that is entirely responsible for the rather unusual behavior of pinned orbitally quantized DW systems. On neglecting the smaller contribution from the 2D Landau levels, we arrive at

$$\tilde{\Phi} \approx g_{DW}\left(\frac{\hbar\nu}{\beta\gamma}\right)^2\left(\frac{1}{2B_0}\frac{\partial E_x}{\partial x} + \frac{q}{m}\Delta B_z\right)^2. \tag{18.17}$$

Thus, the substitution of (18.13) into (18.16) enables us to express the stored energy in terms of the perturbations ΔB_z and $\partial E_x/\partial x$. If ρ_{xx}, ρ_{yy} and ρ_{zz} remain finite, we can assume that $\partial E_x/\partial x \to 0$ in a slowly varying magnetic field. On taking the second derivative with respect to ΔB_z, we then obtain

$$\frac{\partial \Delta\tilde{M}_z}{\partial B_z} = -\frac{\partial^2\tilde{\Phi}}{\partial B_z^2} = -2g_{DW}\left(\frac{\hbar\nu e}{\beta\gamma m}\right)^2 \tag{18.18}$$

with the non-equilibrium susceptibility given by

$$\tilde{\chi} = \frac{\partial \Delta\tilde{M}_z}{\partial H_z} = \left(\left(\mu_0\frac{\partial \Delta\tilde{M}_z}{\partial B_z}\right)^{-1} - 1\right)^{-1}. \tag{18.19}$$

For the purposes of making numerical estimates of $\tilde{\chi}$, it is convenient to express (18.19) in terms of more familiar parameters

$$\tilde{\chi} \approx \left(\left(2\mu_0\eta g_{2D}\left(\frac{\pi\varepsilon_F\mu_m}{4}\right)^2\right)^{-1} - 1\right)^{-1}, \tag{18.20}$$

where $\varepsilon_F \approx \hbar e\nu B_0/m \approx N_{2D}/g_{2D}$ is the Fermi-energy scale of the 2D pocket, $g_{2D} = m/\pi\hbar^2 c$ is its density of states and $\mu_m = e\tau/m$ is its carrier mobility.

For $(TMTSF)_2ClO_4$, where $\eta = 1$ (since the DW and Landau quantization involve the same band), $m \approx 0.1\ m_e$, $c \approx 14$ Å, $g_{2D} \approx 2 \times 10^{45}\ m^3J^{-1}$, $\varepsilon_F \approx 23$ meV and $\mu_m \approx 1\ T^{-1}$ [12, 23, 81], we obtain $\tilde{\chi} \approx -0.04$. For α-(BEDT-TTF)$_2$KHg(SCN)$_4$, where $\eta \approx 0.5$, $m \approx 2\ m_e$, $c \approx 20$ Å (usually

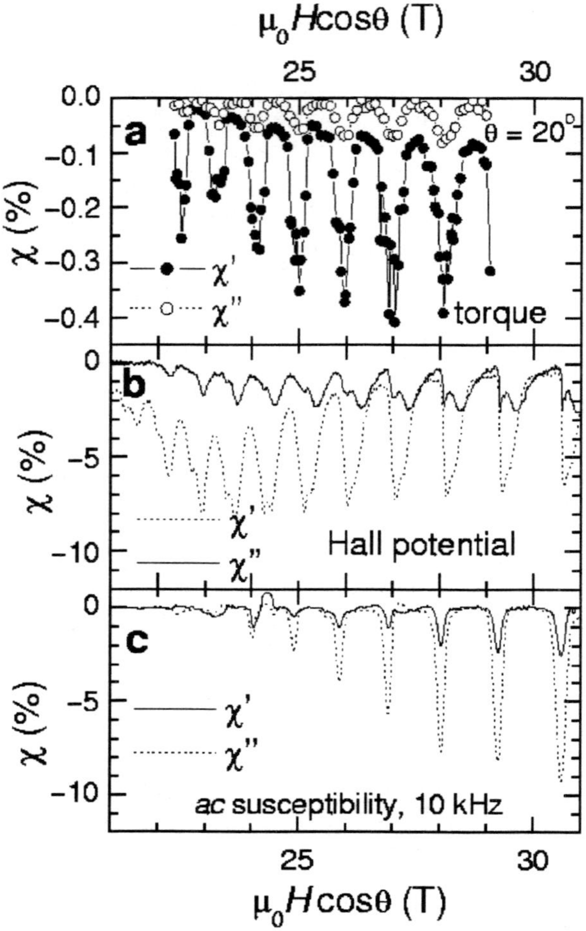

Fig. 18.5. Differential susceptibility of α-$(BEDT\text{-}TTF)_2KHg(SCN)_4$ obtained using different methods. (**a**) From the maximum rate of change in magnetic torque with field on measuring many hysteresis loops like those in Fig. 18.3 at different fields and ∼30 mK [9]. (**b**) From oscillations in the Hall voltage measured between the edge and the center of the uppermost surface layer of a crystal with the excitation provided by means of an ac magnetic field of ∼400 Hz superimposed on a static magnetic field at ∼0.5 K. (**c**) Using the conventional ac susceptibility method in a static background magnetic field at ∼0.5 K. In all cases, $\theta < 20°$

referred to as b), $g_{2D} \approx 3 \times 10^{46}$ m^{3}J^{-1}, $\varepsilon_F \approx 38$ meV and $\mu_m \approx 0.5$ T^{-1} [6], we obtain an estimate of $\tilde{\chi} \approx -0.1$ that is slightly larger. The latter estimate for $\tilde{\chi}$ is also of comparable magnitude to the estimates made from magnetic torque and ac susceptibilty measurements shown in Fig. 18.5 [9].

18.4 Magnetotransport

The magnetotransport of orbitally quantized DW states in equilibrium has been examined throughout the literature. The consequences of non-equilibrium thermodynamic effects are more difficult to predict. The experimental magnetotransport has proven to be rather anomalous compared to ordinary organic conductors. For example, in the case of α-(BEDT-TTF)$_2$KHg(SCN)$_4$, an induced Hall potential can be detected by slowly sweeping the field of a Bitter magnet (shown in Fig. 18.6). This result can be partly interpreted in terms of an unusually large Hall ratio ρ_{xy}/ρ_{xx} of ∼180 [9]. The development of quantized Hall plateaux in ρ_{xy}, a large Hall ratio and reduced longitudinal resistivity ρ_{xx} at low temperatures are partly the natural consequence of the opening of a Landau gap, as demonstrated in (TMTSF)$_2$AsF$_6$ [12]. Such effects appear to be particularly severe in α-(BEDT-TTF)$_2$KHg(SCN)$_4$, however, with the change in resistivity as a function of temperature on entry into the field-induced CDW state resembling that of a phase transition into a dissipationless conducting state (see Fig. 18.7 [25]).

There are several ways in which one can attempt to understand such anomalous data. One is to assume scattering to be an essentially random process, resulting in uniformly broadened Landau levels throughout the bulk, and hence a uniform β. Another is to assume that scattering occurs mostly at spatially extended defects, causing the relative position of the Landau-level eigenvalues to vary throughout the bulk. Alternatively, β itself can vary. In this case β will depend on the local position of μ with respect to the Landau gap as well as on the local concentration of impurities. Another is to consider more radical changes in the potential landscape introduced by the critical state, leading to the possibility of a bulk chiral metal. Below we consider each of these in turn.

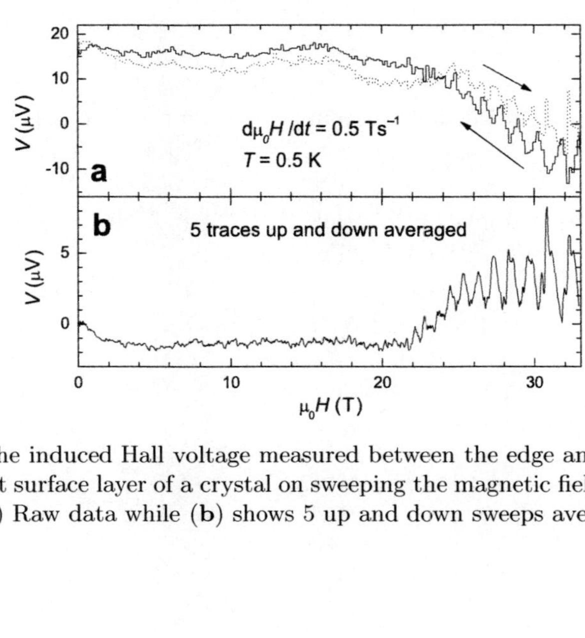

Fig. 18.6. The induced Hall voltage measured between the edge and the center of the uppermost surface layer of a crystal on sweeping the magnetic field up and down at ∼0.5 K. (**a**) Raw data while (**b**) shows 5 up and down sweeps averaged

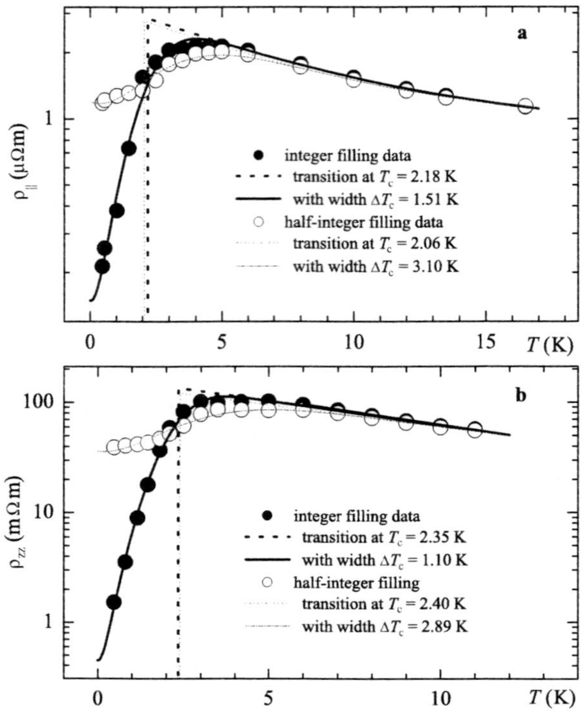

Fig. 18.7. Temperature-dependent intra-layer $\rho_{\parallel} \approx \rho_{xx}$ (**a**) and inter-layer ρ_{zz} (**b**) resistivity of α-(BEDT-TTF)$_2$KHg(SCN)$_4$ at integer (*filled circles*) and half-integer (*open circles*) filling factors at $\mu_0 H \sim 30\,\mathrm{T}$, together with fits of an error function at T_c of width ΔT_c (*solid lines*) renormalized by the extrapolated normal-state resistivity. The *dotted lines* are obtained on setting $\Delta T_c = 0$ in the fitted function

18.4.1 Uniform Transport

Even in the simplest case of a uniform sample (with a uniform β), the simple action of applying a transport current $j_{y,\mathrm{f}}$ within the bulk perturbs the system from equilibrium. A uniform current density, like that depicted in Fig. 18.8a for example, gives rise to a gradient in the magnetic field H_z along $\hat{\mathbf{x}}$ in Fig. 18.8b, which through (18.13) gives rise to a gradient $\nabla\Delta\tilde{\mu}$ in $\Delta\tilde{\mu}$. This gradient exacts an additional force

$$\tilde{\mathbf{F}}_{\mathrm{DW}} = -\frac{2\eta}{\gamma\beta}\nabla\Delta\tilde{\mu} \tag{18.21}$$

per 2D carrier, causing the cyclotron orbits to drift orthogonal to this force. The resulting magnetic current is

$$j_{y,\mathrm{m}} = \frac{N_{\mathrm{2D}}}{qB_0}\left(\frac{2\eta}{\gamma\beta}\frac{\partial\Delta\tilde{\mu}}{\partial x}\right), \tag{18.22}$$

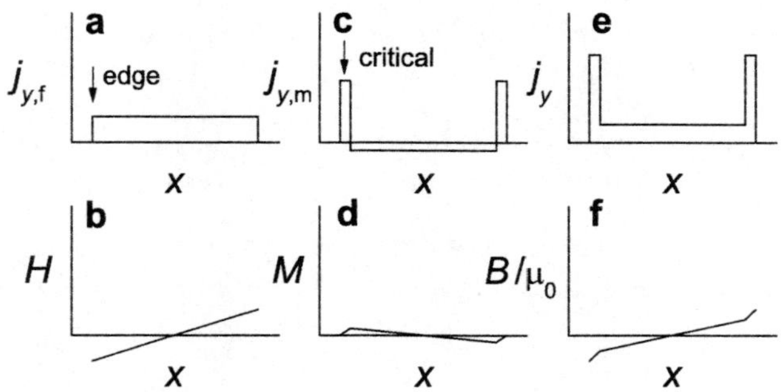

Fig. 18.8. Current and magnetic field distributions across the width of a hypothetical bar-like sample. (**a**) A uniform applied current giving rise to the variation in H in (**b**). (**c**) The induced current due to $\tilde{\chi}$ giving rise to no net magnetization M in (**d**) across the sample owing to the critical region at the edge. (**e**) and (**f**) The total current and magnetic induction B

where $N_{2D} = \varepsilon_F g_{2D}$ is the number density of 2D carriers. This introduces an anomalous contribution to the Hall effect within the bulk, such that the modified Hall conductivity becomes

$$\sigma'_{xy} = \sigma_{xy}(1 + \tilde{\chi})^{-1}, \tag{18.23}$$

where

$$\sigma_{xy} = \frac{2\nu e^2}{h} \tag{18.24}$$

has its usual quantized value. The finite susceptibility in which $\partial B_z/\partial H_z = \mu_0(1+\tilde{\chi})$ effectively results in a back-flowing current $j_{y,\mathrm{m}} = \tilde{\chi} j_{y,\mathrm{f}}$ in response to the applied (free) current $j_{y,\mathrm{f}}$, such that the total current is given by

$$j_y = j_{y,\mathrm{f}} + j_{y,\mathrm{m}} = (1 + \tilde{\chi}) j_{y,\mathrm{f}}. \tag{18.25}$$

On inverting the conductivity tensor in the usual manner, we obtain

$$\begin{aligned} \rho'_{yy} &= \frac{(1+\tilde{\chi})^2 \sigma_{xx}}{\sigma_{xy}^2 + (1+\tilde{\chi})^2 \sigma_{xx}\sigma_{yy}} \\ \rho'_{xy} &= \frac{(1+\tilde{\chi}) \sigma_{xy}}{\sigma_{xy}^2 + (1+\tilde{\chi})^2 \sigma_{xx}\sigma_{yy}}. \end{aligned} \tag{18.26}$$

Here, we assume that $\sigma_{yx} = -\sigma_{xy}$ and that the DW is in equilibrium prior to the application of a current. Given that $|\tilde{\chi}| \lesssim 0.1$, (18.26) can only explain small changes in the Hall ratio and resistivity.

The induced current given by (18.22) is not expected to directly modify the net resistance of a uniform bar-like sample, since the confinement of the current to the sample causes its net value to average to zero as shown in Fig. 18.8c. $\Delta\tilde{\mu}$ and M cannot be finite beyond the sample surface, giving rise to a surface magnetic current that compensates $j_{y,\mathrm{m}}$ within the bulk.

In accordance with the critical state model, this current can only flow at the maximum critical density

$$j_{\mathrm{c}} = \frac{N_{\mathrm{2D}}}{qB_0}|\mathbf{F}_{\mathrm{p,max}}| \tag{18.27}$$

that the DW pinning force can sustain, causing a critical region of finite width to propagate into the interior of the sample, instead if being confined to the edge.

18.4.2 Inhomogeneous Transport

A non-uniform $\beta(\mathbf{r})$ has the potential to radically transform the magneto-transport. If, for example, $\beta(\mathbf{r})$ varies spatially in a random fashion such that $\bar{\beta}v = \int_v \beta(\mathbf{r})\mathrm{d}x\mathrm{d}y\mathrm{d}z$ is the average over the volume v, there can be regions of finite size where both $\beta(\mathbf{r}) \to 1$ and $\beta(\mathbf{r}) \to 0$ are realized. Regions where $\beta(\mathbf{r}) \to 0$ will contribute most significantly to the total conductance, and these are the same regions where locally $\tilde{\chi}(\mathbf{r}) \to -1$, yielding $\rho_{yy} \to 0$ and $\rho_{xy} \to 0$ in (18.26). Hence, we would expect the overall conductance to become filamentary or percolative in nature analogous to an inhomogeneous superconductor [25–30]. While the form and distribution of $\beta(\mathbf{r})$ throughout the sample remains unknown, the electrical resistivity in Fig. 18.7 has already been shown to be fitted by the expression,

$$\rho(T) = \frac{1}{2}\left(\mathrm{erf}\left[\frac{T - T_{\mathrm{c}}}{\Delta T_{\mathrm{c}}}\right] + \mathrm{erf}\left[\frac{-T - T_{\mathrm{c}}}{\Delta T_{\mathrm{c}}}\right] + 2\right)\rho_{\mathrm{n}}(T), \tag{18.28}$$

to astonishing accuracy. This is the same expression that fits broadened transitions in inhomogeneous superconductors [25–30], albeit with a rather wide transition width ΔT_{c} in the present case. Here T_{c} is the midpoint of the transition while ρ_{n} is the extrapolated 'normal-state' resistivity.

Given that superconductivity was the original prediction for systems with CDW-like modulations [25, 31], it is somewhat ironic that the resistivity can be fitted by a model that normally applies to superconductors. The physical mechanism discussed above is nonetheless very different. Another consequence of percolative or filamentary conductance is that the current will take indirect (i.e. meandering) paths throughout the sample. This gives rise to 'current jetting' effects in which the net polarity of the voltage between voltage terminals can switch with respect to that of the net current between current terminals during a four wire transport measurement, as observed in α-(BEDT-TTF)$_2$TlHg(SCN)$_4$ [16].

18.4.3 Ballistic Transport in a Bulk Chiral Metal

Once the critical state is fully established after sweeping the magnetic field over an interval that exceeds double the characteristic coercion field $2H^*$ in

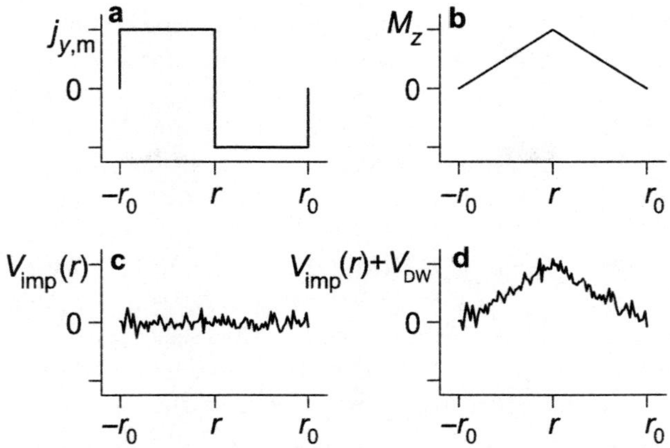

Fig. 18.9. Irreversible current (**a**) and magnetization (**b**) distributions across a hypothetical cylindrical sample of radius r_0 for which the magnetization is fully saturated. The pinning force of the DW introduces an additional effective potential $V_{DW}(r)$ that adds to the impurity potential $V_{imp}(\mathbf{r})$, shown schematically in (**c**), to arrive at the total potential (**d**)

Fig. 18.3b, the profile of $j_{y,m}$ and the magnetization M will develop the form depicted in Fig. 18.9. A bulk chiral metal may occur if the net difference in the pinning potential $\Delta V_{DW} = V_{DW}(r_0) - V_{DW}(0)$ (see Fig. 18.9) exceeds the overall width of the random impurity potential $V_{imp}(\mathbf{r})$. In such a situation, all of the cyclotron orbits will drift continuously about the sample axis ($r = 0$), giving rise to the possibility of ballistic transport should backscattering become inhibited [15].

The average width of the impurity potential is usually given by $\Delta V_{imp} = \hbar/2\tau$, although the actual spatial variation is system dependent. In α-(BEDT-TTF)$_2$KHg(SCN)$_4$, where $\tau^{-1} \approx 0.2 \times 10^{12}\,\mathrm{s}^{-1}$, this corresponds to $\Delta V_{imp} \approx 0.07\,\mathrm{meV}$ [33]. ΔV_{DW} for the situation, depicted in Fig. 18.9d, can be estimated from the experimental data by using the approximate formula

$$\Delta V_{DW} \approx r_0|\mathbf{F}_p| = \frac{j_c r_0 B_0}{N_{2D}} \approx \frac{3M_{sat}B_0}{N_{2D}}, \tag{18.29}$$

where M_{sat} is the saturation magnetization averaged over the volume (assuming a cylindrical approximation). This yields ΔV_{DW} values ranging from $\approx$1.03 meV for $M_{sat} \approx 300\,\mathrm{A\,m^{-1}}$ in [18] to $\approx$0.15 meV for $M_{sat} \approx 50\,\mathrm{A\,m^{-1}}$ in [9]. The former estimate exceeds that of ΔV_{imp} by a factor of $\approx$15, implying that Fig. 18.9d provides a realistic model.

Should ballistic transport become a realistic factor in orbitally quantized DW phases [9,34,35], screening of ΔB_z by currents near the surface of a sample will prevent an energetically costly change in $\Delta\mu$ from developing within the bulk. Setting $\Delta\mu = 0$ in (18.12) then yields

$$\lambda^2\nabla^2 V - V = 0 \quad \text{with} \quad \lambda = [m/(2\mu_0 e^2 N_{2D})]^{1/2}, \tag{18.30}$$

where $\mathbf{E} = \nabla V$, while λ is a penetration depth of similar form to that in superconductors. One consequence of ballistic transport is that the Hall effect becomes 'ideal,' so that $\Delta B_z = -\mu_0 e N_{2D} V / B_0$. On estimating λ for α-(BEDT-TTF)$_2$KHg(SCN)$_4$, we obtain $\lambda \approx 400$ nm [9]. This quantity is too small to have been measured directly, but has been proposed as an explanation for inductive currents in a $\sim$1 mm^2 cross-section sample with a decay time exceeding 10 μs that are heavily weighted towards its edge [9].

The spatial translation of the orbitally quantized DW with respect to the edge in the critical state region in Fig. 18.9 as B_z is swept could be another factor in inducing surface-weighted currents. As consecutive wave fronts of the DW slide past the edge they carry with them magnetic flux, causing currents that were previously pinned to become free near the surface. Loss of pinning should lead to the spontaneous generation of a Hall voltage (or redistribution of charge near the surface) to counter the Lorentz force. A similar spontaneous Hall voltage should occur if the critical state collapses due to an increase in the temperature, since j_c is strongly temperature dependent [18]. This would provide a new mechanism for pyroelectric currents, that are presently the realm of ferroelectric materials [36].

18.5 Future Directions

Orbitally quantized DW systems have become widely known for producing a mechanism by which one can observe a variant of the quantum Hall effect in a bulk material. While this is firmly established experimentally in TMTSF-based salts [11, 12], large Landau filling-factors of $\nu \gtrsim 20$ have proven to be something of an impediment in observing a quantized Hall resistance in α-(BEDT-TTF)$_2M$Hg(SCN)$_4$ salts [16]. Meanwhile, the degree of involvement of orbital quantization in (Per)$_2$Pt(mnt)$_2$ [10] and in α-(BEDT-TTF)$_2M$Hg(SCN)$_4$ when $\mathbf{H}$ is aligned within the layers [8] remains to be established.

In the present chapter, we identify another intrinsic property of orbitally quantized DW systems, which is their ability to store energy inductively in a manner analogous to the vortex state, albeit in the presence of a large static background magnetic field. Such effects are prevalent in layered organic conductors because of their narrow electronic bandwidths, causing them to be somewhat easier to manipulate with a magnetic field than conventional CDW and SDW materials [13]. The magnitude of these persistent current effects are nevertheless rather small compared to those in the vortex state, with dissipationless conduction requiring considerable improvements in sample quality. Sample-dependence has already appeared to be a major factor. Although the in-plane resistivity of some TMTSF-based salts shows a significant drop at low temperatures [12], as might be expected to accompany the quantum Hall effect, this is not universally observed throughout the literature. The same is also true with α-(BEDT-TTF)$_2M$Hg(SCN)$_4$. Samples grown by Tokumoto et al. [18, 25] consistently exhibit the most remarkable drops in resistivity on entry into the high magnetic field DW phase, by as much as a factor of 100, closely followed by those of Kurmoo et al. [35]. The present variability of such effects suggests that there is considerable room for sample improvement, potentially yielding exotic new physical effects. On the other hand, samples that do not exhibit a significant drop in resistivity would be expected to support larger electric effects that would ultimately yield a more conventional non-linear DW conduction to be observed, particularly when μ is no longer within a Landau gap [37].

Acknowledgements

This work is supported by US Department of Energy, the National Science Foundation and the State of Florida.

References

1. E.H. Brandt, Rep. Prog. Phys. **58**, 1465 (1995)
2. D.C. Tsui, H.L. Stormer, A.C. Gossard, Phys. Rev. Lett. **48**, 1559 (1982)
3. L.P. Gor'kov, A.G. Lebed, J. Phys. Lett. (France) **45**, L433 (1984); M. Ribault, J. Cooper, D. Jerome, D. Mailly, A. Moradpour, K. Bechgaard, J. Phys. Lett. (France) **45**, L935 (1984); M. Heritier, G. Montambaux, P. Lederer, J. Phys. Lett. (France) **45**, L943 (1984)
4. F.I.B. Williams, P.A. Wright, R.G. Clark, E.Y. Andrei, G. Deville, D.C. Glattli, O. Probst, B. Etienne, C. Dorin, C.T. Foxon, J.J. Harris, Phys. Rev. Lett. **66**, 3285 (1991)
5. S. Hill, S. Uji, M. Takashita, C. Terakura, T. Terashima, H. Aoki, J.S. Brooks, Phys. Rev. B **58**, 10778 (1998)
6. N. Harrison, L. Balicas, J.S. Brooks, J.S. Sarrao, Z. Fisk, Phys. Rev. B **61**, 14299 (2000)
7. J.E. Hoffman, E.W. Hudson, K.M. Lang, V. Madhavan, H. Eisaki, S. Uchida, J.C. Davis, Science **295**, 466 (2002)
8. D. Andres, M.V. Kartsovnik, P.D. Grigoriev, W. Biberacher, H. Muller, Phys. Rev. B **68**, 201101 (2003)
9. N. Harrison, J. Singleton, A. Bangura, A. Ardavan, P.A. Goddard, R.D. McDonald, L.K. Montgomery, Phys. Rev. B **69**, 165103 (2004)
10. D. Graf, E.S. Choi, J.S. Brooks, M. Matos, R.T. Henriques, M. Almeida, Phys. Rev. Lett. **93**, 076406 (2004); R.D. McDonald, N. Harrison, J. Singleton, A. Bangura, P.A. Goddard, A.P. ramirez, X. Chi, Phys. Rev. Lett. **94**, 106404 (2005)
11. S.T. Hannahs, J.S. Brooks, W. Kang, L.Y. Chiang, P.M. Chaikin, Phys. Rev. Lett. **63**, 1988 (1989)
12. P.M. Chaikin, J. Phys. I (France) **6**, 1875 (1996)
13. G. Grüner, in *Density Waves in Solids, Frontiers in Physics*, vol. 89 (Addison-Wesley, Reading, MA, 1994)
14. S. Senoussi, J. Phys. III (France) **2**, 1041 (1992)

15. L. Balents, M.P.A. Fisher, Phys. Rev. Lett. **76**, 2782 (1995)
16. N. Harrison, M.M. Honold, M.V. Kartsovnik, J. Singleton, S.T. Hannahs, D.G. Rickel, N.D. Kushch, Phys. Rev. B **55**, R16005 (1997)
17. N. Harrison, Phys. Rev. B **66**, 121101 (2002)
18. N. Harrison, L. Balicas, J.S. Brooks, M. Tokumoto, Phys. Rev. B **62**, 14212 (2000)
19. M.J. Naughton, J.S. Brooks, L.Y. Chiang, R.V. Chamberlin, P.M. Chaikin, Phys. Rev. Lett. **55**, 969 (1985)
20. A.H. MacDonald, T.M. Rice, W.F. Brinkman, Phys. Rev. B **28**, 3648 (1983)
21. N. Harrison, Phys. Rev. Lett. **83**, 1395 (1999)
22. N. Harrison, R. Bogaerts, P.H.P. Reinders, J. Singleton, S.J. Blundell, F. Herlach, Phys. Rev. B **54**, 9977 (1996)
23. T. Ishiguro, K. Yamaji, *Organic Superconductors* (Springer, Berlin Heidelberg New York, 1990).
24. A.G. Lebed, H.-I. Ha, M.J. Naughton, Phys. Rev. B **71**, 132504 (2005)
25. N. Harrison, C.H. Mielke, J. Singleton, J.S. Brooks, M. Tokumoto, J. Phys. Cond. Matt. **14**, L389 (2001)
26. L.R. Testardi, Phys. Lett. **35A**, 33 (1971)
27. J. Maza, F. Vidal, Phys. Rev. B **43**, 10560 (1991)
28. J.A. Veira, F. Vidal, Physica C **159**, 468 (1989)
29. S.W. Hsu, S.Y. Tsaur, H.C. Ku, Phys. Rev. B **38**, 856 (1988)
30. J.G. Bednorz, K.A. Muller, Z. Phys. B **64**, 189 (1986)
31. H. Fröhlich, Proc. R. Soc. Ser. A **223**, 296 (1954)
32. B. Huckestein, Rev. Mod. Phys. **67**, 357 (1995)
33. N. Harrison, A. House, I. Deckers, J. Caulfield, J. Singleton, F. Herlach, W. Hayes, M. Kurmoo, P. Day, Phys. Rev. B **52**, 5584 (1995)
34. S. Hill, P.S. Sandhu, J.S. Qualls, J.S. Brooks, M. Tokumoto, N. Kinoshita, Y. Tanaka, Phys. Rev. B **55**, R4891 (1997)
35. M.M. Honold, N. Harrison, J. Singleton, H. Yaguchi, C. Mielke, D. Ricket, I. Deckers, P.H.P. Reinders, F. Herlach, M. Kurmooo, P. Day, J. Phys. Cond.: Matt. **9**, L533 (1997)
36. K. Lijima, Y. Tomita, R. Takayama, I. Ueda, J. Appl. Phys. **60**, 361 (1986)
37. T. Fujita, T. Sasaki, N. Yoneyama, N. Kobayashi, J. Phys. Soc. Jpn. **73**, 1525 (2004)

19

Unconventional Density Waves in Organic Conductors and in Superconductors

K. Maki, B. Dóra, and A. Virosztek

Unconventional density waves (UDW) are one of the ground states in metallic crystalline solids and have been speculated already in 1968. However, more focused studies on UDW started only recently, perhaps after the identification of the low temperature phase in α-(BEDT-TTF)$_2$KHg(SCN)$_4$ as unconventional charge density wave (UCDW) in 2002. More recently, the metallic phase of Bechgaard salts (TMTSF)$_2$X with X = PF$_6$ and ReO$_4$ under both pressure and magnetic field appears to be unconventional spin density wave (USDW). The pseudogap regime of high T_c superconductors LSCO, YBCO, Bi2212, and the one in CeCoIn$_5$ belongs to d-wave density waves (d-DW).

In these identifications, the angular-dependent magnetoresistance and the giant Nernst effect have played the crucial role. These are the simplest manifestations of the Landau quantization of quasiparticle energy in UDW in the presence of magnetic field (the Nersesyan effect). Also we speculate that UDW will be most likely found in α-(BEDT-TTF)$_2$I$_3$, α-(BEDT-TTF)$_2$I$_2$Br, κ-(BEDT-TTF)$_2$Cu(NCS)$_2$, κ-(BEDT-TTF)$_2$Cu(CN)$_2$Br, λ-(BEDT)$_2$GaCl$_4$, and in many other organic compounds.

19.1 Introduction

Until very recently, the electronic ground state in crystalline solids was considered to belong to one of the four canonical ground states in quasi-one-dimensional systems: s-wave superconductors, p-wave superconductors, (conventional) charge density waves, and spin density waves [1–4]. Indeed many condensates discovered since 1972 have appeared to be accommodated comfortably in this scheme. For example, the two CDWs in NbSe$_3$ and SDW in (TMTSF)$_2$PF$_6$ are well known examples [3, 4].

In all these systems, the elementary excitations are of Fermi liquid nature á la Landau [5–7] but have the energy gap Δ and the quasiparticle (condensate) density decrease exponentially as $\exp(-\Delta/T)$ for $T \leq T_c/2$, where T_c is the transition temperature. Therefore, only a small portion of the quasiparticles will be left below $T_c/2$.

The thermodynamics of these systems is practically the same as those in s-wave superconductors as described by the theory of Bardeen, Cooper, and Schrieffer (the BCS theory [8]).

However, since the discovery of heavy fermion superconductors $CeCu_2Si_2$ [9], organic superconductors in Bechgaard salts $(TMTSF)_2PF_6$ [10], high T_c cuprate superconductors [11], Sr_2RuO_4 [12], etc., this simple scheme has to be necessarily modified. First of all, most of these superconductors are unconventional and nodal [13–17]. Parallel to this development, there is a surge of studies on CDW whose quasiparticle energy gap $\Delta(\mathbf{k})$ has line or point nodes at the Fermi surface [18–21]. We may characterize this as the paradigm shift from one dimensional to higher dimensional physics [22]. However, such paradigm shift has been already anticipated. The quasiparticle excitations are of Fermi liquid type and not of Tomonaga–Luttinger type [4,23,24]. Note that we are not interested in the high temperature behavior $T > 200$ K of Bechgaard salts currently discussed in literature [25]. In higher dimensions, the imperfect nesting is inevitable. Further, in a magnetic field, the quasiparticle spectrum is quantized á la Landau in both SDW [26] and CDW [27]. This gives rise to field induced spin density wave (FISDW) with integral quantum Hall effect [4,28,29].

Unconventional density waves (UDW) were first speculated by Halperin and Rice [30] as a possible ground state in the excitonic insulator. Unlike conventional DW, there is no X-ray or spin signal due to the zero average of the gap over the Fermi surface ($\langle\Delta(\mathbf{k})\rangle = 0$). Therefore, UDW is often called the state with hidden order parameter. However, certain confusion has spread around in recent literature with the use of "hidden order." Therefore, it is much wiser to specify what one means. In this sense we prefer to use the notation UDW. Also unlike many other people [18,19,21,31,32], we do not consider the discretized lattice and UDW with minuscule loop current or Z_2 symmetry breaking common to the descendent of the flux phase [31,32].

It is easy to see that such a two-dimensional solution is unstable in three-dimensional environment. In contrast, our UDW has the $U(1)$ gauge symmetry as in conventional DW. Therefore, our UDW can slide and exhibit the nonohmic conductance [33,34] as in conventional DW and generate the phase vortices.

As another difference compared to conventional DW, the nodal excitations persist down to $T = 0$ K, giving rise to an electronic specific heat $\sim T^2$. Indeed, the thermodynamics of UDW is identical to the one in d-wave superconductors [20,35]. The transition from the normal metallic state to UDW is metal to metal. As an example, the T^2 dependence of the electric resistance in the normal state typical to Fermi liquids changes to T linear dependence for $T \leq T_c/2$, which is to be contrasted to the exponentially activated behavior in the conventional counterpart.

In light of the earlier discussion, many of the so-called non-Fermi liquids can be UDW and in fact Fermi liquids as defined by Landau, as we shall see. The change in the exponents distinguishing between various phases stems from the change in the excitation spectrum and not from the different nature of the elementary excitations. We have shown already that the pseudogap phases in both high T_c cuprates like LSCO, YBCO, and Bi2212 [36, 37] and $CeCoIn_5$ [38, 39] are d-DW based on the giant Nernst effect and the angle dependent magnetoresistance observed in these systems [40–44]. Through the angle dependent photoemission spectra, the $d_{x^2-y^2}$ symmetry of the energy gap in the pseudogap region of high T_c cuprates has been established [45]. Also the presence of Fermi arc around the (π, π) direction [46] is consistent with d-DW [47]. Furthermore, the mysterious relation $\Delta(0) \simeq 2.14T^*$ found in the pseudogap region of LSCO, YBCO, and Bi2212 [48–51] can readily be interpreted in terms of UDW. Here $\Delta(0)$ is the maximal energy gap determined by STM and T^* is the pseudogap crossover temperature, which may be identified as T_c in d-DW. The 2.14 gap maximum-transition temperature ratio is the weak-coupling theory value for d-DW and for d-wave superconductors [20, 35].

19.2 Mean-Field Theory

In the following we limit ourselves to quasi-one- and quasi-two-dimensional systems, since for UDW

(a) The Fermi surface nesting plays an important role in the realization of the phase
(b) We do not have yet any well established examples in three dimensional systems

We consider the effective (low energy) Hamiltonian given by

$$H = \sum_{\mathbf{k},\sigma} \xi(\mathbf{k}) c^+_{\mathbf{k},\sigma} c_{\mathbf{k},\sigma} + \frac{1}{2} \sum_{\substack{\mathbf{k},\mathbf{k}',\mathbf{q} \\ \sigma,\sigma'}} V(\mathbf{k},\mathbf{k}',\mathbf{q}) c^+_{\mathbf{k}+\mathbf{q},\sigma} c_{\mathbf{k},\sigma} c^+_{\mathbf{k}'-\mathbf{q},\sigma'} c_{\mathbf{k}',\sigma'}, \qquad (19.1)$$

where $c^+_{\mathbf{k},\sigma}$ and $c_{\mathbf{k},\sigma}$ are the creation and annihilation operators of electrons with momentum $\mathbf{k}$ and spin σ, $\xi(\mathbf{k})$ is the kinetic energy of electrons measured from the Fermi energy in the normal state, and $V(\mathbf{k}, \mathbf{k}', \mathbf{q})$ is the interaction between two particles. In the following we represent $V(\mathbf{k}, \mathbf{k}', \mathbf{q})$ by a separable potential

$$V(\mathbf{k}, \mathbf{k}', \mathbf{q}) = \langle |f(\mathbf{k})|^2 \rangle^{-1} f(\mathbf{k}) f(\mathbf{k}') \delta(\mathbf{q} - \mathbf{Q}), \qquad (19.2)$$

where $\mathbf{Q}$ is the nesting vector. This kind of interaction is readily obtained from the extended Hubbard model with at most nearest-neighbor interaction (on site + nearest neighbor Coulomb interaction, exchange, pair hopping, and assisted hopping terms [20]). For quasi-one-dimensional systems with most conducting or chain direction parallel to the x axis, we consider $f(\mathbf{k}) = \cos(bk_y)$ or $\sin(bk_y)$. For quasi-two-dimensional systems, we consider only d-wave DW with $d_{x^2-y^2}$ and d_{xy} symmetry. At present, these four UDW appears to exhaust all known cases. The quasi-one-dimensional description is applicable to both α-(BEDT-TTF)$_2$KHg(SCN)$_4$ and (TMTSF)$_2$PF$_6$ [52–55], while quasi-two-dimensional UDW applies to the pseudogap phase in high T_c cuprates and in CeCoIn$_5$ [36–39].

Within the mean-field approximation, the Hamiltonian in (19.1) is reduced to quadratic form as

$$H = \sum_{\mathbf{k},\sigma} \left(\xi(\mathbf{k}) c^+_{\mathbf{k},\sigma} c_{\mathbf{k},\sigma} + \Delta(\mathbf{k}) c^+_{\mathbf{k},\sigma} c_{\mathbf{k}-\mathbf{Q},\sigma} + \overline{\Delta}(\mathbf{k}) c^+_{\mathbf{k}-\mathbf{Q},\sigma} c_{\mathbf{k},\sigma}\right) - \sum_{\mathbf{k}} \frac{|\Delta(\mathbf{k})|^2}{V\langle |f(\mathbf{k})|^2\rangle} \tag{19.3}$$

and $\Delta(\mathbf{k})$ obeys the self-consistency equation

$$\Delta(\mathbf{k}) = V f(\mathbf{k}) \sum_{\mathbf{k}',\sigma} f(\mathbf{k}') \langle c^+_{\mathbf{k}'-\mathbf{Q},\sigma} c_{\mathbf{k}',\sigma}\rangle. \tag{19.4}$$

The Hamiltonian is further rewritten as

$$H = \sum_{\mathbf{k},\sigma} \Psi^+_\sigma(\mathbf{k}) \left[\tilde{\xi}(\mathbf{k})\rho_3 + \eta(\mathbf{k}) + \Delta(\mathbf{k})\rho_1\right] \Psi_\sigma(\mathbf{k}) - \sum_{\mathbf{k}} \frac{|\Delta(\mathbf{k})|^2}{V\langle |f(\mathbf{k})|^2\rangle}, \tag{19.5}$$

where $\tilde{\xi}(\mathbf{k}) = (\xi(\mathbf{k}) - \xi(\mathbf{k}-\mathbf{Q}))/2$ and $\eta(\mathbf{k}) = (\xi(\mathbf{k}) + \xi(\mathbf{k}-\mathbf{Q}))/2$. In the following we shall take the tilde off $\tilde{\xi}(\mathbf{k})$, and $\eta(\mathbf{k})$ is the imperfect nesting term. From now on we limit ourselves to UCDW for simplicity. For USDW we have to involve the Pauli spin matrices as well, though the parallel treatment is possible. Also $\Psi^+_\sigma(\mathbf{k})$ and $\Psi_\sigma(\mathbf{k})$ are spinor fields conjugate to each other, $\Psi^+_\sigma(\mathbf{k}) = (c^+_{\mathbf{k},\sigma}, c^+_{\mathbf{k}-\mathbf{Q},\sigma})$. Finally, the single particle Green's function or the Nambu–Gor'kov Green's function is given by

$$G^{-1}(\omega,\mathbf{k}) = \omega - \xi(\mathbf{k})\rho_3 - \eta(\mathbf{k}) - \Delta(\mathbf{k})\rho_1. \tag{19.6}$$

Here ρ_i's are the Pauli matrices operating on the Nambu spinor space [56]. Then the poles of $G(\omega,\mathbf{k})$ determine the quasiparticle energies as

$$\omega = \eta(\mathbf{k}) \pm \sqrt{\xi(\mathbf{k})^2 + \Delta(\mathbf{k})^2}. \tag{19.7}$$

From this, the quasiparticles density of states follows as [20, 35]

$$\frac{N(E)}{N_0} = \mathrm{Re}\left\langle \frac{|E-\eta(\mathbf{k})|}{\sqrt{(E-\eta(\mathbf{k}))^2 - \Delta^2(\mathbf{k})}}\right\rangle, \tag{19.8}$$

where $\langle\dots\rangle$ means the average over the Fermi surface, N_0 is the quasiparticle density of states in the normal state at the Fermi energy. When $\eta(\mathbf{k}) = 0$, all UDW so far discussed acquire the same density of states $N(E) = N_0 g(E/\Delta)$ with

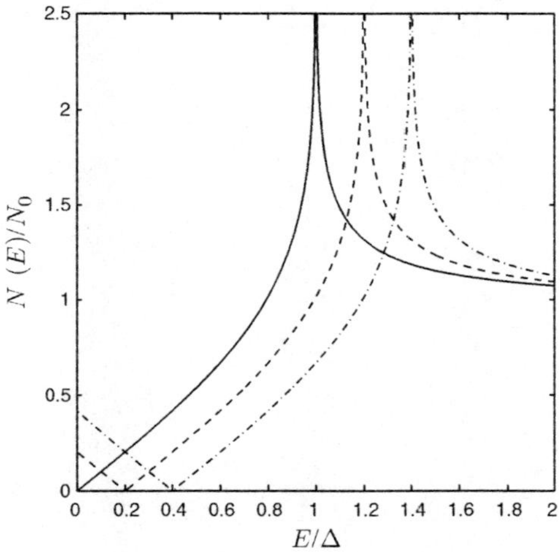

Fig. 19.1. The quasiparticle density of states of UDW is shown for $\mu/\Delta = 0$ (*solid line*), 0.2 (*dashed line*), and 0.4 (*dashed-dotted line*)

$$g(x) = \begin{cases} \frac{2}{\pi} x K(x) \text{ for } x < 1 \\ \frac{2}{\pi} K(x^{-1}) \text{ for } x > 1 \end{cases}, \tag{19.9}$$

where $K(x)$ is the complete elliptic integral of the first kind [35]. For d-wave density waves, in most cases $\eta(\mathbf{k}) = \mu$, i.e., the inclusion of a finite chemical potential as imperfect nesting suffices. Then we obtain

$$N(E) = N_0 g\left(\frac{E-\mu}{\Delta}\right). \tag{19.10}$$

For nonzero chemical potential, $N(0)/N_0 = g(\mu/\Delta) \neq 0$. This is shown in Fig. 19.1. Therefore, the chemical potential produces the Fermi arc or the Fermi pockets detected in ARPES [46]. In quasi-one-dimensional systems, the difference between the different gap functions $f(\mathbf{k})$ is most readily seen in the optical conductivity $\sigma_{yy}(\omega)$ (i.e., the one with electric current parallel to the Fermi surface) [57]. The universal electric conduction (in analogy to the universal heat conduction in nodal superconductors [17]) implies that $\sigma_{yy}(0) \to 0$ as Γ and $T \to 0$ for $\Delta(\mathbf{k}) \sim \sin(bk_y)$, while $\sigma_{yy}(0) \to 2e^2 N_0 v_y^2/\pi\Delta(0)$ for Γ and $T \to 0$ for $\Delta(\mathbf{k}) \sim \cos(bk_y)$ [58]. Here Γ is the quasiparticle scattering rate in the normal state. The Γ dependence of σ_{yy} is shown in Fig. 19.2. Note that the electric conductivity increases with quasiparticle scattering, which is very counter-intuitive. But this has a similar origin to the universal heat conduction in nodal superconductors. In quasi-two-dimensional systems, the question of $d_{x^2-y^2}$-wave density wave vs. d_{xy}-wave density wave is most easily decided by the angle dependent magnetoresistance, when the magnetic field is rotated within the conducting plane as we shall discuss in Sect. 19.4 [37].

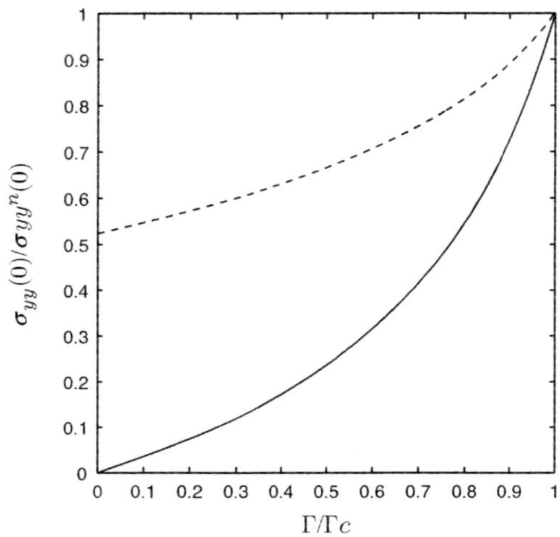

Fig. 19.2. The dc conductivity normalized by its normal state value at $T = 0\,\mathrm{K}$ is plotted in the unitary scattering limit for $\Delta(\mathbf{k}) \sim \sin(bk_y)$ (*solid line*) and $\cos(bk_y)$ (*dashed line*). Γ_c is the critical scattering rate, where the DW phase vanishes. Similar curves are obtained for the Born limit

19.3 Landau Quantization

In the presence of a magnetic field, the quasiparticle energy in UDW is quantized as first discussed by Nersesyan et al. [59, 60]. The quasiparticle spectrum in the presence of magnetic field is obtained from

$$(E - \xi(\mathbf{k} + e\mathbf{A})\rho_3 - \eta(\mathbf{k} + e\mathbf{A}) - \Delta(\mathbf{k} + e\mathbf{A})\rho_1)\Psi(\mathbf{r}) = 0, \tag{19.11}$$

where we have introduced the magnetic field through the vector potential $\mathbf{A}(\mathbf{B} = \nabla \times \mathbf{A})$. It is readily recognized that (19.11) has the same mathematical structure as the Dirac equation in a magnetic field studied in 1936 [61, 62]. Since the Landau quantization in quasi-one-dimensional UDW has been throughoutly discussed in [22] , here we consider the Landau quantization in d-DW [37]. Without loss of generality, we consider here d_{xy}-wave DW or $\Delta(\mathbf{k}) = \Delta\sin(2\phi)$. We assume that the nodal lines are perpendicular to the conducting plane and run parallel to the z axis at $(\pm k_\mathrm{F}, 0, 0)$ and $(0, \pm k_\mathrm{F}, 0)$. Also we take $\xi(\mathbf{k}) = v(k_\parallel - k_\mathrm{F}) + \frac{v'}{c}\cos(ck_z)$, where $k_\parallel$ is the radial vector within the a-b plane, v and v' are the respective Fermi velocities in the a-b plane and parallel to the c-axis.

Let us assume that a magnetic field lies in the $z - x'$ plane and tilted by an angle Θ from the z-axis, $x' = \hat{x}\cos\phi + \hat{y}\sin(\phi)$:

$$\mathbf{B} = B(\cos(\Theta)\hat{z} + \sin(\Theta)(\cos(\phi)\hat{x} + \sin(\phi)\hat{y})). \tag{19.12}$$

We shall focus on the quasiparticle spectrum in the vicinity of Dirac cones at $(\pm k_\mathrm{F}, 0, \pm\pi/2c)$ and $(0, \pm k_\mathrm{F}, \pm\pi/2c)$. Then it is convenient to choose the vector potential as

$$\mathbf{A} = -B(\hat{z}\sin(\Theta) + \cos(\Theta)(\hat{x}\cos(\phi) + \hat{y}\sin(\phi)))(y\cos(\phi) - x\sin(\phi)). \quad (19.13)$$

Then in the vicinity of Dirac cones, (19.11) is recasted as

$$[E - \mu + eB(x\sin(\phi) - y\cos(\phi))(v\cos(\Theta)\cos(\phi) \pm v'\sin(\Theta))\rho_3 - v_2(-\mathrm{i}\partial_y)\rho_1]\,\Psi(\mathbf{r}), \quad (19.14)$$

where $v_2 = 2\Delta/k_\mathrm{F}$. From this, we obtain

$$(E_{1n\pm} - \mu)^2 = 2neBv_2|\cos(\phi)(v\cos(\Theta)\cos(\phi) \pm v'\sin(\Theta))| \quad (19.15)$$

$$(E_{2n\pm} - \mu)^2 = 2neBv_2|\sin(\phi)(v\cos(\Theta)\sin(\phi) \pm v'\sin(\Theta))| \quad (19.16)$$

and $n = 0, 1, 2, 3 \ldots$.

Except for the $n = 0$ state, which is nondegenerate, all other states are doubly degenerated. Also unlike in UDW in quasi-one-dimensional systems, the Landau spectrum consists of four different branches. Furthermore, the chemical potential $\mu(\neq 0)$ is crucial in interpreting the magnetotransport of quasiparticles [37,39]. Finally, (19.15) gives the particle and hole spectrum

$$E_{1n-} = \mu \pm \sqrt{2neBv_2|\cos(\phi)(v\cos(\Theta)\cos(\phi) - v'\sin(\Theta))|}. \quad (19.17)$$

A similar formula can be worked out for (19.16). From these Landau spectra, the thermodynamics as well as the transport properties are readily obtained. In the following, we shall consider the angle dependent magnetoresistance, the nonlinear Hall constant, and the giant Nernst effect as the three hallmarks of UDW.

19.4 Angle Dependent Magnetoresistance (ADMR)

19.4.1 α-(BEDT-TTF)$_2$MHg(SCN)$_4$ Salts with M = K, Rb, and Tl

The nature of the low temperature phase (LTP) in the quasi-two-dimensional organic conductor α-(BEDT-TTF)$_2$MHg(SCN)$_4$ salts with M = K, Rb, and Tl has not been understood until recently [63]. Although the phase transition is clearly seen in magnetotransport measurements, neither charge nor magnetic order has been established [64, 65]. Moreover, the destruction of the LTP in an applied magnetic field suggests a kind of CDW rather than SDW. But the threshold electric field associated with the sliding motion of the density wave turns out to be very different from the one in CDW, but somewhat similar to that observed in SDW [66]: the threshold field increases smoothly with temperature and does not diverge at T_c. Indeed, the threshold electric field in the LTP of α-(BEDT-TTF)$_2$KHg(SCN)$_4$ [33,34,67] is well described in terms of imperfectly nested UCDW. More recently, striking angular dependent magnetoresistance has been detected in this material [68–71]. There have been many unsuccessful attempts to interpret this phenomenon in terms of the reconstruction of the Fermi surface.

We find that Landau quantization of the quasiparticle spectrum in UDW as described in Sect. 19.3 plays the crucial role [52,53]. Also the giant Nernst effect found in LTP of α-(BEDT-TTF)$_2$KHg(SCN)$_4$ can be described in terms of Landau quantized UCDW [72,73]. Therefore, these findings lead us to the conclusion that the LTP in α-(BEDT-TTF)$_2$MHg(SCN)$_4$ with M = K, Rb, and Tl is UCDW.

Before proceeding, it is useful to check the Fermi surface of α-(BEDT-TTF)$_2$MHg(SCN)$_4$ salts as shown in Fig. 19.3 [4,63].

It consists of a quasi-one-dimensional Fermi surface perpendicular to the a axis (the most conducting direction) and small two-dimensional pockets (quasi-two-dimensional Fermi surface) at the corners of the Brillouin zone. Most of the de Haas van Alphen oscillations come from the quasi-two-dimensional Fermi surface, while UCDW appears on the quasi-one-dimensional one. The magnetic field configuration with respect to the conducting plane is also shown in Fig. 19.3, which was used in ADMR measurement.

The magnetoresistance is given in terms of the quasiparticle energy E_n as [22,37]

$$R^{-1} = \sum_n \sigma_n \text{sech}^2\left(\frac{\beta E_n}{2}\right), \tag{19.18}$$

where σ_n's are the level conductivities weakly depending on temperature and magnetic field, $E_n = \sqrt{2nv_a\Delta e|B\cos(\theta)|}$. In the low temperature region, where $\beta E_1 \gg 1$, (19.18) is approximated as

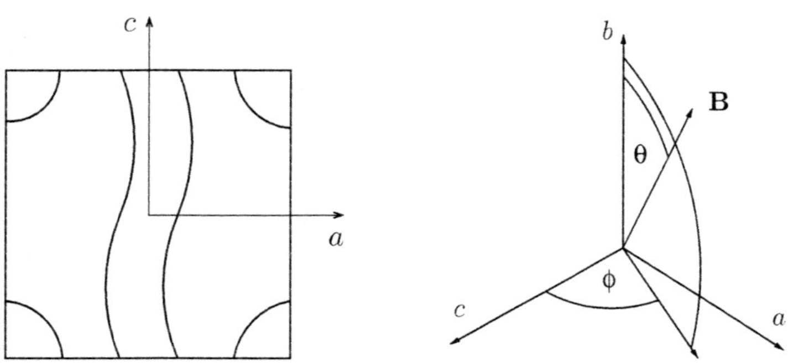

Fig. 19.3. The Fermi surface of α-(BEDT-TTF)$_2$KHg(SCN)$_4$ is shown schematically in the left panel. In the right one the geometrical configuration of the magnetic field with respect to the conducting plane is plotted

$$\begin{aligned}
R^{-1} &= 2\sigma_0 \mathrm{sech}^2\left(\frac{\beta E_0^1}{2}\right) + \sigma_1 \left[\mathrm{sech}^2\left(\frac{\beta(E_1 + E_1^{(1)})}{2}\right)\right. \\
&\quad + \mathrm{sech}^2\left(\frac{\beta(E_1 - E_1^{(1)})}{2}\right) + \mathrm{sech}^2\left(\frac{\beta(E_1 + E_1^{(2)})}{2}\right) \\
&\quad \left. + \mathrm{sech}^2\left(\frac{\beta(E_1 - E_1^{(2)})}{2}\right)\right] \\
&= \frac{4\sigma_0}{1+\cosh(\zeta_0)} + 4\sigma_1 \left[\frac{1+\cosh(x_1)\cosh(\zeta_0)}{(\cosh(x_1)+\cosh(\zeta_0))^2}\right. \\
&\qquad \left. + \frac{1+\cosh(x_1)\cosh(\zeta_1)}{(\cosh(x_1)+\cosh(\zeta_1))^2}\right], \qquad (19.19)
\end{aligned}$$

where $\zeta_0 = \beta E_0^{(1)}$, $\zeta_1 = \beta E_1^{(2)}$, and $x_1 = \beta E_1$,

$$E_0^{(1)} = E_1^{(1)} = \sum_m \varepsilon_m \exp(-y_m), \qquad (19.20)$$

$$E_1^2 = \sum_m \varepsilon_m (1 - 2y_m)\exp(-y_m). \qquad (19.21)$$

Here $E_0^{(1)}$ and $E_1^{(1,2)}$ are corrections to Landau level energies from imperfect nesting (i.e., the warping of the quasi-one-dimensional Fermi surface):

$$\eta(\mathbf{k}) = \sum_n \varepsilon_n \cos(2\mathbf{b}'_n\mathbf{k}). \qquad (19.22)$$

From this we find $y_n = v_a b'^2 e|B\cos(\theta)|[\tan(\theta)\cos(\phi - \phi_o) - \tan(\theta_n)]^2/\Delta c$, $\tan(\theta_n)\cos(\phi - \phi_0) = \tan(\theta_0) + nd_0$, $\tan(\theta_0) \simeq 0.5$, $d_0 \simeq 1.25$, $\phi_0 \simeq 27°$ [22]. A typical fitting of ADMR of α-(BEDT-TTF)$_2$KHg(SCN)$_4$ at $T = 1.4$ K and $B = 14$ T is shown in Figs. 19.4 and 19.5.

Broad peaks centered at $\theta = 0°$ come from E_1, while the series of dips stem from the imperfect nesting term. From these fittings, we can extract Δ (17 K), $v_a \sim 6\times10^6\,\mathrm{cm\,s^{-1}}$, $b' \sim 30$ Å, $\varepsilon_0 \sim 3$ K, and σ_2/σ_1 of the order of 1/10. Finally, we show in Fig. 19.6 a series of ADMR data when the magnetic field is rotated in the plane as well. Clearly, the theory reproduces the general features of the ADMR. Of course, we notice that some interesting details are still missing from our model. Nevertheless, we may conclude that UCDW describes many features of ADMR observed in α-(BEDT-TTF)$_2$KHg(SCN)$_4$ [52, 53].

19.4.2 Bechgaard Salts (TMTSF)$_2$X

The first organic superconductor (TMTSF)$_2$PF$_6$, discovered in 1980 [10], is, perhaps, one of the most fascinating electron system studied so far. The Bechgaard salts are well known for the variety of their ground states. They exhibit

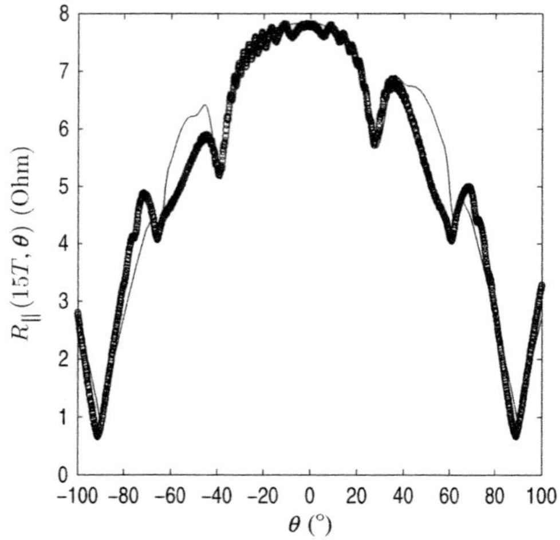

Fig. 19.4. The angular dependent magnetoresistance is shown for current parallel to the a-c plane at $T = 1.4\,\mathrm{K}$, $B = 15\,\mathrm{T}$. The open circles belong to the experimental data, the solid line is our fit based on (19.19)

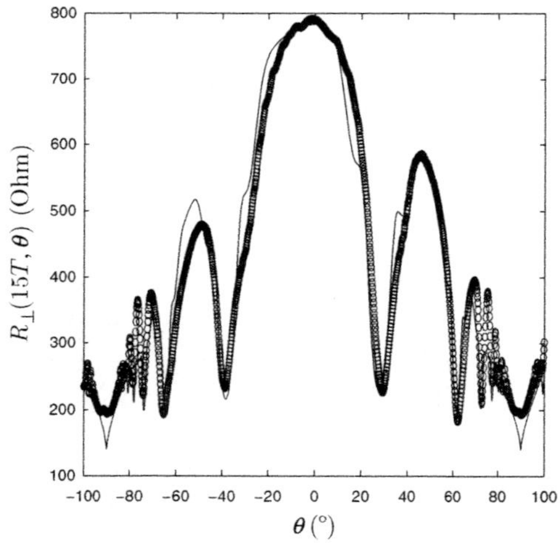

Fig. 19.5. The angular dependent magnetoresistance is shown for current perpendicular to the a-c plane at $T = 1.4\,\mathrm{K}$, $B = 15\,\mathrm{T}$. The open circles belong to the experimental data, the solid line is our fit from (19.19)

spin density wave at ambient-to-moderate pressure, field induced spin density wave, and triplet superconductivity at high pressure ($p > 8\,\mathrm{kbar}$) [2, 4, 74] as shown in Fig. 19.7. We have discovered recently the coexistence of SDW and USDW at $T = T^* \sim 4\,\mathrm{K}$ in $(\mathrm{TMTSF})_2\mathrm{PF}_6$ during the analysis of ADMR across T^* [75]. Also in the large area of the P-B phase diagram, where both SDW and superconductivity are suppressed by pressure and magnetic field (the so-called metallic state), surprising ADMR has been observed

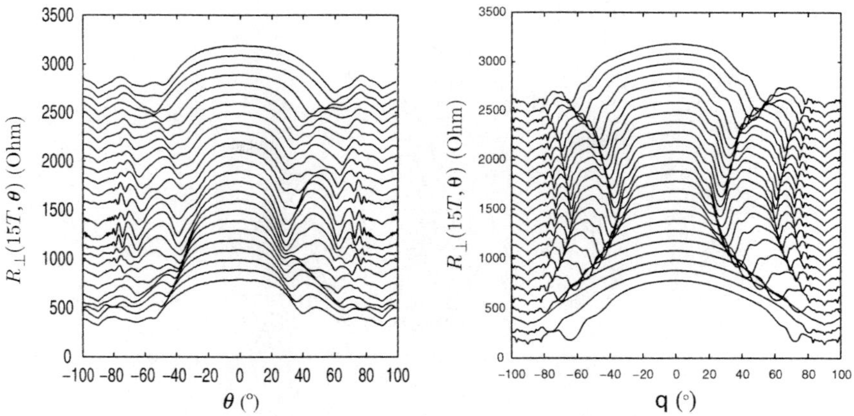

Fig. 19.6. ADMR is shown for current perpendicular to the a-c plane at $T = 1.4\,\mathrm{K}$ and $B = 15\,\mathrm{T}$ for $\phi = -77°, -70°, -62.5°, -55°, -47°, -39°, -30.5°, -22°, -14°, -6°, 2°, 10°, 23°, 33°, 41°, 48.5°, 56°, 61°, 64°, 67°, 73°, 80°, 88.5°, 92°$, and $96°$ from bottom to top. The left (right) panel shows experimental (theoretical) curves, which are shifted from their original position along the vertical axis by $n \times 100$ Ohm; $n = 0$ for $\phi = -77°$, $n = 1$ for $\phi = -70°, \ldots$

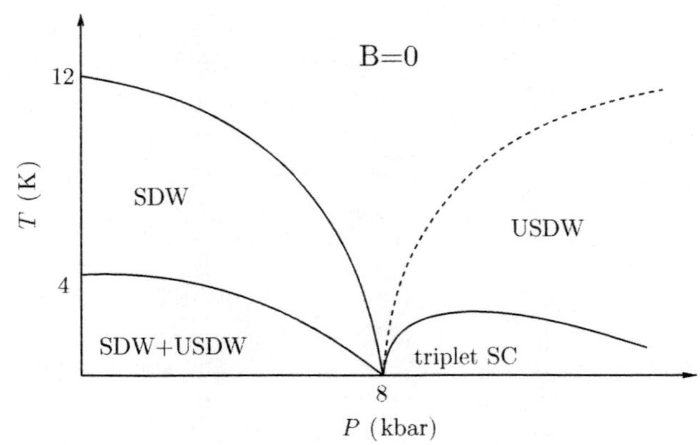

Fig. 19.7. The schematic P-T phase diagram of Bechgaard salts

about a decade ago [76]. These are of three kind: first, when the magnetic field is rotated within the c^*-b' plane perpendicular to the a axis, R_{xx} and R_{zz} exhibits broad peaks with a number of dips [77–80]. These dips were interpreted in terms of Lebed resonance [81, 82]. More recently, very similar ADMR has been seen in the ReO_4 and ClO_4 compounds as well, though the one in the latter is more complex [54]. Second, when the magnetic field is rotated within the a-c^* plane, R_{xx} show very different ADMR, sometimes called "Danner resonance" [83]. Finally, the third angular dependence appears as a kink when the magnetic field is rotated within the a-b' plane.

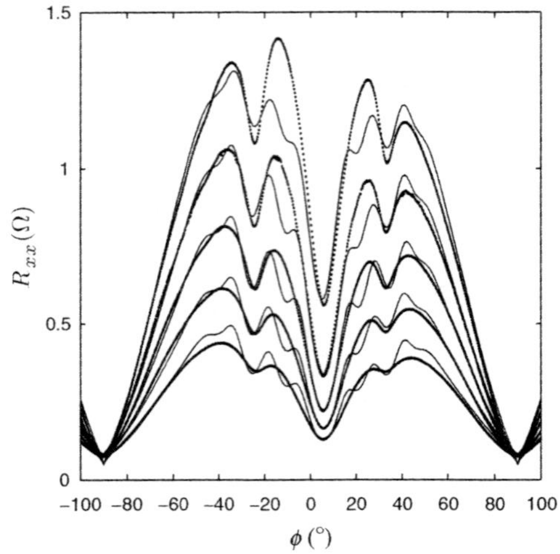

Fig. 19.8. The angular dependent magnetoresistance (R_{xx}) of $(TMTSF)_2PF_6$ at $T = 1.55\,K$ is shown for magnetic fields from 8–4 T from top to bottom. The dots denote the experimental data from [54], the solid line is our fit based on (19.19)

Here we shall concentrate on the case when the magnetic field is rotated within the c^*-b' plane. We propose that the metallic phase should be USDW [55]. When **B** in the c^*-b' plane is tilted by ϕ from the c^* axis, the obtained energy spectrum turns out to be very similar to the one discussed in (19.17). In particular, we get $E_1 = \sqrt{2nv_a\Delta e|B\cos(\phi)|}$, $y_m = v_a d'^2 e|B\cos(\phi)|(\tan(\phi) - \tan(\phi_m))^2/\Delta b$ and $\tan(\phi_m) = \frac{pb\sin(\gamma)}{qc\sin(\beta)\sin(\alpha^*)} - \cot(\alpha^*)$, where b, c, β, and γ are lattice parameters in real space, α^* is a lattice parameter in reciprocal space, p and q are small integers [55]. Then the experimental data [54] of R_{xx} on $(TMTSF)_2PF_6$ and on $(TMTSF)_2ReO_4$ at $T = 1.55\,K$ are shown in Figs. 19.8 and 19.9, respectively. Except for the fact that the theoretical curves exhibit more structures than the data, we think the fitting is excellent. From these, we extract $\Delta = 20\,K$ and $45\,K$ and $v_a = 10^7\,cm\,s^{-1}$ and $3\times10^7\,cm\,s^{-1}$ for the PF_6 and ReO_4 compound, respectively. The Fermi velocities deduced here are also very consistent with known values. Also from the above Δ's, we expect that USDW persists to $T = 9\,K$ and $20\,K$ for the respective compounds. Also the giant Nernst effect in these systems can be crucial to strengthen the case of USDW in Bechgaard salts [84].

19.4.3 κ-$(ET)_2$ Salts, $CeCoIn_5$, and YPrCO

There are three κ-$(ET)_2$ salts with very similar properties: κ-$(ET)_2Cu(NCS)_2$, κ-$(ET)_2Cu[NCN]_2Br$, and κ-$(ET)_2Cu[NCN]_2Cl$ with relatively high superconducting transition temperature $T_c \gtrsim 10\,K$. So far no evidence for unconventional density wave in these systems has been reported, although we have many reasons to believe that the pseudogap phase in high quality crystals

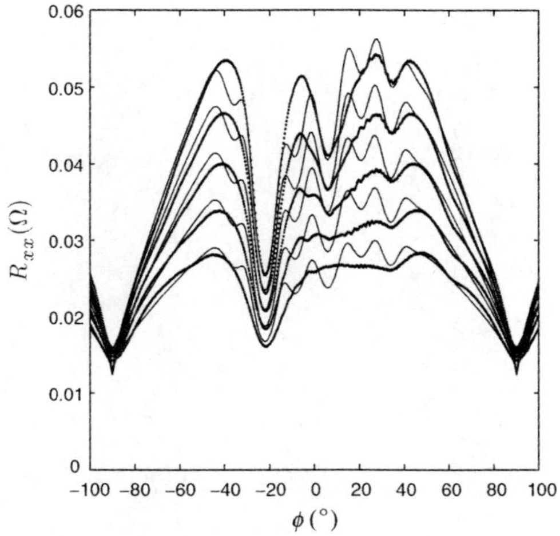

Fig. 19.9. The angular dependent magnetoresistance (R_{xx}) of $(TMTSF)_2ReO_4$ at $T = 1.55\,K$ is shown for magnetic fields from 8–4 T from top to bottom. The dots denote the experimental data from [54], the solid line is our fit based on (19.19)

should be d-DW. First of all, there are many parallels between these organic superconductors and high T_c cuprate superconductors and the heavy fermion compound $CeCoIn_5$:

quasi-two-dimensionality (or layered structure),
the proximity of antiferromagnetic order,
d-wave superconductivity.

Of course d-wave superconductivity is established only for κ-$(ET)_2Cu(NCS)_2$ [85], but this suggests strongly that the other two κ-$(ET)_2$ salt superconductors should be d-wave as well. Second, the giant Nernst signal and ADMR observed in the pseudogap phase of high T_c cuprates [40–43] and in $CeCoIn_5$ [39, 44] are successfully interpreted in terms of d-DW [37]. These suggest strongly the presence of d-DW in κ-$(ET)_2$ salts.

Here we are going to show the fitting of ADMR in the pseudogap phase of $CeCoIn_5$ [39] and in high T_c cuprate $Y_{0.68}Pr_{0.32}Ba_2Cu_3O_7$ [37, 43] in Figs. 19.10 and 19.11, respectively.

From these fittings, we can extract $\Delta = 45\,K$ and $360\,K$, μ (the chemical potential) $= 8.4\,K$ and 40–$60\,K$, v (the planar Fermi velocity) $= 3.3 \times 10^6$ $cm\,s^{-1}$ and $2.3 \times 10^7\,cm\,s^{-1}$ for $CeCoIn_5$ and YPrCO, respectively. These values are very consistent with known parameters of these systems.

19.5 Giant Nernst Effect

Since 2001, the giant Nernst effect has been established as the hallmark of the pseudogap phase in high T_c cuprates [40–42]. As shown in [22, 36], the giant negative Nernst effect follows directly from the Landau quantization of

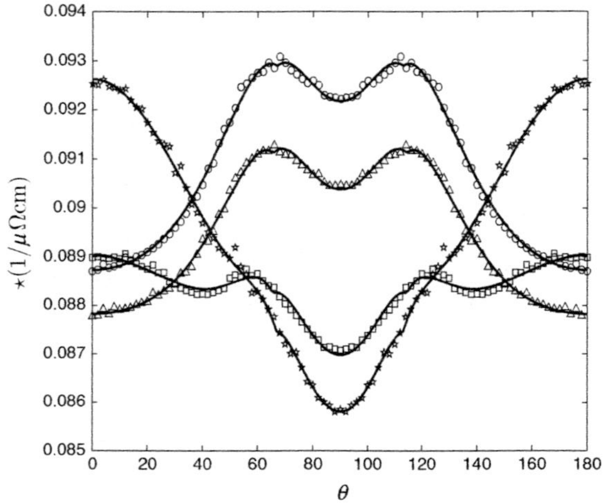

Fig. 19.10. The angle dependent conductivity of $CeCoIn_5$ is shown for $T = 6\,K$ and for $B = 4\,T$ (*circle*), $5\,T$ (*triangle*), $8\,T$ (*square*), $10\,T$ (*star*)

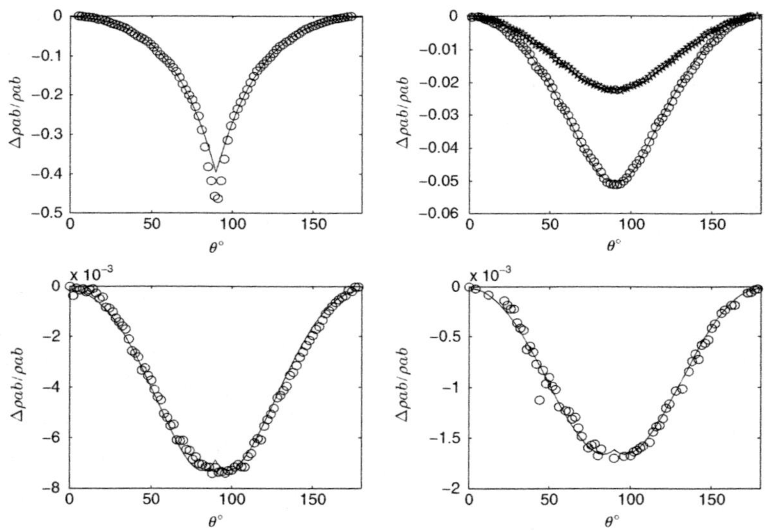

Fig. 19.11. The relative change of the in-plane magnetoresistance of $Y_{0.68}Pr_{0.32}Ba_2Cu_3O_7$ [43] is plotted as a function of angle θ at $B = 14\,T$ for $T = 52\,K$ (*top left*), $60\,K$ and $65\,K$ (*top right*), $75\,K$ (*bottom left*) and $105\,K$ (*bottom right*). The solid line is fit based on (19.19)

the quasiparticle energy spectrum in UDW. When an electric field is applied in the conducting plane in addition to the perpendicular magnetic field, all quasiparticles drift with a drift velocity $\mathbf{v}_D = \mathbf{E} \times \mathbf{B}/B^2$. This gives rise to a heat current $\mathbf{J}_{\mathrm{h}} = TS\mathbf{v}_{\mathrm{D}}$, where S is the quasiparticle entropy given by

$$S = eB\sum_n \left\{\ln\left(1+\mathrm{e}^{-\beta E_n}\right) + \beta E_n\left(1+\mathrm{e}^{\beta E_n}\right)^{-1}\right\}, \tag{19.23}$$

where the sum has to be carried out over all the Landau levels. For d-DW as in high T_{c} cuprates and for $\mathbf{B} \parallel \mathbf{c}$ and $\beta E_1 \gg 1$, the entropy is well approximated as

$$S = 8eB\left[\ln\left(1+\mathrm{e}^{-\zeta_0}\right) + \zeta_0\left(1+\mathrm{e}^{\zeta_0}\right)^{-1} + \ln\left(2\left(\cosh(x_1)+\cosh(\zeta_0)\right)\right) - \right.$$
$$\left. - \frac{\zeta_0\sinh(\zeta_0) + x_1\sinh(x_1)}{\cosh(x_1)+\cosh(\zeta_0)}\right], \tag{19.24}$$

where $\zeta_0 = \beta\mu$, $x_n = \beta\sqrt{2neBv_2v}$ with $v_2/v = \Delta/E_{\mathrm{F}}$. Then the Nernst coefficient is given by

$$S_{xy} = -\frac{S}{B\sigma} = \frac{1}{\sigma}\left\{\frac{L_{2D}}{1+(B/B_0)^2} - 8e\left(\frac{1}{2}\left[\ln(2(1+\cosh(\zeta_0)) - \frac{\zeta_0\sinh(\zeta_0)}{1+\cosh(\zeta_0)}\right]\right.\right.$$
$$\left.\left. + \ln(2(\cosh(x_1)+\cosh(\zeta_0))) - \frac{\zeta_0\sinh(\zeta_0)+x_1\sinh(x_1)}{\cosh(x_1)+\cosh(\zeta_0)}\right)\right\}. \tag{19.25}$$

Here σ has been already defined in (19.18), and $L_{2\mathrm{D}}$ and B_0 comes from quasiparticles not in d-DW state. Here we present our fitting to the data taken on α-(BEDT-TTF)$_2$KHg(SCN)$_4$ [72] in Fig. 19.12.

We have also similar fittings for high T_{c} cuprates [36] and CeCoIn$_5$ [38]. The third hallmark of UDW is the nonlinear Hall coefficient. In UDW in an applied magnetic field, the number of quasiparticles decreases exponentially with decreasing temperature. Therefore, the Hall conductivity is given by [47]

$$\sigma_{xy} = -\frac{2e^2\cos^2(\Theta)}{\pi}n(B,T) \tag{19.26}$$

with

$$n(B,T) = \tanh(\zeta_0/2) + \sinh(\zeta_0)\left[\frac{1}{\cosh(x_1)+\cosh(\zeta_0)}\right.$$
$$\left. + \frac{1}{\cosh(x_2)+\cosh(\zeta_0)}\cdots\right], \tag{19.27}$$

where Θ is the angle the magnetic field makes from the c axis. We have no possibility to compare the theoretical expression (19.26) with available experimental data. But clearly the Hall coefficient will indicate the reduction of the number of quasiparticles due to the opening of the energy gap.

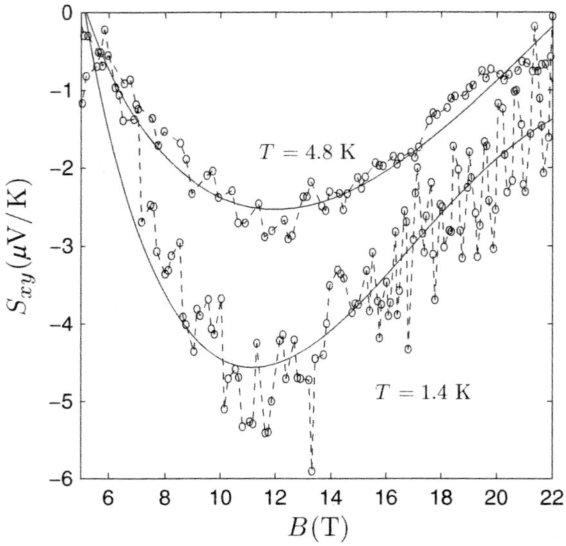

Fig. 19.12. The Nernst signal for heat current along the a direction is shown for $T = 1.4$ K and $T = 4.8$ K (*from bottom to top*), the dashed lines with circles denote the experimental data from [72], the solid line is our fit based on (19.25)

19.6 Concluding Remarks

In the last few years we have experienced a major paradigm shift from conventional condensates to unconventional condensates. These are still the mean-field ground states and described in terms of the generalized BCS theory. For superconductivity, at least d-wave superconductivity in κ-$(ET)_2$ salts and triplet superconductivity in $(TMTSF)_2PF_6$ (most likely chiral f-wave) have been established [17, 86]. As to density waves, more and more previously unidentified condensates appear to belong to UDW. But this game has just started. Also there appears to be still unexplored areas where two of the mean-field order parameters can coexist. Of course, the classic example is $NbSe_3$ where the CDWs, CDW_1 and CDW_2, coexists. A more exotic case will be in Bechgaard salts $(TMTSF)_2PF_6$ where SDW and USDW appears to coexist below $T^* \sim T_c/3$ [75]. However, Gossamer superconductivity appears to be more widespread, where a nodal superconductor coexists with a nodal density wave [47,87]. Actually this is our interpretation of the concept of Gossamer superconductivity introduced by Bob Laughlin. As discussed in [47], the case of Gossamer superconductivity in high T_c cuprate superconductors $CeCoIn_5$ and very likely κ-$(ET)_2$ salts is clear [88].

In other words all the electronic ground states in crystalline solids belong to one of the three possibilities:

(a) Unconventional superconductivity
(b) Unconventional density wave
(c) Coexistence of two of them

These ground states have been expected from the renormalization group analysis of 2D and 3D fermion systems. In these circumstances, the identification of the character and the symmetry of the order parameter is the first step. Now with ADMR, the giant Nernst effect, and the nonlinear Hall coefficient, the exploration becomes much easier.

Acknowledgments

We have benefitted from collaborations and discussions with Carmen Almasan, Philipp Gegenwart, Tao Hu, Bojana Korin-Hamzić, Peter Thalmeier, and Silvia Tomić. B. Dóra acknowledges the hospitality and support of the Max-Planck Institute for Chemical Physics of Solids, Dresden, where part of this work was done. B. Dóra was supported by the Magyary Zoltán postdoctoral program of Magyary Zoltán Foundation for Higher Education (MZFK). This work was supported by the Hungarian Scientific Research Fund under grant numbers OTKA K72613, TS049881, and T046269.

References

1. J. Sólyom, Adv. Phys. **28**, 201 (1979)
2. D. Jerome, H. Schultz, Adv. Phys. **31**, 299 (1982)
3. G. Grüner, *Density Waves in Solids* (Addison-Wesley, Reading, 1994)
4. T. Ishiguro, K. Yamaji, G. Saito, *Organic Superconductors* (Springer, Berlin, 1998)
5. L.D. Landau, Sov. Phys. JETP **3**, 920 (1957)
6. L.D. Landau, Soviet Phys. JETP **5**, 101 (1957)
7. L.D. Landau: Sov. Phys. JETP **8**, 104 (1958)
8. J. Bardeen, L.N. Cooper, J.R. Schrieffer, Phys. Rev. **108**, 1175 (1957)
9. F. Steglich, J. Aarts, C.D. Bredl, W. Lieke, D. Meschede, W. Franz, H. Schafer, Phys. Rev. Lett. **43**, 1892 (1979)
10. D. Jerome, A. Mazaud, M. Ribault, K. Bechgaard, J. Phys. Lett. **41**, L95 (1980)
11. J.G. Bednorz, K.A. Müller, Z. Phys. B **64**, 189 (1986)
12. A.P. Mackenzie, Y. Maeno, Rev. Mod. Phys. **75**, 657 (2003)
13. M. Sigrist, K. Ueda, Rev. Mod. Phys. **63**, 239 (1991)
14. C.C. Tsuei, J.R. Kirtley, Rev. Mod. Phys. **72**, 969 (2000)
15. A. Damascelli, Z. Hussain, Z.-X. Shen, Rev. Mod. Phys. **75**, 473 (2003)
16. H. Won, S. Haas, D. Parker, K. Maki, Phys. Stat. Sol. B **242**, 363 (2005)
17. H. Won, S. Haas, D. Parker, S. Telang, A. Ványolos, K. Maki, in *Lectures in the Physics of Highly Correlated Electron Systems IX*, ed. by A. Avella, F. Mancini (AIP Conference Proceedings 789, Melville, 2005), p. 3
18. E. Cappelluti, R. Zeyher, Phys. Rev. B **59**, 6475 (1999)
19. L. Benfatto, S. Caprara, C. Di Castro, Eur. Phys. J. B **17**, 95 (2000)
20. B. Dóra, A. Virosztek, Eur. Phys. J. B **22**, 167 (2001)
21. S. Chakravarty, R.B. Laughlin, D.K. Morr, C. Nayak, Phys. Rev. B **63**, 094503 (2001)
22. B. Dóra, K. Maki, A. Virosztek, Mod. Phys. Lett. B **18**, 327 (2004)
23. G. Mihály, I. Kézsmárki, F. Zamborszky, L. Forro, Phys. Rev. Lett. **84**, 2670 (2000)

24. A.T. Zheleznyak, V.M. Yakovenko, Eur. Phys. J. B **11**, 385 (1999)
25. T. Giamarchi, Chem. Rev. **104**, 5037 (2004)
26. B. Korin-Hamzic, M. Basletic, K. Maki, Europhys. Lett. **59**, 298 (2002)
27. R.D. McDonald, N. Harrison, J. Singleton, A. Bangura, P.A. Goddard, A.P. Ramirez, X. Chi, Phys. Rev. Lett. **94**, 106404 (2005)
28. T. Vuletic, C. Pasquier, P. Aubin-Senzier, S. Tomic, D. Jerome, K. Maki, K. Bechgaard, Eur. Phys. J. B **21**, 53 (2005)
29. V.M. Yakovenko, H.-S. Goan, J. Phys. I **6**, 1917 (1996)
30. B.I. Halperin, T.M. Rice, in *Solid State Physics*, vol. 21, ed. by F. Seitz, D. Turnbull, H. Ehrenreich (Academic Press, New York, 1968), p. 115
31. I. Affleck, J.B. Marston, Phys. Rev. B **37**, 3774 (1988)
32. P. Thalmeier, Z. Phys. B **95**, 39 (1994)
33. M. Basletić, B. Korin-Hamzić, M.V. Kartsovnik, H. Müller, Synth. Met. **120**, 1021 (2001)
34. B. Dóra, A. Virosztek, K. Maki, Phys. Rev. B **64**, 041101(R) (2001)
35. H. Won, K. Maki, Phys. Rev. B **49**, 1397 (1994)
36. K. Maki, B. Dóra, A. Virosztek, A. Ványolos, Curr. Appl. Phys. **4**, 693 (2004)
37. B. Dóra, K. Maki, A. Virosztek, Europhys. Lett. **72**, 624 (2005)
38. B. Dóra, K. Maki, A. Virosztek, A. Ványolos, Phys. Rev. B **71**, 172502 (2005)
39. T. Hu, H. Xiao, V. Sandu, C.C. Almasan, K. Maki, B. Dóra, T.A. Sayles, M.B. Maples, Phys. Rev. B **73**, 134509 (2006)
40. Y. Wang, Z.A. Xu, T. Kakeshita, S. Uchida, S. Ono, Y. Ando, N.P. Ong, Phys. Rev. B **64**, 224519 (2001)
41. C. Capan, K. Behnia, J. Hinderer, A.G.M. Jensen, W. Lang, C. Marcenat, C. Marin, J. Flouquet, Phys. Rev. Lett. **88**, 056601 (2002)
42. Y. Wang, N.P. Ong, Z.A. Xu, T. Kakeshita, S. Uchida, D.A. Bonn, D. Liang, W.N. Hardy, Phys. Rev. Lett. **88**, 257003 (2002)
43. V. Sandu, E. Cimpoiasu, T. Katuwal, S. Li, M.B. Maple, C.C. Almasan, Phys. Rev. Lett. **93**, 177005 (2004)
44. R. Bel, K. Behnia, Y. Nakajima, K. Izawa, Y. Matsuda, H. Shishido, R. Settai, Y. Onuki, Phys. Rev. Lett. **92**, 217002 (2004)
45. H. Ding, T. Yokoya, J.C. Campusano, T. Takahashi, M. Randeira, M. Norman, T. Mochiku, K. Kadowaki, J. Giapintzakis, Nature **302**, 51 (1996)
46. J.C. Campusano, H. Ding, M.R. Norman, M. Randeira, Phys. B **259–261**, 517 (1999)
47. H. Won, S. Haas, K. Maki, D. Parker, B. Dóra, A. Virosztek, Phys. Stat. Sol. B **243**, 37 (2006), K. Maki, B. Dora, A. Vanyolos, A. Virosztek, Physica C **460–462**, 286 (2007), H. Won, Y. Morita, K. Maki, Phys. Stat. Sol B **244**, 4377 (2007)
48. T. Nakano, M. Momono, M. Oda, M. Ido, J. Phys. Soc. Jpn. **67**, 2622 (1998)
49. M. Oda, R. Kubota, K. Hoya, C. Manabe, M. Momono, T. Nakano, N. Ido, Phys. C 135 (1999)
50. M. Kugler, O. Fischer, C. Renner, S. Ono, Y. Ando, Phys. Rev. Lett. **86**, 4911 (2001)
51. M. Sutherland, D.G. Hawthorn, R.W. Hill, F. Ronning, S. Wakimoto, H. Zhang, C. Proust, E. Boaknin, C. Lupien, L. Taillefer, R. Liang, D.A. Bonn et al., Phys. Rev. B **67**, 174520 (2003)
52. B. Dóra, K. Maki, B. Korin-Hamzić, M. Basletić, A. Virosztek, M.V. Kartsovnik, H. Müller, Europhys. Lett. **60**, 737 (2002)
53. K. Maki, B. Dóra, M.V. Kartsovnik, A. Virosztek, B. Korin-Hamzić, M. Basletić, Phys. Rev. Lett. **90**, 256402 (2003)

54. W. Kang, H.-Y. Kang, Y.J. Jo, S. Uji, Synth. Met. **133–134**, 15 (2003)
55. B. Dóra, K. Maki, A. Ványolos, A. Virosztek, Europhys. Lett. **67**, 1024 (2004)
56. Y. Nambu, Phys. Rev. **117**, 648 (1960)
57. B. Dóra, K. Maki, A. Virosztek, Phys. Rev. B **66**, 165116 (2002)
58. B. Dóra, A. Ványolos, K. Maki, A. Virosztek, Phys. Rev. B **71**, 245101 (2005)
59. A.A. Nersesyan, G.E. Vachnadze, J. Low Temp. Phys. **77**, 293 (1989)
60. A.A. Nersesyan, G.I. Japaridze, I.G. Kimeridze, J. Phys. Condens. Matter **3**, 3353 (1991)
61. W. Heisenberg, H. Euler, Z. Phys. **98**, 714 (1936)
62. V.S. Weisskopf, Kongelige Danske Videns. Selskab Mathematisk-fsiske Meddelelser **6**, 14 (1936)
63. J. Singleton, Rep. Prog. Phys. **63**, 1161 (2000)
64. P. Christ, W. Biberacher, M.V. Kartsovnik, E. Steep, E. Balthes, H. Weiss, H. Müller, JETP Lett. **71**, 303 (2000)
65. D. Andres, M.V. Kartsovnik, W. Biberacher, H. Weiss, E. Balthes, H. Müller, N. Kushch, Phys. Rev. B **64**, 161104(R) (2001)
66. S. Tomic, J.R. Cooper, W. Kang, D. Jerome, K. Maki, J. Phys. I. (Paris) **1**, 163 (1991)
67. B. Dóra, A. Virosztek, K. Maki, Phys. Rev. B **65**, 155119 (2002)
68. J. Caulfield, J. Singleton, P.T.J. Hendriks, J.A.A.J. Perenboom, F.L. Pratt, M. Doporto, W. Hayes, M. Kurmoo, P. Day, J. Phys. Condens. Matter **6**, L155 (1994)
69. A.E. Kovalev, M.V. Kartsovnik, R.P. Shibaeva, L.P. Rozenberg, I.F. Schegolev, N.D. Kushch, Solid State Commun. **89**, 575 (1994)
70. T. Sasaki, N. Toyota, Phys. Rev. B **49**, 10120 (1994)
71. N. Hanasaki, S. Kagoshima, N. Miura, G. Saito, J. Phys. Soc. Jpn. **65**, 1010 (1996)
72. E.S. Choi, J.S. Brooks, J.S. Qualls, Phys. Rev. B **65**, 205119 (2002)
73. B. Dóra, K. Maki, A. Vańyolos, A. Virosztek, Phys. Rev. B **68**, 241102(R) (2003), B. Dóra, K. Maki, A. Ványolos, A. Virosztek, Europhys. Lett. **67**, 3024 (2004), M. Basletic, B. Korin-Hamzic, K. Maki, S. Tomic, Phys. Rev. B **75**, 052409 (2007)
74. I.J. Lee, S.E. Brown, W.G. Clark, M.J. Strouse, M.J. Naughton, W. Kang, P.M. Chaikin, Phys. Rev. Lett. **88**, 017004 (2002)
75. M. Basletić, B. Korin-Hamzić, K. Maki, Phys. Rev. B **65**, 235117 (2002)
76. E.I. Chashechkina, P.M. Chaikin, Phys. Rev. B **65**, 012405 (2002)
77. T. Osada, A. Kawasumi, S. Kagoshima, N. Miura, G. Saito, Phys. Rev. Lett. **66**, 1525 (1991)
78. M.J. Naughton, O.H. Chung, M. Chaparala, X. Bu, P. Coppens, Phys. Rev. Lett. **67**, 3712 (1991)
79. W. Kang, S.T. Hannahs, P.M. Chaikin, Phys. Rev. Lett. **69**, 2827 (1992)
80. I.J. Lee, M.J. Naughton, Phys. Rev. B **58**, R13343 (1998)
81. A.G. Lebed, JETP Lett. **43**, 174 (1986)
82. A.G. Lebed, P. Bak, Phys. Rev. Lett. **63**, 1315 (1989)
83. G.M. Danner, W. Kang, P.M. Chaikin, Phys. Rev. Lett. **72**, 3714 (1994)
84. W. Wu, I.J. Lee, P.M. Chaikin, Phys. Rev. Lett. **91**, 056601 (2003)
85. K. Izawa, H. Yamaguchi, T. Sasaki, Y. Matsuda, Phys. Rev. Lett. **88**, 027002 (2002)
86. A.D. Folgueras, K. Maki, Physica C **480–482**, 643 (2007)
87. R.B. Laughlin, cond-mat/0209269
88. M. Pinteric, S. Tomic, K. Maki, Phys. C **408–410**, 75 (2004)

20

Charge Density Waves in Strong Magnetic Fields

A. Bjeliš and D. Zanchi

In this chapter we review theoretical results and experimental evidences related to the effects of magnetic field H on the charge density wave (CDW) ordering. The theoretical considerations are mostly based on authors' works, while the review of experimental situation comprise recent measurements on α-ET and perylene compounds, and on blues bronzes. The general background for the theoretical part can be found in a chapter of D. Jérome and C. Bourbonnais and in chapters of A.G. Lebed and M. Heritier, whereas the experimental aspects are covered by M.V. Kartsovnik, N. Harrison et al., J. Singleton et al., and S. Uji, and J.S. Brooks.

20.1 Introduction

After more than 30 years of investigations on numerous quasi-one-dimensional electronic systems with a sufficient degree of nesting between two open Fermi surfaces (schematically shown in Fig. 20.1) it could be stated that the phenomena of charge (C) and spin (S) density waves (DWs) are well understood in many aspects. Particularly interesting is behavior of DWs under external magnetic field, intensely studied in the recent period together with the development of new techniques enabling ultrahigh fields of few tenths Teslas. The general rule, based on the Pauli band splitting shown in Fig. 20.1, is that magnetic field **H** destabilizes CDW order by, e.g., decreasing its critical temperature, and favors SDWs, either by increasing the critical temperature [1] or by inducing the SDW order through the Gorkov–Lebed [2] one dimensionalization of imperfectly nested Fermi surface. This rule can be easily understood after introducing S(C)DW operators

$$M_i = \frac{1}{2}\Psi^\dagger \rho_+ \sigma_i \Psi \qquad\qquad i = 1, 2, 3, 4, \tag{20.1}$$

where the four-component fermion field $\Psi^\dagger = \left(\Psi^\dagger_{\uparrow+}, \Psi^\dagger_{\uparrow-}, \Psi^\dagger_{\downarrow+}, \Psi^\dagger_{\downarrow-}\right)$ denotes four Fermi surfaces from Fig. 20.1, ρ's and σ_i are corresponding Pauli matrices

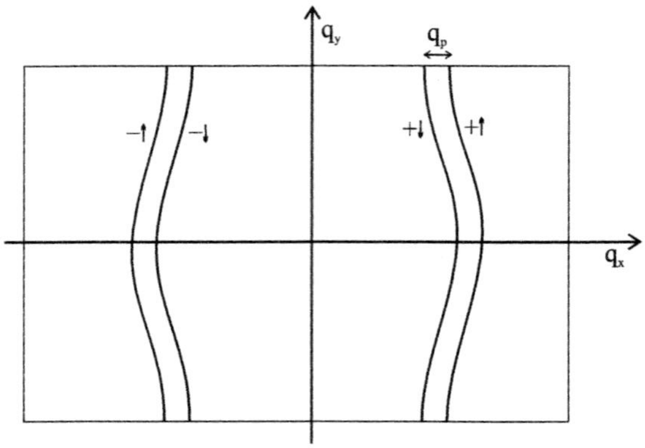

Fig. 20.1. Pauli splitting of left and right corrugated Fermi surfaces

in left–right and spin space with the usual meaning of indices (spin quantization axis being in the magnetic field direction), and $\sigma_4 \equiv I$. The first three components ($i = 1, 2, 3$) in (20.1) define the complex SDW vector amplitude $\mathbf{M}$, while the fourth component M_4 is the complex CDW scalar amplitude. Obviously Pauli splitting does not affect SDW components $M_{1,2}$ since, as is seen in Fig. 20.1, the original longitudinal nesting wave number $2k_\mathrm{F}$ is still preserved for all participating left–right pairs with opposite spin orientations of electrons and holes. Thus, by choosing its orientation perpendicularly to $\mathbf{B}$, SDW avoids the degradation of nesting introduced by Pauli splitting, and in addition strengthens its order through the orbital coupling to the magnetic field. The spin up and spin down DW components, given by $(M_3 \pm M_4)/2$, have respective parallel nesting wave numbers $2k_\mathrm{F} \pm 2q_\mathrm{P}$ where $q_\mathrm{P} \equiv \mu_\mathrm{B}H/v_\mathrm{F}$ denotes Pauli splitting of corresponding Fermi surfaces. This effective degradation of $2k_\mathrm{F}$ coherence is the reason for the weakening of any DW order composed by M_3 and M_4. In particular it leads to the decrease of the CDW critical temperature [3] and the hybridization of the corresponding component M_4 with the SDW component M_3 induced by the Pauli splitting [4].

From the other side, the orbital coupling to the magnetic field still favors the stabilization of these DW components as well, and may even compensate or overwhelm the effects of Pauli splitting. The ratio of corresponding energetic scales for two mechanisms is $\eta \equiv \omega_\mathrm{o}/\omega_\mathrm{P}$, where $\omega_\mathrm{o} = v_F ebH \cos\theta$ is the cyclotron energy and $\omega_\mathrm{P} = \mu_\mathrm{B}H$ is the Pauli splitting energy, with v_F, b, θ being, respectively, Fermi velocity, transverse lattice constant and the inclination of the magnetic field from the second, least conductive, transverse c-direction in the (b, c) plane. In real materials the parameter η may attain, particularly by changing θ, a wide range of values, including the intermediate values of the order of unity. Due to the opposing competition of Pauli and orbital mechanisms, CDWs are thus expected to show richer phase diagrams in magnetic field than SDWs.

This richness was indeed confirmed in measurements performed in the last decade on few CDW materials with critical temperatures of the order of above magnetic energies. In this chapter we review these data, and show that they can be to a great extent interpreted, at least qualitatively, within the random phase approximation (RPA) treatment of the standard theory for DWs in chain compounds [5,6]. We also discuss recent theoretical progress along the same lines.

20.2 Theoretical Background

Let us start from the Hamiltonian for the standard model written in the form convenient for the further analysis

$$\begin{aligned}\mathbf{H}_{\mathrm{o}} = \frac{b}{2\pi}\frac{1}{2}\int \mathrm{d}q_y \int \mathrm{d}x\, \Psi^{\dagger}(x,\, q_y)\big[\mathrm{i}v_{\mathrm{F}}\rho_3\partial_x + 2t_b\rho_3 \sin(q_y b - q_{\mathrm{o}}x) \\ + 2t_b' \cos 2(q_y b - q_{\mathrm{o}}x) - \sigma_3\mu_{\mathrm{B}}H\big]\Psi(x,\, q_y) \\ + \frac{1}{2}\int \mathrm{d}x \sum_{\mathbf{R}_\perp} [-U_s\mathbf{M}^{\dagger}(\mathbf{R})\cdot\mathbf{M}(\mathbf{R}) + U_{\mathrm{c}}M_4^{\dagger}(\mathbf{R})M_4(\mathbf{R})].\end{aligned} \tag{20.2}$$

The transverse band dispersion in the y-direction is parameterized by the direct interchain hopping t_b, while the effective [7] next neighboring hopping t_b' measures the deviation from the perfect nesting. The SDW and CDW coupling constants in (20.2) are related to the usual backward (g_1) and forward (g_2) electron–electron coupling constants by $U_s \equiv 2\pi v_{\mathrm{F}} g_2$ and $U_{\mathrm{c}} \equiv 2\pi v_{\mathrm{F}}(2g_1 - g_2)$. It is useful for further considerations to introduce the ratio of two couplings, $\nu \equiv U_s/(-U_{\mathrm{c}})$. For example, in the simple Hubbard model $U_s = U_{\mathrm{c}} = U$ (i.e. $\nu = -1$), and SDW and CDW orders are favored for the repulsive ($U > 0$) and attractive ($U < 0$) on-site interaction, respectively [8]. We shall concentrate further discussion mostly to the regime $U_{\mathrm{c}} < 0$ and $\nu \geq 0$, relevant for the usual CDW systems with predominant backward electron–phonon interaction.

As was already pointed out, the transverse DW ordering is within RPA fully decoupled from the longitudinal one. The critical temperature for the former, defined by the components $M_{1,2}$, follows from the Stoner criterion $1 - U_s\chi_{\mathrm{o}}(\mathbf{q}) = 0$, where

$$\chi_{\mathrm{o}}(\mathbf{q}) = \sum_{l=-\infty}^{\infty} P(q_x - lq_{\mathrm{o}})I_l^2(q_y), \tag{20.3}$$

is the RPA static susceptibility, $P(q_x - lq_{\mathrm{o}})$ being the corresponding one-dimensional bubble, while the sum of Bessel functions

$$I_l(q_y) = \sum_{l'} J_{l-2l'}\left(\frac{4t_b}{v_{\mathrm{F}}q_{\mathrm{o}}}\sin\frac{q_y b}{2}\right) J_{l'}\left(\frac{2t_b'}{v_{\mathrm{F}}q_{\mathrm{o}}}\cos q_y b\right) \tag{20.4}$$

introduces orbital effects due to the finite corrugation of the Fermi surface.

The Stoner criterion for the longitudinal DWs follows from the diagonalization of the 2×2 susceptibility matrix for $M_{3,4}$ components. The hybridized susceptibilities are given by

$$\chi_{\pm}^{-1} = \chi_g^{-1} \left[\sqrt{1+\delta^2} + \frac{1+\nu}{2} U_{\rm c}\chi_g \mp \sqrt{\left(\frac{1-\nu}{2} U_{\rm c}\chi_g\right)^2 + \delta^2} \right], \qquad (20.5)$$

with $\chi_g^2 \equiv \chi_{\rm o}(q_x + 2q_{\rm P}, q_y)\chi_{\rm o}(q_x - 2q_{\rm P}, q_y)$ and $\delta \equiv [\chi_{\rm o}(q_x + 2q_p, q_y) - \chi_{\rm o}(q_x - 2q_p, q_y)]/2\chi_g$. The critical temperature and the wave vector of the unstable branch follows from the condition $\chi_-^{-1}(\mathbf{q}, T_{\rm c}) = 0$, with the corresponding hybridized DW amplitude $M_- = (\Delta M_4 + \delta M_3)/\sqrt{\Delta^2 + \delta^2}$, where

$$\Delta \equiv -\frac{1-\nu}{2} U_{\rm c}\chi_g + \sqrt{\left(\frac{1-\nu}{2} U_{\rm c}\chi_g\right)^2 + \delta^2}. \qquad (20.6)$$

The above expressions are RPA results in the closed form. Although straightforward, further, mostly numerical, steps toward the specification of phase diagrams for the transition from the metallic phase to hybridized DW states lead to a variety of possibilities. The reason is that the CDW–SDW hybridization in a finite magnetic field essentially depends on three independent ingredients, originating from the geometry (η), nature of electron–electron interaction (ν), and the deviation of the transverse band dispersion from the perfect nesting (measured by $t_b'/t_b'^*$ where $t_b'^* \sim T_{\rm c}^{\rm o}(t_b' = 0)$ is the value of t_b' for which the CDW critical temperature at $H = 0$ disappears). In the further discussion we shall cover the most interesting regimes within the realm of experimental progress made in the last decade.

20.3 Discussion of Specific Regimes

20.3.1 Regime of Perfect Nesting

The phase diagram in the case of perfect nesting ($t_b \neq 0, t_b' = 0$) is shown in Fig. 20.2. For weak magnetic fields the DW order starts with a pure CDW state, denoted by CDW$_{\rm o}$, having the perfect nesting wave vector $\mathbf{Q}_{\rm o} = (2k_{\rm F}, \pi/b)$ (i.e., $q_x = q_y = 0$ in the above notation), but with the decreasing critical temperature [3] due to the Pauli band splitting from Fig. 20.1. As the reduced magnetic field $h \equiv \mu_{\rm B}H/(2\pi T)$ passes a critical value $h_{\rm c}$, equal to 0.304 for $\nu = 0$ and weakly depending on ν and η, this state is replaced either by CDW$_x$ or by CDW$_y$, with the corresponding continuous shifts of the wave vector in one of two longitudinal ($\pm q_x \neq 0, q_y = 0$) or transverse ($q_x = 0, \pm q_y \neq 0$) directions.

Since δ vanishes for $q_x = 0$, CDW$_y$ is a pure CDW state (and therefore independent on parameter ν), appearing due to an effective imperfect nesting

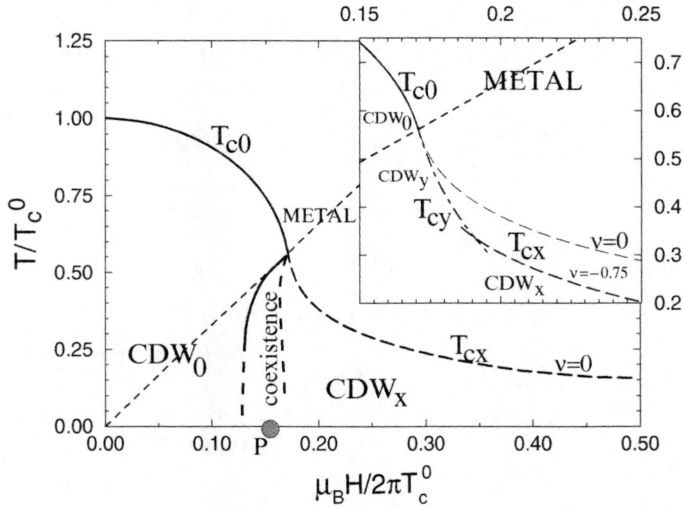

Fig. 20.2. The (H,T) diagram for $t_b' = 0$. The *straight dashed line* $h = \mu_B H/(2\pi T) = 0.304$ defines the critical value h_c. The inset is an enlarged regime with CDW_y. Reproduced from Zanchi et al. [6]

generated by the finite band splitting q_P in χ_g. Indeed, the corresponding Stoner criterion then reads $1 + U_c\chi_o(2q_P, q_y) = 0$, with the orbital quantization entering through (20.3) and (20.4) even for $t_b' = 0$. CDW_y is stable only for $\nu < 0$, and appear in the rather narrow range of values of h between CDW_o and CDW_x (as shown in the inset of Fig. 20.2). The CDW_x state which, as is shown in Fig. 20.2, covers much wider range of large ($h > h_c$) values of magnetic field, is a hybrid of original CDW and SDW components with a finite value of coefficient δ in (20.6). As far as $t_b' = 0$ neither transverse corrugation of the Fermi surface t_b nor the orbital parameter η influence CDW_x stabilization, i.e., it is the result of competing perfect nesting correlations of spin-up and spin-down Fermi sheets in Fig. 20.1.

While all transitions from the metallic phase to low temperature phases are of the second order, usual arguments suggest that the transitions between different CDW phases in the less analyzed low temperature range would have to be of the first order, with ranges of metastability which widen by decreasing temperature. For example, the transition line between the phases CDW_o and CDW_x, shown in Fig. 20.2, is expected to end at the magnetic field $\mu_B H \approx T_c^o$. More detailed mean-field calculation [9] of the upper part of that line suggests that ordered phases could rapidly acquire a nonsinusoidal modulations, with a considerable participation of higher harmonics of the fundamental q_x or q_y wave number at lower temperatures, resembling to soliton lattice modulations. Note in this respect that both CDW_x and CDW_y states have double degenerate values of wave numbers, which remain rather close in the reciprocal space in the whole temperature range down to $T = 0$. Such situations, including that of strictly one-dimensional CDWs with Pauli splitting [10] which is a simplified version of the present problem, are well-known genuine candidates for the formation of regular domain wall patterns with dilute lattices of domain walls (e.g., solitons) [11]. Obviously, even more complex patterns could be realized in the regime of CDW_x-CDW_y coexistence with four almost degenerate coupled CDW components at wave vectors $(\pm q_x, 0)$ and $(0, \pm q_y)$.

20.3.2 Regime of Imperfect Nesting

It was already emphasized that even in the limit of perfect nesting ($t_b' = 0$) the Pauli band splitting in Fig. 20.1 introduces an effective imperfect nesting which in turn activates the orbital mechanism in the unstable susceptibility branch (20.5), and, as was already stressed, affects the phase CDW_y, but still not the phases CDW_o and CDW_x. With finite values of t_b' standard band imperfect nesting additionally influences all CDW phases, and the phase diagram is further enriched. Since the full discussion of various possibilities allowed by variation of three parameters would be a too ambitious task here, we highlight in the further discussion only those aspects and types of behavior which are very probably related to recent experimental observations.

As is well known at $H = 0$ the CDW critical temperature decreases by increasing t_b' (usually by increasing pressure), and finally for $t_b' = t_b'^* \approx T_c^o$, where T_c^o is the critical temperature for the perfect nesting, CDW order is fully eliminated. By increasing magnetic field one gradually restitutes CDW through a series of field-induced phases with the well-defined discrete values of longitudinal wave number $2k_F + Nq_o$, N being an descending series of integers down to $N = 0$, the latter representing the standard CDW phase. However, in contrast to analogous SDW situations, the critical temperature for the ultimate $N = 0$ phase now does not tend toward a perfect nesting $H = 0$ asymptotic limit, but, being affected by the Pauli splitting, passes through a maximum value and then slows down, while the corresponding CDW changes further its longitudinal wave number q_x, showing also additional corresponding degeneracies like the phase CDW_x from Fig. 20.2. This behavior, as well as the evolution of the phase diagram as t_b' passes through the whole range of values, is shown in Fig. 20.3 for the case of pure Hubbard model with an attractive on-site interaction ($\nu = -1$), and in the regime of rather strong orbital effects ($\eta = 2.5$). The same type of behavior is realized for other values of ν, including those covering usual electron–phonon CDW systems with $0 \leq \nu \ll 1$. Note that this cascade of field induced CDWs, which we shall call FICDW-I, appears for low enough magnetic fields, the rough estimation being $\omega_0/8t_b' < 1$.

The next interesting regime covers the right-hand side of the phase diagram in Fig. 20.3. At relatively large values of the magnetic field the critical temperature is again small. Since the critical temperatures for FICDWs are generally very low, much lower than the characteristic temperatures for $H = 0$ and $t_b' = 0$ [12], both original and effective deviations from nesting of two pairs

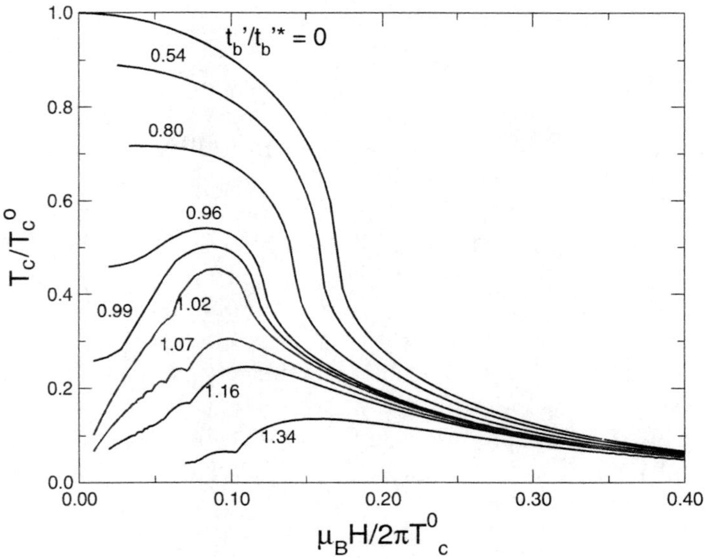

Fig. 20.3. Phase diagram with $T_c(H)$ transition lines for a series of values of $t_b'/t_b'^*$ and $\nu = -1$. Reproduced from Zanchi et al. [5]

of Fermi surfaces might again become significant. More precisely, at high magnetic fields the equal participation of spin-up and spin-down sub-bands in the CDW_o phase is replaced by the overwhelming ordering on one or another sub-band, with the corresponding longitudinal wave numbers $2k_F \pm q_x(H)$, with $q_x(H)$ tending to $2q_P$ as H increases. Two CDW_x phases are degenerate within the simplified model (20.2), but only one is realized within a given monodomain. Further ordering induced by orbital effects is then expected in the sub-band which is less involved in the CDW_x order, with lock-ins at a series of quantum levels as the wave number $q_x(H)$ varies with H. Following the qualitative arguments of Kartsovnik et al. [13], the phase diagram might contain a new cascade of FICDW sub-phases, called here FICDW-II, in which longitudinal wave numbers $2k_F \pm q_x(H) + Nq_o$ appear in the ascending order $N = 0, 1, 2, \ldots$ as magnetic field increases, just opposite to the series of FICDW-Is in the range of low magnetic fields shown in Fig. 20.3. The largest orbital quantum number in this series, N_{max}, is determined from the condition $N_{max} \leq q_x/q_o \leq 2\eta^{-1}$, i.e., such cascade is possible provided $\eta \geq 2$. Note that again this cascade may be realized even for $t_b' = 0$ when the orbital quantization occurs only due to the partial nesting of spin-up and spin-down sub-bands. Furthermore, in contrast to the FICDW-I cascade at low magnetic fields, the present one takes place in the regime of strong CDW–SDW mixing, so that the first-order transitions between neighboring sub-phases are also characterized by a quantization of hybridization coefficients, as follows from (20.6).

20.4 Experiments

First experiments on CDWs in magnetic field were done more than 30 years ago by Tiedje et al. [14] on the organic Q1D system TTF-TCNQ. They found that the critical temperature decreases quadratically with magnetic field. The fast progress starts in 1990s when two large magnets, those in Toulouse and Tallahassee, became available. Among first investigated materials were α-ET compounds [α-(BEDT-TTF)$_2$MHg(SCN)$_2$, M = K, Rb, ...] in which the existence of DW condensate was well documented, but the definite proof that one has the CDW order was the phase diagram in magnetic field. Since these compounds were most extensively experimentally studied, and their CDW properties in magnetic field are now well understood, we shall discuss them in some detail from the point of view of the present theoretical knowledge. We shall also give our interpretations of CDW phase diagrams for other two families of CDW systems with remarkable properties in magnetic field: perylene compounds and blue bronzes.

20.4.1 α-ET Compounds

α-ET compounds have a layer structure. The extremely high anisotropy of the conductivity $\sigma_{\|}/\sigma_{\perp} \simeq 10^5$ is a sign that these materials are essentially two-dimensional. The relevant fermiology [15–18] of α-ETs in the metallic phase comprises one open and strongly corrugated quasi-one-dimensional Fermi surface with Fermi energy of about 50 meV as measured from the bottom of the band, and another, two-dimensional cylindrical hole-like Fermi surface, the upper edge of the band being at about 150 meV. Both bands are less than half filled with a fractional number of electrons and holes, respectively. These compounds have the CDW as the ground state [19], with a wave vector corresponding to the nesting of two quasi-one-dimensional Fermi sheets [20].

Critical temperature for the CDW ordering at ambient pressure and zero magnetic field is 8.5 K for M = K and 12 K for M = Rb. The parameters relevant for the phase diagram in magnetic field, $\eta_o \equiv \eta(\theta = 0)$, ν, and $t_b'^*$, can be estimated from a number of experiments. The parameter ν is certainly larger than $-1/3$ because no superconductivity was observed at zero magnetic field. Its upper limit can be roughly estimated knowing that no reentrant metallic behavior exists in the magnetic filed. According to phase diagram obtained by Grigoriev and Lyubshin [9], and after taking $t_b' = 0.7t_b'^*$ from the comparison of theoretical and experimental curves, we estimate that the upper limit for ν is 0.3. The parameter η_o can be estimated following the arguments and experimental results of Kartsovnik et al. [13]. By comparing the quantized level spacing $G = ebB_z$ with the total Zeeman splitting, one gets $\eta_o \simeq 2.7$.

The following discussion of the behavior of CDW in the magnetic field is mostly based on recent experiments by Kartsovnik and co-workers [13,21,22].

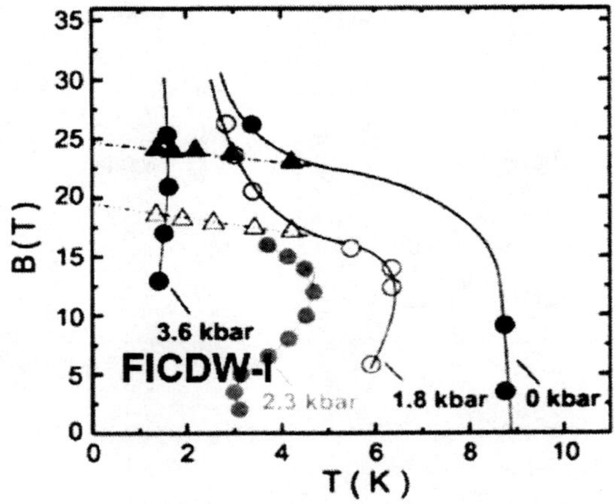

Fig. 20.4. Critical temperature for the CDW ordering in α-(BEDT-TTF)$_2$KHg(SCN)$_2$ for several values of pressure, as a function of the magnetic field oriented along the least conducting direction, $\theta = 0$. Reproduced from Andres et al. [21]

We start with the phase diagram in perpendicular magnetic field and at ambient pressure, shown in Fig. 20.4. As the field increases the critical temperature decreases, with an inflection point at $\mu_B H = 1.8T_{co}$, slightly higher than $1.1T_{co}$ predicted by Zanchi et al. [5]. The critical temperature at the inflection point is $T_{co}/2$, just as predicted in [5]. The magneto-resistance and torque experiments within the ordered state show clearly the first-order transition line that joins the critical temperature at the inflection point. It is the well-known kink transition. Comparison with the theory [5] indicates that this line is in fact the Pauli effect driven transition from CDW$_o$ to CDW$_x$ or CDW$_y$. The fact that the critical temperature never falls to zero for magnetic field up to 30 T, indicates that the parameter ν is not large, in accordance with the calculation in [9]. The phase diagram remains essentially the same for all angles θ up to about 60°: only one kink transition is observed. According to [5] the CDW state above the kink field is a CDW–SDW hybrid. Although the imperfect nesting parameter for this material is nonzero, we expect that this hybridization is well estimated by the expression [5]

$$M_3(\mathbf{q}) = \frac{\delta(\mathbf{q})}{\sqrt{1 + \delta^2(\mathbf{q}) + \nu U_c \chi_g(\mathbf{q})}} M_4(\mathbf{q}), \tag{20.7}$$

which follows from (20.5), (20.6). Namely, if we assume that ν is negative or slightly positive, as long as the imperfect nesting t'_b is lower than about $0.8t'^*_b$ the phase diagram behaves as in the case of the perfect nesting provided that $\eta > 2$. The effect of the imperfect nesting is then just a slight reduction of the critical temperature T_c.

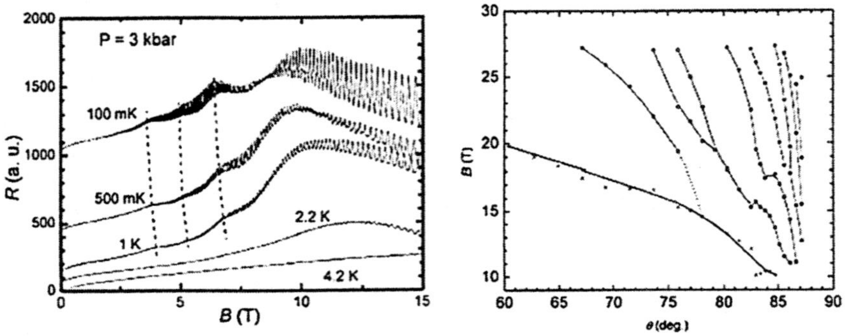

Fig. 20.5. *Left*: Magneto-resistance profiles in α-(BEDT-TTF)$_2$KHg(SCN)$_2$ for several temperatures at $p = 3$ kbar, for magnetic field along the least conducting direction ($\theta = 0$). Multiple kink features in the magneto-resistance are signatures of the first-order phase transitions between the sub-phases of the FICDW-I cascade. *Right*: Positions of the transitions between the sub-phases of the FICDW-II cascade at ambient pressure. Reproduced from Kartsovnik et al. [13]

Let us now discuss the case of α-ET at ambient pressure, but when η becomes smaller than 2. This is achieved by simply turning the sample to angles θ lager than about 60°. First, another kink transition appears (at about 27 T), and than more and more kinks occur as the angle approaches 90°. The experimental data are shown in Fig. 20.5. These multiplication of kink transitions is consistent with the conjecture that we have here the FICDW-II cascade as the cooperative effect of the Pauli and orbital coupling of electrons with the magnetic field as explained in the previous section. The transitions between the FICDW-II sub-phases are of the first order and the quantum number of the phase is expected to decrease as the magnetic field increases, in accordance with this explanation [13]. Another explanation, proposed by Biskup et al. [19] was that the kink transition splitting should be the triple point CDW-CDW$_x$-CDW$_y$. This explanation is based on perfectly nested model that cannot explain the entire cascade of transitions.

We now turn the pressure on, and keep $\eta > 2$ (implying just one kink transition at high magnetic fields and zero pressure). As we increase the pressure, t_b' increases, and the critical temperature for CDW at zero magnetic field decreases following the well-known profile [23]. As one increases the magnetic field, all phase diagram profiles follow remarkably well the predicted curves (see Figs. 20.3 and 20.4). At pressures above about 1 kbar the critical temperature first increases with the magnetic field, and then decreases after attaining some maximum. The initial enhancement of the critical temperature is the FICDW-I. It originates from the orbital quantization within the carrier pockets which are created by the imperfect nesting, just as it is the case with the FISDW [24]. Indeed, within the FICDW-I phase, several sub-phases with first-order transitions between them were observed, revealing the cascade-like structure of the FICDW-I as predicted in Fig. 20.3.

Finally, let us mention that signatures of superconductivity in FICDW-I were reported several times [25–27], and that the Fröhlich collective transport in CDW_x was also invoked [28].

20.4.2 Perylene Compounds

The perylene compounds with generic formula $(Per)_2M(mnt)_2$ have a very anisotropic, but three-dimensional, fermiology. The dispersions are well represented by a four-band model with a Fermi surfaces warping of about 200 K. Thus, a complicated phase diagram is expected a priori. Experimentally it was found that $T_c = 12$ K for M = Au and $T_c = 8$ K for M = Pt. The low values of CDW critical temperatures imply that these materials are good candidates for exploring magnetic field effects. Indeed, the phase diagram in magnetic field is extremely rich as shown in Fig. 20.6. First, one observes the decreasing of T_c, as in the case of α-ET compounds at ambient pressure. After partial or complete suppression of the CDW (depending on field orientation and sample) the FICDW cascade emerges. The low field phase is clearly the CDW_o. The $T_c(H)$ profile decreases in a typical Pauli-like way. This suppression is

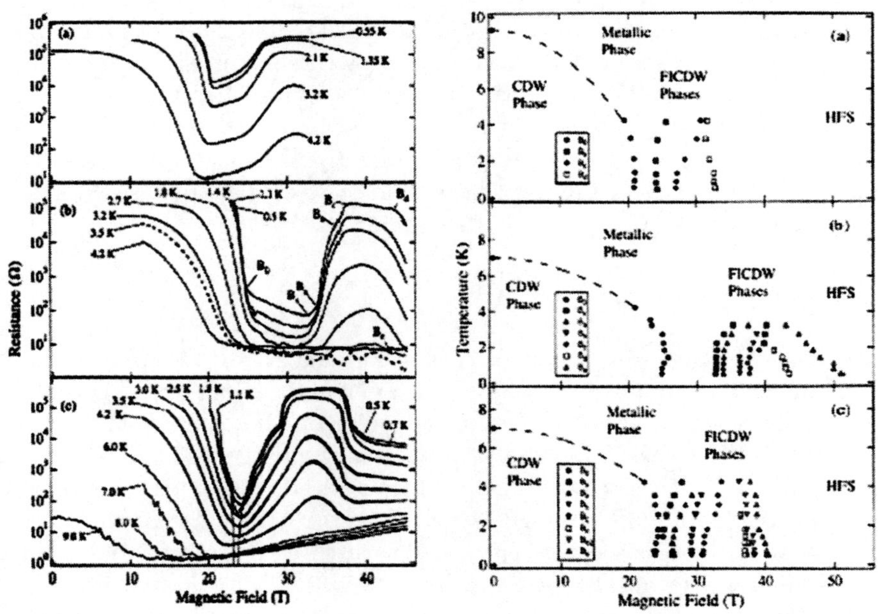

Fig. 20.6. *Left*: Magnetoresistance sweeps at several temperatures for approximate magnetic field orientations H parallel to intermediate conductivity axis (**a**), to the best conduction direction (**b**), to the least conduction direction (**c**). Kinks correspond to the phase transitions between the sub-phases of the FICDW-II cascade. Corresponding phase diagrams on the right panel are to be compared with the Fig. 20.5. Reproduced from Graf et al. [29]

independent of the orientation of the magnetic field, as expected for purely Pauli effect. For all reported directions of magnetic field the CDW_o phase becomes completely suppressed at about 24 T.

It seems puzzling that the metallic state is stable down to $T = 0$ between the CDW_o state and the FICDW cascade. The simplest explanation would be to suppose that both phases CDW_o and FICDW originate from the same band with a poor nesting and an anisotropic three-dimensional dispersion, and that the interaction ratio ν is positive, allowing the total suppression of the CDW_o by magnetic filed, as predicted in [9]. Now, if η is small enough (<2) we can identify the cascade as the FICDW-II phenomenon. The finite threshold field for the FICDW at zero temperature could be a consequence of the three-dimensional (and not two-dimensional) nature of the system [24], with the high field cascade being the FICDW-II. Cascade exists for any direction of H for two reasons. First, since the Fermi surface warping is in both low conducting directions, the pseudoclosed orbits are possible for any orientation of H. However, one would expect no cascade-like orbital effects at all for H parallel to the best conducting direction. That seems not to be the case. The reason is probably the difficulty to orientate precisely the sample in magnetic field, so that we are always left with some small but finite θ, making the parameter η small, which favors the FICDW-II cascade, just as in the case of α-ET shown in Fig. 20.5. An indication of small value of η is the large number of cascades. Also, the fact that perylene is three-dimensional rather than two-dimensional makes it necessary to introduce two different η parameters, one associated with orbits in *ab* plane and another associated with the orbits in *ac* plane (a being the best conducting direction). In fact, any quantitative analysis of the FICDW-II in perylene should start with an analysis within the three-dimensional model which is still lacking.

20.4.3 Blue Bronzes

Blue (or molybdenum) bronzes of general formula $A_xMo_yO_z$, where A is an alkali metal or thallium, are systems which in many aspects can be considered as low dimensional. We shall here concentrate on the compound KMo_6O_{17} with the CDW phase which was studied in magnetic field [30]. The particular fermiology of these compounds is characterized by the presence of three pairs of almost one dimensional Fermi sheets. All these Fermi sheets combine to form three closed Fermi surfaces with three nesting vectors of equal length disposed in an axially symmetric star [31]. Consequently, the CDW in KMo_6O_{17} has three degenerate modulations, corresponding to these three nesting wave vectors. The critical temperature of 105 K is independent on magnetic field up to 40 T within experimental error of about 10%.

Remarkable field-induced phase transitions, however, exist within the range of ordered CDW as shown in Fig. 20.7. These transitions are visible as kinks in magneto-resistance, just like in α-ETs. First two kinks present no hysteresis implying that the transitions are not of the first order, while

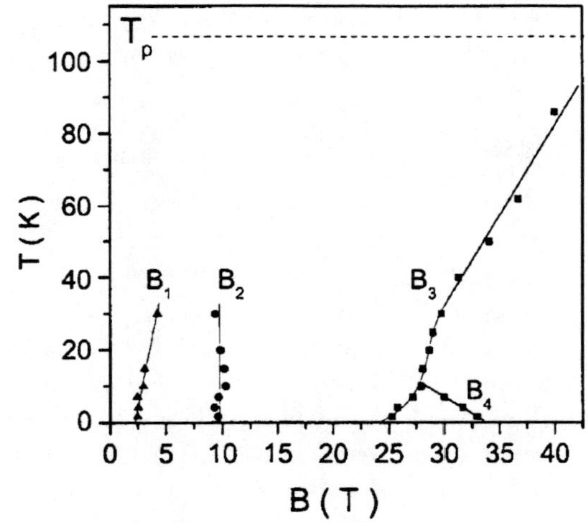

Fig. 20.7. The phase diagram of the moybdenum bronze KMo_6O_{17}. Reproduced from Balaska et al. [30]

second two kinks present strong hysteretic behavior, i.e., they correspond to the first-order transitions. The transition lines of all four kinks depend only slightly on the field direction. Following the model [5], this indicates that the dominant mechanism for all of them is the Pauli coupling, with orbital mechanisms affecting only weakly the critical field. The transitions are of $\mathrm{CDW_o} \rightarrow \mathrm{CDW}_{x/y}$ type and the modulation wave vector is consequently a monotonous function of H, without cascades. The fact that four transitions are present instead of one (or two) as predicted by the theory is most probably the result of the complex fermiology consisting of three quasi-two-dimensional nested Fermi surfaces.

20.5 Conclusions

In the last decade we have been witnessing considerable development in the understanding of the behavior of CDWs in a wide range of values (up to about 30 T) of external magnetic fields. Both, theoretical and experimental evidences show that (T, H) phase diagrams are as a rule more complex than those for SDWs. The reasons are at least twofold. First, unlike SDWs, CDWs cannot avoid an additional deviation from the perfect nesting of Fermi surfaces caused by the Pauli splitting. The consequence is that the spin-up and spin-down Fermi surfaces may nonequally participate in the CDW stabilization. All CDW phases with this property are in fact CDW–SDW hybrids. Second, due to this hybridization, as well as to the activation of both, Pauli and orbital couplings to the magnetic field, two additional relevant parameters, enter into play, namely the relative strength of interaction coupling constants responsible for CDW and SDW correlations (ν) and the ratio of two energetic scales associated to the magnetic field (η). The latter parameter is particularly interesting, since it can be varied experimentally through the wide range of values just by changing the orientation of **H**.

Seemingly just the variants of CDW ordering with a finite CDW–SDW hybridization have been detected in all available CDW materials showing sensibility to the external magnetic field. As argued in Sect. 22.4.2, newly observed CDW phases are, beside FICDW-I which is a direct counterpart of analogous FISDW phases, CDW_x and FICDW-II. Still, the understanding of these phases, particularly of their magneto-transport and collective properties, is far from complete. The further set of open questions deals with the low temperature range of (T, H) phase diagrams. While initial attempts were based on the lowest order terms in the harmonic expansion of sinusoidal CDWs [9], it remains to analyze the possibility of stabilizing nonsinusoidal, possibly soliton lattice-like patterns. The latter are natural candidates for stable low temperature CDW modulations due to the closeness of degenerate or nearly degenerate wave vectors of various CDW phases considered in this review.

References

1. A. Bjeliš, K. Maki, Phys. Rev. **B42**, 10275 (1990)
2. L.P. Gorkov, A.G. Lebed, J. Phys. Lett. (Paris) **45**, L433 (1984)
3. W. Dietrich, P. Fulde, Z. Phys, **265**, 239 (1973)
4. A. Bjeliš, D. Zanchi, Phys. Rev. B **49**, 5968 (1994)
5. D. Zanchi, A. Bjeliš, G. Montambaux, Phys. Rev. B **53**, 1240 (1996)
6. D. Zanchi, A. Bjeliš, G. Montambaux, J. Phys. IV France **9**, Pr10-203 (1999)
7. K. Yamaji, J. Phys. Soc. Jpn. **54**, 1034 (1985)
8. Although one-dimensional systems in the range $U_c < 0$, $\nu < -1/3$ are singlet superconductors, they can be nevertheless included in the present considerations since at high enough external magnetic fields superconducting ground state is suppressed due to orbital effects [5]
9. P.D. Grigoriev, D.S. Lyubshin, Phys. Rev. B **72**, 195106 (2005)
10. S.A. Brazovskii, S.I. Matveenko, Zh. Eksp Teor. Fiz. **87**, 1400 (1984) [Sov. Phys. **60**, 804 (1984)]
11. A. Bjeliš, S. Barišić, in *Electronic Properties of Organic Materials with Quasi-One-Dimensional Structure*, ed. by H. Kamimura (Riedel, New York, Dordecht, 1985), pp. 49–122; J. Phys. C **15**, 5607 (1985)
12. A.G. Lebed, JETP Lett. **78**, 138 (2003) [Pisma ZhETF, **78**, 170 (2003)]
13. M.V. Kartsovnik, D. Andres, P.D. Grigoriev, W. Biberacher, H. Müller, Phys. B **346**, 368 (2004)
14. T. Tiedje, J.F. Carolan, A.J. Berlinsky, Can. J. Phys. **53**, 1593 (1975)
15. T. Ishiguro, K. Yamaji, G. Saito, *Organic Superconductors II* (Springer, Berlin Heidelberg New York, 1998)
16. J. Singleton, Rep. Prog. Phys. **78**, 111 (2000)
17. N. Harrison, E. Rzepniewski, J. Singleton, P.J. Gee, M.M. Honold, P. Day, M. Kurmoo, J. Phys: Condens. Matter **11**, 7227 (1999)

18. A.E. Kovalev, S. Hill, J.S. Qualls, Phys. Rev. B **66**, 134513 (2002)
19. N. Biškup, J.A. Perenboom, J.S. Qualls, J.S. Brooks, Solid State Commun. **107** 503 (1998)
20. P. Foury-Leylekian, S. Ravy, J.-P. Pouget, H. Muller, Syn. Metals **137**, 1271 (2003)
21. D. Andres, M.V. Kartsovnik, W. Biberacher, H. Weiss, E. Balthes, H. Müller, N. Kushch, Phys. Rev. B **64**, 161104(R) (2001)
22. D. Andres, M.V. Kartsovnik, P.D. Gregoriev, W. Biberacher, H. Müller, Phys. Rev. B **68**, 201101 (2003)
23. G. Montambaux, Phys. Rev. B **38**, 4788 (1988)
24. G. Montambaux, M. Heritier, P. Lederer, Phys. Rev. Lett. **55**, 2078 (1985)
25. M. Kartsovnik, D. Andres, W. Biberacher, P.D. Grigoriev, E.A. Shuberth, H. Müller, J. Phys. IV France, **114**, 191 (2004)
26. D. Andres, M.V. Kartsovnik, W. Biberacher, K. Neumaier, E. Schuberth, H. Müller, Phys. Rev. B **72**, 174513 (2005)
27. H. Ito, H. Kaneko, T. Ishiguro, H. Ishimoto, K. Kono, S. Horiuchi, T. Komatsu, G. Saito, Solid State Commun. **85**, 1005 (1993)
28. C.H. Mielke, N. Harrison, A. Ardavan, P. Goddard, J. Singleton, A. Narduzzo, L.K. Montgomery, L. Balicas, J.S. Brooks, M. Tokumoto, Syn. Metals **133**, 99 (2003)
29. D. Graf, E.S. Choi, J.S. Brooks, M. Matos, R.T. Henriques, M. Almeida, Phys. Rev. Lett. **93**, 076406 (2004)
30. H. Balaska, J. Dumas, H. Guyot, P. Mallet, J. Marcus, C. Schlenker, J.Y. Veuillen, D. Vignolles, Solid State Sci. **7**, 690 (2005)
31. J. Dumas, C. Schlenker, Int. J. Mod. Phys. B **93**, 4045 (1993); Gweon et al., Phys. Rev. B **55**, R13353 (1997)

21

Unconventional Electronic Phases in $(TMTSF)_2X$: The Case of $(TMTSF)_2ClO_4$

S. Haddad, M. Héritier, and S. Charfi-Kaddour

21.1 Introduction

The announcement in 1979 of the discovery of superconductivity in the quasi-one-dimensional (q-1D) organic conductor $(TMTSF)_2PF_6$ has been a turning point in the history of the Bechgaard salts family $(TMTSF)_2X$ (denoted hereafter as $(TM)_2X$). These compounds, which attracted the attention of physicists and chemists exhibit a variety of electronic states [1, 2], in particular field-induced spin density wave (FISDW) phases, which are one of the most spectacular features of these materials [3]. Moreover, the proximity of the spin density wave (SDW) and the superconducting (SC) phases in the temperature–pressure phase diagram is a key property of $(TM)_2X$ family, which is reminiscent of the high temperature superconductors. Whether there is a competition or a collaboration between these two phases is a question of great interest. Vuletic et al. [4] reported a macroscopic coexistence of SDW and SC domains in $(TMTSF)_2PF_6$, which has been confirmed experimentally [5] and interpreted theoretically [4, 6].

However, a proper understanding of the interplay between superconductivity and antiferromagnetism in $(TM)_2X$ salts is still lacking. Furthermore, the nature of the pairing in the SC state is an important challenge of these compounds.

To further investigate the properties of the SC phase and its correlation to the SDW state, extensive studies have been carried out on the $(TMTSF)_2ClO_4$ compound since, unlike the other $(TM)_2X$ salts, it becomes superconductor at ambient pressure and is therefore much more widely studied. However, this compound has raised more questions, in particular, those concerning the FISDW phases.

Unlike the $(TMTSF)_2ClO_4$, the overall behavior of the FISDW cascade in the case of $(TMTSF)_2PF_6$ can be interpreted within the quantized nesting model (QNM) [3, 7–13], except the presence of the high temperature nonquantized phases [14], which are consistent with a recent model proposed by Lebed [15]. The $(TMTSF)_2ClO_4$ shows striking deviations from the

QNM's predictions [16–23]. In particular, experiments reported a complex temperature–field phase diagram instead of the simple cascade of FISDW phases as expected by QNM. This enigmatic behavior has inspired several groups to introduce modifications on the QNM in order to account for these deviations. Although the proposed models have different frameworks, they all agree on the origin of the puzzling properties of the $(TMTSF)_2ClO_4$, which has been ascribed to the ClO_4^- anion ordering (AO) transition that occurs at $T_{AO} \sim 24\,K$ when the sample is slowly cooled (relaxed).

In this paper, we go through the main properties of the $(TMTSF)_2ClO_4$ salt by stressing on its unconventional electronic ordered phases. We emphasize on the relaxed state where SC phase appears.

21.2 Structural Properties of the ClO_4^- Anion Ordering

Anion ordering takes place in the case of $(TMTSF)_2X$ salts with noncentrosymmetric anions X such as ReO_4^-, ClO_4^-, BF_4^-, NO_3^-..., which due to their lack of inversion symmetry can take two equivalent positions in the structure. This is in contrast with the centrosymmetric anions (X = PF_6^-, AsF_6^-, Sb_6^-, Br^-...) which have a unique position.

In the relaxed $(TMTSF)_2ClO_4$ with a cooling rate less then $0.1\,K\,min^{-1}$, the ClO_4^- anions order by alternating their orientation in the **b** direction at a temperature below $T_{AO} = 24\,K$ but keeping a uniform ordering in the other two directions. This order, which occurs for entropy reasons, produces a superlattice potential $V(y) = V \cos \pi/by$, with a wave vector $\boldsymbol{Q} = (0, \pi/b, 0)$. This potential leads to inequivalent TMTSF stacks giving rise to two bands crossing the Fermi level and to a four sheets of Fermi surfaces instead of two as in $(TMTSF)_2PF_6$ (Fig. 21.1).

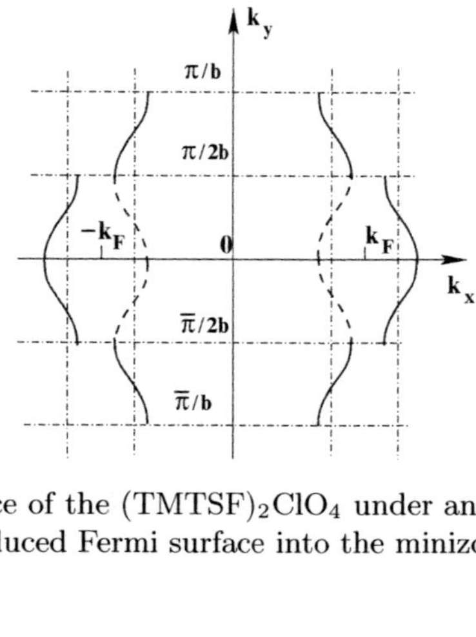

Fig. 21.1. Fermi surface of the $(TMTSF)_2ClO_4$ under anion ordering. The dashed curves represent the reduced Fermi surface into the minizone $-\pi/2b \leq k_y \leq \pi/2b$

The orientation ordering of the anions is achieved when the anion bypasses by thermal excitation a potential barrier induced by the anion surrounding molecules [24]. The AO of the $(TMTSF)_2ClO_4$ was originally pointed out by Gubser et al. [25] and interpreted by Pouget et al. [26].

The electronic properties of $(TMTSF)_2ClO_4$ are dramatically influenced by the cooling rate. By decreasing the temperature, the relaxed sample undergoes an SC transition at 1.2 K. However, when the sample is rapidly cooled through T_{AO} at a rate of at least $50\,K\,min^{-1}$, the anions are frozen with random orientations and a SDW phase appears below 6 K. Intermediate states samples are characterized by a mixed phase where macroscopic-sized regions of SC and SDW phases coexist.

21.2.1 AO vs. Cooling Rate

An X-ray scattering study of the AO transition has revealed the presence of domains of ordered ClO_4 anions in the relaxed sample [27]. As the cooling rate decreases, the average domain size increases and small domain may merge into large one, but a long-range order of ClO_4 is never achieved. This study has also shown that, whatever the cooling rate, there is always a sizable fraction of the ordered ClO_4 in the quenched sample. By slowly heating (thermal annealing) the quenched state, the size of the ordered regions grows by collapsing ordered domains.

The details of the microstructure of the ordered and disordered ClO_4 domains in the sample are substantial for the stability of the low temperature electronic phases of the $(TMTSF)_2ClO_4$. In particular, the Ginzburg Landau coherence length of the superconducting state, which appear in the relaxed state, is limited by the extent of the ordered ClO_4 domains [27].

To discuss the low temperature electronic ground state of the $(TMTSF)_2ClO_4$, it is quite important to seek the dependence of the AO on hydrostatic pressure, which is a crucial parameter in the physics of Bechgaard salts.

21.2.2 AO under Pressure

From the temperature dependence of the resistance measurements under pressure, Guo et al. [28] have argued that the AO transition is robust against pressure and was observed up to 1.1 GPa. They found that the transition temperature increases from 24 K at ambient pressure to 41 K at 1.1 GPa. Early measurements [28] have given lower pressure limits where the AO is still present. The relatively high value proposed in [28] is due to an extreme careful control of the cooling rate since the higher the pressure, the slower the dynamics of anion orientation [24].

21.2.3 Discussion about the Anion Potential Value: Is it Small or Large?

The two-dimensional dispersion relation in the presence of the anion gap can be described by [8]

$$\epsilon(\boldsymbol{k}) = v_{\mathrm{F}}(|k| - k_{\mathrm{F}}) \pm \sqrt{4t_b^2 \cos^2 k_\perp b + V^2} - 2t_b' \cos 2k_\perp b, \tag{21.1}$$

where $k\,(k_\perp)$ is the electron momentum along (across) the chains of the TMTSF molecules, v_{F} is the Fermi velocity and k_{F} is the Fermi point. t_b and t_b' are the effective transfer integrals in the **b** direction to the first and second neighbors, respectively.

The value of V has been the subject of a lot of controversies, since it cannot be extracted from a direct determination. However, a general consensus seems to have been reached recently [29]. Basically, the analysis of the angular dependence of the magnetoresistance (ADM) was the main issue to access to the value of V [30]. Early measurements [31,32] suggested a value of $V \sim 50\,\mathrm{K}$. Yoshino et al. proposed a rather larger value of the order of $0.08t_a$ where t_a is the hopping parameter along the chain. This value was substantially the key argument in the theories assuming a large anion gap [33–37]. More recently, this value has been renewed by the same group who found a much smaller value of the order of $0.023t_a$ [38].

Lebed et al. have ascribed the angular magnetoresistance oscillations in quasi-1D conductors to the 1D-2D crossover occurring at some commensurate directions of the magnetic field. The authors suggested to make use of this effect to determine the band parameters of quasi-1D conductors such as the anisotropy ratio t_a/t_b and the anion gap V [39]. Ha et al. [40] reported ADM measurements on the $(\mathrm{TMTSF})_2\mathrm{ClO}_4$. The analysis of the magnetoresistance oscillations based on the model proposed in [39] gives rise to a value of $V = 0.2t_b$ for the anion potential. ADM measurements show conclusively that V should be much smaller than the interchain hopping parameter t_b. This is supported by structural studies [26] and quantum Hall effect measurements [16,19,41]. The latter cannot be interpreted assuming a large V value [29,34,37].

It is worth stressing that Cooper and Yakovenko have presented recently a model to probe directly the anion gap in $(\mathrm{TMTSF})_2\mathrm{ClO}_4$ based on a geometrical interpretation of the angular magnetoresistance oscillations obtained with an in-plane magnetic field [42].

The outcome of all these studies is the absence of a direct unambiguous determination of V. However, their general trend converges to the conclusion that V should be *small* compared to t_b and may not exceed $50\,\mathrm{K}$. The value of V is crucial for the stability of the low temperature ordered states. In the next section we focus on the main features of the relaxed state namely the SC phase and the FISDW cascade.

21.3 Relaxed State Properties

Superconductivity as a Ground State

The evidence of the SC phase in slowly cooled sample has been established not only by the drop of the resistivity below 1.2 K [43, 44] but also by specific heat measurements [45] and Meissner effect [46, 47].

By decreasing pressure, the SC transition temperature decreases as in the case of other $(TM)_2X$ compounds. However, the dependence of T_C on pressure is more dramatic in $(TMTSF)_2ClO_4$ as argued by Jérome [2] due to the sensitivity of the AO against pressure (see Sect. 21.2.1). The SC state is also systematically dependent on the cooling rate as we will show in the next section.

Superconductivity and SDW: Coexistence or Competition?

Early specific heat measurements [45] have revealed that the SC is strongly suppressed by quenching the sample. These results have been corroborated by magnetoresistance and Meissner signal measurements at different cooling rates [44, 48, 49].

More recently, Joo et al. [50] reinvestigated the dependence of the SC transition temperature on the cooling rate. The authors have found that for rapid cooling rates ($\sim$14 K min^{-1}), the drop of the resistivity, which is the signature of the SC transition, is preceded by an upturn of the resistivity. This feature has been observed in early magnetoresistance measurements [44] and has been ascribed to the onset of a SDW phase in the intermediate state (moderate cooling rates). This remarkable property was assigned to the presence in the sample of isolated domains of SDW phase where the anions are not ordered. These domains separate regions of ordered ClO_4 anions in which SC phase is stabilized in agreement with X-ray measurements discussed in Sect. 21.2.1. The SDW domains act as inhomogeneities whose proportion increases with increasing cooling rate and thereby reducing the SC transition temperature and broadening the transition [44, 45, 47, 49, 50]. As the cooling rate increases, the SDW islands get wider at the expense of the SC domains, which is highlighted by the upturn of the resistivity in the intermediate state. The latter is then characterized by a mixture of the SDW and the SC phases, which *coexist in different regions of the sample* [44, 45, 47, 49, 50]. As pointed out by Schwenk et al. [44], in the intermediate state, *there is not a microscopical coexistence* of the two ground states: it is rather an inhomogeneous mixture. By increasing the cooling rate, the SDW regions merge into larger ones. Eventually, the SDW takes over the SC phase, which is completely suppressed in the quenched state.

The increase of the cooling rate has been described in [51] within a simple model as a decreasing anion potential V, which is in agreement with the collapse in the quenched state of the anion gap [52–55]. The main result

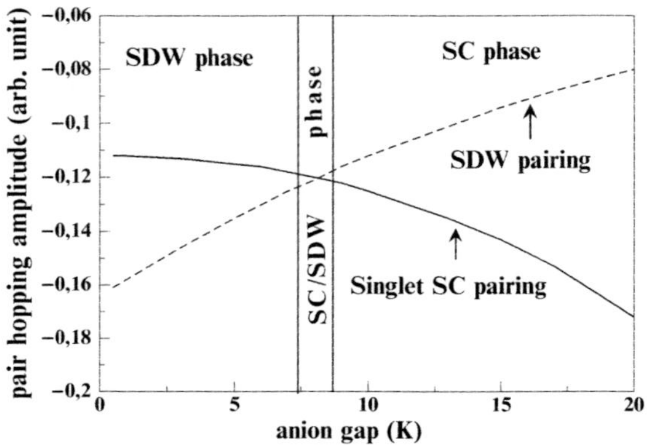

Fig. 21.2. Sc and SDW pair hopping amplitudes vs. anion gap

of this model based on renormalization group calculations is summarized in Fig. 21.2 showing the most divergent pairing amplitudes in the charge density wave (CDW), SDW, and SC channels. In the large V regime (relaxed state), the system is in a singlet SC phase while the ground state of the quenched sample (small V) is a SDW phase. The striking feature of Fig. 21.2 is the presence of a domain of cooling rate where both SC and SDW coexist, which corresponds to the intermediate state. This is reminiscent of the experimental results of [50].

X-ray scattering measurements show that the size of the disordered anions domains increases with the cooling rate. These domains can be regarded as defects or nonmagnetic scattering centers whose concentration is enhanced as the cooling rate is increased [48, 50]. Therefore, the sensitivity of T_c to the anion disorder mimics the effect of nonmagnetic impurities at least for slow cooling rates. The decrease of T_c with the cooling rate has been taken as an evidence of the non-s-wave character of the SC [49, 50]. However, it is interesting to note that anion disorder induced by the cooling rate cannot be considered simply as impurity defects even if they have similar effects on the SC phase. This disorder is in close connection with the electronic mean free path, which depends on the size of the ordered domains [56].

Superconductivity and Nonmagnetic Impurities

The nature of low temperature electronic ground states of the $(TMTSF)_2ClO_4$ is extremely sensitive to the introduction of substituents, which may concern the organic molecule (TMTSF) or the ClO_4 anions. The former may be replaced by the (TMTTF) molecule leading to the formation of the solid solution $[(TMTSF)_{1-x}(TMTTF)_x]_2ClO_4$ [57]. Basically, the substitution of the ClO_4 anions with ReO_4 ones is the most studied case since these anions have the same symmetry. This alloying leads to the solid solution $(TMTSF)_2ClO_{4(1-x)}ReO_{4x}$, which has been first studied by Tomić et al. [59]

and reinvestigated more recently by Joo et al. [50] for higher ReO_4^- concentrations. The structural properties of the AO in $(TMTSF)_2ClO_{4(1-x)}ReO_{4x}$ has been the subject of an intensive X-ray studies [57].

It is worth noting that the ReO_4^- doped samples were free from any magnetic impurity as found by an EPR study [50, 58]. Tomić et al. [59] have shown that actually the SC transition temperature is substantially affected by alloying. A recent study [50, 58] revealed that the suppression of the SC transition temperature is well fitted by a digamma function as in the case of nonmagnetic impurities in unconventional superconductors [60]. The symmetry gap of the SC state in the $(TMTSF)_2ClO_4$ is then expected to be *anisotropic* with an order parameter that changes sign over the Fermi surface. Nevertheless, the study of the influence of nonmagnetic impurities on the SC phase cannot bring to a sudden end the discussion about the possibility to have a singlet (d) or a triplet (p) pairing.

Defects may also be induced instead of alloying by X-ray irradiation. However, the defects are in this case created on the conducting chain of the organic molecules [48]. The influence of such defects on the SC state has been studied by Park et al. who have shown that irradiation makes less effect on the superconductivity, compared to the anion disorder [48]. However, the suppression of the SC phase by irradiation cannot be taken as an evidence of an unconventional pairing, since defects produced by irradiation can be magnetic [2].

Few Comments on the Symmetry of the Superconducting Gap

In spite of exhaustive studies dealing with the SC state in $(TMTSF)_2ClO_4$, there are still controversies on the type of the gap symmetry. Takigawa et al. suggested by NMR experiment that the SC gap is anisotropic with line nodes on the Fermi surface [61]. However, this conclusion was contested since the NMR measurements were done at not sufficiently low temperature.

Based on thermal conductivity measurements, Belin and Behnia [62] argued that the gap should be nodeless. However, the authors did not rule out the possibility of an unconventional order parameter.

To reconcile the NMR measurements and the thermal conductivity ones, Shimahara has proposed a model of nodeless d-wave superconductivity under anion ordering where the anion gap opens precisely at the nodes of the d-wave gap giving rise to a vanishing electronic thermal conductivity [63].

Magnetoresistance [64] and magnetization [65] measurements showed that the upper critical field H_{c2} reaches 5 T, which exceeds the Pauli paramagnetic limit imposed on a singlet SC namely $H_p = 1.84\,T_c^0$ where T_c^0 is the SC transition temperature at zero magnetic field. Nevertheless, violation of the Pauli limit has to be considered with great care because the field induces, as discussed by Jérome [2], a localization of the electronic motion that may affect the SC properties. Furthermore, the enhancement of the critical field may be due, as suggested by Shimahara [66], to an inhomogeneous superconductivity where singlet and relatively weak triplet pairings coexist: at zero field the singlet pairing is dominant but the triplet component becomes mixed to the singlet one as the magnetic field increases. As a result, the upper critical field is strongly enhanced even for weak triplet pairing. The violation of the Pauli limit may also be explained by the recent model proposed by Lebed for type IV superconductors with a singlet gap at zero magnetic field but where a mixed singlet–triplet phase appears when the field exceeds a critical limit H_{C4} [67].

One comes down to the conclusion that the SC gap in $(TMTSF)_2ClO_4$ is unconventional but no clear evidence about the symmetry of the pairing can be deduced from the experimental results, which are still controversial.

21.3.1 Field Induced 3D-2D Crossover

Behnia et al. have carried out magnetoresistance measurements in the R-state of the $(TMTSF)_2ClO_4$ with the magnetic field aligned along the b' direction [68]. The authors reported that the system remains metallic in the **a** direction down to low temperature. More recently, Joo et al. [69] investigated the magnetoresistance of the R-state along the interplane c^* axis and with an in-plane field in the $\mathbf{b}'$ direction. In contrast with the behavior along the chains, the system showed, above a critical field, an insulating character in the interlayer direction, suggesting that the electrons are confined in the ab plane in agreement with early measurements [70]. We have proposed a simple model to interpret this localization by calculating the interlayer conductivity, which was found to be drastically suppressed as the field is increased above a critical value [69, 71]. The latter, as should be expected, is substantially dependent on the interlayer hopping parameter t_c. Independently, Lebed has also obtained an in-plane localization using a rather different model [72].

21.3.2 FISDW Phases: Some Key Features

The temperature–field phase diagram of the $(TMTSF)_2ClO_4$ cannot be explained within the QNM. Despite intensive theoretical studies [29, 33–37, 73–76], many properties are not fully interpreted, in particular the so-called rapid oscillations (RO) of the magnetoresistance. In the following, we briefly go through the main features of this phase diagram.

Unconventional FISDW Phases: Experiments

Early studies of the temperature–field phase diagram of the relaxed $(TMTSF)_2$ ClO_4 salt has already raised a controversy about the nature of the high FISDW phases. Naughton et al. reported the suppression of the last FISDW phase at high field ($H > 27$ T) and the reentrance of the metallic state [77], while Yu et al. argued that the very high field state is rather insulating [78], which has been confirmed by Kang et al. [18]. More striking features have been

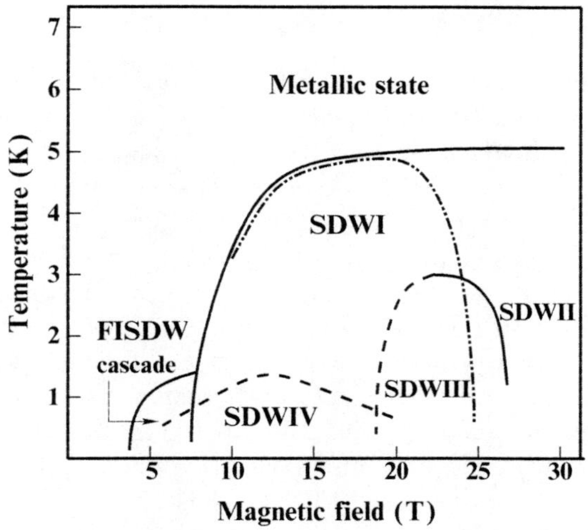

Fig. 21.3. Temperature–field phase diagram of $(TMTSF)_2ClO_4$. Solid lines are the transition boundaries obtained by McKernan et al. [19], dashed ones are the transition lines found by Chung et al. [23], while the dashed-dotted line is the phase boundary obtained by Naughton et al. [77]

obtained by McKernan et al. [19], who have established a temperature–field phase diagram substantially different from the previous ones. They found an original FISDW state with a transition temperature essentially field independent of about 5.5 K above 15 T. Surprisingly, a first-order transition occurs by decreasing the temperature and a second FISDW phase appears *inside* the original one at 3.5 K, but no reentrant metallic state has been reported. The presence of this inner phase has been confirmed by recent magnetoresistance measurements [22, 23, 41], which show that the phase boundary of this new phase collapses at 17 T. From their results, Matsunaga et al. come down to the conclusion that the inner phase and the insulating high field state correspond to *different* FISDW states [41]. Chung et al. have reported the existence of other subphases inside the original one at low field regime [23]. Actually, all the inner phases may be ascribed to the $N = 1$ FISDW state, since the $N = 1$ plateau reported from Hall effect measurements at 8 T survives up to 27 T [19]. However, the high field insulating state (SDWII in Fig. 21.3) may correspond to the $N = 0$ phase as it is characterized by a nonquantized Hall resistance.

To account for all these complicated properties of the FISDW subphases, several theoretical studies have been proposed. In the next section, we give a brief summary of the different models and focus on the one that we proposed in [29, 79].

Unconventional FISDW: Theoretical Approaches

Basically, the theoretical models can be classified into two groups: small V models ($V \ll t_b$) and lagre V ones ($V \sim t_b$). First, Lebed and Bak [73] have proposed a model to deal with the so-called reentrant metallic state obtained by Naughton et al. [77]. Then, by calculating the noninteracting spin susceptibility $\chi_0(\boldsymbol{q})$, Osada et al. argued that the odd-N FISDW states for which the longitudinal component of the nesting vector is $q_x = 2k_{\mathrm{F}} + NG$ are not affected by the anion gap. However, even-N states are strongly suppressed. The nesting vector in these phases is $q_x = 2k_{\mathrm{F}} + NG \pm 2\delta/v_{\mathrm{F}}$, where the effective gap $\delta = VJ_0(4t_b/\omega_{\mathrm{c}})$, J_0 is the Bessel function and $\omega_{\mathrm{c}} = v_{\mathrm{F}}G$ is the magnetic energy. Here $G = eHb$ is the magnetic wave vector, H being the amplitude of the magnetic field and b is the interchain distance in the **b** direction. The reentrance of the metallic state was interpreted as the result of the suppression of the last $N = 0$ FISDW phase.

However, these models fail to explain the presence of the inner phase and the high field insulating state. This failure has been ascribed to the assumption of a small V value. Other models have been proposed where V is supposed to be of the order of t_b [33–37]. These models are based on the scenario of *separate* SDW transitions, which has been already suggested by McKernan et al. [19]. During the first SDW transition only one band of the Fermi surface is gapped while the other remains metallic. By decreasing the temperature, a second SDW transition occurs and the whole Fermi surface becomes gapped. Nevertheless, all the above mentioned models do no account for the first order transition line between the original SDW state and the inner one. Furthermore, it is not obvious within these models how to label the different SDW subphases by definite quantum indexes in agreement with the quantum Hall effect measurements.

In the next part, we give the outline of the model we have proposed in [29, 79] to deal with the high FISDW phases. The main findings include the competition between a SDW phase with a single interband nesting vector and an original SDW phase originating from two intraband nesting processes. The latter do not compete but cooperate to stabilize the corresponding phase. Contrary to the other models, we suggest that in the original SDW phase, the two bands of the Fermi surface are *simultaneously* gapped and two intraband order parameters coexist. As the temperature decreases, this coexistence destabilizes the corresponding phase and eventually a first order transition takes place to the SDW phase with the interband nesting process. This new phase appears *inside* the original one. Within this model, we are able to label the different SDW subphases in accordance with quantum Hall measurements.

As discussed in Sect. 21.2.1, the anion potential V should be small compared to t_b and may then be considered as a perturbation. Osada et al. found that in this case the energy spectrum of the $(\mathrm{TMTSF})_2\mathrm{ClO}_4$ in the presence of a magnetic field reduces to two linearized bands [74]:

$$E_{\boldsymbol{k}}^{m} = v_{\mathrm{F}} \left(|k_x| - k_{\mathrm{F}}^{m} \right) ,$$

where k_{F}^{m} is the Fermi point of the m band ($m = A, B$) written as $k_{\mathrm{F}}^{A} = k_{\mathrm{F}} - \frac{\Delta}{v_{\mathrm{F}}}$ and $k_{\mathrm{F}}^{B} = k_{\mathrm{F}} + \frac{\Delta}{v_{\mathrm{F}}}$. Here Δ is the effective anion gap given by: $\Delta = V J_0\left(\frac{4t_b}{v_F G}\right)$. One may then define two nesting processes: two intraband nestings with wave vectors $q_A^0 = 2k_{\mathrm{F}}^{A}$ and $q_B^0 = 2k_{\mathrm{F}}^{B}$ and an interband nesting with a single wave vector $q_1 = 2k_{\mathrm{F}}$. Osada et al. [74] have shown that the even-N subphases originate from intraband nesting while the odd-N ones are due to interband nesting.

Keeping in mind that the high FISDW phases correspond to the smallest quantum numbers, we will focus on the $N = 0$ and the $N = 1$ phases, which is consistent with the Hall effect data in the high field regime. In [29, 79] we have, contrary to [74], taken into account the electron–electron interactions. We have derived a generalized Stoner criterion for the $N = 0$ phase, which reads as

$$\frac{1}{g_0} = \frac{1}{2}\chi_m^0\left(\boldsymbol{Q}_m^0, T\right) + \frac{1}{2}\frac{g_b}{g_0}\chi_m'\left(\frac{4\Delta}{v_F}, T\right), \tag{21.2}$$

where $\chi_m^0\left(\boldsymbol{Q}_m^0, T\right)$ is the bare susceptibility corresponding to the SDW instability at the wave vector $\boldsymbol{Q}_m^0(q_0^m = 2k_{\mathrm{F}}^m, \pi/b)$, which takes place in the m (m = A, B) band inducing a large gap at k_{F}^m and a much smaller one on the other band ($\bar{m}$) but outside the Fermi level $k_{\mathrm{F}}^{\bar{m}}$. The χ_m' term in (21.2) describes the minority SDW component that forms on the m band outside k_{F}^m and originating from the SDW instability taking place on the $\bar{m}$ band. The g_0 and g_b constants are the intraband amplitudes of two particles scattering processes [80].

The Stoner criterion for the $N = 1$ phase is similar to that derived in the case of a single band model as in the case of $(\mathrm{TMTSF})_2\mathrm{PF}_6$:

$$\frac{1}{g_{\mathrm{f}}} = \frac{1}{2}\left[1 + \frac{g_{\mathrm{t}}}{g_{\mathrm{f}}}\right]\chi_1\left(\boldsymbol{Q}_1, T\right), \tag{21.3}$$

where $\chi_1\left(\boldsymbol{Q}_1, T\right)$ is the susceptibility of the SDW phase with a wave vector $\boldsymbol{Q}_1(q_1 = 2k_{\mathrm{F}} + G, q_{\perp}^1)$. g_{t} and g_{f} are interband coupling constants.

To determine the most stable phase, one has to compare the free energies of the $N = 0$ and the $N = 1$ one, since the total free energy is increased if both phases coexist [29]. Because of the presence of two nesting vectors in the $N = 0$ phase, two intraband order parameters Δ_0^A and Δ_0^B are required, while only one interband order parameter Δ_1 is involved in the $N = 1$ phase, since the latter depends on only a single interband nesting process. The corresponding free energy are respectively given by [29]

$$\begin{aligned} F_0 - F_{\mathrm{norm}} = {} & \frac{a_0}{2}\left(\Delta_0^A\right)^2 + \frac{a_0}{2}\left(\Delta_0^B\right)^2 + \frac{b_0}{2}\left(\Delta_0^A\right)^4 \\ & + \frac{b_0}{2}\left(\Delta_0^B\right)^4 + c_0\left(\Delta_0^A\right)^2\left(\Delta_0^B\right)^2 \end{aligned} \tag{21.4}$$

$$F_1 - F_{\text{norm}} = a_1 \left(\Delta_1\right)^2 + b_1 \left(\Delta_1\right)^4 . \tag{21.5}$$

The substantial meaning of the F_0 expression is that the $N = 0$ phase is due to the two intraband order parameters that do not compete but coexist to stabilize this phase. When the temperature is lowered, a metal–SDW second-order phase transition occurs first, giving rise to the formation of the $N = 0$ phase, because, at large field, it has a higher susceptibility. However, the coupling term $c_0 \left(\Delta_0^A\right)^2 \left(\Delta_0^B\right)^2$ has a dramatic influence on the stability of the $N = 0$ phase since the c_0 coefficient is found to be positive, which enhances the free energy and therefore tends to destabilize the corresponding phase. The larger the order parameters Δ_0^A and Δ_0^B, the stronger the effect. The comparison of the minimized free energies $[F_0]_{\text{min}}$ and $[F_1]_{\text{min}}$ in respectively the $N = 0$ phase and the $N = 1$ state shows that, below a critical temperature T_1^*, $[F_1]_{\text{min}}$ is lowered compared to $[F_0]_{\text{min}}$. At this temperature, a first order transition takes place from the $N = 0$ phase to the $N = 1$ one, which appears inside the $N = 0$ state. In the inner phase the Hall conductivity will be marked by a $N = 1$ plateau as found experimentally.

In Fig. 21.4, we depicted the phase diagram obtained numerically. At T_0, a second-order transition takes place from the metallic state to the $N = 0$ phase. Simultaneously, two gaps Δ_0^A and Δ_0^B open and the whole Fermi surface becomes gapped. By decreasing the temperature, a first order transition occurs at T_1^* from the $N = 0$ to the $N = 1$ phase. The behavior of the T_1^* phase boundary is consistent with experiments [22,23,41]. According to our model, it is possible to label the different high FISDW subphases of the $(TMTSF)_2ClO_4$ (Fig. 21.3): the inner phases SDWIII and SDWIV are nothing but the $N = 1$ state whereas the high field insulating state (SDWI and SDWII) is the $N = 0$ phase.

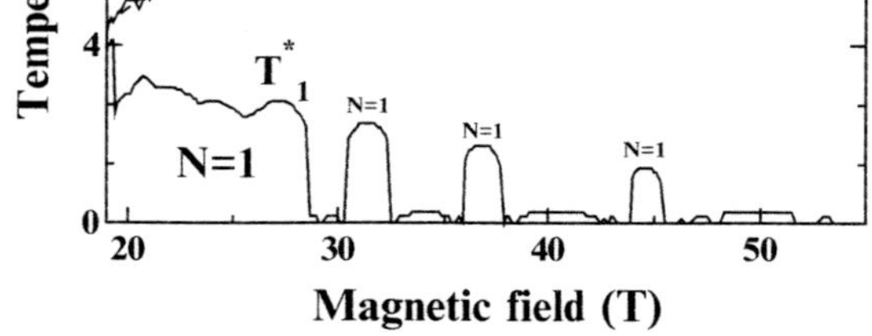

Fig. 21.4. Temperature phase diagram of $(TMTSF)_2ClO_4$ obtained for $t_b = 300\,\text{K}$, $t_b' = 20\,\text{K}$, and $V = 20\,\text{K}$. The dashed line is a guide for eyes

The proposed model of [29] gives a coherent interpretation of the overall features of the high field phase diagram of the $(TMTSF)_2ClO_4$. However, we do not claim to describe its whole properties, in particular the rapid oscillations of the magnetoresistance whose origin is still controversial [81].

21.4 Concluding Remarks

$(TM)_2X$ compounds are under intense scrutiny for their exclusive properties giving rise to new aspects in the physics of low-dimensional conductors. These materials have proved to be excellent candidates to probe the interplay between superconductivity and magnetism. In this brief account, we described some unconventional electronic ground states of this family, in particular, those appearing when the anions X undergo a structural ordering transition. Several issues have been addressed such as the possibility of coexistence of a superconducting state and a SDW one, the dimensional crossover induced by a magnetic field and new FISDW phases. The outcome of this short review is that many questions are still open, essentially the nature of the superconducting pairing, but the present state of the art seems to be like night before the dawn.

Acknowledgment

This short review is a unique opportunity to warmly thank fruitful discussions with Prs. D. Jérome, A.G. Lebed, C. Bourbonnais, T. Osada, P.M. Chaikin, K. Murata, M.J. Naughton, H. Yoshino, K. Maki, K. Bechgaard, and J.P. Pouget and Drs. V. Yakoveneko, N. Matsunaga, N. Dupuis, C. Nickel, G. Abramovici, D. Zanchi, C. Pasquier, P. Auban-Senzier, and N. Joo.

References

1. T. Ishiguro, K. Yamaji, G. Saito, *Organic Superconductors* (Springer-Verlag, Berlin, 1998). For a recent review see the volume of Chem. Rev. **104** (2004)
2. D. Jérome, Chem. Rev. **104**, 5565 (2004)
3. See the contributions of M. Héritier, A.G. Lebed, V. Yakovenko (this book).
4. T. Vuletić, P. Auban-Senzier, C. Pasquier et al., Eur. Phys. J. B **25**, 319 (2002)
5. I.J. Lee, S.E. Brown, W. Yu et al., cond-mat/0504205 (unpublished)
6. D. Podolsky, E. Altman, T. Rostunov et al., Phys. Rev. Lett. **93**, 246402 (2004)
7. L.P. Gor'kov, A.G. Lebed, J. Phys. (France) Lett. **45**, L433 (1984)
8. M. Héritier, G. Montambaux, P. Lederer, J. Phys. (France) Lett. **45**, L943 (1984)
9. M. Héritier, G. Montambaux, P. Lederer, J. Phys. C **19**, L293 (1986)
10. K. Yamaji, Synth. Met. **13**, 29 (1986)
11. P.M. Chaikin, Phys. Rev. B **31**, 4770 (1985)
12. D. Poilblanc, M. Héritier, G. Montambaux et al., J. Phys. C **19**, L321 (1986)

13. K. Maki, Phys. Rev. B **33**, 4826 (1986); A. Virosztek, L. Chen, K. Maki, Phys. Rev. B **34**, 3371 (1986); A.G. Lebed, Sov. Phys. JETP **62**, 595 (1985)
14. A.V. Kornilov, V.M. Pudalov, Y. Kitaoka et al., Phys. Rev. B **65**, 060404R (2002)
15. A.G. Lebed, Phys. Rev. Lett. **88**, 177001 (2002)
16. M.J. Naughton, R.V. Chamberlin, X. Yan et al., Phys. Rev. Lett. **61**, 621 (1988); R.V. Chamberlin, M.J. Naughton, X. Yan et al., Phys. Rev. Lett. **60**, 1189 (1988)
17. R.C. Yu, L. Chiang, R. Upasani et al., Phys. Rev. Lett. **65**, 2458 (1990)
18. W. Kang, S.T. Hannahs, P.M. Chaikin, Phys. Rev. Lett. **70**, 3091 (1993)
19. S.K. McKernan, S.T. Hannahs, U.M. Scheven et al., Phys. Rev. Lett. **75**, 1630 (1995)
20. U.M. Scheven, S.T. Hannahs, C. Immer et al., Phys. Rev. B **56**, 7804 (1997)
21. S. Uji, J.S. Brooks, M. Chaparala et al., Phys. Rev. B **55**, 14387 (1997)
22. J. Moser, H. Kang, W. Kang, D. Jérome (unpublished); J. Moser, Ph.D Thesis, Orsay (unpublished)
23. O.-H. Chung, W. Kang, D.L. Kim et al., Phys. Rev. B **61**, 11649 (2000)
24. J.P. Pouget, J. Phys. IV **10**, Pr3-43 (2000)
25. D. Gubser, W. Fuller, T. Poehler et al., Mol. Cryst. Liq. Cryst. **79**, 225 (1982)
26. J.P. Pouget, G. Shirane, K. Bechgaard et al., Phys. Rev. B **27**, 5203 (1983)
27. J.P. Pouget, S. Kagoshima, T. Tamegal et al., J. Phys. Soc. Jpn. **59**, 2036 (1990)
28. F. Guo, K. Murata, H. Yoshino et al., J. Phys. Soc. Jpn. **67**, 3000 (1998) and references therein
29. S. Haddad, S. Charfi-Kaddour, M. Héritier et al., Phys. Rev. B **72**, 085104-1 (2005)
30. See the contribution of H. Yoshino, K. Murata (this book)
31. T. Osada, S. Kagoshima, N. Miura, Phys. Rev. Lett. **77**, 5261 (1996)
32. S. Uji, T. Terashima, H. Aoki et al., Phys. Rev. B **53**, 14399 (1996)
33. M. Miyazaki, K. Kishigi, Y. Hasegawa, J. Phys. Soc. Jpn. **68**, 313 (1999) and references therein
34. Y. Hasegawa, K. Kishigi, M. Miyazaki, J. Phys. Soc. Jpn. **67**, 964 (1998)
35. K. Sengupta, N. Dupuis, Phys. Rev. B **65**, 35108 (2002)
36. D. Zanchi, A. Bjeliš, Eur. Phys. Lett. **56**, 596 (2001)
37. D. Radić, A. Bjeliš, D. Zanchi, Phys. Rev. B **69**, 014411 (2004)
38. H. Yoshino, S. Shodai, K. Murata, Synth. Met. **133–134**, 55 (2003)
39. A.G. Lebed, M.J. Naughton, Phys. Rev. Lett. **91**, 187003 (2003); A.G. Lebed, N.N. Bagmet, M.J. Naughton, Phys. Rev. Lett. **93**, 157006 (2004); A.G. Lebed, H.-I. Ha, M.J. Naughton, Phys. Rev. B **71**, 132504 (2004)
40. H.-I. Ha, A.G. Lebed, M.J. Naughton, Phys. Rev. B **73**, 033107 (2006)
41. N. Matsunaga, A. Ayari, P. Monceau et al., Phys. Rev. B **66**, 024425 (2002)
42. B.K. Cooper, V.M. Yakovenko, cond-mat/0509039
43. S.S.P. Parkin, M. Ribault, D. Jérome et al., J. Phys. Colloq. **14**, L445 (1981); M.Y. Choi, P.M. Chaikin, P. Haen et al., Solid State Commun. **41**, 225 (1982); K. Murata, T. Ukachi, H. Anzai et al., J. Phys. Soc. Jpn. **51**, 695 (1982)
44. H. Schwenk, K. Andres, F. Wudl, Phys. Rev. B **29**, 500 (1984)
45. P. Garoche, R. Brusetti, K. Bechgaard, Phys. Rev. Lett. **49**, 1346 (1982)
46. D. Maily, M. Ribault, K. Bechgaard, J. Phys. **44**, C3:1037 (1983)
47. H. Schwenk, K. Andres, F. Wudl, Schwenk et al., Phys. Rev. B **27**, 5846 (1983)
48. J.H. Park, S.J. Choi, W. Kang, Synth. Met. **103**, 2121 (1999)

49. N. Matsunaga, A. Ishikawa, A. Hoshikawa et al., J. Low Temp. Phys. **117**, 1735 (1999)
50. N. Joo, P. Auban-Senzier, C. Pasquier et al., Eur. Phys. Lett. **72**, 645 (2005)
51. S. Haddad, S. Charfi-Kaddour, M. Héritier et al., J. Low Temp. Phys. **142**, 461 (2006)
52. K. Murata, priviate communication
53. K. Bechgaard, priviate communication
54. J.P. Pouget, priviate communication
55. N. Matsunaga, priviate communication
56. D. Jérome, priviate communication
57. V. Ilakovac, S. Ravy, K. Boubekeur et al., Phys. Rev. B, **56**, 13878 (1997) and references therein
58. N. Joo, C. Pasquier, P. Auban-Senzier et al., Eur. Phys. J. B **40**, 43 (2004)
59. S. Tomić, D. Jérome, D. Mailly et al., J. Phys. **44**, C3:1075 (1983)
60. A.I. Larkin, JETP Lett. **2**, 130 (1965); K. Maki, H. Won, S. Haas, Phys. Phys. B **69**, 012502 (2004)
61. M. Takigawa, H. Yasuoka, G. Saito, J. Phys. Soc. Jpn. **56**, 873 (1987)
62. S. Belin, K. Behnia, Phys. Rev. Lett. **79**, 2125 (1997)
63. H. Shimahara, Phys. Phys. B **61**, R14936 (2000)
64. I.J. Lee, A.P. Hope, M.J. Leone et al., Appl. Supercond. **2**, 7538 (1994); Synth. Met. **70**, 747 (1995)
65. J. Oh, M. Naughton, Phys. Rev. Lett. **92**, 067001 (2004)
66. H. Shimahara, J. Phys. Soc. Jpn. **69**, 1966 (2000); Phys. Phys. B **62**, 3524 (2000)
67. A.G. Lebed, Phys. Rev. Lett. **96**, 037002 (2006)
68. K. Behnia, L. Balicas, W. Kang et al., Phys. Rev. Lett. **74**, 5272 (1995)
69. N. Joo, P. Auban-Senzier, C.R. Pasquier et al., Eur. Phys. J. B **52**, 337 (2006)
70. G.M. Danner, N.P. Ong, P.M. Chaikin, Phys. Rev. Lett. **78**, 983 (1997)
71. S. Haddad, N. Belmechri, S. Charfi-Kaddour et al., submitted to Phys. Rev. B
72. A.G. Lebed, Phys. Rev. Lett. **95**, 247003 (2005)
73. A.G. Lebed, P. Bak, Phys. Rev. B **40**, 11433 (1989)
74. T. Osada, S. Kagoshima, N. Miura, Phys. Rev. Lett. **69**, 1117 (1992)
75. L.P. Gor'kov, A.G. Lebed, Phys. Rev. B **51**, 3285 (1995)
76. A.G. Lebed, Phys. Rev. B **55**, 1299 (1997)
77. M.J. Naughton, R.V. Chamberlin, X. Yan et al., Phys. Rev. Lett. **61**, 621 (1988); R.V. Chamberlin, M.J. Naughton, X. Yan et al., Phys. Rev. Lett. **60**, 1189 (1988)
78. R.C. Yu, L. Chiang, R. Upasani et al., Phys. Rev. Lett. **65**, 2458 (1990)
79. S. Haddad, S. Charfi-Kaddour, C. Nickel et al., Phys. Rev. Lett. **89**, 087001 (2002)
80. M. Fabrizio, Phys. Rev. B **48**, 15838 (1993)
81. S. Uji, S. Yasuzuka, T. Konoike, et al. Phys. Rev. Lett. **94**, 0772061 (2005) and references therein

Part V

Unconventional Superconducting Properties

22

Mott Transition and Superconductivity in Q2D Organic Conductors

K. Kanoda

Superconductivity emerging from strongly correlated electrons has been expected to have something novel and exotic because interesting phases compete with the superconducting phase in many cases. Superconductivity in quasi-two-dimensional ET compounds with the maximum T_c of 14.2 K, which is the highest among the organic superconductors to date, appears in the vicinity of Mott transition. The symmetry of Cooper pair, which reflects the mechanism of superconductivity, is of keen interest. In this chapter, we first review Mott transition in $(ET)_2X$ families as a key phenomenon for the emergence of superconductivity. Then, we describe the experimental status of the problem of pairing symmetry by reviewing NMR as well as other measurements probing the superconductive nature. This chapter is related to several other chapters in this book: History of organic superconductors is surveyed by D. Jerome; theoretical and experimental reviews on the unconventional nature of superconductivity are given by A.G. Lebed and Si Wu, by R.W. Cherng, W. Zhang, and Sa de Melo, the problems of electron correlations are reviewed by C. Bourbonnais and D. Jerome and by T. Giamarchi.

22.1 Introduction to Quasi-Two-Dimensional Organic Conductors

Superconductivity in organic conductors was first discovered in the quasi-one-dimensional system $(TMTSF)_2PF_6$ and subsequently found to compete with the one-dimensional Fermi-surface instability, namely spin–density wave [1] (see also Chap. 1). To avoid this instability and enhance superconducting transition temperature, T_c, chemists have made great efforts to synthesize systems with higher electronic dimensions. Now, various kinds of quasi-two-dimensional superconductors are available and the highest T_c among the organics to date is attained in this class of materials. In this chapter, we shed light on the physics of superconductivity in quasi-two-dimensional organics.

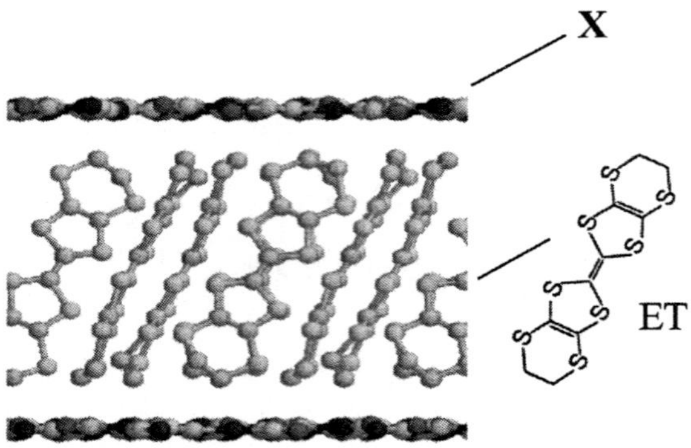

Fig. 22.1. Structure of κ-$(ET)_2X$

The materials dealt with here, κ-$(ET)_2X$, β-$(ET)_2X$ and β'-$(ET)_2X$, which are abbreviated as κ-X, β-X, and β'-X hereafter, are layered materials composed of conducting ET layers with 1/2 hole per ET and insulating layers of various monovalent anions, X^{-1}. Figure 22.1 shows the structure of κ-$(ET)_2X$. The anion, X^{-1}, forms a closed shell, which has no contribution to electronic conduction or magnetism. In the conducting layer, ET molecules form dimers, which are arranged in a checkerboard pattern. Because the dimer carries a hole, the electronic band constructed by the antibonding orbital of the dimer is effectively half filled. Interestingly, the lattice of the dimers is modeled to an isosceles triangular lattice characterized by two inter-dimer transfer integrals, t and t' on the order of 50 meV, whose anisotropy, t'/t, depends on the anion X [2,3]. Thus, the effect of spin frustration is among the interests in this family. The β-X and β'-X have anisotropic triangular lattice with the three sides all inequivalent.

22.2 Mott Transition

These dimeric compounds show a variety of electronic phases including metallic, superconducting and insulating phases, and can be put in a conceptual phase diagram, as shown in Fig. 22.2. The β-I_3 [4,5], κ-$Cu(NCS)_2$ [6], and κ-$Cu[N(CN)_2]Br$ [7] are metals with superconducting transition at low temperatures, and κ-$Cu[N(CN)_2]Cl$ [7] and β'-ICl_2 [8,9] are insulators. The fully deuterated version of κ-$Cu[N(CN)_2]Br$, which is abbreviated as κ-$d[4,4]$-$Cu[N(CN)_2]Br$, is situated on the border of the metal(superconductor)–insulator transition [10]. Considering that the bandwidth of κ-X is varied by replacement of X, it is likely that the Mott transition occurs across κ-$(ET)_2X$. In fact, a rough estimate of the electron–electron repulsive energy U on a dimer gives values of the order of the bandwidth, W, as follows [11,12].

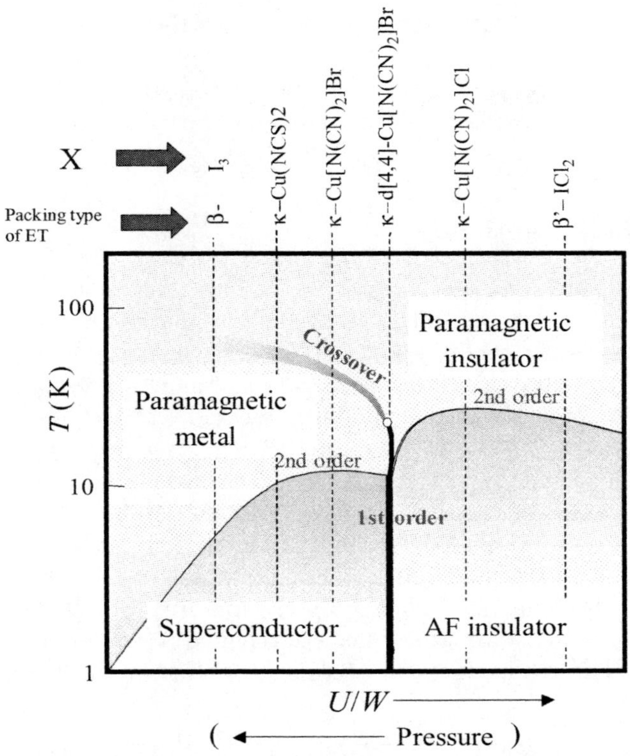

Fig. 22.2. Conceptual phase diagram of κ-$(ET)_2X$ and related dimeric compounds

The two-particle (hole) energy level in the dimer Hubbard model is given by

$$E_2 = 2\epsilon_0 + \frac{U_{\mathrm{ET}}}{2}\left[1 - \sqrt{1 + \left(\frac{4t_{\mathrm{dimer}}}{U_{\mathrm{ET}}}\right)^2}\right], \tag{22.1}$$

where ϵ_0 and U_{ET} are HOMO level and on-site Coulomb energy of ET molecule, respectively, and t_{dimer} is the intradimer transfer energy. Because the one-particle (hole) ground-state level is $E_1 = \epsilon_0 - t_{\mathrm{dimer}}$, the on-dimer Coulomb energy U is $E_2 - 2E_1$, which yields $2t_{\mathrm{dimer}}$ in the limit of $U_{\mathrm{ET}} \gg t_{\mathrm{dimer}}$ [11]. In κ-X, $U_{\mathrm{ET}} \sim 1\,\mathrm{eV}$ and $t_{\mathrm{dimer}} \sim 0.25\,\mathrm{eV}$ [13]. Applying this formula to κ-X, U is estimated at 0.5 eV. On the other hand, the bandwidth, W, is determined by the interdimer transfer integral, t', and roughly given by $2z(t'/2)$ where z is the number of nearest-neighbor dimers and $t'/2$ is the transfer integral between the neighboring antibonding orbitals. For κ-X, $t' \sim 0.1\,\mathrm{eV}$ and $z = 6$ yield $W \sim 0.6\,\mathrm{eV}$. So, U/W is of the order of unity. Although this is an order of estimate neglecting other effects such as intersite Coulomb repulsion V and screening effects, this suggests that κ-X is likely in the critical region of Mott transition [11,14].

22.3 Material Dependence of Normal-State Properties

If material variation is regarded as U/W variation, the critical behavior of Mott transition is extracted from the comparative study of these salts at ambient pressure.

22.3.1 Spin Susceptibility

Figure 22.3 shows spin susceptibility, χ_{spin}, with the diamagnetic core contribution subtracted. Although, at high temperatures, all the salts show temperature–insensitive behaviors with nearly the same values of 4.2–4.5×10^{-4} emu mol^{-1} f.u., χ_{spin} branches off at low temperatures. The metal, β-I_3, situated far from the Mott boundary (see Fig. 22.2) exhibits no remarkable change in χ_{spin} even at low temperatures although a slight decrease is observed below 30 K. However, κ-X shows marked decreases below 50 K and the decrease is steeper as the system is closer to the Mott boundary from κ-$Cu(NCS)_2$ to κ-d[4,4]-$Cu[N(CN)_2]Br$ [10, 15, 16]. It is noted that χ_{spin} of the Mott insulator, κ-$Cu[N(CN)_2]Cl$, is close to those of the metallic members. This means no remarkable anomaly in the uniform ($\boldsymbol{q} = 0$) spin fluctuation across the Mott transition where a marked change occurs in the charge degrees of freedom. The abrupt increase of χ_{spin} at 30 K in κ-$Cu[N(CN)_2]Cl$ and at

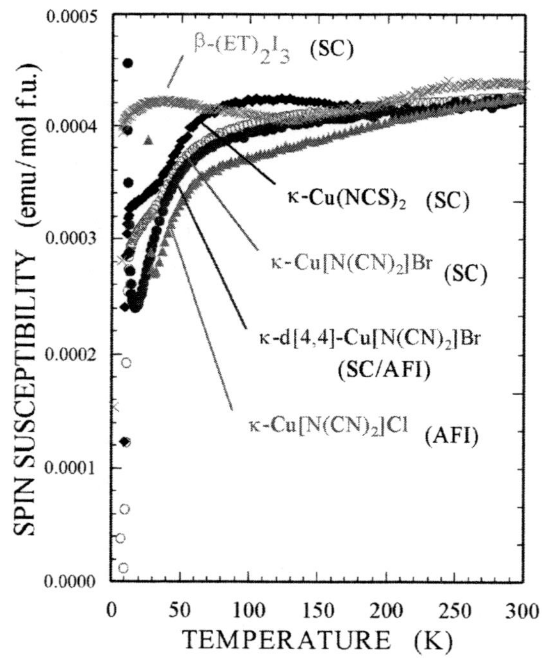

Fig. 22.3. Spin susceptibility of κ-$(ET)_2X$ and β-$(ET)_2I_3$

15 K in κ-d[4,4]-Cu[N(CN)$_2$]Br is due to canting of the spins that undergo antiferromagnetic transition at such temperatures [10, 15].

22.3.2 NMR Relaxation Rate

The spin fluctuations at finite wave numbers ($\boldsymbol{q} \neq 0$) are probed by NMR relaxation rate, $1/T_1$. To probe the fluctuations with high sensitivity, the central carbon with hyperfine coupling several tens times larger than that at the ^{1}H site in ET is a desirable site for NMR. The powder-averaged ^{13}C nuclear spin–lattice relaxation rate divided by temperature, $(T_1T)^{-1}$, is plotted in Fig. 22.4. The Mott insulator, κ-Cu[N(CN)$_2$]Cl, shows a divergent peak at 27 K, which indicates antiferromagnetic transition [17]. The temperature dependence above 27 K in κ-Cu[N(CN)$_2$]Cl salt shows the critical magnetic fluctuations. Remarkably, the superconducting salts κ-Cu(NCS)$_2$ and κ-Cu[N(CN)$_2$]Br trace just this curve above 60 K, which means that the antiferromagnetic spin fluctuations survive even in the superconducting salts at high temperatures [18]. The turnabout of $(T_1T)^{-1}$ at 55 K for κ-Cu(NCS)$_2$ and 45 K for κ-Cu[N(CN)$_2$]Br is an indication of the suppression

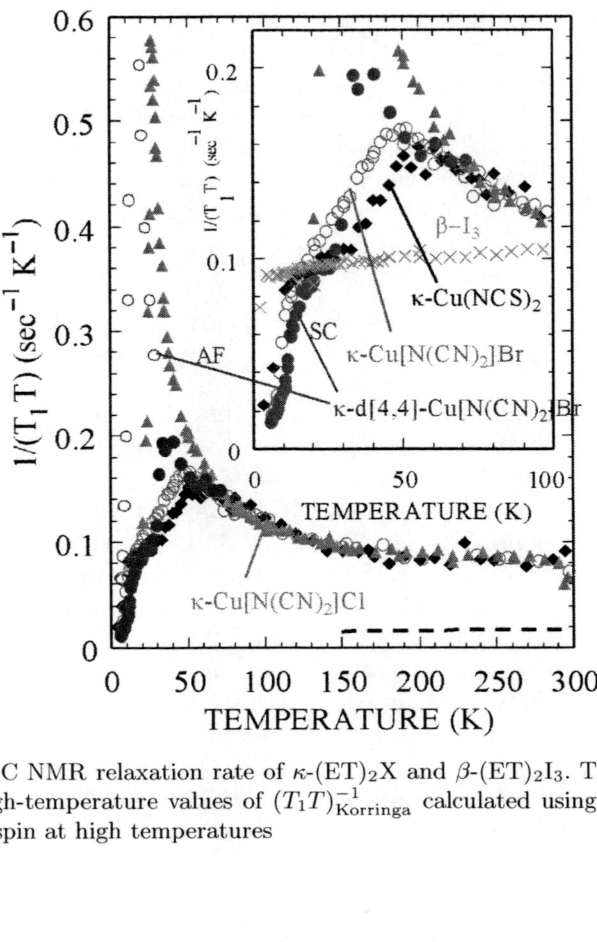

Fig. 22.4. ^{13}C NMR relaxation rate of κ-(ET)$_2$X and β-(ET)$_2$I$_3$. The *broken line* shows the high-temperature values of $(T_1T)^{-1}_{\text{Korringa}}$ calculated using the material-independent spin at high temperatures

of spin fluctuations. These coincide with the inflection point in the temperature dependence of resistivity; namely, the change of spin fluctuations is associated with insulator–metal crossover. In the low-temperature metallic phase, there is an interesting systematic in $(T_1T)^{-1}$. The metal, β-I_3, shows a temperature–insensitive $(T_1T)^{-1}$, a typical behavior of metals. However, as the system approaches the Mott boundary from the metallic side in the order of β-I_3, κ-Cu(NCS)$_2$, κ-Cu[N(CN)$_2$]Br, and κ-d[4,4]-Cu[N(CN)$_2$]Br, the slope of $(T_1T)^{-1}$ at low temperatures becomes higher although the $(T_1T)^{-1}$ values just above T_c are nearly the same for all the salts. This shows that the antiferromagnetic fluctuations at $\boldsymbol{q} \neq 0$ are anomalously depressed on cooling near the Mott transition. Considering the low-temperature decrease in χ_{spin}, the depression of such fluctuations occurs over all $\boldsymbol{q}$ vectors. The sudden decrease of $(T_1T)^{-1}$ below ~10 K for κ-Cu(NCS)$_2$ and κ-Cu[N(CN)$_2$]Br is due to superconducting transition.

The NMR enhancement factor, K_α, defined by the ratio of the observed relaxation rate to the expected one from the uncorrelated Korringa relation, namely $K_\alpha = (T_1T)^{-1}_{\text{obs}}/(T_1T)^{-1}_{\text{Korringa}}$, assesses the relative $\boldsymbol{q}$ profile of spin fluctuations. If $K_\alpha < 1$, spin fluctuations are enhanced at $\boldsymbol{q} = 0$ relative to those at other $\boldsymbol{q}$ vectors; in contrast, if $K_\alpha > 1$, fluctuations are enhanced at finite $\boldsymbol{q}$ vectors. The former is ferromagnetic while the latter is antiferromagnetic. The powder-averaged $(T_1T)^{-1}_{\text{Korringa}}$ is calculated using the observed χ_{spin} and the ^{13}C-site hyperfine coupling tensors, which are assumed not to depend on X [18–20]. The obtained K_α is 5–10 and there is a clear tendency that it is increased as the system approaches Mott transition. This indicates that the finite-$\boldsymbol{q}$ spin fluctuations are prominent compared with the $\boldsymbol{q} = 0$ ones near Mott transition although spin fluctuation intensity is depressed as seen in χ_{spin} and $(T_1T)^{-1}$.

22.3.3 Systematic Variation of Low-Temperature Properties with U/W

The material dependence of χ_{spin}, $(T_1T)^{-1}$, and electronic specific heat coefficient, γ, in the low-temperature limit are shown in Fig. 22.5. This is regarded as the U/W dependence of the quantities. As for χ_{spin} and $(T_1T)^{-1}$, their temperature dependences above T_c are lineally extrapolated to 0 K. The γ is the value under high magnetic fields which destroy superconductivity [21]. According to the Brinkman and Rice's picture of Mott transition, electron correlation is renormalized to the quasiparticle mass of the Fermi liquid, which would be enhanced in a divergent manner [22]. However, it is not the case in the present quasi-two-dimensional systems as seen in the figure. The γ, which is directly related to effective mass, tends to decrease near Mott transition. The effective masses of κ-Cu(NCS)$_2$ and κ-Cu[N(CN)$_2$]Br determined from Shubnikov–de Haas oscillation [13, 23, 24] are consistent with this tendency, but decreases under pressure [25], which is understood in terms of a

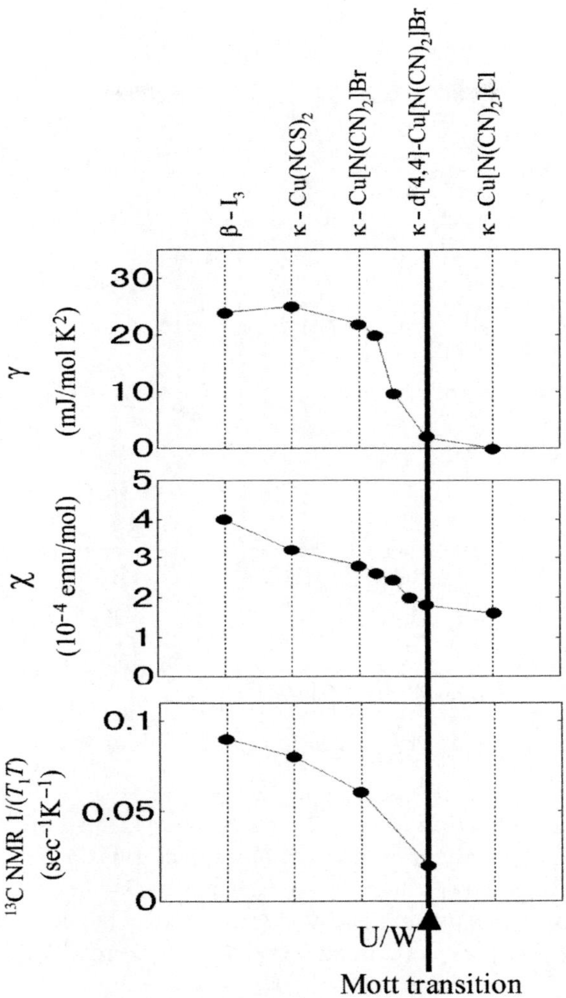

Fig. 22.5. Material dependence of electronic specific heat coefficient, γ, paramagnetic susceptibility, χ, and NMR relaxation rate divided by temperature, $(T_1T)^{-1}$, in the low temperature limit. To be free from the effect of the superconductivity γ is obtained under magnetic fields in excess of H_{c2}, and linear portions in χ and $(T_1T)^{-1}$ above T_c are extrapolated to 0 K. The *vertical bold line* stands for the metal–insulator transition

band widening far from the Mott transition. The $\chi_{\rm spin}$ and $(T_1T)^{-1}$ show the similar tendencies although they include the spin–fluctuation-related factors.

The material dependence of the ground-state properties is shown in Fig. 22.6, where the superconducting transition temperature T_c, Neel temperature T_N and the AF moment are plotted. T_c is enhanced near the Mott transition but is not correlated to the density of states probed by γ. In this sense, the original BCS picture seems not relevant to the present systems.

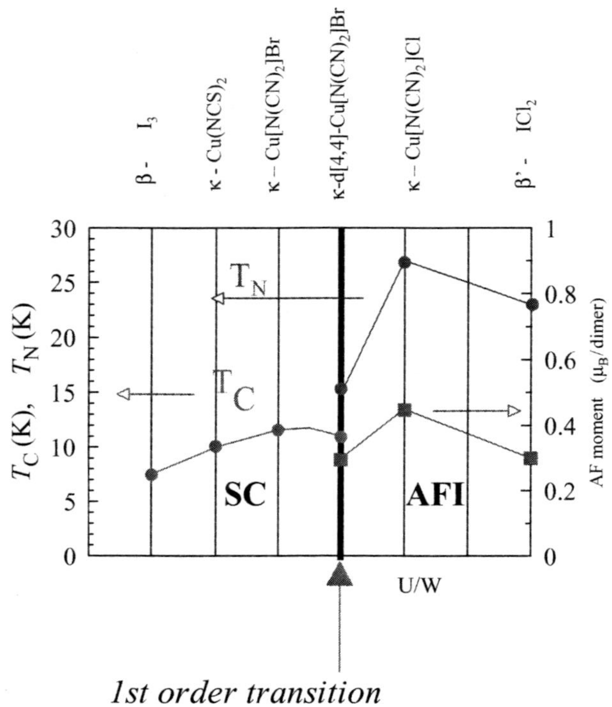

Fig. 22.6. Material dependence of T_c, T_N, and AF moment

Instead, an apparent correlation between K_α and T_c suggests that the relative weight of the AF components in the whole spin fluctuations may be relevant to the occurrence of superconductivity in this family. Such mechanisms for superconductivity have been proposed in several contexts of theoretical frameworks and predicted the d-wave symmetry in Cooper pairing [26–34].

22.3.4 Pressure Study of Mott Transition

To precisely examine the phase diagram not in a conceptual manner, experiments on a fixed material at various pressures and temperatures are needed. The Mott transition is a MI transition without symmetry breaking like the gas–liquid transition, which should be first-order transition or crossover. Therefore, the first-order transition at low temperatures suggested above can have a critical point to the crossover. The pressure study of κ-Cu[N(CN)$_2$]Cl gives the pressure–temperature phase diagram as shown in Fig. 22.7 and revealed detailed nature of Mott transition; namely the existence of critical endpoint of the first-order transition [35, 36] and unconventional criticality [37], which may be addressed as manifestation of marginal quantum critical nature [38]. Possible relationship between the anomalous criticality of Mott transition and the emergence of superconductivity is an intriguing future

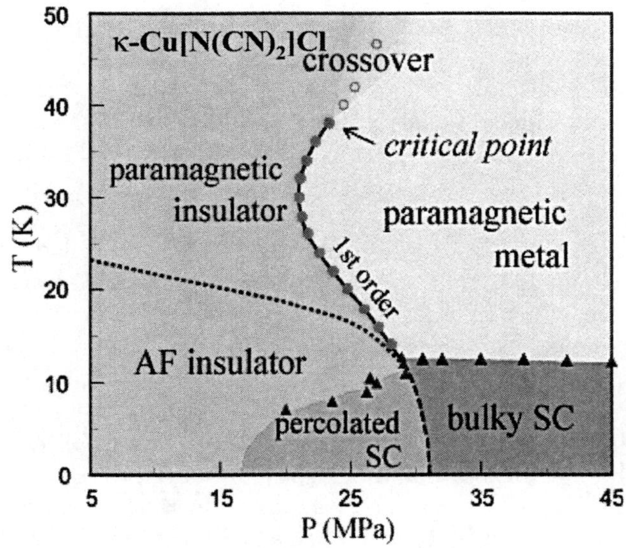

Fig. 22.7. Pressure–temperature phase diagram of κ-Cu[N(CN)$_2$]Cl. The metal–insulator boundary is defined by anomalies in resistivity

problem. Another Mott insulator, β'-ICl$_2$, has a quite high Mott critical pressure of 8.2 GPa, where superconductivity with the highest T_c (14.2 K) among organics appears [39].

22.4 Nature of Superconductivity

The pairing symmetry of Cooper pair, which is closely related to the mechanism of superconductivity, is probed by several experiments such as the temperature and field dependences of penetration depth, electronic specific heat and thermal conductivity, and angular dependence of scanning tunneling spectroscopy (STS) and NMR. This section is mainly devoted to the description of NMR and specific heat studies on superconductivity.

22.4.1 NMR

The first NMR experiment on superconductivity in quasi-two-dimensional compounds was ^{1}H NMR of β-I$_3$ [40]. The result was striking: the relaxation rate, $1/T_1$, formed a highly enhanced peak well below T_c. In the subsequent ^{1}H NMR studies of κ-Cu(NCS)$_2$ and κ-Cu[N(CN)$_2$]Br, the enhancement of $1/T_1$ amounted to more than ten times the value at T_c [41,42]. The feature is different from what is expected in the conventional s-wave pairing, where the $1/T_1$ enhancement is about two at most and occurs just below T_c. The results tempt one to imagine something novel in the pairing mechanism. Before going on to discuss the physical implications, however, one has to clarify what local field the nuclei are in.

There are two kinds of local fields at the nuclear sites in the superconducting state under magnetic field. One is from the thermally activated quasiparticles and the other is from vortices. The latter field is further divided into core-induced quasiparticle contribution and the spatial/temporal field fluctuations of flux quanta. Because the information on pairing symmetry is in the former contribution, the vortex contribution to shift and relaxation rate needs to be eliminated. In layered superconductors modeled by the Josephson-coupled superconducting multilayers like the present materials, this is attained by applying the magnetic field parallel to the layers. Figure 22.8a shows temperature dependence of $1/(T_1T)$ at three different sites (central ^{13}C, outer ^{13}C and ^{1}H sites) in ET as shown in Fig. 22.8b. The applied magnetic field is 1.5 T perpendicular to the layer. The value of $1/(T_1T)$ is normalized to the values of the normal metallic state above T_c, where the relaxation is wholly due to quasiparticles. Therefore, even in the superconducting state, the quasiparticle contributions to the normalized $1/(T_1T)$ should be the same at all three sites. The observed site-dependent variation of the behaviors below T_c is an indication of the other contribution, namely the field fluctuations of flux quanta, and its strong site-dependence. In the order of the central ^{13}C, outer ^{13}C and ^{1}H, the hyperfine coupling to the quasiparticles is weaker, while the flux fluctuations contribute to $1/(T_1T)$ at all sites equally. Therefore, the excess contributions in $1/(T_1T)$ normalized to quasiparticle contribution is due to the dynamics of flux quanta. This contribution is seen to have a peak around 2 K, where the vortex liquid–solid transition is considered to occur. This is the origin of the huge enhancement of ^{1}H $1/T_1$ in the earlier studies [40–42].

In the central ^{13}C sites, the primary contribution to $1/T_1$ is the quasiparticle relaxation, which is shown by shaded region in Fig. 22.8a, although the vortex fluctuation contribution is appreciable as a small peak. Thus, the NMR experiment at the central ^{13}C sites is the best way to probe the pairing symmetry. In this field configuration, however, most of quasiparticles are in the vortex core. To remove the core-induced contribution, the magnetic field should be aligned parallel to the layers so that the Josephson vortices are locked in between the layers without in-plane pancake cores. Because ^{1}H NMR relaxation rate is sensitive to the vortices as seen above, ^{1}H NMR can be used to find the lock-in field condition. Figure 22.9 shows the angular dependence of T_1 at ^{1}H site at 5 K for κ-Cu(NCS)$_2$. The T_1 shows a sharp peak in a narrow range of angle within $\pm 0.12°$ around the parallel field geometry, where the pancake vortices are considered to disappear. According to the simple model, where the lock-in state is realized when the perpendicular component of the external filed is lower than H_{c1}, the condition of the lock-in state should be $H_{c1} > H \sin\theta$, where θ is angle between external field and superconducting layers. The H_{c1} value of 15 Oe [43] yields $\theta_L = 0.15°$. The $2\theta_L$ ($= 0.3°$) is in good agreement with the observed value of 0.25°.

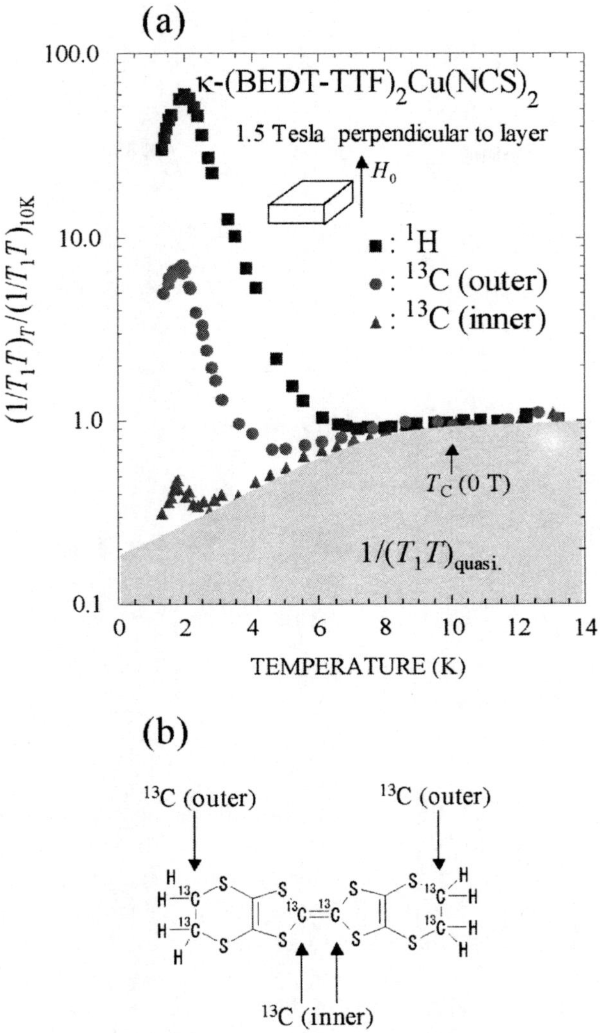

Fig. 22.8. (**a**) Temperature dependence of $1/(T_1T)$ of a single-crystal κ-Cu(NCS)$_2$ at selective ^{13}C and ^{1}H sites (see (**b**)) under fields perpendicular to the conducting layer. The shaded part expresses a quasiparticles contribution

The ^{13}C $1/T_1$ for κ-Cu(NCS)$_2$ and κ-Cu[N(CN)$_2$]Br under magnetic fields of 15 and 23 kOe, respectively, adjusted to the lock-in condition are shown as a function of reduced temperature, T/T_c, in Fig. 22.10. In general, $1/T_1$ is proportional to γ_n^2. The γ_n of ^{1}H is about four times as large as that of ^{13}C, so that T_1 at the ^{1}H site should be 16 times shorter than that at the ^{13}C site under the same field fluctuations. Using the value of $^1T_1 \sim 250$ s, the vortex motion component of $1/^{13}T_1$ at 4.5 K is expected to be 0.00025 s^{-1}, which is negligible compared with the observed $1/^{13}T_1$ value of 0.01 s^{-1} at 4.5 K, which

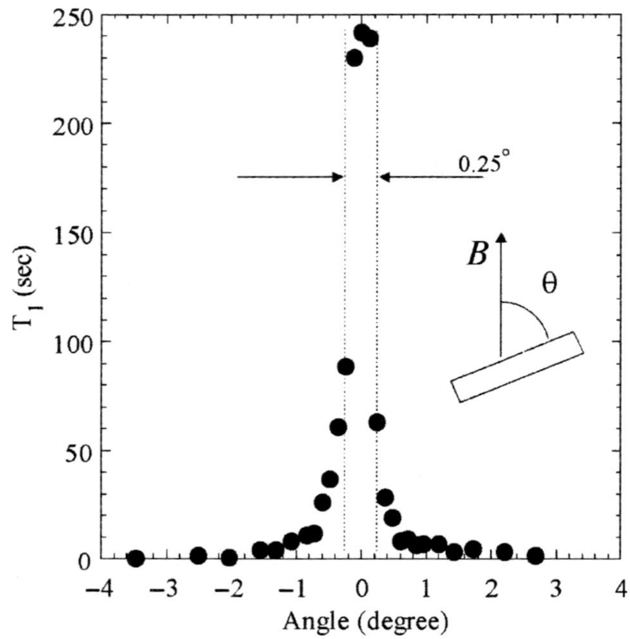

Fig. 22.9. Angular dependence of nuclear spin–lattice relaxation time at ^{1}H site of κ-Cu(NCS)$_2$ under a field of 3.5 kOe and at 5 K

supports that the observed $1/T_1$ is governed by the thermally excited quasiparticles. The conventional s-wave pairing has two distinctive features in NMR relaxation rate; which are an increase just below T_c called the Hebel-Slichter coherence peak, and a decrease at low temperatures following an exponential function of temperature due to the gap opening in the quasiparticle excitation. The experimental data have neither of the characteristics; $1/T_1$ decreases immediately below T_c without any enhancement and follows a T^3-law at low temperatures. Both features strongly suggest the existence of line nodes in gap parameter on the Fermi surface. The superconducting state is likely of non-s-wave in nature, although the strongly anisotropic s-wave state with gap minima smaller than the thermal energy at the lowest temperature (1.5 K), cannot be ruled out. As seen in Fig. 22.11, there is no difference in the profile of $1/T_1$ between the two salts. The κ-Cu[N(CN)$_2$]Br was also studied at higher fields up to 78 kOe [19] and at a lower field of 6 kOe [20], but the results are nearly field-independent in this field range.

The spin susceptibility is probed by Knight shift, namely the position of NMR spectra. Figure 22.11a shows the temperature dependence of ^{13}C NMR spectra of κ-Cu(NCS)$_2$. The distribution of absorption lines above T_c comes from nonequivalent carbon sites with different Knight shifts in the normal state. Below T_c, they move and merge around 120 ppm, which is

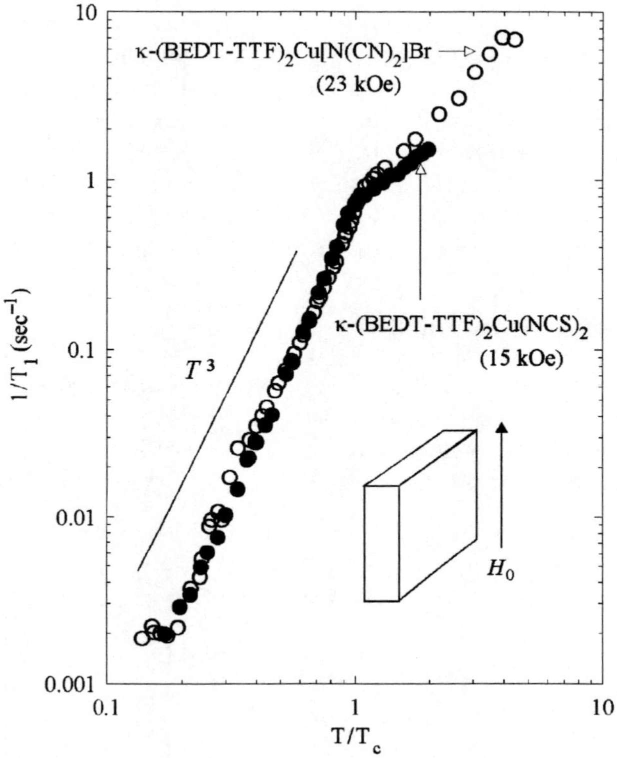

Fig. 22.10. Temperature dependence of $1/T_1$ at the ^{13}C site for single-crystal κ-Cu[N(CN)$_2$]Br and κ-Cu(NCS)$_2$ plotted as a function of T/T_c

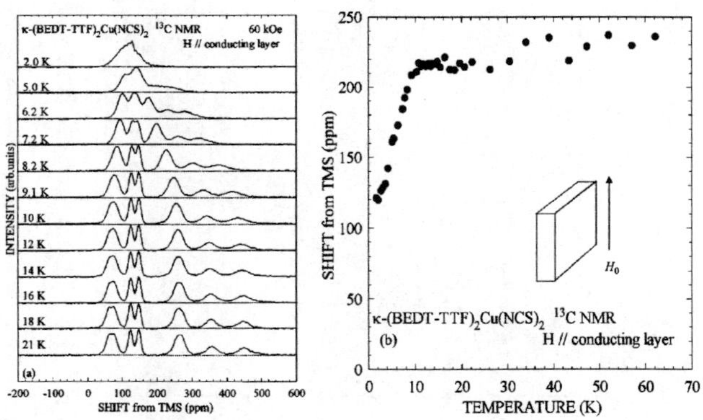

Fig. 22.11. Temperature dependence of (**a**) ^{13}C NMR spectra and (**b**) NMR shift defined by its first moment for a single-crystal κ-Cu(NCS)$_2$ under a magnetic field of 60 kOe parallel to the layer

nearly the chemical shift as observed for the neutral ET molecules. The temperature dependence of the first moment of the spectra is shown in Fig. 22.11b. This indicates that the spin susceptibility decreases below T_c and is almost vanishing at low temperatures, providing a clear evidence for the spin-singlet superconducting state. The similar Knight shift profile is seen in κ-Cu[N(CN)$_2$]Br [19, 20, 44]. All these NMR results suggest that the superconductivity is of d-wave nature with no controversy between the results by three independent research groups in France, US, and Japan [19, 20, 44].

22.4.2 Specific Heat and Other Experiments

Because Cooper pairs carry no entropy, specific heat probes the quasiparticle density of states, which is sensitive to pairing symmetry. Figure 22.12 shows temperature dependence of C_{el}/T of κ-Cu[N(CN)$_2$]Br, where C_{el} is the electronic contribution to specific heat [45]. If the superconducting state

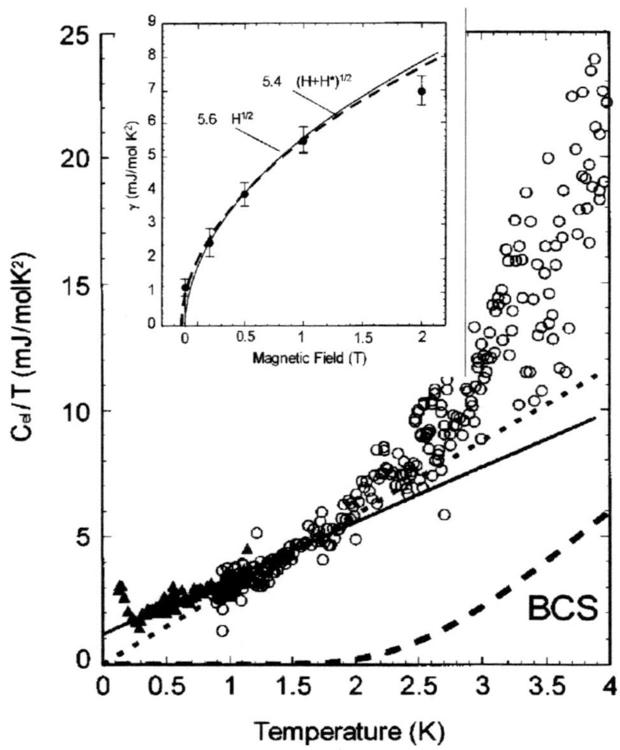

Fig. 22.12. Temperature dependence of electronic specific heat divided by temperature for κ-Cu[N(CN)$_2$]Br. Inset shows field dependence of the low-temperature g term

has the isotropic gap of $2\Delta/k_{\mathrm{B}}T_{\mathrm{c}} = 3.52$ around the Fermi level as is predicted by the BCS-weak coupling theory, C_{el}/T would show the activation type of temperature dependence as shown by the broken line. It is obvious that the observed C_{el}/T values are far larger than the BCS prediction and vary linearly even at low temperatures below 2 K: C_{el} obeys a quadratic temperature dependence. The form of $C_{\mathrm{el}} = \alpha T^2$ is expected in unconventional superconductivity with line nodes in gap function and $\alpha = 3.3 k_{\mathrm{B}}\gamma_{\mathrm{n}}/\Delta_{\max}$ is predicted in the clean limit with $\Delta_{\max}$ the maximum of the gap and γ_{n} the electronic specific-heat coefficient in the normal state [46]. Using the experimental value of $\gamma_{\mathrm{n}}(= 22\,\mathrm{mJ\,mol^{-1}K^{-2}})$ and assuming the weak-coupling formula of $2\Delta/k_{\mathrm{B}}T_{\mathrm{c}} = 3.52$, α yields $3.5\,\mathrm{mJ\,mol^{-1}K^{-3}}$, which is somewhat larger than the observed value ($\alpha = 2.2\,\mathrm{mJ\,mol^{-1}K^{-3}}$ for the solid line or $\alpha = 2.8\,\mathrm{mJ\,mol^{-1}K^{-3}}$ for the dashed line) but would fall into the reasonable range in the strong coupling regime with $2\Delta/k_{\mathrm{B}}T_{\mathrm{c}} > 3.52$. The additional feature of the data is the presence of finite γ value ($\gamma_{\mathrm{res}} = 1.2\,\mathrm{mJ\,mol^{-1}K^{-2}}$) in the low temperature limit. This is attributable to the disorder effect in the context of unconventional superconductivity [47].

Inset of Fig. 22.12 shows the recovery of γ under elevated fields [45]. The experimental values follow a square root of field, as fitted by the solid and broken lines. This is consistent with the line-nodal superconductivity, which is predicted to show the field dependence of $\gamma(H) = k\gamma_{\mathrm{n}}(H/H_{\mathrm{c2}})^{1/2}$ with k of order unity [48]. Using the values of $H_{\mathrm{c2}} = 10\,\mathrm{T}$ and $\gamma_{\mathrm{n}} = 22\,\mathrm{mJ\,mol^{-1}K^{-2}}$ for κ-Cu[N(CN)$_2$]Br, the fitting of the data gives $k = 0.8$, which is in agreement with the prediction.

It should be mentioned that there are several controversial results on specific heat, which are reported to be consistent with conventional gapped superconductivity [49–51]. Many reports on penetration depth measurements suggest line-nodal superconductivity in κ-(ET)$_2$X [52–57], but some others suggest gapped one [58–60]. The STS [61], thermal conductivity [62, 63], and millimeter-wave transmission experiments are all consistent with nodal superconductivity. The mechanism of the electron pairing has long been discussed to be molecular vibration mediated, lattice–phonon mediated, or purely electronic [1]. Paying attention to the fact that the superconductivity appears in compounds containing molecules with the TTF structure, some specific molecular vibration was proposed to be responsible for the occurrence of superconductivity [1, 64]. The analysis of the optical data suggests that the former two couplings are enhanced in ET compounds [65]. On the other hand, the fact that the superconducting phase abuts on the antiferromagnetic phase suggests the possible involvement of antiferromagnetic spin fluctuations in the superconductivity. Theoretical investigations along this line have been intensive [26–34]. Although one cannot make a decisive assessment on the superconducting mechanism at the present stage, many experimental results seem compatible with the consequence of the spin fluctuation-mediated model.

22.5 Pseudogap Behavior

The κ-d[4,4]-Cu[N(CN)$_2$]Br situated just on the Mott boundary exhibits a phase separation into metallic and insulating fractions below 30 K, as clearly identified by the NMR spectral separation [15, 66, 67]. Therefore, the relaxation rates of both phases across the Mott line can be measured separately. Figure 22.13 shows the temperature dependence of $(T_1T)^{-1}$ measured with the applied field parallel to the conducting layers of a single crystal of κ-d[4,4]-Cu[N(CN)$_2$]Br [15]. The $1/(T_1T)$ of the insulating phase continuously increases below 30 K, implying that this phase is a low-temperature continuation of the high-temperature single phase, and forms a divergent peak at 15 K indicating the magnetic order. The whole feature of $1/(T_1T)$ is the same as for the Mott insulator, κ-Cu[N(CN)$_2$]Cl, except for the Neel temperature. The $1/(T_1T)$ of the metallic phase exhibits a sudden drop around 30 K, indicating the suppression of antiferromagnetic spin fluctuations. It is also remarkable that $1/(T_1T)$ shows an anomalous decrease below 30 K, as seen in the inset of Fig. 22.13. Since the superconducting transition temperature of κ-d[4,4]-Cu[N(CN)$_2$]Br is 11 K [16], it turns out that the decrease in $1/(T_1T)$ begins at much higher temperatures than T_c and becomes steeper as

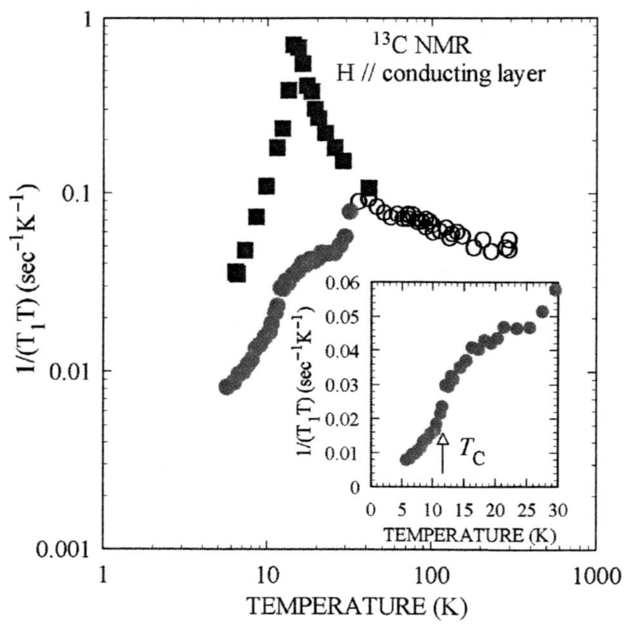

Fig. 22.13. Temperature dependence of $1/(T_1T)$ at the ^{13}C site of the superconducting (*closed circles*) and antiferromagnetic (*closed squares*) phases for the single-crystal κ-d[4,4]-Cu[N(CN)$_2$]Br below 30 K and of the high-temperature phase (*open circles*) above 30 K. Inset shows temperature dependence of $1/(T_1T)$ of the superconducting phase at low temperatures in linear scales

temperature approaches T_c. This behavior is reminiscent of the so-called pseudogapped behavior observed in the underdoped high-T_c cuprates [68]. It is interesting that the bandwidth controlled organics and the band-filling controlled cuprates share one of the most mysterious features in the vicinity of the Mott transition.

The superconducting and insulating phases in κ-(ET)$_2$X meet on the boundary material, κ-d[4,4]-Cu[N(CN)$_2$]Br. At this border, the pseudogapped superconducting phase with T_c of 11 K abuts on the commensurate antiferromagnetic insulating phase with T_N of 15 K. The transition between superconductor and antiferromagnet is of the first order with T_c and T_N not so different at the border.

22.6 Perspectives

As forthcoming issues on quasi-two-dimensional organic superconductors, we raise two topics. One is the effect of spin frustration on superconductivity. As mentioned in Sect. 25.1, the materials in question are modeled to anisotropic triangular lattices. Noticeably, the ambient-pressure Mott insulator κ-Cu$_2$(CN)$_3$ has a nearly isotropic triangular lattice, where spins are highly frustrated. The ^{1}H and ^{13}C NMR experiments show no indication of magnetic ordering down to 30 mK. The spins are likely in the quantum liquid state [69]. Under pressure, it undergoes Mott transition to a Fermi liquid, which exhibits superconductivity at low temperatures [70]. Figure 22.14 shows

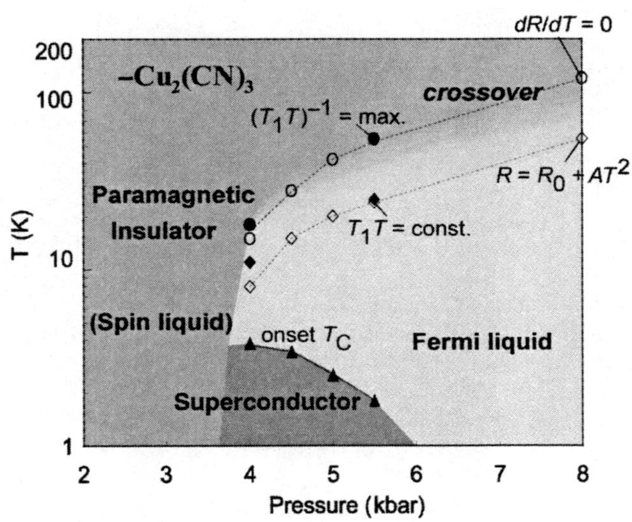

Fig. 22.14. Pressure–temperature phase diagram of κ-Cu$_2$(CN)$_3$. The metal–insulator boundary is defined by anomalies in resistivity and NMR relaxation rate consistently

the pressure–temperature phase diagram of κ-$Cu_2(CN)_3$ constructed on the basis of transport and NMR data. Superconductivity neighbors on spin liquid instead of antiferromagnetic order. The superconductivity emergent from spin liquid is not known so far and is attracting interest.

The other issue is hole doping. The carrier doping, which has proved to be an useful method for finding novel phases in the study of oxides, does not do well in organics. However, there are some exceptions, among which are κ-$(ET)_4Hg_{3-\delta}Z_8$ [Z = Br, Cl] [71]. The ET molecules form the κ-type arrangement. If δ were 0, the band would be half filled and the Mott insulator is expected by the estimate of much larger values of U/W than those of other members in κ-X. In reality, the δ values are 0.11 and 0.22 for Z = Br and Cl, respectively, and contribute to hole doping to the ET layers, consistent with the metallic and superconducting ground states observed. In particular, the Br salt with the t'/t value of nearly unity turns out to be a doped triangular lattice. Although the δ values determined by chemistry of the compounds cannot be varied, these materials provide an opportunity to study the doping effect and, under pressure, the physics of the electron correlation in a comprehensive manner from the strong coupling t–J region to the weakly coupled Hubbard region. The anomalous low-temperature enhancement in NMR relaxation rate and electronic specific heat coefficient γ is observed at ambient pressure [72]. The superconductivity appearing there is quite intriguing.

Acknowledgments

The author is greatly indebted to his colleagues, who contributed to the works presented here: K. Miyagawa, A. Kawamoto, Y. Nakazawa, H. Taniguchi Y. Shimizu, F. Kagawa, and Y. Kurosaki. These works were supported by CREST-JST, MEXT, and JSPS.

References

1. T. Ishiguro, K. Yamaji, G. Saito, *Organic Superconductors* (Springer, Berlin Heidelberg New York 1990)
2. H. Kino, H. Fukuyama, J. Phys. Soc. Jpn. **64**, 2726 (1995)
3. K. Kanoda, J. Phys. Soc. Jpn. **75**, 051007 (2006)
4. E.B. Yagubskii et al., JETP Lett. **39**, 12 (1984)
5. K. Murata et al., J. Phys. Soc. Jpn. **54**, 1236 (1985)
6. H. Urayama et al., Chem. Lett. 55 (1988)
7. J.M. Williams et al., Science **252**, 1510 (1991)
8. H. Kobayashi et al., Chem. Lett. 89 (1986)
9. M. Tokumoto et al., Synth. Met. **19**, 215 (1987)
10. A. Kawamoto et al., Phys. Rev. B **55**, 14140 (1997); J. Am. Chem. Soc. **120**, 10984 (1998)
11. K. Kanoda, Hyperfine Interact. **104**, 235 (1997); Physica C **287**, 299 (1997)
12. K. Miyagawa et al., Chem. Rev. **104**, 5635 (2004)

13. K. Oshima et al., Phys. Rev. B **38**, 938 (1988)
14. McKenzie, Science **278**, 820 (1997)
15. K. Miyagawa et al., Phys. Rev. Lett. **89**, 017003 (2002)
16. H. Taniguchi et al., Phys. Rev B **67**, 014510 (2003); Phys. Rev. B **59**, 8424 (1999)
17. K. Miyagawa et al., Phys. Rev. Lett. **75**, 1174 (1995)
18. A. Kawamoto et al., Phys. Rev. Lett. **74**, 3455 (1995); Phys. Rev. B **52**, 15522 (1995)
19. H. Mayaffre et al., Europhys. Lett. **28**, 205 (1994)
20. S.M. Soto et al., Phys. Rev. B **52**, 10364 (1995)
21. Y. Nakazawa et al., Phys. Rev. B **61**, R16295 (2000)
22. W.F. Brinkman, T.M. Rice, Phys. Rev. B **2**, 432 (1970)
23. T. Sasaki et al., Solid State Commun. **76**, 507 (1990)
24. C.H. Mielke et al., Phys. Rev. B **56**, R4309 (1997)
25. J. Caulfield et al., J. Phys. Condens. Matter **6**, 2911 (1994)
26. H. Kino, H. Kontani, J. Phys. Soc. Jpn. **67**, 3691 (1998)
27. K. Kondo, T. Moriya, J. Phys. Soc. Jpn. **67**, 3695 (1998)
28. J. Schmalian et al., Phys. Rev. Lett. **81**, 4232 (1998)
29. M. Vojta, E. Daggoto, Phys. Rev. B **59**, R713 (1999)
30. K. Kuroki et al., Phys. Rev. B **65**, 100516 (2002); Phys. Rev. B **60**, 3060 (1999)
31. Y. Yanase et al., Phys. Rep. **387**, 1 (2003)
32. J. Liu et al., Phys. Rev. Lett. **94**, 127003 (2005)
33. J.B. Powel, R.H. McKenzie, Phys. Rev. Lett. **94**, 047004 (2005)
34. T. Watanabe et al., Physica C **426**, 289 (2005)
35. P. Limelette et al., Phys. Rev. Lett. **91**, 016401 (2003)
36. F. Kagawa, K. Miyagawa, K. Kanoda, Phys. Rev. B **69**, 064511 (2004); Phys. Rev. Lett. **93**, 127001 (2004)
37. F. Kagawa, K. Miyagawa, K. Kanoda, Nature **436**, 534 (2005)
38. M. Imada, J. Phys. Soc. Jpn. **73**, 1851 (2004); Phys. Rev. B **72**, 075113 (2005)
39. H. Taniguchi et al., J. Phys. Soc. Jpn. **72**, 468 (2003)
40. F. Creuzet et al., Physica B **143**, 363 (1986)
41. T. Takahashi et al., Physica C **153–155**, 487 (1988)
42. T. Takahashi, K. Kanoda, G. Saito, Physica C **185–189**, 366 (1991)
43. P.A. Mansky, P.M. Chaikin, R.C. Haddon, Phys. Rev. B **50**, 15929 (1994)
44. K. Kanoda, K. Miyagawa, A. Kawamoto, Y. Nakazawa, Phys. Rev. B **54**, 76 (1996)
45. Y. Nakazawa, K. Kanoda, Phys. Rev. B **55**, R8670 (1997)
46. K.A. Moler et al., Phys. Rev. Lett. **73**, 2744 (1994)
47. M. Prohammer et al., Phys. Rev. B **47**, 15152 (1993)
48. G.E. Volovik, JETP Lett. **58**, 469 (1993)
49. H. Elsinger, J. Wosnitza, S. Wanka, J. Hagel, D. Schweitzer, W. Strunz, Phys. Rev. Lett. **84**, 2000 6098 (2000)
50. A.E. Kovalev, T. Ishiguro, J. Yamada, S. Takasaki, H. Anzai, J. Exp. Theor. Phys. **92**, 1035 (2001)
51. J. Muller, M. Lang, R. Helfrich, F. Steglich, T. Sasaki, Phys. Rev. B **65**, 140509 (2002)
52. K. Kanoda, K. Akiba, K. Suzuki, T. Takahashi, G. Saito, Phys. Rev. Lett. **65**, 1271 (1990)

53. A. Carrington, I.J. Bonalde, R. Prozorov, R.W. Giannetta, A.M. Kini, J. Schlueter, H.H. Wang, U. Geiser, J.M. Williams, Phys. Rev. Lett. **83**, 4172 (1999)
54. M. Pinteric, S. Tomic, M. Prestrer, D. Drobac, O. Milat, K. Maki, D. Schweitzer, I. Heinen, W. Strunz, Phys. Rev. B **61**, 7033 (2000)
55. M. Pinteric, S. Tomic, M. Prester, D. Drobac, K. Maki, Phys. Rev. B **66**, 174521 (2002)
56. L.P. Le, G.M. Luke, B.J. Sternlieb, W.D. Wu, Y.J. Uemura, J.H. Brewer, T.M. Riseman, C.E. Stronach, G. Saito, H. Yamochi, H.H. Wang, A.M. Kini, K.D. Carlson, J.M. Williams, Phys. Rev. Lett. **68**, 1923 (1992)
57. D. Achkir, M. Poirer, C. Bourbonnas, G. Qurion, C. Lenoir, P. Batail, D. Jerome, Phys. Rev. B **47**, 11595 (1993)
58. D.R. Harshman, A.T. Fiary, R.C. Haddon, M.L. Kaplan, T. Pfiz, E. Koster, I. Shinkoda, D.Ll. Williams, Phys. Rev. B **49**, 12990 (1994)
59. M. Dressel, O. Klein, G. Gruner, K.D. Carlson, H.H. Wong, J.M. Williams, Phys. Rev. B **50**, 13603 (1994)
60. M. Lang, N. Toyota, T. Sasaki, H. Sato, Phys. Rev. Lett. **69**, 1443 (1992)
61. T. Arai, K. Ichimura, K. Nomura, S. Takasaki, S. Nakatsuji, H. Anzai, Phys. Rev. B **63**, 104518 (2001)
62. S. Belin, K. Behnia, A. Deluzet, Phys. Rev. Lett. **81**, 4728 (1998)
63. K. Izawa, H. Yamaguchi, T. Sasaki, Y. Matuda, Phys. Rev. Lett. **88**, 27002 (2002)
64. K. Yamaji, Solid State Commun. **61**, 413 (1987)
65. A. Girlando, M. Masino, A. Brillante, R.G. Della Valle, E. Venuti, Phys. Rev. B **66**, 100507 (2002)
66. T. Sasaki et al., Phys. Rev. Lett. **92**, 227001 (2004)
67. T. Nishi et al., Solid State Commun. **134**, 189 (2005)
68. For example, T. Timusk, B. Statt, Rep. Prog. Phys. **62**, 61 (1999)
69. Y. Shimizu et al., Phys. Rev. Lett. **91**, 107001 (2003)
70. Y. Kurosaki et al., Phys. Rev. Lett. **95**, 177001 (2005)
71. R.N. Lyubovskaya et al., JETP Lett. **42**, 468 (1985); JETP Lett. **46**, 188 (1987)
72. A. Naito et al., Phys. Rev. B **71**, 054514 (2005)

23

Triplet Scenario of Superconductivity vs. Singlet One in $(TMTSF)_2X$ Materials

A.G. Lebed and S. Wu

We discuss experimental data, obtained on organic quasi-one-dimensional $(TMTSF)_2X$ ($X = PF_6$ and ClO_4) superconductors, which reveal unconventional nature of their superconducting phases. Our theoretical interpretation of the data supports a triplet scenario of superconductivity in $(TMTSF)_2PF_6$ material. In particular, we suggest a careful theoretical analysis of the recently measured anisotropic upper critical fields.

On a basis of this analysis, we suggest that the triplet superconducting order parameter is described by a **d**-vector with $d_a(\mathbf{k}) \neq 0$ and $d_b(\mathbf{k}) = d_c(\mathbf{k}) = 0$, which defines a spin part of the order parameter but does not uniquely define its orbital part, $d_a(\mathbf{k})$. We consider two major possible orbital parts of this triplet order parameter: p_x-wave and p_y-wave ones. In the end of the chapter, we discuss in a brief a d-wave singlet scenario of superconductivity, which we consider as a possibility in $(TMTSF)_2ClO_4$ material.

For more detailed discussions of the experimental data, see the chapters by Brown, Chaikin, Naughton, and Jerome. For a different theoretical analysis, see the chapters by Cherng, Zhang, and Sa de Melo.

23.1 Introduction

Early experiments [1–6], performed on quasi-one-dimensional (Q1D) organic superconductors $(TMTSF)_2X$ ($X = PF_6$ and ClO_4) by Chaikin's, Jerome's, and other experimental groups, gave hints on their unconventional nature. In particular, it was found that superconductivity is destroyed by nonmagnetic impurities [1–5] and that the Hebel–Slichter peak is absent in the NMR relaxation rate data [6]. The former feature was interpreted by Abrikosov [7] in terms of possible triplet superconductivity, whereas Hasegawa and Fukuyama [8] suggested several possible unconventional order parameters, including singlet d-wave-like one, which are in agreement with all abovementioned experiments [1–6]. Later, Gor'kov and Jerome [9] provided some more general arguments in favor of a triplet electron pairing in $(TMTSF)_2X$ compounds.

Another unusual feature of $(TMTSF)_2X$ superconductors – a possibility to exceed both the upper critical field, H_{c2}, and the Clogston paramagnetic limit, H_p [10], in the case of a triplet pairing – was discussed in [11–19] in terms of the so-called reentrant superconductivity (RS) phenomenon. The RS phenomenon was first suggested by Lebed [11] and developed by Dupuis et al. [12], Lebed and Yamaji [13], and some others [14–19]. In particular, Sa de Melo et al. [16–18] showed that the RS phase is stable in the presence of two-dimensional fluctuations.

Note that, although the above-mentioned early experiments and theoretical results provided important arguments in favor of unconventional nature of superconductivity in $(TMTSF)_2X$ materials, they did not allow to distinguish between singlet d-wave and triplet superconducting order parameters (for the discussion, see [8]).

Nowadays, interest in a possible triplet pairing in $(TMTSF)_2X$ superconductors has been renewed due to measurements of the upper critical fields in $(TMTSF)_2PF_6$ and $(TMTSF)_2ClO_4$ materials by Naughton's and Chaikin's groups [20–23] and their theoretical analysis by Lebed [14]. These experiments demonstrate that superconductivity significantly exceeds the Clogston paramagnetic limiting field, H_p, in both materials for a magnetic field, applied perpendicular to the conducting Q1D chains, $\mathbf{H} \parallel \mathbf{b}$. It is important [14] that, at low enough magnetic fields, $H \lesssim 1$–$1.5\,\mathrm{T}$, superconducting phases in $(TMTSF)_2PF_6$ [20–22] and $(TMTSF)_2ClO_4$ [23] materials are paramagnetically limited for a magnetic field parallel to the conducting chains, $\mathbf{H} \parallel \mathbf{a}$.

Lebed [14] and Lebed, Machida, and Ozaki (LMO) [15] theoretically analyzed the above-mentioned experimental data. In particular, they showed that, in Q1D superconductors, the paramagnetically limited fields can be much higher than the Clogston limit, $H_p^{\mathrm{Q1D}} \gg H_p$, due to a possible stabilization of the Larkin–Ovchinnikov–Fulde–Ferrell (LOFF or FFLO) phase [10]. Nevertheless the experimentally observed upper critical fields [20–22] exceed all possible paramagnetic limits for singlet superconductivity for $\mathbf{H} \parallel \mathbf{b}$ in $(TMTSF)_2PF_6$ material [14, 15].

Extra experimental arguments in favor of a triplet nature of superconductivity in $(TMTSF)_2PF_6$ have been recently provided by Brown's, Chaikin's, and Naughton's experimental groups [24–26], which have demonstrated no change of the Knight shift in a superconducting phase in relatively high magnetic field, $H \simeq 1.5\,\mathrm{T}$. On the other hand, very recent Knight shift measurements by Brown's and Jerome's groups [27] indicate that low magnetic field, $H \leq 1$–$1.5\,\mathrm{T}$, superconducting phase in $(TMTSF)_2ClO_4$ compound may be of a singlet origin.

It is important that high values of resistive upper critical fields, $\mathbf{H}_{c2} \parallel \mathbf{b}$ [20–23], have been confirmed by thermodynamic (torque) measurements, performed by Oh and Naughton in $(TMTSF)_2ClO_4$ [28]. Destroying of superconductivity by nonmagnetic impurities has also been recently confirmed by Jerome's group [29, 30]. As we discuss below, all above-mentioned experimental data, obtained on $(TMTSF)_2PF_6$ compound, are in agreement with the triplet order parameter suggested by LMO [14, 15]. On the other hand, in $(TMTSF)_2ClO_4$ compound, *d*-wave-like singlet order parameter is likely to exist at low magnetic fields [27].

23.2 Our Goals

From the above, we conclude that the triplet scenario [7] of unconventional superconductivity in $(TMTSF)_2PF_6$ material is more likely than a *d*-wave-like singlet one [8, 31]. In this case, the most fundamental question is to determine the wave functions of the triplet Cooper pairs which, in principle, may correspond to different triplet superconducting phases.

In the review, we directly address this question about a spin part of the triplet Cooper pairs' wave function in $(TMTSF)_2PF_6$ material at low enough magnetic fields. We recall that, in a general theory of unconventional superconductivity, a triplet wave function is described by **d**-vector order parameter [32, 33]:

$$\begin{aligned} &\Delta_{\uparrow\uparrow}(\mathbf{k}) = -d_x(\mathbf{k}) + \mathrm{i}d_y(\mathbf{k}), \quad \Delta_{\downarrow\downarrow}(\mathbf{k}) = d_x(\mathbf{k}) + \mathrm{i}d_y(\mathbf{k}), \\ &\Delta_{\uparrow\downarrow}(\mathbf{k}) = \Delta_{\downarrow\uparrow}(\mathbf{k}) = d_z(\mathbf{k}), \end{aligned} \tag{23.1}$$

where $\Delta_{i,j}(\mathbf{k})$ is a momentum-dependent superconducting gap and z is a spin quantization axis. Therefore, the problem is to obtain some information about the **d**-vector components (23.1).

Below, we argue that the LMO suggestion [15] that, at zero and low magnetic fields, spins of Cooper pairs are confined in $(\mathbf{b}, \mathbf{c})$-plane [34] due to spin–orbital coupling [32, 33] is in agreement with all existing experimental data [1–6, 20–26, 29, 30], obtained on $(TMTSF)_2PF_6$ material. This suggestion corresponds to the following triplet superconducting order parameter [15]:

$$\mathbf{d}(\mathbf{k}) = [d_a(\mathbf{k}) \neq 0, d_c(\mathbf{k}) = 0, d_b(\mathbf{k}) = 0]. \tag{23.2}$$

It is important that the triplet superconducting phase (23.2) is paramagnetically limited in a magnetic field, directed along the conducting chains, $\mathbf{H} \parallel \mathbf{a}$, whereas it does not have any paramagnetic limitations perpendicular to the chains' magnetic field, $\mathbf{H} \parallel \mathbf{b}$ [14, 15]. At high enough magnetic fields, $H \geq 1$–$1.5\,\mathrm{T}$, as stressed in [15], **d**-vector may become a free rotatable one, which corresponds to the absence of the paramagnetic limits for all directions of a magnetic field.

Our review is organized in the following way. In Sect. 23.3, we discuss paramagnetically limited critical fields in the absence of the orbital destructive effects against superconductivity.

In Sect. 23.3.1, we calculate the paramagnetically limited field, H_p^{Q1D}, for a Q1D singlet superconductor. This field is shown [14] to be significantly higher than the Clogston paramagnetic limit, H_p, due to a possible appearance of the LOFF phase.

In Sect. 23.3.2, we compare experimental data [20–22], obtained on $(TMTSF)_2PF_6$ superconductor, with our expression for H_p^{Q1D}, obtained in Sect. 23.3.1. We show that, for $\mathbf{H} \parallel \mathbf{b}$, the experimental upper critical fields, H_{c2}^b, exceed the calculated field, H_p^{Q1D}, which is an argument in favor of a triplet superconducting pairing in $(TMTSF)_2PF_6$ material (see Table 23.1).

In Sect. 23.4, we discuss paramagnetically limited critical fields in the presence of the orbital destructive effects against superconductivity.

In Sect. 23.4.1, we take into account the orbital effects and calculate the corresponding paramagnetically limited upper critical fields, $H_p^{Q1D}(\lambda)$ and H_p^b. We show that the orbital effects decrease the value of the paramagnetically limited field in $(TMTSF)_2PF_6$ conductor down to $H_p^b \simeq 1.7\,\mathrm{T}$.

In Sect. 23.4.2, we compare the experimental upper critical fields, H_{c2}^b [20–22], with the calculated theoretical paramagnetic limits, $H_p^{Q1D}(\lambda)$ and H_p^b. We make a conclusion that the experimental values are several times higher than the theoretical paramagnetic limits in $(TMTSF)_2PF_6$ material, which is a strong argument in favor of a triplet superconducting pairing (see Table 23.2).

In Sect. 23.5, we show that, for magnetic fields parallel to the conducting chains, $\mathbf{H} \parallel \mathbf{a}$, experimentally measured upper critical fields [20–22, 28] demonstrate paramagnetic limitations at low enough magnetic fields, $H \lesssim 1\text{–}1.5\,\mathrm{T}$.

In Sect. 23.6, we make a central statement of our review that the LMO triplet superconducting order parameter (23.2) with $\mathbf{d} = [d_a(\mathbf{k}), 0, 0]$ can

Table 23.1. Experimental values of H_{c2}^b [20–22], obtained on $(TMTSF)_2PF_6$ superconductor, are compared with the theoretical results of (23.10) for a singlet superconducting pairing

Compound	Experimental H_{c2}^b (T)	H_p^{Q1D} (T)
$X = PF_6$ [20]	6	3.8
$X = PF_6$ [21]	6	3.8
$X = PF_6$ [22]	9	3.8

Table 23.2. Experimental data [20–22], obtained on $(TMTSF)_2PF_6$ material, are compared with the theoretical results [14, 15] for a singlet superconducting pairing

Compound	Experimental H_{c2}^b (T)	$H_p^{Q1D}(\lambda = 1)$ (T)	H_p^b (T)
$X = PF_6$ [20]	6	3	1.7
$X = PF_6$ [21]	6	3	1.7
$X = PF_6$ [22]	9	3	1.7

account for all existing experimental data, obtained on $(TMTSF)_2PF_6$ material. We also describe the major physical properties of the corresponding superconducting phase.

In Sect. 23.7, the most spectacular phenomenon, related to the absence of the paramagnetic limit in $(TMTSF)_2PF_6$ superconductor for $\mathbf{H} \parallel \mathbf{b}$ – the so-called RS phase [11] – is considered. The RS phase is expected to be stable in magnetic fields, $\mathbf{H} \parallel \mathbf{b}$, which are much higher than both the upper critical field, $H_{c2}^{\perp}$, and the Clogston paramagnetic limit, $H_{\rm p}$.

In Sect. 23.8, we discuss in brief a singlet scenario of superconductivity in $(TMTSF)_2ClO_4$ material in low magnetic fields, which we consider as a most likely possibility.

23.3 Paramagnetic Limit in Q1D Case: $H_{\rm p}^{\rm Q1D}$

In this section, we calculate a paramagnetically limited upper critical field for a singlet Q1D superconductor and compare it with the existing experimental data.

23.3.1 Theoretical Calculations of $H_{\rm p}^{\rm Q1D}$

In this section, we derive a gap equation, which defines transition temperature for a singlet superconducting order parameter in the presence of Pauli paramagnetic destructive effects against superconductivity. Our goal is to calculate paramagnetically limited upper critical field in a Q1D superconductor, $H_{\rm p}^{\rm Q1D}$, which, as shown in [14], is significantly higher than the textbook Clogston paramagnetic limit, $H_{\rm p}$ [10].

The physical origin of high values of $H_{\rm p}^{\rm Q1D}$ is shown [14] to be due to a possible appearance of the nonuniform LOFF phase, corresponding to nonzero total momenta of Cooper pairs. Below, we demonstrate that the LOFF phase is much more stable in high magnetic fields in a Q1D superconductor than the traditional BCS state and, therefore, $H_{\rm p}^{\rm Q1D} \gg H_{\rm p}$.

In this section, we completely disregard the orbital destructive effects against superconductivity. Below, we follow the result of [14], where $H_{\rm p}^{\rm Q1D}$ is calculated for a typical Q1D electron spectrum, corresponding to two open sheets of Fermi surface (FS):

$$\begin{aligned} \epsilon^{\pm}(\mathbf{p}) &= \pm v_{\rm F}[p_x \mp p_x^0(p_y)] - 2t_c \cos(p_z c^*) + \epsilon_{\rm F}, \\ p_x^0(p_y) &= p_{\rm F} + \frac{2t_b}{v_{\rm F}} \cos(p_y b^*), \end{aligned} \quad (23.3)$$

where the first term represents the free-electron motion along the conducting a-chains, $v_{\rm F}$, $p_{\rm F}$, and $\epsilon_{\rm F} = \sqrt{2}t_a$ are the Fermi velocity, Fermi momentum, and Fermi energy, respectively, $t_a \simeq 2{,}000\,{\rm K}$, $t_b \simeq 200\,{\rm K}$, and $t_c \simeq 5$–$10\,{\rm K}$ are

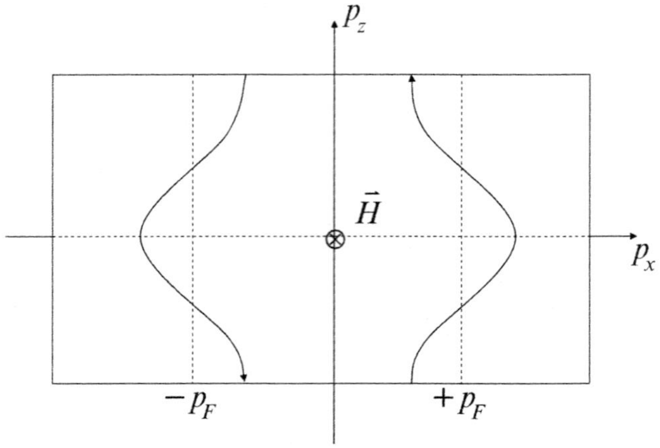

Fig. 23.1. Quasi-one-dimensional (Q1D) electron spectrum of $(TMTSF)_2X$ compounds with two open sheets of Fermi surface (FS). In a magnetic field, $\mathbf{H} = (0, H, 0)$, electrons move along open FS in the extended Brillouin zone

the overlapping integrals of electron wave functions in a tight-binding model, sign $+(-)$ stands for the right (left) sheet of the FS (see Fig. 23.1), and $\hbar \equiv 1$.

Note that, unlike [11–18, 35], we account of a nonlinearity of an electron energy dispersion law along the chains and, thus, treat the Pauli paramagnetic effects for the case of a nonzero value of a Q1D anisotropy ratio, $t_b/t_a \neq 0$. (The previous statements [12, 35] about the absence of the paramagnetic limit in a Q1D case are shown [14] to be incorrect.)

In the presence of a magnetic field, perpendicular to the chains, $\mathbf{H} = (0, H, 0)$, electron Green functions of a Q1D conductor (23.3), in the gauge $\mathbf{A} = (0, 0, -Hx)$, can be written as [14]

$$G^{\pm}_{\mathrm{i}\omega_n}(x, x_1; y, z; s) = \exp[\pm \mathrm{i} p^s_x(p_y)(x - x_1) + \mathrm{i} p_y y + \mathrm{i} p_z z] g^{\pm}_{\mathrm{i}\omega_n}(x, x_1; p_y, p_z; s),$$
$$p^s_x(p_y) = \left\{ p_{\mathrm{F}} + \frac{2t_b}{v_{\mathrm{F}}} \cos(p_y b) + \frac{2s\mu_{\mathrm{B}} H}{v_{\mathrm{F}}} [1 - \alpha \cos(p_y b)] \right\}, \tag{23.4}$$

where $\alpha = \sqrt{2} t_b/t_a \ll 1$ and $s = \pm 1/2$ is a z-projection of an electron spin.

Differential equations for $g^{\pm}_{\mathrm{i}\omega_n}(x, x_1; p_y, p_z; s)$ are obtained by means of the Peierls substitution method, $p_z \to p_z - (e/c)Hx$, using the procedure [36]:

$$\left[\mp \mathrm{i} v_{\mathrm{F}} \frac{\mathrm{d}}{\mathrm{d}x} - 2t_c \cos\left(p_z c^* - \frac{\omega_{\mathrm{c}} x}{v_{\mathrm{F}}} \right) - \mathrm{i}\omega_n \right] g^{\pm}_{\mathrm{i}\omega_n}(x, x_1; p_y, p_z; s) = \delta(x - x_1). \tag{23.5}$$

(Here, $\omega_{\mathrm{c}} = e v_{\mathrm{F}} c^* H/c$ is a cyclotron frequency of electron motion along open FS (23.3), ω_n is the Matsubara frequency [37], μ_{B} is the Bohr magneton, and c is the velocity of light.)

The solutions of (23.5) are

$$g^{\pm}_{\mathrm{i}\omega_n}(x, x_1; p_y, p_z; s) = \frac{-\mathrm{i}\,\mathrm{sgn}(\omega_n)}{v_{\mathrm{F}}} \exp\left[\mp \frac{\omega_n(x - x_1)}{v_{\mathrm{F}}}\right]$$
$$\times \exp\left(\pm \mathrm{i}\lambda \sin\left[\frac{\omega_{\mathrm{c}}(x - x_1)}{2v_{\mathrm{F}}}\right] \cos\left[p_z c^* - \frac{\omega_{\mathrm{c}}(x + x_1)}{2v_{\mathrm{F}}}\right]\right), \qquad (23.6)$$

where $\pm\omega_n(x - x_1) > 0$ and $\lambda = 4t_c/\omega_c$.

A linearized gap equation, determining superconducting transition temperature, $T_c(H)$, is derived by means of Gor'kov equations [37] for nonuniform superconductivity [14]. As a result, we obtain

$$\Delta(x) = \frac{g}{2} \int_{|x-x_1|>a} \mathrm{d}x_1 \frac{2\pi T}{v_{\mathrm{F}} \sinh\left(\frac{2\pi T|x-x_1|}{v_{\mathrm{F}}}\right)}$$
$$\times J_0\left[\frac{2\sqrt{\alpha^2 + q^2}\mu_{\mathrm{B}} H(x - x_1)\beta}{v_{\mathrm{F}}}\right] \cos\left[\frac{2\mu_{\mathrm{B}} H(x - x_1)\beta}{v_{\mathrm{F}}}\right]$$
$$\times J_0\left(2\lambda \sin\left[\frac{\omega_{\mathrm{c}}(x - x_1)}{2v_{\mathrm{F}}}\right] \sin\left[\frac{\omega_{\mathrm{c}}(x + x_1)}{2v_{\mathrm{F}}}\right]\right) \Delta(x_1), \qquad (23.7)$$

where g is a dimensionless coupling constant, $\beta = 1$ for paramagnetically limited singlet and some paramagnetically limited triplet phases and $\beta = 0$ for those triplet phases, which do not have any paramagnetic limits, and $q = (t_b b/\mu_{\mathrm{B}} H)Q$. (Note that the superconducting gap in (23.7) is chosen in the form $\Delta(x, y) = \exp(\mathrm{i}Qy)\Delta(x)$.)

We stress that, at $\beta = 0$, (23.7) coincides with the gap equation [11, 12], describing the RS phenomenon [11]. Therefore, all previous results [11–19] on the reentrance of superconductivity in high magnetic fields in a triplet Q1D superconductor without paramagnetic limit are unchanged. On the other hand, at $\beta = 1$, (23.7) does not coincide with the corresponding gap equation of [11–18, 35] for the case of a singlet s-wave pairing of electrons. Indeed, (23.7) contains an extra Bessel function, $J_0[2\sqrt{\alpha^2 + q^2}\mu_{\mathrm{B}} H(x - x_1)/v_{\mathrm{F}}]$, which comes from a nonlinearity of the energy dispersion law (23.3). This function results in a convergence of the integral (23.7) as $T \to 0$, in contrast to a divergence of the corresponding integrals in [11–18, 35]. (Note that the convergence of the integral (23.7) indicates about instability of the LOFF state at $T = 0$ in high enough magnetic fields.)

Here, we investigate (23.7) at zero temperature, $T = 0$, in the case where the paramagnetic effects are important (i.e., at $\beta = 1$). We calculate the LOFF paramagnetically limited critical field, $H_{\mathrm{p}}^{\mathrm{Q1D}}$, in the absence of the orbital effects (i.e., at $\lambda = 0$). In this case, (23.7) is simplified and it is possible to represent its solutions in the form $\Delta(x) = \Delta \exp(\mathrm{i}Kx)$.

Then (23.7) can be rewritten as

$$1 = g \int_a^{\infty} \frac{\mathrm{d}x}{x} \cos\left[\frac{2\mu_{\mathrm{B}} H x}{v_{\mathrm{F}}}\right] J_0\left[\frac{2\sqrt{\alpha^2 + q^2}\mu_{\mathrm{B}} H x}{v_{\mathrm{F}}}\right] \cos(Kx). \qquad (23.8)$$

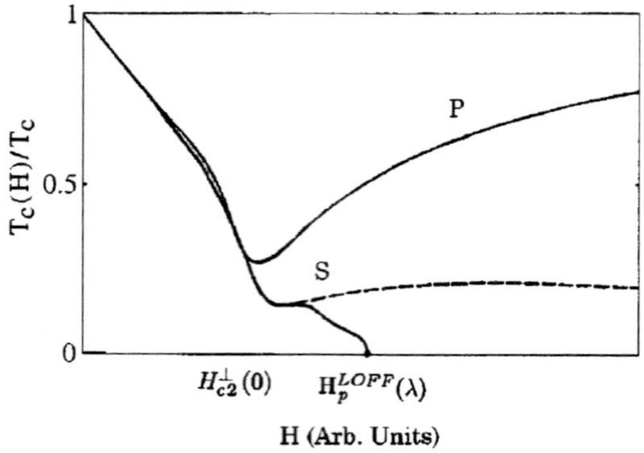

Fig. 23.2. Curve P stands for the reentrant superconducting (RS) phase [11] in a Q1D LMO triplet superconductor (23.2) with the electron spectrum (23.3), which is expected to exist for $\mathbf{H} \parallel \mathbf{b}$ in $(TMTSF)_2PF_6$ superconductor (see Sect. 23.7). *Solid curve* S stands for a Q1D singlet superconductor, which is characterized by a paramagnetically limited upper critical field, $H_{\rm p}^{\rm Q1D} \equiv H_{\rm p}^{\rm LOFF}$ (see (23.10))

If we introduce the Clogston paramagnetically limited field, $H_{\rm p} = \Delta/(\sqrt{2}\mu_{\rm B})$ [10], we can rewrite (23.8) as follows

$$\ln\left[\frac{\sqrt{2}H_{\rm p}^{\rm Q1D}}{H_{\rm p}}\right] = \int_0^\infty \frac{{\rm d}x}{x}\cos(x)[J_0(\sqrt{\alpha^2+q^2}x)\cos(kx) - 1], \tag{23.9}$$

where $k = (v_{\rm F}/2\mu_{\rm B}H)K$.

After taking integral (23.9) analytically [38], it is possible to make sure that the LOFF paramagnetically limited field in a Q1D conductor (23.3) takes its maximum value:

$$H_{\rm p}^{\rm Q1D} \equiv H_{\rm p}^{\rm LOFF} \simeq 0.6\sqrt{t_a/t_b}H_{\rm p}, \tag{23.10}$$

at the LOFF wave vector with the components $K \simeq 2\mu_{\rm B}H/v_{\rm F}$ and $Q = q = 0$ (see Fig. 23.2).

23.3.2 Experimental Exceeding of $H_{\rm p}^{\rm Q1D}$

Let us discuss and compare (23.10) with the existing experimental data [20–22], obtained on $(TMTSF)_2PF_6$ superconductor. Note that Q1D anisotropy ratio, t_a/t_b, is very well known for $(TMTSF)_2X$ materials, $t_a/t_b \simeq 8.5$, from the theoretical analysis [39] of the experimentally observed Lebed magic angle oscillations. If we take $t_a/t_b \simeq 8.5$ and $H_{\rm p} \simeq 2.2\,{\rm T}$ (corresponding to superconducting transition temperature $T_c \simeq 1.2\,{\rm K}$), we find that $H_{\rm p}^{\rm Q1D} \simeq 3.8\,{\rm T}$ from (23.10).

This result, first obtained in [14], is an argument against singlet d-wave scenario [8, 31] of superconductivity in $(TMTSF)_2PF_6$ material, where superconductivity is experimentally found to be stable in magnetic fields of

$H^b_{c2} = 6\,\mathrm{T}$ [20], $H^b_{c2} = 6\,\mathrm{T}$ [21], and $H^b_{c2} = 9\,\mathrm{T}$ [22] ($\mathbf{H} \parallel \mathbf{b}$). For comparisons of (23.10) with the experiments, performed on $(\mathrm{TMTSF})_2\mathrm{PF}_6$ material [20–22], see Table 23.1.

23.4 Paramagnetic Limits in the Presence of the Orbital Effects: $H_\mathrm{p}^{\mathrm{Q1D}}(\lambda)$ and H_p^b

In this section, we calculate the paramagnetically limited upper critical fields for a singlet Q1D superconductor and compare them with the existing experimental data. Here, we take into account the destructive orbital effects against superconductivity.

23.4.1 Theoretical Calculations of $H_\mathrm{p}^{\mathrm{Q1D}}(\lambda)$ and H_p^b

In this section, we calculate the paramagnetically limited upper critical fields for a singlet Q1D superconductor in the presence of the orbital destructive effects against superconductivity, $H_\mathrm{p}^{\mathrm{Q1D}}(\lambda)$ and H_p^b [14, 15].

At first, let us consider (23.7) in the case of weak orbital effects (i.e., at $\lambda \lesssim 1$). At $\lambda \lesssim 1$, we can solve the integral (23.7) by means of a variational method using the trial function, $\Delta_t(x) = \cos(2\mu_\mathrm{B} Hx/v_\mathrm{F})$. As a result, the paramagnetically limited field at $T = 0$, $q = 0$, and $\lambda \leq 1$ is given by [14]

$$H_\mathrm{p}^{\mathrm{Q1D}}(\lambda) \simeq H_\mathrm{p}^{\mathrm{Q1D}} \left[\frac{\alpha}{\beta\sqrt{|4-\beta^2|}} \right]^{\lambda^2/8} , \tag{23.11}$$

where $\beta = \omega_\mathrm{c}/2\mu_\mathrm{B}H(|\beta - 2| \gg \alpha \ll 1)$.

We stress that Q1D$\rightarrow$2D crossover [11] of an orbital electron motion is expected to eliminate the orbital destructive mechanism against superconductivity in $(\mathrm{TMTSF})_2\mathrm{X}$ materials at rather high magnetic fields, $H \simeq 15$–$20\,\mathrm{T}$ (see [11–19]). Therefore, the actual value of $H_\mathrm{p}^{\mathrm{Q1D}}(\lambda)$, in low enough magnetic fields, has to be reduced by the orbital effects, as seen from (23.11). If we take the known value of $v_\mathrm{F} \simeq 2 \times 10^7\,\mathrm{cm\,s^{-1}}$, we find that $\beta \simeq 1.5$ and estimate the reduced paramagnetically limited field at $\lambda \simeq 1$ using (23.11), $H_\mathrm{p}^{\mathrm{Q1D}}(\lambda \simeq 1) \leq 3\,\mathrm{T}$.

Note that the value $\lambda \simeq 1$ at $H \simeq 3.8\,\mathrm{T}$ corresponds to $t_c \simeq 2\,\mathrm{K}$ in (23.3). Our numerical analysis [15] of (23.7) shows that, for the realistic values of $t_c \simeq 5$–$10\,\mathrm{K}$ [40], the region of a stability of the LOFF phase completely disappears and that superconductivity is destroyed by low magnetic field, $H_\mathrm{p}^b \simeq 1.7\,\mathrm{T}$ (see Fig. 23.3).

23.4.2 Experimental Exceeding of $H_\mathrm{p}^{\mathrm{Q1D}}(\lambda)$ and H_p^b

In this section, we strengthen our arguments (see Sect. 23.3.2) that the experimental upper critical fields in $(\mathrm{TMTSF})_2\mathrm{PF}_6$ material for $\mathbf{H} \parallel \mathbf{b}$, H^b_{c2} [20–22],

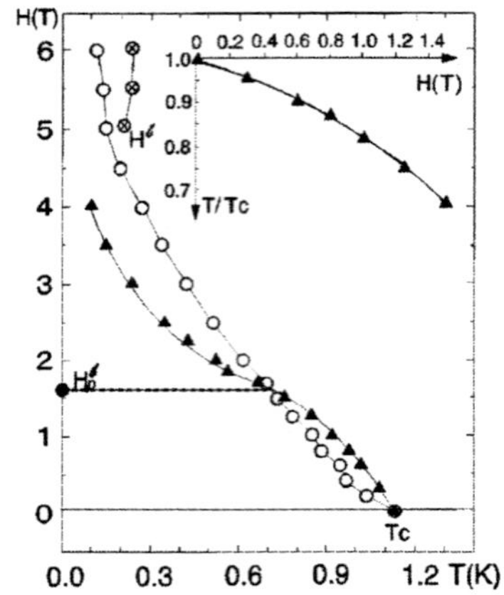

Fig. 23.3. *Circles and triangles* stand for the experimental upper critical fields [20] for **H** ∥ **b** and **H** ∥ **a**, respectively. $H_p^b = 1.7\,\mathrm{T}$ is a theoretical paramagnetic limit for a singlet Q1D superconductor (23.3), reduced by the orbital effects for **H** ∥ **b**. *Inset*: *triangles* stand for the experimental points [20] for **H** ∥ **a**, *solid line* represents our theoretical calculations [15] of the paramagnetically limited upper critical field for the LMO triplet phase (23.2) for **H** ∥ **a**, based on (23.7)

significantly exceed the paramagnetic limit for a singlet Q1D superconductor. As mentioned in Sect. 23.3.2, this exceeding is a major argument in favor of a triplet nature of superconductivity in the above-mentioned material.

For this purpose, let us compare the experimental critical fields [20–22], measured on $(TMTSF)_2PF_6$ superconductor, with our calculations of the paramagnetic limits, $H_p^{Q1D}(\lambda \simeq 1)$ and H_p^b (see Table 23.2 and Fig. 23.3). As follows from Table 23.2, the upper critical fields in $(TMTSF)_2PF_6$ conductor are four to five times higher than the theoretically calculated critical field, $H_p^b = 1.7\,\mathrm{T}$, expected in the case of a singlet Q1D conductor in the presence of the orbital effects.

In conclusion of this section, we discuss other possible explanations of the weakness of the Pauli paramagnetic pair-breaking effects in $(TMTSF)_2X$ superconductors. Note that spin–orbital scattering cannot be responsible for this phenomenon since $(TMTSF)_2X$ materials are very clean superconductors [9] with $1/(T_c\tau) \simeq 0.1$. The ratio $\Delta(0)/(k_b T_c) \simeq 1.9$, found in $(TMTSF)_2ClO_4$ material (see [40]), as well as the small value of superconducting transition temperature, $T_c \simeq 1.2\,\mathrm{K}$, are in favor of a weak coupling of electrons. Therefore, it is very unlikely that possible small deviations from a weak coupling model are responsible for the above-discussed high experimental values of H_{c2}^b.

From the above (see also a caption to Fig. 23.3), we conclude that a triplet pairing of electrons in $(TMTSF)_2PF_6$ material is more probable than a singlet one.

23.5 Paramagnetic Limitations for H || a

Here, we discuss experimental data [20–23] and theoretical work [15] demonstrating that, in low enough magnetic fields parallel to the conducting **a**-axis, $H \leq 1$–$1.5\,\mathrm{T}$, superconducting phases in $(\mathrm{TMTSF})_2\mathrm{X}$ materials are paramagnetically limited.

Let us first consider experimental data, obtained on $(\mathrm{TMTSF})_2\mathrm{PF}_6$ superconductor [20–22]. In [15], the temperature-dependent upper critical field for $\mathbf{H} \parallel \mathbf{a}$, $H^a_{c2}(T)$, is fitted using a combination of the orbital and paramagnetic destructive effects against superconductivity. As it follows from Fig. 23.3 (see the inset), this fit is almost ideal for low enough magnetic fields, $H \leq 1$–$1.5\,\mathrm{T}$. We consider this fact as an argument in favor of the LMO triplet phase (23.2), which is paramagnetically limited for $\mathbf{H} \parallel \mathbf{a}$. Although a quantitative analysis of the upper critical field, measured in [21, 22], has not been done yet, all experimental curves [21, 22] for $\mathbf{H} \parallel \mathbf{a}$ show the characteristic paramagnetically limited behavior at $H \leq 1$–$1.5\,\mathrm{T}$.

At higher magnetic fields, $H \gtrsim 1$–$1.5\,\mathrm{T}$, the experimental curves [20–22] significantly deviate from the paramagnetically limited scenario which may indicate that, at high fields, the **d**-vector order parameter (23.2) becomes a free rotatable one. It is important that a general feature of the anisotropic upper critical fields, measured in [20–22], is that $H^a_{c2} < H^b_{c2}$ at low temperature, whereas the Ginzburg–Landau slopes $(\mathrm{d}H^a_{c2}/\mathrm{d}T)_{T_c} \gg (\mathrm{d}H^b_{c2}/\mathrm{d}T)_{T_c}$. This feature (i.e., an inversion of H^a_{c2}/H^b_{c2} anisotropy) is an additional strong argument in favor of the paramagnetic limitation of H^a_{c2} for $\mathbf{H} \parallel \mathbf{a}$.

Note that $(\mathrm{TMTSF})_2\mathrm{ClO}_4$ superconductor is characterized by even lower experimental upper critical field along **a**-axis, $H^a_{c2}(T=0) \simeq 3.5\,\mathrm{T}$ [23]. This indicates that the above-discussed paramagnetic destructive effects against superconductivity for $\mathbf{H} \parallel \mathbf{a}$ are even stronger in $(\mathrm{TMTSF})_2\mathrm{ClO}_4$ superconductor than in $(\mathrm{TMTSF})_2\mathrm{PF}_6$ one.

23.6 Physical Properties of $\mathbf{d}(\mathbf{k}) = [\mathbf{d}_a(\mathbf{k}), 0, 0]$ Triplet Superconducting Phase

In this section, we describe the major physical features of the LMO triplet superconducting phase with the vector order parameter (23.2), suggested in [15]. First of all, let us demonstrate that this superconducting phase is not paramagnetically limited for $\mathbf{H} \parallel \mathbf{b}$ and is paramagnetically limited for $\mathbf{H} \parallel \mathbf{a}$, which are in accordance with the experimental data for $(\mathrm{TMTSF})_2\mathrm{PF}_6$, discussed in the previous sections (see also Table 23.2).

According to a general theory [32, 33], the electron spin susceptibility tensor, χ_{ij}, for a triplet phase at $T = 0$, is given by

$$\chi_{ij} = \chi_0 \left\langle \delta_{ij} - \frac{d_i^*(\mathbf{k}) d_j(\mathbf{k})}{\mathbf{d}^*(\mathbf{k}) \cdot \mathbf{d}(\mathbf{k})} \right\rangle_k , \qquad (23.12)$$

where $\delta_{ij} = 1$ if $i = j$ and $\delta_{ij} = 0$ if $i \neq j$; $\langle |\mathbf{d}(\mathbf{k})|^2 \rangle_k = 1$, $\langle \cdots \rangle_k$ means an averaging over the FS; and χ_0 is a spin susceptibility of a metallic phase. (Here, we consider only unitary triplet phases [32, 33], i.e., $d_a^*(\mathbf{k}) d_c(\mathbf{k}) = d_a(\mathbf{k}) d_c^*(\mathbf{k})$.) From (23.12), it directly follows that

$$\chi_{bb} = \chi_{cc} = \chi_0 \tag{23.13}$$

and that

$$\chi_{aa} = 0. \tag{23.14}$$

An important fact is that the electron spin susceptibility in a superconducting phase is the same as in a metallic one for $\mathbf{H} \parallel \mathbf{b}$ and $\mathbf{H} \parallel \mathbf{c}$ (see (23.13)). This means that, for such directions of a magnetic field, the LMO triplet superconducting phase (23.2) is not paramagnetically limited. On the other hand, as seen from (23.14), the susceptibility in the superconducting phase is equal to zero for $\mathbf{H} \parallel \mathbf{a}$. Therefore, for $\mathbf{H} \parallel \mathbf{a}$, the LMO superconducting phase (23.2) is characterized by the same paramagnetic limit as a singlet one. All these features are in agreement with the experiments [20–22] and, therefore, we conclude that a spin part of the triplet order parameter (23.2) accounts for all existing experimental data obtained on $(TMTSF)_2PF_6$.

We point out that the LMO order parameter (23.2) defines only a spin part of a triplet vector order parameter and does not uniquely define its orbital part. Therefore, there are infinite number of possibilities, including three p-wave orbital order parameters: p_x-wave, p_y-wave, and p_z-wave ones. Here, we discuss the most probable p_x-wave and p_y-wave orbital order parameters from our point of view.

According to a general theory [32, 33], the triplet superconducting order parameter (23.2) can be rewritten in a simple form:

$$\Delta(\mathbf{p}) = d_a(\mathbf{p}), \tag{23.15}$$

where (23.15) suggests that two qualitatively different situations may occur.

The first main scenario, a fully gapped (i.e., p_x-wave) one, corresponds to the superconducting orbital order parameter:

$$\Delta(\mathbf{p}) = \Delta \, \mathrm{sgn}(p_x), \tag{23.16}$$

where $\Delta(\mathbf{p})$ is characterized by different signs on the left and right sheets of the Q1D FS (23.3) and never zero on the FS (see Fig. 23.4). It is important that the Bogoliubov quasiparticle spectrum for p_x-wave order parameter is fully gapped:

$$\epsilon(\mathbf{p}) = \sqrt{\Delta^2 + v_F^2 (p_x - p_F)^2}. \tag{23.17}$$

The second main scenario corresponds to the following orbital order parameter

$$\Delta(\mathbf{p}) = \Delta \sin(p_y b^*), \tag{23.18}$$

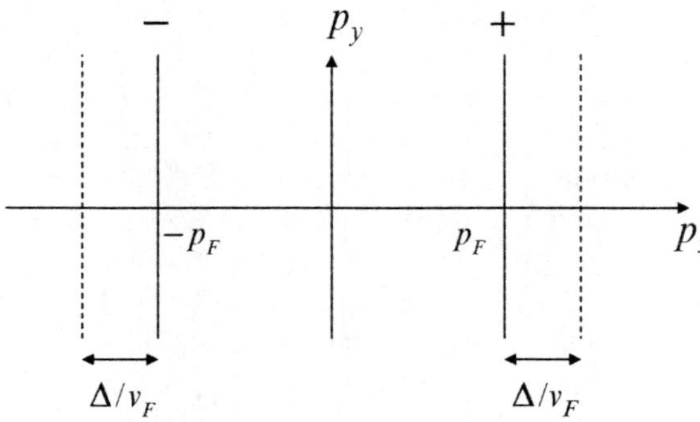

Fig. 23.4. Triplet p_x-wave order parameter (23.16) is positive on a *right sheet* of the Q1D FS (23.3) and negative on a *left one.* Electron spectrum is a fully gaped with the distances in p-space between FS and Bogoliubov quasiparticle levels being Δ/v_F

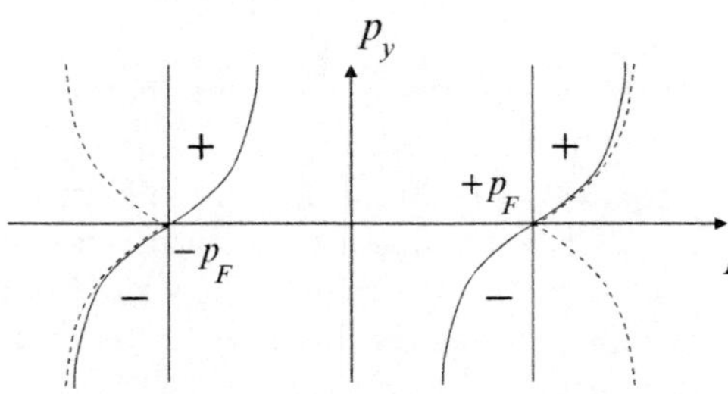

Fig. 23.5. *Solid line*: triplet orbital order parameter p_y changes its sign on each of two sheets of the Q1D FS (23.3). *Dashed line*: a superconducting gap in the Bogoliubov quasiparticle spectrum is characterized by lines of zeros at $p_y = 0$

which is called p_y-wave one and is characterized by zeros of a superconducting gap on the Q1D sheets of the FS (see Fig. 23.5). A spectrum of the Bogoliubov quasiparticles in this case is

$$\epsilon(\mathbf{p}) = \sqrt{\Delta^2 \sin^2(p_y b^*) + v_F^2 (p_x - p_F)^2}. \tag{23.19}$$

In our opinion, a question about the orbital part of the triplet LMO order parameter (23.2) in $(TMTSF)_2X$ conductors is still open. Some of the existing experimental data such as thermopower [41] and specific heat jump measurements [40, 42] are in favor of the fully gapped p_x-wave scenario (see (23.16) and (23.17)) for the superconducting orbital order parameter $\Delta(\mathbf{p})$ (see Fig. 23.4). On the other hand, the existing NMR data [6, 27] are claimed to be in favor of a p_y-wave scenario (see (23.18) and (23.19); see Fig. 23.5).

23.7 Reentrant Superconductivity Phenomenon

In this section, we consider the most spectacular phenomenon, related to the absence of the paramagnetic limit in $(TMTSF)_2PF_6$ superconductor for $\mathbf{H} \parallel \mathbf{b}$ – the so-called RS phase. The RS phase is shown [11] to be stable in magnetic fields, $\mathbf{H} \parallel \mathbf{b}$, which are much higher than both the quasiclassical upper critical field, $H^{\perp}_{c2}$, and the Clogston paramagnetic limit, H_{p}. The most unusual feature of the RS phase is that, at high enough magnetic fields, $H \geq H^*$, superconducting transition temperature may increase with increasing magnetic field, $\mathrm{d}T_c/\mathrm{d}H > 0$ (see Fig. 23.6).

To demonstrate the existence of the RS phase, let us rewrite a common expression (23.7) for the case $\mathbf{H} \parallel \mathbf{b}$, where the paramagnetic limit for the LMO triplet phase (23.2) is absent (i.e., for $\beta = 0$ in (23.7)) [11]:

$$\Delta(x) = \frac{g}{2} \int_{|x-x_1|>a} \mathrm{d}x_1 \frac{2\pi T}{v_{\mathrm{F}} \sinh\left(\frac{2\pi T|x-x_1|}{v_{\mathrm{F}}}\right)}$$
$$\times J_0\left(2\lambda \sin\left[\frac{\omega_{\mathrm{c}}(x-x_1)}{2v_{\mathrm{F}}}\right] \sin\left[\frac{\omega_{\mathrm{c}}(x+x_1)}{2v_{\mathrm{F}}}\right]\right) \Delta(x_1). \qquad (23.20)$$

From the integral equation (23.20), it directly follows that the LMO triplet superconducting phase (23.2) is stable in an arbitrary magnetic field. Indeed, a choice of a periodic solution for superconducting order parameter in (23.20), $\Delta(x_1) = \Delta(x_1 + 2\pi v_{\mathrm{F}}/\omega_{\mathrm{c}})$, makes the integral (23.20) to be logarithmically divergent at low temperatures. This means that (23.20), which defines the superconducting transition temperature, always has periodic solutions at low enough temperatures. At high magnetic fields, these periodic solutions correspond to the RS phase, which is always stable and is qualitatively different from the BCS and LOFF phases.

From physical point of view, a stability of the RS phase in an arbitrary magnetic field is a consequence of 3D → 2D crossover for electron spectrum

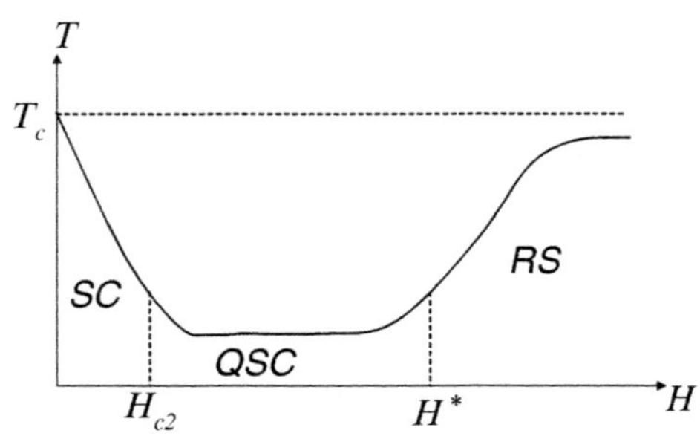

Fig. 23.6. *Textbook theory*: superconductivity is destroyed by the upper critical field H_{c2}. *Refined theory*, suggested in [11]: superconductivity is stable above the upper critical field H_{c2}, with transition temperature being a rising function of a magnetic field at magnetic field higher than some crossover field, $H > H^*$

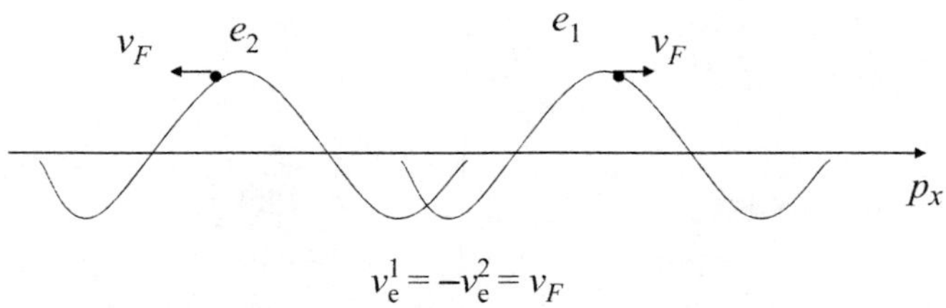

Fig. 23.7. Electron motions in a magnetic field, $\mathbf{H} \parallel \mathbf{b}$, become 2D, since electron trajectories along y-axis become periodic and restricted (see another review by Lebed and Si Wu)

Table 23.3. Experimental values of H_{c2}^b [20–22, 27] are compared with the predictions of the standard Ginzburg–Landau–Abrikosov–Gorkov theory, which disregards the reentrant superconductivity (RS) effects

Compound	Experimental H_{c2}^b (T)	H_p^{Q1D} (T)
X = PF_6 [20]	6	2
X = PF_6 [21]	6	2.5
X = PF_6 [22]	9	2.5
X = ClO_4 [27]	5	3

(23.3) in a magnetic field, perpendicular to the conducting chains [11]. In other words, in a magnetic field, $\mathbf{H} \parallel \mathbf{b}$, electron motion and electron wave functions become 2D (see Fig. 23.7). 2D superconductivity is known to be stable in a parallel magnetic field, which explains the stability of the described above RS phase.

In conclusion, we discuss the existing experimental data, obtained on $(TMTSF)_2PF_6$ [20–22] and $(TMTSF)_2ClO_4$ [23, 28] compounds. Note that the expected quasiclassical values of the upper critical fields for these superconductors in a perpendicular to the conducting chains' magnetic field, $\mathbf{H} \parallel \mathbf{b}$, are of the order of 2–3 T. From Table 23.3, it follows that the experimental upper critical fields are three to four times higher than expected in a standard theory. These facts support the RS scenario [11, 12] of superconductivity in $(TMTSF)_2X$ compounds.

On the other hand, the RS phase with $dT_c/dH > 0$ has not been detected yet. It is expected to occur in high magnetic fields, $H \geq$ 15–20 T [11], in clean samples, which demonstrate a large quadratic magnetoresistance in a metallic phase, $\Delta\rho/\rho \sim H^2 \gg 1$. To our surprise, magnetoresistance in the existing samples of $(TMTSF)_2PF_6$ and $(TMTSF)_2ClO_4$ conductors saturates in much lower magnetic fields, $H \simeq$ 5–10 T, which may reflect intrinsic sample imperfections.

23.8 Singlet Scenario of Unconventional Superconductivity

Very recently, Brown's and Jerome's experimental groups [27] have discovered nonzero Knight shifts for both **a**- and **b**-directions of a magnetic field in $(TMTSF)_2ClO_4$ superconductor at $H \lesssim 1\,T$. Although these shifts are less than in a conventional singlet superconductor, their existence indicates about a possibility that $(TMTSF)_2ClO_4$ superconductor is a singlet *d*-wave one in low magnetic fields, $H \leq 1$–$1.5\,T$. In this case, the high values of the upper critical fields, discussed in the review, may be explained by a triplet nature of superconductivity in high fields. This question about singlet–triplet interplay or coexistence is very far even from a preliminary discussion. We just mention that, according to type-IV superconductivity phenomenon, suggested by Lebed [43], a triplet component of superconducting order parameter always appears in a vortex phase of a singlet superconductor.

Acknowledgments

The author is thankful to N.N. Bagmet (Lebed), S.E. Brown, P.M. Chaikin, D. Jerome, L.P. Gorkov, and J. Singleton for fruitful discussions. This work was partially supported by the NSF grant DMR-0705986.

References

1. M.Y. Choi, P.M. Chaikin, S.Z. Huang et al., Phys. Rev. B **25**, 6208 (1982)
2. R.L. Green, P. Haen, S.Z. Huang et al., Mol. Cryst. Liq. Cryst. **79**, 183 (1982)
3. S. Bouffard, M. Ribault, R. Brusetti et al., J. Phys. C **15**, 2951 (1982)
4. S. Tomic, D. Jerome, D. Mailly et al., J. Phys. (Paris) Colloq. **44**, C3-1075 (1983)
5. C. Coulon, P. Delhaes, J. Amiell et al., J. Phys. (Paris) **43**, 1721 (1982)
6. M. Takigawa, H. Yasuoka, G. Saito, J. Phys. Soc. Jpn. **56**, 873 (1987)
7. A.A. Abrikosov, J. Low Temp. Phys. **53**, 359 (1983)
8. Y. Hasegawa, H. Fukuyama, J. Phys. Soc. Jpn. **56**, 877 (1987)
9. L.P. Gor'kov, D. Jerome, J. Phys. (Paris) Lett. **46**, L643 (1985)
10. A.A. Abrikosov, *Fundamentals of the Theory of Metals* (Elsevier Science Publishers, Amsterdam, 1988)
11. A.G. Lebed, JETP Lett. **44**, 114 (1986)
12. N. Dupuis, G. Montambaux, C.A.R. Sa de Melo, Phys. Rev. Lett. **70**, 2613 (1993)
13. A.G. Lebed, K. Yamaji, Phys. Rev. Lett. **80**, 2697 (1998)
14. A.G. Lebed, Phys. Rev. B **59**, R721 (1999)
15. A.G. Lebed, K. Machida, M. Ozaki, Phys. Rev. B **62**, R795 (2000)
16. R.D. Duncan, C.D. Vaccarella, C.A.R. Sa de Melo, Phys. Rev. B **64**, 172503 (2001)
17. C.D. Vaccarella, C.A.R. Sa de Melo, Phys. Rev. B **64**, 212504 (2001)

18. C.D. Vassarella, C.A.R. Sa de Melo, Phys. Rev. B **63**, 180505(R) (2001)
19. D. Jerome, Nature (London) **387**, 235 (1997)
20. I.J. Lee, M.J. Naughton, G.M. Danner, P.M. Chaikin, Phys. Rev. Lett. **78**, 3555 (1997)
21. I.J. Lee, P.M. Chaikin, M.J. Naughton, Phys. Rev. B **65**, 180502(R) (2002)
22. I.J. Lee, P.M. Chaikin, M.J. Naughton, Phys. Rev. B **62**, R14669 (2002)
23. H.-I. Ha, M.J. Naughton (unpublished)
24. I.J. Lee, P.M. Chaikin, M.J. Naughton, Phys. Rev. B **65**, 180502 (2002)
25. I.J. Lee, S.E. Brown, W.G. Clark et al., Phys. Rev. Lett. **88**, 017004 (2002)
26. I.J. Lee, D.S. Chow, W.G. Clark et al., Phys. Rev. B **68**, 092510 (2003); I.J. Lee, S.E. Brown, W. Yu et al., Phys. Rev. Lett. **94**, 197001 (2005)
27. J. Shinagawa, Y. Kurosaki, F. Zhang, C. Parker, S.E. Brown, D. Jerome, J.B. Christensen, K. Bechgaard, Phys. Rev. Lett. **98**, 147002 (2007)
28. J.I. Oh, M.J. Naughton, Phys. Rev. Lett. **92**, 067001 (2004)
29. F. Tsobnang, F. Pesty, P. Garoche, Phys. Rev. B **49**, 15110 (1994)
30. N. Joo, P. Auban-Senzier, C.R. Pasquier et al., Europhys. Lett. **72**, 645 (2005)
31. H. Shimahara, Phys. Rev. B **61**, R14936 (2000)
32. V.P. Mineev, K.V. Samokhin, *Introduction to Unconventional Superconductivity* (Gordon and Breach, Amsterdam, 1999)
33. M. Sigrist, K. Ueda, Rev. Mod. Phys. **63**, 239 (1991)
34. In this review, we consider an orthorhombic model for a unit cell in $(TMTSF)_2X$ compounds
35. N. Dupuis, Phys. Rev. B **50**, 9607 (1994)
36. L.P. Gor'kov, A.G. Lebed, J. Phys. (Paris) Lett. **45**, L-433 (1984)
37. A.A. Abrikosov, L.P. Gor'kov, I.E. Dzyaloshinsi, *Methods of Quantum Field Theory in Statistical Physics* (Dover, New York, 1963)
38. I.S. Gradshteyn, I.M. Ryzhik, *Tables of Integrals, Series, and Products* (Academic, New York, 1994)
39. A.G. Lebed, M.J. Naughton, Phys. Rev. Lett. **91**, 187003 (2003)
40. T. Ishiguro, K. Yamaji, G. Saito, *Organic Superconductors*, 2nd edn. (Springer, Berlin Heidelberg New York, 1998)
41. S. Belin, K. Benia, Phys. Rev. Lett. **79**, 2125 (1997)
42. P. Garoche, R. Brussetti, D. Jerome et al., J. Phys. (Paris) Lett. **43**, L-147 (1982)
43. A.G. Lebed, Phys. Rev. Lett. **96**, 037002 (2006)

24

Triplet Superconductivity in Quasi-One-Dimensional Conductors

R.W. Cherng, W. Zhang, and C.A.R. Sá de Melo

We review some recent theoretical efforts to understand the spin structure and the orbital symmetry of the order parameter in quasi-one-dimensional conductors. A group theoretical analysis of possible order parameter candidates is presented along with characteristic spectroscopic and thermodynamic properties including density of states, specific heat, superfluid density, and spin susceptibility. In addition, we describe a test for triplet-order parameter symmetries based on the Josephson effect between two triplet superconductors and discuss the possibility of density-induced quantum phase transitions via electrostatic tuning of carrier density. Lastly, we study the competition and coexistence between spin density waves and triplet superconductivity by analyzing the pressure versus temperature phase diagram of quasi-one-dimensional conductors.

24.1 Introduction

After the discovery of quasi-one-dimensional organic superconductors [1], many interesting aspects were studied [2, 3]. The magnetic field versus temperature phase diagram for $(\mathrm{TMTSF})_2\mathrm{PF}_6$ under pressure was revisited by Lee et al. [4]. They found that the upper critical fields along the usual $\mathbf{a}$, $\mathbf{b}'$, and $\mathbf{c}^*$ directions were highly anisotropic and that the upper critical field along $\mathbf{a}$ and $\mathbf{b}'$ displayed a strong positive curvature at lower temperatures. The slope $[-\mathrm{d}H_{\mathrm{c}_2}/\mathrm{d}T]_{T_\mathrm{c}}$ for $\mathbf{H} \parallel \mathbf{b}'$ was smaller than that along the $\mathbf{a}$ direction, i.e., $H_{\mathrm{c}_2}^{(\mathbf{a})} > H_{\mathrm{c}_2}^{(\mathbf{b})}$ at higher temperatures, while, at low temperatures $H_{\mathrm{c}_2}^{(\mathbf{b})}$ exceeded $H_{\mathrm{c}_2}^{(\mathbf{a})}$, after an unusual anisotropy inversion at $H^* = 1.6\,\mathrm{T}$. These results suggested the existence of triplet superconductivity (TSC) in this system. An additional boost for the triplet scenario in $(\mathrm{TMTSF})_2\mathrm{PF}_6$ was given by NMR experiments [5, 6]. Lee et al. [5, 6] found that there is no ^{77}Se Knight shift change in $(\mathrm{TMTSF})_2\mathrm{PF}_6$ for fields $\mathbf{H} \parallel \mathbf{a}$ ($P \approx 7\,\mathrm{kbar}$) [5], and $\mathbf{H} \parallel \mathbf{b}'$ ($P \approx 6\,\mathrm{kbar}$) [6]. These results indicate that $\chi_{b'} \approx \chi_N$, and $\chi_a \approx \chi_N$, where χ_N is the normal state susceptibility. Furthermore, a sharp and narrow (possibly Hebel–Slichter [7]) peak was observed just below T_c in the ^{77}Se NMR relaxation rate $1/T_1$, for $\mathbf{H} \parallel \mathbf{a}$ ($H = 1.43\,\mathrm{T}$), and $P \approx 7\,\mathrm{kbar}$ [5]. The combined work of Lee et al. [4–6] implies the existence of a triplet superconducting phase in $(TMTSF)_2PF_6$. However, ^{77}Se NMR measurements in $(TMTSF)_2PF_6$ should be contrasted with earlier proton NMR results in $(TMTSF)_2ClO_4$ [8] which indicated the absence of the coherence peak just below $T_c = 1.03\,\mathrm{K}$ at $H = 0$, and a T^3 behavior between $T_c/2$ and T_c. Based on their experiments, Takigawa et al. [8] argued that the superconducting state of $(TMTSF)_2ClO_4$ has an anisotropic order parameter vanishing along lines on the Fermi surface, although they were not able to distinguish between singlet and triplet states.

The temperature versus magnetic field phase diagram of $(TMTSF)_2ClO_4$ at ambient pressure was also measured by Lee et al. [9,10], for $\mathbf{H} \parallel \mathbf{b}'$. Their results showed that the Pauli paramagnetic limit is also exceeded in this compound. Furthermore, Belin and Behnia [11] (BB) reported measurements of the thermal conductivity in the superconducting state of $(TMTSF)_2ClO_4$, indicating that their data are inconsistent with the existence of gap nodes at the Fermi surface as suggested by Takigawa et al. [8]. BB's argument was that the T^3 behavior of the proton $1/T_1$ [8] was limited only to $T > T_c/2$, thus, it could not be considered convincing evidence for nodes in the gap because the temperature was not low enough. Even for conventionally gapped superconductors, the exponential behavior of $1/T_1$ occurs only at very low temperatures ($T \ll T_c$). Therefore, these recent experimental results combined [9–11] seem to suggest the existence of a fully gapped triplet superconducting state in $(TMTSF)_2ClO_4$. However, detailed ^{77}Se NMR experiments seem to be lacking for $(TMTSF)_2ClO_4$.

Abrikosov was the first to suggest the possibility of TSC in Bechgaard salts [12,13] based on the behavior of $(TMTSF)_2ClO_4$ and $(TMTSF)_2PF_6$ under X-ray bombardment. These systems exhibited a strong suppression of their critical temperature in the presence of radiation induced nonmagnetic defects [14,15]. Gorkov and Jerome [16] also suggested the possibility of TSC based on a theoretical extrapolation of the semiclassical upper critical fields calculated near the zero field critical temperature. In addition, Lebed [17] pointed out that TSC would manifest itself in Bechgaard salts through a remarkable reentrant phase in high magnetic fields. Later, Dupuis, Montambaux and Sá de Melo (DMS) [18] studied the field versus temperature phase diagram and the vortex lattice structure of quasi-one-dimensional superconductors. DMS concluded that these systems could be either in an inhomogeneous singlet state like the Larkin–Ovchinnikov–Fulde–Ferrel (LOFF) state [19,20] or in a triplet state. Further studies of the LOFF state were performed by Dupuis [21], while an equal spin triplet pairing state (ESTP) was discussed further by Sá de Melo [22].

After the upper critical field measurements of Lee et al. [4], Lebed [23] was able to show that the LOFF state was Pauli paramagnetically limited, thus giving further support for the triplet scenario (at least at high magnetic fields).

The possibility of a magnetic field induced singlet (low fields) to triplet (high fields) transition was considered by Sá de Melo [24,25] and Vaccarella and Sá de Melo [26], but currently there is no experimental evidence of a kink in the upper critical field of these systems. Furthermore, reentrant superconductivity was still expected for the triplet state at high magnetic fields [23,26]. All these previous theories neglected the effects of fluctuations at high magnetic fields. A first attempt to incorporate fluctuation effects was made by Vaccarella and Sá de Melo [27] who showed that phase fluctuations can suppress the reentrant phase at high magnetic fields in the ESTP state.

Lebed, Machida, and Ozaki (LMO) [28] suggested the possibility of a triplet phase with a $\mathbf{d}$ vector with zero-component along the $\mathbf{b}'$-axis and a finite component along the $\mathbf{a}$-axis. A direct consequence of LMO's proposal is the existence of an anisotropic spin susceptibility at zero temperature and low magnetic fields: $\chi_{b'} = \chi_N$, and $\chi_a \ll \chi_N$, where χ_a corresponds to $\mathbf{H} \parallel \mathbf{a}$, $\chi_{b'}$ corresponds to $\mathbf{H} \parallel \mathbf{b}'$, and χ_N corresponds to the normal state susceptibility. A fully gapped singlet "d-wave" order parameter for $(TMTSF)_2ClO_4$ was proposed by Shimahara [29], while gapless triplet "f-wave" superconductivity for $(TMTSF)_2PF_6$ was proposed by Kuroki, Arita, and Aoki (KAA) [30]. Duncan, Vaccarella, and Sá de Melo (DVS) [31,32] performed a group theoretical analysis and identified the triplet "p_x-wave" state as a good candidate for superconductivity in Bechgaard salts. This symmetry is characterized by a fully gapped quasiparticle excitation spectrum, and by spin susceptibilities $\chi_a \approx \chi_N$, and $\chi_{b'} \approx \chi_N$ [24,25,31,32]. All the choices of order parameters mentioned above are consistent with the expectation of weak atomic spin–orbit effects, given that the heaviest element in these systems is Se, however further experimental work is needed to identify the correct symmetry.

This chapter is organized as follows. In Sect. 24.2, we discuss the Hamiltonian and order parameter symmetries consistent with weak spin–orbit effects and TSC. In Sect. 24.3, we review spectroscopic and thermodynamic properties including the density of states, specific heat, superfluid density and spin susceptibility. In Sect. 24.4, we review a test for the order parameter symmetry based on the Josephson effect between two triplet superconductors. In Sect. 24.5, we analyze density induced quantum phase transitions via electrostatic tuning of carrier density. In Sect. 24.6, we discuss the coexistence of TSC and spin–density wave (SDW) order in the pressure versus temperature phase diagram. Lastly, in Sect. 24.7, we summarize the most important results.

24.2 Hamiltonian and Order Parameter Symmetries

In this section, we classify possible order parameter symmetries for quasi-one-dimensional superconductors. Different order parameter symmetries correspond to pairing instabilities generated by attractive electron–electron interactions in different channels. The allowed channels are dictated by the underlying symmetries of the system and given by powerful group theoretical techniques. We will focus on time reversal invariant triplet states assuming weak spin–orbit coupling and refer to [31, 32] for a more general analysis of singlet, strong spin–orbit coupling, and time-reversal symmetry breaking symmetries.

Although the Bechgaard salts have a triclinic structure, we consider an orthorhombic crystal with the crystallographic point group D_{2h} and we assume that the normal state does not break the crystal translation group. The heaviest element in $(TMTSF)_2PF_6$ is ^{77}Se which suggests spin–orbit coupling is weak and rotations in spin space and real space decouple. At zero magnetic field, there is also no experimental evidence for time-reversal symmetry breaking. These considerations give a symmetry group for the normal state of the form $G_\mathrm{n} = SO(3) \times G_\mathrm{c} \times U(1) \times T$, where $SO(3)$ is the group of rotations in spin space, $\mathrm{G_c} = D_{2h}$ is the crystal space group, $U(1)$ is the gauge group and T corresponds to time reversal symmetry.

Having identified the symmetry group, we now turn to the Hamiltonian

$$H = H_\mathrm{kin} + H_\mathrm{int}, \tag{24.1}$$

where H_kin corresponds to the kinetic energy, and H_int to the interaction. We study quasi-one-dimensional systems with a single band in an orthorhombic lattice where

$$H_\mathrm{kin} = \sum_{\mathbf{k},\alpha} (\epsilon_\mathbf{k} - \mu) c^\dagger_{\mathbf{k},\alpha} c_{\mathbf{k},\alpha} \tag{24.2}$$

is expressed in terms of the dispersion

$$\epsilon_\mathbf{k} = -|t_x| \cos(k_x a) - |t_y| \cos(k_y b) - |t_z| \cos(k_z c) \tag{24.3}$$

relative to the chemical potential μ. The transfer integrals $|t_x| \gg |t_y| \gg |t_z|$ reflect the quasi-one-dimensionality of the system. The interaction part is

$$H_\mathrm{int} = \frac{1}{2} \sum_{\mathbf{k}\mathbf{k}'\mathbf{q}} \sum_{\alpha\beta\gamma\delta} V_{\alpha\beta\gamma\delta}(\mathbf{k},\mathbf{k}') b^\dagger_{\alpha\beta}(\mathbf{k},\mathbf{q}) b_{\gamma\delta}(\mathbf{k}',\mathbf{q}), \tag{24.4}$$

when expressed in terms of the pairing operators

$$b^\dagger_{\alpha\beta}(\mathbf{k},\mathbf{q}) = c^\dagger_{-\mathbf{k}+\mathbf{q}/2,\alpha} c^\dagger_{\mathbf{k}+\mathbf{q}/2,\beta}, \tag{24.5}$$

where the labels α, β, γ, and δ are spin indices and the labels $\mathbf{k}$, $\mathbf{k}'$, and $\mathbf{q}$ represent linear momenta.

The interaction tensor $V_{\alpha\beta\gamma\delta}(\mathbf{k},\mathbf{k}')$ must be invariant under the normal state symmetry group G_n. Each spin (momentum) index transforms as an irreducible representation of the $SO(3)$ rotation group (D_{2h} orthorhombic group) and all indices must be combined in an invariant way. In the case of weak spin–orbit coupling and triplet pairing, the model interaction tensor can be chosen to be

$$V_{\alpha\beta\gamma\delta}(\mathbf{k},\mathbf{k}') = \Gamma_{\alpha\beta\gamma\delta} V_\Gamma \phi_\Gamma(\mathbf{k}) \phi^*_\Gamma(\mathbf{k}'). \tag{24.6}$$

Here $\Gamma_{\alpha\beta\gamma\delta} = \mathbf{v}_{\alpha\beta} \cdot \mathbf{v}^{\dagger}_{\gamma\delta}/2$ with $\mathbf{v}_{\alpha\beta} = (\mathrm{i}\sigma\sigma_y)_{\alpha\beta}$ and $\phi_\Gamma(\mathbf{k})$ corresponds to the irreducible representation Γ of the orthorhombic group which are all one-dimensional [33]. This fixes the interaction up to an overall energy scale V_Γ for each representation Γ.

Using either the equation of motion method [34] or the functional integration method [35, 36] we obtain the familiar order parameter and number equations

$$\Delta_{\alpha\beta}(\mathbf{k}) = -\sum_{\mathbf{k}'} V_{\beta\alpha\gamma\delta}(\mathbf{k},\mathbf{k}') \frac{\Delta_{\gamma\delta}(\mathbf{k}')}{2E_{\mathbf{k}'}} \tanh\left(\frac{E_{\mathbf{k}'}}{2T}\right), \tag{24.7}$$

$$N \equiv \sum_{\mathbf{k}} n_{\mathbf{k}} = \sum_{\mathbf{k}} \left[\left(1+\frac{\xi_{\mathbf{k}}}{E_{\mathbf{k}}}\right) f(E_{\mathbf{k}}) + \left(1-\frac{\xi_{\mathbf{k}}}{E_{\mathbf{k}}}\right)(1-f(E_{\mathbf{k}}))\right]. \tag{24.8}$$

where $f(x)$ is the Fermi distribution. These two equations must be solved self-consistently away from the strict BCS limit to accommodate particle–hole asymmetries. Strong coupling effects tend to shift the chemical potential substantially away from the Fermi energy and modify dramatically the previous self-consistency equations near T_c. However, two equations are correct even in the strong coupling (or low density) regime provided that $T \ll T_\mathrm{c}$ [35, 36].

The triplet superconducting state is described by the order parameter $\Delta_{\mathbf{k}}$ and quasiparticle excitation energy $E_{\mathbf{k}}$ given, respectively, by

$$\Delta_{\mathbf{k}} = \mathrm{i}\Delta_\Gamma \mathbf{d}(\mathbf{k}) \cdot \sigma\sigma_y, \quad E_{\mathbf{k}} = \sqrt{\xi_{\mathbf{k}}^2 + \Delta_{\mathbf{k}}^2}, \tag{24.9}$$

where the three-dimensional vector $\mathbf{d}(\mathbf{k})$ is antisymmetric under $\mathbf{k} \to -\mathbf{k}$ and transforms according to the irreducible representations of the orthorhombic group D_{2h}. We also obtain the anomalous $F_{\alpha\beta}(\mathbf{k}, \mathrm{i}\omega_\mathrm{n})$ and single particle $G_{\alpha\beta}(\mathbf{k}, \mathrm{i}\omega_\mathrm{n})$ Green's functions

$$F_{\alpha\beta}(\mathbf{k}, \mathrm{i}\omega_\mathrm{n}) = \frac{\Delta_{\alpha\beta}(\mathbf{k})}{\omega_\mathrm{n}^2 + E_{\mathbf{k}}^2}, \quad G_{\alpha\beta}(\mathbf{k}, \mathrm{i}\omega_\mathrm{n}) = -\frac{\mathrm{i}\omega_\mathrm{n} + \xi_{\mathbf{k}}}{\omega_\mathrm{n}^2 + E_{\mathbf{k}}^2}\delta_{\alpha\beta}. \tag{24.10}$$

Table 24.1 gives the state nomenclature, residual symmetry group, and order parameter $\mathbf{d}(\mathbf{k})$. The notation f_{xyz}, p_x, p_y, p_z is meant to be suggestive of analogous spherical harmonics and will be used throughout this review. In general, the direction of $\mathbf{d}(\mathbf{k})$ is arbitrary, but weak spin–orbit coupling in the Bechgaard salts should lead to the weak pinning of the $\mathbf{d}$ vector.

Table 24.1. Time reversal invariant triplet states in an orthorhombic crystal, assuming weak spin–orbit coupling

State		Residual group	Order parameter $\mathbf{d}(\mathbf{k})$
f_{xyz}	$^3A_{1u}(a)$	$D_\infty(C_\infty) \times D_2 \times I(E) \times T$	$(0,0,1)XYZ$
p_z	$^3B_{1u}(a)$	$D_\infty(C_\infty) \times D_2(C_2^z) \times I(E) \times T$	$(0,0,1)Z$
p_y	$^3B_{2u}(a)$	$D_\infty(C_\infty) \times D_2(C_2^y) \times I(E) \times T$	$(0,0,1)Y$
p_x	$^3B_{3u}(a)$	$D_\infty(C_\infty) \times D_2(C_2^x) \times I(E) \times T$	$(0,0,1)X$

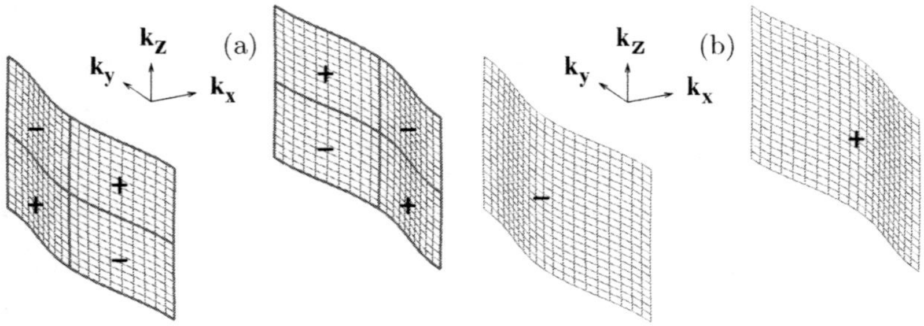

Fig. 24.1. Three-dimensional views of the sign of the orbital component of the **d** vectors on the Fermi surface for (**a**) f_{xyz} and (**b**) p_x. *Thick lines* correspond to order parameter nodes and zeros of the quasiparticle excitation spectrum

Based on recent ^{77}Se NMR experiments [5, 6] it seems that the **d** vector in $(TMTSF)_2PF_6$ is not pointing along **a** or **b**′ direction at least for magnetic fields beyond the 1 T range. Therefore, we choose the weak spin–orbit coupling **d** vectors to be pointing along the $\mathbf{c}^*$ direction.

We emphasize that the basis functions $X(\mathbf{k}), Y(\mathbf{k})$, and $Z(\mathbf{k})$ transform like k_x, k_y, and k_z, but must be periodic in momentum space. This is because the dispersion $\epsilon_\mathbf{k}$ defined in (24.3), where $|t_x| \gg |t_y| \gg |t_z|$ gives a normal state Fermi surface with two disconnected sheets (for weak attractive interaction or high densities) that intersects the Brillouin zones. The minimal basis set must be periodic and may be chosen to be $X(\mathbf{k}) = \sin(k_x a_x)$, $Y(\mathbf{k}) = \sin(k_y a_y)$ and $Z(\mathbf{k}) = \sin(k_z a_z)$, where a_x, a_y, and a_z are lattice constants. In Fig. 24.1, we plot the sign and location of nodes of the order parameter on the Fermi surface for f_{xyz} and p_x symmetries. From (24.9), we observe that gapless quasiparticle excitations occur at the intersection of the Fermi surface and order parameter nodes. In particular, from Fig. 24.1, it is clear that the p_x symmetry has no nodes and is fully gapped while the f_{xyz} symmetry has lines of first-order nodes and isolated second-order nodes and is gapless. The p_y (p_z) symmetries which are not shown have lines of nodes on the $k_y = 0, \pm\pi$ ($k_z = \pm\pi$) planes which intersect the Fermi surface and are gapless. This nodal structure is extremely important for spectroscopic and thermodynamics quantities to be discussed in Sect. 24.3.

24.3 Spectroscopic and Thermodynamic Quantities

The group theoretical analysis of Sect. 24.2 classified the allowed time-reversal triplet states with weak spin–orbit coupling. Qualitative differences among these states, such as quasiparticle excitation spectrum, have observable consequences on measurable spectroscopic and thermodynamic quantities discussed in this section. To compare the symmetries, we solve self-consistently the number and order parameter equations [(24.7) and (24.8), respectively]. We assume that the critical temperature is $T_c = 1.5\,\mathrm{K}$, for a quarter-filled band (with filling factor $\widetilde{N} = 0.5$), and we use parameters $|t_x| = 5{,}800\,\text{K}$, $|t_y| = 1{,}226\,\text{K}$, and $|t_z| = 48\,\text{K}$ to determine V_Γ. Below T_c, we self-consistently solve (24.7) and (24.8) to calculate the quasiparticle density of states (QDOS), specific heat, superfluid density, and spin susceptibility tensor. These quantities can help to distinguish among the allowed symmetries. We will argue that the p_x symmetry is a promising candidate based on current experimental evidence. Extended results and details of our calculations can be found in [31, 32, 37].

The bulk quasiparticle density of states probes the quasiparticle excitation spectrum and can be obtained from the single particle Green's function as

$$\mathcal{N}(\omega) = -\frac{1}{\pi}\mathrm{Tr}\sum_{\mathbf{k}} \mathrm{Im}\, G_{\alpha\beta}(\mathbf{k}, \mathrm{i}\omega_\mathrm{n} = \omega + \mathrm{i}\delta), \tag{24.11}$$

where $G_{\alpha\beta}(\mathbf{k}, \mathrm{i}\omega_\mathrm{n})$ is defined in (24.10). Features of the QDOS shown could be measured, in principle, during STM or photoemission experiments. However, the unconventional superconductivity for triplet states may have surface bound states [38–40]. The QDOS that we discuss is a bulk property, thus the contribution of these surface states is not reflected in our calculated QDOS. As an example, we plot the QDOS for the fully gapped p_x symmetry in Fig. 24.2a. The QDOS vanishes for energies below the gap $|\Delta_{p_x}|$ and exhibits a square root behavior near the threshold. However, the gapless symmetries p_y, p_z have a linear behavior in frequency beginning at zero frequency due to lines of nodes directly on the Fermi surface, while the f_{xyz} symmetry has a logarithmic singularity of the form $\omega \log(\Delta_{f_{xyz}}/\omega)$ due to the presence of isolated second-order nodes.

The temperature dependence of the fermionic contribution to the specific heat is

$$C_\Gamma = \frac{2}{T^2}\sum_{\mathbf{k}} P(E_\mathbf{k}) \left[E_\mathbf{k}^2 + T\xi_\mathbf{k}\frac{\partial \mu}{\partial T} - \frac{T}{2}\frac{\partial |\mathbf{d}(\mathbf{k})|^2}{\partial T} \right], \tag{24.12}$$

where $P(E_\mathbf{k}) = f(E_\mathbf{k})\,[1 - f(E_\mathbf{k})]$. Notice that C_Γ also probes the quasiparticle excitation spectrum (collective mode contributions are not discussed

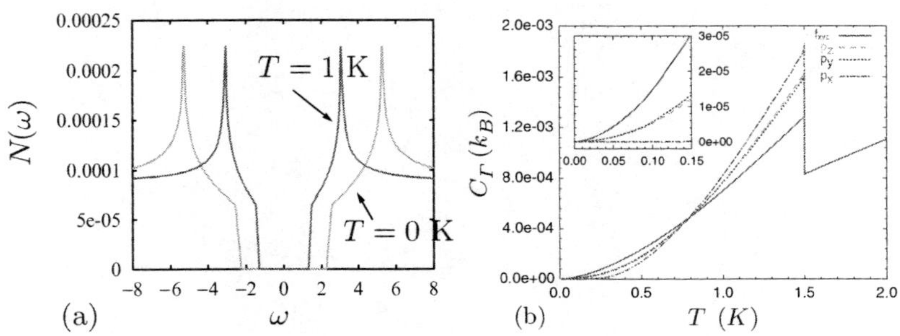

Fig. 24.2. (**a**) Frequency dependence of the QDOS $N(\omega)$ for p_x symmetry and (**b**) temperature dependence of specific heat C_Γ in units of k_B. Inset shows low T behavior of C_Γ

here). This is most evident at low T where the specific heat is directly related to the quasiparticle density of states (in the strict BCS limit) $C_\Gamma \approx 2T\int_0^\infty dww^2\text{sech}^2(w)\mathcal{N}(2Tw)$. We find thermally activated behavior $C_{p_x} \approx (T_{p_x}/T)^{1/2}\exp[-\omega_{p_x}/T]$ where $T_{p_x}, \omega_{p_x} \approx \Delta_{p_x}$ for the p_x symmetry reflecting the presence of a full gap with a power-law prefactor arising from the square root singularity at threshold in the QDOS. For the gapless p_y, p_z symmetries we find power-law behavior $C_{p_{y,z}} \approx (T/T_{p_{y,z}})^2$ where $T_{p_{y,z}} \approx \Delta_{p_{y,z}}$ due to presence of lines of order parameter nodes. The f_{xyz} symmetry exhibits logarithmic behavior $C_{f_{xyz}} \approx (T/T_{f_{xyz}})^2\log(\omega_{f_{xyz}}/T)$ where $T_{f_{xyz}} \approx \Delta_{f_{xyz}}$ due to isolated second-order nodes. The temperature dependence of the specific heat for p_x, p_y, p_z, and f_{xyz} symmetries is ploted in Fig. 24.2b.

The temperature dependence of the superfluid density tensor

$$\rho_{ij}(T) = \frac{1}{V}\sum_{\mathbf{k}}\left[n_{\mathbf{k}}\partial_i\partial_j\xi_{\mathbf{k}} - Y_{\mathbf{k}}\partial_i\xi_{\mathbf{k}}\partial_j\xi_{\mathbf{k}}\right] \qquad (24.13)$$

can also be used to distinguish different weak spin–orbit phases for triplet pairing. Here $n_{\mathbf{k}}$ is the momentum distribution, and $Y_{\mathbf{k}} = (2T)^{-1}\text{sech}^2\left(E_{\mathbf{k}}/2T\right)$ is the Yoshida distribution. This tensor is directly associated with phase twists of the $U(1)$ phase of the $\mathbf{d}$ vector and can be measured in penetration depth experiments. Only the diagonal components are nonvanishing and are highly anisotropic. The low temperature behavior of $\Delta\rho_{xx} \equiv [\rho_{xx}(T)/\rho_{xx}(0) - 1]$ is thermally activated for p_x, power-law for p_y and power-law with logarithmic corrections for f_{xyz} (see [37]). This is due to the interplay between the order parameter nodes and the zeros of ξ_k term in (24.13). In the particular case of the f_{xyz} symmetry, this effect leads to logarithmic corrections only for $\Delta\rho_{xx}$ and pure power-law behavior for $\Delta\rho_{yy}$ and $\Delta\rho_{zz}$. We plot the largest component ρ_{xx} in Fig. 24.3a.

The thermodynamic quantities discussed so far are only sensitive to the magnitude of $\mathbf{d}(\mathbf{k})$. Now we turn to the spin susceptibility tensor χ_{mn} which is sensitive to both the direction and magnitude of $\mathbf{d}(\mathbf{k})$ and can be probed in Knight shift and NMR relaxation rate experiments. We obtain the uniform

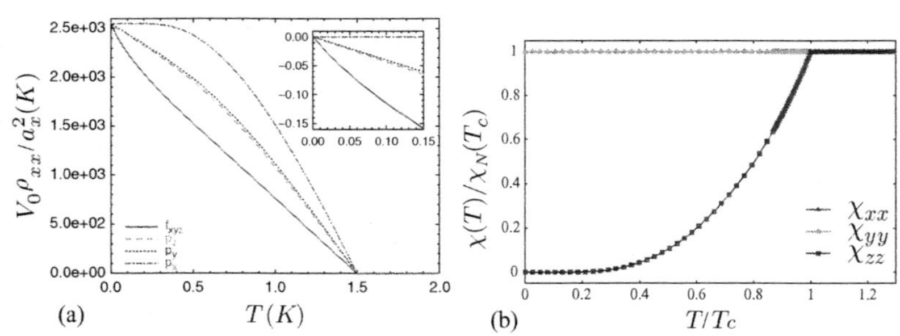

Fig. 24.3. (**a**) Temperature dependence of $V_0\rho_{xx}/a_x^2$ in thermal units (K), where V_0 (a_x) is the unit cell volume (length). Inset shows the low T behavior of $\Delta\rho_{xx}$ (see text). (**b**) Uniform spin susceptibility tensor components χ_{11} (*triangles*); χ_{22} (*circles*); χ_{33} (*squares*) at low temperatures for the p_x symmetry

limit ($\omega \to 0$, $q \to 0$) of χ_{mn} as

$$\chi_{mn}(\mathbf{0},0) = \sum_{\mathbf{k}} \left[\chi_{mn,1}(\mathbf{k}) + \chi_{mn,2}(\mathbf{k})\right], \tag{24.14}$$

where the **k**-dependent tensors have the forms

$$\chi_{mn,1}(\mathbf{k}) = \tilde{g}_m \tilde{g}_n \chi_{\parallel}(\mathbf{k}) \mathrm{Re}\ \hat{d}_m^*(\mathbf{k}) \hat{d}_n(\mathbf{k}), \tag{24.15}$$

$$\chi_{mn,2}(\mathbf{k}) = \tilde{g}_m \tilde{g}_n \chi_{\perp}(\mathbf{k}) \left(\delta_{mn} - \mathrm{Re}\ \hat{d}_m^*(\mathbf{k}) \hat{d}_n(\mathbf{k})\right), \tag{24.16}$$

with $\hat{d}_n(\mathbf{k}) = d_n(\mathbf{k})/|d_n(\mathbf{k})|$, $\tilde{g}_\ell = g_\ell/2$ are the scaled-gyromagnetic factors taking into account the possibility of anisotropies due to spin–orbit coupling. Here, the parallel component is

$$\chi_{\parallel}(\mathbf{k}) = -2\mu_B^2 \frac{\partial f(E_{\mathbf{k}})}{\partial E_{\mathbf{k}}},$$

while the perpendicular component is

$$\chi_{\perp}(\mathbf{k}) = 2\mu_B^2 \frac{\mathrm{d}}{\mathrm{d}\xi_{\mathbf{k}}} \left[\frac{\xi_{\mathbf{k}}}{2E_{\mathbf{k}}}\left(1 - 2f(E_{\mathbf{k}})\right)\right].$$

We plot the temperature dependence of the uniform susceptibility χ_{mn} scaled by the normal state value for the p_x symmetry in Fig. 24.3b. Notice that χ_{xx} and χ_{yy} components are the same as the normal state while χ_{zz} is suppressed at low temperatures. Since the p_x symmetry exhibits weak spin–orbit coupling, a weak magnetic field **H** larger than the spin–orbit coupling pinning field H_f can always rotate $\mathbf{d}(\mathbf{k})$ to be perpendicular to **H** so that the observed susceptibility is always that of the normal state. This pinning field was estimated to be $H_f \approx 0.22\,\mathrm{T}$ (see [32]).

Now we wish to relate our results to current experimental evidence available for quasi-one-dimensional superconductors and propose that the p_x state is a good candidate for the order parameter symmetry. Thermal conductivity measurements [11] suggest the absence of nodes on the Fermi surface and are consistent with the presence of a full gap for the p_x symmetry. Future STM or photoemission experiments probing the QDOS should exhibit the behavior shown in Fig. 24.2a with square root behavior near a finite energy threshold if the p_x is indeed the correct order parameter symmetry. Furthermore, thermodynamic quantities such as the specific heat in Fig. 24.2b and superfluid density tensor in Fig. 24.3a should exhibit thermally activated behavior at low temperatures. Knight shift experiments show that there is no observable change when $\mathbf{H} \parallel \mathbf{a}$ [5] or $\parallel \mathbf{b}'$ [6] at lower temperatures. This is consistent with the p_x symmetry's normal state susceptibility for any direction $\hat{n}$ of **H** provided that the magnitude of the applied field is larger than the small spin–orbit pining field $H_f^{(\hat{n})}$.

Further tests of the symmetry of the superconducting state require experiments that are sensitive to the phase and direction of the order parameter. Thus, we discuss next the Josephson effect between quasi-one-dimensional superconductors as a self-test for the triplet state.

24.4 Josephson Effect

In this section we review a possible test for the order parameter symmetry of Bechgaard salts which is based on the Josephson effect between two triplet superconductors (TS) separated by an insulating (I) barrier [41, 42]. Our suggestion differs from previous proposals invoking tests of the symmetry of the order parameter in the context of triplet heavy fermion superconductors [43,44], which relied on the Josephson effect between a *singlet* and a *triplet* superconductor. We show that the Josephson effect between two TS exists in the simplest case where the tunneling matrix element is spin-conserving (with or without time-reversal invariance), and that it depends on the relative orientation of the **d** vectors of the TS. In addition, we point out that the temperature and angular dependence of the Josephson effect can also help distinguish between different triplet states. We include coherent and incoherent tunneling processes and the effects of surface bound states (within the framework of nonlocal lattice Bogoliubov–de Gennes (BdG) equations).

In anticipation of the existence of surface bound states in particular geometries of unconventional superconductors [38, 39], we prefer to use a real space representation of the *left* (L) and *right* (R) superconductors Hamiltonian

$$H_j = \mathcal{H}_{0,j} + \sum_{\mathbf{r}_j,\mathbf{r}'_j} \Big[\Delta_{j,\alpha_j,\beta_j}(\mathbf{r}_j,\mathbf{r}'_j) c^\dagger_{j,\alpha_j}(\mathbf{r}_j) c^\dagger_{j,\beta_j}(\mathbf{r}'_j) + \text{H.C.}\Big], \tag{24.17}$$

where $\mathcal{H}_{0,j} = \sum_{\mathbf{r}_j,\mathbf{r}'_j} \mathcal{H}(\mathbf{r}_j,\mathbf{r}'_j) c^\dagger_{j,\alpha_j}(\mathbf{r}_j) c_{j,\alpha_j}(\mathbf{r}'_j)$, and indices α_j and β_j (with j = L, R) are spin labels, $\mathbf{r}_j, \mathbf{r}'_j$ are position labels. Repeated greek indices indicate summation. The first term of H_j contains $\mathcal{H}(\mathbf{r}_j,\mathbf{r}'_j) = \Big[t_j(\mathbf{r}_j,\mathbf{r}'_j) - \mu\delta_{\mathbf{r}_j,\mathbf{r}'_j}\Big]$, where $t_j(\mathbf{r}_j,\mathbf{r}'_j)$ are transfer integrals confined to nearest neighbors and μ_j are the chemical potential. The second term of H_j contains the order parameter matrix

$$\Delta_{j,\alpha_j,\beta_j}(\mathbf{r}_j,\mathbf{r}'_j) = V_{j,\alpha_j,\beta_j,\delta_j,\gamma_j}(\mathbf{r}_j,\mathbf{r}'_j)\langle c_{j,\delta_j}(\mathbf{r}'_j) c_{j,\gamma_j}(\mathbf{r}_j)\rangle,$$

where $V_{j,\alpha_j,\beta_j,\delta_j,\gamma_j}(\mathbf{r}_j,\mathbf{r}'_j)$ is the two-body pairing interaction. The tunneling Hamiltonian connecting L and R superconductors is

$$H_{\mathrm{T}} = \sum_{\mathbf{r}_{\mathrm{L}},\mathbf{r}_{\mathrm{R}}} \Big[T_{\alpha_{\mathrm{L}}\alpha_{\mathrm{R}}}(\mathbf{r}_{\mathrm{L}},\mathbf{r}_{\mathrm{R}}) c^\dagger_{\mathrm{L},\alpha_{\mathrm{L}}}(\mathbf{r}_{\mathrm{L}}) c_{\mathrm{R},\alpha_{\mathrm{R}}}(\mathbf{r}_{\mathrm{R}}) + \text{H.C.}\Big], \tag{24.18}$$

with $T_{\alpha_{\mathrm{L}}\alpha_{\mathrm{R}}}(\mathbf{r}_{\mathrm{L}},\mathbf{r}_{\mathrm{R}}) = T_s(\mathbf{r}_{\mathrm{L}},\mathbf{r}_{\mathrm{R}})\delta_{\alpha_{\mathrm{L}},\alpha_{\mathrm{R}}} + \mathbf{T}_v(\mathbf{r}_{\mathrm{L}},\mathbf{r}_{\mathrm{R}})\cdot\sigma_{\alpha_{\mathrm{L}},\alpha_{\mathrm{R}}}$ being the tunneling amplitude, where the first term T_s is spin-conserving and the second term $\mathbf{T}_v$ is spin-dependent.

The general form of the order parameter matrix is

$$\Delta_{j,\alpha_j,\beta_j}(\mathbf{r}_j,\mathbf{r}'_j) = \mathrm{i}\Delta_{si,j}(\mathbf{r}_j,\mathbf{r}'_j)\left[\sigma_2\right]_{\alpha_j,\beta_j} + \mathrm{i}\mathbf{d}_j(\mathbf{r}_j,\mathbf{r}'_j)\cdot\left[\sigma\sigma_2\right]_{\alpha_j\beta_j},$$

where the first term corresponds to the singlet (pseudosinglet) state and the second to the triplet (pseudotriplet) state in the case of weak (strong) spin–orbit coupling. The order parameter matrix $\Delta_{j,\alpha_j,\beta_j}(\mathbf{r}_j,\mathbf{r}'_j)$ satisfies the Pauli exclusion principle, since the function $\Delta_{si,j}(\mathbf{r}_j,\mathbf{r}'_j)$ which represents singlet pairing is symmetric under the exchange $\mathbf{r}_j \leftrightarrow \mathbf{r}'_j$, and the vector $\mathbf{d}_j(\mathbf{r}_j,\mathbf{r}'_j)$ which represent triplet pairing is antisymmetric under the exchange $\mathbf{r}_j \leftrightarrow \mathbf{r}'_j$. The reduced Hamiltonians on both sides of the junction can be separately diagonalized via the lattice BdG transformation $c_{j,\alpha_j}(\mathbf{r}_j) = \sum_{N_j}\left[u_{N_j\alpha_j}(\mathbf{r}_j)\gamma_{N_j} + v^*_{N_j\alpha_j}(\mathbf{r}_j)\gamma^\dagger_{N_j}\right]$, where $u_{N_j\alpha_j}$ and $v_{N_j\alpha_j}$ are two component spinors for each value of α_j and satisfy the corresponding nonlocal BdG equation

$$E_{N_j}u_{N_j\alpha_j}(\mathbf{r}_j) = \sum_{\mathbf{r}'_j}\mathcal{H}(\mathbf{r}_j,\mathbf{r}'_j)u_{N_j\alpha_j}(\mathbf{r}'_j) + \sum_{\mathbf{r}'_j}\Delta_{j,\alpha_j,\beta_j}(\mathbf{r}_j,\mathbf{r}'_j)v_{N_j\beta_j}(\mathbf{r}'_j), \tag{24.19}$$

$$-E_{N_j}v_{N_j\alpha_j}(\mathbf{r}_j) = \sum_{\mathbf{r}'_j}\mathcal{H}^*(\mathbf{r}_j,\mathbf{r}'_j)v_{N_j\alpha_j}(\mathbf{r}'_j) + \sum_{\mathbf{r}'_j}\Delta^*_{j,\alpha_j,\beta_j}(\mathbf{r}_j,\mathbf{r}'_j)u_{N_j\beta_j}(\mathbf{r}'_j). \tag{24.20}$$

This set of equations must be solved self-consistently together with the order parameter equation (see [41, 42]).

From now on we will assume that the superconductors on either side of the junction are in (a) a triplet unitary state (time reversal invariant state in the bulk), (b) characterized by weak spin–orbit coupling, (c) the tunneling matrix element $T_{\alpha_\mathrm{L}\alpha_\mathrm{R}}(\mathbf{r}_j,\mathbf{r}'_j)$ is spin conserving, i.e., the tunnel barrier preserves spin, and thus it is not magnetically active, and (d) the $\mathbf{d}_j$-vector is weakly locked to the $\mathbf{c}^*_j$-axis by the weak spin–orbit coupling, a choice that is consistent with Knight shift experiments of Lee et al. [5]. All these assumptions seem to be applicable to the Bechgaard salts. Next, we discuss the Josephson effect in two situations, first neglecting, and second including surface bound state effects.

The Josephson effect neglecting surface bound states: In this approximation only the contributions from scattering states with well-defined momenta are included. Under this assumption the BdG amplitudes become $u_{N_j\alpha_j}(\mathbf{r}_j) = \exp(i\mathbf{k}_j\cdot\mathbf{r}_j)\tilde{u}_{\mathbf{k}_j,n_j}/\sqrt{V}$; $v_{N_j\alpha_j}(\mathbf{r}_j) = \exp(i\mathbf{k}_j\cdot\mathbf{r}_j)\tilde{v}_{\mathbf{k}_j,n_j}/\sqrt{V}$, where quantum indices N_j are represented by momentum $\mathbf{k}_j$ and discrete index $n_j = 1_j, 2_j$. For the weak spin–orbit p_x, p_y, p_z, and f_{xyz} states, where the direction of the $\mathbf{d}$ vectors is independent of momentum, a simple expression for $J_s(T)$ at zero bias can be obtained

$$J_s^{(\lambda)}(T) = \widetilde{J}^{(\lambda)}(T)\times(\hat{\eta}_\mathrm{L}\cdot\hat{\eta}_\mathrm{R})\times\sin(\bar{\phi}), \tag{24.21}$$

where $\widetilde{J}^{(\lambda)}(T) = (2e/\hbar)\Delta_\mathrm{L}(T)\Delta_\mathrm{R}(T)S^{(\lambda)}_\mathrm{LR}(T)$, $\bar{\phi} = \phi_\mathrm{R} - \phi_\mathrm{L}$, and the label λ identifies incoherent (λ = inc) or coherent (λ = coh) processes. Here, $S^{(\lambda)}_\mathrm{LR}(T) = \sum_{\mathbf{k}_\mathrm{L},\mathbf{k}_\mathrm{R}} Q^{(\lambda)}(\mathbf{k}_\mathrm{L},\mathbf{k}_\mathrm{R})\psi^{\Gamma_\mathrm{L},u}(\mathbf{k}_\mathrm{L})\psi^{\Gamma_\mathrm{R},u}(\mathbf{k}_\mathrm{R})P(\mathbf{k}_\mathrm{L},\mathbf{k}_\mathrm{R})$, where

$P(\mathbf{k}_{\rm L},\mathbf{k}_{\rm R}) = 4T\sum_{m=0}^{\infty}(\hbar^2\nu_m^2 + E_{k_{\rm L}}^2)^{-1}(\hbar^2\nu_m^2 + E_{k_{\rm R}}^2)^{-1}$ and $\psi^{\widetilde{\Gamma},u}(\mathbf{k}_{\rm L}) = X, Y, Z$ or XYZ.

Notice that $\hat{\eta}_{\rm L}\cdot\hat{\eta}_{\rm R} = \cos(\theta_{\rm LR})$, where $\cos(\theta_{\rm LR})$ is the angle between the two $\mathbf{d}$ vectors, and describes a polarization angle just like the Malus's law of electric polarization. The tunneling current $J_s(T)$ depends crucially on the tunneling matrix elements $Q(\mathbf{k}_{\rm L},\mathbf{k}_{\rm R})$. For purely incoherent processes the *only* terms that contribute to $J_s^{(\rm inc)}(T)$ come from $Q^{(\rm inc)} = 2T_s^{1_{\rm L}1_{\rm R}}T_s^{\widetilde{\Gamma}_{\rm L}\widetilde{\Gamma}_{\rm R}} - 2T_s^{1_{\rm L}\widetilde{\Gamma}_{\rm R}}T_s^{\widetilde{\Gamma}_{\rm L}1_{\rm R}}$, where $\widetilde{\Gamma}_j$ are appropriate odd representations of the $\mathbf{d}_j$-vectors. For the p_x symmetry $\psi^{\widetilde{\Gamma}_{\rm L},u}(\mathbf{k}_{\rm L}) = X_{\rm L}$, and $\psi^{\widetilde{\Gamma}_{\rm R},u}(\mathbf{k}_{\rm R}) = X_{\rm R}$, thus (a) $Q_{p_xp_x}^{(\rm inc)}(\mathbf{k}_{\rm L},\mathbf{k}_{\rm R}) = \widetilde{Q}_{p_xp_x}X_{\rm L}X_{\rm R}$ produces a nonvanishing $J_s(T)$, while (b) $Q_{p_xp_x}^{(\rm inc)}(\mathbf{k}_{\rm L},\mathbf{k}_{\rm R}) = \widetilde{Q}_{p_xp_x}1_{\rm L}1_{\rm R}$ produces a trivially vanishing $J_s(T)$. Similarly for the p_y symmetry $\psi^{\widetilde{\Gamma}_{\rm L},u}(\mathbf{k}_{\rm L}) = Y_{\rm L}$, and $\psi^{\widetilde{\Gamma}_{\rm R},u}(\mathbf{k}_{\rm R}) = Y_{\rm R}$, thus (c) $Q_{p_yp_y}^{(\rm inc)}(\mathbf{k}_{\rm L},\mathbf{k}_{\rm R}) = \widetilde{Q}_{p_yp_y}Y_{\rm L}Y_{\rm R}$ produces a nonvanishing $J_s(T)$, while (d) $Q_{p_yp_y}^{(\rm inc)}(\mathbf{k}_{\rm L},\mathbf{k}_{\rm R}) = \widetilde{Q}_{p_yp_y}1_{\rm L}1_{\rm R}$ produces a trivially vanishing $J_s(T)$. For comparison purposes, we analyze next four coherent tunneling processes allowed by group theory corresponding to matrix elements (a) $Q_{p_xp_x}^{(\rm coh)}(\mathbf{k}_{\rm L},\mathbf{k}_{\rm R}) = \widetilde{Q}_{p_xp_x}X_{\rm L}X_{\rm R}\delta_{\mathbf{k}_{\parallel_{\rm L}},\mathbf{k}_{\parallel_{\rm R}}}$, and (b) $Q_{p_xp_x}^{(\rm coh)}(\mathbf{k}_{\rm L},\mathbf{k}_{\rm R}) = \widetilde{Q}_{p_xp_x}1_{\rm L}1_{\rm R}\delta_{\mathbf{k}_{\parallel_{\rm L}},\mathbf{k}_{\parallel_{\rm R}}}$, for the p_x symmetry; and to matrix elements (c) $Q_{p_yp_y}^{(\rm coh)}(\mathbf{k}_{\rm L},\mathbf{k}_{\rm R}) = \widetilde{Q}_{p_yp_y}Y_{\rm L}Y_{\rm R},\delta_{\mathbf{k}_{\parallel_{\rm L}},\mathbf{k}_{\parallel_{\rm R}}}$, and (d) $Q_{p_yp_y}^{(\rm coh)}(\mathbf{k}_{\rm L},\mathbf{k}_{\rm R}) = \widetilde{Q}_{p_yp_y}1_{\rm L}1_{\rm R},\delta_{\mathbf{k}_{\parallel_{\rm L}},\mathbf{k}_{\parallel_{\rm R}}}$, for the p_y symmetry, where the momentum parallel to the junction is conserved.

For coherent processes $\mathbf{k}_{\parallel_{\rm L}} = \mathbf{k}_{\parallel_{\rm R}}$ and the parallel momenta are related by $k_{y_{\rm R}} = \cos(\alpha_{\rm LR})k_{y_{\rm L}} - \sin(\alpha_{\rm LR})k_{z_{\rm L}}$ and $k_{z_{\rm R}} = \sin(\alpha_{\rm LR})k_{y_{\rm L}} + \cos(\alpha_{\rm LR})k_{z_{\rm L}}$, where $\alpha_{\rm LR}$ is the angle between $\mathbf{k}_{\parallel_{\rm L}}$ and $\mathbf{k}_{\parallel_{\rm R}}$. In Fig. 24.4 we show both the temperature and angular dependence of $J_s^{(\lambda)}$ for the case of $\mathbf{a}$-axis tunneling, assuming that $\mathbf{d}_j \parallel \mathbf{c}_j^*$ $(j = L, R)$. The normalized Josephson current $J_s^{(\lambda)}(T)/J_s^{(\lambda)}(0)$ is shown for $\alpha_{\rm LR} = 0$, where *all* L-and R-axes coincide. In the inset, we show $J_s^{(\lambda)}(0,\alpha_{\rm LR})/J_s^{(\lambda)}(0,0)$ for $T = 0$. For $\mathbf{a}$-axis tunneling $\theta_{\rm LR} = \alpha_{\rm LR}$. Notice in the inset of Fig. 24.4 that the Josephson current does *not* change sign for the p_y with $Q^{(\rm coh)} \propto 1_{\rm L}1_{\rm R}$ (double-dotted line). This is a direct manifestation of the *polarization* effect of the $\mathbf{d}$ vector. It is important to emphasize that both the temperature and the angular dependencies of $J_s(T,\alpha_{\rm LR})$ can help distinguish different symmetries, and different (coherent or incoherent) tunneling processes (see Fig. 24.4), as seen experimentally for $\mathbf{c}^*$-axis tunneling of cuprate superconductors [45]. The approach used for $\mathbf{a}$-axis tunneling is good only in the cases of p_y or p_z symmetry but not for p_x or p_{xyz}, since the $\mathbf{d}$ vector changes sign upon reflection at the interface, and leads to low energy surface bound states to be discussed next.

The Josephson effect including surface bound states: To obtain $J_s(T)$ for the p_x symmetry in the case of $\mathbf{a}$-axis tunneling it is crucial to solve the BdG and the order parameter equations self-consistently. We solve the nonlocal lattice BdG equations described above and compare our results with standard local quasiclassical continuum approximations [46–48]. First, we

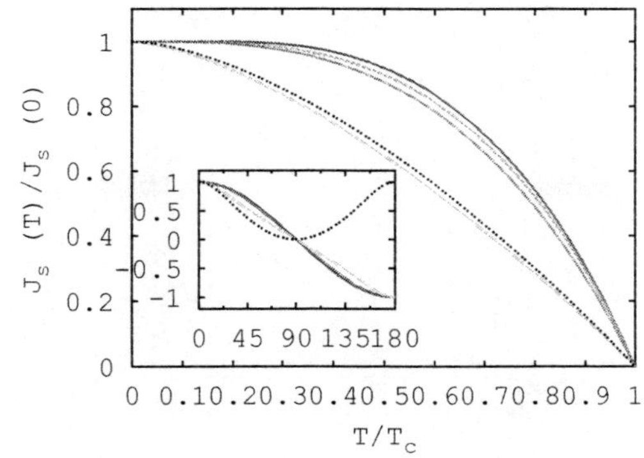

Fig. 24.4. Plot of $J_s(T)/J_s(0)$ versus T/T_c. Inset shows $J_s(0,\alpha_{LR})/J_s(0,0)$ versus α_{LR} at $T=0$. Several **a**-axis tunneling processes are illustrated for p_x and p_y symmetries. *Incoherent processes*: p_x with (**a**) $Q^{(inc)} \propto X_L X_R$ (*solid line*) and (**b**) $Q^{(inc)} \propto 1_L 1_R$ (trivially zero); p_y with (**c**) $Q^{(inc)} \propto Y_L Y_R$ (*dashed line*) and (**d**) $Q^{(inc)} \propto 1_L 1_R$ (trivially zero). *Coherent processes*: p_x with (**a**) $Q^{(coh)} \propto X_L X_R$ (*dotted line*) and (**b**) $Q^{(coh)} \propto 1_L 1_R$ (trivially zero); p_y with (**c**) $Q^{(coh)} \propto Y_L Y_R$ (*dot-dashed line*) and (**d**) $Q^{(coh)} \propto 1_L 1_R$ (*double-dotted line*). The L and R superconductors are assumed to be identical quarter-filled systems, with $T_c = 1.5\,\text{K}$, and parameters $|t_x| = 5{,}800\,\text{K}$, $|t_y| = 1{,}226\,\text{K}$, $|t_z| = 48\text{K}$

take advantage of the translational invariance along the direction parallel to the interface and write the BdG amplitudes as $u_{N_j\alpha_j}(\mathbf{r}_j) = \exp(\mathrm{i}\mathbf{k}_{j\parallel}\cdot\mathbf{r}_{j\parallel})\,\tilde{u}_{\mathbf{k}_{j\parallel},\tilde{N}_j,n_j}(\mathbf{r}_\perp)/\sqrt{V}$; $v_{N_j\alpha_j}(\mathbf{r}_j) = \exp(\mathrm{i}\mathbf{k}_{j\parallel}\cdot\mathbf{r}_{j\parallel})\tilde{v}_{\mathbf{k}_{j\parallel},\tilde{N}_j,n_j}(\mathbf{r}_\perp)/\sqrt{V}$, where quantum indices N_j are represented by momentum $\mathbf{k}_j$ and discrete index $n_j = 1_j, 2_j$. After a Fourier transformation in the parallel coordinates the lattice BdG equations (described above) become one-dimensional in the perpendicular coordinates $\mathbf{r}_\perp, \mathbf{r}'_\perp$. We choose hard boundary conditions, where the BdG amplitudes vanish when $\mathbf{r}_{\parallel L} = \mathbf{r}_{\parallel_R} = 0$ (at the center of the insulating barrier). Second, only the triplet channel component $V_j^{(t)}(\mathbf{r}_j,\mathbf{r}'_j)$ of the general weak spin–orbit coupling interaction $V_{j,\alpha_j,\beta_j,\delta_j,\gamma_j}(\mathbf{r}_j,\mathbf{r}'_j)$ is considered. The triplet component is $V_j^{(t)}(\mathbf{r}_j,\mathbf{r}'_j) = I_{1,j}(\mathbf{r}_j,\mathbf{r}'_j) + I_{2,j}(\mathbf{r}_j,\mathbf{r}'_j)$, and it is assumed that $V_j^{(t)}(\mathbf{r}_j,\mathbf{r}'_j)$ has an on-site ($\mathbf{r}'_j = \mathbf{r}_j$) repulsion U_j and nearest neighbor attractions ($\mathbf{r}'_j = \mathbf{r}_j + \mathbf{a}_\eta$) $V_{j,x}$, $V_{j,y}$, and $V_{j,z}$ only. In addition, we assume that the direction of the $\mathbf{d}(\mathbf{r},\mathbf{r}')$ is independent of position and points along the $\mathbf{c}^*$-axis. Furthermore, we choose the spin quantization axis to be along the direction of $\mathbf{d}(\mathbf{r},\mathbf{r}')$.

Under the previous conditions the order parameter matrix $\Delta_{j,\alpha_j,\beta_j}(\mathbf{r}_j,\mathbf{r}'_j)$ for the p_x symmetry vanishes identically when $\mathbf{r}'_j = \mathbf{r}_j$, and has only off-diagonal matrix elements

$$\Delta_{j,\uparrow_j,\downarrow_j}(\mathbf{r}_j,\mathbf{r}'_j) = V_{j,x}\left[\langle c_{j,\uparrow_j}(\mathbf{r}'_j)c_{j,\downarrow_j}(\mathbf{r}_j)\rangle + \langle c_{j,\downarrow_j}(\mathbf{r}'_j)c_{j,\uparrow_j}(\mathbf{r}_j)\rangle\right]/2$$

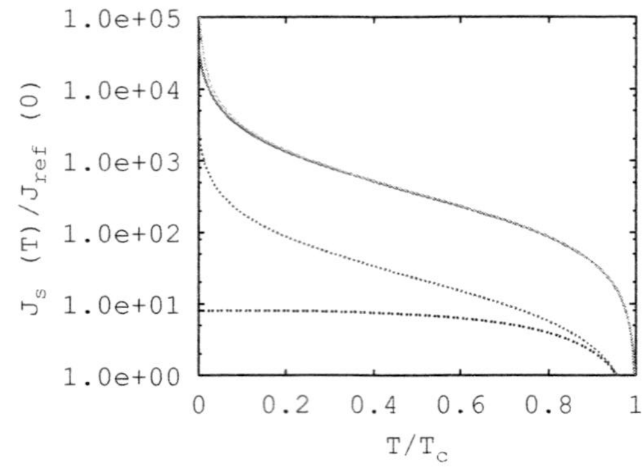

Fig. 24.5. Plots of $J_s(T)/J_{\rm ref}(0)$ versus $T/T_{\rm c}$ for coherent **a** ($\mathbf{c}^*$) axis tunneling where the order parameters are of p_x (p_z) symmetry. Here, $J_{\rm ref}(0)$ is the Landau critical current for a singlet s-wave superconductor with the same $T_{\rm c}$. Quasiclassical results are indicated by *dashed* (*dotted*) *lines* for p_x (p_z), while the nonlocal BdG results are represented by *solid* (*double-dotted*) *lines* for p_x (p_z), for the parameters of Fig. 24.4

and $\Delta_{j,\downarrow_j,\uparrow_j}(\mathbf{r}_j, \mathbf{r}'_j) = \Delta_{j,\uparrow_j,\downarrow_j}(\mathbf{r}_j, \mathbf{r}'_j)$ when $\mathbf{r}'_j = \mathbf{r}_j + a_j\hat{\mathbf{a}}$. The Josephson current $J_s(T)$ is calculated numerically using $T_{\alpha_{\rm L}\alpha_{\rm R}}(\mathbf{r}_{\rm L}, \mathbf{r}_{\rm R}) = T_s(\mathbf{r}_{\rm L}, \mathbf{r}_{\rm R})\delta_{\alpha_{\rm L}\alpha_{\rm R}}$. For definiteness and simplicity we also assume that $T_s(\mathbf{r}_{\rm L}, \mathbf{r}_{\rm R}) = \widetilde{T}\delta_{\mathbf{r}_{\rm L\parallel}, \mathbf{r}_{R\parallel}}$ for $r_{\rm L\perp} = -d/2$ or $r_{\rm R\perp} = +d/2$; and $T_s(\mathbf{r}_{\rm L}, \mathbf{r}_{\rm R}) = 0$, otherwise. Here, d is the separation between the last layer of the L and the first layer of the R superconductor, and corresponds to the *thickness* of the insulating layer. This simple choice guarantees that the parallel momentum $\mathbf{k}_\parallel$ is conserved across the junction, and thus corresponds to a coherent process. Since the tunneling matrix elements connect mostly states that have appreciably large BdG amplitudes near the junction, we use a mixed representation expressed in terms of the parallel momenta $\mathbf{k}_{\parallel_{\rm L}} = \mathbf{k}_{\parallel_{\rm R}} = \mathbf{k}_\parallel$ and the perpendicular coordinates $\mathbf{r}_{\perp_{\rm L}}$ and $\mathbf{r}_{\perp_{\rm R}}$ to calculate numerically $J_s(T)$. In Fig. 24.5 we compare our results for $J_s(T)$ using the nonlocal lattice BdG equations and the local quasiclassical approaches [46–48]. The local quasiclassical approximation is only strictly valid when $k_{\rm F_\perp}\xi_\perp \to \infty$, but it leads to *zero* energy bound states in the case where the quasiclassical order parameter $\Delta(\mathbf{k}, \mathbf{r})$ changes sign upon $\mathbf{r} \to -\mathbf{r}$ [46–48] (as in the case of the **a**-axis tunneling for the p_x symmetry). However, in the nonlocal lattice BdG equations there are only *finite* energy bound states.

The nonexistence of *zero* energy bound states in the nonlocal lattice BdG equations can be understood as follows. Because the order parameter is nonlocal, it can be described in terms of the center of mass $\mathbf{R} = [\mathbf{r} + \mathbf{r}']/2$ and relative $\mathbf{r}_{\rm rel} = [\mathbf{r}' - \mathbf{r}]$ coordinates. Near the surface these two coordinates are entangled and thus lift the *zero* energy bound states degeneracy found in the local theory. This produces, perturbatively, finite energy bound states of the order $|\Delta_0|/\gamma_\perp$ where $\gamma_\perp = \xi_\perp/a_\perp$, with $\xi_\perp$ and $a_\perp$ being the coherence and unit cell lengths, respectively. The finiteness of the energy of the bound states cuts off the low temperature $1/T$ divergence of $J_s(T)$ calculated in the quasiclassical approach at small transparencies [46–48]. For **a**-axis tunneling and the p_x symmetry this amounts to a small correction to the quasiclassical results, as $\gamma_x = \xi_x/a_x \approx 106$ for the Bechgaard salts and $\Delta_{0,p_x} = 3.73\,\mathrm{K}$, leading to the lowest bound state energy to be approximately $T^*_{p_x} = 35.2\,\mathrm{mK}$. Thus, for $T < T^*_{p_x}$ the quasiclassical results are not reliable (see Fig. 24.5). The quasiclassical results fail in a more dramatic way, when we consider $\mathbf{c}^*$ axis tunneling for the p_z symmetry. In this case $\xi_z = 21.5\,\text{Å}$, $a_z = 13.5\,\text{Å}$, $\gamma_z = \xi_z/a_z \approx 1.59$ and $\Delta_{0,p_z} = 3.21\,\mathrm{K}$, resulting in $T^*_{p_z} = 2.02\,\mathrm{K}$, which is larger than the critical temperature $T_\mathrm{c} = 1.5\,\mathrm{K}$ used here. Thus, perturbative corrections to quasiclassical *zero* energy bound states are enormous, and the quasiclassical approximation can not be used except for $T \approx T_\mathrm{c}$ (see Fig. 24.5).

To conclude this section, we emphasize that the Josephson effect may be used to test the orientation and symmetry of the order parameter in quasi-one-dimensional triplet superconductors provided that good tunneling junctions become experimentally available.

24.5 Density Induced Quantum Phase Transitions

We have developed the formalism to study triplet pairing in the quasi-one-dimensional superconductors and applied it to distinguish among possible order parameter symmetries in the previous sections. In this section, we use this framework to address the possibility of topological quantum phase transitions in these systems tuned by the carrier density. Experiments show that Strontium Ruthenate [49] is very sensitive to chemical doping and that $(TMTSF)_2PF_6$ [5] is sensitive to both external pressure and chemical doping. Electrostatic tuning of carrier density has already been demonstrated in cuprate superconductors [50] and amorphous Bismuth [51] and it is conceivable that similar experiments may be carried out for the Bechgaard salts. We find zero-temperature quantum critical lines, where the order parameter does not change symmetry but the ground state topology changes. The elementary excitation spectrum also changes from gapless to gapped. Further details are contained in [52].

We solve the number and order parameter equations ((24.7) and (24.8), respectively) at $T = 0$ with varying filling factors $\widetilde{N}$ and interaction strengths V_Γ using the same values for the transfer integrals t_x, t_y, and t_z as in the spectroscopic and thermodynamic calculations. We characterize three distinct phases based on Fermi surface topology and excitation spectrum. Intuitively, the "normal state" Fermi surface can change topology by touching the boundaries of the first Brillouin zone (BZ) upon increasing $\widetilde{N}$. At the same time, gapless quasiparticle excitations determined by the intersection of the Fermi surface and order parameter nodes can appear or disappear. Trading the filling factor $\widetilde{N}$ for the chemical potential μ, we define the following characteristic

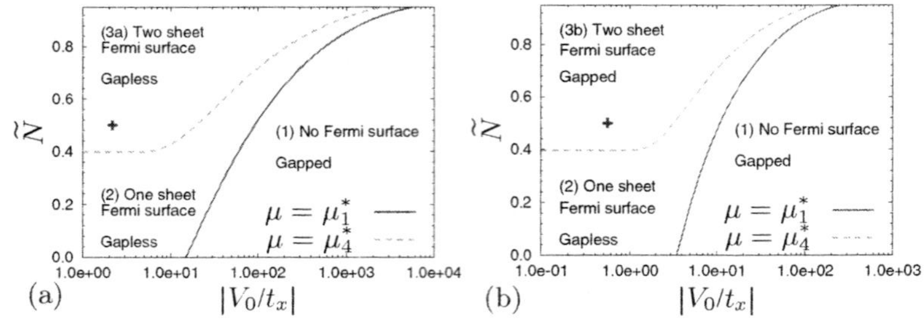

Fig. 24.6. Phase diagrams for (**a**) f_{xyz} and (**b**) p_x symmetries based on Fermi surface connectivity and quasiparticle excitation spectrum. The phase diagrams are symmetric (not shown) around half filling ($\widetilde{N} = 1$) due to particle–hole symmetry. The small cross indicates the parameters compatible with $(TMTSF)_2PF_6$

values $\mu_1^* \equiv t_x + t_y + t_z$, $\mu_2^* \equiv t_x + t_y - t_z$, $\mu_3^* \equiv t_x - t_y + t_z$, $\mu_4^* \equiv t_x - t_y - t_z$. Recall that order parameter nodes occur on the planes $k_i = 0, \pi$ for the p_i symmetry and on the union of the planes $k_i = 0, \pi$ for the f_{xyz} where $i = x, y, z$. The three phases are: (1) no Fermi surface and fully gapped $E(\mathbf{k})$ for all symmetries ($\mu < \mu_1^*$), (2) one-sheet Fermi surface and gapless for all symmetries ($\mu_1^* < \mu < \mu_4^*$), (3a) two-sheet Fermi surface and gapless for f_{xyz}, p_z, p_y ($\mu_4^* < \mu$), and (3b) two-sheet Fermi surface and fully gapped for p_x ($\mu_4^* < \mu$). Phase diagrams for the f_{xyz} and p_x symmetries appear in Fig. 24.6 with the p_y and p_z phase diagrams qualitatively similar to that of f_{xyz}. Since $V_\Gamma/t_x \sim 1$ for the Bechgaard salts, $\widetilde{N}$ only tunes between the (2) and (3) phases. However, larger values of V_Γ/t_x are of greater theoretical interest since $\widetilde{N}$ tunes between all three phases may thus be of interest for future systems exhibiting unconventional superconductivity.

We now turn our attention to thermodynamic quantities that provide signatures of the momentum space topological changes discussed above. In the following calculations we fix the interaction strength to force $\mu = \mu_1^*$ at $\widetilde{N} = 0.5$ (quarter filling). The $T = 0$ electronic compressibility $\kappa = \widetilde{N}^{-2}(\partial\widetilde{N}/\partial\mu)_{T,V}$ is

$$\kappa = \frac{2}{\widetilde{N}^2}\sum_k \left[\frac{1}{2E_k}\left(1 - \frac{\xi_k^2}{E_k^2}\right)\right], \tag{24.22}$$

which we plot in Fig. 24.7a. There are clear anomalies (nonanalyticities) at the phase boundaries. As $\widetilde{N}$ increases, μ crosses μ_j^* where $\mathrm{d}\kappa/\mathrm{d}\widetilde{N}$ decreases discontinuously. These nonanalyticities in κ at $T = 0$ are indicative of a quantum phase transitions between topologically distinct phases discussed above. Finite temperature smears the sharp cusps in κ but clear peaks are still present in the quantum critical region. The measurement of the electronic compressibility may be achieved in a field effect geometry [50,51] through the relation

$$\kappa = VC_\mathrm{d}/Q^2, \tag{24.23}$$

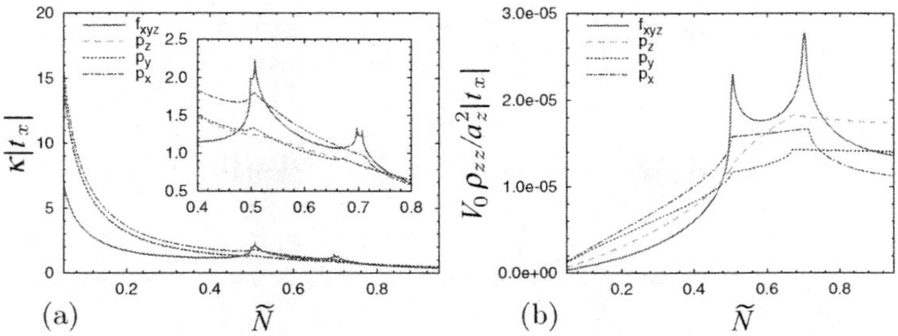

Fig. 24.7. Dependence on filling factor ($\widetilde{N}$) for various symmetries of (**a**) dimensionless electronic compressibility $\kappa|t_x|$, where the inset shows the region $0.4 < \widetilde{N} < 0.8$ and of (**b**) dimensionless zz component of superfluid density tensor $V_0\rho_{zz}/a_z^2|t_x|$, where V_0 (a_z) is the unit cell volume (length)

where $C_\mathrm{d} = [\partial Q/\partial V_e]_{T,V}$ is the differential capacitance, V_e is the applied voltage, Q is the absolute value of the total charge of carriers, and V is the sample volume.

The zero-temperature superfluid density tensor

$$\rho_{ij} = \frac{1}{V}\sum_{\mathbf{k}} [n_{\mathbf{k}}\partial_i\partial_j\xi_{\mathbf{k}}], \tag{24.24}$$

also probes the quantum phase transitions. Here, $n_{\mathbf{k}}$ is the momentum distribution. For all symmetries, both ρ_{xx} and ρ_{yy} appear to be smooth functions of $\widetilde{N}$, but we find ρ_{yy} has a broad, rounded peak in $0.5 < \widetilde{N} < 0.8$ corresponding to $\mu_1^* < \mu < \mu_4^*$. However, as can be seen in Fig. 24.7b, the zz component is very interesting since it exhibits clear anomalies (cusps) such as those seen in κ as a function of $\widetilde{N}$. These nonanalyticities again indicate the existence of a quantum phase transition as μ or $\widetilde{N}$ are tuned electrostatically.

Lastly, fluctuation effects lead to a phase only action ΔS from which phase-only collective mode frequencies can be extracted via $\mathrm{i}\omega_\mathrm{n} \to \omega + \mathrm{i}\delta$. In this case $\omega(\mathbf{q}) = \sqrt{(c_x^2 q_x^2 + c_y^2 q_y^2 + c_z^2 q_z^2)}$, where $c_x^2 = \rho_{xx}/A$, $c_y^2 = \rho_{yy}/A$, and $c_z^2 = \rho_{zz}/A$. Here $A = \widetilde{N}^2\kappa/V$ is proportional to the compressibility. These collective mode frequencies inherit the anomalies of κ and ρ_{ij} and also characterize the quantum phase transition as well. We note that the states we discuss either have a full gap or are in the underdamped regime so that we do not take into account Landau damping which cause collective modes to decay into the two-particle continuum. The anisotropy of $\omega(\mathbf{q})$ reflects the orthorhombic lattice with $\omega(q_x,0,0) = c_x q_x$, $\omega(0,q_y,0) = c_y q_y$, and $\omega(0,0,q_z) = c_z q_z$, where c_i is the speed of sound along the ith direction. However, these modes may be plasmonized in a charged superfluid.

In addition to the possibility of electrostatically tunning transitions between topologically distinct triplet ground states just discussed, we hope that a more detailed study of the pressure versus temperature phase diagram of quasi-one-dimensional superconductors would reveal new quantum phases as well. Thus, we discuss next the possibility of a coexistence phase between TSC and SDW where pressure and temperature are the tunning parameters.

24.6 Coexistence of Triplet Superconductivity and Spin–Density Wave

In this section, we discuss theoretically the pressure versus temperature phase diagram of $(TMTSF)_2PF_6$ indicating TSC, SDW, and TSC/SDW coexistence phases, and show that the TSC and SDW order parameters are both nonuniform in the coexistence region. Furthermore, we investigate the effects of an external magnetic field and suggest that a canting transition of the SDW order parameter may occur and alter the nature of the TSC state in the coexistence region. Further details can be found in [53].

While in the previous sections we were concerned only with TSC, recent experiments [54–56] may have revealed a region of macroscopic coexistence between TSC and SDW, where both orders are nonuniform. This coexistence region can be related to existing theoretical proposals. For instance, strictly one-dimensional theories invoking SO(4) symmetry [57], or negative interface energies [58] have allowed for coexisting TSC and SDW. However, these previous theories are not directly applicable to three-dimensional but highly anisotropic superconductors like the Bechgaard salts, where the SO(4) symmetry is absent, and negative interface energies are not necessary conditions for the coexistence.

The Bechgaard salt $(TMTSF)_2PF_6$ can be described approximately by an orthorhombic lattice with dispersion shown in (24.3). We use natural units ($\hbar = k_B = c = 1$) and work with Hamiltonian $\mathcal{H} = \mathcal{H}_0 + \mathcal{H}_{\text{int}}$, where the noninteracting part is $\mathcal{H}_0 = \sum_{\mathbf{k},\alpha} \xi_{\mathbf{k}} c^\dagger_{\mathbf{k},\alpha} c_{\mathbf{k},\alpha}$, and $\xi_{\mathbf{k}} = \epsilon_{\mathbf{k}} - \mu$ is the dispersion shifted by the chemical potential, which may include a Hartree shift. The interaction part is

$$\mathcal{H}_{\text{int}} = \sum_{\mathbf{kk}'\mathbf{p}} \sum_{\alpha\beta\gamma\delta} V(\mathbf{k},\mathbf{k}') \mathbf{d}^\dagger_{\alpha\beta}(\mathbf{k},\mathbf{p}) \cdot \mathbf{d}_{\gamma\delta}(\mathbf{k}',\mathbf{p}) + \sum_{\mathbf{kk}'\mathbf{q}} \sum_{\alpha\beta\gamma\delta} J(\mathbf{q}) \mathbf{s}^\dagger_{\alpha\beta}(\mathbf{k},\mathbf{q}) \cdot \mathbf{s}_{\gamma\delta}(\mathbf{k}',\mathbf{q}), \tag{24.25}$$

where the first and second terms describe interactions in TSC and SDW channels, respectively.

The order parameter for TSC is $\mathbf{D}(\mathbf{p}) = \langle \sum_{\mathbf{k},\alpha\beta} V \phi_\Gamma(\mathbf{k}) \mathbf{d}_{\alpha\beta}(\mathbf{k},\mathbf{p}) \rangle$, while the SDW order parameter is $\mathbf{N}(\mathbf{q}) = J(\mathbf{q}) \langle \sum_{\mathbf{k},\alpha\beta} \mathbf{s}_{\alpha\beta}(\mathbf{k},\mathbf{q}) \rangle$. With these definitions, the effective Hamiltonian is

$$\mathcal{H}_{\text{eff}} = \mathcal{H}_0 + \mathcal{H}_{\text{TSC}} + \mathcal{H}_{\text{SDW}}, \tag{24.26}$$

where the TSC contribution is $\mathcal{H}_{\text{TSC}} = \sum_{\mathbf{p}} [\mathbf{D}^\dagger(\mathbf{p}) \cdot \sum_{\mathbf{k},\alpha\beta} \phi_\Gamma(\mathbf{k}) \mathbf{d}_{\alpha\beta}(\mathbf{k},\mathbf{p}) + \text{H.C.}] - \sum_{\mathbf{p}} \mathbf{D}^\dagger(\mathbf{p}) \cdot \mathbf{D}(\mathbf{p})/V$, and the SDW term is $\mathcal{H}_{\text{SDW}} = \sum_{\mathbf{q}} [\mathbf{N}(-\mathbf{q}) \cdot \sum_{\mathbf{k},\alpha\beta} \mathbf{S}_{\alpha\beta}(\mathbf{k},\mathbf{q}) + \text{H.C.}] - \sum_{\mathbf{q}} \mathbf{N}(-\mathbf{q}) \cdot \mathbf{N}(\mathbf{q})/J(\mathbf{q})$. The effective action of this

Hamiltonian as a function of $\mathbf{D}(\mathbf{p})$ and $\mathbf{N}(\mathbf{q})$ is obtained by integrating out the fermions. The quadratic terms are

$$S_2^{\mathrm{TSC}} = \sum_{\mathbf{p}} A(\mathbf{p})\mathbf{D}^\dagger(\mathbf{p}) \cdot \mathbf{D}(\mathbf{p}),$$
$$S_2^{\mathrm{SDW}} = \sum_{\mathbf{q}} B(\mathbf{q})\mathbf{N}(-\mathbf{q}) \cdot \mathbf{N}(\mathbf{q}), \tag{24.27}$$

with coefficients found in [53].

Next, we make the following assumptions. First, we assume that the saddle point TSC order parameter is dominated by the zero center of mass momentum component $\mathbf{D}_0 \equiv \mathbf{D}(\mathbf{p} = 0)$, and that $\mathbf{D}_0$ is unitary. Second, we assume that the saddle point SDW order parameter $\mathbf{N}$ is a real vector in $\mathbf{r}$-space, and that it has Fourier components determined by Fermi surface nesting vectors $\mathbf{q} = \mathbf{Q}_i = (\pm Q_a, \pm Q_b, \pm Q_c)$ [3]. In this case, the coefficients $B(\mathbf{Q}_i)$ are identical for all $\mathbf{Q}_i$s, since the lattice dispersion in invariant under reflections and inversions compatible with the D_{2h} group. In addition, the coefficients of all higher order terms involving $\mathbf{N}(\mathbf{Q}_i)$s share the same properties. Given that $\mathbf{N}(\mathbf{r})$ is real, and that we have periodic boundary conditions, we can choose a specific reference phase where $\mathbf{N}(\mathbf{Q}_i)$ are real and identical. Thus, we define $\mathbf{N}_0 \equiv \mathbf{N}(\mathbf{Q}_i)$ for all i, and the quadratic terms are dominated in the long wavelength limit by $S_2^{\mathrm{TSC}} \approx A(0)|\mathbf{D}_0|^2$ and $S_2^{\mathrm{SDW}} \approx (m/2)B(\mathbf{Q}_1)|\mathbf{N}_0|^2$, respectively. Here, m is the number of nesting vectors, and $\mathbf{Q}_1 = (Q_a, Q_b, Q_c)$ is chosen for definiteness.

Notice that the two order parameters $\mathbf{D}(\mathbf{p})$ and $\mathbf{N}(\mathbf{q})$ do not couple to quadratic order, because TSC and SDW are instabilities in particle–particle and particle–hole channels, respectively. Thus, the two orders are independent to this order, and their corresponding vector order parameters are free to rotate. However, this freedom is lost when fourth-order terms are included.

The coupling between $\mathbf{D}$ and $\mathbf{N}$ in fourth-order is given by

$$S_4^{\mathrm{C}} = (C_1 + C_2/2)|\mathbf{D}_0|^2|\mathbf{N}_0|^2 - C_2|\mathbf{D}_0 \cdot \mathbf{N}_0|^2, \tag{24.28}$$

where the coefficients are found in [53]. Notice that the second term in (24.28) can be parametrized as $-C_2\cos^2(\theta)|\mathbf{D}_0|^2|\mathbf{N}_0|^2$, where $\cos^2\theta \equiv |\mathbf{D}_0 \cdot \mathbf{N}_0|^2/|\mathbf{D}_0|^2|\mathbf{N}_0|^2 \leq 1$ is independent of $|\mathbf{D}_0|$ and $|\mathbf{N}_0|$. Since $\mathbf{D}_0$ is unitary, its global phase can be eliminated in S_4^{C}, and θ can be regarded as the angle between $\mathbf{D}_0$ and $\mathbf{N}_0$. The coefficient C_2 for $(\mathrm{TMTSF})_2\mathrm{PF}_6$ is positive, indicating that $\mathbf{D}_0$ and $\mathbf{N}_0$ tend to be aligned ($\theta = 0$) or antialigned ($\theta = \pi$).

Additional fourth-order terms are

$$S_4^{\mathrm{TSC}} = D_1|\mathbf{D}_0|^4, \quad S_4^{\mathrm{SDW}} = D_2|\mathbf{N}_0|^4, \tag{24.29}$$

with coefficients given in [53]. This leads to the effective action

$$S_{\mathrm{eff}} = S_0 + S_2 + S_4, \tag{24.30}$$

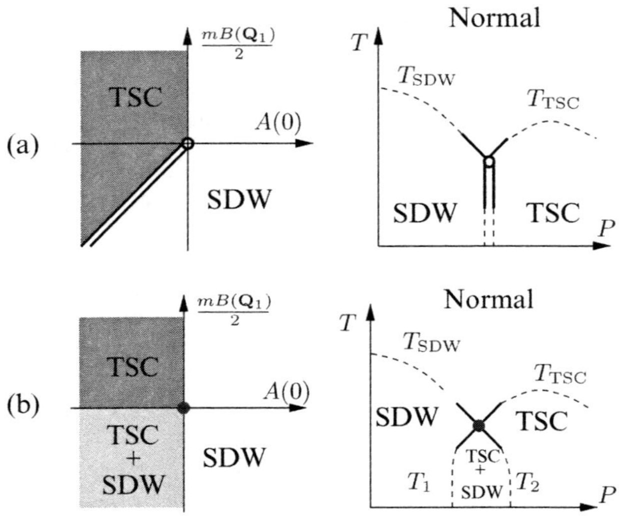

Fig. 24.8. P–T phase diagrams indicating (**a**) first-order transition line with no coexistence phase and (**b**) two second-order lines with a coexistence region between TSC and SDW phases

where S_0 is the normal state contribution, $S_2 = S_2^{\text{TSC}} + S_2^{\text{SDW}}$, and $S_4 = D_1|\mathbf{D}_0|^4 + D_2|\mathbf{N}_0|^4 + C(\theta)|\mathbf{D}_0|^2|\mathbf{N}_0|^2$ with $C(\theta) = C_1 + C_2/2 - C_2\cos^2\theta$. The phase diagram that emerges from this action leads to either bicritical or tetracritical points as illustrated in Fig. 24.8. When $R = C^2(0)/(4D_1D_2) > 1$ the critical point $(P_{\text{c}}, T_{\text{c}})$ is bicritical and there is a first-order transition line at $(m/2)B(\mathbf{Q}_1) = A(0)$ when both $B(\mathbf{Q}_1) < 0$ and $A(0) < 0$, as shown in Fig. 24.8a. However when $R < 1$, $(P_{\text{c}}, T_{\text{c}})$ is tetracritical and a coexistence region for TSC and SDW orders occurs when both $B(\mathbf{Q}_1) < 0$ and $A(0) < 0$, as shown in Fig. 24.8b. The action S_{eff} obtained in three dimensions is not SO(4) invariant, and SO(4) symmetry based theories [57] can only be applied to one-dimensional systems, but not to the highly anisotropic but three-dimensional Bechgaard salts.

The ratio $R \approx 0.12 < 1$ for the Bechgaard salt $(\text{TMTSF})_2\text{PF}_6$ around $(P_{\text{c}}, T_{\text{c}})$, when the interaction strengths V, J are chosen to give the same $T_{\text{c}} = 1.2\,\text{K}$ at quarter filling for parameters $|t_x| = 5{,}800$ K, $|t_y| = 1{,}226$ K, $|t_z| = 58$ K, used in combination with $\phi_\Gamma(\mathbf{k}) = \sin(k_x a_x)$ (p_x-symmetry for TSC) and the nesting vectors $\mathbf{Q} = (\pm\pi/2a_x, \pm\pi/2a_y, 0)$ ($m = 4$). This shows that $(\text{TMTSF})_2\text{PF}_6$ has an TSC/SDW coexistence region as suggested by experiments [54–56]. To investigate the TSC/SDW coexistence region the effective action (24.30) [with $\mathbf{Q} = (\pm Q_a, \pm Q_b, 0)$] is Fourier transformed into real space to give the Ginzburg–Landau (GL) free energy density

$$\mathcal{F} = \mathcal{F}_{\text{n}} + \mathcal{F}_{\text{TSC}} + \mathcal{F}_{\text{SDW}} + \mathcal{F}_{\text{C}}, \tag{24.31}$$

where $\mathcal{F}_{\rm n}$ is the normal state contribution, and $\mathcal{F}_{\rm C} = C(\theta)|\mathbf{N}(\mathbf{r})|^2|\mathbf{D}(\mathbf{r})|^2$ is the coupling term of the two-order parameters. The TSC contribution is $\mathcal{F}_{\rm TSC} = A(0)|\mathbf{D}(\mathbf{r})|^2 + D_1|\mathbf{D}(\mathbf{r})|^4 + \sum_{ij}\gamma_{\rm TSC}^{ij}[\partial_i\mathbf{D}(\mathbf{r})]\cdot[\partial_j\mathbf{D}(\mathbf{r})]$, where $\gamma_{\rm TSC}^{ij}$ is obtained from a small $\mathbf{p}$ expansion of $A(\mathbf{p})$. The SDW contribution is $\mathcal{F}_{\rm SDW} = \int d\mathbf{r}'\,[B(\mathbf{r},\mathbf{r}')\mathbf{N}(\mathbf{r})\cdot\mathbf{N}(\mathbf{r}')] + (D_2/m^2)|\mathbf{N}(\mathbf{r})|^4$, where $B(\mathbf{r},\mathbf{r}')$ is the Fourier transform of $B(\mathbf{q})$ in (24.27). For Bechgaard salt parameters, the prefactor $C(0)$ of the coupling term $\mathcal{F}_{\rm C}$ is positive, and hence represents a local repulsive interaction between the TSC and SDW order parameters. As a consequence, the TSC order parameter is nonuniform in the TSC/SDW coexistence region, and has a modulation induced by the SDW order parameter. Since $R \ll 1$ for $(\mathrm{TMTSF})_2\mathrm{PF}_6$, the coupling term $\mathcal{F}_{\rm C}$ is small in comparison with the other fourth-order coefficients D_1, D_2, and a perturbative solution is possible for $|\mathbf{D}(\mathbf{r})|$ and $|\mathbf{N}(\mathbf{r})|$. At assumed zero TSC/SDW coupling $C(0) = 0$, the saddle point modulation for the SDW order parameter is $\mathbf{N}(\mathbf{r}) = m\mathbf{N}_0\cos(\mathbf{Q}_1\cdot\mathbf{r})$, with $|\mathbf{N}_0| = [-mB(\mathbf{Q}_1)/3D_2]^{1/2}$, while the saddle point for the TSC order parameter is $\mathbf{D}(\mathbf{r}) = \mathbf{D}_0$ with $|\mathbf{D}_0| = [-A(0)/2D_1]^{1/2}$. Including the coupling $\mathcal{F}_{\rm C}$ the new solution for the magnitude of TSC order parameter is

$$
\begin{aligned}
|\mathbf{D}(\mathbf{r})| - |\mathbf{D}_0| = -v\frac{|N_0|^2}{|D_0|}R^{1/2}\Big[&\frac{\cos(2Q_a x)}{4+8\xi_x^2Q_a^2} + \frac{\cos(2Q_b y)}{4+8\xi_y^2Q_b^2} \\
&+ \frac{\cos(2Q_a x)\cos(2Q_b y)}{4+8\xi_x^2Q_a^2+8\xi_y^2Q_b^2} + \frac{1}{4}\Big],
\end{aligned}
\qquad (24.32)
$$

which shows explicitly $2Q_a$ and $2Q_b$ modulations along the **a**- and **b**′-axes, respectively. Here, $\xi_i = \sqrt{|\gamma_{\rm TSC}^{ii}/A(0)|}$ represents the TSC coherence length along the i direction, and $v = (6D_2/D_1)^{1/2}$. Notice that the modulation in $|\mathbf{D}(\mathbf{r})|$ disappears as the SDW order goes away $|\mathbf{N}_0| \to 0$. The qualitative behavior of $|\mathbf{D}(\mathbf{r})|$ is shown in Fig. 24.9a. The new solution for the SDW order parameter to the first-order correction is $|\mathbf{N}(\mathbf{r})| = \sum_i(1 - R^{1/2}|\mathbf{D}_0|^2/4v|\mathbf{N}_0|^2)|\mathbf{N}_0|\cos(\mathbf{Q}_i\cdot\mathbf{r})$, and can be seen in Fig. 24.9b. Notice that

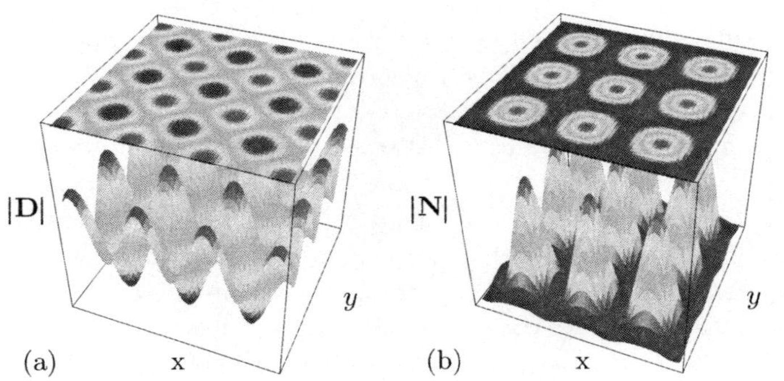

Fig. 24.9. Magnitude of (**a**) TSC and (**b**) SDW order parameters in the x–y-plane, within the coexistence region

the maxima of $|\mathbf{D}(\mathbf{r})|$ coincide with the minima of $|\mathbf{N}(\mathbf{r})|$ indicating that the TSC and SDW orders try to be locally excluded. Since the TSC and SDW modulations are out of phase, experiments that are sensitive to the spatial distribution of the spin–density or Cooper pair charge density may reveal the coexistence of these inhomogeneous phases.

Next, we analyze the effect of magnetic fields on this coexistence region. A uniform magnetic field $\mathbf{H}$ couples with charge via the Peierls substitution $\mathbf{k} \to \mathbf{k} - |e|\mathbf{A}$ in the dispersion relation given in (24.3), where $\mathbf{A}$ is the vector potential, and couples with spin via the paramagnetic term $\mathcal{H}_{\mathrm{P}} = -\mu_0 \mathbf{H} \cdot \sum_{\mathbf{k},\alpha\beta} c^{\dagger}_{\mathbf{k}\alpha} \boldsymbol{\sigma}_{\alpha\beta} c_{\mathbf{k}\beta}$, where μ_0 is the effective magnetic moment. Thus, the effective Hamiltonian becomes

$$\mathcal{H}_{\mathrm{eff}} = \mathcal{H}_0(\mathbf{k} \to \mathbf{k} - |e|\mathbf{A}) + \mathcal{H}_{\mathrm{TSC}} + \mathcal{H}_{\mathrm{SDW}} + \mathcal{H}_{\mathrm{P}}. \tag{24.33}$$

Upon integration of the fermions, the corresponding effective action is

$$S_{\mathrm{eff}}(\mathbf{H}) = S_0(\mathbf{H}) + S_2(\mathbf{H}) + S_4(\mathbf{H}), \tag{24.34}$$

where $S_0(\mathbf{H}) = S_0 + |\mathbf{H}|^2/8\pi - \chi_{\mathrm{n}}|\mathbf{H}|^2/2$, χ_{n} is the uniform electronic spin susceptibility of the normal state, $S_2(\mathbf{H})$ is obtained from S_2 by the Peierls substitution, and $S_4(\mathbf{H}) = S_4 + (E_1 + E_2/2)|\mathbf{H}|^2|\mathbf{D}_0|^2 - E_2|\mathbf{H}\cdot\mathbf{D}_0|^2 + (F_1 - F_2/2)|\mathbf{H}|^2|\mathbf{N}_0|^2 + F_2|\mathbf{H}\cdot\mathbf{N}_0|^2$. The coefficients E_i and F_i can be found in [53]. A detailed calculation shows that the coefficient $E_1 = -E_2/2$, hence the coupling of $\mathbf{H}$ to $\mathbf{D}$ can be described in the more familiar form $F_{\mathrm{M}} - \sum_{\mu\nu} H_\mu \chi_{\mu\nu} H_\nu/2$, where $\chi_{\mu\nu} = \chi_{\mathrm{n}}\delta_{\mu\nu} + E_2 D^*_\mu D_\nu$.

For Bechgaard salts, the coefficients $E_2 < 0$ and $F_2 > 0$ indicating that $\mathbf{D}$ and $\mathbf{N}$ prefer to be perpendicular to the magnetic field $\mathbf{H}$. These conditions, when combined with $C_2 > 0$ in (24.28), show that $\mathbf{D}$ and $\mathbf{N}$ prefer to be parallel to each other, but perpendicular to $\mathbf{H}$. However, the relative orientation of these vectors in small fields is affected by spin anisotropy effects which were already observed in $(\mathrm{TMTSF})_2\mathrm{PF}_6$, where the easy axis for $\mathbf{N}$ is the $\mathbf{b}'$ direction [59]. Such an anisotropy effect can be described by adding a quadratic term $-u_N N^2_{b'}$ with $u_N > 0$, which favors $\mathbf{N} \parallel \mathbf{b}'$. Similarly, the $\mathbf{D}$ vector also has anisotropic effect caused by spin–orbit coupling, and can be described by adding a quadratic term $-u_D D^2_i$, where i is the easy axis for TSC. (Quartic TSC and SDW terms also become anisotropic.)

However, a sufficiently large $\mathbf{H} \parallel \mathbf{b}'$ can overcome spin anisotropy effects, and drive the $\mathbf{N}$ vector to flop onto the $\mathbf{a}$–$\mathbf{c}^*$-plane. This canting (flop) transition was reported [59] in $(\mathrm{TMTSF})_2\mathrm{PF}_6$ for $H \approx 1$ T at zero pressure and $T = 8$ K. If such a spin-flop transition persists near the TSC/SDW critical point $(P_{\mathrm{c}}, T_{\mathrm{c}})$ as suggested in our discussion, then the flop transition of the $\mathbf{N}$ vector forces the $\mathbf{D}$ vector to flop as well, and has potentially serious consequences to the superconducting state. Schematic phase diagrams are shown in Fig. 24.10a, b. For $P < P_{\mathrm{c}}$, if a flop transition occurs for $H_{\mathrm{F}} < H_1(0)$ (see Fig. 24.10a), then $\mathbf{N}$ flops both in the pure SDW and in the TSC/SDW coexistence phases, in which case it forces $\mathbf{D}$ vector to flop as well. If the flop

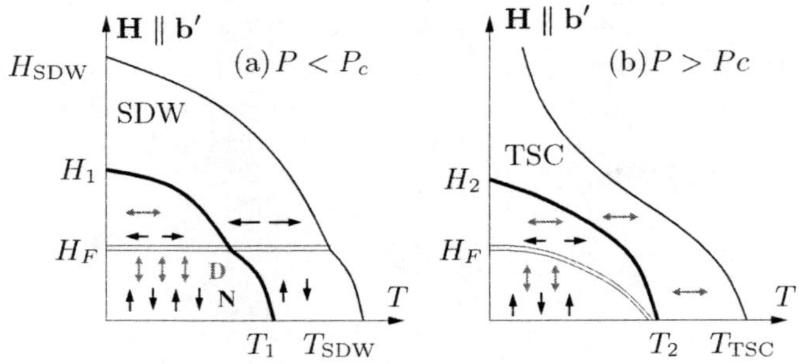

Fig. 24.10. H–T phase diagrams showing the TSC/SDW coexistence region (*thick solid line*) and canting transitions (*double line*) for (**a**) $P < P_c$ and (**b**) $P > P_c$

transition occurs for $H_{\rm SDW}(0) < H_{\rm F} < H_1(0)$ (not shown) then only the pure SDW phase is affected. This situation is qualitatively different for $P > P_{\rm c}$. In the zero (weak) spin–orbit coupling limit the **D** vector is free to rotate in a magnetic field and tends to be perpendicular to **H** to minimize its magnetic free energy $F_{\rm M}$. Thus, for $\mathbf{H} \parallel \mathbf{b}'$ and $|\mathbf{H}| > H_2$, the **D** vector lies in the **a**–**c*** plane since there is no SDW order. However, at lower temperatures and small magnetic fields when TSC and SDW orders coexist, the spin anisotropy field forces **N** to be along **b**′ and **N** forces **D** to flop from the **a**–**c*** plane to **b**′ direction. This canting transition occurs at $H_{\rm F} < H_2(0)$ (see Fig. 24.3b), when **N** flops in the TSC/SDW coexistence phase, and forces the **D** vector to flop as well.

In this section we showed that the TSC and SDW order parameters can coexist in the P–T phase diagram of quasi-one-dimensional organic conductors. In the coexistence region the TSC order parameter is nonuniform, and its modulation is induced via the SDW order parameter. We also pointed out that theories based on the SO(4) symmetry are strictly valid only in the one-dimensional limit, and cannot be applied to these highly anisotropic three-dimensional systems. Furthermore, we discussed that a magnetic field induced canting transition of the SDW order parameter affects dramatically the TSC order parameter and the phase diagram of the coexistence region, both below and above the critical pressure.

24.7 Summary

In this review, we have discussed properties of triplet quasi-one-dimensional superconductors, with emphasis on symmetries consistent with weak spin–orbit coupling effects. We have reviewed spectroscopic and thermodynamic properties for various symmetries and indicated that the triplet p_x state is an excellent candidate for the order parameter. We also discussed the Josephson effect between two triplet superconductors as a possible test of the symmetry and direction of the vector order parameter. Furthermore, we investigated the existence of several quantum phase transitions when the charge carrier density is electrostatically tuned. Lastly, we analyzed the coexistence of TSC and SDW order in the pressure versus temperature phase diagram of quasi-one-dimensional superconductors. We would like to thank NSF (DMR-0304380) and the NDSEG program for support.

References

1. D. Jerome, A. Mazaud, M. Ribault, K. Bechgaard, J. Phys. Lett. **41**, L95 (1980)
2. J.M. Williams, J.R. Ferraro, R.J. Thorn, K.G. Carlson, U.U. Geiser, H.H. Wang, A.M. Kini, M.H. Whangboo, *Superconductors (Including Fullerenes), Synthesis, Structure, Properties and Theory*, (Prentice Hall, Englewood, 1992)
3. T. Ishiguro, K. Yamaji, G. Saito, *Organic Superconductors*, 2nd edn (Springer-Verlag 1998)
4. I.J. Lee, M.J. Naughton, G.M. Danner, P.M. Chaikin, Phys. Rev. Lett. **78**, 3555 (1997)
5. I.J. Lee, S.E. Brown, W.G. Clark, M.J. Strouse, M.J. Naughton, W. Kang, P.M. Chaikin, Phys. Rev. Lett. **88**, 017004 (2002)
6. I.J. Lee, D.S. Chow, W.G. Clark, M.J. Strouse, M.J. Naughton, P.M. Chaikin, S.E. Brown, Phys. Rev. B **68**, 092510 (2003)
7. L.C. Hebel, C.P. Slichter, Phys. Rev. **113**, 1504 (1959)
8. M. Takigawa, H. Yasuoka, G. Saito, J. Phys. Soc. Jpn. **56**, 873 (1987)
9. I.J. Lee, A.P. Hope, M.J. Leone, M.J. Naughton, Appl. Supercond. **2**, 753 (1994)
10. I.J. Lee, A.P. Hope, M.J. Leone, M.J. Naughton, Synth. Met. **70**, 747 (1995)
11. S. Belin, K. Behnia, Phys. Rev. Lett. **79**, 2125 (1997)
12. A.A. Abrikosov, JETP Lett. **37**, 503 (1983)
13. A.A. Abrikosov, J. Low Temp. Phys. **53**, 359 (1983)
14. M.Y. Choi, P.M. Chaikin, S.Z. Huang, P. Haen, E.M. Engler, R.L. Greene, Phys. Rev. B **25**, 6208 (1982)
15. S. Bouffard, M. Ribault, R. Brussetti, D. Jerome, K. Bechgaard, J. Phys. C: Solid State Phys. **15**, 2951 (1982)
16. L.P. Gorkov, D. Jerome, J. Phys. Lett. **46**, L643 (1985)
17. A.G. Lebed, JETP Lett. **44**, 89 (1986)
18. N. Dupuis, G. Montambaux, C.A.R. Sá de Melo, Phys. Rev. Lett. **70**, 2613 (1993)
19. A.I. Larkin, Y.N. Ovchinnikov, JETP Lett. **20**, 762 (1965)
20. P. Fulde, R.A. Ferrel, Phys. Rev. **135**, A550 (1964)
21. N. Dupuis, Phys. Rev. B **51**, 9074 (1995)
22. C.A.R. Sá de Melo, Physica C **260**, 224 (1996)
23. A.G. Lebed, Phys. Rev. B **59**, R721 (1999)
24. C.A.R. Sá de Melo, in *The Superconducting State in Magnetic Fields: Special Topics and New Trends*, Chap. 15, ed. by C.A.R. Sá de Melo (World Scientific, Singapore, 1998), pp. 296–324
25. C.A.R. Sá de Melo, J. Supercond. **12**, 459 (1999)
26. C.D. Vaccarella, C.A.R. Sá de Melo, Physica C **341–348**, 293 (2000)
27. C.D. Vaccarella, C.A.R. Sá de Melo, Phys. Rev. B **63**, R180505 (2001)
28. A.G. Lebed, K. Machida, M. Ozaki, Phys. Rev. B **62**, R795 (2000)

29. H. Shimahara, Phys. Rev. B **61**, R14936 (2000)
30. K. Kuroki, R. Arita, H. Aoki, Phys. Rev. B **63**, 094509 (2001)
31. R.D. Duncan, C.D. Vaccarella, C.A.R. Sá de Melo, Phys. Rev. B **64**, 172503 (2001)
32. R.D. Duncan, R.W. Cherng, C.A.R. Sá de Melo, Physica C **391**, 98 (2003)
33. M. Tinkham, *Group Theory and Quantum Mechanics*, (McGraw-Hill, New York, 1964)
34. V.P. Mineev, K.V. Samokhin, *Introduction to Unconventional Superconductivity* (Gordon and Breach, Amsterdam, 1999)
35. C.A.R. Sáde Melo, M. Randeria, J.R. Engelbrecht, Phys. Rev. Lett. **71**, 3202 (1993)
36. J.R. Engelbrecht, M. Randeria, C.A.R. Sá de Melo, Phys. Rev. B **55**, 15153 (1997)
37. R.W. Cherng, C.A.R. Sá de Melo, Phys. Rev. B **67**, 212505 (2003)
38. L.J. Buchholtz, G. Zwicknagl, Phys. Rev. B **23**, 5788 (1981)
39. C.R. Hu, Phys. Rev. Lett. **72**, 1526 (1994)
40. K. Sengupta, I. Zutic, H.J. Kwon, V.M. Yakovenko, S. Das Sarma, Phys. Rev. B **63**, 144531 (2000)
41. C.D. Vaccarella, R.D. Duncan, C.A.R. Sá de Melo, cond-mat/0302080 (2003)
42. C.D. Vaccarella, R.D. Duncan, C.A.R. Sa de Melo, Physica C **391**, 89 (2003)
43. J.A. Pals, W. van Haeringen, Phys. Rev. B **15**, 2592 (1977)
44. V.B. Geshkenbein, A.I. Larkin, JETP Lett. **43**, 395 (1986)
45. Q. Li, Y.N. Tsay, M. Suenaga, R.A. Klemm, G.D. Gu, N. Koshizuka, Phys. Rev. Lett. **83**, 4160 (1999)
46. Y. Tanaka, S. Kashiwaya, Phys. Rev. B **53**, R11957 (1996)
47. Y.S. Barash, H. Burkhardt, D. Rainer, Phys. Rev. Lett. **77**, 4070 (1996)
48. T. Lofwander, V.S. Shumeiko, G. Wendin, Supercond. Sci. Technol. **14**, 43 (2001)
49. A.P. Mackenzie, Y. Maeno, Rev. Mod. Phys. **75**, 657 (2003)
50. C.H. Ahn, J.M. Triscone, J. Mannhart, Nature **424**, 1015 (2003)
51. K.A. Parendo, K.H. Sarwa B. Tan, A. Bhattacharya, M. Eblen-Zayas, N.E. Staley, A.M. Goldman, Phys. Rev. Lett. **94**, 197004 (2005)
52. R.W. Cherng, C.A.R. Sá de Melo, Phys. Rev. B **74** 212505 (2006)
53. W. Zhang, C.A.R. Sá de Melo, Phys. Rev. Lett. **97**, 047001 (2006)
54. T. Vuletic, P. Auban-Senzier, C. Pasquier, S. Tomic, D. Jerome, M. Heritier, K. Bechgaard, Eur. Phys. J. B **25**, 319 (2002)
55. A.V. Kornilov, V.M. Pudalov, Y. Kitaoka, K. Ishida, G.Q. Zheng, T. Mito, J.S. Qualls, Phys. Rev. B **69**, (2004)
56. I.J. Lee, S.E. Brown, W. Yu, M.J. Naughton, P.M. Chaikin, Phys. Rev. Lett. **94**, 197001 (2005)
57. D. Podolsky, E. Altmam, T. Rostunov, E. Demler, Phys. Rev. Lett. **93**, 246402 (2004)
58. W. Zhang, C.A.R. Sá de Melo, J. Appl. Phys. **95**, 10B108 (2005)
59. K. Mortensen, Y. Tomkiewicz, T.D. Schultz, E.M. Engler, Phys. Rev. Lett. **46**, 1234 (1981)

25

Theory of the Fulde–Ferrell–Larkin–Ovchinnikov State and Application to Quasi-Low-dimensional Organic Superconductors

H. Shimahara

The Fulde–Ferrell–Larkin–Ovchinnikov (FFLO or LOFF) state is a superconducting state of Cooper pairs with finite center-of-mass momenta which is stabilized by the Zeeman energy at high magnetic fields in clean type II superconductors. In this chapter, we review recent developments of the theory of the FFLO state in quasi-low-dimensional (QLD) systems, and its application to QLD organic superconductors. For the FFLO state to occur, the orbital pair-breaking effect needs to be sufficiently weak that superconductivity survives up to the Pauli paramagnetic limit. This condition is satisfied in layered superconductors with small interlayer electron hopping energy when the magnetic field is precisely aligned parallel to the most conductive layers. The FFLO state is favored in QLD systems because of the Fermi surface effect, analogous to the nesting effects for charge density waves (CDW) and spin density waves (SDW), arising in the FFLO state due to the finite center-of-mass momenta of the Cooper pairs. In particular, QLD organic superconductors are good candidates because of their low-dimensionality, narrow electron band, and possible anisotropy of the gap function, all of which contribute to the stabilization of the FFLO state. Experimentally, upper critical field curves exhibit upturns at low temperatures in some quasi-two-dimensional organic superconductors, which agrees with the theoretical prediction of the FFLO state. In addition, sensitivity to misalignment of the magnetic field and impurities has been observed, which agrees with theoretical predictions. These experimental and theoretical results suggest the possibility of the appearance of the FFLO state in organic superconductors. We discuss the crossover from vortex states to the FFLO state in QLD systems. It is known that the FFLO state is nothing but the vortex state with the infinite Landau level index.

25.1 The FFLO State

25.1.1 Basis of the FFLO State

In normal-state metals in magnetic fields, the Fermi surfaces of up and down spin electrons split due to the Zeeman energy. In contrast, the BCS ground state for singlet pairing is not spin-polarized, as is seen in the vanishing spin susceptibility. However, it is possible that Cooper pairs are formed on the split Fermi surfaces in a magnetic field. Such Cooper pairs have a finite center-of-mass momentum $\mathbf{q}$ due to the displacement of the Fermi surfaces. The resultant superconducting state has a lower Zeeman energy than the BCS state with $\mathbf{q} = 0$, but a smaller condensation energy. Therefore, the state can occur at high magnetic fields with $H \gtrsim \Delta_0/\mu_\mathrm{B}$, where Δ_0 and μ_B denote the zero field BCS gap and the Bohr magneton, respectively. This state is called the Fulde–Ferrell–Larkin–Ovchinnikov (FFLO or LOFF) state, since the possibility of such a state was proposed independently by Fulde and Ferrell [1] and by Larkin and Ovchinnikov [2]. The FFLO state has been examined by many authors [3]. It is easily verified by energy considerations that when it occurs the area of the FFLO state in the phase diagram should be between the normal and BCS states on the temperature and magnetic field plane.

The strength of the magnetic field Δ_0/μ_B is of the order of the Pauli paramagnetic limit H_P. For example, weak coupling theory gives $H_\mathrm{P} = \Delta_0/\sqrt{2}\mu_\mathrm{B}$ in an isotropic superconductor. Therefore, a necessary condition is that the orbital pair-breaking effect is suppressed so that superconductivity survives up to the Pauli paramagnetic limit [4]. The orbital effect is suppressed, for example, by the low dimensionality of a system and a large effective mass.

Therefore, if observed upper critical field is larger than a value of H_P estimated from observed zero-field transition temperature $T_\mathrm{c}^{(0)}$, it may suggest the presence of the FFLO state, although there may be some other possible mechanisms. Here, the formula $H_\mathrm{P} = \Delta_0/\sqrt{2}\mu_\mathrm{B} \approx 1.86\,(\mathrm{T\,K^{-1}}) \times T_\mathrm{c}^{(0)}\,(\mathrm{K})$ is convenient for a crude estimation, but we need to be careful that this formula is based on some assumptions, such as weak coupling, isotropic pairing, absence of spin–orbit coupling and impurities, and so on.

Because of the finite center-of-mass momentum, $\mathbf{q} \neq 0$, the FFLO state has several characteristics, such as phase or sign changes of the order parameter in real space and sensitivity to the Fermi surface structure, pairing anisotropy, and impurities, as we shall review later.

25.1.2 Spatial Structure of the Order Parameter

In this section, we summarize the spatial structure of the order parameter of the FFLO state in the limit where the orbital pair-breaking effect is negligible. The state first proposed by Fulde and Ferrell [1] is described by an order parameter of the form

$$\Delta(\mathbf{r}) = \Delta_1 \mathrm{e}^{\mathrm{i}\mathbf{q}\cdot\mathbf{r}}. \tag{25.1}$$

In general, because of the symmetry of the system there are several equivalent $\mathbf{q}$'s that give the same upper critical field. We write these $\mathbf{q}$'s as $\mathbf{q}_m$ with $m = 1, 2, \ldots, M$, where M denotes the number of equivalent $\mathbf{q}$'s. For example, in an isotropic system $M = \infty$, while $M = 4$ or 8 in a fourfold symmetric system. It is easily verified that the $\mathbf{q}_m$'s have the same length. Any superpositions of the form

$$\Delta(\mathbf{r}) = \sum_m \Delta_m \mathrm{e}^{\mathrm{i}\mathbf{q}_m \cdot \mathbf{r}} \tag{25.2}$$

have the same upper critical field within the second-order phase transition. Even when $M > 6$, only $1, 2, 3, 4$, or $6 \Delta_m$'s can be nonzero at the same time in the summation, if we require the resultant state to be periodic. The degeneracy is removed below the upper critical field, because the gap equation becomes nonlinear in that case, and states with the lowest free energy are present. Near the upper critical field it can be examined with the Ginzburg–Landau (GL) expansion of the free energy [2, 5].

Examining periodic solutions of the form of (25.2), Larkin and Ovchinnikov [2] showed that the state described by the linear combination

$$\Delta(\mathbf{r}) = \Delta_1(\mathrm{e}^{\mathrm{i}\mathbf{q}\cdot\mathbf{r}} + \mathrm{e}^{-\mathrm{i}\mathbf{q}\cdot\mathbf{r}}) = 2\Delta_1 \cos(\mathbf{q} \cdot \mathbf{r}) \tag{25.3}$$

has the lowest free energy near the upper critical field in a three-dimensional (3D) system with spherical Fermi surfaces within the fourth-order GL expansion.

In two-dimensional (2D) systems with cylindrical Fermi surfaces, it has been shown that the states with 2D oscillations with the order parameters

$$\begin{aligned}
&\Delta(\mathbf{r}) \propto \cos(qx) + \cos(qy), \\
&\Delta(\mathbf{r}) \propto \exp(\mathrm{i}\mathbf{q}_1 \cdot \mathbf{r}) + \exp(\mathrm{i}\mathbf{q}_2 \cdot \mathbf{r}) + \exp(\mathrm{i}\mathbf{q}_3 \cdot \mathbf{r}), \\
&\Delta(\mathbf{r}) \propto \cos(\mathbf{q}_1 \cdot \mathbf{r}) + \cos(\mathbf{q}_2 \cdot \mathbf{r}) + \cos(\mathbf{q}_3 \cdot \mathbf{r})
\end{aligned} \tag{25.4}$$

have the lowest free energy at low temperatures, depending on the temperature and the pairing anisotropy, where $\mathbf{q}_1 = (q, 0, 0)$, $\mathbf{q}_2 = (-q/2, \sqrt{3}q/2, 0)$, and $\mathbf{q}_3 = (-q/2, -\sqrt{3}q/2, 0)$ [5]. We call these states square, triangular, and hexagonal states, respectively. Here, more complicated superpositions of a larger number of terms $\mathrm{e}^{\mathrm{i}\mathbf{q}_m\cdot\mathbf{r}}$ have not been considered because they do not have a periodic structure in real space [6, 7]. In 2D systems, the fourth-order GL expansion of the free energy does not break down for s-wave pairing. For d-wave pairing, it does not break down in almost all temperature region, but it does in a very limited temperature region, for which the possibility of first-order transition has been suggested [5].

In contrast, in 3D systems, the fourth-order GL expansion breaks down for the cubic state $\Delta(\mathbf{r}) \propto \cos(qx) + \cos(qy) + \cos(qz)$ [2] and for the triangular and square states [5], which suggests the possibility of first-order transitions to these states. Mora and Combescot showed by a numerical calculation beyond the GL expansion that the transition to the FFLO state from the normal state is of first order and the transition temperature is slightly higher than that obtained where the second-order transition is assumed. They also showed that states with 2D and 3D oscillations have lower free energies than the state of (25.3) at low temperatures [8] as in 2D systems.

When the magnetic field decreases below the upper critical field, the order parameter has many plane wave components of $\mathbf{q}$'s with different lengths $|\mathbf{q}|$.

In the presence of the orbital pair-breaking effect, the order parameter structure is determined taking into account the coexistence of the FFLO state and the vortex states [9–11]. We shall review this in Sect. 25.3.

25.1.3 Exotic Superconductors and the FFLO State

In spite of the theoretical proposal, no experimental indication of the FFLO state had been obtained until exotic superconductors, such as organic, cuprate, and heavy fermion superconductors, were discovered.

The reason why the FFLO state has not been observed in conventional alloy type-II superconductors is explained as follows. First, the FFLO state is unstable against nonmagnetic impurities due to the order parameter oscillations in contrast to the isotropic BCS state with $\mathbf{q} = 0$ as examined by Takada [12]. Therefore, for the occurrence of the FFLO state, sample needs to be a clean type-II superconductor, in which $l \gg \xi_0$ and $\kappa \gg 1$ are satisfied, where l is the electron mean free path, ξ_0 the superconducting coherence length, and κ the Ginzburg–Landau parameter. Second, orbital pair-breaking effect needs to be sufficiently weak that $H_{c20} \gtrsim H_{\mathrm{P}}$ is satisfied, for the occurrence of the FFLO state, as explained in Sect. 25.1.1 and reviewed in Sect. 25.3, where H_{c20} denotes the pure orbital limit, i.e., the upper critical field in the absence of the Pauli paramagnetic effect. These conditions, $l \gg \xi_0$, $\kappa \gg 1$, and $H_{c20} \gtrsim H_{\mathrm{P}}$, are difficult to be satisfied in conventional alloy type-II superconductors.

In contrast, the exotic superconductors can be clean type-II superconductors which is strongly Pauli limited [13–17]. The orbital pair-breaking effect is weak when the electron effective mass is large and when the magnetic field is applied parallel to the conductive layer in layered compounds. In addition, the quasi-low-dimensionality and possible pairing anisotropy may contribute to stabilization of the FFLO state due to a Fermi surface effect in some of the exotic superconductors as we shall review in Sect. 25.2. Therefore, the FFLO state has been studied extensively in connection with the exotic superconductors.

25.1.4 Lower Critical Field

The lower critical field of the FFLO state due to the transition to the BCS state with $\mathbf{q} = 0$ is a subject of interest. Some results have been obtained assuming an FFLO state that oscillates only in a single direction and no orbital pair-breaking effect. Under these assumptions, Burkhardt and Rainer examined the isotropic 2D system [15]. They showed that the transition between the FFLO state and the uniform BCS state is of second order, and it occurs such that the period of the spatial oscillation increases as the magnetic field decreases and diverges at the lower critical field. This result does not essentially change in the isotropic 3D system [18], if the orbital pair-breaking effect is ignored, although it is present in actuality in the 3D system. In the presence of the orbital pair-breaking effect, however, it was found that the transition at the lower critical field is of first order [17]. Observation of the phase transition at the lower critical field can be experimental evidence of the presence of a new superconducting state such as the FFLO state.

25.2 Nesting Effect for the FFLO State

Because of the finite center-of-mass momenta $\mathbf{q}$'s of the Cooper pairs, the stability and the structure of the FFLO state strongly depend on the Fermi surface structure and the pairing anisotropy. For example, the FFLO upper critical field diverges when $T \to 0$ in 1D systems [19–21], while it remains finite in 2D systems [15,16,22,23] and in 3D systems [1,2]. The Fermi surface effect on the stability of the FFLO state can be examined by introducing a concept analogous to the Fermi surface nesting for charge density waves (CDW) and spin density waves (SDW) [16,24,25].

It is well known that CDW or SDW are stable at low temperatures when the Fermi surface and the surface shifted by $\mathbf{Q}$ touch over a finite area, which is called the nesting effect and the vector $\mathbf{Q}$ is called the nesting vector. Analogously, the FFLO state is more stable if the Fermi surfaces of up spin electrons and down spin electrons touch over a larger area, when the latter Fermi surface is inverted and shifted by $\mathbf{q}$, i.e., $\mathbf{k} \to -\mathbf{k}+\mathbf{q}$, although it can be stable without such an effect. These considerations on the nesting condition for the FFLO state can be mathematically verified by examining the gap equation.

In general, there are several $\mathbf{q}$'s which give the best nesting condition. We write them as $\tilde{\mathbf{q}}_1, \tilde{\mathbf{q}}_2, \ldots, \tilde{\mathbf{q}}_{\tilde{M}}$, where $\tilde{M}$ denotes the number of such $\mathbf{q}$'s. For example, in a cylindrically symmetric system $\tilde{M} = \infty$, while $\tilde{M} = 4$ or 8 in a fourfold symmetric system. In the absence of the orbital effect, a state expressed by (25.2) with $\mathbf{q}_m = \tilde{\mathbf{q}}_m$ and $M = \tilde{M}$ occurs at $T = 0$. At low temperatures, the $\mathbf{q}_m$'s differ due to a temperature effect, satisfying $\mathbf{q}_m \parallel \tilde{\mathbf{q}}_m$ and $|\mathbf{q}_m| < |\tilde{\mathbf{q}}_m|$. Depending on the pairing anisotropy and Fermi surface structures, $\mathbf{q}_m$'s may jump at a finite temperature [9,26–28].

The pairing anisotropy needs to be taken into account when the nesting condition for the FFLO state is examined [24–26]. The parts of the Fermi surfaces near the momenta at which the gap function is maximum contribute more to the gap equation than those near the nodes of the gap function. For example, in a system with cylindrical Fermi surfaces, a nesting vector $\tilde{\mathbf{q}}_m$

can point in any direction in the p_xp_y-plane for s-wave pairing due to the symmetry, while the $\tilde{\mathbf{q}}_m$'s are oriented in the directions of the $\pm p_x$ and $\pm p_y$ axes for $d_{x^2-y^2}$-wave pairing [9,27,28].

In one-dimension, the Fermi surfaces are flat in 3D momentum space. Therefore, the Fermi surfaces touch over a finite area by the transformation $\mathbf{k} \to -\mathbf{k} + \tilde{\mathbf{q}}_m$ with nesting vectors $\tilde{\mathbf{q}}_1 = -\tilde{\mathbf{q}}_2$, which satisfies $|\tilde{\mathbf{q}}_m| = 2h/v_{\mathrm{F}} \equiv \tilde{q}$, where $h \equiv \mu_B|\mathbf{H}|$ and v_{F} denote the Zeeman energy and the Fermi velocity, respectively. In this case, the upper critical field diverges when $T \to 0$, as mentioned earlier.

In contrast, the upper critical field of the FFLO state becomes finite in a 3D system with spherical Fermi surfaces, because the Fermi surfaces touch only at a point by the same transformation with $|\mathbf{q}| = 2h/v_{\mathrm{F}}$. To be precise, the length of the optimum $\mathbf{q}$'s is about 20% longer than $2h/v_{\mathrm{F}}$, because vector $\mathbf{q}$'s that cause the Fermi surfaces to cross on a line give better nesting conditions than those that cause them to touch at a point.

In a 2D system with cylindrical Fermi surfaces, the upper critical field is finite but enhanced compared to a 3D system with spherical Fermi surfaces. This is because in the cylindrical system the Fermi surfaces of up and down spin electrons touch on a line by the transformation $\mathbf{k} \to -\mathbf{k} + \tilde{\mathbf{q}}_m$ with $|\tilde{\mathbf{q}}_m| = 2h/v_{\mathrm{F}}$.

Therefore, it may appear that 1D systems are the best candidates for an FFLO superconductor. In such systems, however, the nesting instability to CDW or SDW occurs for realistic energy parameters. Hence, 2D systems are the best practical candidates, because CDW and SDW are suppressed, with the FFLO state being favored. Here, the 2D systems include quasi-one-dimensional (Q1D) systems in which the nesting instabilities are suppressed. In Q1D systems, due to imperfect nesting of Fermi surface [29], the upper critical field becomes convergent, but the FFLO critical field is much higher than H_{P} as in the isotropic 2D system. Furthermore, we assume the presence of interlayer interactions, so that the superconducting long-range order is stabilized. Therefore, precisely speaking, such systems should be called quasi-low-dimensional (QLD) systems. In QLD systems, the critical field as a function of the temperature T exhibits an upturn, i.e., a downward convex curve ($\mathrm{d}^2H_{c2}(T)/\mathrm{d}T^2 > 0$), at low temperatures.

It appears that if the Fermi surfaces have flat parts, the upper critical field of the FFLO state is more enhanced by the nesting effect for the FFLO state, as in 1D systems. This should be the case in certain systems, but in actuality the situation is more complicated because the curvatures of the Fermi surfaces of the up and down spin electrons may differ if their numbers change in the magnetic field. It also appears that the $\tilde{\mathbf{q}}_m$'s become perpendicular to the flattest parts of the Fermi surface, because they are the shortest nesting vectors and cause the slowest spatial variation of the order parameter, which appears to be energetically favored. In actuality, however, this does not necessarily occur. For example, in the square lattice tight-binding model with various electron concentrations n_{e}, the FFLO state is most stable due to the

Fermi surface effect near $n_e \approx 0.63$, for which the Fermi surfaces do not have a nearly flat area [25]. Also, the $\tilde{\mathbf{q}}_m$'s are not perpendicular to the flattest parts of the Fermi surfaces. Similar results have been obtained for a model with Fermi surfaces more realistic for organic superconductors [24].

In the above, we have considered the limit where the orbital pair-breaking effect is negligible. In this limit, for $d_{x^2-y^2}$-wave pairing, the directions of $\mathbf{q}_m$'s jump from $\theta_{\mathbf{q}} = m'\pi/2$ to $\theta_{\mathbf{q}} = \pi/4 + m'\pi/2$ at a temperature $T_1 \approx 0.06 T_c^{(0)}$, where $\theta_{\mathbf{q}}$ denotes the angle between $\mathbf{q}$ and the p_x-axis, and $m' = 0, 1, 2, 3$. Therefore, the upper critical field exhibits a kink at $T = T_1$, and increases sharply with an upturn at low temperatures $T \leq T_1$. In contrast, in the presence of a substantial orbital effect, it is appropriate to assume a model in which $\mathbf{q} \parallel \mathbf{H}$, as we shall discuss later. In this case, the kink due to the jump of $\mathbf{q}$ does not appear, and H_{c2} is enhanced due to the nesting effect when a magnetic field is applied so that $\mathbf{H} \parallel \tilde{\mathbf{q}}_m$.

The upturn of $H_{c2}(T)$ is a characteristic of the FFLO state in QLD systems, which vanishes rapidly when the perpendicular component of the magnetic field increases and when the sample is doped with impurities. There are some mechanisms that exhibit similar upturn behaviors [30] other than the FFLO state, but their signatures are rather different from that of the FFLO state. In the mechanism based on the FFLO state, the upturn occurs only at low temperatures. In contrast, the experimental data and results for some theories of high-T_c superconductors exhibit upturns for all temperatures $T < T_c^{(0)}$ [30]. Dimensional crossover is another mechanism producing an upturn behavior similar to that of the FFLO state. This mechanism was examined by Lebed originally in quasi-one-dimension [31], and subsequently by Dupuis et al. [32], and extended to quasi-two-dimensions by Lebed and Yamaji [33]. It was predicted for layered superconductors in a parallel magnetic field that the superconducting transition temperature decreases due to the orbital effect as the magnetic field increases, and at higher fields it recovers to the value in the absence of the orbital pair-breaking effect. In this mechanism, however, the upper critical field does not exceed the Pauli paramagnetic limit unless triplet pairing is considered.

25.3 Vortex States and the FFLO State

The orbital pair-breaking effect is present to a greater or lesser extent in real materials, and vortex states are expected to occur. In film systems, the orbital effect can be avoided by applying the magnetic field in an exactly parallel direction. However, this is practically difficult in films made from metal, because the misalignment of the magnetic field θ_H from the film needs to be less than $\sim 10^{-4}$ deg. for the orbital effect to be suppressed. This requirement is due to the fact that the Fermi velocity and the coherence length are large [9]. Furthermore, surface scattering may destroy the FFLO state in film systems.

In contrast, in layer systems, such as QLD organic superconductors, surface scattering is negligible if the sample size is of macroscopic scale. Since the coherence length is much shorter in organic superconductors than in metals, the misalignment θ_H only needs to be less than 0.1–1 deg., as we shall discuss later. However, even in these systems, the orbital effect exists in practice, because of interlayer electron hopping and the finite component of the magnetic field perpendicular to the most conductive plane. In experiments, the perpendicular component cannot be eliminated completely, and even a very small perpendicular component may give rise to appreciable strength of the orbital effect since the intralayer conductivity is usually large.

25.3.1 Coexistence of the FFLO State and Vortex States in 3D Systems

In the presence of the orbital pair-breaking effect, the spatial variation of $\Delta(\mathbf{r})$ perpendicular to the magnetic field is described by a vortex lattice state. Hence, FFLO oscillation can only occur parallel to the magnetic field, i.e., only the FFLO state with $\mathbf{q} \parallel \mathbf{H}$ can coexist with the vortex state [4]. Therefore, at low temperatures, the FFLO state is most stable when the magnetic field is applied parallel to any one of the $\tilde{\mathbf{q}}_m$'s.

For example, for spherically symmetric systems, Gruenberg and Gunther found that the FFLO state occurs when the orbital pair-breaking effect is sufficiently weak that $H_{c20}/H_P \gtrsim 1.28$ [4]. In terms of the Maki parameter $\alpha \equiv \sqrt{2}H_{c20}/H_P$, the condition is written as $\alpha \gtrsim 1.81$. This result holds qualitatively in other 3D systems if the orbital pair-breaking effect is substantial, although the critical value of H_{c20}/H_P is different.

25.3.2 Tilted Magnetic Field in 2D Systems

When the system has a layer structure and the interlayer electron hopping energy is small, the orbital pair-breaking effect is weak, if the magnetic field is applied parallel to the conductive layer. As a result, the upper critical field reaches a value near the Pauli paramagnetic limit H_P, and thus the FFLO state may occur.

We note that the magnetic field cannot be exactly aligned parallel to the conductive layer experimentally in real materials. Therefore, even if the interlayer electron hopping energy is negligible in the gap equation, the orbital pair-breaking effect is present because of the finite perpendicular component $H_\perp$ of the magnetic field. Since the solutions of the linearized gap equation in the conductive layer are completely described by Abrikosov functions, it appears that there is no possibility for FFLO oscillation when $H_\perp$ is finite.

Bulaevskii examined a 2D system in tilted magnetic fields at $T = 0$ [22]. In the result of numerical calculation, the upper critical field of the vortex state tends to approach that of the FFLO state when $H_\perp$ decreases, as physically expected. Extensions to finite temperatures have also been made [9, 34]. Buzdin and Brison assumed a gap function of the form

$$\Delta(\mathbf{r}) \propto \exp[-im\varphi]\exp[-\rho^2/2]\rho^m, \tag{25.5}$$

where $\rho = r\sqrt{eH/c\hbar}$ and the integer m corresponds to the quantum number of the angular momentum [34]. It appears that a large supercurrent flows around $r = 0$ for the gap function of (25.5) when m increases. Shimahara and Rainer have obtained complete solutions of the linearized gap equation

$$\Delta(\mathbf{r}) \propto (-1)^n \mathrm{e}^{\mathrm{i}kx} H_n\left[\sqrt{2\kappa_\perp}\left(y - \frac{k}{\kappa_\perp}\right)\right] \exp\left[-\frac{\kappa_\perp}{2}\left(y - \frac{k}{\kappa_\perp}\right)^2\right] \tag{25.6}$$

at the second-order upper critical field [9], where H_n denotes the Hermite polynomial of the Landau level index n, and we have defined $\kappa_\perp = 2|\mathrm{e}|H_\perp/c$. We define $n(\kappa_\perp)$ as the Landau level index n that gives the highest upper critical field for each given value of $\kappa_\perp \propto H_\perp$. We have proven that $n(\kappa_\perp)$ diverges in the limit $\kappa_\perp \to 0$, i.e., in the limit of an exactly parallel magnetic field, $H_\perp \to 0$, below $T = T^* \approx 0.56T_\mathrm{c}$, so that the limit

$$q = \lim_{\kappa_\perp \to 0} \sqrt{2\kappa_\perp n(\kappa_\perp)} \tag{25.7}$$

is finite, where q is equal to $|\mathbf{q}|$ of the FFLO state in the limit of a parallel magnetic field. The relation (25.7) connects $\mathbf{q}$ and n, which characterize the FFLO state and the Abrikosov vortex states, respectively. It indicates that the FFLO state is nothing but the vortex state with the infinite Landau level index. In fact, the gap functions with large n oscillate in space with the wave number given by (25.7). As an example, the gap function with $n = 20$ is shown in Fig. 25.1. The spatial oscillations of these states and the FFLO state are of the same physical origin, i.e., due to the Pauli paramagnetic effect. The nodes of the order parameter in real space are present so that the spin polarization energy is lowered for both oscillations. The theory can be extended to d-wave pairing [9]. Therefore, the vortex states with very high Landau level indexes can be regarded as the FFLO state in practice.

It is easily verified[1] that the index m in [34] is different from the Landau level index n. For example, taking the limit $H_\perp \to 0$ where the FFLO

[1] If we express the magnetic field $H_x = H_y = 0$, $H_z = H$ by the vector potential $A_\varphi = H\rho/2$, $A_z = A_\rho = 0$, the Schrödinger equation for a particle with charge $-e$ and mass m becomes $\frac{-\hbar^2}{2m}\left[\frac{1}{\rho}\frac{\partial}{\partial\rho}\left(\rho\frac{\partial\psi}{\partial\rho}\right)\right] - \frac{\mathrm{i}}{2}\hbar\omega\frac{\partial\psi}{\partial\varphi} = E\psi$, where $\omega = |\mathrm{e}|H/mc$. The energy eigenvalue is obtained as $E = \hbar\omega[n_\rho + (|m| + m + 1)/2]$ with non-negative integer n_ρ, and the corresponding general solution is expressed as $\psi(\rho, \varphi, z) \propto R_{n_\rho m}(\rho)\mathrm{e}^{\mathrm{i}m\varphi}\mathrm{e}^{\mathrm{i}p_z z/\hbar}$ with $R_{n_\rho m}(\rho) \propto \exp(-\rho^2/4a_H^2)\rho^{|m|}F(-n_\rho, |m|+1, \rho^2/2a_H^2)$, where $a_H = \sqrt{\hbar/m\omega}$ and F denotes the confluent hypergeometric function. Therefore, the Landau level index n is expressed as $n_\rho + (|m| + m)/2$, but not as m.

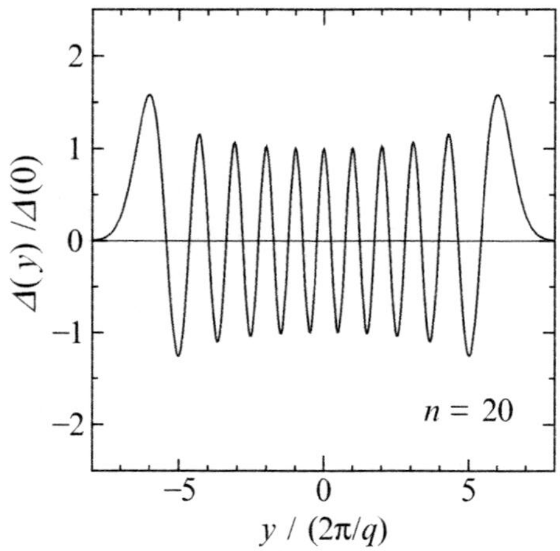

Fig. 25.1. Spatial dependence of the gap function calculated from (25.6) with $n = 20$ and $k = 0$

state occurs, it seems that the angular momentum m must vanish since the supercurrent vanishes, while n diverges.

The vortex lattice structure in a tilted magnetic field is obtained by minimizing the free energy among the superconducting states expressed by linear combinations of (25.6) [10, 11]. Klein has completed this work and revealed the detailed behavior of the order parameter structure in the tilted magnetic field, notably when n becomes infinitely large, i.e., for a nearly parallel magnetic field [11]. It was shown that the results of [5] are recovered in the limit of a parallel magnetic field, as expected.

The strength ratio of the orbital magnetic effect and the Pauli paramagnetic effect is expressed by the parameter

$$r_{\rm m} \equiv \sin\theta_H / z_{\rm m} \tag{25.8}$$

with

$$z_{\rm m} = \frac{g}{4} \frac{T_{\rm c}^{(0)}}{\Delta_0} \frac{m^*}{m} \frac{1}{p_{\rm F}\xi_0}, \tag{25.9}$$

where g, m, m^*, $p_{\rm F}$, ξ_0, and θ_H denote the g-factor, bare electron mass, electron effective mass, Fermi momentum, BCS coherence length, and the tilt angle (i.e., the angle between the magnetic field and the conductive layer), respectively. The parameter $z_{\rm m}$ is proportional to the Maki parameter α.

For quasi-two-dimensional (Q2D) organic superconductors, we obtain $z_{\rm m} \sim 1/20$ as a rough estimate [9]. It was found that the vortex states with Landau level $n \geq 1$ occur when $r_{\rm m} \lesssim 0.8$, and that the critical field is very close to that of the FFLO state when $r_{\rm m} \lesssim 0.1$ [9]. The former and the latter conditions are realized when $\theta_H \lesssim 2.3$ (deg.) and $\theta_H \lesssim 0.3$ (deg.), respectively, if $z_{\rm m} \sim 1/20$. In contrast, for metal superconductors, we obtain $z_{\rm m} \sim 10^{-4}$.

Therefore, it is practically difficult to apply a parallel magnetic field with sufficient accuracy that the high-n vortex states, which are practically regarded as the FFLO state, can be observed in metallic film superconductors.

25.3.3 Orbital Pair-Breaking Effect in Q2D Systems

In this section, we discuss the FFLO state in superconductors with layered structures in which the interlayer electron hopping energy $t_\perp$ is finite.

When $t_\perp$ is very small and the magnetic field is parallel to the conductive layers, an approximation where the orbital pair-breaking effect is neglected becomes appropriate. Therefore, the states of the form of (25.2) with $\mathbf{q}_m \parallel \tilde{\mathbf{q}}_m$ and $M = \tilde{M}$ occur at low temperatures near the upper critical field.

When $t_\perp \neq 0$, the situation is essentially the same as that in Sect. 25.3.1. The gap function can be written in terms of a linear combination of the Abrikosov functions near the second-order transition point, with respect to the coordinates perpendicular to the magnetic field. Hence, the vector $\mathbf{q}$ of the FFLO state is oriented parallel to the magnetic field as far as $t_\perp \neq 0$, however small it is. This fact appears to contradict the values of $\mathbf{q}$ at $t_\perp = 0$, where the $\mathbf{q}$'s are determined from the Fermi surface structure and the pairing symmetry, because it is unlikely that the limit of $t_\perp \to 0$ is not continuously recovered.

We note again that the FFLO state is nothing but the vortex state with the infinite Landau level index, as reviewed in Sect. 25.3.2, where we discussed the limit of $\theta_H \to 0$ with $t_\perp = 0$ [9]. Also, in the limit of $t_\perp \to 0$ with $\theta_H = 0$, the FFLO oscillation perpendicular to $\mathbf{H}$ should be recovered continuously. Therefore, the Landau level index n should diverge in this limit. This behavior is easily verified by extending the theory of [9] to Q2D systems. It is also conjectured that in Q2D systems with sufficiently small $t_\perp$, internal phase transitions between states with different Landau level indexes occur below the upper critical field.

For example, we have shown that the FFLO state can be expressed by a linear combination of three, four, or six plane waves as expressions in (25.4) in 2D systems at low temperatures. From this limit, we can deduce that states with higher Landau level indexes occur in Q2D systems with finite but small $t_\perp$ if the magnetic field is applied parallel to the highly conductive layers. In this case, the gap function oscillates in space not only in the direction of $\mathbf{H}$, but also perpendicular to $\mathbf{H}$, since the Abrikosov functions with large n oscillate in space as explained in Sect. 25.3.2.

We have considered only systems with an interlayer hopping energy $t_\perp$ of strength such that the mean field theory is applicable. This condition is written as $t_\perp \gg k_\mathrm{B}T_\mathrm{c}^{(0)}$, because the energy scale of the thermal fluctuation is proportional to $k_\mathrm{B}T_\mathrm{c}^{(0)}$. On the other hand, in order to omit $t_\perp$ in the resultant mean field equation, it needs to be satisfied that $t_\parallel \gg t_\perp$ where $t_\parallel$ denotes an intralayer hopping energy. These conditions are consistently satisfied if

$t_{\parallel} \gg t_{\perp} \gg k_{\rm B}T_{\rm c}^{(0)}$. A similar consideration holds for pairing interactions. Lebed and Yamaji proposed a theory which quantitatively accounts the orbital effect in layered superconductors in parallel fields [33].

25.4 Candidate Organic Superconductors

Quasi-low-dimensional organic superconductors have some features that favor the FFLO state as explained in Sect. 25.1.3: (1) they can be clean type II superconductors due to their narrow electron band, (2) the nesting effect due to the quasi-low-dimensionality contributes to the stabilization of the FFLO state, (3) the orbital pair-breaking effect is strongly suppressed for magnetic fields parallel to the highly conductive layers. Based on these features, the possibility of the FFLO state occurring in organic superconductors has been suggested by many authors [9, 14–16, 24, 35–37].[2]

25.4.1 κ-(BEDT-TTF)$_2X$

Enhancement of the upper critical field of the FFLO state at low temperatures due to the nesting effect can be expected in Q2D organic superconductors away from the SDW and CDW instabilities. We have examined the Fermi surface effect based on a realistic Fermi surface structure of a κ-(BEDT-TTF)$_2X$ system [24]. We found that the upper critical field of the FFLO state can become 1.5–2.5 times the Pauli paramagnetic limit of the BCS state with $\mathbf{q} = 0$. Nam et al. have obtained the T–H phase diagram of the compound κ-(BEDT-TTF)$_2$Cu(NCS)$_2$ [38]. They found that the upper critical field exceeds the Pauli paramagnetic limit $H_{\rm P}$ and exhibits an upturn at low temperatures [38]. Singleton et al. argue from their experimental data of resistance and magnetic measurements that a transition from the superconducting mixed state into an FFLO state occurs in this compound [39]. Ohmichi et al. also found a similar upturn in κ-(BEDT-TTF)$_2$Cu(NCS)$_2$ [40]. In the compound κ-(BEDT-TTF)$_2$Hg$_{2.89}$Br$_8$, the critical field exceeds $H_{\rm P}$, but does not exhibit an upturn [37,41]. It can be explained that the FFLO state is destroyed by random scattering due to the random distribution of the anions. Manalo and Klein compared theoretical results and experimental data of $H_{\rm c2}$ of the FFLO state in κ-(BEDT-TTF)$_2$Cu(NCS)$_2$ and found agreement both with regard to the angular and the temperature dependence of $H_{\rm c2}$ [42].

[2] Organic superconductors, such as (TMTSF)$_2$X and (DMET)$_2$X, are traditionally called Q1D superconductors, but in the sense that SDW is suppressed and the Fermi surfaces are sufficiently warped in 2D momentum space, they can be classified as being Q2D. They are good candidates for FFLO superconductors, unless the superconductivity is of triplet pairing.

25.4.2 λ-(BETS)$_2X$

The Q2D organic compound λ-(BETS)$_2$FeCl$_4$ exhibits insulating antiferromagnetic long-range order at low magnetic fields, while it is metallic at high fields. In the high field region around $B \approx 33\,\mathrm{T}$, field-induced superconductivity (FISC) has been observed [43–45]. FISC is attributed to the Jaccarino–Peter mechanism [46, 47]. At $B \approx 33\,\mathrm{T}$, the Zeeman energy is completely canceled by the energy due to the exchange field created by the localized spins on the anion FeCl$_4$. The maximum transition temperature $T \approx 4.2\,\mathrm{K}$ of the FISC at $B \approx 33\,\mathrm{T}$ [43, 44] is close to the zero field transition temperature 4.7 K in the isostructural material λ-(BETS)$_2$GaCl$_4$ [48]. Therefore, the orbital effect is expected to be very weak in λ-(BETS)$_2$FeCl$_4$. The lower and higher critical fields of the FISC show upward and downward convex curves, respectively, as functions of temperature, which can be explained as being due to the FFLO state in a Q2D system [44,49,50]. However, the critical fields of the FISC calculated by a traditional model of the FFLO state are very different from the experimental data [44]. The curve of the experimental critical field can be fit very well if we assume a very weak p-wave attractive interaction [49, 51], which must exist in real materials to some extent even in singlet superconductors [52, 53]. It has also been attempted to attribute the discrepancy to the 2D in-plane anisotropy [50], although the resultant tricritical temperature seems too low to fit the experimental data.

Tanatar et al. have obtained the H–T phase diagram of the compound λ-(BETS)$_2$GaCl$_4$ by thermal conductivity experiments, and argue that the experimental data suggest the FFLO state by comparing the results of clean and dirty samples, and also by the dependence on the field inclination from the exactly parallel direction [48].

The above two compounds can be continuously connected by replacing magnetic Fe atoms with nonmagnetic Ga atoms. Uji et al. have examined the compounds λ-(BETS)$_2$Fe$_x$Ga$_{1-x}$Cl$_4$ systematically and obtained a global phase diagram [54]. They have compared the experimental phase diagrams of FISC and those based on Jaccarino–Peter–Fisher theory [47], which does not take into account the FFLO state. For every value of x, there are areas of superconductivity that cannot be covered only by the BCS state with $\mathbf{q} = 0$. As mentioned earlier, the FFLO state seems to be most plausible explanation of those areas.

It is worthwhile to compare λ-(BETS)$_2$FeCl$_4$ with κ-(BETS)$_2$FeBr$_4$, in which both FISC and low field superconductivity have been observed by Konoike et al. [55]. The phase diagram of FISC at high fields is well reproduced by the Jaccarino–Peter–Fisher theory, which does not take into account the FFLO state [55], in contrast to λ-(BETS)$_2$FeCl$_4$. The absence of the FFLO state in κ-(BETS)$_2$FeBr$_4$ at high fields is considered to be due to rather strong orbital pair-breaking effect, whose presence can be verified if we consider that the maximum transition temperature of the FISC ($T_\mathrm{c} \approx 0.6$–$0.7\,\mathrm{K}$ at $H \approx 13\,\mathrm{T}$) is much smaller than the zero field transition temperature

($T_c^{(0)} \approx 1.4\,\mathrm{K}$). This is confirmed more quantitatively by a theory explained later [56].

The upper critical field of low field superconductivity is not reproduced by Jaccarino–Peter–Fisher theory, which does not take into account the coexisting antiferromagnetic long-range order at low fields. We have extended the Jaccarrino–Peter mechanism to antiferromagnetic superconductors [57, 58], and have applied the theory to κ-$(BETS)_2FeBr_4$ [56]. We have determined model parameters from the curves of the critical fields of FISC and shown that the curve of upper critical fields of low field superconductivity in the canted antiferromagnetic phase is well reproduced with those parameters and without any additional fitting parameters. It was also shown that in contrast to FISC, low field superconductivity can exhibit the FFLO state.

25.5 Other Exotic Superconductors

It is worthwhile reviewing studies on the FFLO state in other exotic superconductors for future researches of organic superconductors. Exotic superconductors, such as heavy fermion and cuprate superconductors, can be candidates for FFLO superconductors, as well as organic superconductors, as explained in Sect. 25.1.3 [13–17].

In particular, Q2D heavy fermion superconductors are good candidates for FFLO superconductors, because the orbital effect is suppressed due to both the large effective mass and the low-dimensionality, and, besides, the low-dimensionality is advantageous for the occurrence of the FFLO state due to the nesting effect as explained in Sect. 25.2. The compound $CeCoIn_5$ has large effective mass [59] and Q2D Fermi surfaces [60]. Izawa et al. found that the upper critical field at low temperatures was of first order, which indicates strongly Pauli limited superconductivity [61]. Murphy et al. observed the upper critical field which depended strongly on the orientation of field, and argue that it is consistent with the FFLO state [62].

Bianchi et al. [63] and Radovan et al. [64] have obtained the phase diagram including the lower critical field by specific heat measurements. They found that along the upper critical field curve the transition changed from second order at high temperatures to first order at low temperatures, and the transition at the lower critical field is of second order. They argue that the new superconducting phase between the upper and lower critical fields is possibly the FFLO state. The phase diagram and quasi-particle density of states have been examined theoretically [65, 66].

Following these discoveries, much experimental evidence of the FFLO state has been obtained. The results of ultrasonic measurements suggest a softening of the vortex lattice due to the node perpendicular to the vortex line due to the FFLO state [67]. NMR data suggest that the new superconducting phase consists of the BCS state and the normal state, which is consistent with the FFLO state [68]. Thermal conductivity [69] and penetration depth [70] measurements have also been made and their results support the FFLO state.

25.6 Conclusion and Future Prospects

We have reviewed some theoretical predictions of characteristic behavior of the FFLO state and observations in organic superconductors, such as the upturn and the sensitivity to the misalignment of the magnetic field and impurities of the upper critical field and the existence of the lower critical field with a tricritical point near $T \sim 0.56T_c^{(0)}$. Based on these experimental and theoretical facts, it seems plausible that the FFLO state occurs in some organic superconductors, although the possibility of explanations other than the FFLO state cannot be completely excluded.

The theoretical predictions follow from a straightforward extension of BCS mean field theory to include the possibility of $\mathbf{q} \neq 0$. At present, there is no indication that the candidate organic superconductors are not in the range where the FFLO state may occur. It may be considered a priori that the fluctuations characteristic to the FFLO state may make mean field theory inappropriate, but to the best of our knowledge no theory shows that the FFLO state is unstable. The effects of low dimensional fluctuations have been examined in a phenomenological Ginzburg–Landau model [71], showing that the FFLO state is stable to the same extent as the BCS state unless the system is rotationally symmetric. Ohashi has confirmed a part of the results on the basis of a microscopic model [72].

In future researches, ultrasonic and NMR measurements of candidate organic superconductors, like those performed in $CeCoIn_5$, may help identify the FFLO state. Yang and Agterberg have proposed an experiment using the Josephson effect to detect the existence of the FFLO state [73]. They argue that the Josephson current in a Josephson junction between an FFLO state and the BCS state is suppressed, but may be recovered by applying a magnetic field in the junction. Bulaevskii et al. have proposed an experiment to detect the intrinsic pinning of vortices due to the nonuniformity of the FFLO state [74]. They have shown that the interlayer critical current and the conductivity have peaks when the magnetic field is perpendicular to the in-plane wave vector of the FFLO state and when the period of the Josephson vortex lattice induced by the magnetic field is commensurate with the FFLO period. It is also of importance to determine the spatial structure of the order parameter directly by scanning tunneling microscope (STM) measurements. Neutron scattering experiments may provide information on the spatial structure. These measurements, which are sensitive to the spatial structure and the phase of the order parameter, may give solid evidence of the existence of the FFLO state. Specific heat measurements may also give solid evidence of the transition between the FFLO state and the BCS state with $\mathbf{q} = 0$ at the lower critical fields. To perform these experiments, it is important to synthesize large and clean samples of organic superconductor.

Acknowledgements

The author wishes to thank U. Klein and Y. Matsuda for useful discussions.

References

1. P. Fulde, R.A. Ferrell, Phys. Rev. **135**, A550 (1964)
2. A.I. Larkin, Yu.N. Ovchinnikov, Zh. Eksp. Teor. Fiz. **47**, 1136 (1964); [translation: Sov. Phys. JETP **20**, 762 (1965)]
3. For review, see R. Casalbuoni, G. Narduli, Rev. Mod. Phys. **76**, 263 (2004); Y. Matsuda, H. Shimahara, J. Phys. Soc. Jpn. **76**, 051005 (2007)
4. L.W. Gruenberg, L. Gunther, Phys. Rev. Lett. **16**, 996 (1966)
5. H. Shimahara, J. Phys. Soc. Jpn. **67**, 736 (1998)
6. H. Shimahara, Czech. J. Phys. **46**(Suppl. 2), 561 (1996), (Proceedings of the 21st International Conference on Low Temperature Physics, Prague, August 8–14, 1996)
7. C. Mora, R. Combescot, Europhys. Lett. **66**, 833 (2004); R. Combescot, G. Tonini, Phys. Rev. B **72**, 094513 (2005)
8. C. Mora, R. Combescot, Phys. Rev. B **71**, 214504 (2005)
9. H. Shimahara, D. Rainer, J. Phys. Soc. Jpn. **66**, 3591 (1997)
10. U. Klein, D. Rainer, H. Shimahara, J. Low Temp. Phys. **118**, 91 (2000)
11. U. Klein, Phys. Rev. B **69**, 134518 (2004)
12. S. Takada, Prog. Theor. Phys. **43**, 27 (1970); see also S. Takada, T. Izuyama, Prog. Theor. Phys. **41**, 635 (1969)
13. K. Gloos, R. Modler, H. Schimanski, C.D. Bredl, C. Geibel, F. Steglich, A.I. Buzdin, N. Sato, T. Komatsubara, Phys. Rev. Lett. **70**, 501 (1993)
14. G. Yin, K. Maki, Phys. Rev. B **48**, 650 (1993)
15. H. Burkhardt, D. Rainer, Ann. Physik **3**, 181 (1994)
16. H. Shimahara, Phys. Rev. B **50**, 12760 (1994)
17. M. Tachiki, S. Takahashi, P. Gegenwart, M. Weiden, M. Lang, C. Geibel, F. Steglich, R. Modler, C. Paulsen, Y. Ōnuki, Z. Phys. B **100**, 369 (1996)
18. S. Matsuo, S. Higashitani, Y. Nagato, K. Nagai, J. Phys. Soc. Jpn. **67**, 280 (1998)
19. K. Machida, H. Nakanishi, Phys. Rev. B **30**, 122 (1984)
20. Y. Suzumura, K. Ishino, Prog. Theor. Phys. **70**, 654 (1983)
21. A.I. Buzdin, V.V. Tugushev, Zh. Eksp. Teor. Fiz. **85**, 73 (1983); [Sov. Phys. JETP **58**, 428 (1983)]
22. L.N. Bulaevskii, Zh. Eksp. Teor. Fiz. **65**, 1278 (1973); [translation: Sov. Phys. JETP **38**, 634 (1974)]
23. K. Aoi, W. Dieterich, P. Fulde, Z. Physik **267**, 223 (1974)
24. H. Shimahara, J. Phys. Soc. Jpn. **66**, 541 (1997)
25. H. Shimahara, J. Phys. Soc. Jpn. **68**, 3069 (1999)
26. H. Shimahara, K. Moriwake, J. Phys. Soc. Jpn. **71**, 1234 (2002); H. Shimahara, S. Hata, Phys. Rev. B **62**, 14541 (2000)

27. K. Maki, H. Won, Czech. J. Phys. **46**, 1035 (1996)
28. K. Yang, S.L. Sondhi, Phys. Rev. B **57**, 8566 (1998)
29. A.G. Lebed, Phys. Rev. B **59** R721 (1999)
30. A.S. Alexandrov, V.N. Zavaritsky, W.Y. Liang, P.L. Nevsky, Phys. Rev. Lett. **76**, 983 (1996); Y.N. Ovchinnikov, V.Z. Kresin, Phys. Rev. B **54**, 1251 (1996); A.A. Abrikosov, Phys. Rev. B **56**, 5112 (1997); V.B. Geshkenbein, L.B. Ioffe, A.J. Millis, Phys. Rev. Lett. **80**, 5778 (1998)
31. A.G. Lebed, Pis'ma Zh. Eksp. Teor. Fiz. **44**, 89 (1986); [JETP Lett. **44**, 114 (1986)]
32. N. Dupuis, G. Montambaux, C.A.R. Sá de Melo, Phys. Rev. Lett. **70**, 2613 (1993)
33. A.G. Lebed, K. Yamaji, Phys. Rev. Lett. **80**, 2697 (1998)
34. A.I. Buzdin, J.P. Brison, Europhys. Lett. **35**, 707 (1996)
35. N. Dupuis, Phys. Rev. B **51**, 9074 (1995); N. Dupuis, G. Montambaux, Phys. Rev. B **49**, 8993 (1994)
36. I.J. Lee, M.J. Naughton, G.M. Danner, P.M. Chaikin, Phys. Rev. Lett. **78**, 3555 (1997)
37. T. Ishiguro, J. Phys. IV France **10**, Pr3-139 (2000)
38. M.-S. Nam, J.A. Symington, J. Singleton, S.J. Blundell, A. Ardavan, J.A.A.J. Perenboom, M. Kurmoo, P. Day, J. Phys. Condens. Matter **11**, L477 (1999)
39. J. Singleton, J.A. Symington, M.-S. Nam, A. Ardavan, M. Kurmoo, P. Day, J. Phys. Condens. Matter **12**, L641 (2000); J.A. Symington, J. Singleton, M.-S. Nam, A. Ardavan, M. Kurmoo, P. Day, Physica B **294–295**, 418 (2001)
40. E. Ohmichi, T. Ishiguro, J. Yamada, H. Anzai, T. Osada, Synth. Met. **133–134**, 245 (2003)
41. E. Ohmichi, T. Ishiguro, T. Sakon, T. Sasaki, M. Motokawa, R.B. Lyubovskii, R.N. Lyubovskaya, J. Supercond. **12**, 505 (1999)
42. S. Manalo, U. Klein, J. Phys. Condens. Matter **12**, L471 (2000)
43. S. Uji, H. Shinagawa, T. Terashima, T. Yakabe, Y. Terai, M. Tokumoto, A. Kobayashi, H. Tanaka, H. Kobayashi, Nature **410**, 908 (2001)
44. L. Balicas, J.S. Brooks, K. Storr, S. Uji, M. Tokumoto, H. Tanaka, H. Kobayashi, A. Kobayashi, V. Barzykin, L.P. Gor'kov, Phys. Rev. Lett. **87**, 067002 (2001)
45. O. Cépas, R.H. McKenzie, J. Merino, Phys. Rev. B **65**, 100502(R) (2002)
46. V. Jaccarino, M. Peter, Phys. Rev. Lett. **9**, 290 (1962)
47. O. Fisher, Helv. Phys. Acta **45**, 331 (1972)
48. M.A. Tanatar, T. Ishiguro, H. Tanaka, H. Kobayashi, Phys. Rev. B **66**, 134503 (2002)
49. H. Shimahara, J. Phys. Soc. Jpn. **71**, 1644 (2002)
50. M. Houzet, A. Buzdin, L. Bulaevskii, M. Maley, Phys. Rev. Lett. **88**, 227001 (2002)
51. H. Shimahara, Phys. Rev. B **62**, 3524 (2000)
52. H. Shimahara, J. Phys. Soc. Jpn. **69**, 1966 (2000)
53. H. Shimahara, M. Kohmoto, Phys. Rev. B **65**, 174502 (2002)
54. S. Uji, T. Terashima, C. Terakura, T. Yakabe, Y. Terai, S. Yasuzuka, Y. Imanaka, M. Tokumoto, A. Kobayashi, F. Sakai, H. Tanaka, H. Kobayashi, L. Balicas, J.S. Brooks, J. Phys. Soc. Jpn. **72**, 369 (2003)
55. T. Konoike, S. Uji, T. Terashima, M. Nishimura, S. Yasuzuka, K. Enomoto, H. Fujiwara B. Zhang, H. Kobayashi, Phys. Rev. B **70**, 094514 (2004)

56. Y. Suginishi, H. Shimahara, Phys. Rev. B **74**, 024518 (2006); Phys. Rev. B **75**, 099902(E) (2007)
57. H. Shimahara, J. Phys. Soc. Jpn. **71**, 713 (2002)
58. H. Shimahara, J. Phys. Soc. Jpn. **73**, 2635 (2004)
59. C. Petrovic, P.G. Pagliuso, M.F. Hundley, R. Movshovich, J.L. Sarrao, J.D. Thompson, Z. Fisk, P. Monthoux, J. Phys. Condens. Matter **13**, L337 (2001)
60. H. Shishido, R. Settai, D. Aoki, S. Ikeda, H. Nakawaki, N. Nakamura, T. Iizuka, Y. Inada, K. Sugiyama, T. Takeuchi, K. Kindo, T.C. Kobayashi, Y. Haga, H. Harima, Y. Aoki, T. Namiki, H. Sato, Y. Ōnuki, J. Phys. Soc. Jpn. **71**, 162 (2002)
61. K. Izawa, H. Yamaguchi, Y. Matsuda, H. Shishido, R. Settai, Y. Onuki, Phys. Rev. Lett. **87**, 057002 (2001)
62. T.P. Murphy, Donavan Hall, E.C. Palm, S.W. Tozer, C. Petrovic, Z. Fisk, R.G. Goodrich, P.G. Pagliuso, J.L. Sarrao, J.D. Thompson, Phys. Rev. B **65**, 100514(R) (2002)
63. A. Bianchi, R. Movshovich, C. Capan, P.G. Pagliuso, J.L. Sarrao, Phys. Rev. Lett. **91**, 187004 (2003)
64. H.A. Radovan, N.A. Fortune, T.P. Murphy, S.T. Hannahs, E.C. Palm, S.W. Tozer, D. Hall, Nature **425**, 51 (2003)
65. H. Adachi, R. Ikeda, Phys. Rev. B **68**, 184510 (2003)
66. H. Won, K. Maki, S. Haas, N. Oeschler, F. Weickert, P. Gegenwart, Phys. Rev. B **69**, 180504(R) (2004)
67. T. Watanabe, Y. Kasahara, K. Izawa, T. Sakakibara, Y. Matsuda, C.J. van der Beek, T. Hanaguri, H. Shishido, R. Settai, Y. Onuki, Phys. Rev. B **70**, 020506 (2004)
68. K. Kakuyanagi, M. Saitoh, K. Kumagai, S. Takashima, M. Nohara, H. Takagi, Y. Matsuda, Phys. Rev. Lett. **94**, 047602 (2005)
69. C. Capan, A. Bianchi, R. Movshovich, A.D. Christianson, A. Malinowski, M.F. Hundley, A. Lacerda, P.G. Pagliuso, J.L. Sarrao, Phys. Rev. B **70**, 134513 (2004)
70. C. Martin, C.C. Agosta, S.W. Tozer, H.A. Radovan, E.C. Palm, T.P. Murphy, J.L. Sarrao, Phys. Rev. B **71**, 020503(R) (2005)
71. H. Shimahara, J. Phys. Soc. Jpn. **67**, 1872 (1998); Physica B **259–261**, 492 (1999)
72. Y. Ohashi, J. Phys. Soc. Jpn. **71**, 2625 (2002)
73. K. Yang, D.F. Agterberg, Phys. Rev. Lett. **84**, 4970 (2000)
74. L. Bulaevskii, A. Buzdin, M. Maley, Phys. Rev. Lett. **90**, 067003 (2003)

Part VI

Electron Correlations in Organic Conductors

26

SO(4) Symmetry in Bechgaard Salts

D. Podolsky, E. Altman, and E. Demler

Recent experiments give compelling evidence that superconductivity in Bechgaard salts involves spin-triplet Cooper pairing. The proximity of triplet superconductivity and antiferromagnetism in these compounds prompts us to study the interplay between the two orders in quasi-one-dimensional electron systems. We find that the two orders are unified in a natural way through an SO(4) symmetry group, and we show that SO(4) is a good approximate symmetry of the system near the phase transition between the two orders, without the need for fine-tuning of microscopic parameters. In this chapter we study the experimental consequences of SO(4) symmetry, including predictions for the phase diagram and for the low-energy excitation spectrum of Bechgaard salts.

26.1 Competing Orders in Strongly Correlated Electron Systems: Emergence of Higher Symmetries

Symmetry is a powerful principle in elucidating the properties of a complex system. The symmetry group of a quantum mechanical Hamiltonian contains information regarding the degeneracy of states, and provides a scheme by which to organize the spectrum of excitations. In situations where the symmetry of the Hamiltonian is spontaneously broken with the onset of long range order, the pattern of symmetry breaking determines the spectrum of gapless Goldstone modes uniquely [1].

Although the symmetry of the Hamiltonian is explicit in most cases, there are situations in strongly correlated electron systems where, for special values of parameters, the Hamiltonian displays a higher symmetry than manifest. This typically occurs at the phase boundary between two seemingly unrelated orders, where the extra symmetry generators rotate the degenerate order parameters into one another. The addition of new Goldstone modes leads to a suppression of the critical temperature. In addition, the enhanced symmetry group constrains the form of the Ginzburg–Landau free energy for the competing orders, leading to strong constrains on the topology of the phase diagram near the phase transition.

These ideas are illustrated by the Hubbard model on a bipartite lattice [2,3],

$$\mathcal{H} = -t \sum_{\langle ij \rangle \sigma} c^{\dagger}_{i\sigma} c_{j\sigma} + U \sum_{i} \left(n_{i\uparrow} - \frac{1}{2} \right) \left(n_{i\downarrow} - \frac{1}{2} \right) - \mu \sum_{i\sigma} n_{i\sigma}. \quad (26.1)$$

The model is invariant under spin SO(3) rotations, $[\mathcal{H}, S^{\alpha}] = 0$, generated by the total spin operators, $S_{\alpha} = \frac{1}{2} \sum_i c^{\dagger}_{is} \sigma^{\alpha}_{ss'} c_{is'}$, and under the SO(2) group of phase rotations, $[\mathcal{H}, Q] = 0$, which are generated by the total charge $Q = \frac{1}{2} \sum_{i\sigma} (n_{i\sigma} - \frac{1}{2})$. Exactly at half filling ($\mu = 0$), the charge group is enlarged by the inclusion of two new generators, η^{-} and $\eta^{+} \equiv (\eta^{-})^{\dagger}$, where $\eta^{-} = \sum_i (-1)^i c_{i\downarrow} c_{i\uparrow}$, as can be verified through explicit calculation,

$$[\mathcal{H}, \eta^{\pm}] = \mp 2\mu \eta^{\pm}. \quad (26.2)$$

The "pseudospin" operators $\eta_x = \frac{1}{2}(\eta^{+} + \eta^{-})$, $\eta_y = \frac{1}{2i}(\eta^{+} - \eta^{-})$, and $\eta_z = Q$ satisfy an SO(3) algebra. The η are spin-singlet operators, and therefore commute with the S^{α}, $[S^{\alpha}, \eta^{a}] = 0$. Hence, the total symmetry group is $\mathrm{SO}(4) \approx \mathrm{SO}(3)_{\mathrm{spin}} \times \mathrm{SO}(3)_{\mathrm{pseudospin}}$.

In the negative-U model, electrons in the ground state like to form on-site spin-singlet pairs. These pairs may condense to form a singlet superconductor, or they may instead prefer to arrange themselves into a checkerboard of CDW. In fact, these two types of order are connected by the pseudospin group,

$$[\eta^{a}, \Delta^{b}] = \mathrm{i}\epsilon^{abc} \Delta^{c}, \quad (26.3)$$

where $\Delta^{x} = \mathrm{Re}(\sum_i c_{i\downarrow} c_{i\uparrow})$, $\Delta^{y} = \mathrm{Im}(\sum_i c_{i\uparrow} c_{i\downarrow})$, and $\Delta^{z} = \frac{1}{2} \sum_{i\sigma} (-1)^i n_{i\sigma}$. Thus, by symmetry, singlet superconductivity ($\Delta_{x,y}$) and checkerboard CDW (Δ_z) are actually degenerate at half filling.

Since the two types of order are degenerate, it is possible to construct a new class of low energy states by alternating regions with local SC and CDW orders that slowly twist into one another. For short-ranged interactions, the energy of such states is made arbitrarily small by making the twist arbitrarily slow. Thus, enhanced symmetry leads to gapless Goldstone modes associated with the pseudospin operators, η, which generate the rotations between the two orders. In the quantum system, such low energy modes are created by acting with η on the ground state, as shown in (26.2). Away from $\mu = 0$, the excited state is a massive pseudo-Goldstone mode whose gap softens as the SO(4) symmetric point $\mu = 0$ is approached to yield a true Goldstone mode.

The dynamics of these low energy modes can also be studied by a quantum rotor model [4], which is obtained by coarse-graining the lattice into small clusters, and then projecting onto the low energy Hilbert space on each cluster. By symmetry, the remaining degrees of freedom must decompose into well-defined representations of SO(4). In this case, they are mapped into a local three-dimensional vector order parameter $\boldsymbol{\Delta}_i$ on cluster i, subject to the local constraint $|\boldsymbol{\Delta}_i|^2 = 1$. The quantum rotor model then consists of a kinetic energy term for each rotor, in addition to a ferromagnetic coupling for rotors on adjacent clusters.

The presence of new low energy modes leads to enhanced fluctuations of the order parameters, and therefore to a reduction in critical temperature T_c. For instance, in two dimensions, it is possible to have a finite temperature phase transition into a long-range checkerboard CDW (which is a discrete order parameter), or into a quasi-long-range-ordered superconductor. However, the negative-U Hubbard model at half filling has an SO(3)-symmetric order parameter, which, by the Mermin–Wagner theorem, cannot be spontaneously broken at finite temperature. Hence, enhanced fluctuations suppress T_c all the way to zero in this case. As we will see below, a transition can change from second order to first order near a point of enhanced symmetry due to increased fluctuations.

The Hubbard model (26.1) provides an example of enhanced symmetry where all generators commute exactly with the microscopic Hamiltonian. However, this model is highly fine tuned. In particular, inclusion of a more general hopping or interaction term, or a chemical potential generically destroys the symmetry. Below, we will argue that quasi-one-dimensional Bechgaard salts have an approximate SO(4) symmetry at the transition between AF and TSC phases. This comes from an exact microscopic symmetry of Luttinger liquids, *without fine tuning of parameters*. Hopping between the chains breaks the symmetry, but only weakly. This SO(4) symmetry differs from the pseudospin of the Hubbard model in the generators of the group and the order parameters that they act upon.

26.2 SO(4) Symmetry in Quasi-One-Dimensional Systems

26.2.1 Order Parameters and Generators at Half filling

At half filling, AF order is described by a real Néel order parameter $\boldsymbol{N}$. Hence, the Néel and triplet superconducting order parameters are

$$
\begin{aligned}
N_\alpha &= \frac{1}{2}\sum_{kss'}\left(a^\dagger_{+,ks}\sigma^\alpha_{ss'}a_{-,ks'} + a^\dagger_{-,ks}\sigma^\alpha_{ss'}a_{+,ks'}\right), \\
\Psi^\dagger_\alpha &= \frac{1}{i}\sum_{kss'} a^\dagger_{+,ks}(\sigma_2\sigma^\alpha)_{ss'}a^\dagger_{-,-ks'},
\end{aligned}
\tag{26.4}
$$

where $a^\dagger_{\pm,ks}$ creates right/left moving electrons of momentum $\pm k_f + k$ and spin s. Here, we assume the orbital component of a triplet superconducting order parameter $\boldsymbol{\Psi} \propto p_x$, where x is the intrachain direction.

The total spin and charge are

$$S_\alpha = \frac{1}{2} \sum_{r,kss'} a^\dagger_{r,ks} \sigma^\alpha_{ss'} a_{r,ks'},$$

$$Q = \frac{1}{2} \sum_{ks} \left(a^\dagger_{+,ks} a_{+,ks} + a^\dagger_{-,ks} a_{-,ks} - 1 \right). \tag{26.5}$$

The spin operators S_α form an $\mathrm{SO(3)_{spin}}$ algebra of spin rotations, $[S_a, S_b] = \mathrm{i}\epsilon^{abc} S_c$, under which N_α and Ψ_α transform as vectors. On the other hand, the charge Q rotates the real and imaginary parts of Ψ_α into one another, while maintaining the real vector N_α invariant. What is missing for a unified description of AF and TSC orders is an operator $\Theta^\dagger$, which rotates $\boldsymbol{\Psi}$ into $\boldsymbol{N}$. Examination of the order parameters shows that such an operator must have momentum $\pm 2k_f$, charge ± 2, and spin 0,

$$\Theta^\dagger = \sum_k \left(a^\dagger_{+,k\uparrow} a^\dagger_{+,-k\downarrow} - a^\dagger_{-,k\uparrow} a^\dagger_{-,-k\downarrow} \right). \tag{26.6}$$

The operators Θ and $\Theta^\dagger$ are combined with the charge to define the "isospin" generators, $I_x = \frac{1}{2}(\Theta^\dagger + \Theta)$, $I_y = \frac{1}{2\mathrm{i}}(\Theta^\dagger - \Theta)$, $I_z = Q$, which satisfy an isospin $\mathrm{SO(3)_{iso}}$ algebra $[I_a, I_b] = \mathrm{i}\epsilon^{abc} I_c$. These spin and isospin algebras are independent of one another, $[I_a, S_\alpha] = 0$, so that the total group defined by these generators is $\mathrm{SO(4)} \approx \mathrm{SO(3)_{spin}} \times \mathrm{SO(3)_{isospin}}$. The vectors $\boldsymbol{\Psi}$ and $\boldsymbol{N}$ are then combined naturally into a single tensor

$$\hat{Q} = \begin{pmatrix} \mathrm{Re}\, \Psi_x & \mathrm{Im}\, \Psi_x & N_x \\ \mathrm{Re}\, \Psi_y & \mathrm{Im}\, \Psi_y & N_y \\ \mathrm{Re}\, \Psi_z & \mathrm{Im}\, \Psi_z & N_z \end{pmatrix}. \tag{26.7}$$

The rows (columns) of $\hat{Q}$ transform as a vector under the spin (isospin) SO(3) algebra, $[S_\alpha, Q_{b\beta}] = \mathrm{i}\epsilon^{\alpha\beta\gamma} Q_{b\gamma}$ ($[I_a, Q_{b\beta}] = \mathrm{i}\epsilon^{abc} Q_{c\beta}$). Hence $\hat{Q}$ transforms in the (1,1) representation of the SO(4) algebra.

On general grounds, the total charge Q is a good quantum number of any closed system. Furthermore, for systems without magnetic fields or spin–orbit coupling, the total spin S_α also commute with the Hamiltonian. Microwave absorption experiments in $\mathrm{(TMTSF)_2AsF_6}$ measured the anisotropy in the exchange couplings to be 10^{-6} [5], which is unlikely to play a significant role in the competition between AF and TSC phases. Hence, we conclude that both Q and S_α are good symmetries of the Hamiltonian everywhere in the phase diagram. Below, we will show that at the transition between AF and TSC phases, the isospin generators $\Theta^{(\dagger)}$ are also good symmetries of the system.

26.2.2 Symmetries of a Luttinger Liquid

Incommensurate Case: $\mathrm{SO(3)_{spin}} \times \mathrm{SO(4)_{isospin}}$ Symmetry

For now, we assume that the system has incommensurate filling, so that Umklapp processes are not allowed. At incommensurate filling, there is no mechanism to pin a SDW in a clean Luttinger liquid. Hence, the SDW order parameter is a complex vector,

$$\Phi_\alpha = \sum_{kss'} a^\dagger_{+,ks} \sigma^\alpha_{ss'} a_{-,ks'} . \tag{26.8}$$

Hence, there are four vector order parameters at incommensurate filling, $\mathrm{Re}\,\boldsymbol{\Phi}(\equiv \boldsymbol{N})$, $\mathrm{Im}\,\boldsymbol{\Phi}$, $\mathrm{Re}\,\boldsymbol{\Psi}$, and $\mathrm{Im}\,\boldsymbol{\Psi}$. To connect all four, we need to enlarge the charge $\mathrm{SO}(3)_{\mathrm{isospin}}$ group defined by Q, Θ, and $\Theta^\dagger$ into an $\mathrm{SO}(4)_{\mathrm{isospin}}$ group, which has the additional generators Λ_z, Λ, and $\Lambda^\dagger$:

$$\Lambda_z = \frac{1}{2} \sum_{ks} \left(a^\dagger_{+,ks} a_{+,ks} - a^\dagger_{-,ks} a_{-,ks} \right),$$

$$\Lambda^\dagger = \sum_k \left(a^\dagger_{+,k\uparrow} a^\dagger_{+,-k\downarrow} + a^\dagger_{-,k\uparrow} a^\dagger_{-,-k\downarrow} \right).$$

Note that Λ_z is the difference between the charge carried by the right and left movers, which is a symmetry of the Luttinger Hamiltonian due to the absence of Umklapp.

The phase diagram for interacting electrons in one dimension was obtained from bosonization and renormalization group analyses has been discussed extensively (see, e.g., [6,7]). This system has a phase boundary between SDW and TSC phases when the Luttinger parameter in the charge sector, K_ρ, equals 1. Along the entire phase boundary $\mathrm{SO}(3)_{\mathrm{spin}} \times \mathrm{SO}(4)_{\mathrm{isospin}}$ is an exact symmetry of the Luttinger Hamiltonian, as shown explicitly in [8]. Here we argue for the existence of such symmetry by analogy with the spin sector.

In the absence of backward scattering ($g_1 = 0$), the Luttinger parameter in the spin sector K_σ equals 1, and the spin Hamiltonian does not have a sine-Gordon term. Thus, the spin Hamiltonian for $g_1 = 0$ has the same form as the charge Hamiltonian at the TSC/AF boundary ($K_\rho = 1$) and incommensurate filling (no Umklapp). When $g_1 = 0$, the spin of right and left movers is conserved separately, so that the full spin-symmetry in this case is $\mathrm{SO}(3) \times \mathrm{SO}(3) \approx \mathrm{SO}(4)_{\mathrm{spin}}$. Thus, we conclude that the charge sector at $K_\rho = 1$ and in the absence of Umklapp must have an $\mathrm{SO}(4)_{\mathrm{isospin}}$ symmetry. In fact, the generators Q, Θ, and Λ can be obtained by bosonizing the spin operators of right and left movers, converting spin variables into charge variables, and refermionizing.

Half Filling: SO(4) Symmetry

In Bechgaard salts $(\mathrm{TMTSF})_2\mathrm{X}$, three out of every four conduction states are occupied. At quarter filling, Umklapp processes involving interactions of four electrons are allowed. Such interactions are weak, and furthermore are irrelevant in the RG sense for $K_\rho > 1/4$ (whereas we are interested in the regime near the SDW/TSC boundary, where $K_\rho \approx 1$) [9]. On the other hand, due to structural dimerization in Bechgaard salts [10], a gap splits the conduction band into a completely filled lower band and a half-filled upper band. Hence, Bechgaard salts are half-filled systems, and the Hamiltonian is modified to include two-electron Umklapp scattering g_3. The value of g_3 is small, since it is proportional to the dimerization, which is less than 1% in $(TMTSF)_2PF_6$ [10].

Analysis of the phase diagram of Luttinger liquids at half filling reveals that there is still a direct transition between AF and TSC orders at $K_\rho = 1$. The Umklapp term allows scattering of two right moving electrons into two left moving ones, and vice versa. Thus, it does not commute with the operator $\Lambda_z = Q_+ - Q_-$, which leads to breaking of the $\mathrm{SO}(4)_{\mathrm{isospin}}$ symmetry. We can understand this effect at the level of the Ginsburg–Landau (GL) free energy (which is discussed in more detail in the following section). To linear order in g_3, Umklapp modifies the quadratic part of the GL free energy

$$\Delta F = \frac{g_3}{2L}\left((\mathrm{Re}\,\boldsymbol{\Phi})^2 - (\mathrm{Im}\,\boldsymbol{\Phi})^2\right), \tag{26.9}$$

while keeping all quartic terms invariant. The new term pins the SDW and breaks the $\mathrm{SO}(4)_{\mathrm{isospin}}$ symmetry down to the subgroup $\mathrm{SO}(3)_{\mathrm{isospin}}$, which is generated by Q and $\Theta^{(\dagger)}$, and discussed in Sect. 26.2.1. The three remaining vector order parameters, $\boldsymbol{N}$, $\mathrm{Re}\,\boldsymbol{\Psi}$, and $\mathrm{Im}\,\boldsymbol{\Psi}$, are assembled into a tensor order parameter $\hat{Q}$ (26.7).

Unlike the $\mathrm{SO}(3) \times \mathrm{SO}(4)$ symmetry discussed in the incommensurate case, the SO(4) symmetry at half filling is not a rigorous symmetry of the system. The generators of this group do not commute with the Hamiltonian of the system exactly. However, the main emphasis of our work is to understand the finite temperature phase diagram of $(TMTSF)_2PF_6$. This is obtained from the classical GL free energy, which at the AF/TSC phase boundary has SO(4) symmetry if we retain Umklapp processes to linear order in g_3. In addition, with regards to quantum properties, SO(4) symmetry is a good starting point to study the collective modes of the system when g_3 is small. For small g_3, modes found assuming SO(4) symmetry will have a finite overlap with the actual excitations of the system. In particular, the quantum numbers of the Θ mode discussed in Sect. 26.2.1, including charge two and center of mass momentum $2k_f$, are not affected by Umklapp. These properties determine which experimental probes couple to Θ. We must keep in mind, however, that the explicit breaking of SO(4) due to higher order corrections in g_3, and also due to interchain coupling, may give a small energy gap and finite broadening to Θ, even at the AF/TSC phase boundary.

26.3 Competition of Spin–Density Wave Order and Triplet Superconductivity in Bechgaard Salts

In this section we investigate the consequences of SO(4) symmetry for true finite temperature phase transitions, when three-dimensional fluctuations of the order parameter are important. One may be concerned that SO(4) symmetry is immediately destroyed by interchain coupling. As shown in Ref. [8], even in the case where the interchain coupling t_b is large enough to make the system into a highly anisotropic Fermi liquid, approximate SO(4) symmetry prevails in the Ginzburg–Landau (GL) free energy. This implies that our analysis of the phase diagram, based on classical SO(4) symmetry, is valid even when the normal state is described by a highly anisotropic Fermi liquid rather than a collection of weakly coupled Luttinger liquids.

The GL free energy for competing AF and TSC orders near the AF/TSC phase transition is strongly constrained by SO(4) symmetry

$$\begin{aligned} F = \frac{1}{2}\left(\nabla Q_{a\alpha}\right)^2 + \bar{r}Q_{a\alpha}^2 + \delta r(Q_{z,\alpha}^2 - Q_{x,\alpha}^2 - Q_{y,\alpha}^2) \\ + \tilde{u}_1 Q_{a\alpha}^2 Q_{b\beta}^2 + \tilde{u}_2 Q_{a\alpha} Q_{a\beta} Q_{b\alpha} Q_{b\beta} . \end{aligned} \quad (26.10)$$

We assume that temperature and pressure control the quadratic coefficients $\bar{r}$ and δr. When $\delta r = 0$ the model has full SO(4) symmetry. Away from this line it only has spin and charge SO(3) × SO(2) symmetry. The quartic terms in (26.10) are the only ones allowed by SO(4) symmetry. We also note that a derivation of the GL energy for weakly interacting quasi-one-dimensional electrons following the usual approach yields the model in (26.10) with $\tilde{u}_1 = 21\zeta(3)/16\pi^2 v_f T^2$ and $\tilde{u}_2 = -7\zeta(3)/8\pi^2 v_f T^2$ [8].

The sign of $\tilde{u}_2$ in model (26.10) determines whether the triplet superconductor is unitary ($\mathrm{Re}\,\boldsymbol{\psi} \propto \mathrm{Im}\,\boldsymbol{\psi}$) or nonunitary ($\mathrm{Re}\,\boldsymbol{\psi} \times \mathrm{Im}\,\boldsymbol{\psi} \neq 0$). The unitary case, $\tilde{u}_2 < 0$, is of experimental relevance to $(TMTSF)_2PF_6$, and in the remainder we concentrate exclusively on this case. The mean field diagram is then composed of an AF phase separated from a TSC phase by a first-order phase transition, and a disordered (normal) phase separated from the two other phases by second-order lines, see Fig. 26.1

An analysis of the $d = 4 - \epsilon$ renormalization group (RG) is done to understand the role of thermal fluctuations in model (26.10), and in slightly perturbed models where the quartic coefficients do not lie exactly on the SO(4) symmetric manifold. We find that the RG equations have only two fixed points: a trivial Gaussian fixed point $\bar{r} = \delta r = \tilde{u}_1 = \tilde{u}_2 = 0$ and

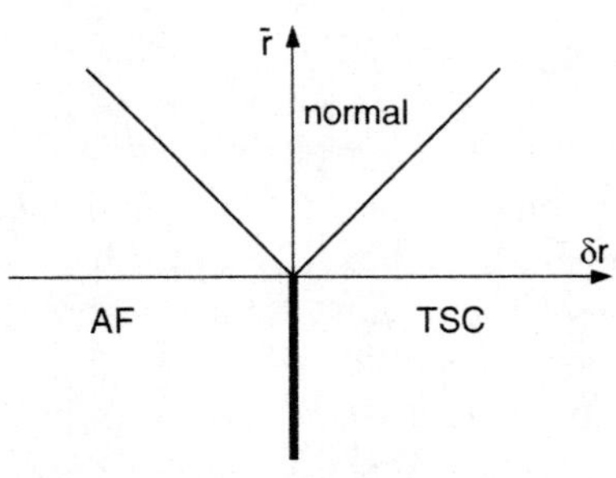

Fig. 26.1. Mean field phase diagram of (26.10) in the unitary case $\tilde{u}_2 < 0$. There is a first-order transition (*thick line*) between AF and TSC phases

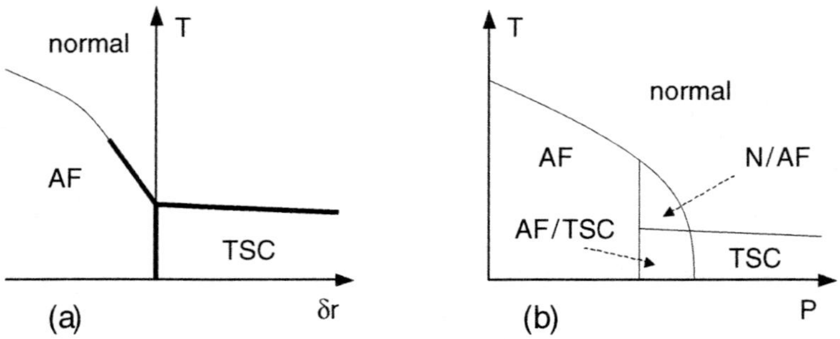

Fig. 26.2. (**a**) Phase diagram of model (26.10) in the large N limit. *Thick lines* represent first-order phase transitions. The direct transition between AF and TSC phases, tuned by δr, is first order even at mean field level for $N = 3$. Enhanced fluctuations due to SO(4) symmetry make the AF/normal transition first order near the AF/TSC phase boundary. (**b**) Temperature–pressure phase diagram of $(TMTSF)_2PF_6$ [14, 15]. N/AF and AF/TSC correspond to coexistence regimes of the appropriate phases

an SO(9) Heisenberg point $\bar{r} \neq 0$, $\delta r = 0$, $\tilde{u}_1 \neq 0$, $\tilde{u}_2 = 0$. All RG flows starting with $\tilde{u}_2 \neq 0$ are runaway flows. This analysis can be generalized to order parameters $\boldsymbol{N}$ and $\boldsymbol{\Psi}$ that are N-component vectors, in which case the SO(4) $\approx$ SO(3) $\times$ SO(3) symmetry becomes SO(3) $\times$ SO(N). However, we find that even in the large N limit, all flows with $\tilde{u}_2 < 0$ are runaway flows, indicating the absence of fixed points with unitary TSC in $4 - \epsilon$ dimensions.

The absence of a fixed point in the RG flow could be due to a fluctuation-induced first-order phase transition, which would preclude a multicritical point in the phase diagram. To inspect this possibility, we study model (26.10) directly in $d = 3$ dimensions in the large N limit. The results are shown in Fig. 26.2a. We find a first-order transition between AF and TSC phases along the SO(4) symmetric line $\delta r = 0$, as predicted by mean field theory. However, the large N results differ from the mean field theory in two important aspects: the transition between normal and TSC phases is first order and, close to the SO(4) symmetric line, fluctuations induce a first order transition between AF and normal phases.

The large-N first order N/TSC transition was first discussed by Bailin et al. in the context of ^{3}He [11]. For $N = 3$, the order of this transition remains an open question [12,13]. However, for large N the N/TSC transition remains first order arbitrarily far from the SO(4) symmetric point $\delta r = 0$, and is therefore not a consequence of SO(4) symmetry. On the other hand, the AF/N transition is first order only close to the SO(4) symmetric line, and is a direct consequence of the enhanced fluctuations due to SO(4) symmetry.

Assuming that the experimentally controlled pressure changes an extensive variable conjugate to δr, such as the volume of the system, the first-order transition broadens into a coexistence region of TSC and AF. This is consistent with the experimental phase diagram of $(TMTSF)_2PF_6$ shown in Fig. 26.2b.

26.4 Collective Modes

The collective modes near the AF/TSC boundary give possible direct experimental tests of SO(4) symmetry. We introduce a quantum rotor model to study the spectrum of low energy collective excitations in the neighborhood of the SO(4) symmetric point

$$\mathcal{H}_\mathrm{r} = \frac{1}{2\chi_1}\sum_i \boldsymbol{S}_i^2 + \frac{1}{2\chi_2}\sum_i \boldsymbol{I}_i^2 - J\sum_{\langle ij\rangle a\alpha} Q_{i,a\alpha}Q_{j,a\alpha} + \tilde{u}_1 \sum_{iab\alpha\beta} Q_{i,a\alpha}^2 Q_{i,b\beta}^2$$
$$+ \tilde{u}_2 \sum_{iab\alpha\beta} Q_{i,a\alpha}Q_{i,a\beta}Q_{i,b\alpha}Q_{i,b\beta} + \delta r \sum_{i\alpha}(Q_{i,z\alpha}^2 - Q_{i,x\alpha}^2 - Q_{i,y\alpha}^2). \quad (26.11)$$

In this lattice model each site has spin and isospin vector operators $\boldsymbol{S}_i$ and $\boldsymbol{I}_i$, and an SO(4) tensor order parameter $Q_{i,a\alpha}$. They satisfy the commutation relations $[S_{i,\alpha}, S_{j,\beta}] = \mathrm{i}\delta_{ij}\epsilon_{\alpha\beta\gamma}S_{i,\gamma}$, $[I_{i,a}, I_{j,b}] = \mathrm{i}\delta_{ij}\epsilon_{abc}I_{i,c}$, $[S_{i,\alpha}, Q_{j,a\beta}] = \mathrm{i}\delta_{ij}\epsilon_{\alpha\beta\gamma}Q_{j,a\gamma}$, $[I_{i,a}, Q_{j,b\alpha}] = \mathrm{i}\delta_{ij}\epsilon_{abc}Q_{j,c\alpha}$, and $[Q_{i,a\alpha}, Q_{j,b\beta}] = 0$. In (26.11) the unit length constraint of the rigid rotors is replaced by quartic interaction terms $\tilde{u}_1$ and $\tilde{u}_2$. The collective mode spectrum is obtained from the Heisenberg equations of motion for order parameters and symmetry generators. These equations are linearized in the ordered states, the ensuing eigenvalue problem is solved to yield the spectrum of excitations. In addition to the Goldstone (massless) modes due to the exact SO(3)×SO(2) symmetry, there are pseudo-Goldstone (massive) modes due to the approximate SO(4) symmetry. The latter soften as we approach the SO(4) symmetric point $\delta r = 0$.

The symmetry breaking pattern in each ordered phase determines the low energy spectrum, see Fig. 26.3. In the AF phase there is a (degenerate) doublet of massless spin waves, and a doublet of massive isospin waves. Their dispersion about $k = 2k_f$ is $\omega_{\mathrm{AF,S}}(k) = N\sqrt{\frac{J}{2\chi_1}}|\Delta k|$ and

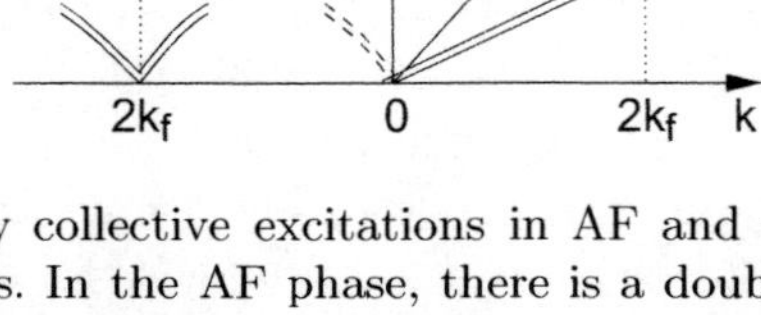

Fig. 26.3. Low energy collective excitations in AF and unitary TSC phases for short range interactions. In the AF phase, there is a doublet of spin waves S and a doublet of massive isospin waves I. Due to translational symmetry breaking, the $2k_f$ modes also appear near $k = 0$ (*dashed lines*). In the TSC phase, there is a phase mode ϕ, a doublet of spin waves S, and a massive Θ mode. The massive modes in both phases soften as the phase transition is approached

$\omega_{\mathrm{AF,I}}(k) = N\sqrt{\frac{4|\delta r|}{\chi_2} + \frac{J}{2\chi_2}(\Delta k)^2}$, where $\Delta k = k - 2k_f$. Of these modes, only the spin waves couple to neutron scattering. In the TSC phase we find a doublet of massless spin waves, a massless phase mode ϕ, and a massive Θ mode. In the presence of long range Coulomb interactions, the ϕ mode is shifted to the plasma frequency. The spin modes are centered at $k = 0$, $\omega_{\mathrm{TSC,S}}(k) = \psi\sqrt{\frac{J}{2\chi_1}}|k|$, whereas the Θ mode has a minimum at $k = 2k_f$, $\omega_{\mathrm{TSC},\Theta}(k) = \psi\sqrt{\frac{4\delta r}{\chi_2} + \frac{J}{2\chi_2}(\Delta k)^2}$.

The intensity of spin polarized neutron scattering is proportional to the dynamic spin structure factor, $\chi''_{zz}(q,\omega) = \sum_n |\langle n|S^z_q|0\rangle|^2 \delta(\omega - \omega_{n0})$ at low temperatures. Here $|0\rangle$ is a ground state and the sum over n runs over all excited states. The contribution of the Θ excitation to the spin structure factor at $\boldsymbol{Q} = (2k_f, \pi, \pi)$ is

$$\chi''_{\Theta}(\boldsymbol{Q},\omega) \propto |\langle 0|\Psi^z|0\rangle|^2 \delta(\omega - \omega_\Theta). \tag{26.12}$$

(due to interchain hopping, the center of mass momentum of the Θ excitation in quasi-one-dimensional systems is $(2k_f, \pi, \pi)$). Hence, the Θ excitation appears as a resonance in inelastic neutron scattering [16–18] and its intensity is proportional to the square of the pairing amplitude.

From (26.6) we observe that the Θ excitation is a collective mode in the particle–particle channel (i.e., it has charge $Q = 2$). Deep in the normal phase it cannot be probed by conventional methods, such as electromagnetic waves or neutron scattering, as these only couple to particle–hole channels (e.g., spin or density). The situation changes when the system becomes superconducting. In the presence of a condensate of Cooper pairs charge is not a good quantum number and particle–particle and particle–hole channels mix. This makes the Θ excitation accessible to neutron scattering experiments in the TSC phase. For quasi-one-dimensional $(TMTSF)_2PF_6$ we expect strong pairing fluctuations even above T_c. Hence, precursors of the Θ resonance should be visible in the normal state, with strong enhancement of the resonant scattering intensity appearing when long range TSC order develops [19].

The most striking feature of the Θ-resonance, which identifies it as a generator of the SO(4) symmetry, is the pressure dependence of the resonance energy inside the TSC phase. When the pressure is reduced and the system is brought closer to the phase boundary with the AF phase, we predict the energy of the Θ-resonance to be dramatically decreased. Mode softening is not expected generically at first-order phase transitions and provides a unique signature of the SO(4) quantum symmetry.

Another possible approach to detecting the Θ excitation involves tunneling experiments with a (singlet superconductor)/$(TMTSF)_2ClO_4$ junction shown in Fig. 26.4 (analogous experiments in the context of π excitations in the high T_c cuprates are discussed in [20]). A singlet superconductor provides a reservoir of Cooper pairs that can couple to Θ pairs in $(TMTSF)_2ClO_4$. One needs to overcome, however, the momentum mismatch between the two

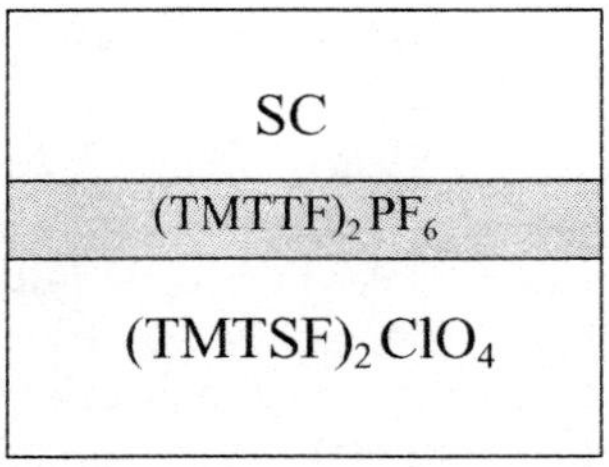

Fig. 26.4. Tunneling experiment for detecting the Θ excitation in $(TMTSF)_2ClO_4$ material. A singlet superconducting material with a higher transition temperature than $(TMTSF)_2ClO_6$ provides a reservoir of Cooper pairs that can couple resonantly to Θ pairs. Momentum mismatch between the Cooper pairs in SC and Θ pairs in $(TMTSF)_2ClO_4$ is compensated by scattering of electrons in a layer of the SP material $(TMTTF)_2PF_6$

types of pairs. This can be overcome by the use of an intermediate layer of the quasi-one-dimensional material $(TMTTF)_2PF_6$. This salt is quarter filled and displays spin-Peierls (SP) order. The modulations of the SP order thus have a periodicity of four TMTTF sites, matching the $(2k_f, \pi, \pi)$ wave vector of $(TMTSF)_2ClO_4$. The small mismatch between the two wave vectors, due to differences in the lattice constant in these compounds, can be compensated by a parallel magnetic field [21]. We expect peaks in the current–voltage characteristics of the junction when the voltage bias compensates the energy difference between Cooper and Θ pairs $2eV = \omega_\Theta$. Peaks in I–V should be present even above the superconducting transition temperature of $(TMTSF)_2ClO_4$ and only require the other material to be superconducting. The choice of $(TMTSF)_2ClO_4$ is made as this material is likely to be close to the AF/TSC transition at ambient pressure [22]. This eliminates the need for pressure cells, which would make the experiments more difficult.

References

1. E. Demler, W. Hanke, S.C. Zhang, Rev. Mod. Phys. **76**, 909 (2004)
2. C.N. Yang, Phys. Rev. Lett. **63**, 2144 (1989)
3. S.C. Zhang, Phys. Rev. Lett. **65**, 120 (1990)
4. E. Demler, S.C. Zhang, N. Bulut, D.J. Scalapino, Int. J. Mod. Phys. B **10**, 2137 (1996)
5. J.B. Torrance, H.J. Pedersen, K. Bechgaard, Phys. Rev. Lett. **49**, 881 (1982)
6. T. Giamarchi, H.J. Schulz, Phys. Rev. B **39**, 4620 (1989)
7. J. Sólyom, Adv. Phys. **28**, 201 (1979)
8. D. Podolsky, E. Altman, T. Rostunov, E. Demler, Phys. Rev. B **70**, 224503 (2004)
9. T. Giamarchi, Phys. B **230–232**, 975 (1997)
10. N. Thorup, G. Rindolf, H. Soling, K. Bechgaard, Acta Cryst. B **37**, 1236 (1981)
11. D. Bailin, A. Love, aM.A. Moore, J. Phys. C **10**, 1159 (1977)

12. D. Podolsky, E. Altman, T. Rostunov, E. Demler, Phys. Rev. Lett. **93**, 246402 (2004)
13. M. de Prato, A. Pelissetto, E. Vicari, eprint `cond-mat/0312362` (2003)
14. A.V. Kornilov, V.M. Pudalov, Y. Kitaoka, et al., Phys. Rev. B **69**, 224404 (2004)
15. T. Vuletic, P. Auban-Senzier, C. Pasquier, et al., Euro. Phys. J. B **25**, 319 (2002)
16. H.F. Fong, B. Keimer, P.W. Anderson, et al., Phys. Rev. Lett. **75**, 316 (1995)
17. H. Mook, M. Yethiraj, G. Aeppli, et al., Phys. Rev. Lett. **70**, 3490 (1993)
18. J. RossatMignod, L.P. Regnault, C. Vettier, et al., Physica (Amsterdam) B **180**, 383 (1992)
19. E. Demler, S.C. Zhang, in *Proceedings of the Conference on High Temperature Superconductivity*, ed. by S. Barnes, J. Ashkenazi, J. Cohn, F. Zou (AIP, 1999)
20. Y.B. Bazaliy, E. Demler, S.C. Zhang, Phys. Rev. Lett. **79**, 1921 (1997)
21. D.J. Scalapino, Phys. Rev. Lett. **24**, 1052 (1970)
22. C. Bourbonnais, F. Creuzet, D. Jérome, et al., J. Phys. (Paris), Lett. **45**, 755 (1984)

27

From Luttinger to Fermi Liquids in Organic Conductors

T. Giamarchi

This chapter reviews the effects of interactions in quasi-one-dimensional systems, such as the Bechgaard and Fabre salts, and in particular the Luttinger liquid physics. It discusses in details how transport measurements both d.c. and a.c. allow to probe such a physics. It also examines the dimensional crossover and deconfinement transition occurring between the one-dimensional case and the higher-dimensional one resulting from the hopping of electrons between chains in the quasi-one-dimensional structure.

27.1 Introduction

Organic conductors, such as TMTTF and TMTSF compounds offer unique challenges. Indeed from the theoretical point of view most of our understanding of interacting electronic problems is based on Landau's Fermi liquid (FL) theory [1–3]. However it is well known that the effects of interactions can be greatly enhanced by reduced dimensionality. In one dimension, interactions destroy the Fermi liquid and lead to a quite different state known as a Luttinger liquid (LL) [4]. For commensurate systems such as the organic conductors, interactions can also lead to a Mott insulating (MI) state, another state showing clearly the effects of strong correlations.

The organic conductors are thus natural candidates to search for the existence and properties of such states. However because of their three-dimensional nature, they provide not a single one-dimensional electron gas, but a very large number of such one-dimensional systems coupled together. This allows for a unique new physics to emerge where the system is able to crossover from a one-dimensional behavior to a more conventional three-dimensional one [5]. This richness is also a drawback, since it is now an important issue to know whether MI or LL physics can be realized at all in these systems. Although for the members of the TMTTF family it was soon undisputable that they are indeed Mott insulators and that interaction effects were important [6], the situation was far from being clear for the TMTSF compounds that were behaving as good metals with many characteristics of a nice Fermi liquid. Despite important efforts no convincing experimental case could be made for LL behavior and the nature of the normal phase and of the crossover scale between a one-dimensional and higher-dimensional behavior of these compounds remained hotly debated. Understanding the nature of the normal phase and the effect of interactions in these compounds, in addition of being important in its own right, was of course potentially crucial in connection with the low temperature ordered phase and in particular of the superconducting one.

Theoretical progress in computing the transport properties [7, 8] and corresponding experimental measurements of the optical conductivity [9, 10] allowed for a solution of this dilemma and proved the LL properties of the above-mentioned organic conductors. In addition, this led to a definite reexamination [5, 8] of what was commonly believed as the main reason [11] for the Mott insulating nature of the parent compounds, namely the dimerization of the organic chain. It also allowed to clearly determine the crossover scale between the one- and higher-dimensional behavior in these systems, leading to a very consistent understanding of the physics of these materials as well as the one of quarter-filled compounds [6, 12, 13].

Thus, in this chapter, I will focus on the issue of the transport in quasi-one-dimensional organic conductors and how it can be used to probe for the MI and LL physics, and more generally on the question of dimensional crossover and deconfinement between the low-dimensional Luttinger liquid or Mott insulator and a more conventional high-dimensional metal (HDM). I mostly concentrate here the specific applications to the organics and refer the reader to the review of C. Bourbonnais and D. Jerome [6] for a general introduction, specific experimental data and references on the quasi-one-dimensional organics and to previous literature for further details on the derivations [4, 7, 8], further theoretical issues [5, 8] and references.

The plan of this chapter is as follows. In Sect. 27.2, I review the basic questions and concepts for a system of coupled one-dimensional chains. In Sect. 27.3, I discuss the transport properties of isolated chains and how one can use them to probe for MI and LL physics, as well as various characteristics of the interactions in these systems. In Sect. 27.4, I discuss effects specifically due to the coupling between the chains, such as the deconfinement transition and some transverse transport properties. Finally conclusions and perspectives are presented in Sect. 27.5.

27.2 General Ideas

The main ideas and challenges in connection with the quasi-one-dimensional nature of the organics and the observation of LL behavior are summarized here.

The chains are characterized by intra- ($t_{\|}$) and interchain ($t_{\perp}$) single particle hopping. The main effect of the interchain single particle hopping is to induce a dimensional crossover between a one-dimensional situation and a higher-dimensional one. In the absence of interactions such dimensional crossover is easy to understand. In Fourier space the kinetic energy becomes

$$\epsilon(k_{\|}, k_{\perp}) = -2t_{\|} \cos(k_{\|}a) - 2t_{\perp} \cos(k_{\perp}b), \tag{27.1}$$

where b denotes a perpendicular direction. If the perpendicular hopping $t_{\perp}$ is much smaller than the parallel one $t_{\|}$, which is the relevant case for the quasi-one-dimensional organics, then (27.1) leads to the open Fermi surface of Fig. 27.1.

If one is at an energy scale (for example the temperature T or the frequency ω) larger than the warping of the Fermi surface then the warping is washed out, which means that no coherent hopping can take place between the chains. In that case the system is indistinguishable from one with a flat Fermi surface and can thus be considered as a one-dimensional system. On the other hand if the temperature or energy is much smaller than the warping of the Fermi surface all correlation functions are sensitive to the presence of the warping, and the system is two- or three-dimensional. Since I considered free electrons in the above example, this crossover occurs at an energy scale of the order of the interchain hopping, as is summarized in Fig. 27.1. In presence of interactions and commensurability, the problem is of course much more complicated and interactions affect drastically this behavior compared to the noninteracting case. Indeed, for the organics $t_{\perp}$ is much smaller than the intrachain characteristic energy scales such as the kinetic energy

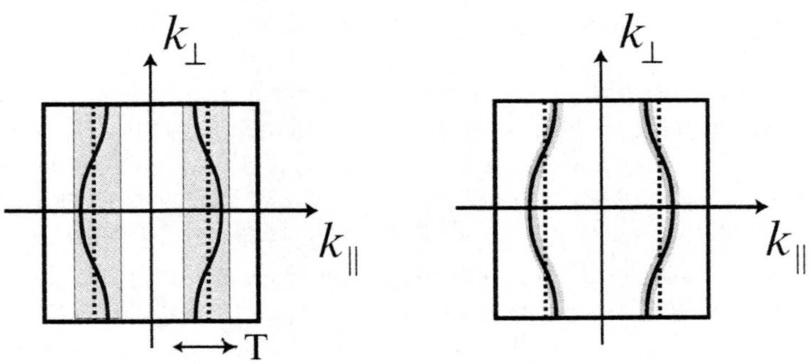

Fig. 27.1. Dimensional crossover for noninteracting electrons. $k_{\|}$ is the momentum along the chains and $k_{\perp}$ the one perpendicular to the chains. (*left*) If the temperature T (or any other external energy scale, represented by the *gray area*) is larger than the warping of the Fermi surface due to interchain hopping the system cannot feel the warping. It is thus behaving as a one-dimensional system. (*right*) At a lower temperature/energy the system feels the two- (or three-) dimensional nature of the dispersion and thus behaves as a full two- (or three-) dimensional system. There is thus a dimensional crossover as the temperature/energy is lowered

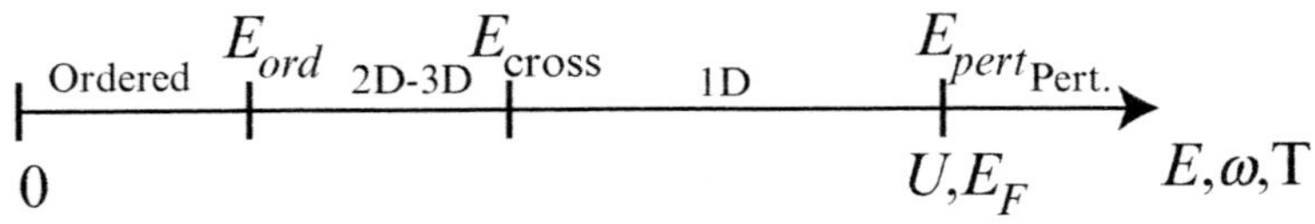

Fig. 27.2. Separation of energy scales if the interchain hopping $t_\perp$ is much smaller than the intrachain one $t_\parallel$. For energies larger than the intrachain hopping (or equivalently the Fermi energy E_{F+}) and interactions (denoted generically U), simple perturbation theory is valid. Below this scale the system is in an interacting one-dimensional regime. The interchain hopping couples the chains at an energy E_{cross} and destroys the one-dimensional physics. For noninteracting particles $E_{\text{cross}} \sim t_\perp$ but this scale is renormalized by interactions into $t_\perp^\nu$ in a LL. For commensurate systems (Mott insulators) the Mott gap can suppress the single particle hopping and drive E_{cross} to zero. In all cases the system can have a transition to an ordered state at an energy E_{ord}. If the dimensional crossover takes place before (as is the case shown in the above figure), this transition should be described from the two- or three-dimensional interacting theory

and the interactions. In that case the chains can experience the full effect of the interaction in a one-dimensional regime before the processes due to interchain hopping can spoil this pure one-dimensional physics. The dimensional crossover scale must thus be computed from the one-dimensional interacting theory. This is summarized in Fig. 27.2.

For the case of a commensurate system, the situation is even more complicated and a phase diagram as a function of the temperature (or another energy scale probing the system) and interchain hopping $t_\perp$ is shown in Fig. 27.3 If the chains are uncoupled, $t_\perp = 0$, one recovers the behavior of one-dimensional interacting electrons. For an incommensurate system this gives the well-known Luttinger liquid behavior, which I will discuss in more details in Sect. 27.3. The organics are commensurate systems, and thus at low temperature one can expect to have a Mott insulator behavior. An important question, which I will discuss further in Sect. 27.3, concerns the origin of this Mott insulating behavior. Indeed the material has a quarter-filled band (of holes) [6], but a small dimerization of the order of $\Delta_d \sim 100\,\text{K}$ makes the band effectively half filled. One has thus to determine which commensurability is important for the Mott behavior. Regardless of this point, one must get an insulating behavior for an isolated chain at low temperature, characterized by a Mott gap in the charge excitation spectrum Δ_ρ. For temperatures $T \ll \Delta_\rho$, one thus sees the Mott insulating behavior. This is the region represented as (MI) in the phase diagram of Fig. 27.3. For temperatures larger than the Mott gap $T \gg \Delta_\rho$ the Mott behavior is not observable and one has a crossover to the Luttinger liquid behavior in the isolated chain. This is denoted as (LL) in the phase diagram of Fig. 27.3.

The interchain coupling brings the system from the one-dimensional behavior to a higher-dimensional one. However, if each chain develops a gap, it means that the single-particle Green's function decays *exponentially*. The

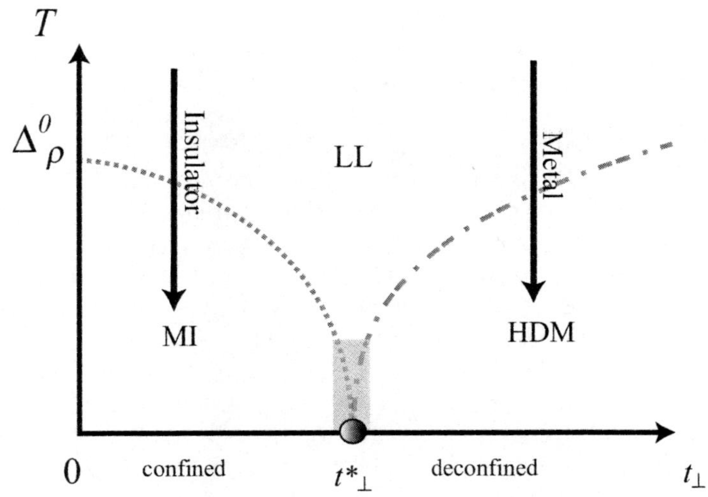

Fig. 27.3. Phase diagram expected for coupled one-dimensional commensurate chains as a function of the temperature T (or an energy E) and the interchain hopping $t_\perp$. In the absence of $t_\perp$ the system is a one-dimensional Mott insulator with a Mott gap Δ_ρ^0. If $t_\perp$ is weak the ground state is still a Mott insulator (MI). For temperature larger than the Mott gap $\Delta(t_\perp)$ (*dotted line*) one observes a crossover to a Luttinger liquid (LL) regime. Beyond a critical value $t_\perp^*$ the system has a deconfinement transition toward a high-dimensional metal (HDM). Additional complications can occur near this point as schematically represented by the *gray box* (see text). Above a certain crossover scale T^* (*dash-dotted line*) the LL behavior is recovered since no coherent hopping can take place between chains. These two crossovers can be observed in different materials or upon application of pressure by lowering the temperature as indicated by the *two arrows*. The two confined and deconfined regions correspond, respectively, to apparent insulating and metallic behavior when the temperature is lowered. After [5, 14]

single-particle hopping is now an irrelevant variable. The formation of a gap is thus in direct competition with the interchain hopping. For small interchain hopping the system thus remains an insulator but with a smaller gap $\Delta_\rho(t_\perp)$ than for an isolated chain. As far as single particle hopping is concerned the system is thus essentially one-dimensional. This is the regime corresponding to the insulating part of Fig. 27.3. As shown in Fig. 27.2, the system can still in this regime undergo a transition to an ordered state if particle–hole (density–density) or particle–particle (Josephson coupling) interactions are present between the chains. Such interactions are in any case generated to second order in the single particle hopping.

The competition between the Mott physics and the interchain hopping means that by increasing the interchain hopping to a critical value one can break the one-dimensional Mott gap. Thus at $T = 0$ a quantum phase transition occurs for $t_\perp = t_\perp^*$ above which the insulating Mott state is destroyed and the system becomes an HDM. This transition is known as a deconfinement transition where both the nature of the state as well as the effective dimensionality of the system change. The properties and consequences of this transition will be discussed in Sect. 27.4. Beyond this critical value of $t_\perp$ the low temperature properties are the ones of the HDM. What is the nature of such an HDM (and in particular whether it is a Fermi liquid or not) and how it is affected by the fact that it stemmed from coupled one-dimensional chains is of course an important question. As the temperature is increased above a temperature $T^*(t_\perp)$ one can expect a crossover toward a one-dimensional LL behavior again. Indeed at high temperature coherent hopping between the chains cannot take place. This dimensional crossover energy scale is drastically affected by both the interactions present in the chains *and* the commensurability, and is thus quite different from the one for noninteracting electrons.

This generic phase diagram is directly relevant for the Q1D organics [6] due to their commensurate nature. In particular they are very good realization of quasi-one-dimensional systems with hopping integrals of the order of $t_a \simeq 3{,}000\,\mathrm{K}$, $t_b \simeq 300\,\mathrm{K}$, $t_c \simeq 20\,\mathrm{K}$, leading to relatively well separated energy scales in which one is indeed dominated by the intrachain hopping. The band is quarter filled, with a small dimerization along the chains giving some half-filled character to the system as well. The various parameters ($t_\parallel$, $t_\perp$, and the dimerization) can be tuned either by changing the chemistry of the compound or by applying external pressure, so the phase diagram of Fig. 27.3 can be roughly seen as a temperature (or energy) – pressure phase diagram. I will come back to the role of pressure in Sect. 27.3. Experimentally, at ambient pressure, the $(TMTTF)_2PF_6$ compound displays an insulating behavior (MI). A transition to a metallic phase is found, with increasing pressure and the properties of the TMTTF compounds evolve toward those of the compounds of the TMTSF family, which are good conductors. This evolution is clear from the a-axis resistivity measurements (see, e.g., Fig. 1.5 of [6]). Such an insulating behavior is well consistent with what one would expect for a one-dimensional Mott insulator. The minimum of the resistivity (followed by an activated law as temperature is lowered) defines the onset of the MI regime in Fig. 27.3. It is thus clear that the interactions play a crucial role in the TMTTF family even at relatively high energies. For the TMTSF, the question is more subtle in view of the metallic behavior at ambient pressure, and it was even suggested that such compounds could be described by a FL behavior with weak interactions [15]. On the contrary, interpretations of deviations of $1/T_1$ in NMR [16] or magnetoresistance [17] as due to a one-dimensional behavior would suggest that one-dimensional effects would persist to temperatures as low as $20\,\mathrm{K}$, a much too low scale compared to the naive one given by the bare interchain hopping $t_b \sim 300\,\mathrm{K}$. TMTTF and TMTSF families thus prompt for very fundamental questions in connection with one-dimensional physics:

1. Are interactions also important for the metallic members of the family or can they be simply regarded as a Fermi liquid with an anisotropic Fermi surface?
2. If indeed one can identify Luttinger liquid behavior, what are the Luttinger parameters?
3. If the system is a Mott insulator, is this mostly due to the dimerization of the band or is the quarter filling commensurability sufficient?
4. What are the deconfinement scale $t_\perp^*$ and beyond that point the crossover scale T^* below which the system is not one-dimensional any more? What is the nature of the high-dimensional metallic phase?

I will now show how transport measurement, especially optical conductivity, has proven to be a key tool in addressing and, to a large extent, answering these important questions.

27.3 Mott Insulators and One-Dimensional Transport

Let me first examine the properties of the system in a regime where the coherent hopping between the chains can be neglected. This is the regime corresponding to the LL and MI parts of Fig. 27.3. In that regime the properties are essentially the ones of isolated chains, and, in particular, the transport properties along the chains can be computed from a pure one-dimensional limit.

27.3.1 Theory of Transport

Here, I recall only the salient points on transport in connection with the quasi-one-dimensional organics and refer the readers to [4, 5, 7, 8] for more details on the transport in one dimension and references.

If the filling is not commensurate, all excitations of a one-dimensional system are sound waves of density and spin–density. A convenient basis to describe such a system is provided by the so-called bosonization technique [4]. The energy of these excitations is given by a standard elastic-like Hamiltonian. The Hamiltonian of the system is the sum of a part containing only charge excitations and one containing only spin excitations.

$$H = H_\rho + H_\sigma, \tag{27.2}$$

where H_ν ($\nu = \rho, \sigma$) is of the form

$$H = \frac{1}{2\pi} \int \mathrm{d}x \; [u_\nu K_\nu (\pi \Pi_\nu(x))^2 + \frac{u_\nu}{K_\nu} (\nabla \phi_\nu(x))^2], \tag{27.3}$$

where ϕ_ν and Π_ν are conjugate variables $[\phi_\nu(x), \Pi_\nu(x')] = \mathrm{i}\hbar\delta(x - x')$. The fields ϕ_ν are related to the long wavelength distortions of the charge $\rho(x)$

and spin $\sigma(x)$ electron density by $\rho(x), \sigma(x) = -\sqrt{2}\nabla\phi_{\rho,\sigma}(x)/\pi$. $u_{\rho,\sigma}$ are the velocities of these collective excitations. In the absence of interactions, $u_\rho = u_\sigma = v_{\mathrm{F}}$. Interactions, of course, renormalize the velocities of charge and spin excitations, as in higher dimensions. $K_{\rho,\sigma}$ are dimensionless parameters depending on the interactions. For systems with spin rotation symmetry $K_\sigma = 1$ (for repulsive interactions), while the spin excitations are gapped for attractive interactions. $K_\rho = 1$ in the absence of interactions and quite generally $K_\rho < 1$ for repulsive ones. The three parameters u_ρ, u_σ, and K_ρ completely characterize the low energy properties of a one-dimensional system. They can be computed for a given microscopic model as a function of the interactions [18–23], but as was shown by Haldane [24–26], the form (27.2) is the generic low energy form. This means that (27.2) and the parameters u_ρ, u_σ, K_ρ play a role similar to the one of the Landau Fermi liquid Hamiltonian (and Landau parameters) in higher dimensions. To have again in one dimension a concept equivalent to the Fermi liquid, i.e., a generic description of the low energy physics (for energies lower than E_{pert} of Fig. 27.2) of the interacting problem, is of course extremely useful. This removes part of the caricatural aspects of any modelization of a true experimental system and allows to easily deal with extensions such as the commensurability with the lattice.

The form (27.2) immediately shows that an excitation that is looking like a free electron (i.e., that carries both charge and spin) cannot exist. This is a very important difference between a Luttinger and a Fermi liquid since in the latter, in addition to collective modes of charge and spin, individual excitations (quasiparticles) carrying both charge and spin and looking essentially like a free electron do exist [1–3]. In addition, the correlation functions in a LL display nonuniversal power laws with exponents dependant on the interactions via the Luttinger parameter K_ρ. For example the single particle correlation function decays with distance or time with an exponent $\zeta = \frac{1}{4}[K_\rho + K_\rho^{-1}] + \frac{1}{2}$. The fact that it decays faster than $1/r$ which is the case for free electrons or a Fermi liquid shows directly that single particle excitations do not exist in one dimension. Similarly spin–spin or density–density correlations have a $2k_{\mathrm{F}}$ oscillating part decaying with an exponent $K_\rho + 1$. Probing such power laws is thus a direct proof of the LL behavior.

For commensurate systems one has to modify the Hamiltonian (27.2) to take the commensurability with the lattice into account. Such commensurability is at the root of the Mott transition. Although one can of course work out the Mott transition from microscopic models such as the Hubbard model [27], the Luttinger liquid theory provides an excellent framework to take into account the effects of a lattice and describe the Mott transition. It is particularly well adapted for the case of the organics since, as we will see, the Mott gap is smaller than E_{pert} and thus the LL theory is indeed a suitable starting point at these energies. To incorporate the Mott transition in the Luttinger liquid description one must take into account that in presence of a lattice the wavevector is in fact defined modulo a vector of the reciprocal lattice (that is, in one dimension a multiple of $2\pi/a$ with a the lattice spacing). Thus, in addition to the interaction processes that truly conserve momentum $k_1 + k_2 = k_3 + k_4$ one can now have umklapp processes [28] such that $k_1 + k_2 - k_3 - k_4 = Q$ where Q is a vector of the reciprocal lattice. Since umklapps do not conserve momentum they are the only ones that can lead to a finite resistivity, and are responsible for the T^2 law in a Fermi liquid [29]. The umklapp process is also responsible for the Mott transition in one dimension. In order for such process to be efficient at the Fermi level as it is necessary to have $4k_{\mathrm{F}} = 2\pi/a$ namely $k_{\mathrm{F}} = \pi/2$ or one electron per site (half filling). This corresponds to the case where two electrons are scattered from one side of the Fermi surface ($-k_{\mathrm{F}}$) to the other side ($+k_{\mathrm{F}}$). This is indeed the most standard case for having a Mott insulator. But in fact, umklapps are not restricted to one particle per site [30,31], but occur for any commensurate fillings. Indeed, if $2pk_{\mathrm{F}} = 2\pi q/a$ (where p and q are integers) then one can show that an additional term must be added to (27.2). For even commensurabilities ($p = 2n$), that corresponds to case of the quasi-one-dimensional organics) this term is [7,8,30]

$$H_{\frac{1}{2n}} = g_{\frac{1}{2n}} \int \mathrm{d}x \, \cos(n\sqrt{8}\phi_\rho(x)), \tag{27.4}$$

where n is the order of the commensurability ($n = 1$ for half filling – one particle per site; $n = 2$ for quarter filling – one particle every two sites and so on). The coupling constant $g_{1/2n}$ is the umklapp process corresponding to the commensurability n.

If the bosonization representation can give the universal form of the Hamiltonian and the umklapp term, the amplitude of the umklapp coefficients $g_{\frac{1}{2n}}$ depends on the precise microscopic interaction. At half filling, for a Hubbard model, $g_{\frac{1}{2}}$ is of the order of the interaction U. Higher commensurability umklapps can be estimated perturbatively. For a quarter-filled band such that $8k_{\mathrm{F}} = 2\pi/a$ (this corresponds to $n = 2$ in the above notations), to produce an umklapp one needs to transfer *four* particles from one side of the Fermi surface to the other to get the proper $8k_{\mathrm{F}}$ momentum transfer. This can be done in higher-order perturbation terms by doing three scatterings as shown in Fig. 27.4. For weak interactions the amplitude of such a process would thus be of order $U(U/W)^2$, where W is the bandwidth. In addition to the above process, there is an additional one for the Bechgaard or Fabre salt family. Indeed in these systems the stack is slightly dimerised [6,32]. This dimerization opens a gap in the middle of the band as indicated in Fig. 27.4. Thus although the system is originally quarter filled the dimerization turns the system into a half-filled band. This means that even if the system is quarter filled, a nonzero $g_{1/2}$ exists in addition to $g_{1/4}$. If Δ_d is the dimerization gap the strength of such umklapp is $g^d_{1/2} = U(\Delta_d/W)$. Note that contrary to what happens in a true half-filled system the umklapp coefficient is now much smaller than the typical interaction U. This allows to get a small Mott gap even if the interactions are large. In the presence of dimerization a quarter-filled system can thus be a Mott insulator, either because of the half-filling umklapp (that

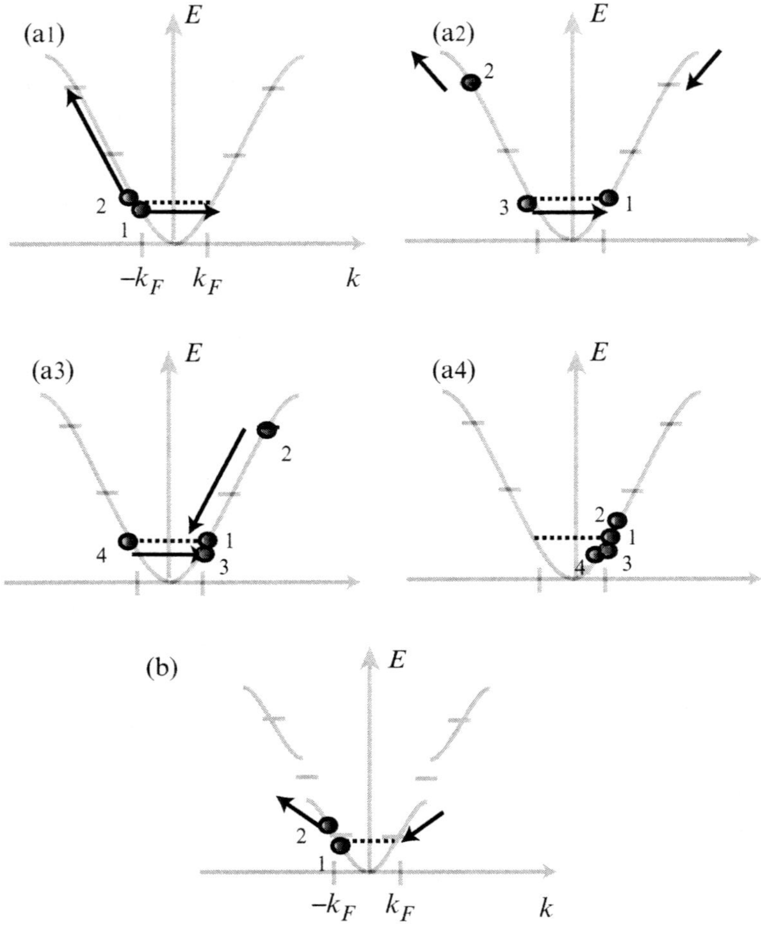

Fig. 27.4. Umklapp processes important for the organic conductors. (**a1–a4**) A quarter-filled umklapp can be constructed from a third-order perturbation theory in the interaction U. It consists in transferring four particles from one side of the Fermi surface to the other. The sequence of scattering due to the interaction is shown in figures (**a1–a4**). Three electrons are transferred from $-k_\mathrm{F}$ to $+k_\mathrm{F}$ by three successive interaction processes, while the momentum difference is absorbed by a fourth electron until it reaches the opposite side of the Fermi surface. The processus needs two intermediate states of high-energy W of the order of the bandwidth. Thus for small interactions the amplitude for such a process is of order $U(U/W)^2$. (**b**) Dimerization opens a gap in the band. Because of this gap the quarter-filled band becomes effectively half filled. This reduces the zone boundary. The dimerization gap Δ_d thus creates even for a quarter-filled system an half-filling umklapp where two particles can be transferred from one side of the Fermi surface to the other. The amplitude of such a process is proportional to the dimerization gap and thus of the order of $U\Delta_d/W$. After [5]

exists now because of the dimerization) or because of the quarter-filled one. Which process is dominant depends of course on the strength of the dimerization and of the interactions, and has important consequences on the physics of the system [8].

From (27.4) all the properties of the Mott transition and transport in a one-dimensional system can be worked out. The system is a Mott insulator for $K_\rho < K_\rho^* = 1/n^2$ where n is the order of the commensurability. The larger the commensurability the smaller K_ρ needs to be for the system to become insulating. For a commensurability $n = 1$, that is, half filling the critical value is $K_\rho = 1$. This means that, contrarily to the higher-dimensional case, *any* repulsive interactions turn the system into an insulator. For a quarter-filled band ($n = 2$) the critical value is $K_\rho = 1/4$. To get the insulator one needs both pretty strong interactions *and* interactions of a finite range, since the minimum value of K_ρ for a local interaction is $K_\rho = 1/2$ [33]. This is physically obvious to stabilize a structure in which there is a particle every two sites one cannot do it with purely local interactions. The range of the interactions in addition of their strength and thus the precise chemistry of the compound controls the range of values of K_ρ that one is able to explore.

Of course the Mott transition and the Luttinger physics have drastic consequences on the transport properties and one can expect quite different properties than for Fermi liquid. Thus transport can be used as an efficient probe. As we will see it allows to probe both the single particle behavior (or absence thereof) and the Luttinger liquid collective excitations [7,8,30,34–37]. A schematic plot of the a.c. conductivity (at $T = 0$) is shown in Fig. 27.5. In the Mott insulator σ is zero until ω is larger than the optical gap $2\Delta_\rho$. For frequencies larger than the Mott gap, interactions dress the umklapps and give a nonuniversal (i.e., interaction-dependent) power law-like decay. Such a power law can be described by renormalization group calculations coupled to a memory function formalism [7, 8]. The results of this approach have been subsequently confirmed by form factor calculations [35]. If one ignores the renormalization of K_ρ by the umklapp (for the effect of the renormalization of K_ρ see [7]) one gets for the a.c. conductivity, for frequencies larger than the Mott gap

$$\sigma(\omega) \sim \omega^{4n^2 K_\rho - 5}, \tag{27.5}$$

where n is the order of commensurability.

The d.c. conductivity can be computed by the same methods [7, 8, 37] and is shown in Fig. 27.6.

Here again the dressing of umklapps by the other interactions results in a nonuniversal power law dependence for temperatures larger than the Mott gap Δ_ρ. Within the same approximations than for (27.5) one obtains

$$\rho(T) \sim T^{4n^2 K_\rho - 3}. \tag{27.6}$$

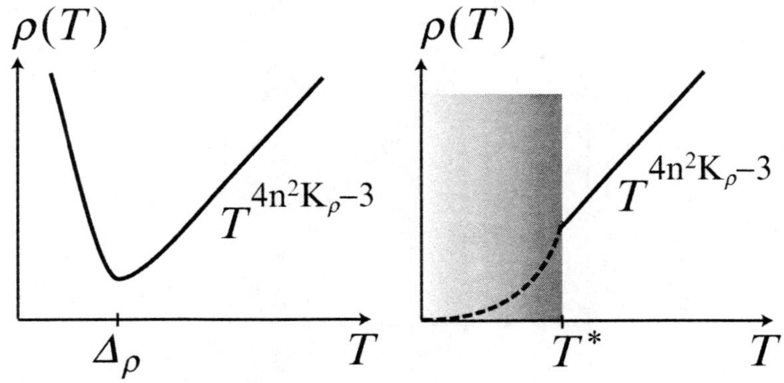

Fig. 27.6. D.c. conductivity along the chains as a function of the temperature T for a commensurability of order n. Δ_ρ is the Mott gap. *Left*: confined region. Above the Mott gap the d.c. transport shows an exponent $4n^2K_\rho - 3$ characteristic of the Luttinger liquid. Below the Mott gap the number of carriers is exponentially small, and any scattering will give an exponentially small conductivity. *Right*: deconfined region. The Mott gap scale does not exist any more. Above the dimensional crossover scale T^* the temperature dependence is essentially identical to the one on the left and shows LL behavior. Below the scale T^* the system must be described by a two- or three-dimensional theory and one can expect a temperature dependence much more conventional (it would be T^2 for a simple Fermi liquid)

be done on the d.c. transport and gives also good agreement [6, 32, 38] with the predicted power law (27.6). Recent d.c. transport data on $(TMTSF)_2PF_6$ [39] are also shown in Fig. 27.8. From the data of Fig. 27.8 one sees that $(TMTTF)_2PF_6$ shows a quite consistent behavior with the above theoretical description. For temperatures T larger than about 100 K one recovers a power law $\rho_a \propto T^{0.56}$ quite compatible, using (27.6) with the value of $K_\rho = 0.22$–0.23 obtained from the optics.

Additional experiments both in optics [40–43] and on d.c. transport [6, 32, 39] have confirmed the LL nature, and thus the upper part of the phase diagram of Fig. 27.3.

Quite remarkably, optical conductivity is in fact an excellent probe of the LL behavior. Indeed if the control parameter is the temperature several limitations are present. First the compounds are quite compressible, and it is important to make pressure corrections to compare the theoretical prediction (at constant volume) with the experimental observations [6,32,38]. Not taking into account this volume change with temperature was initially responsible for the seemingly T^2 behavior observed till room temperature. However the pressure correction is hard to perform with great accuracy and more importantly the range of temperature between T^* and room temperature, which is in practice about the maximum range available, is very limited (typically 100–300 K at best in PF_6) making difficult a quantitative and convincing test of a power law regime. Optics does not suffer from those limitations. Since the parameter

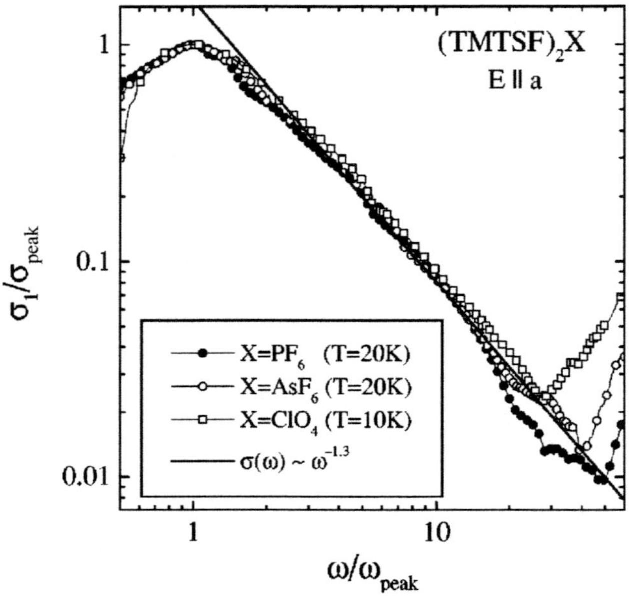

Fig. 27.7. Optical conductivity along the chain axis in the TMTSF family. The conductivity is rescaled by the gap in various samples. A power law behavior is clearly observed. The optical conductivity thus allows to show that even in the metallic (deconfined) regime LL behavior is still present. Because of the wide frequency range accessible on which the power law is seen it also allows to extract the LL parameter K_ρ reliably. From [10] (copyright 1998 by the American Physical Society)

varied is the frequency, no such expansion correction is needed. The range of energy that can be explored is also much larger and limited only by the bandwidth of the system. This allows for a fit of the power law on more than a decade. It is important to note that such an analysis of a.c. transport has also been used with success in other types of one-dimensional materials [44]. It is thus a method of choice to probe for the physics of these systems.

Indeed quite importantly the a-axis optical measurements described above even allow for a quantitative determination [10] of the LL parameter K_ρ and a better understanding of the mechanism behind the Mott transition in these materials. A fit of the frequency dependence of the longitudinal conductivity (see Fig. 27.7) can be performed using (27.5). A commensurability of order one ($n = 1$) does not allow for a consistent fit of both the exponent and the Mott gap [10]. Such a commensurability would lead to $K_\rho \sim 1$, a nearly noninteracting system and then to a gap much smaller than the one observed experimentally. This indicates that, at least for the TMTSF members of the family, the dominant umklapp comes from the quarter-filled nature of the band. Formula (27.5) with $n = 2$ thus yields $K_\rho \simeq 0.23$, indicating quite strong electron–electron interactions. Moreover this indicates that the finite

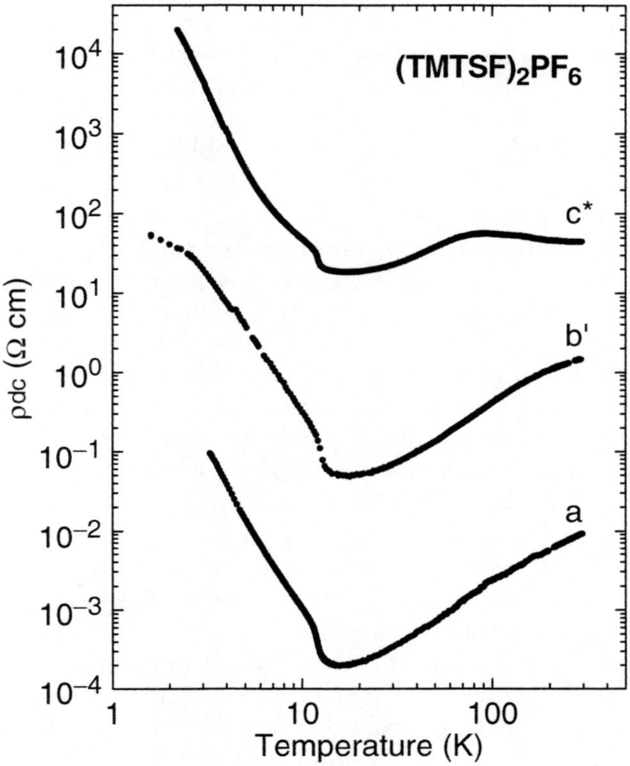

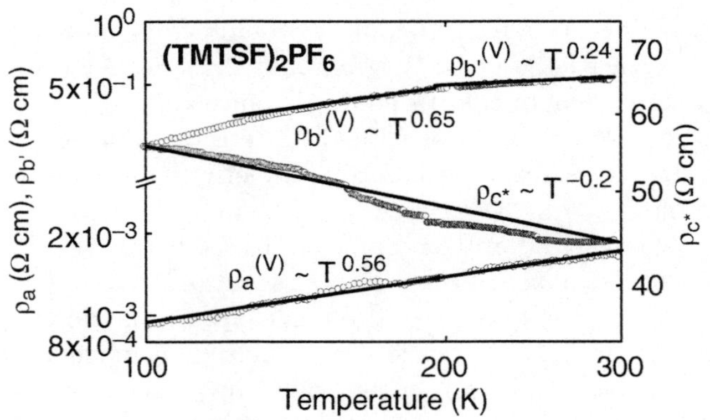

Fig. 27.8. (*top*) Conductivity of $(TMTSF)_2PF_6$ at constant pressure. The difference between the a- and c-axis at high temperature is directly visible and is a signature of the LL behavior. The crossover temperature scale is here slightly below 100 K as indicated by the change of behavior of $\rho_c(T)$. (*bottom*) The conductivity, corrected to be the constant volume one is shown above the crossover scale T^*. Both the $\rho_a(T)$ and $\rho_c(T)$ temperature dependence are well compatible with the value of the LL parameter $K_\rho \sim 0.23$ extracted from the optics. The crossover in the $\rho_b(T)$ dependence is obviously much broader but the same qualitative tendency than in $\rho_c(T)$ can be seen. From [39] (copyright 2005 by the American Physical Society)

range nature of the interactions should be taken into account, with interactions extending at least to nearest neighbors. A modelization of the organics thus should not be done with a purely local Hubbard model, with an interaction U, but also take into account at least the nearest neighbor interaction V. The optical data is thus consistent with an interpretation of the insulating state as a quarter-filled Mott insulator, suggesting that contrarily to what was commonly believed [11] the dimerization plays little role at least in the TMTSF family. In the TMTTF family, dimerization is larger and it is unclear there which process is dominant. Note that because of the anions other transitions can exist such as a ferroelectric transition [45–47]. It is important to note that this also suggests [5, 10], that a very important effect of pressure, is not so much to affect the dimerization, but to affect the hoppings and thus reduce the interaction versus kinetic energy ratio. This makes the system effectively less interacting with pressure. This prediction has been recently confirmed on optical measurements under pressure where a consistent decrease of the Luttinger parameter K_ρ has been observed with increasing pressure [43]. Given the fact that the value of K_ρ is very close to the critical value $K_\rho = 0.25$ for which the quarter-filled umklapp becomes irrelevant, such a variation of K_ρ can trigger a rapid variation of the Mott gap upon application of pressure, or when going from the TMTTF members to the TMTSF members. The importance of the quarter-filled umklapp in this family of compounds has been also clearly confirmed by properties of parents compounds with a structure similar to the Bechgaard salts but that do not have dimerization [12, 13, 48, 49]. These compounds turned out to be Mott insulators [6]. Under pressure they share most of the features of the Bechgaard and Fabre salts, indicating that the same physics is at hand [13, 49, 50]. It would be of course very interesting to further investigate the phase diagram and the transport properties under pressure of these compounds. In addition, since they share the same basis microscopic features, it would be specially interesting to assert whether these quarter-filled systems also exhibit superconductivity under pressure as in the Bechgaard salts.

Finally the last question that can be addressed by the transport along the chains is one of the values of the crossover scales $\Delta_\rho(t_\perp)$ or $T^*(t_\perp)$ as shown in Fig. 27.3. $\Delta_\rho(t_\perp)$ is easily seen from the upturn of the resistivity along a-axis or directly from the optics for the insulating members of the family. In the metallic regime $T^*(t_\perp)$ can be estimated by the crossover between a T^2 behavior at low temperature to the nonuniversal power law (27.6) corresponding to the LL at high temperatures. For example this suggests a crossover scale of about $T^* \sim 100\,\mathrm{K}$ for $(\mathrm{TMTSF})_2\mathrm{PF}_6$ as shown in Fig. 27.8. However a much more precise determination of this scale is provided by a measure of the transverse transport [51] that I now examine.

27.4 Coupled Chains

Let us now investigate the effects that are direct consequences of the coupling between the chains. There are of course the deconfinement transition at $t_\perp^*$ and the two crossover scales $\Delta(t_\perp)$ and $T^*(t_\perp)$. But the very fact that many chains are presents means that transverse transport effects can be probed as well, even in the high energy (LL) of the phase diagram of Fig. 27.3. Such transverse transport is sensitive on how electrons can tunnel from one of the chain to the other. It thus reflects directly how well single particle excitations can exist, and is therefore also a way to probe the LL nature of the system. It is of course also a very sensitive way to address the question of the dimensional crossover since one can expect a drastically different type of transverse transport depending on whether single particle excitations exists in the chain and thus can hop or not.

Let me first discuss the LL region in Fig. 27.3. In that region the hopping is incoherent between the chains. Thus the transverse conductivity can be computed in the high temperature or high frequency regime by an expansion in the perpendicular hopping [52]. One finds a power law in frequency of temperature controlled by the single particle Green's function exponent. At finite temperatures

$$\sigma_\perp(T \gg \omega) \propto T^{2\alpha-1} \tag{27.7}$$

for $k_\mathrm{B}T \gg E_\mathrm{cross}$, while at high frequency ($\hbar\omega \gg E_\mathrm{cross}$), one gets

$$\sigma_\perp(\omega \gg T) \propto \omega^{2\alpha-1}, \tag{27.8}$$

where $\alpha = \zeta - 1 = \frac{1}{4}(K_\rho + K_\rho^{-1}) - \frac{1}{2}$ is the exponent in the single particle density of states. E_cross is the scale at which this expansion breaks down, either $\Delta_\rho(t_\perp)$ or $T^*(t_\perp)$ as given by the dotted and dashed lines in Fig. 27.3. Note that in the regime where chains are in the LL state if one takes $K_\rho \sim 0.23$ as given by the measurement of the intrachain transport, the transverse conductivity *decreases* with decreasing temperature or frequency. One has thus a very different behavior of the intra- and interchain transport. This is to be contrasted from a normal Fermi liquid regime, where one can expect similar temperature dependencies in both directions. The change of behavior can thus be used to detect the dimensional crossover scale T^*. This is quite clear in Fig. 27.8.

Observed optical conductivity along c-axis [42] is compatible with the power law growth of (27.8) and a value of $K_\rho \sim 0.23$, as determined by the in-chain transport. But clearly much more experiments would be needed since the measurement is extremely difficult and the data not at the same level of accuracy than the intrachain transport. At the price of the pressure correction the d.c. transport can also be used. For $(\mathrm{TMTTF})_2\mathrm{PF}_6$, as shown in Fig. 27.8, the temperature dependence of the in-chain conductivity gives, with (27.6), a value of $K_\rho = 0.22$–0.23 well compatible with the one from the optics. In the same way $\rho_c \propto T^{-0.2}$ and (27.7) gives $\alpha = 0.69$ again well compatible with $K_\rho = 0.22$. Note that the fact that the d.c. transport is at least qualitatively reproduced with the *same* value of K_ρ gives strong credence to a LL physics interpretation of the data. The interpretation of ρ_b is more complex since no simple power law is seen over the entire temperature range. However a change in the anisotropy behavior takes place above $T^* \sim 100\,\mathrm{K}$ in Fig. 27.8 and the fit to a power law shows a clear downturn of the exponent. A possible interpretation of the data is thus as being in a crossover regime between the low-temperature Fermi liquid one and the high-temperature Luttinger liquid. Note that it is reasonable to expect a much larger crossover region for the b'-axis transport, than for the c^* axis given the much higher value of the transfer integral in this direction. Despite this general agreement, a quantitative understanding of the transport along b clearly requires further work both for transport [39] and for optics [42].

The temperature dependence below $T \approx 100\,\mathrm{K}$ is the same for the a and b' directions implying a similar transport mechanism, and the anisotropy ratio corresponds roughly to the expected band structure value. Below this scale, the d.c. resistivity follows a power law $\rho_a, \rho_b \propto T^2$, as expected for a Fermi liquid. Note that although the system is now two-dimensional because the hopping in the b direction is now coherent, the temperature is still much larger than the hopping in the c direction $T \gg t_c$. Thus one can still use formula (27.7) but putting $\alpha = 0$ as would befit a Fermi liquid. This gives $\rho_c \propto T$ which is effectively consistent with the experimental data. The experiments in $(TMTSF)_2PF_6$ thus directly confirms the above theoretical analysis and the Luttinger liquid behavior. The crossover scale can be experimentally determined to be $T^* \sim 100\,\mathrm{K}$ for the case of PF_6. For the case of $(TMTSF)_2ClO_4$ a similar analysis suggests a much higher value $T^* \geq 200\,\mathrm{K}$ [39]. However several problems remain with this compound, in particular concerning the high temperature resistance anisotropy which is much smaller than normally expected. More experimental data are clearly needed in that case. The crossover scale can also be followed under pressure (see Fig. 1.8 of [6]). Another way to determine the crossover scale is provided by the optical b-axis conductivity since coherent hopping between the chains manifests itself as the appearance of a Drude peak in the b-axis conductivity [42].

Another measurement that can in principle probe the nature of the Luttinger liquid is of course the Hall effect. Indeed in a Fermi liquid the Hall effect is essentially a measure of Fermi surface properties. At low temperatures (in the HDM part of the diagram) the Hall effect can be quite successfully described in this framework [53–55]. At temperatures larger that T^* one could expect the Hall effect to reflect again the nature of the interactions in the LL phase. However both the theory and the experiments concerning this quantity are more complicated. Experiments with the magnetic field along the c-axis [56, 57] observe a weak temperature dependence while field along a leads to an essentially temperature independent Hall effect [58]. On the theoretical side, quite surprisingly it was shown that in the absence of scattering along the chains the Hall effect in a Luttinger liquid does not show any trace of the interactions and is equal to the band value [59,60] $R_{\rm h}^0$. Including the scattering along the chains was done recently [61,62] for the case of the half-filled umklapp and magnetic field perpendicular to the chains, leading to a Hall effect behaving as

$$R_{\rm h} = R_{\rm h}^0 \left[1 - A\left(\frac{g_{\frac{1}{2}}}{\pi v_{\rm F}}\right)^2 \left(\frac{T}{W}\right)^{3K_\rho - 3}\right], \tag{27.9}$$

where A is a dimensionless constant and W is the bandwidth of the material. This shows that some temperature dependence is to be expected in the Luttinger regime, in qualitative agreement with the observations [56,57].

As one can obtain the crossover temperature T^*, one can also determine the deconfinement critical value $t_\perp^*$ from the transport measurements. This can be done for example by monitoring the occurrence of metallic behavior in the a-axis (see Fig. 1.8 of [6]). In addition, a measure of the gap extracted from the optical conductivity shows that the change of nature between insulating to metallic behavior occurs when the observed gap is roughly of the order of magnitude of the interchain hopping [40] (see Fig. 27.9).

On the theoretical side understanding quantitatively the deconfinement transition and even the crossover scale T^* for deconfined systems is a major theoretical challenge. In the absence of commensurability, the crossover scale between a LL and the HDM can be determined by looking at the renormalization of the interchain hopping [63–69]. If one neglects the renormalization

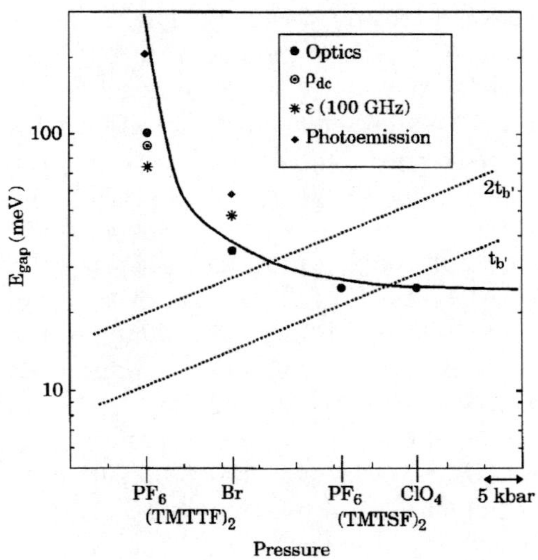

Fig. 27.9. A comparison of the measured gap in the optical conductivity with the interchain hopping. The change of behavior from insulator to metallic occurs when the two quantities are of the same order of magnitude showing that the difference between the various members of the TM families is indeed linked to a deconfinement transition. From [41] (copyright 2000 by EDP Sciences)

of α by the interchain hopping, then one has [64]

$$T^* \sim W \left(\frac{t_\perp}{W}\right)^{1/1-\alpha} . \tag{27.10}$$

For the noninteracting case $\alpha = 0$ and one recovers $E^* \sim t_\perp$. Since α for an interacting system is always positive, the scale at which the dimensional crossover takes place is always *smaller* than for free fermions. Interactions thus tend to make the system more one-dimensional. This reduction of the crossover scale comes again from the fact that in a Luttinger liquid *single particle excitations* are strongly suppressed. For the commensurate case, the scale T^* must be computed in presence of the umklapp term, and is thus dependent also on the deconfinement scale. Unfortunately the RG study, although it provides the scale at which the LL is unstable cannot carry easily through the low temperature HDM. What is the nature of this phase is thus still a major challenge. Even if it is a Fermi liquid, since this Fermi liquid stems from the high-temperature non-Fermi liquid phase, its features are certainly quite special. In particular the quasiparticle residue Z and lifetime of the quasiparticles could in principle retain the memory of the strong correlations that existed in the one-dimensional phase [5, 52]. In addition since the strength of the hopping depends on the transverse momentum $k_\perp$ these quantities could be varying on the Fermi surface and lead to the presence of hot spots [70].

Besides the RG analysis various methods have been tried to tackle the deconfinement transition. This is a difficult problem and much less is known than for the dimensional crossover. Both scaling arguments [8] and study of two chain systems [71–75], showed the importance of the energy scale T^* in comparison with the Mott gap Δ_ρ^0 in the absence of $t_\perp$. A rule of thumb to get the position of this deconfinement transition is to compare the two scales T^* and Δ_ρ^0. Thus, roughly if $T^* > \Delta_\rho^0$ one is deconfined, whereas for $T^* < \Delta_\rho^0$ the gap wins and the chains are confined, only allowing for two particles hopping. Of course, this is only a rule of thumb and one should, in principle, solve the full coupled problem to obtain the critical value $t_\perp^*$ at which deconfinement occurs. No full solution of this problem exists so far. An RG analysis properly incorporating the umklapp terms and the interchain coupling up to the deconfinement transition is quite difficult to realize in a controlled way since both phase correspond to strong coupling fixed points. An RPA treatment [76] of the hopping does produce an insulator–metal transition via the formation of pockets on the Fermi surface. It however neglects any feedback of the hopping on the one-dimensional gap itself and thus grossly overestimate the position of the transition. It also cannot give a full deconfinement with an open Fermi surface, since the one-dimensional gap never closes. A quite promising method is a mean field approach (ch-DMFT) treating the chains as an effective bath [14, 52, 77–80]. This method shows clearly the deconfinement transition and gives access to some of the properties of the HDM phase beyond the transition. An analysis has been performed for the half-filled case and I refer the reader to [14,79] for more details. The full analysis of the quarter-filled band, relevant for the quasi-one-dimensional organics still remains to be done. A caricature of this case, corresponding roughly to a very large on-site interaction U and a moderate nearest neighbor interaction V can however be performed by considering an half-filled band of *spinless* fermions [80]. In that case a strong depletion of the Mott gap with increasing $t_\perp$ has been observed and the deconfinement transition has been analyzed. In such an approach deconfinement occurs first through formations of pockets at a first critical value t_{c1} but then at a slightly larger value t_{c2} a full open Fermi surface is recovered. In the HDM phase, effects of the interactions can still be felt, as in particular the presence of hot spots on the Fermi surface [80]. Despite these progresses, more work, both theoretical and experimental, is needed to completely understand the deconfinement transition. On the experimental side, it would of course be particularly interesting to have information on single particle excitations. Unfortunately photoemission or STM tunneling experiments on the organic conductors seems to be difficult due to the ionic nature of the systems and the surface problems it entails. Some results consistent with Mott physics and Luttinger liquids were observed, in particular the Mott gaps [41,81,82]. But given the very large energy scales at which for example a depletion of the density of state has been observed, interpretation of these results in terms of LL should be taken with a grain of salt.

The physics below the dimensional crossover scale also remains to be fully understood. However we at least now know reliably from the above-mentioned transport experiments the crossover scale to the HDM. This allows to sort out effects that can be attributed to the one-dimensional behavior from those that one has to understand in a more conventional high-dimensional system. For the TMTSF members the crossover scale being at least of the order of $\sim$100 K (for PF_6) one can expect that a Fermi liquid approach should be a useful starting point much below this temperature. Indeed, Fermi liquid theory has been quite successful in explaining many of the *low temperature* ordered phase and properties of these compounds (see, e.g., the chapters on the field induced spin–density waves in this book). On the other hand some correlation effects beyond simple Fermi liquids still persists even way below T^*. This is clear both from the above-mentioned theoretical calculations and, from the experimental side, by the existence of anomalies such as the ones in NMR [16,83]. Such anomalies, being below T^* cannot be attributed to Luttinger liquid behavior. What is their explanation is still an open and a challenging issue.

27.5 Conclusions and Perspectives

I have presented in this chapter the main concepts and questions relevant to tackle the normal phase physics of quasi-one-dimensional systems. The most important ones for isolated chains are the Luttinger liquid theory and the

Mott insulating physics, which are quite special in one dimension. For quasi-one-dimensional systems an extremely rich physics stems from the coupling between the chains. Its most spectacular expression is the presence of a deconfinement transition between a one-dimensional insulator and an HDM. The quasi-one-dimensional organic conductors provide wonderful systems to investigate these phenomena. But this question is pertinent for many experimental systems and is in particular now investigated in systems such as cold atomic gases as well [84–86].

I have shown here how a good theoretical understanding of the transport properties allows to probe the unconventional physics of these systems. In particular the transport has allowed to prove the Luttinger liquid properties of the quasi-one-dimensional organics. It was also instrumental in determining the crossover scale (about 100 K for PF_6) between the one-dimensional LL properties and the more conventional high-dimensional metallic ones. It also forced to a reexamination of the mechanism underlying the insulating behavior in this compounds showing clearly that the quarter-filled nature of the system is enough in itself to lead to an insulating behavior. These findings were confirmed by the existence of nondimerized compounds with Mott insulating properties. Thanks to these recent progress we have now a consistent description of the relevant properties and energies scales in the quasi-one-dimensional organics. This framework now provides a solid reference to focuss on the important still unsolved questions.

Of course the remaining challenges are numerous. The low energy phase properties are still not largely understood, and if some are strongly reminiscent of the ones of a Fermi liquids, some deviate markedly from them, such as the NMR. Even if one has a Fermi liquid it is unlikely to be a plain vanilla one, since it will remember that it stemmed from a low-dimensional highly interacting system. This can manifests itself, as is apparent on, e.g., the mean field solution, by the variation of the Fermi liquid parameters along the Fermi surface. Much work thus remains to be done to understand this phase, and possibly have a clue on the consequences on the ordered phases, such as the superconducting one. In a similar way the field opened by the new nondimerized compound must be explored. In particular it is important to determine whether they have indeed the same properties than the TMTTF and TMTSF salts under suitable pressure. There are no doubts that transporting both a.c. and d.c. will prove a useful tool to tackle these systems too. Finally of course, if any for the theorist, since for now no chemist has managed to dope such systems, it would be crucial to complete the phase diagram by adding the doping axis, since most of the questions raised here become even more crucial for doped Mott insulators.

Acknowledgements

Many people have contributed, directly via enjoyable collaborations or indirectly via scientific discussions, to the work presented here or more generally to my own understanding of the field. The list of the persons I would like to thank would be too long to be given here, but I would like to specially mention L. Degiorgi, M. Dressel, A. Georges, D. Jérome, and H.J. Schulz. Part of this work was supported by the Swiss national fund for research under MaNEP and division II.

References

1. L.D. Landau, Sov. Phys. JETP **3**, 920 (1957)
2. P. Nozieres, *Theory of Interacting Fermi Systems* (Benjamin, New York, 1961)
3. A.A. Abrikosov, L.P. Gorkov, I.E. Dzyaloshinski, *Methods of Quantum Field Theory in Statistical Physics* (Dover, New York, 1963)
4. T. Giamarchi, *Quantum Physics in One Dimension* (Oxford University Press, Oxford, 2004)
5. T. Giamarchi, Chem. Rev. **104**, 5037 (2004)
6. C. Bourbonnais, D. Jerome, *The Physics of Organic Superconductors and Conductors* (Springer, Berlin Heidelberg New York)
7. T. Giamarchi, Phys. Rev. B **44**, 2905 (1991)
8. T. Giamarchi, Physica B **230–232**, 975 (1997)
9. M. Dressel, A. Schwartz, G. Grüner, L. Degiorgi, Phys. Rev. Lett. **77**, 398 (1996)
10. A. Schwartz et al., Phys. Rev. B **58**, 1261 (1998)
11. V.J. Emery, R. Bruinsma, S. Barisic, Phys. Rev. Lett. **48**, 1039 (1982)
12. R. Kato, H. Kobayashi, A. Kobayashi, J. Am. Chem. Soc. **111**, 5224 (1989)
13. K. Heuzé et al., Adv. Mater. **15**, 1251 (2003)
14. S. Biermann, A. Georges, A. Lichtenstein, T. Giamarchi, Phys. Rev. Lett. **87**, 276405 (2001)
15. L.P. Gorkov, Physica B **230–232**, 970 (1997)
16. C. Bourbonnais et al., J. Phys. (Paris) Lett. **45**, L755 (1984)
17. K. Behnia et al., Phys. Rev. Lett. **74**, 5272 (1995)
18. A. Luther, I. Peschel, Phys. Rev. B **12**, 3908 (1975)
19. H.J. Schulz, Phys. Rev. Lett. **64**, 2831 (1990)
20. M. Ogata, M.U. Luchini, S. Sorella, F.F. Assaad, Phys. Rev. Lett. **66**, 2388 (1991)
21. F. Mila, X. Zotos, Europhys. Lett. **24**, 133 (1993)
22. N. Kawakami, S.K. Yang, Phys. Lett. A **148**, 359 (1990)
23. N. Kawakami, S.K. Yang, Phys. Rev. B **44**, 7844 (1991)
24. F.D.M. Haldane, J. Phys. C **12**, 4791 (1979)
25. F.D.M. Haldane, J. Phys. C **14**, 2585 (1981)
26. F.D.M. Haldane, Phys. Rev. Lett. **45**, 1358 (1980)
27. E.H. Lieb, F.Y. Wu, Phys. Rev. Lett. **20**, 1445 (1968)
28. I.E. Dzyaloshinskii, A.I. Larkin, Sov. Phys. JETP **34**, 422 (1972)
29. J.M. Ziman, *Electrons and Phonons* (Clarendon, Oxford, 1962)
30. T. Giamarchi, A.J. Millis, Phys. Rev. B **46**, 9325 (1992)
31. H.J. Schulz, in *Strongly Correlated Electronic Materials: The Los Alamos Symposium 1993*, ed. by K.S. Bedell et al. (Addison–Wesley, Reading, MA, 1994), p. 187
32. D. Jérome, Chem. Rev. **104**, 5565 (2004)

33. H.J. Schulz, Phys. Rev. Lett. **64**, 2831 (1990)
34. T. Giamarchi, Phys. Rev. B **46**, 342 (1992)
35. D. Controzzi, F.H.L. Essler, A.M. Tsvelik, Phys. Rev. Lett. **86**, 680 (2001)
36. E. Jeckelmann, F. Gebhard, F.H.L. Essler, Phys. Rev. Lett. **85**, 3910 (2000)
37. A. Rosch, N. Andrei, Phys. Rev. Lett. **85**, 1092 (2000)
38. D. Jérome, *Organic Superconductors: From* $(TMTSF)_2PF_6$ *to Fullerenes*, Chap. 10 (Marcel Dekker, New York, 1994), p. 405
39. M. Dressel et al., Phys. Rev. B **71**, 075104 (2005)
40. V. Vescoli et al., Science **281**, 1191 (1998)
41. V. Vescoli et al., Eur. Phys. J. B **13**, 503 (2000)
42. W. Henderson et al., Eur. Phys. J. B **11**, 365 (1999)
43. D. Pashkin, M. Dressel, C.A. Kuntscher, 2006, cond-mat/0608242
44. Y.-S. Lee, K. Segawa, Y. Ando, D.N. Basov, Phys. Rev. Lett. **94**, 137004 (2005)
45. P. Monceau, F. Nad, S. Brazovskii, Phys. Rev. Lett. **86**, 4080 (2001)
46. S. Brazovskii, J. Phys. IV **114**, 9 (2004)
47. S. Brazovskii, *The Physics of Organic Superconductors and Conductors* (Springer, Berlin Heidelberg New York)
48. K. Hiraki, K. Kanoda, Phys. Rev. B **54**, R17276 (1996)
49. T. Itou et al., Phys. Rev. Lett. **93**, 216408 (2004)
50. T. Itou et al., Phys. Rev. B **72**, 113109 (2005)
51. J. Moser et al., Eur. Phys. J. B **1**, 39 (1998)
52. A. Georges, T. Giamarchi, N. Sandler, Phys. Rev. B **61**, 16393 (2000)
53. V.M. Yakovenko, Synth. Metal **103**, 2202 (1999)
54. V.M. Yakovenko, A.T. Zheleznyak, Synth. Metal **120**, 1083 (2001)
55. V.M. Yakovenko, *The Physics of Organic Superconductors and Conductors* (Springer, Berlin Heidelberg New York)
56. J. Moser et al., Phys. Rev. Lett. **84**, 2674 (2000)
57. B. Korin-Hamzić et al., Phys. Rev. B **67**, 014513 (2003)
58. G. Mihaly, I. Kezsmarsky, F. Zamborsky, L. Forro, Phys. Rev. Lett. **84**, 2670 (2000)
59. A. Lopatin, Phys. Rev. B **57**, 6342 (1997)
60. A. Lopatin, A. Georges, T. Giamarchi, Phys. Rev. B **63**, 075109 (2001)
61. G. Leon, T. Giamarchi, J. Low Temp. Phys **142**, 315 (2006)
62. G. Leon, C. Berthod, T. Giamarchi, 2006, cond-mat/0608427
63. S. Brazovskii, V. Yakovenko, J. Phys. (Paris) Lett. **46**, L111 (1985)
64. C. Bourbonnais, L.G. Caron, Physica **143B**, 450 (1986)
65. X.G. Wen, Phys. Rev. B **42**, 6623 (1990)
66. C. Bourbonnais, L.G. Caron, Int. J. Mod. Phys. B **5**, 1033 (1991)
67. V.M. Yakovenko, JETP Lett. **56**, 510 (1992)
68. D. Boies, C. Bourbonnais, A.-M.S. Tremblay, Phys. Rev. Lett. **74**, 968 (1995)
69. H.J. Schulz, in *Correlated Fermions and Transport in Mesoscopic Systems*, ed. by T. Martin, G. Montambaux, J. Tran Thanh Van (Editions frontières, Gif sur Yvette, France, 1996), p. 81
70. A.T. Zheleznyak, V.M. Yakovenko, Synth. Metal **70**, 1005 (1995)
71. Y. Suzumura, M. Tsuchiizu, G. Grüner, Phys. Rev. B **57**, R15040 (1998)
72. M. Tsuchiizu, Y. Suzumura, T. Giamarchi, Prog. Theor. Phys. **101**, 763 (1999)
73. M. Tsuchiizu, Y. Suzumura, Phys. Rev. B **59**, 12326 (1999)
74. M. Tsuchiizu, P. Donohue, Y. Suzumura, T. Giamarchi, Eur. Phys. J. B **19**, 185 (2001)

75. K. Le Hur, Phys. Rev. B **63**, 165110 (2001)
76. F.H.L. Essler, A.M. Tsvelik, Phys. Rev. B **65**, 115117 (2002)
77. E. Arrigoni, Phys. Rev. Lett. **83**, 128 (1999)
78. E. Arrigoni, Phys. Rev. B **61**, 7909 (2000)
79. S. Biermann, A. Georges, T. Giamarchi, A. Lichtenstein, in *Strongly Correlated Fermions and Bosons in Low Dimensional Disordered Systems*, ed. by I.V. Lerner et al. (Kluwer, Dordrecht, 2002), p. 81, cond-mat/0201542
80. C. Berthod, T. Giamarchi, S. Biermann, A. Georges, 2006, cond-mat/0602304
81. B. Dardel et al., Europhys. Lett. **24**, 687 (1993)
82. F. Zwick et al., Phys. Rev. Lett. **79**, 3982 (1997)
83. W. Yu et al., Int. Rev. Mod. Phys. B **16**, 3090 (2002)
84. A.F. Ho, M.A. Cazalilla, T. Giamarchi, Phys. Rev. Lett. **92**, 130405 (2004)
85. T. Stöferle et al., Phys. Rev. Lett. **92**, 130403 (2004)
86. M.A. Cazalilla, A.F. Ho, T. Giamarchi, Phys. Rev. Lett. **95**, 226402 (2005)

Index

Springer Series in

MATERIALS SCIENCE

Editors: R. Hull R. M. Osgood, Jr. J. Parisi H. Warlimont

50 **High-Resolution Imaging and Spectrometry of Materials**
Editors: F. Ernst and M. Rühle

51 **Point Defects in Semiconductors and Insulators**
Determination of Atomic and Electronic Structure from Paramagnetic Hyperfine Interactions
By J.-M. Spaeth and H. Overhof

52 **Polymer Films with Embedded Metal Nanoparticles**
By A. Heilmann

53 **Nanocrystalline Ceramics**
Synthesis and Structure
By M. Winterer

54 **Electronic Structure and Magnetism of Complex Materials**
Editors: D.J. Singh and D. A. Papaconstantopoulos

55 **Quasicrystals**
An Introduction to Structure, Physical Properties and Applications
Editors: J.-B. Suck, M. Schreiber, and P. Häussler

56 **SiO_2 in Si Microdevices**
By M. Itsumi

57 **Radiation Effects in Advanced Semiconductor Materials and Devices**
By C. Claeys and E. Simoen

58 **Functional Thin Films and Functional Materials**
New Concepts and Technologies
Editor: D. Shi

59 **Dielectric Properties of Porous Media**
By S.O. Gladkov

60 **Organic Photovoltaics**
Concepts and Realization
Editors: C. Brabec, V. Dyakonov, J. Parisi and N. Sariciftci

61 **Fatigue in Ferroelectric Ceramics and Related Issues**
By D.C. Lupascu

62 **Epitaxy**
Physical Principles and Technical Implementation
By M.A. Herman, W. Richter, and H. Sitter

63 **Fundamentals of Ion-Irradiated Polymers**
By D. Fink

64 **Morphology Control of Materials and Nanoparticles**
Advanced Materials Processing and Characterization
Editors: Y. Waseda and A. Muramatsu

65 **Transport Processes in Ion-Irradiated Polymers**
By D. Fink

66 **Multiphased Ceramic Materials**
Processing and Potential
Editors: W.-H. Tuan and J.-K. Guo

67 **Nondestructive Materials Characterization**
With Applications to Aerospace Materials
Editors: N.G.H. Meyendorf, P.B. Nagy, and S.I. Rokhlin

68 **Diffraction Analysis of the Microstructure of Materials**
Editors: E.J. Mittemeijer and P. Scardi

69 **Chemical–Mechanical Planarization of Semiconductor Materials**
Editor: M.R. Oliver

70 **Applications of the Isotopic Effect in Solids**
By V.G. Plekhanov

71 **Dissipative Phenomena in Condensed Matter**
Some Applications
By S. Dattagupta and S. Puri

72 **Predictive Simulation of Semiconductor Processing**
Status and Challenges
Editors: J. Dabrowski and E.R. Weber

73 **SiC Power Materials**
Devices and Applications
Editor: Z.C. Feng

Springer Series in

MATERIALS SCIENCE

Editors: R. Hull R. M. Osgood, Jr. J. Parisi H. Warlimont

74 **Plastic Deformation in Nanocrystalline Materials**
By M.Yu. Gutkin and I.A. Ovid'ko

75 **Wafer Bonding**
Applications and Technology
Editors: M. Alexe and U. Gösele

76 **Spirally Anisotropic Composites**
By G.E. Freger, V.N. Kestelman, and D.G. Freger

77 **Impurities Confined in Quantum Structures**
By P.O. Holtz and Q.X. Zhao

78 **Macromolecular Nanostructured Materials**
Editors: N. Ueyama and A. Harada

79 **Magnetism and Structure in Functional Materials**
Editors: A. Planes, L. Mañosa, and A. Saxena

80 **Micro- and Macro-Properties of Solids**
Thermal, Mechanical and Dielectric Properties
By D.B. Sirdeshmukh, L. Sirdeshmukh, and K.G. Subhadra

81 **Metallopolymer Nanocomposites**
By A.D. Pomogailo and V.N. Kestelman

82 **Plastics for Corrosion Inhibition**
By V.A. Goldade, L.S. Pinchuk, A.V. Makarevich and V.N. Kestelman

83 **Spectroscopic Properties of Rare Earths in Optical Materials**
Editors: G. Liu and B. Jacquier

84 **Hartree–Fock–Slater Method for Materials Science**
The DV–X Alpha Method for Design and Characterization of Materials
Editors: H. Adachi, T. Mukoyama, and J. Kawai

85 **Lifetime Spectroscopy**
A Method of Defect Characterization in Silicon for Photovoltaic Applications
By S. Rein

86 **Wide-Gap Chalcopyrites**
Editors: S. Siebentritt and U. Rau

87 **Micro- and Nanostructured Glasses**
By D. Hülsenberg and A. Harnisch

88 **Introduction to Wave Scattering, Localization and Mesoscopic Phenomena**
By P. Sheng

89 **Magneto-Science**
Magnetic Field Effects on Materials: Fundamentals and Applications
Editors: M. Yamaguchi and Y. Tanimoto

90 **Internal Friction in Metallic Materials**
A Reference Book
By M.S. Blanter, I.S. Golovin, H. Neuhäuser, and H.-R. Sinning

91 **Time-dependent Mechanical Properties of Solid Bodies**
By W. Gräfe

92 **Solder Joint Technology**
Materials, Properties, and Reliability
By K.-N. Tu

93 **Materials for Tomorrow**
Theory, Experiments and Modelling
Editors: S. Gemming, M. Schreiber and J.-B. Suck

94 **Magnetic Nanostructures**
Editors: B. Aktas, L. Tagirov, and F. Mikailov

95 **Nanocrystals and Their Mesoscopic Organization**
By C.N.R. Rao, P.J. Thomas and G.U. Kulkarni

96 **Gallium Nitride Electronics**
By R. Quay

97 **Multifunctional Barriers for Flexible Structure**
Textile, Leather and Paper
Editors: S. Duquesne, C. Magniez, and G. Camino

98 **Physics of Negative Refraction and Negative Index Materials**
Optical and Electronic Aspects and Diversified Approaches
Editors: C.M. Krowne and Y. Zhang